MATHEMATICAL REVIEW

Area of a circle of radius R	$A = \pi R^2$
Circumference of a circle	$C = 2\pi R$
Surface area of a sphere	$A = 4\pi R^2$
Volume of a sphere	$V = \frac{4}{3}\pi R^3$
Area of a triangle	$A = \frac{1}{2}bh$
Volume of a circular cylinder of length l	$V = \pi R^2 l$
Pythagorean Theorem	$C^2 = A^2 + B^2$

$\sin\theta = A/C$

$\cos\theta = B/C$ $\tan\theta = \dfrac{\sin\theta}{\cos\theta}$

$\tan\theta = A/B$

Quadratic Equation: Where $ax^2 + bx + c = 0$

$$x = \frac{-b \pm \sqrt{b^2 - 4ac}}{2a}$$

PHYSICAL CONSTANTS

Quantity	Symbol	Value
Gravitation constant	G	$6.672\,59 \times 10^{-11}$ N·m²/kg²
Speed of light in vacuum	c	$2.997\,924\,58 \times 10^8$ m/s
Electron charge	e	$1.602\,18 \times 10^{-19}$ C
Planck's Constant	h	$6.626\,076 \times 10^{-34}$ J·s
		$4.135\,669 \times 10^{-15}$ eV·s
Universal gas constant	R	$8.314\,510$ J/mol·K
Avogadro's number	N_A	$6.022\,137 \times 10^{23}$ mol⁻¹
Boltzmann Constant	k_B	$1.380\,66 \times 10^{-23}$ J/K
		$8.617\,39 \times 10^{-5}$ eV/K
Coulomb force constant	k_0	$8.987\,55 \times 10^9$ N·m²/C²
Permittivity of free space $(1/\mu_0 c^2)$	ϵ_0	$8.854\,19 \times 10^{-12}$ C²/N·m²
Permeability of free space	μ_0	$1.256\,64 \times 10^{-6}$ T·m/A
Permeability constant	$\mu_0/4\pi$	10^{-7} T·m/A
Electron mass	m_e	$9.109\,39 \times 10^{-31}$ kg
Electron rest energy	$m_e c^2$	$0.510\,999$ MeV
Electron magnetic moment	μ_e	$9.284\,77 \times 10^{-24}$ J/T
Electron charge/mass ratio	e/m_e	$1.758\,82 \times 10^{11}$ C/kg
Electron Compton wavelength	λ_c	$2.426\,31 \times 10^{-12}$ m
Proton mass	m_p	$1.672\,623 \times 10^{-27}$ kg
		$1.007\,276$ u
Proton rest energy	$m_p c^2$	938.272 MeV
Proton magnetic moment	μ_p	$1.410\,608 \times 10^{-26}$ J/T
Neutron mass	m_n	$1.674\,929 \times 10^{-27}$ kg
		$1.008\,66$ u
Neutron rest energy	$m_n c^2$	939.566 MeV
Neutron magnetic moment	μ_n	$9.662\,37 \times 10^{-27}$ J/T
Bohr magneton	μ_B	$9.274\,015 \times 10^{-24}$ J/T
Stefan-Boltzmann Constant	σ	$5.670\,51 \times 10^{-8}$ W/m²·K⁴
Rydberg constant	R	$1.097\,373 \times 10^7$ m⁻¹
Bohr radius	r_1	$5.291\,77 \times 10^{-11}$ m
Faraday constant	F	$9.648\,53 \times 10^4$ C/mol

PHYSICS

VOLUME TWO

To Ca, b. w. l.

V O L U M E   T W O
PHYSICS
Calculus

EUGENE HECHT
Adelphi University

Brooks/Cole Publishing Company

An International Thomson Publishing Company I**T**P

Pacific Grove • Albany • Bonn • Boston • Cincinnati • Detroit •
London • Madrid • Melbourne • Mexico City • New York • Paris •
San Francisco • Singapore • Tokyo • Toronto • Washington

Brooks/Cole Publishing Company
A Division of Wadsworth, Inc.

Sponsoring Editor: *Harvey Pantzis*
Marketing: *Connie Jirovsky & Margaret Parks*
Editorial Associate: *Beth Wilbur*
Production Editor: *Ellen Brownstein*
Production Service: *HRS Electronic Text Management*
Manuscript Editor: *Monique Condon*
Art Coordinator: *HRS Electronic Text Mgmt.*
Interior Illustration: *Precision Graphics, Carl Brown, LM Graphics, Matrix Communications, and HRS Electronic Text Management*
Cover Photo: *Tom Skrivan*

Permissions Editor: *May Clark & Jennifer Burke*
Interior & Cover Design: *E. Kelly Shoemaker & Vernon T. Boes*
Photo Coordination & Digital Photo Design: *Larry Molmud & HRS Electronic Text Mgmt.*
Photo Researcher: *Stuart Kenter & Carolyn Hecht*
Typesetting: *Beacon Graphics*
Cover Printing: *Lehigh Press Lithographers*
Printing and Binding: *Quebecor/Hawkins*

For more information, contact:

BROOKS/COLE PUBLISHING COMPANY
511 Forest Lodge Road
Pacific Grove, CA 93950
USA

International Thomson Publishing Europe
Berkshire House 168-173
High Holborn
London WC1V 7AA
England

Thomas Nelson Australia
102 Dodds Street
South Melbourne, 3205
Victoria, Australia

Nelson Canada
1120 Birchmount Road
Scarborough, Ontario
Canada M1K 5G4

International Thomson Editores
Campos Eliseos 385, Piso 7
Col. Polanco
11560 México D. F. México

International Thomson Publishing GmbH
Königswinterer Strasse 418
53227 Bonn
Germany

International Thomson Publishing Asia
221 Henderson Road
#05-10 Henderson Building
Singapore 0315

International Thomson Publishing Japan
Hirakawacho Kyowa Building, 3F
2-2-1 Hirakawacho
Chiyoda-ku, Tokyo 102
Japan

Printed in the United States of America

10 9 8 7 6 5 4 3 2 1

Library of Congress Cataloging-In-Publication Data
Hecht, Eugene.
　PHYSICS: calculus / Eugene Hecht.
　　p.　cm.
　Includes index.
　ISBN 0-534-34157-8
　1. Physics.　2. Calculus.　3. Mathematical physics.　I. Title.
QC21.2.H43　1996　　　　　　　　　　　　95-37110
530–dc20　　　　　　　　　　　　　　　　CIP

Preface

Physics is the study of the material Universe—all there *is*. And that's a bold and wonderful agenda. The Universe is incredibly awesome and tantalizingly mysterious, and we, after all, are just beginning to understand it. Almost 3000 years in the making, physics—incomplete as it is—stands as one of the great creations of the human intellect. It has been a privilege and an unending joy to have spent much of my life studying physics, and it is out of gratitude and admiration that this book takes its form. If this work, while insightfully teaching basic physics, transmits a sense of the grandeur, unity, and vitality of the subject, it will have met my primary objectives.

TO THE STUDENT

This book has been designed for the calculus-based Introductory Physics course, and it contains the standard range of material from kinematics to quantum mechanics. It is predicated, however, on the belief that it's time to return to fundamentals; today's texts have become too mathematical and too advanced. By contrast, this work limits the associated math to basic calculus and very basic vector analysis. It omits a variety of obscure high-level topics and instead uses its facilities to transmit a deeper understanding of the fundamental concepts of modern-day physics. It covers all the grand insights, but with a self-restraint that stops short of examining every possible side issue. As a result the book can progress at a slower pace and provide much more support in the process.

The text assumes that the student comes with only a modest knowledge of algebra, geometry, and trigonometry. It presumes that the reader is now studying or has already studied calculus but retains little more than a cloudy memory of that experience. Of course, the derivative of x^2 equals $2x$, even though one may not remember exactly what a derivative is. In short, ***whatever mathematics is required will be retaught, in place, as the need arises*** and is further elaborated in an extensive **tutorial appendix**. It is strongly suggested that the math appendix be reviewed before starting the course.

TO THE PROFESSOR

Back to Basics

Over the last five decades the Introductory Physics course and the texts that support it have undergone a dramatic transformation. Prior to the Second World War a typical text dealt almost entirely with the discussion of concepts and principles; there were comparatively few equations, little analysis, and a small selection of straightforward problems. The idea of a vector quantity was introduced, but only in a qualitative way that allowed for the addition of forces. The mathematics was algebra with a touch of trigonometry—no calculus, no vector analysis. The war brought physics into a new social prominence, and even before it ended, calculus was already finding its way, if only tentatively, into the introductory course. By the mid-1950s the distinction had been established between so-called College Physics (algebra/trig. based) and University Physics (calculus based). Standard treatises (e.g., *University Physics* by Sears and Zemansky, 2nd edition, 1955) fully embraced calculus but contained no rigorous discussion of vectors. The rudiments of vector analysis entered the introductory discourse only in the early 1960s.

Driven by a seemingly endless supply of high-quality students of engineering and science, University Physics became increasingly more sophisticated mathematically, even as it remained philosophically naive and minimally developed conceptually. The tension between competing texts (*Physics for Students of Science and Engineering* by Halliday and Resnick, 1960) and a boundless optimism in the teaching community reshaped the next generation of books and the courses that used them. By 1970 serious difficulties were already cropping up and there was a token reduction in the "level of sophistication" (*Fundamentals of Physics* by Halliday and Resnick, 1st edition), but even the revised works were doing vector calculus (with **i**, **j**, **k** vectors) by Chapter 3. Texts ranged well beyond the once-traditional bounds of the curriculum, treating numerous subtleties (e.g., the Poynting vector, the Maxwell–Boltzmann distribution, and the power associated with a wave on a string) that had previously always been dealt with in higher level courses.

Today's established texts are the end product of that rush toward analytic and computational prowess that rolled on, untempered and unchallenged, across the 1970s and 1980s. A quick thumb through any of these works will reveal a splendid but dauntingly unintegrated minicompendium of contemporary physics. Scattered among the foundational concepts (and undifferentiated from them) one can find a plethora of such unlikely items as the gradient operator in all its glory, a derivation of the partial differential wave equation, the relativistic Doppler effect, an analysis of forced oscillations, and the quantum mechanical treatment of a particle in a well. Any first-year graduate student should be pleased to have mastered the scope and depth of the material now proffered as Introductory Physics.

This state of affairs is only acceptable for a core cadre of highly motivated gifted students of the physical sciences. The clientele of the 1990s is much broader than that and these people are not at all well served by the standard texts of the day. Indeed, for many students these books are simply overwhelming, and that puts inordinate demands on the professor. From their earliest chapters these texts require an extensive familiarity with vector analysis (based on **i**, **j**, **k** vectors); something few undergraduates have at the time. Thus, at the very beginning of the experience, when students are most vulnerable, and when they should be concentrating on learning *how* to learn physics, they are made to deal with the unnecessary burden of mastering mathematical machinery, albeit elegant machinery, that they can easily do without. Moreover, the standard texts seem to make the assumption that someone who is taking, or has taken, a course in calculus has learned calculus. With little or no additional preparation, they swiftly go on to incorporate vector calculus, something quite unheard of, even for physics majors, thirty or forty years ago.

There should be a realistic alternative, especially one that is thoroughly modern, philosophically mature, and pedagogically effective. And that's what *this* book is all about. ***Where appropriate, it restrains the mathematics and limits the range of the discourse so that it can focus on, elaborate, and teach the fundamentals of physics***. Calculus and vector analysis are both painstakingly developed as tools and then only used insofar as they illuminate the physics. This text is not embarrassed to go slowly; to justify where it *is* going; to stop to take stock of where it has been; to point out the marvelous unity of the subject; to skip obscure topics; and above all, to seem simple, indeed occasionally to *be* simple. Informed by an integrated twentieth-century perspective guided by a commitment to providing a conceptual overview of the discipline, ***this book is a return to basics***.

Pedagogy

The present work comprises an elaborate pedagogical system that has assimilated the important educational research findings of the last several decades. Its many innovations (e.g., in notation, graphics, topic order, problem handling, and photography) have been extensively tested in colleges and universities across North America and Europe.

The text is rich in small but important teaching advances that are difficult to notice without actually using the book. For example, to take advantage of the student's familiarity with motion, a familiarity predicated on automobiles with odometers and speedometers (neither of which ever shows a negative value), Chapter 2 first treats **motion along a curved path**. This small shift in sequencing has proven to be quite significant pedagogically. Locating an object as it moves along some axis involves dealing with its *displacement* from the origin, and that's a far more subtle matter than treating the overall *distance* it traveled. The standard texts begin kinematics, as they have for over half a century, with the special case of one-dimensional motion (something that would be unrealistic even on a flat Earth), and talk about negative distances and negative velocities, which, to the modern traveler, is counterexperiential. They go on to completely muddle the difference between speed and velocity (for example, talking about velocity while inexplicably using the lightfaced letter *v* to represent it). Far better to begin with motion specified by the scalar path length along a curve (using distance and speed) and once that's mastered to move on to the vector concept of displacement, even in one dimension (using displacement and velocity).

This raises some minor, but confusing, linguistic issues that are associated with the standard treatments—issues that are generally glossed over, although they should have been dealt with years ago. In the name of student well-being, this work addresses these annoyances (e.g., one cannot have a negative speed if speed is the *magnitude* of the velocity, and so what shall we call a quantity like $v = -10$ m/s? Since *acceleration* is always defined as a vector, how shall we refer to -9.81 m/s^2? It's not really the acceleration and it's not the magnitude of the acceleration).

The concept of **significant figures** represents another major annoyance and point of confusion for the uninitiated. It's amazing that the leading texts, though they pay lip service to the concept, still have not taken the pains to follow through and actually correct all their wrong numerical results. Cautious students who correctly use significant figures, rightly complain (year after year) that the process of checking their answers with the back of the book is extremely frustrating. A professor may choose to be rigorous or not on the matter, but a text that points out the correct procedures must scrupulously adhere to them—this book (p. 13) does just that, *throughout*.

Another important pedagogical improvement is in **notation**. Consider the needless confusion engendered by the ill-conceived use of symbols in most introductory texts. How many different symbols are used for force—*F*, *f*, *w*, *W*, *N*, *T*, *R*—in any one book? How could that not be confusing to the uninitiated? This text uses one symbol for force, *F*. There are tensile forces F_T, normal forces F_N, reaction forces F_R, weights F_W, friction forces F_f, elastic forces F_e, electrical forces F_E, and so on. When a student sees F representing a physical quantity anywhere in this book, it is force. In the same way, W is work, nothing else.

Pacing

The question of pacing is a good example of the kinds of concerns that have made this book a highly successful teaching device.

The Introductory Physics course makes a wide range of challenging intellectual demands on the student, and the experienced instructor knows that the first few weeks of the semester can be crucial. It is then that some students are likely to become overwhelmed. With that in mind, the material of kinematics has been rearranged in order to introduce the physics more gradually and allow time for the ideas to be assimilated. Consequently, Chapter 2 deals only with the physical concepts of speed, displacement, and velocity. The explanations are elaborate, and there are many examples, graphs, and illustrations. Acceleration comes only later, in Chapter 3.

Anticipating that a successful development of calculus requires a substantial mathematical foundation, Chapter 1 contains an extensive preparatory section (p. 17) on graphs and functions. The basic idea of differentiation is introduced in Chapter 2 (p. 34) and methodically developed. A selection of problems (p. 63) and an extensive appendix (p. A-13) support the effort. The concept of vectors appears for the first time as it relates to displacement (p. 38). It is justified, made logically appealing, and very carefully extended and applied to velocity. With the visual aid of multiframe drawings (e.g., pp. 51–54), vector addition is next applied to relative velocities.

Only then, after the student has presumably worked out dozens of problems and has begun to learn how to learn physics, do we turn to Chapter 3 and acceleration. By the time the equations of uniform acceleration are reached (p. 75), the typical student is much better able to deal with the logical complexities involved. In contrast to standard texts, which roar through integral calculus in the first fifty pages or so (one manages it in twenty-eight), *this book derives the equations of constant acceleration without recourse to calculus and then leisurely introduces integration (p. 101) after the student has had ample opportunity to assimilate the physics*.

With the mathematics of trigonometry, rudimentary vector algebra, and calculus in place, and with a more realistic understanding of the demands of the experience, the serious student is then ready to move ahead more rapidly.

The Utility of History

The history of physics is very occasionally incorporated into the standard texts, and when it is, it's more as decoration than anything else. By contrast, this book selectively incorporates historical materials for a variety of pedagogical reasons. For instance, when details of the lives of any of the great scientists are given, it is with an eye toward making these larger-than-life figures less intimidating (e.g., p. 84 on Galileo, or p. 116 on Newton) and their work a little more approachable. In a similar vein, *the book is responsive to the significant contribution made by women in physics*. It highlights the accomplishments of such outstanding twentieth-century scientists as Amalie Noether, Maria Goeppert Mayer, and Lise Meitner, among many others.

Most importantly, the text follows a historical approach whenever doing so allows the physics to unfold more clearly. This book treats the **history of ideas** in order to make those ideas more immediately accessible. Once the student learns what Buridan (1330) was thinking about when he conceived the notion of momentum, the concept instantly becomes understandable (p. 127): *mv* makes sense. The brilliant idea of *Conservation of Momentum* came out of Descartes's metaphysical musings in a way that's perfectly reasonable (p. 139). Having read why Huygens was unhappy with Descartes's momentum (p. 325) and why he squared the *v* as an alternative (thereby discovering a new conserved quantity), the concept of kinetic energy comes alive—that's effective pedagogy.

A Modern Approach

The central glory of twentieth-century physics is the discovery of an overarching unity in nature. There is revealed an internal simplicity that is well confirmed, even if its complete comprehension is, as yet, just beyond our reach. All matter is composed of myriad identical clones of a small number of fundamental interacting particles. Everything physical is presumably understandable within that context. Thus, to treat the subject as if it were an encyclopedic collection of unrelated ideas is to miss the whole point of the twentieth century. *From the beginning of this text to its end we will study the unity of natural phenomena, the various manifestations of matter interacting with matter via the fundamental Four Forces*. Even as we explore concepts conceived centuries ago, we will bring to bear the perspective of contemporary physics.

An Atomic Perspective

Bulk matter is atomic, and to truly comprehend its behavior (e.g., mechanical, thermal, electrical, magnetic, acoustical, and optical), we must appreciate how atoms interact to produce the phenomena of everyday life. *This book presents a great variety of physical concepts from an atomic perspective* (e.g., friction, p. 182; Young's Modulus, p. 381; and heat, p. 568). Accordingly, optics begins with a discussion of atomic scattering (p. 936). The treatment dispenses with Huygens's rather contrived principle (which originated in the aether theory) and instead, derives the laws of reflection and refraction in a much more satisfying way from atomic scattering.

It's no longer appropriate to wait until the end of the second semester to learn that the objects of our world are composed of atoms and that the phenomena of nature are understandable in that context. Nor is it necessary to teach the first semester as if nineteenth-century thought was still valid in all regards—it is not.

Symmetry and Conservation

The discoveries of this century have radically changed our perception of every aspect of the physical universe. And yet Classical Physics is often taught as though the last hundred years of revelation had little or no effect on our thinking. This is certainly not the case.

One of the most far-reaching insights of twentieth-century physics (one still not discussed in books at this level) is the theoretical importance of the relationship between symmetry and conservation. That relationship, which culminates in the theory of the electroweak interaction, is a leitmotif that runs throughout this text. Students are fascinated to learn that symmetries of space and time manifest themselves in the conservation laws. The relationship is first discussed, albeit briefly, in Chapter 1, with an introduction to the concept of symmetry. It recurs in Chapter 4 with Conservation of Linear Momentum, in Chapter 8 with Conservation of Angular Momentum, in Chapter 9 with Conservation of Energy, and so on all the way up to gauge symmetry in Chapter 33. The **symmetry-conservation diagram**, which appears ten times (e.g., pp. 7, 143, 258, 300, 347, etc.), in successively developing versions, speaks to the centrality of the conservation laws. But it also serves two other important functions. *It marks the student's progress through the year's work and underscores the unity of the entire discipline.*

Modern insights are introduced whenever they are relevant and as early as possible. Chapter 1, for example, briefly summarizes and considers the interrelationships of the various theoretical formalisms (Classical Physics, Relativity, and Quantum Mechanics). It is important for the student to have an overview of the discipline and to know that this is an ongoing, unending study. Chapter 1 also introduces some of the vocabulary and ideas that will be considered throughout the development. In this way, the concept of quantization will be a familiar one long before it is dealt with formally in Chapter 30. Likewise, atoms, electrons, protons, neutrons, photons, gravitons, and quarks are all introduced early in the book.

The primary role played by the **Four Forces** of nature is an ever-present theme, and an effort is made to see everyday phenomena from this unifying perspective. Students are made aware from the outset that contact forces, friction, cohesion, and adhesion, for example, are results of electromagnetic interaction, although they will have to wait until Chapter 17 to study the details.

Practical Applications

A driving force that shapes this text is the belief that students are more easily motivated to learn about what directly affects their lives and concerns. The perspective of the book is therefore largely practical: How does a rocket work (p. 137)? Why did the Nimitz Freeway collapse (p. 461)? How does an airplane really fly (p. 424)? How do we breathe (p. 405)? The text is especially rich in applications of physics to the life sciences, but it strives to be broadly interdisciplinary, drawing from and reflecting back upon biology (e.g., p. 427), geology (e.g., p. 466), astronomy (e.g., p. 117), architecture (e.g., p. 221), medicine (e.g., p. 381), and meteorology (e.g., p. 591). This course has been blind to the universality of physics for far too long: *all* physical processes are the purview of contemporary physics.

The careful reader will likely detect that the author, who earned an undergraduate degree in engineering physics and was a practicing engineer for several years, still takes endless delight in exploring how things work. The text has a subtle engineering sensibility (e.g., Table 6.1, p. 199) that also embraces bio-engineering concerns (e.g., p. 203 and p. 385).

Problem Solving and Independent Study

Examples

This text contains a great many sample calculations systematically worked out in detail, in order to guide students through the processes of analysis and numerical problem solving. All such examples are followed by a **Quick Check**, which teaches a wide variety of verification methods and helps to establish the habit of checking one's computations.

Problem-Solving Techniques

Every chapter includes a section called **Suggestions on Problem Solving** that explores techniques applicable to the particular problems that will be confronted. It may also contain approximation methods as well as discussions of the pitfalls and common errors specific to the material at hand. For example, a common error in kinematics is to compute the average speed of a uniformly accelerating object using $v_{av} = \frac{1}{2}(v - v_0)$ rather than $v_{av} = \frac{1}{2}(v + v_0)$; students are appropriately cautioned.

Core Material

Every chapter includes a distilled review of the core material for that chapter.

Discussion Questions

All chapters contain a selection of discussion questions designed to develop and extend the conceptual understanding of the material.

Multiple Choice Questions

A group of roughly 20 multiple choice questions similar to those found on national medical (MCAT) and optometry (OATP) school entrance exams is included in each chapter. Among other types, these comprise single-concept calculational questions as well as probing conceptual questions. More and more universities are using multiple choice questions on examinations, and *this is the only text at this level that provides the student with a sample of the genre.*

Numerical Problems

An extensive selection of problems, most built on real data and referring to actual situations, is provided at the end of each chapter. An instructor will find a wealth of choices from which to assign homework. The problem sets are arranged in three levels of difficulty (indicated by the symbols [I], [II], and [III], respectively) and always include a large selection of single-concept problems that explore one idea at a time (from several perspectives) to help students establish a strong foundation of competence and confidence with the basics. Each set of problems is deployed as a carefully developed and integrated unit that will carry the student from one level of mastery to the next. All students working diligently should soon be able to do grade-I problems without much trouble. Grade-II problems require a more developed competence, and a student who is comfortable with them should do well in the course. Grade-III problems are designed to be still more challenging.

Problem solving, like playing an instrument, can only be mastered by a combination of practice and guiding examples. Thus, ***approximately 15 percent of the end-of-chapter problems are worked out succinctly at the back of the book in order to encourage and strengthen independent study.*** (These problems are indicated by a boldface numeral.) The pedagogical scheme, which has proven to be highly successful, is as follows: as much as possible, problems are presented in small groupings according to type. For each such group a representative example is solved, very concisely, in the back of the book. This allows the student working outside of the classroom to have somewhere to turn when stymied by a problem, rather than just give up. No other Introductory Physics text provides this kind of support. Another 10 percent of the solutions are provided in a student solutions manual, which is available at the instructor's request to the bookstore. (These problems are indicated by an italic numeral.)

Wherever appropriate, a chapter will contain a selection of calculus-based problems whose type and level are indicated via the designations [c], [cc], or [ccc], respectively. ***The student is carefully guided through the more demanding problems by a variety of*** Hints. This is an important and unique feature of this book. It allows someone who is less well prepared to nonetheless experience the methodology of calculus.

Organization

A good working text has responsibilities to both the student and the instructor. It should, as much as possible, provide a complete resource to students who are studying on their own at home. At the same time, it must offer an instructor a broad range of conceptual material from which to fashion a course that meets his or her specific requirements.

It is impossible to lecture in class on every topic covered in a text that deals with the whole of Introductory Physics. The book has been organized in a way that addresses this problem and is therefore a more effective teaching instrument. All section headings are coded in three colors: green, black, and red. Topics bearing green headings are intended to be read and understood by students working on their own prior to coming to class. This information can then be assumed as basic communal knowledge and thereafter requires little or no classroom time. Topics bearing **black headings** represent the basic material from which the instructor can fashion the specific course. Topics with red headings are enrichment sections that can be used to enhance and strengthen the primary discourse.

Units

Units play a special role in physics and so represent an issue that requires considerable attention. Without exception, the proper units to be used in physics are those of the Systéme International (SI). Thus, SI units will be used almost exclusively throughout this text. The commitment to the Systéme International must, however, be flexible enough to contend with three outstanding challenges: (1) The United States is the only major country in which there still lingers an antique system of units. Consequently, the intuitive experiences of most students are framed in terms of feet, miles, and pounds. Moreover, the data of everyday life, which must be brought into the discussion, is invariably given in U.S. Customary units. Apparently, baseball pitchers hurl balls in miles per hour exclusively. (2) In order to extend our analyses into the domains of other disciplines, we must deal realistically with their idiosyncratic unit preferences. Like it or not, the pressure in the lungs is measured in centimeters of water, and for good reason. (3) There is a vast body of extant scientific data in a variety of units; the well-trained student must have access to that treasure. Thus, this text will gradually, compassionately, and yet unswervingly move toward the total adoption of SI units. Even as it does, it will attempt to deal effectively with the substantive issues considered here.

Art and Photography Program

The graphics and text for this book were conceived and developed from the very start as an integrated whole; every photograph and every drawing is justified in its inclusion. The result is an unparalleled level of teaching effectiveness. There are about 1,400 fresh, insightful, and uncompromisingly accurate illustrations that make it possible for students to visualize a diversity of physical phenomena. Many of these are unique multiframe sequential drawings that allow the reader to apprehend the temporal unfolding of complex events (e.g., Figures 2.32, 2.33, 13.38, and 13.39). There are nearly 500 photographs (many of them overlaid with explanatory graphics) that form an important part of the pedagogical program.

A Complete Ancillary Package

Accompanying the text is a complete ancillary package, which includes the following instructional aids.

Instructor's Solutions Manual
Includes answers to all discussion questions, answers to all multiple choice questions, and solutions to all problems.

Calculus Problem Workbook

A student guide that provides step-by-step strategies for solving calculus-based physics problems. It contains an extensive collection of exercises, solutions, and answers.

Student's Solutions Manual

Includes answers to selected odd-numbered discussion questions, answers to odd-numbered multiple choice questions, and solutions to selected odd-numbered problems (approximately 10 percent) not already solved in the book.

Transparencies

Approximately 60 full-color transparencies illustrating key concepts.

Test Bank

Includes more than 1,000 multiple choice and short-answer questions.

Electronic Testing (EXP-Test, Chariot MicroTest III)

All test items are available in electronic format for DOS, Windows, and Macintosh platforms.

BCX-Physics Software

An electronic study guide for the Macintosh, DOS, and Windows platforms, consisting of multiple choice and fill-in-the-blank questions with hints, selected from the text.

Acknowledgments

Over the several years during which this work has developed, there have been many people who were kind enough to contribute to the project. Accordingly, I take this opportunity to express my appreciation to them all, especially to Professors R. Neiman of Illinois Wesleyan University, Z. Hlousek of California State University, Long Beach, and V. Saxena of Purdue University. I also thank Professors G. N. Rao and J. Dooher of Adelphi University for the delightful time we have spent talking about physics. Anyone wishing to exchange ideas is welcome to write to the author c/o Physics Department, Adelphi University, Garden City, NY 11530.

The design of the book was carried out by Kelly Shoemaker, whose skill and uncompromising commitment to excellence will always be appreciated. It actually became a book under the careful guidance of Lorraine Burke of HRS. Her expertise, good humor, and patience have earned my admiration and respect. I thank my editor Harvey Pantzis for his vision, understanding, and continuing support without which this project would not have happened. Jamey A. Hecht, Kurt Norlin, and Ed Burke each generously brought to bear a special talent and I thank them. Lee Young, May Clark, Elizabeth Rammel, Ellen Brownstein, and Vernon Boes all kindly contributed to the effort as did Beth Wilbur, who was in charge of the author's mental health, and stood a constant vigil. Her good offices kept everything on track and calmed the tempest. I am especially grateful for the friendship of Connie Jirovsky; her keen insight, understanding, and wisdom played a formative role in shaping this effort from the very beginning.

Finally, I bow appreciatively to my personal editor, Carolyn Eisen Hecht (*eshet chayil*). She tracked down hundreds of photographs, read thousands of pages, and made superb suggestions that have greatly enhanced the clarity of the exposition. Knowing full well the sacrifices demanded by this project, I gratefully thank her for going through it all, one more time.

Eugene Hecht

Brief Contents

Contents

37.5 m/s
83.9 mi/h

31.8 m/s
71.2 mi/h

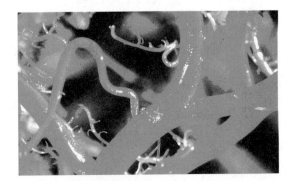

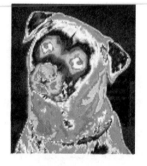

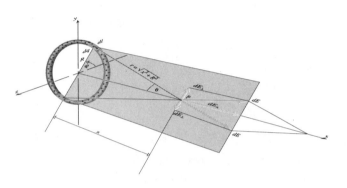

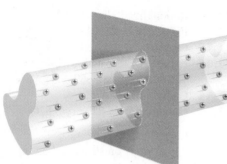

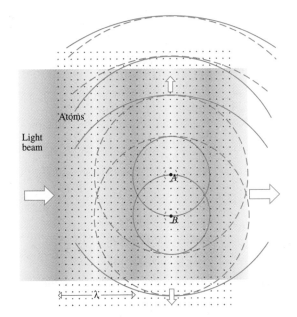

24 Radiant Energy: Light 905

25 The Propagation of Light: Scattering 935

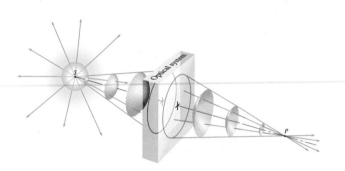

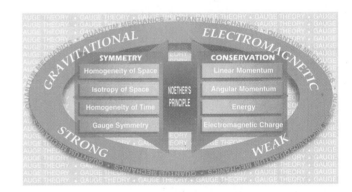

NOTE: The Table of Contents is color-coded as per p. xiv in the Preface.

17

Electrostatics: Forces

THUS FAR, WE HAVE concentrated on the gravitational interaction and the associated property of matter called *mass*. Certainly, the electromagnetic interaction was always there in our considerations, but in the guise of such macroscopic notions as friction, cohesion, elasticity, contact force, and so forth. Now we focus on it and on the associated characteristic of matter called *charge*. The electromagnetic interaction binds matter in all its observable forms (Chapter 10). It is responsible for holding the subatomic particles together as atoms, for holding the atoms together as molecules, for holding the molecules together as objects, indeed, for holding your nose onto your

A chunk of amber rubbed on fur picks up small pieces of colored paper.

face. All biological processes are governed by the interaction of charges: seeing, feeling, moving, thinking, living—all of it.

The story begins in ancient times with amber, a yellow-brown material (fossilized pine-tree resin) used for jewelry for thousands of years. Probably while polishing amber, the Greeks noticed that it had an extraordinary quality: rubbed with cloth or fur, it attracts small bits of lightweight matter—tufts of lint, hair, etc. Even Plato wrote about the "marvels concerning the attraction of amber." The Greek word for amber is *elektron,* and by the mid-seventeenth century it was already being suggested that a substance, activated by rubbing, possessed some sort of "amber stuff," or *electricity.* And people came to refer to the *charge* of electricity imparted to an object in the same sense as an "amount" (much like a charge of gunpowder).

When rubbed with woolen cloth, a chunk of amber or a plastic comb can pick up a small piece of Styrofoam™ broken from a coffee cup. We are led by our understanding of the laws of mechanics to conclude that there is yet another force, the *electric force.* Something must be causing it, and so we call the physical attribute responsible for that interaction, electric or more precisely, **electromagnetic charge**. (Electric and magnetic phenomena are one and the same—both are manifestations of charge.)

The subject of electromagnetism is usually arranged into conceptual subdivisions, one of which is **electrostatics**, *the study of charges at rest.*

ELECTROMAGNETIC CHARGE

Charge gives rise to electric force, and we are only now beginning to figure out how it manages to do that. Charge is fundamental and cannot be described in terms of simpler, more basic concepts. We know it by what it does, not by what it is—if you like, it is what it does, and that's that.

If we vigorously stroke a plastic pen on a woolen glove and hold the two apart but close, fibers on the glove will stand on end, straining to reach up to the pen, making the attraction obvious. It was a French botanist, C. Dufay, who first studied the *repulsive* interactions of electricity. He found that objects of the same material electrified in the same way repelled one another. Two pieces of glass rubbed with silk will repel each other, just as two chunks of amber rubbed with fur will (Fig. 17.1). Yet the charged glass will attract both the silk and the charged amber, and vice versa. Sometime around 1734, Dufay concluded "that there are two distinct Electricities"—two kinds of electric charge—and he was quite right. Summarized in contemporary terms: **like charges repel, unlike charges attract** (Fig. 17.2).

17.1 Positive and Negative Charge

Today we follow Benjamin Franklin's lead, arbitrarily calling the two kinds of charge *positive* and *negative*. In ordinary circumstances, we deal with the electrical behavior of solids (plastic combs, nylon sweaters, TV screens, etc.) and there the positive charges are locked up in the nuclei of the essentially stationary atoms, while some negative charges are more or less free to be transferred. There are two kinds of charge but, in the vast majority of cases, only one is mobile.

An object that contains the same amount of positive as negative charge in close proximity attracts and repels an external charge equally, and thereby cancels its own ability to exert a net force. It behaves as though it had no charge at all and is electrically **neutral**. That ability to combine charges to produce a null response allows us

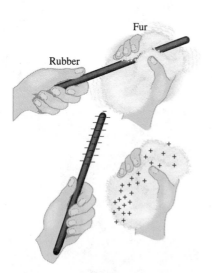

Fur

Rubber

Figure 17.1 Charging by contact. Electrons are transferred from fur to rod. The hard-rubber rod becomes negatively charged, the fur becomes positively charged.

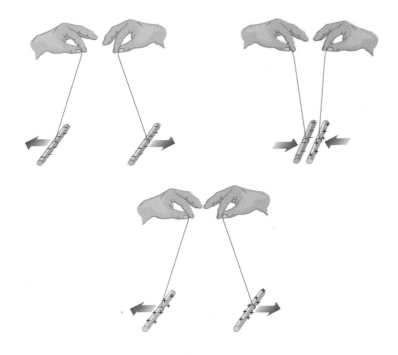

Figure 17.2 Attraction and repulsion of charges. Like charges repel; unlike charges attract.

to create a kind of electrical algebra: 10 units of positive charge and 10 units of negative charge add up to a net of zero units of charge, in the sense that together they produce no observable external electric force. Any macroscopic object, yourself included, contains a vast number of minute individual charges, but on the whole it is usually neutral. That's why you are not noticeably pulled on by the electric force as you are by the gravitational force.

17.2 Charge Is Quantized and Conserved

Charge is a property of most of the subatomic entities that make up the material Universe. An electron is negatively charged; it repels other electrons and it attracts protons. The latter are positively charged. There is no way to discharge an electron, no way to peel off its charge and make it naked, neutral. If the electron is truly a fundamental particle, and most physicists believe it is, charge is an inseparable aspect of the thing in itself.

The amount of the charge of the electron, q_e, has been determined experimentally. Moreover, all electric charge comes in whole number multiples of that basic amount (the numerical value of which we will deal with shortly). Whether positive or negative, **charge is quantized**, it appears in certain specific amounts—every subatomic entity that has been observed to date has had a charge of either 0, $\pm q_e$, or $\pm 2q_e$. Since matter on the subatomic scale comes in specific lumps—only a small number of fundamental particles exist—it's not surprising that charge comes in lumps as well. Still, it is curious that the proton has a charge ($+q_e$) equal in size to that of the electron ($-q_e$), since these particles are otherwise quite different. Modern experiments have shown that the magnitudes of the charges of the electron and proton are so nearly alike that if their ratio does differ from 1 at all, it does so by less than 10^{-20}. That's fascinating, especially since the proton seems to have a complex structure. Contemporary theory maintains that most heavy subatomic particles are actually composite systems (Ch. 33) made up of several varieties of smaller fun-

Electricity is of two kinds, positive and negative. The difference is, I presume, that one comes a little more expensive, but is more durable; the other is a cheaper thing, but the moths get into it.

STEPHEN LEACOCK (1869–1944)
Canadian humorist

This old AM radio represents so much practical physics that we'll come back to it several times throughout our study of Electromagnetic Theory.

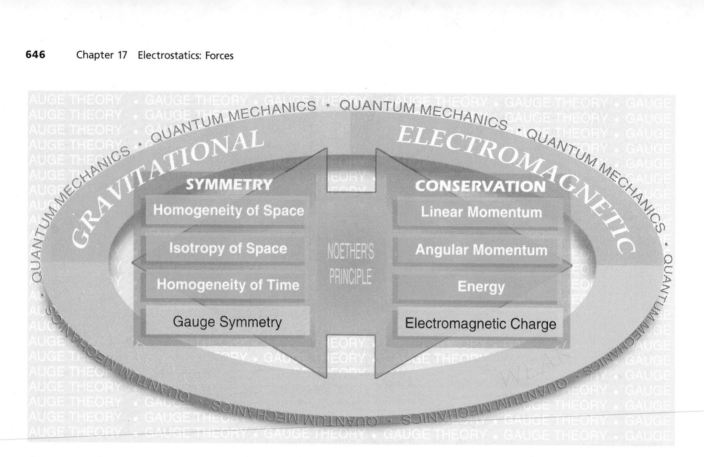

Figure 17.3 This chapter is the first of several that deal with the effects of the electromagnetic force. Thus far, we have recognized Conservation of Linear Momentum, Angular Momentum, and Energy. Now we add Conservation of Electromagnetic Charge. Later, on p. 700, we will relate charge conservation to symmetry.

damental entities called *quarks* (Fig. 1.2, p. 6). These, in turn, are supposed to have charges of $\pm\frac{1}{3}q_e$ and $\pm\frac{2}{3}q_e$. It is believed, however, that quarks cannot ordinarily exist in the free state (that is, independently), and so the observable unit of charge is indeed q_e.

A neutral "chargeless" object is electrified by either losing or gaining charge; the glass rod loses electrons, the silk used to rub it gains them. Unlike mass (which can be converted into energy, Ch. 29), electromagnetic charge is scrupulously conserved. The net charge (the difference between the amount of positive and negative charge) within an isolated system is always constant. That concept is the **Law of Conservation of Charge** (Fig. 17.3). Whenever a positive charge appears, somewhere in the vicinity there will appear an equal negative charge. As far as we can tell, the net amount of charge in the Universe is constant, which does not mean charge cannot be created; it just means that, if it is, as much positive as negative will be produced. This constant net charge of the Universe may well equal zero (that's a pretty thought), but clearly no one knows.

17.3 Charging by Rubbing

We are surrounded by charges that form a subtle electrostatic environment that we generally don't pay much attention to. For example, everyone has probably felt the sparks that often punctuate a short shuffle across a carpet on a dry winter's day. By contrast, in the potentially explosive anesthetic-rich environment of a hospital operating room, even a minute spark of this sort could be disastrous. Most of us have stroked plastic foodwrap so it adheres to some container, rubbed a balloon on a shirt to stick it on a wall, used a copy machine, or fussed with clinging clothes—the electrostatic effects of charge are everywhere.

Atoms are neutral; they possess as many electrons as protons. But the outer electrons are the least strongly bound and they are most easily shed. The process whereby these electrons pass from one object to another is not entirely understood.

TABLE 17.1   The triboelectric sequence

Asbestos Fur (rabbit) Glass Mica Wool Quartz Fur (cat) Lead Silk Human skin, aluminum Cotton Wood Amber Copper, brass Rubber Sulfur Celluloid India rubber	On contact between any two substances shown in the column, the one appearing above becomes positively charged, the one listed anywhere below it becomes negatively charged.

Electrical action-reaction before *the* **Principia**. *It is commonly believed, that Amber attracts the little Bodies to itself; but the Action is indeed mutual, not more properly belonging to the Amber, than to the Bodies moved, by which it also itself is attracted; . . ."*

MAGALOTTI (1665)
Florentine Academy

Nonetheless, we do have a good overall idea of what's happening. Different materials have different affinities for electrons. When two substances are put in contact, one of them may give up some of its loose electrons while the other draws them into itself. For example, when a sheet of plastic is pressed down onto a metal plate, electrons will be transferred from the donor plastic to the grabber metal. The plastic, having lost electrons, now contains a number of immobile positive ions (atoms missing negative charge) on its surface and, as a whole, has become charged. The positive plastic attracts the negative metal and the two cling to one another.

In much the same way, when a hard-rubber rod is stroked with a piece of fur, the rod draws off electrons, becoming negatively charged, and the donor fur assumes an equal, positive charge. *The rubbing seems to do little beyond increasing the area brought into intimate contact.* Moreover, there are degrees of grabbers, and a substance that can snatch electrons away from one material may well find itself serving as donor to a still more potent grabber. Glass rubbed with asbestos will draw off electrons from that fibrous material, becoming negative, but if stroked with some persistence against silk or flannel, the glass will emerge positively charged, having lost electrons (Fig. 17.4). We can bring some order to the whole business by arbitrarily defining glass-rubbed-on-silk to be *positive,* and then any charged object that is attracted to it is *negative.*

By comparing the behavior of various materials, a listing (Table 17.1) known as the *triboelectric* (*tribo,* meaning friction) *sequence* has been formulated. The materials toward the top of the list tend to lose electrons easily, the ones near the bottom tend to gain them effectively. The farther apart on the list the two are, the more intense the resulting electrification, which is why rabbit fur and hard rubber are still the mainstays of electrostatic demonstrations (asbestos is carcinogenic and should be avoided).

17.4 The Transfer of Charge

A negatively charged object contains an excess of electrons that can move about, somewhat at least, and that repel each other. When such an object is placed in con-

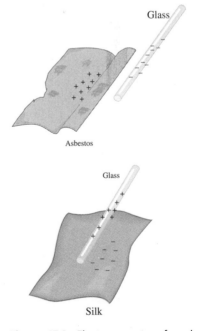

Figure 17.4 Electrons are transferred in contact from the asbestos to the glass, and from the glass to the silk.

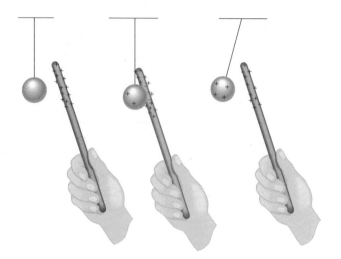

Figure 17.5 Charging a sphere of pith positively by direct transfer of electrons from ball to rod. This leaves the ball positively charged, and it is immediately repelled from the rod.

A new and extremely sensitive technique for revealing fingerprints has been developed at Los Alamos National Laboratory. It makes use of the fact that unlike charges attract. Negatively charged gold particles adhere to the positively charged proteins, which are always left behind when fingers touch a surface.

tact with a neutral body, some of those electrons are forced over onto that body, charging it negatively. Similarly (Fig. 17.5), a positively charged body has a deficiency of electrons or, equivalently, an excess of positive ions. When placed in contact with a neutral body, it attracts and draws off electrons, becoming less positive itself, while causing the now electron-diminished once-neutral body to become positive. Only electrons are transferred, but the system behaves exactly as though positive charge flowed from the charged object into the neutral one. That being the case, it's common to talk about the transfer of positive charge even though no positive charge actually moves.

The arrangement in Fig. 17.2 suggests a straightforward scheme for determining the polarity of any distribution of charge. Thus, suppose we want to find the sign of a charged elephant and don't have a modern electrometer handy. Just suspend two identical, lightweight neutral objects—two Ping-Pong balls, say, or two pith balls.* Now touch a rod of glass-rubbed-on-silk to one of the pith balls, thereby essentially transferring positive charge to it. Next, back the electrified pachyderm into the second pith ball, which will take on some of the unknown elephant-charge. If the two pith balls then attract, the elephant *may* be negative: if they repel, it surely is positive (see Discussion Question 15).

17.5 Insulators and Conductors

Any astute observer might ask, "Is it possible that our elephant got charged only at one end?" The answer is "Yes." If you rub the middle of a plastic comb with a piece of wool, that region will have the ability to pick up tissue tufts, but the comb's ends

*Pith is the pulpy, soft stuff in the center of certain dried plant stems. It was the Styrofoam of generations past, used in work with electricity for centuries and still to be found in physics storage rooms around the world.

will not—they remain neutral, even though the middle is electrified. This behavior represents a class of materials (such as wood, plastic, glass, air, hair, cloth, leather—and dry elephant), all of which are variously known as **insulators, nonconductors**, or **dielectrics**. Charges in a nonconductor have limited mobility and will only move when their mutual repulsion is great enough to overcome the tendency to be held in place by the host atoms. *When an insulator receives a charge, it retains that charge, confining it within the localized region in which it was introduced.**

A **conductor** *allows charge introduced anywhere within it to flow freely and redistribute.* Metals (copper, gold, aluminum, etc.) are among the best conductors at ordinary temperatures. Incidentally, the fact that there are both good and bad conductors of electricity was first realized in 1729 by Stephen Gray, who carried out a series of brilliant experiments, despite the severe handicap of being a pensioner in a London poorhouse.

The distinction between conductors and nonconductors (and it's not always a clear one) arises from the relative mobility of charge within the material. The atoms of metals hold their own outermost electrons weakly and so a bulk sample contains a tremendous number of free electrons, roughly one per atom. Any added electrons join that sea, which moves among the unmoving positive ions. Alternatively, the atoms of a nonconductor hold fast to their own electrons and will even latch on to excess ones introduced on them (Fig. 17.6). Despite this, *no material is a perfect insulator;* all allow some redistribution of charge. Thus, human skin is a respectable nonconductor compared to copper, although it is a fairly decent conductor when compared to glass. Pure water is a modest insulator, but a pinch of some dissolved impurity such as table salt will provide enough ions to turn it into a good conductor (*ion* means *goer* or *traveler*).

Air is a good insulator, particularly dry air, even though it contains some 300 ions per cubic centimeter. Nevertheless, if enough negative charge builds up on an object, electrons under the influence of their mutual repulsion can be propelled into the surrounding gas. The air will have some of its own electrons ripped off, becoming ionized and creating a temporary conductive pathway along which the bulk of the charge then flows. Collisions with the gas increase its temperature and cause some of the atoms to emit light. The result is the familiar glowing trail known as a **spark**.

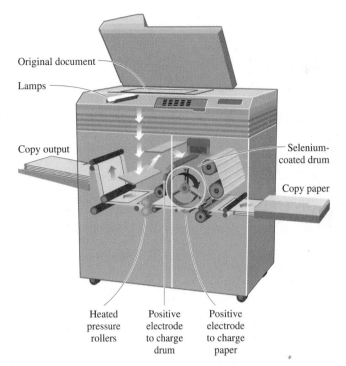

Original document

Lamps

Copy output

Selenium-coated drum

Copy paper

Heated pressure rollers

Positive electrode to charge drum

Positive electrode to charge paper

Figure 17.6 The drum is the centerpiece of both the electrostatic (photo) copier and the laser printer. It's an aluminum cylinder with a coating of photoconducting selenium, which is an insulator in the dark and a conductor in the light. The entire surface is first positively charged, then the image of the document is projected on the drum. Wherever light falls, the charge is conducted away. The dark regions remain positive and attract negatively charged toner powder. A sheet of highly positively charged paper then picks up the toner, to which it is fixed by heated rollers.

Charged Conductors

The free flow of electrons in a conductor is responsible for several interesting properties. First, unlike an insulator, where charge pretty much stays where it's put, a cluster of identical charges introduced on a conductor experiences a mutual repulsion that sends them all scurrying. Constantly pushed apart, they move until they can separate no farther and are as distant from one another as possible (Fig. 17.7). Charges tossed onto a metal sphere will very quickly stream around until they are

---

*The great bane of electrostatic demonstrations is dampness. Warm the objects being charged, work in a dry place, and keep metal, glass, and plastics clean by occasionally wiping them with alcohol. Airborne water molecules are electrically polarized; the hydrogen "Mickey Mouse ears" are positive and tend to attract electrons and so inevitably discharge apparatus.

Here, a bolt of lightning just misses a *Space Shuttle*. Years earlier, a similar bolt destroyed an unmanned rocket.

uniformly distributed and at rest on the outer surface. ***No matter what the shape of the conductor, excess charge always resides on its outer surface***. Charges simply push each other to the very extremities of the object. With a nonspherical conductor the charge distribution will be nonuniform, bunching up somewhat in the remote regions (Fig. 17.8). Each charge is impelled to get as far away from as many others as possible, even if that means getting closer to a few charges in the process. That's why charge tends to concentrate on the sharp protrusions of a conductor, a fact well known to people who worry about sparks.

Another manifestation of the free flow of electrons in a conductor is the manner in which charge is transferred. Envision a negative conductor made to touch an uncharged metal body. Electrons are propelled onto the neutral body by their mutual repulsion, which depends on how densely packed they were to begin with. The charge flows much as a fluid flows from a filled chamber into a connecting empty chamber of arbitrary shape. That gravity-powered flow continues until the liquid levels are the same, the pressures equalize, and equilibrium is reached. As we will see in Sect. 18.7 (p. 696), a very similar balance determines the amount of charge transported. Evidently, if a total excess charge Q is placed on one of two identical metal spheres (Fig. 17.9) and those spheres are brought into contact and then separated, a charge of $\frac{1}{2}Q$ will end up on each of them.

In 1786 Rev. A. Bennet introduced a device that, until the modern era of electronics, was the premier electrostatic indicator. The **gold-leaf electroscope** is basically two extremely thin metal leaves hanging parallel to each other from a conducting wire, all surrounded by a protective glass enclosure (Fig. 17.10). A

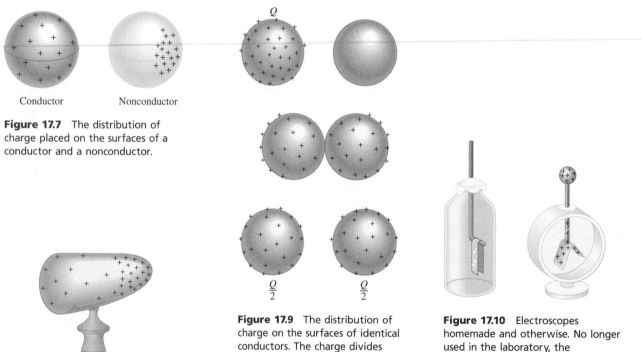

Figure 17.7 The distribution of charge placed on the surfaces of a conductor and a nonconductor.

Figure 17.8 Charge tends to bunch up on the pointed regions of a conductor.

Figure 17.9 The distribution of charge on the surfaces of identical conductors. The charge divides evenly on the two spheres.

Figure 17.10 Electroscopes homemade and otherwise. No longer used in the laboratory, the electroscope is now primarily a teaching device. To make one, remove the thin aluminum foil from the wrapper on a stick of gum. Hang it on a thick wire, lower it into a bottle, and seal the bottle with wax or clay.

charge introduced onto the metal knob on top spreads out; some travels down the wire and divides equally onto each of the leaves. Repelling one another, the leaves spring apart at an angle proportional to the net charge. And there they stay, leaves apart, until the charge is deliberately removed or until it leaks off into the air.

THE ELECTRIC FORCE

In the mid-eighteenth century, the question of the nature of the force acting between charges was a major scientific issue. Naturally, speculation had a decidedly Newtonian bent, relying heavily on discoveries regarding gravity. As early as 1760, Daniel Bernoulli confirmed that his own rather crude experiments were at least consistent with an inverse-square law for electrical attraction and repulsion.

Benjamin Franklin observed that pith balls lowered inside a metal cup appeared to be unaffected by whether the cup was charged or not. We know that excess charge repels itself to the outer extremities of the cup, leaving the inner surface chargeless. That fact is easy to confirm with a *proof-plane,* a small metal disk on an insulating handle (Fig. 17.11). If the proof-plane is touched to the outside of the electrified cup, an electroscope shows that the proof-plane is thereafter charged. Yet touching the inside wall of the cup leaves the proof-plane neutral—***there is no charge on the inside wall of a hollow electrified conductor***. That's clear, but it does not explain why there is no electric force inside the hollow—the pith ball is not attracted to a nearby wall even though there are charges just on the other side of that wall. Franklin communicated these findings to his English friend Joseph Priestley, the discoverer of oxygen.

Priestley made a not altogether surprising guess at the cause, proposing that the force between charges varied inversely with the square of the distance separating them. Newton had already shown (p. 246) that a particle would experience a zero gravitational force when placed inside the cavity of a uniform spherical shell of mass. And Priestley assumed that since gravity varied as $1/r^2$, the electric force did as well. Though he was correct, the analogy is really a poor one: gravitationally, the null effect is a special case where the spherical geometry works to balance the $1/r^2$ drop-off of the force. By contrast, the zero electric force inside a hollow conductor is independent of the shape of the shell, primarily because of the way the free charges distribute themselves on the outer surface.

There had been some direct experimental tests of the inverse-square law by Robison (1769) and later much more completely by Cavendish (1775), but the latter, a rather strange fellow (p. 242), never made his results public. Consequently, by the time Charles Augustin de Coulomb published his definitive study (1785), it was already widely suspected that the electrical force between two *point-charges* varied inversely with the square of their separation.

17.6 Coulomb's Law

Coulomb's apparatus (Fig. 17.12) consisted of a lightweight insulating rod hung horizontally from its middle on a long, fine silver wire. On one end of the rod was a small pith ball covered with gold leaf; on the other, counterbalancing it, was a paper disk (a damping vane). Positioned behind the suspended ball was another identical one fixed in place so that initially the two were in contact. They were then both electrified simultaneously upon being touched by a charged rod. Immediately, the suspended ball swung away from the fixed one, twisting the silver wire, and coming to rest some distance away (Fig. 17.13).

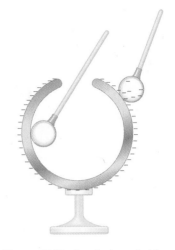

Figure 17.11 Proof-planes inside and outside a hollow charged conductor.

Charles Coulomb (1736–1806).

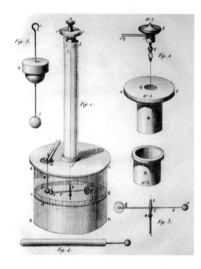

Figure 17.12 C. A. Coulomb's torsion balance (from *Histoire et Mémoires de L'Académie Royale des Sciences,* 1785). A light horizontal rod has a small sphere (*a*) on its left end and a disk counterweight (*g*) on its right end. An identical sphere (*t*) is inserted at the left through a hole (*m*). The arrangement is shown in Fig. 3.

Figure 17.13 A detail of Coulomb's device. Once charged, the two spheres repel and move apart. That behavior twists the wire until the force it exerts matches the electrostatic repulsion.

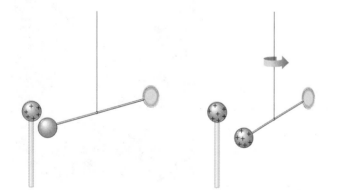

Coulomb had already determined how much force it took to twist the wire through any angle. Now with the movable sphere pushed aside, he slowly twisted the suspension wire by hand, forcing it back closer to its original position near the fixed sphere, against the balls' mutual repulsion. While doing that, he measured the separations and determined the forces holding the two spheres at each location. His results were clear: "The repulsive force between two small spheres charged with the same type of electricity is inversely proportional to the square of the distance between the centers of the two spheres." The electric force (F_E) of interaction between the spheres (and, by Newton's Third Law, the forces are equal in magnitude and opposite in direction even if the charges are unequal) is proportional to $1/r^2$, where r is the center-to-center distance

$$F_E \propto \frac{1}{r^2}$$

just as is the gravitational interaction. Modern experiments confirm that if the power to which r is raised differs from 2, it must deviate by *less* than 3×10^{-16}.

Provided the two interacting spheres are small compared to their separations, we can expect the charge distributions on each of them to remain fairly, though not precisely, uniform. The presence of either sphere causes a shift of charge on the other sphere (that's part of the mechanism of interaction), just as a rock sags when a butterfly lands on it. When the spheres are very small, the repulsive forces between charges on either one of them will be powerful enough to keep the distribution almost uniform, especially if the other disturbing sphere is far away. With this restriction in mind, *we assume uniform charge distributions on each of the interacting spheres.*

Newton showed that a particle of mass, located outside a thin spherical shell, will be drawn gravitationally toward the sphere's center as if all the shell's mass is concentrated at that point (p. 245). In an analogous way, we can conclude that *a uniformly charged conducting sphere produces the same electric force as if all its charge is concentrated at a single point at its center.* The force law can be assumed to be precise for two point-charges separated by a distance r (and nearly so for two small charged spheres).

Coulomb ingeniously determined how the force law varied with the magnitude of the two interacting charges q_1 and q_2. By touching the charged sphere with an identical uncharged sphere (Fig. 17.9), he was able to exactly halve the charge on it, even without knowing what that charge was. In the same way, he could neatly reduce it to a quarter, an eighth, etc., of its initial value simply by repeating the touch-and-halve procedure with two spheres. He found for the two interacting spheres that halving the charge on one halved the force, while halving the charge on both quartered it. Coulomb concluded that the force of interaction was proportional to the product of the charges q_1q_2 and so

$$F_E \propto \frac{q_1q_2}{r^2}$$

The similarity with the gravity formula is obvious and striking, though quite unexplained theoretically even to this day.

The right side of this relationship could be made identical to the left side (not just proportional to it) by defining the unit of charge appropriately: one unit of charge on each sphere would produce one unit of force when the separation was one unit of length. That definition was formulated in the early days in the system that used *centimeters, grams,* and *seconds.* There, the unit of force was the *dyne* (1 dyne $=10^{-5}$ N) and the unit of charge was the now long-forgotten *franklin.* But the force law is really not a very good way to measure charge—the distributions on the spheres are not actually uniform. So nowadays the unit of charge is defined via the more precisely measured interactions of electrical currents (Sect. 21.10). Accordingly, there is no reason for the right side of the above proportionality to equal the left side; in fact, there is no reason to even expect that the units will be equivalent. If charge is defined in some other way, the force law must contain a constant of proportionality (k) that also has units. The electric force between two point charges q_1 and q_2 in some medium (air, vacuum, whatever), separated by a distance r is

$$F_{\mathrm{E}} = k\frac{q_1 q_2}{r^2} \qquad (17.1)$$

The force acts along the line connecting the charges; like charges repel whereas unlike charges attract. This is the modern formulation of **Coulomb's Law.** The internationally accepted (SI) unit of charge is the *coulomb* (C), and k is a constant specific to the medium in which the charges are imbedded. In the special case of vacuum $k = k_0 = 8.98755179 \times 10^9$ N·m^2/C$^2 \approx 9.0 \times 10^9$ N·m^2/C^2. The value of k in air is smaller than k_0 by only 0.06%, which is generally quite negligible. When the charges are imbedded in other media such as water, glass, or oil, we will have to use appropriate constants for each. For example, the value of k for water is measured to be about 1/80 that for vacuum, making the forces between charges that much smaller, but we need not worry about that here.

Until 1983, k_0 was determined by measurements, much as is the constant G. For reasons we will see later, k_0 turns out to be numerically precisely equal to 10^{-7} times the speed of light (c) in vacuum squared, and since c is now fixed by definition (p. 10), so too is k_0. Notice that the units of k are whatever it takes to cancel out all the units on the right of Eq. (17.1) and still leave N, the unit of force.

Two small spheres, each carrying 1 coulomb of charge and separated in vacuum by 1 meter, will each experience a force of about 9×10^9 N. That is a tremendous force (equal to the weight of more than 2400 jumbo jets), and it arises because a coulomb is a tremendous amount of charge. Since the best modern experimental value of the ***charge of the electron*** is $-1.60217733 \times 10^{-19}$ C (within an error of $\pm 0.00000049 \times 10^{-19}$ C), 1 coulomb corresponds to a charge of more than six million million million electrons.

Example 17.1 A hydrogen atom consists of an electron of mass $m_{\mathrm{e}} = 9.1094 \times 10^{-31}$ kg, moving about a proton of mass $m_{\mathrm{p}} = 1.6726 \times 10^{-27}$ kg at an average distance of 0.53×10^{-10} m. Determine the ratio of the electrical and gravitational forces acting between the two particles.

Solution: [Given: $\pm q_{\mathrm{e}} = \pm 1.60 \times 10^{-19}$ C, $m_{\mathrm{e}} = 9.1094 \times 10^{-31}$ kg, $m_{\mathrm{p}} = 1.6726 \times 10^{-27}$, and $r = 0.53 \times 10^{-10}$ m. Find: F_{E} and F_{G}.] From Coulomb's Law

$$F_{\mathrm{E}} = k\frac{q_{\mathrm{e}} q_{\mathrm{p}}}{r^2}$$

$$F_{\mathrm{E}} = (9.0 \times 10^9 \text{ N·m}^2/\text{C}^2)$$

$$\times \frac{(-1.60 \times 10^{-19} \text{ C})(+1.60 \times 10^{-19} \text{ C})}{(0.53 \times 10^{-10} \text{ m})^2}$$

and $\qquad F_{\mathrm{E}} = -8.2 \times 10^{-8}$ N

The minus sign here simply indicates that the force is

(continued)

(continued)

attractive, and that usage is widely adhered to in electrostatics. The sign of the force tells us only if it is attractive ($-$) or repulsive ($+$). The force on the proton due to the electron is equal in magnitude and opposite in direction to the force on the electron due to the proton, but each equals -8.2×10^{-8} N. Had we introduced this usage earlier, the Law of Universal Gravitation [Eq. (7.2)] would have had a minus sign in front of it since it's always attractive and the masses are always positive; hence

$$F_G = -G\frac{m_p m_e}{r^2} = -(6.67 \times 10^{-11} \text{ N·m}^2/\text{kg}^2)$$

$$\times \frac{(1.67 \times 10^{-27} \text{ kg})(9.11 \times 10^{-31} \text{ kg})}{(0.53 \times 10^{-10} \text{ m})^2}$$

$$F_G = -3.6 \times 10^{-47} \text{ N}$$

And so $\boxed{F_E/F_G = 2.3 \times 10^{39}}$; the electrical force is far stronger than the gravitational force.

▶ **Quick Check:** $F_E/F_G = kq_e q_p/Gm_e m_p = 2.3 \times 10^{39}$.

The electrostatic generator charges each strand of this young woman's hair; thereafter they separate as far as possible.

By human standards, the coulomb is definitely a great deal of charge. We are accustomed to taking our charge in far smaller doses, usually in the form of sparks that carry much less than a microcoulomb (1 μC = 10^{-6} C). Usually rubbing will build up charge on an ordinary-sized object with a density of up to 10 nanocoulombs per square centimeter (1 nC = 10^{-9} C), which is a practical limit beyond which there will tend to be discharging into the air.

The electrical force is a vector quantity. It has been confirmed experimentally that when several charges are present, each exerts a force given by Eq. (17.1) on every other charge. *The interaction between any two charges is independent of the presence of all other charges.* Thus, *the net force on any one charge is the vector sum of all the forces exerted on it due to each of the other charges interacting with it independently.* Only in the special case where the charges are arrayed along a straight line will the force vectors be colinear and can they be added or subtracted as scalars (provided we keep track of their directions).

Example 17.2 Figure 17.14a shows three tiny uniformly charged spheres. Determine the net force on the middle sphere due to the other two.

Solution: [Given: $q_1 = +5.0$ μC, $q_2 = -4.0$ μC, $q_3 = +10.0$ μC, $r_{12} = 2.0$ cm, and $r_{23} = 6.0$ cm. Find: F_2.] Watch out for units—we want everything in SI. The net force on charge 2 is the vector sum of the force exerted on 2 by 1, namely, $\mathbf{F}_{21}$; and the force exerted on 2 by 3, namely $\mathbf{F}_{23}$:

$$\mathbf{F}_2 = \mathbf{F}_{21} + \mathbf{F}_{23}$$

The next step is to sketch the problem out pictorially showing the directions of the forces, as is done in Fig. 17.14b. Because the charges have opposite polarities, the forces are both attractive and act in opposite directions. Coulomb's Law provides the numerical values of those forces; accordingly

$$F_{21} = k\frac{q_1 q_2}{r_{21}^2} = (9.0 \times 10^9 \text{ N·m}^2/\text{C}^2)$$

$$\times \frac{(+5.0 \times 10^{-6} \text{ C})(-4.0 \times 10^{-6} \text{ C})}{(2.0 \times 10^{-2} \text{ m})^2} = -450 \text{ N}$$

(continued)

(continued)

and

$$F_{23} = k\frac{q_2 q_3}{r_{23}^2} = (9.0 \times 10^9 \text{ N·m}^2/\text{C}^2)$$

$$\times \frac{(-4.0 \times 10^{-6} \text{ C})(+10.0 \times 10^{-6} \text{ C})}{(6.0 \times 10^{-2} \text{ m})^2} = -100 \text{ N}$$

These minus signs (attraction) have been integrated into the solution via the directions of the forces in the diagram and are no longer of any concern—forget about them! The two vectors in Fig. 17.14c must now be added together. As ever, when treating colinear vectors, we take the direction of the x-axis to be positive. Hence

$$F_2 = F_{21} + F_{23} = (-450 \text{ N}) + (+100 \text{ N}) = -350 \text{ N}$$

The resulting force is $\boxed{3.5 \times 10^2 \text{ N acting to the left}}$.

▶ **Quick Check:** From Coulomb's Law, $F_{23} = (2/3^2)F_{21}$; hence, $F_2 = (7/9)F_{21} = (7/9)(-450)$.

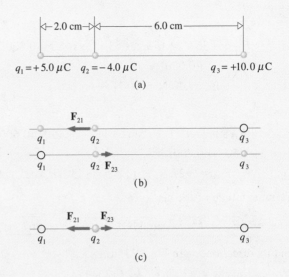

Figure 17.14 Opposite charges attract; like charges repel. Notice how strongly the distance $(1/r^2)$ affects the force: $F_{23} < F_{21}$, even though $q_3 > q_1$.

In theory, if we know the details of some charge distribution, no matter how complicated, we can compute the net force it exerts on a single external charge q. Practically, the summation of all the individual Coulomb interactions can be accomplished either via calculus (approximating the situation as if the charge were distributed continuously) or, more directly, using a computer to carry out the large number of separate applications of Eq. (17.1) that would be necessary. Our analysis initially focuses on the simplest cases of just a few charges.

Example 17.3 Figure 17.15a depicts three small charged spheres at the vertices of a 3-4-5 right triangle. Calculate the force exerted on q_3 by the other two charges.

Solution: [Given: $q_1 = +50$ μC, $q_2 = -80$ μC, $q_3 = +10$ μC, $r_{12} = 50$ cm, $r_{13} = 30$ cm, and $r_{23} = 40$ cm. Find: F_3.] To start, determine if the forces are attractive or repulsive and then draw them acting center-to-center on q_3, as in Fig. 17.15b. Since these two force vectors are not colinear, we must, as usual, resolve them into perpendicular components and sum those individually. Figure 17.15c provides the appropriate geometry. But first we need to compute the magnitudes of the two force vectors; consequently

$$F_{31} = k\frac{q_1 q_3}{r_{31}^2} = (9.0 \times 10^9 \text{ N·m}^2/\text{C}^2)$$

$$\times \frac{(+50 \times 10^{-6} \text{ C})(+10 \times 10^{-6} \text{ C})}{(30 \times 10^{-2} \text{ m})^2} = +50 \text{ N}$$

which is positive because like charges repel. Furthermore

$$F_{32} = k\frac{q_2 q_3}{r_{32}^2} = (9.0 \times 10^9 \text{ N·m}^2/\text{C}^2)$$

$$\times \frac{(-80 \times 10^{-6} \text{ C})(+10 \times 10^{-6} \text{ C})}{(40 \times 10^{-2} \text{ m})^2} = -45 \text{ N}$$

which is negative because unlike charges attract. Hence, from Fig. 17.15c, forgetting the attractive-minus sign,

(continued)

(continued)
we have

$$F_{3x} = F_{32} \cos 36.9° + F_{31} \cos 53.1° = (45 \text{ N})(0.800)$$
$$+ (50 \text{ N})(0.600)$$

and
$$F_{3x} = 36 \text{ N} + 30 \text{ N} = 66 \text{ N}$$

whereas

$$F_{3y} = -F_{32} \sin 36.9° + F_{31} \sin 53.1° = -(45 \text{ N})(0.600)$$
$$+ (50 \text{ N})(0.800)$$

and
$$F_{3y} = -27 \text{ N} + 40 \text{ N} = +13 \text{ N}$$

The magnitude of the net force is

$$F_3 = \sqrt{F_{3x}^2 + F_{3y}^2} = \sqrt{(66 \text{ N})^2 + (13 \text{ N})^2} = \boxed{67 \text{ N}}$$

and its direction, as in Fig. 17.15d, is

$$\theta = \tan^{-1}\frac{F_{3y}}{F_{3x}} = \tan^{-1}\left(\frac{13 \text{ N}}{66 \text{ N}}\right) = \boxed{11°}$$

▶ **Quick Check:** There is no different easy way to arrive at the above results, but since the force between two $\pm 10 \ \mu\text{C}$ charges 10 cm apart is $\approx \pm 10$ N, these values are at least the right order-of-magnitude. Repelled by q_1 and attracted to q_2, the force on q_3 must be in either the first or fourth quadrants.

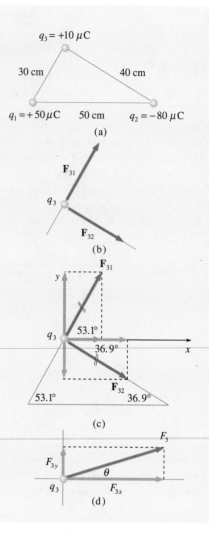

Figure 17.15 The force F_3 on a charge q_3 due to two other charges q_1 and q_2.

Figure 17.16 Inducing a charge on an electroscope. The negative rod repels electrons down into the leaves. The two equally charged leaves repel each other and move apart.

Storing even 1 coulomb is a formidable task, and yet the Earth appears to be carrying a horrendous charge of roughly $-400\,000$ C. In fact, in storm-free regions, the ground leaks about 1500 C every second to the atmosphere. That flow is returned by lightning bolts (up to 20 C each). This ability of the Earth to store vast quantities of charge (because of its great size) makes it an ideal dumping ground for our excess charges. Any conductor in good contact with the Earth (like the water pipes in a building) will carry off all the charge one could possibly want to get rid of. That is precisely what is meant by the phrase to **ground** something (or in Britain, to "earth" it). Incidentally, the pictorial symbol for ground is ⏚.

17.7 Electrostatic Induction

It is not necessary for a charged object to physically touch an electroscope in order that the leaves respond to its presence (Fig. 17.16). A negatively charged object, such as a hard-rubber wand stroked with fur, located anywhere near the top of the electroscope will repel free electrons within the conducting knob and support wire.

These will be forced down into the leaves, which will spring apart and stay that way as long as the charged object remains nearby to maintain the imbalance. Figure 17.17 is a closeup of what happens in a metal under these conditions. The nearer the wand comes, the greater the Coulomb repulsion, and the more electrons enter the leaves, which makes the leaves more negative, and they spread farther apart. When the wand is removed, the displaced electrons immediately flow back, being mutually repelled by their brethren as well as attracted by the fixed positive ions of the metal (knob, wire, etc.). The leaves hang vertically, and the electroscope, which was neutral in total, reverts to its normally unsegregated charge distribution.

Instead of being transferred to the electroscope, the negative charge on the wand has **induced** a negative charge on the leaves. We could have equally well imagined that the negative wand attracted positive charges toward it, although that is not what happened. Alternatively, a positively charged wand would have attracted electrons upward and thus induced a positive charge on the leaves.

Suppose that the knob is now grounded and the game is repeated, bringing a negatively charged ball nearby (Fig. 17.18). Under the influence of the ball, mobile electrons in the metal (knob, support wire, and leaves), interacting in mutual repulsion, get a lot farther away from each other and the ball by flowing into ground. The scope becomes positively charged, and the leaves stand apart. Moving the ball away allows electrons to return up from ground, and the system relaxes back to its original neutral condition.

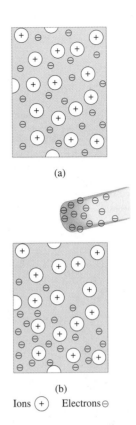

Ions ⊕   Electrons ⊖

Figure 17.17 (a) The conductor is neutral and its electrons are uniformly distributed. (b) A charged rod repels electrons downward. The top of the conductor becomes positive, the bottom negative.

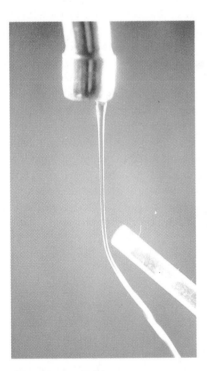

Water is composed of polarized molecules. When a charged rod is brought near the stream, these molecules align themselves so that they are drawn toward the rod. Do you think nonpolar liquids such as gasoline will show the same effect?

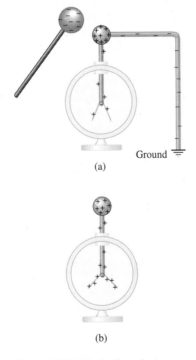

Figure 17.18 By letting electrons flow off to ground, the wire allows the electroscope to become permanently charged.

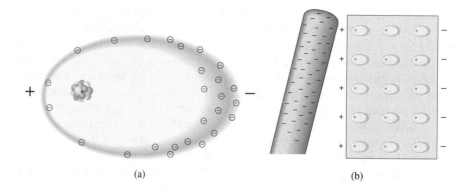

(a)   (b)

Figure 17.19 (a) A polarized atom. Electrons shifted to the right expose the nucleus. In this polarized state, the right end is negative and the left end is positive. (b) An electrically polarized dielectric.

Now bring the ball back and hold it close. The scope is again made positive, the leaves again part, but this time we go one step further and disconnect the ground lead, isolating the device. That done, the scope is shy a clutch of electrons and is more or less permanently positively charged with the leaves standing apart. Again charge was induced—the electroscope was never touched by the ball.

By contrast, electrons within a dielectric, such as a tuft of tissue paper, are far less mobile and ordinarily remain atom-bound. When a negatively charged rod is brought near such a nonconductor, it repels the electron clouds surrounding the atoms and in effect distorts or *polarizes* them (Fig. 17.19). Each nucleus, now slightly exposed, represents the positive end of the elongated atom, and the electron cloud bunched up on the opposite side constitutes the negative end. The result is that one side of the dielectric is electrified positively, the opposite side negatively. A charge has been induced, and it shows up on the surfaces of the specimen.

This is just what happens when you negatively charge a comb with a piece of wool or your clean hair and then use it to pick up bits of lint or tissue paper (Fig. 17.20). The region of the tissue closest to the comb takes on a positive charge; the region farthest away becomes negative. Since by Coulomb's Law the force drops off with distance squared, the mutual attraction of comb and tissue exceeds their slightly weaker repulsion because of the greater distance to the negative end, and the two objects experience a net tug compelling them toward one another. That's why, on a dry day, a piece of tissue will stick to the face of the cathode-ray tube (CRT) of a computer and why your TV screen is always covered with dust. It's also why a charged balloon sticks to a wall: it induces an opposite surface charge on the wall, and the two cling together until the excess electrons leak off the balloon.

THE ELECTRIC FIELD

Here we are once more, this time wand in hand lifting lint and pulling pith and yet not actually touching either—action-at-a-distance again, the fundamental puzzle of affecting motion without apparent contact. Physicists are still struggling to learn how one speck of matter reaches across the "void" to influence another. Today, the most prevalent theoretical picture is based on the concept of the *field,* which was introduced to help visualize the distribution of forces in the space surrounding a material object (p. 257). Inevitably, it itself became the medium for an explanation of action-at-a-distance.

Just as there is a gravitational force, there is an electromagnetic force; just as we envisioned a gravitational field, we envision an electromagnetic field. A **field of**

Figure 17.20 A negatively charged comb picking up bits of tissue paper.

force exists in a region of space when an appropriate object placed at any point therein experiences a force. We can imagine a gravitational force field surrounding any object of mass m, and in the same way we can imagine an electric force field surrounding an object of charge q.

Consider a small sphere carrying a uniform positive charge. This is the ***primary charge distribution***, and we want to study its influence. As a probe, we use a tiny pith ball on a thread (Fig. 17.21). By tradition, our detector (called a **test-charge**) is always positive. Accordingly, it will be repelled from the charged sphere no matter where it is positioned. At every point in the region surrounding the sphere (or primary charge distribution), the detector (of charge q_0) will sense a force of a specific strength and direction, and so at every point we assign a corresponding force vector (Fig. 17.22). It should be noted that if we actually drew a vector at *every* point, the picture would be solid black and useless; the illustration is a compromise. That, then, is the vector force field associated with this particular charged object. It happens to be radial and uniformly outward in all directions because the object is a uniformly charged sphere.

To appreciate the central creation in all of this—the electromagnetic field—we go back more than 100 years to the man who gave it form, the consummate experimentalist Michael Faraday. In 1812, at the age of 21, Faraday was given free tickets to the evening lectures on chemistry by Sir Humphry Davy at the Royal Institution in London. The young man was so dazzled by the experience that he sent Davy a leather-bound copy of his meticulous notes, accompanied by a request for a job as an assistant. A while later Davy fired his lab man for brawling and, remembering Faraday's flattering gesture, offered him the job of bottle washer. The bright young man accepted and quickly rose from lackey to protégé to rival.

Faraday was the first to introduce a visual representation of the electric force field. His scheme, using so-called *lines-of-force,* is a more convenient alternative to our field of force vectors. Faraday's version is equivalent to merging successive vectors into a continuous line that is tangent to all of them. ***The force experienced by a positive test-charge at any point in space is in the direction tangent to the line of force at that point.*** Notice that the lines-of-force flow radially outward from a positive point-charge, as they do for a uniformly charged sphere (Fig. 17.23).

The lines diverge, getting farther apart as they extend out, just as did the gravitational field, treated earlier. The same number of lines pass out through a small imaginary surrounding sphere centered on the charge as pass through a larger concentric sphere whose surface is more distant (Fig. 17.24). As the area ($A = 4\pi r^2$) of the encompassing sphere increases, the density of lines decreases in proportion to $1/A$ or, equivalently, to $1/r^2$. Happily, Coulomb's Law tells us that the force exerted by a point-charge q on a test-charge q_0 also drops off as $1/r^2$. It follows that the density or concentration of the lines-of-force corresponds to the strength of the force field. Since this is true for a single point-charge, it is true for the superposition of the force fields of many point-charges. ***The more lines drawn in a region—that is, the denser the concentration of lines—the greater the field they represent; the farther apart the lines, the weaker the field***.

To Faraday, these intricate patterns of lines came to stand for an invisible physical reality. For him, the field pervading space became an entity that reached from puller to pulled. An electrified object sends out its field into space, and the pith ball detector immersed within that web interacts with the field in proportion to its own charge. Charge-field-charge—that concept was Faraday's answer to the magic of action-at-a-distance. The field was not only an illustrative device; it was a reality

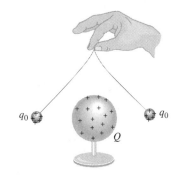

Figure 17.21 A primary charge Q affects a test-charge q_0, repelling it radially.

Figure 17.22 The vector force field surrounding a positive primary charge q.

I can hardly imagine anyone who knows the agreement between observation and calculation, based on action at a distance, to hesitate an instant between this simple and precise action on the one hand and anything so vague and varying as lines of force on the other.

SIR GEORGE B. AIRY (1801–1892)
British astronomer and mathematician

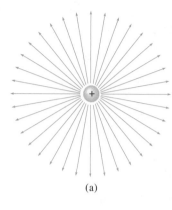

(a)

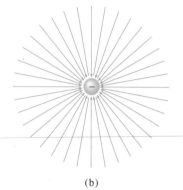

(b)

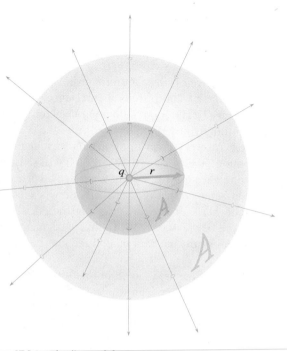

Figure 17.24 The lines-of-force stream through surrounding spheres. The area (A) of a sphere is $4\pi r^2$ and as r increases the density of lines, the number per unit area decreases as $1/r^2$, which is in accord with Coulomb's Law.

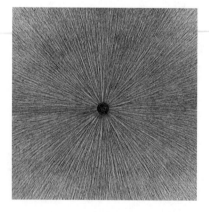

(c)

Figure 17.23 (a) Lines-of-force emanating from a positive point-charge. (b) Lines-of-force converging to a negative point-charge. (c) Fine rayon fibers suspended in oil tend to align themselves when in the vicinity of a charged object. The fiber patterns can be thought of as revealing the lines-of-force.

(though no less mysterious than the ghostly action-at-a-distance it now replaced). After all, what was the field made of, if anything?

One obvious disadvantage of concentrating on force is that its magnitude at every point in space depends not only on the primary charge distribution, but also on the size of the test-charge q_0. What we really want is a map showing the field of a primary body independent of the detector, a map that could be used to compute the force at every point in space when any size charge is placed there. Thus, regardless of its source, we define the **electric field (E)** *at a point in space to be the electric force experienced by a positive test-charge at that point divided by that charge*:

$$\mathbf{E} = \frac{\mathbf{F}}{q_0} \qquad (17.2)$$

Electric field has the SI units of newtons per coulomb (N/C). The tangent to a line-of-force at any point is the direction of the E-field. The custom is to draw the same pattern of lines, but to call them *electric field lines*. Only in the special case when the E-field is uniform (the field lines are parallel) is $\mathbf{E}$ a constant. More commonly $\mathbf{E}$ varies from point to point in space depending on the charge distribution.

Conversely, knowing $\mathbf{E}$ at any location in space (whatever the source) we can calculate the force $\mathbf{F}$ that would arise on any point-charge q placed at that location; accordingly

$$\mathbf{F} = q\mathbf{E} \qquad (17.3)$$

Notice that $\mathbf{F}$ and $\mathbf{E}$ point in the same direction when q is positive.

Example 17.4   At a particular moment the electric field at a point 30 cm above an electric blanket is 250 N/C, straight up. Compute the force that acts on an electron at that location, at that moment.

Solution: [Given: $q_e = -1.60 \times 10^{-19}$ C, $E = 250$ N/C. Find: **F**.]

$$F = q_e E = (-1.60 \times 10^{-19} \text{ C})(250 \text{ N/C})$$

and

$$\boxed{F = -4.01 \times 10^{-17} \text{ N}}$$

The minus sign arises from the charge and tells us that the force on the electron is in the opposite direction to the E-field; namely, downward.

▶ **Quick Check:** $E = F/q = (-4 \times 10^{-17} \text{ N})/ (-1.6 \times 10^{-19} \text{ C}) = 2.5 \times 10^2$ N/C.

The electrically responsive cells in a shark allow it to detect the weak electric fields created by the operation of the muscles of its prey.

Prior to the introduction of electrical technology in the nineteenth century, the strongest electric field most humans were likely to encounter was the static atmospheric field of about 120 N/C to 150 N/C in fair weather, and up to 10 000 N/C during thunderstorms (Table 17.2).

Figure 17.23a shows the direction in which a positive test-charge q_0 would experience a force and so tend to accelerate. It is also a mapping of the electric field of the primary point-charge $+q$. Alternatively, when the primary charge is negative, the field lines converge inward toward it, as in Fig. 17.23b. The special thing about a point-charge source is that we can determine a formula for its E-field using Eq. (17.2). Thus, substituting for F from Coulomb's Law, we obtain

$$E = \frac{F}{q_0} = k \frac{qq_0}{r^2} \frac{1}{q_0}$$

and *the scalar value of the electric field of a point-charge* q is

$$E = k \frac{q}{r^2} \tag{17.4}$$

Michael Faraday (1791–1867) was one of 10 children of a blacksmith in London. As a youngster, with little formal education, Michael was apprenticed to a bookbinder. But he longed to enter "the service of Science" and ultimately became one of the greatest experimentalists of all time.

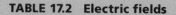

TABLE 17.2   Electric fields

Source	Field strength (N/C)
Background radiation in space	3×10^{-6}
In-house wires	10^{-2}
Radio waves	$\approx 10^{-1}$
Outside an electrified building	$\approx 10^{-1}$
Center of typical living room	≈ 3
In a fluorescent tube	10
30 cm from electric clock	15
30 cm from stereo	90
Laser beam (low power)	10^2
Atmosphere (fair weather)	≈ 150
30 cm from electric blanket	250
Built up by splashing water in a shower	800
Sunlight (average)	10^3
Atmosphere (thunderstorm)	10^4
Van de Graaff accelerator	2×10^6
Breakdown of air	3×10^6
X-ray tube	5×10^6
At cell membrane	10^7
Created by pulsed laser system	5.7×10^{11}
At electron in hydrogen atom	6×10^{11}
Surface of a pulsar	$\approx 10^{14}$
Surface of uranium nucleus	2×10^{21}

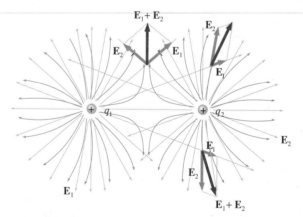

Figure 17.25   The two fields E_1 and E_2 superimpose, coexisting without interacting. The net field E is the vector sum of the two. In this diagram, at each point in space, $E = E_1 + E_2$. Lines of E never intersect one another.

This result is extremely important: it allows us to calculate the **E**-field anywhere in space due to a known distribution of point-charges (electrons, protons, whatever). If there are just a few such charges, the calculation can be done directly by hand. Each charge contributes a field whose strength at any point is given by Eq. (17.4) and whose direction is radially inward (when q is negative) or outward (when q is positive), as in Fig. 17.25. These contributions add vectorially to produce the net **E**.

Example 17.5 Figure 17.26 shows two point-charges (or tiny charged spheres) each of $+10$ nC separated in air by a distance of 8.0 m. Compute the electric field at points A, B, and C.

Solution: [Given: $q_1 = q_2 = +10 \times 10^{-9}$ C. Find: **E** at points A, B, and C.] Starting at A, make a sketch of the layout (Fig. 17.26b) and then draw in vectors for the field $\mathbf{E}_1$ due to q_1 and $\mathbf{E}_2$ due to q_2. To do that, imagine a positive test-charge at A. The force on it due to q_1 acts along the center-to-center line, is repulsive, and so points to the right. That means the field $\mathbf{E}_1$ at A is to the right along the axis. Similarly, the force due to q_2 on our imaginary test-charge is to the left, as is $\mathbf{E}_2$. Next, calculate E_1 and E_2 via Eq. (17.4) and add them vectorially. We are spared that last effort because $E_1 = E_2$ and so the two cancel and the field at A is zero. A test-charge exactly at A would just sit there force-free in unstable equilibrium.

At point B, the two E-fields act as drawn in Fig. 17.26c and we must find their components. First, let's calculate E_1 and E_2. Since the charges and distances happen to be the same, the magnitudes of the two contributing fields are equal; consequently

$$E_1 = E_2 = k\frac{q}{r^2} = (9.0 \times 10^9 \text{ N·m}^2/\text{C}^2)$$

$$\times \frac{(+10 \times 10^{-9} \text{ C})}{(4.0/\sin 45°)^2} = 2.81 \text{ N/C}$$

Now for the vector components. From the geometry, the fields are seen to act at 45°. This time, because the charges are equal and point B is on the midline, the horizontal field components are equal and opposite, and cancel. Only the vertical components contribute

$$E_B = E_1 \sin 45° + E_2 \sin 45° = 2(2.81 \text{ N/C})(0.707)$$

and $\boxed{E_B = 4.0 \text{ N/C}}$

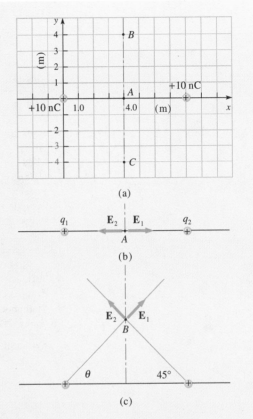

Figure 17.26 The electric fields of two charges q_1 and q_2 at points A, B, and C.

directed straight up in the positive y-direction. The same value applies to point C, where $\boxed{E_C = 4.0 \text{ N/C}}$ directed downward in the negative y-direction.

▶ **Quick Check:** As we will see presently, these results agree with the diagram (Fig. 17.27) for the field of two equal charges. The field of a 1-nC positive charge at 1 m is ≈ 9 N/C, so these answers are the right order-of-magnitude.

17.8 Rationalized Units

An English electrical engineer, Oliver Heaviside, pointed out sometime around 1892 that most of the equations for the E-fields of different charge distributions contained a factor of 4π. He suggested that the constant k in Coulomb's Law be written in terms of $1/4\pi$ and some new constant ε, so that

$$k = \frac{1}{4\pi\varepsilon} \tag{17.5}$$

TABLE 17.3 The permittivity (ε) and relative permittivity* ($\varepsilon/\varepsilon_0$) of some common substances

Substance	Permittivity ($C^2/N \cdot m^2$)	Relative permittivity ($\varepsilon/\varepsilon_0$)
Vacuum	8.85×10^{-12}	1.000 00
Air	8.85×10^{-12}	1.000 54
Body tissue	71×10^{-12}	8
Glass	44×10^{-12}–89×10^{-12}	5–10
Mica	27×10^{-12}–53×10^{-12}	3–6
Nylon	31×10^{-12}	3.5
Paper	18×10^{-12}–35×10^{-12}	2–4
Polyethylene	20×10^{-12}	2.3
Polystyrene	23×10^{-12}	2.6
Rubber	18×10^{-12}–27×10^{-12}	2–3
Silicone oil	19×10^{-12}–25×10^{-12}	2.2–2.8
Sodium chloride	50×10^{-12}	5.6
Teflon	19×10^{-12}	2.1
Ethanol (25°C)	2.2×10^{-10}	24.3
Methanol (20°C)	3.0×10^{-10}	33.6
Water (20°C)	7.1×10^{-10}	80

*Also called the *dielectric constant*.

This would have the effect of causing 4π to appear in expressions for the E-field only in situations where the charge distribution was spherically symmetric, and 2π to appear when it was axially symmetric. Since this was a purely formal device, the system of units it engendered was said to be *rationalized*. The SI system embraced the practice, and in 1960 it became official. The constant ε is called the **permittivity** (from the Latin *permittere,* to let go through). In the case of vacuum, $k_0 = 1/4\pi\varepsilon_0$, where ε_0 is referred to as the **permittivity of free space**. It follows from the value of k_0 that

$$\varepsilon_0 = 8.854\,187\,8 \times 10^{-12} \ C^2/N \cdot m^2$$

Table 17.3 lists the permittivities of several common materials. It also provides the corresponding **relative permittivity** ($\varepsilon/\varepsilon_0$) for each. Often called the **dielectric constant** (K_e), this ratio is unitless and applies to all systems of units. In the early days when there were several competing systems, tabulating ratios was often the only sensible way to present experimental data. The users simply multiplied K_e by the appropriate value of ε_0 in the system of units they happened to be working with.

Example 17.6 A point-charge of 10 μC is surrounded by water with a dielectric constant of 80. Calculate the magnitude of the electric field 20 cm away.

Solution: [Given: $q = 10$ μC, $r = 0.20$ m and $K_e = 80$. Find: E.] The field of a point-charge is $E = kq/r^2 = q/4\pi\varepsilon r^2$, where $\varepsilon = K_e\varepsilon_0$. Hence

(continued)

(continued)

$$E = \frac{10 \times 10^{-6} \text{ C}}{4\pi 80(8.85 \times 10^{-12} \text{ C}^2/\text{N·m}^2)r^2}$$

and

$$\boxed{E = 28 \text{ kN/C}}$$

▶ **Quick Check:** In vacuum $E \approx (9 \times 10^4/0.04)$ N/C $\approx 2.3 \times 10^6$ N/C. With water, it will be a factor of 80 smaller; that is, $E \approx 28$ kN/C.

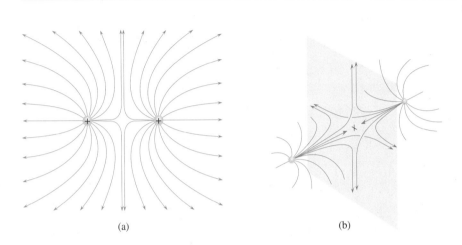

(a)

(b)

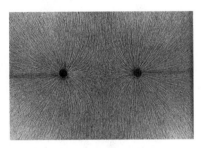

(c)

Figure 17.27 (a) The electric field lines due to two positive charges of equal value. (b) The E-field is three-dimensional. (c) Rayon fibers suspended in oil align with electric field lines.

17.9 Field Lines

Because electric field lines diverge away from a positive point-charge and converge in toward a negative point-charge, we say the former is the *source* of the field and the latter the *sink*. *In electrostatic situations, field lines always begin on positive charge and end on negative charge*. Both ends of the field lines are not always pictured, but they terminate somewhere on a charge (Fig. 17.28), even if that charge is on the walls of the room or beyond.

Figure 17.27 shows the E-field of two equal positive point-charges (each q). The pattern is three-dimensional (Fig 17.27b), and it actually corresponds to the picture we would get if we rotated the drawing (Fig. 17.27a) about the line connecting the charges. As was found analytically (Fig. 17.26), the field is zero at the very center. Very far from the pair (at a distance much greater than their separation), the field will appear as if it were due to a single positive charge ($2q$).

Lines of the net field never cross. If they did, the field would have two different values at that point, which is silly since two superimposed fields combine to yield a single net field. A positive test-charge at that point must experience a single net force, and that is the direction of **E**.

Figure 17.29 shows a positive and negative charge of the same magnitude, a configuration that is special enough to have its own name—it's called a **dipole**. Because the charges have opposite signs and equal magnitudes, the field is not zero anywhere nearby (as it was in Fig. 17.27). Still, at greater and greater distances, the lines get increasingly sparse, and the field rapidly decreases. It must, since as seen from very far away the charges will essentially coalesce and cancel. When the charges are unequal, as in Fig. 17.30, at very great distances the field should resemble the field associated with a single net charge.

Imagine two large, parallel horizontal metal plates each equally charged—the upper one positive, the lower negative (Fig. 17.31). The charge distributes itself more or less uniformly over the inner faces of the plates; the mutual attraction draws most

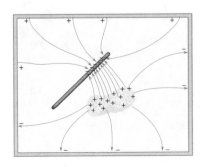

Figure 17.28 A hard-rubber rod rubbed on fur. Many of the field lines begin or end on the walls. *The net enclosed charge is zero.*

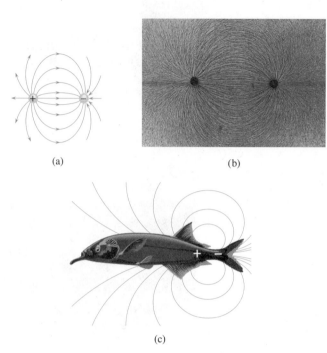

(a)

(b)

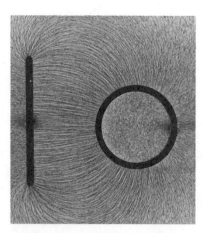

(c)

Figure 17.29 (a) A dipole field set up by two equal opposite charges. (b) Rayon fibers suspended in oil reveal the pattern of electric field lines. (c) The Elephant Gnathonemus produces a dipole electric field and detects nearby objects by their effects on that field.

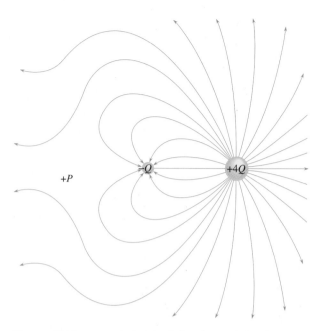

Figure 17.30 A rough sketch of the electric field of two charges $+4Q$ and $-Q$. Notice that the field is zero at point *P*.

The external *E*-field is perpendicular to a conductor. Moreover, there is no field inside a conductor unless it surrounds a net charge.

of the opposite charges as close together as possible (leaving just a few stray charges on the outer surfaces). A positive test-charge placed anywhere in the gap (away from the ends) would be repelled from the top plate, attracted to the bottom plate, and down the charge would go. The *E*-field between the plates is down and *uniform,* the field lines are parallel, and *E* is constant. This is the most convenient way to produce a uniform field and so is used in many devices, including cathode-ray tubes.

17.10 Conductors and Fields

Imagine a neutral conductor, solid or hollow, in a field-free region of space. Now suppose we add some electrons to it, either on the surface or inside. The mutual repulsion between free charges causes them to redistribute. The charges are initially bunched up, there is a transient *E*-field, and very quickly they move apart, coming to rest on the outer surface. The process generally only takes a fraction of a second, depending on the physical details of the conductor. Once settled, the distribution is such that each free charge experiences a zero net force—if it did not, the charge would accelerate until it could move no more, and equilibrium would ultimately be established. If none of the charges experiences a net electric force, none of them is in an electric field. Since field lines either begin or end on charge, the field could only extend inside the conductor if there were a remaining excess of free charges there and that's impossible. ***The electrostatic field inside a charged conductor, anywhere beneath the surface, is zero*** (provided it does not encompass a space in which there is an isolated charge). Faraday dramatically proved the point by constructing a room within a room, covering the inner enclosure with tinfoil. He sat inside this *Faraday cage,* as it has come to be called, with an electroscope at hand, while the entire structure was charged by an electrostatic generator—no field could be detected inside, even while sparks were flying outside.

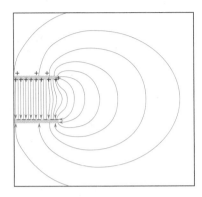

(a)

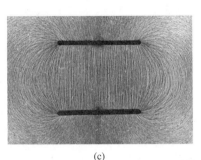

(b)

(c)

Figure 17.31 (a) The E-field of a parallel plate capacitor. (b) A closeup of edge effects. (c) The field as illustrated by rayon fibers suspended in oil.

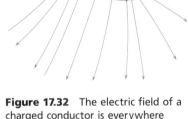

Figure 17.32 The electric field of a charged conductor is everywhere perpendicular to the surface. Charge concentrates on regions with a small radius of curvature.

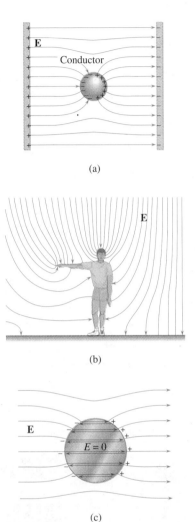

(a)

(b)

(c)

Figure 17.33 Neutral conductors immersed in an applied E-field. (a) A conducting sphere in the uniform field of a parallel plate capacitor. (b) A person in the E-field of the out-of-doors environment. The presence of a conductor distorts the field. (c) The external field polarizes the conductor, making one side positive and the other negative, which creates a self-field that cancels the applied field and leaves a zero internal electric field.

As for the electrostatic field external to the body of a conductor—*it must always be perpendicular to the surface*. It does not matter whether we are talking about a charged object (Fig. 17.32) and its own field or a neutral conductor in an external field (Fig. 17.33). If the field were not perpendicular, it would have a component parallel to the surface. There would be forces on the free surface charges, and they would move. Static equilibrium, which is ordinarily observed to set in quickly, can only be reestablished when there are no tangential field components.

Unlike gravity, we can shield against electric fields simply by surrounding the region to be isolated with a closed conductor. This is true as well in the case of non-static fields because the electrons can redistribute themselves very quickly. Electronic components are often encased in metal cans, and wires (for example, the leads for your stereo amplifier and cable TV) are surrounded by braided copper sheathing to keep out stray electric fields. If a portable radio is placed in a closed metal pot, the E-field signal will not be able to reach the antenna. That's why a car radio's reception is so poor in a metal-encased tunnel or on a steel bridge.

The electrostatic field within the material of a conductor is always zero, but it is possible to have a field in the hollow of a conductor if we put an isolated charge in-

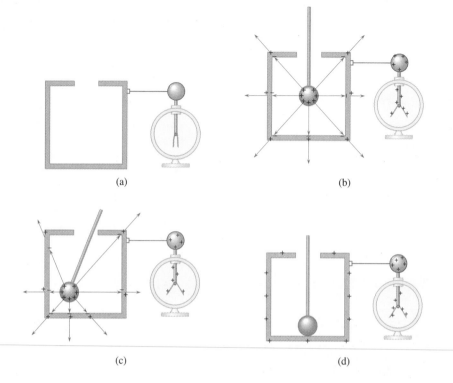

(a)

(b)

(c)

(d)

Figure 17.34 The Faraday ice-pail experiment. A charge held inside the conductor causes an equal and opposite charge to move to the outer and inner surfaces.

side it. Imagine that we introduce a positively charged ball into a nearly closed conductor attached to an electroscope (Fig. 17.34), just as Faraday did in 1843 using a small ice-pail. Once inside, it doesn't matter where the ball is, the outside will be charged positively, and the leaves of the electroscope will stand apart at an unchanging angle. If the ball touches the inside surface, it discharges, becoming neutral, but the leaves do not move at all—the same charge was induced on the outside as was carried by the ball. Notice that the net charge inside the conductor is zero both before and after the ball touched the inner surface.

During the Cold War, the CIA had the windows of the United States Embassy in Moscow screened with wire mesh and the rooms covered with metallic wallpaper. The idea was to keep out the intense electric fields (microwaves) being beamed at the building by their counterparts in the Soviet secret service. The United States had been scanning much of Moscow's microwave communications,

A tube in the old radio pictured on p. 645. It's surrounded by a metal shield that reduces interference caused by stray electric fields. Transistors and sensitive integrated circuits are commonly shielded by encasing them in metal.

The man in the Faraday cage on the left sits safely as sparks from a large Van de Graaff generator flash all around him.

using instruments in the Embassy, and now the Russians were jamming the eavesdropping. The field was not static, but the mesh and wallpaper still isolated the staff from the flood of radiation that had already made several of them ill. In effect, the Embassy was turned into a Faraday cage.

17.11 E-fields of Continuous Charge Distributions

The tremendous analytic power of calculus can be used to compute electric fields for a variety of charge distributions, provided the charge can be approximated as continuous. The game is played by dividing a distribution into minuscule charge elements dq and summing the effects of these (force, **E**-field, whatever) via integration. In principle, each element dq must become vanishingly small in dimension and at the same time correspond to a vanishingly small amount of charge—something that only makes sense if the charge is continuous and so endlessly divisible. In actuality, the pointlike electrons and protons are discrete, but there are so many of them (billions upon billions), and they are so close together, that when a charged object is viewed from outside at even a modest distance, the discrete nature of the phenomenon blurs and the charge might just as well be continuous.

Ultimately, we will have to integrate the contributions to the net **E**-field from each element dq, over the spatial extent of the charged body. That means that we must relate dq with the space variables denoting the body. To accomplish this, let's first suppose that the charge is distributed uniformly and define several *charge densities*. If the body is essentially one-dimensional, that is, long, narrow, and thin (like a wire of length l), and carries a net charge Q, the **linear charge density** λ is defined as

$$\lambda = \frac{Q}{l} \tag{17.6}$$

with units of C/m. If the body is two- or three-dimensional and the charge is distributed over its *extended* surface area A, the **surface charge density** σ is defined as

$$\sigma = \frac{Q}{A} \tag{17.7}$$

with units of C/m². Finally, if the charge is distributed throughout a three-dimensional region of volume V, the **volume charge density** ρ is defined as

$$\rho = \frac{Q}{V} \tag{17.8}$$

Even more generally, we can write the differential charge elements in any one of three forms:

$$dq = \lambda \, dl \qquad dq = \sigma \, dA \qquad dq = \rho \, dV$$

where λ, σ, and ρ are usually found from Eqs. (17.6) to (17.8) but might also be functions of position when the charge is located nonuniformly.

All of this is not to say that we can now easily calculate anything we like. However wonderful calculus is, computing electric fields can nonetheless be quite daunting for all but the simplest charge distributions, which have certain symmetries that can be taken advantage of. We *will* be able to determine lots of important field configurations, especially after mastering Gauss's Law, but finding the **E**-field of something as seemingly simple as a charged banana is well beyond the mathematical level of this treatment.

Example 17.7 Figure 17.35 depicts a narrow ring carrying a uniformly distributed net charge Q. Find the electric field it produces at a point P on the central axis an arbitrary distance x from the plane of the ring. [Hint: This is one of those situations that is greatly simplified because of the symmetry of the charge distribution.]

Solution: [Given: a ring-shaped charge distribution, Q. Find: $\mathbf{E}$.] The field of a point-charge is $E = kq/r^2$ and so each charge element produces a tiny contribution $dE = k\,dq/r^2$ at P. It will turn out that the problem simplifies so much that the charge element can be left as dq, but that's a rarity and since we don't know as much at this point, let's go ahead and write $dq = \lambda\,dl$. Use $\lambda = Q/2\pi R$, inasmuch as each segment of the ring is essentially one-dimensional. Because of the symmetry, for every segment on one side of the ring there will be an identical segment directly opposite it, equidistant from P. Each will contribute an identical field component dE_x along the x-axis, and these add to one another. They also each produce a component perpendicular to the x-axis, and these cancel one another. So we need only worry about the x-components. Thus $dE_x =$

$\cos\theta\,dE$ where $\cos\theta = x/r = x/\sqrt{x^2 + R^2}$ and

$$dE_x = \frac{k\,dq}{r^2}\cos\theta = \frac{k\lambda\,dl}{(x^2 + R^2)}\frac{x}{\sqrt{x^2 + R^2}}$$

Now integrate both sides of the equation. At P, x is constant, as is everything else—k, λ, and R—and all of it can be taken out of the integral, leaving only the integral of dl around the ring, which equals the circumference $2\pi R$:

$$E_x = \frac{k\lambda x}{(x^2 + R^2)^{3/2}}\int_{\text{ring}} dl = \left[\frac{k\lambda x}{(x^2 + R^2)^{3/2}}\right]2\pi R$$

Since $\lambda = Q/2\pi R$,

$$E_x = \frac{kQx}{(x^2 + R^2)^{3/2}}$$

▶ **Quick Check:** For P very far away $x \gg R$; $(x^2 + R^2)^{3/2} \to (x^2)^{3/2} = x^3$ and $E \to kQ/x^2$, which is the field of a point charge Q. And that's what we would expect since the ring would indeed look like a point from very far away. (See Problem 27 for an alternative solution.)

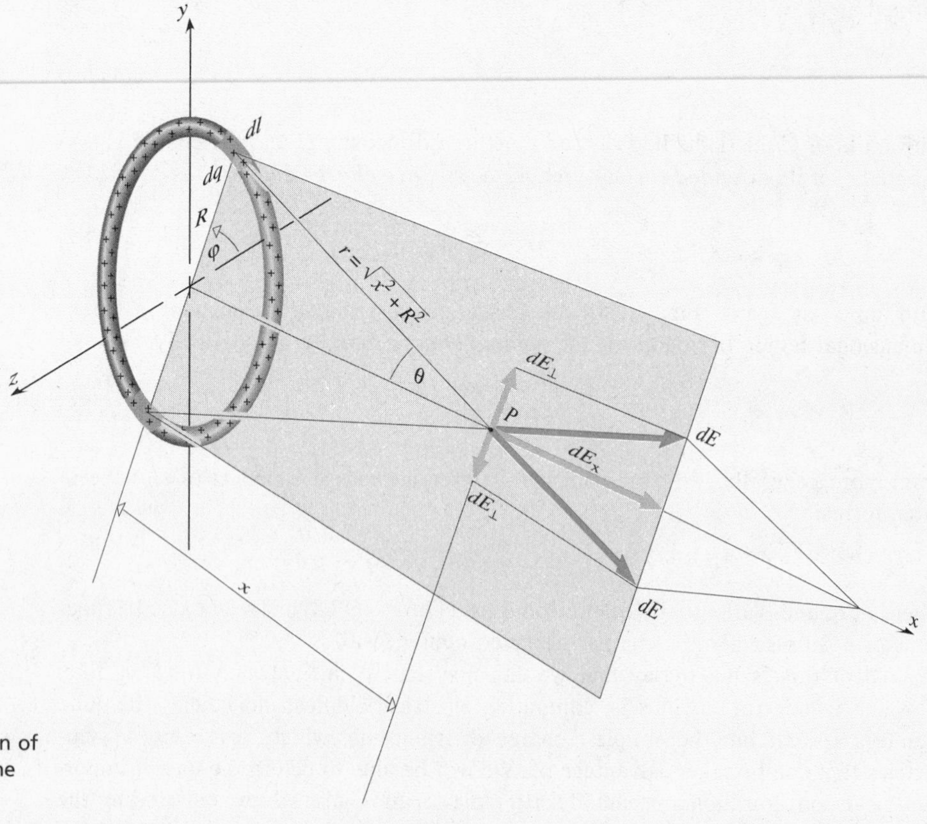

Figure 17.35 The axial E-field produced by a uniform distribution of positive charge distributed over the surface of a narrow ring.

17.12 Gauss's Law

We can determine the E-field of any extended charge distribution directly using Coulomb's Law, but the process is often very difficult. There is another calculational approach that, in certain situations, is relatively simple. It's due to J. K. F. Gauss, the great nineteenth-century German mathematician and astronomer. Gauss's Law is about the relationship between the flux of the electric field and the sources of that flux, charge. The ideas derive from fluid dynamics, where both the concepts of field and flux were introduced. The flow of a fluid, as represented by its velocity field, is depicted via streamlines, much as the electric field is pictured via field lines. Figure 11.30 (p. 417) shows a moving fluid within which there is a region isolated by an imaginary closed surface. The surface is a *tube of flow* bounded by flat area discs (A_1 and A_2) at either end, as pictured in Fig. 17.36. The discharge rate, or *volume flux* (Av), is the volume of fluid flowing past a point in the tube per unit time as given by Eq. (11.13): $J = Av = \Delta V/\Delta t$.

Let's first be a bit more precise about this notion of flux. Consider a fluid flowing perpendicularly through a rectangular wire frame surrounding an area A, as in Fig. 17.36a. When the frame is tilted up at an angle θ, some of the flow misses the frame, passing beneath rather than through it. The frame's effective surface area now corresponds to $A_\perp = A \cos \theta$ with as many flow lines passing through the tilted A as through $A_\perp$. This is a small point, but it's one of the basic notions necessary to really understanding flux; we'll revisit it several times.

As can be seen in Fig. 17.36b, the same amount of fluid passes per second through the different cross-sectional areas A and A' even though $Av \neq A'v$. It therefore makes sense that the flux through either area, perpendicular or not, should be the same and we will have to define it so it is. For the tilted area A', the flux is determined by either the component of the velocity perpendicular to A' times A', that is, $v_\perp A'$, or by v times the component of A' perpendicular to v, that is, $vA'_\perp$. Since $v_\perp = v \cos \theta$ and $A'_\perp = A' \cos \theta$, in either case, the flux is the same, namely, $vA' \cos \theta$.

Consider the tube of flow, that is, the bent pipe-shaped surface, closed at its ends by area caps A_1 and A_2, in Fig. 17.36c. Fluid enters and leaves the tube, and we will need to distinguish the flux-in from the flux-out. Thus, construct area vectors (**A**) normal to the surfaces and pointing *outward*. The flux is then defined as $vA \cos \theta$, where θ (as before) is the angle between **v** and **A**. In other words, $J = \mathbf{v} \cdot \mathbf{A}$. On the input face, the cosine of the angle between **v** and **A** is $\cos 180° = -1$, and on the output side $\cos 0° = +1$; the flux-in is $-v_1 A_1$, the flux-out is $+v_2 A_2$, and no flux enters or leaves through the curved side where the angle between **v** and **A** is everywhere 90°. From the Continuity Equation, the volume flux through both end surfaces is equal in magnitude—what flows in per second, flows

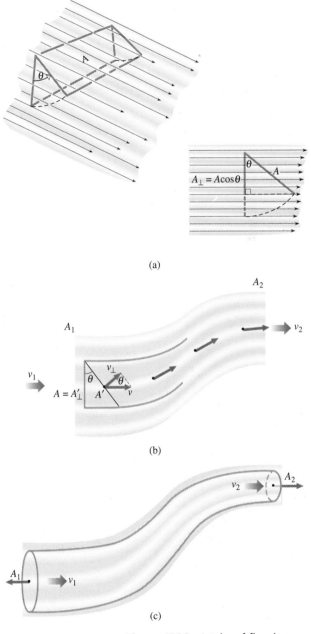

(a)

(b)

(c)

Figure 17.36 A tube of flow in a moving fluid. (a) and (b) The flux of the velocity field is defined as $v_\perp A = vA_\perp$. (c) To compute the total flux into and out of a region of space, the area vectors **A**$_1$ and **A**$_2$ are defined as pointing outward.

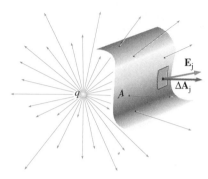

Figure 17.37   An imaginary area A in the E-field of an arbitrary charge q. The flux through the tiny area element is $\mathbf{E}_j \cdot \Delta\mathbf{A}_j$.

out per second. ***The net fluid flux (into and out of the closed area) summed over all the surfaces equals zero***. If, however, a small pipe is inserted into the region either sucking out fluid (a sink) or delivering fluid (a source), the net flux would then be nonzero—a situation we will contend with presently.

To apply these ideas to the electric field, consider an imaginary area A placed in some arbitrary electric field, as depicted in Fig. 17.37. The tiny area element ΔA_j is small enough so that the E-field at that location, let it be E_j, *can be assumed to be constant over its extent*. In accord with the above discussion, we define the **electric flux** $(\Delta\Phi_E)$ through ΔA_j as

$$\Delta\Phi_{Ej} = E_j\,\Delta A_j \cos\theta = E_{\perp j}\,\Delta A_j = \mathbf{E}_j \cdot \Delta\mathbf{A}_j$$

and hence the total electric flux through the entire surface is the sum of all such contributions:

$$\Phi_E = \Sigma\,\Delta\Phi_{Ej} = \Sigma\,E_{\perp j}\,\Delta A_j$$

The assumption that E is constant over each area element improves as $\Delta A_j \to 0$, and the sum thereupon transforms into an integral evaluated over the surface:

$$\Phi_E = \int_{\text{surface}} \mathbf{E} \cdot d\mathbf{A} = \int_{\text{surface}} E_\perp\,dA \tag{17.9}$$

This then is the defining expression for the **electric flux**. The obvious question is: "What does it mean physically?" Fluid flux is the amount (volume) of liquid passing through an area per second; there is an actual flow of something. As for the electric flux, classically speaking, nothing is flowing; the **E**-field is established and static, but it does nonetheless pass through the area. The value of E at any and every point in space is the strength of the field at that point, whereas the flux can be thought of as a measure of the "amount" of field traversing A. About a hundred years ago physicists picturesquely equated the flux with the number of lines of force passing through the area, but that's a rather simplistic view that gives the lines a distinct and separate reality, which is misleading, at best. In Section 17.12 we'll see that contemporary quantum theory maintains that streams of virtual photons are at the heart of the electromagnetic interaction. Yet, physicists have been slow to augment the nineteenth-century continuous field-line imagery, with the twentieth-century virtual-photon flux. Whatever flux *is*, it is a potent idea that will be useful practically, both here and in the study of magentism.

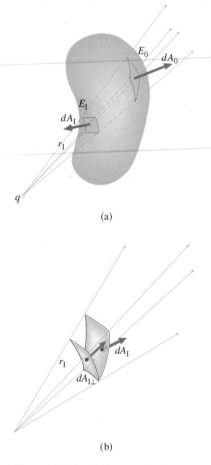

Figure 17.38   (a) Flux passing into and out of a closed surface. (b) The projection of area element dA_I down onto the surface of a sphere produces an element $dA_{I\perp}$ perpendicular to r_I.

Figure 17.38a shows a charge q outside of a *closed* surface. Flux enters on one side through an input area element dA_I and emerges on the other through an output area element dA_O. Notice (Fig. 17.38b) that the projection of dA_I perpendicular to r_I is the element $dA_{I\perp}$, which is part of a sphere. The same flux that traverses $dA_{I\perp}$ also traverses dA_I, and that's true as well for $dA_{O\perp}$ and dA_O. The projected area elements are parts of concentric spheres centered on q. Although the area element increases as the radius squared ($dA_{O\perp} > dA_{I\perp}$), the E-field simultaneously drops off as the radius squared. The result is that the flux-in ($-E_I dA_I$) is equal in magnitude but opposite in sign to the flux-out ($+E_O dA_O$) and they cancel one another. Integrating over the entire *closed* surface, we get the net flux:

$$\Phi_E = \oint \mathbf{E} \cdot d\mathbf{A} = \oint E_\perp\,dA \tag{17.10}$$

which, in this case, equals zero. The $1/r^2$ dependence of E is crucial to this entire analysis—it links Gauss's Law to Coulomb's Law. ***When there are no sources or sinks of the field within the region encompassed by the closed surface, the net flux through the surface equals zero***—that much is a general rule for all such fields.

Example 17.8 A uniform electric field of 2.1 kN/C passes through a rectangular area that measures 22 cm by 28 cm. The field makes an angle of 30° with the normal to the area. Determine the electric flux through the rectangle.

Solution: [Given: area 0.22 m × 0.28 m, $\theta = 30°$, and $E = 2.1$ kN/C. Find: Φ_E.]

$$\Phi_E = EA \cos \theta = (2.1 \text{ kN/C})(0.062 \text{ m}^2)(0.866)$$
$$\Phi_E = 0.11 \text{ kN·m}^2/\text{C}$$

▶ **Quick Check:** Flux has the units of E-field (N/C) times area (m^2), or N·m^2/C, so the units are all right. The area is 0.22 m × 0.28 m = 0.06 m^2 and 0.866 times that is ≈0.05 m^2, so the flux should be ≈1/2 m^2 × 0.21 kN/C.

In order to find out what would happen in the presence of internal sources and sinks, consider a spherical surface of radius r centered on and surrounding a positive point-charge q (Fig. 17.39). The E-field is everywhere outwardly radial and at any distance r is entirely perpendicular to the surface: $E = E_\perp$ and

$$\Phi_E = \oint E_\perp \, dA = \oint E \, dA \qquad (17.11)$$

Moreover, since E is constant over the surface of the sphere, it can be taken out of the integral:

$$\Phi_E = E \oint dA$$

The sum of all the area elements over a sphere of area A equals $4\pi r^2$; that is,

$$\Phi_E = E 4\pi r^2 \qquad (17.12)$$

But we know from Eqs. (17.4) and (17.5) that the point-charge (in a medium of permittivity ε) has an electric field given by

$$E = \frac{1}{4\pi\varepsilon} \frac{q}{r^2}$$

and so Eq. (17.12) becomes

$$\Phi_E = \frac{q}{\varepsilon}$$

This is the electric flux associated with a single point-charge q within the closed surface. Had we used a closed surface more complicated than a sphere, the same kind of argument that was applied to the analysis of Fig. 17.38 would show that the net flux is independent of the shape of the surface—any surface effectively projects down to a sphere because the flux that traverses $dA_{1\perp}$ traverses dA_1.

Since all charge distributions are made up of point-charges, it is reasonable that *the net flux due to a number of charges contained within any closed area is*

$$\Phi_E = \frac{1}{\varepsilon} \Sigma q$$

The sources (positive charges) provide an outwardly directed flux; the sinks (negative charges) draw the flux inward; and the difference, in or out of the surface, is the net flux associated with the charge distribution.

If there are an equal number of positive and negative charges inside the closed surface ($\Sigma q = 0$), the field begins on the positive ones, ends on the negative ones, and no net electric flux emerges (Fig. 17.40). Combining this last equation for Φ_E with Eq. (17.11) yields

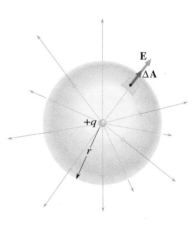

Figure 17.39 A spherical Gaussian surface surrounding a spherical primary charge distribution q at a distance r.

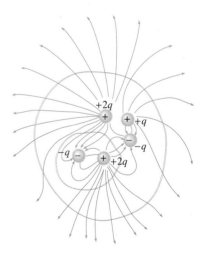

Figure 17.40 A schematic representation of the E-field of a distribution of charge. From far away it should look like the field of $+3q$. Given that there are eight lines per $\pm q$, surround any region by a closed area and subtract the number of lines in from those out.

$$\oint \mathbf{E} \cdot d\mathbf{A} = \frac{1}{\varepsilon}\Sigma q \qquad (17.13)$$

This is **Gauss's Law**, and it turns out to be even more general than Coulomb's Law, since it is applicable at any instant to both stationary and moving charges.

Because of the complexity of carrying out the calculation on the left side of Eq. (17.13), we shall limit our use of it to situations where the imaginary surface, the so-called *Gaussian surface*, leads to especially simple results. Thus, *we will usually construct a closed surface out of faces that are either parallel to the field* ($E_\perp = 0$) *or are perpendicular to the field* ($E_\perp = E$), *in which case the field is then constant over the surface selected.* With conductors, it's often useful to have part of the surface inside the material where $E = 0$. In most cases, we will stick with simple Gaussian surfaces, such as spheres and cylinders.

Example 17.9 A long, straight wire of length L, in air, carries a uniform positive charge Q. Use Gauss's Law to find the electric field at a perpendicular distance R from the wire at a point that is far from the conductor's ends. To operate in a field region free of the complications of end effects, limit the analysis to the domain where $L \gg R$.

Solution: [Given: total charge Q, distance R, and wire length L. Find: E.] First, determine the field configuration with an eye to finding its symmetries. From the facts that the charge is uniform and the field lines must begin on positive charges and be perpendicular to the conductor's surface, we can assume that *the field is radially outward* (Fig. 17.41). Because the wire is very long, the field in the middle region cannot have a component toward either end—either end is equally remote, neither is special; the field must be radial. Considerations like these are the first step in the application of Gauss's Law. Thus, given the cylindrical symmetry of

the field, surround the wire with a cylindrical Gaussian surface of arbitrary radius R and length l, closed with flat end caps of area A_1 and A_3. Apply Eq. (17.13) over the closed imaginary surface made up of A_1, A_2, and A_3, which will involve the field at a distance of R from the wire—just what we are after. Accordingly, processing the left side of the equation yields

$$\oint E_\perp \, dA = E_{\perp 1}A_1 + E_{\perp 2}A_2 + E_{\perp 3}A_3 = \frac{1}{\varepsilon}\Sigma q$$

But $E_{\perp 1}$ and $E_{\perp 3}$ are both zero because the field is parallel to the ends. Since the wire is surrounded by air $\varepsilon \approx \varepsilon_0$. Furthermore, the entire field passes perpendicularly through the curved surface and also has a constant value over that surface; that is, $E = E_{\perp 2}$, hence

$$EA_2 = E(2\pi Rl) = \frac{1}{\varepsilon_0}\Sigma q$$

where $A_2 = 2\pi Rl$. Now for the right side: If the total wire of length L carries a charge Q then the charge per

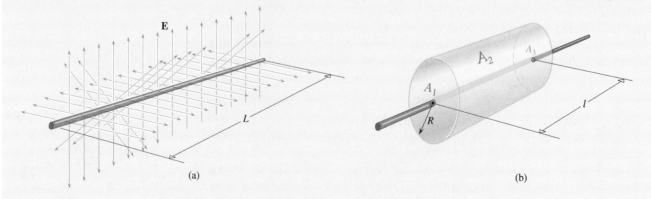

Figure 17.41 (a) The field of a line charge. (b) The corresponding Gaussian surface.

(continued)

(continued)
unit length (λ) is Q/L and the charge inside the Gaussian surface (Σq), which we made to be of length l, is λl. Consequently

$$E(2\pi Rl) = \frac{1}{\varepsilon_0}\lambda l$$

and

[*straight charged wire*] $$\boxed{E = \frac{\lambda}{2\pi R\varepsilon_0}}$$ (17.14)

Since R is unspecified, we really have a formula for E everywhere beyond the wire where the fringing at the ends is negligible.

▶ **Quick Check:** This result should have the units of F/q. Since λ is charge over length, it follows from Coulomb's Law that the units are all right. We can expect that at $R = 0$, $E = \infty$, and at $R = \infty$, $E = 0$, and both are the case. Moreover, the 2π is there because of the cylindrical symmetry.

The $1/R$ dependence of the field determined in Example 17.9 comes from the fact that as the point P moves farther from the wire, more charge contributes effectively to establishing the radial field. That happens because when P is close to the wire only the charges nearby contribute much to E. Charges way down on the wire to the right or left produce fields that are almost parallel to the wire. These are oppositely directed and nearly antiparallel so that much of their contribution cancels in the vicinity of the line of charge. Consequently, as a detector at P moves away, instead of the field falling off as $1/R^2$, it "sees" more and more charge, and the field diminishes more slowly, namely, at $1/R$. Note too that the field is independent of the Gaussian surface used to derive it. That's reasonable since there are an infinite number of possible surfaces that can be used (most of them with great difficulty). Thus l, which related only to the surface, cancels out of the analysis.

As another application of Gauss's Law consider a *very large flat sheet of charge* immersed in a medium having a permittivity ε, as shown in Fig. 17.42. Its E-field (not far from the middle), by symmetry, must be perpendicular, outward, and *uniform*. Thus, a Gaussian surface in the shape of a cylinder with end faces of area $A = A_1 = A_2$ encompasses a charge of σA. As a result

$$\oint E_\perp \, dA = E_{\perp 1}A_1 + E_{\perp 2}A_2 + E_{\perp 3}A_3 = \frac{1}{\varepsilon}\Sigma q$$

and since $E = E_{\perp 1} = E_{\perp 3}$ and $E_{\perp 2} = 0$, we have

$$EA + EA = \frac{\sigma A}{\varepsilon}$$

And, finally, *the E-field of a large sheet of charge is*

[*large sheet of charge*] $$E = \frac{\sigma}{2\varepsilon}$$ (17.15)

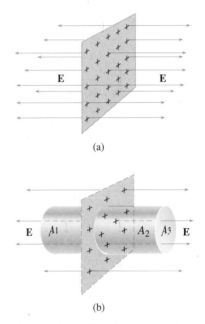

Figure 17.42 (a) A large sheet of charge. (b) The Gaussian surface (a cylinder perpendicular to the sheet) used to calculate **E**.

Figure 17.43 pictures a pair of parallel, equally and oppositely charged metal plates. The two fairly uniform sheets of opposite (excess) charge mutually attract and hold each other in place on the inner faces of the plates. Except for fringing at the edges (Fig. 17.31), the net field everywhere should be the vector sum of the two overlapping *uniform* fields set up by the two opposite sheets of charge. Outside this *parallel plate capacitor*, the two uniform fields are oppositely directed and cancel—ideally, there is no E-field outside the plates. Inside, the two fields, each given by Eq. (17.15), are in the same direction and add: $E = 2(\sigma/2\varepsilon)$ and *the field be-*

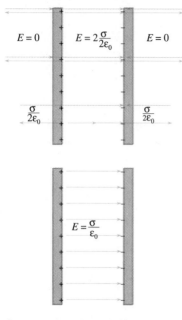

Figure 17.43 The *E*-field between two charged sheets. The ability of a parallel plate configuration to produce a uniform field region is used in all sorts of devices; for example, see Fig. 17.44.

tween the plates of a parallel plate capacitor is

[*charged parallel plate capacitor*]
$$E = \frac{\sigma}{\varepsilon} \qquad (17.16)$$

It's left as Problem 43 to show that this same result follows directly from Gauss's Law.

17.13 Quantum Field Theory

A charged particle interacts with other charged particles. It creates a web of interaction around itself that extends out into space and corresponds to what has been called action-at-a-distance. That imagery leads to the concept of the electric field, which is a representation of the way the electromagnetic interaction reveals itself on a macroscopic level. The static electric field is, in effect, a spatial conception summarizing the interaction among charges. Classically, we say that one charge sets up an *E*-field in space and another charge immersed in that field interacts directly with it, and vice versa. The picture is straightforward, but many questions come to mind. Does the *E*-field have a physical reality? Is anything actually flowing? How does the field produce a force on a charge? Does it take time to exert its influence?

As early as 1905, Einstein already considered the classical equations of Electromagnetic Theory to be descriptions of the average values of the quantities being considered. Classical theory beautifully accounted for everything being measured, but it was oblivious to the exceedingly fine granular structure of the phenomenon. Using thermodynamic arguments, Einstein proposed that electric and magnetic fields were quantized, that they are particulate rather than continuous. After all, classical theory evolved decades before the electron was even discovered. If charge (the fundamental source of electromagnetism) is quantized, shouldn't the theory reflect that in some fundamental way?

Today, we are guided by Quantum Mechanics, a highly mathematical theory that provides tremendous computational and predictive power but is nonetheless disconcertingly abstract. Contemporary physics holds that all fields are quantized; that each of the fundamental Four Forces is mediated by a special kind of field particle. These *messenger particles* are continuously absorbed and emitted by the interacting material particles (electrons, protons, etc.). It is this ongoing exchange that *is* the interaction. The mediating particle of the electric field is the **virtual photon**. This

Figure 17.44 One type of ink jet printer fires charged droplets of ink at the paper. The droplets pass through a set of parallel plates, which carries a charge proportional to the control signal. The resulting *E*-field steers the beam of ink vertically as the paper moves by.

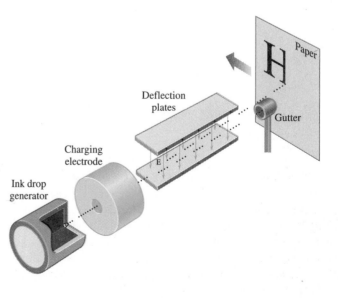

massless messenger travels at the speed of light and transports momentum and energy. When two electrons repel one another, or an electron and proton attract, it is by emitting and absorbing virtual photons and thereby transferring momentum from one to the other, that transfer being a measure of the action of force. The messenger particles of the electromagnetic force are called *virtual* photons because they are bound to the interaction. Virtual photons can never escape to be detected directly by some instrument, however unsettling that is philosophically and however hard that makes it to establish their existence. Indeed, virtual photons exist only as the means of interaction.

On a macroscopic level, messenger particles can manifest themselves as a continuous field provided they can group in very large numbers. Fundamental particles have an intrinsic angular momentum, or *spin* (p. 291), that determines their grouping characteristics. Quantum Theory tells us that the desired field behavior can only occur if forces are mediated by messenger particles having angular momenta equal to integer multiples of $h/2\pi$ (that is, $0, 1h/2\pi, 2h/2\pi, 3h/2\pi, \ldots$). The angular momentum of the virtual photon is $1(h/2\pi)$ and it's therefore referred to as a **spin-1 particle**. For reasons we shall study later, the exceedingly important class of interactions that have spin-1 messengers are known as **gauge forces** and the electromagnetic force is the model for all the gauge forces we shall encounter.

Today, the magic of action-at-a-distance is understood via the no-less mysterious exchange of virtual particles, but at least now there is a highly predictive mathematical theory in place that describes the phenomenon.

Core Material

Charge is the property of matter that gives rise to electric force. *Like charges repel, unlike charges attract.* Charge is conserved. The electrostatic force between two charges is

$$F_E = k\frac{q_1 q_2}{r^2} \qquad [17.1]$$

which is **Coulomb's Law**. The unit of charge is the *coulomb* (C), and k is a constant specific to the medium in which the charges are imbedded. In the special case of vacuum, $k_0 = 8.98755179 \times 10^9$ N·m²/C² $\approx 9.0 \times 10^9$ N·m²/C². The *charge of the electron* is $-1.60217733 \times 10^{-19}$ C. The net force on any one charge is the vector sum of all the forces exerted on it due to each of the other charges interacting with it independently (p. 654).

The **electric field** (**E**) *at any point in space is the electric force experienced by a positive test-charge at that point divided by that charge:*

$$\mathbf{E} = \frac{\mathbf{F}}{q_0} \qquad [17.2]$$

in units of newtons per coulomb (N/C). The field is represented by field lines that begin and end on charges. The denser the concentration of lines, the greater the field. The force **F** that would arise on any point-charge q placed in the field is

$$\mathbf{F} = q\mathbf{E} \qquad [17.3]$$

The *electric field of a point-charge q* is

[*point-charge*] $\qquad E = k\dfrac{q}{r^2} \qquad [17.4]$

In terms of the **permittivity** (ε)

$$k = \frac{1}{4\pi\varepsilon} \qquad [17.5]$$

The *permittivity of free space* is a constant defined as $\varepsilon_0 = 8.8541878 \times 10^{-12}$ C²/N·m².

The field inside a charged conductor in the steady state is zero. External field lines must be perpendicular to the surface of any conductor.

The **electric flux** is

$$\Phi_E = \int_{\text{surface}} \mathbf{E} \cdot d\mathbf{A} = \int_{\text{surface}} E_\perp \, dA \qquad [17.9]$$

Gauss's Law (p. 674) is

$$\oint \mathbf{E} \cdot d\mathbf{A} = \frac{1}{\varepsilon}\Sigma q \qquad [17.13]$$

We define the *linear charge density* as $\lambda = Q/L$, the *surface charge density* as $\sigma = Q/A$, and the *volume charge density* as $\rho = Q/V$. The electric field of a long, straight charged wire is then

[*straight wire*] $\qquad E = \dfrac{\lambda}{2\pi R\varepsilon_0} \qquad [17.14]$

The *E*-field of a large sheet of charge is

[*large sheet*] $$E = \frac{\sigma}{2\varepsilon}$$ [17.15]

The field between the plates of a parallel plate capacitor is

[*parallel plates*] $$E = \frac{\sigma}{\varepsilon}$$ [17.16]

Suggestions on Problem Solving

1. Remember that the signs arising from the application of Coulomb's Law just relate to whether the force is attractive or repulsive. The forces between two point-charges are equal in magnitude and opposite in direction, and act along their center-to-center line. Draw a diagram, sketch in the directions of the forces, and then forget those signs. As ever, take to the right and up as the positive directions.

2. Given several point-charges, you may need to find the net force acting *on one of them*. Draw the direction of each force acting on the charge. Next, determine the magnitude of each force via Coulomb's Law. Resolve each force vector into perpendicular components and find the resultant of all the components in the usual way—that is, the net force.

3. The electric field *at a point* in space due to several point-charges is calculated much as is the force but using Eq. (17.4) instead of Coulomb's Law. Imagine a positive test-charge at that point and draw in the field vectors accordingly.

4. When applying Gauss's Law, first find the direction of the *E*-field so you can take advantage of its symmetries. Remember there is no field inside a conductor and embedding all or part of the Gaussian surface inside the conductor puts it in a zero-field region. The fields of all of the easy configurations have long since been figured out using Gauss's Law. Once you have learned how to make the standard dozen-or-so calculations, you have pretty much mastered the thing. Most professionals do not remember the field inside a charged coaxial cable; they just quickly derive it when needed using Gauss's Law, and that's the best reason to learn the method.

Discussion Questions

1. A common demonstration is to toss some small fragments of paper onto a highly charged conducting sphere. The paper initially sticks to the sphere and then after a little while pops off as if shot into the air. In 1676, Newton sent a note to the Royal Society describing a similar experiment where he charged a glass disk and held it horizontally above "some little fragments of paper." These he observed "sometimes leaping up to the glass, and resting there awhile; then leaping down and again resting." Explain what was happening.

2. Imagine a spherical dielectric shell—one made of glass, for example. A charge of $+Q$ is carefully distributed uniformly over the entire outer surface. Is there an electric field anywhere inside the shell? Explain.

3. Figure Q3 shows a dipole in a room (which is represented by a large grounded conductor). Make a rough sketch of its *E*-field and explain your reasoning.

4. Static electric buildup can be troublesome as, for example, in a photographic processing laboratory where film easily becomes charged (usually positively), attracting dust, and even creating sparks. To control the effect, there are static eliminators, one variety of which uses radioactive polonium-210. This material pours out a constant stream of positively charged alpha particles. How would that help?

5. William Gilbert in the sixteenth century devised a simple instrument to detect electrification, which he called a *versorium*. To make one, just balance a length of any material "lightly pivoted on a needle." A piece of a drinking straw stuck on a tack set in clay will work nicely once it's made free to rotate (Fig. Q5). Or fold a narrow strip of paper into a long upside-down V-shape and balance it horizontally on a tack. Now rub something like a comb to electrify it. Use the versorium to see if you have succeeded—that is, wave the comb near the detector. Put your versorium near a TV screen and see if it can detect when you turn the electron beam on. What happens and why?

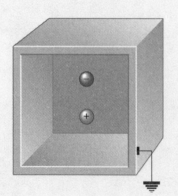

Figure Q3

Figure Q5

6. Go into a dark closet with a fluorescent bulb. Rub it vigorously with a piece of cloth until it warms up a bit and starts to glow. Now stroke it with just your dry hand. In fact, simply squeeze the bulb and then quickly open your hand—the lamp will light rather brightly. Now wet it with a little water. What happens? Charge up a wand (my favorite is a clear plastic tube that a set of windshield wipers came in) and wave it around. Does its effect on the bulb change when the lamp is wet?

7. Stephen Gray attached a lead ball to the end of a glass tube with a 3-ft length of moistened "parcel string." Under the ball he placed some fragments of brass leaf (very thin foil). When he rubbed the tube, the leaf was attracted up to the ball. Explain what was happening. (By the way, Gray later redid the experiment using 80 feet of string, and "the Tube being rubbed, the Ball attracted the Leaf-Brass.")

8. Why do long strips of ordinary plastic tape freshly pulled from the roll usually crumple and stick to themselves and everything nearby? Go in a dark closet and pull off a length of transparent plastic tape—sparks will blaze brilliantly at the point of contact with the roll.

9. An electric typewriter on a desk not far from a TV set disturbs the picture, generating colored bands across the screen. The roof antenna is attached to the set via an ordinary flat twin-lead TV wire. What would you suggest as a possible solution to the problem and why?

10. Figure Q10 shows a hollow charged conductor with two proof-planes touching each other inside of it. The scheme is used to determine that no E-field is inside a charged conductor. Explain how that might be accomplished.

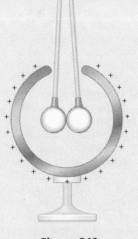

Figure Q10

11. A neutral, banana-shaped solid conductor is grounded to the plumbing via a wire. A negatively charged sphere is brought near it, the ground lead is disconnected, and the sphere removed. Describe the resulting E-field and surface charge density on the metal banana, if any.

12. Why are electrostatic phenomena, like sparks when you take off a sweater in the dark, more common in winter than in summer?

13. During the time Newton was president of the Royal Society, its curator of experiments was Francis Hauksbee. The following (Fig. Q13) is a demonstration Hauksbee performed that seems a very early hint of Faraday's lines-of-force. Onto a semicircular wire frame, Hauksbee fastened "several pieces of Woollen Thread . . . so as to hang down at pretty nearly equal distances." At the center, he placed a glass "Tube," which was subsequently charged by rubbing. The threads immediately swung about so that they pointed directly at the center of the glass. Explain what was happening and discuss the implications regarding field lines.

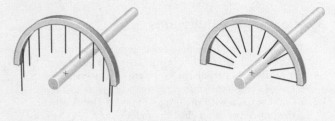

Figure Q13

14. Figure Q14 shows a variation of the most widely used electrostatic generator, a device that carries the name of its inventor, Robert J. Van de Graaff (1902–1967). The two pulleys are covered with different materials so that when they are contacted by the motor-driven belt, the belt acquires a negative charge from the bottom pulley and a positive charge from the top one. In some research models, electrons are literally sprayed on the belt at the base. Figure out how the generator works and explain the crucial role of the hollow conductor dome. What, if anything, limits the amount of charge that can be built up on the dome? How would that compare to the case where the belt delivered the charge to the outside surface?

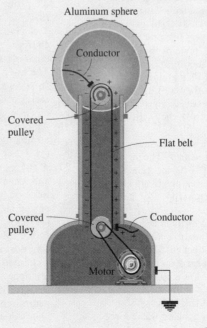

Figure Q14

15. Suppose that you approach an elephant with two known oppositely charged pith balls, in order to determine if the beast is itself charged. The first ball (+) is attracted, the second (−) repelled. What does that tell you? Which ball provides the unambiguous evidence? Explain.

16. What, if anything, might happen to the leaves of a charged electroscope if you waved a lit match around near it? Explain.

Multiple Choice Questions

1. The electrostatic force between a negative electron and a neutral neutron is   (a) negative and attractive   (b) positive and repulsive   (c) zero   (d) sometimes attractive and sometimes repulsive   (e) none of these.

2. By comparison with the force of gravity, the electrical attraction between an electron and a proton   (a) is just about the same size   (b) is very much stronger   (c) is very much weaker   (d) cannot be compared   (e) none of these.

3. When rubbed with a piece of wool, sulfur and glass will become charged   (a) positively and negatively, respectively   (b) negatively and positively, respectively   (c) both positively   (d) both negatively   (e) none of these.

4. When the center-to-center separation between two small charged spheres is doubled, the electric force between them   (a) is halved   (b) doubles   (c) is quartered   (d) is quadrupled   (e) none of these.

5. The SI units of electric flux are   (a) N/C^2 (b) $N \cdot m/C$ (c) $N \cdot m^2/C$ (d) $C/N \cdot m$ (e) none of these.

6. Some cars and trucks (especially those carrying combustible fuels) have straplike bands attached to their bottoms that drag along the ground in a somewhat vain attempt to   (a) discharge static electricity   (b) get rid of moisture   (c) build up negative charge   (d) protect them in case they get hit by lightning   (e) none of these.

7. Suppose we have three identical conducting spheres and one of them carries a charge of Q. If they are all brought into contact and then separated   (a) they will each have a charge of $Q/3$ (b) they will each have a charge of Q (c) only one will be charged with Q (d) they will all be discharged   (e) none of these.

8. A metal sphere is grounded through a switch, and a positively charged balloon is brought near it. The switch is opened and the balloon taken away. The sphere is now   (a) neutral   (b) negatively charged   (c) positively charged   (d) charged, but we cannot know its polarity   (e) none of these.

9. A metal sphere is grounded through a switch, and a positively charged balloon is brought near it. The balloon is then taken away and the switch is opened. The sphere is now   (a) neutral   (b) negatively charged   (c) positively charged   (d) charged, but we cannot know its polarity   (e) none of these.

10. If the charge on each of two identical tiny spheres is doubled while their separation is also doubled, their force of interaction will   (a) double   (b) become halved   (c) be quartered   (d) stay unchanged   (e) none of these.

11. If you are stretched out in a bathtub full of water, you are likely   (a) not to be grounded because the water is in a tub   (b) to be grounded if the tub has metal feet   (c) not to be grounded because you are inside a conductor   (d) to be grounded because the water connects you electrically to the pipes   (e) none of these.

12. The E-field of a point-charge 4.0 m away is measured to be 100 N/C. The field 2.0 m away from that charge is   (a) 400 N/C   (b) 50 N/C   (c) 100 N/C   (d) 800 N/C   (e) none of these.

13. A nonconductor is charged and then brought near a conductor. Consequently   (a) the two electrostatically repel each other   (b) the two electrostatically attract each other   (c) only the nonconductor is repelled   (d) there is no electrostatic interaction at all   (e) none of these.

14. Three charges ($q_1 = 1$ nC, $q_2 = 2$ nC, and $q_3 = 5$ nC) are separated in space (by distances of $r_{12} = 1$ m, $r_{23} = 2$ m, and $r_{13} = 3$ m). The ratio of the magnitudes of the forces on q_3 due to q_1 and to q_2 is   (a) 2/9   (b) 5/2   (c) 5/9   (d) 9/2   (e) none of these.

15. Two charges in vacuum attract each other. The same two charges, with the same separation, are now immersed in ethanol, which has a relative permittivity ($\varepsilon/\varepsilon_0$) of 25. The interaction is now   (a) increased by a factor of $\sqrt{25}$ (b) decreased by a factor of $\sqrt{25}$ (c) decreased by a factor of 25   (d) increased by a factor of 25   (e) none of these.

16. An electron in an electric field of 100 N/C experiences a force of   (a) 1.6×10^{-10} N   (b) 1.6×10^{-21} N   (c) 3.2×10^{-17} N   (d) 1.6×10^{-17} N   (e) none of these.

17. A charge $+q$ is placed at the origin of a coordinate system, and a charge of $+Q$ is located at $+a$ on the x-axis. The force on $+Q$ is found to be **F**. A third charge $-q$ is now placed at $+2a$ on the x-axis and the force on $+Q$ is now   (a) zero   (b) **F**   (c) $\frac{1}{2}F$ (d) 2**F**   (e) none of these.

18. In Fig. MC18, which drawing depicts the net field of a tiny positively charged conducting sphere in the gap of a charged parallel plate capacitor?   (a)   (b)   (c)   (d)   (e) none of these.

19. In Fig. MC18, which drawing depicts the net field of a tiny neutral conducting sphere in the gap of a charged parallel plate capacitor?   (a)   (b)   (c)   (d)   (e) none of these.

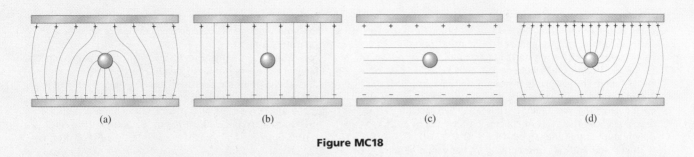

Figure MC18

Problems

THE ELECTROMAGNETIC CHARGE
THE ELECTRIC FORCE

1. [I] By roughly how much does the mass of a copper object change when, upon being stroked with a piece of woolen cloth, it acquires an excess charge of $-1.0\ \mu C$?

2. [I] How many electrons are needed to produce a charge of -1.0 C?

3. [I] Two tiny spheres carrying the same charge are 1.0 m apart in vacuum and experience an electrical repulsion of 1.0 N. What is their charge?

4. [I] A very small conducting sphere in air carries a charge of 5.0 picocoulombs and is 0.20 m from another such sphere carrying a charge Q. If each sphere experiences a mutual electrical repulsion of 2.0 μN, find Q.

5. [I] Two protons are fired directly at each other in a vacuum chamber. What is the force of repulsion at the instant they are 1.0×10^{-14} m apart?

6. [I] Two point-charges of $+0.50\ \mu C$ are 0.10 m apart. Determine the electric force they each experience in air.

7. [I] Two equally charged small spheres repel each other with an electric force of 1.0 N when 0.50 m apart, center-to-center, in air. What is the charge on each sphere?

8. [I] Compute the gravitational attraction between two electrons separated by 1.0 mm in vacuum, and compare that with the electrical repulsion they experience.

9. [I] An equilateral triangle with sides of 2.0 m is inscribed within a circle. A tiny charged sphere carrying $(+10\ \mu C)$ is then fixed at each vertex, and one of $-25\ \mu C$ is placed at the center of the circle. What is the net force acting on that central charge (magnitude and direction)?

10. [I] Two charged spheres each containing a quantity of excess electrons equal in number to Avogadro's number are separated by 1000 km in vacuum. Compute to two significant figures the electrical interaction between the spheres.

11. [I] Three small spheres immersed in air, each carrying a charge of $+25$ nC, are located at the vertices of a right isosceles triangle with a hypotenuse of 1.414 m, lying along the x-axis. Find the net electric force on the charge opposite the hypotenuse, located above it on the positive y-axis.

12. [I] Point-charges of $+1.0$ nC, $+3.0$ nC, and -3.0 nC are located, one each, at the three corners of an equilateral triangle of side-length 30 cm. Find the net electrostatic

force exerted on the 1.0-nC charge. Assume the surrounding medium is vacuum.

13. [I] A square nonconducting framework is set up in space with a tiny metal sphere mounted at each corner. The spheres at either end of the 45° diagonal are charged equally with $+45$ nC each, while the other two spheres are both given charges of -45 nC. What is the net force on a charge of 10 nC at the very center of the square?

14. [c] A glass rod of length L lies along the positive x-axis with one end at $x = 0$. Given that it is charged positively in such a way that its linear charge density is $\lambda(x) = (30.0\ \mu C/m^3)x^2$, find the total charge on the rod.

15. [c] An explosion takes place, and for an instant a spherical ball of positive charge is formed. It has a radius R and a volume charge density that increases radially, in a linear way, out from the very center where it is zero. Write an expression for the total charge of the sphere.

16. [II] How many electrons are there in a tablespoon (15 cm³) of water? What is the net charge of all of these electrons?

17. [II] Three small negatively charged metal spheres in vacuum are fixed on a horizontal straight line, the x-axis. One $(-12.5\ \mu C)$ is at the origin, another $(-5.0\ \mu C)$ is at $x = 2.0$ m, and the third $(-10.0\ \mu C)$ is 1.0 m beyond that at 3.0 m. Compute the net electric force on the last sphere due to the other two.

18. [II] Two charges of $+4.0$ nC and -1.0 nC are fixed to a baseline at a separation of 1.0 m. Where on the baseline should a third charge of $+2.0$ nC be placed if it is to experience zero net electric force?

19. [II] Figure P19 shows four point-charges fixed at the corners of a rectangle, in vacuum. Please compute the net electrostatic force acting on the 100-μC charge.

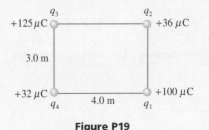

Figure P19

20. [II] Three very small charged spheres are located in a plane in air as follows: $q_1 = +15$ μC at point $(0,0)$, $q_2 = -20$ μC at point $(2,0)$, and $q_3 = +10$ μC at point $(2,2)$. Find the net force acting on the last charge.

21. [II] Redo Problem 19 where now q_4 is -32 μC.

22. [II] A cubic framework is inscribed within a sphere of 1.0-m radius. Tiny conducting spheres are then fixed at each corner and subsequently charged. There are four diagonals, and the pairs of spheres on the ends of each diagonal are equally charged with $+10$ nC, -10 nC, $+20$ nC, and -20 nC, respectively. What would be the resulting electric force on a $+100$-nC point-charge located at the center of the sphere? The surrounding medium is air.

23. [II] Two 2.0-g pith balls hang in air on cotton threads 50 cm long from a common point of support. The balls are then equally charged, and they spring apart, each making an angle of $10°$ with the vertical. Find the magnitude of the charge on each ball.

24. [cc] Return to the rod in Problem 14 where now the positive charge distribution is uniform. Determine the force on a test charge q_0 placed at a distance l from the end of the rod on the x-axis. [Hint: The test charge is at $x = (L + l)$, and take dq to be at x. The distance from dq to q_0 is then $r = (L + l - x)$.]

25. [cc] Figure P25 shows a wire carrying a charge Q uniformly distributed along its length L. Prove that the force acting on a charge q placed at point P right in the middle at a distance h is given by

$$F = \frac{kqQ}{h\sqrt{(L/2)^2 + h^2}}$$

Set up the analysis in terms of θ—we will revisit this geometry in Problem 59 where the hints will encourage a somewhat different approach. [Hint: You may need $d(\tan\theta)/d\theta = \sec^2\theta$ and $\sec^2\theta = 1 + \tan^2\theta$. Integrate over half the wire and double the answer.]

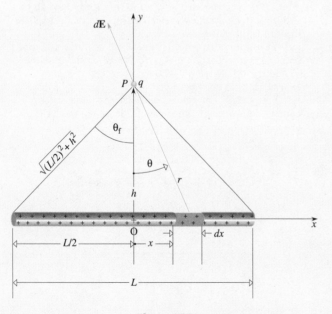

Figure P25

26. [cc] How is the answer to the previous problem affected if the wire, which still carries a net charge of Q, is now infinitely long?

27. [cc] Calculate the force on a tiny positively charged sphere (q) placed in front of a uniformly charged ring (Q) at a distance x along the central symmetry axis. The ring has a radius R and lies in the yz-plane. As a change of pace, do the analysis in terms of φ, which is the angle locating a differential charge element dQ in the plane of the ring. [Hint: See Fig. 17.35 and use $d\varphi$ rather than dl. Note that the charge per radian is $Q/2\pi$ and so $dQ = (Q/2\pi)\,d\varphi$.]

28. [III] Two charges $+q$ and $-q$ reside in vacuum on the y-axis at locations of $-\frac{1}{2}d$ and $+\frac{1}{2}d$, respectively. Determine the force on a third charge $+Q$ located at a distance of $+x$ from the origin on the x-axis.

29. [III] Three free charges (two of which are $+Q$ and $+2Q$, separated by a distance d) are in equilibrium. Find the size, polarity, and location of the third charge.

THE ELECTRIC FIELD

30. [I] Determine the electric force acting on an electron placed in a uniform north-to-south E-field of 8.0×10^4 N/C in vacuum.

31. [I] A test-charge of $+5.0$ nC placed at the origin of a coordinate system experiences a force of 4.0×10^{-6} N in the positive y-direction. What is the electric field at that location? Assume the medium is vacuum.

32. [I] A $+10$-μC test-charge at some point beyond a charged sphere experiences an attractive force of 40 μN. Please compute the value of the E-field of the sphere at that point in a vacuum.

33. [I] A very small conducting sphere carrying a charge of -20 nC is attached to a force gauge and lowered into a uniform electric field. A force of 2.0 nN due east keeps the sphere in equilibrium. Describe the E-field, assuming air is the medium.

34. [I] A small positively charged object is placed, at rest, in a uniform electric field in vacuum. Write an equation giving its speed v after a time t in terms of its mass m and charge q.

35. [I] Determine the magnitude and direction of an E-field if an electron placed in it, in vacuum, is to experience a force that will exactly cancel its weight at the Earth's surface.

36. [I] What is the magnitude and direction of the electric field of an electron at a point 1.0 m away in vacuum?

37. [I] Consider a hydrogen atom to be a central proton around which an electron circulates at a distance of 5.3×10^{-11} m. Find the electric field at the electron due to the proton.

38. [I] Two point-charges of $+10$ nC and -20 nC lie on the x-axis at points $x = 0$ and $x = +10$ m, respectively. Find the electric field on the axis at point $x = +5.0$ m. Assume vacuum.

39. [I] Positive point-charges of $+20$ μC are fixed at two of the vertices of an equilateral triangle with sides of 2.0 m, located in vacuum. Determine the magnitude of the E-field at the third vertex.

40. [I] Redo Problem 39, this time with charges of $+20$ μC and -20 μC at either end of the baseline (creating the field at the remaining vertex).

41. [I] Two point-charges of $+50$ nC each are separated in air by 1.414 m. What is the value of the net electric field they produce at a point that is 1.0 m away from both of them?

42. [I] Determine a formula for the electric field of a point-charge Q in vacuum using Gauss's Law.

43. [I] Using Gauss's Law, determine the electric field in the air gap of a charged parallel plate capacitor. Use a cylindrical Gaussian surface with one endface embedded in the metal of one of the plates.

44. [I] Two flat metal plates of area 2.0 m² are placed parallel to each other and both are then charged, one with $+10$ μC and the other with -10 μC. Determine the electric field in the air gap anywhere far from the edges.

45. [I] A pair of flat, horizontal parallel aluminum plates 100 cm × 50 cm has a 1.0-mm vacuum gap between them. How should they be charged if there is to be a uniform upward E-field of 1000 N/C in the gap?

46. [I] Two flat metal plates of area 2.0 m² are placed parallel to each other and both are then charged, one with $+10$ μC and the other with -10 μC. Determine the electric field inside anywhere far from the edges when the gap is completely filled with mica (use a permittivity of 41×10^{-12} C²/N·m²). How does the field now compare to the case where the gap was empty?

47. [c] A very long glass rod carries a uniform positive linear charge density of λ. The rod lies on the x-axis with one end at the origin and the other far off in the positive x-direction. Find the electric field at point P located at $x = -x_0$. [Hint: Start with Coulomb's Law using $dq = \lambda \, dx$ and integrate over the wire. The distance from P to dx is $(x_0 + x)$. Take the limits to be 0 and ∞.]

48. [c] An electric field having a strength given by $E = Cy^2$ is directed upward parallel to the z-axis. The constant C has units of N/C·m². A wire rectangle H wide and L long lies in the xy-plane so that the z-axis passes through its center and the long sides are parallel to the y-axis. Determine the electric flux through the rectangle. [Hint: Divide the rectangle into rectangular segments dy wide by H long. Find the differential flux through one such area element and then integrate over the entire length L.]

49. [II] A section of an advertising sign consists of a long tube filled with neon gas having electrodes inside at both ends. An electric field of 20 kN/C is set up between the electrodes, and neon ions accelerate along the length of the tube. Given that the ions each have a mass of 3.35×10^{-26} kg and are singly ionized, determine their acceleration.

50. [II] Three point-charges are placed at the corners of an isosceles triangle. At the left and right, end points of the base are $+1.0$ μC and $+1.0$ μC, respectively, and at the vertex $+3.0$ μC. The base of the triangle is 40 cm long, and the altitude is 30 cm high. Find the E-field at the midpoint of the baseline.

51. [II] Use Gauss's Law to determine a formula for the electric field outside of a long, uniformly charged cylinder of radius R with a positive surface charge density of σ.

52. [II] Use Gauss's Law to prove that there can be no net charge within a hollow conductor.

53. [II] Use Gauss's Law to find a formula for the electric field very close to any charged conducting surface in vacuum.

54. [II] With the previous problem in mind, compute the surface charge density on a conducting surface in the immediate vicinity of a point (in vacuum) where the electric field due to the charge is 5.0×10^5 N/C.

55. [II] An electron is placed in a uniform electric field of 1.5×10^4 N/C. Please determine its acceleration.

56. [II] Two point-charges of $+10$ nC and -20 nC lie on the x-axis at points $x = 0$ and $x = +10$ m, respectively. Find a point where the net electric field is zero, if such a point exists.

57. [II] Consider an idealized uniform spherical shell of charge $+Q$ surrounded at a distance by a concentric idealized shell of uniform charge $-Q$. Determine the E-fields between and beyond the shells. Assume vacuum.

58. [II] Consider the case of a small metal sphere suspended at the center of the cavity within a large hollow metal sphere. The central conductor is electrified with a charge of $+Q$. Write expressions for the E-field both inside the gap and outside the large sphere. Assume vacuum everywhere.

59. [cc] Determine the electric field at some point P a perpendicular distance x from a straight wire of length L having a uniform positive charge density λ. The wire lies along the y-axis with its center at the origin, and because $L \gg x$, the wire is essentially infinitely long. [Hint: As a change of pace take the following approach: With r the distance from dy to P and with θ the angle between r and x, use the fact that $y = x \tan \theta$ to get $dy = x \, d\theta/\cos^2\theta$. Compare your answer with Eq. (17.14).]

60. [cc] A large flat square sheet of plastic having sides L, covered uniformly with a positive charge Q, is in the vertical yz-plane. Find the electric field at a perpendicular distance h from the center of the sheet along the x-axis, where $h \ll L$. [Hint: Divide the sheet into narrow vertical strips of area $L \, dz$ each with a charge of $\sigma L \, dz$ and a charge per unit length of $\lambda = \sigma L \, dz/L$. Use this and the results of the previous problem to show that the field at P of a single differential strip is $dE = 2k\sigma \, dz/\sqrt{z^2 + h^2}$.] You can leave your answer in terms of inverse tangents.

61. [cc] Show that the electric field at point P (Fig. P61), a perpendicular distance x away from the center of a flat uniformly charged disc of radius R carrying a net positive charge Q, is given by

$$E_x = 2\pi k\sigma \left[1 - \frac{x}{(x^2 + R^2)^{1/2}} \right]$$

[Hint: Use a ring-shaped charge element and establish that $dq = \sigma(2\pi r \, dr)$. Then utilize the results of Example 17.7, p. 670.]

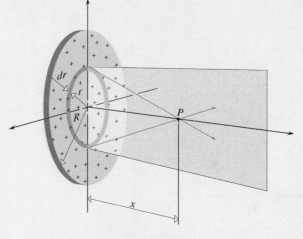

Figure P61

62. [cc] Considering the last two problems, determine the electric field for each one as the size of the charged object gets so large that it might well be infinite.

63. [cc] A thin glass ring of radius R lies in the xy-plane with the origin, O, of a coordinate system at its center. The upper half of the ring is given a uniform positive charge per unit length λ. Determine the magnitude and direction of the electric field at O. [Hint: Measure θ to either side of the y-axis.]

64. [cc] Redo the previous problem, but now let the linear charge density be given by the function $\lambda(\theta) = \lambda_0 \cos \theta$, where the angle is measured from the y-axis, as before, and has maximum values of $\pm\pi/2$.

65. [cc] A very crude model of the atom pictures it as a minute nucleus of charge $+Q$ surrounded by a negative spherically symmetric cloud of charge ranging out to a radial distance $r = R$. Suppose that the volume charge density of the cloud, for values of r equal to or less than R, is given by

$$\rho(r) = -\frac{5\rho_0}{2}\left[1 - \frac{r^2}{R^2}\right]$$

where $\rho_0 = Q/\frac{4}{3}\pi R^3$ and ρ is defined out to $r = R$. Find the electric field inside and out of the atom. What is the net charge inside a radius of $r = R$? [Hint: First find an expression for $q(r)$ by integrating $\rho(r)$, then use Gauss's Law.]

66. [cc] According to Example 17.7 the electric field at a point P on the axis of a uniformly charged ring is given by

$$E = \frac{kQx}{(x^2 + R^2)^{3/2}}$$

Now consider the fact that when P is near the ring, the elemental E-field contributions from segments on opposite sides of it are nearly antiparallel to one another and add weakly to $E = E_x$. On the other hand when P is far away the addition is more effective along the x-axis, but the field is dropping off. All of this suggests that there is a point where the axial field is a maximum—find it.

67. [III] A uniform ball of charge (Q) has a radius R. Determine the electric field inside the ball.

68. [III] With Problem 67 in mind, determine a formula for the field outside the sphere and draw a curve of E versus r.

69. [III] If a proton is considered a uniform ball of charge of radius 1.0×10^{-15} m, what is the E-field just beyond its surface?

18

Electrostatics:
Energy

AN UNDERSTANDING OF ELECTROSTATICS is necessary before we study a number of important topics, from electric currents to atomic theory. This chapter continues the development, elaborating the primary idea of *electric potential*, the concept that corresponds in informal language to the notion of *voltage*. Of course, everybody knows what voltage is—batteries come in 1.5- and 9-volt varieties and household wall outlets deliver electricity at a voltage of 110 volts—but what does that really mean?

Smoke particles are removed before they leave the stacks at this plant. The gray structures at the bases of the smokestacks are the electrostatic precipitators. (See Fig. Q21.)

ELECTRIC POTENTIAL

Chapter 17 dealt with electric *force* and its more convenient equivalent, electric field. This chapter focuses on *energy* as it relates to electrostatics. Both **F** and **E** are vector quantities and therefore sometimes a little complicated to deal with. Energy is a scalar, and that will make the analysis a lot simpler.

18.1 *Electrical*-PE

It takes a force to raise a mass in the Earth's gravitational field (Fig. 18.1a). We do work on the object to overcome the downward force field. And the mass goes up from an initial value of *gravitational*-PE to a higher one. The Coulomb force has the same mathematical form as the gravitational force and so it, too, must be conservative (p. 321); potential energy is associated with each of these interactions. When a charge is made to move *against* the influence of an electric field, as for instance in Fig. 18.1b, it will experience a change in its *electrical*-PE. Such motion requires the application of an external force and the expenditure of energy by the agency providing that force (Fig. 18.2).

The opposite occurs when the field does positive work on the charge and propels it to a new lower-energy location. Thus, if an electron is released in the vicinity of a positively electrified object, it is drawn toward the plus charges; work is done *by the field* on the electron. As it "descends" in the field, the electron loses electrical potential energy (PE$_E$) while it gains speed, and its KE increases equivalently. The electron "falls" to the positive object (or away from a negative one), just as a proton "falls" to a negative object (or away from a positive one), just as a rock falls to the Earth.

A piano held in midair next to a pea has more *gravitational*-PE with respect to the Earth than does its little green companion because it has more mass. Similarly, a highly positively charged pith ball immersed in an *E*-field will experience a greater force and have a larger *electrical*-PE than will a single proton at that point in the field, because of its greater charge.

18.2 Electric Potential

Imagine a charged body of some sort. We'll refer to it as the primary charge because we want to know about *its* influence on the surrounding scene. In Chapter 17, we initially pictured the force field of a primary charge. But that was dependent on the test-charge used to make the survey, so we then eliminated that dependence by dividing out the test-charge (q_0). The result was the **E**-field, which depends only on the primary charge and which tells us all about the distribution of force (acting on *any* charge placed in the field) in the surrounding space.

What we want now is an energy measure that is again independent of the test-charge being used for the survey. Accordingly, we find the *electrical*-PE of a test-charge at all points in space within the field of the primary charge and divide each such value by q_0. This operation associates a number (positive or negative) with ev-

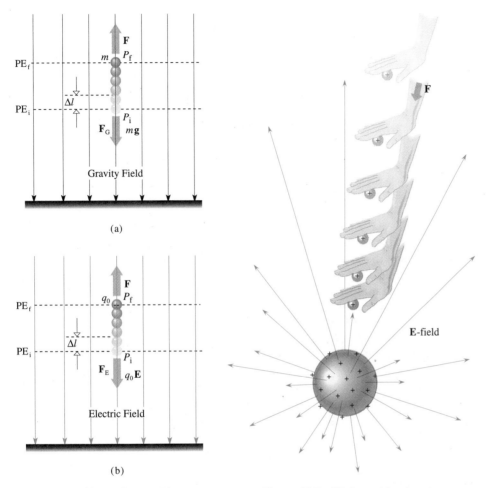

Figure 18.1 (a) An object, with a mass m, being raised in a uniform downward gravitational field by a force **F** experiences a downward gravitational force $\mathbf{F}_G = m\mathbf{g}$. (b) An object, with a charge q_0, being raised in a uniform downward electric field by a force **F** experiences a downward electric force $\mathbf{F}_E = q_0\mathbf{E}$.

Figure 18.2 Work must be done in moving a charge against an electric field. Here a positive charge is forced toward a positively charged sphere. It's being pushed against the E-field.

ery point in space and yields a map of the *potential energy per unit charge*. That scalar quantity is known as the **electric potential** (V) or just the *potential* for short:

$$\text{Electric potential} = \frac{\textit{electrical}\text{-PE}}{\text{charge}} \qquad (18.1)$$

The units of potential are joules per coulomb (J/C) and in honor of Alessandro Volta, 1 J/C is defined to be 1 volt (1 V). It is common to speak of the electric potential as the ***voltage***.

Example 18.1 The most commonly used unit of electric field is the volt per meter. Show that this is consistent with the units introduced thus far.

Solution: We want to show the equivalence between N/C and V/m. The brute force way is simply to convert everything into the most basic form possible in terms of meters, kilograms, seconds, and coulombs:

$$1 \frac{N}{C} = 1 \frac{kg \cdot m/s^2}{C}$$

while

$$1 \frac{V}{m} = 1 \frac{J/C}{m} = 1 \frac{N \cdot m/C}{m} = 1 \frac{N}{C} = 1 \frac{kg \cdot m/s^2}{C}$$

and the two are, indeed, identical.

▶ **Quick Check:** We could have stopped one step back and not made the first and last conversions, but we didn't know it was going to work out that way. Alternatively, multiply the top and bottom of 1 N/C by m: 1 N·m/C·m = 1 (J/C)/m = 1 V/m.

We now examine the difference in electric potential ΔV between two points, P_i and P_f, in an arbitrary E-field (Fig. 18.1b). Before we begin, review the analysis in the section called "Work Done By a Changing Force" (p. 318). Take any path from P_i to P_f, divide it into n segments; and represent the jth one as Δl_j, as was done in Fig. 9.7. Suppose a positive test-charge q_0 is displaced through the small distance Δl_j in an electrostatic field $\mathbf{E}$. There is a corresponding change in its *electrical*-PE, in the amount of ΔPE_{Ej}. We take Δl_j to be so small that the E-field does not change appreciably over its extent—that's the crucial assumption. A force $\mathbf{F}$ is applied and the charge is brought from rest at its initial location to rest at its final location; $\Delta KE = 0$. The work done in moving the charge the distance Δl_j against the field is $\Delta W_j = F_{\parallel j} \Delta l_j$, where $F_{\parallel j}$ is the component of $\mathbf{F}$ in the direction parallel to the displacement. Because $F_{\parallel j}$ and $E_{\parallel j}$ are in opposite directions, $F_{\parallel j} = -q_0 E_{\parallel j}$.

The work done on the charge by the applied force $F_{\parallel j}$ is

$$\Delta W_j = F_{\parallel j} \Delta l_j = -q_0 E_{\parallel j} \Delta l_j$$

and that work goes into changing the potential energy:

$$\Delta W_j = \Delta PE_{Ej} = -q_0 E_{\parallel j} \Delta l_j$$

By definition, the change in potential is

$$\Delta V_j = \frac{\Delta PE_{Ej}}{q_0} = -E_{\parallel j} \Delta l_j \tag{18.2}$$

The net change in electric potential in going from P_i to P_f is gotten by taking the sum of all the contributions along each tiny displacement Δl_j. Accordingly

$$\Delta V = \sum_{j=1}^{n} (-E_{\parallel j} \Delta l_j)$$

When the field varies from place to place the approximation that $E_{\parallel j}$ is constant over the extent of Δl_j must be improved by letting $\Delta l_j \to 0$ and $n \to \infty$, which transforms the sum into the line integral

$$\Delta V = -\int_{P_i}^{P_f} E_{\parallel} \, dl = -\int_{P_i}^{P_f} \mathbf{E} \cdot d\mathbf{l} \tag{18.3}$$

This is the most general statement of the relationship between the electric potential difference at two points in space and the electric field permeating that space, which gives rise to the difference.

Potential Difference

The quantity ΔV is referred to as the **difference in potential** (or the **potential difference**) between the initial (P_i) and final (P_f) points in the field and is often written as

$$\Delta V = V_{P_f} - V_{P_i} \tag{18.4}$$

The potential difference between two points P_i and P_f numerically equals the work done against the field in moving a unit positive charge from P_i to P_f with no appreciable acceleration:

$$V_{P_f} - V_{P_i} = \frac{W_{P_i \to P_f}}{q_0} \tag{18.5}$$

The work done can be greater than zero, less than zero, or zero; the same is true for ΔV.

Example 18.2 A small sphere carrying a positive charge of 10.0 μC is moved against an E-field through a potential difference of +12.0 V. How much work was done by the applied force in raising the potential of the sphere?

Solution: [Given: $q_0 = 10.0\ \mu$C and $\Delta V = +12.0$ V.

Find: W.] The defining relationship between ΔV, q_0, and W is Eq. (18.5); hence,

$$W = \Delta V\, q_0 = (+12.0\ \text{V})(10.0 \times 10^{-6}\ \text{C}) = \boxed{120\ \mu\text{J}}$$

▶ **Quick Check:** $q_0 = W/\Delta V = (120\ \mu\text{J})/(12\ \text{V}) = 10\ \mu$C.

Remember that energy is a relative quantity. There is no signpost floating in space that says, "This is the absolute zero of PE." Only changes in *electrical*-PE, and therefore *only changes in potential, are important* (Table 18.1).

A positive charge released in an electric field accelerates in the direction of the field. *It will not travel along a path that corresponds to a field line if the line is curved* (after a time Δt, $\mathbf{p}_f = \mathbf{p}_i + \Delta\mathbf{p}$, but $\mathbf{p}_i$ is not parallel to $\Delta\mathbf{p}$, which is parallel to $\mathbf{E}$). The positive charge naturally descends from a high potential to a low potential, losing PE_E and gaining KE. Here, $\Delta V < 0$, but in general *the change in potential equals the negative of the work done per unit charge by the field on the positive charge. Work done against the field always increases* PE_E; *work done by*

The man straddling the 138 000-V power line landed there in a parasail accident. The brief surge of charge that brought him up to 138 kV burned his hands and feet, but otherwise he was fine.

TABLE 18.1 **Common potential differences**	
Biochemical	1 mV–100 mV
Dry cell	1.5 V
Automobile battery	12 V
Household electricity	
U.S.A.	110 V–120 V
Much of Europe and Asia	240 V–250 V
Voltage induced in 600-mile Alaska pipeline by solar storm	1000 V
Power plant generator	24 000 V
Transmission lines	
Local	4 400 V
Cross country	120 000 V
Extra-high-voltage	500 kV–1 MV
Van de Graaff generator (1940)	4.5 MV
Folded tandem generator	25 MV
Lightning	10^8 V–10^9 V

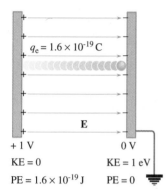

$q_e = 1.6 \times 10^{-19}$ C

E

+1 V   0 V
KE = 0   KE = 1 eV
PE = 1.6×10^{-19} J   PE = 0

Figure 18.3 A proton falling through a potential difference of 1 volt. It follows the E-field being pushed along by the field and loses an amount of potential energy equal to 1 unit of charge times 1 volt, or 1 eV.

the field always decreases the PE_E. The change in PE equals the negative of the work done *by* a conservative force field.

A negative charge released in an E-field moves spontaneously, experiencing a drop in *electrical*-PE. But it travels from a low potential to a higher potential ($\Delta V > 0$). That's evident from Eq. (18.2), where ΔPE_E is negative and q_0 is negative, so that ΔV is positive. **When a negative charge moves from a point of high potential in a field to a point of low potential, it increases its** PE_E.

The work done on a charge in moving it from point P_i to point P_f is independent of the path taken, just as the work done in climbing a mountain is independent of the path. If it were not, we could make a perpetual-motion machine, going from P_i to P_f along a small-change-in-potential path and coming back along a larger one, thereby gaining a bit of energy on each cycle.

When a particle (Fig. 18.3) with a charge equal to $+q_e$ moves from one point in a field to another, dropping 1 volt in potential, it will thereby decrease its *electrical*-PE by (p. 383) 1 electron volt (eV):

$$1 \text{ eV} = 1.6 \times 10^{-19} \text{ J}$$

Example 18.3 Figure 18.4 depicts the basic elements of a cathode-ray tube, the kind found in computer terminals and oscilloscopes. Electrons are "boiled" off a heated cathode and emerge through a pinhole, being drawn toward the first anode, which is at a relatively small positive potential above the cathode. A second anode that is 8000 V to 20 000 V above the cathode (depending on the design of the tube) accelerates the beam up to speed. Determine the change in the potential energy of each electron on traversing the gun, given that the second anode has a voltage of 20 kV above the cathode. Assuming that an electron has negligible motion at the cathode, write a general expression for its final speed in terms of ΔV. Find its specific final speed in this case.

Solution: [Given: $v_i = 0$, and $\Delta V = 20$ kV. Find: v_f.] Any one electron traverses a potential difference of 20×10^3 V and so drops in potential energy by

$$\Delta\text{PE}_E = q_e \Delta V$$

This result is true regardless of the details of the field existing within the gun. Remember that a negative charge loses PE_E as it *increases* in potential (however weird that may sound).

$$\Delta\text{PE}_E = (-1.6 \times 10^{-19} \text{ C})(20\,000 \text{ V})$$

and $\boxed{\Delta\text{PE}_E = -3.2 \times 10^{-15} \text{ J}}$

The electron's energy is conserved and so

$$\text{KE}_i + \text{PE}_i = \text{KE}_f + \text{PE}_f$$

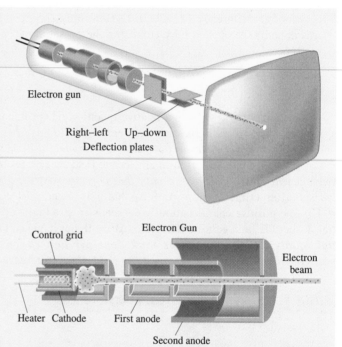

Electron gun

Right–left   Up–down
Deflection plates

Control grid   Electron Gun

Electron beam

Heater Cathode   First anode

Second anode

Figure 18.4 A cathode-ray tube (see Fig. Q19). A beam of electrons passes between two perpendicular sets of plates. These are appropriately charged, and the beam is deflected to any desired point on the face of the tube.

and since $\text{KE}_i = 0$

$$\text{KE}_f = \text{PE}_i - \text{PE}_f = -\Delta\text{PE}$$

and $$\tfrac{1}{2} m_e v_f^2 = -q_e \Delta V$$

(continued)

(continued)
Consequently

$$v_f = \left[\frac{-2q_e\,\Delta V}{m_e}\right]^{\frac{1}{2}} = \left[\frac{-2(-3.2 \times 10^{-15}\ \text{J})}{9.1 \times 10^{-31}\ \text{kg}}\right]^{\frac{1}{2}}$$

and $\boxed{v_f = 8.4 \times 10^7\ \text{m/s}}$

▶ **Quick Check:** The high voltage and tiny mass reasonably result in a tremendous speed of $\approx\frac{1}{4}$c. As we will see in Chapter 28, this rapid motion is more accurately treated using the Special Theory of Relativity. Note that the electron experiences an energy drop of 20 keV = 3.2×10^{-15} J = ΔKE.

18.3 Potential in a Uniform Field

Let's now derive an expression for the potential difference between two points in a **uniform electric field**, *one in which the strength of the field is constant everywhere and the field lines are straight and parallel.* Consider the positive charge in Fig. 18.5. As we have already seen in Eq. (18.3), the potential difference between any two points in an *E*-field is

$$\Delta V = -\int_{P_i}^{P_f} E_{\parallel}\,dl = -\int_{P_i}^{P_f} \mathbf{E} \cdot d\mathbf{l} = -\int_{P_i}^{P_f} E\cos\theta\,dl$$

where $E_{\parallel} = E\cos\theta$. Because the field is constant and θ is constant, the net potential difference encountered in traversing the straight-line distance D from point $P_i = A$ to point $P_f = B$ is

$$V_B - V_A = -\int_A^B E\cos\theta\,dl = -E\cos\theta \int_A^B dl = -D(E\cos\theta)$$

It follows that

$$V_B - V_A = -ED\cos\theta$$

Using $\pm D\cos\theta = d$, this becomes

[*a uniform E-field*] $$V_B - V_A = \pm Ed$$

(18.6)

The potential difference is $+$ when the displacement has a component that is opposite to the field and $-$ when it has a component parallel to the field. ***The potential difference only depends on the displacement parallel or antiparallel to the field.***
Observe in Fig. 18.5a that

$$V_C - V_A = -Ed$$

and so $$V_B - V_A = V_C - V_A$$

Points B and C are at the same electric potential. Another way to see this is to realize that if the points B and C are such that there is a path from one to the other that is every-

Figure 18.5 A positive charge moving diagonally across a uniform electric field. In (a) there is a component of displacement parallel to the field, and the potential drops in going from *A* to *B*. In (b) there is a component of displacement against the field, and the potential rises in going from *A* to B.

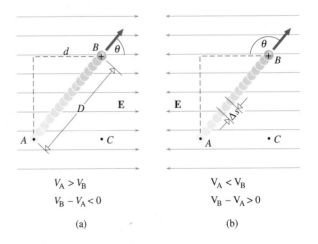

$V_A > V_B$
$V_B - V_A < 0$

(a)

$V_A < V_B$
$V_B - V_A > 0$

(b)

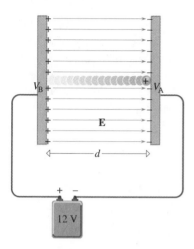

Figure 18.6 A battery connected across two parallel metal plates. There is a uniform E-field between the plates, and a positive charge experiences a drop in potential $V_A - V_B = -Ed$ upon traveling from the left plate to the right plate.

where perpendicular to the field, $E_\parallel = 0$, $F_\parallel = 0$, and $\Delta W = 0$. No work is done by or against the field in moving from B to C, which is the case gravitationally when a mass is moved horizontally on the Earth's surface. Thus, since $\Delta W = q\,\Delta V$, $\Delta V = 0$, and the two points are at the same potential.

The simplest situation, that of moving a positive charge along (or opposite to) a uniform E-field, is depicted in Fig. 18.6. Then $\theta = 0$ (or 180°) and $V_B - V_A = +Ed$, while $V_A - V_B = -Ed$. There are a number of devices, such as batteries, generators, fuel cells, solar cells, thermoelectric cells, that can provide electrical energy and sustain a fairly constant potential difference across their two output terminals. In Fig. 18.6, ***the positive terminal of the battery is 12 V higher than the negative terminal***. These terminals are connected to the parallel plates by excellent conductors, along which it is assumed there will be no appreciable loss in electrical energy and no measurable decrease in potential—that's a little unrealistic, but we will get back to that point in Chapter 19 when resistance is considered. This battery will maintain a constant 12-V difference across the parallel plates and sustain a constant field in the gap.

Example 18.4 If the separation of the plates in Fig. 18.6 is 2.0 mm, determine the magnitude of the electric field in the air gap.

Solution: [Given: $d = 2.0$ mm and $V = 12$ V. Find: E.] The plus side is at a higher potential than the minus side: $V_B > V_A$ by 12 V. We would like an expression relating E, V, and d, and so we use

$$V_B - V_A = +Ed \qquad [18.6]$$

whereupon

$$12\ \text{V} = +E(2.0 \times 10^{-3}\ \text{m})$$

Hence

$$\boxed{E = 6.0\ \text{kV/m}}$$

▶ **Quick Check:** This is a fair-sized field (Table 17.2), but that's reasonable since the spacing over which the voltage drops 12 V is only 2 mm. That corresponds to a drop of 6 V/mm, or 6 kV/m.

18.4 Potential of a Point-Charge

Earlier, we calculated the E-field of a point-charge and saw how that could be used to determine the fields for all sorts of charge distributions. The same is true for the potential; that is, once we have the potential of a point-charge we can use it to determine the potentials of other more complicated systems. To that end, return to p. 332 where we computed the change in *gravitational*-PE when a test-mass m changed position in the field of a spherical mass M. Because the gravitational field of a point-mass and the electric field of a point-charge both vary inversely with r^2, finding the corresponding change in PE is a bit complicated. The change in PE is the work done, which is the force times the displacement, but the force is a variable. Still, the mathematical aspect of the problem has already been solved in the process of arriving at Eq. (9.20):

$$\Delta\text{PE}_G = GmM\left(\frac{1}{R} - \frac{1}{r}\right)$$

and we would gain little by doing it again. This is the change in *gravitational*-PE that occurs when a mass m is moved from R *out* to r *against* the *attractive* $1/r^2$-gravitational field produced by a sphere of mass M. It follows that

$$\Delta PE_E = kq_0 Q\left(\frac{1}{r_B} - \frac{1}{r_A}\right)$$

This is the change in *electrical*-PE that occurs when a positive charge q_0 is moved from r_A *in* to r_B *against* the *repulsive* $1/r^2$-electric field produced by a tiny sphere of positive charge Q, as shown in Fig. 18.7. In that case $r_A > r_B$ and $\Delta PE_E > 0$. Similarly, when a positive charge q_0 is moved from r_A out to r_B by the repulsive $1/r^2$-electric field, $r_A < r_B$ and $\Delta PE_E < 0$. Inasmuch as

$$\Delta V = \frac{\Delta PE_E}{q_0}$$

[+ *point-charge Q*] $$V_B - V_A = kQ\left(\frac{1}{r_B} - \frac{1}{r_A}\right) \tag{18.7}$$

This result is the sought-after potential difference encountered in moving from point A to point B in the electric field of a tiny positively charged sphere or, equivalently, a point-charge Q.

Again, as we saw in the case of gravity, it is often useful to take the zero of potential, which is totally arbitrary, at infinity (where the force is zero). To that end, let $r_A \to \infty$, whereupon $1/r_A \to 0$ and

$$V_B - 0 = kQ\left(\frac{1}{r_B} - 0\right)$$

represents the electric potential at the finite point B with respect to zero at infinity. To simplify matters, we might as well drop the subscript B since the equation applies to any point in the region beyond the sphere of positive charge. Accordingly,

[+ *point-charge; V = 0 at* ∞] $$V = kQ\left(\frac{1}{r}\right) \tag{18.8}$$

This is the electric potential at a distance r from an isolated positive point-charge Q measured with respect to the zero at ∞. Alternatively, we can say that this is *the work done per unit charge against the E-field of a positive point-charge Q in bringing a test-charge from infinity to a distance r from Q.* The potential at a point can be positive, negative, or zero depending on Q, and, of course, it is a directionless quantity, a scalar. Figure 18.8 shows the PE_E of a point-charge placed in the **E**-field of a point-charge; it's significant because it corresponds to the electron-proton interaction of a hydrogen atom.

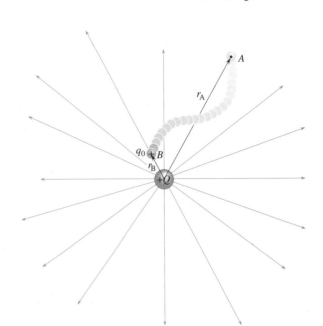

Figure 18.7 Here, a positive charge q_0 is moved inward toward Q, going from a distance r_A to a distance r_B. The potential difference ($V_B - V_A$) is positive because q_0 is pushed downward against the field. Its PE_E increases and, if released, q_0 will be pushed out and away.

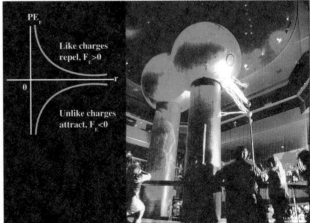

Figure 18.8 The potential energies of pairs of interacting point-charges. Here, a positive point charge is located at the origin, and the other charge is at r. Like charges repel and the $PE_E > 0$; unlike charges attract and the $PE_E < 0$. In each case, the force is the negative of the slope of the PE_E curve.

Example 18.5 What is the potential difference encountered in going from point B to point A in Fig. 18.7 if the tiny sphere carries a charge of $Q = +10 \ \mu C$, $r_A = 20$ cm, and $r_B = 10$ cm?

Solution: [Given: $Q = +10 \ \mu C$, $r_A = 20$ cm, and $r_B = 10$ cm. Find: ΔV.] Going from point B to point A, a positive test-charge moves out along the field and so drops in potential. Using Eq. (18.7)

$$V_A - V_B = -(V_B - V_A) = -kQ\left(\frac{1}{r_B} - \frac{1}{r_A}\right)$$

$$V_A - V_B = -(9.0 \times 10^9 \ \text{N·m}^2/\text{C}^2)(+10 \times 10^{-6} \ \text{C})$$
$$\times (10 \ \text{m}^{-1} - 5.0 \ \text{m}^{-1})$$

and
$$\boxed{V_A - V_B = -0.45 \ \text{MV}}$$

▶ **Quick Check:** From Eq. (18.8), the potentials at distances of r_B and r_A are $+0.90$ MV and $+0.45$ MV, respectively. Hence, going from point B to point A there is a change in potential of -0.45 MV.

18.5 Equipotentials

Return to Fig. 18.5 and observe that we can move a positive test-charge from point B to point C so that the displacement is everywhere perpendicular to the uniform E-field and no work is done. The potential does not change in the process, and we say that *the line from B to C, which is everywhere perpendicular to the field, is an equipotential line*—a line along which the potential energy of a test-charge remains unchanged.

Envision a positive point-charge or, equivalently, a tiny charged conducting sphere (Fig. 18.9). The potential at any fixed distance R beyond the charge is given by

$$V = kQ\left(\frac{1}{R}\right) \qquad [18.8]$$

which is constant in all directions. In other words, the **equipotential surfaces** of a point-charge are a series of concentric spheres, everywhere perpendicular to the E-field—this is an important point.

These ideas should bring to mind the concept of a conductor. Since we found that the electric field in the vicinity of the surface is always perpendicular to a conductor regardless of its shape, the conductor's surface must also be an equipotential. That's true whether the conductor is charged or not. Another way to appreciate this concept is to realize that there is no E-field within the substance of a conductor in electrostatic equilibrium. As a consequence, a test-charge transported from one point within a conductor to another will not have to be moved against an electric force, will not have work done on it, and will therefore experience no change in potential. By analogy, although a ball will roll down a hill by itself, if it were placed on the smooth surface of an idealized spherical planet, it would not spontaneously move at all. The ball, by itself, can only drop to a lower *gravitational*-PE, and yet everywhere on the surface it has the same PE or, if you like, all points on the surface are at the same gravitational potential.

The entire body of a conductor, devoid as it is of any E-field, must be an **equipotential volume**. Indeed, provided there are no encompassed isolated charges (as in Fig. 17.34), the total region contained within a conductor, hollow or not, is at the same potential. Because the Earth itself is a conductor, its surface is an equipotential and, by custom, it is often taken to be the zero of potential. That's just what is done in the electrical system in your home, where the "hot" terminal in each of

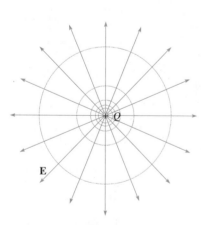

Figure 18.9 Field lines and equipotentials for a small positive charge. The equipotential surfaces are concentric spheres centered on Q.

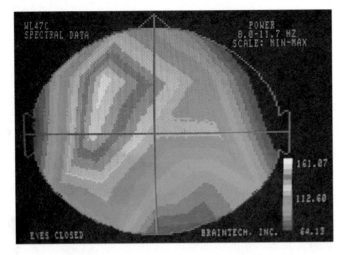

A map of the equipotentials in the brain of a person with epilepsy. This picture was made about 0.1 s after the person received a stimulus.

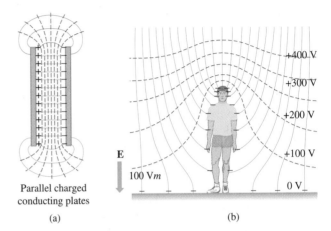

Figure 18.10 Electric fields and equipotentials. (a) A pair of charged conducting parallel plates. The person in (b) is grounded and so is at 0 potential.

the outlets oscillates between ±120 V with respect to the zero-potential terminal. The latter is physically connected to ground (p. 885).

Figure 18.10 shows both the *E*-field lines and the associated equipotentials for several configurations. In each case, the equipotentials are drawn successively at fixed-voltage intervals, ΔV. When the field is uniform, as in Fig. 18.11, the equipotentials are evenly spaced. The inclusion pictorially of equipotentials beautifully complements the field-line diagrams. A drawing of field lines immediately reveals the direction in which a charge will experience a force when placed anywhere in the region, and that corresponds to the direction in which it will accelerate as well. Moreover, if the field is uniform, the charge will move along a field line. A glance at a diagram containing equipotentials reveals the energy change that will occur when a charge moves from one point to another along *any* path. Together, the field lines and equipotentials provide a complete picture of the influence of the primary charge distribution everywhere in space.

18.6 The Potential of Several Charges

Potential is a scalar quantity. Thus, if there are several charges present, the sum of their superimposed potentials anywhere in the surrounding space is equal to the algebraic sum of the individual contributions. We can make use of this fact provided we stick with point-charges because we already have, via Eq. (18.8), the potential function for a point-charge. Therefore, *if there are two or more point-charges, the net potential at any location will be the scalar sum of the potentials at that location due to each charge.* Consequently, it is certainly possible that at some point in space a positive potential due to one charge can be canceled by a negative potential due to another charge. If we imagine a positive test-charge brought to that zero-potential point from infinity, in the process it will be repelled by one charge as much as it is attracted by the other, and no net work will be done by or against the field. *There can be a net E-field at a point even though the potential at that point is zero, and conversely, there can be a nonzero potential at a point where the net field is zero.*

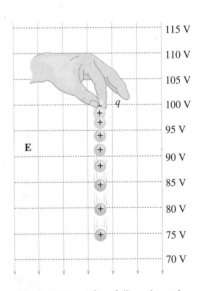

Figure 18.11 After falling through a potential difference of 25 V, the charge has lost an amount of potential energy 25*q* equal to its gain in kinetic energy.

Example 18.6 Figure 17.26, p. 663, shows two charges each of +10 nC at (0, 0) and (8.0, 0). Determine the net potential at point (4.0, 0). Then let one of the charges be −10 nC and recompute the potential at that same point.

Solution: [Given: $Q_1 = Q_2 = \pm 10$ nC at (0, 0) and (8.0, 0), respectively. Find: V at point A, (4.0, 0).] For each charge, there is a contribution given by

$$V = kQ\left(\frac{1}{r}\right) \qquad [18.8]$$

and so at point A

$$V_A = kQ_1\left(\frac{1}{r_1}\right) + kQ_2\left(\frac{1}{r_2}\right)$$

Thus, with both charges positive

$$V_A = \frac{(9.0 \times 10^9 \text{ N·m}^2/\text{C}^2)(+10 \times 10^{-9} \text{ C})}{4.0 \text{ m}}$$
$$+ \frac{(9.0 \times 10^9 \text{ N·m}^2/\text{C}^2)(+10 \times 10^{-9} \text{ C})}{4.0 \text{ m}}$$

and $\boxed{V_A = 45 \text{ V}}$

Although the E-field is zero at that point (see Fig. 17.27), the potential is not; it is positive. Work must be done against the fields of both charges to haul a test-charge in from infinity.

With either charge negative, we have a dipole and

$$V_A = \frac{(9.0 \times 10^9 \text{ N·m}^2/\text{C}^2)(+10 \times 10^{-9} \text{ C})}{4.0 \text{ m}}$$
$$+ \frac{(9.0 \times 10^9 \text{ N·m}^2/\text{C}^2)(-10 \times 10^{-9} \text{ C})}{4.0 \text{ m}}$$

and $\boxed{V_A = 0 \text{ V}}$

▶ **Quick Check:** In the field of a dipole, as a test-charge is moved in from infinity, the positive charge repels it, and work must be done to overcome that repulsion. At the same time, the negative charge attracts the test-charge, doing negative work on it. The result of the push-pull is that no net work needs to be done on the test-charge to bring it to A, and the potential there is the same as it was at the start of the journey at infinity—namely, zero.

Figure 18.12 shows the fields and potentials for two equal charges. When both charges are positive (Fig. 18.12a), the field lines emerge from the system, and as we move out along them, the positive potential drops off, approaching zero at infinity. At the very center, there is a finite positive potential. Note how the equipotentials combine into a single surface that becomes increasingly more like a sphere; from very far away, the system looks like a single positive charge.

By contrast, in the case of the dipole (Fig. 18.12b), the field extends out the positive side and in the negative side. The equipotentials surrounding the positive charge are positive and those surrounding the negative charge are negative. There is a zero-potential plane down the middle to which the field is everywhere perpendicular. A test-charge can be brought from infinity by traveling within the plane and therefore along a path perpendicular to the field. No work is done, and no change from zero occurs in the potential. Of course, since the E-field is conservative, we can actually come from infinity to any point on the plane via any path, and the change in potential would still be zero.

18.7 Potential of a Continuous Charge Distribution

Suppose we have some charged object, a well-rubbed balloon will do. Of course the charge isn't continuous—

A complex distribution of charges results in a complex pattern of field lines and equipotentials. See if you can figure out what's happening here.

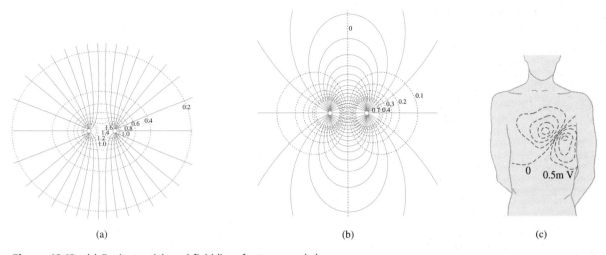

Figure 18.12 (a) Equipotentials and field lines for two equal charges. (b) Equipotentials and field lines for two opposite charges of equal magnitude (a dipole). (c) Equipotentials (in millivolts) at some instant across the chest due to heart activity.

nothing substantial is continuous—but if it were, we could use the calculus to determine the potential at any point (provided we had a fucntion that adequately described the charge distribution). With that as the justification, let's assume that charge is continuous (see Section 17.11). A point-charge results in a potential of

$$V = kQ\left(\frac{1}{r}\right)$$ [18.8]

and therefore a differential element of charge dq, which is essentially a point-charge, will result in a potential of

$$dV = \frac{kdq}{r}$$

For some distribution of charge

[V = 0 at ∞] $$V = k\int \frac{dq}{r}$$ (18.9)

where r is the distance from the charge element dq to the point P, where we want the potential.

Example 18.7 Find the electric potential at P on the central axis of the ring-shaped charge distribution discussed in Example 17.7 and pictured in Figure 17.35 on p. 670.

Solution: [Given: A uniform ring of positive charge. Find: V on the axis.] Consider the charge element dq remembering that potential is a scalar and all the differ-

ential contributions (dV) simply add algebraically. Using Eq. (18.9) and the fact that $r = \sqrt{x^2 + R^2} = $ constant,

$$V = k\int \frac{dq}{r} = k\int \frac{dq}{\sqrt{x^2 + R^2}} = \frac{k}{\sqrt{x^2 + R^2}}\int dq$$

and the integral simply equals the net charge Q. Therefore

(continued)

(continued)

$$V = \frac{k}{\sqrt{x^2 + R^2}}Q$$

▶ **Quick Check:** The units are those of k times charge over distance, which agrees with Eq. (18.8). The potential at $x = 0$, namely $V = kQ/R$, is also reasonable since there (where $r = R$ for all dq) the contributions from all the charge elements are equal and add algebraically. Furthermore, when $x \gg R$, $\sqrt{x^2 + R^2} \to \sqrt{x^2}$ and $V = kQ/x$, which is the potential of a point charge, and that too is appropriate when we are so far away that the size of the ring is inconsequential. The best check, however, is to confirm that $E_x = -dV/dx$, but that will have to wait until the next section.

18.8 Determining *E* from *V*

For any charge distribution the resulting *E*-field is everywhere perpendicular to the equipotential lines. In fact, specifying the electric potential completely specifies the field—they are complementary formulations, one from the force perspective, the other from the energy perspective. We already know how to find V from E; now let's study the converse process.

The relationship between the potential difference and the electric field sustaining that difference was stated earlier in integral form as

$$\Delta V = -\int_{P_i}^{P_f} E_{\parallel} \, dl = -\int_{P_i}^{P_f} \mathbf{E} \cdot d\mathbf{l} \qquad [18.3]$$

With dV being the potential difference between two points separated by an infinitesimal distance dl, we can rewrite this integral as

$$\int dV = -\int E_{\parallel} \, dl$$

from which it follows that

$$dV = -E_{\parallel} \, dl$$

where $E_{\parallel}$ is the component of the field parallel to the path. Rearranging things yields

[*E parallel to path*]
$$E_{\parallel} = -\frac{dV}{dl} \qquad (18.10)$$

When a positive charge moves in the direction of increasing V, that is, increasing PE_E, it moves against **E**, and that's why there is a minus sign in the equation. Where the *E*-field is strong, V changes rapidly with distance, and vice versa; *E equals the negative gradient of V.*

We can find E (in units of V/m) via Eq. (18.10) provided we know the function $V(l)$. For instance, if we have $V(r)$, that is, if there is spherical symmetry and the potential only depends on r, the path length l refers to r and therefore $E_r = -dV/dr$; the field is radial. Whereas if we have $V(x, y, z)$ the path length l, in turn, refers to either x, y, or z, and the derivative is taken holding the other two coordinates constant. To be precise, the field components are given by the *partial derivatives:* $E_x = -\partial V/\partial x$, $E_y = -\partial V/\partial y$, and $E_z = -\partial V/\partial z$. In the cases we'll generally encounter, $V(l)$ is a function of only one space variable, and ordinary derivatives will do nicely.

Example 18.8 Use Eq. (18.8) for the potential of a point-charge, namely, $V(r) = kQ/r$, to determine the electric field at some distance r from Q.

Solution: [Given: A point-charge and $V(r) = kQ/r$. Find: The E-field.] Since the potential only depends on r, so too will E; the field will be purely radial. Hence

$$E_r = -\frac{dV}{dr} = -\frac{d}{dr}(kQr^{-1}) = -\left(-k\frac{Q}{r^2}\right)$$

and

$$\boxed{E_r = k\frac{Q}{r^2}}$$

The sign of Q determines whether the field points radially inward or outward.

▶ **Quick Check:** By now you should know that this is, indeed, the field of a point-charge: Eq. (17.4).

18.9 Potential and the Arrangement of Charge

In 1672, Otto von Guericke devised a machine that greatly reduced the effort it took to build up charge by rubbing. He produced a large sphere of sulfur (because that material was easily electrified) and mounted it on a crank-driven mechanism that whirled it around so it could be stroked as it spun. Spinning within his cupped hand, it built up sizable quantities of "electric virtue" that sparked away impressively. Before long, all sorts of revolving rubbing machines were devised, and everything in sight from milk buckets to chickens was being charged by these *electrostatic generators*.

Imagine von Guericke's sulfur sphere. As it becomes increasingly more charged, it takes increasingly more work to further charge it. Clearly, there is a relationship between the charge, the geometry, and the work needed to bring yet another electron to the sphere. With a given charge Q distributed over a *large* sphere, the net force exerted on a new incoming electron (since the Coulomb force drops off as $1/r^2$) is smaller than if that same Q was on a tiny sphere. After all, on a large sphere, the individual charges that constitute Q are far apart from one another and from the new charge. Thus, *the potential of a charged sphere increases, both as its radius R decreases and as its net charge increases*. Indeed, the more tightly packed the surface charge, the more externally supplied work must be done against it, and the higher the potential. We already know that for a charged sphere of radius R

$$V = kQ\left(\frac{1}{R}\right) \qquad\qquad [18.8]$$

just as one might anticipate. Furthermore, since $Q = 4\pi R^2 \sigma$

[sphere] $$V = 4\pi kR\sigma \qquad\qquad (18.11)$$

If we have two different-size spheres, each with the same σ, the larger would hold more charge, more work would be needed to charge it, and so its potential would be higher.

An object like von Guericke's sphere cannot be charged endlessly; neither the potential nor σ keeps going up as the thing is rubbed indefinitely. The sphere, which initially strongly drew electrons from the hand, becomes highly negatively charged. The charges are increasingly crowded together until a point is reached, depending on the size of the sphere, where they strongly repel and block any further arrival of electrons. The sphere reaches a maximum potential. In contrast, the Van de Graaff generator avoids that physical limitation by having charge introduced inside a conductor, where there is no repelling field (Fig. Q14, p. 679). Its maximum potential can be so great that it is ultimately limited by the surrounding air. The very high electric field at the surface of the generator will ionize the air, causing it to be-

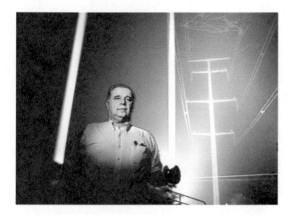

Power lines in rural areas are often operated at several hundred thousand volts. The voltage drop per meter from the line down to the zero of ground can be quite high. For this farmer, it's enough to light the fluorescent bulbs he's holding in his hands.

Figure 18.13 We have seen this diagram five or six times before. Now note that Conservation of Electromagnetic Charge is related to gauge symmetry, and that's the first clue to the importance of Gauge Theory.

come a conductor—a phenomenon known as *dielectric breakdown.* Air breaks down, ionizing and supporting sparks when the voltage across it is ≈30 kilovolts per centimeter. A little spark one-eighth inch long, the kind that comes from rug shuffling, corresponds to a potential difference of almost 10 kV.

Suppose we run an electrostatic generator at a sustained potential *V* and wish to transfer some charge from it to a neutral conducting object. Bringing the two in contact allows charge to flow from the generator to the conductor. *That flow continues until the object reaches the same potential as the generator.* Similarly, suppose a neutral conductor is brought in contact with a charged conductor. Charge will flow to the formerly neutral body *until both reach the same potential.*

When there is a potential difference between two bodies, Δ*V* equals the work that must be done per unit of positive charge to transfer charge from one body to the other. If the two are brought into contact, charge will spontaneously flow "downhill," and the potential will rapidly equalize at a value that may be positive, negative, or zero.

18.10 Conservation of Charge

Conservation of Charge had been known for over a century before physicists, guided by Noether's Principle, began to search for a corresponding symmetry associated with classical Electromagnetic Theory. What they found was an invariance arising from the arbitrariness of the electric and magnetic (which we are not going to discuss) potentials. That mathematical behavior is called *gauge symmetry,* and many now believe that it is the fundamental characteristic shared by all correct theories

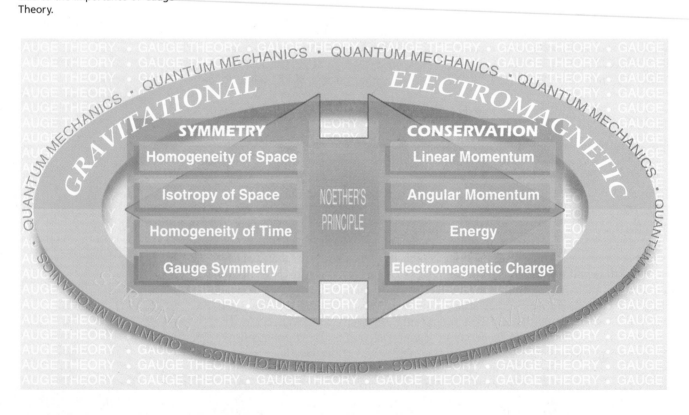

(Fig. 18.13). Conservation of Charge is related to the fact that the proper formulation of electromagnetism is gauge symmetric. That notion sounds complicated, and we will not attempt to explain gauge symmetry until Chapter 33, but we can get a sense of the central idea very simply. Let's look at just the electric-potential part of the symmetry. There is no observable effect that depends on an absolute value of electric potential. We will show that the invariance of phenomena with respect to electric potential is consistent with Conservation of Charge.

Imagine a laboratory immersed in a uniform *E*-field (Fig. 18.14). We, outside the lab, fix the arbitrary zero-potential anywhere we like and then bring a positive charge up to the lab from that $V = 0$ level. We do a precise amount of work on the test-charge and it has a corresponding precise ΔPE_E, as far as we are concerned. Yet no experiment performed *inside* the laboratory can measure the PE_E of the charge *with respect to the chosen $V = 0$ level*. In fact, the laboratory can be translated up, down, or sidewise in the uniform field, and the experimenter inside at the new location will not observe any change—that's due to the arbitrariness of the potential.

What would happen if it were possible for a charge to vanish while all the remaining laws of physics—Conservation of Energy, Conservation of Momentum, etc.—were precisely in effect? The energy possessed by the charge would have to be liberated to the system in the lab in some way if energy is to be conserved. (In whatever manner that energy is transferred, momentum must also be conserved.) The experimenter in the lab and we outside are at rest with respect to each other and must see the same amount of liberated energy. That's paradoxical—the amount of potential energy the charge possesses is arbitrary, and yet the amount of energy that would be liberated if the charge were to vanish is not. Furthermore, by measuring the energy given out, observers in the laboratory could presumably determine the charge's absolute potential, which is impossible: the $V = 0$ level can be anywhere. Hence, **a single charge cannot vanish**. On the other hand, if an equal pair of positive and negative charges was brought to the lab, *no* net work would be done against the field. They certainly could annihilate each other, or some other pair, still conserving charge, energy, and momentum. *In this Universe, energy is relative and charge must be conserved.*

CAPACITANCE

Sometime around 1706, Hauksbee replaced von Guericke's sulfur ball with a large glass globe, thereby creating a friction electric machine that was to become popular throughout Europe. By mid-century the science of electricity was drawing eager crowds, who came to see the wonders performed by itinerant lecturers with cartloads of mysterious paraphernalia. In London, one of Hauksbee's Influence Machines was even installed at the so-called Temple of Health, that it might provide a therapeutic environment around the "magnetico-electrico" bed, ostensibly to aid in matters connubial.

In these early days, rubbing machines could develop prodigious voltages but only tiny trickles of charge—the pressing problem was to store the charge, so that it could be built up and then dumped in a single powerful blast. It was soon recognized that when a conductor was electrified, the size of the conductor determined the amount of charge it could store. In the beginning, bars of metal were used, gun barrels and the like, and sometimes even people themselves, but more ambitious practitioners suspended massive cannons, charging them to tremendous levels. Any such charge-storing device was dubbed a *condenser* by Volta, but that term has now been replaced by the word **capacitor**.

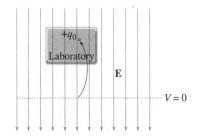

Figure 18.14 A charge in a uniform *E*-field is raised from an arbitrary zero-*V* level up to a laboratory. No experiment can be performed in the lab that will reveal the original level from which the charge was raised.

Subatomic events played out in a liquid-hydrogen bubble chamber. A gamma-ray photon (no track) descends from the top of the picture. It strikes an atom and knocks out an electron (long green track). The remainder of the photon's energy goes into creating an electron positron pair. Because of the applied magnetic field, these two low-energy particles move along tight spirals. Another invisible photon creates a second electron-positron pair further down in the picture. This time, all of the photon's energy goes into the pair. Each particle has more KE and therefore moves along a path that is nearly straight. Note how charge is conserved.

Alessandro Giuseppe Antonio
Anastasio Count Volta (1745–1827).

It was Volta who introduced the expression "electrical capacity" in analogy with the concept of heat capacity. At a given potential (V), the amount of charge (Q) that can be stored by a body depends on its physical characteristics, all of which we lump together under the name **capacitance** (C). The more charge, the greater the capacitance; while the less voltage that is needed to accomplish the feat, the greater the capacitance. In other words, C must vary directly with Q and inversely with V:

$$C = \frac{Q}{V} \tag{18.12}$$

The unit of capacitance is coulombs per volt, and to honor Faraday, it is called a *farad* (F): 1 farad $= 1$ F $= 1$ C/V. *Keep in mind that capacitance is always a positive quantity, and using the proper signs for Q and V will ensure that.*

One farad is a rather large capacitance; microfarad (1 μF $= 10^{-6}$ F) and pico-farad (1 pF $= 10^{-12}$ F) capacitors smaller than the size of a grain of rice are commonly used in radios and TV sets. But fairly hefty capacitors (about the size of a soup can) can still be found in air conditioners where large amounts of charge are dumped into the compressor motor to get it going.

To find the capacitance of a metal sphere of radius R, suppose it is carrying a charge Q. Its potential (p. 699) is then

$$V = \frac{kQ}{R}$$

and so

$$C = \frac{Q}{V} = \frac{Q}{\left(\dfrac{kQ}{R}\right)} = \frac{R}{k}$$

Hence, the capacitance of a sphere is

$$C = 4\pi\varepsilon R \tag{18.13}$$

(see Problem 78). If the sphere is surrounded by air ($k \approx k_0 = 1/4\pi\varepsilon_0$)

$$C = 4\pi\varepsilon_0 R$$

The larger the sphere, the greater the capacitance, but ($4\pi\varepsilon_0$) is very small ($\approx 10^{-10}$), and even a large sphere will have only a modest capacitance. That's true as well for other shapes that would be much harder to calculate directly, such as your body. You are a capacitor that can store enough charge at a high enough voltage to produce observable sparks. In fact, the capacitance of a human (measured while standing on 5 cm of insulation) is roughly from 100 pF to 110 pF (for people 68 kg to 105 kg, respectively). Compare that to a cow, which is a lot larger and comes in a fairly standard 200-pF model. These values reflect a good deal of interaction with the Earth and, as we will see shortly, the presence of another conductor will appreciably increase the capacitance of any object. Thus, an isolated person, several meters above the Earth, has a capacitance of only about 50 pF.

Example 18.9   Determine the capacitance of an isolated metal sphere 50 cm in diameter and immersed in vacuum.

Solution: [Given: $R = 25$ cm. Find: C.] From Eq. (18.13)

$$C = 4\pi\varepsilon_0 R = \frac{R}{k_0} = \frac{0.25 \text{ m}}{(9.0 \times 10^9 \text{ N·m}^2/\text{C}^2)} = \boxed{28 \text{ pF}}$$

▶ **Quick Check:** This sphere has an area of ≈ 0.8 m^2 as compared to the 2-m^2 area of a person whose capacitance is ≈ 50 pF.

The breakthrough in storing charge was made almost by accident in 1745 by G. von Kleist and again independently by P. van Musschenbroek. Both of these European gentlemen were experimenting with electricity and happened to insert the conductor being charged into a hand-held jar; they were probably attempting to collect "electric fluid." Van Musschenbroek dangled a brass wire, attached to a gun barrel that was being charged, into a flask "partly filled with water." It discharged, and all at once his body convulsed "as if it had been struck by lightning; . . . I thought it was all up with me," he recounted.

They had unknowingly constructed a device in which a conductor (the wire) was separated from another grounded conductor (a sweaty hand) by an insulating medium (the glass). Ordinarily, the isolated conductor being charged would rapidly reach the potential of the generator and thereafter repel any further charge. The new arrangement of conductor-insulator-conductor forestalled that cutoff. The charge put on one conductor, the central wire, induced an equal and opposite charge on the other conductor, the moist hand. That induced charge, having the opposite polarity and being relatively nearby, acted to reduce the wire's repulsion of additional charge. The result was a considerable increase in the charge stored before the device, which came to be called a *Leyden jar,* reached the potential of the generator (Fig. 18.15).

In Europe and the Colonies, anyone interested in electricity was likely to have a Leyden jar—sparks were flying everywhere (Fig. 18.16). The Abbé Nollet had 180 of Louis XV's fearless guards join hands in a circle, or *circuit.* With the first man holding the outer terminal of a charged Leyden jar, the last victim gleefully touched the central wire and shocked the whole assembly.

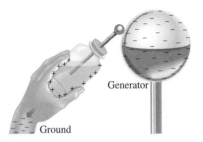

Figure 18.15 Charging a Leyden jar with the help of a generator and a grounded experimenter.

18.11 The Parallel Plate Capacitor

The renowned tamer of lightning and writer of racy prose, Ben Franklin, had his Leyden jars, too, but he went one step further. Franklin was among the first to use a new and more convenient configuration consisting of flat metal plates separated by sheets of window glass. That simple arrangement allowed for a dramatic increase in the size of the conductor-insulator-conductor sandwich. Franklin's flattened Leyden jar is a **parallel plate capacitor**.

At the end of the eighteenth century, Volta carried out a series of measurements of the potentials of large charged objects. He attached the test object via a conductor to a grounded electroscope. The leaves would then spring apart in proportion to the charge impressed on them. That, in turn, is proportional to the potential of the scope, which equals the potential of the object. In other words, the angle of the leaves of the electroscope is effectively proportional to the difference in potential between them and the grounded case. Suppose the test object is a positively charged flat plate as in Fig. 18.17. If an identical grounded plate is brought nearby, the leaves gradually descend as it approaches. In effect, the negative charge induced on the new plate will draw up some of the positive charge from the leaves, holding it fixed on the near side of the positive plate. But that means that introducing the second (now negatively charged) plate drops the potential of the first plate, which can be restored to its original value by further increasing the charge. *Thus, the second plate considerably enhances the ability of the parallel-plate capacitor to store charge at a given voltage.*

We now determine the capacitance of the parallel plate capacitor (Fig. 18.18) as a function of its physical characteristics. If the plates carry opposite charges of $\pm Q$ and have a difference of potential ΔV between them, then $C = Q/\Delta V$. We can calculate this potential difference in terms of the E-field between the plates starting with

Figure 18.16 A Leyden jar. The improved version is lined with metal foil. A small chain connects the central wire with the inner metal surface.

Figure 18.17 The operation of the capacitor as a charge storer. (a) The charges on a single plate experience a repulsive force, which contributes to establishing the potential. In (b), we bring a grounded neutral plate close to the positive plate. A group of negative charges are drawn onto this second plate, essentially neutralizing much of the positive charge, and reducing the potential.

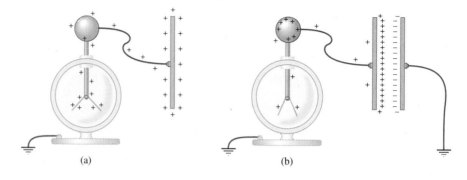

(a)   (b)

$$\Delta V = Ed$$

from Eq. (18.6), where d is the plate separation. Recall that

$$E = \frac{\sigma}{\varepsilon} = \frac{Q}{A\varepsilon} \qquad [17.16]$$

wherein A is the area of *each* plate and $\sigma = Q/A$. Hence

$$\Delta V = Ed = \frac{Qd}{A\varepsilon}$$

and so

$$C = \frac{Q}{\Delta V} = \frac{Q}{Qd/A\varepsilon}$$

which simplifies to

$$C = \frac{\varepsilon A}{d} \qquad (18.14)$$

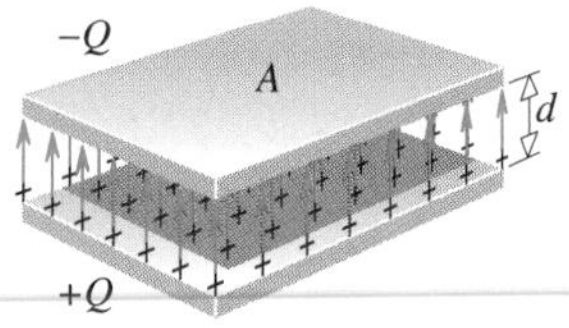

Figure 18.18 A charged parallel plate capacitor. Each plate has an area A and stores a charge Q.

To produce as large a capacitance as possible, we must make A large, d small, and use a material in the gap with a large permittivity (Table 17.3).

Example 18.10 Determine the size of a 1.00-F parallel plate capacitor if the plates are square and separated by 1.00 mm of air. How would things change if the gap were filled with a sheet of glass having a relative permittivity, or dielectric constant, of 10?

Solution: [Given: $C = 1.00$ F, $d = 1.00$ mm, $\varepsilon = $ either ε_0 or $10\varepsilon_0$, and $A = L \times L$. Find: L.] This problem involves the physical characteristics of a parallel plate capacitor and one equation should come to mind immediately:

$$C = \frac{\varepsilon A}{d} \qquad [18.14]$$

Hence, with air

$$A = L^2 = \frac{dC}{\varepsilon_0} = \frac{(1.00 \times 10^{-3} \text{ m})(1.00 \text{ F})}{(8.85 \times 10^{-12} \text{ C}^2/\text{N·m}^2)}$$

$$A = 0.113 \times 10^9 \text{ m}^2$$

and $\boxed{L = 10.6 \text{ km}}$. The plates are gigantic, about 6.6 miles on a side. When glass replaces air, ε replaces ε_0, and since $\varepsilon = 10\varepsilon_0$, the area is 10 times smaller and $L = (10.6 \text{ km})/\sqrt{10} = 3.35$ km.

▶ **Quick Check:** $C \approx (10^{-11} \text{ C}^2/\text{N·m}^2)(10^8 \text{ m}^2)/(10^{-3} \text{ m}) \approx 1$ F.

A dielectric has a larger value of permittivity than vacuum and therefore a smaller internal field ($E = \sigma/\varepsilon$), as is evident in Fig. 18.19. The dielectric becomes polarized in the external field and takes on a surface charge. As a result, a small

Figure 18.19 The effect of the dielectric in a parallel plate capacitor. It becomes polarized and thereby reduces the internal field. (a) Polarization of the dielectric. (b) The *E*-field due to the polarized dielectric is opposite the *E*-field due to the charge on the plates. (c) The internal *E*-field is reduced, and more charge can be stored at a given potential difference.

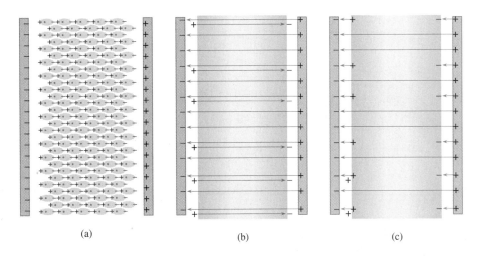

(a)   (b)   (c)

internal *self-field* opposes the applied field within the dielectric. The effect is a weaker field in the gap and a lower voltage ($\Delta V = +Ed$) across it. Slipping a sheet of insulation into a charged capacitor (Fig. 18.20) effectively cancels some of the charge on the plates via the opposite polarity-induced charge on the surfaces of the dielectric.

Modern parallel plate capacitors come in different forms. The most common is a sandwich of metal foil (aluminum), dielectric (waxed paper, mylar, etc.), and metal foil, rolled up into a tight little cylinder and sealed (Fig. 18.21). It's represented symbolically in diagrams by two parallel lines of equal length.

A computer keyboard. When a key is depressed, the spacing of the plates in an air capacitor changes, thereby changing the capacitance and registering the keystroke.

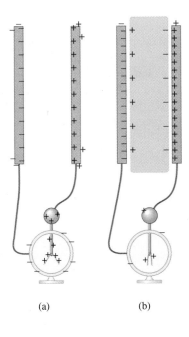

(a)   (b)

Figure 18.20 (a) A charged capacitor attached to an electroscope that indicates the potential difference. (b) Inserting a dielectric essentially neutralizes some of the charge on the plates and lowers the potential difference.

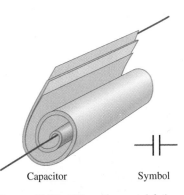

Capacitor   Symbol

Figure 18.21 When the metal foil, dielectric, metal foil sandwich is rolled up and sealed, it forms a parallel plate capacitor with a fixed capacitance.

Capacitors in the circuit of the AM radio shown on p. 645.

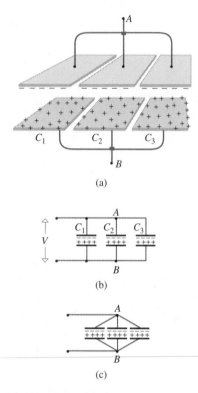

Figure 18.22 (a) Three capacitors in parallel represented in several effectively identical ways. (b) The equivalent capacitance is $C = C_1 + C_2 + C_3$. The circuits in (a), (b), and (c) are all electrically the same.

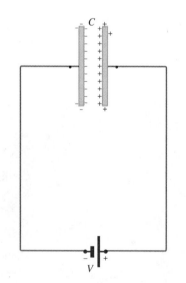

Figure 18.23 A capacitor across a battery. The capacitor's plates become charged as shown, and there is a voltage V across them.

18.12 Capacitors in Combination

Inasmuch as the early interest in capacitors was purely for charge storage, it's not surprising that Leyden jars were wired to one another to increase that ability. To the same end, Franklin connected almost a dozen parallel plate capacitors together to produce great blasts of electricity.

The Parallel Circuit

Franklin, like others, used two basic wiring schemes to create two different results. Figure 18.22 shows the so-called **parallel** arrangement of three capacitors, though any number of them could be connected that way. The distinguishing characteristic of the parallel scheme is that one terminal (or plate) from each capacitor is connected to the same common wire, and all the remaining plates are connected to a second common wire. All the top plates are attached and must be at the same potential, and all the bottom plates are attached and must be at the same potential. **When a wire is used to connect two points together, those two points, and every point along the wire, are assumed to be at the same potential.** The result here is essentially one large capacitor whose two plates are each segmented into three connected parts.

Figure 18.23 shows a capacitor attached across a battery, which is represented by the standard symbol, ⊣⊢. The longer lighter line corresponds to the higher potential and is labeled +. It is V volts higher than the other terminal, labeled −. Negative charge flows out of the − terminal of the battery in the form of electrons. These charges gradually build up on the negative low-potential plate of the capacitor. They, in turn, repel an equal number of electrons off the other plate, leaving it positive, and this charge circulates back to the battery. The path is called a **circuit** and electrons continue to flow, as if around the circuit, for the brief time until the capacitor reaches the same potential difference as the battery. At that point, any further charge is repelled by the capacitor as forcefully as it is propelled by the battery. There is a field traversing the gap, and the voltage difference across the capacitor is given by Eq. (18.6), $\Delta V = Ed$. This difference increases during the charging until it finally equals the battery's voltage, at which time the charging ceases.

It is customary to write the potential difference (ΔV) across any circuit element (capacitors, batteries, resistors, etc.) simply as V. In Fig. 18.22, where the capacitors are in parallel, the voltage across the combination V must equal the voltage across each one:

$$V = V_1 = V_2 = V_3$$

The three capacitors will reach exactly the same state if they are charged to voltage V all as a single unit, or if each is charged separately and then attached together.

An air capacitor. This variable capacitor is from the tuning circuit of the radio shown on p. 645. Turning the knob of the radio causes a set of movable metal plates to slide between a set of stationary plates, thereby changing the capacitance and the station.

The net charge stored (Q) is the sum of the individual amounts stored on each capacitor, and so

$$Q = Q_1 + Q_2 + Q_3$$

Hence

$$CV = C_1V_1 + C_2V_2 + C_3V_3$$

and

[*in parallel*] $$C = C_1 + C_2 + C_3 \qquad (18.15)$$

where we just add on more terms if there are more capacitors. The **equivalent capacitance of several capacitors in parallel is the sum of all the individual capacitances**. A single capacitor C, given by Eq. (18.15), would be electrically indistinguishable from the parallel array of *smaller* capacitors C_1, C_2, and C_3.

There are many kinds of circuit elements, and they can all be in parallel with each other provided both terminals of one are connected to both terminals of another. *A place where three or more leads come together is called a* **node**. The two regions in Fig. 18.22 where the three terminals are connected, top and bottom, constitute two nodes (A and B). *When elements are in parallel, any number of leads or circuit branches can converge at a node*.

Example 18.11 Figure 18.24 shows two capacitors attached to a 12-V battery. Determine the equivalent capacitance and the charge it would carry. What is the charge on each of the capacitors in the figure?

Solution: [Given: $C_1 = 20$ μF, $C_2 = 30$ μF, and $V = 12$ V. Find: C, Q, Q_1, and Q_2.] First, notice that the top terminals of both capacitors are connected, as are the bottom terminals—the capacitors are in parallel. You might also want to redraw the diagram, as is done in part (b). Since the potential across each capacitor is 12 V

$$V = \frac{Q_1}{C_1} = \frac{Q_2}{C_2}$$

$$Q_1 = (12 \text{ V})C_1 \quad \text{while} \quad Q_2 = (12 \text{ V})C_2$$

Thus

$$Q_1 = (12 \text{ V})(20 \times 10^{-6} \text{ F}) = 2.4 \times 10^{-4} \text{ C}$$

and

$$Q_2 = (12 \text{ V})(30 \times 10^{-6} \text{ F}) = 3.6 \times 10^{-4} \text{ C}$$

Since the capacitors are in parallel

$$C = C_1 + C_2 = (20 \text{ } \mu\text{F}) + (30 \text{ } \mu\text{F}) = \boxed{50 \text{ } \mu\text{F}}$$

The charge it would carry is gotten from $C = Q/V$; namely

$$Q = CV = (50 \times 10^{-6} \text{ F})(12 \text{ V}) = \boxed{6.0 \times 10^{-4} \text{ C}}$$

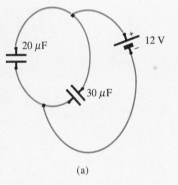

(a)

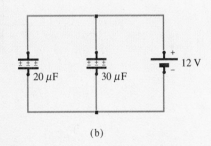

(b)

Figure 18.24   (a) A circuit of two capacitors and a battery. (b) A redrawn equivalent.

▶ **Quick Check:** For parallel capacitors we must have $Q = Q_1 + Q_2 = 2.4 \times 10^{-4} \text{ C} + 3.6 \times 10^{-4} \text{ C} = 6.0 \times 10^{-4} \text{ C}$.

The Series Circuit

Another basic way to connect circuit elements is known as **series**, and it's illustrated in Fig. 18.25. Here *one and only one terminal of a circuit element is connected to one and only one terminal of an adjacent circuit element. Two elements will not be in series if any other branch in the circuit connects to the point at which the two are attached.* Imagine that the three series capacitors are put across a battery of voltage V, as in Fig. 18.25c. Electrons travel from the battery to the negative plate of C_3, giving it a charge of $-Q$. These, in turn, repel an equal quantity of charge ($-Q$) from the positive plate of C_3 to the negative plate of C_2. As the negative plate of C_2 charges up to $-Q$, it repels an equal number of electrons from its positive plate to the negative plate of C_1. An amount of electrons equivalent to $-Q$ is repelled back to the positive terminal of the battery. Hence

$$Q = Q_1 = Q_2 = Q_3$$

Notice that the net charge stored by the three capacitors is effectively $+Q$ on the left-most plate and $-Q$ on the right-most plate; the rest of the charge on the remaining plates ($-Q$ and $+Q$ and $-Q$ and $+Q$, respectively) cancels out—only Q's worth of electrons went in on the right and out on the left. Thus, the equivalent capacitor will also have a charge Q (which is *not* the sum of the charges on the individual capacitors).

The equivalent capacitor will be across the battery, just as is the series string of C_1, C_2, and C_3. Hence, the voltage across C is V. If we go from point A on the left to point D on the right, we will, by necessity, drop in potential by an amount V; the battery is connected to points A and D. Yet, in going from point A to point B, we drop V_1; in going from B to C, we drop an additional V_2; in going from C to D, we drop another amount V_3; hence

$$V = V_1 + V_2 + V_3$$

Using the definition, Eq. (18.10), $V = Q/C$, and

$$\frac{Q}{C} = \frac{Q_1}{C_1} + \frac{Q_2}{C_2} + \frac{Q_3}{C_3}$$

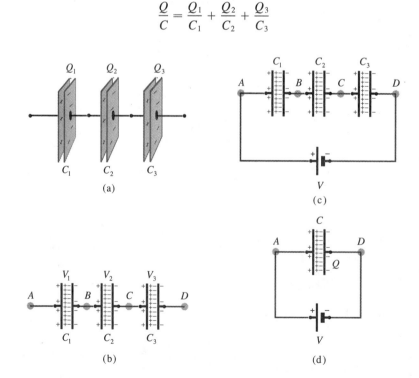

Figure 18.25 (a) Three capacitors in series and (b) their graphic representation. (c) The same series capacitors across a battery. (d) The equivalent circuit wherein $\frac{1}{C} = \frac{1}{C_1} + \frac{1}{C_2} + \frac{1}{C_3}$.

But all the charges are equal and therefore

$$\frac{1}{C} = \frac{1}{C_1} + \frac{1}{C_2} + \frac{1}{C_3}$$ (18.16)

and we just keep adding on terms if there are more capacitors.

Example 18.12 The circuit shown in Fig. 18.26a consists of a 12-V battery and three capacitors. Determine both the voltage across and charge on each capacitor after the switch S is closed and electrostatic equilibrium is established. Find the equivalent capacitance of the network.

Solution: [Given: $C_1 = 2.0$ μF, $C_2 = 2.0$ μF, $C_3 = 5.0$ μF, and $V = 12$ V. Find: C, V_1, V_2, V_3, Q_1, Q_2, and Q_3.] First, redraw the circuit as in Fig. 18.25b to make things a bit clearer. The two 2.0-μF capacitors are in series, and their equivalent capacitance, Eq. (18.16), is

$$\frac{1}{C} = \frac{1}{C_1} + \frac{1}{C_2}$$

which is easy to compute with a calculator. However, if you would rather do it in your head, remember the equivalent form (which is left as a problem to prove)

$$C = \frac{C_1 C_2}{C_1 + C_2} = \frac{(2.0\ \mu\text{F})(2.0\ \mu\text{F})}{2.0\ \mu\text{F} + 2.0\ \mu\text{F}} = 1.0\ \mu\text{F}$$

As shown in Fig. 18.26c, this 1.0-μF equivalent of the series pair is itself in parallel with the 5.0-μF capacitor, and so they can be combined into a single capacitor via Eq. (18.15); namely,

$$C = 5.0\ \mu\text{F} + 1.0\ \mu\text{F} = \boxed{6.0\ \mu\text{F}}$$

which is the equivalent capacitance of the whole network. Now, working back from the simplified diagram of Fig. 18.26c, $V = 12$ V and so

$$Q_3 = C_3 V_3 = (5.0\ \mu\text{F})(12\ \text{V}) = \boxed{60\ \mu\text{C}}$$

There are 12 V across the combination of the two 2.0-μF capacitors and, hence, there must be a potential difference of $\boxed{6.0\ \text{V}}$ across each one. Therefore

$$Q_1 = Q_2 = (2.0\ \mu\text{F})(6.0\ \text{V}) = \boxed{12\ \mu\text{C}}$$

▶ **Quick Check:** The charge on the equivalent capaci-

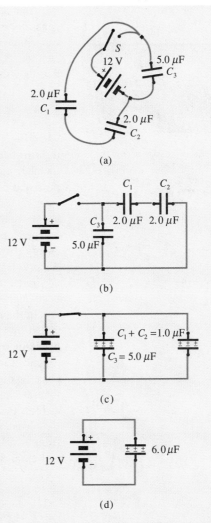

Figure 18.26 (a) A capacitive circuit and (b) a redrawn version. (c) A simplified configuration and (d) the simplest equivalent circuit.

tor is $Q = CV = 72$ μC. This must equal the net charge on the two series capacitors (that is, the charge on either one—namely, 12 μC) plus $Q_3 = 60$ μC, and, happily, it does.

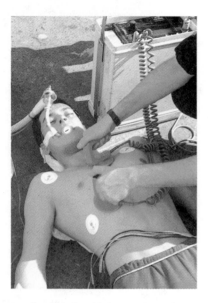

Sending a pulse of charge across the chest can start the heart beating with a steady rhythm. The necessary energy is first built up and stored in a capacitor.

Until now, we have assumed that a capacitor placed across a voltage source quickly reaches electrostatic equilibrium, which is actually something of an idealization. A real capacitor leaks charge across its dielectric spacer at a rate dependent on the geometry, the insulating material, and the voltage across it. A high-voltage mica capacitor might have over 1000 times more resistance (p. 733) to this flow of charge than does a low-voltage paper capacitor, even if they have the same capacitance. When two such capacitors are connected in series with a source of several hundred volts, some charge will continue to flow, electrostatic equilibrium will not be established, and the voltage across the mica one will be proportionately larger than that across the paper one. Thus, at substantial applied voltages, the charges are generally *not* equal on each different series capacitor even when they have the same capacitance. As a result, Eq. (18.16) does not hold in such cases. We limit our discussion to electrostatic equilibrium and will not worry about high voltages and leakage, but it's good to remember that this treatment is restricted.

18.13 Energy in Capacitors

Charging a parallel plate capacitor requires that work must be done on the charges to bring them from wherever they are to the plates. Each electron is forced over to the plate against the repulsive action of all the other electrons that preceded it. Suppose that we wish to electrify a capacitor with a total charge Q to a potential difference V, starting from a potential difference of zero. The end result is two oppositely charged plates in close proximity. The amount of energy stored in the process is independent of the details of how the capacitor got charged, just as the amount of energy stored via a boulder on a mountaintop is independent of how it got up there. Accordingly, let's work with the simplest scenario. Assume that we carry an amount of charge $+Q$ from one plate across the gap to the other, thereby leaving the first plate charged with $-Q$ (Fig. 18.27).

In the beginning, when the voltage is near zero and each new charge is only repelled by the few previous arrivers, the work done is small. As the charge on the plates increases, the voltage increases, the repulsive force increases, and the work expended increases; this process goes on until a potential difference of V is reached. Once again, we are faced with determining the work done in a process where the force is changing, but this time things are simple. We can treat this sort of problem as if all the charge were transported at once against an average potential difference.

Because $V = Q/C$, the variation in voltage as charge is built up is linear, going from 0 to V. Hence, the average potential difference is just the sum of the initial 0 and final V divided by two (remember Fig. 3.6 and the Mean-Speed Theorem, p. 77): $V_{av} = (0 + V)/2 = \frac{1}{2}V$. Hence, the work done in charging the capacitor, $W = QV_{av}$, is

$$W = Q\tfrac{1}{2}V$$

and if we think of this energy as being stored as *electrical*-PE, we have

$$\mathrm{PE_E} = \tfrac{1}{2}QV \tag{18.17}$$

Equivalently

$$\mathrm{PE_E} = \tfrac{1}{2}CV^2 = \tfrac{1}{2}Q^2/C \tag{18.18}$$

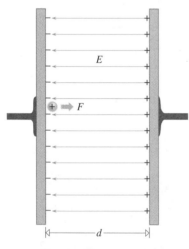

Figure 18.27 Carrying a charge through a distance d against a field requires an external force and the doing of work.

If we want to know how PE_E varies with, say, Q, we must pick the expression that contains Q and no other quantity depending on Q; thus, $PE_E = \frac{1}{2}Q^2/C$ tells us that doubling Q quadruples PE_E.

Example 18.13   How much energy is stored in each of the capacitors of Fig. 18.26 in the process of charging them?

Solution: [Given: $C_1 = 2.0$ μF, $C_2 = 2.0$ μF, $C_3 = 5.0$ μF, and $V = 12$ V. Find: the potential energy stored in each capacitor.] Using Eq. (18.17) and the results of Example 18.12, namely, $Q_1 = Q_2 = 12$ μC and $Q_3 = 60$ μC, we get

Capacitor 1 $PE_E = \frac{1}{2}QV = \frac{1}{2}(12 \; \mu C)(6.0 \; V)$

$$\boxed{PE_E = 36 \; \mu J}$$

Capacitor 2 $PE_E = \frac{1}{2}QV = \frac{1}{2}(12 \; \mu C)(6.0 \; V)$

$$\boxed{PE_E = 36 \; \mu J}$$

Capacitor 3 $PE_E = \frac{1}{2}QV = \frac{1}{2}(60 \; \mu C)(12 \; V)$

$$\boxed{PE_E = 0.36 \; mJ}$$

for a grand total of 0.43 mJ.

▶ **Quick Check:** If the equivalent capacitor is truly equivalent to all the capacitors in the system, it should store the same amount of energy. Using Eq. (18.18), it follows that

$$PE_E = \frac{1}{2}CV^2 = \frac{1}{2}(6.0 \; \mu F)(12 \; V)^2 = 432 \; \mu J$$

Figure 18.28 shows a charged capacitor and therefore one with a potential difference across its plates. Each electron, given the opportunity, would spontaneously descend in potential energy. Thus, when a wire—a so-called *short circuit*—is connected across the capacitor, electrons immediately move from the $-$ plate to the $+$ plate, canceling the charge and reducing the potential difference across the capacitor to zero.

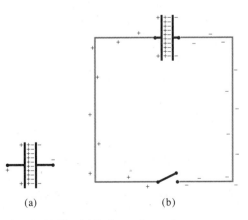

(a)   (b)

Figure 18.28   (a) A charged capacitor. (b) Closing the switch would allow charge to flow and cancel, leaving a net charge of zero.

Example 18.14 The two capacitors C_1 and C_2 in Fig. 18.29 were first put across different batteries so that they took on voltages of $V_1 = 12$ V and $V_2 = 6.0$ V, respectively. They were then attached as shown. Compute the charge and energy stored in each capacitor once the switch is closed and electrostatic equilibrium restored.

Solution: [Given: $C_1 = 4.0$ μF, $C_2 = 2.0$ μF, $V_1 = 12$ V, and $V_2 = 6.0$ V. Find: the potential energy stored and the charge on each capacitor once the switch is closed.] A wire connects the two positive plates even before the switch is closed, but no redistribution of charge can occur because the charges are effectively bound in place on each capacitor by the oppositely charged opposing plate. Even after the switch is closed, electrons on the negative plates cannot get to the positive plates, and although the charge will redistribute itself, the net amount $Q_1 + Q_2$ is fixed. Let's find the two final charge distributions. Two unknowns will require two equations, so begin by determining the net charge. Before they were attached to one another

$$Q_1 = C_1 V_1 = (4.0 \ \mu\text{F})(12 \ \text{V}) = 48 \ \mu\text{C}$$

and

$$Q_2 = C_2 V_2 = (2.0 \ \mu\text{F})(6.0 \ \text{V}) = 12 \ \mu\text{C}$$

The net charge is

$$Q_1 + Q_2 = 60 \ \mu\text{C}$$

Once connected, the charge redistributes itself until any potential difference between the two positive plates, and between the two negative plates, vanishes. Each capacitor has the same final voltage V across it (they're in parallel), and

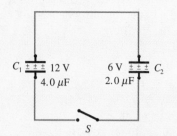

Figure 18.29 Two charged capacitors in a circuit containing a switch S.

$$V = \frac{Q_1}{C_1} = \frac{Q_2}{C_2}$$

whereupon

$$Q_1 = \frac{Q_2(4.0 \ \mu\text{F})}{2.0 \ \mu\text{F}} = 2Q_2$$

But the total charge is 60 μC; hence $\boxed{Q_1 = 40 \ \mu\text{C}}$ and $\boxed{Q_2 = 20 \ \mu\text{C}}$. To find the potential energy we use Eq. (18.18), and thus for C_1

$$\text{PE}_{\text{E1}} = \frac{\frac{1}{2}Q_1^2}{C_1} = \frac{\frac{1}{2}(40 \ \mu\text{C})^2}{4.0 \ \mu\text{F}} = \boxed{0.20 \ \text{kJ}}$$

for C_2

$$\text{PE}_{\text{E2}} = \frac{\frac{1}{2}Q_2^2}{C_2} = \frac{\frac{1}{2}(20 \ \mu\text{C})^2}{2.0 \ \mu\text{F}} = \boxed{0.10 \ \text{kJ}}$$

▶ **Quick Check:** The larger capacitor, which has twice the capacitance of the other, stores twice the charge and twice as much energy in reaching the same voltage.

The energy stored in a capacitor can be imagined as associated with the electric field that exists in the gap. We do work charging up the capacitor or, equivalently, establishing the field. Looking at the region between the plates, $C = \varepsilon_0 A/d$ and $V = Ed$, thus

$$\text{PE}_{\text{E}} = \frac{1}{2}CV^2 = \frac{1}{2}\left(\frac{\varepsilon_0 A}{d}\right)(Ed)^2 = \frac{1}{2}\varepsilon_0(Ad)E^2 \tag{18.19}$$

The electrical energy stored by the E-field is proportional to the square of the strength of the field (E^2). In this sense, the electromagnetic field (as we shall see in Chapter 24) can be treated as a tenuous form of matter carrying energy and momentum.

One good way to provide a tremendous blast of electrical energy is to store charge in a giant capacitor. This capacitor bank at the Lawrence Livermore National Laboratory supplies 600 MJ of energy to the Nova laser, the most powerful in the world.

Core Material

The **electric potential** V, or just the *potential,* is defined as

$$\text{Electric potential} = \frac{\textit{electrical}\text{-PE}}{\text{charge}} \qquad [18.1]$$

where 1 J/C = 1 V. The most general statement of the relationship between the potential difference at two points in space and the electric field permeating that space is

$$\Delta V = -\int_{P_i}^{P_f} E_\parallel dl = -\int_{P_i}^{P_f} \mathbf{E} \cdot d\mathbf{l} \qquad [18.3]$$

The potential difference between two points P_i and P_f numerically equals the work done against the field in moving a unit positive charge from P_i to P_f with no acceleration:

$$\Delta V = V_{P_f} - V_{P_i} = \frac{W_{P_i \to P_f}}{q_0} \qquad [18.5]$$

In a *uniform electric field,* the potential difference is

$$V_B - V_A = \pm Ed \qquad [18.6]$$

The potential difference is + when the displacement has a component that is opposite to the field and − when it has a component parallel to the field.

When a positive charge is moved from r_A to r_B against the $1/r^2$-electric field of a *sphere* of positive charge Q

$$V_B - V_A = kQ\left(\frac{1}{r_B} - \frac{1}{r_A}\right) \qquad [18.7]$$

The potential at a distance r from a positive point-charge, measured with respect to the zero at ∞, is given by

$$V = kQ\left(\frac{1}{r}\right) \qquad [18.8]$$

The potential of a pointlike charge element dq is

$$[V = 0 \text{ at } \infty] \qquad V = k\int \frac{dq}{r} \qquad [18.9]$$

The E-field can be found from the potential using

$$[E \text{ parallel to path}] \qquad E_\parallel = -\frac{dV}{dl} \qquad [18.10]$$

Capacitance (C) is a measure of *the capacity to store charge:*

$$C = \frac{Q}{V} \qquad [18.12]$$

the units of which are coulombs per volt, where 1 farad = 1 F = 1 C/V. For a parallel plate capacitor

$$C = \frac{\varepsilon A}{d} \qquad [18.14]$$

When two or more capacitors are connected, the equivalent capacitance for them in **parallel** is

$$C = C_1 + C_2 + C_3 + \cdots \qquad [18.15]$$

and in **series**

$$\frac{1}{C} = \frac{1}{C_1} + \frac{1}{C_2} + \frac{1}{C_3} + \cdots \qquad [18.16]$$

The energy stored in the process of charging a capacitor is

$$\text{PE}_E = \tfrac{1}{2}QV = \tfrac{1}{2}CV^2 = \tfrac{1}{2}Q^2/C \qquad [18.18]$$

Suggestions on Problem Solving

1. When calculating the PE_E of a charge, be careful with signs. An electron at a positive potential has a negative potential energy. Similarly, when a charge "falls" through a potential difference, thereby losing potential energy, the corresponding change in KE is positive. A negative charge "falls" (in the opposite direction to the E-field) to a *higher potential;* a positive charge "falls" (in the same direction as the E-field) to a lower potential.

2. If a parallel-plate capacitor has charges on its two plates of $+Q$ and $-Q$, the charge on the capacitor is Q, and that value is indicated in the defining equation $C = Q/V$. Similarly, the area A is essentially the area of overlap, generally the area of *either* plate.

3. When looking at equations, read C as capacitance and Q as charge. But C is a coulomb, which is the unit of charge. Don't confuse them.

4. Keep the following quantities handy: $\varepsilon_0 = 8.85 \times 10^{-12}$ C^2/N·m^2, $1/\varepsilon_0 = 1.129 \times 10^{11}$ N·m^2/C^2, $4\pi\varepsilon_0 = 1.113 \times 10^{-10}$ C^2/N·m^2, and $1/4\pi\varepsilon_0 = k_0 = 8.988 \times 10^9$ N·m^2/C^2.

5. Remember that Eq. (18.14) yields an equivalent capacitance for a series string of capacitors, which is always less than the smallest capacitance in the group. When applying this equation, be wary. Don't substitute into the right side, carry out the math, and present the result as the answer, forgetting that you have actually computed $1/C$ and not C. For only two capacitors you might find the following form more convenient

$$C = \frac{C_1 C_2}{C_1 + C_2}$$

In any event, you could use it for a quick check to see that the magnitude of your answer is right.

6. To remember the set of equations for a point-charge Q, memorize only Coulomb's Law: force is kqQ/r^2. By definition, divide that quantity by charge to get the field, kQ/r^2; in effect, multiply force by distance to get work and PE, kqQ/r; by definition, divide that quantity by q to get potential, kQ/r. Although simplistic and not rigorous, this scheme is helpful as a memory device.

Discussion Questions

1. A negative charge in an electric field moves from a point where the potential is zero to a point where it is -100 V. Discuss the energy change and the work done.

2. Imagine a hollow, spherical, positively charged conductor of radius R that is far away from any other bodies. Draw a graph of its potential V, as a function of r out from its center. Discuss your results.

3. With Question 2 in mind, suppose we place a neutral hollow conducting sphere in the vicinity of the charged sphere. Draw a rough graph of the potential along the center line of both spheres and compare it to the previous potential without the neutral conductor. Discuss your results.

4. With Question 3 in mind, suppose we now ground the neutral conductor. Describe the potential along the center line. Discuss your results if they are any different from those of Question 3.

5. Using equipment like that depicted in Fig. 17.34, Faraday showed that when a neutral conductor touched the inside of a hollow charged conductor, it remained neutral. Use that result to justify the conclusion that *provided there is no external field, a neutral conductor assumes the potential of the region of space in which it is introduced.* What happens to the potential of the neutral conductor (the little sphere) if the charge on the surrounding body in Fig. 17.34 is increased?

6. Describe the potential of a negative point-charge as a function of distance r. Explain your thinking.

7. A small, neutral spherical conductor is placed between the plates of a large, charged parallel plate capacitor. Describe the E-field and the equipotentials in the gap.

8. A positive point-charge is located a short distance above a large conducting horizontal plane. Describe the field lines and equipotentials. Compare your results with that of an electric dipole. Explain your conclusions.

9. Imagine a square with point-charges at each corner. At the ends of one diagonal, the charges are $+q$; at the ends of the other they are $-q$. This arrangement is called an *electric quadrupole.* Make a rough sketch of the field lines and equipotentials and explain your reasoning.

10. Figure Q10 shows a positive point-charge located at the corner of two intersecting conducting planes. Describe the E-field and the equipotentials. Compare your answer to that of the previous question. Explain your observations.

Figure Q10

11. Is it correct to maintain that *when a positive charge is deposited on a body it raises the potential of that body and, moreover, it raises the potential of the entire vicinity, including that of any other bodies in the vicinity?* Contrarily, *does the introduction of a negative charge lower the potential of every point in the neighboring region?* Explain your answer in detail. How might you prove your conclusion experimentally?

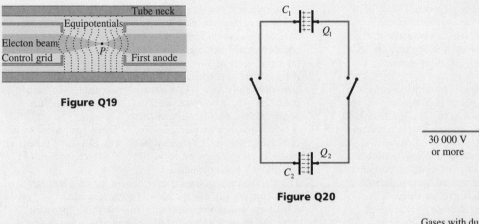

Figure Q19

Figure Q20

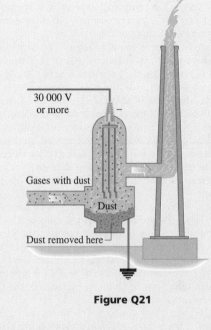

Figure Q21

12. Is it correct to maintain that, in general, *the potential of a conductor, which certainly depends on its own charge, also depends on the distribution of charge everywhere else nearby?* Under what circumstances, if any, does the potential of a charged conductor only depend on its charge, structure, and the medium it is in?

13. We know that points infinitely far from all charge are at zero potential and yet, as a practical matter, we take the Earth, which may well be charged, to be at zero-potential. How can this inconsistency be resolved?

14. Water has a rather large dielectric constant $(\varepsilon/\varepsilon_0)$. How might that characteristic contribute to its ability to keep substances such as table salt, once dissolved, in solution?

15. Given a region that is an equipotential volume, what can you say about the possibility that there is a net charge within it?

16. Given that a particular equipotential volume (different from its surroundings) is bounded by an equipotential surface, what can you say about the charge, if any, on that surface?

17. It is sometimes desirable, especially in high-voltage applications, to replace a capacitor by an equivalent string of several capacitors in series. Discuss the possible reasons for this. How do the sizes of the several series capacitors compare to the original single one?

18. Envision a parallel plate capacitor across the terminals of a battery. When a dielectric is inserted between the plates, the capacitance increases, as does the charge, and more energy is stored by the device. Where is that additional energy stored?

19. Figure Q19 is a profile view showing the equipotentials in the space between the control grid and the first anode of an electron gun (Fig. 18.4). The holes in the ends of the metal surfaces of the grid and anode cause a distortion of the equipotentials, and the arrangement serves as an *electron lens*. The beam is brought to a focus at the "crossover" point *P*. Explain how this process happens. (Incidentally, another such lens in the second anode focuses the beam onto the screen).

20. The two identical capacitors in Fig. Q20 are charged at different voltages such that $Q_1 > Q_2$. What will be the charge on each capacitor after the two switches are closed? Explain your answer completely.

21. The Cottrell precipitator is shown mounted to a chimney in Fig. Q21. How do you think it works to remove 99% of the ash and dust that passes through it?

Multiple Choice Questions

1. The quantity 1 C·V is equivalent to   (a) 1 V/m   (b) 1 N·m   (c) 1 C/N   (d) 1 V/N   (e) none of these.

2. In the case of a nonconductor   (a) its surface must be at a single potential   (b) its entire volume, except for the surface, is at a constant potential   (c) different regions may well be at different potentials   (d) the potential must be zero everywhere within it   (e) none of these.

3. Electric field lines always point toward   (a) ground   (b) a region of higher potential   (c) a region of lower potential   (d) positive charge   (e) none of these.

4. The potential as we get closer and closer to a point-charge

(a) approaches $\pm\infty$ (b) is zero   (c) is indeterminate   (d) is exceedingly small, but not zero   (e) none of these.

5. Given a group of nearby charges whose net value is nonzero, the equipotential surface at a very great distance is   (a) nearly a plane   (b) nearly a sphere   (c) quite indeterminate   (d) nearly a cylinder   (e) none of these.

6. Any closed equipotential surface that does not surround a net charge must   (a) be at zero-potential   (b) be a sphere   (c) enclose an equipotential volume   (d) be infinitely small   (e) none of these.

7. Suppose we examine a charged metal cup with a device

that measures potential with respect to ground. What will happen as the probe from the device that touches the cup changes from contact with the outside to contact with the inside? (a) the reading will ascend (b) the reading will descend partway (c) the reading will go to zero (d) nothing (e) none of these.

8. A volume of space is found to have a constant potential everywhere within it. It follows that in that region (a) the E-field is zero (b) the potential is zero (c) the E-field is finite and uniform (d) the potential gradient is a nonzero constant (e) none of these.

9. When two charged metal objects are connected to each other by a conducting wire, the one that gains electrons or, equivalently, loses positive charge is said, in comparison to the other object, to have had (a) a greater electrical potential energy (b) a lower capacitance (c) a lower dielectric constant (d) a higher potential (e) none of these.

10. Electrical potential determines the flow of positive charge just as (a) pressure determines the flow of fluid, and temperature the flow of thermal energy (b) kinetic energy determines the flow of matter, and charge the flow of electricity (c) temperature determines the flow of entropy, and entropy the flow of heat (d) power determines the flow of fluid, and efficiency the flow of work (e) none of these.

11. Generally, when any conductor is connected to ground (a) nothing happens (b) charge flows so that the conductor takes on a potential above zero (c) charge flows so that the conductor takes on a potential below zero (d) charge flows so that the conductor takes on the potential of the ground (e) none of these.

12. The movement of charge in an electric field from one point to another at a constant speed without the expenditure of work, by or against the field (a) is impossible (b) can only occur along a field line (c) can only occur along an equipotential (d) can only occur in a uniform field (e) none of these.

13. A body that when grounded takes on electrons is said to have originally had (a) a negative potential (b) zero-potential (c) a positive potential (d) an original net negative charge (e) none of these.

14. A neutron somehow picks up 10 eV. That's equivalent to it increasing its (a) charge by 10 C (b) electrical potential by 10 V (c) energy by 16×10^{-19} J (d) capacitance by 10 μF (e) none of these.

15. The electrostatic potential everywhere inside a hollow conductor is (a) always zero (b) never positive (c) always a nonzero constant (d) constant, provided there are no enclosed isolated charges (e) none of these.

16. The capacitance of a parallel plate capacitor is (a) independent of the plate separation (b) dependent on the charge (c) dependent on the voltage (d) independent of the plate area (e) none of these.

17. We can increase the capacitance of a parallel plate capacitor by (a) cooling the plates (b) bringing the plates closer together (c) decreasing the permittivity of the medium in the gap (d) increasing the voltage (e) none of these.

18. If the voltage across a capacitor is doubled, the amount of energy it can store (a) doubles (b) is halved (c) is quadrupled (d) is unaffected (e) none of these.

19. If the charge on a capacitor is halved, its stored energy (a) is halved (b) is quartered (c) is unchanged (d) is doubled (e) none of these.

Problems

ELECTRIC POTENTIAL

1. [I] A tiny sphere carrying a charge of −25.0 nC is moved 100 cm in a uniform electric field with no acceleration. It goes from a location at a potential of zero to a point where the potential is 100 V. How much work is done on it by the applied force? What is the significance of the sign of ΔW?

2. [I] A stream of singly ionized gold atoms pouring from a small oven impinges on a metal target. The target is attached to the positive terminal of a 12-V battery, and the oven is attached to the negative terminal. How much kinetic energy do the ions pick up in the process of crossing over to the target?

3. [I] What voltage should be put across a pair of parallel metal plates 10.0 cm apart if the field between them is to be 1.00 V/m?

4. [I] Two charged parallel metal plates, inside the evacuated cathode-ray tube of a radar system, are separated by 1.00 cm and have a potential difference of 25.0 V. What is the value of the electric field in the gap?

5. [I] Figure P5 shows two hollow concentric metal spheres of radii R_a and R_b. If the inner one is charged with $+Q$ and

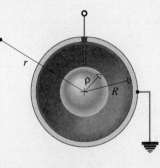

Figure P5

the outer surface is grounded, what is the potential at any point P outside the larger sphere? Explain your thinking.

6. [I] A 10.0-cm diameter metal sphere carries a charge of +0.100 μC. What is the potential 10.0 m away in the surrounding air?

7. [I] Two small spheres carrying charges of $+30.0 \ \mu C$ and $-50 \ \mu C$ are 100 cm apart in air. What is the potential at a point on the center-to-center line midway between them?

8. [I] Figure P8 depicts two metal objects. Sketch in some field lines and equipotentials. Discuss your answer.

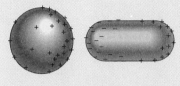

Figure P8

9. [I] An electron is to be accelerated from rest at a grounded cathode, to a metal plate at $+500$ V. Express in electron volts how much kinetic energy it will gain. How much electrical potential energy will it lose, if any?

10. [I] It is fairly easy to strip the two electrons off a helium atom leaving a bare nucleus of two neutrons and two protons. That so-called *alpha particle* is to be accelerated, essentially from rest, up to a KE of 100 keV by having it "fall" through a potential difference. What is the necessary voltage difference?

11. [c] As a rather simple model for the nucleus of an atom consider a uniform spherical distribution of positive charge Q, having a radius R. Figure P11 is a graph of the electric field; without integrating use it to find the difference in electric potential between the surface of the nucleus at $r = R$ and any point inside; $V_R - V_r$. [Hint: $\int E_r \, dr$ is the area under the curve.]

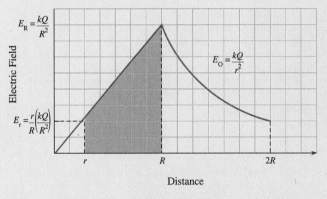

Figure P11

12. [c] Redo the previous problem using integration and the fact that the field *inside* the sphere ($r < R$) is

$$E_r = kQ \frac{r}{R^3}$$

13. [c] With Problem 11 in mind, what is the potential anywhere outside of the sphere (taking the zero to be at infinity)? [Hint: Solve this using Gauss's Law to find E and then integrate.] Knowing that **the electric potential**

must always be continuous, what is the value of V at the surface? Check your answer with the results of Problem 11.

14. [c] Considering the previous three problems, what is the potential inside the charged sphere?

15. [c] Starting with the electric potential for a point charge, show that $E_x = kQx/r^3$. [Hint: Use the Pythagorean Theorem in three dimensions.]

16. [c] The electric potential outside a long uniformly charged conducting cylinder of radius R at a perpendicular distance r is

$$V = \frac{\lambda}{2\pi\varepsilon_0} \ln \frac{R}{r}$$

where λ is the linear charge density. Determine the electric field outside the cylinder. [Hint: Remember that $\frac{d}{dx} \ln u = \frac{1}{u} \frac{du}{dx}$.]

17. [c] The electric potential on the symmetry axis of a uniformly charged ring of radius R at a distance x is

$$V(x) = k \frac{Q}{\sqrt{x^2 + R^2}}$$

where Q is the total positive charge it carries. Compute the axial electric field.

18. [c] Two conducting plates are held horizontally parallel to each other (and to the xy-plane) at a fixed separation. The bottom plate, which is grounded, carries a charge of $-Q$ while the top plate carries a charge of $+Q$. At a point P between the plates, an arbitrary distance y up from the bottom one, the potential is $V(y) = Cy$ above ground ($V = 0$). Here C is a constant with units of V/m. Determine the three components of the electric field. Explain your results. Why doesn't the charge on the lower plate run off into ground?

19. [II] An electron initially at rest in an X-ray tube crosses a potential difference of 30 kV and crashes into a target that then emits radiation. Compute the KE of the electron (in joules) at impact. Determine its maximum speed.

20. [II] A proton is released from rest in a uniform electric field of 500 V/m. How fast will it be moving after traveling 40 cm in and parallel to the field?

21. [II] A metal target sphere of 20-cm diameter suspended out in space is given a charge of $+1.00$ nC. How much work must be done on a proton in taking it from very far away (essentially infinity) to the surface of the sphere?

22. [II] A hydrogen atom consists of a proton around which circulates an electron at an average distance of 0.053 nm. Determine the potential at that distance due to the proton and find the potential energy of the electron (in joules).

23. [II] With Problem 21 in mind, through what voltage must the proton be accelerated from rest (by some sort of space weapon) if it is to arrive at the sphere at a speed of 8.5×10^4 m/s? Neglect any gravitational effects.

24. [II] Refer to Fig. P5. If the inner sphere is charged with $+Q$ and the outer surface is no longer grounded, how does the charge distribute itself, and what is the potential in the region beyond the spheres in terms of ρ, R, and Q? Explain your answer. Write an expression for the field beyond the spheres.

25. [II] Two horizontal parallel metal plates 10.0 cm apart in a vacuum chamber are to be used to suspend an electron in "midair." What voltage must be put across the plates?

26. [II] Tiny conducting spheres carrying charges of $+60\ \mu C$ are fixed at each of the four corners of a square 20 cm on a side. What is the potential at the very center? What is the electric field at the center? Assume the medium is vacuum.

27. [II] Two very small metal spheres carrying charges of $+10\ \mu C$ and $-25\ \mu C$ are located at coordinates $(0, 0)$ and $(3.0\ m, 0)$, respectively, in air. How much work would have to be done to bring a third sphere with a charge of $-10\ \mu C$ from very far away to the point $(0, 4.0\ m)$?

28. [II] Point-charges of $+300$ nC, -700 nC, $+500$ nC, and -100 nC are located in sequence at the corners of a square 40 cm on a side. Determine the potential at the center.

29. [II] The fluid within a living cell is rich in potassium chloride, while the fluid outside it predominantly contains sodium chloride. The membrane of a resting cell is far more permeable to ions of potassium than sodium, and so there is a transport out of positive ions, leaving the cell interior negative. The result is a voltage of about -85 mV across the membrane, called the *resting potential*. The membrane (about 50 atom-layers) is roughly 8 nm thick. Assuming the E-field across the cell membrane is constant, determine its magnitude.

30. [cc] What is the electric potential difference between two points at perpendicular distances of r_1 and r_2 (where $r_1 < r_2$) from the middle of a very long straight wire having a uniform linear charge density of λ? When is your answer negative and what does that mean? [Hint: We have already determined an expression for the electric field at a perpendicular distance r; use it.]

31. [cc] Show that the electric potential at point P in front of the uniformly charged disc pictured in Fig. P61 (Chapter 17, p. 683) is

$$V = \frac{kQ}{R^2} 2[\sqrt{x^2 + R^2} - x]$$

The total charge on the disc is $+Q$. [Hint: Divide the disc into charge elements and integrate. You might find it useful to note that $2r\,dr = d(r^2) = d(x^2 + r^2)$.]

32. [cc] A uniformly charged rod lies along the x-axis with one end at the origin and the other at $x = L$, as pictured in Fig. P32. It carries a total charge of $+Q$. Determine the electric potential at a point P on the y-axis at $y = h$. [Hint: $\int dx/\sqrt{x^2 \pm p^2} = \ln(x + \sqrt{x^2 \pm p^2})$.]

33. [cc] With the previous problem in mind, find the electric potential at a point a distance h away on the perpendicular bisector.

34. [cc] A long thin plastic rod is uniformly sprayed with positive charge so that it has a linear charge density of $+800$ nC/m. The rod is positioned on the x-axis where it extends from $x = 0$ to $x = 100$ cm. Prove that

$$V = k\lambda \ln\left[\frac{h + L}{h}\right]$$

where L is the length of the rod and h is the distance from its end at 100 cm to point P which lies on the x-axis at $x \geqslant 100$ cm. Determine the electric potential on the axis at $x = 150$ cm.

35. [cc] Starting with the electric potential of the uniformly charged disc (see Problem 31), find the axial electric field.

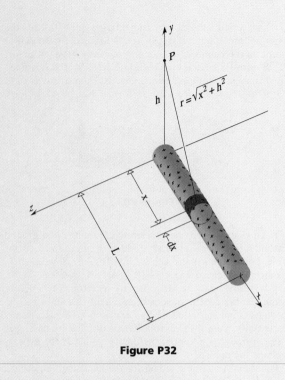

Figure P32

Show that very close to the disc the field approaches that of an infinitely large charged sheet.

36. [cc] The electric potential in a certain region of space is given by

$$V(x, y, z) = (6.00\ \text{V/m})x + (4.00\ \text{V/m}^2)y^2 + (0.00\ \text{V/m})z$$

Determine the several components of the electric field everywhere and its magnitude in particular at the point $(6.00\ m, 1.00\ m, 5.00\ m)$.

37. [cc] Figure 18.12b depicts the equipotentials and field lines for a dipole consisting of two equal charges of opposite sign $(\pm Q)$, separated by a distance l, lying on the x-axis, centered on the origin. The electric potential of such a dipole is

$$V = kp\frac{\cos\theta}{r^2}$$

where θ is the angle, measured up from the x-axis, made by a line drawn from the origin to the point P and $p = Ql$ is the *dipole moment*. Calculate the x- and y-components of the electric field. [Hint: Take $r \gg l$ and rewrite $\cos\theta$ in terms of x and y.]

38. [III] Two very small conducting spheres surrounded by transformer oil are charged with $+50.0\ \mu C$ and $-40.0\ \mu C$, respectively. Determine the point on the line connecting them where the potential is zero, if indeed such a point exists. The center-to-center separation is 1.00 m.

39. [III] A total of eight tiny conducting spheres, each carrying a charge -100 nC, are placed one each at the corners of a cube 1.0 m on a side. Find the potential at the center. What is the electric field at the center?

40. [III] If the inner sphere in Fig. P5 is charged with $+Q$ and

the outer surface is grounded, show (without using calculus) that

$$\Delta V = kQ\left(\frac{1}{\rho} - \frac{1}{R}\right)$$

is the expression for the potential difference across the gap. (see Problem 78).

CAPACITANCE

41. [I] A 100-pF capacitor is charged by putting it across a 1.5-V battery. What is the charge on its plates?

42. [I] A 48.0-μF capacitor, with an impregnated paper dielectric, is placed across the terminals of a 12-V battery. How much charge flows from the battery to the capacitor?

43. [I] Cathode-ray tubes used in computers and TV sets are coated inside and out with a conducting graphite paint called *Aquadag*. The inside layer is connected to the second anode (see Fig. 18.4) and helps to accelerate the beam and keep it narrow. The outer layer is connected to the chassis, which is ground. This arrangement creates an Aquadag-glass-Aquadag capacitor (which is connected across the power supply providing the high voltage for the anode, and tends to smooth out voltage variations). The capacitance is generally only about 500 pF, but all such tubes always carry a warning that even though the set has been turned off, the CRT must be discharged, otherwise you risk a dangerous shock. Explain why this warning is necessary and compute the charge for a tube voltage of 20 kV.

44. [I] Estimate the capacitance of the Earth. Its radius is 6371 km. Give the answer to two significant figures.

45. [I] What is the radius of a conducting sphere in air if it has a capacitance of 10 pF (or as it used to be called, $\mu\mu$F)?

46. [I] As a very rough estimate, assume you have about the same capacitance as a conducting sphere 3/4 m in diameter. How much charge can you store at a potential of 100 V?

47. [I] A parallel plate capacitor immersed in transformer oil carries a charge of $+20\ \mu$C on one plate and $-20\ \mu$C on the other, when there is a voltage of 4.0 V across it. What is its capacitance?

48. [I] What is the capacitance of two parallel metal plates each with an area of 100 cm^2 separated by 1.0 mm of air?

49. [I] What is the capacitance of two parallel metal plates each with an area of 100 cm^2 separated by 1.0 mm if the gap is filled with glass having a dielectric constant of 10?

50. [I] Thus far, we have used units for ε that relate back to Coulomb's Law. Accordingly, prove that 1 C^2/N·m^2 is equivalent to 1 F/m.

51. [I] A little flat ceramic capacitor consists of two circular plates 0.50 cm in diameter separated by a dielectric 1/3 mm thick with a relative permittivity of 4.8. Compute its capacitance.

52. [I] Show that capacitance has the dimensions of $[Q^2T^2]/[ML^2]$. ($Q{\rightarrow}$charge, $T{\rightarrow}$time, $M{\rightarrow}$mass, and $L{\rightarrow}$length.)

53. [I] Given three capacitors of values 60 pF, 30 pF, and 20 pF, determine the net capacitance when they are placed successively in series and then in parallel.

54. [I] What is the equivalent capacitance of the circuit between the terminals A and B, indicated in Fig. P54?

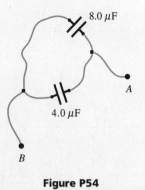

Figure P54

55. [I] Suppose we make an alternating stack of aluminum foil and paper (numbering each sheet of metal successively): 101 sheets of foil, 100 of paper, each 12 cm by 50 cm. The paper has a thickness of 0.22 mm and a relative permittivity of 4.1. If all the odd-numbered foil sheets are connected to a common wire and all the even-numbered sheets are connected to a second common wire, what will be the capacitance measured between the two wires? [Hint: There's one sheet of dielectric per capacitor.]

56. [I] What is the equivalent capacitance of the circuit between the terminals A and B, indicated in Fig. P56?

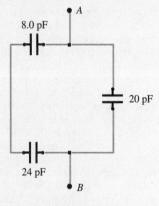

Figure P56

57. [I] What is the equivalent capacitance of the circuit between the terminals A and B, indicated in Fig. P57?

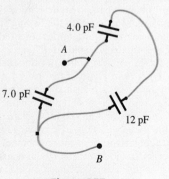

Figure P57

58. [I] What is the equivalent capacitance of the circuit between the terminals *A* and *B*, and again between *A* and *C*, indicated in Fig. P58?

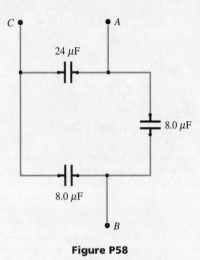

Figure P58

59. [I] A 20-pF capacitor with a mica insulating sheet $\frac{1}{2}$ mm thick is put across a 12-V battery. How much energy will it store?

60. [I] A bolt of lightning corresponding to about 20 C descends through a potential difference of upwards of 150 MV. How much energy is involved? Incidentally, at any one time there might be 2000 thunderstorms rolling over the Earth with 100 lightning bolts flashing every second!

61. [II] Figure P61 depicts a neuron, or nerve cell, and shows the long signal-carrying axon (which, from spine to fingers can be more than a meter in length). As discussed in Problem 37, the axon membrane is usually positive on the outside and negative on the inside. The dielectric constant of the membrane has been measured to be about 7. Given that the membrane wall is a mere 6.0 nm thick and that the axon radius is 5 μm, determine its capacitance per unit area. Explain any assumptions you make.

62. [II] The Leyden jar shown in Fig. 18.16 has a 22-cm-diameter base and the metal foil (inside and out) goes up to a height of 35.5 cm. The inside metal is attached to the central conductor by a wire. The glass has a thickness of 2.25 mm and a dielectric constant of 7.2. Estimate its capacitance.

63. [II] With Problem 61 in mind, determine the surface charge density of the axon. In the resting state, when there is no signal being transmitted, the potential difference across the membrane is about 70 mV.

64. [II] With the previous problem in mind, determine the capacitance per unit length of the axon.

65. [II] When a pulse is propagated down an axon, there is a shift of ions across a segment of membrane that causes a reversal of polarity and an overall change in potential of about 100 mV (see Fig. P65). This voltage spike (and the associated repolarization of the membrane) propagates along from one region to the next and constitutes the signal, just as a flame propagates down a length of fuse. After a few milliseconds, this so-called *action potential*

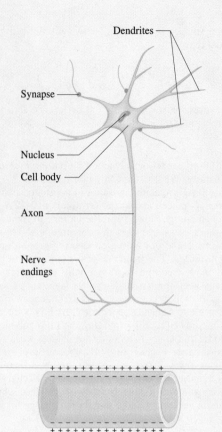

Dendrites

Synapse

Nucleus

Cell body

Axon

Nerve endings

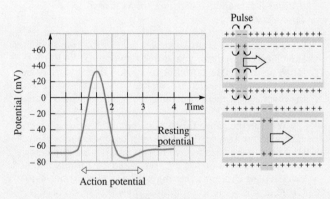

Figure P61

Figure P65

pulse passes any given point on the axon, which then returns to its resting potential (−70 mV). How much energy is required to recharge a 1-m length of axon in the wake of a pulse, so that it will be ready to transmit the next pulse?

66. [II] What is the equivalent capacitance of the grouping shown in Fig. P66?

67. [II] What is the equivalent capacitance of the circuit between terminals *A* and *B* in Fig. P67?

68. [II] What is the equivalent capacitance of the circuit between terminals *A* and *B* in Fig. P68?

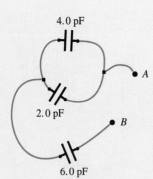

Figure P66

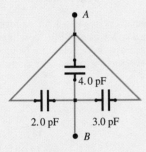

Figure P67

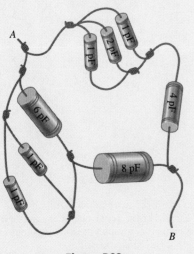

Figure P68

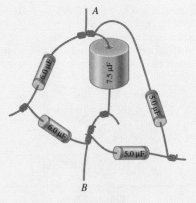

Figure P69

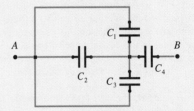

Figure P70

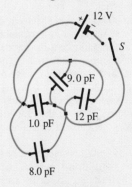

Figure P71

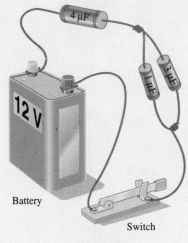

Figure P72

69. [II] What is the equivalent capacitance of the circuit between terminals A and B in Fig. P69?

70. [II] What is the equivalent capacitance of the circuit between terminals A and B in Fig. P70, if each of the capacitors is 2.0 pF?

71. [II] What is the voltage across the 9.0-pF capacitor in the circuit shown in Fig. P71 after the switch is closed? Determine the equivalent capacitance across the battery. How much charge is drawn from the battery?

72. [II] Determine the voltage across the 4.0-μF capacitor after the switch is closed in the circuit shown in Fig. P72.

73. [II] Two capacitors of 100 μF and 50 μF are separately charged to 250 μC and 100 μC, respectively. They are then attached in parallel so the + plate of one goes to the − plate of the other, and vice versa. Determine the final voltage across the two.

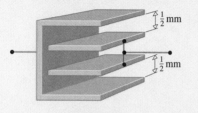

Figure P74

74. [II] Figure P74 shows four metal plates with the outer pair and inner pair wired together. Given that there is air in the gaps, which are $\frac{1}{2}$ mm wide, and that the area of each plate is 0.01 m², what is the capacitance of the device?

75. [II] Return to Fig. P71 and, overlooking losses, find the net energy provided by the battery when the switch is closed.

76. [II] In fair weather, the constant atmospheric E-field near the surface of the Earth varies from 120 V/m to 150 V/m. Compute the corresponding range of energies stored per cubic meter in the field.

77. [cc] A very long coaxial cable of length L consists of a central copper wire of radius R_a surrounded by the empty space enclosed by an outer concentric conducting cylindrical shell, of inner radius R_b. If the wire carries a net charge of $+Q$ and the shell a charge of $-Q$, find the capacitance of the cable via a direct analysis using calculus. [Hint: Neglect end effects and assume the internal field is radial. Use Gauss's Law to find E and then compute ΔV across the cable. Remember that C is always positive.]

78. [cc] A thin-walled spherical metal ball of radius R_a is surrounded by an outer thin spherical metal shell of radius R_b. The central ball (held in place by a narrow spacer that has a negligible effect) carries a charge $+Q$ while the outer sphere has a charge of $-Q$. Determine its capacitance directly using calculus. From your previous work find the capacitance of a single charged sphere. [Hint: Remember that $C = Q/V$ and, moreover, that C is always positive.]

79. [III] Figure P79 shows three plates each 50 cm by 60 cm and separated by 1.0 mm. The plates are immersed in transformer oil with a dielectric constant of 4.5. Find the total capacitance of the system.

Figure P79

80. [III] What is the difference in potential between points A-D, B-D, and C-D in Fig. P80?

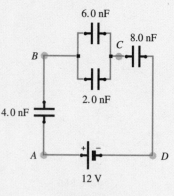

Figure P80

81. [III] If a 12-V battery is placed across terminals A and B in the circuit shown in Fig. P81, how much energy will be stored by the capacitive network? Notice that there are two places where one wire goes over another without touching it.

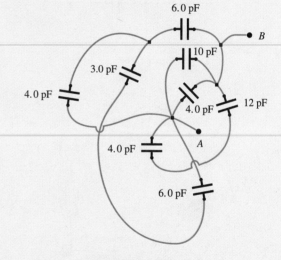

Figure P81

82. [III] Fig. P5 shows two hollow metal concentric spheres of radii R_a and R_b. If the inner one is charged with $+Q$ and the outer surface is grounded, show (without using calculus) that the capacitance is

$$C = \frac{R_a R_b}{k(R_b - R_a)}$$

Moreover, if the gap (d) is very small and A is the area of either inside surface ($4\pi R_a^2 \approx 4\pi R_b^2$), show that

$$C \approx \frac{\varepsilon A}{d}$$

as with Eq. (18.12). Compare your answer with that of Problem 78.

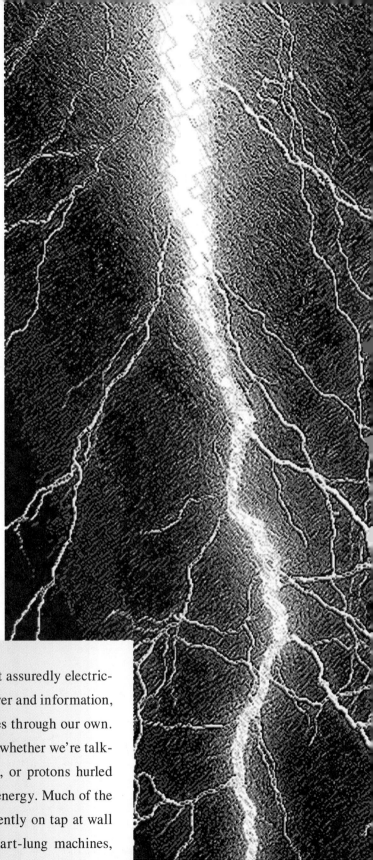

19

Direct
Current

IF THERE IS A life's blood in our technology, it's most assuredly electricity—electricity coursing along wire veins, delivering power and information, trickling through a metal nervous system just as it trickles through our own. **The ordered flow of charge is called electric current,** whether we're talking about electrons propelled down a wire by a battery, or protons hurled through space by an exploding star. And currents carry energy. Much of the energy we "consume" is delivered by electricity conveniently on tap at wall outlets everywhere: refrigerators, VCRs, computers, heart-lung machines, all plug into the electric stream and draw energy from it.

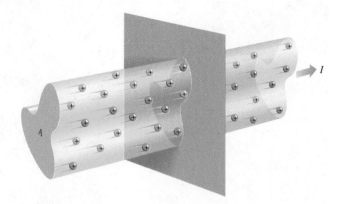

Figure 19.1 A beam of positive particles of cross-sectional area *A* constituting a current *I*. During each interval of time Δt, an amount of charge Δq passes through the plane, such that $I = \Delta q / \Delta t$.

This chapter is about electric currents, the batteries that sustain them, and the resistances that impede them.

CURRENTS

To quantify **electric current** (*I*), envision a stream of positive charge (Fig. 19.1). Now picture an imaginary reference plane cutting across the flow at some point of observation and determine the net charge Δq traversing the plane in a time Δt. The ratio $\Delta q / \Delta t$ is the average rate at which charge passes the point of observation during that interval, the *average current*. Often, the flow is constant, in which case

$$I = \frac{\Delta q}{\Delta t} \tag{19.1}$$

When the current changes from moment to moment, we define the ***instantaneous current*** as

$$I = \lim_{\Delta t \to 0} \left(\frac{\Delta q}{\Delta t} \right) = \frac{dq}{dt} \tag{19.2}$$

The SI unit of electric current is the *ampere* (A) or simply the *amp*, where 1 amp corresponds to a flow of 1 coulomb of charge per second; thus

$$1 \text{ A} = 1 \text{ C/s}$$

Minute currents in the microamp range (10^{-6} A = 1 μA) are commonplace, even within the human body. Microamp currents are generated in bone and connective tissue during exercise and seem to play a vital role in sustaining the health of these structures.

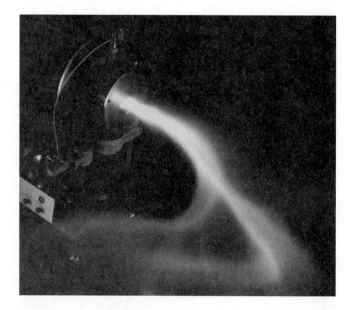

A beam of electrons (60 kiloamps, 3 MeV) deflected by both air scattering and a magnetic field. This is as much an electrical current as any that ever negotiated a toaster.

Normally, a current in a metal wire is a stream of free electrons. But, the **mobile charge carriers** constituting a current can be positive, negative, or both. The latter may be the case, for example, in a semiconductor or a plasma. Stars, streetlights, and fluorescent lamps contain plasmas wherein streams of oppositely charged carriers can be made to flow past each other in opposite directions.

The mobile charge carriers in Fig. 19.1 happen to be positive—just the sort of picture Ben Franklin had in mind for a current. Because of him, *the direction of flow of positive charge is traditionally taken to be the direction of current, regardless of the actual sign of the participating carriers.* Since electrons are the carriers in ordinary wires, this custom can be a little awkward at times, though it's easy enough to live with. *A flow of negative carriers to the left is equivalent to an equal flow of positive carriers to the right* (Fig. 19.2).

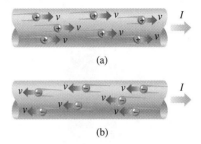

(a)

(b)

Figure 19.2 (a) The direction of an electric current is the direction of flow of positive charge. (b) Thus, it is opposite to the flow of negative charge.

Example 19.1 A constant downward electron beam transports 3.20 μC of negative charge in 200 ms across the vacuum chamber of an electron microscope. Determine the beam current and the number of electrons traversing the chamber per second.

Solution: [Given: $\Delta q = 3.20 \times 10^{-6}$ C, and $\Delta t = 200 \times 10^{-3}$ s. Find: I and the number of electrons per second.] From Eq. (19.1)

$$I = \frac{\Delta q}{\Delta t} = \frac{3.20 \times 10^{-6} \text{ C}}{200 \times 10^{-3} \text{ s}} = \boxed{16.0 \ \mu\text{A}}$$

The current is *upward* and equal to 1.60×10^{-5} C/s.

The number of electrons transported per second, each with a charge of -1.60×10^{-19} C, is

$$\frac{-1.60 \times 10^{-5} \text{ C/s}}{-1.60 \times 10^{-19} \text{ C}} = \boxed{1.00 \times 10^{14} \text{ electrons/s}}$$

Again, we have to be careful with the signs since the current transfers -1.6×10^{-5} C/s.

▶ **Quick Check:** 16 μA flowing for 200 ms transports 3.2 μC. Since 1 A corresponds to (1 A)/(1.6 $\times$ 10^{-19} C) = 6.2 $\times$ 10^{18} electrons/s, 16 μA is about 10 $\times$ 10^{13} electrons/s.

Example 19.2 An electron gun fires a pulse of charge lasting 2.0 μs. The average current of the burst is 1.0 μA. How many electrons are there in the pulse?

Solution: [Given: $\Delta t = 2.0 \ \mu$s, $I = 1.0 \ \mu$A. Find: number of electrons.] $I = \Delta q/\Delta t$, so if we first find the amount of charge in a pulse, we can then, knowing the charge on the electron, find the number of particles in the pulse; accordingly

$$\Delta q = I\Delta t = (1.0 \ \mu\text{A})(2.0 \ \mu\text{s}) = 2.0 \text{ pC}$$

The number of electrons is then

$$\frac{2.0 \times 10^{-12} \text{ C}}{1.602 \times 10^{-19} \text{ C/electron}} = \boxed{12 \times 10^{6} \text{ electrons}}$$

▶ **Quick Check:** $\Delta q \approx 10^{-6}$ s $\times$ 10^{-6} A $\approx 10^{-12}$ C: 10^{-12} C/10^{-19} C/electron $\approx 10^{7}$ electrons.

A current (either a flow of electricity in a wire or water in a pipe) is usually impeded in some way by the environment through which it progresses. That *resistance* inevitably results in the expenditure of energy from the flow. To be sustained, a current must be driven by an external source of energy. In the case of electricity, a nonelectrostatic source must continuously supply energy to the mobile charge carriers, pushing them along. It's not obvious how to perform such a process; indeed,

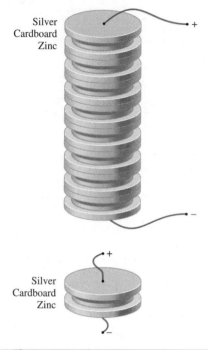

Silver
Cardboard
Zinc

Silver
Cardboard
Zinc

Figure 19.3 The voltaic pile, in which the elementary cell was a zinc-cardboard-silver sandwich. Volta wrongly thought that it was the zinc-silver contact that provided the effect.

Volta demonstrating his "pile" to Napoleon in a painting by a contemporary, A. E. Fragonard. Some years later, Volta gave Faraday a battery of this sort.

sustained currents were unknown up until the end of the 1700s. It was only then that a strange series of events led to the development of the electric battery, thereby plunging the world into the Age of Electricity.

19.1 The Battery

By the late 1700s, many researchers working with Leyden jars had experienced muscle spasms due to accidental shocks. Fascinated by the relationship between electricity and life, they began to study the electrical excitation of muscles in animals. The frog, whose muscular legs were a popular delicacy, became a logical martyr to the cause. In a short time, dissected frog's legs were twitching and convulsing on command from one end of Europe to the other. Even Faraday kept a froggery in the basement of the Royal Institution.

In 1780, Luigi Galvani, a noted Italian anatomist, made the first of several accidental discoveries that would arouse a storm of controversy. Each of his little green participants was humanely terminated, being impaled through the spinal cord on a sharp bronze hook. Thus prepared, the frogs waited their turns hanging on an iron trellis in the garden. To his surprise, Galvani noticed that every now and then the disembodied legs would spastically convulse for no apparent reason. He concluded that he had discovered "in the animal itself" a natural source of electricity.

Alessandro Giuseppe Antonio Anastasio Volta, already renowned as an "electrician," soon became Galvani's staunchest critic. It seemed to Volta that the source of the electricity was not the animal, but the two different metals brought in contact. The frog was simply a current meter. Recognizing that his tongue was a highly sensitive and convenient muscle, Volta became his own guinea pig (or, in this case, frog). He placed different pairs of metals on his tongue and brought them into contact. Expecting to feel a contractive spasm, he was surprised when the arrangement produced a metallic taste that lasted as long as the two metals were in contact.* This implied the incredible notion that "the flow of electricity from one place to another is continuing without interruption." He had created a source of sustained current.

Volta soon produced two devices that were the forerunners of the modern electric battery. He replaced the hook, frog, and trellis with something much more convenient. The **voltaic pile** was a stack of small disks of zinc, brine-soaked cardboard, and silver, layered in order: zinc-cardboard-silver, and so on (Fig. 19.3). Each zinc-cardboard-silver sandwich formed a unit (much like the aluminum, wet tongue, and silver) called a *cell*. These were repeated about 20 times (with the cells effectively in series) to build up the voltage. The result was a device that produced an appreciable continuous current. The other variation on the theme was the *voltaic wet cell*. It consisted of a drinking glass filled with brine or a dilute acid into which two metal strips were immersed, one of zinc, the other of copper. By putting several of these cells in series (Fig. 19.4), Volta created the world's first *electric battery*. **A battery is two or more cells electrically attached to one another.** By 1838, the Great Western Railway already had a voltaic pile powering a telegraph link. The granddaddy of the cheap flashlight battery, the *carbon-zinc cell* was conceived by Bunsen in 1842. In

*Volta was not the first to perform this little experiment, nor did he ever fully understand it. He used tin and silver or gold, but aluminum foil and sterling silver work well. Try it.

1859, Planté gave the world the lead-plate-dipped-in-sulfuric-acid storage device that has come to be known as the automobile battery.

How Batteries Work

Almost any two different solid conductors immersed in a variety of active solutions, known as *electrolytes,* function more or less as a battery. Chemical energy stored in the interatomic bonds (typically at less than about 3 eV) is converted into *electrical*-PE as the solution and one or both of the conducting plates, the *electrodes,* become involved in the chemical reaction. In the voltaic wet cell, the acid attacks the copper (Cu), and some of its positive Cu^{++} ions go into solution, leaving behind a negatively charged plate. Similarly, zinc (Zn) ions go into solution, too, but zinc is more soluble than copper and the zinc plate becomes even more negative—with even more excess electrons. The copper is slightly lower in potential than the acid; the zinc is a lot lower than the acid. The result is a difference in potential between the electrodes, with the copper higher and so positive, and the zinc lower and negative (Fig. 19.5).

The *electromotive series* (Table 19.1) is a list of metals in decreasing order of the tendency of each to become ionized by losing an electron. If we construct a simple cell with two metal electrodes immersed in a uniform electrolyte, the cell's voltage can be approximated from the values in the table. Accordingly, the potential of copper is +0.34 V, whereas zinc is −0.76 V (each with respect to hydrogen as the reference). Thus, copper is (0.34 V) − (−0.76 V) = +1.1 V higher than zinc, and this value is the largest voltage we can expect to produce across the terminals of a simple copper-zinc cell.

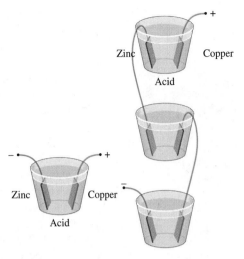

Figure 19.4 A single voltaic wet cell and three wired in series to form a battery.

Figure 19.5 A voltaic wet cell composed of a copper and zinc plate immersed in a solution of dilute sulfuric acid (H_2SO_4). This device is the kind the young poet Percy Shelley was playing with at Oxford, burning holes in the carpet of his apartment with the splashed acid.

TABLE 19.1	Electromotive series for several metals*	
Substance	**Electrode potential (V)**
Lithium	−3.0
Potassium	−2.9
Sodium	−2.7
Aluminum	−1.7
Zinc	−0.76
Iron	−0.44
Tin	−0.14
Lead	−0.13
Copper	+0.34
Mercury	+0.80
Silver	+0.80
Platinum	+1.2
Gold	≈+1.3

*These values correspond to the potentials required to ionize the corresponding metal atoms. The signs are a matter of convention referenced to hydrogen (and measured at 25°C).

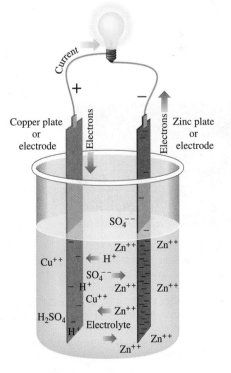

A little dry cell torn open to show the central carbon rod. The zinc casing is corroding in spots.

*A potential difference that can be used to supply energy and thereby sustain a current in an external circuit is an **electromotive force**, or emf (pronounced ee em ef), although that's a misnomer—it's not force at all. Practically, **the emf is the voltage measured across the terminals of a source when no current is being drawn from or delivered to it**.*

A strip of copper (some wire or a penny) and a strip of zinc (a few galvanized nails) immersed in salt water easily puts out ≈0.7 mA at ≈0.6 V. Add some vinegar and the emf will go up to perhaps 0.9 V. A small speaker from an old radio makes a fine current detector. Attach one wire, or *lead,* from the cell to the speaker and then tap the other lead from the cell on the remaining terminal. The speaker will clatter with each touch. Stick a carbon rod (a pencil lead will do) into a lemon and surround it with a bunch of galvanized nails. The resulting cell will deliver ≈0.5 mA at a voltage of roughly 0.7 V. Replace the carbon with something as ordinary as a paper clip, and the lemon cell will still generate an emf, though down now to ≈0.2 V. Try a grapefruit or even some sauerkraut.

A given kind of cell will generate a voltage difference that is determined by its chemical makeup, independent of size. The size determines the total current a cell can deliver, not the voltage; the greater the quantity of each substance chemically reacting, the more charge is liberated. An ordinary flashlight "battery," a *dry cell* (Fig. 19.6a), has an emf of 1.5 V. A *mercury cell,* one of those little button-sized

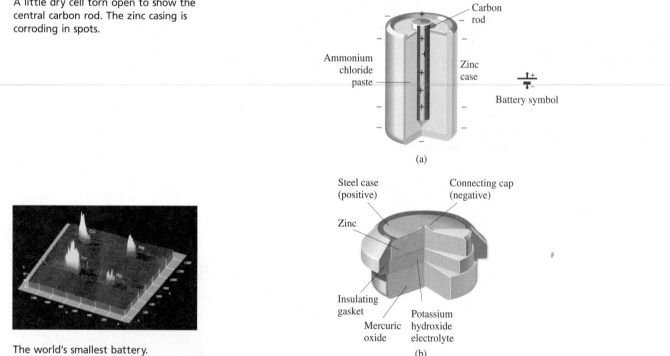

Figure 19.6 (a) The basic dry cell, essentially as Leclanché developed it in 1868. The modern version has a so-called *depolarizing* region of manganese dioxide around the carbon rod. That region controls the buildup of hydrogen ions that would otherwise reduce the emf. (b) A mercury cell used in calculators and watches.

The world's smallest battery. Measuring 70 nm long and made of fewer than 5×10^5 atoms, it is 1/100 the diameter of a red blood cell. It consists of tiny pillars of copper and silver deposited on a graphite surface. When immersed in a copper sulfate solution, it produces an emf of 20 mV for about 45 minutes.

calculator, watch, and hearing-aid "batteries," has an emf of about 1.4 V, while the *lead storage cell* of auto fame is a 2-V device. One of its great virtues is that it can be recharged by the generator in the car. The *nickel-cadmium cell* used in recharge-able computer battery packs has an emf of 1.2 V.

Cells in Series and Parallel

To boost the potential difference provided, cells are connected in series. The crucial point is that *the voltage across the series-connected battery is the sum of the voltages across each constituent cell*. Refer to Fig. 19.7a. Point *B* is 1.5 V higher than *A*, and point *D* is 4.5 V higher than *A*. This sort of series stacking is just what you are doing when you load two, three, or four D-cells, top (+) to bottom (−), into a flashlight or a portable radio in order to get the 3.0 V, 4.5 V, or 6.0 V needed to operate the device. It's also the way the cells are wired in a car battery to yield 12 V.

If a wire is connected across the terminals of a cell (as in Fig. 19.8a), a current will move around the closed conducting path. In practice, this is a bad idea since a wire having little resistance corresponds to a *short circuit* that will drain the cell. It could also be dangerous if enough current flows to melt the wire. As a rule, the bat-tery provides power to a *load*—for example, a motor, radio, or light bulb (Fig. 19.8b). **A steady-state current can only exist in a closed circuit** and *the same current flows in and out of the load*. **Current is never used up by a circuit element**. Because the polarity of the cell is fixed, the direction of the flow of charge is constant, and we say that *I* is a **direct current** (dc). If the same thing is done with a battery comprising several cells in series (Fig. 19.9), the same amount of current will pass through each cell. In series the voltages add, while the current remains un-altered as it passes in and out of each element (reminiscent of the way series capac-itors behave where the voltages add and the net charge is the charge on any one). By contrast, cells attached in parallel (Fig. 19.10a) form a battery whose voltage is the same as the individual voltages but whose current capacity is the sum of the individ-ual current outputs (again matching the behavior of capacitors). If you want a lot of current at a low voltage, put the cells in parallel; if you want both a large current and a large voltage, stack the cells in parallel and then put the stacks in series.

Battery manufacturers provide (though not often right on the battery where it ought to be) a crude measure of the current capacity in the form of an **amp-hour rating**. A little 1.5-V AA-cell, the common penlight finger-sized "battery," is rated at about 0.6 amp-hour, whereas the larger 1.5-V D-cell can deliver as much as 3 amp-hours. Presumably, one can draw a steady 0.3 amps for 10 hours or 0.1 amp for

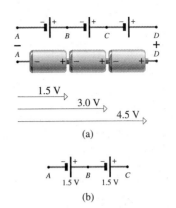

Figure 19.7 Cells in series. As connected in (a), the voltages add up, and point *D* is 4.5 V above point *A*. As connected in (b), the voltages subtract and *A* and *C* are at the same potential.

But when I took the frog into a closed room, laid it on an iron plate and began to press the hook that was fixed in its spinal cord against the plate, lo and behold, the same contractions and the same kicks!

GALVANI

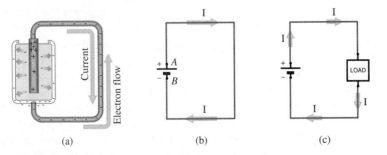

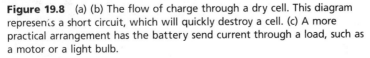

Figure 19.8 (a) (b) The flow of charge through a dry cell. This diagram represents a short circuit, which will quickly destroy a cell. (c) A more practical arrangement has the battery send current through a load, such as a motor or a light bulb.

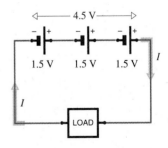

Figure 19.9 When cells in series form a battery, the same current *I* passes through each cell: *I* circulates around the circuit and is undiminished as it enters and leaves each element.

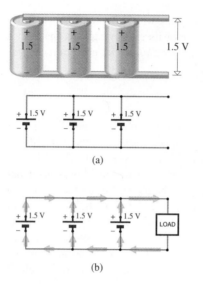

Figure 19.10 (a) Three 1.5-V cells in parallel. The voltage across the battery is the voltage across each cell. (b) Currents in the various segments of a parallel array of cells with a load across the terminals. The current through the load is the sum of the currents provided by each cell.

An old 45-V battery made up of thirty 1.5-V dry cells in series.

30 hours before discharging a D-cell. It will never be able to put out 300 amps for 0.01 hour, but you get the point. Most operating batteries tend to produce positive hydrogen ions which, if unchecked, build up and rapidly degrade the operation of the cell—a process known as *polarization.* Modern dry cells contain a region of manganese dioxide around the anode that eliminates the hydrogen, gradually *depolarizing* the cell. The discharge rate is limited by the need to have the depolarization keep pace with the liberation of H^+ ions. Dry cells are designed to function intermittently so they have a chance to recuperate as the depolarizer works. Have you ever noticed how a dim flashlight will come on strong again after a rest? Mercury cells are self-depolarizing, removing the need for rest periods. They are especially well suited for the continuous operation of low-current solid-state commercial, scientific, and medical devices—cameras, hearing-aids, alarm systems, and so on.

What you want from a car battery is a lot of current to power the starter motor that turns the engine until it gets going on its own. A heavy-duty 12-V truck battery with an amp-hour rating of near 160 can provide a tremendous current; even 10 amps is a great deal—but of course it's only at a meager 12 V. Plenty of charge is available to flow from such a battery, but there's little push to propel it, so it's fairly safe. That's like having all the water of the Atlantic Ocean behind a dam one foot deep; lots of water stored, but little pressure. Typically a lead-acid automobile battery stores about 0.5 kW·h (that is, 0.5×10^3 W·h $\times$ 60 min/h $\times$ 60 s/min = 1.8 MJ) of energy. By comparison, the batteries that power diesel submarines while submerged weigh as much as several hundred tons and consist of row upon row of interconnected rechargeable cells that can retain 10 000 times the energy of a car battery.

Example 19.3 Design a battery to be constructed of D-cells (each rated at 3 amp-hours) that will provide a maximum operating current of 5.0 A with an emf of 4.5 V. One restriction is that no cell be required to supply in excess of 1.0 A. What can you expect will be the lifetime of your battery while delivering maximum current?

Solution: The 1.0-A restriction per cell demands we put five cells in parallel to form a 5.0-A unit, as in Fig. 19.11a. The maximum current through each cell would then be 1.0 A as the unit provides the required 5.0 A with an emf of 1.5 V. To bring the voltage up to 4.5 V, we put three such units in series, as in Fig. 19.11b. With a 3.0-amp-hour rating, operating at a maximum of 1.0 A, we anticipate a lifetime of $\boxed{3\ \text{h}}$ for the battery.

▶ **Quick Check:** The 5.0-A current entering an adjacent unit splits up into five separate 1.0-A currents, one of which passes through each cell. These reunite as a 5.0-A stream on leaving the unit.

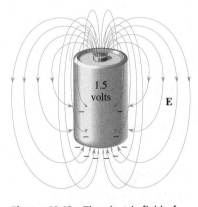

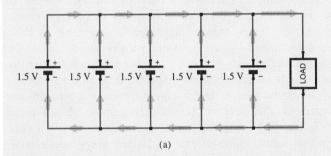

(a)

Figure 19.11 A battery made up of D-cells each rated at 3 amp-hours. (a) The voltage across this battery is the voltage across each of its cells—namely, 1.5 V. Thus, the voltage across the load is 1.5 V. (b) By putting three strings of cells in series, the voltage across the load is now 4.5 V.

(b)

19.2 Electric Fields and the Drift Velocity

The electric field produced by an isolated battery rises out of the positive terminal (*anode*) and returns to the negative terminal (*cathode*), looking more or less like the field of a dipole (Fig. 19.12). A negative mobile charge carrier in the vicinity of the anode would be drawn toward it, and one near the cathode would be repelled. Now suppose we attach a few meters of ordinary copper hookup wire across the terminals of the battery. Electrons, forced away from the cathode, stream into the wire, repelling each other and spreading out to the wire's surface as they progress along its length. The surface becomes nonuniformly charged, although there is no net charge on the wire as a whole. The copper resists the flow of charge (p. 733), and so there must be a sustained driving force (that is, an *E*-field within the wire maintained by a battery, generator, or power supply). Once the surface charges accumulate in a region, they naturally inhibit any further lateral flow of charge and the bulk of the current progresses down the length of the wire. Thus, a surface charge quickly builds up that creates a uniform axial field along the length of the wire (and a field outside it as well). Open the circuit, the surface charges redistribute, the field

Figure 19.12 The electric field of a dry cell. The field lines are distributed symmetrically around the central axis.

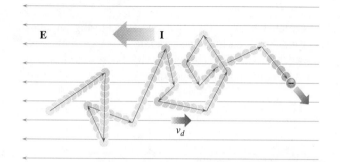

E   **I**

v_d

Figure 19.13 The zigzag path of an electron colliding with impurities and imperfections. It has an overall drift velocity, which (because of its negative charge) is opposite to the E-field.

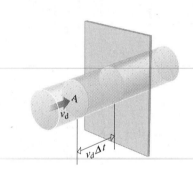

A

v_d

$v_d \Delta t$

Figure 19.14 A wire carrying a current. Positive charge carriers drift along at an axial speed v_d and move a distance $v_d \Delta t$ in a time Δt.

vanishes, and the current ceases; an *E*-field cannot be maintained inside a conductor under electrostatic conditions.

If there are any kinks or turns in the wire, a few additional electrons initially bunch up on the surface at those locations until the resulting internal steady-state *E*-field they contribute to is completely axial. After a very short time, a stable distribution of surface charges is in place, a uniform axial *E*-field exists, and a constant current progresses around the closed circuit. That current may be imagined as continuously emerging from the positive terminal of the battery, moving along the wire (where it loses energy due to the resistance), and coming back into the battery (where it's pumped up in energy), only to be sent out and around again. In the steady state there is current everywhere in the closed circuit and no further buildup of charge anywhere; **current neither bunches up nor gets used up**.

Figure 19.13 depicts the zigzag path of a typical electron as it makes its way along an ordinary metal conductor. Pushed by an externally sustained *E*-field, the electron advances but quickly scatters every-which-way off thermally vibrating metal ions, impurities (that is, foreign atoms), and imperfections (that is, lattice flaws and crystal grain boundaries). In copper at room temperature, the electron travels at about 10^6 m/s, collides roughly every 10^{-14} s, and gets little farther than 10^{-8} m (or several hundred atom lengths) between impacts. Yet the field relentlessly forces it to move axially along the wire in the opposite direction to **E** with an **average drift speed**, v_d, regardless of the detours. Accordingly, suppose that a wire carries a current as shown in Fig. 19.14. A group of mobile charge carriers will sweep past the plane of observation during a time interval Δt. Indeed, every mobile carrier within a little cylindrical volume of length $v_d \Delta t$ and cross-sectional area A will traverse the plane during that interval. If η is the **number of carriers per unit volume** in the wire, and $v_d \Delta t A$ is the volume of carriers swept past the plane, then $\eta v_d \Delta t A$ is the number of mobile charges passing a point of observation in a time Δt. Given that each carrier has a charge q_e, the net amount of charge (Δq) transported in a time Δt is

$$\Delta q = (\eta v_d \Delta t A) q_e$$

It follows from Eq. (19.1) that the current is

$$I = \frac{\Delta q}{\Delta t} = \eta v_d A q_e \qquad (19.3)$$

Example 19.4 Determine the order-of-magnitude of v_d in copper, assuming a current of 1 A, a cross-sectional area of $1.0 \text{ mm}^2 = 1.0 \times 10^{-6} \text{ m}^2$, and the availability of one conduction electron per atom. Discuss the notion that this is only an order-of-magnitude calculation.

Solution: [Given: $I = 1$ A, $A = 1.0 \times 10^{-6}$ m^2, and 1 electron per Cu atom. Find: v_d.] Equation (19.3) relates I, A, and v_d and immediately suggests that we find η. It

follows from Chapter 10 that η equals the number of moles of copper per cm^3 multiplied by Avogadro's number of atoms per mole—that is, the number of atoms per cubic centimeter. Copper has an atomic mass of 63.5 u and a density of 8.9 g/cm^3, and so the number of moles per cm^3 is (8.9 g/cm^3)/(63.5 g/mol). Thus η equals

$$\frac{(1 \text{ electron/atom})(6.0 \times 10^{23} \text{ atom/mol})(8.9 \text{ g/cm}^3)}{63.5 \text{ g/mol}}$$

(continued)

(continued)

or $\eta = 8.4 \times 10^{22}$ electrons/cm³

and $\eta = 8.4 \times 10^{28}$ electrons/m³

From Eq. (19.3)

$$v_d = \frac{I}{\eta A q_e}$$

$$v_d = \frac{1\ \text{A}}{(8.4 \times 10^{28}\ \text{electrons/m}^3)(1.0 \times 10^{-6}\ \text{m}^2)(1.6 \times 10^{-19}\ \text{C})}$$

and so

$$\boxed{v_d = 0.7 \times 10^{-4}\ \text{m/s}} \approx 0.1\ \text{mm/s}$$

Notice that had we used a current of 10 A, the speed would have been 1 mm/s, so this should be considered only an order-of-magnitude result.

▶ **Quick Check:** The size of our computed η compares well with those given in Table 10.2, p. 364. This relatively low-drift speed is much like a wind powered by a pressure difference: the molecules are randomly zigzagging around at about 1 km/s, while the organized breeze is much slower.

Electrons at room temperature make progress down a wire very slowly, typically at a mere 1 mm/s or so. How, then, can the telephone manage to transmit electrical signals along its clutter of wires at close to the speed of light? The answer is simple: the electron we push on in New York is not the one that tickles the phone in San Francisco. The starting electron might take 16 minutes to travel the first meter of the journey; it may not even be out the door before the message is over! The situation is like a long pipe filled with water. You push inward at this end, and a pulse of water carrying energy spurts out the other end. A disturbance of the water rapidly propagates down the pipe even though any given sample of liquid hardly moves. It's the electric field that can be thought of as traveling down the wire at near the speed of light, carrying the signal, setting the electrons in motion before it. When we buy electrical energy, we don't buy electrons, there are plenty of those in the wires we already have—we buy additional amounts of electron motion. *The power company pushes around our electrons and sends us a bill for how much work they did in the process.*

RESISTANCE

There is a great diversity in the ability of materials to conduct electricity, and Georg Simon Ohm set himself the task of finding order in the seeming chaos. His experiments had to be modest in scale. Ohm was a high school teacher in Cologne and never far from poverty. He had been inspired to the particular challenge by the recent publications of Fourier, whose work established that the rate of flow of heat along a conducting rod was proportional to the temperature difference between its ends. Ohm wondered whether the rate of flow of charge along a conducting rod was likewise proportional to the voltage difference between its ends. Remember that in the steady state there is a uniform E-field along the current-carrying rod and so $V_B - V_A = \pm Ed$. As we move a distance d from point A to point B in the direction of **E**, that is, in the direction of I in Fig. 19.15, there will be a voltage drop, $V = -Ed$. If we attach a length L of almost any kind of hookup wire across a battery, there will be a total voltage drop of $-EL$ along the wire and that, in turn, is determined by the voltage the battery sustains across its terminals.

Suppose we take a sample of metal wire and attach it successively to the terminals of different batteries, thereby applying different known voltages across it. In

Georg Simon Ohm (1787–1854).

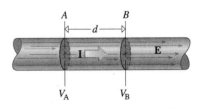

Figure 19.15 Moving along the uniform E-field from A to B a distance d, there is a voltage drop $V = -Ed$.

each case, we could measure the resulting current passing through the sample with a device known as an *ammeter* (Ohm used a kind of magnetic torsion balance to accomplish the same thing). What we would find is that the current the battery could force through a specimen depends linearly on the applied voltage, $I \propto V$; doubling the voltage doubles the current. And this relationship is true for a variety of different conducting materials—gold, copper, brass, etc.—though each behaves in a characteristic way. Ohm suggested that every sample manifested a **resistance** (R) to the flow of charge. The greater the resistance (symbolized diagrammatically by -⋀⋀-), the less current any battery could push through it. The current in the circuit, the current through the sample, varies directly with the applied V and inversely with R. In 1826, Ohm published his results:

$$I = \frac{V}{R} \qquad (19.4)$$

or
$$V = IR \qquad (19.5)$$

Ohm's Law, as this relationship is called, deals with a rather limited, rather special set of circumstances, and yet it is of tremendous practical value. It applies to conductors (at a constant temperature), notably the common metals and several non-metallic conductors as well. A plot of V versus I, Fig. 19.16, is a straight line *passing through the origin* with a slope of R. Put slightly differently, R is independent of I and V for these important materials. Moreover, for a circuit element to be *ohmic*, reversing the potential difference across it must simply reverse the current through it. A cup of copper sulphate solution with copper electrodes in it obeys Ohm's Law and is an ohmic conductor, though not one you are likely to find in a typical circuit. There are many materials and devices that are *nonohmic*—an ionized gas is just one (Fig. 19.17b). So Ohm's "Law" is a useful practical statement that applies to an important class of materials, but it's nothing like the grand fundamental pronouncements of Coulomb's Law or the Law of Universal Gravitation.

To honor Ohm—recognition for his work was depressingly late in coming—the unit of resistance was named the *ohm* and symbolized by the Greek capital letter omega, Ω. Ohm's Law serves to define resistance as $R = V/I$, and so 1 ohm = 1 volt per ampere.

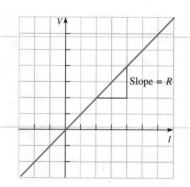

Figure 19.16 An ohmic circuit element has a linear V versus I curve. The slope of the line equals R.

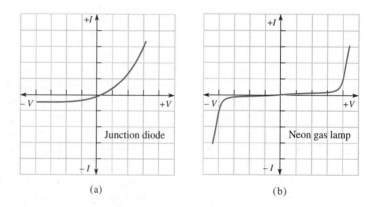

(a)   (b)

Figure 19.17 Many electrical devices, like the two shown here, are nonohmic.

Example 19.5 A small ohmic light bulb is placed in series with two D-cells, as shown in Fig. 19.18. The ammeter in series with the bulb reads the current in the circuit (0.50 A) without introducing any appreciable voltage drop across its own terminals. The voltmeter attached to the terminals of the bulb reads the voltage across it (3.0 V) without introducing any appreciable change in the current through the bulb. What is the resistance of the bulb?

Solution: [Given: at the bulb, $V = 3.0$ V, and $I = 0.50$ A. Find: R.] Note that the total voltage across the bulb is the net voltage produced by the batteries, 1.5 V + 1.5 V. From Ohm's Law

$$R = \frac{V}{I} = \frac{3.0 \text{ V}}{0.50 \text{ A}} = \boxed{6.0 \ \Omega}$$

▶ **Quick Check:** $V = IR = (0.50 \text{ A})(6.0 \ \Omega) = 3.0$ V. The low resistance draws a sizable current even at a low voltage.

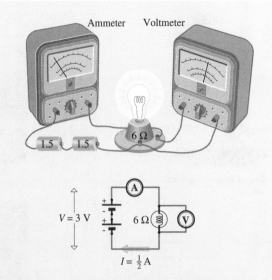

Figure 19.18 A circuit consisting of two 1.5-V cells and a lamp, all in series. The ammeter -Ⓐ- is placed in the arm of the circuit through which the desired current passes. The voltmeter -Ⓥ- is placed across the two points where a potential difference is to be determined.

All kinds of electrical devices, including the very wires that connect them, have resistance. Even so, there are specific circuit elements called **resistors** whose primary function is to introduce a certain known resistance and in so doing, control the currents and voltages in a circuit. They range from fractions of an ohm to millions of ohms (megohms, MΩ). Used in almost all electronic devices from radios to computers, they regulate the flow of charge (Fig. 19.19).

Example 19.6 Suppose someone falling out of a tree grabs an overhead power line. The wire has a resistance of 60 microhms per meter and is carrying a dc current of 1000 amps. With hands a meter apart, what is the voltage across him? Will the unfortunate soul get much of a shock?

Solution: [Given: $R/L = 60 \ \mu\Omega$/m, $I = 1000$ A, and $d = 1.0$ m. Find: V.] The voltage drop across 1.0 m of the line is

$$V = IR = (1000 \text{ A})(60 \ \mu\Omega/\text{m} \times 1.0 \text{ m})$$

hence
$$V = 0.060 \text{ V} = \boxed{60 \text{ mV}}$$

You know you can handle the terminals of a 1.5-V dry cell without feeling any electricity, so 60 mV is much too small a voltage to push a detectable amount of current through a human body (see Sect. 23.6).

▶ **Quick Check:** $V = IR = (10^3 \text{ A})(6 \times 10^{-5} \ \Omega) = 6 \times 10^{-2}$ V.

19.3 Resistivity

The resistance of a piece of wire, or anything else for that matter, is specific to that sample—it tells us nothing about the material that the wire is made of. Alternatively, it would be nice to have some measure of how each kind of material behaves independent of geometry. We did the same sort of thing when we expressed the

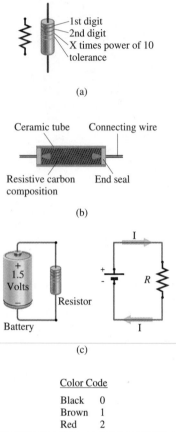

A close-up of the circuitry in the radio pictured on p. 645. The gray cylinder is a 1.5-kΩ wire-wound resistor. Such resistors are made by wrapping wire of some exotic alloy such as nichrome, manganin, or constantan around an insulating core. They are more precise and have better temperature stability than the less expensive carbon resistors.

Color Code	
Black	0
Brown	1
Red	2
Orange	3
Yellow	4
Green	5
Blue	6
Violet	7
Gray	8
White	9
Gold	± 5%
Silver	± 10%

Figure 19.19 (a) Although resistors come in many forms, the most common is the little striped brown cylinder. These are carbon composition resistors, as shown in (b). The stripes are a color code indicating the resistance. (c) shows the schematic representation.

spring constant (which is specific to each spring) in terms of Young's Modulus (p. 381), which is characteristic of the material making up the spring. Thus, we follow Ohm's lead, and for the very simplest geometry, we can guess at what he found experimentally. How does the resistance of a rod vary with its shape and composition? The analogy with water flowing through pipes suggests that the resistance is directly proportional to the length of the conductor (L)—the longer the rod, the more scattering the electrons will experience in traversing it. Similarly, the narrower the pipe, the more the resistance, and so we can anticipate that electrical resistance will also vary inversely with the cross-sectional area, $R \propto 1/A$. From Eq. (19.3), $I \propto Av_d$, where the drift velocity should depend on E, and E in turn depends on V. Hence, $I \propto AV$, and so from Ohm's Law, $R \propto 1/A$. In any event, Ohm found experimentally that $R \propto L/A$ and introduced a material-dependent constant of proportionality ρ, the **resistivity**. To make the statement an equation:

$$R = \rho \frac{L}{A} \tag{19.6}$$

A long extension cord (large L) should have heavy-gauge wire (large A) to keep R small. This result is similar to Eq. (15.8), which describes the heat current ($\Delta Q/\Delta t$) driven by a temperature difference (ΔT): the ratio of driving influence to resulting thermal current is proportional to the sample's length over its cross-sectional area.

Table 19.2 lists the resistivities, in units of ohm-meters ($\Omega \cdot$m), for a number of important materials. Substances with resistivities of less than about 10^{-5} $\Omega \cdot$m, such as silver and copper, are called **conductors**. **Insulators** such as glass, rubber, and

TABLE 19.2 Resistivities*	
Substance	**Resistivity (ρ) (in $\Omega \cdot$m)**
Aluminum	2.8×10^{-8}
Brass	$\approx 8 \times 10^{-8}$
Constantan (60% Cu, 40% Ni)	$\approx 44 \times 10^{-8}$
Copper	1.7×10^{-8}
Iron	$\approx 10 \times 10^{-8}$
Manganin ($\approx$84% Cu, $\approx$12% Mn, $\approx$4% Ni)	44×10^{-8}
Mercury	96×10^{-8}
Nichrome ($\approx$59% Ni, $\approx$23% Cu, $\approx$16% Cr)	100×10^{-8}
Platinum	10×10^{-8}
Silver	1.6×10^{-8}
Tungsten	5.5×10^{-8}
Carbon	3.5×10^{-5}
Germanium	0.46
Silicon	$100-1000$
Glass	$10^{10}-10^{14}$
Neoprene	10^{9}
Polyethylene	$10^{8}-10^{9}$
Polystyrene	$10^{7}-10^{11}$
Porcelain	$10^{10}-10^{12}$
Teflon	10^{14}
Sodium chloride (saturated solution)	0.044
Blood	1.5
Fat	25

*Values determined at or near 20°C.

teflon typically have resistivities greater than about 10^5 $\Omega \cdot$m. Between 10^{-5} and 10^5 $\Omega \cdot$m are the so-called **semiconductors**, like silicon and germanium.

The plastic polyacetylene is ordinarily a semiconductor, but when doped with iodine, it becomes a conductor. (A material is said to be *doped* when small amounts of a foreign substance are introduced into it.) An iodine atom removes an electron from a carbon atom in the polymer chain, leaving behind a "hole" that behaves like a positively charged particle. The holes advance in the direction of the E-field as if they were a flow of positive charge, a current. This metallic-looking plastic is, ounce-for-ounce, twice as conductive as copper and, though still in the developmental stage, it promises a new age of inexpensive plastic electronic devices.

Example 19.7 A length of nichrome ribbon with a rectangular cross section of 0.25 mm $\times$ 1.0 mm is to be used as the heating element in a toaster. How long should it be if it's to have a total resistance of 1.5 Ω at room temperature?

Solution: [Given: cross section 0.25 mm $\times$ 1.0 mm, nichrome ribbon; $R = 1.5$ Ω. Find: L.] The resistance, cross section, length, and type of wire are related via Eq. (19.6). The cross-sectional area is

$$A = (0.25 \times 10^{-3})(1.0 \times 10^{-3}) = 0.25 \times 10^{-6} \text{ m}^2$$

Using Table 19.2, we have

$$L = \frac{RA}{\rho} = \frac{(1.5 \ \Omega)(0.25 \times 10^{-6} \text{ m}^2)}{100 \times 10^{-8} \ \Omega \cdot \text{m}} = \boxed{0.38 \text{ m}}$$

▶ **Quick Check:** The resistance of this wire per meter ($L = 1$ m) is $\rho l/A = 4$ Ω/m. Hence, $1.5/4 = L/1$, and $L = 0.38$ m.

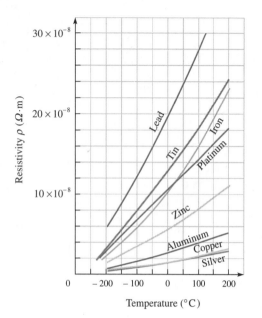

Figure 19.20 The resistivities of several metals over a range of temperatures. All rise with T and all are nearly linear.

The Temperature Dependence of Resistivity

When the temperature of a conductor increases, the corresponding increase in the random vibrations of its atoms and ions increases the scattering of electrons, impeding their progress and elevating the resistivity of the material. For example, during the fraction of a second while the tungsten filament in a light bulb rises roughly 2000°C as it becomes incandescent, its resistance increases by a factor of about 10. Experiments show that ρ usually varies *almost* linearly with modest changes in temperature (ΔT), as shown in Fig. 19.20. Accordingly, we write an expression for ρ at any temperature relating it to a known value (ρ_0) at some reference temperature:

$$\rho \approx \rho_0 (1 + \alpha_0 \Delta T) \qquad (19.7)$$

where values of ρ_0 appear in Table 19.2. Values of α_0, the **temperature coefficient of resistivity**, are given in Table 19.3, and ΔT is the difference in temperature (either in °C or K) from the reference value. Because α_0 also varies somewhat with temperature, it has been subscripted to indicate that it, too, must be correlated to the reference temperature. For precise work, there are handbooks of physical data that supply values of both ρ_0 and α_0 at different temperatures. The approximation sign in Eq. (19.7) is a reminder that the formula holds best in the vicinity of the reference temperature at which α_0 and ρ_0 are given. Notice that for most pure metals $\alpha_0 \approx 1/273$ K^{-1} (see Problem 79).

The temperature coefficients of the semiconductors in Table 19.3 are negative; they become less resistive when the temperature increases. As with conductors, in-

The tungsten helical filament of an ordinary incandescent light bulb. It has a high resistance and becomes white hot at 110 V.

TABLE 19.3 Temperature coefficients of resistivity*

Substance	α_0 (K^{-1})
Aluminum	0.003 9
Brass	0.002
Constantan (60% Cu, 40% Ni)	0.000 002
Copper	0.003 93
Iron	0.005 0
Manganin ($\approx$84% Cu, $\approx$12% Mn, $\approx$4% Ni)	0.000 000
Mercury	0.000 89
Nichrome ($\approx$59% Ni, $\approx$23% Cu, $\approx$16% Cr)	0.000 4
Platinum	0.003 927
Silver	0.003 8
Tin	0.004 2
Tungsten	0.004 5
Carbon	$-0.000 5$
Germanium	-0.05
Silicon	-0.075
Sodium chloride (saturated solution)	-0.005

*Values determined at or near 20°C.

creasing T increases the scattering of charge carriers. The density of charge carriers in a semiconductor increases strongly with T—carriers that were initially not free to move can be set loose after absorbing thermal energy. The resulting increase in η has the effect of increasing I for a given V and, hence, decreasing R, via Equations (19.3) and (19.4).

Example 19.8 One of the most useful devices for measuring temperature is the platinum-resistance thermometer. Typically, about 2.0 m of pure platinum wire 0.1 mm in diameter is formed into a coil with a resistance at 0°C of 25.5 Ω. Taking the temperature coefficient of resistivity to be 0.003 927 K^{-1}, determine the change in resistance corresponding to a 1.00°C change in temperature. At what temperature is the thermometer if its resistance is 35.5 Ω?

Solution: [Given: $R_0 = 25.5$ Ω, and $\alpha_0 = 0.003\,927$ K^{-1}. Find: ΔR corresponding to $\Delta T = 1.00$°C, and T corresponding to $R = 35.5$ Ω.] Since

$$\rho \approx \rho_0(1 + \alpha_0 \Delta T) \qquad [19.7]$$

it follows from Eq. (19.6) that

$$R \approx R_0(1 + \alpha_0 \Delta T) \qquad (19.8)$$

We want the change in resistance $(R - R_0)$ that occurs when $\Delta T = \pm 1$, which can either be in °C or K; thus

$$R - R_0 = R_0 \alpha_0 \Delta T = (25.5\ \Omega)(0.003\,927\ \text{K}^{-1})(1.00\ \text{K})$$

and $\qquad\qquad R - R_0 = 0.100\ \Omega$

The thermometer changes resistance by

$\boxed{0.100\ \Omega\ \text{per 1 K}}$ or 1°C; hence, a change of $+10.0\ \Omega$ from its 0°C reading means a temperature of $\boxed{+100°C}$.

▶ **Quick Check:** For $\Delta T = 100$ K, $R - R_0 = (25.5\ \Omega)(0.003\,927\ \text{K}^{-1})(100\ \text{K}) = 10.0\ \Omega$.

19.4 Superconductivity

Resistance, although helpful in controlling currents, is the bane of electrical technology; it limits the operation of almost everything. If we could get rid of it, we could put nuclear power plants safely away from populated areas, cut the cost of electricity, float cars frictionlessly on magnetic fields, make tiny powerful motors, and improve computers. We could revolutionize the entire technology. Today, the goal of eliminating resistance seems just over the horizon in the realm of superconductivity.

Soon after H. Kamerlingh Onnes succeeded in liquefying helium in 1908, he began a study of the temperature dependence of the dc resistance of metals. The use of liquid helium as a refrigerant allowed him to routinely operate down to about 1 K (−458°F). Onnes, working first with platinum, found that its resistance dropped as the temperature descended, though it leveled off at a fairly constant value below roughly 4 K. He next selected mercury to examine because it could be obtained at ultrahigh purity and he wrongly believed that the resistance of any pure metal would vanish as it approached 0 K—something certainly suggested by Fig. 19.20. The resistance of mercury did slowly decrease with T, but at 4.2 K it inexplicably plunged to an unmeasurably small value. "Mercury has passed into a new state," wrote Onnes, "which on account of its extraordinary electrical properties may be called the superconducting state."

This total absence of dc resistivity below a **critical temperature** (T_c) is known as **superconductivity**. Onnes was lucky in selecting mercury: only 27 elements become superconducting under ordinary pressure, and many of those do so at critical temperatures well below 4.2 K (see Table 19.4). Incidentally, platinum does not make the transition to superconductivity, nor do the other good conductors, copper, silver, and gold. Still, well over a thousand alloys and compounds undergo this remarkable transformation.

TABLE 19.4 Critical temperature of some superconducting elements

Element	T_c (K)
Aluminum	1.175
Beryllium	0.026
Cadmium	0.52
Gallium	1.083 3
Indium	3.405
Lead	7.23
Mercury (α)	4.154
Molybdenum	0.916
Niobium	9.25
Osmium	0.655
Protactinium	1.4
Tantalum	4.47
Tin	3.721
Titanium	0.39
Tungsten	0.015 4

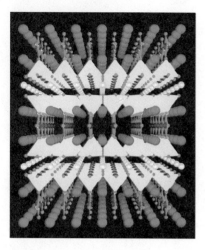

A computer model of a high-temperature, superconducting compound. The atoms are yttrium (gray), barium (green), copper (blue), and oxygen (red).

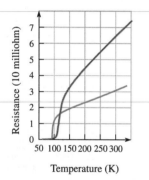

Figure 19.21 Resistance versus temperature of a thallium compound (blue) and a europium compound (red). Both become superconducting below around 100 K. These are typical of high-temperature superconductors. (Adapted from *Physics Today,* April 1988.)

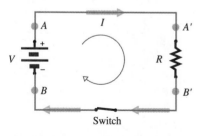

Figure 19.22 Points A and A', and B and B', are at the same potential.

Although a substance in its normal state has resistance and so must have an emf across it and an *E*-field within it to sustain a current, no such electric field exists in a superconductor. A current once initiated will continue on its own in a closed superconducting loop, perhaps indefinitely. One such experiment ran for over $2\frac{1}{2}$ years with no observed diminution in the circulating supercurrent. By contrast, a current circulating in a normal resistive material without any driving force would last for less than a second. Careful measurements indicate that the decay time for a supercurrent is at least 10^5 years, implying that if there is any resistance at all in the superconducting state, it's at least 10^{-12} times that of the normal state.

The basic understanding of low-temperature superconductivity is known as the BCS Theory after its Nobel-laureate (1972) creators John Bardeen, Leon Cooper, and Robert Schrieffer, who proposed it in 1957. Theirs is a quantum mechanical formulation maintaining that, unlike a conductor in the normal state where electrons behave independently, the electrons in a superconductor are paired. There is a long-range attractive interaction between paired electrons that is mediated by the electrons' interaction with the vibrating lattice. In effect, the densely packed electrons are all linked together in a kind of overlapping web of two-by-two interactions.

The situation is like a dance hall strewn with tables and chairs and yet crowded with people dancing independently—that's the normal state of a conductor. To make the transition to the superconducting state, partners are designated and each couple is provided a long scarf. The two people in every pair go on dancing far apart, but each holds an opposite end of their scarf. While bound together in pairs with an attractive scarf-interaction, every dancer repels all the other people on the floor. If the group is now impelled to drift in one direction—each person being pushed somehow (for example, by tilting the floor)—no single dancer can collide with an obstacle in the room and bounce away from the collective flow. The group acts as a coherent unit restraining all the participants to move together. Likewise, no single electron can be scattered by some imperfection from the collective motion of a supercurrent—no scattering, no resistance.

A breakthrough occurred in 1986, when it was discovered that an exotic ceramic compound of barium, lanthanum, copper, and oxygen had an unprecedentedly high critical temperature of 35 K. The race was on. By the beginning of 1987, physicists had prepared a ceramic (substituting yttrium for lanthanum) for which T_c was 98 K ($-283°$F), well above the temperature of liquid nitrogen (77 K). And by early 1988, a thallium compound (Tl-Ca-Ba-Cu-O) with a rather balmy critical temperature of 125 K had been produced (Fig. 19.21). Though still far from room temperature, values of T_c continue to creep upward and a variety of new high-temperature superconducting devices now exists.

19.5 Voltage Drops and Rises

Figure 19.22 depicts an idealized battery (having no internal resistance of its own) connected to a resistor by two ideal zero-resistance leads (generally, heavy-gauge copper wire will do nicely, but you can imagine these to be superconductors). The voltage across the battery (V) is its emf, and since the hookup wires have no resistance, there is no drop in potential along them—points A and A' are at the same voltage, as are points B and B'. In this simple circuit, the voltage across the terminals of the battery equals its emf, equals the voltage across the resistor.

In the steady state, charges introduced from the battery into that idealized wire coast along freely ($E = 0$) until they encounter the resistor. If we suppose that the mobile charge carriers are positive, they effortlessly traverse the ideal wire and arrive at the resistor (at point A'). Impeded in its progress, positive charge accumulates in the vicinity of that point, repelling positive charges out and away from the other end of the resistor, which becomes negatively charged. An E-field now exists within the resistor that has an axial component along its length. With the potential across R equal to its maximum value, namely that of the battery (V), the E-field drives a steady-state current I through the resistor. From point B' the carriers coast at a constant potential back to the battery. Thus, there are surface charges distributed around the circuit with concentrations ($+$) at A and A', and ($-$) at B and B'.

Note that I leaves the high potential (positive) side of the battery and enters the high potential (positive) side of the resistor. In passing through the resistor, the mobile charge carriers are scattered somewhat, imparting random KE (that is, thermal energy) to the atoms of the resistive medium. As a result, energy is transferred from the current to the resistor—each positive charge q drops in potential V and thus in potential energy qV, on traversing the resistor. The carriers return to the negative terminal of the battery, which raises their potential by V, thus pumping them up in energy qV, and sending them out again. Clearly, the battery supplies electrical energy, and the resistor dissipates that exact amount as thermal energy (in accord with the Law of Conservation of Energy). Current will circulate until the battery supplies all the energy it can and inevitably runs down. **The current is the transporter of energy**.

No point in the circuit is grounded, and so no point has a potential referenced to the zero of ground. The circuit is said to *float*, and what is important is the voltage of one point with respect to another: ***voltage differences are the practical quantities***.

Figure 19.23 depicts a variable resistor, or *potentiometer,* a pot for short. This device is usually a long coil of metal wire (its end points being A and C) with a central slide pressing against it, making contact with the wire (at a variable point B). When placed across an ideal battery, the emf (V) equals V_{AC} and V_{AB} is any desired fraction thereof. This sort of *voltage divider* is very useful. When you turn the knob on the volume control of a radio or stereo, you're moving the slide on just such a pot.

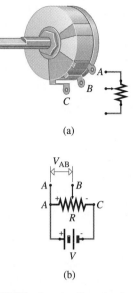

(a)

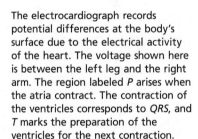

(b)

Figure 19.23 A potentiometer is a variable resistor. It is commonly found in radios, stereos, amplifiers, and tape decks.

19.6 Energy and Power

A current in a circuit is like a moving fluid capable of transporting energy from the source of emf (a battery, generator, solar cell, etc.) to some device (a toaster or a TV), where it can subsequently be utilized. The transport of energy by a current is one of the fundamental features of electricity, one that we must quantify.

Suppose that a small amount of charge Δq traverses a circuit element and moves through a constant potential difference of V. It will, in doing so, change its electrical potential energy (ΔPE_E) by an amount $\Delta q V$. The rate at which this happens, the rate

The electrocardiograph records potential differences at the body's surface due to the electrical activity of the heart. The voltage shown here is between the left leg and the right arm. The region labeled P arises when the atria contract. The contraction of the ventricles corresponds to QRS, and T marks the preparation of the ventricles for the next contraction.

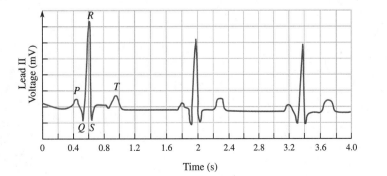

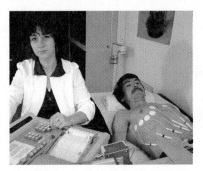

of transfer of energy, is

$$\frac{\Delta PE_E}{\Delta t} = \frac{\Delta q}{\Delta t} V$$

which, by definition (p. 322) is the **power**, P (either delivered or consumed), over a time Δt. Since $\Delta q/\Delta t$ is the current, the power at any instant is

$$P = IV \qquad (19.9)$$

The unit of electrical power is an ampere·volt where

$$1 \text{ A·V} = (1 \text{ C/s})(1 \text{ J/C}) = 1 \text{ J/s} = 1 \text{ W}$$

In Fig. 19.22, the power delivered by the battery to the current I in raising its potential V is IV. Similarly, the power delivered thermally to the resistor by the current I while dropping in potential by V is also IV. If several batteries are connected in a circuit, it is possible that a current may *enter* one of them at its positive terminal. Such a current would emerge after a drop in potential—that is, after depositing energy in that battery. It is by supplying energy to a battery in this fashion that it can be recharged.

Using Ohm's Law, $V = IR$, **the power dissipated in a resistance R can be expressed as**

$$P = I(IR) = I^2 R \qquad (19.10)$$

or

$$P = \left(\frac{V}{R}\right)V = \frac{V^2}{R} \qquad (19.11)$$

J. P. Joule (1841) first showed experimentally that the "heating-power" of an electric current through a resistance had the form of Eq. (19.9). To recognize his accomplishment, we speak of the thermal energy produced in this way as **joule heat**.

Example 19.9 The circuit in Fig. 19.24, containing a battery of unknown voltage, carries a current of 5.0 A. Determine the power either dissipated or provided by each circuit element.

Solution: [Given: a 12-V battery, an unknown battery, $I = 5.0$ A, and $R = 10$ Ω. Find: power for each.] The same 5.0-A current passes through both elements in going from A to B. Hence, the power *delivered to* (since I enters on the + side) the 12-V battery is

$$P = IV = (5.0 \text{ A})(12 \text{ V}) = \boxed{60 \text{ W}}$$

It's receiving energy from the current. Resistors only dissipate power, and here that is in the amount of

$$P = I^2 R = (5.0 \text{ A})^2 (10 \text{ Ω}) = \boxed{0.25 \text{ kW}}$$

The net power dissipated is therefore 0.060 kW +

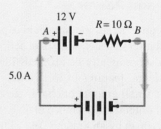

Figure 19.24 Each circuit element either provides or dissipates power.

0.25 kW = 0.31 kW and that's provided by the second battery—current leaves its + terminal.

▶ **Quick Check:** The voltage of the second battery is $V_{AB} = 12 \text{ V} + IR = 12 \text{ V} + 50 \text{ V} = 62 \text{ V}$. The power it delivers is therefore P = IV = (5.0 A)(62 V) = 0.31 kW, which is the power dissipated.

Example 19.10 Calculate the price of electrical energy in dollars per kilowatt-hour as supplied by a 50¢ D-cell having a 3.0 amp-hour rating. Compare that with the ≈10¢ per kW·h for electricity on tap from a wall outlet.

Solution: [Given: 50¢ for 3.0 amp-hours. Find: price per kW·h.] First, 1.0 kW equals 1000 J/s; hence, 1.0 kW·h = (1000 J/s)(3600 s/h), or 3.6 MJ. The cell delivers 3.0 amp-hours at 1.5 V; $IV = P$, which is the power delivered, and that multiplied by the time during which it is delivered is the energy; consequently

$$Pt = (3.0 \text{ amp-hours})(1.5 \text{ V}) = 4.5 \text{ W·h}$$

or 4.5×10^{-3} kW·h. At 50¢ per cell, that's (50¢)/$(4.5 \times 10^{-3}$ kW·h) or $\boxed{\$111 \text{ per kW·h}}$.

▶ **Quick Check:** The energy of the cell is 4.5 W·h divided by the operating voltage, which yields the number of amp-hours: 3.

As a rule, for a given type of cell, the bigger it is, the higher the amp-hour rating, the more energy it stores. That's why portable devices that need a lot of energy, such as heavy-duty flashlights or large speakers, operate on D-cells rather than A-cells, even though both have the same emf of 1.5 V.

19.7 Current Density and Conductivity

When a uniform conducting wire, having a substantial resistance, is put across the terminals of a battery, a current flows. The battery establishes a nonelectrostatic E-field within the wire that runs along the circuit parallel to the walls of the conductor; it propels the charge. The fairly uniform pattern of field lines produces a similar pattern of current flow lines that follow the geometry of the circuit. Unlike the electrostatic case, the surface of the conductor is no longer an equipotential. Now the equipotentials are perpendicular to the wire, and the E-field within it, and the voltage drop (equal to the battery's emf) occurs uniformly with distance along the wire's entire length.

If the conductor is a spoon (or a careless electrician) it will have a nonuniform cross section and a current that varies in density from place to place accordingly, even though $V = IR$. To better deal with that situation consider the **current density** J in a uniform wire of cross-sectional area A, defined as

$$J = \frac{I}{A} \tag{19.12}$$

As a rule current flows parallel to **E** so we write the current density as a vector **J** in the direction in which a positive charge would move. The most general formulation of current flow is then

$$I = \int \mathbf{J} \cdot d\mathbf{A} \tag{19.13}$$

where I is the net current passing through the surface described by the integral and **J** can vary from point to point in space, as can the cross-sectional area. Put a potential difference across the spoon and a constant I will progress that is determined by the resistance as a whole. By contrast, J varies along the length as the cross-sectional area varies; I is the gross flow specific to a given conductor, J is the detailed flow specific to the geometry.

Just as $V = IR$ is the macroscopic statement of Ohm's Law for a conductor (where there's no regard for what it's made of or its configuration), there is an equivalent microscopic version in terms of J, ρ, and E that depends on the resistivity of the material and the geometry. Accordingly, consider a wire of length L and cross-sectional area A. Put into Ohm's Law the alternative forms: $V = EL$, $I = JA$, and $R = \rho L/A$, whereupon

$$V = IR$$

$$EL = (JA)(\rho L/A)$$

in which case

$$E = J\rho \qquad (19.14)$$

This is Ohm's Law recast. It is customary to introduce the **conductivity** $\sigma = 1/\rho$ and write

$$\mathbf{J} = \sigma \mathbf{E} \qquad (19.15)$$

Example 19.11   A copper wire 1.00 mm in diameter carries 15.0 A. Determine the electric field inside the wire.

Solution: [Given: $D = 1.00$ mm, copper wire, and $I = 15.0$ A. Find: E.] We are given the physical features of the conductor and required to find the internal E-field, which should bring to mind the microscopic version of Ohm's law:

$$E = \frac{J}{\sigma} = \rho\frac{I}{A} = \rho\frac{I}{\pi R^2}$$

Using Table 19.2

$$E = \frac{(1.7 \times 10^{-8}\ \Omega \cdot \text{m})(15.0\ \text{A})}{\pi(0.500 \times 10^{-3}\ \text{m})^2} = 0.32\ \text{V/m}$$

▶ **Quick Check:** In this expression for E the units are $(\text{A} \cdot \Omega)/\text{m} = \text{A/m} \times \text{V/A} = \text{V/m}$, which is correct. Notice that the field is fairly small and we expect this because copper is a good conductor.

Core Material

The ordered flow of charge is called **electric current** (I):

$$I = \frac{dq}{dt} \qquad [19.2]$$

The SI unit of current is the *ampere:* $1\ \text{A} = 1\ \text{C/s}$. A nonelectrostatic potential difference that sustains a current in an external circuit is called an **electromotive force**, or emf (p. 728).

The current in a wire of cross-sectional area A, in terms of the **average drift speed**, v_d, the number of carriers per unit volume, η, and the charge per carrier, q_e, is

$$I = \frac{\Delta q}{\Delta t} = \eta v_d A q_e \qquad [19.3]$$

The current passing through a resistor varies directly with the applied V and inversely with R:

$$V = IR \qquad [19.5]$$

which is **Ohm's Law**, and it applies to a limited range of mate-

rials. The unit of resistance is the *ohm* (Ω). The resistance of a wire can be expressed in terms of its length L, cross-sectional area A, and a material-dependent constant of proportionality ρ, the **resistivity**, as

$$R = \rho\frac{L}{A} \qquad [19.6]$$

Resistivity is temperature-dependent (p. 738); accordingly

$$\rho \approx \rho_0(1 + \alpha_0 \Delta T) \qquad [19.7]$$

α_0 is called the **temperature coefficient of resistivity**. The total absence of dc resistivity below some **critical temperature** (T_c) is known as **superconductivity** (p. 739).

The power either provided or dissipated by a circuit element is

$$P = IV \qquad [19.9]$$

The unit of electrical power is the *watt*. The power *dissipated* in a resistance R is

$$P = I(IR) = I^2R \qquad [19.10]$$

or

$$P = \left(\frac{V}{R}\right)V = \frac{V^2}{R} \qquad [19.11]$$

Figure 19.25 summarizes the diagrammatic representations of the circuit elements introduced thus far.

The microscopic form of Ohm's Law is

$$E = J\rho \qquad [19.14]$$

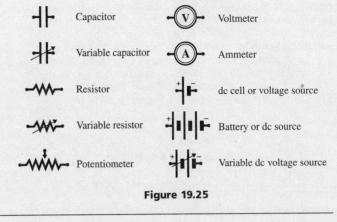

Capacitor		Voltmeter	
Variable capacitor		Ammeter	
Resistor		dc cell or voltage source	
Variable resistor		Battery or dc source	
Potentiometer		Variable dc voltage source	

Figure 19.25

Suggestions on Problem Solving

1. An important type of derivation involves determining the current, given a flow of charge or vice versa. For example, a detector measures N particles, each carrying a charge q, arriving *per second per unit of its surface area*—what current strikes the instrument if its area is A? The basic definition is $I = \Delta q/\Delta t$, but to use this equation you must first find Δq, the amount of charge that strikes the detector during the time Δt. We are given N and so Nq is the amount of charge arriving *per second* per unit of surface area. Consequently, NqA is the total amount of charge arriving on the detector per second and is equal to I. We obtained I without finding Δq explicitly because N had the time built into it to begin with. This kind of analysis of flow is very important in physics—we've seen it before and we'll see it again.

2. Remember that all parts of an ideal hookup wire in a circuit are at the same potential. Drops or rises in potential occur only across circuit elements (resistors, batteries, etc.), and in the case of a resistor, that means only when a current is passing through it (take a look at Multiple Choice Question 9). **A steady-state current cannot exist in a length of conductor unless it forms part of a closed path.**

3. Whenever you have a circuit diagram with a battery in it, place + and − signs at the appropriate terminals. Current, being imagined as composed of positive charge, always emerges from the positive terminal (provided there are no other batteries bucking it) and re-enters at the negative one. In a simple circuit where you know the direction of the current, follow it around, labeling all the resistors with + signs where current enters and − signs where it leaves. This will show the voltage drops and rises, something that will be explored further in the Chapter 20.

4. The voltage of any point in a circuit is only known relative to some other point in that circuit. Thus, one side of a battery might be 12 V higher than the other side (though it could be 10 000 V higher than your nose). For example, the entire chassis of an automobile serves as a common conductor called "ground" in the trade, though it isn't usually grounded and so actually floats—the + terminal of the battery is 12 V higher than the engine block. Often a point in a circuit is grounded, as is done with stereo systems and computers. Its potential is then taken as zero, and all voltage drops or rises are referenced with respect to it.

Discussion Questions

1. Ohm once remarked that the laws of electricity "are so similar to those given for the propagation of heat . . . that even if there existed no other reasons, we might with perfect justice draw the conclusion that there exists an intimate connection between these natural phenomena." What was he alluding to? Discuss the nature of electrical and thermal conductivity for metals.

2. Figure Q2 shows several 9-V batteries opened up so we can look inside. Explain what you see.

3. Suppose you wish to set up a circuit with a battery and some resistors so that you can measure currents and voltages and confirm the ideas of this chapter. How should you select the hookup wire, or doesn't it matter?

4. What happens when you turn the key in the ignition of your automobile? Should you start your car on a cold rainy night with the wipers, lights, heater, and defroster all on? Explain.

5. Figure Q5 shows a 1.5-V dry cell attached to a 1.5-V

Figure Q2

flashlight bulb via two single-pole double-throw knife switches. Each switch attaches the central post (or pole) to either the right or left terminal on the device—the blade is only thrown to a horizontal position. You probably have an arrangement of this kind in your house, especially if it has a long flight of steps that are lighted. What's it for? Make a simplified drawing of the circuit and discuss its operation. As shown, is the light on? What happens when either switch *A* or switch *B* is thrown?

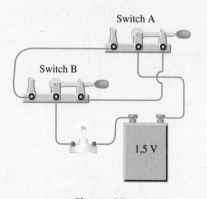

Figure Q5

6. Suppose you parked your car and left the lights on all night. Why would you have trouble starting it the next day? Would it be more bothersome to start on a really cold winter's day? After several tries, the engine just makes some disheartening clicking noises and you've had it—what has happened? A friend comes to your aid and "jump starts" your car. How is that done and what does it accomplish?

7. With Question 6 in mind, it's assumed that once the engine is running, all will return to normal. Explain. In other words, how does your engine keep running once the jumper cables are removed, given that the battery is still exhausted? Is it then a good idea to let the engine run or drive for a while without turning on any unnecessary electrical devices? Why? What's happening in the electrical system while you drive the car with the radio playing?

8. Figure Q8 shows an ordinary metal-body flashlight. Explain how it works and discuss the significance of the arrangement of the dry cells.

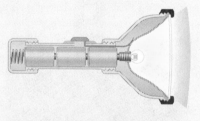

Figure Q8

9. Describe Volta's tin-tongue-silver cell. Why don't you taste anything until the two metals touch? Volta thought that the current of "electric fluid" in his pile arose from a "contact force" between the two metals. It was not until about 25 years later that the chemical action taking place was recognized (by C. A. Becquerel, A. De La Rive, G. F. Parrot, and others).

10. Figure Q10 shows a vertical arrangement of pipes with water flowing around the circuit at a rate that keeps the levels constant. This system is a liquid analogy to an electrical circuit. Draw the corresponding electrical diagram and discuss the relationship between the various parameters. Consider the system from an energy perspective. What does the pump "pump" besides water? No wonder nineteenth-century scientists liked the notion of "electric fluid." Have you ever heard a modern electrician mutter things like "turn on the juice"?

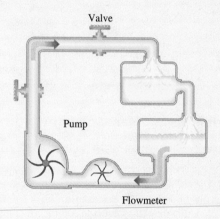

Figure Q10

11. Discuss everything of significance you know about the dry cell—that is, the ordinary flashlight "battery." Would it still work if it was actually dry? Ask a few elderly people if they remember how leaky old-fashioned D-cells were when they were kids—explain. Why do expensive electronic devices recommend that they not be stored with the batteries in them?

12. Question 10 deals with a fluid system using gravity. Figure Q12 represents a horizontal variation that is independent of gravity. Relate the liquid and electrical parameters. Considering fluid friction, talk about the influence of the narrow-bore tube.

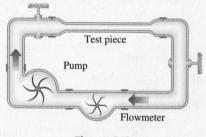

Figure Q12

13. When an ordinary tungsten light bulb burns out, it's likely to flash bright blue-white for an instant. Explain what is happening. Suggest a possible reason these lamps usually burn out soon after being turned on.

14. Suppose that you are watching a demonstration performed by your instructor using a small battery (anywhere from 4.5 V to 9 V) and a conductor with a high resistance (a small-diameter strand of steel wool will serve nicely). A single filament of steel is stretched across the terminals, and within seconds

a region of it glows red hot and bursts into a shower of molten beads. (*Don't try this—it's dangerous!*) Why did the wire melt? I once saw someone accidentally drop a wrench across the terminals of a 12-V car battery. There was a tremendous sparking flash, and molten beads of metal flew all over the place. Explain. Why did the wrench only melt where it touched the battery?

15. When lightning hits the ground, it often spreads out radially as it penetrates downward. Thus, it is possible for a large four-legged animal standing with its bodyline radial with the bolt to be electrocuted. Explain. Considering ground currents, what is the best posture for a human out in a thunderstorm—lying down, squatting, or standing up? Read the newspaper article on the right (Fig. Q15).

16. High-voltage cables are generally uninsulated, and a bird standing on one of them may have its feathers puffed up. Why? Why aren't our little feathered friends electrocuted on their perches? Could a chicken walk safely on the third rail of an electrified railroad?

Figure Q15

Multiple Choice Questions

1. If 10 amperes circulate in a closed circuit, how much charge passes any point therein in 2 s? (a) 10 C (b) 5 C (c) 20 C (d) 200 C (e) none of these.

2. The difference in potential between the electrodes of a voltaic cell when there is no current being drawn is (a) zero (b) 1.5 V (c) its emf (d) the power (e) none of these.

3. The emf of a voltaic cell (a) is independent of the chemical interactions taking place within it (b) is dependent on the size of the cell (c) is dependent on its amp-hour rating (d) is independent of the plate size (e) none of these.

4. Usually, the larger a voltaic cell is, the more (a) voltage it can supply (b) current it can supply (c) potential it can develop (d) emf it can sustain (e) none of these.

5. The moving nonconducting belt of a Van de Graaff generator carries 10 μC of charge up to the metal sphere (Fig. Q14, Chapter 17) each second. The steady-state potential difference between the sphere and the grounded source of charge is 3×10^6 V. What, if any, is the emf of the generator? (a) 10 μV (b) 3 μV (c) 3 MV (d) 30 μV (e) none of these.

6. A statue of a chicken made of pure gold has a resistance of 0.10 mΩ between the beak and the tail. An exact duplicate of the piece is made in an alloy that has a resistivity 10 times greater than gold. The resistance between the same two points will now be (a) 0.10 mΩ (b) 1.00 mΩ (c) 0.01 mΩ (d) 10.0 mΩ (e) none of these.

7. A metal wire has a resistance of 1.0 Ω. What will be the resistance of a wire made of the same material but twice as long and with half the cross-sectional area? (a) 0.40 Ω (b) 2.00 Ω (c) 0.02 Ω (d) 40.0 Ω (e) none of these.

8. The resistance of a superconducting material drops, essentially to zero, rather suddenly when (a) the sample is heated above T_c (b) the sample is exposed to a magnetic field (c) the sample is cooled below T_c (d) the sample experiences a current I_c (e) none of these.

9. Referring to Fig. MC9, the voltage at point B is (a) +12 V (b) −12 V (c) 0 V (d) −6 V (e) none of these.

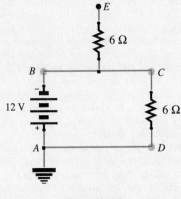

Figure MC9

10. Referring to Fig. MC9, the voltage at point D is (a) +12 V (b) −12 V (c) 0 V (d) −6 V (e) none of these.

11. Referring to Fig. MC9, the voltage at point C is (a) +12 V (b) −12 V (c) 0 V (d) −6 V (e) none of these.

12. Referring to Fig. MC9, the voltage at point *E* is
    (a) +12 V   (b) −12 V   (c) 0 V   (d) −6 V   (e) none
    of these.
13. An 8.00-Ω speaker is connected across the output terminals
    of a power amplifier that delivers 64.0 W to it. The current
    supplied to the speaker is   (a) 2.83 A   (b) 8.00 A
    (c) 2.00 A   (d) 64.0 A   (e) none of these.
14. Figure MC14 shows the voltage across a tungsten light bulb
    filament plotted against the current through it. The curve
    bends upward because   (a) the filament is getting hot and
    running out of electricity   (b) it takes more voltage to
    propel the same current than it would if the device were
    ohmic because the resistance increases   (c) the resistance
    goes down, so we get more voltage for a given current
    (d) the resistance stays constant, but the current decreases
    (e) none of these.
15. A copper wire has a resistance of 10 Ω. What will be its
    new resistance if the wire is shortened by cutting it in
    half?   (a) 20 Ω   (b) 10 Ω   (c) 5 Ω   (d) 1 Ω   (e) none
    of these.
16. When the temperature of a length of aluminum wire
    is lowered, its resistance   (a) increases slightly
    (b) decreases correspondingly   (c) stays the same
    (d) increases as ΔT (e) none of these.
17. If a potential difference of 12 V across a resistor results
    in a current of $\frac{1}{2}$ A, how much power is dissipated?
    (a) 48 W   (b) 12 W   (c) 4 W   (d) 6 W   (e) none
    of these.
18. Imagine a length of ordinary insulated hookup wire. The
    wire's resistance is not dependent on   (a) the conductor's

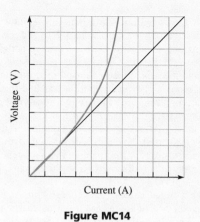

Figure MC14

length   (b) the conductor's radius   (c) the material
making up the insulator   (d) the material making up the
conductor   (e) none of these.
19. The units of $\Omega \cdot A^2$ correspond to   (a) current   (b) energy
    (c) power   (d) voltage   (e) none of these.
20. One microvolt (1 μV) corresponds to   (a) 10^6 V
    (b) 10^{-6} V   (c) 1000 V   (d) 1/1000 V   (e) none of these.
21. A resistor operated at 100 V generates joule heat at a
    rate of 20 W. When placed across a 50-V source, it will
    draw   (a) 0.10 A   (b) 5 A   (c) 20 A   (d) 4.0 A
    (e) none of these.

Problems

CURRENTS
1. [I] The moving nonconducting belt of a Van de Graaff
   generator carries 10 μC of charge up to the metal sphere
   (Fig. Q14, Chapter 17) each second. Determine the
   corresponding current.
2. [I] A beam of positrons carries 1.4 C past a point in space
   in 2.0 s. What current does that correspond to?
3. [I] If a quantity of singly ionized sodium ions (Na$^+$) equal
   to Avogadro's number streams past a point in 1000 s, what
   is the current?
4. [I] The photo on p. 724 shows an electron beam
   representing a current of 60 kA. How many electrons flow
   out of the device per second?
5. [I] The starter motor in an automobile is a small but
   powerful electrical device that "turns over" the main
   gasoline engine, moving it through a cycle to get it started.
   Typically, it will draw about 180 A from the battery for
   perhaps 2.0 s. How much charge flows through the circuit?
6. [I] A particle accelerator contains two beams flowing
   side-by-side in opposite directions; the beams will
   ultimately be made to collide. One is a stream of protons
   (each with a charge of $+1.60 \times 10^{-19}$ C), the other a
   stream of antiprotons (each with a charge of
   -1.60×10^{-19} C). Given that either beam can deliver
   1.0×10^{14} particles per second to the colliding region,

what is the net current in the machine as the two streams
race past each other?
7. [I] A synchrotron accelerates protons up to nearly the
   speed of light, imparting energies to them of 500 MeV. If
   the beam current is 1.0 mA, how many protons will hit a
   target in 0.10 s?
8. [I] With the previous problem in mind, how many protons
   are there in each 1.0-cm-long segment of the beam?
   Assume a uniform particle density throughout the beam.
9. [I] A wire is connected across the terminals of a battery
   for precisely 60.0 s during which time a constant current
   of 2.00 A circulates around the loop. What was the net
   charge that flowed past any point on the wire?
10. [I] A portable tape recorder is powered by six 1.5-V
    AA-cells in series. What is its operating voltage?
11. [I] A torpedo, a giant saltwater ray that can develop a
    voltage of 220 V, is covered with cells known as
    electroplaques, each of which produces a potential
    difference of about 0.15 V. Several thousand rows (each
    made up of a series-connected array of cells) are then
    connected in parallel to build up a sizeable current. How
    many cells would you guess form each row? Why do
    freshwater electric fish in general develop higher voltages
    than saltwater ones?

12. [I] A NiCd cell has a voltage of 1.2 V and an amp-hour rating of 34. It is a sealed storage cell with a nickel anode and a cadmium cathode immersed in an alkaline electrolyte. How long can it operate when providing 2.0 A to some load?

13. [I] A battery is rated at 10 amp-hours. How much charge does that correspond to?

14. [I] The positive terminal of a 1.5-V dry cell is attached to ground via a length of heavy hookup wire. The negative terminal is then connected, via the same kind of wire, to a neutral brass ball. What is the potential of the ball? Describe its state of charge. What is the potential of the central carbon electrode?

15. [c] A time-varying current flowing along a copper wire delivers charge according to the function $q = (2.00 \text{ C/s}^3)t^3 + (4.00 \text{ C/s}^2)t^2$. Determine the instantaneous current—write it in amps.

16. [c] An ion beam having a cross-sectional area of 1.00 mm^2 transports charge according to the expression $q = (1.00 \text{ C/s}^3)t^3 + (6.00 \text{ C/s}^2)t^2 - (2.00 \text{ C/s})t$. Determine the current in amps at $t = 2.00$ s.

17. [c] A beam of electrons corresponding to a time-varying current $I(t) = (3.00 \text{ A/s}^2)t^2 - (2.00 \text{ A/s})t + 3.00 \text{ A}$ impacts on a detector. How much charge is delivered to the detector during the interval from $t = 1.00$ s to $t = 2.00$ s?

18. [c] The current in a circuit diminishes from $I(0)$, which is its value at $t = 0$, according to the expression

$$I(t) = I(0)e^{-t/\tau}$$

where τ is called the time constant (the exponent must be unitless). Find an expression for the net amount of charge that has flowed past any point in the circuit after the first t seconds.

19. [c] Determine an expression for the instantaneous current density when a time-varying current flowing along a gold rod, having a cross-sectional area of 2.00 mm^2, delivers charge according to the function $q = (4.00 \text{ C/s}^3)t^3 - (4.00 \text{ C/s})t$. What is the current density, in amps per meter squared, at $t = 1.00$ s?

20. [c] A long rod of conducting material lies on the x-axis (Fig. P20). It is placed between the terminals of a battery, and a current passes along its length L. The material is made in layers and is therefore nonuniform top to bottom. Thus the current density depends on the position across the square ($l \times l$) cross section of the rod, being greatest at the bottom and decreasing with y toward the top according to the expression

$$J = C_1(1 - C_2 y)$$

wherein C_1 and C_2 are constants. Determine the total current along the length of the rod. Check the units of your answer.

21. [II] A pure gold wire with a 1.00 mm $\times$ 1.00 mm cross section carries a flow of electrons having a current density (I/A) equal to 1.0 MA/m^2. How long will it take for an amount of electrons equal to Avogadro's number to pass a point on the wire?

22. [II] An ion generator used to clean room air puts out a stream of negatively charged molecules that attach themselves to airborne pollutants, which are then collected

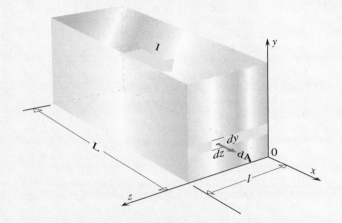

Figure P20

electrostatically. Oxygen molecules tend to pick up electrons, thereby becoming negative oxygen ions. According to one company's literature, at 1.0 m from a particular generator, a detector would record the arrival of "168 million ions/sec./cm^2." Assuming that each ion is singly charged, what current impinges on a 10.0-cm^2 target at 1.0 m from the device?

23. [II] Figure P23 is a diagram of a portion of the electrical system of an automobile. List which switches—A, B, C, D, E, and F—must be closed in order to (a) blow the horn; (b) turn on the headlights; (c) turn on the tail lights; (d) turn on only the parking lights; (e) activate the inside dome light. When do the side marker lights go on? *Note that the whole body of the car, including the engine block, is generally wired together as a common "ground."*

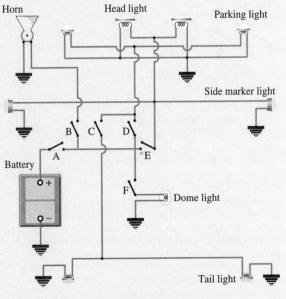

Figure P23

24. [II] A particular model of automobile battery, when fully charged, can deliver roughly 4.0×10^5 C before becoming completely run down. What is the amp-hour rating of such a battery?

25. [II] We wish to make up a panel of silicon solar cells that will provide at least 440 mA at 9.0 V when placed in an appropriately illuminated region. Given that each cell provides 22 mA at 0.45 V, design the panel.

26. [II] A rectangular aluminum wire with a cross section 1.0 mm × 2.0 mm carries an electron current of 0.10 A. Given that aluminum has 6.0×10^{22} atoms per cm^3 and each contributes roughly one free electron, determine the electron's drift speed.

27. [II] If in Problem 26 the current is kept constant while the cross-sectional area of the wire is halved, how long will it take a typical electron to travel down a 1.0-m length of the aluminum?

28. [II] The drift velocity of electrons in a pure copper wire carrying a current of 1 A and having a cross-sectional area of 1.0 mm^2 was found in Example 19.4 to be ≈0.1 mm/s. How would that change if the current was increased to 20 A, all else kept as is? What would happen if the current was doubled and the cross-sectional area was halved?

29. [cc] In Chapter 15 we wrote the following expression for the rate at which heat (Q) is conducted along a rod of uniform cross section:

$$\frac{dQ}{dt} = -k_T A \frac{dT}{dx}$$

Here heat is driven by a temperature gradient. Now show that in the electrical case

$$\frac{dq}{dt} = -\sigma A \frac{dV}{dx}$$

where the current is driven by the potential difference. Of course, both phenomena depend on electron transport. [Hint: Use the facts that $I = JA$, $J = \sigma E$, and $E = -dV/dx$.]

30. [cc] Suppose that the rod in Fig. P20 is made of a material such that the current density within it is given as

$$J = J_0 \sin \frac{\pi y}{l} \sin \frac{\pi z}{l}$$

Where is the current density zero? What is its maximum value and where does it occur? What is the total current through the rod? [Hint: When doing the double integral here first hold y constant and integrate to form a differential strip dy wide and l long.]

31. [III] In the process of recharging a rundown automobile battery, it's attached to an electronic 12-V charger. As soon as the device is turned on, an ammeter shows that the battery draws 7.0 A, but as it revives, the current slowly drops until after 6.0 hours it's down to 3.0 A. Assuming the current decreased linearly with time, how much charge passed through the battery?

RESISTANCE

32. [I] A small high-torque variable-speed dc motor requires an input of 10 mA at 3.0 V. Determine the electrical resistance of the motor.

33. [I] A 100-V electric heater draws 10 A. What is its resistance?

34. [I] If electric contacts are placed on the scalp, time-varying differences in potential will be observed.

These can be recorded by an electroencephalograph. Voltage differences of ≈0.5 mV will appear across resistances of ≈10 kΩ. What size currents are involved?

35. [I] A wooden stick in contact with the metal sphere of a 100-kV Van de Graaff generator carries a current of 2.0 μA down to ground. Calculate the stick's resistance.

36. [I] If the current in a 10-Ω resistor is 500 mA, what is the voltage across its terminals?

37. [I] Given a solid cube of metal, under what circumstances, if any, will $R = \rho$?

38. [I] A 1.0-m-long wire of pure silver at 20°C is to have a resistance of 0.10 Ω. What should be its diameter?

39. [I] Considering a length of metal wire, show that Ohm's Law can be written as

$$I = \frac{E}{\rho/A}$$

40. [I] A nichrome wire with a cross-sectional area of 1.5×10^{-6} m^2 is to be used in a heater. If the design calls for a 3.0-Ω coil, what length of wire will be needed?

41. [I] A 5.0-m-long wire has a cross-sectional area of 2.0 mm^2 and a resistance of 40 mΩ. What is the resistivity of the material constituting the wire?

42. [I] According to the American Wire Gauge system, No. 0000 wire, which is the heaviest, has a diameter of 11.7 mm. What would be the resistance of 100 m of copper AWG No. 0000 at 20°C?

43. [I] Copper telegraph wire has a resistance of about 10 Ω per mile. What's its diameter in millimeters?

44. [I] A wire of pure gold is drawn through a die so that it is stretched out to twice its original length. Given that its volume is unchanged in the process and its new cross-sectional area is constant, compare the new with the original resistance.

45. [I] A narrow rod of pure iron has a resistance of 0.10 Ω at 20°C. (a) What is its resistance at 50°C? (b) A narrow rod of manganin has a resistance of 0.10 Ω at 20°C. What is its resistance at 50°C?

46. [I] A carbon rod used to generate the bright light in a movie-theater projector has a resistance of 110 Ω at 20.0°C. What will be its resistance at 520°C? (Incidentally, the hottest point on a functioning carbon arc, and the point of greatest luminosity, is typically at about 3500°C.)

47. [I] A portable tape recorder has a plate on its underside indicating that it uses 1 W at 9 V dc. What net current does it draw from its battery?

48. [I] Referring to Problem 47, what is the net resistance of the tape recorder?

49. [I] A high-torque dc motor designed to drive cassette decks has a no-load speed of 7400 rpm at 9 V with a torque of 9.6 in.·oz and a no-load current of 16 mA. How much power will it draw from a 9-V battery when turning freely?

50. [I] What is the maximum current that should be passed through a 100-Ω, 10-W resistor?

51. [I] The little speaker in a portable radio is labeled 8 Ω, 0.2 W. What current does that correspond to?

52. [I] A windmill with 6-ft propellers generates dc at 12 V. In a strong wind it will produce up to 200 W of electrical power, which is fed to a 230-amp-hour battery. What's the maximum current the windmill will deliver?

53. [I] An automobile starter motor will draw about 180 A from the 12-V battery of a car for perhaps 2.0 s in the

process of starting the gasoline engine. How much power does it use?

54. [II] Figure P54 shows a variable resistance (total 100 Ω) across which is a voltage drop of 12 V supplied by a battery. What must be the resistance of that portion of the resistor between A and B (namely, R_{AB}) if the voltage V_{AB} is to be   (a) 12 V;   (b) 6.0 V;   (c) 3.0 V?

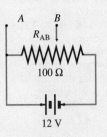

Figure P54

55. [II] A length of copper wire 1.0-m long with a cross-sectional area of 1.0 mm² is part of a circuit carrying 1.0 A. What is the value of the steady-state electric field within the wire?

56. [II] A variable slidewire resistor with a total of 50 Ω is made by wrapping a single layer of 200 turns of varnished copper wire around an insulating cylindrical core. When it carries a current of 3.0 A, what is the voltage drop across each turn of wire?

57. [II] It is said that the carbon filaments in the early incandescent light bulbs that Edison produced lost more than two-thirds of their resistance soon after they were turned on. Explain this occurrence and compute their approximate operating temperature. The latter was actually about 1900°C.

58. [II] Imagine that you are going to connect a remote speaker to your stereo system. The speaker has a resistance of 4.0 Ω, and so you want to use hookup wire whose total resistance is small by comparison, say, 0.25 Ω. If the speaker is to be 15 m away, what diameter copper wire should be used? [Hint: You'll need two leads.]

59. [II] A platinum resistance thermometer made up of a coil of wire with a resistance of 10 Ω at 20°C is placed in a chamber at 420°C. What will be its new resistance if α_0 is fairly constant at 0.003 9 K⁻¹?

60. [II] Show that if a conductor has its temperature changed, it will experience a fractional change in its resistivity given by

$$\frac{\Delta\rho}{\rho_0} = \alpha_0 \Delta T$$

What assumption must be made here?

61. [II] An iron wire at 20°C is heated until its resistance doubles. At what temperature will that occur? Assume the temperature coefficient is constant over that temperature range.

62. [II] When currents are transported at very high voltages, it's desirable to keep the electric fields surrounding the conductors down to levels that will not break down the

surrounding air, thereby controlling sparking. Accordingly, conductors at power plants and high-voltage laboratories are often large-diameter pipes. What is the resistance per meter of a copper pipe 2.5 cm thick with an inside diameter of 15.0 cm?

63. [II] A powerhouse near a waterfall has a large dc generator that produces electricity for a factory 0.50 mile away. The energy is transmitted over two cables each with a resistance of 0.25 Ω/mile. Given that the factory requires 45 kW at a voltage of 110 V to run its equipment, what must be the output of the powerhouse?

64. [II] A 0.8-in.-square silicon solar cell delivers 90 mA at 0.45 V when illuminated by sunlight (at 100 mW/cm²). Suppose 10 such cells are connected in series; how much power could the panel deliver?

65. [II] For greater flexibility, electrical wires often consist of several fine strands twisted together rather than being constructed of one thick lead. A copper wire is made up of 10 fine fibers, each with a resistance of 2.0 mΩ. When placed across a voltage difference, a total current of 0.12 A traverses the wire.   (a) What is the net resistance of the length of wire?   (b) What voltage exists across it?   (c) How much power is dissipated by each strand? [Hint: How much current does each strand carry?]

66. [II] A credit card–sized calculator uses two tiny 1.5-V cells in series that provide a normal operating power of 0.000 18 W. Determine the current passing through each cell when the device is in use.

67. [II] An old-fashioned trolley car draws 12 A from an overhead wire at +500 V (the rails are grounded). What power is delivered to the motor? If the motor is 86% efficient, what power does it develop in propelling the car?

68. [II] When put across the terminals of two D-cells in series, a small flashlight bulb draws 330 mA. How much power does it consume? How much energy does it take from the cells in 1.0 minute of operation?

69. [II] The moving nonconducting belt of a Van de Graaff generator carries 10 μC of charge up to the metal sphere (Fig. Q14, Chapter 17) each second. The steady-state potential difference between the sphere and the grounded source of charge is 3 MV. What minimum power must be supplied to the generator to sustain its operation?

70. [II] A stereo tuner-amplifier with a maximum power output of 50 W per channel (that is, 50 W to each of two speakers) has its right channel connected as shown in Fig. P70. Since the speaker will be destroyed if it receives more than 36 W, a fuse is installed to limit the current entering it. What should be the rated maximum current of the fuse?

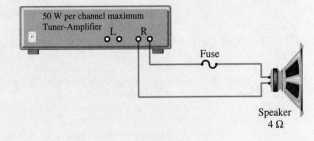

Figure P70

71. [II] When turned on, a flashlight bulb draws 1/3 A at 3.9 V. How much power does it require? Incidentally, more than 95% of that power appears in the form of thermal energy and not light. What is the resistance of the filament in the lamp?

72. [II] How much current does an ideal 1.00-hp dc electric motor draw when operating off a portable 100-V generator?

73. [II] A NiCd battery has an amp-hour rating of 10 and consists of five cells connected in series. What is the maximum power the battery will deliver if it is to operate for 5.0 h?

74. [cc] A piece of resistive material of resistivity ρ is formed in the shape of a hollow cylinder of length L. It has an inner radius r_i and an outer radius r_o. Show that if a voltage is put between the inner and outer surfaces the resistance will be

$$R = \frac{\rho}{2\pi L} \ln \frac{r_o}{r_i}$$

[Hint: Generalize the expression $R = \rho L/A$ to get $dR = \rho dL/A$ where dL is in the direction of the current and A is the area across which it flows.] Compare your answer to that of Problem 81.

75. [cc] Consider a hollow spherical shell of resistive material having an inner radius of r_i, an outer radius of r_o, and a resistivity of ρ. Determine its resistance when a potential difference is applied between the inner and outer surfaces.[Hint: Generalize the expression $R = \rho L/A$ to get $dR = \rho dL/A$ where dL is in the direction of current flow and A is the area across which it flows. Incidently, each differential element of resistance is in series with the next and their resistances add (see Section 20.2).]

76. [III] A silicon solar cell 5 mm × 4 mm produces a current of 5 mA at 0.45 V under sunlight illumination of 100 mW/cm². Determine its efficiency.

77. [III] A modern incandescent lamp has a tungsten filament with a melting point of 3400°C. Ordinarily, with the bulb evacuated, it's operated at about 2200°C. The efficiency can be improved almost threefold by raising the temperature to 2800°C, but that causes the tungsten to evaporate and shortens the life appreciably. The alternative (an idea introduced by Langmuir) is to fill the bulb with nitrogen or argon to suppress evaporation of the tungsten. What is the fractional change in the resistance of the filament when raised from 2200°C to 2800°C? Use the value of α_0 found in Table 19.3. Given the following values of α_0 for tungsten (from the *Handbook of Chemistry and Physics*)—0.004 5 at 18°C, 0.005 7 at 500°C, 0.008 9 at 1000°C—how good was the calculation you just made? Discuss your answer and make a very rough estimate of α_0 at 2500°C. Now estimate the fractional change in resistance using this new value.

78. [III] Determine the power rating in kilowatts of an electric heater that will raise the temperature of 10 liters of water from 25°C to 85°C in 15 minutes, assuming no loss of thermal energy. Given that the heater coil has a resistance of 10 Ω, how much current does it draw?

79. [III] Figure 19.20 shows how the resistivities of various pure metals each seem to be heading toward zero at some Celsius temperature $-T'$. Make a plot of R versus T (in °C), assuming it to be a straight line. Is that reasonable? Taking $T = 0°C$ as the reference (that is, the line crosses the R-axis at $R = R_0$ and $T = 0$), show that for any point (T, R) on the line

$$R = R_0 \frac{T' + T}{T'}$$

and that $\alpha_0 = 1/T'$. How does this result compare with the values of resistivity given in Table 19.3? Notice how most values approximate $1/273 = 3.7 \times 10^{-3}$.

80. [ccc] The metal rod in Fig. P80 has a constant thickness, but it tapers down from a height of H to a height of h over its length L. If a voltage V is put across its two small rectangular faces what current will traverse the rod? [Hint: Cut it into rectangular differential slices. You will need to relate x and y via an equation.]

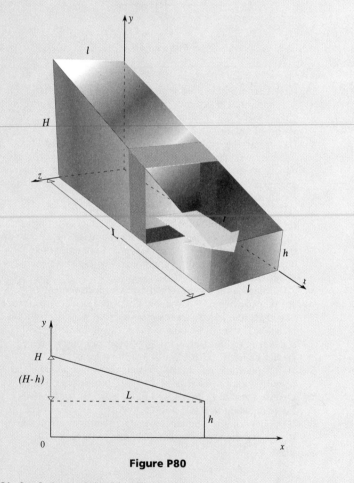

Figure P80

81. [ccc] A coaxial cable of length L consists of a metal wire in contact with and surrounded by a cylinder of resistive material with an inner radius r_i and an outer radius r_o that has a resistivity ρ. It, in turn, is held in place by a tight-fitting outer conducting cylinder. If a voltage V is applied between the wire and cylinder, find   (a) the electric field at some arbitrary distance r in the resistive material,   (b) the current flowing radially between the two conductors,   (c) the resistance of the material. Compare your answers to that of Problem 74.

20

Circuits

NOW THAT WE ARE familiar with the basics of dc—batteries, resistors, capacitors, and Ohm's Law—we can apply this knowledge to the treatment of circuits, where several elements are attached together. For example, we can study the electrical system of an automobile or design an experimental setup to measure the response of pigeons to the sight of popcorn. Today, virtually every field of science uses electric circuits in its research.

A variety of sensors, or *input transducers,* convert nonelectrical signals into electrical ones—the stereo cartridge on a record player and the microphone are familiar examples. Any up-to-date hospital intensive care unit can

A circus poster from 1879. A steam engine powered a generator that supplied current to the arc lamps that lit this circus "day and night."

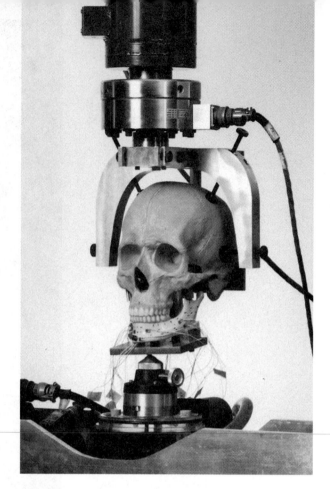

A testing device applies axial loading to the jaw. Here the mandible was fractured and repaired with a surgical plate assembly. Strain gauges around the jaw convert the resulting strain distribution into electrical signals.

electronically monitor a patient's body temperature, blood pressure, and respiration, measuring infusions and drainage while recording electrocardiograms, all with transducers. One kind of strain gauge is made from fine alloy wire that changes its resistance as it's stretched and compressed. Such gauges are widely used to monitor mechanical apparatus and in medical studies of bones, joints, and the behavior of muscles. Clearly, there's more to resistive devices and sources than we have considered thus far. Learning to deal with simple circuits lays the groundwork for understanding a wide range of practical electrical systems.

CIRCUIT PRINCIPLES

A circuit, in simple form, is a few elements (for example, a resistor and a battery) joined together to make at least one closed current path (usually with some purpose in mind beyond merely confounding students). Often a system converts electrical energy into light via an incandescent lamp or into thermal energy via a toaster. There might be an input transducer, a microphone, at one end and an output transducer, a speaker, at the other end. In any event, we are generally interested in determining a variety of circuit parameters, such as voltage drops, currents, power dissipated, and so on—you can't send a 100-W signal out to a 20-W speaker, at least not twice. The only limitation in this chapter is that the currents and voltages are

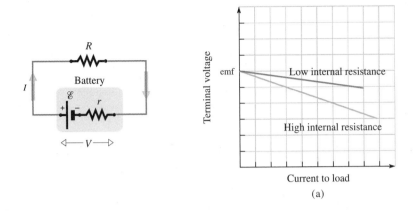

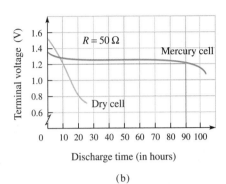

Figure 20.1 The terminal voltage of a battery will change as more and more current is drawn from it. (a) The effect is small with a fresh battery. (b) Most cells suffer a drop in terminal voltage when they are in prolonged use.

dc—they may rise and fall, but they do not fall below zero; they do not reverse direction, the sources do not change polarity.

20.1 Sources and Internal Resistance

So far, the battery has been taken to be an ideal constant-voltage source. That assumes that its terminal voltage remains fixed regardless of the resistive load across it or the length of time it provides power. Alas, that assumption is generally not the case. Although mercury cells approximate this behavior, they are small and usually supply very little current. Most other batteries show a marked decrease in terminal voltage as the current supplied by them increases (Fig. 20.1). Start your car with the headlights on. As the starter motor draws hundreds of amps, the battery's terminal voltage drops and the lights become noticeably dimmer.

When a voltaic cell provides current to an external circuit, there is a transport of charge from one electrode to the other across the electrolyte, and that does not happen in an unimpeded way. The cell resists current traversing it in either direction. This **internal resistance** (r) is an inseparable aspect of any real battery, cell, or other power source. Insofar as the plot in Fig. 20.1a is linear, r is ohmic, and the battery can be represented most simply by a resistor in series with an ideal emf ($\mathscr{E}$), as in Fig. 20.2. The + and − terminals of the source are the points A and B, respectively, and it doesn't matter on which side the internal resistor is drawn as long as it is between those terminals.

When the battery supplies current to an external circuit, as in Fig. 20.3a, I leaves the positive terminal A, traverses the circuit, returns to the battery at terminal B and passes internally to A. (Label the resistor with + and − signs, indicating the ends at which current enters and leaves, respectively.) If we trace from B to A, there will be a voltage drop (−) across the internal resistor of $-Ir$ followed by a rise (+) equal to the emf, and so with $V = V_A - V_B$, we have

$$V = \mathscr{E} - Ir \qquad (20.1)$$

The terminal voltage V, *the potential difference measured*

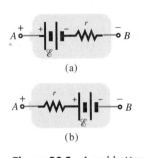

Figure 20.2 A real battery can be represented by an emf in series with an internal resistance r. The order doesn't matter since (a) and (b) are equivalent. The voltage difference between A and B is the terminal voltage of the battery.

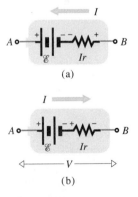

Figure 20.3 (a) When the battery supplies current to an external circuit (not shown here), there is a voltage drop across r, and $V < \mathscr{E}$. (b) When the battery receives current from an external source (not shown here), $V > \mathscr{E}$.

by an ideal voltmeter across the battery, is less than the emf when the current goes from *B* to *A* and the battery provides power.

When the external circuit supplies current to the battery, as in Fig. 20.3b, *I* enters at the positive terminal *A* and passes internally to terminal *B*. The battery is having power supplied to it—it's being charged. Tracing from *B* to *A*, we move first from − to + across the internal resistor, which corresponds to a voltage rise (+) of +*Ir*, followed by another rise (+) equal to the emf. Hence

$$V = \mathcal{E} + Ir \tag{20.2}$$

and the terminal voltage exceeds the emf.

Example 20.1 A voltage source with an emf of 12 V and an internal resistance of 0.40 Ω supplies 1.5 A to an external load. If a high-quality voltmeter is placed across the terminals of the source, what will it read? What will it read if a switch in series with the source open-circuits the system?

Solution: [Given: $\mathcal{E} = 12$ V, $r = 0.40$ Ω, and $I = 1.5$ A. Find: *V* when $I = 1.5$ A and when $I = 0$.] First, draw a circuit diagram with the source functioning as in Fig. 20.3a. The terminal voltage when current exists is given by Eq. (20.1),

$$V = \mathcal{E} - Ir = 12 \text{ V} - (1.5 \text{ A})(0.40 \text{ Ω})$$

and $V = 11.4$ V or, to two significant figures, $\boxed{V = 11 \text{ V}}$. Opening the switch causes *I* to become zero, whereupon $\boxed{V = \mathcal{E} = 12 \text{ V}}$.

▶ **Quick Check:** The drop across *r* is $Ir = (1.5 \text{ A}) \times (0.40 \text{ Ω}) \approx 1$ V. $V \approx 12$ V − 1 V ≈ 11 V.

The internal resistance of a fresh battery is small: ≈0.05 Ω or so for a D-cell and only ≈2 mΩ for a 12-V lead storage battery. As a battery ages, its internal resistance increases while its emf gradually decreases. Still, it's possible to have an old 1.5-V dry cell with an internal resistance as high as several ohms and an emf only a few percent less than normal. As a consequence, using a meter to measure the terminal voltage of a worn D-cell with no load on it ($I = 0$) may yield a value near its normal emf. Clearly, that's an unreliable indication of the condition of the cell. Measuring *V* while there is a load on the cell (for example, while there is a flashlight bulb across it) is a better test. For most applications, if *V* does not exceed at least 1.1 V, the cell should be replaced.

Example 20.2 A 5.9-Ω load resistor is placed across the terminals of a battery that has an emf of 12 V and an internal resistance of 0.10 Ω. Write a general expression for the current in the circuit and then compute its specific value. What's the terminal voltage of the battery?

Solution: [Given: $R = 5.9$ Ω, $r = 0.10$ Ω, and $\mathcal{E} = 12$ V. Find: *I* in general and in particular, and *V*.] First draw a circuit diagram (Fig. 20.4). Current leaves the + side of the battery. We can find *I* using Ohm's Law if we know the voltage across *R*; namely, the terminal voltage. So, the first thing is to find *V*. From Eq. (20.1), $V = \mathcal{E} - Ir$, which in turn equals *IR*, so

$$V = IR = \mathcal{E} - Ir$$

and

$$I(R + r) = \mathcal{E}$$

so that

$$\boxed{I = \frac{\mathcal{E}}{R + r}} \tag{20.3}$$

(continued)

(continued)

Specifically

$$I = \frac{12 \text{ V}}{5.9 \ \Omega + 0.10 \ \Omega} = \boxed{2.0 \text{ A}}$$

As for the terminal voltage

$$V = \mathscr{E} - Ir = 12 \text{ V} - (2.0 \text{ A})(0.10 \ \Omega) = 11.8 \text{ V}$$

or to two significant figures $\boxed{V = 12 \text{ V}}$.

▶ **Quick Check:** $V = IR = (2.0 \text{ A})(5.9 \ \Omega) = 11.8 \text{ V}$.

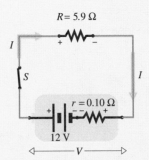

Figure 20.4 A battery with an emf of 12 V and internal resistance r supplies a current I to the resistor R when the switch S is closed. The current (a flow of positive charge) leaves the positive terminal of the battery and returns, with less energy, to the negative terminal.

If an ammeter and voltmeter are added to the circuit of Fig. 20.4, as indicated in Fig. 20.5, it becomes an easy matter to determine r. Note that *the ammeter is always placed in the branch of the circuit through which the required current passes*, whereas *the voltmeter is always placed across the two points whose potential difference is to be measured*.

Before we move on, let's standardize some of the terminology: a **branch** is one or more circuit elements (in series) carrying a single current. A **node** is a place (often a single point) where three or more branches come together. Nodes are interconnected by branches, and branches begin and end on nodes. A circuit that has no nodes can be thought of as a single branch that closes on itself. Any closed current path is a **loop**. The circuit in Fig. 20.4 is a single loop, and the ammeter is positioned in that loop. The two points where the voltmeter attaches to the circuit (Fig. 20.5) are nodes, and the voltmeter constitutes one of three branches.

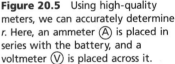

Figure 20.5 Using high-quality meters, we can accurately determine r. Here, an ammeter Ⓐ is placed in series with the battery, and a voltmeter Ⓥ is placed across it.

20.2 Resistors in Series and Parallel

Suppose we have a circuit with several branches and lots of resistors and sources and we want to compute the voltage across, and the current through, each element. The computation is likely to be difficult, but two special configurations are particularly easy to deal with. When the resistors are in *series* or in *parallel* or a combination of both, they can be replaced by a single *equivalent resistor* (as was done earlier with capacitors, p. 706). Then Ohm's Law can be applied and the problem solved step-by-step. How do we arrive at this **equivalent resistance** (R_e)?

Figure 20.6a shows an ideal voltage source in series with two resistors R_1 and R_2. *The same current I enters and leaves each element in series*. The difference in potential between points A and B is the voltage of the source V. If we go from B to A across the battery, we encounter a voltage rise of $+V$. Similarly, tracing from B to A across the resistors, we encounter a rise of V_2 followed by another rise of V_1, which must also equal V; thus

$$V = V_1 + V_2$$

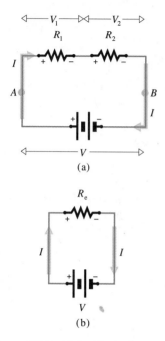

Figure 20.6 (a) An ideal voltage source in series with two resistors. (b) The equivalent circuit where $R_e = R_1 + R_2$.

and using Ohm's Law
$$V = IR_1 + IR_2$$

We want an equivalent resistance that will replace the load, as indicated in Fig. 20.6b, so that the same current I circulates. Hence

$$V = IR_e$$

and so
$$IR_e = IR_1 + IR_2$$

Consequently

[*in series*]
$$R_e = R_1 + R_2 \qquad (20.4)$$

The equivalent resistance is simply the sum of the series resistances, no matter how many there happen to be. To see how this works, suppose $R_1 = 5.0\ \Omega$, $R_2 = 1.0\ \Omega$, and $V = 12$ V. From Eq. (20.4), $R_e = 6.0\ \Omega$, and using Ohm's Law for Fig. 20.6b, we find the current to be $I = V/R_e = 2.0$ A. Returning to the original circuit of Fig. 20.6a, we see that $V_1 = IR_1 = 10$ V and $V_2 = IR_2 = 2.0$ V and, as expected, $V_1 + V_2 = 12$ V. The analysis of the circuit in Fig. 20.4 can now be reinterpreted in light of the above discussion. Accordingly, $R_e = R + r$, and since $I = V/R_e$, $I = V/(R + r)$, which is Eq. (20.3).

Figure 20.7a shows two resistors in parallel with a constant voltage dc source, though there could equally well have been ten of them. The (as yet undetermined) current I supplied by the source comes to node A and splits into two branch currents I_1 and I_2, which are (as we will find presently) inversely proportional to the resistances of the two branches. These currents recombine into I at node B:

$$I = I_1 + I_2$$

The voltage across each resistor is V, and therefore Ohm's Law gives

$$I = \frac{V}{R_1} + \frac{V}{R_2}$$

But in the equivalent circuit of Fig. 20.7b $I = V/R_e$, and so

$$\frac{V}{R_e} = \frac{V}{R_1} + \frac{V}{R_2}$$

Canceling V yields

[*in parallel*]
$$\frac{1}{R_e} = \frac{1}{R_1} + \frac{1}{R_2} \qquad (20.5)$$

and the summation would just continue on were there additional resistors in the circuit.

Equation (20.5) for two resistors can be rewritten as

$$R_e = \frac{R_1 R_2}{R_1 + R_2} \qquad (20.6)$$

Figure 20.7 (a) An ideal voltage source across two resistors in parallel. Notice how the current I splits at node A and recombines at B. (b) The equivalent circuit where $1/R_e = 1/R_1 + 1/R_2$.

which makes it apparent that when either resistor is much larger than the other, say, $R_1 \gg R_2$, then $R_1 + R_2 \approx R_1$ and $R_e \approx R_2$; the equivalent resistance approximates the *smaller* of the two. ***The equivalent parallel resistance is always less than either of the contributing resistances***. All of the circuits in Fig. 20.8 are electrically identical, and all have the same equivalent resistance between A and B.

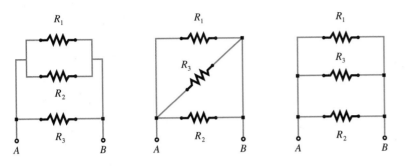

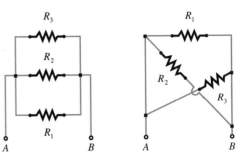

Figure 20.8 Each of these circuits has the same equivalent resistance between A and B. In each version, R_1, R_2, and R_3 are in parallel.

Example 20.3 Figure 20.9a represents three light bulbs with resistances of 2.0 Ω, 4.0 Ω, and 8.0 Ω attached across a source with an emf of 6.0 V and an internal resistance of 1.0 Ω. Find the current through each bulb.

Solution: [Given: $R_1 = 2.0$ Ω, $R_2 = 4.0$ Ω, $R_3 = 8.0$ Ω, $r = 1.0$ Ω, and $\mathscr{E} = 6.0$ V. Find: I_1, I_2, and I_3.] Draw a diagram putting in all the required quantities. The three bulbs are in parallel so their equivalent (R) is gotten from Eq. (20.5):

$$\frac{1}{R} = \frac{1}{2.0 \text{ Ω}} + \frac{1}{4.0 \text{ Ω}} + \frac{1}{8.0 \text{ Ω}} = 0.875 \text{ Ω}^{-1}$$

and $R = 1.14$ Ω, which is the equivalent resistance of the load (Fig. 20.9b). R and r are in series, and so using Eq. (20.4), the equivalent resistance of the entire circuit (R_e) is ($R + r$) = 2.14 Ω. With R_e across the emf in Fig. 20.9c, the current provided by the source (via Ohm's Law) is

$$I = \frac{\mathscr{E}}{R_e} = \frac{6.0 \text{ V}}{2.14 \text{ Ω}} = 2.8 \text{ A}$$

Going back to Fig. 20.9b, we can find the terminal voltage across R and then each branch current. The terminal voltage follows from Eq. (20.1); namely, a drop of $-Ir = -2.8$ V followed by a rise of $\mathscr{E} = +6.0$ for a total of $V = +3.2$ V. The branch currents are then

$$I_1 = \frac{V}{R_1} = \frac{3.2 \text{ V}}{2.0 \text{ Ω}} = \boxed{1.6 \text{ A}}$$

$$I_2 = \frac{V}{R_2} = \frac{3.2 \text{ V}}{4.0 \text{ Ω}} = \boxed{0.80 \text{ A}}$$

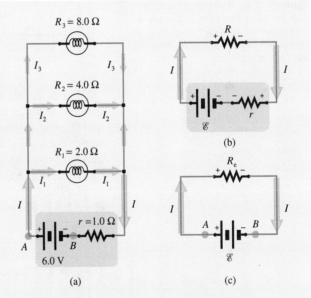

Figure 20.9 (a) Three light bulbs in parallel with a real battery. (b) The bulbs have resistances that are in parallel such that $1/R = 1/R_1 + 1/R_2 + 1/R_3$. (c) The equivalent resistance "seen" by the emf between points A and B is R_e.

and $I_3 = \dfrac{V}{R_3} = \dfrac{3.2 \text{ V}}{8.0 \text{ Ω}} = \boxed{0.40 \text{ A}}$

▶ **Quick Check:** Note that R is less than any of the constituent resistors, which is as it should be. $I = I_1 + I_2 + I_3 = (1.6 \text{ A}) + (0.80 \text{ A}) + (0.40 \text{ A}) = 2.8$ A, which is appropriate. The smaller the resistance of the branch, the more current it carries.

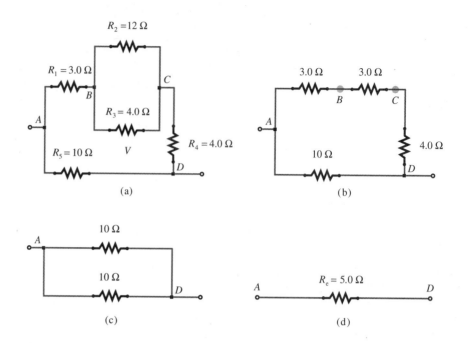

Figure 20.10 A cluster of resistors and the successive combinations and simplifications leading to a single equivalent resistance R_e between points A and D.

Figure 20.10a depicts a cluster of attached resistors. There is no source in the circuit and so no current, but we can still think about adding one, say, across points A and D, and consequently still be concerned with the equivalent resistance of the network. The node points are labeled to keep track of the process as the circuit is successively contracted. Resistors R_2 and R_3 are in parallel, and the result of adding those (namely, 3 Ω) is in series with R_1 and R_4 for a total of 10 Ω. This resistance is in parallel with R_5, since both resistors are attached at A and D, thus yielding $R_e = 5.0$ Ω between points A and D. This means that the whole cluster behaves in precisely the same way as would a single 5-Ω load. Put a battery across A and D, and the same current will be delivered to either configuration. This value is *not* necessarily the equivalent resistance between A and C, or A and B (Problems 14, 15, and 16)—each of those is quite distinct.

Example 20.4 Imagine a wire square and label its corners A, B, C, and D going clockwise. Now place a larger square surrounding (and with sides parallel to) the first, and label its corners E, F, G, H clockwise, with E closest to A. Insert a 1.0-kΩ resistor in each arm of each square (eight in total) and attach another between A and E. Place a 12-V DC source between C and G. (a) If possible, simplify the circuit and determine an equivalent resistance between C and G. (b) What current is provided by the source? (c) What is the voltage across points G and E? (Perhaps it should be said that this is about as complicated a circuit as you are likely to see.)

Solution: [Given: nine resistors, $R = 1.0$ kΩ each, and $V = 12$ V. Find: R_e, I, and V between G and E.]

Figure 20.11a is the circuit diagram. We can simplify it as is or, alternatively, imagine the wires to be flexible and lift up the inside square, with the resistor and source attached, and place it outside $E–F–G–H$, as in Fig. 20.11b. Branches $A–B–C$ and $A–D–C$ are in parallel, as are $E–F–G$ and $E–H–G$, and each has a resistance of 1.0 kΩ + 1.0 kΩ = 2.0 kΩ (Fig. 20.11c). Thus, the resistance of each square $E–F–G–H$ and $A–B–C–D$ reduces to

$$\frac{1}{2} \text{ k}\Omega + \frac{1}{2} \text{ k}\Omega = \frac{1}{R}$$

and $R = 1.0$ kΩ. The three 1.0-kΩ resistors in Fig. 21.11d are in series with the source. (a) The equiva-

(continued)

(continued)

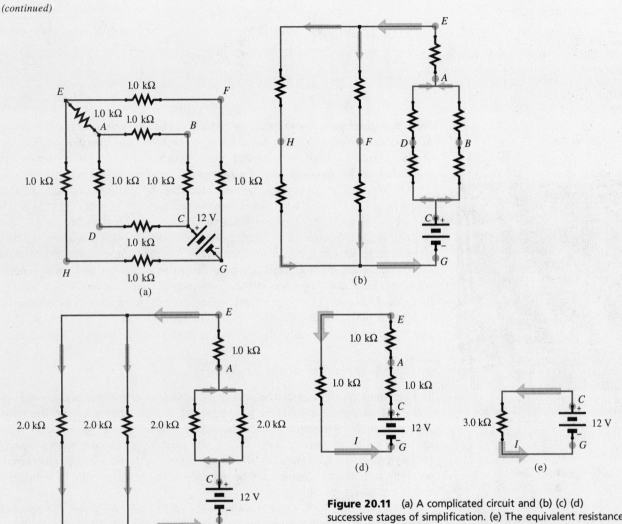

Figure 20.11 (a) A complicated circuit and (b) (c) (d) successive stages of simplification. (e) The equivalent resistance between points C and G is 3.0 kΩ, and that's across the 12-V source.

lent resistance in Fig. 20.11e is $\boxed{3.0 \text{ k}\Omega}$. (b) Since $V = IR_e$, $\boxed{I = 4.0 \text{ mA}}$. (c) A current of 4.0 mA leaves the battery and splits at C. Because the two branches C–D–A and C–B–A have the same resistance, the current divides into two equal streams of 2.0 mA each. The voltage drop in going from C to A is given by $V_{AC} = IR = (2.0 \text{ mA})(1.0 \text{ k}\Omega + 1.0 \text{ k}\Omega) = 4.0$ V. In

going from A to E, there is another drop of $V_{EA} = IR = (4.0 \text{ mA})(1.0 \text{ k}\Omega) = 4.0$ V. C is 12 V above G, A is 8.0 V above G, and $\boxed{E \text{ is } 4.0 \text{ V above } G}$.

▶ **Quick Check:** 4.0 mA leaves A, enters E, and splits so that 2.0 mA goes through both branches E–F–G and E–H–G. The voltage drop from E along either branch to G is (2.0 mA)(2.0 kΩ) = 4.0 V.

The resistors in Fig. 20.6a can be slid around so that the battery is between them and nothing changes. ***The same current passes through every resistor in a given***

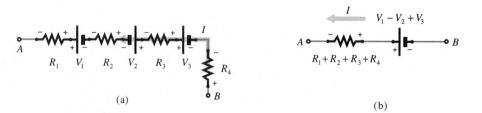

Figure 20.12 A series of sources separated by resistors is equivalent to a single source having the net voltage and a single resistor having the combined resistance.

(a)

(b)

An electric car powered by storage batteries can travel about 100 miles on a single charge.

branch, regardless of the presence of sources in that branch, and the resistors are in series even though they are not directly connected to one another. In the process of simplifying a circuit, sources and resistors in series can be interchanged, provided they are all returned after the currents are computed (Fig. 20.12).

Often power is provided to a number of devices by putting them each in parallel across the source. That's what you do when you "plug in" a toaster (p. 885) or a TV set—it is placed across the two leads that supply electrical energy to the home (Fig. 20.13). All household appliances are in parallel—in that way each can be designed to operate at a specific voltage ($\approx$120 V in the United States, $\approx$220 V most everywhere else). That voltage can be provided to each appliance independent of whatever else is drawing power in the circuit, which would not be the case were they in series. Moreover, if any one device open-circuits (for example, as when a light bulb burns out), the rest are unaffected. The same is true if you have several telephone extensions—they are all in parallel, supplied with the same low-voltage dc.

20.3 Maximum Power and Impedance

The amount of power that can be provided from a source to a load is a practical concern. Accordingly, we examine an important insight known as the **Maximum Power Transfer Theorem**: *A maximum amount of power will be transferred from any source, with an internal resistance r, to a load of resistance R when R equals r.* For example, there are several different output taps on the back of a stereo amplifier precisely so that its resistance can be matched to the resistance of the speakers for maximum power transfer.

Return to Fig. 20.5, where a source with an internal resistance r is connected across a load R. Remembering that

$$I = \frac{\mathscr{E}}{R + r} \qquad [20.3]$$

Figure 20.13 Household wiring, using alternating current, puts all appliances and lights in parallel across $\approx$110 V.

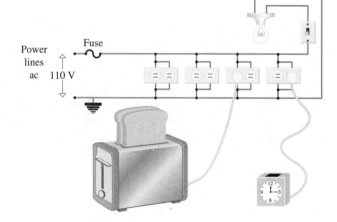

the power delivered to the load is

$$P = I^2 R = \frac{\mathscr{E}^2}{(R + r)^2} R \qquad (20.7)$$

Notice that when $R \gg r$, the denominator is essentially R^2, and the power drops as $1/R$ to a low value. When $R \ll r$, the denominator is essentially r^2, but the power, which varies with R, must also be small. Somewhere between these extremes, P has a maximum. A plot of P versus R, Fig. 20.14, makes the same point.

The process of transferring energy from one system to another is fundamental, and the above ideas apply to a wide range of physical phenomena beyond electricity (for example, light, sound, mechanical vibrations, etc.). It is found that by matching a certain characteristic of the provider-of-energy to that of the receiver-of-energy, a maximum transfer will take place. That characteristic is a generalized notion that corresponds to the resistance offered to the transport of energy, and it's known as the *impedance* (p. 878). Figure 12.39 shows this effect for mechanical waves on discontinuous ropes—maximum transmission occurs when the two ropes have equal densities (that is, equal impedances). Similarly, the kinetic energy transferred from a moving object to a stationary one (as the result of an elastic collision) has the same form where the impedance corresponds to the measure of the resistance to the change in motion—namely, the mass. When the masses are equal (that is, when the impedances are matched), the energy transferred is a maximum. One can contemplate the impedance of the ear, the impedance of a musical instrument, the impedance of an antenna, and so on, wherever energy is transferred.

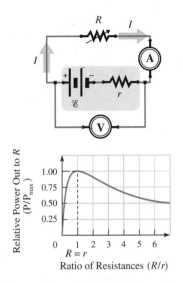

Figure 20.14 The ratio of the power out to a load resistor R, divided by the maximum power supplied to it, plotted against the ratio of R to the internal resistance of the source r. When $R = r$, the power P equals P_{max} and $P/P_{max} = 1$.

20.4 Ammeters and Voltmeters

Electrical meters of all kinds come in two varieties, analog and digital. An analog meter usually has a pointer that moves across a scale, whereas a digital meter provides direct numerical values of whatever is being measured. An analog meter is constructed around a moving coil galvanometer (p. 815). The latter is a coil of wire with a resistance r of perhaps 200 Ω or less, having a pointer fixed to it, placed between the poles of a magnet. Passing a current through the coil causes it to twist proportionately and the pointer swings around. The digital meter is an electronic device that takes a reading from the circuit, which it amplifies, digitizes, and displays. Digital meters tend to be more accurate, sturdier, and more expensive than analog meters, but they are both used in the same way.

Figure 20.15a represents the basic analog detector, a galvanometer comprising a resistor in series with an ideal rotating coil. The input that causes a maximum deflection of the pointer is called the **full-scale current**, and for an inexpensive device, that's about 1 mA. To use the galvanometer as an **ammeter**, a current meter, we must direct a small fixed fraction of the main stream of current through the galvanometer. That's accomplished by placing a small *shunt* resistor ($R_s \ll r$) across the galvanometer (Fig. 20.15b). The resulting instrument is an ammeter, and it is always positioned within the branch where the current I is to be measured. The branch current I entering the ammeter splits with part I_g going through the galvanometer and most of it, I_s, being shunted through the small resistor R_s. The specific value of R_s depends on the full-scale current sensitivity of the galvanometer. Thus, if the coil has a full-scale current of $I_g = 1.0$ mA and we want the ammeter to read $I = 1.0$ A at full scale, then, since $I = I_g + I_s$, $I_s = 0.999$ A. Because the voltage drops across R_s, and $r = 200$ Ω are equal

$$I_g r = I_s R_s$$

hence
$$R_s = \frac{I_g r}{I_s} = \frac{(1.0 \text{ mA})(200 \text{ Ω})}{0.999 \text{ A}} = 0.20 \text{ Ω}$$

Notice that the analog dc ammeter has a specific polarity; the current must enter at the + terminal or the meter may be damaged. One nice feature of the digital ammeter is that we usually needn't worry about polarity.

A **voltmeter**, Fig. 20.15c, is a galvanometer in series with a resistance R. The voltmeter is always placed *across* the two points whose potential difference is to be measured. To ensure that little current passes through the coil, R is very large (usu-

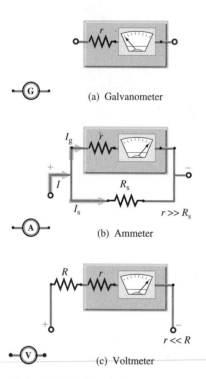

(a) Galvanometer

(b) Ammeter

(c) Voltmeter

Figure 20.15   (a) A galvanometer having an internal resistance r. (b) When the galvanometer is placed in parallel with a low-resistance shunt R_s, we have an ammeter. (c) When the galvanometer is in series with a high-resistance R, we have a voltmeter.

ally in the high kilohm to megohm range). For example, suppose the voltmeter is to measure from 0 to 100 V, using our cheap galvanometer with a full-scale current of $I_g = 1.0$ mA and $r = 200\ \Omega$. That current must exist when the voltage is a maximum of 100 V; thus

$$V = I_g(R + r) = 100\ \text{V} = (1.0\ \text{mA})(R + 200\ \Omega)$$

and $R = 10 \times 10^4\ \Omega$. This meter is not a very good one because R is not really large enough. If it were placed across a several-kilohm resistor in a circuit, it would appreciably affect the current and produce an erroneous reading. By contrast, a fine electronic voltmeter might have a resistance of $10^8\ \Omega$ or more. Notice that the dc analog voltmeter also has a specific polarity and must be connected with its + terminal at the higher potential.

20.5   *R-C* Circuits

Any circuit containing a capacitor also has some resistance, even if it's only due to the wiring. Such **R-C circuits** are both commonplace and important—especially if you happen to be wearing a cardiac pacemaker. And that points up one of the most interesting features of *R-C* circuits. The transient states of the resistive circuits considered thus far are extremely brief, and so voltages and currents are taken to be constant at their steady-state values. By contrast, *R-C* circuits have time-dependent voltages and currents. That makes them very useful for creating a variety of circuits that produce time-varying signals (in a pacemaker, for example).

Imagine a capacitor with its plates charged at $\pm Q_i$ as shown in Fig. 20.16. There is an initial voltage $V_i = Q_i/C$ across the capacitor, but no charge flows while the switch S is open. At $t = 0$, the switch is closed, the voltage across the resistor is immediately V_i, and the charge on the capacitor begins to redistribute itself under the influence of the Coulomb force. For an instant there is an initial current through the circuit of $I_i = V_i/R = Q_i/CR$. But as the charge on the capacitor decreases ($\Delta Q < 0$), the voltage decreases, and that decreases the current: $I = -\Delta Q/\Delta t$. At any moment, the voltage across the resistor IR equals the voltage across the capacitor Q/C, such that

$$IR = -\frac{\Delta Q}{\Delta t}R = \frac{Q}{C}$$

and

$$\frac{\Delta Q}{\Delta t} = -\frac{1}{RC}Q \qquad (20.8)$$

which is a rather special and yet wonderfully common situation. ***Whenever a physical quantity changes*** $(\Delta Q/\Delta t)$ ***at a rate that depends on the quantity itself*** (Q), ***the quantity varies exponentially*** (Appendix A-3). If we plot Q versus t (Fig. 20.17a), *the slope of the curve is proportional to the value of the curve:* when the charge is large, the slope $(\Delta Q/\Delta t)$ is negative and large; when Q is small, the slope is small. In other words,

$$\frac{dQ}{dt} = -\frac{1}{RC}Q$$

$$\int_{Q_i}^{Q}\frac{dQ}{Q} = -\frac{1}{RC}\int_{0}^{t}dt$$

$$\ln\frac{Q}{Q_i} = -\frac{t}{RC}$$

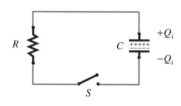

Figure 20.16   A charged capacitor C in series with a resistor and an open switch S. The charge on each plate is $\pm Q_i$.

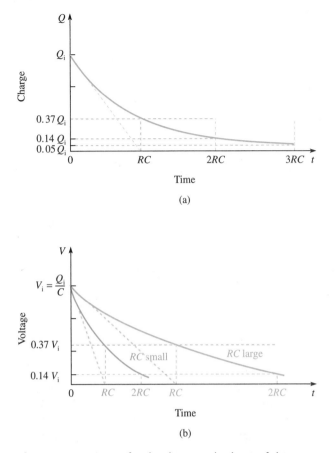

Figure 20.17 Curves for the charge and voltage of the capacitor in Fig. 20.16 once *S* is closed. (a) After each interval of time equal to *RC*, the curve of *Q* versus *t* drops to 36.787 9% of the previous value. (b) The voltage also decays exponentially.

and exponentiating both sides, using the fact that $e^{\ln x} = x$,

$$Q(t) = Q_i e^{-t/RC} \qquad (20.9)$$

Since $e^0 = 1$, at $t = 0$, $Q = Q_i$, which makes sense. At $t = \infty$, $e^{-\infty} = 0$ and $Q = 0$, which also makes sense; the capacitor is discharged. Notice from Eq. (20.8) that at $t = 0$ the slope is $-Q_i/RC$. The initial slope intersects the time axis at $t = RC$. At that time, $Q = Q_i e^{-1} = 0.37 Q_i$. The quantity *RC* is called the **time constant** because it influences the temporal behavior of the circuit. That's evident in the plots of the voltage across the capacitor in Fig. 20.17b for two different values of *RC*. The voltage drops 63% of the way down toward zero in one time constant, and the smaller *RC* is, the faster that drop happens. The smaller is *R*, the less the resistor retards the flow of charge; and the smaller is *C*, the faster the capacitor discharges.

We now examine the inverse process of charging a capacitor, but the treatment will just be qualitative; a more detailed analysis is left for Problems 49, 50, and 51. The circuit in Fig. 20.18 shows an uncharged capacitor in series with a resistor and an ideal battery. When the switch is closed, charge flows and a time-dependent current begins to circulate. At that moment, the charge on the capacitor *Q* is essentially zero as is the voltage drop across it; hence, $\mathscr{E} = I_i R$. At any instant thereafter

$$\mathscr{E} = IR + Q/C$$

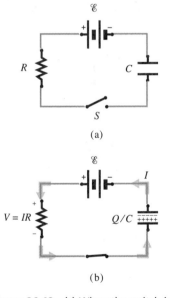

Figure 20.18 (a) When the switch in the circuit is closed, a transient current appears. (b) Positive charge goes out of the battery, across the resistor, and builds up on one plate of the capacitor. An equal amount of charge is repelled off the other plate and returns to the negative terminal of the battery.

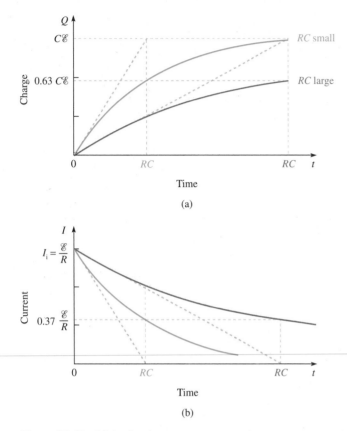

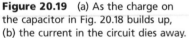

Figure 20.19   (a) As the charge on the capacitor in Fig. 20.18 builds up, (b) the current in the circuit dies away.

Since $\mathscr{E}$ is constant, as the charge builds up (Fig. 20.19a), the current correspondingly dies off (Fig. 20.19b). As the current approaches zero, $\mathscr{E}$ approaches Q/C, and the maximum charge on the capacitor approaches $C\mathscr{E}$. The current never actually reaches zero; after 20 time constants ($t = 20RC$), it's down to $I = I_i e^{-20} = 2 \times 10^{-9} I_i$, which isn't much.

R-C circuits are often used as timers to control the periodic activation of other devices. For example, the intermittent action of an automobile windshield wiper is controlled by an *R-C* circuit. A cardiac pacemaker circuit with a time constant RC of 0.8 s can be used to reach a certain triggering voltage at a frequency of $1/(0.8 \text{ s}) = 1.25$ times per second, thereupon activating a signal to the heart at a rate of 75 pulses per minute.

NETWORK ANALYSIS

The first approach to analyzing any circuit is to reduce it to a manageable equivalent scheme, but there are many configurations that can't be simplified. Figure 20.20 shows a representative selection wherein the elements are neither in series nor in parallel. Complicated circuits are often called *networks*—Fig. 20.21 is an example of the genre. But before we get into the systematic approach to the solution of such circuits, it's appropriate to look at some special cases that are both important and remarkably simple. For example, although the circuit in Fig. 20.21 appears to be a small horror, there are aspects of it that are easily solved. What is the current through, and the voltage drop across, the 6.0-Ω resistor (R_1)? Happily, the voltage drop across the whole uppermost branch is apparent—it's 12 V because the two resistors (6.0 Ω and 4.0 Ω in series) are directly across the terminals of the 12-V source. All the complex circuitry in the middle has no effect on the voltage across the upper (10-Ω) branch. Thus, whatever the current may be, the drop across R_1 is 6/10 of 12 V, or 7.2 V. The current is then $I = V_1/R_1 = (7.2 \text{ V})/(6.0 \text{ }\Omega) = 1.2$ A or, alternatively, $I = V/(R_1 + R_2) = 1.2$ A.

Figure 20.22 depicts another rather formidable network, but there's something special here, too. What is the voltage drop across, and the current through, the 6.0-Ω resistor? Notice that when the current reaches node *A*, it "sees" two identical paths and therefore splits into equal parts. The voltage drop across R_1 is equal to the drop across R_3, and so points *C* and *D* are at the same potential. There is no voltage difference across R_5 and no current through it; R_5 has no effect on this so-called *bridge circuit!* It's as if the branch from *C* to *D* were removed.

20.6   Kirchhoff's Rules

The systematic analysis of networks proceeds from two basic rules set forth by the German physicist Gustav Robert Kirchhoff in the late 1800s. There's really nothing new about these ideas; what *is* new is the application. Kirchhoff's first rule is

The algebraic sum of the voltage rises and drops encountered in going around any closed path formed by any portion of a circuit must be zero.

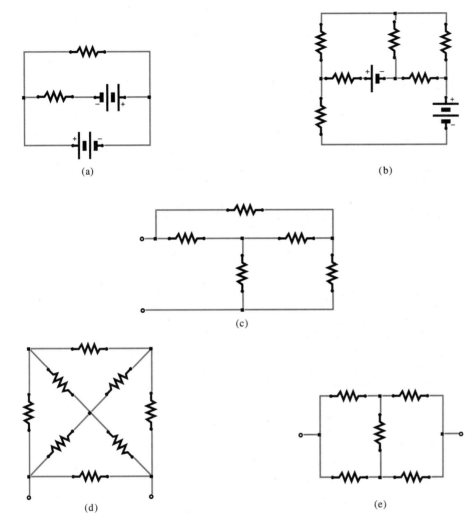

An electronic photographic flash unit. The rechargeable battery is a stack of four cells at the upper left. The large cylinder is the capacitor that stores charge and energy. To fire the flash, the capacitor dumps its charge, via the red and blue leads, into the lamp. After a short wait, determined by *RC*, it's recharged and ready to go again.

Figure 20.20 Several circuits in which the elements are neither in series nor in parallel.

Figure 20.21 We can easily calculate the current through R_1 since the voltage across the upper branch is 12 V.

This **Loop Rule** maintains that if we start at some arbitrary point in a circuit at a given potential and follow any path that returns to that same point, there can be no net change in potential.

In exactly the same way, there is no change in *gravitational*-PE if you return to the same spot after meandering up and down a mountainside. Thus, the sum of the voltage rises (taken as positive) and the voltage drops (taken as negative) must equal zero. For a charge traversing a loop and returning to its starting point, the net energy gained must equal the net energy lost. The Loop Rule is a restatement of the Law of Conservation of Energy.

To apply the Loop Rule, let's determine the potentials at points *A*, *B*, *C*, *D*, *E*, and *F* in Fig. 20.23. This circuit has only one loop, and therefore only one branch, and one un-

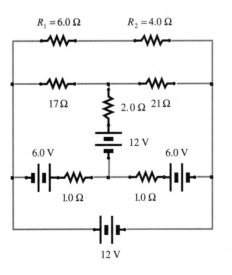

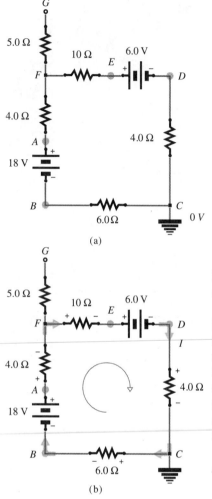

(a)

(b)

Figure 20.23 The analysis of a circuit using the Loop Rule.

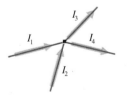

Figure 20.24 The Node Rule states that the sum of the currents entering a node must equal the sum of the currents leaving the node.

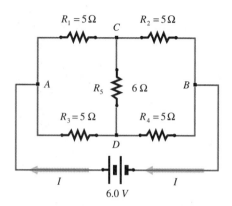

Figure 20.22 By symmetry, the voltage at C equals the voltage at D, so $V_{CD} = 0$. Therefore, no current passes through R_5, and R_1 and R_2 are in series, as are R_3 and R_4.

known branch current, I. Kirchhoff's Rules must be applied (thereby producing an equation) as many times as there are unknowns in the system.

The next step is to guess at the directions of the branch currents and draw them in. Assume that the larger battery dominates and the current emerges from its positive terminal. But not to worry, if ever the guess is wrong, the solution for I will simply come out negative, indicating that it's really in the opposite direction. Now label all the rises and drops with $+$ and $-$ signs depending on the direction of the current. **The signs of the emfs are independent of the current.** At this point, we are ready to apply the Loop Rule. Start at any point, say A, and go around the loop in either direction, writing the loop equation as you go. Let's do it clockwise: from A to F, there's a drop of $-4.0I$; from F to E, a drop of $-10I$; from E to D, a drop of -6.0 V; from D to C, a drop of $-4.0I$; from C to B, a drop of $-6.0I$; and from B to A, a rise of $+18$ V. Thus

$$-4.0I - 10I - 6.0 - 4.0I - 6.0I + 18 = 0$$

and

$$-24I + 12 = 0$$

hence

$$I = \tfrac{1}{2} \text{ A}$$

The current is positive, and therefore all the drops are correct as indicated. Since C is grounded and at a potential of zero, B is at -3.0 V. That is, to go from B to C, we rise $(\tfrac{1}{2}$ A$)(6.0$ $\Omega)$: A is at $+15$ V, F is at $+13$ V, E is at $+8.0$ V, and D is $+2.0$ V. Notice that the potential at any point—for example, E—is indeed the same whether we go from C to D to E or from C to B to A to F to E.

Kirchhoff's second rule is known as the **Node Rule**:

The sum of all the currents entering any node in a circuit must equal the sum of all the currents leaving that node.

In the steady state, there is neither an accumulation nor a diminution of charge anywhere in the circuit. Hence, at a node, the net amount of charge flowing in must equal the net amount flowing out—that's the Principle of Conservation of Charge. Figure 20.24 illustrates the point: $I_1 + I_2 = I_3 + I_4$.

To see how all of this comes together in practice, let's apply Kirchhoff's Rules to the network depicted in Fig. 20.25a, determining the current through every element. First, notice that the circuit has only two nodes—*C* and *E*—and therefore there are three branches *C–B–A–E*, *C–D–E*, and *C–E*. Assign a current arbitrarily to each branch, as shown in Fig. 20.25b and label with signs all the voltage drops and rises.

Three branches means three unknown currents, which requires three equations. But when it comes to picking those three equations, we have to be cautious. There are three loops: the lower one *A–B–C–E–A*, the upper one *A–B–C–D–E–A*, and the little one *C–D–E–C*, as well as two nodes *C* and *E*. Thus, there are five possible equations, *but not any three will do*—we need three independent equations. To that end, *use a mix of loop and node equations and always avoid using all the equations of either kind*. For instance, here the two node equations are

$$I_1 + I_2 = I_3 \quad \text{at } C$$

and

$$I_3 = I_1 + I_2 \quad \text{at } E$$

These are identical and not independent—only one of them can be used. *If two nodes involve the same set of currents, they will not produce independent equations* (and you should not waste time playing with them both). The three loop equations are also not independent. Consequently, we utilize any two loop equations and either node equation.

Using loop *A–B–C–E–A* and going around clockwise starting at *A*, we have

$$-10I_1 - 20I_1 - 30I_3 + 40 = 0$$

or

$$-30I_1 - 30I_3 + 40 = 0 \qquad \text{(i)}$$

Similarly, going clockwise around loop *C–D–E–C* starting at *C*, yields

$$-20 + 10I_2 + 30I_3 = 0 \qquad \text{(ii)}$$

These two equations can be solved simultaneously if we use the node equation to get rid of one of the variables, say, $I_2 = I_3 - I_1$. Hence, Eq. (ii) becomes

$$-20 + 10(I_3 - I_1) + 30I_3 = 0$$

and

$$-10I_1 + 40I_3 - 20 = 0 \qquad \text{(iii)}$$

Multiplying this by 3 and subtracting it from Eq. (i) yields

$$-150I_3 + 100 = 0$$

and $I_3 = 2/3$ A. It follows from Eq. (i) that $I_1 = 2/3$ A. And, finally, something of a surprise: $I_2 = 0$. As a *quick check,* we see that the voltage difference in going from *C* to *E* is −20 V via either branch and, moreover, the sum of the drops and rises is indeed zero around each loop.

As a last point, observe that loop *A–B–C–D–E–A* produces (going clockwise from *A*) the equation

$$-10I_1 - 20I_1 - 20 + 10I_2 + 40 = 0$$

which is

$$-30I_1 + 10I_2 + 20 = 0 \qquad \text{(iv)}$$

If we subtract this result from Eq. (i), we get Eq. (ii) and vice versa, proving that we could not solve the problem with just those three—they're not independent.

If there are N nodes, use N−1 node equations along with as many loop equations as is necessary to form a total number of equations equal to the number of unknowns. Every element of the circuit must appear in at least one loop equation.

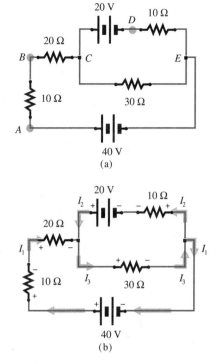

Figure 20.25 A network to be analyzed using Kirchhoff's Rules. Each branch has a current assigned to it.

Example 20.5 Determine the power delivered to the circuit by the 12-V source in Fig. 20.26.

Solution: [Given: circuit. Find: P delivered by 12-V source.] What's needed is the current through the 12-V source: $P = IV$. Let's first simplify the circuit as much as possible—Fig. 20.26b, c, and d show the series and parallel combinations. There are two nodes in the final circuit configuration A and B and three branches B–F–A,

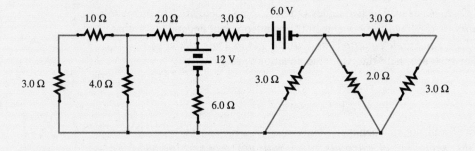

(a)

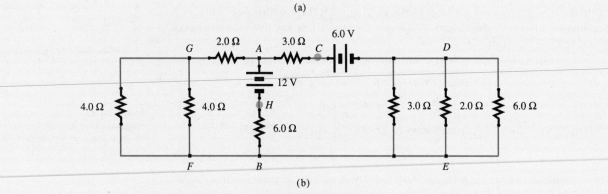

(b)

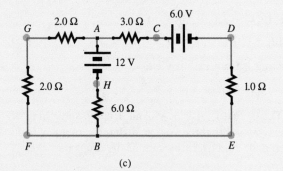

(c)

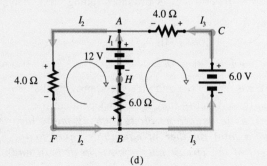

Figure 20.26   A circuit to be analyzed. It is first simplified as much as possible.

(d)

(continued)

(continued)

B–H–A, and B–C–A. Hence, we assign three branch currents. The same currents are involved at A and B so only one node equation will be used; accordingly

$$I_1 + I_3 = I_2 \qquad \text{(i)}$$

Starting at B and going clockwise around loop B–F–A–H–B, the Loop Rule yields

$$+ 4.0I_2 - 12 + 6.0I_1 = 0 \qquad \text{(ii)}$$

Starting at B and going clockwise around loop B–H–A–C–B, the Loop Rule yields

$$- 6.0I_1 + 12 + 4.0I_3 - 6.0 = 0$$

or

$$- 6.0I_1 + 4.0I_3 + 6.0 = 0 \qquad \text{(iii)}$$

We just want I_1, so use Eq. (i) to get rid of I_2 in Eq. (ii), which then becomes

$$10I_1 + 4.0I_3 - 12 = 0 \qquad \text{(iv)}$$

Subtracting Eq. (iv) from Eq. (iii) results in

$$-16I_1 + 18 = 0$$

and $I_1 = +1.125$ A. Current passes through the source from $-$ to $+$; hence, power *is* delivered by the source. Consequently

$$\text{P} = IV = (1.125 \text{ A})(12 \text{ V}) = 13.5 \text{ W}$$

or to two figures $\boxed{\text{P} = 14 \text{ W}}$.

▶ **Quick Check:** From Eq. (ii), $I_2 = 1.31$ A and the potential difference from B to A via F is $+5.24$ V, which must equal the rise from B to A via H; namely, $-(6.0 \, \Omega) \times (1.125 \text{ A}) + 12 = +5.25$ V.

Core Material

The voltage (V) across the terminals of a source of emf ($\mathscr{E}$) having an internal resistance (r) is

$$V = \mathscr{E} - Ir \qquad [20.1]$$

when the source provides power, and

$$V = \mathscr{E} + Ir \qquad [20.2]$$

when power is provided to it (p. 755).

The **equivalent resistance** of two or more resistors in series is

$$R_e = R_1 + R_2 + \cdots \qquad [20.4]$$

The equivalent resistance of two or more resistors in parallel is

$$\frac{1}{R_e} = \frac{1}{R_1} + \frac{1}{R_2} + \cdots \qquad [20.5]$$

For two resistors, this equation can be expressed as

$$R_e = \frac{R_1 R_2}{R_1 + R_2} \qquad [20.6]$$

In an R-C circuit the rate at which the charge on a capacitor decays is given by

$$Q(t) = Q_i e^{-t/RC} \qquad [20.9]$$

The **Maximum Power Transfer Theorem** states that *a maximum amount of power will be transferred from any source with an internal resistance r to a load of resistance R when r equals R* (p. 762).

Kirchhoff's **Loop Rule** is *the algebraic sum of the voltage rises and drops encountered in going around any closed path formed by any portion of a circuit must be zero.* Kirchhoff's **Node Rule** is *the sum of all the currents entering any node in a circuit must equal the sum of all the currents leaving that node.*

Suggestions on Problem Solving

1. A common error is to use Eq. (20.5) for the equivalent resistance of a parallel string of resistors, computing and adding $1/R_1$ and $1/R_2$ and then presenting that as R_e rather than $1/R_e$; look out for this one. Note, too, that the reciprocal of the sum is *not* equal to the sum of the reciprocals: $1/(R_1 + R_2) \neq 1/R_1 + 1/R_2$. Equation (20.6), $R_e = (R_1 R_2)/(R_1 + R_2)$, is very useful for getting a quick sense of parallel combinations without a calculator. The reciprocal function on the calculator is wonderfully helpful in computing Eq. (20.5) directly. For example, to compute R where $1/R = 1/4 + 1/3$, hit the following sequence of keys on your calculator: [4] [1/x] [+] [3] [1/x] [=] [1/x] to get 1.7.

The expression $R_1 R_2 R_3/(R_1 + R_2 + R_3)$ is *not* the correct formula for three resistors in parallel!

2. When analyzing a circuit, study it carefully. Work with a drawing of the network. If the original version is not neatly diagrammed, redraw it before proceeding. Is there anything extraordinary about the circuit (as, for example, in Fig. Q3)? Are there any obvious shortcuts to finding the required quantities (as, for example, in Fig. 20.21)? Is the circuit symmetrical in such a way that some sort of bridge is present so that elements might be redundant (as, for example, in Fig. 20.22)? Remember, if no current traverses a particular resistor, it can be removed from the

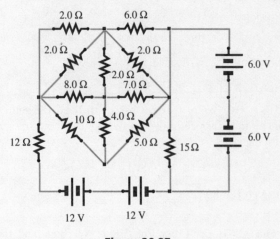

Figure 20.27

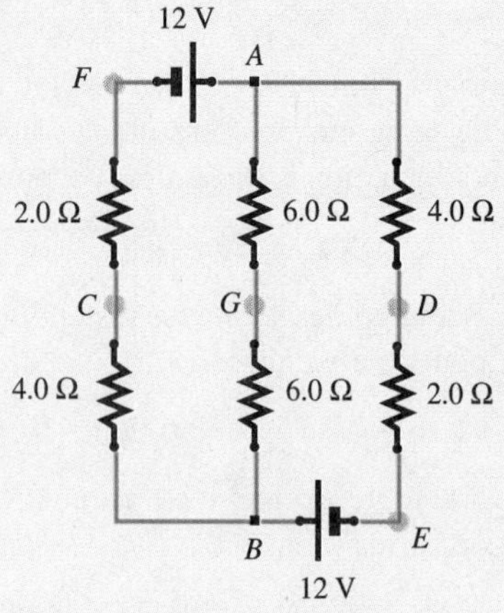

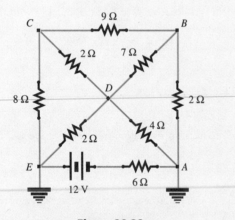

Figure 20.28

circuit. All right, then, what current passes through the 8-Ω resistor in Fig. 20.27? Zero! All the sources buck each other—there's no current anywhere. Try Fig. 20.28. What's the current in the 6-Ω resistor? By inspection, it must be 2 A to the left. Both ends of the bottom branch are grounded, so the net potential difference between E and A is zero. Going from A to E, there must be a drop of 12 V across the resistor since there is a rise of 12 V across the battery. In Fig. 20.28, what is the current in the 9-Ω resistor?

Figure 20.29

Again, it's zero because there is no voltage across E–A. All the current from the battery ($I = 2$ A) goes from E into the zero-resistance path back to A via ground. When ground connections are indicated, it should be assumed that all such leads are wired to a common line even if not shown.

Referring to Fig. 20.29, symmetry demands that current only circulate in the outer elements. Points A and B are at the same potential because the circuit is symmetrical. Therefore, no current can go across the resistors in that branch. The current through both batteries is 2 A.

3. The first step in an analysis is to simplify the circuit by combining series and parallel groupings wherever possible. Redraw the circuit whenever needed, keeping track of what you have done so that you can work your way back to the original. (a) If the system reduces to one source across an equivalent resistance, use Ohm's Law to find what is unknown and work backwards reconstructing the original circuit step-by-step. (b) If the circuit cannot be further simplified, apply Kirchhoff's Rules to what you have. (c) Line up the several variables, essentially making columns of them so you can see what needs to be done at a glance. For example

$$-3I_2 - 2I_3 \qquad\quad + 12 = 0 \qquad\qquad \text{(i)}$$

$$\qquad\quad + 2I_3 - 3I_4 - 6 = 0 \qquad\qquad \text{(ii)}$$

$$-3I_2 \qquad\qquad - 3I_4 + 6 = 0 \qquad\qquad \text{(iii)}$$

Discussion Questions

1. Figure 20.5 depicts a battery with an internal resistance of r in series with a load R. What experimental procedure might be used to determine r?

2. What is meant by a complete or closed circuit and what is the relationship between such a circuit and a steady-state direct current? When the switch in Fig. 20.18 is closed, a transient current circulates in the circuit. Explain how this could happen in spite of the gap in the capacitor that stops charge from literally crossing it.

3. Figure Q3 is a reproduction of a circuit taken from a simulated MCAT exam created by a well-known company that prepares students for that test. The question asks for the current I and the answer given is "9 amps toward G." What's wrong with that and why do you think they give 9 A as the answer?

4. In light of Question 2, discuss the statement: *In the steady state, a capacitor acts as an open circuit to dc*.

5. The circuit of Fig. Q5 contains a gas-filled lamp—the sort of flashing light often used at construction sites. At low voltages, the lamp has nearly infinite resistance, but at a certain breakdown voltage, which is less than the terminal voltage of the

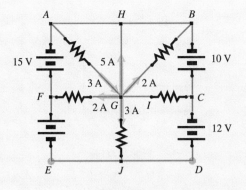

Figure Q3

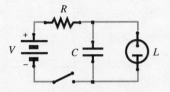

Figure Q5

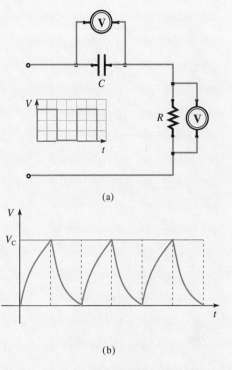

(a)

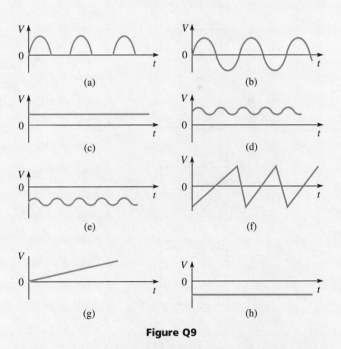

(b)

Figure Q7

battery, it becomes a very good conductor. A blast of charge through the lamp causes it to flash brightly. What happens once the switch is closed?

6. Suppose we have a fluid voltaic cell across which we place a load resistor. Next we measure the current from, and voltage across, the source with an ammeter and a voltmeter, respectively. It can be observed that when the plates are moved apart, the current diminishes. Moreover, when the plates are partly lifted out of the electrolyte, the current again drops, although in both cases the voltage remains unchanged. What's happening to the internal resistance of the cell? Relate your conclusions to the fluid-flow analogy of current.

7. *R-C* circuits are often used to change the shape of a signal. The arrangement of Fig. Q7 shows a network being fed a square-wave signal comprising a series of rectangular dc voltage pulses rising to $+V$ and falling to 0. Part (b) shows the corresponding voltage across the capacitor. Explain its shape. What can you say about the value of RC as compared to the width of each pulse? Draw a curve of the output signal as seen across the resistor and explain its features. Write a general expression relating V, V_R, and V_C.

8. Explain how a flashlight bulb works. Draw a picture showing the relationship between a bulb's filament and its two terminals. Why does the brightness depend on the current through the bulb? What effect would increasing the voltage across a bulb have on its brightness? Incidentally, light bulbs are not ohmic.

9. Figure Q9 depicts several voltage-versus-time curves. These might be the outputs of a generator, transducer, etc. Which of them corresponds to dc? Explain.

10. Draw a flashlight bulb and show where the current usually enters and leaves. Draw a D-cell and label its two terminals. Now draw a diagram showing how to connect the bulb and cell with two lengths of wire. How might you connect them with one wire? Show the path of the current in each case.

11. Figure Q11 shows three arrangements of identical flash-

Figure Q9

light bulbs along with an ideal 6-V source. Compare the brightness of all the bulbs, listing them in descending order. Explain.

12. Compare the brightness of all the bulbs in Fig. Q12, listing them in descending order. The battery is ideal. Explain.

13. Compare the brightness of each identical bulb in Fig. Q13. Explain.

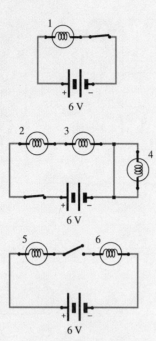

Figure Q11

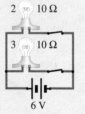

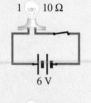

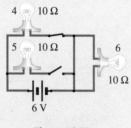

Figure Q12

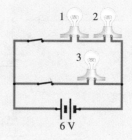

Figure Q13

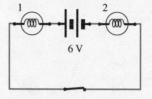

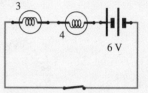

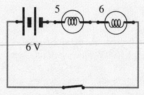

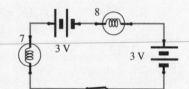

Figure Q14

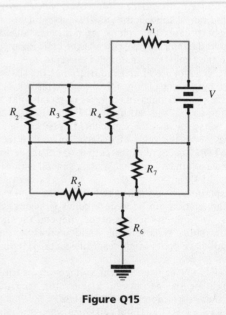

Figure Q15

14. All the bulbs in Fig. Q14 are identical. Discuss the relative brightness of each and explain your answer.

15. If all the resistors in Fig. Q15 are identical, rate them in order of the amount of current passing through each one. Explain.

Multiple Choice Questions

1. The equivalent resistance of a 4-Ω resistor in parallel with a 12-Ω resistor is   (a) 48 Ω (b) 8 Ω (c) 16 Ω (d) 3 Ω (e) none of these.

2. An ideal source of emf is connected in series with a resistor in a closed loop, and a current of 0.50 A circulates. An additional 6.0-Ω resistor is added in series, and the current drops to 0.30 A. The original resistance was   (a) 9.0 Ω (b) 6.0 Ω (c) 15 Ω (d) 3.0 Ω (e) none of these.

3. In the previous question, what was the emf?   (a) 9.0 V   (b) 12 V   (c) 15 V   (d) 4.5 V   (e) none of these.

4. A 24-W, 12-V ohmic lamp is placed in a circuit with 6 V across it, at which time it draws a current of   (a) 2 A   (b) 1 A   (c) 0   (d) $\frac{1}{2}$ A   (e) none of these.

5. The device in Fig. MC5 consists of a rotating switch, five different resistors, and a galvanometer. What is the apparatus?   (a) an ohmeter   (b) a multirange voltmeter   (c) a multirange ammeter   (d) a multirange wattmeter   (e) none of these.

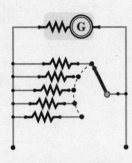

Figure MC5

6. The circuit in Fig. MC6 contains four identical light bulbs in series with an ideal battery. A wire is then clipped to points B and E and   (a) lamps 3 and 4 become brighter   (b) lamp 1 gets dimmer   (c) lamps 2, 3, and 4 remain unaffected   (d) lamp 1 brightens   (e) none of these.

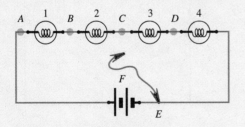

Figure MC6

7. In Fig. MC6, a wire is clipped to points B and E and   (a) lamps 2, 3, and 4 become brighter   (b) lamps 1 and 2 get dimmer   (c) lamps 2, 3, and 4 go out altogether   (d) lamp 1 goes out   (e) none of these.

8. In Fig. MC6, a wire is clipped to points F and E and   (a) lamps 1, 2, 3, and 4 become brighter   (b) lamp 1 gets slightly dimmer   (c) lamps 2, 3, and 4 remain unaffected   (d) lamps 1, 2, 3, and 4 get slightly dimmer   (e) none of these.

9. In Fig. MC6, a wire is clipped to points C and E and   (a) more current is drawn from the battery   (b) less current is drawn from the battery   (c) the same current is drawn from the battery   (d) not enough information is given to tell what will happen to the current   (e) none of these.

10. If Fig. MC6, a wire is clipped to points C and E and   (a) the voltage across lamps 1 and 2 increases   (b) the voltage across lamps 3 and 4 increases   (c) the voltage across only lamp 1 increases   (d) the voltage across lamp 2 decreases   (e) none of these.

11. In Fig. MC6, a wire is clipped to points C and E and   (a) the power delivered by the battery remains unchanged   (b) the power delivered by the battery increases   (c) the power dissipated by the circuit decreases   (d) the power dissipated by the circuit is halved   (e) none of these.

12. Suppose the four lamps in Fig. MC6 are rewired so they are in parallel across the battery: (a) each lamp will put out the same amount of light as before   (b) each lamp will put out more light than before   (c) each lamp will put out less light than before   (d) there is not enough information to tell what will happen   (e) none of these.

13. The two circuits (i) and (ii) shown in Fig. MC13 are each composed of identical resistors connected to a battery. The resistors in (i) and the resistors in (ii), respectively, are hooked up in   (a) series and series   (b) parallel and

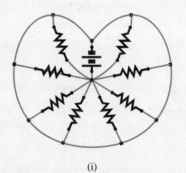

(i)

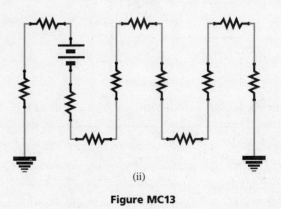

(ii)

Figure MC13

parallel (c) series and parallel (d) parallel and series (e) none of these.

14. A battery of emf 12 V and internal resistance 0.2 Ω is placed across a variable resistor. The resistor is adjusted until it dissipates a maximum amount of power, at which point the current through it is (a) 1 A (b) 0.4 A (c) 30 A (d) 60 Ω (e) none of these.

15. A battery attached to a load supplies 2 A with a terminal voltage of 12.0 V. If the battery dissipates 0.4 W, its emf is (a) 11.8 V (b) 12 V (c) 12.2 V (d) 12.4 V (e) none of these.

16. Two resistors of 5 Ω and 20 Ω are connected in parallel across an ideal source of 20 V. The current supplied by the source is (a) 4 A (b) 5 A (c) 20 A (d) 1 A (e) none of these.

17. Figure MC17 shows a portion of a network. The current *I* is (a) +2 A (b) +4 A (c) not enough information to tell (d) −4A (e) none of these.

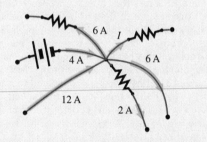

Figure MC17

18. What is the current in any one of the 4-Ω resistors in the circuit of Fig. MC18? (a) 12 A (b) 1.2 A (c) 3 mA (d) 0 (e) none of these.

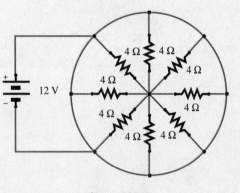

Figure MC18

19. In Fig. MC19, (a) the battery with an emf of 12 V is

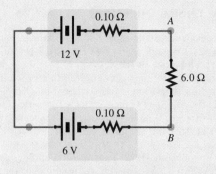

Figure MC19

being charged (b) the battery with an emf of 6 V is being charged (c) the battery with an emf of 6 V is being discharged (d) neither battery is being charged (e) none of these.

20. In Fig. MC19, the higher-voltage battery (a) internally dissipates more power via joule heating than the lower-voltage battery (b) internally dissipates less power via joule heating than the lower-voltage battery (c) dissipates more power than the 6-Ω resistor (d) has a terminal voltage of 14 V (e) none of these.

21. What is the equivalent resistance between A and B in Fig. MC21? (a) slightly more than 1 kΩ (b) 4 kΩ (b) slightly more than 1 Ω (c) slightly more than 17 kΩ (d) slightly less than 1 kΩ (e) none of these.

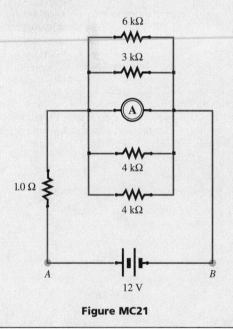

Figure MC21

Problems

CIRCUIT PRINCIPLES

1. [I] What is the current in the circuit of Fig. P1 before and after the wire is clipped on at *C* and *B*? What happens to the 12-Ω bulb after the wire is attached? The battery has a negligible internal resistance.

2. [I] A dc source with an internal resistance of 0.10 Ω is

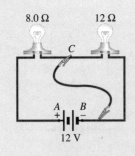

Figure P1

connected across a length of nichrome wire having a resistance of 20 Ω. If a voltmeter across the nichrome indicates a drop of 10 V, what is the emf of the source?

3. [I] Each lamp in Fig. P3 has a resistance of 20 Ω. How much current is drawn from the battery before and after the switch is closed? How does the power supplied to the circuit change when the switch is closed? The battery has a negligible internal resistance.

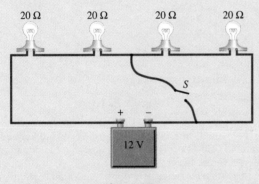

Figure P3

4. [I] A dc source with a terminal voltage of 100 V internally dissipates 40 W as it delivers 2.0 A. What is its emf?

5. [I] Two 2.0-Ω resistors are in parallel. What is their equivalent resistance?

6. [I] Three resistors with values of 2.0 Ω, 3.0 Ω, and 6.0 Ω are connected in parallel. What is the equivalent resistance?

7. [I] A portable generator in a field hospital produces dc at 100 V. Five 100-W lamps are attached in parallel and the string placed across the terminals of the generator. How much current must the source provide for the lamps to operate as designed?

8. [I] Three resistors with values of 1/2 Ω, 1/3 Ω, and 1/6 Ω are connected in parallel. What is the equivalent resistance?

9. [I] Show that if there are a number N of identical resistors in parallel, each with a value R, then $R_e = R/N$.

10. [I] If in Fig. P10 the ideal ammeter reads 3.20 A, what will the ideal voltmeter read?

11. [I] What is the equivalent resistance of the circuit in Fig. P11 between terminals A and B? Note that the wires cross but do not make contact at the center.

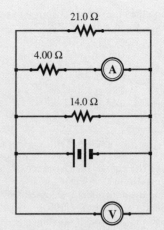

Figure P10

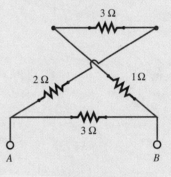

Figure P11

12. [I] Determine the equivalent resistance of the circuit between points A and B in Fig. P12.

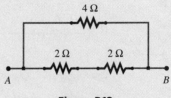

Figure P12

13. [I] Determine the equivalent resistance of the circuit between points A and B in Fig. P13.

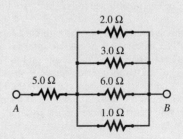

Figure P13

14. [I] Determine the equivalent resistance of the circuit between points A and B in Fig. P14.

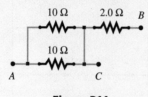

Figure P14

15. [I] Determine the equivalent resistance of the circuit between points A and C in Fig. P14.
16. [I] Determine the equivalent resistance of the circuit between points B and C in Fig. P14.
17. [I] Determine the equivalent resistance of the circuit between points A and B in Fig. P17.

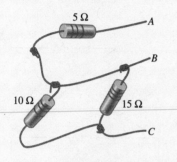

Figure P17

18. [I] Determine the equivalent resistance of the circuit between points C and B in Fig. P17.
19. [I] Figure P19 shows a portion of a circuit, the rest of which contains both sources and resistors. If the ammeter reads 9 A, what current passes through each resistor shown?

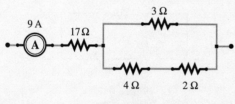

Figure P19

20. [I] Given that the ammeter in Fig. P20 reads 1.0 A, what is the emf of the ideal dc source as indicated by the voltmeter?
21. [I] How much current passes through each lamp in Fig. P21? The battery has negligible internal resistance.
22. [I] A 10-Ω resistor is attached at one end to the + terminal of a 20-V dc source whose − terminal is grounded. The resistor's other end is attached to the + terminal of a 40-V dc source whose − terminal is grounded. What current traverses the resistor?
23. [I] A 20-V lamp designed to dissipate 80 W is placed in series with a resistor R and a 60-V dc source. What value should R be for the lamp to operate properly?

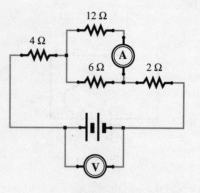

Figure P20

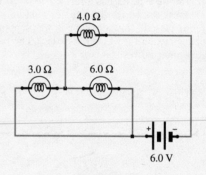

Figure P21

24. [I] If the total resistance of a voltmeter is 6.0 kΩ and it contains a 50-Ω coil movement, describe the other circuit element in the meter.
25. [I] An ammeter movement consists of a 50.0-Ω coil. If 0.10% of the current entering the meter is to pass through the coil, how big must the shunt resistor be?
26. [I] Show that the time constant of an R-C circuit has the correct units.
27. [I] An electronic flash fires a blast of energy from a 800-μF capacitor into a xenon lamp. It recharges through a series resistor of 5.0 kΩ. How long will it take to recharge 63% of its maximum charge?
28. [I] A charged 600-μF capacitor is in series with an open switch and a 4.0-kΩ resistor. If the switch is closed, how long will it take for the charge on the capacitor to decay down to 37% of its original value?
29. [I] An uncharged 1.0-μF capacitor is in series, through a switch, with a 2.0-MΩ resistor and a 12.0-V battery (with negligible internal resistance). The switch is closed at $t = 0$ and a current I_i immediately appears. Determine I_i. How long will it take for the current in the circuit to drop to $0.37I_i$?
30. [I] A 3.0-μF capacitor is put across a 12-V ideal battery. After an hour, it is disconnected and put in series, through a switch, with a 200-Ω resistor. (a) What is the initial charge on the capacitor? (b) What is the initial current when the switch is closed?
31. [II] Determine the equivalent resistance of the circuit between points A and B in Fig. P31.

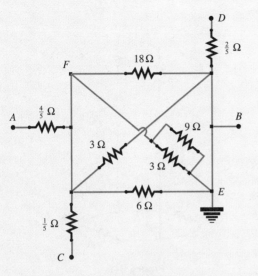

Figure P31

32. [II] Two resistors R_1 and R_2 are in parallel with each other and with an ideal source ($r = 0$) having a terminal voltage V. Show that the branch currents are given by

$$I_1 = I\left(\frac{R_2}{R_1 + R_2}\right) \quad \text{and} \quad I_2 = I\left(\frac{R_1}{R_1 + R_2}\right)$$

where the larger current goes through the smaller resistor. (These are good relationships to remember.)

33. [II] Figure P33 shows a portion of a circuit, the rest of which contains both sources and resistors. If the ammeter reads 9 A, what current passes through each resistor in the diagram?

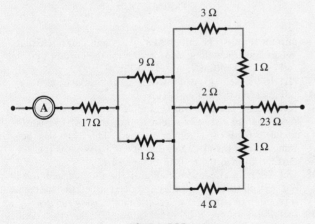

Figure P33

34. [II] Imagine that N is the number of identical cells in series, each having an internal resistance r and emf $\mathcal{E}$. Derive an expression for the current through a load resistor R placed across the terminals of the battery.

35. [II] What is the equivalent resistance between points A and B of the circuit shown in Fig. P35? How much power would be dissipated by this circuit if a constant 20-V dc

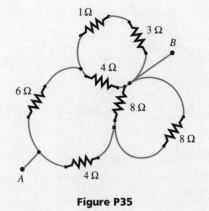

Figure P35

source with a 0.10-Ω internal resistance was placed across A and B?

36. [II] How much current passes through each of the 5.0-Ω resistors in Fig. P36? How much power is delivered by the dc source?

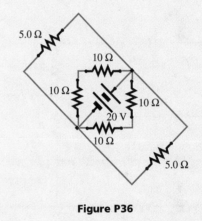

Figure P36

37. [II] Given the circuit in Fig. P37, calculate the current in each resistor. What power is delivered by the battery? What is the potential difference between A and C?

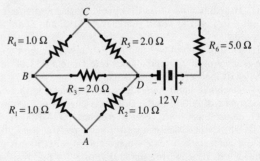

Figure P37

38. [II] The battery in Fig. P38 has an emf of 9.0 V and an internal resistance of 0.50 Ω.   (a) What current does it supply?   (b) What is the total power dissipated by the

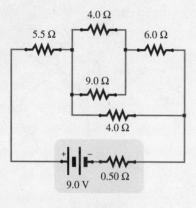

Figure P38

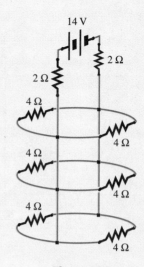

Figure P40

entire circuit? (c) What is the terminal voltage of the battery?

39. [II] What is the current provided by the battery in Fig. P39, given that its internal resistance is 0.50 Ω? What is its terminal voltage?

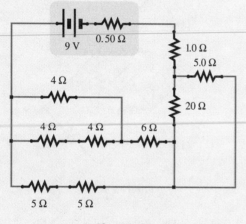

Figure P39

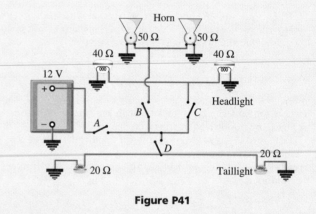

Figure P41

current through the circuit? Draw a rough plot of current versus time. How long does it take for the current to drop to 37% of its initial value?

46. [II] In Problem 45, what is the charge on the capacitor 6.0 ms after the switch is closed?

47. [II] A 12-V battery with negligible internal resistance is placed across a series combination of a 1.0-MΩ resistor and a 12.0-μF capacitor for 10 hours. The battery is then removed and the circuit closed at $t = 0$. What is the current through the resistor at $t = 24$ s?

48. [cc] We have already studied a discharging capacitor in an R-C circuit (see Fig. 20.17a) and determined the charge as a function of time. Now derive an expression for $I(t)$.

49. [cc] Return to the R-C circuit in Fig. 20.18b, which depicts a capacitor being charged, and derive an expression for $I(t)$. What is the initial value of the current, I_i, at $t = 0$? [Hint: Begin with the sum of the voltage rises and drops around the loop. Taking the derivative of that relationship will provide an equation that can be solved for $I(t)$.]

50. [cc] Regarding the R-C circuit in Fig. 20.18b, which depicts a capacitor being charged, determine an expression for its charge as a function of time, $Q(t)$. Write your answer in terms of Q_f the final or maximum charge that

40. [II] Find the current supplied by the source in Fig. P40. The resistors are mounted around a cylindrical form.

41. [II] How much power is dissipated by the automobile circuit in Fig. P41 when switches A, B, C, and D are all closed?

42. [II] Design a voltmeter using a galvanometer with a coil resistance of 100 Ω and a full-scale current of 1.00 mA that will measure 100 V full scale.

43. [II] Design a shunt such that a galvanometer with a coil resistance of 100 Ω and a full-scale current of 1.00 mA can be used as an ammeter to measure up to 1.00 A.

44. [II] The coil of a galvanometer has a resistance of 20 Ω, and it deflects full scale when a current of 0.50 mA passes through it. By shunting the coil with a 2.0-mΩ resistor, it becomes an ammeter. What full-scale current will it now read?

45. [II] A 6.0-μF capacitor is charged up to 12 V and subsequently connected through a switch to a 100-Ω resistor. At $t = 0$, the switch is closed. What is the initial

occurs at $t = \infty$. [Hint: Use the results of the previous problem.]

51. [cc] Considering the previous problem, suppose the capacitor in Fig. 20.18 is fully charged when the switch is opened, the battery removed, and the switch closed. How much energy will be dissipated as joule heat in the resistor during the capacitor's subsequent discharge? [Hint: Use the discharge current determined in Problem 48 and integrate the power over time.]

52. [III] Imagine a wire cube with identical resistors R in each arm (Fig. P52). What is the equivalent resistance between points A and B?

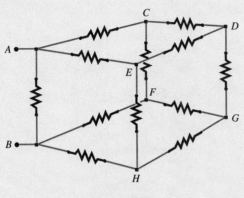

Figure P52

NETWORK ANALYSIS

53. [I] Use Kirchhoff's Loop Rule to solve for the currents in branches $A-D-C$ and $A-B-C$ of the circuit in Fig. P53. Then use the Node Rule to find the current in branch $A-C$.

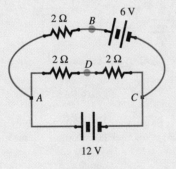

Figure P53

54. [I] Find the current through each element of the circuit in Fig. P54.

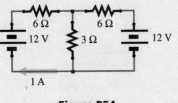

Figure P54

55. [I] Apply Kirchhoff's Rules to the circuit of Fig. P55, and solve for the three branch currents. Next, simplify the network, determine all the currents, and check your answers.

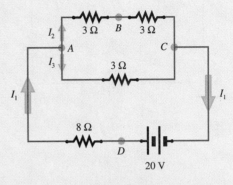

Figure P55

56. [I] Solve for the unknown source voltage and the power delivered by the 12-V battery in Fig. P56.

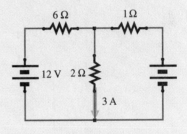

Figure P56

57. [I] The transistor circuit of Fig. P57 is to be checked for proper operation. According to specifications, the collector voltage (point C) should be at a constant $+6$ V with respect to ground. If that's the case, what should be the voltage measured at point D? [Hint: Read Discussion Question 5 to learn that a capacitor is an open circuit to dc.]

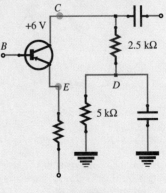

Figure P57

58. [I] If the ammeter and voltmeter in Fig. P58 read 2.0 A and 10 V, respectively, what current passes through the 1.0-Ω resistor?

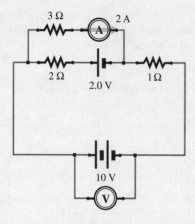

Figure P58

59. [I] The ammeter in Fig. P59 reads 2.0 A; what will the voltmeter read?

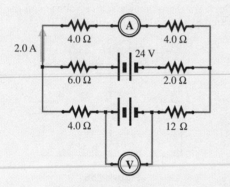

Figure P59

60. [I] Referring to the bridge circuit of Fig. P60, if $I = 6$ A, $I_2 = 4$ A, and $I_3 = 0$, find I_1, I_4, and I_5.

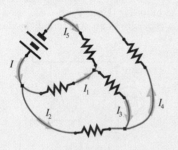

Figure P60

61. [I] Determine the current in each branch of the circuit of Fig. P61.
62. [I] Find the current in each resistor of the circuit in Fig. P62 using Kirchhoff's Rules. Then simplify the circuit and compare your results.
63. [I] The switch in the circuit of Fig. P63 is closed, and a steady state is established. What is the charge on the capacitor?

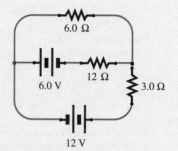

Figure P61

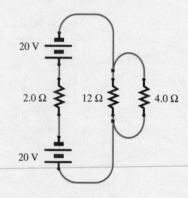

Figure P62

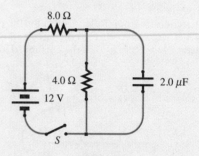

Figure P63

64. [I] Figure P64 shows a 200-V dc generator (the pictorial symbol used is another fairly common one) supplying 100 A to a load via a two-lead cable having a resistance of 0.20 Ω per length of conductor. What is the voltage across the load?

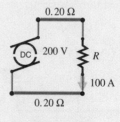

Figure P64

65. [II] Find the values of R_1 and $\mathcal{E}_1$ in Fig. P65.

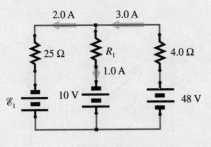

Figure P65

66. [II] The variable resistor in Fig. P66 is adjusted to 20 Ω, whereupon the ammeter (which has a negligible internal resistance) reads zero. Use Kirchhoff's Rules to determine the power provided by the sources. What is the voltage at points B, C, and D?

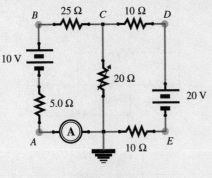

Figure P66

67. [II] The variable resistor in Fig. P67 is adjusted until the ammeters read (#1) 120 mA and (#2) 80 mA, with the directions of the currents as shown. Find the value of R.

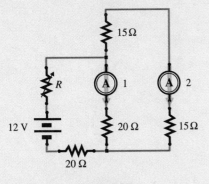

Figure P67

68. [II] The potentiometer in Fig. P68 has a total resistance of 200 Ω. At what position must the slider be placed so that the 40.0-Ω coil of wire receives 1.00 A? Check your results using Kirchhoff's Loop Rule.

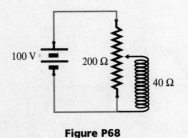

Figure P68

69. [II] With only switch S_1 closed in Fig. P69, (a) what is the steady-state reading of the voltmeter? (b) What is the charge on the 3.0-μF capacitor?

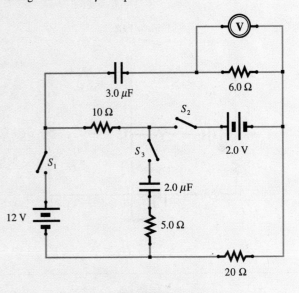

Figure P69

70. [II] With Problem 69 in mind, how much power does the 12-V battery supply in the steady state a few minutes after all the switches are closed? What's the charge on the 2.0-μF capacitor?

71. [II] Find the values of R, V, and all the unknown branch currents in the network of Fig. P71, given that $I_3 = 1.0$ A.

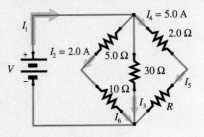

Figure P71

72. [II] Solve for the currents in each branch of the circuit in Fig. P72.

73. [II] Given that 5.0 A passes along the branch from C to B in Fig. P73, what is the voltage of points A, D, E, F, and G?

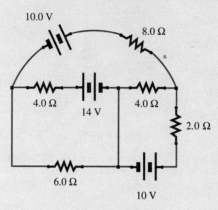

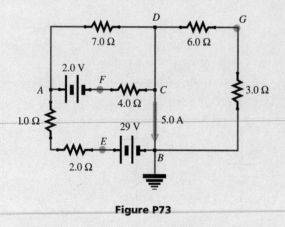

Figure P72

Figure P73

74. [III] Solve for the currents in each branch of the circuit in Fig. P74.

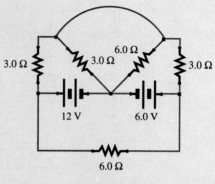

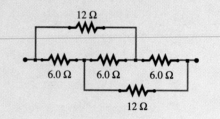

Figure P74

75. [III] Determine the equivalent resistance between the terminals of the group of resistors shown in Fig. P75 using Kirchhoff's Rules.

Figure P75

21

Magnetism

OUR UNDERSTANDING OF MAGNETISM has evolved into a sophisticated picture of whirling electrons, of fields and currents. Yet, the modern era of magnetism only began in 1820, and it's not surprising that the theory remains incomplete—we still can't determine theoretically the magnetic characteristics of the proton or the neutron.

The strange power of the **lodestone** to cling to iron tools was discovered in ancient times. The lodestone is an oxide of iron (Fe_3O_4) known as *magnetite*. That rich iron ore, some of it permanently magnetized, occurs in many parts of the world. The word *magnet* comes from the Greek (*magnes*),

A lodestone with a few steel nails clinging to it. It is a chunk of magnetite, a common iron ore, magnetized by the Earth's magnetic field.

which probably derives from the ancient colony of Magnesia, where the ore was mined 2500 years ago.

Chinese legend has it that Emperor Hwang-ti (ca. 2600 B.C.) was guided in battle through dense fog by a small pivoting figure that always pointed south, a lodestone embedded in its outstretched arm. By about 1100 A.D., the magnetic compass had come to the West, and the lodestone's ability to align itself with the north-south axis of the "Universe" was awesome.

For all the centuries from Hwang-ti to the 1800s, the only practical source of magnetism (apart from the gradual realization that the Earth itself was the mother of all magnets) was the dark lodestone. The surge of activity and innovation that began in the nineteenth century followed the discovery by Oersted (p. 795) that electric currents give rise to magnetic force. For centuries, electricity and magnetism had been taken as two distinct powers, and now they were connected. *Charges generate electric fields. Charges in motion, in addition, generate magnetic fields.* The two fields are different manifestations of a single phenomenon—**electromagnetism**.

MAGNETS AND THE MAGNETIC FIELD

The earliest scholar to study the lodestone was probably Thales (ca. 590 B.C.). Almost 150 years later, the philosopher Socrates dangled soft iron rings clinging to one another beneath a lodestone. That's the same game most of us have played with a magnet and a box of paper clips. Just as a charged comb can induce charge on scraps of paper, a magnet can magnetize nearby pieces of iron.

There are many variations on what passes for a chunk of iron, and they all behave differently magnetically. Cast iron is a hard, brittle material rich in carbon (from 2% to about 7%), whereas pure iron is rather soft. Between these two extremes is the carbon-iron alloy *steel*. Soft iron retains its magnetized condition only so long as the inducer (the lodestone) is kept nearby. When removed, the specimen quickly demagnetizes, just as a charged piece of paper quickly depolarizes. Low-carbon soft steel, the stuff of paper clips and common nails, demagnetizes a bit more gradually. In contrast, a piece of hard steel, once magnetized, retains much of its power, and we speak of it as a **permanent magnet**, although that's an exaggeration.

The Chinese were versed in the art of making permanent magnets by the beginning of the second century A.D. A manuscript from the period suggests stroking an iron rod or needle from end to end along a lodestone, repeatedly, and *always in the same direction*. Try it with a magnet and a screwdriver made of hard tool steel.

21.1 Poles

On a summer's day in 1269 a French engineer, Peter de Maricourt—alias Peter Peregrine, Peter the Pilgrim—sat down to write a long letter to a friend. (Peter was whiling away time in the trenches with an army that was laying siege to a city in Italy.) In the letter, he described his researches on magnetism: military engineers were concerned about the compass, often being required to construct long tunnels leading under fortress walls.

Peregrine was the first to introduce the concept of the **magnetic pole**. He had several lodestones ground into the shape of a sphere to resemble the Earth and called them *terrellas*. A steel needle placed anywhere on the surface of

ON THE LODESTONE

The stone not only attracts iron rings but also imparts to them a similar power whereby they attract other rings.

Plato recounting the words of Socrates (ca. 400 B.C.)

A magnetite crystal found in a bacterium.

Similar crystals, this one about a millionth of an inch long, have been found in the human brain.

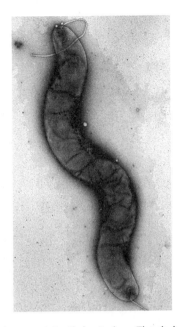

A magnetotactic bacterium. The dark line of dots is a chain of magnetite crystals that functions as a compass needle.

a terrella aligned itself in a particular way (Fig. 21.1). By drawing lines on the stone in the directions assumed by the needle, he determined that they all crossed at two opposing points, just as "all the meridian circles of the Earth meet in the two opposite poles of the world." If a piece of a needle was placed in contact with the lodestone, it would stand straight upright at, and only at, the poles.

Most magnets have two poles, where the force is clearly strongest. A straight bar magnet is the simplest two-pole configuration (known as a **dipole**), and Peregrine made iron bar magnets as well. Beyond that, it is possible for a magnet to have any number of poles, odd or even, provided it's two or more. Some modern flexible magnets are made in long strips with hundreds of poles.

Peregrine next put a terrella in a wooden bowl and set it afloat in a large vessel of water. As soon as it was released, it spun around, bowl and all, the *north-seeking* or **north pole** always pointing northward, the *south-seeking* or **south pole** always pointing southward. Of course, this was a compass, not all that different from the crude floating-needle instruments already in widespread use. Holding another terrella, whose poles were determined and marked, he approached the floating stone. When the north pole of one was brought near the south pole of the other, the little boat lunged toward the hand-held stone, but when either two north poles or two south poles were positioned near each other, the boat was pushed away. Peregrine had discovered the basic mutual interaction of all magnets: **like magnetic poles repel and unlike magnetic poles attract** (Fig. 21.2).

Naturally, Peregrine tried to isolate a single **monopole**, a piece of magnet that was simply and only north polar or south polar. And what more obvious way to do that than to split a magnet in two (Fig. 21.3)? Surprise! No matter how we break a magnet, the fragments are always bipolar—*the monopole cannot be isolated.* It's as if a magnet were composed of a succession of microscopic bar magnets with opposite

Figure 21.1 Several small steel needles attracted to the surface of a spherical magnet. Only at the poles do the needles stand straight up. A compass pointer pivoted so that it can swing vertically shows the same behavior in the Earth's magnetic field.

Figure 21.2 The attraction of unlike poles and the repulsion of like poles can be observed easily by suspending one bar magnet on a thread. The suspended magnet will swing away from the like pole of a second, stationary magnet and swing toward the stationary magnet's unlike pole.

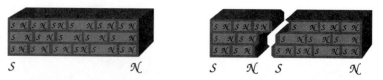

Figure 21.4 A magnet behaves as if it were composed of tiny bipolar units, tiny bar magnets, or dipoles.

Figure 21.3 The fragments of a bar magnet always have two poles.

poles touching and neutralizing each other everywhere but at the ends. When the magnet is broken, the appropriate poles appear (Fig. 21.4). That behavior will be understood only after we learn that **the electron itself is the fundamental dipole magnet**.

21.2 The Magnetic Field

A high point in the study of magnetism was provided in 1600 by William Gilbert (p. 238), then physician to Queen Elizabeth I. Gilbert compared the Earth to a large spherical lodestone, maintaining that the compass needle was drawn to the planet's magnetic pole, not to the heavens as everyone else had thought. It was simply a matter of one magnet pulling on another.

Dr. Gilbert probed the region surrounding a magnet with a little compass and concluded that "Rays of magnetick virtue spread out in every direction in an orbe." The statement seems almost identical in spirit to the nineteenth-century vision of Faraday's *lines-of-force of the magnetic field*. One need only connect the little compass arrows with smooth arcs to transform the imagery from an "orbe of virtue" into a "magnetic field." Less than half a century after Gilbert, Descartes carried the mapping process one step further. He sprinkled iron filings around a magnet, and they aligned themselves like minute compass needles to form curved continuous filaments, which even more potently *suggest* lines-of-force, lines of *magnetic field* (Fig. 21.5).

Again we are faced with the mystery of action-at-a-distance, and again our answer is to assume that every magnet establishes, in the space surrounding it, a **magnetic field** (*B*). As before, we say a field exists in a region of space when an appropriate object placed at any point therein experiences a force. The fields span the space and communicate the interaction—the fields mediate the Third Law's action-reaction. Remember that we probed the gravity field with a test-mass, just as we probed the electric field with a test-charge. Now we will map the magnetic field, not with a monopole, but with the next best thing, a dipole, a tiny test-compass.

A compass needle, able to turn freely, placed near the south pole of a bar magnet simultaneously has its own south pole repelled and its north pole attracted (Fig. 21.6a). It experiences a net torque and twists around into a new equilibrium orientation such that the torque vanishes (Fig. 21.6d). That's why a compass needle aligns itself with the local field. The strength of a magnetic field (*B*) at every point in space can be determined from the torque tending to realign the test-compass. For the moment this approach will suffice, although a more practical one will be forthcoming. The SI unit for *B* is the *tesla* (after Nikola Tesla), abbreviated T (Table 21.1). The compass is a tiny bar magnet that will settle tangent to the field, and *we arbitrarily take the arrow from its south to its north pole as the direction of the field in which it is immersed.*

Once again the concentration of field lines, the number per unit cross-sectional area, is proportional to the strength of the field. The field lines in Fig. 21.7 are more densely concentrated near the poles, where *B* is largest. By convention *the field*

A magnet and a box of paper clips. Each clip is magnetized, becoming a small temporary magnet. The clips pack in densely where the field is strong. They form bridges that arch from pole to pole, crudely suggesting a pattern of field lines.

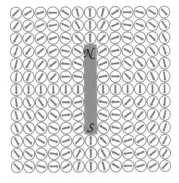

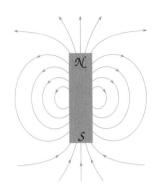

Figure 21.5 The force field around a bar magnet, as revealed by an array of small compasses. This drawing shows what is happening in only one plane. The field is three-dimensional. The photo shows iron filings lining up in the vicinity of a small bar magnet. By tradition, we say the field lines emerge from the north pole, curve around, and enter the south pole. But by just looking at this photo, we can't tell which pole is which.

TABLE 21.1 Magnetic fields

Source	Field (T)
Nucleus (at surface)	10^{12}
Neutron star (at surface)	$\approx 10^{8}$
Highest yet attained in laboratory	
explosive compression ($\approx 10^{-6}$ s)	1.5×10^{3}
pulsed coils ($\approx 10^{-3}$ s)	100
constant (dc, superconducting, 1993, MIT)	37.2
constant (dc, room temperature)	23.5
No acute effects on bacteria, mice, or fruit flies	14
Large laboratory electromagnet	5
Sunspot (within)	0.3
Human exposure limit (full body, dc, for minutes)	≈ 0.2
Small ceramic magnet (nearby)	≈ 0.02
Small bar magnet (near pole)	10^{-2}
Sun (at surface)	10^{-2}
Hair dryer (60 Hz, nearby)	$1 \times 10^{-3} - 2.5 \times 10^{-3}$
Can opener (60 Hz, nearby)	$0.5 \times 10^{-3} - 1 \times 10^{-3}$
Jupiter (at poles)	8×10^{-4}
Blender (60 Hz, nearby)	$10^{-4} - 0.5 \times 10^{-3}$
Earth (dc, at surface)	0.5×10^{-4}
Color TV (60 Hz, nearby)	10^{-4}
Transmission line (maximum under, 765 kV, 4 kA)	$\approx 0.5 \times 10^{-4}$
Toaster (60 Hz, nearby)	$0.1 \times 10^{-4} - 1 \times 10^{-4}$
Sunlight (rms)	3×10^{-6}
Refrigerator (60 Hz, nearby)	10^{-6}
Mercury (at surface of planet)	2×10^{-7}
Human body (produced by)	$\approx 310^{-10}$
Interstellar space	$\approx 10^{-10}$
Earth (50–60 Hz, at surface)	10^{-12}
Shielded region (smallest value measured)	1.6×10^{-14}

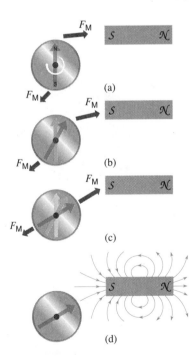

Figure 21.6 A compass needle mounted so that it can turn freely experiences a torque in the *B*-field of a magnet and swings around until that torque vanishes.

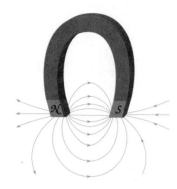

Figure 21.7 The horseshoe magnet concentrates its field in the immediate region between its poles.

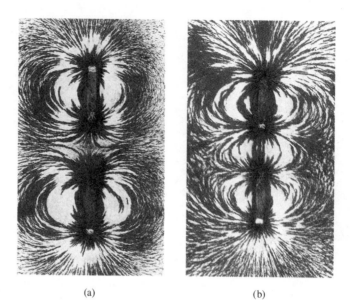

(a)   (b)

If we suppose that each iron filing is a compass needle, then the pattern they form reveals the magnet's *B*-field lines. In (a), the two magnets have like poles facing each other. In (b), unlike poles face each other. Of course, these views are in only one plane; the field is three-dimensional.

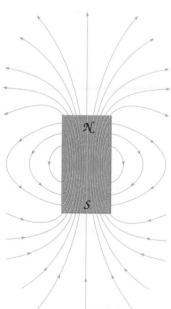

B-field

Figure 21.8 The magnetic field inside and outside of a permanent magnet. Electric field lines begin on positive charges and end on negative charges. Here there are no magnetic charges (monopoles), so magnetic field lines are closed. They must turn back on themselves. The field in three dimensions is like a system of concentric misshapen doughnuts.

extends out from the north pole and into the south pole. The north end of the test-compass is attracted to the south pole of the magnet responsible for *B* (and the south end is attracted to its north pole). As usual, *field lines never intersect, nor should they begin or end on anything but sources and sinks:* if there were monopoles clustered at both ends of a bar magnet, the field lines would begin and end there. But there are not, and **the lines of B form closed loops** (Fig. 21.8).

21.3 The Earth's Magnetism

The Chinese of the eleventh century were already aware that the compass needle does not align itself everywhere along a true north-south direction. That knowledge came to Europe at least as early as 1436, but it's still a surprise to some to learn that a magnetic compass rarely points true north. While a compass in Chicago does align itself almost due north, one in Los Angeles tilts about 15° to the east, and one in New York leans almost 15° west of north.

Gilbert was right, at least to the extent that the Earth behaves as if it contained, as its main field source, a dipole. It acts as if a relatively short bar magnet were embedded at its center (Fig. 21.9), tilted over at about 11.5° along the 70° W meridian. But matters are much more complicated than that.

The core of the planet is too hot to be totally solid and even too hot ($\approx$2500 K) for magnetic materials to remain magnetic—we are not dealing with a buried bar magnet. Besides that, the field changes in time, both in size and direction. There is also compelling evidence that the field has reversed itself some 300 times in the past 170 million years and may have done so just 30,000 years ago.

As a further complication, the *dip poles*, where the field points straight up and down, not only wander, but are 500 miles from the *geomagnetic poles* (the poles of Gilbert's magnet toward which a compass points). Moreover, neither the geomagnetic nor the dip poles are at the geographic poles, as fixed by the planet's spin axis. The final irony is linguistic: the Earth's northern magnetic pole is actually a *south* pole, and the southern magnetic pole is—you guessed it—a *north* pole. That's why the north-seeking pole of a compass needle points north—it's attracted toward the Earth's south magnetic pole.

21.4 Monopoles

Unfortunately, in 1785, Coulomb showed that a workable magnetic force law could be stated in an identical form to the electrostatic force law, provided one introduced the notion of magnetic charges, or *monopoles*. That theory was so reasonable it took more than 100 years before it was discredited. *Magnetic charges are not the source of magnetism.*

Even so, in 1931, P. A. M. Dirac presented a theoretical argument resurrecting the magnetic monopole. Besides establishing a symmetry between electric $(+, -)$ and magnetic (N, S) particles, the existence of the monopole would also provide a ready explanation for the quantization of charge. Today, there are competing versions of the so-called Grand Unification Theory (Ch. 33), which attempts to unify all the forces of nature. These formulations maintain that monopoles were created in high-energy collisions during the Big Bang that birthed the Universe. If they still survive at all, it's not likely there are many around and what few there are, are flying through space (certainly they're not clustered at the ends of magnets). The slowest, and by far the most massive, of all the elementary particles (weighing about as much as a bacterium), monopoles would be fascinating little creatures, each much smaller in volume than a proton. For one thing, they would exert an attraction between their opposite numbers almost 5000 times greater than the attraction that exists between the electron and proton.

At present, there are at least 35 major monopole hunts in progress worldwide. To date, not a single monopole has been spotted. But of course that incredible shyness may spring from the fact that monopoles are exceedingly scarce, or they just don't exist anymore, or perhaps they never did.

21.5 Magnetism on an Atomic Level

As we shall see, charge in motion produces magnetic force, and in particular a current moving in a circular path is magnetically identical to a dipole (p. 799). Moreover, electrons behave in ways that suggest they are perpetually spinning.* It follows that *the electron itself corresponds to a circulating charge and is the ultimate subatomic dipole magnet* (Fig. 21.10). The observed magnetic response of electrons is consistent with each having a purely dipole field down to a radius of 10^{-12} m. The orbitlike motion of an electron about the nucleus of an atom also constitutes a current and produces an additional magnetic field. Atomic nuclei can generate dipole fields too, but these are typically a thousand times weaker. Together, the two elec-

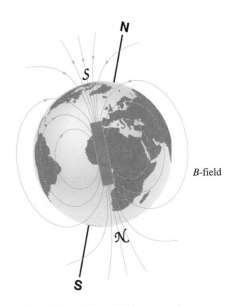

Figure 21.9 The Earth's magnetic influence resembles that of a tilted bar magnet. A compass needle aligns itself with the field and points roughly toward the north geographic pole, which is not far from the Earth's south magnetic pole. The field extends thousands of kilometers out into space and is rotationally symmetrical around the axis of the hypothetical bar magnet.

B-field

*Although physicists commonly talk about spinning subatomic particles in the same way they talk about spinning basketballs, in their heart of hearts they know better. Quantum Mechanics has made it clear that the old notion of spin needs a modern interpretation. By spin, we mean a fundamental quality (like mass and charge, whatever they are) that is associated with the manifestation of angular momentum, but not necessarily with the existence of spinning, with turning round and around. The latter picture produces relativistic inconsistencies and must be rejected. Thus, an electron has spin though it is not spinning in the usual sense of the word. Whatever it's doing, we are confident that an electron has an intrinsic angular momentum and an intrinsic magnetic dipole field.

Having made many and divers compasses . . . I found continuallie that after I had touched the yrons with the stone, that presentlie the north point thereof wouilde bend or decline downwards under the horizon in some quantitie.

ROBERT NORMAN
The Newe Attractive (1581)

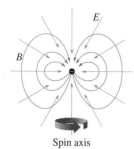

Spin axis

Figure 21.10 The electric and magnetic fields of an electron. The *E*-field is radially inward. The *B*-field forms a system of concentric dough-nut shapes about the spin axis. The magnetic field is that of a minute dipole.

I have seen Samothracian iron rings even leap up and at the same time iron filings move in a frenzy inside brass bowls, when this Magnesian mineral was placed beneath.

LUCRETIUS (First century B.C.)

The magnetic-domain structure of a polycrystalline sample of cobalt samarium. Each little domain has its own random magnetic field orientation. The sample as a whole is said to be unmagnetized.

tron mechanisms (spin and orbit) account for the magnetic behavior of all the various forms of bulk matter.

Usually, electrons in atoms come in oppositely moving pairs and their magnetic fields cancel—the vast majority of atoms and molecules have no net magnetic field. Still, no material exists that will not respond magnetically to an applied *B*-field. Most substances (water, glass, copper, lead, salt, rubber, diamond, wood, and so on), the majority of gases (such as nitrogen, carbon dioxide, and hydrogen), and millions of organic compounds (for example, plastics) are very weakly magnetic in what might at first seem a startling way. They are *repelled* by the pole of a strong magnet. Faraday was the first to observe this peculiar phenomenon (1845), and he called it **diamagnetism**. It's still amusing to swing a small piece of glass dangling on a thread into a field almost 100 000 times stronger than the Earth's, only to have the glass pop right out and stay out at some gravity-defying angle. Humans are mostly water and should be able to feel this repulsion. But that requires a tremendously strong *B*-field (you feel nothing even at 4 T, which is quite large), and very few people have actually experienced the repulsion.

Diamagnetism is associated with the orbital motions of atomic electrons. Turning on a *B*-field changes their angular momenta, and that added motion produces a field that opposes the applied field (p. 836). Accordingly, diamagnetism is present in all substances, although it is observable only when not swamped out by other much stronger effects.

When an atom has an odd number of electrons or a structure in which not all its electrons are paired, the atom will have a net magnetic dipole field of its own. In a substance made up of countless such atoms, these dipoles *en masse* produce only a feeble response because the ordinary thermal agitation of the atoms keeps them disoriented. Substances of this kind are called **paramagnetic** and they include the elements aluminum, oxygen, sodium, platinum, and uranium, among others. If placed near a powerful magnet, they will be drawn in toward a pole, but only weakly.

The last of the three major magnetic classes of substances (there are others that are less important) is known as **ferromagnetic**. This group, to which newly concocted materials are always being added, includes magnetite, a troop of alloys such as steel and Alnico, and a number of elements. Iron, cobalt, and nickel have been known for centuries to be "magnetic" at room temperature, and there are a half-dozen other elements that become ferromagnetic at low temperatures. On average, ferromagnetic materials have a greater number of unpaired, spin dipoles per atom. But more importantly, *these dipoles enter into large-scale cooperative alignments.* In other words, the uncompensated spin dipoles of each atom interact strongly with the dipoles of adjacent atoms, locking together in a parallel orientation that tends to persist even at room temperature.* Such substances are strongly attracted to the poles of a magnet and are themselves easily magnetized.

An interesting application arises when we place a thin layer of a ferromagnetic material on a plastic substrate. Typically, that's done with extremely fine particles of γ-Fe_2O_3. These needlelike grains can conveniently be magnetized in microscopic patterns, thereby "permanently" recording information. Of course, what you have then is magnetic recording tape, computer memory disks, etc. (See Discussion Question 8.)

Magnetic Domains

Ferromagnetic substances are composed of very many microscopic **domains**—islands of order—throughout each of which tremendous numbers of atomic spin dipoles

*That powerful coupling is a quantum-mechanical effect, first explained by Werner Heisenberg in 1928.

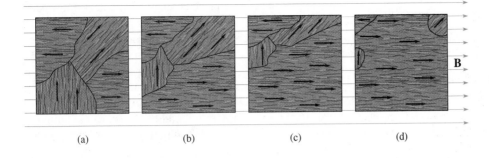

Figure 21.11 The growth of domains aligned in the direction of an applied magnetic field. As the applied B-field increases, the pattern goes from (a) to (b) to (c) to (d). Regions parallel to the field grow at the expense of those not parallel.

are aligned parallel to one another. Each domain is a tiny (roughly 5×10^{-5} m across) magnet that can be viewed under a microscope. In an unmagnetized specimen, the orientations of these domains are random, and their fields cancel (Fig. 21.11). That configuration is a compromise between total order of the spin dipoles and total disorder, and exists because it is energetically the most economical.

When a ferromagnetic sample is brought into a magnetic field, its domain structure can be drastically altered in two basic ways. The low-energy process (which occurs even with fairly weak fields) is one in which *the many domains that happen to be already aligned with the applied field grow at the expense of the domains that are misaligned at the start and subsequently shrink*. This is what occurs when soft iron is placed in a field and becomes magnetized (Fig. 21.12). This state with aligned domains and a resulting induced magnetic field (in which there is a good deal of energy) is unstable. Without support from outside, the induced field collapses and the iron spontaneously demagnetizes, rearranging domains.

The other magnetizing process, which requires a higher applied B-field, results in the *irreversible reorientation of the domains* (Fig. 21.13). All the electron dipoles coupled together within each domain can literally be rotated into alignment with the applied field, just like a compass needle. This mechanism prevails in substances that become permanently magnetized. Domains in materials with irregular internal structures, such as steel, cannot easily change shape. The domains rotate instead, and having once been forced to do so against a kind of internal friction, they tend to stay put. When a steel knife blade (some stainless steel is "nonmagnetic") is stroked with a strong magnet, the effect is to rotate the domains into that direction, to "comb" them into alignment.

Permanent magnets and ferromagnetic materials are used in VCRs, TVs, stereo headsets, automobiles, speakers, tape decks, motors, and telephones. They float in space in a thousand different satellites, are on the backs of millions of credit cards, and in the inks on countless dollar bills and personal checks. (Try holding a dollar bill near a powerful magnet.)

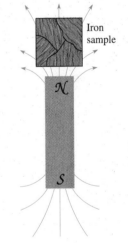

Figure 21.12 An iron sample placed in the magnetic field of a bar magnet. The domains of the sample are influenced, and they align with the applied field. The sample becomes magnetized and is drawn toward the bar magnet.

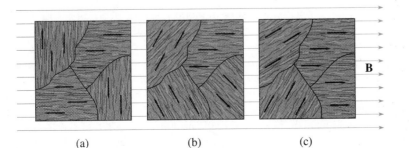

Figure 21.13 The reorientation of domains, bringing about an alignment with the applied field. As the applied magnetic field increases from (a) to (b) to (c), the electrons in the domains rotate so that their fields align with the applied field. The domains do not change shape as they did in Fig. 21.12. This realignment mechanism occurs in materials that become permanently magnetized.

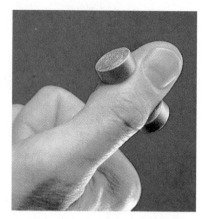

These two extremely powerful magnets are made of a new (1985) alloy of 14 parts iron, 2 parts neodymium, and 1 part boron. The alloy is usually powdered, the grains aligned in a strong B-field, and then the material is heated until it fuses.

Anything that reorients the domains of a piece of magnetized steel diminishes its supposedly permanent field. Banging on a magnet with a hammer will do just that, by knocking regions out of alignment—providing the domains with enough energy to unpin themselves. If the magnet's temperature is made high enough, the vibrating atoms will jostle themselves out of alignment and disrupt the long-range order. Pierre Curie realized that there is a limiting temperature, now called the **Curie temperature**, for each substance beyond which ferromagnetism vanishes and the material becomes paramagnetic. In 1894, Curie found that this critical temperature for iron is 770°C, though it had been known for a long time that red-hot iron has no magnetic power. The Curie temperature for magnetite is roughly 575°C, and for the other important iron ore, hematite, it's about 675°C. Interestingly, in all three cases, the Curie temperature is well below the melting point.

Permeability

The presence of a dielectric in an applied electric field has the effect of producing a weaker net internal E-field (p. 705). Similarly, when a diamagnetic material is placed in an applied magnetic field, it decreases the net B-field within the medium, but the effect is very small. On the other hand, the presence of a paramagnetic material will very slightly enhance the field within the medium. By contrast, on being immersed in a B-field, a ferromagnetic material becomes strongly magnetized. That, in turn, contributes an additional field component that adds to the original field and tends to cause the new net field to follow the contours of the metal (Fig. 21.14a). Lord Kelvin called this property **permeability**, observing that iron is hundreds of times more permeable than air. Special alloys that are millions of times more permeable are used to magnetically shield things such as delicate wristwatches and color TV picture tubes.

Just the opposite effect occurs in superconductors, which are both perfectly conducting and perfectly diamagnetic. Below a certain applied field strength, a superconductor will maintain itself in a state where, internally, $\mathbf{B} = 0$. When the sample (immersed in a magnetic field) is cooled below its critical temperature, it becomes superconducting. As in Fig. 21.14c, it then almost completely expels the B-field from its interior and is therefore said to be *perfectly diamagnetic*. To be precise, an external field does penetrate a superconductor, but only in a very thin surface layer 10^{-5} cm to 10^{-6} cm thick. Only if the applied field is made to exceed the so-called *critical field* will it again penetrate the body of the specimen. The diamagnetic behavior of superconductors was discovered experimentally in 1933 and is known as the **Meissner Effect**.

ELECTRODYNAMICS

A charged particle, whether at rest or in motion, has an electric field $\mathbf{E}$, which is not much more than saying that charges interact electrically. Now, suppose that the particular charged particle has no intrinsic magnetism of its own. Nonetheless, when such a charge moves in space, it exerts magnetic forces and possesses a magnetic field. This *magnetism arises out of motion, and motion is relative*. A beam of protons is a current I and as such exerts both electric and magnetic forces. And yet, if we run along with the flow, essentially causing I to become zero, the magnetic force vanishes. It must vanish, because there is no longer a source of the B-field.

The Special Theory of Relativity provides the realization that current-generated magnetism is a facet of electricity. A modification of the electrical interaction arising from relative motion appears as a force, the magnetic interaction. On a theoretical level well beyond our needs, Coulomb's Law can be modified to include the effects of relative motion. Though that modification is typically quite small, its consequences are far-reaching, and we call them *magnetism*. Indeed, if the speed of light were infinite, currents would not generate magnetic fields. Electricity and magnetism are the two sides of a single phenomenon—*electromagnetism*—that looks different to observers in relative motion.

The study of the electromagnetic interaction in all its manifestations is known as **electrodynamics**, a word prophetically coined by Ampère, who initiated the unification over 150 years ago when he linked the source of magnetism to currents.

21.6 Currents and Fields

Soon after Volta's invention of the battery, there were renewed attempts to find some relationship between electricity and magnetism. Evidence that a connection existed had been at hand for decades. The 1735 volume of the *Philosophical Transactions of the Royal Society of London* carried a paper entitled "Of an Extraordinary Effect of Lightning in Communicating Magnetism." It was the report of a bolt of lightning that struck a tradesman's house, blasting apart a box full of knives and forks, hurling them "all over the room . . . but what was most remarkable" was that they were all strongly magnetized afterwards!

On July 21, 1820, Hans Oersted, professor of physics at Copenhagen University, delivered a lecture on electricity to some advanced students. By chance, a wire leading to a voltaic pile was nearly parallel to and above a compass that happened to be on the table along with other paraphernalia. When the circuit was closed, the needle swung around almost perpendicular to the current-carrying wire as if gripped by a powerful magnet (see photo below).

The news of Oersted's discovery reached Paris on September 4, when Dominique F. J. Arago reported it to a skeptical gathering of the Paris Academy of Sciences. A young professor, André Marie Ampère attended the talk, and within two weeks he completed a series of experiments of his own. Ampère showed that the magnetic force experienced by a compass needle in the vicinity of a current-

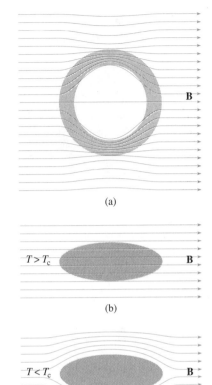

Figure 21.14 (a) When a material with a high permeability is placed in a *B*-field, most of the field lines pass through the material, effectively shielding the region it surrounds. (b) By contrast, a substance (such as lead) in its nonsuperconducting or normal state allows the *B*-field to pass through it. (c) When the temperature drops and the material becomes superconducting, it completely expels the applied *B*-field via the Meissner Effect.

(a)   (b)

Oersted's demonstration. (a) With no current in the wire, the compass needle points north. (b) When a current exists, the needle swings so that it almost aligns with the new field created by the current. The Earth's field causes a small northerly deflection of the needle.

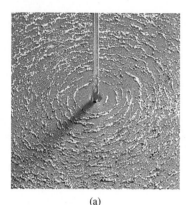

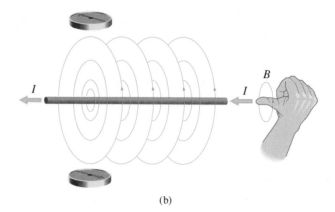

(a)

(b)

Figure 21.15 (a) The circular magnetic field surrounding a current-carrying wire, as revealed in an iron-filing pattern. (b) When the thumb on the right hand points in the direction of the current (I), the fingers curl in the direction of the circular magnetic field (B).

carrying straight wire acted at right angles to the current along a series of concentric circles (Fig. 21.15). *Point the thumb of the right hand in the conventional direction of the current and the direction in which a compass will point, the direction of the B-field, anywhere in the surroundings, is given by the direction of the fingers curling around the wire at that location.* (Call this the **Right-Hand-Current Rule**.) That rather surprising perpendicular character of the force explains why it took so long to discover.

The Current-Carrying Wire

A straight current-carrying wire generates a circular, or more accurately, a cylindrical magnetic field in the space surrounding it. *That field is constant at any given perpendicular distance from the wire and gets weaker as that distance increases.* Experiments, principally by Jean-Baptiste Biot and Félix Savart (1820), established that the B-field near a long straight wire in air is directly proportional to the current I and inversely proportional to the perpendicular distance r from the wire: $B \propto I/r$. The next logical step is to introduce a constant of proportionality that balances the units, yielding teslas on both sides, and thereby produce an equality. Since the SI system is "rationalized," the constant is defined so that $1/2\pi$ shows up in the equations for fields when there is axial symmetry, as there is here (that is, a cross section of the field over a plane perpendicular to the symmetry axis is the same wherever the plane is located). Furthermore, it was found that the B-field depends on the magnetic behavior of the medium in which the wire is immersed. Hence, we follow tradition and introduce the constant, $\mu/2\pi$. The Greek letter mu (μ) represents the medium-dependent constant that, naturally enough, is called the **permeability**.*

The magnetic field at any point outside a long straight current-carrying wire (but not so far from it that distortions due to the ends of the wire show up) is then

$$B = \frac{\mu}{2\pi} \frac{I}{r} \tag{21.1}$$

In vacuum, the value of the permeability is, by definition

$$\mu_0 = 4\pi \times 10^{-7} \text{ T·m/A}$$

Hans Christian Oersted (1777–1851). Besides his work in electricity and magnetism, Oersted was the first to prepare pure metallic aluminum (1825).

*The constant $\mu/2\pi$ happens to have the μ on top because the permeability was originally defined so that Coulombs's Law for magnetic charge (no longer of any interest) would have a factor of $1/4\pi\mu$ in front to match the $1/4\pi\varepsilon$ in front of Coulomb's Law for electrical charge.

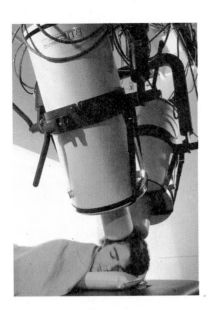

Today, superconducting quantum interference devices (SQUIDS) are being used to measure the minute ($\approx 10^{-13}$ T) magnetic fields generated by currents in the brain and heart. The magnetoencephalograph can locate the source of nerve signals within the brain to within a few milimeters.

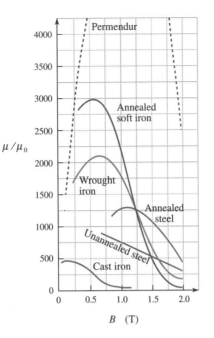

Figure 21.16 The permeability of magnetic materials is not constant, but instead changes as the field affects the material. Notice how the permeability initially rises, peaks, and then falls off as the material reaches saturation. Permendur, which is half iron and half cobalt, keeps a high μ even at large values of B.

An expression related to Eq. (21.1) will be used later to define the unit of current, and that's where this value of μ_0 comes from. It might be helpful to think of μ_0 as a scale factor for the units rather than a reflection of some physical characteristic of free space. Equation (21.1) works fine provided μ is constant, which it is for diamagnetic and paramagnetic media. It is not constant for ferromagnetic media, where μ changes as the B-field changes (Fig. 21.16), and Eq. (21.1) should then be used with care. For ferromagnetic media, μ is usually much larger than μ_0. Typically, in media that are likely to be of concern to us such as air or water, $\mu \approx \mu_0$. For diamagnetic media, $\mu/\mu_0 < 1$, whereas for paramagnetic media $\mu/\mu_0 > 1$, but in both cases $\mu/\mu_0 \approx 1$. The differences are small. In air $\mu_{air}/\mu_0 = 1 + 3.6 \times 10^{-7}$; whereas $\mu_{water}/\mu_0 = 1 - 0.88 \times 10^{-5}$. Thus, we can rewrite Eq. (21.1), **the magnetic field of a long straight wire in vacuum**, as

[*long straight wire*]
$$B = \frac{\mu_0}{2\pi}\frac{I}{r}$$
(21.2)

A. M. Ampère (1775–1836). Legend has it that Ampère was the classic absent-minded professor, who once even forgot to attend a dinner with the Emperor Napoleon.

This equation provides the strength of the field (Fig. 21.17) of any line of current—a straight beam of protons transporting I coulombs per second through space will produce a B-field given by Eq. (21.2).

Example 21.1   The overhead power cable for a street trolley is strung horizontally 10 m above the street. A long straight section of it carries 100 amps dc due east. Describe the magnetic field produced by the current and determine its value at ground level just under the wire. Compare that to the strength of the Earth's field.

Solution: [Given: $r = 10$ m and $I = 100$ A. Find: B.] First, draw a diagram. Fig. 21.15 will suffice here with the current assumed heading east. At ground level (at a

point beneath the easterly current), the Right-Hand-Current Rule tells us that **B** points due north. Using Eq. (21.2)

$$B = \frac{(4\pi \times 10^{-7} \text{ T·m/A})(100 \text{ A})}{2\pi(10 \text{ m})} = \boxed{2.0 \times 10^{-6} \text{ T}}$$

which, from Table 21.1, is only 4% of the Earth's field.

▶ **Quick Check:** $\mu_0 = 1.2566 \times 10^{-6}$ T·m/A, $B \approx (10^{-6} \text{ T·m/A})(10^2 \text{ A})/60 \text{ m} \approx 2 \times 10^{-6}$ T.

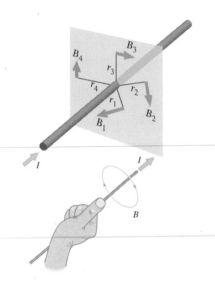

Figure 21.17 The B-field surrounding a current-carrying straight wire lies in a plane perpendicular to the wire. At any point on that plane, the B-field is perpendicular to the line from the wire to the point, and decreases inversely with the distance from the wire to the point.

The Current Loop

During those first weeks of excitement, Ampère had another lovely idea. Since a straight current-carrying wire is surrounded by concentric rings of magnetic force, bending the conductor into a loop should concentrate that force (as does bending a bar magnet into a horseshoe magnet). Using the field imagery, we can imagine the lines of B that extend far out into space on one side of a straight wire, being carried around and crowded within the small region then encompassed by the loop (Fig. 21.18). The field inside the loop is much stronger than the field outside. Again the Right-Hand-Current Rule provides the direction of B. The field will effectively vanish if the loop is squashed such that the current doubles back on itself—oppositely directed fields then overlap and cancel. The current-circle brings to mind the picture of a dipole field. Recognizing this similarity, Ampère boldly suggested that currents were the basic cause of magnetism (and to the degree that the intrinsic magnetism of an electron is due to the motion of its charge, he was right).

Biot and Savart determined experimentally that the field at the very center of a current loop points axially outward (Fig. 21.19) along the z-axis. They found B to be directly proportional to I and inversely proportional to the radius R of the loop:

[*circular loop, center*] $$B_z = \frac{\mu_0 I}{2R} \qquad\qquad (21.3)$$

In this case, the field's cross section over a plane perpendicular to the central z-axis is different at different values of z, and so the factor of 2π is not present.

Stacking several loops in parallel results in the overlapping of their individual fields, providing a proportionately increased net effect. Each loop simply adds (vectorially) its field to the fields of all the others. Thus, a tight short coil (Fig. 21.19) composed of N closely wrapped turns of wire each carrying a current I has a field at its center of

[*circular coil, center*] $$B_z = N\frac{\mu_0 I}{2R} \qquad\qquad (21.4)$$

This sort of coil, with a negligible length compared to its diameter, resembles a stubby disk magnet. If the coil is delicately pivoted so it can rotate freely about a

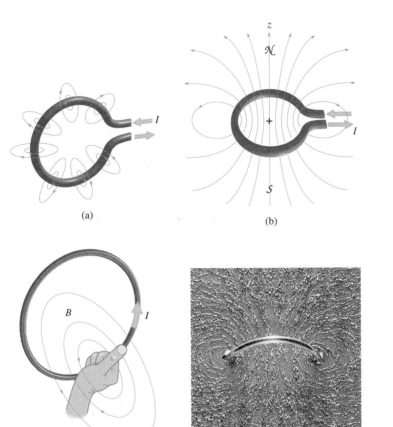

(a)

(b)

(c)

(d)

Figure 21.18 (a) Each segment of a current-carrying loop is surrounded by a circular *B*-field. (b) These combine to produce a dipole field very much like the field of a bar magnet. Remember that the field is 3-dimensional and more or less axially symmetrical around the central *z*-axis. (c) The Right-Hand-Current Rule gives the direction of **B**. (d) The field pattern, as revealed with iron filings, in a plane perpendicular to the loop.

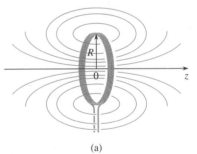

(a)

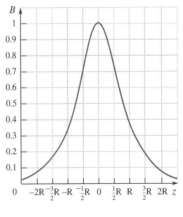

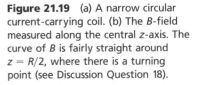

(b)

Figure 21.19 (a) A narrow circular current-carrying coil. (b) The *B*-field measured along the central *z*-axis. The curve of *B* is fairly straight around $z = R/2$, where there is a turning point (see Discussion Question 18).

diameter, it will swing into alignment with an applied field just as a compass would. Ampère observed as much in 1820. This twisting behavior is the basis of both the moving coil galvanometer and the electric motor (p. 815).

The Solenoid

Carrying the coil concept one step further, Ampère wound wire into a long helix (Fig. 21.20) or *solenoid* (from the Greek, *solen* meaning "tube") and found that, with a current passing through it, it acted like a bar magnet. Within the space encompassed by a long, narrow (at least 10 times longer than it is wide), tightly wound solenoid, the *B*-field is strong and quite uniform, especially in the middle and around the central *z*-axis. The solenoid is one of the most useful magnetic devices — a typical home has dozens of them operating bells, chimes, and speakers. The solenoid is the central component of the relay that mechanically controls equipment such as washing machines, dishwashers, clothes driers, and furnaces. As a circuit element, the solenoid is in radios, TVs, and computers.

A solenoid is helical and not quite the same as stacking a bunch of separate flat loops; here, the current progresses from one end of the coil to the other and that adds a small additional contribution to *B* (p. 804). On the other hand, if the solenoid is wound with overlapping turns, an even number of layers will bring the end wire back to the beginning and cancel the longitudinal current.

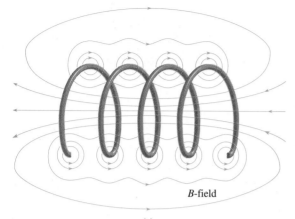

(a)

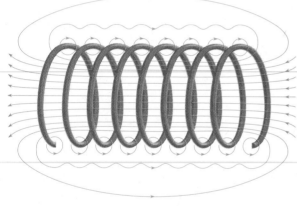

(b)

Figure 21.20 The solenoid. (a) The magnetic field of a loosely wound current-carrying coil. (b) When the coil is wound tighter and there are more loops, the field inside becomes larger and more uniform. The Right-Hand-Current Rule provides the direction of **B**.

Figure 21.21 (a) The *B*-field of a finite solenoid carrying a current *I*. (b) The value of the field at any point *P* on the central *z*-axis is called B_z, and it varies with θ_l and θ_r. At the very center of the coil ($z = 0$), the field is a maximum equal to $\mu_0 nI$.

It's reasonable to assume that the field inside a solenoid increases directly with *I*. Moreover, the field should increase with the number of current loops contributing, but here there's the additional concern of packing the loops as close together as possible. Carrying the same current, 100 turns spread over a length of a meter will produce a much weaker inside field than 100 turns distributed over a centimeter. What's needed is a measure of the density of the winding, namely the *number of turns per unit length* (*n*) of solenoid. Hence, if *L* is the length of the coil and *N* the total number of turns, $n = N/L$. Then $B \propto nI$, and it's found experimentally (Fig. 21.21) that in air

[*solenoid, center,
axial field*]

$$B_z \approx \mu_0 nI \qquad (21.5)$$

The field at any axial point P is actually given by

$$B_z = \tfrac{1}{2}\mu_0 nI(\cos\theta_r - \cos\theta_1)$$

where θ_r and θ_1 are the angles made with the right and left edges of the coil. As $\theta_r \to 0$ and $\theta_1 \to 180°$, as they would for an infinitely long coil, the field approaches that of Eq. (21.5). For a coil 10 times longer than its diameter, Eq. (21.5) yields results that are only about 0.5% too large, which is good enough for our purposes.

If such a coil is imagined cut across the middle into two equal-length solenoids carrying the same current as the original, we can expect that the fields at their ends (which add up to $\mu_0 nI$ in the middle of the uncut coil) should be

[solenoid, ends] $B_z \approx \tfrac{1}{2}\mu_0 nI$

About half the field "leaks out" from between the windings of a real solenoid.

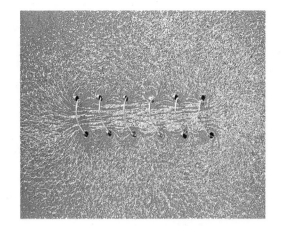

The magnetic field lines produced by a current-carrying coil. The pattern is again formed using iron filings.

Example 21.2 A 20-cm-long solenoid with a 2.0-cm inside diameter is tightly wound on a hollow quartz cylinder. There are several layers with a total of 20×10^3 turns per meter of a niobium-tin wire. The device is cooled below its critical temperature and becomes superconducting. Since the wire is then without resistance, it can easily carry 30 A and not develop any I^2R losses. Compute the approximate field inside the solenoid near the middle. What is its value at either end?

Solution: [Given: a solenoid where $n = 20 \times 10^3$ m^{-1}, and $I = 30$ A. Find: B_z.] The solenoid is long and narrow and will obey the approximations that lead to Eq. (21.5), thus

$$B_z \approx \mu_0 nI = (1.257 \times 10^{-6}\ \text{T·m/A})$$
$$\times (20 \times 10^3\ \text{m}^{-1})(30\ \text{A})$$

and $\boxed{B_z \approx 0.75\ \text{T}}$

which is a formidable field, over 10^4 times that of the Earth. The field at either end is about half this, $\boxed{0.38\ \text{T}}$.

▶ **Quick Check:** $B_z \approx \mu_0 nI \approx (10^{-6}\ \text{T·m/A}) \times (2 \times 10^4\ \text{m}^{-1})(3 \times 10^1\ \text{A}) \approx 0.6$ T.

Sometime around 1825, W. Sturgeon wrapped 18 turns of bare wire around a varnished iron bar and sent a current through the coil; in so doing, he created the first powerful *electromagnet*. The field set up by the current aligned the domains within the iron to produce a combined magnetic field of unprecedented strength (Fig. 21.22). The American physicist Joseph Henry heard about the feat and set about to better it. Legend has it that he tore apart his wife's petticoats so that he could insulate his wires with the silk stripping. By using many turns of insulated wire, Henry enhanced the field while keeping I relatively low. In 1831, he produced a modest-sized device powered by an ordinary battery that could lift more than a ton of iron. When the current was interrupted, the soft iron core almost completely demagnetized spontaneously, and the load was released.

21.7 Confirming Ampère's Hypothesis

By the beginning of the twentieth century, Ampère's hypothesis—that all magnetism is due to currents—was widely held, though it certainly had not been con-

Talking about Ampère. He further deduced from this analogy the consequence that the attractive and repulsive properties of magnets depend on electric currents which circulate about the molecules of iron and steel.

D. F. J. ARAGO (1820)
French physicist

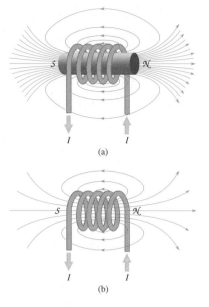

Figure 21.22   The field of a current-carrying coil (a) with and (b) without a ferromagnetic core. The *B*-field set up by the current in the coil magnetizes the core, aligning its domains, and that increases the field even more.

Steel nails align themselves to reveal the *B*-field configuration surrounding the opposite poles of two powerful superconducting magnets. Frost around the necks of the stainless steel Dewar flasks is a clue to the fact that they're operating at liquid helium temperature (around 4.2 K).

Joseph Henry (1797–1878). To honor him for his many original contributions, the International Electrical Congress of 1893 named the unit of inductance the henry.

firmed. And that was despite the fact that decades earlier, the great physicist James Clerk Maxwell had proposed several experimental ways of resolving the issue. One of his suggestions was to see whether or not a magnet behaved like a gyroscope. If the magnetism of a bar magnet is due to hidden circulating currents, these whirling charges should have angular momentum and display gyroscopic behavior.

In 1909, S. J. Barnett demonstrated the inverse effect: simply *rotating a ferromagnetic specimen magnetizes it* precisely as if it had been placed in a uniform external *B*-field directed along the axis of rotation. Barnett assumed there were microscopic current whirls initially at random orientations throughout the sample. (This was decades before the introduction of the notion of the intrinsic spin of the electron, and ferromagnetism was understood in terms of atomic current loops.)

To model the arrangement, imagine several small gyroscopes (mounted so they can swivel in any direction) attached to the horizontal platform of a centrifuge. With the little gyros spinning in random directions, the centrifuge is turned on and rotates about a vertical axis. All of the gyros now revolving with the platform will align themselves vertically, having the same sense of rotation as the centrifuge. This is quite similar to the way a gyrocompass aligns itself with the planet's spin axis. In Barnett's demonstration, the rotation of the sample as a whole caused the tiny current loops to swing around into alignment with the bulk spin axis. That alignment was subsequently revealed by the magnetization of the specimen. The argument applies equally well to spinning electrons, which is the way it is interpreted today.

Maxwell's idea was finally confirmed in 1915, by no less a personage than Albert Einstein, who worked in conjunction with W. J. de Haas. Their experiment verified that the magnetization of a sample causes it to rotate. Since this work was done before the appreciation of electron spin, their theoretical arguments were somewhat in error, and their numerical result turned out to be off by a factor of 2, but that didn't seem to bother Einstein. In the experiment, a ferromagnetic cylinder was hung vertically inside a solenoid whose *B*-field was periodically reversed. Einstein and de Haas explained that magnetizing the sample aligns current whirls, which, in turn, must align their angular momentum vectors, thus increasing the overall internal angular momentum of the sample. But the Law of Conservation of Angular Momentum demands that the net change in **L** must be zero; the cylinder as a whole must therefore rotate in the opposite direction. And rotate it did. "We have given firm proof," wrote Einstein, "of Ampère's molecular currents . . . ."

The speaker of the radio shown on p. 645. You can see the permanent magnetic housing at the rear of the speaker. The paper cone is attached to a current-carrying coil that causes the cone to vibrate. (See Fig. Q9, p. 821.)

21.8 Calculating *B*-Fields

There are two equivalent schemes for calculating magnetic fields due to currents, one devised by Laplace based on the data of Biot and Savart, the other by Ampère. We will first study *Ampère's Law,* which is simpler to use, provided the geometry is simple. We cannot *derive* the law from basics (that is, from the properties of the electron)—physicists simply don't know enough yet to do that. It can, however, be *deduced* from experimental observations. In its general nature Ampère's Law relates the net current and the associated *B*-field, much as Gauss's Law relates the net charge and the associated *E*-field. Like Gauss's Law, Ampère's Law will supply us with a wide range of useful results that are easily computed provided we stick to situations where symmetry can be exploited. By contrast, the Biot-Savart Law makes use of infinitesimal current elements to calculate B, just as Coulomb's Law (in the form $dE = k\,dq/r^2$) uses infinitesimal charge elements to calculate E.

Ampère's Law

Ampère's Law is a little obscure physically—it will take a bit of doing to justify it, but it's worth it, as we'll see in future chapters. Accordingly, imagine a straight current-carrying wire and the circular *B*-field surrounding it. We know from experiments that $B = \mu_0 I/2\pi r$, which is Eq. (21.2). Now, suppose we put ourselves back in time to the nineteenth century when it was common to think of magnetic charge (q_m). Let's define this monopole charge so that it experiences a force when placed in a magnetic field B equal to $q_m B$ in the direction of B, just as an electric charge q_e experiences a force $q_e E$. Suppose we carry this north-seeking monopole around a closed circular path perpendicular to and centered on a current-carrying wire and determine the work done in the process. Since the direction of the force changes, because **B** changes direction, we will have to divide the circular path into tiny segments (Δl) and sum up the work done over each. Work is the component of the force parallel to the displacement times the displacement: $\Delta W = q_m B_\parallel \Delta l$, and the total work done by the field is $\Sigma\, q_m B_\parallel \Delta l$. In this case, **B** is everywhere tangent to the path, so that $B_\parallel = B = \mu_0 I/2\pi r$, which is constant around the circle. With both q_m and B constant, the summation becomes

$$q_m \Sigma\, B_\parallel \Delta l = q_m B \Sigma\, \Delta l = q_m B 2\pi r$$

where $\Sigma\, \Delta l = 2\pi r$ is the circumference of the circular path.

If we substitute for B the equivalent current expression, which varies inversely with r, the radius cancels—the work is independent of the circular path taken. Since no work is done in traveling perpendicular to **B**, the work must be the same if we move along a radius from one circular segment to another as we go around. Indeed, W is independent of path altogether—the work will be the same for any *closed path* encompassing the current. Putting in the current expression for B and canceling the

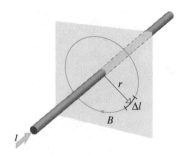

Figure 21.23 Using Ampère's Law. The circle of radius r surrounding the current-carrying wire is the Ampèrian path we have chosen. The B-field is everywhere parallel to the path elements Δl.

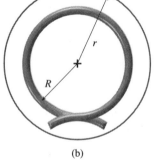

(a)

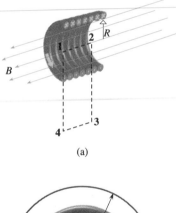

(b)

Figure 21.24 Using Ampère's Law. A helical coil carrying a current (out of the page and to the right at the bottom and into it and to the left at the top). (a) To find the field inside the solenoid, use path 1–2–3–4. (b) To find the field outside, use a circle of radius r.

"charge" we get the rather remarkable expression

$$\Sigma B_{\parallel} \Delta l = \mu_0 I$$

which is to be summed over any closed path surrounding I. The magnetic charge has disappeared, which is nice, since we no longer expect to be able to perform this little thought experiment with a monopole. Still, the physics was consistent, and the equation should hold, monopoles or no. Moreover, if there are several current-carrying wires encompassed by the closed path, their fields will superimpose and add, yielding a net field. The equation is true for the separate fields and must be true as well for the net field. Hence

$$\Sigma B_{\parallel} \Delta l = \mu_0 \Sigma I$$

As $\Delta l \to 0$ the sum becomes an integral *around a closed path*:

$$\oint \mathbf{B} \cdot d\mathbf{l} = \mu_0 \Sigma I \qquad (21.6)$$

Today this equation is known as **Ampère's Law**, though at one time it was commonly referred to as the "work rule."

Ampère's Law can be used to compute B generated by some pattern of currents (without actually having to rely on calculus) when the configuration of the field is known and is not very complicated. For example, consider the simplest case of a straight wire carrying a current I. *Wherever possible, we want **B** either perpendicular ($B_{\parallel} = 0$) or parallel ($B_{\parallel} = B$) to the encompassing path.* The field here (Fig. 21.23) is known to be circular, so we choose a circular Ampèrian path, around which $B_{\parallel} = B$. Hence, $B \Sigma \Delta l = B 2\pi r$ and because the net current is I,

[*long straight wire*] $$B = \frac{\mu_0 I}{2\pi r} \qquad [21.2]$$

As another and last example, let's find the field inside a very long solenoid (Fig. 21.24). Its asymmetry requires a bit of tricky maneuvering. As an Ampèrian loop that follows the B-field, with segments either parallel or perpendicular to it, we construct the path 1–2–3–4–1. The summation $\Sigma B_{\parallel} \Delta l$ is carried out over the four straight segments. Since the field lines close on themselves and since the system is symmetrical, there cannot be a radial component of the field. Thus B must be perpendicular to segments 2–3 and 4–1, and they make no contribution. Even if there were a radial field, in going around the loop, the contribution along 2–3 would be equal and opposite to that from 4–1, and they would cancel. Now for segment 3–4: the field outside an infinitely long solenoid is zero, and the field outside a finite but long solenoid must be small, axial, and drop off with r. In any event, we take 3–4 so far from the solenoid that the field there is negligible and that segment makes no appreciable contribution. What remains is segment 1–2, over whose length L the field is parallel (that is, $B_{\parallel} = B_z$). Thus, if the number of turns of wire encompassed by the path is N, the net current is NI, and Ampère's Law yields

$$B_z L = \mu_0 N I$$

but N/L is the number of turns per unit length n and so

[*solenoid, center, axial field*] $$B_z = \mu_0 n I \qquad [21.5]$$

Note that the field inside is independent of its distance from the central z-axis and must be uniform. It's also independent of the cross-sectional shape of the coil.

If the solenoid is wound in a single helical layer, charge is transported from one end to the other, and there is a net current I in the z-direction that produces a circular external field. Applying Ampère's Law using the path shown in Fig. 21.24b, we find that the field has a component B_ϕ around the solenoid, given by

[*solenoid, outside*] $$B_\phi = \mu_0 I/2\pi r$$

This is the field of a straight wire, and it's very small compared to the axial field inside the solenoid, where n might be 10^4 turns per meter. (See Discussion Question 11.)

The Biot-Savart Law

J. B. Biot and his young colleague F. Savart announced the results of their experiment less than two months after Arago reported Oerstead's discovery. They had very carefully and methodically measured the forces exerted on a tiny magnet everywhere in the vicinity of a straight current-carrying wire. This they accomplished by determining the frequency of oscillation of the magnet, which was delicately suspended on a silk cocoon thread. Their result was essentially Eq. (21.1), and that was a splendid piece of work even though their theoretical interpretation of it was, alas, uninspired. Nonetheless the great physicist, astronomer, and mathematician Pierre-Simon Laplace took that observation and showed that the force on the magnet due to each current element varied as the inverse-square of their separation.

Taking guidance from Coulomb's Law one might well ask, "What is the magnetic equivalent of a charge element dq that could be integrated over a circuit to yield B?" Charges in motion produce magnetic fields, and we might guess that a tiny contribution to the field, dB, produced by a minute clump of charge moving along some path (l) is proportional to both the amount of charge, dq, and its speed $v = dl/dt$. Thus $dB \propto dq\,(dl/dt)$, or if the current, dq/dt, is constant $dB \propto I\,dl$. The little energized circuit fragment $I\,dl$, which has the units of A·m, is rather loosely called a **current element**. Moreover, we can define the vector $d\mathbf{l}$ to be along the path in the direction of the current (Fig. 21.25), in which case the field $d\mathbf{B}$ at any point P, due to the current element $I\,d\mathbf{l}$, is found to be in the direction of the vector cross product, $d\mathbf{l} \times \mathbf{r}$. The field contribution is always perpendicular to the plane defined by $d\mathbf{l}$ and $\mathbf{r}$. **All B-field lines close on themselves and encircle the current that is their source.**

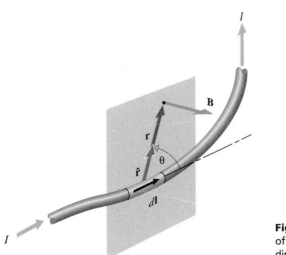

Figure 21.25 The magnetic field of a current element $I\,d\mathbf{l}$ is in the direction of dl × r.

To make things look a little neater it's customary to introduce a ***unit vector*** in the direction of **r** (such a vector is, in effect, pure direction; it has a magnitude equal to one and no physical units). Accordingly, define the unit position vector that points in the direction from $d\mathbf{l}$ to P, as $\mathbf{r}/r \equiv \hat{\mathbf{r}}$ (the cap distinguishes it from an ordinary vector). We can now write a more inclusive vector equation beginning with the observation that $d\mathbf{B} \propto I d\mathbf{l} \times \hat{\mathbf{r}}$ (p. 207). The cross product provides the direction of $d\mathbf{B}$, which is known from experiments to be correct, but it also implies that $dB \propto \sin\theta$, where θ is the angle between $d\mathbf{l}$ and $\hat{\mathbf{r}}$, and that is one of the insights Laplace deduced. The closer $\hat{\mathbf{r}}$ approaches to the direction of $d\mathbf{l}$, the smaller is θ and the smaller is the **B**-field contribution.

The next issue to settle is the dependence of the field contribution, dB, on the distance from $I d l$ to P. It was natural for Laplace to suspect that dB varied as $1/r^2$; after all, the Law of Universal Gravitation and Coulomb's Law were both inverse-square laws.

The last missing item is a constant of proportionality that will take care of the units on both sides of the equation. Here we jump to the contemporary formulation in terms of the permeability. Putting it all together, a modern-day expression of Laplace's conclusion is

$$d\mathbf{B} = \frac{\mu_0}{4\pi} \frac{I d\mathbf{l} \times \hat{\mathbf{r}}}{r^2} \tag{21.7}$$

or in scalar form

$$dB = \frac{\mu_0}{4\pi} \frac{I dl \sin\theta}{r^2} \tag{21.8}$$

either of which has come to be known as the **Biot-Savart Law** (pronounced *bee'yo-s'vahr*).

The whole point of this exercise is to be able to compute **B**-fields, not for a mathematically idealized current element, but for real macroscopic currents, usually flowing in some sort of closed circuit. Consequently, relying on the fact that the individual differential field contributions can be added to one another, we have

$$B = \int \frac{\mu_0}{4\pi} \frac{I dl \sin\theta}{r^2} \tag{21.9}$$

As always, we will have to get around the inherent vector nature of the problem by focusing on symmetrical situations. As a consequence, the scalar integral will do for our purposes.

The Biot-Savart Law doesn't tell us anything new about the nature of the field or its origins. Indeed, because it relies on the notion of a current element, which has no separate steady-state reality of its own, we might even suspect that we are dealing with the manifestations of something far more subtle. This then is one of those practical relationships that continues to derive its reason to exist from the simple fact that it works. When we finish calculating **B**-fields for a number of current arrangements, we'll be finished with the Biot-Savart Law; not so for Ampère's Law, which is more fundamental and will later illuminate the nature of electromagnetic waves.

Example 21.3   Use the Biot-Savart Law to confirm that the *B*-field near the middle of an *essentially* infinitely long straight wire, in air, carrying a constant current *I*, is given by $B = \mu_0 I / 2\pi r$.

Solution: [Given: straight wire and *I*. Find: *B* nearby.] Notice in Fig. 21.26 that point *P* is a constant perpendicular distance *x* from the wire. Moreover, every contribution ($d\mathbf{B}$) from any current element $I d\mathbf{l}$ is in the same direction at *P* and all can be added algebraically. That means that the scalar form of the Biot-Savart Law will suffice. Since $dl = dy$

$$B = \frac{\mu_0}{4\pi} I \int \frac{dy \sin \theta}{r^2}$$

There are two variables θ and *y*, and they are not independent of each other. Consequently, we will have to express one of them in terms of the other. Below the axis *y* is negative and $\theta < 90°$, whereas above the axis *y* is positive and $\theta > 90°$. It follows that $y = -x/\tan\theta = -x \, \text{ctn}\, \theta$, in which case $dy = -x \, d(\text{ctn}\, \theta) = -x(-\csc^2\theta)\, d\theta = x \, d\theta/\sin^2\theta$. Furthermore, $\sin \theta = x/r$ and the integral becomes

$$B = \frac{\mu_0}{4\pi} I \int \frac{x \, d\theta}{r^2 \sin \theta} = \frac{\mu_0}{4\pi} I \int \frac{x \, d\theta}{\left(\dfrac{x}{\sin \theta}\right)^2 \sin \theta}$$

and so

$$B = \frac{\mu_0}{4\pi} \frac{I}{x} \int_0^\pi \sin \theta \, d\theta$$

Notice that the limits go from the extremes of $\theta = 0$ far below the origin to $\theta = \pi$ far above the origin

$$B = \frac{\mu_0}{4\pi} \frac{I}{x}[-\cos \theta]_0^\pi = \frac{\mu_0}{2\pi} \frac{I}{x}$$

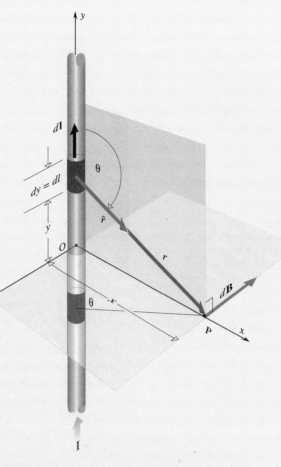

Figure 21.26   The *B*-field contribution from a current element on a long straight wire. The vector **dB** lies in the xz-plane in the direction fixed by *I* d**l** × **r**.

▶ **Quick Check:** This is identical to Eq. (21.2) except that we now use *x* instead of *r* for the distance.

MAGNETIC FORCE

Oersted demonstrated that a current exerted a force on a compass needle. To establish that this was a purely magnetic interaction, Ampère did away with the iron needle altogether. He passed a current through two parallel wires, one of which was suspended so that it could swing in response to the *B*-field of the other, and swing it did. Inasmuch as currents exert forces on magnets, it follows from Newton's Third Law that magnets ought to exert forces on currents. As we will see, Ampère's two-wire experiment proves the point, as do a number of other elegant arrangements, including the electric motor (p. 817) and generator (p. 842).

A TV special—"Monty Hall meets the magnet." The B-field of this fairly strong horseshoe magnet exerts forces on the electron beam and distorts the TV picture. This demonstration can be done with a black-and-white set with no risk of harm to it, but it's not advisable to try it on a color picture tube.

We know now that these interactions, which are describable on a macroscopic level in terms of currents, are fundamentally due to a magnetic force experienced by mobile charge carriers. Nowadays it's easy enough to send a beam of charged particles (for example, in a cathode-ray tube) through a known field and observe the effects firsthand. Several conclusions are forthcoming: a charge q moving through a magnetic field **B** with a velocity **v** experiences a force $\mathbf{F}_M$, which, reasonably enough, is proportional to q, v, and B; that is, $F_M \propto qvB$. Thus, no relative motion ($v = 0$), no magnetic force. Further, the two vectors **v** and **B** determine a plane, and the force is perpendicular to that plane, as shown in Fig. 21.27. *If a particle with an opposite charge is introduced, the force reverses*—the sign of q affects the sign of F_M.

The magnitude of the force depends on the angle θ between **v** and **B**: in particular, *when $\theta = 0$ or $180°$ and the particle is moving along or opposite to the field, the force is zero. When $\theta = 90°$ or $270°$ and the particle is moving perpendicular to the field, the force is a maximum (qvB)*. In other words, $F_M \propto \sin \theta$, and putting it all together

$$F_M = qvB \sin \theta \qquad (21.10)$$

There is no constant of proportionality here because the equation is basically the one that will be used to define B. The zero-force line can be defined as the direction of the magnetic field. From the formula, a 1-N force will be exerted on a 1-C charge moving at 1 m/s at 90° to a 1-T magnetic field.

Not surprisingly, the force, the field, and the velocity are related by a right-hand rule. If you put the fingers of your *right* hand in the direction of **v** and curl them through the smallest angle to **B**, your thumb will point in the direction of $\mathbf{F}_M$, as in Fig. 21.28. All of this information can be represented in a single vector equation using the cross product (p. 207)

$$\mathbf{F}_M = q\mathbf{v} \times \mathbf{B} \qquad (21.11)$$

To keep the Earth's B-field from exerting forces on the electron beam in a color TV set, the large end of the picture tube is shrouded in a casing of a high-permeability material that shields the internal region.

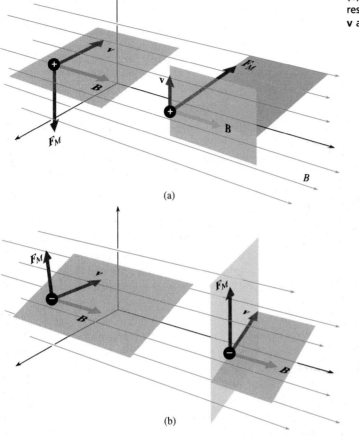

(a)

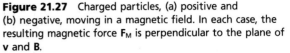

Figure 21.27 Charged particles, (a) positive and (b) negative, moving in a magnetic field. In each case, the resulting magnetic force F_M is perpendicular to the plane of **v** and **B**.

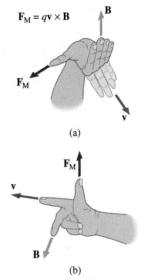

$F_M = q\mathbf{v} \times \mathbf{B}$

(a)

(b)

Figure 21.28 (a) Point the fingers of the right hand in the direction of **v**. Close them (through the smallest angle) toward **B**, and the extended thumb points in the direction of F_M. (b) A positive charge moving with a velocity perpendicular to a magnetic field experiences a force perpendicular to both.

(b)

Example 21.3 A conventional water-cooled electromagnet produces a 3.0-T uniform magnetic field in the 4-inch gap between its flat pole pieces. The field is aligned horizontally pointing due north. A proton is fired into the field region at a speed of 5.0×10^6 m/s. It enters traveling in a vertical north-south plane, heading north and downward at 30° below the horizontal. Compute the force vector acting on the proton at the moment it enters the field.

Solution: [Given: a proton with $v = 5.0 \times 10^6$ m/s, at 30° below the horizontal in the northerly direction, $B = 3.0$ T, north. Find: F_M.] First, make a drawing—Fig. 21.29. The proton has a *positive* charge of $+1.60 \times 10^{-19}$ C and so $\mathbf{v} \times \mathbf{B}$ is due east, F_M is due east. From Eq. (21.10)

$$F_M = q_e vB \sin \theta = (+1.6 \times 10^{-19} \text{ C})(5.0 \times 10^6 \text{ m/s})$$
$$\times (3.0 \text{ T})(\sin 30°)$$

and $\boxed{F_M = 1.2 \times 10^{-12} \text{ N}}$

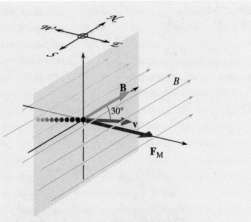

Figure 21.29 A proton traveling in the plane enters the field region at an angle of 30° below the horizontal.

▶ **Quick Check:** $F_M = q_e vB \sin \theta \approx (10^{-19} \text{ C}) \times (10^7 \text{ m/s})(1 \text{ T})(1) \approx 10^{-12}$ N.

The *B*-field coils, or yoke, which surround the neck of a TV picture tube. They steer the electron beam so that it sweeps across the screen.

Because $\mathbf{F}_M$ is perpendicular to $\mathbf{v}$, it's purely a *deflecting* force; it changes the direction of $\mathbf{v}$ without altering v. Since there can never be a component of magnetic force along the motion, there will be no tangential acceleration. Thus, **no work will be done on a moving charge by a B-field, and no change in its energy can occur in the process**.

21.9 The Trajectory of a Free Particle

Imagine a positively charged particle q entering perpendicularly into a uniform magnetic field (Fig. 21.30). Because the magnetic force is always perpendicular to the velocity, an otherwise free particle will experience a centripetal acceleration that is also always perpendicular to the motion; that is, radial. The particle will be forced to move along a circular arc. If the field is strong enough and the particle stays in it long enough, and moreover doesn't lose any energy, it will swing out a complete circle. In reality, free accelerating charges radiate electromagnetic energy and spiral inward.

Here, a positive particle is carried into a counterclockwise trajectory; had the particle been negative, it would have simply swung the other way into a clockwise circle, as in the photo on p. 701. We know that an object will move in a circle of radius R provided there is a centripetal force on it whose magnitude is

$$F_C = \frac{mv^2}{R} \tag{5.11}$$

Here, with $\theta = 90°$

$$qvB = \frac{mv^2}{R}$$

and

$$R = \frac{mv}{qB} \tag{21.12}$$

For a given q, the radius R of the path is determined by the momentum mv and the field. If we somehow impart energy to the particle (for example, via an applied E-field), thereby increasing its momentum, and if at the same time we increase B proportionately, the radius of the orbit can be kept fixed. This is the central feature of all modern ring-shaped particle accelerators. To get the largest possible $mv = qBR$, we need both the largest available field and orbital radius. Today, the approach is to construct a great doughnut-shaped hollow chamber miles in diameter, pump out all

Figure 21.30   (a) A positive particle traveling through a uniform *B*-field experiences a force that causes it to move in a curved path. (b) The force is always perpendicular to the velocity, so the particle will ideally move in a circle. (It actually radiates, loses energy, and spirals inward.)

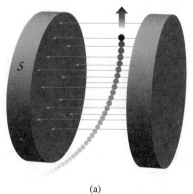

(a)

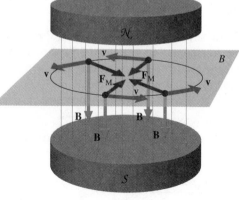

(b)

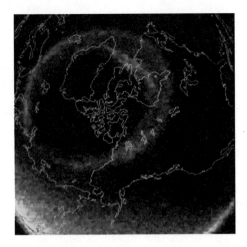

A high-altitude view of the aurora taken by a satellite about three Earth-radii away from the planet. A computer graphic shows the location of land masses. The U.S.A. is at the right of center.

The aurora borealis seen from the surface of the Earth.

the air within, and surround it with powerful superconducting electromagnets, so B can be as large as possible. A beam of, say, protons is fired in tangentially and, by adjusting mv and B for the fixed R of the chamber, the particles are brought up to tremendous energies (Ch. 33).

The *deflection yoke* on the neck of a TV picture tube consists of a pair of coils that create a set of crossed magnetic fields, one for vertical deflection and one for horizontal deflection. The current through each coil creates a variable magnetic field that steers the beam across the face of the tube.

(a)

(a) A beam of electrons in an old cathode-ray tube. (b) The beam is bent using a horseshoe magnet. (c) A beam of electrons bent into a circular orbit by the magnetic field of a set of large Helmholtz coils. The small amount of gas in the tube glows when it's ionized by collisions with the electrons, making the trajectories of the electrons visible.

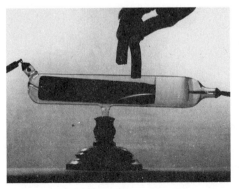

(b)

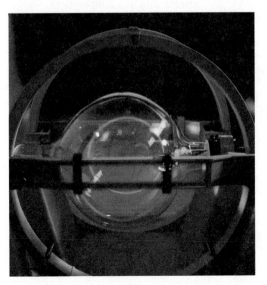

(c)

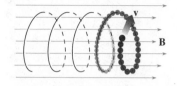

Figure 21.31 A particle, which has an initial component of **v** in the direction of **B**, moving along a spiral.

Auroras and Radiation Belts

On a grand scale, the magnetic field acts on charged particles as a kind of cosmic accelerator provoking and guiding a range of violent occurrences from X-ray emissions and stellar flares to the blazing northern and southern lights—the auroras—in the Earth's atmosphere.

When a charged particle enters a uniform magnetic field with an initial component of velocity parallel to **B**, it progresses as it spirals, following a helical path (Fig. 21.31). Things get much more complicated in a nonuniform *B*-field. Figure 21.32 depicts a little dipole initially traveling to the right in a field that gets stronger in that direction. The magnetic forces on each end are different. The dipole is pushed to the left, decelerates, and may ultimately reverse direction if the field is strong enough. Something similar happens to a spiraling charge, which is a tiny current (Fig. 21.32b). The charge, too, behaves like a dipole and will be decelerated and reversed as if reflected from a so-called *magnetic mirror*. If two such regions exist, the particle may be trapped, bouncing back and forth in what is called a *magnetic bottle,* Fig. 21.32c. In the laboratory such arrangements are used, for example, in the study of controlled thermonuclear fusion, to retain immensely hot plasmas that would otherwise be destroyed by contact with the walls of an ordinary container.

In space, particles known as *cosmic rays* come streaming toward the Earth (mostly from the Sun) and encounter the planet's *B*-field. Some are deflected away, others spiral down along the field (Fig. 21.33). Some of these particles become trapped in a form of magnetic bottle pinched off at its ends near the poles where the field increases rapidly. These form the Van Allen radiation belts that surround the planet. When some of those fast-moving charges collide with air molecules, the latter emit light in a wonderful display of color known as the aurora borealis. When a nuclear bomb was detonated high above the Johnston Islands in the Pacific (1960s), particles spiraling along the Earth's magnetic field caused an auroral display that was seen as far south as Hawaii.

21.10 Forces on Wires

Free charges experience forces when they traverse a magnetic field, something that would happen whether they were in a beam in vacuum or traveling as an ordinary current down a copper wire. When constrained to move within a conductor, the charges impart an average force to the conductor.

Return for a moment to Fig. 19.14 where a current traveling along a wire is viewed in terms of the drift of mobile charge carriers. A little segment of the wire of length *l* contains a total number of carriers equal to its volume (*Al*) times the number of mobile charges per unit volume (η): $Al\eta$. Moving at a speed equal to the drift speed (v_d) through a *B*-field, all these charges experience a net force

Figure 21.32 (a) A magnetic dipole moving in the direction of a nonuniform *B*-field. That field may be thought of as arising from an unseen south pole at the right. The south end of the dipole is pushed back more strongly than the north end is pulled forward. The net force is F_M. (b) A spiraling positive charge is the equivalent of a dipole, and it, too, experiences a backward force. (c) When the field is properly shaped, it behaves like a magnetic bottle and can capture charges that spiral back and forth within it.

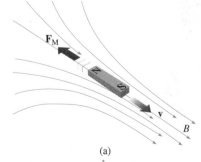

(a)

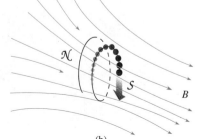

(b)

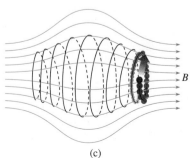

(c)

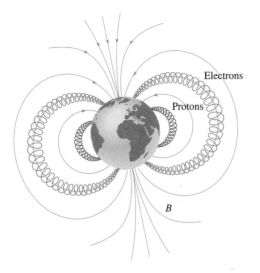

Figure 21.33 The Van Allen belts and the charged particles spiraling within them. Shown here is a cross-sectional view; the belts surround the planet.

$$F_M = (q_e v_d B \sin \theta)(Al\eta)$$

But the current is

$$I = \eta v_d A q_e \qquad [19.3]$$

Hence, for a wire segment of length l

$$F_M = IlB \sin \theta \qquad (21.13)$$

and the direction of the force on the wire is the same as the direction of the force on the individual mobile carriers (always taken as positive charges). In other words, the direction of the force is given by the direction of $\mathbf{v} \times \mathbf{B}$, as shown in Fig. 21.34 (or if you like, index finger in the direction of $\mathbf{v}$, middle finger in the direction of $\mathbf{B}$, and thumb in the direction of $\mathbf{F}$). Ampère used the arrangement of Fig. 21.35 to establish that no matter how he applied a field, the force on the current-carrying wire segment was always perpendicular to it.

When $\mathbf{B}$ is not uniform, or the wire isn't straight so that θ changes from place to place, Eq. (21.13) must be written as a differential

$$d\mathbf{F}_M = I d\mathbf{l} \times \mathbf{B} \qquad (21.14)$$

which is integrated over the conductor (see Problem 80).

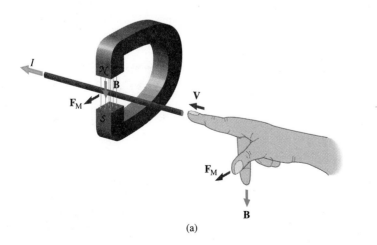

(a)

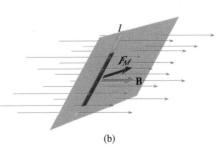

(b)

Figure 21.34 (a) The force on a current-carrying wire in a magnetic field. (b) Positive charges moving in the direction of I. $\mathbf{F}_M$ is then in the direction of $\mathbf{v} \times \mathbf{B}$.

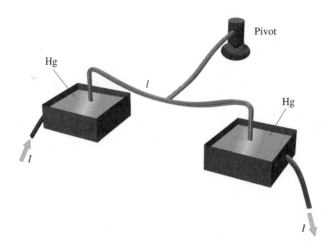

Figure 21.35 An arrangement used by Ampère to prove that F_M is always perpendicular to l. A segment of wire was set so as to freely rotate about a pivot. No matter what the direction of the applied magnetic field was, the wire never rotated because there was never a force tangent to the wire.

Example 21.5 A flat, horizontal rectangular loop of wire is positioned, as shown in Fig. 21.36, in a hypothetical 0.10-T uniform vertical magnetic field. The sides of the rectangle are F–C equal to 30 cm and C–D equal to 20 cm. Determine the total force acting on the loop when it carries a current of 1.0 A.

Solution: [Given: $B = 0.10$ T, F–$C = 30$ cm, C–$D = 20$ cm, and $I = 1.0$ A. Find: F_M.] Current travels from the positive terminal of the battery clockwise around the circuit. The directions of the forces on each segment are arrived at via $\mathbf{v} \times \mathbf{B}$ and are indicated in the diagram.

Because the forces on segments F–C and D–E are equal and opposite, they cancel. The total force is that acting on segment C–D. Using Eq. (21.13), we have

$$F_M = IlB \sin \theta = (1.0 \text{ A})(0.20 \text{ m})(0.10 \text{ T})(\sin 90°)$$

and

$$\boxed{F_M = 0.020 \text{ N}}$$

▶ **Quick Check:** A 1-A current in a 1-m wire at 90° to a 1-T field experiences a force of 1 N. Here, the length is 0.2 of that value and the field is 0.1 of it, so we can expect a force 0.02 times smaller.

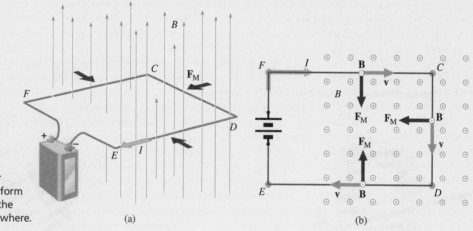

Figure 21.36 (a) A rectangular loop of wire in an idealized, uniform vertical field. (b) The forces on the wire segments are inward everywhere.

(a)

(b)

The Torque on a Current Loop

Imagine a lightweight current-carrying coil in the shape of a rectangle (Fig. 21.37) supported in a vertical plane so it turns easily about a vertical axis. The lengths of its horizontal and vertical sides are l_h and l_v, respectively. When placed in a uniform horizontal B-field, the forces on the top and bottom wires are oppositely directed

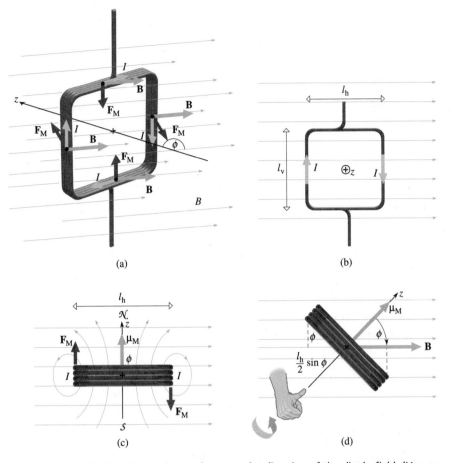

Figure 21.37 (a) Here the z-axis was chosen as the direction of the dipole field. (You can find the direction by using the Right-Hand-Current Rule.) (b) The coil swings around to align its dipole field with the applied external field. (c), (d) Views looking down from the top. The vector **u** is defined below.

and act parallel to the axis of rotation and so have no effect on the allowed motion of the coil. But the forces on the vertical segments act at a distance from that axis and produce a torque that tends to twist the coil so that it becomes perpendicular to the field. The coil is equivalent to a little dipole with the z-axis (Right-Hand-Current Rule) pointing from the south to the north pole. Like a compass, it tends to swing around such that the z-axis aligns with the B-field and $\phi = 0$.

The forces on the two vertical segments are equal and constant, independent of the angle between the z-axis and **B** because they are always perpendicular to the field. For each segment formed of N wires

$$F_{\mathrm{M}} = NIl_{\mathrm{v}}B$$

On the other hand, the torque (τ) changes because the moment-arm (the perpendicular distance from the line-of-action of the force to the axis of rotation) for each force varies with ϕ as $\frac{1}{2}l_{\mathrm{h}}\sin\phi$. At any orientation, the torque due to either vertical segment is $F_{\mathrm{M}}\frac{1}{2}l_{\mathrm{h}}\sin\phi$, as shown in Fig. 21.37d. Hence, the total torque on the coil is twice this or

$$\tau = F_{\mathrm{M}}l_{\mathrm{h}}\sin\phi = NIl_{\mathrm{v}}l_{\mathrm{h}}B\sin\phi$$

Moving coil meters are used in a tremendous variety of gauges. Transducers provide electrical signals that are proportional to the physical quantities being measured, and these signals are displayed by galvanometer gauges.

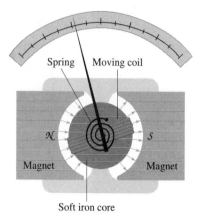

Figure 21.38 A current-carrying coil hung so that it can turn in a permanent *B*-field. The variation of torque with ϕ complicates matters, but it can be removed. If a stationary cylindrical ferromagnetic core is placed within the coil, the field lines can be concentrated and made almost radial, thereby rendering ϕ effectively equal to 90° regardless of the position of the coil.

and since $l_v l_h$ is the area A

$$\tau = NIAB \sin \phi \qquad (21.15)$$

The fact that this formula depends on A rather than on the details of the geometry suggests that it will be the same for any shaped coil, and that is confirmed by a more general treatment. If the coil is hung on a fine wire (Fig. 21.38), it becomes the movement of a galvanometer.

The Magnetic Dipole Moment

Equation (21.15) looks very much like the torque formula of Eq. (6.3), suggesting a vector cross product. Accordingly, we define IA to be the magnitude of the **magnetic dipole moment** ($\boldsymbol{\mu}_M$) of a single current loop. For a coil of N turns, the magnitude of the dipole moment is NIA, and it has the units A·m². Using the Right-Hand-Current Rule, we assign a direction to $\boldsymbol{\mu}_M$ such that it points along the extended thumb of the right hand when the fingers curl in the direction of I (Fig. 21.37d). The torque on any current loop [Eq. (21.15)] then becomes

$$\boldsymbol{\tau} = \boldsymbol{\mu}_M \times \mathbf{B} \qquad (21.16)$$

The magnetic moment tends to align with the B-field just like a compass needle; indeed, the compass needle is a little magnetic dipole.

Example 21.6 The Bohr model depicts the hydrogen atom as an electron circulating around a proton in an orbit with a radius of 0.0529 nm at a speed of 2.2 × 10⁶ m/s. Compute the orbital magnetic moment of the electron μ_B, also called the *Bohr magneton*.

Solution: [Given: $r = 0.0529$ nm and $v = 2.2 \times 10^6$ m/s. Find: μ_B.] By definition $\mu_M = IA$, so we need the current and the area of the orbit. The current is due to an electron of charge q_e moving past a point once in an interval of time equal to one period T: $I = q_e/T$. The

circumference is $2\pi r$ and therefore $T = 2\pi r/v$, whereas $A = \pi r^2$ and so

$$\mu_B = \left(\frac{q_e v}{2\pi r} \right)(\pi r^2) = \tfrac{1}{2} q_e vr$$

$$\mu_B = \tfrac{1}{2}(1.60 \times 10^{-19}\ \text{C})(2.2 \times 10^6\ \text{m/s})(0.0529\ \text{nm})$$

and $\boxed{\mu_B = 9.3 \times 10^{-24}\ \text{A·m}^2}$

▶ **Quick Check:** The units of $\tfrac{1}{2} q_e vr$ are C(m/s)m = (C/s)m² = A·m².

The intrinsic magnetic moment of the electron μ_e has a magnitude of very nearly 1 Bohr magneton. The electron is the archetypal dipole magnet. Because electrons in an atom tend to pair up with opposing spins in closed shells, the magnetic moments of atoms are due to unpaired electrons and are typically equal to just a few Bohr magnetons.

The DC Motor

Now suppose we again suspend a pivoted electromagnet, or *armature,* in a *B*-field and send a current through it, as in Fig. 21.39a. That coil swings around into alignment with the fixed magnetic field; north and south poles come together (c). The coil has inertia and will slightly overshoot the mark. If at the instant (d) it sails past alignment, the polarity of the armature is inverted (by reversing the current through it), and the coil will be violently rotated all the way around until it is once more horizontal (f). But it again overshoots, and if the current is reversed for a second time (g) as it passes the pole, the armature will swing around back to the original position, and so on, round and around.

The timed current reversals are easy to accomplish automatically whenever the armature rotates through 180°. Current is fed into the coil via a split-ring arrangement known as a *commutator*. In the diagram, two metal strips rest on either side of the commutator (each making contact with one-half of the split-ring), but in commercial motors that's done with spring-loaded carbon blocks known as *brushes*. Whenever they pass over the split, the brushes essentially change places and the current through the armature reverses. This continuous whirling machine is a rudimentary version of the ever popular dc motor—the great mover that has powered a range of devices from the trolley car and golf cart to automobile starters, toy trains, electric cars, and so on.

Two Parallel Wires—the Ampère

Ampère passed a current through two long parallel wires and found that they experienced equal and opposite forces. The idea was to demonstrate that a current-carrying wire interacts with the B-field it is immersed in, regardless of whether that field is due to a permanent magnet or another current.

Figure 21.40a shows how the field of I_2 extends over to I_1, a distance d away, where it has a value given by Eq. (21.2) of

$$B_2 = \frac{\mu_0 I_2}{2\pi d}$$

and points directly to the left. If l is some length of the upper wire, Eq. (21.10) provides the force on it where $\theta = 90°$. Thus, the force per unit length on wire 1 due to the field of wire 2 is

$$\frac{F_M}{l} = I_1 B_2 = \frac{\mu_0 I_1 I_2}{2\pi d} \qquad (21.17)$$

The same force acts on wire 2 due to the field of wire 1 (the equation is symmetrical in the currents), except that B_1 is to the right and the force (in the direction of $\mathbf{v} \times \mathbf{B}$) is upward in the plane. Newton's Third Law makes the same prediction. The wires attract each other. Reversing either one of the currents reverses its field and the forces both reverse, becoming outwardly directed and therefore repulsive.

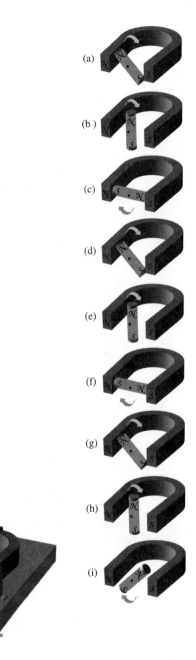

(a)

(b)

(c)

(d)

(e)

(f)

(g)

(h)

(i)

(j)

Figure 21.39 A simple electric motor. (a) The coil swings around with like poles repelling each other. (b) Then unlike poles attract and the rotation continues. (c) Just as the magnetic poles are about to align, the direction of the current reverses and the process continues.

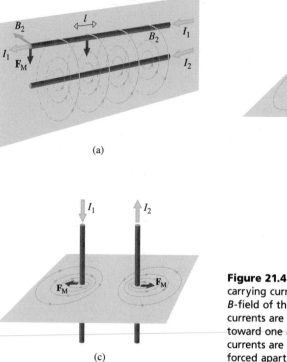

Figure 21.40 (a) Two parallel wires carrying current. Each wire is in the B-field of the other. (b) When the currents are parallel, they are forced toward one another. (c) When the currents are antiparallel, they are forced apart.

Example 21.7 What is the force per unit length experienced by each of two extremely long parallel wires carrying equal 1.0-A currents in opposite directions while separated by a distance of 1 m in vacuum?

Solution: [Given: $I_1 = I_2 = 1.0$ A and $d = 1$ m. Find: F_M.] From Eq. (21.17)

$$\frac{F_M}{l} = \frac{\mu_0 I_1 I_2}{2\pi d} = \frac{(4\pi \times 10^{-7} \text{ T·m/A})(1.0 \text{ A})(1.0 \text{ A})}{2\pi(1 \text{ m})}$$

and so

$$\boxed{\frac{F_M}{l} = 2 \times 10^{-7} \text{ N/m, repulsion}}$$

▶ **Quick Check:** $B_2 = \mu_0 I_2/2\pi r = (2 \times 10^{-7}$ T); $F_M/l = I_1 B_2 = 2 \times 10^{-7}$ N/m.

The ideas in Example 21.7 form the basis for the SI definition of the ampere:

One ampere is the constant current that, if present in two infinitely long parallel straight wires one meter apart in vacuum, produces a force of exactly 2×10^{-7} newtons per meter of length.

Practically, we can't use infinitely long wires, but very precise measurements can be made with a delicate instrument called a *current balance*. It turns out to be much more reasonable to define the ampere first and use it to define the coulomb:

One coulomb is that amount of charge transported in one second through any cross section of a wire carrying a current of one ampere.

Along with the kilogram, the meter, and the second, the ampere is considered a basic unit.

Core Material

Like magnetic poles repel and unlike magnetic poles attract. The SI unit for B, the magnetic field, is the *tesla* (T). The field of a magnet extends out from the north pole and into the south pole. The field lines form closed loops (p. 790).

A straight current-carrying wire generates a circular magnetic field in the space surrounding it given by

$$[straight\ long\ wire] \qquad B = \frac{\mu_0}{2\pi}\frac{I}{r} \qquad [21.2]$$

where μ is the permeability and for vacuum $\mu_0 = 4\pi \times 10^{-7}$ T·m/A. For a current loop of radius R

$$[circular\ loop,\ center] \qquad B_z = \frac{\mu_0 I}{2R} \qquad [21.3]$$

If there are N turns of wire

$$[circular\ coil,\ center] \qquad B_z = N\frac{\mu_0 I}{2R} \qquad [21.4]$$

For a long coil with n turns per unit length

$$[solenoid,\ center] \qquad B_z \approx \mu_0 nI \qquad [21.5]$$

$$[solenoid,\ ends] \qquad B_z \approx \tfrac{1}{2}\mu_0 nI$$

Ampère's Law

$$\oint \mathbf{B} \cdot d\mathbf{l} = \mu_0 \Sigma I \qquad [21.6]$$

provides a means of computing magnetic fields (p. 803).

The **Biot-Savart Law** is

$$d\mathbf{B} = \frac{\mu_0}{4\pi}\frac{I d\mathbf{l} \times \hat{\mathbf{r}}}{r^2} \qquad [21.7]$$

or in scalar form

$$dB = \frac{\mu_0}{4\pi}\frac{I dl \sin\theta}{r^2} \qquad [21.8]$$

and so

$$B = \int \frac{\mu_0}{4\pi}\frac{I dl \sin\theta}{r^2} \qquad [21.9]$$

The magnitude of the force on a charge moving in a B-field depends on the angle (θ) between $\mathbf{v}$ and $\mathbf{B}$:

$$F_M = qvB \sin\theta \qquad [21.10]$$

and $\qquad \mathbf{F_M} = q\mathbf{v} \times \mathbf{B} \qquad [21.11]$

The force on a current-carrying wire of length l is

$$F_M = IlB \sin\theta \qquad [21.13]$$

The torque on a coil of N turns with an area A, making an angle ϕ with the field, which is

$$\tau = NIAB \sin\phi \qquad [21.15]$$

With the orbital **magnetic moment** defined as NIA, the torque becomes

$$\tau = \boldsymbol{\mu_M} \times \mathbf{B} \qquad [21.16]$$

The force per unit length exerted on each other by two current-carrying straight wires is

$$\frac{F_M}{l} = \frac{\mu_0 I_1 I_2}{2\pi d} \qquad [21.17]$$

One ampere is the constant current that, if present in two infinitely long parallel straight wires one meter apart in vacuum, will produce a force of exactly 2×10^{-7} newtons per meter of length.

Suggestions on Problem Solving

1. This chapter has several equations that depend directly on q and that produce reversed results when the sign of the charge changes. As soon as you see you are dealing with negative charges, *be extra careful of directions.*

2. It's helpful to remember that $\mu_0/4\pi = 10^{-7}$ N/A^2 and so $\mu_0/2\pi = 2 \times 10^{-7}$ N/A^2, where $\mu_0 = 1.257 \times 10^{-6}$ N/A^2.

3. Physicists don't usually memorize all the equations for the fields arising from the various current configurations. Instead, many just remember Ampère's Law and derive what they need. When applying Ampère's Law to a new situation, first es-

tablish the direction of the B-field. That's usually done via symmetry arguments and the realization that the field lines are closed. Since there are no monopoles in magnets, B-field lines are not going to be radial, as they were with the E-fields of a small spherical charge or a line charge. Select an Ampèrian loop that is everywhere perpendicular or parallel to $\mathbf{B}$. Any lengths or distances that are specific to your Ampèrian path and not arbitrary (as is the radius of the loop in Fig. 21.23) must cancel out of the final equation for B.

Discussion Questions

1. Assuming that space is isotropic, discuss the symmetry of the electric field of a point charge, both at rest and in uniform motion. What can you expect for the symmetry of the magnetic field of a charge moving with a constant velocity (that is, a current)?

2. Why is a chunk of iron attracted to either pole of a magnet? What does this tell you about the reliability of attraction as a test of whether something is permanently magnetized or not? A classic puzzle involves two seemingly identical rods, one steel

and magnetized (with poles at its ends), the other soft iron and not magnetized. How can you tell which rod is which, using nothing else and not bending or breaking either rod?

3. James Clerk Maxwell confirmed experimentally that the *B*-field of a long straight current-carrying wire drops off inversely with distance. His apparatus is pictured in Fig. Q3 and it consists of a lightweight disk, free to rotate, on which rest four bar magnets. No matter how large the current through the central wire, there was never any rotation of the disk. Given that the poles are at distances of R_S and R_N from the wire, use the idea of fictitious magnetic charges and the torques they would experience to explain the experiment.

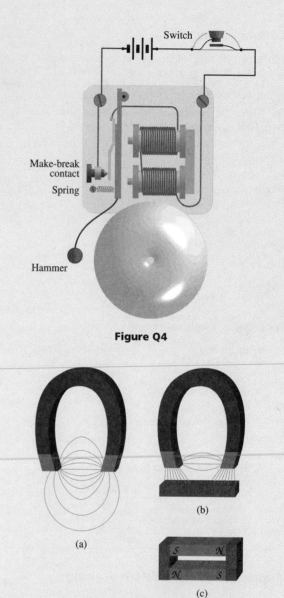

Figure Q4

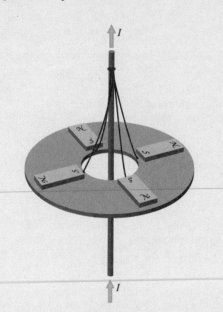

Figure Q3

4. An electric door bell is basically an automatically interrupted electromagnet. Referring to Fig. Q4, describe how it works.

5. Figure Q5 shows a horseshoe magnet with and without a piece of iron, called a *keeper,* across its poles. Explain what happens to the field. To understand the virtues of a keeper, realize that an ordinary bar magnet produces a field that extends externally from N to S and then continues through the magnet to form closed loops. But we can think of the poles at the bar's ends setting up an opposing internal field (N to S), which tends to demagnetize the magnet. The keeper essentially cancels these bare poles via induction. Figure Q5c shows the arrangement (with two soft-steel keepers) usually used to package a pair of bar magnets—explain how it works.

6. Suppose you placed a candle flame between the poles of a powerful electromagnet so that it was in a nonuniform high-field region. What, if anything, would happen to the flame? Why? What would happen to a soap bubble filled with smoke?

7. A good way to demagnetize the heads of a tape recorder (or anything else) is to send an alternating current—one that reverses direction periodically—through a nearby coil, which is then slowly moved away from the heads, decreasing *B*. Discuss how this process will demagnetize an object.

Figure Q5

8. Figure Q8 is a schematic diagram of a setup for tape-recording the output of a microphone. Explain how it works.

9. A speaker consists of a coil fixed to the back of a flexible cone, as in Fig. Q9. The coil is mounted in an assembly that is a cylindrical permanent magnet and a soft steel core. Describe how the device works.

10. A lovely way to shield against the Earth's magnetic field was used in a monopole experiment at Stanford University. A deflated lead-foil balloon is cooled below its critical temperature. Inflating the balloon then provides a field-free region inside of its hollow. How does that work?

11. Figure Q11 shows a straight section of wire carrying a downward current surrounded by a spiral of wire carrying the same current but upward. What will happen to these wires (they are pivoted and free to rotate as a unit), when a current-carrying

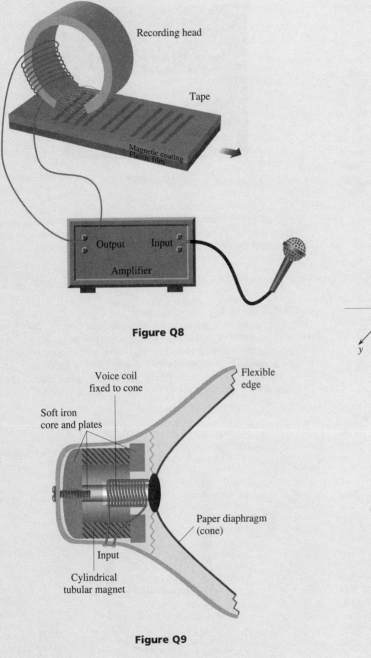

Recording head

Tape

Magnetic coating
Plastic film

Output   Input

Amplifier

Figure Q8

Voice coil
fixed to cone

Flexible
edge

Soft iron
core and plates

Paper diaphragm
(cone)

Input

Cylindrical
tubular magnet

Figure Q9

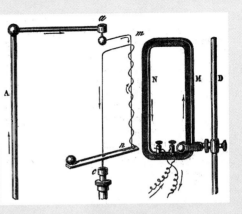

Figure Q11

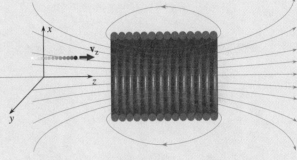

Figure Q12

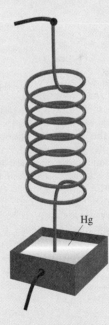

Hg

Figure Q13

coil or a bar magnet is brought nearby? Discuss the situation from the perspective of Ampère's Law. How does it relate to the field outside of a solenoid as considered in Fig. 21.24b?

12. It's been known since the late nineteenth century that a short coil can act like a lens, deflecting charged particles toward the central symmetry axis (Fig. Q12). Accordingly, explain how the electron initially traveling parallel to the z-axis, but displaced from it, ends up moving in toward that axis.

13. A very flexible helical coil is suspended (Fig. Q13) so that its lower end just dips into a cup of mercury. What will happen when a current is sent through the coil? Incidentally, this

is called *Roget's Spiral*. What would be the effect of putting an iron rod up the middle of the spiral?

14. Most radiators, steel wastepaper baskets, and metal garbage cans in the world are magnetized. Explain how this might

happen naturally, and figure out the likely polarity. Steel ships are also magnetized, which prompted someone to wrap ac current-carrying coils around ships during World War II in order to foil magnetic mines. W. Gilbert makes the following observation (in his book of 1600) concerning a "glowing mass of iron": "Let the smith be standing with his face to the north, his back to the south. Let him always, while he is striking the iron direct the same point of it toward the north and let him lay down that end toward the north [during cooling]." What will happen to the iron?

15. In 1821, Faraday devised a primitive motor he called a *rotator.* Actually, he produced two versions of it in a single unit (Fig. Q15). One had a pivoted wire carrying a sizable current swinging around a vertical, fixed bar magnet immersed in mercury. The other had a pivoted magnet rotating around a fixed current-carrying wire, also in mercury. Explain how they worked.

Figure Q17

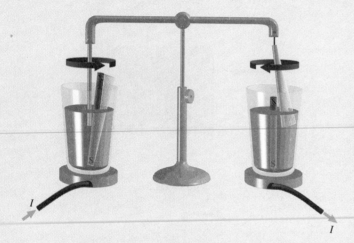

Figure Q15

16. Describe what you think will happen to a long glass tube filled with very fine iron filings when it is placed in a strong magnetic field, shaken, and then gently removed. What effect will subsequently shaking it have? Compare this with a solid bar of iron from the perspective of domains.

17. Figure Q17 shows a small cylindrical permanent magnet floating above a superconducting tin disk bathed in liquid helium at ≈ 1.2 K. The magnet was placed on the disk, and the latter was cooled below its transition temperature, at which point the magnet spontaneously jumped into the air. Explain what happened. This same kind of magnetic levitation is being applied via high-temperature superconductors to produce frictionless magnetic bearings for gyroscopes, computer disk drives, and the like.

18. Since the field of a narrow coil changes linearly with z at a distance of around $R/2$, as shown in Fig. 21.19b, Helmholtz

put two such coils (each with N turns) parallel to one another a distance R apart. The result is a fairly uniform B-field over the large central region given by

$$B_z \approx \frac{0.72 \mu_0 NI}{R}$$

and depicted in Fig. Q18. Discuss the advantages of these Helmholtz coils as a provider of uniform field (see the photo on the bottom right of p. 811). In what direction does the current progress in the coils in the diagram? How much current would it take to provide a region in which the Earth's field was canceled for coils of 1.0-m diameter with 200 turns each? (See Problem 45 for a derivation of the above equation.)

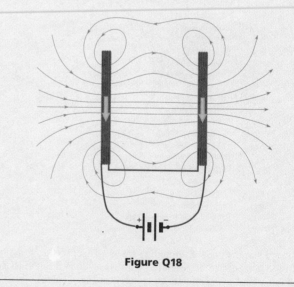

Figure Q18

Multiple Choice Questions

1. The field-line pattern around the two bars in Fig. MC1 shows that (a) neither bar is a permanent magnet (b) both bars must be permanent magnets with like poles adjacent to each other (c) either both bars are permanent magnets with like poles adjacent, or one is permanent and one is a soft iron bar (d) both must be identical permanently magnetized bars with opposite poles adjacent to each other (e) none of these.

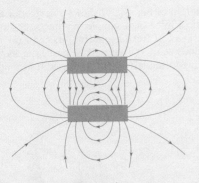

Figure MC1

2. The field-line pattern around the two bars in Fig. MC2 shows that   (a) neither bar is a permanent magnet   (b) both bars must be permanent magnets with like poles adjacent to each other   (c) either both bars are permanent magnets with like poles adjacent, or one is permanent and one is a soft iron bar   (d) both bars must be permanently magnetized with opposite poles adjacent to each other   (e) none of these.

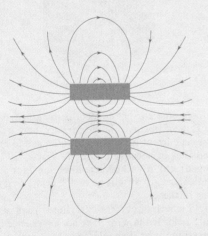

Figure MC2

3. Iron filings are sprinkled around the two bars in Fig. MC3 making a pattern showing that   (a) neither bar is a permanent magnet   (b) both bars must be permanent magnets with like poles adjacent to each other   (c) the top bar is soft iron, the bottom is a permanent magnet   (d) both must be identical permanent bar magnets with opposite poles adjacent to each other   (e) none of these.

4. A long, magnetized needle is floated vertically with its north pole up. It is held in place, and a bar magnet is brought near, as shown in Fig. MC4. When released, the floating needle will   (a) stay exactly where it is   (b) rush toward the north pole of the magnet in a straight line   (c) swing in an arc away from the north pole and over to the south pole   (d) move in a straight line to the south pole   (e) none of these.

5. The net force on a magnetic dipole in a uniform magnetic field is   (a) toward the north pole   (b) toward the south

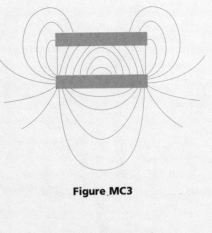

Figure MC3

Figure MC4

pole   (c) zero   (d) not enough information given   (e) none of these.

6. At the instant shown in Fig. MC6 (assuming no interaction between them), which of the little magnets experiences a net downward magnetic force?   (a) 1, 2, and 5   (b) 1 and 5   (c) 2 and 4   (d) 1, 3, 5, and 6   (e) none of these.

7. Figure MC7a shows the B-field of a long current-carrying wire immersed in a uniform magnetic field. Figure MC7b depicts the resultant field. The wire experiences   (a) zero force   (b) a force to the right   (c) a downward force   (d) a force to the left   (e) none of these.

8. Referring to Fig. MC8 (and taking "up" as out of the plane)   (a) particle 1 experiences an upward force while particle 3 experiences a downward force   (b) particle 1 experiences an upward force while particle 4 experiences a force due north   (c) particle 3 experiences an upward force while particle 2 experiences no force at all   (d) particle 4 experiences an upward force while particle 2 experiences a downward force   (e) none of these.

9. A charge can move through a magnetic field and not experience a force by   (a) moving quickly   (b) traveling parallel to B (c) moving perpendicular to B (d) traveling very slowly   (e) none of these.

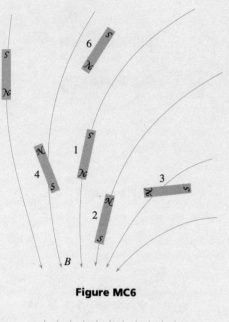

Figure MC6

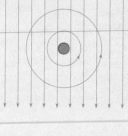

Figure MC7

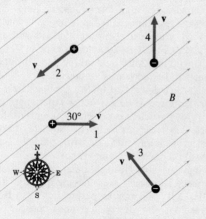

Figure MC8

10. The diagram of Fig. MC10 shows a device for measuring the vertical force on a sample in a magnetic field. If we hang a chunk of fresh apple from its end   (a) it will be drawn downward   (b) it will be pushed upward   (c) it will swing to the left   (d) nothing will happen   (e) none of these.

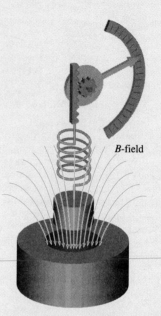

Figure MC10

11. If the force on a current element is imagined as arising from a flow of positive charge carriers, how will it change if the same current is considered to be due to an oppositely directed flow of negative carriers?   (a) it will be unchanged   (b) it will reverse direction   (c) it will be perpendicular   (d) it will be at 45° in the first quadrant   (e) none of these.

12. A straight length of current-carrying wire is in a uniform magnetic field. If the wire does not experience a force, (a) everything is as it should be   (b) it must be parallel to **B**   (c) we have an impossible situation   (d) the wire must be perpendicular to **B**   (e) none of these.

13. A circular flat coil of N turns and enclosed area A, carrying a current I, has its symmetry z-axis parallel to a uniform B-field in which it is immersed. The torque on the coil is   (a) zero   (b) $NIBA$ (c) NBA (d) IBA (e) none of these.

14. Figure MC14 shows a compass with several turns of wire wrapped around it. Such a device can best be used in a circuit to   (a) measure power   (b) indicate the presence of resistance   (c) measure a voltage difference   (d) indicate the presence of current   (e) none of these.

15. Just wrap 150 turns of a heavy (#22) insulated wire around an iron rod (for example, a door hinge pin), attach the leads to a 1.5-V D-cell via a switch and you have   (a) a radio   (b) an electromagnet   (c) a galvanometer   (d) an ammeter   (e) none of these.

16. The American physicist Henry Rowland (1876) placed some charges (in a fixed location) on a nonconducting disk, which he then rotated at high speed about its central axis

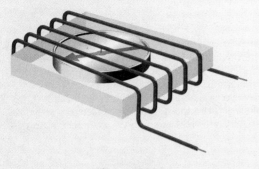

Figure MC14

near a delicate compass.   (a) the current created a B-field that deflected the compass   (b) there was no E-field, so nothing happened to the compass   (c) the charges attracted the compass, which moved toward the disk   (d) the charges repelled the compass via Coulomb's Law   (e) none of these.

17. Which way will the bar move in Fig. MC17 once the current is established?   (a) straight up   (b) to the left   (c) to the right   (d) it will not move at all   (e) none of these.

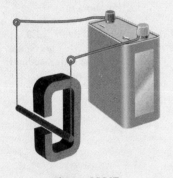

Figure MC17

18. The horseshoe-shaped iron bar in Fig. MC18 is wrapped with a wire, and once the dc current is turned on   (a) it becomes demagnetized   (b) the right end becomes a south pole and the left a north pole   (c) the left end becomes a south pole and the right a north pole   (d) both ends become north poles and the bottom becomes a south pole   (e) none of these.

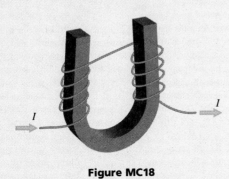

Figure MC18

19. If the current through a long solenoid is doubled while the coil's length is also doubled, keeping the total number of turns constant, the magnetic field at a point inside near the axis is   (a) four times larger   (b) half the original size   (c) unchanged   (d) one-quarter the size   (e) none of these.

20. The two (noninteracting) samples (1 and 2), which have swung into alignment in the B-field, as shown in Fig. MC20, may be, respectively,   (a) ferromagnetic and paramagnetic   (b) diamagnetic and ferromagnetic   (c) paramagnetic and diamagnetic   (d) paramagnetic and ferromagnetic   (e) none of these.

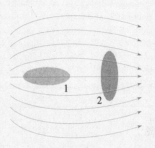

Figure MC20

21. The movable loop in Fig. MC21 and the coil carry currents in the directions indicated. The loop will   (a) looking down from above, rotate counterclockwise   (b) experience no force   (c) looking down from above, rotate clockwise   (d) experience an upward force   (e) none of these.

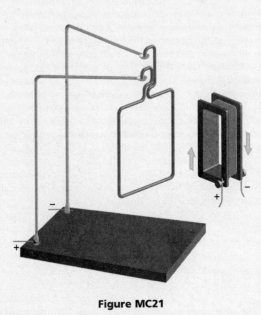

Figure MC21

Problems

MAGNETS AND THE MAGNETIC FIELD
ELECTRODYNAMICS

1. [I] Show that the units T·m/A and N/A^2 for μ are equivalent.

2. [I] The most common pre-SI unit of magnetic field, one still to be found in use, is the *gauss* (1 tesla = 10^4 gauss). The magnetic field in intergalactic space is around 10^{-6} gauss, as compared with the 8-kilogauss field of a powerful samarium-cobalt permanent magnet. Express these quantities in teslas.

3. [I] Determine the B-field 50 cm from a long narrow wire carrying a current of 10 A immersed in air.

4. [I] A magnetic field of 6.0 μT is to be produced 10 cm from a single long straight wire in air. How much current must it carry?

5. [I] The return stroke of a bolt of lightning typically carries a peak current of 20 kA up from the ground. What is the maximum magnetic field associated with the bolt 1.0 m away?

6. [I] A long straight wire carrying a current of 4.00 A is suspended in air and the field around it is measured with a magnetometer. It is found that at some point P the field is 0.660×10^{-5} T. How far from the wire is P?

7. [I] A straight wire 1.00 m long having a resistance of 1.2 Ω is attached to a 12-V battery. What is the magnitude of its B-field 2.0 cm away in air?

8. [I] A long, somewhat wiggly copper wire carries a current of 20 A. What is the B-field at a distance of 50 cm away, given that this dimension is much greater than those of any of the bends in the wire? (See Discussion Question 11.)

9. [I] A single flat loop of superconducting wire 50 cm in diameter carries a current of 25 A. What is the magnitude of the magnetic field at its center?

10. [I] A narrow flat circular coil with a diameter of 20 cm consists of 100 turns of wire. What is the magnetic field at its center when it carries a current of 5.0 A?

11. [I] A beam of protons is made to travel in a nearly circular orbit by a perpendicularly applied external B-field. Determine the magnetic field at the center of a 20-cm-diameter orbit produced by a 0.10-mA proton beam.

12. [I] A flat circular coil having a diameter of 25 cm is to produce a B-field at its center of 1.00 mT. If it has 100 turns, how much current must be provided to it?

13. [I] We wish to make a hollow solenoid 10 cm long and 1.5 cm in diameter having 200 turns of wire. How much current must we send through it to produce a field of roughly 0.50 mT inside the coil?

14. [I] An air-filled solenoid is to be 80 cm long and carry a current of 20 A. How many turns of a superconducting wire should it have if the field inside it near the middle is to be roughly 2.0 T?

15. [I] An air-core solenoid has 100 turns per centimeter and a resistance of 60 Ω. Determine the magnetic field inside it near its middle when it is connected across a 12-V battery.

16. [I] A 5.0-cm-diameter solenoid 50 cm long is made by wrapping four layers each of 1000 turns of wire on a thin hollow plastic core. Calculate the approximate B-field generated near the coil's ends when it carries a current of 1.5 A.

17. [c] Find the B-field at the very center of a circular loop of wire of radius R carrying a current I. Compare your answer with that of the more general case depicted in Problem 39. What is the field at the center of a flat circular coil if it is composed of N tightly wound turns of wire?

18. [c] A point charge Q is moving at a constant (fairly low) velocity $\mathbf{v}$. Use the Biot-Savart Law to show that the B-field it produces at a point P a distance r away is

$$\mathbf{B} = \frac{\mu_0}{4\pi} Q \frac{\mathbf{v} \times \mathbf{r}}{r^3}$$

19. [II] Suppose we remove the magnetic deflection yoke from the neck of a TV picture tube so the electron beam travels straight down the central axis. For the brightness level given, there are 6.0×10^{12} electrons arriving at the screen per second. Determine the magnetic field (magnitude and direction) caused by the beam at a radial distance of 1.5 cm from it.

20. [II] Two long horizontal straight parallel wires are 28.28 cm apart and each carries a current of 2.0 A in the same direction, namely due south. What is the **B**-field at a point that is a perpendicular distance of 20 cm from both wires?

21. [II] A superconducting niobium wire will return to the normal state when the magnetic field at its surface exceeds 0.100 T. If the wire has a diameter of 2.00 mm, what is the *critical current;* that is, what maximum current can it carry without quenching the superconducting state?

22. [II] Two long current-carrying wires are depicted in Fig. P22. What is the value of the B-field at point P, 1.0 m away from the crossing point?

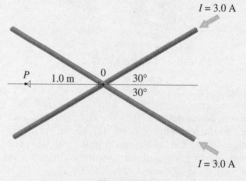

Figure P22

23. [II] Considering Problem 22, what would be the magnitude of B at P if one of the currents was reversed?

24. [II] Show that the dimensions of $\varepsilon_0\mu_0$ correspond to those of 1 over speed squared: $[T^2]/[L^2]$. This point will come up again later when we see that $\varepsilon_0\mu_0 = 1/c^2$. Indeed, since the speed of light is now defined to be exactly 299 792 458 m/s, both ε_0 and μ_0 are also exact.

25. [II] A long straight vertical wire carries a vertically upward current of 15 A. A small horizontal compass is placed 10 cm away due north of the wire. At that point, the Earth's B-field has a horizontal component of 0.50×10^{-4} T directed due north. Determine the equilibrium direction of the compass needle.

26. [II] A rather simplistic model of the hydrogen atom has a single electron revolving around a nuclear proton with an orbital radius of 0.53×10^{-10} m at a speed of 2.19×10^6 m/s. Determine the magnetic field at the proton due to the electron.

27. [II] The field along the central z-axis of a current-carrying loop is

$$B_z = \frac{\mu_0 I R^2}{2(R^2 + z^2)^{\frac{3}{2}}}$$

where $z = 0$ is at the center of the circular loop of radius R. Show that this expression is consistent with Eq. (21.3). What is the approximate form of the equation at great distances away along z?

28. [II] A solenoid 30 cm long and 2.0 cm in diameter is wrapped around an aluminum core. The coil contains 500 turns of fine insulated wire carrying 5.0 A. If the permeability of aluminum is 1.257×10^{-6} T·m/A, what is the value of B produced at the center of the coil?

29. [II] Consider the Earth (radius 6.4×10^3 km) from the perspective of the Dynamo Theory, which maintains that the B-field is due to currents in the spinning planet. Given that its liquid iron outer core has a mean radius of about 2.3×10^3 km and that the field at either pole is about 0.6×10^{-4} T, approximate the net current that would produce the dipole magnetic field. (Make your model as simple as possible, but discuss its shortcomings.) Draw a rough sketch of the planet and the currents. In what direction does I progress? (Take a look at Problem 27 and Fig. 21.19.)

30. [II] Consider the solenoid of Problem 28, this time with a silicon-iron core having a relative permeability of 6000 times that of free space. Determine the approximate field at the center of the coil.

31. [II] Figure P31 shows a toroidal coil of N turns carrying a current I. Determine the B-field inside the coil. How does it depend on position? What is the relationship between the B-field inside the torus and that of a single straight wire along the axis of symmetry of the coil carrying a current of NI?

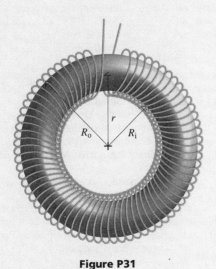

Figure P31

32. [II] In 1962, scientists at the Geophysical Observatory in Tulsa, Oklahoma, measured a magnetic field of 1.5×10^{-8} T being generated by a large tornado about 9.0 km away. Use Ampère's Law to approximate the current swirling downward within the funnel.

33. [II] With Problem 31 in mind, use Ampère's Law to prove that the B-field lies entirely within the confines of the torus.

34. [II] Why is it that a long solenoid can be regarded as a toroidal coil with an infinite radius?

35. [II] Figure P35 shows a portion of an infinite conducting sheet carrying a current (traveling in the z-direction) per unit of width (along the x-axis) given by i amps per meter. Imagine the sheet as if it were made up of an infinite number of adjacent parallel wires. Use symmetry arguments and the Right-Hand-Current Rule to establish that **B** is a constant everywhere (independent of distance) above and below the sheet. Further, show that B points in the negative x-direction above the sheet and in the positive x-direction below the sheet. Using Ampère's Law, show that

$$B = \tfrac{1}{2}\mu_0 i$$

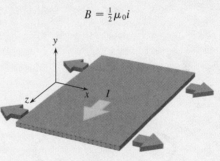

Figure P35

36. [cc] Determine the B-field inside a very long straight inhomogeneous rod of radius R, carrying a current that increases out from the center according to the expression $J = Cr$. [Hint: Give your answer in terms of R, r, and I. Neglect any end effects.]

37. [cc] Determine the B-field inside a very long straight inhomogeneous rod of radius R, carrying a current that decreases out from the center according to the expression $J = C/r$. Give your answer in terms of I. Neglect any end effects.

38. [cc] Use the Biot-Savart Law to determine the B-field at a nearby point P on the bisector of a long straight wire, in air, carrying a constant current I. [Hint: Referring to Fig. 21.26, **r** and the wire make an external angle of $\theta = \theta_a$ above the origin and an internal angle of $\theta = \theta_b$ below the origin. Write your answer as a function of θ_a, θ_b, I, and x.]

39. [cc] Use the Biot-Savart Law to determine the magnetic field at a point P anywhere on the central axis of a circular loop of wire of radius R carrying a counterclockwise current I. [Hint: See Fig. P39 for the geometry and check your answer against Problem 27.]

40. [cc] Figure P40 shows a wire bent into the form of a section of a circle centered at point P. If a current I flows along the wire, what is the magnitude of the resulting B-field at P? Do not overlook the straight lengths of wire. [Hint: Study the diagram. Write your answer in terms of R, I, and φ using the fact that $l = R\varphi$.]

41. [cc] The bent wire in Fig. 41 carries a current I. Write

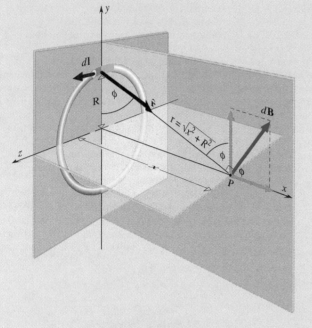

Figure P39

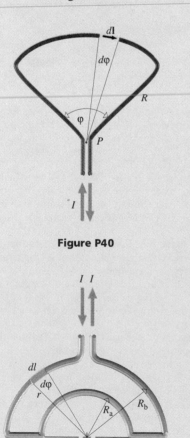

Figure P40

Figure P41

an expression for the resulting B-field at P. [Hint: Use the quantities shown in the diagram. Watch out for the directions of the contributing B-fields.]

42. [cc] Determine the magnetic field at a point P a perpendicular distance h from one end of a straight wire of length L carrying a constant current I. [Hint: Put the wire on the x-axis such that $dl = dx$. You may need to remember that $\sin \theta = \sin(180° - \theta)$.]

43. [cc] The bent wire depicted in Fig. P43 carries a constant current of I as shown. Compute the **B**-field it produces at point P. [Hint: Give your answer in terms of I, L, R, and h where R is measured from P to the curved wire. Watch out for the directions of the contributing B-fields.]

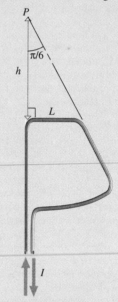

Figure P43

44. [cc] Determine the B-field at a perpendicular distance h near the middle of a very long straight inhomogeneous rod of radius R, carrying a current that increases out from the center according to the expression $J = Cr$. Take $h > R$. [Hint: Give your answer in terms of R and C.]

45. [cc] Return to the Helmholtz coils (each having N turns) discussed in Discussion Question 18 and prove that the equation given for the B-field is correct. [Hint: Compute the field due to both coils at the center of the arrangement where $x = R/2$.]

46. [III] A long straight hollow cylindrical conductor with an inner radius of R_i and an outer radius of R_o carries a current I, which is distributed uniformly over its cross section and which propagates along its very long length. Given that it is immersed in air, determine the magnetic fields in the inner and outer air-filled regions, as well as in the conductor. Draw a graph of B against radial distance from the central axis.

47. [III] Imagine two horizontal parallel infinite planes, one above the other. If these are each thin conductors carrying equal currents, the top one traveling west to east, the bottom one east to west, find the B-field between them if the current per unit width is i. Take a look at Problem 35. What is the field outside the sheets?

MAGNETIC FORCE

48. [I] A small ball of plastic carries a positive charge and is shot at high speed along the x-axis in the horizontal x–y plane. It passes through a uniform horizontal B-field, which makes an angle with the x-axis of $+23.5°$. In what direction is the magnetic force it will experience?

49. [I] A negatively charged projectile is fired upward in the vertical z–y plane at an angle of $+20°$ to the y-axis. It travels through a uniform magnetic field directed downward in the negative z-direction. What is the direction of the resulting magnetic force?

50. [I] A proton travels downward in the vertical x–z plane at an angle of $-40°$ with respect to the x-axis. It passes through a uniform horizontal B-field directed in the $+y$-direction. In what direction is the resulting magnetic force?

51. [I] A proton in a research facility is propelled at 2.0×10^6 m/s perpendicular to a uniform magnetic field. If it experiences a force of 4.0×10^{-13} N, what is the strength of the field?

52. [I] Show that 1 T $= 1$ kg/s·C.

53. [I] Show that 1 T $= 1$ V·s/m².

54. [I] Show that 1 T $= 1$ N/A·m.

55. [I] An electron of mass 9.11×10^{-31} kg travels in a circular orbit within a large evacuated chamber. The orbit has a 2.0-mm radius and is perpendicular to a B-field of 0.050 T. What's the electron's speed?

56. [I] An electron moving at 2.5×10^6 m/s within a cathode-ray tube (CRT) makes an angle of $45°$ with a uniform 1.2-T magnetic field. What is the magnitude of the force it will experience?

57. [I] A beam of protons of various speeds enters a region where it encounters perpendicular electric and magnetic fields both at right angles to **v**. Show that only particles for which $v = E/B$ will pass through undeflected. Accordingly, such an arrangement is called a *velocity selector*.

58. [I] A proton moving at 3.00×10^6 m/s through a uniform magnetic field experiences a maximum force of 5.2×10^{-12} N, straight upward when it's traveling due east. Determine **B**.

59. [I] A straight wire carrying an upwardly directed current is in the x–z plane making an angle of $+20°$ with the horizontal x-axis. The wire is immersed in a uniform B-field in the positive z-direction. What is the direction of the force on the wire?

60. [I] A 10-m-long wire is strung more or less horizontally in an east-west direction, and a current of 10 A is sent from west to east along it. Assuming the Earth's B-field at that location has a magnitude of 0.5×10^{-4} T and is horizontal and due north, find the magnitude of the force on the wire.

61. [I] A 1.50-m-long straight wire experiences a maximum force of 2.00 N when in a uniform 1.333-T B-field. What current must be passing through it?

62. [I] A wire is positioned in a uniform magnetic field so that the force on it—4.00 N—is a maximum. If the wire is 20 cm long, and carries 10.0 A, what is the magnitude of the B-field?

63. [I] A straight current-carrying ($I = 6.0$ A) wire makes an angle of $31.2°$ with a 0.01-T uniform B-field. What is the magnitude of the force exerted on a 1.0-cm length of the wire?

64. [I] A narrow flat coil wound on a square frame has 200 turns and sides of 20 cm. It carries a current of 1.25 A and is positioned in a 0.50-T magnetic field. What is the maximum torque that can be exerted by the field on the coil?

65. [I] A circular flat coil of wire encompassing an area of 1.3×10^{-3} m² has 20 turns and carries a current of 1.5 A. If its rotational symmetry axis makes an angle of $32°$ with a B-field of 0.90 T, what is the torque acting on it?

66. [I] Two parallel wires 50.0 m long are 25.0 cm apart and each carries a current of 10 A in the same direction. What is the force (magnitude and direction) between them?

67. [II] An electron having a velocity of 5.0×10^6 m/s along the x-axis enters a region where there is a uniform 5.0-T B-field making an angle of $60°$ to the x-axis. Determine the magnitude of its acceleration.

68. [II] A cosmic ray proton traverses a uniform B-field perpendicularly at a speed of 2.0×10^7 m/s in the positive x-direction, experiencing an acceleration of 4.0×10^{12} m/s² in the positive y-direction. Determine **B**.

69. [II] A doubly charged positive helium ion, one missing two orbital electrons, has a mass of 6.7×10^{-27} kg and is accelerated through a potential difference of 10 kV. It then enters a region in which there is a uniform perpendicular B-field of 1.50 T. What is its subsequent path in the field?

70. [II] A tiny plastic sphere of mass 0.10 mg is shot horizontally at a speed of 200 m/s in the x–y plane. Given that it carries a charge of -10 μC and travels along the positive y-axis, determine the nature of the smallest uniform magnetic field that will keep the particle moving horizontally despite the Earth's gravity.

71. [II] Write an expression for the momentum of a particle of charge q and mass m moving in a circular orbit of radius R in a uniform magnetic field B.

72. [II] A proton is sailing through the outer region of the Sun at a speed of 0.15c. It traverses a locally uniform magnetic field of 0.12 T at an angle of $25°$. What is the radius of its helical orbit? [Hint: $v_{\parallel}$ and $v_{\perp}$ can be considered separately.]

73. [II] The American physicist E. H. Hall discovered (1879) that when a current travels along a conducting plate of width l, which is perpendicular to a magnetic field, a potential difference V appears across the plate as shown in Fig. P73. Prove that

$$V = vBl$$

The Hall probe makes a very convenient magnetometer. Discuss the difference you might expect if the probe is made of copper in one case, where the charge carriers are negative, and germanium, where the charge carriers are positive "holes." The Hall effect reveals a difference between positive charge moving to the right and negative charge moving left. Explain.

74. [II] A single power line 50.0 m long is stretched more or less horizontally at an angle of $20°$ east of north at a location where the Earth's field is 0.50×10^{-4} T at $5.0°$ west of north. What is the total force on the wire when it carries 1.0 kA?

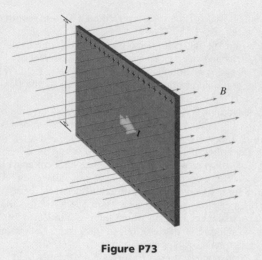

Figure P73

75. [II] With Problem 73 in mind, suppose we want to use the Hall effect to measure the blood's flow rate (v) in an artery. If we apply a transverse magnetic field, it will cause the positive and negative ions in the blood to separate, thereby producing a voltage across the artery. Given that the artery has an inner diameter of 4 mm, if a uniform 0.50-mT field produces a voltage difference of 1.0 μV, what is the flow rate?

76. [II] A flat rectangular coil of 10 turns is suspended vertically from one arm of a beam balance that is brought into horizontal alignment with a few weights on the other pan. The bottom section of the coil, which is 10 cm long, is next exposed to a uniform perpendicular magnetic field of 0.60 T. A current of 300 mA is sent through the coil, which is pulled downward. How much weight must now be added to the other pan to restore balance?

77. [II] In a moving-coil galvanometer, the coil, which has N turns, surrounds a steel core so that the B-field always makes an angle of $\phi \approx 90°$ with the symmetry axis (the z-axis). If the current through the coil is reduced by 25%, how does the torque on it change?

78. [II] A single circular loop of wire 10 cm in radius is hung on a fine silver ribbon so that the plane of the loop is parallel to the magnetic field in a uniform region between the pole pieces of a large electromagnet. If the torque on the coil is 0.100 N·m when 5.0 A passes through the coil, what is the magnitude of the B-field?

79. [II] A long wire is stretched out horizontally to a 1.0-m-diameter tree, around it and back. Both ends are connected to the terminals of a battery and 10 A is drawn by the wire. If both lengths are parallel and the Earth's field is negligible, what force exists between each meter of the wires?

80. [cc] A rigid wire is bent into the shape of a half circle of radius R and placed flat down on a horizontal table. Two straight leads of length L are then rigidly attached, each to one of the ends, making a U-shaped path with the outward curve pointing south and the two straight wires leading to the north. A battery is then placed across the two leads so that a current I circulates in the closed circuit. The table is pushed between the poles of a large vertical magnet, and a downward B-field fills and surrounds the U-shaped region.

The current flows in such a way as to produce an outward horizontal force on the U; determine that force.

81. [cc] A long bar magnet lies flat on a horizontal table (along the $-y$-axis with its north pole uppermost a short distance $-h$ beneath the origin). A thin metal rod of length L is also lying on the table along the x-axis with its center at the origin. The magnet and rod form a symmetrical T-shape. The rod is then connected to a battery and a constant current I passes through it. Determine the resulting magnetic force, if any, on the rod. [Hint: Assume the field of the magnet is radial and of the form $B = C/r^2$. Both B and θ change along the rod.]

82. [III] A positive particle of charge q and mass m is accelerated through a potential V and is subsequently bent into a circular orbit of radius R by a uniform perpendicular magnetic field B. Write an expression for R^2 in terms of V, B, and m/q; the latter is called the charge-to-mass ratio. A modern mass spectrometer allows R to be measured directly, and from that m is determined with a precision of about 1 part in 10^7.

83. [III] A positively charged particle is moving in a circle under the influence of a perpendicular magnetic field. Write an expression for the frequency of its orbital motion and show that it is independent of both radius and speed. Overlook the fact that it radiates.

84. [III] Suppose we have two tiny spheres 1.00 m apart, each carrying 1.00 C. If they now move along parallel straight paths (1.00 m apart) at 1.00 m/s, compare the electric force to the magnetic force they experience.

85. [ccc] A nonconducting sphere of radius R carries a uniformly distributed charge $+Q$ on its surface. If it is spinning at a rate ω what is its magnetic moment? [Hint: Divide the surface into a series of rings or loops (as in Fig. P85) each of which has a differential magnetic moment $d\mu_M$. Show that the current corresponding to a revolving loop is given by

$$dI = \frac{Q}{4\pi} \omega \cos\theta \, d\theta$$

Then use $\mu_M = IA$.]

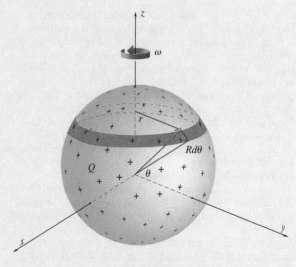

Figure P85

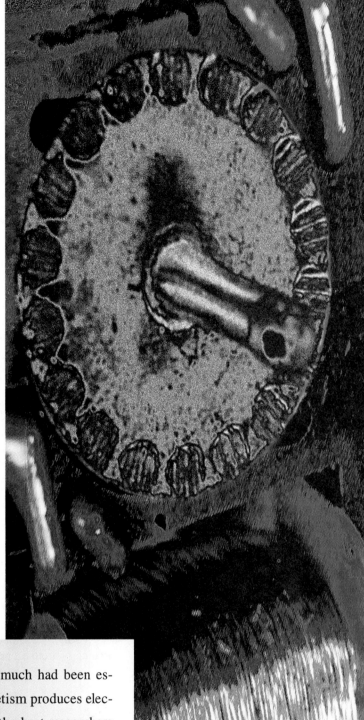

22

Electromagnetic Induction

ELECTRICITY PRODUCES MAGNETISM; THAT much had been established in the early 1820s, and so the converse—magnetism produces electricity—seemed a reasonable thing to expect. And yet, the best researchers of the day could only come up with results that were ambiguous and unconvincing. Still, the agenda was obvious enough: a charge can electrify a nearby object by induction, and a magnet can magnetize a nearby piece of iron by induction; it was only reasonable to expect that a current should induce a current in a nearby conductor.

Since a steady current generates a steady magnetic field, should not a steady magnetic field generate a steady current? However logical that was, it was wrong— currents *were* being induced, right before their eyes, but these were transient currents, something no one expected and no one was ready to "see." A steady magnetic field does not impart energy to free charges, it does no work on them, and yet for a current to exist it must get energy from somewhere. A constant current in a wire sitting next to and at rest with respect to another wire will not induce a current in that second wire. But a changing magnetic field is something very different. It can impart energy to charges and it can produce currents.

ELECTROMAGNETICALLY INDUCED EMF

In 1821, Ampère conducted an experiment during which he observed the momentary effects of what has come to be known as **electromagnetic induction**, but that was not the main thrust of his research and he did not appreciate what was at hand. "Convert magnetism into electricity" was the brief remark Faraday jotted in his notebook in 1822, a challenge he set himself with an easy confidence that made it seem so attainable. Two years later, Arago by chance observed that the needle of a fine compass tended, after it was jolted, to oscillate for a shorter-than-normal time when it was in a housing with a copper bottom (p. 846). Arago then performed the inverse experiment and rotated a copper disk beneath a compass needle and found, inexplicably, that the needle followed the disk around.

After several years doing other research, Faraday returned to the problem of electromagnetic induction in 1831. His first apparatus made use of two coils mounted on a wooden spool. One, called the *primary,* was attached to a battery and a switch; the other, the *secondary,* was attached to a galvanometer (Fig. 22.1a). Initially, the results were totally negative until he installed a much more powerful battery composed of a hundred cells. Even then, the effects were faint and transient, and yet their mere existence was enough for him. Faraday found that the galvanometer deflected in one direction just for a moment whenever the switch was closed, returning to zero almost immediately, despite the constant current still in the primary. Whenever the switch was opened, interrupting the primary current, the galvanometer in the secondary circuit momentarily swung in the opposite direction and then promptly returned to zero.

Using a ferromagnetic core to concentrate the "magnetic force," Faraday wound two coils around opposing sections of a soft iron ring (Fig. 22.1b). Now the effect

Figure 22.1 (a) A current in one coil produces a time-varying magnetic field that couples to the other coil, inducing a transient current in it. (b) An iron core becomes magnetized by the field of the primary current, increasing the field and improving the coupling to the secondary current.

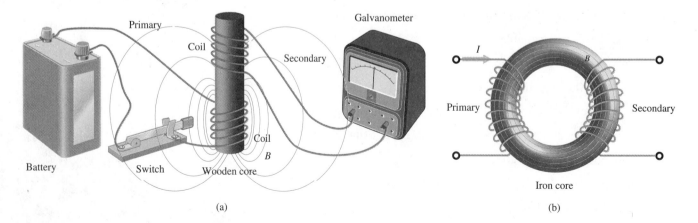

(a)

(b)

was unmistakable—a changing magnetic field generated a current. Indeed, as he would continue to discover, *change* was the essential aspect of electromagnetic induction. Within a few weeks, the modest, untutored genius had explored the phenomenon and was ready to present his first paper on the subject.

Joseph Henry in New York didn't hear of Faraday's achievement until April, and when he did, the news was a bitter disappointment. He had performed much the same experiment as Faraday well over a year before (1830) though the results went unpublished. Like Faraday, Henry appreciated the implications of the discovery even if he felt no urgency to publish them. When he finally dashed off an article in response to the news from Europe, the experiment he described was somewhat more sophisticated than that performed in London, and it even contained an important feature Faraday had missed (p. 847).

Mr. Faraday (he declined knighthood) went on to use his new knowledge to build an electric generator (p. 842), one of the single most far-reaching accomplishments of the era. Later, when asked by Prime Minister Gladstone if his research had any practical value, Faraday quipped, "Why, sir, you will soon be able to tax it." Electromagnetic induction is now at the very heart of the generation, delivery, and utilization of electrical energy—and they do indeed tax it.

22.1 Faraday's Induction Law

Figure 22.2 depicts a modern version of an experiment performed by Faraday to explore the dependence of electromagnetic induction on variations in B. The oscilloscope displays, as a function of time, the voltage—otherwise known as the **induced emf** ($\mathscr{E}$)—across the coil's terminals. The exact shape of the resulting curve is not important; it depends on the details of the motion of the magnet. Observe that thrusting a south pole toward the coil produces a positive emf and yanking it away produces a negative emf (for this particular winding). Approaching the coil with a north pole reverses the polarity of the induced emf. Futhermore, the amplitude of the emf depends on how rapidly the magnet is moved; when $v = 0$, the emf = 0. In this arrangement, *the induced emf depends on the rate of change of B* through the coil and not on B itself. A weak magnet moved rapidly can induce a greater emf than a strong magnet moved slowly.

When the same changing B-field passes through two different wire loops, as in Fig. 22.3, the induced emf is larger across the terminals of the larger loop. In other

Figure 22.2 (a) As the S-pole approaches the right end of the coil, a current circulates from left to right through the coil, and the right end of the coil becomes an S-pole. The right side of the coil becomes positive, the left side negative, and it acts like a battery, sending current clockwise through the external circuit. (b) When the magnet is stationary, the emf is zero. (c) Removing the magnet induces a north pole at the right end of the coil, which attracts the receding S-pole of the magnet. (d) No motion, no flux change, and $\mathscr{E} = 0$.

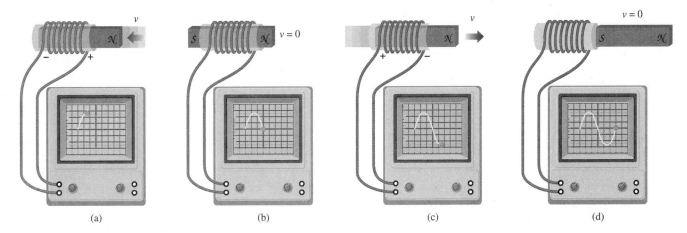

(a)   (b)   (c)   (d)

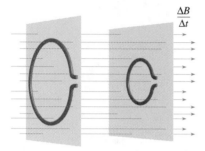

$\frac{\Delta B}{\Delta t}$

Figure 22.3 A time-varying magnetic field passes through both loops. It induces a larger emf in the larger loop, which embraces a greater perpendicular area.

words, here where the *B*-field is changing, *the induced emf is proportional to the area A of the loop penetrated perpendicularly by the field.* If the loop is successively tilted over, as in Fig. 22.4, the area presented perpendicularly to the field ($A_\perp$) varies as $A \cos \theta$ and, when $\theta = 90°$, the induced emf is zero because no amount of *B*-field then penetrates the loop: when $\Delta B/\Delta t \neq 0$, emf $\propto A_\perp$. The converse also holds: *when the field is constant, the induced emf is proportional to the rate-of-change of the perpendicular area penetrated.* Thus, if a coil is twisted or rotated or even squashed (Fig. 22.5) while in a constant *B*-field so that the perpendicular area initially penetrated is altered, there will be an induced emf $\propto \Delta A_\perp/\Delta t$ and it will also be proportional to *B*. In summary, then, when $A_\perp =$ constant, emf $\propto A_\perp \Delta B/\Delta t$ and, when $B =$ constant, emf $\propto B\Delta A_\perp/\Delta t$.

All of this suggests that the emf depends on the rate-of-change of both $A_\perp$ and *B*, that is, on the rate-of-change of their product (see the Product Rule in Appendix F). This brings to mind the notion of the flux of the field (p. 671)—the product of field and area where the penetration is perpendicular. Accordingly, we define the **flux of the magnetic field** as

$$\Phi_M = B_\perp A = BA_\perp = BA \cos \theta$$

or more generally [remembering Eq. (17.9) for Φ_E] as

$$\Phi_M = \int \mathbf{B} \cdot d\mathbf{A} \qquad (22.1)$$

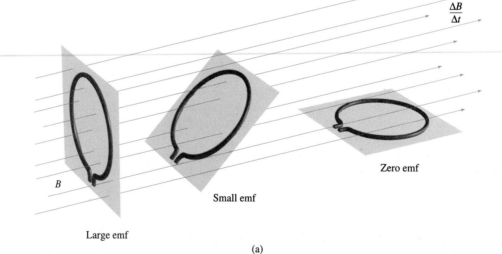

$\frac{\Delta B}{\Delta t}$

B

Large emf

Small emf

Zero emf

(a)

Figure 22.4 (a) The induced emf is proportional to the perpendicular area intercepted by the field. (b) The area of the ring ($A = \pi r^2$) projects as an ellipse. Vertically *r* projects to $r \cos \theta$; horizontally *r* projects to *r*. The area of the ellipse is $A_\perp = \pi(r \cos \theta)r = A \cos \theta$.

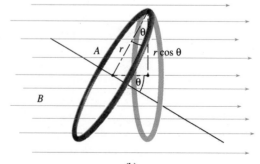

B

(b)

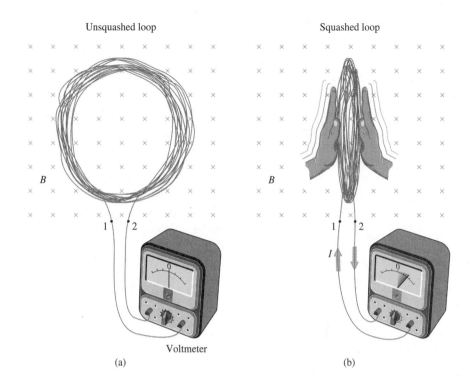

Figure 22.5 (a) A downward B-field perpendicular to a wire loop. The coil starts at point 1, winds clockwise through a number of turns, and ends at point 2. (b) Squashing the loop reduces $A_\perp$, changes the flux, and, therefore, induces an emf. The voltmeter will show a momentary jump in voltage, after which it will return to zero.

The units of magnetic flux are tesla-meter^2, which is called a **weber** (Wb) to honor one of the early workers in magnetism, Wilhelm Weber: 1 Wb = 1 T·m^2. Clearly, B must be measured in Wb/m^2, which is the old unit that was replaced by the tesla: 1 Wb/m^2 = 1 T. It's common in the literature to find B referred to as **flux density** (flux per unit area).

The results of the above observations for a one-turn loop of wire can be stated in terms of an average induced emf:

$$\mathscr{E}_{av} = -\frac{\Delta\Phi_M}{\Delta t}$$

where the significance of the minus sign will be considered presently. If we have a functional representation for the moment-by-moment flux it's useful to define the instantaneous emf as

$$\mathscr{E} = -\frac{d\Phi_M}{dt} \tag{22.2}$$

A single loop of wire will experience an induced voltage (in volts) that equals the time rate-of-change of magnetic flux through it at any given instant (in SI units the constant of proportionality is unity). If there are N turns of wire in a coil, each has an induced voltage that is in series with the others so that the net *induced emf* is

$$\mathscr{E} = -N\frac{d\Phi_M}{dt} \tag{22.3}$$

which is one of the *fundamental equations of electromagnetism.* It is a summary of observations and not derivable from any previous formulas. The equation is generally known as **Faraday's Induction Law**, though Faraday himself never wrote it.

Example 22.1 A circular flat coil of 200 turns of wire encloses an area of 100 cm². The coil is immersed in a uniform perpendicular magnetic field of 0.50 T that penetrates the entire area. If the field is shut off so that it drops to zero in 200 ms, what is the average induced emf? Given that the coil has a resistance of 25 Ω, what current will be induced in it?

Solution: [Given: $N = 200$, $A = 100$ cm², $B_i = 0.50$ T, $B_f = 0$, $\Delta t = 200$ ms, and $R = 25$ Ω. Find: emf and I.] The B-field links the entire area perpendicularly; hence, the initial flux is

$$\Phi_M = BA = (0.50\ T)(0.010\,0\ \text{m}^2) = 0.005\,0\ \text{T·m}^2$$

and so $\Delta\Phi_M = -0.005\,0$ T·m² since the final flux is zero. It follows from Eq. (22.3) that

$$\mathscr{E}_{av} = -N\frac{\Delta\Phi_M}{\Delta t} = -200\frac{(-0.005\,0\ \text{T·m}^2)}{0.200\ \text{s}} = \boxed{5.0\ \text{V}}$$

(Don't worry about the signs here; we'll deal with that part presently.) From Ohm's Law

$$I = \frac{V}{R} = \frac{(5.0\ \text{V})}{(25\ \Omega)} = \boxed{0.20\ \text{A}}$$

▶ **Quick Check:** Remember that a change of flux of 1 T·m² through 1 turn in 1 s produces 1 V. Here, $N\Delta\Phi_M$ (which is the change in the total flux linking the coil) equals one, so 5.0 V is reasonable.

Equation (22.3) makes the point that the emf depends on the rate-of-change of flux linking all the turns of wire, and it's useful to formalize the notion by speaking about the **flux linkage** in the coil—namely, $N\Phi_M$. Thus, **the induced emf is the negative of the time rate-of-change of the flux linkage**.

22.2 Lenz's Law

The negative sign in Eq. (22.3) relates the polarity of the induced emf to the flux change, which can occur in a variety of ways: the field can be increased, decreased, or moved; the loop area can be turned, squashed, or yanked out of the field, and yet there is always a consistent, reproducible sense in which the emf appears. It was a physicist working in Russia—Heinrich Friedrich Emil Lenz (pronounced *lents*)— who first (1834) published an elegant statement of the phenomenon that has come to be known as **Lenz's Law**:

> **The induced emf will produce a current that always acts to oppose the change that originally caused it.**

In Fig. 22.6a, a south pole approaches a coil, inducing an emf across its terminals. The emf will be such as to cause an induced current, which in turn creates an induced magnetic field (B_I). This field opposes the *change* of flux linking the coil. The instigating effect is an *increasing B-field directed to the right* (toward the approaching magnet's south pole). The opposing induced field, tending to cancel an increasing field to the right, will be to the left, and the current in the coil circulates appropriately. *The polarity of the emf indicates the direction in which it drives current in the external circuit* and would be measured by a voltmeter across the terminals of the coil. In other words, the approach of a south pole induces a current that causes the coil to manifest a south pole at its near end, which opposes the advance of the magnet and thereby the change of the field. If the magnet is withdrawn, as in Fig. 22.6b, the change—*a decreasing field to the right*—is opposed by an induced field to the right. The current direction is now reversed, as is the emf. A north pole is induced at the coil's right end, which attracts the magnet tending to oppose its motion away.

Change is at the heart of the phenomenon of induction, just as it underlies the concept of energy—change is a manifestation of energy. Not surprisingly, Lenz's

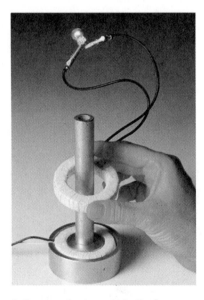

A time-varying current in the lower coil produces a time-varying magnetic flux that passes through the upper coil. That, in turn, induces a current in the upper coil, and the bulb lights.

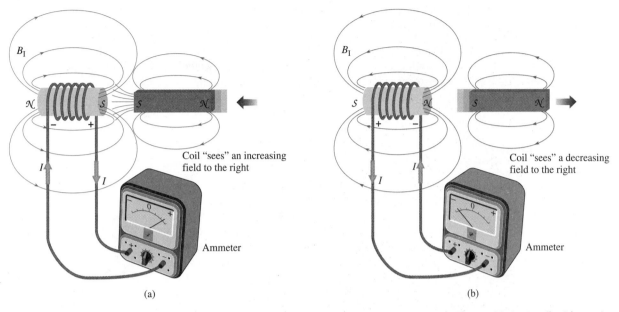

(a)   (b)

Figure 22.6 A coil with an air core, an ammeter, and a bar magnet. (a) The approach of an S-pole induces an emf (+ on the right, − on the left) that causes a current to circulate clockwise in the external circuit. (b) Removing the S-pole induces a counterclockwise current and the reversal of the emf, driving it. Note the polarity of the connections on each meter and the sign of its reading.

Law can be appreciated as a result of the Law of Conservation of Energy. For example, when an external agent moves the magnet in Fig. 22.6a, it must overcome the counterforce immediately exerted by the coil. The work done in the process provides the electrical energy needed to build up and sustain the induced current. The magnetic field is the intermediary between the external mover pushing the magnet and the resulting current. Were that not the case, the induced current would be created at no cost of mechanical work, and energy would not be conserved. Thus, no matter how the flux changes, work must be done on the system and the induced current, which is a manifestation of that work, must oppose the change—it doesn't necessarily stop it, but it always opposes it. *The work done by the magnet on the current* (W_{mc}) *must equal in size the work done by the current on the magnet* (W_{cm}), *and that work is negative:* $W_{mc} = -W_{cm}$.

Example 22.2 The loop in Fig. 22.5a has an area of 0.25 m² and is immersed in a uniform perpendicular 0.40-T downward magnetic field. If the coil has 200 turns, a resistance of 5.0 Ω, and is squashed to a zero area in 100 ms, what average current will be induced in it during the collapse? Discuss the direction of the current and the polarity of the emf. What happens to the current?

Solution: [Given: $A_i = 0.25$ m², $A_f = 0$, $B = 0.40$ T, $N = 200$, $R = 5.0$ Ω, and $\Delta t = 0.100$ s. Find: I.] The initial flux is

$$\Phi_M = BA = (0.40 \text{ T})(0.25 \text{ m}^2) = 0.100 \text{ T·m}^2$$

and so $\Delta\Phi_M = -0.100$ T·m² since the final flux linking the loop is zero. It follows from Eq. (22.3) that the average emf is

$$\mathscr{E}_{av} = -N\frac{\Delta\Phi_M}{\Delta t} = -200\frac{(-0.100 \text{ T·m}^2)}{0.100 \text{ s}}$$

and $\quad\quad \mathscr{E}_{av} = 2.0 \times 10^2$ V

Hence, from Ohm's Law, the average current is

$$I = \frac{V}{R} = \frac{(200 \text{ V})}{(5.0 \text{ Ω})} = \boxed{40 \text{ A}}$$

The current circulates clockwise around the loop so as to induce a B-field within its confines, which will tend to counter the decrease in downward flux. Thus, terminal 2 is positive with respect to terminal 1.

▶ **Quick Check:** The sides of the coil as it is crushed will experience a force via Eq. (21.13)—a force on a current-carrying wire in a B-field. That force will everywhere be in the direction of **v** × **B**, where **v** is in the

(continued)

(continued)

direction of motion of the positive current. For a clock-wise current, the force is to the left on the left side and to the right on the right side. The induced force will oppose the external force that is crushing the coil. Be-cause of resistance, the energy associated with the current is dissipated as joule heat, and the current stops as the emf $\rightarrow 0$; it would continue to circulate if the coil was superconducting.

Figure 22.7 A high-speed maglev. Superconducting coils on the bottom of the car induce opposing B-fields in aluminum coils in the guideway that lift the car.

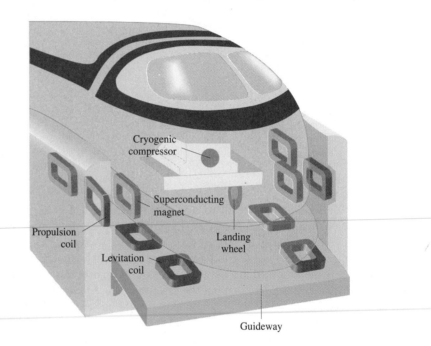

In one version of the magnetically levitated train (Fig. 22.7), there are super-conducting electromagnets in the bottom of the car. These pass above two rows of closed coils mounted in the guideway, and as they do, currents are induced in the coils creating an opposing field (via Lenz's Law) that lifts the vehicle six inches into the air. It is propelled by successively energizing the coils in the sides of the guideway.

22.3 Motional Emf

Suppose a length of ordinary wire is moved perpendicularly across a uniform magnetic field at a speed v (Fig. 22.8). The mobile charge carriers within it are dragged along to the right with the wire at the same speed. Moving through the field, they experience a force

$$\mathbf{F}_M = q\mathbf{v} \times \mathbf{B} \qquad\qquad [21.11]$$

parallel to the wire. In this case, the carriers are electrons and they flow downward. Accordingly, there is an induced emf across the wire—the top is at a higher potential than the bottom. As *within a battery,* negative charge is propelled from the $+$ terminal to the $-$ terminal.

Once a charge begins to move along the wire with some drift speed v_d, it will experience yet another force ($F_\perp = qv_d B$) due to this new motion in the B-field. The charge is then also propelled perpendicularly across the wire, but that lateral motion is restrained by the boundaries of the conductor. In accord with Lenz's Law, the net transverse force on all the mobile charges is opposite to the velocity of the wire and must be overcome by an externally applied force if the motion of the wire is to be sustained. *The external agency that moves the wire supplies energy to the system* (by doing work on the wire against the net transverse force). That energy is imparted to the induced current.

The motional emf (the change in potential) equals the work done on a positive test-charge in bringing it from one end of the rod to the other (that is, the change in its potential energy) per unit charge (p. 687). The force qvB acting parallel to the wire's length l does work on a charge in the amount of $qvBl$, and so the induced emf is

$$\mathscr{E} = vBl \tag{22.4}$$

Here the wire moves perpendicular to the B-field, the angle θ between **v** and **B** is 90°, and sin 90° = 1 in Eq. (21.7). It need not be so, in which case $v_\perp = v \sin \theta$ instead of v must appear in Eq. (22.4).

The downward transit of electrons makes the bottom end of the wire negative, leaving behind a positive upper end. This charge buildup continues until the repulsion thus produced on any subsequent approaching charge matches the driving force qvB, and the current stops. Simply put, the bottom end gets so negatively charged that the induced **motional emf** cannot push any more electrons down to it. If the motion of the wire stopped, the emf would go to zero and electrons would flow back up or, if you like, the hypothetical mobile positive charges would descend from the higher potential and move downward.

Because there is no closed circuit, there cannot be a steady current produced by this wire generator. Instead, at equilibrium, the separated charges will have created an electric field E that opposes any further motion of charge—the induced current is transient. As we saw earlier (p. 691), since emf = El, it follows that the electric field in the wire, which exactly counters the motional emf, is

$$E = vB \tag{22.5}$$

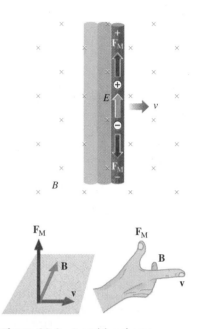

Figure 22.8 A positive charge experiences an upward force $F_M = qv \times B$ as the wire cuts across the B-field.

Example 22.3 A 1.0-meter-long wire held in a horizontal east-west orientation is dropped at a place where the Earth's magnetic field is 2.0×10^{-5} T, due north. Determine the induced emf 4.0 s after release.

Solution: [Given: $l = 1.0$ m, $t = 4.0$ s, and $B = 2.0 \times 10^{-5}$ T. Find: emf.] From Eq. (22.4), we will need the speed, thus

$$v = v_0 + gt = 0 + (9.8 \text{ m/s}^2)(4.0 \text{ s}) = 39.2 \text{ m/s}$$

hence

$$\mathscr{E} = vBl = (39.2 \text{ m/s})(2.0 \times 10^{-5} \text{ T})(1.0 \text{ m})$$

and

$$\boxed{\mathscr{E} = 0.78 \text{ mV}}$$

▶ **Quick Check:** $v \approx (10 \text{ m/s}^2)(4 \text{ s}^2) \approx 40$ m/s; $\mathscr{E} \approx 80 \times 10^{-5}$ V.

Cutting Field Lines

With the moving wire in Fig. 22.8, there is no closed loop and so no flux change; the induced emf arises from the motion of the charges across the field. We might expect that if the wire was kept stationary and the field moved to the left, the same relative

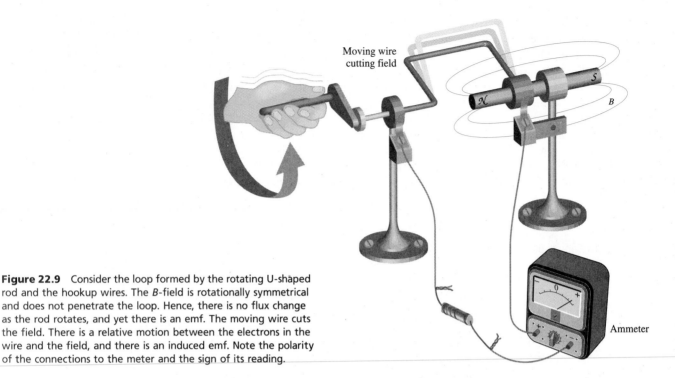

Moving wire
cutting field

Ammeter

Figure 22.9 Consider the loop formed by the rotating U-shaped rod and the hookup wires. The B-field is rotationally symmetrical and does not penetrate the loop. Hence, there is no flux change as the rod rotates, and yet there is an emf. The moving wire cuts the field. There is a relative motion between the electrons in the wire and the field, and there is an induced emf. Note the polarity of the connections to the meter and the sign of its reading.

motion would exist and the same emf would be induced and, indeed, that's what happens experimentally. With this in mind, Faraday provided an alternative model in the spirit of Eq. (22.4); it's pure 19th-century imagery, but it can be helpful, just don't take it literally. He envisioned the emf arising when the wire cut across magnetic field lines. Remember that the strength of a field B can be related to the number of field lines per unit area. Thus, through any perpendicular area A, the number of field lines is defined to be BA, the flux. When the wire in Fig. 22.8 moves with speed v, it sweeps across an area equal to vl per second. The number of field lines "cut" per second is therefore vlB, which, according to Eq. (22.4), equals the emf: *the induced emf equals the number of field lines cut per second by a conductor.*

Figure 22.9 shows the field of a permanent magnet being cut by a revolving wire. The field is rotationally symmetrical, and there is no flux through the loop bounded by the device, and yet if the crank is turned at a steady rate, a steady emf will be induced and a steady current will circulate. Both the emf and the current will vary linearly with the rotation rate.

Now suppose the moving wire of Fig. 22.8 is made to be part of a closed loop by placing it on a conducting U-shaped track (Fig. 22.10) having an appreciable resistance. Closing the loop will allow a current I to circulate, thereby extracting energy from the arrangement, if only as joule heat. The moving wire again produces an emf $= vBl$ much like a battery, and the resulting current provides energy at a rate of $P = I\mathscr{E}$; we have a dc generator. Notice that if the rod moves at a constant speed v in a time Δt it travels a distance $v\Delta t$ and sweeps out an area $\Delta A = v\Delta tl$. It therefore cuts a number of field lines equal to $B\Delta A = Bv\Delta tl$, and it does so at a rate of $B\Delta A/\Delta t = Bvl$, which indeed equals the emf! From Lenz's Law, the induced current progresses counterclockwise and thereby generates a force to the right opposing the applied force that is moving the wire.

Alternatively, we can say that the flux through the loop increases at a rate $B\Delta A/\Delta t = Bv\Delta tl/\Delta t = Bvl = emf$, which is the same result. Again, Lenz's Law main-

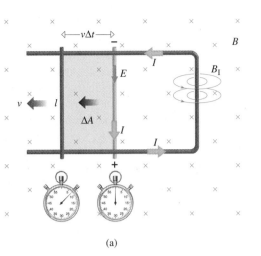

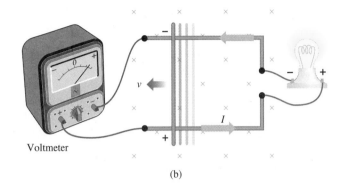

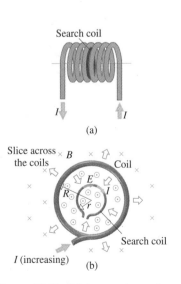

(a)   (b)

Figure 22.10 A straight wire sliding along, and in contact with, a stationary U-shaped conductor. (a) As the wire moves to the left, there is a time-rate-of-change of the flux equal to $B\Delta A/\Delta t = B(lv\,\Delta t)/\Delta t = Bvl$. Notice how the induced field B_I *within* the loop opposes the increase of flux. (b) Current circulates counterclockwise as if the load were placed across a battery with + and − terminals.

tains that the induced current will generate a *B*-field (out of the plane, inside the loop) that tends to decrease the downward flux increase. Inasmuch as the flux through a perpendicular area is BA and the number of field lines through that area is BA, it's not surprising that these two interpretations of the experimental results are *nearly* equivalent. What is remarkable is that we need two distinct notions (Eq's. 22.3 and 22.4) to deal with the whole range of physical possibilities associated with induced emfs. One might wonder whether or not we are missing something.

22.4 Induced Magnetic and Electric Fields

A stationary charge has a constant *E*-field at every surrounding point in space, whereas a moving charge—a current—produces a time-varying *E*-field. But a current is also surrounded by a *B*-field. This suggests a fresh way of looking at things from a field perspective; namely, **a time-varying E-field produces a B-field**. Let's now examine the reciprocal idea that a time-varying *B*-field is always accompanied by an *E*-field.

Picture a long solenoid of radius *R* carrying a current (Fig. 22.11). To probe for an electric field, we place within the solenoid a small coaxial coil of radius *r* having one or more turns, a so-called ***search coil*** (reminiscent of the test-charge). If the primary current is now increased, the flux through the search coil will increase, an emf will be induced across its terminals, and thus an induced *E*-field (E_I) will exist along it. Alternatively, we can envision the increasing current creating additional *B*-field lines that arise out of the wires and move inward toward the center of the primary (decreasing in density away from the wire). These lines cut the search coil, causing a positive charge therein to have an outward relative motion with respect to the field and thus be moved (in the direction of $\mathbf{v} \times \mathbf{B}$) to flow as a clockwise current, as required by Lenz's Law (Fig. 22.11b). At any point in the conductor of the search coil, a positive charge would experience a tangential force and a tangential electric field.

Since the wire loop of the search coil doesn't physically contribute to the *E*-field, it's useful to reconsider the above situation without the search coil in place. Experiments show that a charged particle will be accelerated in a uniformly varying *B*-field precisely as if there was a circular *E*-field present. In general, **an induced electric field accompanies a time-varying magnetic field** (p. 908). There not being any sources or sinks in the form of charges, *the induced E-field lines must close on*

Figure 22.11 (a) A current-carrying coil. As *I* increases, *B* increases, and inside the coil, field lines move in toward the center. (b) A search coil within the solenoid experiences an *E*-field.

themselves. The search coil simply allowed us to detect the presence of an independent *E*-field associated with the time-varying *B*-field. This is exactly what the circular UHF loop antenna you might have on top of your TV set is supposed to do.

Imagine a ring of wire immersed in a time-varying *B*-field. As long as the ring encompasses a time-varying flux $d\Phi_M/dt$, there will be an induced electric field (E_I) in the wire and an associated induced current. We can envision a positive charge circulating around the loop. Work is being done on it, just as if the charge were being pushed against friction—there will be joule heating and energy dissipated. But this is not an electrostatic situation nor, for that matter, is it equivalent to a battery-powered current. *Every point on the ring is identical to every other point*; this is very different from the battery-powered circuit where the battery provides the voltage rise across its terminals and the sum of the rises around the circuit equals the drops. We can go around an ordinary circuit with a voltmeter and measure the voltage (e.g., above ground) at every point and the values gotten will be independent of how the measurement is made; not so with the induced current (see Discussion Question 15). Since the environment at every point on the ring is identical, how could one point have a well-defined potential that is different from some other point?

As long as the time-varying *B*-field is sustained there will be an E_I and current will circulate. There is no static force field due to the presence of some primary charge that can store potential energy in the interaction. Whatever causes the time-varying *B*-field is the energy source; E_I is the agency that mediates the forces acting on the charges in the wire. Unlike the electrostatic *E*-field, which is predicated on an interaction with a primary charge, the induced *E*-field does not impart a specific potential to every point it occupies. **The induced E-field is not conservative.**

Faraday's Induction Law [Eq. (22.3)] can now be recast in terms of the induced *E*-field, which is a bit more concrete than the notion of a time-varying magnetic flux. We already have an expression for the relationship between the *E*-field and the voltage difference [Eq. (18.3)] and it can be carried over to this discussion, namely, $V = \int \mathbf{E} \cdot d\mathbf{l}$. It comes from the idea of the work done moving a point-charge in the field, and that's valid even if we now have to be careful because it's no longer path-independent; the concept of potential energy is inapplicable and the notion of potential is ambiguous. Still, we can use this line integral to compute the emf for any specific path between two points even in an induced *E*-field: $\mathcal{E} = \int \mathbf{E}_I \cdot d\mathbf{l}$. Our concern is with the emf induced in a closed circuit encompassing a time-varying flux. Thus *the line integral of the induced electric field around any closed path equals the time rate-of-change of the magnetic flux through any surface bounded by that path*:

$$[\textit{Faraday's Law}] \qquad \oint \mathbf{E}_I \cdot d\mathbf{l} = -\frac{d\Phi_M}{dt} = -\frac{d}{dt}\int \mathbf{B} \cdot d\mathbf{A} \qquad (22.6)$$

A time-varying B-field is always accompanied by an E-field. If there is a wire loop present it need not actually be touched by the *B*-field; it has only to encompass the region where there is a time-varying flux and it will experience an E_I and an emf. Indeed, if there were no wire loop there, at all, there still would be an encircling E_I. This restatement of Faraday's Law is one of the four pillars of Electromagnetic Theory known as Maxwell's Equations.

GENERATORS

Michael Faraday devised the world's first electric generator in 1831 (Fig. 22.12). It is a seemingly simple thing: a copper disk hand-cranked to rotate between the poles of

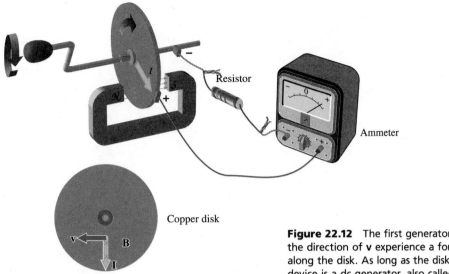

Copper disk

Figure 22.12 The first generator. Positive charge carriers moving in the direction of **v** experience a force $q\mathbf{v} \times \mathbf{B}$ that is radially outward along the disk. As long as the disk is turned in one direction, this device is a dc generator, also called a *dynamo*.

a permanent magnet. Charges in the metal, set in motion across the field as the disk turns, experience a radial force acting along the conductor. A conventional current of positive charges moves outward in the direction of $\mathbf{v} \times \mathbf{B}$. Work done on the mobile charges raises them in potential such that the rim terminal is positive and the axis terminal is negative. As long as the rotation rate is kept constant (which is easy to do), the emf is quite steady, and a continuous current is supplied to a load by this dc generator, or **dynamo**. The emf produced was small and the device impractical (modern versions have been used in electroplating where precise dc is desirable); still, its effect on the world was probably as far-reaching as any other single invention in all of history.

22.5 The AC Generator

Currently, the people of the world use electrical energy at a tremendous rate of about 10^{13} W. Almost all the electric current supplied commercially is generated with induction machines that produce a relative rotation between coils of wire and magnetic fields. The external power to rotate either the coils or the field is usually supplied by steam turbines (steam blasting against fanlike blades that are forced into rotation). The steam generally comes either from burning fossil fuels or from nuclear reactors. A simplified version of an alternating current generator is shown in Fig. 22.13. Usually a coil, of several turns of wire wound on an iron armature, is rotated in the constant field of a magnet. Brushes rubbing against two slip rings attached to the ends of the coil carry off the induced current. As the coil rotates, its two long parallel sides (1–2 and 3–4), each of length l, move through the field in opposite directions. The emf induced across side 1–2, moving at a speed $v_{\perp} = v \sin \theta$ with respect to B, follows from Eq. (22.4):

$$\mathscr{E} = v_{\perp} Bl = Blv \sin \theta$$

At the moment pictured in Fig. 22.13c, $\mathbf{v} \times \mathbf{B}$ points from 2 to 1, so point 2 is at a lower potential than point 1 (work is done in bringing positive charges over to point 1). The same emf appears across the length 4–3 with point 4 lower than point 3. Since no emf is induced along the lengths of 1–4 and 2–3, the net emf around the single loop is $\mathscr{E} = 2(Blv \sin \theta)$ with point 4 negative (lower) and point 1 positive

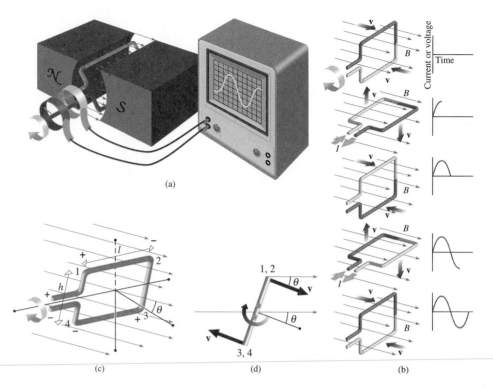

Figure 22.13 (a) A simple ac generator consisting of a loop rotating in a magnetic field. (b) Various stages in the rotation cycle, along with graphs of the corresponding generator outputs. (c) Detail of the conducting loop tilted at an angle θ. (d) The same loop shown in profile.

(higher): a conventional current will circulate from 4 to 1 driven by induction and then out (from the + terminal, near 1) into the external circuit, around and back into the loop (at the − terminal, near 4). When there are N turns of wire forming a coil, the emf of each loop is in series with the next, yielding a net emf of

$$\mathscr{E} = 2NBlv \sin \theta$$

When θ exceeds 180°, the polarity reverses. Since both θ and v can be written in terms of ω, we use the fact that when the angular speed ω is constant, $\theta = \omega t$. Moreover, $v = r\omega$, where here $r = \frac{1}{2}h$. Keeping in mind that lh is the area A, $lv = l \times (\frac{1}{2}h\omega) = \frac{1}{2}A\omega$ and the emf becomes

$$\mathscr{E} = NAB\omega \sin \omega t \qquad (22.7)$$

This same result is derived with a bit more finesse via Faraday's Law in Problem 84.

Regardless of the actual shape of the coil, the emf will be alternating with a frequency $f = \omega/2\pi$, via Eq. (12.2). In the United States and Canada, f is typically 60 Hz, whereas in much of the rest of the world f is 50 Hz. AC voltages can be changed quite easily and so you can plug an American hairdryer (using a converter) into a British wall outlet—the dryer isn't noticeably affected by the frequency difference. On the other hand, you will have a lot more trouble with something requiring timing signals, like playing an American video tape on a French VCR.

The simple generator of Fig. 22.13, with a few horseshoe magnets providing the field, was once commonplace (you've probably seen someone in an old movie cranking a telephone or perhaps a detonator just prior to setting off explosives). In any event, when large voltages (several kilovolts) and currents (of 50 amps and more) are involved, the slip rings and brushes, which tend to spark and deteriorate, become too troublesome. Today, this sort of *revolving-armature* machine (the armature is the component in which the emf is induced) is far less important commercially than it once was. To avoid the difficulties associated with voltage exceeding about 600 V, the coil assembly carrying the induced current is made stationary, and electromagnets are mounted on the turning shaft to produce a *revolving-field* machine. The current creating the revolving *B*-field is comparatively small, and slip rings and brushes can handle the task well. Meanwhile, the large induced current passes through a motionless, heavy, continuous wire. This kind of *alternator* powers the electrical system in most automobiles, where it's turned by the engine via belts and pulleys. The ac output must subsequently be converted to dc in order to charge the battery.

22.6 The DC Generator

There are many applications where dc electrical power is required, even though it's somewhat more difficult to produce. For example, although small motors (for things like hand drills and vacuum cleaners) run nicely on ac, large electrical motors generally don't do quite so well. The really big motors used on electric railway systems (trolley cars, subway trains, etc.) usually operate on dc.

The polarity reversals of the voltage from an ac generator can be eliminated using a split-ring *commutator* (Fig. 22.14), just like that on a simple dc motor. Thus, the negative half of the ac signal is reversed and made positive. The result is a bumpy direct current that rises and falls but never goes negative (Fig. 22.15). The single-coil arrangement produces an emf that is zero twice during each revolution. Moreover, the voltage is relatively small for much of the cycle. By contrast, if the armature comprises two perpendicular coils (Fig. 22.16) and the commutator has four segments, so that the brushes are always in contact with the coil having the greatest emf, the output voltage across the terminals will be appreciably smoother, and the zeros will no longer occur. With 90° between coils, there will be a relative shift in the induced voltages of 90°, which is equivalent to a shift in time of $\pi/2\omega$, as shown in Fig. 22.16b. With three coils 60° apart, three voltages each shifted in time by $\pi/3\omega$ will be induced, yielding a final voltage equal to the envelope of the three curves, as in Fig. 22.17. Still more coils can be added, and the output voltage can thereby be made almost constant with only a slight ripple.

It's curious that the similarity between the dynamo and the dc motor was not recognized until the mid-nineteenth century and not really appreciated until a remarkable accident occurred in 1873. Not only do they look alike, they are actually identical: the action of the dynamo is just the reverse of the action of the dc motor. The dynamo converts mechanical energy into electrical energy—the armature is turned mechanically from the outside, and out comes a current. The motor converts electrical energy into mechanical energy—in comes a current from the outside, and around goes the armature.

The large hall at the 1873 Vienna Exposition was filled with modern gadgets. One of the Gramme dynamos driven by a steam engine was pouring forth electrical

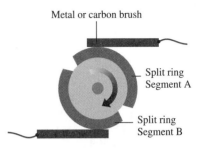

Figure 22.14 A two-segment commutator. At the moment shown, the top brush is touching segment *A* and the bottom one, segment *B*. As the ring turns, segment *A* moves into contact with the bottom brush. Reversing the connection reverses the signal.

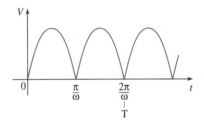

Figure 22.15 The pulsating dc from a single-coil generator with a two-segment commutator. Just as the sinusoidal voltage is about to go negative, the commutator flips it.

Figure 22.16 (a) A two-coil dc generator. Compare this device with the one depicted in Fig. 22.13 and note how the brushes are in contact with one of the loops at maximum voltage. (b) Emf of the two-coil dc generator. Notice how this emf is equivalent to two voltages like the one shown in Fig. 22.15, shifted by $\pi/2\omega = \frac{1}{4}\tau = 90°$ with respect to one another. A single coil contributes for $\frac{1}{4}$ of a cycle until the brushes shift and the other coil kicks in. It then contributes for the next $\frac{1}{4}$ cycle, and so on. At any instant only one coil is providing the voltage.

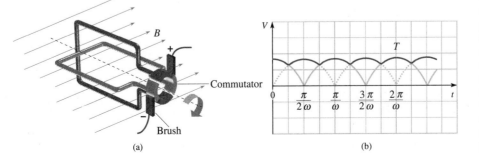

(a) (b)

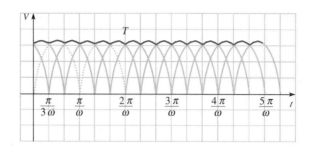

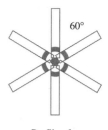

Profile of
a triple coil

Figure 22.17 The emf of a three-coil dc generator. Each component signal is shifted by $\pi/3\omega = \frac{1}{6}\tau = 60°$.

Example 22.4 A simple single-coil dc generator rotates at a constant frequency of 60 Hz in a 0.40-T magnetic field. Given that the coil has 10 turns and encompasses an area of 1200 cm², what will be its maximum emf?

Solution: [Given: $f = 60$ Hz, $A = 1200$ cm², $B = 0.40$ T, and $N = 10$. Find: the maximum emf.] The maximum emf ($\mathscr{E}_m$) is the amplitude of the oscillating voltage given by

$$\mathscr{E} = NAB\omega \sin \omega t \qquad [22.7]$$

namely

$$\mathscr{E}_m = NAB\omega$$

Here $\omega = 2\pi f = 2\pi(60$ Hz$) = 376.99$ rad/s and

$$\boxed{\mathscr{E}_m = 0.18 \text{ kV}}$$

▶ **Quick Check:** $\omega \approx 6(60) \approx 360$ rad/s; $\mathscr{E}_m \approx 10 \times (0.12$ m²$)(0.4$ T$)(360$ rad/s$) \approx 0.17$ kV.

power when a workman unwittingly connected the output leads from another dynamo to the energized circuit. In an instant, the second device began to spark and whine and come alive, whirling around at great speed. The dynamo had become a motor. That impromptu mating of machines gave birth to the future as we have come to know it.

22.7 Eddy Currents

When Arago rotated a copper disk beneath a compass and found that the needle soon began revolving around with the disk, he was actually observing the effects of electromagnetic induction. The rotating disk "sees" a nonuniform time-varying magnetic field. An emf is induced and loops of current are set up in the disk. These currents generate their own counter B-field that opposes, via Lenz's Law, the cause of the induction—the needle is thereby made to rotate such that its relative motion with respect to the disk tends to vanish.

If an extended conductor translates with respect to a B-field, which is not uniform over the entire conductor, or if different portions of the conductor move at different velocities (that is, it rotates) with respect to the field, currents will be induced within it that circulate in closed paths. Instead of the transient current in the uniform-field situation of Fig. 22.8, a closed current loop can now form. For example, with Faraday's disk (open circuited, as shown in Fig. 22.18) the induced current loops, known as **eddy currents**, set up two dipole field regions, each north on one face of the disk and south on the other. The leading dipole, emerging from the external B-field on the left, attracts the poles of the magnet, thus tending to oppose the rotation. The trailing eddy-current dipole, moving into the B-field of the magnet, presents its north to the magnet's north and its south to the magnet's south, thus repelling the magnet and again resisting the motion. Consequently, the eddy currents produce a Lenz's Law drag on the rotating disk.

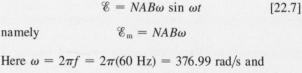

Induced
current

Applied B-field

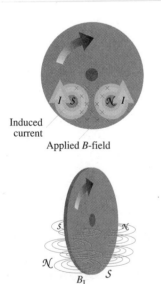

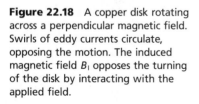

Figure 22.18 A copper disk rotating across a perpendicular magnetic field. Swirls of eddy currents circulate, opposing the motion. The induced magnetic field B_I opposes the turning of the disk by interacting with the applied field.

Similarly, the iron armature of a motor or generator (or even the core of an electromagnet) will harbor eddy currents that waste energy in the form of joule heating. To reduce this effect, most cores are built up of stacks of separate insulated sheets of iron.

Eddy currents caused by the geomagnetic field are commonplace, even if they are rarely of much consequence—they're induced in a metal spoon whenever you stir a cup of coffee, and in a coin when you flip it. In the early days of the Space Age, engineers were surprised to find that many of their low-altitude spin-stabilized satellites were gradually "despinning." The cause was traced to eddy currents induced in the walls of spacecraft, which were rotating in the planet's magnetic field. Not all such effects are detrimental: eddy currents are used to great advantage in heating induction furnaces, in damping the oscillations of delicate devices, and in the operation of a variety of instruments, including the metal detector now familiar to air travelers and beachcombers alike. In the latter, a time-varying magnetic field induces currents in a metal object, and the resulting induced *B*-field reveals its presence to a pickup coil.

SELF-INDUCTION

Whenever a voltage from some external source is placed across the terminals of a coil, the resulting current will produce a magnetic field. But this phenomenon raises an interesting point first considered by Joseph Henry. A coil laced with an increasing field must experience an induced emf—what does it matter that the coil itself is involved in creating the field? Field lines stream out from the center of the wire of the coil "cutting" it, just as they would cut a search coil immediately adjacent to it. Mobile charges within the wire experience a time-varying *B*-field, a perpendicular force, and a resulting induced emf along the length of the conductor. Or, if you prefer, the flux linking the coil will be changing in time, and Eq. (22.3) therefore requires that there be an induced emf. With either interpretation, it follows from Lenz's Law that the induced emf must oppose the cause of itself and so it is called a *back-emf*. Because of this opposition, the steady state is not reached instantaneously; the current builds gradually, as does its associated *B*-field. This process of **self-induction** retards the increase or decrease of current in a coil (and to a lesser extent in other circuit elements, hookup wires, transmission lines, and so on). As such, it is an extremely important aspect of almost all real ac devices from stereos to satellites.

Faraday delivering one of his famous Christmas lectures (1855). On the left in the front row is the Prince Consort between his two princely charges. This room at the Royal Institution in London still exists, and they still give demonstrations there.

22.8 Inductance

To quantify the self-induction of a current-carrying coil, we realize that the flux linkage $N\Phi_M$ is proportional to the current I producing it:

$$N\Phi_M = LI \qquad (22.8)$$

The constant of proportionality L is called the **self-inductance**, or just the *inductance,* for short. In a sense, it is the electrical equivalent of inertia, a resistance to change. The SI unit of inductance is the **henry** (H); a flux linkage of 1 weber is established in a coil by a current of 1 ampere circulating therein when the coil's inductance is 1 henry.

The antenna of the old radio pictured on p. 645. The antenna consists of a coil wrapped around a ferrite core (the gray, cylindrical bar). The time-varying magnetic field component of an incoming electromagnetic wave induces a signal, an emf, in the coil.

As we have seen so many times before, the constant of proportionality L is a composite of several of the system's physical characteristics. Here, it depends on the size and shape of the coil and on the surrounding medium. The flux linkage does not vary linearly with I if the permeability of the medium is not constant—that is, if it depends on B and therefore on I. Consequently, if there is a ferromagnetic material within the coil, or even nearby, L becomes a function of I and though Eq. (22.8) still holds, it is inherently more complicated than it would be were L constant.

In practice, when the precise value of an inductance is needed, it's usually measured, but if the geometry is simple enough, it can be approximated theoretically. As an example, to compute the inductance of a long hollow coil of length l with n turns per unit length ($n = N/l$) and a cross-sectional area A, we recall that

$$B_z \approx \mu_0 n I \qquad [21.5]$$

Assuming B to be constant across the region encompassed by the solenoid (although we know the field actually decreases in toward the center), we should get a useful expression, even if it's likely to be slightly too large. Using Eq. (22.8) the inductance of a long air-core solenoid becomes

$$L = \frac{N\Phi_M}{I} = \frac{NBA}{I} \approx \frac{\mu_0 N^2 A}{l} \qquad (22.9)$$

If some material with a permeability μ is made to fill the hollow, we simply replace μ_0 by μ. When that material is ferromagnetic, the equation still works, provided we know μ at the particular current being used. There are commercial inductors that allow their inductances to be adjusted by sliding a ferrite slug partway into the coil. (See the photo on p. 850.) Ferrites (oxides of magnesium, manganese, zinc, or nickel) are especially useful in suppressing core eddy currents at the high operating frequencies of radio and television circuits.

Example 22.5 A solenoid 3.0-cm long with a cross-sectional area of 0.50 cm^2 and comprising 300 turns of fine copper wire in a single layer is to be used as the antenna for a radio. The magnetic field component of the incoming electromagnetic signal will oscillate within the coil and induce an emf that will then be processed by the rest of the radio, and out will come music. (a) Determine the coil's inductance when the core is air-filled. (b) Approximate the inductance when a ferrite core is used instead, given that its relative permeability at the anticipated current level is 400.

Solution: [Given: $l = 0.030$ m, $A = 0.50 \times 10^{-4}$ m^2, $N = 300$, and $\mu/\mu_0 = 400$. Find: L.] (a) Using Eq. (22.9)

$$L \approx \frac{\mu_0 N^2 A}{l}$$

$$L \approx \frac{(1.26 \times 10^{-6} \text{ T·m/A})(300)^2(0.50 \times 10^{-4} \text{ m}^2)}{(0.030 \text{ m})}$$

and to two significant figures $L \approx 1.88 \times 10^{-4}$ H $\approx$ $\boxed{0.19 \text{ mH}}$. (b) Given $\mu = 400\mu_0$, L is increased by a factor of 400 and $\boxed{L \approx 75 \text{ mH}}$.

▶ **Quick Check:** $N^2 A \approx 9 \times 10^4 (0.5 \times 10^{-4}) \approx 4.5$; $N^2 A/l \approx 150$; $L \approx 1\frac{1}{4}(150) \ \mu\text{H} \approx 190 \ \mu\text{H}$.

22.9 The Back-Emf

Faraday's Induction Law can be applied to the inductance, thereby providing an expression for the back-emf. From Eq's. (22.3) and (22.9) the average self-induced emf is

$$(\mathscr{E}_L)_{av} = -N\frac{\Delta\Phi_M}{\Delta t} = -\frac{\Delta(LI)}{\Delta t}$$

If the inductance is constant, $\Delta(LI) = L\Delta I$ and

$$(\mathscr{E}_L)_{av} = -L\frac{\Delta I}{\Delta t}$$

or more generally, letting $\Delta t \to 0$, in the limit

$$\mathscr{E}_L = -L\frac{dI}{dt} \qquad (22.10)$$

The instantaneous induced emf is proportional to the time rate-of-change of the current in the coil. *The positive direction is that of I.* An inductance of 1 H will induce a back-emf of 1 V when the current through it changes at a rate of 1 A/s. This expression also tells us that 1 V = 1 H·A/s and so 1 H = 1 V·s/A, or 1 H = 1 Ω·s. *If the coil itself has negligible resistance, there will not be an appreciable voltage drop across it due to the current traversing it, and the back-emf will be the voltage measured across its terminals.*

Suppose an externally supplied current is made to *increase* through a coil passing from its terminals A to B (Fig. 22.19a). In response to the changing flux, an induced current (I_1) then moves from B to A, opposing ΔI and reducing I in the circuit, which in effect raises the potential of point A making it positive with respect to B. This back-emf is in the opposite direction to I and is negative—it's a voltage drop. The coil's inductance impedes the original *changing* current, thereby producing a voltage drop across its terminals (as if it now had some kind of "resistance" even though $IR = 0$). While the current is increasing, there will be a voltage across the terminals as shown. If the current begins decreasing, an induced current will pass from A to B to oppose the decrease. Point B will be positive with respect to point A, and the back-emf will reverse, becoming positive (Fig. 22.19b).

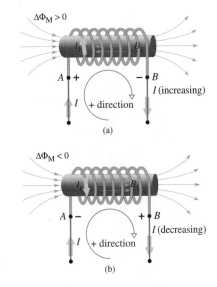

Figure 22.19 (a) I is increasing, an induced current I_1 opposes that increase, and a negative back-emf (a voltage drop) appears across the terminals. (b) I is decreasing, I_1 opposes that decrease, and a positive back-emf (a voltage rise) appears across the terminals.

Example 22.6 The current in a 50-μH coil (for which R is negligible) goes from 0 to 2.0 A in 0.10 s. Determine the average self-induced emf measured across its terminals.

Solution: [Given: $L = 50$ μH, $I_i = 0$, $I_f = 2.0$ A, and $\Delta t = 0.10$ s. Find: $(\mathscr{E}_L)_{av}$.] From Eq. (22.10) the numerical value of the back-emf is

$$(\mathscr{E}_L)_{av} = -L\frac{\Delta I}{\Delta t} = -(50 \times 10^{-6}\text{ H})\frac{2.0\text{ A}}{0.10\text{ s}}$$

and $\boxed{(\mathscr{E}_L)_{av} = -1.0\text{ mV}}$

The negative sign tells us what we presumably knew—namely, that the back-emf opposes the increase in current. The polarity shown in Fig. 22.19 already took the sign into consideration.

▶ **Quick Check:** The current is changing at a moderate rate of 20 A/s, but the inductance is only 50 microhenries, so we can expect an emf in the millivolt range.

An **inductor** is a device designed expressly to introduce inductance into a circuit, something that is done for a variety of reasons. For instance, an inductor (or *choke*) opposes a changing current and will therefore impede the progress of alternating currents while passing, with very little opposition, steady currents. Not sur-

This adjustable ferrite-core inductor is part of the tuning circuit of the AM radio on p. 645. We'll come back to what it does when we consider resonant circuits (p. 881).

prisingly, this impedance (p. 878) to ac increases with both the frequency of the current and the inductance of the inductor. Accordingly, inductors are often used to separate or filter out ac from dc. Typically, an inductor consists of a coil of wire wound on a hollow cylinder that may contain air or some kind of ferromagnetic core. Values usually range from microhenries (μH), used at high frequencies as in radio and television, to several henries, used, for instance, in low-frequency (60 Hz) choke-filtered power supplies. The circuit symbols for an inductor with and without a ferromagnetic core are ‑⁓⁓⁓‑ and ‑⁓⁓⁓‑ , respectively.

Wire-wound resistors are also coils, but in that case having an appreciable inductance is generally not desirable. The solution in practice is to double the wire over on itself before winding the coil. Current then circulates in opposite directions simultaneously, and the flux in these *noninductive* coils is zero.

22.10 The *R-L* Circuit: Transients

Figure 22.20 shows a battery in series with an inductor L and a resistor R. The inductor is imagined to have zero resistance, although we could have just as well lumped its more realistic nonzero resistance into R. As soon as the switch is closed, an increasing current I just begins to circulate, but it's opposed by the back-emf, so it builds only gradually. From Kirchhoff's Loop Rule (p. 766), at any instant

$$V = L\frac{dI}{dt} + IR \qquad (22.11)$$

Remember that V, R, and L are fixed but I varies from moment to moment. Inasmuch as the battery voltage V is constant, it follows from the equation that as the current increases, it must increase ever more slowly. This occurs because as IR increases, the back-emf decreases, and dI/dt decreases. That's shown in the exponential I-versus-t curves of Fig. 22.21a. The slopes do indeed decrease with time. The inductor causes the current to rise slowly, reaching a maximum value of V/R as $dI/dt \to 0$, as $t \to \infty$.

A mathematical expression for $I(t)$ can be gotten by solving Eq. (22.11). First, get all the current terms on one side

$$\frac{V - IR}{dI} = \frac{L}{dt}$$

Then flip it and integrate,

$$\int_0^I \frac{dI}{V - IR} = \int_0^t \frac{dt}{L}$$

which yields the natural log

$$\frac{1}{-R}\ln\left[\frac{V - IR}{V}\right] = \frac{t}{L}$$

Multiply both sides by $-R$ and exponentiate both sides using the fact that $e^{\ln x} = x$:

$$\frac{V - IR}{V} = e^{-Rt/L}$$

and

$$I(t) = \frac{V}{R}(1 - e^{-t/\tau}) \qquad (22.12)$$

where the **time constant** is $\tau = L/R$.

Figure 22.20 An *R-L* circuit with the switch just closed.

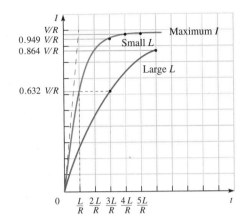 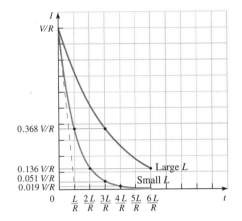

Since

$$L\frac{dI}{dt} = V - IR$$

at the instant the switch is closed ($t = 0$), the current begins to rise from its value of zero and, hence, the emf initially equals V; that is, $IR = 0$ and $L(dI/dt) = V$. Thus, the initial slope of the curve at $t = 0$ is

$$\frac{dI}{dt} = \frac{V}{L}$$

which we could have gotten, alternatively, by taking the derivative of $I(t)$. Had the current continued to rise at its initial rate of increase (that is, with the slope it had immediately after the switch was closed—namely, V/L) it would have reached its maximum value (V/R) at a time L/R (such that V/R divided by L/R equals V/L), which is the **time constant**. After an interval of 1 time constant, the current reaches an amount of 0.632 of its final value (V/R). After an interval 5 time constants long, the current is within 1% of its steady-state value.

Notice that when R is very large, most of the voltage drop will be across the resistor, the back-emf will be small, and the delay it causes in the current buildup will be small. On the other hand, the larger the inductance, the greater the back-emf, the less tilted the I-t curve is initially, the larger the time constant, and the longer it takes for the current to reach maximum [which brings to mind the *R-C* circuit (p. 765), where the time constant is RC].

Figure 22.21 (a) The graph shows I versus t for an *R-L* circuit with a time constant of L/R. For comparison, another curve is included where the time constant is 3 times as large. Note that the curves rise to 63.2% of maximum value after L/R, and 63.2% of the remaining 36.8% after $2L/R$, and 63.2% of the remaining 13.6% after $3L/R$, and so on. (b) When the switch in Fig. 22.22 is opened and the current decays, the current drops 63.2% of the remaining amount during each successive one-time-constant interval. Thus the current falls to 36.8% V/R after L/R, and then 63.2% of 36.8% V/R to 13.6% V/R at $2L/R$, then 63.2% of 13.6% V/R to 5.1% V/R at $3L/R$, and so on.

Example 22.7 Determine the time constants for a series R-L circuit consisting in one case of a 100-Ω resistor and a 10-H inductor, and in the other case of the same resistor with a 1.0-H inductor. Assuming the circuit of Fig. 22.20, compare the two resulting currents.

Solution: [Given: $L = 10$ H and 1.0 H, and $R = 100\ \Omega$. Find: the time constants.] The time constant is L/R; hence

$$\frac{L}{R} = \frac{10\ \text{H}}{100\ \Omega} = \boxed{0.10\ \text{s}}$$

as compared to

$$\frac{L}{R} = \frac{1.0\ \text{H}}{100\ \Omega} = \boxed{10\ \text{ms}}$$

With the larger inductor, we will have to wait 0.10 s for the current to reach 63% of its maximum value, as compared to only 10 ms for the smaller one.

▶ **Quick Check:** The inductances differ by a factor of 10, so the time constants L/R must also: (0.10 s)/ (10 ms) = 10.

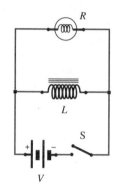

Figure 22.22 An inductor, a battery, and a lamp. The resistance of the lamp (R) is much greater than that of the coil, and the lamp gets little current when S is closed. Remarkably, it glows brightly for a moment after the switch is opened.

Opening the switch once the steady state has been established will cause the current to gradually decay, being sustained by the back-emf (Fig. 22.21b). Have you ever pulled the plug on an appliance that was still turned on and seen a spark at the ends of the prongs? If a lamp with a large resistance, compared to that of the coil, is positioned as in Fig. 22.22 and the switch is closed, it will only glow faintly in the steady state: most of the current will then be passing through the inductor since the back-emf is zero and the resistance low. On opening the switch, the lamp will not immediately go out, but instead will burn even more brightly for a moment and then gradually diminish in intensity. The back-emf at the instant S is opened can be greater than the steady-state voltage across the lamp— that condition will not last long, but it is what makes the lamp flare up initially. The larger the L, the larger the time constant, and the longer the lamp glows on (see Problem 86).

One might well ask at this point, "Where is the energy coming from that powers the lamp when the switch is opened?" Clearly, it's not from the battery and that only leaves the inductor itself. Classical theory maintains that the B-field must possess energy that it imparts to the coil in the form of an induced current when the steady-state current, which sustained the field, is disrupted. The model suggests that when the switch is opened the field returns its energy to the coil. The realization that the field carries energy (and momentum), which are properties of matter, raises a number of interesting questions that will have to be dealt with later. For example, how exactly is energy transferred from the field to a charged particle?

22.11 Energy in the Magnetic Field

In Chapter 18, we talked about the energy stored in a charged capacitor in the process of building up the electric field that spans the gap between its plates:

$$\text{PE}_E = \tfrac{1}{2}CV^2 = \tfrac{1}{2}\varepsilon_0(Ad)E^2 \qquad [18.17]$$

The field in the gap can be imagined as retaining this energy uniformly within the space of volume Ad, which it occupies. Thus, if we introduce the concept of *energy per unit volume of the electric field* (u_E), the above expression becomes

$$u_E = \tfrac{1}{2}\varepsilon_0 E^2 \qquad (22.13)$$

This is the **energy density** of the electric field; wherever there is an E-field in space, this will be the energy per unit volume associated with it.

The analogous situation exists when we establish a current through an inductor, thereby building up a magnetic field. In Fig. 22.20, the battery must do work W against the back-emf if it is to send a current through the inductor. During a tiny interval of time Δt, the battery, which provides power to the inductor at a rate P, does a small amount of work

$$\Delta W = \text{P}\Delta t = I\mathscr{E}\,\Delta t = IL\frac{\Delta I}{\Delta t}\Delta t$$

and so
$$\Delta W = IL\,\Delta I$$

where I is the current at any instant. We want the total amount of work done in increasing the current from $I = 0$ to $I = I_f$:

$$\Sigma\,\Delta W = \Sigma\,IL\,\Delta I$$

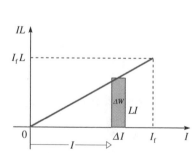

Figure 22.23 A plot of IL versus I. The area under the curve is the work done in building up a current of I_f through an inductor L.

The right side is the area under a plot of IL versus I, as shown in Fig. 22.23. Since L is constant, this plot is a straight line and the total area (W) under it between 0 and I_f is the triangular area of height I_fL and base I_f; namely, $W = \tfrac{1}{2}LI_f^2$. This is the work done on the inductor and it's also the magnetic potential energy PE_M stored. Remembering that we are considering the final current, the subscript f can be dropped for simplicity. The energy stored in an inductor carrying a current I is then

$$PE_M = \tfrac{1}{2} L I^2 \tag{22.14}$$

where we *assume* it's associated with the B-field. (For an alternative derivation that makes more direct use of calculus see Problem 87.) Thus, with a long solenoid of length l, where from Eq. (22.9) $L = NBA/I$, the field existing inside is $B \approx \mu_0 NI/l$, and so

$$PE_M = \tfrac{1}{2} L I^2 = \tfrac{1}{2} NBAI = \tfrac{1}{2} (Al) \frac{B^2}{\mu_0} \tag{22.15}$$

Representing the *energy per unit volume of the magnetic field* as u_M and realizing the Al is the volume occupied by the B-field, we have

$$u_M = \tfrac{1}{2} \frac{B^2}{\mu_0} \tag{22.16}$$

Although this formula was derived for a long solenoid, it turns out to be generally valid. Apparently, it is possible to store electromagnetic energy in electric and magnetic fields in vacuum. As we shall see in Chapter 24, it is even possible to transport this energy across free space—indeed, that's what light is.

So far, we have dealt with a variety of subtle electromagnetic phenomena in terms of classical field theory. It allows us to understand everything we have looked at in terms of electric and magnetic fields using rather picturesque nineteenth-century images of flux and field lines. Indeed, these fields seem to become more substantial with each extension of the analysis. **The classical field appears as a continuous something that can store, transfer, and transport energy** (although it's not obvious exactly how it does that). This picture will have to be profoundly modified when we confront the modern concept that fields are quantized and radiant energy exists only in minute discrete bursts. Usually however, when observed on a macroscopic scale, quantum fields average out to, and become indistinguishable from, classical fields. Though the classical electromagnetic field is an indispensable conceptual tool, as we'll see, the model can no longer be taken quite so literally.

Core Material

The **flux of the magnetic field** is

$$\Phi_M = \int \mathbf{B} \cdot d\mathbf{A} \tag{22.1}$$

with units of **webers**: 1 Wb = 1 T·m². As a result of flux changes, the induced emf is

$$\mathscr{E} = -\frac{d\Phi_M}{dt} \tag{22.2}$$

which is **Faraday's Induction Law** (p. 835). **Lenz's Law** is

The induced emf will produce a current that always acts to oppose the change that originally caused it.

When a wire of length l moves at a speed v perpendicularly through a B-field, the induced **motional emf** (p. 839) is

$$\mathscr{E} = vBl \tag{22.4}$$

Another form of Faraday's Law is

$$[\textit{Faraday's Law}] \quad \oint \mathbf{E}_I \cdot d\mathbf{l} = -\frac{d}{dt} \int \mathbf{B} \cdot d\mathbf{A} \tag{22.6}$$

An ac generator revolving at a rate ω produces an emf given by

$$\mathscr{E} = NAB\omega \sin \omega t \qquad [22.7]$$

The net flux through a current-carrying coil—the flux linkage $N\Phi_M$—is proportional to the current I producing it:

$$N\Phi_M = LI \qquad [22.8]$$

The constant of proportionality, L, is the **self-inductance**, or just the *inductance*. The self-inductance of a long air-core solenoid is

$$L \approx \frac{\mu_0 N^2 A}{l} \qquad [22.9]$$

The induced, or **back-emf**, in a coil is

$$\mathscr{E}_L = -L \frac{dI}{dt} \qquad [22.10]$$

When such a coil is placed in series with a battery of voltage V and a resistor R, we have an R-L circuit where

$$V = L\frac{dI}{dt} + IR \qquad [22.11]$$

The *energy per unit volume of the magnetic field* is

$$u_M = \frac{1}{2}\frac{B^2}{\mu_0} \qquad [22.16]$$

Suggestions on Problem Solving

1. When determining the polarity of an emf induced in a coil, it's useful to imagine a resistor placed across the terminals, forming an external circuit. The current enters the resistor on the (+) high side and exits on the (−) low side, as in Figs. 22.10b, 22.13c, and 22.19. The induced current *inside* the coil travels from (−) to (+) just as it does in a battery, which often seems to be a sticky point for students. The induced emf is the work done, in accord with Lenz's Law, on the charge, per unit charge.

2. Be aware that $\Delta \mathbf{B}$ may point in a different direction from the **B**-field. Thus, if the field is directed to the east and decreasing, $\Delta \mathbf{B}$ points west. If there is an induced current, its induced field will oppose this change and so must point east.

3. Imagine an electrostatic E-field and suppose that we carry a charge from point 1 to point 2. Work will be done and an emf will exist between 1 and 2, an emf that is independent of the path taken. The electrostatic E-field is conservative, but the E-field arising from a changing magnetic flux is *not* conservative. The emf does depend on the path taken between 1 and 2 since different amounts of work may well be needed. Given a time-varying B-field in some localized region of space, it's often possible to go around a closed loop so the path either encompasses the varying flux totally or in part, or even avoids it, thereby leading to different path-dependent values of the emf.

The work done in moving a charge around a closed path in an electrostatic E-field is zero—you come back to the same energy you started with. By contrast, in the case of an induced E-field, if we move a positive charge around a loop, work may be done continuously. *The concept of potential is ambiguous with induced E-fields* (see Discussion Question 15).

4. The sign in Eq. (22.10) for the back-emf is there to remind us that the emf opposes that which causes it. If the current is building at a rate of, say, 2.0 A/s, then dI/dt is positive and the emf is negative (meaning it opposes the buildup). However, without a clear picture of the system, the minus sign by itself doesn't provide a complete picture. In such cases, it's often easier to put aside the sign, remembering that the emf opposes its cause. For example, given that a certain back-emf is measured to be 10 V while the current buildup is 2.0 A/s, find the inductance. Since this situation represents an increase of current, the emf must be put into Eq. (22.10) as −10 V or you will get a negative L, which is impossible. Had the current been decreasing, it would be entered as −2.0 A/s and the emf would then be positive, again yielding a positive L. It's common to just give the numerical value of the emf in a problem (without talking about which terminal is higher or lower), so you must be careful not to come up with a negative L.

Discussion Questions

1. In 1825, long before the successful work by Henry and Faraday on induction, Jean Daniel Colladon attempted to observe the effect by attaching a helical coil to a sensitive galvanometer. To shield the delicate instrument from the direct influence of the moving magnet, which he planned to wave around near the coil, Colladon wired the galvanometer to the rest of the circuit via two long leads so it could be safely located in an adjacent room. Because he had no assistant, whenever he moved the magnet he had to walk over to the galvanometer to observe its response. Since the name of Colladon is never included with that of Faraday and Henry, what do you think he saw and what did he not see? Explain.

2. During a major period of solar activity, magnetic storms of such severity occur that voltages of upwards of 1 kV can appear across the length of the Alaska pipeline from Prudhoe Bay to Valdez. Explain how this happens. (In 1956, a particularly strong magnetic storm severely affected the first transatlantic voice cable. Such effects were a troublesome problem for early telegraph operators.)

3. Using rechargeable batteries, it is possible to power small appliances such as electric toothbrushes with watertight sealed units. Rather than having the customary two exposed terminals that plug directly into a dc power supply, these devices are simply positioned, handle down, in a well within a holder that is continuously attached to ac. During the long periods of nonuse, both the holder well and the handle become quite warm. As the toothbrush is slightly lifted from the base, you can feel a vibration (at what seems to be 60 Hz). How might such a system work? Does any "electricity" actually pass from the base to the batteries? A similar arrangement could be used to power an arti-

ficial heart by passing electromagnetic energy into the chest without the need for wires through the skin. Discuss how this might be done.

4. Imagine a superconductor in its normal state in the shape of a ring immersed in a magnetic field parallel to the central axis (that is, perpendicular to the plane of the ring). Suppose the ring is cooled and becomes superconducting. (a) Describe what happens to the field. (b) If the ring is pulled perpendicularly out of the field, what will happen to the flux in the hole in the "doughnut"? (c) Account for the energy associated with the work done on the ring (if any) in yanking it from the field region.

5. Imagine a superconducting ring supported horizontally so that it can be approached from below by a bar magnet. Describe and explain what will happen as the magnet (north pole upward) is brought near the ring. Since a superconductor has zero resistance, an induced E-field would result in an infinite current. How, then, must a superconductor behave in the presence of a changing B-field in order that this impossibility not occur?

6. A promising future source of energy is the controlled fusion reaction, the same phenomenon that powers the stars. Deuterium and tritium atoms are ionized. The resulting plasma of positive nuclei and electrons is confined to a ring-shaped region within a vacuum chamber (Fig. Q6) by magnetic fields. The plasma is raised up to temperatures in excess of 100 million K, whereupon the nuclei undergo fusion and liberate great amounts of energy. The confinement field, which actually spirals around the toroid, is provided in part by external current-carrying coils wrapped around the chamber. A crucial component of this confinement field is generated by toroidal currents circulating in the plasma itself. Explain how such a toroidally directed current (along the axis of the chamber) could be induced in the plasma. Describe the kind of current required and trace the transfer of energy into the plasma.

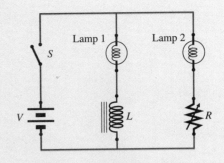

Figure Q8

Figure Q9

Figure Q6

7. Imagine a hand-cranked generator in parallel with a 50-W light bulb and a 100-W light bulb. Suppose each bulb can be switched out of the circuit. Compare the amount of effort it would take to light each bulb steadily with the effort needed to crank the generator in a sustained fashion without a load. Explain what causes the difference if there is one.

8. The switch in the circuit shown in Fig. Q8 is closed, and after a long wait, the resistor is set so the two lamps are equally bright, at which point the switch is again opened. After another long interlude, the switch is closed. Describe what subsequently happens to the two lamps.

9. Figure Q9 shows a 70-kV, 60-Hz power line in a remote area of the countryside. A shifty local resident has erected a large open loop just below the line with the intention of drawing off power. Is this possible and, if so, how would it be transferred? Where would the energy stolen come from? Would any-

one be able to detect the loss? Might it be possible to bug a telephone using the same approach? Explain.

10. Consider several inductors (L_1, L_2, and L_3) connected alternatively in series and then in parallel. What do you think will be the equivalent inductance in each case? Explain your reasoning.

11. A circuit contains an air-core coil of inductance L and resistance R in series with a power supply. Discuss and compare the amounts of energy stored with and without an iron core in place. Compare the final currents established. Where does the difference in energy come from?

12. Figure Q12 illustrates the construction of a so-called variable reluctance microphone. A diaphragm is attached to a light flexible rod made of ferromagnetic material that, in turn, is fixed via a permanent magnet to a C-shaped structure also made of magnetic material. How does it work? Comment on the way the two coils are wound.

13. Motors and generators are the same device, the distinction appearing in the operation rather than in the construction—mechanical power in and electrical out and you have a generator; electrical in and mechanical out, and you have a motor. It's reasonable, therefore, to ask, "Is a motor in some way a generator while it's turning?" Explain. If there is a generated back-current, how might it depend on the load? What do you think limits the speed of a free-turning motor? Describe the operation of the motor with a load attached. Why do many motors have open slots on their sides and little internal fan blades attached to their shafts? If you jam a motor—bind a drill or a blender so it's receiving

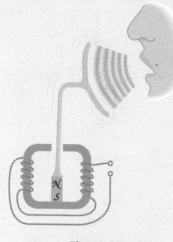

Figure Q12

current but not turning—it won't be long before you smell burning insulation. What's happening, and why?

14. Figure Q14 depicts a strip of recorded magnetic tape (which is like a succession of little magnets) passing under a playback head. The latter is a ferromagnetic C-shaped structure with a coil wrapped around it. How does the playback head read the tape? Comment on the relationship between the fineness of the read and write heads, the speed of the tape, and the density of information.

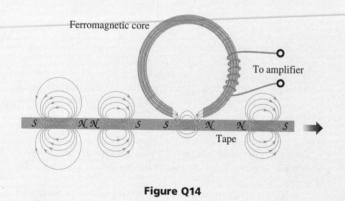

Figure Q14

15. The circuits in Fig. Q15 are adapted from an article by R. H. Romer entitled "What do 'voltmeters' measure?: Faraday's law in a multiply connected region" (*Am. J. Phys.*, **50** no. 12, Dec. 1982, 1089). Part (a) illustrates the nonconservative nature of the induced *E*-field, in this instance surrounding a long solenoid (perpendicular to the page) carrying a time-varying current. The induced current, I_i, passes through both resistors. The meters (which draw negligible current) show different readings and even different polarities. Note that no flux links either circuit 1–3–4–2–10–9–1 or 1–8–7–2–5–6–1, and these must obey Kirchhoff's Loop Rule (p. 766). Each voltmeter actually reads the work done per unit charge in moving charge through

the meter itself. Discuss what's happening. What does the meter on the left in part (b) read and why?

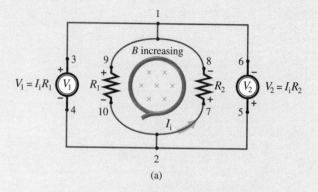

(a)

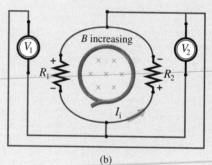

(b)

Figure Q15

16. Figure Q16 shows an electron orbiting between the poles of an electromagnet in a device called a betatron. The field is gradually being increased. How does the machine accelerate the electron? What keeps it in orbit?

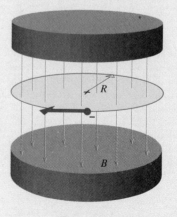

Figure Q16

Multiple Choice Questions

1. A single horizontal loop of wire is moving in a horizontal plane at a constant speed *across* a uniform vertical magnetic field that "fills" and surrounds the loop. The emf induced across its terminals will be   (a) time varying   (b) constant   (c) negative   (d) positive   (e) none of these.

2. Figure MC2 shows a horizontal length of copper wire moving across a horizontal magnetic field. There will be   (a) a negative voltage induced across its ends   (b) a positive voltage induced across its ends   (c) no voltage induced across its ends   (d) a time-varying voltage induced across its ends   (e) none of these.

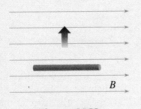

Figure MC2

3. Figure MC3 shows a horizontal length of copper wire moving across a horizontal magnetic field. There will be   (a) a negative voltage induced across its ends   (b) a positive voltage induced across its ends   (c) no voltage induced across its ends   (d) a time-varying voltage induced across its ends   (e) none of these.

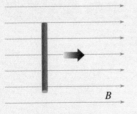

Figure MC3

4. A closed loop moves at a constant speed parallel to a long straight current-carrying wire, as in Fig. MC4.   (a) the induced current in the loop will progress clockwise   (b) there will be no induced current in the loop   (c) the induced current in the loop will progress counterclockwise   (d) the induced current in the loop will vary with the speed at which the loop moves   (e) none of these.

5. The bar magnet in Fig. MC5 is moving at a constant speed toward the coil. The voltage measured across points *A* and *B* is   (a) higher at *B* than *A* and increasing   (b) higher at *A* than *B* and increasing   (c) zero   (d) higher at *A* than *B* and decreasing   (e) none of these.

6. The more rapidly a magnet approaches a coil (as in Fig. MC5), the   (a) lower the current in the coil   (b) greater the resistance of the coil   (c) greater the induced voltage across the coil   (d) more it is attracted   (e) none of these.

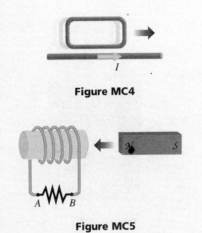

Figure MC4

Figure MC5

7. The wire loop of area 0.050 m^2 in Fig. MC7 is in a uniform downward *B*-field, which is increasing at 0.010 mT/s. The induced emf is such that the potential of   (a) *C* is higher than *A* by 0.50 μV   (b) *A* is higher than *C* by 0.50 μV   (c) *A* is the same as that of *C*   (d) *C* is higher than *A* by 0.50 mV   (e) none of these.

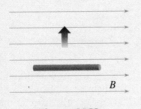

Figure MC7

8. The two coils in Fig. MC8 are wrapped on an iron bar. When the switch is closed   (a) a current momentarily passes through *R* from right to left   (b) a constant current circulates through *R* from right to left   (c) a current momentarily passes through *R* from left to right   (d) a constant current circulates through *R* from left to right   (e) none of these.

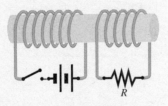

Figure MC8

9. The coil in Fig. MC9 has a core made up of a stack of insulated iron wires. The reason for using such a configuration is to   (a) generate as much thermal energy

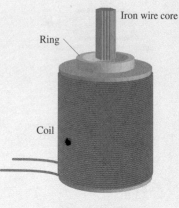

Iron wire core

Ring

Coil

Figure MC9

as possible (b) be able to make it inexpensively (c) reduce eddy current losses (d) make it strong to resist magnetic bending (e) none of these.

10. When a loose metal ring is placed on the coil in Fig. MC9 and the latter is suddenly fed a large current, the ring will (a) pop into the air (b) initially vibrate and then stop (c) initially remain stationary but get very hot (d) have nothing happen to it (e) none of these.

11. The solenoid and battery of Fig. MC11 are moving at a constant speed toward the coil on the left. The voltage measured across points 1 and 2 is (a) zero (b) higher at 1 than 2 and increasing (c) higher at 2 than 1 and increasing (d) higher at 1 than 2 and decreasing (e) none of these.

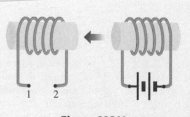

1   2

Figure MC11

12. We wish to produce a clockwise (looking down) current in the loop on the right in Fig. MC12. The variable resistor should be (a) left as is (b) decreased (c) increased (d) reversed (e) none of these.

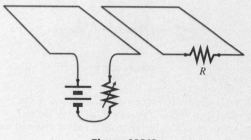

R

Figure MC12

13. The copper ring in Fig. MC13 is in a uniformly increasing magnetic field. The induced electric field within it is (a) clockwise and constant (b) counterclockwise and constant (c) clockwise and changing (d) counterclockwise and changing (e) none of these.

Figure MC13

14. Figure MC14 shows an aluminum ring and the current induced in it by the nearby magnet that is free to move along its central axis. (a) the magnet must be stationary (b) the magnet must be moving to the right (c) the magnet must be moving to the left (d) not enough information to say anything about the magnet (e) none of these.

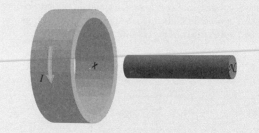

I

Figure MC14

15. A solenoid is physically altered by doubling the number of turns it has while halving the current through it, leaving everything else unchanged. (a) its self-inductance stays the same (b) its self-inductance doubles (c) its self-inductance is halved (d) its self-inductance is four times greater (e) none of these.

16. The coil in Fig. MC16 is rotating at a constant rate about an axis perpendicular to the field. The induced voltage across its terminals will (a) always be zero when $\theta = 0$ (b) always be zero when $\theta = 90°$ (c) sometimes be zero when $\theta = 90°$ (d) never be zero (e) none of these.

17. A small light bulb is in series with an air-core coil and a dc power supply such that the lamp glows brightly. An iron core is then inserted into the coil, and an hour later the lamp (a) is brighter (b) goes out completely (c) grows dimmer (d) is unaffected (e) none of these.

18. A small light bulb is in series with an air-core coil and an ac power supply such that the lamp glows brightly. An iron core is then inserted into the coil, and the lamp (a) grows brighter (b) goes out completely (c) grows dimmer (d) is unaffected (e) none of these.

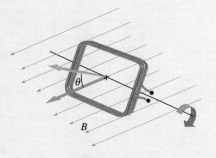

Figure MC16

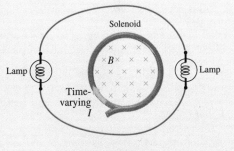

Figure MC20

19. An electric locomotive going uphill draws power from the feeder lines. In its most efficient mode of operation, when it goes downhill, it   (a) must draw even more power   (b) can generate and return power to the lines   (c) will neither draw nor produce power   (d) draws the same amount of power   (e) none of these.

20. Figure MC20 shows an end-view looking down onto a long narrow solenoid carrying an increasing clockwise current. Two identical light bulbs are connected in a circuit that encircles the solenoid.   (a) the bulb on the right glows, the one on the left does not   (b) the bulb on the left glows, the one on the right does not   (c) both bulbs glow equally   (d) neither bulb glows   (e) none of these.

21. Figure MC21 shows an end-view looking down onto a long narrow solenoid carrying an increasing clockwise current. Two identical light bulbs are connected in a circuit that encircles the solenoid. A wire is then attached as shown.   (a) the bulb on the right glows more brightly, the one on the left does not light at all   (b) the bulb on the left glows more brightly, the one on the right does not light at all   (c) both bulbs glow equally   (d) neither bulb glows   (e) none of these.

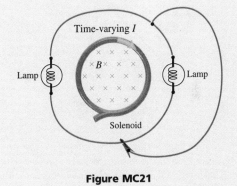

Figure MC21

Problems

ELECTROMAGNETICALLY INDUCED EMF

1. [I] A magnetic field of 1.2 mT passes perpendicularly and uniformly through a region of area 25 cm^2. What is the corresponding flux through that region?

2. [I] A uniform magnetic field of 100 mT passes through a loop of wire having an area of 0.020 m^2. The field makes an angle of 30° with the perpendicular to the plane of the loop. What is the magnetic flux through the loop?

3. [I] A flux of 6.0 mWb passes uniformly along an iron bar of cross-sectional area 50 cm^2. Compute the flux density in the bar.

4. [I] A coil wrapped around a portion of an iron ring generates a flux that passes through another coil also wrapped around the ring. This secondary has 100 turns and is penetrated by a flux of 0.016 Wb. Given that the ring has a cross-sectional area of 8.0 cm^2, what is the magnitude of the magnetic field in the toroid?

5. [I] A changing magnetic field has an initial value $\mathbf{B}_i$, which points due south and has a strength of 0.5 T. If the final value of the field, $\mathbf{B}_f$, is 0.6 T pointing due south, what is the magnitude and direction of the change in the field, $\Delta\mathbf{B}$?

6. [I] A changing magnetic field has an initial value $\mathbf{B}_i$, which points downward and has a strength of 0.6 T. If the final value of the field, $\mathbf{B}_f$, is 0.5 T pointing upward, what is the magnitude and direction of the change in the field, $\Delta\mathbf{B}$?

7. [I] A single loop of wire with an encompassed area of 0.25 m^2 is perpendicular to a 0.40-T uniform magnetic field. The loop is yanked from the field in 200 ms; what is the induced emf?

8. [I] A 100-turn coil of wire is removed laterally from an axial B-field in 20 ms, and an emf of 1.0 V is induced across it in the process. How big was the field if the cross-sectional area of the coil is 4.0 cm^2?

9. [I] A magnet is moved close to a coil of 150 turns in which it introduces a flux change of 30 mWb. If a voltage of 60 V appears across its terminals and if the resistance

of the coil is negligible, how much time elapsed during the flux change?

10. [I] A coil of 100 turns is penetrated by a flux of 0.050 Wb in 0.020 s. Determine the numerical value of the average induced emf.

11. [I] The flux passing through a coil of 200 turns changes uniformly from a constant 0.010 Wb to 0.070 Wb in a time of 1.5 s. Find the numerical value of the emf induced during that interval.

12. [I] A 10-turn coil of wire of area 0.25 m² is in a perpendicular magnetic field that changes in time as illustrated in Fig. P12. Determine the induced voltage across the loop as a function of time.

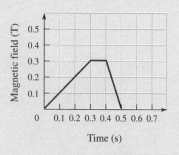

Figure P12

13. [I] A Boeing 747 jumbo jet with a wingspan of 60 m is flying in an area where the Earth's B-field is 0.050 mT. Given that the plane traverses the field perpendicularly at a speed of 200 m/s, what is the potential difference induced between its wing tips?

14. [I] A straight 30-cm length of copper wire is moved at a constant speed of 0.50 m/s perpendicularly across a uniform magnetic field of 1.5 T. Determine the emf appearing across its ends.

15. [I] How could you measure the potential difference in Problem 13? If we attached a small light bulb across the wing tips, would it light?

16. [I] An emf of 0.45 V is induced in a straight conductor having a length of 20 cm moving at right angles to a magnetic field at a constant speed of 600 cm/s. Calculate the value of the field.

17. [I] A straight copper rod 100 cm long is initially held horizontally pointing due east and then released. It falls across the Earth's field of 0.5×10^{-4} T, which is pointing north and 50° below the horizontal. What is the induced emf in the rod when it reaches a speed of 2.8 m/s?

18. [c] Considering the induced emf in a circuit bounding an area A in a magnetic field B, show that in general

$$\mathscr{E} = -(A \cos \theta)\frac{dB}{dt} - (B \cos \theta)\frac{dA}{dt} + BA\omega \sin \theta$$

19. [c] A flat circular loop of radius R is immersed in a uniform perpendicular magnetic field B. The loop is then made to rotate about one of its diameters at a constant rate ω. Determine an expression for the induced emf in the loop. [Hint: First find Φ_M and then take its derivative. Remember that $\theta = \omega t$.]

20. [c] A flat coil of N turns of wire encompassing an area A is placed in the xy-plane. It is immersed in and surrounded by a magnetic field in the z-direction given by the expression $B = B_0 = (1 + t/\tau)$. Determine the induced emf in the coil.

21. [c] A narrow flat coil of area A having N turns rests on a horizontal surface which is placed between the pole pieces of a very large electromagnet. A uniform downward B-field fills and surrounds the coil. If the field is made to decrease such that $B(t) = B_0e^{-Ct}$, find the expression for the induced emf in the coil.

22. [c] A search coil with N turns is placed inside a very long solenoid having n turns per unit length. The solenoid is then connected to an ac power supply that sends a current through it given by $I = I_0 \sin \omega t$. Given that the search coil has a cross-sectional area of A_{sc} and is perpendicular to the axis of the solenoid, find the emf induced in it.

23. [c] The rotating coil in a generator consists of N turns, each one having an area A. If it rotates with a frequency f, use the calculus to find the output emf.

24. [II] A specimen is to be exposed to a controllable magnetic field and is therefore positioned inside a 0.55-m-long narrow air-core solenoid of cross-sectional area 2.0×10^{-4} m² comprising 10 turns per cm. To monitor the field, a small search coil (connected to a voltmeter) is wrapped around the outside of the solenoid at its middle. When the current in the solenoid is increased from zero to 4.9 A in 5.00 ms, what will be the emf across the search coil given that it consists of 240 turns of fine wire?

25. [II] The rectangular loop in Fig. P25 has a resistance R and is moving left with a constant speed v into a region of uniform magnetic field. The field is a little unrealistically assumed to be constant, dropping immediately to zero beyond its rectangular boundaries. Describe the induced current as a function of time.

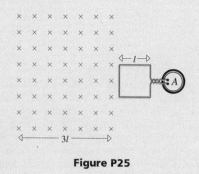

Figure P25

26. [II] A flux of 8.0 mWb generated by a large electromagnet links a coil of 100 turns placed between its poles. Suppose the current in the electromagnet is gradually and continuously reduced to zero and then reversed so that a flux of 8.0 mWb is uniformly reestablished in the opposite direction in a time of 200 ms? What is the emf induced in the coil?

27. [II] Imagine a vertical magnetic field decreasing at a constant rate of 20 mT/s. A flat circular coil of 220 turns

with a 20-cm diameter has a 30-μF capacitor across its terminals. If the coil is placed perpendicular to the field, what is the steady-state charge on the capacitor?

28. [II] A flat circular coil of 10 turns and diameter 10 cm (shown in Fig. P28) is located in a uniform magnetic field of 0.20 T that passes through the coil area at 45°. What will be the voltage, on average, across the terminals if the plane of the coil is smoothly rotated (through 45°) so that it's parallel to the field in a time of 0.10 s? Indicate which terminal has the higher potential. What will the voltage be when the coil comes to a stop?

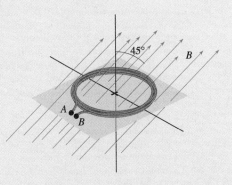

Figure P28

29. [II] A single loop of copper wire in the shape of a square 4.0 cm on each side is lying flat on a horizontal table. A large electromagnet is positioned with its north pole above and to the left a little so that the uniform magnetic field is downward onto the loop, making an angle of 30° with the vertical. Compute the average induced emf across the loop as the field varies linearly from 0 to its final value of 0.500 T in 200 ms. What is the direction of the induced current?

30. [II] A wire of length l is moved at a speed v across a perpendicular magnetic field B by a force F, as shown in Fig. P30. Given that the lamp has a resistance R while the resistance of all the wires is negligible, derive an expression for F.

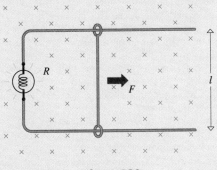

Figure P30

31. [II] With Problem 30 in mind, what is the power supplied to the moving wire by the external agency? Show that this power is equal to I^2R dissipated in the lamp.

32. [II] Imagine a flat circular coil of N turns and resistance R connected to a ballistic galvanometer (that is, one with negligible damping). The flux through the coil (Φ_M) is changed and an induced charge ΔQ flows through the galvanometer, whose pointer is initially displaced by an amount $\theta \propto \Delta Q$, known as the *throw*. Show that

$$\Delta Q = \frac{N\Delta\Phi_M}{R}$$

33. [II] With Problem 32 in mind, suppose a small search coil is attached to a ballistic galvanometer for which $\Delta Q = K\theta$, where K is a measured constant and θ is the initial deflection. The coil is first placed perpendicularly in a uniform B-field that is to be measured. It is then rapidly yanked out, causing the galvanometer to swing. Write an expression for B in terms of θ, R, K, N, and A, the area of the search coil.

34. [II] A flat circular coil of 20 turns with an area of 5.0×10^{-2} m^2 is located in a uniform magnetic field of 10 mT. Initially, the magnetic flux passes through the coil perpendicularly. The coil is then rotated, in 150 ms, so that its central axis makes an angle of 50° with **B**. Determine the average induced emf resulting from the rotation.

35. [II] A small search coil (of area 0.50 cm^2, resistance 0.50 Ω, and having 12 turns) is attached to a ballistic galvanometer. The coil is placed inside, at the center of, and coaxial with, a large solenoid. When the current in the solenoid is rapidly reversed, the galvanometer deflects, indicating a pulse of 2.0 μC of charge. Determine the initial B-field of the solenoid. [Hint: Remember Problems 32 and 33.]

36. [cc] A long straight wire carrying a constant current I lies along the x-axis, which is drawn in chalk on a horizontal laboratory table. Parallel to it, a distance y_0 away, is a rectangular wire circuit that extends from $x = 0$ to $x = l$ and from $y = y_0$ to $y = y_0 + w$. Determine the magnetic flux through the circuit.

37. [cc] A flat coil made of N tightly wound turns of copper wire encompasses an area A and has a resistance R. It is held vertically and then rotated through a small angle φ so that the plane of the coil then makes a fixed angle φ with the yz-plane. An electromagnet is turned on, and the coil is immersed in a magnetic field given by $B_x = B_0 \cos \omega t$. Compute the size of the induced current. (Assume the coil's inductance is negligible and there is an ammeter across its terminals.)

38. [cc] A search coil composed of N turns of wire is placed inside a very long solenoid having n turns per unit length. The solenoid is then connected to an ac power supply that sends a current through it given by $I = I_0 \sin \omega t$. The search coil has a cross-sectional area of A_{sc} and is perpendicular to the axis of the solenoid. By pulling on its leads, the area of the search coil is made to decrease linearly in time according to the equation $A_{sc}(t) = A_0(1 - \alpha t)$. Find the emf induced in it.

39. [cc] A long straight wire carrying a current I runs along the x-axis, as drawn in chalk on a lab work table. The current in the wire is in the $+x$-direction. A thin copper rod of length l is placed on the table parallel to the y-axis, with its lowest end a distance $+y_0$ from the current-carrying wire. The rod is then moved, with a constant

speed v, parallel to the current. Find the steady-state potential difference between the ends of the rod. [Hint: Use Eq. (21.2) and generalize Eq. (22.4) into an integral since $B = B(y)$.]

40. [cc] A copper rod of length L is screwed perpendicularly (down through one end) to the top of a plastic wand. The wand is held vertically in a downwardly directed magnetic field B and spun at a constant rate ω about its symmetry axis so that the copper rod sweeps out a horizontal circle. Determine the induced emf along the length of the rod. Neglect the mass of the charged particles, which would otherwise lead to an additional tiny inertial contribution due to the centripetal force. [Hint: Measure r along the rod from the screw outward. Generalize Eq. (22.4) into an integral since $v(r) = r\omega$.]

41. [cc] Redo the previous problem, but this time imagine the emf to be due to the changing flux through an imaginary time-varying loop swept out by the rod. Since there is no actual loop there, why does this tack work? [Hint: Deal with the area swept out by the moving rod.]

42. [cc] With the last two problems in mind, suppose the copper rod is screwed to the top of the wand at its midpoint rather than at its end. If it is again spun at ω, what will be the induced emf between the center of the rod and either end? What will be the difference in potential between the two ends? Neglect the mass of the charged particles, which would otherwise lead to an additional tiny inertial contribution due to the centripetal force.

43. [cc] A very long tightly wound solenoid of radius R is being supplied an increasing current such that its B-field is increasing linearly with time. Determine the induced electric field at some arbitrary perpendicular distance $r > R$, roughly midway between the ends of the coil. [Hint: Ignore the very small external field.]

44. [III] Imagine two stationary (zero-resistance) vertical conductors with a horizontal wire crossbar (of length l) capable of moving downward freely while staying in contact (Fig. P44). The crossbar (of mass m) is dropped and it falls perpendicular to the uniform field B. Describe its motion completely, writing an equation for the acceleration and the maximum speed, if there is one.

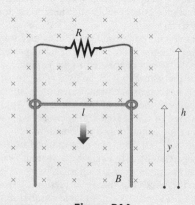

Figure P44

GENERATORS
SELF-INDUCTION

45. [I] A generator, which consists of a flat coil, is rotated about its central axis (Fig. 22.13a) so that the longer portion of it moves perpendicular to a 0.80-T magnetic field. The coil has an active length of 50 cm (that is, there is a total of 50 cm of wire cutting the field normally). What is the emf when the machine turns at a rate such that each conductor moves at 4.0 m/s?

46. [I] We wish to design an ac generator using an armature on which is wound a flat tight rectangular coil (8.00 cm × 20.0 cm) of 150 turns. The generator output is to be sinusoidal with a maximum voltage of 20.0 V and a frequency of 50.0 Hz. How strong should the B-field be and at what angular speed should the coil turn?

47. [I] A 25-turn coil is made to rotate in a uniform 0.35-T magnetic field. Each turn has two 10-cm wire lengths that cut the field at varying angles as the coil rotates. What is the emf when the wire is moving with a speed of 22 m/s at 90° to the field?

48. [I] In the United States, the national power grid supplies ac to ordinary consumers with a maximum voltage of 170 V at 60 Hz (the standard 120-V "average" household service). Given that a generator has a 100-turn coil of area 0.50 m², what strength magnetic field should be used?

49. [I] An ac generator produces an emf that (in SI units) has the form $\mathscr{E} = 100 \sin 376.99t$. What is the maximum voltage? What is the frequency of the output?

50. [I] If a current of 2.0 A in a coil produces a flux linkage of 6.0 mWb, what is the self-inductance?

51. [I] A long solenoid of 500 turns carrying a current of 3.8 A produces within itself a uniform magnetic flux of 2.0 mWb. Compute the self-inductance of the coil.

52. [I] An air-core coil having a self-inductance of 3.0 mH carries a current that creates a magnetic field within it of magnitude B. The coil is slid onto a ferromagnetic rod that has a relative permeability of 2000 when in a field of magnitude B. What will be the new inductance with the core in place?

53. [I] An air-core coil with 500 turns has an inductance of 200 mH. If a current of 2.0 A passes through the coil, what will be the flux within it at that moment?

54. [I] Show that the unit of μ_0 is H/m.

55. [I] What is the number of turns of a 2.5-H coil if, when 1.80 A is being carried, the flux linking it is 1.80 mWb?

56. [I] Determine the inductance of a 0.50-m-long solenoid having 200 turns wound on a hollow core 4.0 cm in diameter.

57. [I] A 40-cm-long air-core solenoid with a cross-sectional area of 4.5 cm² is to be constructed for a radio circuit so that it has an inductance of about 1.0 mH. How many turns of wire should it have?

58. [I] A coil with an inductance of 10 H has a current within it that is increasing at a rate of 4.0 A/s. What is the back-emf induced across the coil?

59. [I] If an emf of 10 V is induced across a coil when the current within it is changing at a constant rate of 2.0 A/s, what is its inductance?

60. [I] A current in a 5.0-H coil drops from 2.5 A to zero uniformly in a time of 10 ms. What is the induced emf?

61. [I] A current in a coil rises from zero to 4.0 A in 0.020 s. If the back-emf is measured to be 500 V, what is the inductance of the coil?

62. [I] A 10.0-H inductor is in series with a 12-V battery, a 6.0-Ω resistor, and a switch. What is the steady-state current attained after the switch is closed?

63. [I] In Fig. 22.21 the current reaches a value of $0.63V/R$ in 1 time constant. Show that $1 - (1/e) = 0.632$, where $e = 2.7183$.

64. [I] How much energy is stored in a 200-mH inductor when a current of 10 A circulates through it?

65. [I] What is the inductance of a coil if it stores 15 J of energy when the current through it is 5.0 A?

66. [I] What is the total energy density in a region where the Earth's surface magnetic field is 0.50×10^{-4} T and its surface electric field is 100 V/m? Notice the difference in the energies stored in the two fields.

67. [II] We wish to make a dc generator that puts out a constant current. Accordingly, a flat copper disk with a radius of 25 cm is placed in a uniform magnetic field so that its axis of rotation is parallel to B. A 20-Ω resistor is connected between the axis and the rim of the disk. The disk is rotated at a rate of 360 rpm and a dc current of 1.25 mA passes through the resistor. Find B. [Hint: Consider a radial strip of conductor and show that the rate at which it sweeps out area is $\frac{1}{2}r^2\omega$. Also, look at Problem 69.]

68. [II] A simple generator, which consists of a flat rectangular coil (Fig. 22.13) with a turning radius of 10 cm, is rotated about its central axis so that it moves perpendicularly across a 0.60-T magnetic field. The coil has an active length of 30 cm (that is, there is a total of 30 cm of wire cutting the field). What is the emf when the machine turns at 10 rev/s?

69. [II] A helicopter hovering in the air has its 5.0-m rotor blades revolving at 240 rpm in a plane that intersects the local B-field of the Earth (0.050 mT) at an angle (with the normal to the plane) of 40°. Determine the induced emf between the hub and the tip of each rotor blade. [Hint: Look at the hint for Problem 67.]

70. [II] A 25-turn flat coil is made to rotate in a uniform 0.35-T magnetic field. Each turn has two 10-cm wire lengths that cut the field at varying angles as the coil rotates. What is the emf when the wire is moving with a speed of 22 m/s at 30° with respect to the field?

71. [II] A simple ac generator consists of a coil, each turn of which has two 20-cm lengths of wire that cut the uniform 50-T B-field as they revolve at a radius of 5.0 cm from the central axis. Write an expression for the emf as a function of time given that the coil has 200 turns and revolves at 6000 rpm.

72. [II] An inductor in the motor circuit of a robot consists of an air-core solenoid of 100 turns with an inductance of 200 mH. If the coil carries a current of 5.0 A, what is the flux linkage? How much energy does the coil store?

73. [II] A 5.0-A current through a 1.5-H coil is reversed, and in the process an average emf of 100 V is induced. Compute the time it took to reverse the current.

74. [II] We wish to make an inductor that will be used in an experiment and must be formed around an iron rod

for which $\mu = 1.5 \times 10^3 \mu_0$. The rod has a diameter of 2.0 cm and is 27 cm long. Given that the inductance is to be 300 mH, approximately, how many turns of wire will be needed? Why should we stress the word *approximately* here?

75. [II] A long narrow solenoid having a 2.0-cm radius is made of copper wire wrapped with two turns per millimeter on a plastic core. How much energy is stored per unit length of coil when 5.0 A circulates through it?

76. [II] Write an expression for the self-inductance of an air-core toroidal coil of N turns, with a mean radius of b and a cross-sectional area A.

77. [II] An experimental setup is comprised of an air-core solenoid (having an inductance of 300 mH and negligible resistance) connected in series to a sample of bone (of 25 Ω), an ammeter, and a 150-V dc source via a switch. The intent is to provide a gradually building current to the sample.   (a) What will be the steady-state current? (b) Determine the time constant of the system. (c) Determine the current after such a time has elapsed.

78. [II] An inductor ($L = 2.00$ H and $R = 4.0$ Ω) is connected in series with an ammeter (of negligible resistance) and a varying voltage source. Figure P78 is a plot of the current in the circuit versus time. Draw a corresponding plot of the voltage across the inductor versus time.

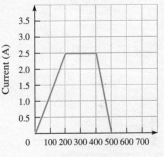

Figure P78

79. [II] A 24-H iron-core inductor of negligible resistance is placed in series with a 12-V battery, a 6.0-Ω resistor, and a switch. Determine: (a) the steady-state current;   (b) the time constant; and   (c) roughly how long it would take for the current to reach within 1% of its maximum value.

80. [II] A 24-H iron-core inductor of negligible resistance is placed in series with a 12-V battery, a 6.0-Ω resistor, and a switch. Once the switch is closed, the current in the circuit is

$$I = \frac{V}{R}\left\{ 1 - \exp\left[\frac{-t}{(L/R)}\right]\right\}$$

Compute the current at a time of 2.0 s after the switch is closed.

81. [II] Inductance transducers, mounted in catheters (narrow tubes inserted into the body), have been used to measure

blood pressure. Blood presses against a diaphragm (a few millimeters in diameter), which bends a proportionate distance backward. On the back side of the diaphragm is a small ferromagnetic shaft (of permeability μ) that is thereby displaced, entering into an air-core coil of length l (a distance d). The resulting change in inductance ΔL is measured electrically and calibrated to correspond to the blood pressure. Show that

$$\Delta L \approx \frac{d(\mu - \mu_0)N^2 A}{l^2}$$

82. [II] The so-called "solenoid" in your car is mounted on the starter motor. When the key is turned, 12 V are applied to the solenoid, which then carries a current that generates a magnetic field. The ferromagnetic plunger (Fig. P82) is drawn into the coil, and that moves a rod that does two things: it closes a heavy-duty switch sending about 100 A

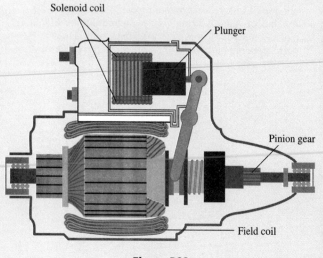

Figure P82

to the starter motor and it engages a gear so that the motor can "turn over" the engine. If the solenoid (which is 200 turns wrapped on a hollow core 5.0 cm in diameter and 9.5 cm long) has a resistance of 1.20 Ω, determine the approximate energy density in the coil, given that the plunger fills the space and has a relative permeability of 500 at that field.

83. [II] Determine the approximate amount of energy stored in the Earth's magnetic field in the first 200 km above the planet's surface. Even though the dipole field drops off as $1/r^3$, this distance is comparatively small, so assume $B = 0.4 \times 10^{-4}$ T throughout. (That value averages-in the variation of B between the equator and pole.) The

mean radius of the Earth is 6371.23 km. Compare your answer to the energy in a gallon of gasoline, $\approx 10^8$ J.

84. [cc] In Section 22.5 we derived the generator equation by considering the motion of the constituent conductors with respect to the B-field. Let's now reanalyze the problem in terms of Faraday's Law. The coil of a generator consists of N turns of wire encompassing an area A, rotating at a rate ω, in a magnetic field B. Find an expression for the induced emf.

85. [cc] A current having a frequency f is sent through a coil with an inductance of L. Direct measurements show that the voltage across the coil varies cosinusoidally with a maximum value of V_0. Find an expression for the current in the coil. What is its maximum value?

86. [cc] In Fig. 22.22 the battery, resistor, and inductor form an R-L circuit carrying a steady-state current I. The switch is thrown open leaving the inductor and resistor in series, and the current begins to decay in accord with Fig. 22.21b. Derive an expression for $I(t)$. [Hint: Begin with Kirchhoff's Loop Rule.]

87. [cc] Using calculus, derive Eq. (22.14), which describes the energy stored in an inductor, by directly computing the work done (by the energy source, e.g., a battery) in driving charge through the inductor against the induced emf. [Hint: The energy involved is associated with the product of the changing emf and the time-varying charge passing through the inductor. Start with $d(PE_M)$.]

88. [cc] An air-filled coaxial cable of length l consists of a central copper wire of radius r_a surrounded by a metal cylindrical sheath with an inner radius of r_b. Determine the self-inductance per unit length of the cable when the two conductors are attached and a steady current I passes up one and back through the other. [Hint: Compute the magnetic flux in the gap. Start by finding its value through a loop of area $l\,dr$.]

89. [cc] With the previous problem in mind, determine the energy stored in the cable and then confirm that your answer agrees with the above-computed expression for L.

90. [III] Derive the expression for the emf of a rotating coil

$$\mathcal{E} = NAB\omega \sin \omega t \qquad [22.7]$$

by considering the work done by a torque acting through an angle.

91. [III] Suppose a 200-Ω resistor is placed in series with a 50-mH inductor, a switch, and a 120-V dc power supply. Compute the rate-of-change of the current when $t = 0$, $t = L/R$, and $I = 1.0$ A. What is the maximum current?

92. [III] An air-core solenoid is attached to a 12-V battery via a switch. The switch is closed, and after a while the current reaches a constant level of 2.0 A. If the current changes at a rate of 12 A/s when $I = 1.0$ A, what is the resistance and inductance of the coil?

93. [III] The current in a long solenoid of 210 turns generates a flux within it of 10 mWb. If that current is gradually reversed in a time of 200 ms, what will be the induced emf?

23

AC and Electronics

HOWEVER RUDIMENTARY, BATTERIES WERE the only practical source of sustained electric current throughout the early 1800s. By the 1870s, the dynamo (p. 845) had made the transition from laboratory curiosity to practical workhorse, and dc was still the preferred form of electricity. As yet undeveloped, alternating current (ac) was widely viewed as inherently inferior.

ALTERNATING CURRENT

Nikola Tesla (1856–1943) was one of the leading figures in the development of alternating current. The spectacular display crackling around Tesla was produced by high-frequency (20,000 Hz) high-voltage ac. The handwritten inscription reads "To my illustrious friend Sir William Crookes of whom I always think and whose kind letters I never answer! June 17, 1901."

Instead of maintaining a fixed polarity, each terminal of an ac generator, though always opposite to the other, alternates between + and −. The electrons that constitute a typical alternating current move first forward then backward, oscillating essentially in place at some given number of cycles per second corresponding to the generator frequency. Remember that the electrons in a dc current drift quite slowly; what moves at nearly the speed of light is the disturbance of the electrons, the front behind which the electrons are moving and beyond which they are not yet affected by the driving voltage source. An alternating current transports energy in the same way as does a direct current—namely, in the form of the organized kinetic energy of mobile-charge carriers. And joule heat arises with ac, just as it does with dc.

When the practical high-resistance, low-current incandescent lamp was introduced by Edison around 1880, each installation came with its own 110-volt generator, and it was dc. Most other lamps of the era were low-resistance devices that had to be put in series so that a high current could pass through each one ($P = I^2R$ and for a required P, a small R means a large I). No one lamp could be shut off without open-circuiting the system. Edison's carbon-filament lamps had a high resistance and could be put in parallel, where the feeder current would be divided into many small branch currents, each powering one bulb that could be turned on and off at will (Fig. 20.13, p. 762)—which, of course, is the way your home is wired.

Meanwhile, ac was quietly being transformed by a brilliant, wildly eccentric, young engineer named Nikola Tesla, who had briefly been associated with (and subsequently came to despise) Mr. Edison. With Tesla's invention of a practical ac induction motor, alternating current became far more appealing, particularly to shrewd visionaries like George Westinghouse who hired Tesla and bought the rights to his motor. Slowly, as the market expanded, there developed a titanic struggle for control of the industry between the hustlers of *high-voltage ac* (primarily Westinghouse and General Electric, the J. P. Morgan combine to which Edison would ultimately sell out) and those of *low-voltage dc* (led rather unscrupulously by Edison).

When central generating stations came into being and especially when they were located at remote energy sources like Niagara Falls (1895), the electrical power produced had to be transmitted over long distances. But the very wires that carry electricity have some resistance, and that poses a major problem. A medium-sized city might easily require ≈10 MW of power ($P = IV$). If that amount is to be provided at a modest 100 V or so, then 100 000 A will have to be supplied. There's the difficulty: the joule heating ($P = I^2R$) in the delivery wires varies as I^2, not just as I. A two-wire line of one-quarter-inch-diameter copper wire has a resistance of ≈1.7 Ω/mile. Carrying 10^5 A, the joule heating losses are ≈1.7×10^{10} W/mile. For each mile, that's ≈1.7×10^7 kW and every hour the line loses 1.7×10^7 kW·h of energy. At a cost of roughly 10¢ per kW·h, sending 10^5 A down the line wastes 1.7 million dollars per hour per mile!

There was no economically feasible way out but to lower the current. Clearly, if the voltage was raised to 100 000 V, the same power could be efficiently delivered by 100 A! Thus, raising the voltage by a factor of 10^3 allows for the lowering of the cur-

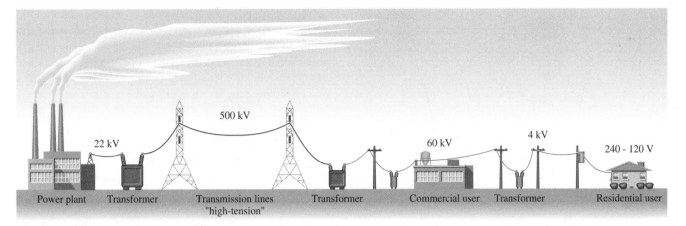

22 kV

500 kV

60 kV

4 kV

240 - 120 V

Power plant   Transformer   Transmission lines
"high-tension"   Transformer   Commercial user   Transformer   Residential user

Figure 23.1 The transmission of ac power. Transformers are used to increase the voltage so power can be efficiently sent long distances via "high-tension" wires. The voltage is then stepped down for commercial and residential use. For simplicity, the ground and other "hot" leads are not shown.

rent by 10^3, and that drops the associated power lost over the same lines by a factor of 10^6. Since there already existed a very simple way to raise and lower the voltage of ac (via transformers, p. 882) but no comparable means for dc (until recently), the contest ultimately went to the high-voltage-ac people.

Electrical power is usually generated at around 22 kV and boosted by step-up transformer to about 500 kV, which is then transmitted long distances via high-tension lines (Fig. 23.1). Using a transformer, the voltage is then reduced to perhaps 60 kV for heavy industrial consumers—usually the more power required, the higher the voltage at which it's delivered. Substations in each area step the voltage down further to about 4 kV for distribution to local communities. This voltage is what's carried by those endless ugly overhead wires that crisscross most suburban towns. Thereafter, the voltage is finally reduced by small pole-mounted transformers, each of which feeds a cluster of buildings. **Line voltage** in the United States and Canada is ordinarily 110 V, 115 V, or 120 V and can be anywhere in that range, varying from time to time.

In the early days, there were a number of different frequencies supplied (at first 125 Hz and 133 Hz were common and later 25 Hz, 35 Hz, 50 Hz, and 60 Hz became popular), depending on the producer. Lower frequencies were more suitable for use with the induction motor, but they still had to be high enough to keep incandescent lamps from flickering. Today, most of Europe and Asia uses 50 Hz (220 V), whereas the United States and Canada have adopted 60 Hz (110 V) as standard.

Alternating current was initially supplied (as was dc) using conventional two-wire lines in what is called a *single-phase* system—the voltage rises and falls sinusoidally (p. 844), as does the current. Under such circumstances, power is inefficiently available in pulses. Tesla was among the few to realize that by simultaneously generating several sinusoidal voltages, each shifted in phase with respect to the other, a transmission line could carry more power and actually do so at a constant, continuous rate. Using three wires instead of two and sending out three (120°-shifted) sinusoidal voltages between them (Fig. 23.2), Tesla was able to triple the average amount of power delivered by the line. Equally as important, this kind of *polyphase* ac was necessary to drive the big new induction motors that required no brushes or commutators and were extremely reliable. Under the irresistible pressure of these benefits, the entire ac power system was overhauled, and the transmission of *three-phase* ac became almost universal. Today, although industrial users are often supplied directly with three-phase ac, the typical domestic consumer is usually provided with two single-phase 120-V ac lines (p. 886).

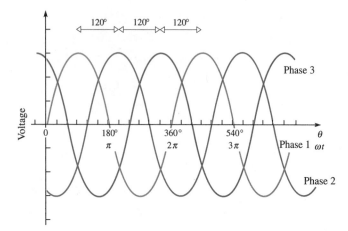

Figure 23.2 The emf produced by a three-phase ac generator. Each signal is shifted by 120°.

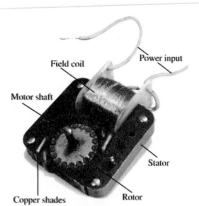

An ac shaded-pole induction motor of the kind that powers devices such as blenders and can openers. The field coil sets up an oscillating B-field in the stator. Currents induced in the copper wire "shades" retard the changing B-field in those regions, producing a primitive rotating field that induces currents in the rotor. These currents result in a torque that sets the rotor turning.

The availability of efficient ac generators and motors, the invention of the ac watt-hour meter (which made the selling of ac energy practicable), the development of efficient transformers, and the economical three-phase transmission system have led to today's vast national power grids. We can simply plug in, sit back, and listen to the music.

23.1 AC and Resistance

As we saw earlier [Eq. (22.7)], the emf produced by an ac generator can be represented by a sine function of angular frequency $\omega = 2\pi f = 2\pi/T$. In addition to the ubiquitous wall outlet, there are electronic devices known as *oscillators* that also provide a harmonically varying potential difference. They're often used in the laboratory, though there are many gadgets that have oscillators built into them—the modern radio receiver is one. Whatever the source, the terminal voltage then has the form

$$v(t) = V_m \sin \omega t = V_m \sin 2\pi f t \tag{23.1}$$

where $v(t)$ is the **instantaneous voltage**. The lowercase letter v is used here to distinguish it from the several different possible practical measures of ac voltage that will be considered presently. A plot of $v(t)$ as in (Fig. 23.3) rises and falls between two peak or **maximum voltage** values, $+V_m$ and $-V_m$, which is what would be seen if we plugged the source into an oscilloscope. Then we could read the maximum value directly from the scale on the screen just as we could read the so-called **peak-to-peak voltage** (V_{pp}), which is simply $2V_m$.

With a purely resistive load R across the terminals of an ac source, a current will circulate that increases and decreases, reversing direction over and over again, precisely in step with the oscillating voltage—the applied emf directly drives the **instantaneous current** i. Ohm's Law takes the form

$$i(t) = \frac{v}{R} = \frac{V_m}{R} \sin \omega t$$

and with the **maximum current** given by $I_m = V_m/R$, it becomes

$$i(t) = I_m \sin \omega t = I_m \sin 2\pi f t \tag{23.2}$$

Both the voltage and the current are zero when $2\pi f t$ equals either 0, π, 2π, etc., since $\sin 0 = \sin \pi = \sin 2\pi = 0$. This occurs when $t = 0$, $t = 1/2f = \pi/\omega$, or $t = 1/f = 2\pi/\omega$, and so on, as shown in Fig. 23.3. For 60-Hz ac, the zeros happen at $t = 0$, $1/120$ s, $1/60$ s, etc. The current reverses itself every $1/120$ s, and the period is $T = 1/f = 1/60$ s.

The early workers with electricity were faced with a practical problem when it came to measuring ac; they couldn't have the pointer on an ammeter or a voltmeter flutter back and forth at 60 Hz. Indeed, what does one even mean by the statement "the ac voltage* is 120 V?" Engineers were accustomed to dealing with dc, where

---

*Despite the obvious awkwardness, such terms as ac current and ac voltage are now part of the jargon. In such cases, the ac should be read as *ay cee* and not alternating current.

it makes sense to speak of a 10-A current, but how does one describe the strength of an alternating current? It was decided to use the *ampere* as the unit of alternating current and define it in such a way as to be, insofar as possible, equivalent to the direct-current ampere. There were three possible methodologies that could be used: chemical, magnetic, or thermal. But only the thermal effect of a current varies as i^2 and is independent of the direction, or sign, of the current. Since both ac and dc produce joule heating, ***the alternating-current ampere was defined as the quantity of current that produces the same amount of heating in a resistor as does a direct-current ampere during the same interval of time***.

Nowadays, ac ammeters are calibrated to read this **effective current** (I_{eff}): 10 amps ac effective will generate the same amount of thermal energy as 10 amps dc. So common is the usage, it is automatically assumed that a device rated at simply 10 amps ac refers to 10 A effective. And the same is true for ac voltmeters, which are, as a rule, calibrated in **effective voltage** (V_{eff}). Moreover, since we are likely to deal only in effective values, the subscripts are often dropped, and it is understood that the I and V mean effective current and voltage, respectively.

The effective values can be related to the maximum values by looking at the power; I_{eff} is defined by its ability to cause heating in a resistor, which is equivalent to its ability to transfer an average level of power. The **instantaneous power** p*(t)* dissipated in a resistor is

$$\text{p} = i^2 R$$

What we actually measure via a calorimeter is the effect of such power averaged over many cycles. Inasmuch as the R is constant

$$P_{\text{av}} = [i^2]_{\text{av}} R \qquad (23.3)$$

and this result by definition is to equal $I^2 R$ or, if you like, $I_{\text{eff}}^2 R$. The average rate at which thermal energy is developed in a resistor by an alternating current is

$$P_{\text{av}} = I_{\text{eff}}^2 R \qquad (23.4)$$

Comparing these two equations, it follows that

$$I = I_{\text{eff}} = \sqrt{[i^2]_{\text{av}}}$$

The effective current equals the root of the mean of the square of the instantaneous current or, for short, the **rms current**. It's not uncommon in the literature to find the symbol I_{rms} used instead of I or I_{eff}. Had we started with p $= v^2/R$, we would have come upon a precisely analogous definition for the **rms voltage**.

Returning to Eq. (23.3) and substituting Eq. (23.2) into it yields

$$P_{\text{av}} = [(I_m \sin \omega t)^2]_{\text{av}} R = [I_m^2 \sin^2 \omega t]_{\text{av}} R$$

Because I_m^2 is constant

$$P_{\text{av}} = I_m^2 [\sin^2 \omega t]_{\text{av}} R \qquad (23.5)$$

The average of any time-varying function taken over some interval is equal to the area under the curve divided by the length of the interval. In Fig. 23.4, the area under the curve equals the area under the constant $\frac{1}{2}$-line since the tops of the peaks can be imagined cut off and moved so they fill the troughs. For any interval ωt, of several periods, the area ($\frac{1}{2} \times \omega t$) divided by the duration (ωt) is just $\frac{1}{2}$: $[\sin^2 \omega t]_{\text{av}} = \frac{1}{2}$. Another way to see this result is to start with the identity $\sin^2 \omega t + \cos^2 \omega t = 1$. Realizing that $\sin^2 \omega t$ and $\cos^2 \omega t$ are identical except for a 90° phase shift, they must both have the same average value over an interval greater than one period. But

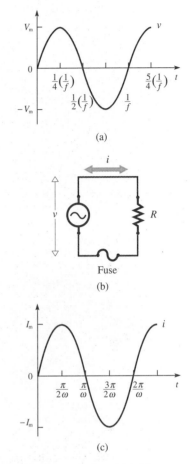

Figure 23.3 A sinusoidal voltage (a) applied to a resistive load (b) results in a sinusoidal in-phase current (c).

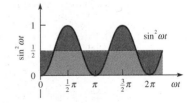

Figure 23.4 The area under the curve $\sin^2 \omega t$ divided by the extent of the curve (ωt) equals the average value; $[\sin^2 \omega t]_{\text{av}} = 1/2$.

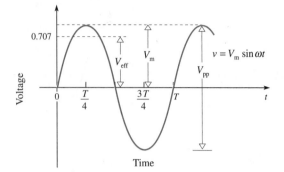

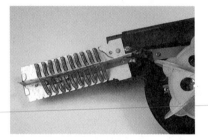

Figure 23.5 A sinusoidal voltage and the corresponding effective, maximum, and peak-to-peak values.

The heating coil in a hair dryer. Current passing through the coil causes it to become red hot ($P = I^2R$) via joule heating. The white plastic star-shaped fan blows air over the coil and out the nozzle.

if we take the average of both sides of the identity, it follows that $[\sin^2 \omega t]_{av} = [\cos^2 \omega t]_{av} = \frac{1}{2}$. Thus, from Eq. (23.5)

$$P_{av} = \tfrac{1}{2} I_m^2 R \qquad (23.6)$$

and since $P_{av} = I_{eff}^2 R$, we have the sought-after expression for the effective current in terms of the maximum current, namely

$$I = I_{eff} = \frac{I_m}{\sqrt{2}} \qquad (23.7)$$

And, similarly

$$V = V_{eff} = \frac{V_m}{\sqrt{2}} \qquad (23.8)$$

The effective current or voltage is just 0.707 times the corresponding maximum value (Fig. 23.5). A wall outlet providing 120 volts effective is actually supplying a sinusoidal emf with a maximum terminal voltage of (120 V)/(0.707) = 170 V, which is a good reason for you to be even more careful with 120-V ac than 120-V dc.

From Ohm's Law, which is valid at any instant, $V_m = I_m R$ and

$$V_{eff} = I_{eff} R \qquad \text{or} \qquad V = IR \qquad (23.9)$$

Moreover

[*resistive load*]
$$P_{av} = I_{eff}^2 R = I_{eff} V_{eff} \qquad (23.10)$$

which is nice and simple.

Example 23.1 According to the little metal plate on a hair dryer, it is rated at 120 V 1200 W. Assuming the load to be purely resistive (it really isn't because of the motor), how much current does it draw and what is its resistance? What is the maximum value of the current in the dryer?

Solution: [Given: $V = 120$ V and $P_{av} = 1200$ W. Find: I, I_m, and R.] From Eq. (23.10), the effective current is

$$I = \frac{P_{av}}{V} = \frac{1200 \text{ W}}{120 \text{ V}} = \boxed{10.0 \text{ A}}$$

The maximum current then follows from Eq. (23.7), as

$$I_m = 1.414 I = \boxed{14.1 \text{ A}}$$

From Ohm's Law

$$R = \frac{V}{I} = \frac{120 \text{ V}}{10.0 \text{ A}} = \boxed{12 \ \Omega}$$

▶ **Quick Check:** $P_{av} = I^2 R = (10.0 \text{ A})^2 (12 \ \Omega) = 1200$ W.

23.2 AC and Inductance

The distinctive aspects of ac become evident when there are inductors or capacitors in the circuit. It's then that the currents through, and the voltages across, these elements are *not in-phase*. As a result of these phase differences (which are different for capacitors and inductors), all sorts of interesting things happen. For example,

Ohm's Law doesn't apply in the same simple way it did with dc: in general, the current in an ac circuit (which is not purely resistive) is not equal to the applied voltage divided by the resistance of the circuit. Moreover, the algebraic sum of the ac voltage drops across each element of a series circuit (as measured by individual voltmeters) may not equal the applied voltage (as it does in a dc circuit). Energy is still conserved and Kirchhoff's Rules apply, but for sinusoidal voltages (and currents) that are out-of-phase, we can't simply add their effective values algebraically.

To study the ac behavior of an inductor, consider the circuit shown in Fig. 23.6a. It consists of an inductor L, having negligible resistance, placed across the terminals of an ac source. At the instant shown in Fig. 23.6b, an increasing clockwise current exists, creating an increasing flux in the coil. Remember (p. 848) that in response, an induced current i_I will appear in the windings, and it will oppose the change in the flux via Lenz's Law. The induced current at that moment emerges from the inductor at point A, and there is a potential difference across the terminals of the inductor, which is $+$ at point A and $-$ at point B. The inductor then behaves much like a battery being charged. Because the current from the source is sinusoidal, there will be a continuously changing flux and a sustained harmonic back-emf. In the steady state, the back-emf is 180° out-of-phase with the emf of the source. In Fig. 23.6, trace around the circuit and when the back emf is a drop, the source voltage will be a rise. In the unattainable case where $R = 0$, the two are numerically equal. At any instant, the sum of the voltage rises and drops around the loop is then zero.

Ideally, the coil has no resistance so that the source emf need not be larger than what it takes to overcome the back-emf in order to sustain current in the circuit. On average there cannot be any energy dissipated by an ideal inductor. However, a real inductor has resistance; consequently, the difference between the source voltage and the back-emf will equal iR [Eq. (22.11)], and some small amount of energy provided by the source will be dissipated as i^2R.

The instantaneous current in the inductor can be determined from the voltages in the circuit because $\varepsilon_L = -L\, di/dt$. The ac source voltage is $v(t) = V_m \sin \omega t$ and taking the sum of the voltage rises and drops around the loop yields

$$v(t) + \varepsilon_L = 0$$

and so $V_m \sin \omega t = L\, di/dt$. To find $i(t)$ put the time-dependent terms on the left and integrate:

$$\int \frac{V_m}{L} \sin \omega t\, dt = \int di$$

Since $-\cos \theta = \sin(\theta - \pi/2)$

$$i(t) = -\frac{V_m}{\omega L} \cos \omega t = \frac{V_m}{\omega L} \sin(\omega t - \pi/2) \qquad (23.11)$$

The current has been written as a sine so that it can more easily be compared with the source voltage.

Figure 23.7 is a plot of this instantaneous current through, and the voltage across, the inductor. Notice how the current peaks at a later time than the voltage: **the current lags the voltage** (or the voltage leads the current) *by one-quarter cycle* (90° = $\pi/2$ rad). The instantaneous voltage depends on the product of the induc-

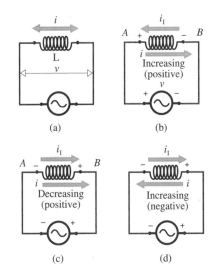

Figure 23.6 (a) An inductor placed across the terminals of an ac source. (b) Note that when i is positive (clockwise) and increasing, the terminal voltage v is positive. (c) When i is positive and decreasing, v is negative, and (d) when i is negative and increasing, v is negative.

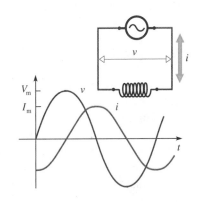

Figure 23.7 An inductor effectively holds back the current so that i (the current through it) lags v (the voltage applied by the source). (See Fig. 23.6.)

tance and the *rate-of-change* of the current. The voltage is zero when the current curve levels off (that is, when its slope is zero) at its maximum values. The voltage is maximum when the current curve is steepest (that is, when its slope is maximum) at points where it crosses the time axis. *The higher the frequency, the tighter the curves are, and the greater the slopes at these points.*

A real inductor inhibits current both because of its resistance and because of its inductance (via the associated back-emf). We can speak of the total effect as representing an ***impedance*** (p. 878), part of which, the ***reactance***, is due purely to the inductive behavior exclusive of any resistance. To elaborate, observe that it follows from Eq. (23.11) that

$$I_m = \frac{V_m}{\omega L} \tag{23.12}$$

and since $I = 0.707 I_m$ and $V = 0.707 V_m$

$$V = \omega L I \tag{23.13}$$

This result is so much like Ohm's Law, with the ωL term impeding the current, that we introduce a new quantity called the **inductive reactance** X_L, defining it as

$$X_L = \omega L \tag{23.14}$$

whereupon $$V = I X_L \tag{23.15}$$

Consequently, 1 volt per 1 amp corresponds to an inductive reactance of 1 ohm (see Problem 37).

In this ideal situation, there is no resistance and the reactance equals the impedance. The greater the inductance, the greater the back-emf, and the more the inductor inhibits or "chokes" the current. Note that in the case of dc ($\omega = 0$), there is no back-emf—the reactance is zero and there is no inhibition of current. An inductor with a large L and small R is usually called a *choke coil* because of its ability to control ac currents without wasting nearly as much power as would a resistor. Because the inductive reactance increases with frequency, a choke can be used to suppress the high end of a mix of frequencies. For example, in a loudspeaker system there are usually at least two drivers, a small high-frequency tweeter and a large low-frequency woofer. By putting an inductor in series with the woofer (Fig. 23.8), high-frequency currents (representing high-pitched sound) are suppressed, allowing the woofer to respond to the signal range it was designed to handle best.

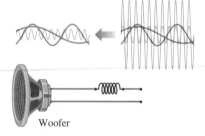

Woofer

Figure 23.8 The inductor's impedance increases with frequency, so it passes low-frequency signals and blocks high-frequency signals.

Example 23.2 A radio circuit contains a 400-mH inductor with a resistance of 0.50 Ω. This is supplied (across its terminals) with an 80-V-effective ac signal of 100 Hz. Determine both the reactance of the coil and its rms current.

Solution: [Given: $L = 0.400$ H, $R = 0.50$ Ω, $V = 80$ V, and $f = 100$ Hz. Find: X_L and I.] Let's first find the inductive reactance:

$X_L = \omega L = 2\pi f L = 2\pi(100 \text{ Hz})(0.400 \text{ H}) = \boxed{251 \ \Omega}$

By comparison, R is negligible and we can treat the coil as if it were a pure inductance. Hence

$$I = \frac{V}{X_L} = \frac{80 \text{ V}}{251 \ \Omega} = \boxed{0.32 \text{ A}}$$

▶ **Quick Check:** $X_L \approx (6)10^2(4 \times 10^{-1})\Omega \approx 0.24$ kΩ; $V = I X_L \approx (0.32 \text{ A})(\frac{1}{4} \text{ kΩ}) \approx 0.08$ kV.

Figure 23.9 is a plot of the instantaneous power ($p = iv$) associated with the ideal inductor. Where i and v are both positive or both negative, p is positive, and where either one (*i* or *v*) is negative, p is negative. Thus, the area under the curve (subtracting what is below the axis from what is above it) per cycle is zero. Energy is stored in the alternating magnetic field, but *the average power supplied per cycle in a purely inductive circuit is zero*. Power is alternately absorbed from the generator (positive portion of the cycle) and then returned to it (negative portion of the cycle).

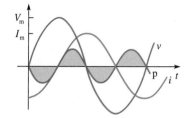

Figure 23.9 The power drawn by a pure inductance averages to zero. Note how p = 0 whenever either *v* or *i* is zero. The relative heights of the curves are unimportant.

23.3 AC and Capacitance

Imagine a capacitor *C* placed across the terminals of a battery. We are already familiar with what happens in the transient state: charge piles up on the plates until the mutual repulsion between the charges stops any more from arriving. A *reverse potential* appears (via $v = q/C$) that increases until it reaches the battery potential, at which point the current stops. Because *C* is constant, $\Delta v = \Delta q/C$; a change in the charge is associated with a change in the voltage. By definition, the current is the time rate-of-change of *q*, which corresponds to a time rate-of-change of *v*: no $\Delta v/\Delta t$, no steady state *i*.

Now suppose the battery is replaced by an ac source (Fig. 23.10). We want to compare *v* across the capacitor to *i* passing through it. As the sinusoidal voltage of the generator rises positively (0° to 90° in the ac cycle), we can imagine positive charges traveling to the upper plate and an equal number repelled away from the lower plate. A large positive current immediately circulates, because at $t = 0$ there is no charge on the plates and no reverse potential to inhibit it. As the impressed voltage rises, the charge on the plates increases in step with it, and the reverse potential increases too, making it harder for more charge to be deposited—the clockwise current dies off.

At the moment the voltage peaks (90°) so that $\Delta v/\Delta t = 0$, the current is zero. But then the voltage begins to decrease (90° to 180° in the ac cycle), whereupon the capacitor starts to discharge: $\Delta v/\Delta t \neq 0$ and $i \neq 0$. Current appears in the negative counterclockwise direction. After the voltage passes through zero, it reverses and increases (180° to 270° in the ac cycle), increasing the charge on the plates in the reverse direction. That again decreases the counterclockwise current until it becomes zero, when the voltage peaks in the negative direction (270°). The negative applied voltage then decreases, and positive charge leaves the capacitor flowing clockwise and constituting a positive current (270° to 360° in the ac cycle). Apparently, **the instantaneous current in a capacitor leads the instantaneous voltage across it by one-quarter of a cycle** (90°).

Figure 23.10 The voltage and current in an ac capacitive circuit. The positive direction is taken to be clockwise.

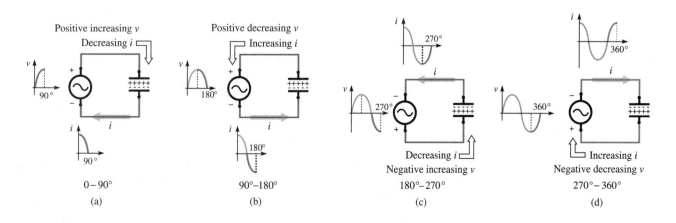

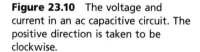

0–90°	90°–180°	180°–270°	270°–360°
(a)	(b)	(c)	(d)

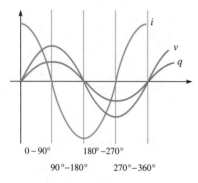

Figure 23.11 The current, charge, and voltage for an ac capacitive circuit.

We can describe this behavior analytically in the following way. Since $v = q/C$ and the source voltage is $v = V_m \sin \omega t$, it follows that

$$q = CV_m \sin \omega t \qquad (23.16)$$

The charge and voltage are in-phase as shown in Fig. 23.11. Now to find the instantaneous current $i(t)$. Since $i = dq/dt$

$$i = \frac{d}{dt}(CV_m \sin \omega t)$$

and so

$$i(t) = \omega CV_m \cos \omega t = \omega CV_m \sin(\omega t + \pi/2) \qquad (23.17)$$

This time, $\omega CV_m = I_m$ and so the *rms* values share the same relationship—namely, $\omega CV = I$. Hence

$$V = \frac{1}{\omega C} I$$

and the form of Ohm's Law suggests that we introduce a **capacitive reactance** X_C, defined as

$$X_C = \frac{1}{\omega C} \qquad (23.18)$$

hence

$$V = IX_C \qquad (23.19)$$

The reactance, in ohms (Problem 36), is a measure of the capacitor's opposition to the passage of alternating current.

When $\omega = 0$, the reactance is infinite, and no finite dc voltage will produce a current through an ideal capacitor—**capacitors block dc**. As the frequency goes up, the capacitive reactance goes down, in contrast to the inductive reactance; a given source voltage results in a higher and higher current as ω is increased. What stops the current is the peaking of the voltage arising from the buildup of charge on the plates. At high frequency, the charge has little opportunity to pile up and thereby decrease the current before the applied voltage is reversed. Alternatively, increasing ω increases the slope of the v curve, which means a larger value for the time-rate-of-change of q, and so a larger current. If the capacitor is getting charged and discharged more rapidly, the current must increase. Raising the capacitance lowers the reactance; a large capacitor can store more charge before the voltage across its plates rises to appreciably diminish the current. Alternatively, since $\Delta q/\Delta t = C\Delta v/\Delta t$, the larger is C, the greater the current.

Figure 23.12 indicates how a capacitor in series with a tweeter in a speaker system will suppress the low frequencies in the input signal; $(X_C \propto 1/\omega)$. It will pass on, with little attenuation, the high-frequency components that the tweeter was designed to convert into sound.

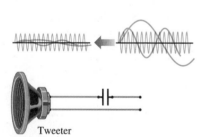

Figure 23.12 The capacitor represses low frequencies and passes high frequencies.

Example 23.3 A 50-μF capacitor is connected across the terminals of an oscillator set to have a sinusoidal output at 50 Hz with a maximum voltage of 100 V. Determine the effective current in the circuit.

(continued)

(continued)

How would this change if the frequency were raised to 5 kHz?

Solution: [Given: $C = 50$ μF, $V_m = 100$ V, and $f = 50$ Hz. Find: I.] Knowing the maximum voltage, we can find the effective voltage and, from that and the reactance, we can find I. So, figure out X_C:

$$X_C = \frac{1}{\omega C} = \frac{1}{2\pi f C} = \frac{1}{2}\pi(50\text{ Hz})(50\text{ }\mu F) = 63.7\text{ }\Omega$$

Now $V = 0.707 V_M = 70.7$ V and therefore

$$I = \frac{V}{X_C} = \frac{70.7\text{ V}}{63.7\text{ }\Omega} = \boxed{1.1\text{ A}}$$

At 5 kHz, the reactance would be lower by a factor of 100 and the current raised by a factor of 100.

▶ **Quick Check:** $V = IX_C = I/\omega C = I/2\pi f C \approx$ (1 A)/6(50 Hz)(50 $\times$ 10^{-6} F) $\approx$ 1/0.015 V $\approx$ 0.07 kV.

As with the inductor, because the instantaneous current and voltage are out-of-phase by 90°, no average power will be dissipated by an ideal capacitor. Energy stored in the electric field between the plates is returned to the source. Thus, *only resistance will dissipate power in an ac circuit, converting electrical energy into thermal energy.*

L-C-R AC NETWORKS

Electronic devices such as radios, televisions, stereos, and so on, utilize inductors, capacitors, and resistors to process electrical signals—that is, currents and voltages. Arrangements of these elements can be used to reshape a signal, to filter out or perhaps accentuate certain frequencies, or to remove any dc that might be present, and so on.

23.4 Series Circuits

Suppose we put an inductor, a capacitor, and a resistor in series across the terminals of an oscillator with an instantaneous voltage v, as shown in Figure 23.13. Now represent the current by either a sine or cosine function. Let it be

$$i(t) = I_m \sin \omega t \qquad (23.20)$$

This same current exists in each element of the circuit. And since the instantaneous voltage across the resistor (v_R) is in-phase with the current, we have

$$v_R(t) = I_m R \sin \omega t \qquad (23.21)$$

By comparison, the instantaneous voltage across the capacitor (v_C) lags the current by 90°, or $\pi/2$ radians; the sine function describing the voltage reaches the same value as $\sin \omega t$ at a later time. Using Eq. (23.18), we obtain

$$v_C(t) = \frac{I_m}{\omega C} \sin\left(\omega t - \frac{\pi}{2}\right) \qquad (23.22)$$

Thus, $\sin \omega t = 0$ at $t = 0$, whereas $\sin(\omega t - \pi/2) = 0$ *later,* when $t = \pi/2\omega$. On the other hand, the instantaneous voltage across the inductor leads the current by 90° or

Figure 23.13 The same current passes through each element in a series circuit. The voltages across each are usually not in-phase and $v = v_L + v_C + v_R$.

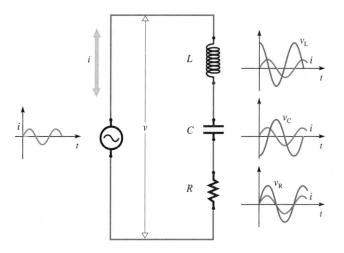

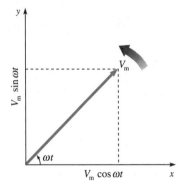

Figure 23.14 As the arrow rotates, the x- and y-components oscillate harmonically.

$\pi/2$ radians, as

$$v_L(t) = I_m\omega L \sin\left(\omega t + \frac{\pi}{2}\right) \qquad (23.23)$$

Thus, $\sin \omega t = 0$ at $t = 0$, whereas $\sin (\omega t + \pi/2) = 0$ *earlier,* when $t = -\pi/2\omega$.

In order for the applied voltage v to drive the current i through the circuit, it must equal the sum of the instantaneous voltages

$$v = v_R + v_C + v_L \qquad (23.24)$$

If we added the three sinusoids via algebra and trigonometry (and the mathematics is laborious), we would end up with a resultant sinusoid

$$v(t) = V_m \sin(\omega t + \theta) \qquad (23.25)$$

Interestingly, the sum of any number of sine functions of the same frequency is a sine function of that same frequency (Fig. 13.5, p. 494). The complete analysis yields expressions for V_m and θ:

$$V_m = \sqrt{(I_m R)^2 + \left(I_m\omega L - \frac{I_m}{\omega C}\right)^2} \qquad (23.26)$$

and

$$\theta = \tan^{-1}\frac{\omega L - 1/\omega C}{R} \qquad (23.27)$$

which, when computed and substituted into Eq. (23.25), gives the potential difference across all three elements. We have not provided the details of that calculation because there is a much simpler practical scheme for arriving at the same results, a method called **phasor addition**.

Figure 12.5 (p. 448) illustrated how a line rotating at a rate of ωt can be projected onto either axis in order to generate harmonic functions. Suppose, then, that we draw an arrow of length V_m and have it rotate counterclockwise at a rate ω, where at any instant it makes an angle ωt. The corresponding sine and cosine components are shown in Fig. 23.14. This revolving "arrow," which looks like a vector and has some of the properties of a vector, is formally a different beast called a *phasor* (designated by boldfaced type). The important thing is that **phasors add like vectors** and, in so doing, we in effect add their components—namely, the sinusoids. Combine the phasors and we combine the sinusoids, which is what we want, and that is easily done graphically.

Since the current and the voltage across the resistor are in-phase, we draw these two phasors one on top of the other in Fig. 23.15. Their lengths are I_m and $V_{Rm} = I_m R$, respectively, and their y-components are the sine functions of Eqs. (23.20) and (23.21). Because all of the phasors will be referenced to the current phasor, it's customary to simplify things a little by working at $t = 0$ (and remembering to put ωt in the expression for the phase when needed), which has the effect of placing the current phasor on the x-axis. Henceforth, all phasors will be drawn with respect to the x-axis. If its phase is $(\omega t + \alpha)$, it is tilted α radians above the x-axis; if its phase is $(\omega t - \beta)$, it is tilted β radians below the x-axis. Figure 23.16 shows how two phasors **A** and **B** are added tip-to-tail like vectors. The projection on the x-axis is then the sum of the individual cosine functions, and the projection on the y-axis is the sum of the sine functions. *Our phase shifts will only be $\pm90°$.*

Figure 23.15 The phasors **i** and **v**$_R$ are in-phase; that is, the ac current is in-phase with the voltage across the resistor.

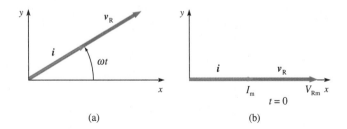

(a)

(b)

Consider a circuit with just an inductor and a resistor (Fig. 23.17). Because the voltage across the inductor leads the current by 90°, $\mathbf{v}_L$ leads $\mathbf{v}_R$ by 90°, and the two phasors (with lengths $V_{Lm} = I_m \omega L$ and V_{Rm}) are drawn as shown in Fig. 23.17b. For simplicity, this configuration is redrawn at $t = 0$ in Fig. 23.17c. If we put these instantaneous voltages across the terminals of a dual-beam oscilloscope, we'll see two sinusoids shifted 90° with respect to each other. The two phasors are next added "vectorially" to produce a resultant phasor $\mathbf{v}$ in Fig. 23.27d. The length of $\mathbf{v}$ corresponds to the maximum voltage (V_m), which will be measured across the combination of R and L. The angle θ (positive if it's a lead, negative if a lag) is the phase angle of the resultant voltage sinusoid with respect to the current through the circuit. And that's it; Eq. (23.25) provides the expression for v once we compute V_m and θ from the diagram. Figure 23.17e shows the individual instantaneous voltages that can be added point-by-point to get v. Notice how v is indeed shifted by some angle (namely, $\theta < 90°$) with respect to v_R. The larger is L, the larger is θ. Similarly, if R is made larger, θ becomes smaller.

Return to the L-C-R circuit of Fig. 23.13. The voltage across the capacitor lags v_R by 90° and its phasor, of length $V_{Cm} = I_m/\omega C$, is added in along with $\mathbf{v}_R$ and $\mathbf{v}_L$ in Fig. 23.18. The two opposing phasors on the y-axis add to yield a single phasor of length $|V_{Lm} - V_{Cm}|$, which may or may not point in the positive y-direction, depending on which is larger in a particular situation. Finally, using the Pythagorean Theorem to find the length of the resultant phasor (Fig. 23.18c), we obtain

$$V_m = \sqrt{(V_{Rm})^2 + (V_{Lm} - V_{Cm})^2} \qquad (23.28)$$

and this is identical to Eq. (23.26). Moreover

$$\tan \theta = \frac{(V_{Lm} - V_{Cm})}{V_{Rm}} \qquad (23.29)$$

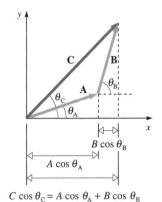

$$C \cos \theta_C = A \cos \theta_A + B \cos \theta_B$$

Figure 23.16 The addition of phasors **A** and **B**. Phasors add like vectors, and their projections onto the x-axis add like cosines.

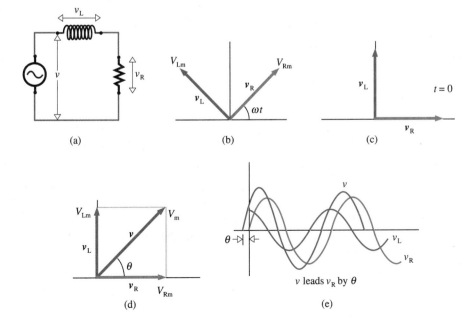

(a) (b) (c)

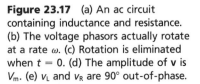

v leads v_R by θ

(d) (e)

Figure 23.17 (a) An ac circuit containing inductance and resistance. (b) The voltage phasors actually rotate at a rate ω. (c) Rotation is eliminated when $t = 0$. (d) The amplitude of **v** is V_m. (e) v_L and v_R are 90° out-of-phase.

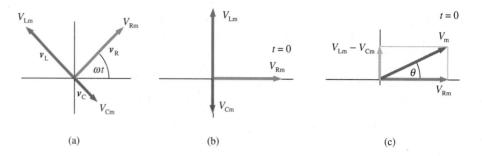

Figure 23.18 (a) The phasor diagram for a circuit containing resistance, inductance, and capacitance. (b) The magnitudes of the voltage phasors at $t = 0$. (c) The magnitude of the voltage across all the elements R, L, and C is V_m.

(a)   (b)   (c)

which is equivalent to Eq. (23.27). Inasmuch as the same expression holds for the effective voltages, Eq. (23.28) tells us that the sum of the separate voltage readings made with a meter across each element will generally exceed the voltage across the source.

Example 23.4 A series circuit contains a 240-Ω resistor, a 3.80-μF capacitor, and a 550-mH inductor. It's placed across the terminals of an oscillator set to 100 Hz. If an ammeter in the circuit reads 250 mA effective, what is the maximum voltage of the oscillator?

Solution: [Given: $R = 240\ \Omega$, $C = 3.80\ \mu$F, $L = 550$ mH, $f = 100$ Hz, and $I = 250$ mA. Find: V_m.] We have an expression for the maximum voltage, namely Eq. (23.26), and so must first determine I_m; thus

$$I_m = 1.414 I = 353.5\ \text{mA}$$

Hence, $I_m R = 84.84$ V, $I_m \omega L = 122.2$ V, and $I_m/\omega C = 148.1$ V. And so

$$V_m = \sqrt{(I_m R)^2 + (I_m \omega L - I_m/\omega C)^2}\quad [23.26]$$

hence

$$V_m = \sqrt{(84.84\ \text{V})^2 + (-25.9\ \text{V})^2}$$

and

$$\boxed{V_m = 88.7\ \text{V}}$$

Notice that *the voltages across the capacitor* (148 V) *and inductor* (122 V) *are higher than the voltage across all three components taken together* (88.7 V).

▶ **Quick Check:** The reactances are $X_L = \omega L = 346\ \Omega$ and $X_C = 1/\omega C = 419\ \Omega$, which are comparable. We can therefore expect the voltage drops V_{Lm} and V_{Cm} to be roughly equal, yielding only a small vertical phasor $V_{Cm} - V_{Lm}$. The net voltage should be a little larger than $V_{Rm} = 84.8$ V, and it is.

Going back to Eq. (23.26), factor out the current and divide both sides by $\sqrt{2}$ to get effective values, whereupon

$$V = I\sqrt{R^2 + \left(\omega L - \frac{1}{\omega C}\right)^2} = I\sqrt{R^2 + (X_L - X_C)^2}$$

The quantity $(X_L - X_C)$ is the **reactance** of the circuit; it's a measure of the net nonresistive influence impeding the current, and it's denoted by X:

$$X = (X_L - X_C) \tag{23.30}$$

It follows that

$$V = I\sqrt{R^2 + X^2}$$

Ohm's Law again suggests that the measure of a circuit's entire ability to restrain ac current (inductively, capacitively, and resistively) can be defined by its **impedance** (Z), where

$$Z = \sqrt{R^2 + X^2} \tag{23.31}$$

given in ohms. And so Ohm's Law survives into ac provided that it's written as

$$V = IZ \qquad (23.32)$$

Notice how the impedance can also be thought of as a kind of vector quantity (one that isn't time-varying) in the sense that Z is equal to the magnitude of the resultant of adding X and R as if they were vectors. The associated diagram (Fig. 23.19) is often called the ***impedance triangle***, and we see immediately that $\tan \theta = X/R$, which is equivalent to Eq's. (23.27) and (23.29).

Figure 23.20 summarizes some of the results for two-element series circuits. Of course, a circuit may contain several resistors, capacitors, or inductors. We already know how to add any number of resistors (p. 757) and capacitors (p. 708) in series; suffice it to say without proof that inductors add as do resistors (see Discussion Question 10 in Chapter 22). Thus, to analyze an ac series circuit with many components, we first combine all of the same kinds of elements, whereupon all the above equations apply, with R, L, and C being the resultant values.

On average, the power drawn by an L-C-R circuit is dissipated by the resistor: $P_{av} = I^2 R$. This process can be expressed in terms of the voltage by noting from Fig. 23.18c that

$$\cos \theta = \frac{V_{Rm}}{V_m} = \frac{I_m R}{I_m Z} = \frac{R}{Z}$$

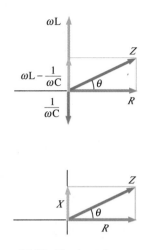

Figure 23.19 The impedance triangle. Reactances and resistance can be added as if they were vectors. The resultant is the impedance (Z) of the circuit. The "vector" $\mathbf{X_C}$ always points down, whereas $\mathbf{X_L}$ always points up. The "vector" $\mathbf{X}$ is the sum of the two, and so $X = X_L - X_C$. When $X_L > X_C$, θ is above the x-axis and positive. When $X_C > X_L$, X is negative, $\mathbf{X}$ points down, and θ is negative; it is beneath the "vector" $\mathbf{R}$, which always points in the positive x-direction.

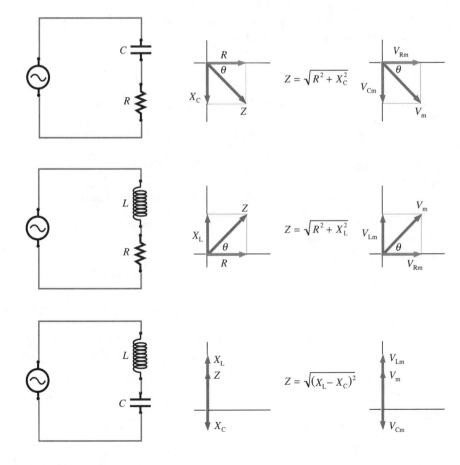

Figure 23.20 A summary of the behavior of various ac series circuits.

Accordingly $$P_{av} = I^2 Z \cos \theta$$

and so $$\boxed{P_{av} = IV \cos \theta} \tag{23.33}$$

This expression is known as the **real**, or *average,* or **dissipated power**. It differs from the corresponding dc equation by the term $\cos \theta$, which is called the **power factor** of the circuit. The power factor is a measure of the relative influence of the resistance in dissipating power. With a purely resistive circuit, $Z = R$, $\cos \theta = 1$, and the real power is IV as expected. For either a purely inductive or capacitive circuit, the phase angle is either $\pm 90°$, and no real power is transferred from the source. A typical circuit has a power factor of less than 1 (or, since the power factor is often given in percent, less than 100%). For example, in a series ac circuit containing a resistor and an inductor for which the resistance equals the inductive reactance, it follows from the impedance triangle that $\theta = 45°$ and $\cos \theta = 0.707$. Small single-phase motors have power factors of around 0.6 to 0.8.

The product IV is called the **apparent power** and is measured in volt-amps (V·A) to distinguish it from real power. Practically, an amount of power equal to IV must be supplied even if a portion of that, $(1 - \cos \theta)IV$, is stored in the fields and returned to the source. A piece of electrical equipment that has an appreciable reactance, such as a transformer, fluorescent lamp, motor, etc., has a power factor much less than 100% and must, in order to operate properly, be supplied more power than it consumes. With a power factor of 80%, a motor that consumes 800 W must be supplied with 1000 V·A in order to operate. A large commercial user pays for the power that has to be provided, even if some of it is returned.

Example 23.5 An oscillator set for 500 Hz puts out a sinusoidal voltage of 100 V effective. A 24.0-Ω resistor, a 10.0-μF capacitor, and a 50.0-mH inductor in series are wired across the terminals of the oscillator. (a) What will an ammeter in the circuit read? (b) What will a voltmeter read across each element? (c) What is the real power dissipated in the circuit?

Solution: [Given: $R = 24.0$ Ω, $C = 10.0$ μF, $L = 50.0$ mH, $f = 500$ Hz, and $V = 100$ V. Find: I, V_L, V_C, V_R, and P_{av}.] (a) Having V, to find the current we'll need the impedance; accordingly

$$X_L = \omega L = 2\pi(500 \text{ Hz})(50.0 \times 10^{-3} \text{ H}) = 157.1 \ \Omega$$

$$X_C = \frac{1}{\omega C} = \frac{1}{2\pi}(500 \text{ Hz})(10.0 \times 10^{-6} \text{ F}) = 31.8 \ \Omega$$

and therefore

$$Z = \sqrt{(24.0 \ \Omega)^2 + (125.3 \ \Omega)^2} = 127.5 \ \Omega$$

Thus $$I = \frac{V}{Z} = \frac{100 \text{ V}}{127.5 \ \Omega} = \boxed{784 \text{ mA}}$$

(b) Across each element, a voltmeter will read

$$V_R = IR = (784 \text{ mA})(24.0 \ \Omega) = \boxed{18.8 \text{ V}}$$

$$V_L = IX_L = (784 \text{ mA})(157.1 \ \Omega) = \boxed{123 \text{ V}}$$

$$V_C = IX_C = (784 \text{ mA})(31.8 \ \Omega) = \boxed{24.9 \text{ V}}$$

(c) To determine the power, we must first compute the power factor:

$$\cos \theta = \frac{R}{Z} = \frac{24.0 \ \Omega}{127.5 \ \Omega} = 0.188$$

$$P_{av} = IV \cos \theta = (0.784 \text{ A})(100 \text{ V})(0.188) = \boxed{14.7 \text{ W}}$$

▶ **Quick Check:** From the fact that $\cos \theta = 0.188$, $\theta = 79.2°$ and, using Eq. (23.29), $\tan \theta = (123 \text{ V} - 24.9 \text{ V})/(18.8 \text{ V}) = 5.22$ and $\theta = 79.2°$.

Series Resonance

An ac series circuit can function in a remarkable way at a specific frequency, $\omega_0 = 2\pi f_0$, known as its **resonant frequency**. The phenomenon is the electrical equivalent of the mechanical concept of resonance considered earlier (p. 458). Figure 23.21 depicts the frequency-dependent behavior of R, X_L, and X_C, and what we see is that, at the resonant frequency, the capacitive and inductive reactances are equal. Consequently, inasmuch as

$$Z = \sqrt{R^2 + (X_L - X_C)^2} \qquad (23.34)$$

at resonance $$Z = R$$

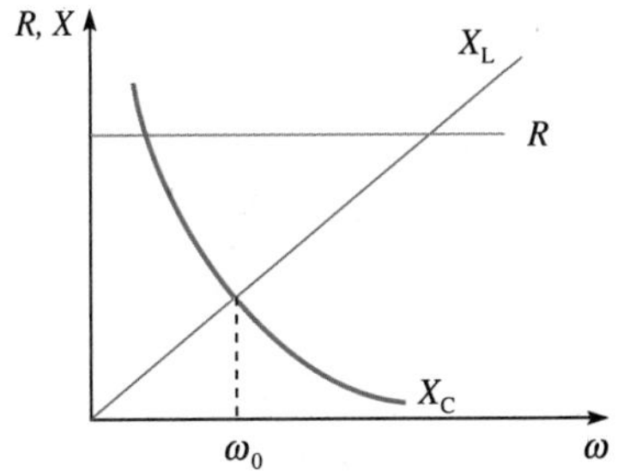

The condition for resonance exists when

$$\omega L = \frac{1}{\omega C}$$

and that occurs at $\omega_0 = 2\pi f_0$, whereupon

[resonance] $$f_0 = \frac{1}{2\pi\sqrt{LC}} \qquad (23.35)$$

Thus, at resonance $\theta = 0$, $Z = R$, and $P_{av} = IV$. Because the impedance is then a minimum, for a given voltage V, the current I is a maximum: $V = IR$. In other words, suppose that a wide range of frequencies is fed into an L-C-R circuit. If we adjust or *tune* L or C or both so that Eq. (23.35) holds at a particular frequency, say 1 kHz, then the circuit will have a peak current at 1 kHz—all other currents at other frequencies will be considerably less. Figure 23.22 indicates how the current curves are affected by the resistance of the circuit.

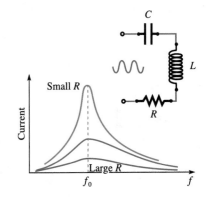

Figure 23.21 At the resonant frequency (ω_0), the inductive and capacitive reactances are equal.

Figure 23.22 At resonance, the current in a series L-C-R circuit reaches a maximum. The smaller the resistance, the sharper the peak, and the narrower the range of frequencies selected by the circuit. Compare this to Fig. 12.24 (p. 460).

Example 23.6 A series circuit contains a 50.0-Ω resistor adjacent to a 200-mH inductor attached to a 0.050-μF capacitor, all connected across an oscillator with a terminal sinusoidal voltage of 150 V effective. (a) What is the resonant frequency? (b) What voltages will be measured by voltmeters across each element at resonance? (c) What is the voltage across the series combination of the inductor and capacitor?

Solution: [Given: $R = 50.0\ \Omega$, $C = 0.050\ \mu$F, $L = 200$ mH, and $V = 150$ V. Find: f_0, V_L, V_R, and V_C.] From Eq. (23.35)

$$f_0 = \frac{1}{2\pi\sqrt{LC}} = \boxed{1.59\ \text{kHz}} \quad \text{and} \quad \omega_0 = 10.0\ \text{krad/s}$$

To find the voltages, we need the reactances and the current:

$$X_L = \omega L = (10.0\ \text{krad/s})(0.200\ \text{H}) = 2.00\ \text{k}\Omega$$

$$X_C = \frac{1}{\omega C} = \frac{1}{(10.0\ \text{krad/s})(0.050 \times 10^{-6}\ \text{F})} = 2.00\ \text{k}\Omega$$

and $Z = R$. Consequently

$$I = V/R = (150\ \text{V})/(50.0\ \Omega) = 3.00\ \text{A}$$

and therefore

$$V_R = IR = (3.00\ \text{A})(50.0\ \Omega) = \boxed{150\ \text{V}}$$

$$V_L = IX_L = (3.00\ \text{A})(2.00\ \text{k}\Omega) = \boxed{6.00\ \text{kV}}$$

$$V_C = IX_C = (3.00\ \text{A})(2.00\ \text{k}\Omega) = \boxed{6.00\ \text{kV}}$$

Notice that although there is 6.00 kV across the inductor

(continued)

(continued)

and 6.00 kV across the capacitor, the corresponding instantaneous voltages are 180° out-of-phase—the voltage across the combination is zero!

▶ **Quick Check:** The fact that the reactances are equal at resonance is a good indication that we haven't messed up the numbers.

AM Radio

A radio broadcasting system essentially converts sound (20 Hz–20 kHz) into electromagnetic waves (p. 834) that travel a lot faster and farther. We could transmit such waves at the same frequencies as the information, the sound, but that would require an antenna of tremendous size and is totally impractical. The solution is to use a convenient high-frequency radiowave (the electromagnetic **carrier wave**) and impress the information on it. In AM, or *amplitude modulation,* the carrier's amplitude is made to vary with the information. Thus, the high-frequency signal of Fig. 23.23 carries all the music or talk in the form of relatively low-frequency changes in height, or strength, of the signal. The valuable information is the envelope of the signal and the carrier itself will ultimately be discarded by the receiver.

Figure 23.24 is a rudimentary AM radio receiver. The antenna picks up a tumult of signals composed of the transmissions from all of the stations reaching it. That hodgepodge is available to the tuning circuit by way of the coupling between L_1 and L_2. When you turn the tuning knob on a radio, you are adjusting the capacitor C_1 to resonate the input circuit at the frequency of, say, WCBS. Only that frequency and its immediate surroundings will then be passed to the next stage, the crystal diode (p. 819). This is a one-way gate that chops off the negative portion of the signal. The resulting positive voltage is applied to a filter formed by C_2 and R. When the diode drops the current to zero, the capacitor discharges through R, but the time constant of this RC circuit is large compared to the period of the carrier; the discharge is slow, and voltage across the filter hardly decreases before the diode passes current again and C_2 recharges. The result is a slightly wiggly, low-frequency voltage across R that otherwise corresponds in shape to the envelope of the carrier. The original information (Fig. 23.23b) oscillated above and below the axis, whereas this signal is only positive—it's been displaced by a constant positive voltage. That dc component is eliminated by a blocking capacitor (C_3). The resulting slowly oscillating voltage corresponds almost exactly to the oscillating sound wave. When it's fed into headphones, out comes a sound wave identical to the one that was heard in the studio at WCBS.

Figure 23.23 An amplitude modulated (AM) signal. (a) A constant amplitude, constant frequency carrier is made to carry information, as in (b), by modulating its amplitude, as in (c).

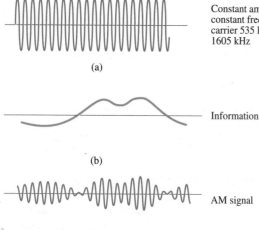

(a)

Constant amplitude, constant frequency carrier 535 kHz to 1605 kHz

(b)

Information

(c)

AM signal

23.5 The Transformer

In 1831, Faraday discovered the principle of electromagnetic induction that underlies the transformer, but it took about 50 years before the latter became a practical instrument. In broad terms, the **transformer** *is an induction device used to convert energy in the form of a large time-varying current at a low voltage into nearly the same amount of energy in the form of a small time-varying current at a high voltage (or vice versa).*

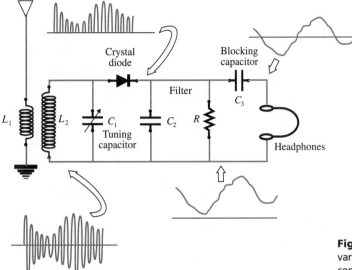

Figure 23.24 An early radio receiver showing a signal at various stages of being processed. The two coupled coils L_1 and L_2 constitute a transformer, a device we'll deal with next.

Imagine two coils wrapped around an iron core, as shown in Fig. 23.25. An ac power source is applied to the ***primary*** winding, and the switch is initially left open in the circuit of the ***secondary*** winding. The time-varying primary current I_p creates a time-varying magnetic flux that circulates through the secondary coil. Because of its high permeability, the iron core enhances the flux generated by the primary current by a factor of about 10 000. It provides an easy path for the field, which is very nearly totally constrained to pass within its volume, effectively coupling the two coils.

It might seem that closing the switch S_1 would essentially short-circuit the source, causing a tremendous input current; after all, the primary winding has a low resistance. Indeed, that's exactly what would happen if we put a constant dc source across the primary. With an ac source, the sizable self-inductance L, due largely to the core, will result in a substantial induced back-emf via Eq. (22.10). This back-emf will oppose the applied voltage, keeping the primary current I_p very small. Since the load on the source is essentially purely inductive, almost no energy is drawn from the source, and with the secondary open-circuited, no energy is transferred.

Ideally, the same time-varying flux passes through all the turns of both coils and so the induced emf on any one turn is the same as that on any other. Hence, the total induced emf on the primary is proportional to the total number of its turns (N_p) just as the total induced emf on the secondary is proportional to its number of turns (N_s). Assuming the windings have negligible resistance and therefore sustain no IR voltage drops, the induced emf across the primary will be numerically equal to the terminal voltage across it (V_p). Similarly, the emf induced across the secondary will equal its terminal voltage (V_s). It follows, then, that the ratio of the effective voltages will equal the ratio of the numbers of turns:

$$\frac{V_p}{V_s} = \frac{N_p}{N_s} \qquad (23.36)$$

The coil with the higher number of turns corresponds to the higher voltage, which could be either the primary or the secondary, depending on how we wire it. *The ratio of the number of turns in the higher-voltage winding to the number of turns in*

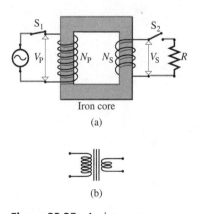

Figure 23.25 An iron-core transformer. (a) The input side is the primary; the output side is the secondary. (b) The symbol for an iron-core transformer.

the lower-voltage winding is called the **turn ratio**. For example, a transformer with a 10:1 turn ratio has 10 times the number of turns on one coil as on the other.

For any given primary voltage, we need only select an appropriate turn ratio in order to produce any desired secondary voltage. When $V_p > V_s$, it's called a **step-down transformer**, and when $V_p < V_s$, it's a **step-up transformer**. The spark coil in an automobile is a commonplace example of the step-up variety. The 12-V dc from the battery is chopped into pulses by a switching device. The "coil" is a transformer that boosts this pulsating 12-V input up to 20 kV, which activates the spark plugs that ignite the gasoline in each of the cylinders (p. 620).

Example 23.7 The ac adapter for a pocket calculator contains a transformer, two diodes, and a capacitor (p. 822). The transformer takes the 120-V ac at the wall down to 15.0-V ac, which the rest of the circuit converts to dc. If the transformer's secondary consists of exactly 50 turns, how many turns must the primary have? What is the turn ratio?

Solution: [Given: $V_p = 120$ V, $V_s = 15.0$ V, $N_s = 50$. Find: N_p and N_p/N_s.] We have the primary and secondary voltages as well as the secondary number of turns, which suggests Eq. (23.36), and consequently

$$N_p = \frac{N_s V_p}{V_s} = \frac{50(120 \text{ V})}{15.0 \text{ V}} = \boxed{400 \text{ turns}}$$

This is a step-down transformer, and so the turn ratio is

$$\frac{N_p}{N_s} = \frac{400}{50} = \boxed{8}$$

▶ **Quick Check:** The transformer is of the step-down variety, and we expect $N_p > N_s$. Further, the ratio of the voltages (120 V)/(15 V) does equal the turn ratio.

Transformers lose energy because of the resistance of the coils and because of eddy currents and residual magnetization of the core. To control resistance losses, the low-voltage, high-current coil is usually made of relatively thick wire. A typical core is a laminate of insulated sheets, thereby removing the possibility of large-scale eddy currents flowing in cross-sectional loops. A well-built modern transformer will have a remarkable efficiency of better than 98%.

With the switch S_2 in Fig. 23.25 closed, a secondary current progresses around the circuit. If we then make the reasonable approximation that energy losses are negligible, the average power-in equals the average power-out, and so from Eq. (23.33)

$$I_p V_p \cos \theta_p = I_s V_s \cos \theta_s \qquad (23.37)$$

It's not at all obvious that the two power factors are equal and will cancel, but that's the way it will turn out. The presence of a secondary current means there will be an induced flux in the secondary coil that is proportional to $N_s I_s$. By Lenz's Law, this induced flux opposes the cause of the secondary current—namely, the alternating core flux due to the small initial ac primary current. In other words, the secondary flux momentarily decreases the primary flux and therefore the induced primary back-emf is decreased, which allows more current to come from the generator that, in turn, increases the back-emf until it again equals the generator voltage. The primary current I_p, which is then no longer negligible, provides the energy supplied to the secondary. The process stabilizes when the increase in the primary flux (which is proportional to $N_p I_p$) equals the induced secondary flux; thus

$$N_p I_p = N_s I_s$$

A speaker and output transformer from the AM radio shown on p. 645. The transformer allows the output amplifier of the radio to efficiently supply power to the speaker. (See Problem 89.)

Combining this equation with Eq. (23.36) yields

$$I_p V_p = I_s V_s \qquad (23.38)$$

and comparing this with Eq. (23.37), it follows that *the power factors are equal.* When the load on the secondary is large and resistive, the current and emf are essentially in-phase. Thus, the secondary power factor is 100% (as is the primary power factor), and Eq. (23.38) corresponds to the equality of the average powers in and out.

Example 23.8 Assume that the little transformer in the ac adapter in Example 23.7 is lossless (which it certainly isn't). (a) If it supplies 450 mA at 15.0-V ac, what is the current in the primary for a line voltage of 120 V effective? (b) If the secondary circuit has a power factor of 80%, what is the average power supplied by the wall outlet?

Solution: [Given: $V_p = 120$ V, $V_s = 15.0$ V, $I_s = 450$ mA, $\cos \theta = 0.80$. Find: I_p and P_p.] (a) From Eq. (23.38)

$$I_p = \frac{V_s I_s}{V_p} = \frac{(15.0\ \text{V})(0.450\ \text{A})}{120\ \text{V}} = \boxed{56.3\ \text{mA}}$$

(b) The wall supplies the same average power as the secondary provides, and so from Eq. (23.33)

$$P_p = I_s V_s \cos \theta_s = (0.450\ \text{A})(15.0\ \text{V})(0.80) = \boxed{5.4\ \text{W}}$$

▶ **Quick Check:** Since the power factors are equal, $P_p = I_p V_p \cos \theta_p = (0.056\,3\ \text{A})(120\ \text{V})(0.80) = 5.4$ W.

Nowadays, transformers can be found around the home in a variety of appliances: fluorescent lamps, television sets, toy-train power supplies, high-intensity desk lights, battery chargers, stereos, door chimes, radios, and so on.

23.6 Domestic Circuits and Hazards

Although three-phase ac is supplied to large commercial users, the ordinary consumer is provided—via three wires down from the pole (Fig. 23.26)—with a pair of single-phase 120-V lines. Two of these wires are "hot" and the third is neutral, grounded back at the transformer. Inside the building, the neutral is usually attached to the entering water pipe and all electrical metal conduits, receptacle boxes, wall brackets, etc. are connected to it. The two hot leads (one red and one black), along with the neutral, go through a meter to a breaker or fuse box. The instantaneous voltages between red and ground (120 V) and black and ground (120 V) are 180° out-of-phase and can be combined to yield 240 V effective (between red and black), for use with heavy-duty air conditioners, water heaters, etc. Emerging from the breaker box are several parallel pairs of leads (black and white, hot and neutral, respectively) that carry power to the rest of the house. This arrangement is what's known as 240-V ac service, and it's fairly standard in North America. When you plug your TV into the wall, you are attaching one of the leads in the power cord to a black (hot) wire in the outlet and the other lead to a white (neutral) wire, across the two of which there is 120-V-effective, single-phase ac.

A typical home is wired internally with several separate lines leading to the main, each attached in series with a fuse or circuit breaker of its own. The idea is to limit the amount of current that can be drawn by any single line—too much current and the wires in the wall can get dangerously hot (via $I^2 R$). In fact, that's how the fuse

A pole-mounted transformer used to drop the voltage from the supply cables. Three leads provide two separate 110-V lines for domestic service. One of the three wires from the transformer is common to both lines. It attaches to a horizontal, uninsulated support cable, around which the other two wires are wrapped. These two wires go off on the right and left to individual houses.

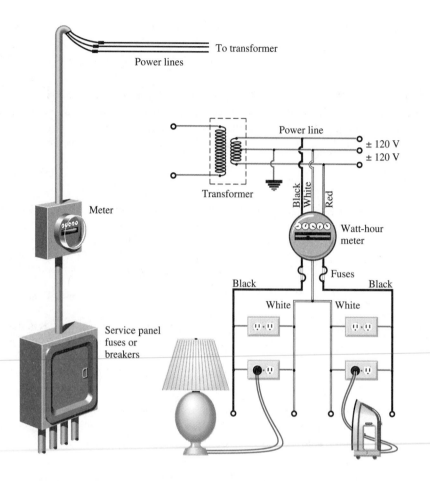

Figure 23.26 Domestic ac service.

The electrical lines enter a house through an energy meter, which records the number of kilowatt hours delivered.

works: current passes through a thin metal element that will melt and open-circuit when I exceeds the designated value. Household wiring is typically rated at 15 or 20 amps effective and fused accordingly. If a line contains too large a fuse and is made to carry too large a current by plugging in too many appliances, there will be an appreciable voltage drop in the wiring itself. The temperature of the line will rise, and its resistance will therefore also rise. The terminal voltages at the outlets may then be appreciably less than 120 V, and lights will dim, TV pictures will shrink, and you will be risking burning the place down. If a line is rated at 20 A, it will provide (120 V)(20 A) = 2400 W, which is just enough power to run a dishwasher, simultaneously make a slice of toast, and allow you to watch it all happen under a 100-W light bulb (Table 23.1).

The newest three-wire system includes an additional ground lead connected to the metal housing of the appliance. Thus, the whole steel body of a washing machine or refrigerator is attached to ground via the round prong on the three-prong plug (Fig. 23.27). Experience has shown that occasionally the hot wire in an electric device can be exposed and the whole appliance become "hot." For example, suppose you move your washing machine to do some cleaning and then while pushing it back against the wall, a sharp corner of the cabinet cuts into the power cord and touches the hot lead. In the old two-wire system, the entire machine would sit there at 120 V waiting for someone (standing on a damp basement floor) to touch it and get a nasty shock. In the new system, as soon as the hot lead makes contact with the grounded case, the current is shorted out and the fuse blows.

TABLE 23.1 Electrical power consumption

Appliance	Typical wattage (W)
Range	12 000–16 000
Clothes drier	5000–8000
Oven	4000–8000
Hair dryer	1000–1300
Dishwasher	1200–1500
Furnace blower	1200
Iron	1100
Toaster	1100
Waste disposal	1000
Oil burner	800
Refrigerator (big, double-door)	800
Vacuum cleaner	600
Washing machine	550
Blender	400
Fan	200
TV	100
Typewriter	90
Humidifier	40
Clock	4
Motors:	
1 hp	1500
1/2 hp	1000
1/4 hp	700
1/6 hp	450

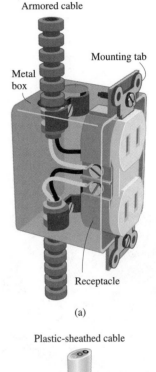

(a)

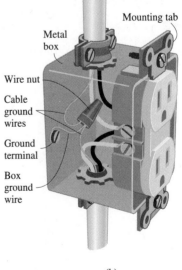

(b)

Figure 23.27 (a) An old-style two-prong outlet. (b) A modern three-prong outlet. The round third prong opening is internally connected to the metal mounting tab.

This raises the interesting question of the effect of electricity on the human body. Over 1000 people a year are accidentally electrocuted in the United States alone. Included are the foolhardy who balance plugged-in appliances on bathroom shelves within reach of the tub, as well as the poor souls who invariably try to retrieve kites from high-voltage lines, usually with long damp sticks (that are fair conductors at high enough voltages). Of course, anyone in a tub of water is grounded via the pipes. In the end "it's the current that kills ya"; the voltage is relevant only to the extent that it must be sufficiently high to force the current through the body. If very little current is available, as in a small electrostatic generator, even tremendous voltages, in excess of 100 kV, will be harmless for short durations.

Usually, we can feel a 1-mA current, and up to about 5 mA, however unpleasant, is as a rule harmless. When a substantial amount of charge flows (>10 mA) through muscles, it causes wrenching spasms. If those muscles are in the hand, the effect may be little more than tiny burns and a deep ache. At above 15 mA or so, one loses voluntary muscle control—rather awkward if you are holding on to a "hot" wire and can't let go. Up to roughly 50 mA, currents will cause considerable pain, but probably no massive malfunction of any crucial body process. More sizable currents across the torso can paralyze the respiratory system and disrupt the steady pumping of the heart. A current of approximately 100 mA sustained for a second or more through the heart will cause it to go into a lethal condition of ventricular fibrillation (irregular beating). The idea is to avoid letting current pass through the body in general and certainly to keep it away from the heart. Electricians working with high voltages (120 V can be lethal, 240 V demands great care), will often position one arm well away from the circuit; they risk blasting the fingers of one hand but avoid creating a hand-to-hand pathway across the chest.

The resistance of the body is determined to a large extent by the contact resistance with the outer layer of the skin. The wet human stuff within each of us is rich in ions and is a fairly good conductor. Therefore, depending on the condition of the skin, the area of contact, and the intimacy of that contact, the resistance (hand-to-hand or head-to-foot) may vary from perhaps 100 kΩ to over 1.5 MΩ dry, and possibly 100 times less, wet. If we assume a body resistance of 200 kΩ and an outlet voltage of $V = 120$ V $= IR$, it follows that $I = 1.2$ mA—far from problematic. If you are dripping wet and $R = 1$ kΩ, $I = 120$ mA and you are in big trouble. Be exceedingly careful with 120-V ac and don't even consider puttering around with 240-V ac. One is not likely to be able to let go of a high-voltage line, and the resulting burns will quickly lower the skin resistance, along with the chance of survival.

ELECTRONICS

Over the past three decades, electronics has become dominated by solid-state components, principally semiconductor diodes and transistors. Each of us is likely to use dozens of intregrated circuits (IC) in the course of an ordinary day; they are in computers, calculators, telephones, pacemakers, TVs, answering machines, radios, VCRs, elevators, wristwatches, cameras, cars, microwave ovens, and so on. An IC microprocessor "chip" (perhaps $\frac{1}{4}$ in. by $\frac{1}{4}$ in.) in the realm of what's called "very large-scale integration" might contain 450 000 minute transistors, along with a multitude of diodes, resistors, and capacitors. A digital watch requires about 5000 transistors; a little pocket calculator utilizes roughly 20 000 transistors; a computer, whose equivalent might once have filled a room with vacuum tubes, now contains a tiny 100 000-transistor chip and comfortably fits on a person's lap. By the end of the century, the 1 000 000-transistor chip will probably be commonplace.

A 4-megabit integrated circuit (IC) memory chip.

23.7 Semiconductors

An isolated atom can exist in any one of a number of distinct energy levels; its electron cloud can only have certain configurations. The atom typically has many electrons (for example, silicon has 14), distributed in closed shells about the nucleus. Only the outer group—the *valence electrons*—are involved in the behavior we are concerned with here. When these outer electrons (in silicon, there are 4) are in their lowest-energy configuration, the atom as a whole is in its lowest-energy state (the one it usually occupies), called the *ground state.* When two atoms are near each other, their interaction causes a slight shift in the allowed levels. When there are a tremendous number of interacting atoms, as in a solid, the shifted levels are so numerous and so close together that they form energy bands. The ground state is called the **valence band**, and the electrons therein are usually held to their respective atoms.

Given enough energy, one of the outer electrons can be ripped from its atom, to move relatively freely through the lattice of atoms. That electron is then said to be in the more energetic **conduction band**. Figure 23.28 shows how, in an insulator, the almost totally empty conduction band is separated from the occupied valence band by a sizable gap (10 eV or greater) from which electrons are prohibited. Very few electrons can pick up enough energy thermally to be propelled across the gap. Thus, very few move freely in the insulator; hence, the high resistivity. By comparison, in a conductor the two bands overlap, and no clear distinction between the valence and conduction bands exists. Valence electrons are free to wander from atom to atom, and the resistance is low. An *elemental* (or *intrinsic*) semiconductor such as silicon or germanium has a small forbidden energy gap (in silicon it's only about 1.1 eV). Still, at room temperature (300 K), the average KE of the electrons ($kT \approx 4.1 \times 10^{-21}$ J) is about 0.026 eV so that very few high-energy electrons can jump the gap—the conduction band will be nearly empty.

Doping

Devices such as transistors and diodes are fabricated using *impurity* semiconductors prepared by adding minute quantities of foreign atoms (just a few parts per million) to an intrinsic semiconductor. The process is known as **doping**, and it affects the availability of mobile-charge carriers, producing two distinct kinds of systems. Figure 23.29 shows how the electrons of a silicon single crystal are shared (in so-called covalent bonds) between the four nearest neighbor atoms. When such a crystal is doped with a five-valence-electron atom (Fig. 23.30) like arsenic, the fifth electron is not locked in place—it does not fit and can move around freely within the crystal. Such an electron resides in an energy level just below the conduction band, into which it can easily be made to jump. Because these mobile-charge carriers are negative, the system is referred to as an **n-type** semiconductor.

If the silicon is doped with a three-valence-electron atom like gallium (Fig. 23.31), there will be a deficiency of one electron; in effect, there will be a **hole** in the negative distribution of electrons. An outer electron from a nearby silicon atom can drop out of its cloud and fill the hole, but this will leave a new hole in the place where the electron originally was. The hole moves about like a bubble in a cup of water—it is the absence of negativity and so behaves as if it were a positive mobile-charge carrier. Because the carriers are positive, this system is known as a **p-type** semiconductor. The presence of the gallium gives rise to a number of empty levels just above the valence band. Electrons can jump into these levels, leaving behind holes in the valence band that can then move around in response to an applied electric field, thus constituting a current.

23.8 The *pn*-Junction and Diodes

One of the most practical arrangements of semiconductors is the **pn-junction**, the interface formed by joining a *p*-type to an *n*-type semiconductor. In practice, a polished slice of single crystalline *p*-type silicon is heated to around 1000°C and exposed to a vapor of arsenic or phosphorus, which diffuses into the surface. The uppermost layer is transformed into *n*-type silicon, which is then coated with a protective insulating layer of silicon dioxide. Because the single crystal structure is continuous across the junction, electrons from the *n*-type region can diffuse across to the *p*-type region, where they fill an equivalent number of holes. The *n*-type region is left positive (because of a deficiency of electrons) and the *p*-type region is negative (because of an excess of electrons). The resulting internal potential difference cuts off further transfer of charge, leaving a central region depleted of carriers. This **depletion layer** is essentially an insulator and the junction then resembles a charged capacitor (Fig. 23.32).

To see how such a *pn*-junction can function as a **diode** (that is, as a one-way gate passing current in one direction and blocking it in the other), examine Figure 23.33.

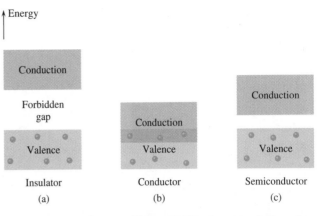

Figure 23.28 A representation of the band structure of solids. (a) The valence band in an insulator is occupied with charge carriers (electrons), but it's separated from the conduction band by a gap, so it is a poor conductor. (b) In the conductor, the two bands overlap and electrons flow easily. (c) A semiconductor has a gap, but it's small.

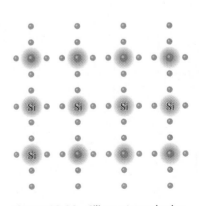

Figure 23.29 Silicon atoms sharing electrons in a crystal.

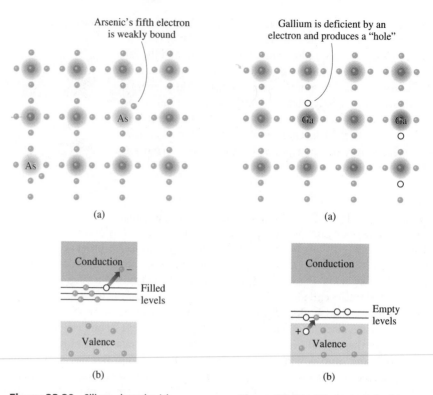

Figure 23.30   Silicon doped with arsenic. The impurity atoms result in filled levels just below the conduction band. Electrons from these levels can reach the conduction band relatively easily. These are the mobile-charge carriers.

Figure 23.31   Silicon doped with gallium. The impurity atoms result in empty levels just above the valence band. Electrons from the valence band can reach these levels relatively easily. The holes left behind in the valence band are the mobile-charge carriers.

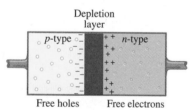

Figure 23.32   Schematic representation of a junction diode.

Here, an external potential difference is applied across the junction diode such that the positive terminal is attached to the *p*-side and the negative to the *n*-side. This opposes the internal potential difference, and the diode is **forward biased**. The positive terminal repels holes into the junction, where they are met by electrons repelled by the negative terminal—at first a tiny current traverses the diode. The depletion layer shrinks as the voltage is raised to about 650 mV (or about 300 mV for germanium), at which point the layer vanishes and the amount of current increases abruptly as carriers flow freely across the diode.

By contrast, when the diode is *reverse biased,* electrons are attracted away from the junction toward the positive terminal just as holes are attracted away toward the negative terminal. The depletion layer broadens, and only an exceedingly small current can traverse the diode.

Although we can take the opportunity to study only the diode, there are a number of other important applications of the *pn*-junction. These include the photovoltaic (solar) cell, the light-emitting diode (LEDs are the bright little red lights in VCR, camera, and stereo displays), and the *diode laser.*

Rectification

The process of converting ac to dc is called **rectification**, and the junction diode is a popular rectifier. The vast majority of electronic devices require some dc, and if

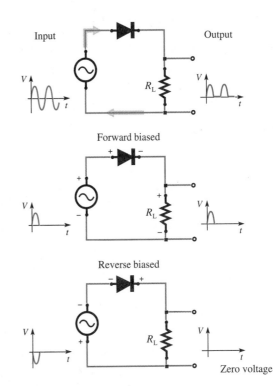

Figure 23.34   Rectification using a diode. Only when it's forward biased does the diode conduct. Think of it as an active switch that opens and closes depending on the polarity of the signal.

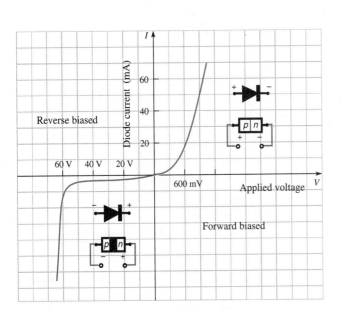

Figure 23.33   Current versus voltage for a junction diode. Observe the different voltage scales. A forward-biased junction diode conducts. A reverse-biased diode does not conduct. The symbol for the diode is an arrow in the direction it conducts a conventional current.

the power source is 60 Hz ac, a rectifier will be needed. As arranged in Fig. 23.34, just the positive portion of the signal is passed by the diode, which will allow only a clockwise current. The output voltage across the *load resistor* R_L is dc, but hardly constant. To smooth out the voltage, we add a capacitor to make an *RC*-filter (Fig. 23.35). The voltage across *C* rises positively, and it becomes charged. As the voltage starts to drop from its maximum, the capacitor discharges into the load resistor (it can't send a current backwards through the diode). With a long-time constant (*RC*), that discharge will be so slow that the next positive voltage peak will interrupt it and recharge the capacitor. The output, which still isn't flat, is said to have a *ripple* and, for obvious reasons, the circuit is known as a **half-wave rectifier**. A **full-wave rectifier** makes use of both the positive and negative peaks, producing a much smoother output with less ripple. One version, which you're likely to find if you cut open the ac adapter for your calculator or portable CD player (p. 892), uses a center-tapped transformer and two diodes (Fig. 23.36).

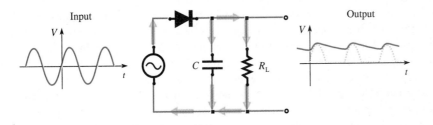

Figure 23.35   Rectification and filtering. The diode passes on only the positive peaks, and the *RC*-filter smooths them out.

Figure 23.36   A full-wave rectifier. Note that the direction of the current through the load resistor is the same regardless of the polarity of the input signal. The output is unfiltered, but a capacitor across the output would smooth the signal appreciably.

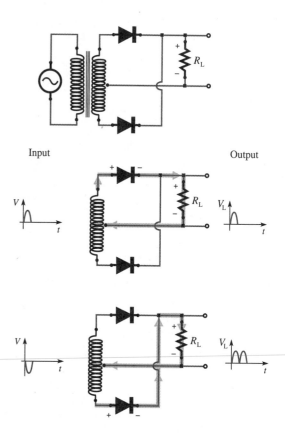

23.9   Transistors

The transistor is effectively two diodes, back-to-back. Usually made of silicon, it is formed in a three-tiered sandwich of either *pnp* or *npn* doped regions. These *epitaxial* layers (from the Greek *epi* meaning *upon* and *taxis* meaning *arranged*) are grown in place so as to preserve the single-crystal structure (Fig. 23.37). One layer, called the *emitter* (labeled *E*), is highly doped. It therefore has a low resistance and is rich in mobile-charge carriers. The middle, very thin layer is the lightly doped *base* (labeled *B*). The remaining layer is the *collector* (labeled *C*) and it, too, is lightly doped. Reasonably enough, **the emitter emits mobile-charge carriers to the collector**. In a *pnp* transistor, the principal carriers are positive holes, and the arrow in the symbolic representation of the transistor is in the direction of traditional current. In an *npn* transistor, the principal carriers are negative electrons, and a positive current is imagined to pass from collector to emitter while the actual electron flow goes from emitter to collector. In either case, within the transistor symbol the emitter-base arrow always points from a *p*-type to an *n*-type region.

The transistor is usually placed in series with a dc source able to provide an appreciable current. It then serves as an electrical control valve, opening fully or partially, and allowing a proportionately large current to pass, or closing and cutting off the current altogether. This it does, in effect, by varying the emitter-collector resistance (indeed, the name *transistor* comes from combining the words *transfer* and *resistor*). The electrical control of the valve arises by way of a tiny current through the base—the base or signal current. Variations in this very small input swing the valve open proportionately and thus control the very much larger emitter-collector current. In other words, the tiny input signal current is *amplified* in the form of an identically shaped, but much larger, output current.

The full-wave rectifier in an ac-to-dc adapter. You can see the transformer (attached to the plug prongs via two yellow wires), two little (black) diodes, and a (blue) filter capacitor. This is one of those black-box plugs that supplies power to calculators, rechargeable flashlights, portable tape recorders, CD players, etc.

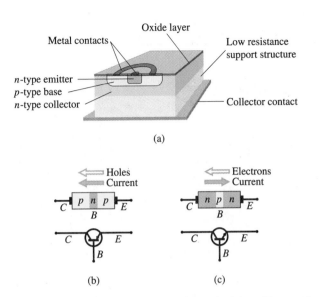

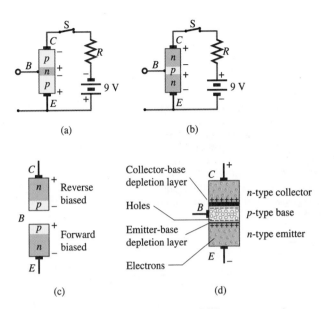

Figure 23.37 (a) The transistor is formed of doped layers of *p*- and *n*-type semiconductors. (b) The mobile-charge carriers in the *pnp* device are holes. (c) Electrons are the mobile-charge carriers in the *npn* transistor.

Figure 23.38 (a) A *pnp* transistor and (b) an *npn* transistor, each with a voltage between collector and emitter. (c) The device functions as if it were two *np* diodes oppositely biased. (d) In the *npn* transistor, electrons flow from the emitter (*n*) to the collector. In the *pnp* transistor, holes flow from the emitter (*p*) to the collector.

Figure 23.38 shows both a *pnp* and an *npn* transistor, each with a battery supplying a voltage between collector and emitter. Consider either transistor—say, the *npn*. It can be imagined as two *np* diodes back-to-back (*np-pn*)—two junctions formed, one on each surface of the base and separated by the layer's small ($\approx$10-μm) thickness. Recall that when the *p*-region is positive with respect to the *n*-region, the junction is forward biased, as in Fig. 23.33. Thus, while the *E-B* junction is forward biased by the battery, the *C-B* junction is reverse biased (and the same is true for the *pnp* transistor). Immediately after the switch is closed, if the forward bias exceeds about 650 mV for silicon (and $\approx$300 mV for germanium), electrons will easily flow from the emitter into the base. The majority of these will continue on, cross the thin base, move into the collector and out into the circuit, driven by the battery. But this current will not continue long. Holes in the base will soon be depleted, and that will change things drastically. The holes migrate across the *E-B* junction toward the negative terminal of the battery, and they can also be lost by recombining with electrons flowing toward the base from the emitter. The total effect is to cause the appearance of a net negative charge on the base that, opposing the electron flow from the emitter, will soon stop that current almost entirely (Fig. 23.38d). A relatively small charge inhibits the progress of a good deal of current that could be provided by the battery. The transistor is then like an opened switch with a nearly infinite resistance, which is identically the case with the *pnp* transistor, except there the principal carriers flowing from *E* to *C* are holes.

This blocked situation can be reduced or eliminated by injecting holes into the base of the *npn* transistor or, equivalently, drawing electrons out of it. By sending a small positive current into the base, a proportionately large positive current will pass from *C* to *E* and around the external circuit (Fig. 23.39). Again, the same thing will happen with a *pnp* transistor, provided electrons are injected into (or holes removed from) its base. ***The presence of a small base current controls the flow of charge from emitter to collector.***

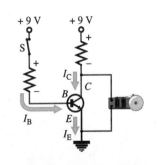

Figure 23.39 An *npn* transistor used as a switch, perhaps in a burglar alarm. Opening the switch *S* rings the bell. Here $I_E = I_B + I_C$. As long as current enters the base, the collector-emitter current is uninhibited. Interrupting I_B will cause the transistor to essentially open the circuit. With the path to ground blocked, current I_C goes through the bell instead, and it rings.

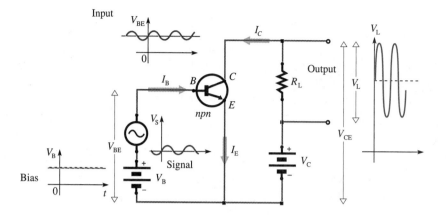

Figure 23.40 The common-emitter amplifier. The signal V_S is super-imposed on a bias voltage V_B so that the input V_{BE} is always positive. The amplified output is taken off the load resistor. This kind of amplifier has a large current gain (50–250), a high voltage gain, and a medium input resistance (≈ 2 kΩ).

Amplifiers

Most transducers, used either in the home or laboratory, rarely generate enough electrical output to directly power an appropriate indicating device. Some high-quality microphones produce voltages of only a few tenths of a microvolt or less. The ac signals from a radio antenna or a tape recorder pick-up head are all a few millivolts and much too weak to be fed directly to a loudspeaker, especially if you want to hear a thundering performance. Such signals are usually amplified several times in successive stages before sending them out to the speakers or some other *load*. The design of an amplifier (often with one or two transistors) depends on the type of input and load, and there is an endless variety of possibilities.

Figure 23.40 depicts one of the most useful basic amplifier configurations, the ***common emitter***. Its name comes from the fact that the emitter is common to both the base and collector circuits. An alternating signal is fed to the base on top of a constant dc bias voltage of the battery (perhaps 1.5 V). That way, the input into the base is never negative, and the base always receives the necessary influx of positive charge. Without the bias, the transistor would clip off the negative portion of the signal.

A small change in base or ***input current*** ΔI_B (typically several microamps) results in a large change in the collector or ***output current*** ΔI_C (perhaps a few milliamps), and we define the ratio $\Delta I_C / \Delta I_B$ as the ***current gain*** of the amplifier. And there is a corresponding definition for the ***voltage gain***. According to the manufacturer, a general purpose *npn* transistor like the 2N3904 has a maximum current gain of 400.

Though individual transistors are still being used to fabricate amplifiers and countless other devices, especially in small production runs, the integrated circuit is dominant in many areas of electronics. A *monolithic IC* is formed on a single crystal wafer, usually silicon, but there are other competing semiconductors as well. Using diffusion techniques (the vapor deposition of layers of materials and etching), hundreds of thousands of components can be formed and "wired" together. A microscopic monolithic resistor is just a tiny sheet of isolated semiconductor. A monolithic capacitor is a little reverse-biased *pn*-junction. Monolithic transistors are usually *npn*, formed by successive diffusions. Although several hundred thousand transistors and all their accompanying resistors and capacitors can be made to fit on a dime, the basic operation of such a technological wonder is not any different from the kinds of circuits we have discussed.

An integrated circuit.

Core Material

A sinusoidal ac source has a terminal voltage of the form

$$v(t) = V_m \sin \omega t = V_m \sin 2\pi ft \qquad [23.1]$$

and with a load resistor R across its terminals

$$i(t) = I_m \sin \omega t = I_m \sin 2\pi ft \qquad [23.2]$$

The **effective** (or *rms*) **current** is given by (p. 870)

$$I = I_{eff} = \frac{I_m}{\sqrt{2}} \qquad [23.7]$$

and the **effective** (or *rms*) **voltage** is

$$V = V_{eff} = \frac{V_m}{\sqrt{2}} \qquad [23.8]$$

Thus the average power dissipated is

$$P_{av} = I_{eff}^2 R = I_{eff} V_{eff} \qquad [23.10]$$

With a purely inductive load, *the current lags the voltage* (or the voltage leads the current) *by one-quarter cycle* (90°). The **inductive reactance** X_L is defined as

$$X_L = \omega L \qquad [23.14]$$

whereupon

$$V = IX_L \qquad [23.15]$$

The average power supplied per cycle in a purely inductive circuit is zero (p. 873).

The instantaneous current in a capacitor leads the instantaneous voltage across it by one-quarter of a cycle (90°). The **capacitive reactance** X_C is defined as

$$X_C = \frac{1}{\omega C} \qquad [23.18]$$

whereupon

$$V = IX_C \qquad [23.19]$$

Only resistance will dissipate power in an ac circuit converting electrical energy into thermal energy (p. 875).

In a series *L-C-R* circuit

$$V_m = \sqrt{(I_m R)^2 + \left(I_m \omega L - \frac{I_m}{\omega C}\right)^2} \qquad [23.26]$$

and

$$\theta = \tan^{-1} \frac{\omega L - 1/\omega C}{R} \qquad [23.27]$$

The quantity $(X_L - X_C)$ is called the **reactance** of the circuit:

$$X = X_L - X_C \qquad [23.30]$$

The **impedance** is

$$Z = \sqrt{R^2 + X^2} \qquad [23.31]$$

in ohms. Ohm's Law for ac is then

$$V = IZ \qquad [23.32]$$

The average power dissipated is

$$P_{av} = IV \cos \theta \qquad [23.33]$$

where the product IV is called the *apparent power* (p. 880).

At **resonance**, where the capacitive and inductive reactances are equal (p. 881), we have

$$f_0 = \frac{1}{2\pi\sqrt{LC}} \qquad [23.35]$$

$\theta = 0$, $Z = R$, and $P_{av} = IV$.

For a transformer, the ratio of the effective voltages will equal the ratio of the numbers of turns, expressed as

$$\frac{V_p}{V_s} = \frac{N_p}{N_s} \qquad [23.36]$$

Moreover

$$V_p I_p = V_s I_s \qquad [23.38]$$

A *pn-junction* is the interface formed by joining a *p*-type to an *n*-type semiconductor. A *pn*-junction can function as a **diode**. The process of converting ac to dc is called *rectification* and the junction diode is a popular rectifier (p. 890). The junction transistor, usually made of silicon, is formed in a three-tiered sandwich of either *pnp* or *npn* doped regions. It's a kind of electrical switch that can be used to control a large current and, in that sense, to amplify a signal.

Suggestions on Problem Solving

1. When converting from effective values to maximum values, or vice versa ($V = 0.707V_m$ or $V_m = 1.414V$), keep in mind that you are simply multiplying by either $1/\sqrt{2}$ or $\sqrt{2}$. Which of these to use is easy to remember: if you are computing V_m, it's larger than V, and you multiply by 1.414 to get a larger number.

2. Remember that ω is in radians per second. Don't mix up degrees and radians when dealing with the phase angle. Accordingly, sin ωt must be computed using the radian setting on your calculator. For example, when $f = 60$ Hz and $t = 0.01$ s, sin $2\pi ft = \sin 3.77$ rad $= -0.59 = \sin 216°$.

3. In effect, an inductor holds back, or chokes, ac current, allowing voltage to develop, and a capacitor holds back ac voltage, allowing current to develop. In series circuits, the phasor diagrams are for voltages (the current is the same for each element). In series, *whenever the phase angle is positive, voltage leads current;* the applied voltage leads the circulating current. *Whenever the phase angle is negative, voltage lags current.*

4. Many of the problems you will confront in this chapter can be checked by recomputing the results using a different approach. That is especially true in series circuits where the impedance triangle usually provides a complementary perspective.

Discussion Questions

1. Heavy wire connects a light bulb to a large air-core coil of several hundred turns (Fig. Q1). The coil goes to a pair of fuses, which connect to a double-pole single-throw switch, and finally the leads plug into the power line via a wall outlet. Both leads contain fuses just to make sure the hot line, which is often unknown, is fused. When the switch is closed, will the lamp light? What, if anything, will happen to the light level after the iron core is inserted?

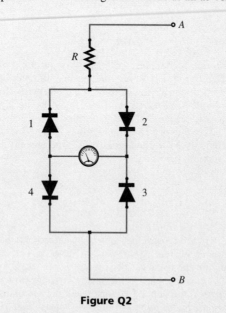

Figure Q1

2. Figure Q2 shows a galvanometer in a circuit with four diodes. Explain how this arrangement works as an ac voltmeter.

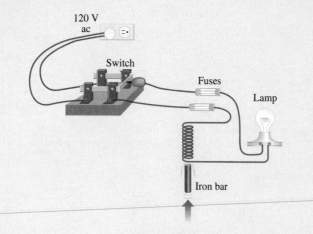

Figure Q2

3. The transmission of electrical power via high-voltage dc is becoming increasingly attractive. List a few reasons why this might be the case.

4. Does the input to a transformer have to be dc for there to be an output?

5. The laboratory setup in Fig. Q5 is used for testing diodes. The oscilloscope is essentially a voltage-measuring in-

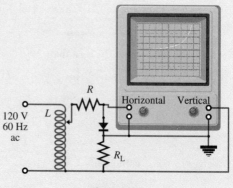

Figure Q5

strument, and that determines how it is used in the circuit. Voltages fed across the horizontal and vertical terminals cause the beam to be deflected proportionately along those directions. Explain what's happening here. What is being measured on the scope?

6. The voltage tester in Fig. Q6 is a little neon discharge lamp that glows brightly when 110 V is across its terminals (with no concern for polarity). Explain what should be observed in each part of the diagram if all is functioning normally. The shorter slot (here on the right) is supposed to be wired "hot" in modern receptacles. What would happen if one lead of the tester was inserted in the right-hand slot and the other touched the faucet on the kitchen sink?

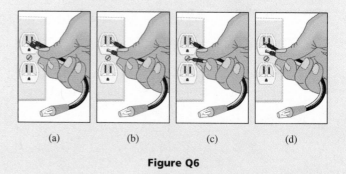

    (a)   (b)   (c)   (d)

Figure Q6

7. The circuit in Fig. Q7 is being studied, using an oscilloscope. The scope is essentially a voltage meter. Here, it's set to scan horizontally and at the same time be deflected vertically by the voltage signal coming in across its two terminals (see Problem 22). Sketch the curves that would result with the scope's probes across points A and B, B and C, and A and C. Discuss.

8. Imagine that you have a faulty wall outlet and wish to replace it. Why would it be wise to shut off the voltage to the outlet? You go to the fusebox and unscrew the fuse and return to the receptacle. Need you be concerned about touching the cover plate, which you want to unscrew and remove? What can you conclude if the tester held as in Fig. Q8 lights up? What do you do next?

9. Imagine that a fuse blows (or a breaker pops) and the lights in a room go out. You remove the toaster and shut off the hair dryer and then screw in a new fuse, but the lights don't go

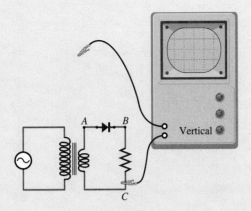

Figure Q7

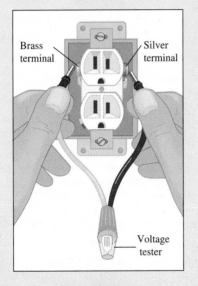

Figure Q8

back on. How can you check the installation of the fuse with a neon tester?

10. Suppose a 240-V air conditioner doesn't seem to be working (at least it doesn't do anything when the on-switch is turned). A check of the outlet (Fig. Q10) reveals that a neon tester lights brightly between terminals B and C, but not between A and B or A and C. What's wrong?

Figure Q10

11. Figure Q11 shows an R-C combination being fed by an ac signal, with the voltage across the capacitor as the output. It is left for Problem 79 to show that the ratio of the output to the input voltage is

$$\frac{V_o}{V_i} = \frac{1}{\sqrt{1 + (2\pi fRC)^2}}$$

What happens to this ratio when $f \to 0$? When $f \to \infty$? Why is it called a low-pass filter?

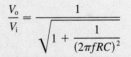

Figure Q11

12. Figure Q12 shows a three-prong adapter. What purpose does it serve and how? What will happen if you just plug in the adapter and don't bother attaching the plug's ground contact to anything?

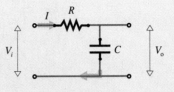

Cover-plate screw

Ground wire

Figure Q12

13. Figure Q13 shows an R-C combination being fed by an ac signal, with the voltage across the resistor as the output. It is left for Problem 78 to show that the ratio of the output to the input voltage is

$$\frac{V_o}{V_i} = \frac{1}{\sqrt{1 + \dfrac{1}{(2\pi fRC)^2}}}$$

What happens to this ratio when $f \to 0$? When $f \to \infty$? Why is it called a high-pass filter?

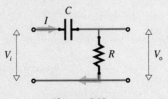

Figure Q13

14. The circuit shown in Fig. Q14 represents the wiring arrangement for a switch and a light fixture. As is customary in such installations, two or more wires are joined (that is, twisted together) by "wire nuts"—those are the small plastic caps wherever there is a splice. The cable is plastic-sheathed and contains two insulated leads and a third bare ground wire. Describe the circuit and draw a simplified diagram. Where does power enter? What does the switch do? What do the gray wires do? Which is the hot lead?

15. What is the device pictured in the circuit diagram in Fig. Q15? How does it work—that is, what is the function of each portion of the circuit?

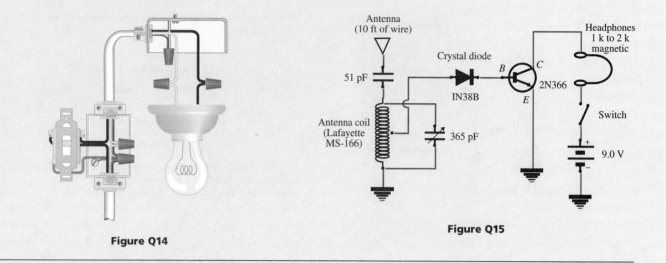

Figure Q14

Figure Q15

Multiple Choice Questions

1. In a sinusoidal-ac circuit, the *rms* current is   (a) $1.414I_m$ (b) $I_m/1.414$ (c) $I_m/0.707$ (d) $I/0.707$ (e) none of these.

2. In a sinusoidal-ac circuit, the peak-to-peak voltage equals   (a) 2 V   (b) 2(0.707) V   (c) 2(1.414) V   (d) 1.414 V/2   (e) none of these.

3. If we double the frequency in a sinusoidal-ac circuit, the effect on a capacitor in that circuit is to   (a) double its reactance   (b) increase its reactance by a factor of four   (c) leave its reactance unchanged   (d) halve its reactance   (e) none of these.

4. When the instantaneous voltage and current in an ac circuit are in-phase, we know that   (a) the total reactance is zero   (b) the capacitive reactance is zero   (c) the inductive reactance is zero   (d) the resistance is zero   (e) none of these.

5. In an *R-L* series ac circuit, increasing *L* and leaving everything else fixed has the effect of   (a) lowering the impedance   (b) lowering the reactance   (c) increasing the capacitive reactance   (d) increasing the power factor   (e) none of these.

6. In an *R-L* series ac circuit, increasing *R* and leaving everything else fixed has the effect of   (a) lowering the impedance   (b) lowering the reactance   (c) increasing the capacitive reactance   (d) increasing the power factor   (e) none of these.

7. In an *L-C-R* circuit, when the inductive and capacitive reactances are equal,   (a) the resistance is zero   (b) the phase factor is zero   (c) the phase angle is zero   (d) the current is zero   (e) none of these.

8. In an *R-L* ac series circuit   (a) the instantaneous current and voltage are in-phase everywhere   (b) the instantaneous current leads the voltage across *L*   (c) the instantaneous voltage across *L* leads the current   (d) the instantaneous voltage across *L* lags the

current by 90°   (e) none of these.

9. A variable capacitor is wired across the terminals of an ac source. Increasing the capacitance   (a) increases the reactance and decreases the current   (b) decreases the reactance and increases the current   (c) decreases the reactance, leaving the current unchanged   (d) decreases the reactance and decreases the current   (e) none of these.

10. In an ac series *L-C* circuit where the resistance is zero, we can expect that the phase angle will   (a) always be zero   (b) never be zero   (c) be + or −90°   (d) be 180°   (e) none of these.

11. Measuring across the two terminals that span an *R-C* ac series circuit, we find that   (a) the instantaneous current leads the voltage by 90°   (b) the instantaneous current lags the voltage by 90°   (c) the instantaneous current and voltage are in-phase   (d) the instantaneous power leads the resistance   (e) none of these.

12. A capacitor with a reactance of 100 Ω is in series with a 100-Ω resistor. The ac voltage across the two   (a) lags the current by 90°   (b) lags the current by 45°   (c) leads the current by 90°   (d) leads the current by 45°   (e) none of these.

13. A pure 1.0-H inductor is in series with a 0.20-μF capacitor across a 110-V, 60-Hz wall outlet. The resulting power factor is   (a) zero   (b) infinite   (c) 1.414   (d) 0.707   (e) none of these.

14. In an ac circuit, the power factor equals the   (a) apparent power   (b) real power   (c) real divided by the apparent power   (d) apparent divided by the real power   (e) none of these.

15. True or average power equals apparent power when   (a) the voltage is equal to the current   (b) the voltage is in-phase with the current   (c) the phase factor equals 0.707   (d) the current is very large   (e) none of these.

Problems

ALTERNATING CURRENT

1. [I] The maximum output voltage of an ac generator is +120 V at a time equal to $\frac{1}{4}$ cycle. What will be the instantaneous terminal voltage   (a) $\frac{1}{4}$ cycle later, and (b) $\frac{1}{8}$ cycle later (at $\frac{3}{8}$ of a cycle)?

2. [I] A sinusoidal oscillator puts out an instantaneous voltage represented by $v = (75\ \text{V})\sin(376.99\ \text{rad/s})t$. What is its frequency and maximum voltage?

3. [I] Write an expression for the instantaneous current delivered by an ac generator supplying 10 A effective, at 50 Hz.

4. [I] The maximum potential difference across the terminals of a 20.0-Hz sinusoidal-ac source is +50.0 V (occurring at $t = \frac{1}{4}$ cycle). If at $t = 0$, $v = 0$, find v at $t = 2.00$ ms.

5. [I] If the *rms* voltage read across a resistor by a voltmeter in a sinusoidal-ac circuit is 100 V, what is the maximum voltage?

6. [I] The effective current passing through a resistor R in a sinusoidal-ac circuit is 2.00 A. What is the maximum voltage drop across the resistor if $R = 100\ \Omega$?

7. [I] Determine the current drawn by a lit 100-W light bulb plugged into a 120-V wall outlet.

8. [I] The instantaneous voltage measured across an ac source is $v = 200 \sin 2\pi 70t$. What is its effective voltage and frequency?

9. [I] A resistor is in series with an ac generator. An ammeter in series with the resistor reads 1.50 A, and a voltmeter across the resistor reads 75.0 V. What average power is being supplied by the source?

10. [I] If the generator in Problem 8 is placed across a 200-Ω resistor, what current will be measured by an ac ammeter in series with the resistor?

11. [I] A 1.0-Ω resistor is attached across an ac generator. An oscilloscope shows that the sinusoidal current in the circuit has a maximum value of 0.50 A. What average power does the resistor dissipate?

12. [I] An 80-μF capacitor C is placed across the terminals of a 60-Hz sinusoidal-ac oscillator. What is the capacitive reactance of C?

13. [I] If a capacitor in series with a sinusoidal oscillator has a reactance of 200 Ω when $f = 50.0$ Hz, what reactance will it have when the frequency is raised to 5.00 kHz?

14. [I] Determine the current that will be drawn by a 45.0-μF capacitor connected across a 240-V, 50.0-Hz source of sinusoidal-ac.

15. [I] A 60-Hz ac generator has a terminal voltage of 120 V. What size capacitor should be placed in series with it so that a current of 1.00 A circulates?

16. [I] Figure P16 depicts a simple circuit for studying the ac behavior of capacitors. What current will the milliammeter read? If either capacitor is disconnected, what will happen to the current?

17. [I] A 0.15-H coil with a negligible resistance has a reactance of 10 Ω when wired into a sinusoidal-ac circuit. Determine the frequency of the source.

18. [I] Determine the inductive reactance of a 10.0-mH coil connected to a 100-Hz ac generator.

19. [I] A high-quality 250-mH inductor (with negligible resistance) is attached across the terminals of a 60-Hz

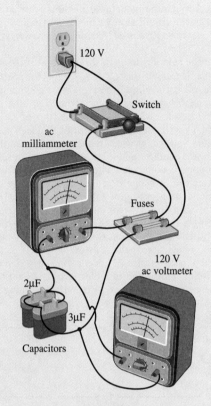

Figure P16

generator having an *rms* output of 125 V. Determine the reactance of the coil.

20. [I] What is the inductance of a coil having negligible resistance if at an angular frequency of 628.3 rad/s, its reactance is 200 Ω?

21. [c] Suppose that the ac power supply in Fig. 23.10 provided a voltage that, rather than being sinusoidal as was the case earlier, is given by $v = V_m \cos \omega t$.   (a) Find the charge on the capacitor.   (b) Find the current in the circuit.

22. [c] An ac power supply is in series with a capacitor. If the current in the circuit is found to be $i(t) = I_m \sin \omega t$, write an expression for the charge on the capacitor as a function of time. Take the charge to be zero at $t = 0$.

23. [c] A capacitor is placed across an ac power supply. What is the voltage across the capacitor if the current in the circuit is found to be $i(t) = I_m \sin \omega t$? Write your answer as a sine (rather than a cosine) function. [Hint: Make use of the results of the previous problem.]

24. [c] A capacitor is attached to the two terminals of an ac power supply and a voltage of $v(t) = V_m \sin(\omega t + \varphi)$ is applied. Determine the current in the circuit.

25. [c] An inductor is placed across the terminals of an ac generator. If the current in the circuit is measured to be $i(t) = I_m \sin \omega t$, what is the voltage across the inductor? Write your answer as a sine function.

26. [c] The two terminals of an ac power supply are attached across an inductor, and a voltage of $v(t) = V_m \sin(\omega t + \varphi)$ is applied. Determine the current in the circuit.

27. [c] An inductor is placed across an ac generator. If the current through the inductor at $t = 0$ is I_0, what is the current in the circuit at any time if the voltage of the source is $v(t) = V_m \cos \omega t$? Write your answer as a sine function.

28. [c] Use the calculus to show directly that if $i(t) = I_m \cos \omega t$, it follows that $I_{eff} = I_m/\sqrt{2}$. [Hint: Start with the integral definition of the average or mean and integrate over ωt as the variable that goes from 0 to 2π.]

29. [II] If at a time of 2.00 ms after the start of a cycle, a sinusoidal oscillator has an output of 95.1% of its maximum voltage, at what frequency is it operating?

30. [II] An ac generator and a load resistor are in series. An oscilloscope is connected across the resistor. The vertical deflection on the scope is caused by the voltage input and corresponds to a setting of 150 mV/cm. The horizontal axis is set for scan and is swept internally by the scope, producing a sinusoidal image on the screen. If the peak-to-peak displacement of the signal is 4.0 cm, what is the *rms* voltage across the resistor?

31. [II] The circuit in Fig. P31 needs your help. Every time the switch is closed the fuse blows. What could be wrong? [Hint: One of the components is malfunctioning.]

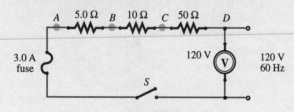

Figure P31

32. [II] Referring to Fig. P32, what will the two meters in the circuit read?

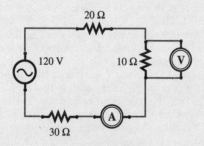

Figure P32

33. [II] Referring to Fig. P33, what will the two meters in the circuit read?

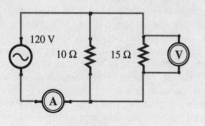

Figure P33

34. [II] A 1200-W hair dryer is plugged into a 120-V line. What's its resistance and how much current does it draw?

35. [II] Figure P35 shows a laboratory setup for studying the behavior of a resistive load (R) in an ac circuit. Here, a transformer inputs the 120 V from the wall outlet and supplies 30 V at 60 Hz, and the milliammeter reads 500 mA effective. Assuming the losses in the wiring to be negligible, what average power does the resistor dissipate? How would that change if the transformer was removed and the circuit plugged directly into the wall outlet?

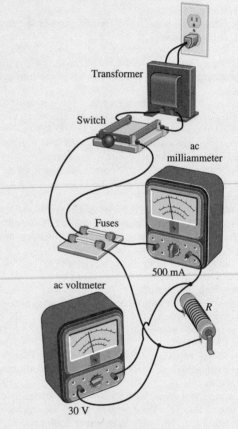

Figure P35

36. [II] Prove that the ohm is the SI unit of capacitive reactance.

37. [II] Prove that the ohm is the SI unit of inductive reactance.

38. [II] A sinusoidal oscillator with a terminal voltage of 125 V at a frequency of 55 Hz is put in series with a capacitor of 100 μF. What effective current will be provided by the oscillator?

39. [II] A high-quality coil with an inductive reactance of 100 Ω and negligible resistance is placed across the terminals of a 150-V ac generator. Determine the maximum current through the coil.

40. [II] Determine the current that would be read by an ammeter in series with a 410-mH inductor of negligible resistance connected to a wall outlet at 120 V, 60 Hz.

41. [II] Figure P41 shows a 60-Hz ac laboratory setup for trouble-shooting the 20-μF capacitor, which has been around for a while and needs to be checked before being used. (a) What can you say about it? Now, suppose we put another suspicious 10-μF capacitor across the first and find that the fuse blows every time the switch is closed. (b) What would you then conclude?

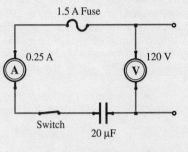

Figure P41

42. [cc] Given a sinusoidal voltage $v(t) = V_m \sin \omega t$, (a) what is the average value over one period? (b) What is its average value over half a period? (c) What is its rms value?

43. [cc] Figure P43 shows the waveform of an ac voltage source. Using the calculus, compute the rms or effective voltage. [Hint: Write an expression for $v(t)$ and then use the definition of the mean or average value.]

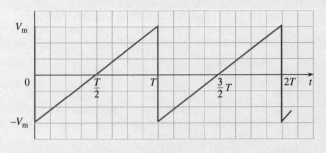

Figure P43

44. [cc] Suppose we add a dc component to the signal in Fig. P43 so that it is entirely positive, rising linearly from zero to a new maximum voltage V_m in a time of one period T. Find the mean voltage and the rms voltage and compare your answer to that of the previous problem.

45. [cc] A voltage signal (shown in Fig. P45) consists of a series of parabolic humps given by the expression $v(t) = V_0[1 - 4(t/T)^2]$. Determine its rms value.

46. [III] A 100-Ω resistor is wired across the terminals of an oscillator, which sends a current through it given by

$$i(t) = 2.40 \cos 180t$$

in SI units. (a) Determine the period of the current oscillation. (b) Find the effective voltage of the source. (c) Write an expression for the instantaneous voltage across the load.

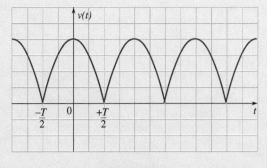

Figure P45

47. [III] Write an expression for the maximum instantaneous power dissipated in a sinusoidal-ac circuit consisting of a source and a resistor. How does this expression compare with the average power?

L-C-R AC NETWORKS

48. [I] A 6.0-μF capacitor in series with a 500-Ω resistor are both placed across a 120-V, 60-Hz ac source. Draw the impedance triangle for the circuit. Check your answer by computing Z.

49. [I] A coil has a resistance of 6.10 Ω and a reactance of 8.05 Ω. What is its impedance?

50. [I] An inductor has a reactance of 8.0 Ω and a resistance of 6.0 Ω. When connected across a 120-V ac outlet, how much current will it draw?

51. [I] A capacitor with a reactance of 160 Ω is in series with a 150-Ω resistor and an ac source. What is the net impedance of the load?

52. [I] A coil has an inductance of 32.0 mH and a resistance of 30.5 Ω. What is its impedance when placed across a 240-V, 200-Hz sinusoidal oscillator?

53. [I] If the total reactance of an L-C-R circuit is 1200 Ω and the total resistance is 500 Ω, what is the impedance?

54. [I] In a series L-C-R circuit, the inductive reactance is 1800 Ω, the capacitive reactance is 2600 Ω, and the resistance is 1000 Ω. What is the impedance?

55. [I] An L-C-R ac series circuit has a value of $(X_L - X_C) = +200$ Ω and a resistance of 400 Ω; what is the phase angle and does the current lead or lag the applied voltage?

56. [I] A 2.0-kΩ resistor and a 2.0-μF capacitor are connected in series across a 60-Hz ac generator. What is the impedance of the circuit?

57. [I] A series circuit contains a capacitor, a resistor, and a 40-V, 240-Hz sinusoidal-ac source. If the net impedance of the load (resistor and capacitor) is 160 Ω, what value of current will be measured by an ammeter in the circuit?

58. [I] A coil with a reactance of 150 Ω and a resistance of 25 Ω is placed across a 120-V ac harmonic source. Determine the phase angle between the instantaneous voltage across the coil and the current.

59. [I] A capacitor with a reactance of 0.20 kΩ is placed in series with a 100-Ω resistor and the combination is attached to the terminals of an ac source. What is the phase angle between the instantaneous current and the instantaneous voltage across the combination?

60. [I] In the previous problem, what is the rms current if the source has a terminal voltage of 0.12 kV?

61. [I] A coil has a reactance of 1.55 kΩ and draws 26 mA from an ac generator with a terminal voltage of 60 V effective. What is its resistance?

62. [I] In an *L-C* series circuit with negligible resistance, the voltage drops across the inductor and capacitor are 100 V and 120 V, respectively. What is the phase angle?

63. [I] A circuit consists of a capacitor with a reactance of 2500 Ω in series with an inductor of negligible resistance and a reactance of 2000 Ω. What is the impedance of the circuit?

64. [I] A choke coil (of negligible resistance) with a reactance of 2.5 kΩ and a 2.5-kΩ resistor are in series across a 120-V ac generator. What average power is dissipated in the circuit?

65. [I] An *L-C-R* series circuit contains a 500-Ω resistor, a 5.0-H choke coil, and a capacitor. What value of capacitance will cause the circuit to resonate at 1000 Hz? What can you say about the practicality of your result?

66. [I] An *L-C-R* series circuit contains a 1.00-μF capacitor, a 5.00-mH coil, and a 100-Ω resistor. What is its resonant frequency?

67. [I] A little radio has a tuning circuit consisting of a variable capacitor and an antenna coil. If the circuit has a maximum current when $C = 300$ pF and $f = 40.0$ kHz, what is the inductance of the coil?

68. [I] A transformer has a 500-turn primary and a 2500-turn secondary. If the primary voltage is 120-V ac, what is the secondary voltage?

69. [I] A typical neon sign operates at a voltage of about 11.5 kV. If a transformer is to raise the 115-V line voltage up to that level, what should be its turn ratio?

70. [I] The primary of a transformer has 600 turns and a voltage across it of 240 V. How many turns should the secondary have if the output voltage is to be 60.0 V?

71. [I] A load resistor is placed across the secondary terminals of a step-down transformer with a turn ratio of 4:1. If the primary voltage and current are 120 V and 1.0 A, respectively, what will be the secondary voltage and current? Assume 100% efficiency.

72. [II] Suppose we want to find the inductance of a coil experimentally. We use an ohmmeter to determine that its resistance is 50 Ω. Next, we attach the coil to an oscillator, which supplies 300 mA to it at 20.0 V and 1.20 kHz. Determine *L*.

73. [II] A 30-mH coil with a resistance of 30 Ω is supplied by a 240-V, 200-Hz sinusoidal-ac source. Calculate the effective current it draws and the phase angle between the instantaneous current and the supply voltage.

74. [II] A 5.2-H choke coil with a resistance of 20 Ω is in series with a 1200-Ω resistor. The circuit is plugged into a transformer that puts out 60-V ac at 60 Hz. What voltage drop will appear across the resistor? Across the inductor?

75. [II] A 400-mH inductor with negligible resistance and a 185-Ω resistor are in series across a 60.0-Hz sinusoidal-ac source. If the voltage drop across the resistor is 370 V, what is the voltage across the inductor? What is the voltage of the source?

76. [II] Imagine yourself in a lab. On the table is an inductor rated at 1.5 H, supposedly with a resistance of 50 Ω. It's in series with a 0.5-A fuse. Every time the circuit is plugged into the 110-V, 60-Hz wall outlet the fuse blows. (a) Determine the current that should be in the circuit if all were well. (b) What would you guess was wrong? How might you confirm that?

77. [II] In an experimental setup, a capacitor and a 20-Ω resistor are in series across a 120-V, 60-Hz sinusoidal-ac source. An ammeter indicates that the circuit carries an *rms* current of 1.20 A. Determine the capacitance.

78. [II] Figure Q13 shows an *R-C* combination known as a high-pass filter (see Discussion Question 13). It's being fed an ac signal composed of a broad range of different frequency information. The voltage across the resistor is the output that is passed on to the next circuit for further processing. Show that the ratio of the output to the input voltage is

$$\frac{V_{\text{o}}}{V_{\text{i}}} = \frac{1}{\sqrt{1 + \dfrac{1}{(2\pi f R C)^2}}}$$

79. [II] Figure Q11 shows an *R-C* combination known as a low-pass filter (see Discussion Question 11). It's being fed an ac signal composed of a broad range of different frequency information. The voltage across the capacitor is the output that is passed on to the next circuit for further processing. Show that the ratio of the output to the input voltage is

$$\frac{V_{\text{o}}}{V_{\text{i}}} = \frac{1}{\sqrt{1 + (2\pi f R C)^2}}$$

80. [II] A series *L-C-R* circuit is connected across a 10.0-kHz source. The 1.2-H inductor and 1000-Ω resistor are fixed whereas the capacitor is variable. (a) At what capacitance will the current in the resistor be a maximum? (b) What will be the voltage drop across the resistor if the source voltage is 50 V? (c) What average power will be dissipated?

81. [II] A transformer is used to power a 6.0-V door chime. The 240-turn primary, which has an inductance of 3.0 H and negligible resistance, is connected across the standard 120-V, 60-Hz ac. (a) What turn-ratio should the transformer have? (b) With the secondary open-circuited, determine the primary current.

82. [II] Draw the impedance triangle for the circuit in Fig. P82 and determine the phase angle and impedance. Check your answers by direct computation. How much current is circulating? What is the resonant frequency of the circuit, and if it were driven at that frequency, how much current would be present?

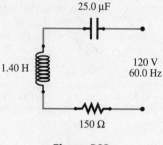

25.0 μF

1.40 H

120 V
60.0 Hz

150 Ω

Figure P82

83. [II] A transformer delivers power at a rate of 45 kW. If it loses 300 W in hysteresis and eddy current effects (the so-called iron loss) and 500 W in joule heat (the so-called copper loss), what is its efficiency?

84. [cc] Consider an ac generator in series with an inductor, capacitor, and resistor (Fig. 23.13). The source voltage is $v(t) = V_m \sin \omega t$. Write out the differential equation governing the circuit in terms of the charge. [Hint: Start with Kirchhoff's Loop Rule.]

85. [cc] Consider an ac generator in series with an inductor, capacitor, and resistor (Fig. 23.13). The source voltage is $v(t) = V_m \cos \omega t$. Write out the differential equation governing the circuit in terms of the current. [Hint: Start with Kirchhoff's Loop Rule.]

86. [cc] Referring to the parallel circuit depicted in Fig. P86, determine the instantaneous current through each element. Where appropriate, write your answers in terms of reactances.

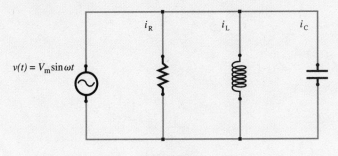

Figure P86

87. [c] Examine the voltage output of the diode in Fig. 23.34. What is the rms current, given that $i(t) = I_m \sin \omega t$ for $0 < t < T/2$ and $i(t) = 0$ for $T/2 < t < T$ where T is the period?

88. [III] A transformer is supplied 8.20 kW by an ac source across its primary. The terminal voltage at the secondary is 220 V and the load is such that there is a power factor of 82%. Determine the current in the secondary circuit.

89. [III] To transfer a maximum amount of power from a source to a load (p. 762), the two should have the same impedance but often don't (audio amplifiers have a high impedance, speakers have a low impedance). A transformer can be used to match impedances (see photo on p. 884). Let Z_s be the load attached to the secondary and V_p and I_p be the voltage and current in the primary when the transformer and load are connected to the source. The source then sees an effective load Z_p; that is, it behaves as though Z_p were attached across its terminals (rather than the transformer and Z_s). Show that

$$Z_p = Z_s \left(\frac{N_p}{N_s} \right)^2$$

What turn-ratio should a transformer have if it is to match a 3.2-kΩ amplifier to an 8.0-Ω speaker?

24

Radiant Energy: Light

"**W**HAT *IS* LIGHT?" IS one of those superbly simple questions we humans have pondered, and struggled with, for well over 2000 years. We are not without answers, marvelous answers that reflect the highest accomplishment of human ingenuity and creativity, but our understanding even now is far from complete, far from satisfactory. This chapter, and the next three as well, are about electromagnetic radiant energy, of which light is only that small part we happen to see. Thus, we begin the study of *Optics,* the study of the behavior of radiant energy.

Light. *Caloric is the name assigned by the new nomenclature to the element of heat. It is the most subtile of all bodies, and exists in nature in very great quantities. . . . Light is occasioned by the transmission of caloric, or the matter of heat, which travels through the air, in one second of time, about 170,000 miles. . . . It is heat projected. The identity is the same; and consists of infinitely small particles thrown off in every direction from a luminous body.*

H. G. SPAFFORD
General Geography and Rudiments of Useful Knowledge (1809)

Sources of Light Waves. *Just as sound waves are disturbances set up in the air by the vibrations of bodies of ordinary dimensions, so light waves are disturbances set up in the aether probably by the vibrations of the minute corpuscles, or* electrons, *of which the atoms of ordinary matter are supposed to be built up.*

ROBERT A. MILLIKAN AND
HENRY G. GALE
A First Course in Physics (1906)

Augustin Jean Fresnel (1788–1827).

THE NATURE OF LIGHT

Among the jumble of ideas that came from the ancient Greeks were the seeds of two significant lines of thought. The Atomists evolved the *emission theory,* which pictured light as a torrent of exceedingly minute, high-speed particles. The other influential conception was advanced by Aristotle. To the four elements of nature (fire, air, earth, and water), he added a fifth—the *aether.** There was no such thing as empty space. All the void was filled with aether and so Aristotle could propose that human vision "arises from a movement, produced by the body we perceive, in the interposed medium. . . ." Light was aethereal motion.

24.1 Waves and Particles

In the 1660s, Robert Hooke studied the color patterns associated with thin transparent films, such as soap bubbles. The repetitive form of those patterns guided him to propose a rudimentary *wave theory* in which light consisted of very rapid vibrations of the aether propagating at tremendous, though finite, speed. The fastest naturally occurring thing known at the time was sound, and light was even faster—the lag between the perception of thunder and lightning proved that. So it seemed reasonable to suppose that only a wave, which is a *self-sustaining disturbance of a medium without the transport of matter,* could travel at such tremendous speed. Furthermore, two beams of light can cross (as they must, for example, when you look through a small hole at two separate sources), and yet they never scatter each other as streams of particles surely would.

Although Newton seems earnestly to have tried to stay out of the wave-versus-particle debate, he nonetheless ultimately favored the corpuscular picture. His rejection of the wave theory came from his inability to explain the observed straight-line propagation of light. To Newton's mind, a wave should spread out markedly as it passes through an aperture (Fig. 24.1), filling almost the whole region beyond. It couldn't possibly produce a narrow beam, and yet such beams could easily be made. Unhappily, neither he nor anyone else of that era knew very much about waves. Waves do spread out, just as Newton had maintained, but will only do so in the way he envisioned when the wavelength is roughly the same size as the hole through which they pass and the wavelength of light is minute (Fig. 13.36, p. 517, makes the point nicely).

Aware of Hooke's work with thin films, Newton proposed a remarkably prophetic solution. Light has a dual nature: it is a stream of material particles, but these are capable of setting up vibrations in the aether. The resulting pattern in the distribution of aether channels the stream of corpuscles—waves of aether guide particles of light.

Although the wave theory did have a few important backers in the eighteenth century, such as Benjamin Franklin and Leonhard Euler, the age was dominated by the works of Newton. His doting disciples soon forgot the master's own doubts and, in his name, light *became* a stream of particles.

In the first few decades of the 1800s, the wave theory was reborn with a new analytic vigor. Resurrected principally and independently by English physician Thomas Young and French civil engineer Augustin Fresnel, the wave conception

*Though no longer accepted, the notion of a material aether has played a tremendously important role in the development of physics. This divine, invisible, elastic "goop" would ultimately become the transmitter of all sorts of influences from gravity and electricity to heat and "life force." The word itself was first used by the poet Hesiod (700 B.C.) to refer to the uppermost tenuous reaches of the Earth's atmosphere.

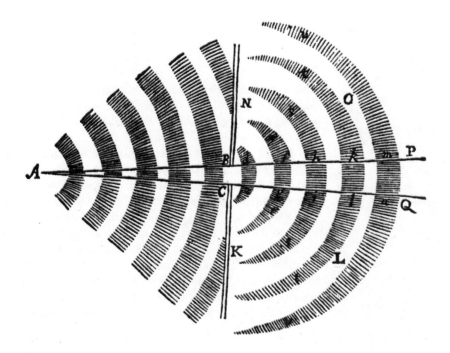

Figure 24.1 Taken from the *Principia*, this is Newton's picture of the diffraction of a wave as it passes through a hole. It's a fairly good representation, provided the hole is smaller than the wavelength.

slowly gained acceptance. By supposing that light was a *transverse* disturbance (p. 465), they could understand everything that had been observed over the centuries. Light was once again a wave—presumably an elastic wave in the aethereal sea.

24.2 Electromagnetic Waves

While this saga was being played out, and quite independent of it, the study of electricity and magnetism was reaching a high point of its own. The masterwork of the age was created by a good-humored, bright young man, one James Clerk Maxwell of Kirkcudbrightshire in the Lowlands of Scotland.

William Thomson (later to become Lord Kelvin) was, in the 1840s, the first to begin to mathematize the notion of *lines-of-force*. Faraday himself was ill at ease with mathematics, never having had any formal training in the subject. One day in 1846, Faraday was asked to step in at the last moment and give the evening lecture at the Royal Institution in place of Charles Wheatstone who had had stagefright and run off minutes before. During the impromptu remarks that followed, Faraday conjectured that light might be a wave of some sort propagated along the lines-of-force. So it was that Faraday's visions and Thomson's lovely though disconnected thoughts became the stimulus for Maxwell, who began to pursue his own researches within months of graduating from Cambridge in 1854. He drew together everything fundamental that was known about electricity and magnetism in a set of four equations. Two of these are credited to Gauss, the third is Faraday's Induction Law, and the fourth, Ampère's Circuital Law—from these Maxwell built his theory.

We saw earlier, in Chapter 22, that a time-varying *B*-field is always accompanied by an induced *E*-field. Thus if the flux through the loop in Fig. 24.2 is changing in time, an emf will be induced in the loop and a nonelectrostatic *E*-field will exist within and beyond the confines of the *B*-field. (We can simplify the notation a little by dropping the subscript I because all the fields of interest here will be induced.) Since the flux causing the *E*-field arises from a time-varying *B*-field, we

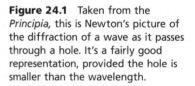

Figure 24.2 A loop enclosing an area *A* in a magnetic field **B**. Here, the area is flat and perpendicular to the field.

A **Letter From Maxwell to Kelvin**

Trin. Coll., Feb. 20, 1854

Dear Thomson

Now that I have entered the unholy estate of bachelor-hood I have begun to think of reading. This is very pleasant for some time among books of acknowledged merit wh one has not read but ought to. But we have a strong tendency to return to Physical Subjects and several of us here wish to attack Electricity.

Suppose a man to have a popular knowledge of electrical show experiments and a little antipathy to Murphy's Electricity, how ought he to proceed in reading & working so as to get a little insight into the subject wh may be of use in further reading?

If he wished to read Ampère Faraday &c how should they be arranged, and at what stage & in what order might he read your articles in the Cambridge Journal?

If you have in your mind any answer to the above questions, three of us here would be content to look upon an embodiment of it in writing as advice. . . .

Propagation of Light by the Aether. *As the atoms of matter vibrate in the aether in which they are immersed, they communicate their vibration to it. The vibrations thus started in the aether are propagated through it in every direction in minute waves and with an inconceivable velocity.*

J. A. GILLET AND W. J. ROLFE
Natural Philosophy for the Use of Schools and Academies (1882)

James Clerk Maxwell (1831–1879).

can rewrite Eq. (22.6) as

$$\oint \mathbf{E} \cdot d\mathbf{l} = -\int \frac{d}{dt}\mathbf{B} \cdot d\mathbf{A} \qquad (24.1)$$

This variation of Faraday's Law is today known as one of **Maxwell's Equations**. In essence, it says that *a time-varying B-field generates an E-field*: When the right side is nonzero, the left side is nonzero.

Figure 24.3 depicts a localized upwardly directed increasing magnetic field **B** surrounded by an induced electric field **E**. Three points are crucial here: **E** and **B** are everywhere perpendicular to each other; the lines of **E**, which do not now begin and end on charges, must therefore close on themselves; and **E** is not restricted to the region actually containing the flux—it extends beyond that.

The second equation that will be of special interest is Ampère's Circuital Law (p. 804). It relates the amount of **B** parallel to a closed path C, with the total current, ΣI, passing within the confines of C, that is, passing through *any* area bounded by C:

$$\oint \mathbf{B} \cdot d\mathbf{l} = \mu_0 \Sigma I \qquad [21.6]$$

This expression says that moving charges are the source of the magnetic field, and although that's true, it's not the whole truth. That much is evident by the fact that, while charging or discharging a capacitor, one can measure a magnetic field in the region between the plates, even though no actual current traverses the device (Fig. 24.4). Moreover, Ampère's Law is not very particular about the area used, provided it's bounded by the curve C, which makes for an obvious problem when charging a capacitor, as shown in Fig. 24.5. If flat area A_1 is used, a net current of I flows through it and there is a B-field along curve C—the right side of the equation is nonzero, so the left side is nonzero. But if area A_2 is used instead to encompass C, no net current passes through it and the field must now be zero, even though nothing physical has actually changed. Something is obviously wrong! Notice that if Q is the charge on either plate and A the area thereof, it follows from Eq. (17.16), $E = \sigma/\varepsilon$, that

$$E = \frac{Q}{\varepsilon_0 A}$$

is the electric field spanning the gap. As the charge builds on the plates, the field changes as well, and taking the derivative of both sides yields

$$\varepsilon_0 A \frac{dE}{dt} = \frac{dQ}{dt}$$

where the expression on the right is the current actually flowing into, and out of, the capacitor. What this result suggests is that charge building up on the plates creates a changing electric field that crosses the gap and is in some way equivalent to a current. Maxwell hypothesized the existence of just such a mechanism, which he called the **displacement current**, $I_D = \varepsilon_0 d\Phi_E/dt$, where $\Phi_E = E_\perp A$ is the flux of the E-field. In this case, where $E_\perp = E$ and $d\Phi_E = A\, dE$, it follows that $I_D = \varepsilon_0 A\, dE/dt$. In other words, a time-varying E-field functions as an effective current I_D, and he added this term into the sum in Eq. (21.6) along with the actual currents: ΣI

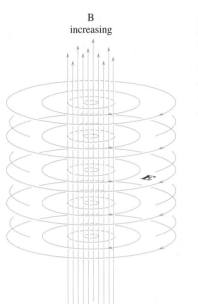

B
increasing

Figure 24.3 Imagine a wire loop surrounding the increasing *B*-field. A current would be induced such that the induced magnetic field **B**ᵢ is downward, telling us the direction of the current and, therefore, the direction of **E**.

Figure 24.4 As a capacitor gets charged, a transient current exists, producing a *B*-field. In the gap where no actual current exists there is nonetheless a *B*-field. It's due to the time-varying *E*-field.

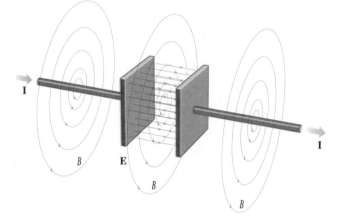

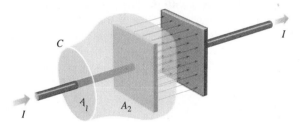

Figure 24.5 Ampère's Law is oblivious to which area A_1 or A_2 is bounded by the path *C*. But a current passes through A_1 and not through A_2, and that means something is very wrong.

henceforth would be $\Sigma(I + I_D)$. So modified, Ampère's Law becomes another of Maxwell's Equations; namely

$$\oint \mathbf{B} \cdot d\mathbf{l} = \mu_0 \sum \left(I + \varepsilon_0 \frac{d\Phi_E}{dt} \right) \tag{24.2}$$

We can write the net flux as an integral via Eq. (17.9), namely, $\Phi_E = \int \mathbf{E} \cdot d\mathbf{A}$. Moreover, in free space, where there are no real currents ($I = 0$), Ampère's Law simplifies to

$$\oint \mathbf{B} \cdot d\mathbf{l} = \mu_0 \varepsilon_0 \frac{d}{dt} \int \mathbf{E} \cdot d\mathbf{A} \tag{24.3}$$

and finally since we are only dealing with changing fields

$$\oint \mathbf{B} \cdot d\mathbf{l} = \mu_0 \varepsilon_0 \int \frac{d}{dt} \mathbf{E} \cdot d\mathbf{A} \tag{24.4}$$

a time-varying **E**-*field generates a* **B**-*field* (Fig. 24.6). This idea is totally new—it had never been proposed or observed prior to Maxwell. It's the crucial creation that allows us to understand electromagnetic waves.

Each form of radiant energy (radiowaves, microwaves, infrared, light, ultraviolet, X-rays, and γ-rays) is a web of oscillating electric and magnetic fields inducing

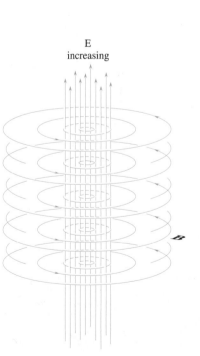

E
increasing

Figure 24.6 The increasing *E*-field shown here has the effect of an upward current. The Right-Hand-Current Rule tells us the direction of the associated *B*-field.

one another. A fluctuating E-field creates a B-field perpendicular to itself, surrounding and extending beyond it. That magnetic field sweeping off to a point further on in space is varying there, and so it generates a perpendicular electric field that moves out and beyond that point as well, and so on: dE/dt creating a perpendicular B, dB/dt creating a perpendicular E, extending out indefinitely in an independent undulation of interlacing fields, an **electromagnetic wave**.

When an electromagnetic wave interacts with bulk matter, the E-field component has a much greater effect on charges than does the B-field. Accordingly, experiments have revealed that it is principally the electric field of the wave that effectuates vision, photochemistry, fluorescence, and so on. We shall usually explicitly consider only the E-field component assuming that the B-field tags along—the two are inseparable.

24.3 Waveforms and Wavefronts

A *progressive* electromagnetic wave is a self-supporting, energy-carrying disturbance that travels free of its source (the light from the Sun sails through space, unleashed and on its own, for 8.3 minutes before arriving at Earth).

Armed with an E-field meter,* we could presumably measure the variations in electric field as a wave swept by. Figure 24.7 attempts to show pictorially what might be seen as the meter-pointer wiggles, first positively then negatively, while the size of the E-field at that point in space fluctuates with the passing wave. The sizes of the drawn arrows correspond to the instantaneous values of the strength of the E-field—*nothing is actually displaced in space*. This phenomenon is not a wave ris-

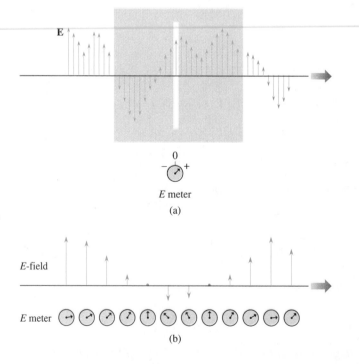

Figure 24.7 The electric field of an electromagnetic wave. (a) If we observe the wave at a single, stationary location, we see an oscillating field as the wave passes. (b) If we observe the wave at many different locations all at the same time, we see the profile of the wave.

*Though these do exist for relatively low-frequency signals, a field meter that would function fast enough to follow the exceedingly rapid swings of the electric field associated with light is still beyond the capabilities of present technology.

ing and falling so many centimeters in the air like a rope. What's being pictured is a succession of readings in time, on a single meter, at one point in space. Alternatively, Fig. 24.7b can also be thought of as showing the many readings from a whole row of meters at different points along a line in the direction of propagation of the wave, all at the same instant. The construct is of a disturbance whose "outline," or **profile**, is irregular.

Figure 24.8 depicts an electromagnetic wave whose profile is sinusoidal. (Take another look at Chapter 12 for a discussion of waveforms and the definitions of period T, wavelength λ, frequency f, and phase speed v.) Here, as with all periodic waves,

$$v = f\lambda \qquad\qquad [12.18]$$

In vacuum, $v = c$, whereas in all other material media, v is usually less than c, though it can be greater under certain circumstances (p. 946).

Example 24.1 In New York City, WNYC broadcasts FM radio signals at 93.9 MHz. Assuming that the speed of propagation of these electromagnetic waves in air is negligibly different from c, please determine the corresponding wavelength.

Solution: [Given: $v = c$ and $f = 93.9$ MHz. Find: λ.] Rewriting Eq. (12.18), we have

$$\lambda = \frac{v}{f} = \frac{2.998 \times 10^8 \text{ m/s}}{93.9 \times 10^6 \text{ Hz}} = \boxed{3.19 \text{ m}}$$

▶ **Quick Check:** $v = f\lambda \approx 300 \times 10^6$ m/s. Incidentally, TV sound and FM radio are VHF (very high frequency) signals ranging in wavelength from 1 m to 10 m.

In Fig. 24.8, along with the vertical E-field is included the ever-present accompanying B-field. Every point on the disturbance experiences a sinusoidal oscillation in time, and the whole wave extends along the line-of-travel as an advancing sinusoid moving with a speed v. The E-field has an *amplitude* of E_0. It oscillates between $\pm E_0$, and so the wave can be represented mathematically as

$$E = E_0 \sin \frac{2\pi}{\lambda}(x - vt) \qquad\qquad (24.5)$$

very much like the mechanical wave of Eq. (12.19). To get a sense of the size of E_0, realize that bright sunlight on Earth might have an E-field amplitude of about 10 V/cm as compared with the tremendous value of 10^{10} V/cm that can occur in the

An Anticipation of the Wave Theory. Just as a stone thrown into water becomes the cause and center of various circles, sound spreads in circles in the air. Thus, every body placed in the luminous air spreads out in circles and fills the surrounding space with infinite likenesses of itself and appears all in all and in every part.

LEONARDO DA VINCI (ca. 1508)

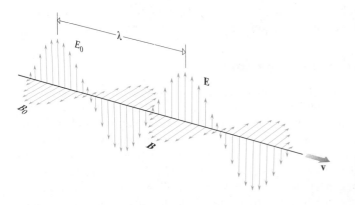

Figure 24.8 A harmonic electromagnetic wave. The coupled E- and B-fields are perpendicular to each other, and the propagation direction is given by **E** × **B**.

focused beam of a high-power laser. The quantity $2\pi/\lambda$ arises so often in the analysis of waves that it's given its own name, the **propagation number** (or the angular *wave number*), and is symbolized (p. 486) by the letter k.

Example 24.2   Write an expression for the profile of the E-field of a harmonic electromagnetic wave propagating in vacuum in the positive x-direction if its frequency is 600 THz (green light) and its amplitude is 8.00 V/cm.

Solution: [Given: $v = c$, $E_0 = 800$ V/m, and $f = 600$ THz. Find: the profile.] Equation (24.5) is what's needed, but we will first have to find $k = 2\pi/\lambda$, and that means finding λ:

$$\lambda = \frac{v}{f} = \frac{2.998 \times 10^8 \text{ m/s}}{600 \times 10^{12} \text{ Hz}} = 500 \times 10^9 \text{ m} = 500 \text{ nm}$$

Hence, $k = 12.6 \times 10^6$ m^{-1} and so

$$E = (800 \text{ V/m}) \sin(12.6 \times 10^6 \text{ m}^{-1})x$$

▶ **Quick Check:** $k = 2\pi/\lambda$ and $\lambda = 2\pi/k$; hence, $2\pi f/k = f\lambda = 2\pi(600 \text{ THz})/(12.6 \times 10^6 \text{ m}^{-1})$ should equal c, and it does.

Refer back to the harmonic electromagnetic wave traveling in vacuum (shown in Fig. 24.8) and notice that the direction of propagation corresponds to the direction of **E** × **B**. The wave is *transverse;* **E** and **B** are perpendicular to each other and to the direction of propagation as well. Such **transverse electromagnetic waves**—so-called **TEM waves**—exist in vacuum, but things usually get much more complicated for waves traversing material media, where the fields may not be totally transverse.

Electromagnetic waves, like sound waves, are three-dimensional. An ideal point source would radiate sinusoidal waves uniformly in all directions, as in Fig. 13.12. Here, *the surfaces of constant phase*—the **wavefronts**—are spherical. More practically, if we examine the light from a street lamp even a few hundred meters away, the portion of the wavefront intercepted by an eye or a small telescope will be essentially spherical. By allowing a spherical wave to expand out (Fig. 13.13) over a great distance, the wavefronts will get larger and flatter, ultimately resembling planes.

It might seem easier to just say that a wavefront corresponds to a surface over which the disturbance has some constant strength or magnitude. Indeed, quite often waves are *homogeneous;* they have constant magnitudes over their wavefronts—but not always! A homogeneous harmonic plane wave traveling in free space can be envisioned as a moving stack of flat surfaces, like a deck of cards, over each of which E and B are constant, arranged so that as we go from one plane to the next, the fields change in a sinusoidal way (Fig. 24.9). The most popular form of laser light resembles plane waves that have a far greater electric field strength at the center (and are much brighter there) than at the edge of the beam—such a wave is *inhomogeneous.*

24.4 The Speed of Propagation: c

Let's now return to Maxwell's Equations and make sure that our wave picture is consistent with them. Accordingly, imagine a sheet or flat pulse of uniform magnetic field **B** pointing in the z-direction laced with a uniform electric field **E** pointing in the y-direction. The combination, resembling a rudimentary plane wave, is traveling through empty space in the x-direction at speed c (Fig. 24.10). To find out what kind of E-field is accompanying the B-field, we need only apply Faraday's Law. Envision an imaginary closed rectangular loop C of width L and indefinite length suspended in the xy-plane. In a time interval dt, the pulse advances a distance

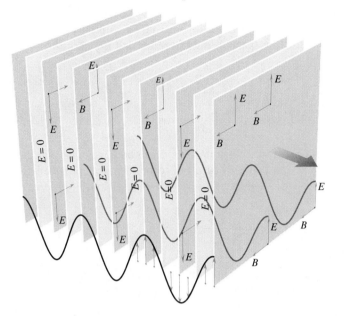

Figure 24.9 A plane harmonic electromagnetic wave. Notice how the fields change from one plane to the next. Over each plane both **E** and **B** are constant.

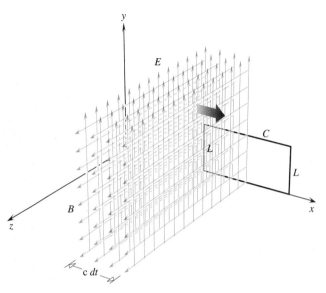

Figure 24.10 An electromagnetic pulse propagating past a loop. The change in magnetic flux through the loop induces a current in the loop and, therefore, an E-field as well.

($c\,dt$) and sweeps out an area of loop equal to ($c\,dt$)L. Hence, the change of flux through the loop is $d\Phi_M = B(cL\,dt)$ and using Faraday's Law

$$\oint \mathbf{E} \cdot d\mathbf{l} = -\frac{d\Phi_M}{dt} = -\frac{d}{dt}\int \mathbf{B} \cdot d\mathbf{A} \qquad [22.6]$$

we obtain

$$\oint \mathbf{E} \cdot d\mathbf{l} = -\frac{d\Phi_M}{dt} = -BcL$$

The minus sign tells us that the induced emf opposes the flux change. Consequently, we will take the reminder, make sure that everything is in its proper direction, and drop the minus sign, which is unnecessary clutter at this point. Now focus on the left side of the equation. Since **E** is upward, it is perpendicular to the top and bottom sides of the loop and therefore the only nonzero contribution to the sum is from the single vertical side, along which $E_{\parallel} = E$. The entire sum equals EL and, as a result

$$E = cB \qquad (24.6)$$

The pulse we fashioned is indeed in accord with Faraday's Law, provided that the strengths of the electric and magnetic fields are related by Eq. (24.6). That relationship turns out to be true in general for electromagnetic waves in space.

It's worth considering what is happening from a slightly different perspective. As the B-field sweeps by, the magnetic flux in the region of space of the loop increases in the z-direction. That, in turn, induces an E-field in that region of space. The direction of this field is the direction of the induced current, which must be upward, so its induced B-field will oppose the increasing flux in the loop. In short, a traveling B-field generates an accompanying E-field, and only such coupled perpendicular fields propagate through space as a wave.

We could now put the loop in the zx-plane and do the whole analysis over for the E-field, using Maxwell's version of Ampère's Law. Instead, let's simply compare the two of Maxwell's Equations for free space, using the result that $E = cB$; thus

$$\oint \mathbf{E} \cdot d\mathbf{l} = -\int \frac{d}{dt}\mathbf{B} \cdot d\mathbf{A} \qquad [24.1]$$

and

$$\oint \mathbf{B} \cdot d\mathbf{l} = \mu_0\varepsilon_0\int \frac{d}{dt}\mathbf{E} \cdot d\mathbf{A} \qquad [24.4]$$

Forgetting about the minus sign, if we substitute cB for E in both formulas and then multiply the second one by c, the two will turn out to be identical, provided that $\varepsilon_0\mu_0 c^2 = 1$; that is

$$c = \frac{1}{\sqrt{\varepsilon_0\mu_0}} \qquad (24.7)$$

and that must be the case—both equations must be satisfied simultaneously. Substituting in the values of the constants, we get

$$c = \frac{1}{\sqrt{(8.85 \times 10^{-12}\ \text{C}^2/\text{N·m}^2)(4\pi \times 10^{-7}\ \text{N·s}^2/\text{C}^2)}} = 3.00 \times 10^8\ \text{m/s}$$

Amazing! This is the speed of light in vacuum.

Maxwell's calculation was far more elegant than the one we just went over. He had begun with the several laws of Electromagnetic Theory and combined them to form an expression that was well known to describe all wave phenomena. And where there should have been a variable associated with the medium, there alternatively was E or B. And where there should have been the speed of the wave, there was $1/\sqrt{\varepsilon_0\mu_0}$. He had in hand these two constants (they had been determined in 1856 by Weber and Kohlrausch), and he had the speed of light (3.15×10^8 m/s) measured by Fizeau with a rotating toothed wheel in 1849. The conclusion was inescapable: *light was "an electromagnetic disturbance in the form of waves" propagated in the aether.* And this description was true, as well, for "radiant heat, and other radiations if any" (light and "radiant heat"—that is, infrared—were known at the time, as was something called chemical rays, now referred to as ultraviolet). Maxwell's analysis, which forms the basis of classical Electromagnetic Theory, stands as one of the greatest theoretical achievements in the history of physics.

All electromagnetic waves propagate in vacuum at exactly

$$c = 2.997\,924\,58 \times 10^8\ \text{m/s}$$

which is a tremendous speed. Light travels 1 ft in $\approx 10^{-9}$ s. As we'll see, this wave speed is a macroscopic manifestation of the fact that photons exist only at the speed c.

24.5 Energy and Irradiance

If we are to deal with light quantitatively, we have to measure the amount of radiant energy arriving on a surface—a photographic plate, a retina, whatever. It would be nice to determine the strength of the E- and B-fields directly, but that's technically impractical at the high frequencies ($\approx 10^{15}$ Hz) of light. The next best thing to find is the average amount of light energy flowing during a convenient interval of time. A surface is illuminated and we measure the amount of energy arriving per second on each square unit of area. Earlier (p. 497), we called that quantity *intensity,* but in modern optics the *average amount of energy-per-unit-area-per-unit-time* is known as the **irradiance** (*I*) and it's specified in J/s·m^2 or W/m^2.

Figure 24.11 shows a beam of light of cross-sectional area *A* impinging on a plane in vacuum. It travels at c and in a time Δt, all the light in the cylindrical section of length (c Δt) and volume $V = (c\,\Delta t)A$ will sweep down onto the plane. If we knew the energy per unit volume *u* in the beam, then the amount of energy striking the surface in the time Δt would just be $uV = u(c\,\Delta t)A$. But we do know *u* from Eqs. (22.13) and (22.16)—it's the energy density in the *E*-field, $u_E = \frac{1}{2}\varepsilon_0 E^2$, plus the energy density in the *B*-field, $u_M = \frac{1}{2}B^2/\mu_0$. Before we press on, notice that for an electromagnetic wave $E = cB$ and $c = 1/\sqrt{\varepsilon_0\mu_0}$, and so $u_M = \frac{1}{2}E^2/c^2\mu_0 = \frac{1}{2}\varepsilon_0 E^2 = u_E$. Not surprisingly, the energy stored in the *B*-field of an electromagnetic wave equals the energy stored in the *E*-field. Thus, $u = \varepsilon_0 E^2$. Since the amount of energy arriving in a time Δt is $u(c\,\Delta t)A$, the energy-per-unit-area-per-unit-time is then $u(c\,\Delta t)A/A\,\Delta t = uc = c\varepsilon_0 E^2$.

Because *E* varies very rapidly, this quantity has to be averaged in time to obtain *I*:

$$I = [c\varepsilon_0 E^2]_{av}$$

Here *E* is a time-varying sinusoidal function that we will write as $E = E_0 \sin \phi$, keeping in mind that ϕ varies in time. Thus

$$I = [c\varepsilon_0 E_0^2 \sin^2 \phi]_{av} = c\varepsilon_0 E_0^2 [\sin^2 \phi]_{av}$$

since $c\varepsilon_0 E_0^2$ is constant. We saw in Fig. 23.4 that $[\sin^2 \phi]_{av} = \frac{1}{2}$ and so

$$I = \tfrac{1}{2}c\varepsilon_0 E_0^2 \qquad (24.8)$$

The irradiance (or intensity), which we measure with a meter, is proportional to the square of the amplitude of the electric field of the wave.

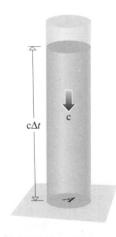

Figure 24.11 Traveling at a speed c, the column of light c Δt high will impinge on the plane in a time Δt. Thus, a volume of the beam equal to Ac Δt will sweep down onto the plane in a time Δt.

Example 24.3 A 1.0-mW laser beam with a frequency of 4.74×10^{14} Hz has a cross-sectional area of 3.14×10^{-6} m^2. Determine (a) the energy arriving in 1.00 second on a screen intercepting the beam perpendicularly; (b) the *radiant flux density,* or irradiance; and (c) the amplitude of the electric field. Take the medium to be vacuum.

Solution: [Given: beam-power P = 1.0×10^{-3} W, cross-sectional area A = 3.14×10^{-6} m^2, and $f = 4.74 \times 10^{14}$ Hz. Find: (a) energy arriving in 1.00 s, (b) irradiance I, and (c) E_0.] (a) The beam-power or *radiant flux* is P; hence, in $\Delta t = 1.00$ s, the energy that arrives is

$$P\Delta t = (1.0 \times 10^{-3} \text{ W})(1.00 \text{ s}) = \boxed{1.0 \times 10^{-3} \text{ J}}$$

(b) The irradiance or energy-per-unit-area-per-unit-time is

$$I = \frac{P}{A} = \frac{1.0 \times 10^{-3} \text{ W}}{3.14 \times 10^{-6} \text{ m}^2} = \boxed{3.2 \times 10^2 \text{ W/m}^2}$$

(c) Once we have *I*, we can get the amplitude from Eq. (24.8); thus

$$E_0^2 = \frac{2I}{c\varepsilon_0} = \frac{2(3.2 \times 10^2 \text{ W/m}^2)}{(3.00 \times 10^8 \text{ m/s})(8.85 \times 10^{-12} \text{ C}^2/\text{N·m}^2)}$$

and $\boxed{E_0 = 0.49 \text{ kV/m}}$

▶ **Quick Check:** $I \approx (1.0 \text{ mW})/(3 \times 10^{-6} \text{ m}^2) \approx 0.3 \text{ kW/m}^2$; $\frac{1}{2}c\varepsilon_0 \approx 1.3 \times 10^{-3} \text{ C}^2/\text{N·m·s}$; $E_0^2 = I/\frac{1}{2}c\varepsilon_0 \approx (0.3 \times 10^3)/(1.3 \times 10^{-3})(\text{V/m})^2 \approx 0.2 \times 10^6 \text{ (V/m)}^2$ and $E_0 \approx 0.5 \times 10^3 \text{ V/m}$.

24.6 The Origins of EM Radiation

Though all forms of electromagnetic (EM) radiation share a single vacuum speed c, they do differ in frequency and wavelength. Still, there really is only one entity, one essence of electromagnetic "stuff." Maxwell's Equations, which are independent of frequency, do not suggest any fundamental differences in kind. Thus, it is natural enough to look for a common basic source-mechanism. What we find is that all the various types of radiant energy seem to have a common origin in that they are all associated somehow with *nonuniformly moving charges*. Of course, classically, we are dealing with waves in the electromagnetic field, and charge *is* that which gives rise to the field.

Free Charge

A stationary charge will have a constant *E*-field, no motional *B*-field and, hence, produce no radiation. (Where would the energy come from if it did radiate, and what would turn it on or off?) A uniformly moving charge has both an *E*- and a *B*-field, but these do not abruptly disengage, the motion is continuous, and, again, there is no radiation. If you moved along with the charge, the current would thereupon vanish; hence, *B* would vanish, and we would be back to the previous case of $v = 0$. That's reasonable since it would make no sense at all if the charge stopped radiating just because you started walking along next to it. Besides, if uniformly moving charges radiated, they would not move uniformly very long (not without an external agent supplying energy) and what then would happen to the Law of Inertia? That leaves *nonuniformly moving charges,* which, assuredly, will be accompanied by time-varying *E*- and *B*-fields.

We know, in general, that *free charges* (those not bound within an atom) emit electromagnetic radiation when accelerated. This much is true for charges sailing around in circles within a synchrotron, moving at a changing speed on a straight line in a linear accelerator, or simply oscillating back and forth in a radio antenna—**if charge accelerates, it radiates**. A portion of its kinetic energy is converted into radiant energy.

The Dipole

Perhaps the simplest electromagnetic wave-producing mechanism to visualize is the oscillating dipole: two charges, one plus and one minus, vibrating to and fro along a straight line. The electric field lines, which begin and end on charges, close on themselves whenever the two charges overlap and thereby effectively vanish (Fig. 24.12). Moving charge constitutes a current (in this instance, an oscillating current), which is accompanied by an oscillating magnetic field whose closed circular lines-of-force are in planes at right angles to the motion. The *E*- and *B*-fields, which are in-phase, rising and falling together, are perpendicular to each other and to the propagation direction as well. The frequency of the radiation is determined by the oscillation

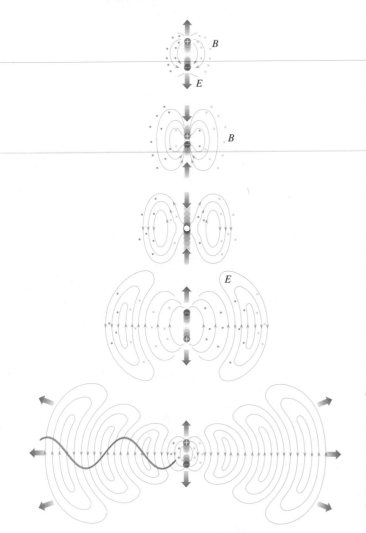

Figure 24.12 A cross-sectional view of the three-dimensional pattern of electromagnetic waves radiated by an oscillating dipole. The loops are electric field lines. The dots and crosses depict the emergence and exit of the circles of magnetic field, which are in planes perpendicular to the page and surround the dipole.

of the current. In brief, far from the dipole there will be developed an outgoing TEM wave. Keep in mind that this pattern is actually three-dimensional, roughly resembling a series of almost spherical concentric tires expanding radially outward.

It's an easy matter to attach an ac generator between two conducting rods and send currents oscillating up and down that transmitting "antenna." The result, Fig. 24.13, is a fairly standard AM radio tower. The antenna will function most efficiently if its length corresponds to the wavelength being transmitted or, more conveniently, to half that wavelength. The radiated wave is then formed at the dipole in sync with the oscillating current producing it. Unhappily, AM radiowaves are typically several hundred meters long. Accordingly, the antenna shown uses the conducting Earth so that it functions as if half of the $\frac{1}{2}\lambda$-dipole were buried in the ground, which allows us to build the thing only $\frac{1}{4}\lambda$ tall.

To receive such a signal, one with a vertically oscillating E-field, we need only stick a straight length of wire up into the air, more or less parallel to E. The electric field strength of the incoming signal might be anywhere from microvolts per meter to millivolts per meter depending on where you are and what you want to receive. A signal with an amplitude of 1.0 mV/m reaching an antenna 2.0 m tall will induce a voltage having an amplitude of 2.0 mV. An old wire coat hanger will work adequately well as a replacement antenna for your car when you are driving around in the city, where the signals could be as strong as 10 mV/m.

Photons

Einstein's Special Theory of Relativity revolutionized Classical Mechanics, completely removing the theoretical need for a stationary aether. Deprived of the necessity for an all-pervading elastic fiction, physicists simply had to get used to the idea that electromagnetic waves could propagate through an aetherless free space. The conceptual emphasis passed from aether to field—*light became an electromagnetic wave propagating in the electromagnetic field.* Aether vanished and field appeared as an entity in itself.

In 1905 Einstein shocked the scientific world with another brilliant insight. He boldly introduced a novel form of corpuscular theory that immediately explained several experimental problems that had developed since the late 1800s (Sect. 30.2). The energy carried by light (and all other forms of electromagnetic radiation) is not smoothly spread out across the wave, but is somehow concentrated at points within it. Einstein asserted that light consists of massless quanta, "particles" of electromagnetic radiation.

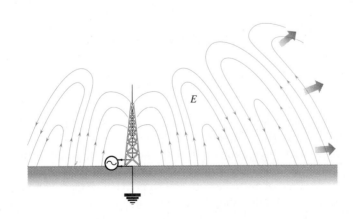

Figure 24.13 The *ground wave* generated by an AM antenna. The wave tends to hug the planet's surface, where most of the people with the radios and TVs are likely to be. Typically, this gives a commercial radio station a range of somewhere between 25 and 100 miles.

24.7 Energy Quanta

Each *quantum of electromagnetic radiation,* or **photon,** as it came to be called in the 1920s, has an energy proportional to its frequency. The constant of proportionality, $h = 6.626 \times 10^{-34}$ J/Hz (or 4.136×10^{-15} eV/Hz), is known as Planck's Constant. The energy of a photon of frequency f is

$$E = hf \tag{24.9}$$

Example 24.4 A helium-neon laser puts out a beam of red light containing a very narrow range of frequencies centered at 4.74×10^{14} Hz. Determine the energy of each photon at that frequency.

Solution: [Given: $f = 4.74 \times 10^{14}$ Hz. Find: E.] The energy per photon follows from Eq. (24.9), where

$$E = hf = (6.626 \times 10^{-34} \text{ J/Hz})(4.74 \times 10^{14} \text{ Hz})$$

and

$$\boxed{E = 3.14 \times 10^{-19} \text{ J}}$$

▶ **Quick Check:** Using h in eV/Hz, we get E = 1.96 eV, and since 1 eV = 1.602×10^{-19} J, E = 3.14×10^{-19} J. Photons of light have an energy of roughly 2 or 3 eV.

The emission and absorption of light (that is, the transfer of electromagnetic energy and momentum from, or to, a material object) takes place in a corpuscular way. Because photons have zero mass, each quantum of light carries very little energy. Thus, even an ordinary flashlight beam must be thought of as a torrent of perhaps 10^{17} photons per second. In general, when we "see" light, what is recorded with eye, detector, or film is the average energy-per-unit-area-per-unit-time arriving at some surface.

Example 24.5 Reconsider the laser beam in Example 24.3. Taking the frequency of the light to be 4.74×10^{14} Hz, determine (a) the *photon flux*—the average number of photons per second impinging on a perpendicular flat target; and (b) the *photon flux density*—the number of photons per second per unit area hitting that target.

Solution: [Given: beam-power P = 1.0×10^{-3} W, cross-sectional area $A = 3.14 \times 10^{-6}$ m^2, and $f = 4.74 \times 10^{14}$ Hz. Find: (a) photon flux and (b) photon flux density.] (a) We have the power or energy per second, 1.0×10^{-3} W; hence, if we divide that quantity by the energy of each photon (from Example 24.4), we will get the photon flux

$$\frac{P}{hf} = \frac{1.0 \times 10^{-3} \text{ W}}{3.14 \times 10^{-19} \text{ J}} = \boxed{3.2 \times 10^{15} \text{ photons/s}}$$

(b) The photon flux density is just the flux per unit area, or

$$\frac{3.18 \times 10^{15} \text{ photons/s}}{3.14 \times 10^{-6} \text{ m}^2} = \boxed{1.0 \times 10^{21} \text{ photons/s·m}^2}$$

▶ **Quick Check:** The photon flux density (1.013×10^{21}) times the energy per photon equals 0.32 kW/m^2, the irradiance.

Light, and all other forms of electromagnetic radiation, interacting with matter in the processes of emission and absorption, behave like streams of particlelike concentrations of energy that exist only at the speed c, and which otherwise propagate in a wavelike fashion. We cannot say whether light *is* particle or wave; it seems to be both and so is likely neither. The "wee beasties" of the microworld, whether they be protons or photons, don't play our game of waves *or* grains—of "either/or." The electromagnetic wave (like the water wave made up of countless molecules) is an illusion of continuity. On the most fundamental level, there is no such thing as a continuous electromagnetic wave. Yet torrents of photons behave exactly as if they were dissolved into a smooth classical TEM wave.

So here we have the latest picture of what seems to defy picturing: powerful in its ability to explain, wonderful in its subtlety—it is the latest theory of light, but probably not the last.

24.8   Atoms and Light

By far the most important mechanism for the emission and absorption of radiant energy—especially of light—is the *bound charge,* electrons confined within atoms. Much of the chemical and optical behavior of a substance is determined by only its outer electrons; the remainder of the cloud is formed into closed, essentially unresponsive shells around and tightly bound to the nucleus. Although it's not clear what exactly happens when an atom radiates, we do know with some certainty that light is emitted during changes in the outer charge distribution of the electron cloud (Sect. 30.5). It is the absorption and emission of light via these electrons that determines almost all optical phenomena in nature.

Each electron is usually in the lowest possible energy state available to it, and the atom as a whole is in its **ground state**. There it will remain, indefinitely, if left undisturbed. Any mechanism that can pump energy into an atom, such as a collision with another atom, with a photon or an electron, will affect this situation. In addition to the ground state, there are specific well-defined higher energy levels, so-called **excited states**.

At low temperatures, atoms (and molecules) tend to be in their ground states but, as the temperature rises, more and more of them become excited through collisions. This process is indicative of a class of relatively gentle excitations (glow discharges, flames, sparks, etc.), which energize only the outermost unpaired valence electrons rather than the far more strongly bound inner ones. For the moment, we will concentrate on these outer-electron transitions, which give rise to the emission of light, infrared, and ultraviolet.

When just the right amount of energy (ΔE) is imparted to an atom (that is, to a valence electron), it can respond by suddenly jumping from a lower to a higher energy level. Thus, the amount of energy that can be absorbed by an atom is quantized (that is, limited to specific, well-defined amounts). This excitation of the atom is a short-lived resonance phenomenon. Usually, after a time of about 10^{-8} s or 10^{-9} s, the excited atom spontaneously relaxes back to a lower state, most often the ground state, losing the excitation energy along the way (Fig. 24.14). This transition can occur by the emission of light or (especially in dense materials) by conversion to thermal energy via interatomic collisions within the medium.

When the downward atomic transition is accompanied by the emission of light, the energy of the photon (hf) exactly matches the quantized energy decrease of the atom (ΔE). Thus, $\Delta E = hf$ and there is a specific frequency associated with both the emitted photon and the atomic transition between the two states. The latter is effectively a *resonance frequency of the atom, one of several at which it very efficiently absorbs and emits energy.*

Even though we don't know exactly what is going on during that 10^{-8} s, it is helpful to imagine the orbital electron somehow making the downward energy transition via a diminishing, oscillatory motion at the specific resonance frequency. Light can therefore be imagined as emitted in a short oscillatory pulse, or wave-

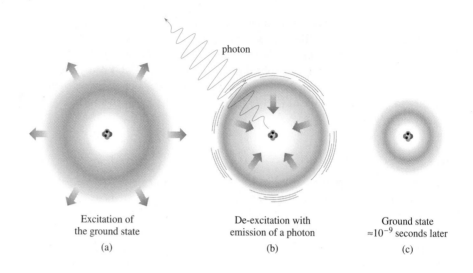

Figure 24.14 A schematic representation of an atomic emission. (a) An atom is raised to an excited state. (b) With the emission of a photon, it (c) spontaneously drops back to the ground state.

photon

Excitation of the ground state
(a)

De-excitation with emission of a photon
(b)

Ground state ≈10^{-9} seconds later
(c)

train, that is somehow representative of the radiated photon. Keep in mind that there are two intertwined models that have evolved to represent radiant energy: the particlelike and the wavelike—and light is both. Without the more complete development of Quantum Theory (which even then leaves something to be desired), we will have to irreverently talk about particles at one moment and waves at another.

24.9 Scattering and Absorption

The process whereby an atom absorbs a photon and emits another photon is known as **scattering**. The transmission of light through a windowpane, the reflection from a mirror or a face, the coloration of the sunset sky, are all governed by scattering.

We see most things not because they are self-luminous but because they absorb part of the incident ambient light and redirect some of the remainder toward our eyes. An atom (one on your cheek, for instance) can react to incoming light in essentially two different ways depending on the incident frequency (color) or, equivalently, on the incoming photon energy ($E = hf$). The atom can absorb the light or, alternatively, its ground state can scatter the light.

If the photon's energy matches that of one of the excited states, the atom will absorb it, making a quantum jump to that higher energy level. In the dense atomic clutter of solids and liquids, it's very likely that the excitation energy will be transferred, via collisions, to the random KE of the atoms rather than being reemitted when the atom returns to the ground state—the photon vanishes, its energy converted into thermal energy. This process (the taking up of a photon and its conversion into thermal energy) is called **dissipative absorption**. Your skin dissipatively absorbs certain frequencies and scatters others, which is what gives it the color it has under white light illumination. A red apple appears red because it has a resonance in the blue and absorbs out the yellow-BLUE-green band, reflecting mostly red.

In contrast to this resonant process, *ground-state,* or **nonresonant elastic scattering**, occurs for incoming light of other frequencies; that is, other than resonance frequencies. Even now, the light passing through your eyes is progressing via elastic scattering. Envision an atom in its lowest state and suppose that it interacts with a photon of frequency *f*, whose energy is too small to cause an excitation up into any of the higher states. Nonetheless, the electromagnetic field of the light can drive the electron cloud into oscillation. There can be no resulting atomic transitions; the atom will remain in its ground state while the cloud vibrates ever so slightly at the frequency of the incident light. The electron, once it begins to oscillate, is, of course, an accelerating charge and so will immediately begin to reemit light of that same frequency, f. This scattered light consists of a photon that sails off in some direction carrying the same amount of energy as did the incident photon. In effect, we are imagining the atom to resemble a little dipole oscillator. When a material (like a piece of glass) is bathed in light, this almost-omnidirectional scattering gives each atom the appearance of being a tiny source of spherical wavelets. As we'll see in the next chapter, it is precisely this nonresonant scattering that accounts for the transmission of light through all transparent material media and for the reflection of light from most surfaces.

THE ELECTROMAGNETIC-PHOTON SPECTRUM

The whole spread of radiant energy, which conceptually ranges in wavelength between zero and infinity, is referred to as the **electromagnetic spectrum**. It is usually subdivided into seven more or less distinct regions. These were delineated originally as much by historical circumstance as by physical necessity, and so there tends to be a good bit of overlap in the categories. Needless to say, light was discovered first, then infrared (1800), ultraviolet (1801), radiowaves (1888), X-rays (1895), gamma rays (1900), and, finally, it was just a technical matter of filling in the microwaves, which was done in the 1930s, primarily with an eye toward radar (Fig. 24.15).

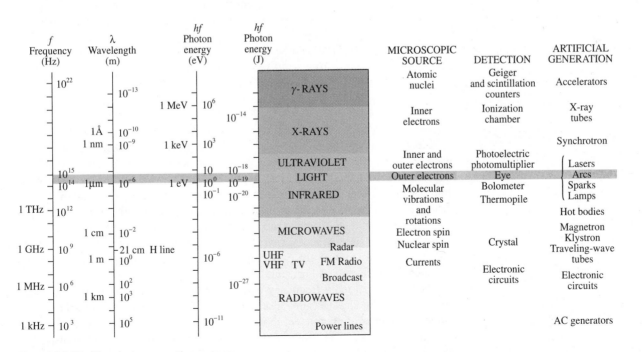

Figure 24.15 The electromagnetic spectrum.

24.10 Radiowaves

Great, slowly rising and falling electromagnetic waves (more than 18 million miles long) have been measured impinging on the Earth, streaming in from the depths of space. These faint cosmic flutters are one extreme of the radiant energy grouping known as **radiowaves**, which extends in wavelength down to about 0.3 meter.

AM radio and television transmissions in the United States are confined to the frequency band from 535 kHz to 1605 kHz (p. 882). The major failing of the AM technique is that it picks up extraneous noise sent out by everything from lightning and electric motors to car ignitions. In fact, electronic calculators transmit radiowaves—not very powerfully and therefore not very far, but far enough to be picked up by a nearby AM radio. Turn the radio's dial to the low-frequency end between stations and punch in a beeping tune on your calculator. You can even watch the effects of lightning—flashing white dashes running horizontally across your TV picture—during a storm. The picture is transmitted AM, the sound, FM. Or hold a telephone receiver next to your ear near a computer and listen to the radiated hum.

Example 24.6 The AM receiver of Fig. 23.24 (p. 883) is tuned to accept a carrier frequency of 1.0 MHz. If the inductance in the tuning circuit has a value of $L = 300$ μH, (a) determine the value of the tuning capacitor. (b) What is the wavelength of the carrier wave? (c) What is the energy of a radio frequency carrier photon in electron volts?

Solution: [Given: $f_0 = 1.0$ MHz and $L = 300$ μH. Find: C, λ, and E.] (a) Using Eq. (23.35), $f_0 = 1/(2\pi\sqrt{LC})$

$$C = \frac{1}{4\pi^2 f_0^2 L} = \frac{1}{4(3.14)^2(1.0 \times 10^6 \text{ Hz})^2(300 \times 10^{-6} \text{ H})}$$

and $\qquad C = 84 \times 10^{-12}$ F $= \boxed{84 \text{ pF}}$

(b) The wavelength follows from $c = f\lambda$; namely

$$\lambda = \frac{c}{f} = \frac{3.0 \times 10^8 \text{ m/s}}{1.0 \times 10^6 \text{ Hz}} = \boxed{3.0 \times 10^2 \text{ m}}$$

(c) $E = hf = (6.63 \times 10^{-34} \text{ J/Hz})(1.0 \times 10^6 \text{ Hz}) = 6.63 \times 10^{-28}$ J. Since 1.00 J $= 6.24 \times 10^{18}$ eV;

$$\boxed{E = 4.1 \times 10^{-9} \text{ eV}}$$

▶ **Quick Check:** Light has a frequency of about 500 THz and an energy of around 1 eV. These radio photons are roughly 0.5×10^9 lower in frequency and should be that much lower in energy, thus $E = (1 \text{ eV})/(0.5 \times 10^9) = 2 \times 10^{-9}$ eV, which is the same order-of-magnitude as above.

Frequency modulation (FM) was introduced primarily to decrease the amount of noise and to extend the frequency content of the information transmitted. The range of FM is from about 88 MHz to 108 MHz, corresponding to wavelengths of from 3.4 m to 2.8 m. The carrier is made to vary in frequency (Fig. 24.16) in proportion to the amplitude of the audio message. Noise, which only affects the amplitude of the signal, is simply cropped off by the receiver since all the useful information is in the frequency distribution. FM signals, and therefore TV as well, are often transmitted with the E-field horizontal, which is why a typical TV receiving antenna has several $\frac{1}{2}\lambda$ horizontal bars.

At 1.0 MHz, a radio frequency photon has an energy of 6.6×10^{-28} J, a very small quantity by any measure. In an ordinary flow of radio emission, there is an incredible number of photons, each individually so weak as to be essentially undetectable on its own. The result is an apparently continuous wavelike transport of energy. Radio photons have such low energies that they are not very likely to encounter atomic resonances, meaning that we can expect nonconductors (such as glass, bricks, and concrete) to be fairly transparent to radiowaves, while metals (with their free electrons) certainly are not. Roll yourself in a ball around a portable radio and see what happens to the reception. Next, put some metal foil around the radio. Note the difference. Incidentally, the human body most effectively serves as a conducting antenna in the range from about 30 MHz to 300 MHz, which is in the FM and VHF regions. Perhaps you have noticed as much while playing with an indoor TV antenna.

24.11 Microwaves

The frequency region extending from about 10^9 Hz (1 GHz) up to roughly 3×10^{11} Hz is the domain of the **microwaves**. In wavelength, that corresponds to the range from approximately 30 cm to 1.0 mm. Electromagnetic radiation in the region from less than 1 cm to about 30 m can penetrate the Earth's atmosphere, and that makes microwaves especially useful for space-vehicle communications and radio astronomy. The ground state of the cesium atom consists of two very closely spaced levels, separated by only 4.14×10^{-5} eV. When the atom drops from one level to the other, the resulting microwave emission has a splendidly precise frequency of $9.192\,631\,77 \times 10^9$ Hz. This is the basis for the cesium clock, the present-day laboratory standard of frequency and time (p. 13).

To understand how the microwave oven works, recall that molecules can absorb and emit energy by altering the state of motion of their constituent atoms. The molecule can be made to vibrate and/or rotate; again, the energy associated with either such motion is quantized, and molecules therefore possess a number of rotational and vibrational energy levels in addition to those due to their electrons. Only when a molecule is polar will it experience forces via an electromagnetic wave that cause it to rotate into alignment with the changing E-field. Because molecules are massive and not able to swing around easily, we can anticipate low-frequency rotational resonances (infrared of 0.1 mm to microwave of 1 cm).

For example, the water molecule is polar—the hydrogen end is positive, the oxygen end is negative. When exposed to an alternating electromagnetic wave, it will swing around, trying to stay lined up with the E-field. Water molecules will very efficiently absorb microwave radiation at or near a resonant frequency, thereupon exhibiting large-amplitude oscillations. The oscillatory KE of these excited molecules is rapidly converted into thermal energy via collisions with other molecules. The microwave oven (12.2 cm, 2.45 GHz) is an obvious application. Clearly, the thing to be heated has to contain water—a dry paper plate will remain quite cool. The diathermy machine, used to warm muscles and joints in order to relieve soreness, works on the same principle, at the same frequency.

Microwaves are now used for everything from carrying phone conversations and interstation TV to guiding planes and catching speeders (via radar) to studying the origins of the Universe and chatting with astronauts. Even though individual photon energies are small (in general, you want to avoid absorbing large numbers of them), people have been killed by massive exposure. In the United States, the sup-

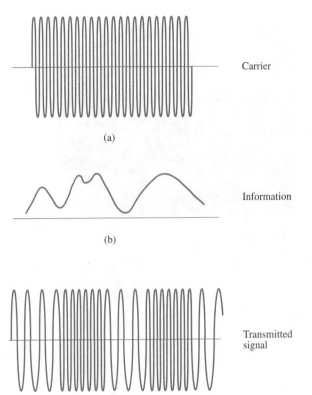

(a)

Carrier

(b)

Information

(c)

Transmitted signal

Figure 24.16 Frequency modulation (FM). (a) A constant-frequency, constant-amplitude carrier is modulated by (b) an information signal such that the amplitude of the information determines the frequency of the FM signal, as shown in (c).

A frequency-modulated signal.

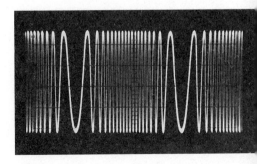

Microwave antennae on the top of the Eiffel Tower in Paris.

A microwave (radar) image of the surface of Venus taken by the *Magellan* spacecraft in 1992.

posedly safe limit to environmental microwave exposure is assumed to be 10 mW/cm^2, which some maintain is far too high.

24.12   Infrared

In 1800, the renowned astronomer (musician and deserter from the Hanoverian Foot Guards) Sir William Herschel made a surprising discovery. Using a prism, he had been studying the different amounts of "heat" conveyed by the various colors of sunlight. To his astonishment, he found that his thermometer registered its greatest increase just "beneath" the red region. Herschel rightly concluded that he was observing the effects of an invisible radiation, now called **infrared** (beneath the red).

The infrared (or IR) band merges with microwaves at around 300 GHz (1.0 mm) and extends to about 385 THz (780×10^{-9} m; that is, 780 nm), where the photons have enough energy (1.6 eV) to break apart certain molecules. Most any material will radiate IR via thermal agitation of its molecules—just heat it up and it will pour forth IR. Infrared is copiously emitted from glowing coals and home radiators—roughly half the radiant energy from the Sun is IR, and an ordinary light bulb puts out far more IR than light. Like all warm-blooded creatures, we ourselves are IR emitters. The human body radiates very weakly, starting practically around 3000 nm; it peaks in the vicinity of 10 000 nm and trails off from there. This fact is exploited by some rather nasty "heat"-sensitive snakes (Crotalidae pit vipers and Boidae constrictors) that are active at night (p. 586).

In addition to rotating, a molecule can vibrate in several different modes. The corresponding vibrational emission and absorption spectra are, generally, in the infrared (1000 nm to 0.1 mm). Many molecules have vibrational and rotational resonances in the IR and are good absorbers, converting radiant energy into thermal energy, which is one reason IR is often misleadingly called "heat waves" (just put your face in the sunshine and you'll warm to the notion).

Heat lamps (1000 nm–2000 nm) used in physical therapy (and in bad restaurants) offer a more penetrating radiation than light. Typically, near-IR (that is, near

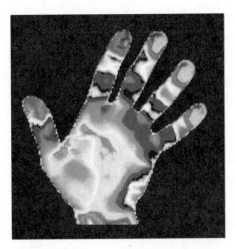

An infrared picture of a normal hand showing a typical temperature distribution. Abnormal blood flow becomes immediately obvious in IR images of this kind.

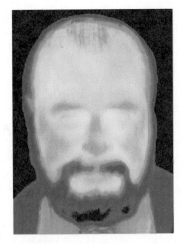

A portrait of the author taken using self-radiated IR. The beard is cool, and there seem to be no abnormal hot spots. The top of the mustache is warm from exhaled air.

the visible) enters the human body to a depth of not much more than about 3 mm below the skin, regardless of its color. If you've ever warmed yourself in sunlight streaming through a window, you already know that ordinary glass passes a good fraction of the incident near-IR; so does the cornea and lens of the human eye, and that's a good reason not to stare at the Sun, especially through cheap sunglasses that deceptively often pass much more IR than light.

There are photographic films sensitive to near-IR (<1300 nm); TV systems that produce continuous infrared pictures known as *thermographs;* there are IR spy satellites that look out for rocket launches, IR satellites that look out for crop diseases, and IR satellites that look out into space; there are "heat-seeking" missiles that are guided by infrared; IR lasers and IR astronomical telescopes peering at the sky. You probably change stations on your TV with an IR-emitting remote control unit. Wherever subtle variations in temperature are of concern, from detecting brain tumors and breast cancer to simply spotting a lurking burglar, IR systems have found practical use.

24.13 Light

The narrow band of the spectrum that we humans "see" is often referred to as *light.* That's a rather inaccurate specification because we can "see" X-ray shadow patterns cast directly on the retina. Furthermore, many of us can see—if only poorly—into both the IR (up to roughly 1050 nm) and ultraviolet (down to about 312 nm). At those extremes, the sensitivity of the eye has dropped by a factor of about a thousand. Accordingly, let's fix the meaning of the word *light* to stand for that tiny range of the electromagnetic spectrum from 780 nm to 390 nm, one octave.

Newton was the first to realize that **white light** is actually a mixture of all the colors of the visible spectrum; that the prism does not create color by changing white light to different degrees, as had been thought for centuries, but simply fans out the light, separating it into its constituent colors. Not surprisingly, the very con-

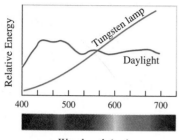

Figure 24.17 The distribution of the various frequencies in the light from a tungsten lamp and in sunlight.

cept of white light seems dependent on our perception of the Earth's daylight spectrum. The phenomenon of "daylight" changes from moment to moment and place to place, but it is nonetheless recognizable as a broad, slightly wiggly frequency distribution that falls off more rapidly in the violet than in the red (Fig. 24.17).

What we perceive as whiteness is a wide mix of frequencies roughly in the same amounts as that of daylight, and that's what we mean when we talk about "white light." Be that as it may, lots of different distributions will still appear more or less "white." We recognize a piece of paper to be white whether it's seen indoors in incandescent light or outside in skylight, even though those whites are quite different. Indeed, many pairs of colored light beams (for example, 656 nm red and 492 nm cyan) will produce the sensation of whiteness, and the human eye cannot always distinguish one white from another—it cannot frequency analyze light into its harmonic components the way the ear can analyze sound (p. 493).

Colors themselves are the subjective human physiological and psychological responses primarily to the various frequency regions extending from about 384 THz for red, through orange, yellow, green, and blue, to violet at about 769 THz (Table 24.1). Thus, a harmonic wave of the single frequency 508 THz will be perceived as yellow, as will 507 THz and 509 THz, etc. Each such single-frequency, single-color light is spoken of as *monochromatic*. But to be monochromatic, the wavetrain must be perfectly sinusoidal, extending from $-\infty$ to $+\infty$. It can have no beginning and no end, and that's impossible. In reality, the best we can hope for is light having a narrow band of frequencies—**quasimonochromatic light**.

Color corresponds to the human perception of photon energy or frequency. It is not a property of the light itself but a manifestation of the electrochemical sensing system: eye, nerves, brain. To be precise, we should not say "yellow light," but instead refer to light that is "seen to be yellow." Actually, a variety of different frequency mixtures can evoke the same color response from the eye-brain sensor. For example, a beam of red light (peaking at, say, 690 THz) overlapping a beam of green light (peaking at, say, 540 THz) will result, believe it or not, in the perception of *yellow* light, even though there are no frequencies present in the yellow band. The eye-brain averages the input and "sees" yellow! Furthermore, the nice reddish-blue or purple known as *magenta* does not even exist as a single frequency—it's not in the white-light spectrum.

The human eye, under daytime illumination levels, is most responsive to yellow-green (one reason that sodium-yellow street lights are so common and why yellow-tinted eyeglasses can be useful). Sensitivity gradually diminishes at both higher (blue, violet, ultraviolet) and lower (orange, red, IR) frequencies. It should be no surprise to learn that the solar spectrum also peaks at around 560 nm (2.2 eV) in the yellow-green so that there are usually plenty of the "right" photons flying around.

The wavelength of even red light at 0.000 000 780 m is rather small—780 nm is roughly 1/100 the thickness of this page. On an atomic scale the wavelengths of light are immense, several thousand times the size of an atom. If a uranium atom were enlarged to the size of a pea, a single wavelength of red light would then be about 54 feet long. This disparity is one of the crucial factors in determining the way light reflects off material objects. Remember that most of what we see is via reflected light. We make it a point not to look directly at bright, self-luminous objects even when they are around (apart from TV sets and campfires, which aren't very bright). Seeing the world almost exclusively in reflected light gives rise to a richly colored, shadow-filled, high-contrast, detailed picture—there's essentially no background noise.

TABLE 24.1 *Approximate frequency and vacuum wavelength ranges for the various colors*

Color	λ_0(nm)	f(THz)*
Red	780–622	384–482
Orange	622–597	482–503
Yellow	597–577	503–520
Green	577–492	520–610
Blue	492–455	610–659
Violet	455–390	659–769

*1 terahertz (THz) = 10^{12} Hz, 1 nanometer (nm) = 10^{-9} m.

In bright sunlight, where more than 10^{17} photons arrive each second on every square centimeter of a surface, the quantum nature of the process can easily be overlooked. Even so, light quanta are energetic enough ($hf \approx 1.6$ eV up to 3.2 eV) to produce effects on a distinctly individual basis. For instance, the human eye can detect as few as 10 photons impinging on it and perhaps as few as 1 arriving at the retina. Quanta of light can break up delicate chemical bonds, and so substances such as aspirin and wine are protected by dark bottles, and light-sensitive photographic films are commonplace. Premature infants sometimes develop jaundice, due to an excess of bilirubin in the blood, a condition that is successfully treated by exposing them to light. Blue-light photons have enough energy to dissociate the bilirubin molecule. Light is a major agency for the Sun → Earth transport of energy. Powered by sunlight, the process of photosynthesis results in the removal of upwards of 200 thousand million tons of carbon yearly from atmospheric carbon dioxide and the subsequent generation of complex organic molecules to the greater good of life on the planet.

24.14 Ultraviolet

A year after IR was discovered, J. Ritter (1801) found yet another invisible radiation. It was well known that white silver chloride would turn black, liberating metallic silver in the presence of light, especially blue light (this reaction was the precursor of photography). Ritter found that silver chloride did its little trick even more efficiently when exposed to the spectral region "beyond" the violet, where there was no visible radiation. This was the discovery of what came to be known as **ultraviolet** (UV) radiation, which corresponds to the range from about 8×10^{14} Hz to 2.4×10^{16} Hz. This is the so-called "black light," which is neither black nor light. Ultraviolet controls certain annoying dermatological conditions, tans the skin, and activates the synthesis of vitamin D within it.

At wavelengths of ≈300 nm and below, at the edge of the solar spectrum, UV can cause sunburn as well as tanning. Interestingly, this is roughly the energy needed (4 eV) to break a carbon-carbon bond. Passage through the atmosphere filters out much of this radiation, especially in the higher latitudes and at the low sun-angles that occur in winter and in the early and late parts of the day, even in summer. Someone in Chicago has to be a lot more determined to get tanned than someone in Florida. But then again, solar UV is the major cause of skin cancer in human beings. Our continued concern for the ozone (O_3) layer stems from the fact that this gaseous envelope absorbs (<320 nm) what would otherwise be a lethal stream of solar UV photons.

Ultraviolet in the wavelength range less than 300 nm will depolymerize nucleic acids and destroy proteins, both of which are strong absorbers, making UV quite incompatible with life on this planet. Extended exposure to UV, in time, causes wrinkles, liver spots, actinic keratosis (precancerous dark blotches), and finally cancer (80% of which is the curable form of basal-cell carcinoma). UV also inhibits the body's immune system, which may explain why some viral diseases, such as fever blisters and chicken pox, get more severe when exposed to sunshine.

Some materials reflect UV much as they reflect light, so that long exposure while playing around on snow or water can be de-

An ultraviolet photo of Venus taken by *Mariner 10*.

ceptively hazardous. The same is true for lying about on a beach on a totally over-cast summer day, since water vapor passes a good deal of UV ($\approx$50% in this case). In contrast, ordinary window glass invariably contains iron oxide contaminants, which make it quite opaque to near-UV. It's a waste of time to attempt to tan your-self behind such a window no matter how warm it gets.

Humans don't see ultraviolet very well because the cornea absorbs it, especially at the shorter wavelengths, while the eye's lens absorbs most effectively beyond 300 nm. Someone who has had a lens removed because of cataracts can see UV (λ > 300 nm). It now seems that in addition to insects such as honeybees, a fair number of creatures can visually respond to UV as well. Pigeons, for one, are quite capable of recognizing patterns illuminated by only UV and likely employ that abil-ity to navigate by the Sun, even on overcast days.

An atom emits a UV photon when the electron makes a long jump down from a highly excited state. For example, the outermost electron of a sodium atom can be raised into higher and higher energy levels until it's simply torn loose altogether at 5.1 eV. The atom is then said to be ionized. Should it subsequently recombine with a free electron, the latter will quickly descend to the ground state, most likely in a series of jumps to ever lower levels, each resulting in the emission of a photon. If, however, the electron makes one long plunge to the ground state, a single 5.1-eV ul-traviolet photon will result. Still more energetic UV photons can be generated when the inner electrons of an atom are excited.

Example 24.7   Suppose that a singly ionized copper atom recombines with an electron. The ionization en-ergy of copper is 7.72 eV. What is the shortest possible wavelength that can be emitted via the recombination?

Solution: [Given: E = 7.72 eV. Find: the minimum λ.] The maximum amount of energy available, correspond-ing to a transition directly to the ground state, equals 7.72 eV. This value, in turn, is associated with a maxi-mum frequency given by E = hf and therefore a mini-mum wavelength where E = hc/λ; hence

$$\lambda = \frac{hc}{E} = \frac{(4.136 \times 10^{-15} \text{ eV/Hz})(3.00 \times 10^8 \text{ m/s})}{7.72 \text{ eV}}$$

and

$$\boxed{\lambda = 161 \text{ nm}}$$

This, the most energetic radiation, is in the UV.

▶ **Quick Check:** 161 nm is very roughly $\frac{1}{4}$ the wave-length of light, and so the energy of the photon should be approximately 4 times the energy of a light photon, which is roughly 1.5 eV. That value yields about 6 eV, which is close to the given value of 7.72 eV. Thus, 161 nm must be at least the right order-of-magnitude.

The unpaired valence electrons of isolated atoms are the source of much colored light. But when these atoms combine to form molecules or solids, those valence electrons can get paired up in the very process of forming the chemical bonds that hold the thing together. As a direct consequence, the electrons are often more tightly bound and their molecular-excited states are higher up; that is, in the UV. This gives rise to selective molecular UV absorption. Molecules in the atmosphere, such as N_2, O_2, CO_2, and H_2O, have such electronic resonances in the UV (which contributes to making the sky blue, p. 936).

Each UV quantum can carry enough energy (from 3.2 eV to 100 eV) to inde-pendently ionize an atom or rip apart a chemical bond. The particlelike aspects of radiant energy begin to become increasingly more evident as the frequency in-creases. At wavelengths less than around 290 nm, UV is germicidal; that is, it kills microorganisms.

24.15 X-rays

When Wilhelm Röntgen discovered X-rays on November 5, 1895, it was quite by accident (p. 1094). Within months, the marvelous rays were at work everywhere. Without the slightest inkling of their inherently dangerous nature, X-rays were used for everything conceivable, from removing facial hair to examining luggage. Too often, the results of that cavalier attitude were horribly tragic. (And, of course, the Victorian ladies were well warned of the only danger anyone seems to have been concerned with: lurking X-ray peeping Toms.) Incredibly, X-rays were even being used as late as the 1950s to treat acne, though it's only recently that the victims began developing cancer, particularly of the thyroid.

Extending in frequency from roughly 2.4×10^{16} Hz to 5×10^{19} Hz, X-rays have exceedingly short wavelengths; most are smaller than the size of an atom. The individual photon energies (100 eV to 0.2 MeV) are so large that X-ray quanta can interact with matter one at a time in a clearly granular fashion, almost like bullets of energy. The primary mechanism for the production of this radiation is the rapid deceleration of high-speed charged particles.

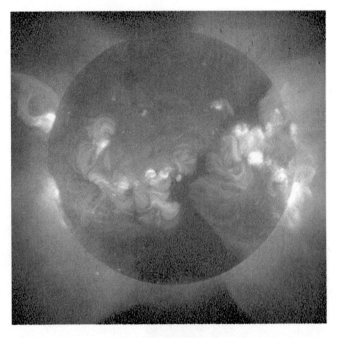

An X-ray photograph of the Sun. Modern focusing methods are producing detailed images of distant celestial sources. Orbiting X-ray telescopes have given us an exciting new view of the Universe.

Example 24.8 An electron flies across an X-ray tube. Traversing a voltage difference of 1.00×10^4 V, it crashes into a metal target. Assuming it makes a rare head-on collision and comes to rest with the emission of a single photon, what is the wavelength of the radiation?

Solution: [Given: $V = 1.00 \times 10^4$ V and $q_e = 1.60 \times 10^{-19}$ C. Find: λ.] An electron that falls through a potential difference of V picks up an amount of energy Vq_e, which then appears as the photon energy $E = hf = hc/\lambda$; hence

$$\lambda = \frac{hc}{E} = \frac{hc}{Vq_e}$$

and
$$\lambda = \frac{(6.626 \times 10^{-34} \text{ J/Hz})(3.00 \times 10^8 \text{ m/s})}{(1.00 \times 10^4 \text{ V})(1.60 \times 10^{-19} \text{ C})}$$

$$\boxed{\lambda = 1.24 \times 10^{-10} \text{ m}}$$

Notice that the calculation could have been done a bit more easily using electron volts (Problem 35)

▶ **Quick Check:** This value of λ is 0.12 nm, which is just about the size of an atom, and we know that's the correct wavelength for X-rays.

Diagnostic X-rays used in medicine have energies from 20 keV to 100 keV. Traditional medical film-radiography of this simple variety produces shadow castings. What arrives at the film is a crude mapping of the absorption that took place as the beam crossed all the interposed tissue.

In the 1970s, the marriage of the X-ray machine and the computer gave rise to a marvelous advance in X-ray technique known as *computed tomography*. Tomography (from the Greek *tomos*, or slice) is the process of creating an image of a cross-sectional region of a three-dimensional object—in this case, a person. The CT (or CAT) scan is done by rotating an X-ray source 360° around the patient and recording the transmission across the body at hundreds of different viewing angles. From that data, the computer then constructs a picture of the traversed region (see the

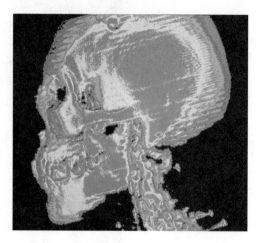

A modern X-ray CAT scan can reveal, slice-by-slice, structural details never before visible in ordinary X-ray photos.

Wilhelm Konrad Röntgen (1845–1923). As a result of his discovery of X-rays in 1895, Röntgen became the instant hero of his age and the first winner of the Nobel Prize in Physics.

photo above), producing a splendidly detailed representation showing all the soft tissue in that "slice" of the body.

24.16 Gamma Rays

Just as the atom is an interacting system of charged particles that exists only in well-defined configurations with well-defined energies, so, too, is the nucleus. Accordingly, the nucleus has a lowest-energy configuration, a ground state to which it will return once excited. A nucleus in one of its several well-defined excited states drops back to its ground state with the emission of a **gamma-ray** photon. Because the nuclear energy states are tightly bound, γ-ray energies range from keV to MeV.

There is no substantive difference between γ-rays and X-rays, other than the superficial historical one that the former were first seen to come from nuclei and the latter from atomic electrons—high-energy photons from a synchrotron come from neither. Thus, the distinction between X-rays and gamma rays vanished with the introduction of modern high-energy machines. Hundreds of modern hospitals are equipped with sophisticated devices such as linear accelerators and betatrons that provide multi-MeV electron beams that generate energetic photons for cancer treatment.

Core Material

Light (and all other forms of radiant energy) is electromagnetic. A *progressive* wave is a self-supporting, energy-carrying disturbance that travels free of its source (p. 907). The most important form of disturbance, from a theoretical perspective, is the **harmonic**, or sinusoidal, wave, which is given by

$$E = E_0 \sin k(x - vt) \qquad [24.5]$$

$k = 2\pi/\lambda$ is the **propagation number**. As with all waves

$$v = f\lambda \qquad [12.18]$$

In particular, for electromagnetic waves in vacuum

$$E = cB \qquad [24.6]$$

and

$$c = \frac{1}{\sqrt{\varepsilon_0 \mu_0}} \qquad [24.7]$$

where $c = 2.997\,924\,58 \times 10^8$ m/s

The energy-per-unit-area-per-unit-time conveyed by a lightwave is its **irradiance** I, where

$$I = \tfrac{1}{2} c \varepsilon_0 E_0^2 \qquad [24.8]$$

Light is wavelike, though unlike traditional waves, electromagnetic energy is not smoothly spread out across the wavefront but somehow concentrated within it. These energy concentrations are known as **photons**. The energy of a photon is

$$E = hf \qquad [24.9]$$

where $h = 6.626 \times 10^{-34}$ J/Hz (or 4.136×10^{-15} eV/Hz) is Planck's Constant.

The **electromagnetic spectrum** ranges from radiowaves to microwaves, infrared, light, ultraviolet, X-rays, and gamma rays (p. 921).

Suggestions on Problem Solving

1. Planck's Constant is given as either 6.6×10^{-34} J/Hz or 4.1×10^{-15} eV/Hz and, depending on how the data of the problem is provided, one or the other is the more convenient. Note that 1 Hz is 1 cycle per second, so the units of h are also J·s.

2. Many of the introductory wave problems make use of two equations ($v = \lambda f$ and $f = 1/T$) in endless variations. That was the case with sound, and it's true here as well.

3. Keep in mind that although $\lambda = c/f$, if we have a wavelength range $\Delta\lambda$, there will be a corresponding frequency range, but $\Delta\lambda \neq c/\Delta f$. We have to calculate $\lambda_1 = c/f_1$ and $\lambda_2 = c/f_2$ and then determine $\Delta\lambda = \lambda_2 - \lambda_1$ and $\Delta f = f_2 - f_1$.

4. When a charge q falls through a potential difference V, the energy change, qV, is in joules, *not* electron volts. That's a common point of confusion.

Discussion Questions

1. Why are fire engines sometimes painted yellow?

2. The Stanford Linear Accelerator imparts energies of up to 20 000 MeV to electrons, accelerating them along its straight two-mile course. Discuss the behavior of these particles as it pertains to electromagnetic radiation.

3. Does your hair dryer radiate radiowaves? How might you determine as much?

4. When NASA sends space probes to bodies beyond Earth, they routinely bathe the vehicles on the launch pad in ultraviolet radiation for several hours. Why? What's the significance of the so-called hole in the ozone layer?

5. People such as Ben Franklin rejected the corpuscular theory of light on the basis of momentum considerations. Devise an argument he might have used.

6. Is the region of space you are in right now crisscrossed by electromagnetic waves? In fact, are you permeated by them even as you read this question? Explain.

7. How does Newton's corpuscular theory of light compare with the modern quantum picture of light?

8. Central to Maxwell's theory of electromagnetic waves is the insight that the electric and magnetic fields generate one another. Explain. In vacuum, electromagnetic waves (including light) are said to be *transverse*. What does that mean?

9. How is it possible to cook a piece of meat on a paper plate in a microwave oven? What would happen to a dry cube of ice immediately after being placed in an operating microwave oven?

10. In his experiments (1780) on animal electricity, Luigi Galvani at one point tied a long wire to a dissected frog via a nerve, and another wire from the feet was grounded down a well. Waiting for a thunderstorm, he observed the legs twitching in rhythm with the lightning. In what sense was this the first frog-radio? What was happening? Incidentally, H. Hertz (1880s) vainly attempted to use a frog to detect the TEM waves he was generating in his laboratory.

11. What is the physical meaning of the idea of whiteness? That is, when do we see a thing as being white? In what way is whiteness a property of the detector (that is, the human eyeball)?

12. One often hears terms such as "ultraviolet light" and "infrared light" (although X-ray light and radio light are never spoken of). Why are such terms at best misleading?

13. Is a pulse a wave, or must a wave be something that "rises and falls" over and over again? What waves in a light-wave?

14. In what regard is it better to say "light that is perceived to be yellow" rather than "yellow light"?

15. Figure Q15 shows the phenomenon of optical levitation. A tiny glass sphere one-thousandth of an inch in diameter is suspended in an upward laser beam. What can you say about the ability of radiant energy to transfer momentum? Might it be possible to travel around in space not far from the Sun using a sail craft?

16. A few decades ago, some doctors were treating tonsillitis with X-rays. What would you expect would be a possible consequence of receiving such therapy?

Figure Q15

Multiple Choice Questions

1. The source of electromagnetic waves is   (a) a constant current   (b) a charge moving only in circles   (c) any accelerating charge   (d) any accelerating particle   (e) none of these.
2. The radio antenna for an AM station is a 75-m-high tower that is equivalent to $\frac{1}{4}\lambda$; another $\frac{1}{4}\lambda$ corresponds to the ground reflection. At what frequency does the station transmit?   (a) 10 kHz   (b) 75 kHz   (c) 1 MHz   (d) 300 kHz   (e) none of these.
3. When sodium atoms are excited (for example, by heating salt in a flame) they emit bright yellow light at two wavelengths of 589.0 nm and 589.6 nm. That emission implies that the sodium atom has two nearby energy levels separated by only   (a) 2 eV   (b) 2×10^{-3} eV   (c) 0.6 eV   (d) 0.6 MeV   (e) none of these.
4. In comparison to UV, light has   (a) wavelengths that are shorter   (b) frequencies that are higher   (c) wavelengths that are equal   (d) frequencies that are equal   (e) none of these.
5. Dental X-ray photos are usually taken while operating the machine at about 50 000 V. The minimum wavelength of such radiation is   (a) 0.025 nm   (b) 0.25 nm   (c) 2.5 nm   (d) 25 nm   (e) none of these.
6. Which statement is not true for a photon?   (a) it is electromagnetic in nature   (b) it always travels at the speed c independent of frequency   (c) it possesses energy independent of frequency   (d) it has wavelike properties   (e) none of these.
7. An X-ray tube produces a beam of photons by bombarding a dense metal target with high-energy electrons. The resulting beam possesses a wide range of frequencies and is terminated at a maximum energy that   (a) depends only on the target material and the temperature   (b) depends on the voltage across the tube and is constant provided the voltage is constant   (c) is not constant and changes even when the voltage is kept constant   (d) depends on the tube length and the shielding   (e) none of these.
8. Which of the following requires a physical medium to travel in?   (a) lightwaves   (b) radiowaves   (c) sound waves   (d) gamma rays   (e) none of these.
9. The propagation number of a wave   (a) varies inversely with wavelength   (b) has to do with the number of pulses in a burst   (c) varies inversely with frequency   (d) depends on the speed of the wave   (e) none of these.
10. Nowadays it is possible to directly measure, using electronic techniques, the frequencies of electromagnetic oscillations ranging up to 500 MHz. That corresponds to a wavelength of   (a) 0.6 m   (b) 0.6 cm   (c) 6.0 m   (d) 6.0 cm   (e) none of these.
11. What is the frequency of 1-GeV gamma rays?   (a) 2.4×10^{20} Hz   (b) 1.5×10^{42} Hz   (c) 3×10^8 Hz   (d) 2.4×10^{23} Hz   (e) not enough information given.
12. For all practical purposes, the light from the following source can best be considered a plane wave   (a) a street light overhead   (b) a nearby desk lamp   (c) a match held in the hand   (d) a star in the constellation Orion   (e) none of these.
13. An oscillating dipole is best described as   (a) two poles moving in a circle   (b) a single pole moving in two directions   (c) two equal and opposite charges moving to and fro along a line   (d) two like charges oscillating   (e) none of these.
14. Light travels 1.000 m in vacuum in   (a) 1.000 s   (b) 3.336×10^{-19} s   (c) 0.334 s   (d) 3.336×10^{-9} s   (e) none of these.
15. A typical AM radiowave is   (a) 1.0 m long   (b) 1.0 cm long   (c) millions of meters long   (d) hundreds of meters long   (e) none of these.
16. Is there a relationship between the electric and magnetic fields of an electromagnetic wave in vacuum?   (a) no   (b) yes, $B = cE$ (c) yes, $E = cB$ (d) yes, $E = B/v$ (e) none of these.
17. An AM radio station transmits a signal whose electric field is received with a strength of 1.5 mV/m. If the antenna on a portable radio at that location is 0.75 m long and straight up in the air, the input voltage will be   (a) 1.5 mV   (b) 1.1 mV   (c) 7.5 mV   (d) 1.5 V   (e) none of these.
18. A typical atomic transition lasts about   (a) 10^{-5} s to 10^{-6} s   (b) 1.0 s to 10.0 s   (c) 0.01 s to 0.001 s   (d) 10^8 s to 10^9 s   (e) none of these.
19. In order of decreasing frequency, the entire electromagnetic spectrum is made up of   (a) radiowaves, microwaves, IR, light, UV, and γ-rays   (b) γ-rays, X-rays, UV, IR, microwaves and radiowaves   (c) radiowaves, microwaves, IR, light, X-rays, and γ-rays   (d) light, IR, and UV   (e) none of these.
20. The ground state of an atom is the state   (a) with the highest energy   (b) nearest the ground   (c) with no energy   (d) with the least energy   (e) none of these.
21. The molecular rotational and vibrational energy states are primarily responsible for the emission of   (a) radiowaves   (b) light and UV   (c) IR and microwaves   (d) X-rays and gamma rays   (e) none of these.

Problems

THE NATURE OF LIGHT
PHOTONS
THE ELECTROMAGNETIC SPECTRUM

1. [I] A lightwave has a wavelength of 500 nm. What is its propagation number?

2. [I] An infrared electromagnetic wave has a propagation number of 2000π m^{-1}. What is its wavelength?
3. [I] The propagation number of a harmonic electromagnetic wave is 6.283×10^{-4} m^{-1}. What is its wavelength?
4. [I] An electromagnetic wave has a profile given by

$$E = E_0 \sin kx$$

If the amplitude of the wave is 20.0 V/m, what is the size of the field at $x = 0$?

5. [I] An electromagnetic wave at $t = 0$ has the profile

$$E = E_0 \sin kx$$

Draw a plot of E versus x showing E at points $x = 0$, $\lambda/4$, $\lambda/2$, $3\lambda/4$, and λ. Remember that $k = 2\pi/\lambda$.

6. [I] Draw a plot of the function

$$E = E_0 \sin (kx - \pi/2)$$

and compare it to your result from Problem 5. How far has the profile advanced?

7. [I] The electric field of a microwave is given by

$$E = (20 \text{ V/m}) \cos \frac{2\pi}{1.00 \text{ mm}} [x - (3.00 \times 10^8 \text{ m/s})t]$$

What is its value at $x = 0$ and $t = 0$?

8. [I] The electric field of a TEM wave has the form

$$E = (5.0 \text{ V/m}) \sin k(x - vt)$$

What is the value of the E-field at $x = \lambda/4$ and $t = 0$?

9. [I] Make a sketch of the profile or shape of the wave $E = (10 \text{ V/m}) \sin [k(x - vt) + \varepsilon]$ when $\varepsilon = 0$ and again when $\varepsilon = \pi/2$. Set $t = 0$ and plot the curve for various values of x (namely, $x = 0, \lambda/4, \lambda/2, 3\lambda/4, \lambda$), remembering that $k = 2\pi/\lambda$. What does the phase-shifted wave look like?

10. [I] Make a sketch of the wave

$$E = E_0 \sin k(x - vt)$$

at $t = 0$ given that $\lambda = 10$ m and $E_0 = 2.0$ V/m.

11. [I] In the article "The Longest Electromagnetic Waves" (*Sci. Am.,* March 1962), J. R. Heirtzler described the detection of waves 18.6×10^6 miles in wavelength. What type of waves were they and what was their period?

12. [I] How far does light travel in vacuum in 10^{-9} s?

13. [I] What is the vacuum wavelength of a 20.0-Hz electromagnetic wave?

14. [I] If light is emitted from an atom in little wavetrains (Fig. P14) lasting up to 10^{-8} s, how long, at most, is such a disturbance in space? If we approximate the wavelength as 500 nm, roughly how many waves long is the train?

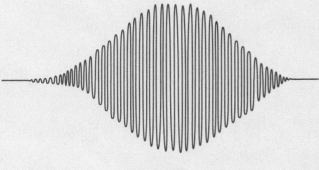

Figure P14

15. [I] A tuning circuit in an FM radio is designed to pick up a station at 100 MHz. If the capacitance of the input circuit is 0.5 pF, how much is the inductance?

16. [I] In 1887, Heinrich Rudolf Hertz succeeded in generating and detecting long-wavelength electromagnetic waves. His transmitter was an induction coil (a device that converted low-voltage dc into high-voltage ac) attached to a loop of wire ending in a spark gap. Across the room, a wire loop with its own open gap served as the receiver. When the circuit discharged, an oscillatory spark flashed across the transmitter's gap, and the event was almost immediately reported by a fainter spark at the distant receiver. If the frequency of the waves was 75 MHz, what was their wavelength?

17. [I] Determine the frequency of a 200-m-long AM radiowave, assuming the conditions to be that of vacuum.

18. [I] On December 12, 1901, Marconi, using a 20-kW transmitter attached to a 200-ft antenna, sent electromagnetic signals (with a 1-km wavelength) across the Atlantic for the first time. Compute the frequency of the emission. Assume the speed in air is the same as that in vacuum.

19. [I] The energy arriving per second on a 2.00-m^2 detector held perpendicular to the light from a bright star is about 2.4×10^{-9} J. Determine the irradiance of the radiation.

20. [I] The irradiance 1 m from a candle flame is just about 1.5×10^{-3} W/m^2. How much energy will arrive in 2.00 s on a disk having a 1.00-cm^2 area held as close to perpendicular as possible 1.00 m from the flame?

21. [I] The maximum irradiance I of solar radiation arriving on the lawn of the White House is 1.05×10^3 W/m^2. What maximum energy will impinge each minute on a flat collector with a 1.00-square-meter area?

22. [I] What is the energy of a tangerine-colored photon where λ_0, the vacuum wavelength, equals 616 nm?

23. [I] Determine the vacuum wavelength, frequency, and energy in joules of a 2.00-eV photon. What type of radiation is it?

24. [II] Show that $\varepsilon_0(\Delta\Phi_E/\Delta t)$ has the units of current.

25. [II] Show that the wave equation in Problem 10 can be written as

$$E = E_0 \sin (kx - \omega t)$$

26. [II] Show that the electric field of a progressive harmonic electromagnetic wave, as given in Problem 10, can be written as

$$E = E_0 \sin 2\pi f \left[\left(\frac{x}{v} \right) - t \right]$$

27. [II] What is the magnitude of the electric field given in SI units by $E = 20 \cos(kx - \omega t + \pi)$ at $x = 0$, when $t = 0$, $t = T/4$, and $t = T$?

28. [II] The electric field of an electromagnetic wave is given by

$$E = 2.0 \times 10^2 \sin[3.0 \times 10^6 \pi(x - 3.0 \times 10^8 t)]$$

where everything is in SI units. What is the wave's frequency? [Hint: Compare this expression with the one in Problem 10.]

29. [II] What is the amplitude of the magnetic field associated with the TEM wave in the previous problem?
30. [II] The antenna shown in Fig. P30 is called a folded dipole. How long should it be if it's to receive FM signals at 90 MHz?

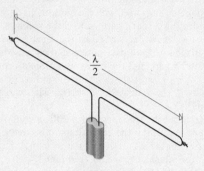

$\dfrac{\lambda}{2}$

Figure P30

31. [II] In 1982, workers at Bell Labs produced optical pulses lasting 30.0 femtoseconds. How many wavelengths of the 620-nm red light correspond to one of these little wavetrains?
32. [II] Light with a frequency of 6.50×10^{14} Hz is traveling through vacuum. How many of these waves (end-to-end) are there per centimeter?
33. [II] A TEM wave given by

$$E = 2.0 \times 10^2 \sin[3 \times 10^6 \pi(x - 3.0 \times 10^8 t)]$$

where everything is in SI units, impinges on a perpendicular surface in space. What is the irradiance?

34. [II] A beam of TEM harmonic waves has an irradiance of 13.3 W/m². What is the amplitude of the electric field?
35. [II] An electron falls through a potential difference of 1.00×10^4 V. What is the maximum energy of a resulting X-ray photon? What is the minimum corresponding wavelength? Do this calculation in electron volts and then compare the results with that of Example 24.8.
36. [III] It takes an energy of 33 keV to remove the innermost electron from an atom of iodine. Show that iodine will be a powerful absorber of X-rays at a frequency of 8.0×10^{18} Hz.
37. [III] If the electric field of an electromagnetic wave traveling in vacuum, pointing in the y-direction, is given in SI units by

$$E_y = 200 \sin (10^7 x - \omega t)$$

find (a) the vacuum wavelength and (b) the frequency. (c) In what direction does the wave progress?
38. [III] Using the results of Problem 37, write an expression for the accompanying B-field.
39. [III] A laser that emits pulses of UV lasting 2.00 ns has a beam diameter of 2.5 mm. If each burst contains an energy of 3.0 J, (a) what is the length in space of each pulse? (b) what is the average energy per unit volume (J/m³), the energy density, in one of these pulses?
40. [III] Given the wave function for an electromagnetic harmonic disturbance expressed in SI units

$$E = 10^2 \sin \pi(3 \times 10^6 x - 9 \times 10^{14} t)$$

find (a) the amplitude, (b) the speed, (c) the frequency, (d) the wavelength, (e) the period, and (f) the direction of propagation.

25

The Propagation of Light: Scattering

OUR PRESENT CONCERN IS with the two basic laws that describe the *reflection* (p. 942) and *refraction* (p. 948) of light. But the underlying questions are: How does light move through bulk matter? And, what happens to it as it does? Each such encounter is a stream of photons sailing through an array of atoms suspended in the void. The details of that marvelous journey determine why the sky is blue and blood is red, why your cornea is transparent and your hand opaque, why snow is white and rain is not.

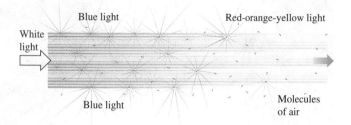

Figure 25.1 A beam of light traversing a region of widely spaced molecules. The light laterally scattered is mostly blue, and that's why the sky is blue. The unscattered light, which is rich in red, is viewed only when the Sun is low in the sky at sunrise and sunset.

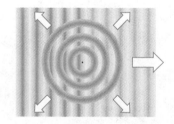

Figure 25.2 Scattering of some of the energy out from a plane wave in the form of a spherical wavelet. The process is continuous and hundreds of millions of photons per second stream out of the scattering atom in all directions.

SCATTERING

At its core, this chapter is about **scattering**, the absorption and prompt re-emission of electromagnetic radiant energy by atoms and molecules. The processes of reflection and refraction are macroscopic manifestations of scattering occurring on a submicroscopic level. Let's first consider the propagation of radiant energy through various homogeneous media.

25.1 Rayleigh Scattering: Blue Skies

Imagine a narrow beam of sunlight, a torrent of photons having a broad range of frequencies, advancing through empty space. As it progresses, the beam spreads out very slightly, but apart from that, all the energy continues forward at c. There is no scattering, and the beam cannot be seen from the side (the photons move straight ahead). Nor does the light tire or diminish in any way. When a star in a nearby galaxy 1.7×10^5 ly away was seen to explode in 1987, the flash of light that reached Earth had been sailing through space for 170,000 years. Photons are timeless.

Now, suppose we mix a wisp of air into the void—some molecules of nitrogen, oxygen, and so forth. Since these molecules have no resonances in the visible (pp. 920 and 928), no one of them can be raised into an excited state by absorbing a quantum of light, and the gas is transparent. Instead, each molecule behaves as a little oscillator whose electron cloud can be driven into a ground-state vibration by an incoming photon. Immediately upon being set oscillating, the molecule initiates the re-emission of light. A photon is absorbed, another photon of the same frequency is emitted; the light is *elastically scattered*. The molecules are randomly oriented, and photons scatter out every which way (Fig. 25.1). Even when the light is fairly dim, the number of photons is immense, and it looks as if the molecules are scattering little classical spherical wavelets (Fig. 25.2)—energy streams out in every direction. Still, the scattering process is quite weak and the gas tenuous, so the beam is very little attenuated unless it passes through a tremendous volume of air.

The amplitudes of these ground-state vibrations, and therefore the amplitudes of the scattered light, increase with frequency because the molecules all have electronic resonances in the UV. The closer the driving frequency is to a resonance, the more vigorously the oscillator responds (p. 460). So, violet light is strongly scattered laterally out of the beam, as is blue to a slightly lesser degree, as is green to a considerably lesser degree, as is yellow to a still lesser degree, and so on. The beam that traverses the gas will thus be richer in the red end of the spectrum, while the light scattered out (sunlight not having much violet in it in the first place) will be richer in blue. That, in part, is why the sky is blue.

Long before Quantum Mechanics, Lord Rayleigh (1871) analyzed scattered sunlight in terms of molecular oscillators and correctly concluded that the intensity of the scattered light was proportional to $1/\lambda^4$, which increases with f^4. Before this work, it was widely believed that the sky was blue because of scattering from minute dust particles. Since that time, *scattering involving particles smaller than a wavelength* has been referred to as **Rayleigh scattering**. A human's blue eyes, a blue jay's feathers, the blue-tailed skinks's blue tail, and the baboon's blue buttocks are all colored via Rayleigh scattering.

Example 25.1 A beam of white light traverses a medium composed of randomly distributed particles that are each much smaller than a typical wavelength. Compare the amount of scattering occurring for the red (710 nm) component with that of the violet (400 nm) component.

Solution: [Given: $\lambda_r = 710$ nm and $\lambda_v = 400$ nm, compare their scattered intensities.] The degree of Rayleigh scattering is proportional to $1/\lambda^4$. But $\lambda_r = 1.775\lambda_v$ and

so $1/\lambda_r^4 = (1/1.775\lambda_v)^4$; hence, violet is scattered $(1.775)^4 = \boxed{9.93}$ times more intensely than red.

▶ **Quick Check:** We can check the order-of-magnitude of the answer by noting that red is very roughly twice the wavelength of violet and the answer must be between $1^4 = 1$ and $2^4 = 16$; at least we pushed the right keys on the calculator.

As we will see in a moment, a dense uniform substance will not appreciably scatter laterally, and that applies to much of the lower atmosphere. Something else beyond Rayleigh scattering must be contributing to making the lower dense regions of the sky blue. What happens in the atmosphere is that thermal motion of the air results in rapidly changing *density fluctuations* on a local scale. And these momentary, fairly random fluctuations cause more molecules to be in one place than another and to radiate more in one direction than another. A theory of scattering from these fluctuations gives very much the same results as Rayleigh obtained for a tenuous gas.

Sunlight streaming into the atmosphere from one direction is scattered in all directions. Without an atmosphere, the daytime sky would be as black as the void of space, as black as is the Moon sky. When the Sun is low over the horizon, its light passes through a great thickness of air (far more so than it does at noon). With the blue-end appreciably attenuated, the reds and yellows propagate along the line-of-sight from the Sun to produce Earth's familiar fiery sunsets.

25.2 Scattering and Interference

In dense media, there are a tremendous number of close-together atoms or molecules contributing an equally tremendous number of scattered electromagnetic wavelets. These wavelets overlap and interfere in a way that does not occur in a tenuous medium. As a rule, the denser the substance through which light advances, the less the lateral scattering.

The phenomenon of interference has already been discussed (p. 514) and will be treated in further detail in Chapter 27; here, the basics suffice. Interference is ***the superposition of two or more waves producing a resultant disturbance that is the sum of the overlapping wave contributions***. Figure 25.3 shows two harmonic waves

Figure 25.3 The superposition of two waves. These combine to form a resultant wave. (a) When the component waves are in-phase, we have *constructive interference*, and the resultant is large. (b) As the phase difference increases, the resultant decreases. (c) When the component waves are 180° out-of-phase, the resultant has its smallest value, and we have *destructive interference*.

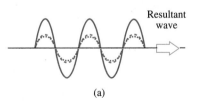

(a)

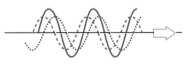

(b)

(c)

················ Wave 1
- - - - - - - - Wave 2
───────── The sum of wave 1 and wave 2

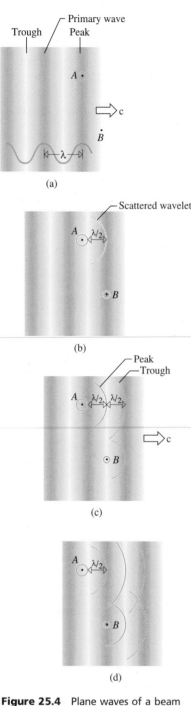

Figure 25.4 Plane waves of a beam of light moving to the right, passing over molecules *A* and *B*. These scatter wavelets that travel in-phase in the forward direction. In actuality, billions of such wavelets combine constructively, scattering light in the forward direction.

of the same frequency traveling in the same direction. When they are in-phase, the resultant at every point is the sum of the two wave-height values (Fig. 25.3a). This extreme case is called *total constructive interference*. When the phase difference reaches 180°, the waves tend to cancel, and we have the other extreme, called *total destructive interference* (Fig. 25.3c).

The simple theory of Rayleigh scattering has the molecules randomly arrayed in space so that the phases of the wavelets scattered off to the side have no particular relationship to one another and there can be no sustained interference between them. That situation will occur when the separation between the molecular scatterers is roughly a wavelength or more, as it is in a tenuous gas. When there are many scatterers, the random hodgepodge of overlapping waves effectively averages away the interference. *Random, widely spaced scatterers driven by an incident wave emit wavelets that are essentially independent of one another in all directions except forward.* Laterally scattered light, unimpeded by interference, streams out of the beam. And this is approximately the situation existing about 100 miles up in the Earth's tenuous high-altitude atmosphere, where a good deal of blue-light scattering takes place.

To see why the forward direction is special, why the wave advances in any medium, consider Fig. 25.4. It depicts a sequence in time showing two molecules *A* and *B*, fairly far apart, interacting with an incoming plane wave—a solid line represents a wave peak (a positive maximum); a dashed line corresponds to a trough (a negative maximum). In (a), the incident wavefront impinges on molecule *A*, which begins to scatter a spherical wavelet. For the moment, suppose the wavelet is 180° out-of-phase with the incident wave. Thus, *A* begins to radiate a trough (a negative *E*-field) in response to being driven by a peak (a positive *E*-field). Part (b) shows the spherical wavelet and the plane wave overlapping, marching out-of-step but marching together. The incident wavefront impinges on *B* and it, in turn, begins to reradiate a wavelet, which must also be out-of-phase by 180°. In (c) and (d), we see the point of all of this, namely, that both wavelets are moving forward with the incident wave—they are in-phase with each other, but out-of-phase with the incident wave. And that condition would be true for all such wavelets regardless of both how many molecules there were and how they were distributed. Because of the asymmetry introduced by the beam itself, ***all the scattered wavelets add constructively with each other in the forward direction***.

25.3 The Transmission of Light through Dense Media

Now, suppose the amount of air in the region under consideration is increased. In fact, imagine that each little cube of air, one wavelength on a side, contains a great many molecules, whereupon it is said to have an appreciable *optical density*. At the wavelengths of light, the Earth's atmosphere at STP has about three million molecules in such a λ^3-cube. The wavelets ($\lambda \approx 500$ nm) radiated by sources so close together (≈ 3 nm) cannot properly be treated as random. The light beam effectively encounters a fairly uniform medium with no discontinuities to destroy the symmetry. Again the scattered wavelets interfere constructively in the forward direction (that much is independent of the arrangement of the molecules), but now destructive interference predominates in all other directions. ***No light ends up scattered laterally or backwards***.

For example, Fig. 25.5 shows the beam moving through an ordered array of close-together scatterers. Some molecule *A* radiates spherically out of the beam, but because of the ordered close arrangement, there will be a molecule *B*, a distance

≈λ/2 away, such that both wavelets cancel in that transverse direction. Here, where λ is thousands of times larger than the scatterers and their spacing, it is very likely that there will always be pairs of molecules that tend to negate each other's wavelets in any given direction. Even if the medium were not perfectly ordered, the net electric field at a point in any direction will be the sum of a great many tiny scattered fields, each somewhat out-of-phase with the next (Fig. 25.6), so that the sum will always be negligibly small. The more dense, uniform, and ordered the medium is, the more complete will be the destructive interference and the less the nonforward scattering. The beam advances undiminished.

The scattering phenomenon on a per-molecule basis is extremely weak. In order to have half its energy scattered, a beam of green light will have to traverse ≈150 km of atmosphere. Since about 1000 times more molecules are in a given volume of liquid than in the same volume of vapor (at atmospheric pressure), we can expect to see an increase in scattering. Still, the liquid is a far more ordered state with much less pronounced density fluctuations, and that should suppress the nonforward scattering appreciably. Accordingly, an increased scattering, per unit volume, is observed in liquids, but it's more like 5 to 50 times as much rather than 1000 times. *Molecule for molecule,* liquids scatter substantially less than gases. Put a few drops of milk in a tank of water and illuminate it with a bright flashlight beam. A faint but unmistakable blue haze will scatter out laterally, and the direct beam will emerge decidedly reddened.

Transparent amorphous solids, such as glass and plastic, will also scatter light, but very weakly. Good crystals, like quartz and mica, with their almost perfectly ordered structures, scatter even more faintly. Of course, imperfections of all sorts (dust and bubbles in the liquids; flaws and impurities in the solids) will serve as scatterers, and when these are small, as in the gem moonstone, the emerging light will be bluish. In the same way, some inexpensive plastic food containers and white garbage-bag plastic look pale blue-white in scattered light and are distinctly orange in transmitted light.

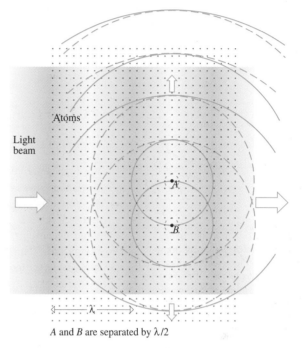

Figure 25.5 A plane wave impinging from the left onto a material composed of many closely spaced atoms. Among countless others, a wavefront stimulates two atoms, *A* and *B*, that are very nearly one-half wavelength apart. The wavelets they emit interfere destructively. Trough overlaps crest and they completely cancel each other in the direction perpendicular to the beam. That process happens over and over again and little or no light is scattered laterally.

REFLECTION

When a beam of light impinges on the surface of a transparent material, such as a sheet of glass, the wave "sees" a vast array of very closely spaced atoms that will somehow scatter it. Remember that the wave may be ≈500 nm long while the atoms and their separations (≈0.2 nm) are thousands of times smaller. In the case of transmission through a dense medium, the scattered wavelets cancel each other in all but the forward direction and just the ongoing beam is sustained. But that can only hap-

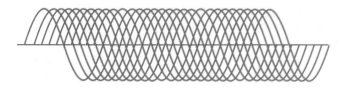

Figure 25.6 When a great many slightly shifted waves arrive at a point in space, there is generally as much positive *E*-field as negative, and the resultant disturbance is very nearly zero.

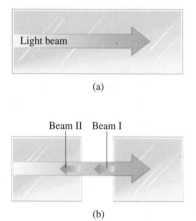

(a)

(b)

Figure 25.7 (a) A light beam traveling through a dense homogeneous medium such as glass. (b) If the block of glass is cut and parted, the light is reflected backward at the two new interfaces. Beam I is externally reflected, and beam II is internally reflected. When the two pieces are pressed back together, the two reflected beams cancel one another.

The cruiser *Aurora,* which played a key role in the Communist Revolution (1917), docked in St. Petersburg. Where the water is still, the reflection is specular. The image blurs where the water is rough.

pen if there are no discontinuities, whereupon the scattering is uniform. This is not the case at an interface between two different transparent media (such as air and glass), which is a jolting discontinuity. When a beam of light strikes such an interface, some light is always scattered backward, and we call this phenomenon **reflection**.

25.4 Internal and External Reflection

Imagine that light is traveling across a large block of glass (Fig. 25.7a). Now, suppose that the block is sheared in half perpendicular to the beam. The two segments are then separated, exposing the smooth flat surfaces depicted in Fig. 25.7b. Just before the cut was made, there was no lightwave traveling to the left inside the glass—we know the beam only advances. Now there must be a wave (beam I) moving to the left, reflected from the surface of the right-hand block. The implication is that a region of scatterers on and beneath the exposed surface of the right-hand block is now "unpaired," and the backward radiation they emit can no longer be canceled. The region of oscillators that was adjacent to these, prior to the cut, is now on the section of the glass that is to the left. When the two sections were together, these scatterers presumably also emitted wavelets in the backward direction that were 180° out-of-phase with, and canceled, beam I. Now they produce reflected beam II. Each molecule scatters light in the backward direction and, in principle, each and every molecule contributes to the reflected wave. Nonetheless, in practice, it is a thin layer ($\approx\lambda/2$ deep) of unpaired atomic oscillators near the surface that is effectively responsible for the reflection. For an air-glass interface, about 4% of the energy of an incident beam falling perpendicularly *in* air *on* glass will be reflected straight back out by this layer of unpaired scatterers. And that's true whether the glass is 1.0 mm thick or 1.00 m thick.

Beam I reflects off the right-hand block, and because light was initially traveling from a less to a more optically dense medium, this is called **external reflection**. Since the same thing happens to the unpaired layer on the section that was moved to the left, it, too, reflects backwards. With the beam incident perpendicularly *in* glass *on* air, 4% must again be reflected, this time as beam II. And this process is referred to as **internal reflection**. If the two glass regions are made to approach one another increasingly closely (so that we can imagine the gap to be a thin film of, say, air—p. 1028), the reflected light will diminish until it ultimately vanishes as the two faces merge and disappear and the block becomes continuous again.

Remember this 180° relative phase shift between internally and externally reflected light—we will need it later on. This kind of reflection is of practical importance when you have a microscope with lots of compound lenses and perhaps a dozen or two interfaces each kicking back $\approx$4% of the incident light (just try looking through 20 or 30 layers of plastic food wrap). To overcome that difficulty, modern high-quality lenses are covered with antireflection coatings (p. 1030).

We know from experience with the common mirror that white light is reflected as white—it certainly isn't blue. To see why, first realize that the layer of scatterers responsible for the reflection is very roughly $\lambda/2$ thick (per Fig. 25.5). Thus, the larger the wavelength, the deeper the

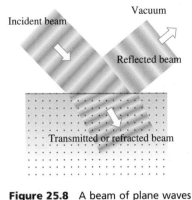

Figure 25.8 A beam of plane waves incident on a distribution of molecules constituting a piece of clear glass or plastic. Part of the incident light is reflected and part refracted.

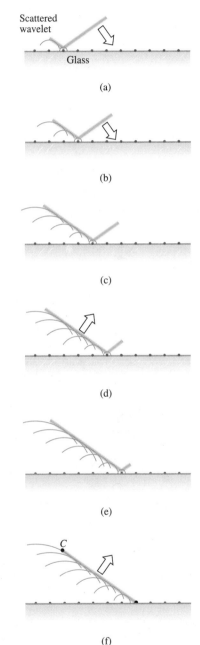

layer contributing, and the more scatterers there are acting together. This tends to balance out the fact that each scatterer is less efficient as λ increases (remember $1/\lambda^4$). The combined result is that *the surface of a transparent medium reflects all wavelengths about equally and doesn't appear colored in any way.* That, as we will see, is why this page looks white under white-light illumination.

25.5 The Law of Reflection

Figure 25.8 shows a beam composed of plane wavefronts impinging at some angle on the smooth flat surface of an optically dense medium (let it be glass). Assume that the surrounding medium is vacuum. Follow one wavefront as it sweeps in and across the molecules on the surface. For the sake of simplicity, in Fig. 25.9 we have omitted everything but one molecular layer at the interface. As the wavefront descends, it excites one scatterer after another, each of which reradiates a stream of photons that can be thought of as a hemispherical wavelet in the incident medium. Because the wavelength is so much greater than the separation between the molecules, the wavelets advance together and add constructively in only one direction, and there is one well-defined *reflected* beam. (That would not be true if the incident radiation was short-wavelength X-rays, in which circumstance there would be several reflected beams. And it would not be true if the scatterers were far apart compared to λ, as they are for a diffraction grating, in which case there would also be several reflected beams.)

In Fig. 25.10, the line $\overline{AB}$ lies along an incoming wavefront while $\overline{CD}$ lies on an outgoing wavefront—in effect, $\overline{AB}$ transforms on reflection into $\overline{CD}$. With Fig. 25.9 in mind, we see that the wavelet emitted from A will arrive at C in-phase with the wavelet just being emitted from D (as it is stimulated by B), so long as the distances $\overline{AC}$ and $\overline{BD}$ are equal. In other words, if all the wavelets emitted from all the surface scatterers are to overlap in-phase and form a single reflected plane wave, it must be that $\overline{AC} = \overline{BD}$. Then, since the two triangles have a common hypotenuse

$$\frac{\sin \theta_i}{\overline{BD}} = \frac{\sin \theta_r}{\overline{AC}}$$

Figure 25.9 An incoming plane wave scattering off a layer of molecules. Because the wavelength is thousands of times larger than the molecular spacings, the reflection can be treated as occurring at the surface.

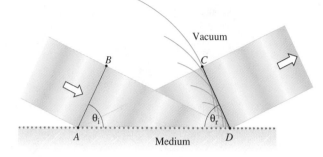

Figure 25.10 Wavefront geometry for reflection. The reflected wavefront $\overline{CD}$ is formed of waves scattered by the atoms on the surface from A to D. Just as the first wavelet arrives at C from A, the atom at D emits, and the wavefront along $\overline{CD}$ is completed.

All the waves travel in the incident medium with the same speed v_i. It follows that in the time Δt it takes for point B on the wavefront to reach point D on the surface, the wavelet emitted from A reaches point C. In other words, $\overline{BD} = v_i \Delta t = \overline{AC}$, and so from the above equation, $\sin \theta_i = \sin \theta_r$, which means that

$$\theta_i = \theta_r \qquad (25.1)$$

The angle of incidence equals the angle of reflection. This equation is the first part of the **Law of Reflection**.

Drawing wavefronts can get things a bit cluttered, so we now introduce another convenient scheme for visualizing the progression of light. The imagery of antiquity was in terms of straight-line streams of light, a notion that got into Latin as "radii" and reached English as "rays." *A ray is a line drawn in space corresponding to the direction of flow of radiant energy*. It is a mathematical device and not a physical entity. In a medium that is uniform (homogeneous), rays are straight. If the medium behaves in the same manner in every direction (isotropic), *the rays are perpendicular to the wavefronts*. Thus, for a point source emitting spherical waves (Fig. 13.12, p. 496), the rays, which are perpendicular to them, point radially outward from the source. Similarly, the rays associated with plane waves are all parallel (Fig. 25.11a). Rather than sketching bundles of rays, we can simply draw one incident ray and one reflected ray (Fig. 25.11b). *All the angles are measured from the perpendicular (or normal) to the surface*, and thus θ_i and θ_r have the same numerical values as before (Fig. 25.10).

The ancient Greeks knew the Law of Reflection—it can be deduced by observing the behavior of a flat mirror, and nowadays that observation can be done most simply with a flashlight or, even better, a laser. The second part of

Figure 25.11 (a) We select one ray to represent the beam of plane waves. Now both the angle of incidence θ_i and the angle of reflection θ_r are measured from a perpendicular drawn to the reflecting surface. (b) The incident ray and the reflected ray define a plane, known as the *plane-of-incidence*, perpendicular to the reflecting surface.

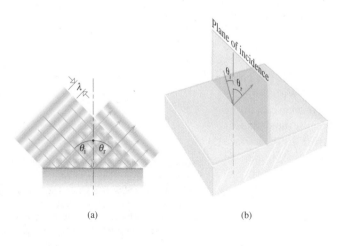

(a)

(b)

the Law of Reflection maintains that *the incident ray, the perpendicular to the surface, and the reflected ray all lie in a plane* called the **plane-of-incidence** (Fig. 25.11b)—this is a three-dimensional business. Try to hit some target in a room with a flashlight beam by reflecting it off a stationary mirror and the importance of this second part of the law becomes obvious!

When a beam is incident upon a reflecting surface that is smooth (one for which any irregularities are small compared to a wavelength), the light re-emitted by millions upon millions of atoms will combine to form a single well-defined beam in a process called **specular reflection**. If, on the other hand, the surface is rough, although the angle of incidence will equal the angle of reflection for each ray, the whole lot of rays will emerge every which way, constituting what is called **diffuse reflection** (Fig. 25.12). Both of these conditions are extremes—the reflecting behavior of most surfaces lies somewhere between them (see the photo below).

25.6 The Plane Mirror

While standing in front of a plane (flat) mirror* (that is, any polished smooth surface), every point on your body that is illuminated by some external source scatters light, some of which heads toward the mirror (Fig. 25.13). Every molecule on your face that is driven by the incident E-field reradiates a more or less spherical wavelet or, if you prefer, sends out a cone of rays. Some portion of these wavelets will reflect from the mirror and subsequently reach your eye. But the eye-brain receiver, accustomed to straight-line propagation, will see the source as if it were behind the mirror. *The rays received by the observer diverge from the image point* and, as such, the image is said to be **virtual**—*it appears behind the mirror* and *cannot be projected onto a screen*. Every point on your hand, for instance, will have its corresponding image point behind the mirror, recreating the appearance of that hand (Fig. 25.14).

Figure 25.12 Diffuse reflection. When the surface roughness is large compared to λ, the scattering molecules are far apart, and there is no longer a direction in which all the wavelets will add constructively to produce a single reflected wave. Energy goes off in a broad range of directions as shown in the photo below.

Specular reflection. A laser beam reflected from a mirror in a well-defined beam. Note that the angle of incidence equals the angle of reflection. The laser is at the upper left in both photos.

Diffuse reflection. Here the surface is rough compared to λ, and the light (unable to interfere effectively) scatters in every direction. There is no well-defined reflected beam.

*Most mirrors in common use outside of the laboratory are back-silvered—they reflect light off a layer of metal *behind* the glass; that's done in order to protect the delicate reflecting coating. For simplicity, we'll draw only front-silvered mirrors of the type used in optics labs.

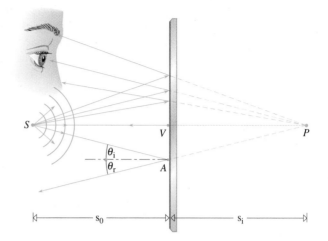

Figure 25.13 The light from an atom (S) on the face of someone looking into a mirror. Rays from S strike the mirror, $\theta_i = \theta_r$, and some reflect back to the viewer's eye. The eye-brain system apprehends the rays as if they were straight and being emitted from P, the image of S behind the mirror.

Figure 25.14 (a) The image of the left hand can be easily traced by using rays from it perpendicular to the mirror ($\theta_i = 0 = \theta_r$). They simply reflect back on themselves. A left hand becomes a life-size right hand. (b) The light from the young man's hand reflects both to his eye and the eye of the woman standing next to him. Both can therefore see the image of his hand behind the mirror.

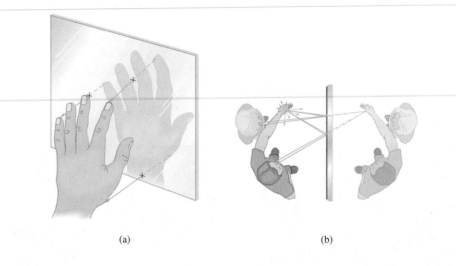

(a)   (b)

Example 25.2 Using Fig. 25.13, prove that the image is located the same distance behind the mirror as the object is in front of it. That is, the **object distance** s_0 equals the **image distance** s_i.

Solution: The exterior angle of triangle SAP equals $\theta_i + \theta_r$, and that equals the sum of the alternate inte-rior angles of that triangle; namely, $\angle VSA + \angle VPA$. But $\angle VSA = \theta_i = \theta_r$ and therefore $\angle VSA = \angle VPA$, which means that triangles VAS and VAP are congruent by angle-angle-side. It follows that $s_0 = s_i$, *the image of the source is the same perpendicular distance behind the mirror as the object is in front of it.*

The fact that the mirror is flat and that the angle of incidence equals the angle of reflection, combine to produce an undistorted, right-side up, life-size image standing just as far behind the surface as the object is in front of it. Realize that the image of a left hand (the outline of which can be determined using rays falling perpendicularly on the mirror as in Fig. 25.14a) is a right hand. Smack your left hand up against a mirror (the reflected fingers must be directly against the actual fingers)—the image must be a right hand. *A single reflection changes a right-handed system into a left-handed one and vice versa.*

Example 25.3   What is the length of the smallest vertical plane mirror in which you can see your entire standing body all at once, and how should it be positioned?

Solution: [Given: that $s_0 = s_i$, determine the minimum height of the mirror.] Refer to Fig. 25.15. A ray from your toe will enter your eye, striking point H somewhere such that $\angle DHC = \angle BHC$. Triangles CHD and CHB are congruent and so $\overline{GH} = \overline{HI} = \frac{1}{2}\overline{BD}$. Similarly, if you are to see the top of your head, $\overline{EF} = \overline{FG} = \frac{1}{2}\overline{AB}$. Thus, a mirror of length $\overline{FH}$ will do the job, where

$$\overline{FH} = \overline{FG} + \overline{GH} = \tfrac{1}{2}\overline{AB} + \tfrac{1}{2}\overline{BD} = \tfrac{1}{2}\overline{AD}$$

A mirror half your height with its upper edge lowered by half the distance between your eye and the top of your head serves nicely.

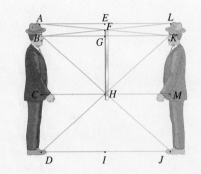

Figure 25.15   A determination of the smallest mirror ($\overline{FH}$) in which a person can see one's entire body.

Generally, the amount of light reflected from a surface increases as the angle of incidence increases. The amount reflected actually approaches 100% at glancing incidence where $\theta_i \approx 90°$. Furthermore, even the ability of a rough surface to reflect specularly improves as the light skims across it at glancing incidence. Hold this book horizontally, right at eye level so that the light from a lamp grazes off a page into your eye—you'll see a bright image of the bulb reflected in the paper.

REFRACTION

We saw earlier that as light propagates through a transparent homogeneous medium, there will be elastic scattering from the atoms. The initial beam, let's call it the *primary wave*, results in the emission of scattered wavelets. These, in turn, cancel each other in all but the forward direction, wherein they combine to form what we shall call the *secondary wave* (Fig. 25.4). The result is two waves (the primary and secondary), not necessarily in-phase with each other, overlapping and propagating together as one net disturbance—the **refracted wave**.

Before we press on, realize that *both waves travel at* c: **photons do not exist at any speed other than** c. *When light traverses any material, it travels in the interatomic void.* And yet, if we were to measure the speed of a macroscopic beam of light in a material, we would generally find some value other than c! All that atomic absorbing and re-emitting has the effect of producing a net speed, which can be either greater than, less than, or equal to c, depending on the medium and the frequency of the radiant energy. This process is marvelously subtle, and not every aspect of it can be fully developed here; still, it's so fundamental that it must be examined, if only superficially.

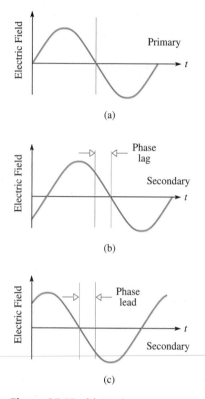

Figure 25.16 (a) A primary wave and two possible secondary waves. In (b), the secondary wave lags behind the primary—it takes longer to reach any given value. In (c), the secondary wave reaches any given value at an earlier time than the primary; that is, it leads the primary.

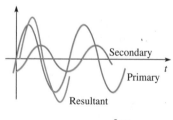

Figure 25.17 If the secondary lags the primary, the resultant will also lag it and vice versa.

25.7 The Index of Refraction

The crucial feature here is that the secondary wave is generally somewhat out-of-phase with the primary wave. What happens is that the primary wave drives the molecular oscillators into a ground-state vibration. But this vibration will not usually follow in step with the driver; there will be a phase difference, as we saw earlier when we studied Slinkies (p. 460). When the driving frequency is much below the oscillator's resonant frequency, there is only a slight phase difference. But that difference increases as the frequency of the incident beam increases toward the resonance. The result is the emission by every driven oscillator, of a scattered wavelet, each of which is out-of-phase by the same amount (from 0° to 180°) with respect to the primary wave. To this must be added another phase shift of 90°, which arises between the oscillators and the reconstituted secondary wave. We needn't worry about all the details; what's of interest is that *for a primary wave with a frequency below resonance, the secondary wave lags behind it, and for a primary wave above resonance, the secondary wave leads it* (Fig. 25.16).

The refracted wave is the sum of the primary wave and this phase-shifted secondary wave. Remember that the phase speed of any wave is the rate at which any point of constant phase, like a crest, moves. But adding the secondary wave to the primary has the effect of changing the phase. Figure 25.17 shows how the phase of the refracted wave can be advanced or retarded, and that is exactly equivalent to increasing or decreasing its average speed by some small amount. Notice that a small secondary wave produces only a small phase shift. When the phase is retarded, a crest will get to some point beyond it in the medium just a little later (at a larger value of t) than it would have otherwise. A phase lag means a time lag (the crest will take longer to get anywhere in the forward direction), and that means a speed $v < c$. Keep in mind that an electromagnetic wave is the averaged macroscopic manifestation of a torrent of photons. Such a wave effectively slows down as it enters the medium, even though the individual photons, without which there is no wave, only travel at c.

X-rays have frequencies in excess of the electron resonances and so generally $v > c$. On the other hand, in transparent media, light frequencies are most often lower than the electron resonances of the medium, and $v < c$. This situation ($v < c$) is usually the case, but it most assuredly is not always the case, especially for colored materials that have resonances in the visible range. *Because of the ongoing process of absorption and re-emission, electromagnetic waves propagate through material media at speeds other than c.* The dependence of the speed of propagation on the frequency of the radiant energy is known as **dispersion**; it's responsible for the separation of white light into its constituent colors via a prism.* The ratio of the speed of an electromagnetic wave in vacuum to that in a medium is defined as the **index of refraction** n; accordingly

$$n = \frac{c}{v} \tag{25.2}$$

*One need not fret about violating Relativity Theory with speeds in excess of c—these are phase speeds, there is no modulation and the waves carry no information. If they were modulated somehow, the signal would travel at a reduced rate known as the *group velocity*.

Keep in mind that *n* always varies somewhat with the frequency of the illumination. For most transparent materials, this variation, although significant, is not very large across the visible spectrum, and so it's common to use a single value of *n* (see Table 25.1) for such a substance illuminated with light of any frequency (Fig. 25.18). However, as *f* approaches that of an atomic resonance, *n* changes drastically, and there is a large increase in dissipative absorption as the amplitudes of the atomic oscillations increase. That's what happens to ordinary glass in the UV and IR (and it's why you can't get a tan behind a window that you can easily see through).

Example 25.4   What is the apparent speed of light (589 nm) in diamond?

Solution: [Given: from Table 25.1, $n_d = 2.42$. Find: v.] From the definition of the index of refraction, Eq. (25.2), we have

$$v = \frac{c}{n_d} = \frac{3.00 \times 10^8 \text{ m/s}}{2.42} = \boxed{1.24 \times 10^8 \text{ m/s}}$$

▶ **Quick Check:** The index of vacuum is 1, and since diamond has an index of around 2.4, the speed in diamond is 2.4 times slower than c ≈ 3 × 10⁸ m/s, or roughly 1 × 10⁸ m/s.

TABLE 25.1	Approximate indices of refraction of various substances*
Air	1.000 29
Ice	1.31
Water	1.333
Ethyl alcohol (C_2H_5OH)	1.36
Fused quartz (SiO_2)	1.458 4
Carbon tetrachloride (CCl_4)	1.46
Turpentine	1.472
Benzene (C_6H_6)	1.501
Plexiglass	1.51
Crown glass	1.52
Sodium chloride (NaCl)	1.544
Light flint glass	1.58
Polystyrene	1.59
Carbon disulfide (CS_2)	1.628
Dense flint glass	1.66
Lanthanum flint glass	1.80
Zircon ($ZrO_2 \cdot SiO_2$)	1.923
Fabulite ($SrTiO_3$)	2.409
Diamond (C)	2.417
Rutile (TiO_2)	2.907
Gallium phosphide	3.50

*Values vary with physical conditions—purity, pressure, etc. These correspond to a wavelength of 589 nm.

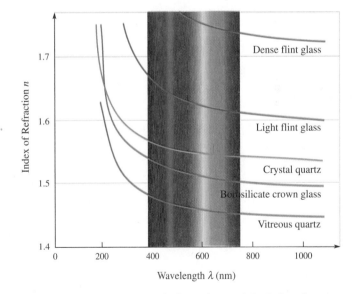

Figure 25.18   The wavelength dependence of the index of refraction for various materials. Transparent substances such as these usually have resonances in the UV (the region below 380 nm) and in the IR (far off to the right), which is why all of the curves here rapidly rise in the UV. That means that glass will dissipatively absorb UV. Can you guess why UV lamps are made of vitreous quartz rather than glass?

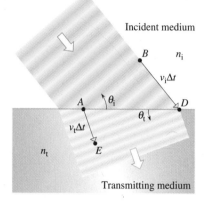

Figure 25.19 The wave picture of refraction. The atoms in the region of the surface of the transmitting medium reradiate wavelets that combine constructively to form a refracted beam.

25.8 Snell's Law

Let's examine what is happening to the transmitted beam in Fig. 25.8. Each molecule radiates wavelets into the glass that expand out at the speed c. These can be imagined as combining into a secondary wave that then recombines with the unscattered remainder of the incoming beam, the primary wave, to finally form the refracted wave. However we visualize it, immediately on entering the transmitting medium, there is a single net field, a single net wave—the refracted, or transmitted, light. As we have seen, this transmitted wave usually propagates with an effective speed $v_t < c$.

Fig. 25.19 picks up where we left off with Fig's. 25.8 and 25.10. The diagram depicts several wavefronts, all shown at a single instant in time. Remember that each wavefront is a surface of constant phase and, to the degree that the phase of the net field is retarded by the transmitting medium, each wavefront is held back, as it were. The wavefronts bend as they cross the boundary because of the speed change. Alternatively, we can envision Fig. 25.19 as a multiple-exposure picture of a single wavefront showing it after successive equal intervals of time. Notice that in the time Δt, which it takes for point B on a wavefront (traveling at speed v_i) to reach point D, the transmitted portion of that same wavefront (traveling at speed v_t) has reached point E. If the glass ($n_t = 1.5$) is immersed in an incident medium that is vacuum ($n_i = 1$) or air ($n_i = 1.000\,3$) or anything else where $n_t > n_i$, $v_t < v_i$ and $\overline{AE} < \overline{BD}$, the wavefront bends. The refracted wavefront extends from E to D, making an angle with the interface of θ_t. As before, the two triangles ABD and AED in Fig. 25.19 share a common hypotenuse ($\overline{AD}$), and so

$$\frac{\sin \theta_i}{\overline{BD}} = \frac{\sin \theta_t}{\overline{AE}}$$

where $\overline{BD} = v_i \Delta t$ and $\overline{AE} = v_t \Delta t$. Hence

$$\frac{\sin \theta_i}{v_i} = \frac{\sin \theta_t}{v_t}$$

Multiply both sides by c, and since $n_i = c/v_i$ and $n_t = c/v_t$

$$n_i \sin \theta_i = n_t \sin \theta_t \tag{25.3}$$

This equation is the first portion of the **Law of Refraction**, also known as **Snell's Law**. At first, the indices of refraction were simply experimentally determined constants of the physical media. Later on, Newton was actually able to derive Snell's Law using his own corpuscular theory. By then, the significance of n as a measure of the speed of light was evident. Still later, Snell's Law was shown to be a natural consequence of Maxwell's Electromagnetic Theory.

It is again convenient to transform the diagram into a ray representation (Fig. 25.20) wherein all the angles are measured from the perpendicular. Along with Eq. (25.3), there goes the understanding that *the incident, reflected, and refracted rays all lie in the plane-of-incidence*. When $n_i < n_t$ (that is, when the light is initially traveling within the lower-index medium), it follows from Snell's Law that $\sin \theta_i > \sin \theta_t$, and since the sine function is everywhere positive between $0°$ and $90°$, then $\theta_i > \theta_t$. Rather than going straight through, *the ray entering a higher-index medium bends toward the normal* (Fig. 25.21a). The reverse is also true; that is, on entering a medium having a lower index, the ray, rather than going straight through, will bend *away* from the normal (Fig. 25.21b). Notice that this im-

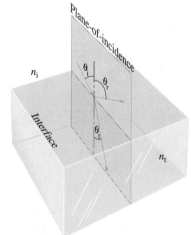

Figure 25.20 The incident, reflected, and transmitted beams all lie in the plane-of-incidence. The incident medium, containing the incoming beam, has an index of refraction n_i. The transmitting medium has an index n_t.

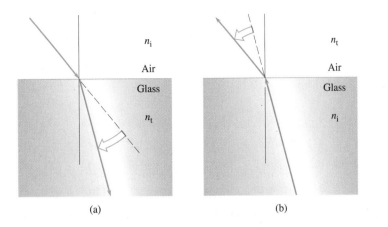

Figure 25.21 (a) When a beam of light enters a more optically dense medium, one with a greater index of refraction ($n_i < n_t$), it bends toward the perpendicular. (b) When a beam goes from a more dense to a less dense medium ($n_i > n_t$), it bends away from the perpendicular.

plies that the rays will traverse the same path going either way, into or out of either medium—the arrows can be reversed and the resulting picture is still true.

Example 25.5 Suppose that the beam of light in Fig. 25.22 travels in air before it's incident on the glass plate ($n_g = 1.5$) at 60°. (a) At what angle will it be transmitted into the block? (b) Show that the beam emerges from the far side parallel to the original incoming light.

Solution: [Given: an air-glass interface and $\theta_i = 60°$. Find: θ_{t1} and θ_{t2}, using Fig. 25.22.] (a) Applying Snell's Law at the first interface

$$n_a \sin \theta_{i1} = n_g \sin \theta_{t1}$$

and

$$\sin \theta_{t1} = \left(\frac{n_a}{n_g}\right) \sin \theta_{i1} = \left(\frac{1.0}{1.5}\right) \sin 60° = 0.577$$

Hence $\theta_{t1} = 35.3°$, that is, $\boxed{35°}$. (b) Applying Snell's Law to the second interface where $\theta_{t1} = \theta_{i2}$, we obtain

$$n_g \sin \theta_{i2} = n_a \sin \theta_{t2}$$

and

$$\sin \theta_{t2} = \frac{n_g}{n_a} \sin \theta_{i2} = \frac{1.5}{1.0} \sin 35.3° = 0.866$$

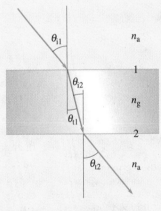

Figure 25.22 Light incident on a parallel plate of glass of index n_g immersed in air of index n_a.

Hence $\theta_{t2} = 60°$, and the light emerges parallel to the incident beam.

▶ **Quick Check:** On crossing the first interface, the ray bent toward the normal, which is what it should have done. It emerged parallel to the incident ray, which it must do if the interfaces are parallel, and that's a check of the calculation in itself.

Fig. 25.19 illustrates the three important changes that occur in the beam traversing the interface. (1) It changes direction. Because the leading portion of the wavefront in the glass slows down, the part still in the air advances more rapidly, sweeping past and bending the wave toward the normal. (2) The beam in the glass has a broader cross section than the beam in the air; hence, the energy is spread thinner. (3) The wavelength decreases because the frequency is unchanged while the speed decreases, $\lambda = v/f$. This latter notion suggests that the color aspect of light is better thought of as its frequency (or energy, $E = hf$) than its wavelength, which changes with the medium through which the light moves. When we do talk about wavelengths and colors, we should always be referring to *vacuum wavelengths* (henceforth to be given as λ_0).

Example 25.6 Someone is wearing a scarlet-red bathing suit that reflects primarily $\lambda_a = 629$ nm in air ($n_a \approx 1.00$). What is the corresponding wavelength in water ($n_w = 1.33$)? Do bathing suits change color underwater? Explain.

Solution: [Given: $\lambda_a = 629$ nm for $n_a = 1.00$. Find: λ_w when $n_w = 1.33$.] $n_a = c/v_a = c/f\lambda_a$; hence, $f = c/n_a\lambda_a = 4.77 \times 10^{14}$ Hz. Using the same reasoning, in water we have

$$\lambda_w = \frac{c}{n_w f} = \frac{3.00 \times 10^8 \text{ m/s}}{1.33(4.77 \times 10^{14} \text{ s}^{-1})} = \boxed{473 \text{ nm}}$$

which, were it in vacuum, would be seen as blue. Of course, the suit still appears red in water.

▶ **Quick Check:** Since wavelength decreases as index increases, $\lambda_w/\lambda_a = n_a/n_w$, and $\lambda_w = \lambda_a(1/1.33) = 473$ nm.

Notice that had we substituted $(\lambda_0 f)$ for c in Example 25.6, we would have gotten

$$\lambda = \frac{\lambda_0}{n} \tag{25.4}$$

for the wavelength in any medium of index n.

The fact that rays leaving a medium of higher index bend away from the perpendicular gives rise to the familiar effect in which distances within liquids appear foreshortened when viewed from above. Figure 25.23 shows a fish underwater a real distance down d_R, which seems to be at an apparent depth d_A. These distances can be related in a particularly simple way if we limit the problem to one where the fish is not far away horizontally from the viewer; that is, x is small compared to the depth. Then θ_i and θ_t are both small. For small angles, the cosine is approximately 1.0 and the tangent then equals the sine:

$$\sin \theta_i \approx \tan \theta_i = \frac{x}{d_R} \qquad \text{and} \qquad \sin \theta_t \approx \tan \theta_t = \frac{x}{d_A}$$

It follows from Snell's Law that

$$\frac{n_i x}{d_R} = \frac{n_t x}{d_A}$$

and

$$\frac{n_i}{n_t} = \frac{d_R}{d_A}$$

which is why a pencil seems to bend when it's dipped into water—the immersed end appears higher than it should in comparison to the portion remaining in the air.

In all the situations thus far treated, it was assumed that the reflected and refracted beams always had the same frequency as the incident beam, and ordinary experience tells us that that is a very reasonable assumption. Light of frequency f impinges on a medium and presumably drives

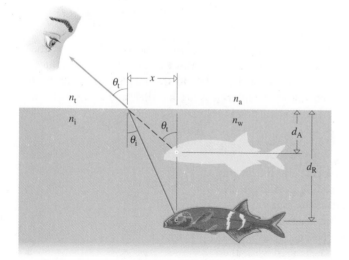

Figure 25.23 When light emerges from water into air ($n_w > n_a$), the ray bends away from the perpendicular. Anything below the surface appears higher in the liquid than it actually is, which makes spear fishing tougher than it might otherwise seem.

the molecules into simple harmonic motion. That's certainly the case when the amplitude of the vibration is fairly small, as it is when the electric field driving the molecules is small. The E-field for bright sunlight is only about 1000 V/m (while the B-field is less than a tenth of the Earth's surface field). This isn't very large compared to the fields keeping a crystal together, which are of the order of 10^{11} V/m, just about the same magnitude as the cohesive field holding the electron in an atom. We can usually expect the oscillators to vibrate in simple harmonic motion and so the frequency will remain constant—the medium will ordinarily respond linearly. That will not be true, however, if the incident beam has an exceedingly large-amplitude E-field, as can be the case with a high-power laser. So driven, at some frequency f the medium can behave in a nonlinear fashion, resulting in reflection and refraction of harmonics ($2f$, $3f$, etc.) in addition to f. Nowadays, second-harmonic generators are available commercially; you shine red light (694.3 nm) into an appropriately oriented transparent nonlinear crystal (of, for example, potassium dihydrogen phosphate, KDP, or ammonium dihydrogen phosphate, ADP) and out will come a beam of UV (347.15 nm).

25.9 Total Internal Reflection

Often a beam of light originates in air (a low-index medium) and impinges on glass or water (a high-index medium). The reverse is also possible. When we look at a swimming fish, the reflected light from the fish originates in the water (a high-index medium) and impinges on the air (a low-index medium). When the light is incident on an interface where $n_i > n_t$, a curious and rather important thing happens. As the angle of incidence is made larger and larger (Fig. 25.24), the transmitted beam bends more and more away from the normal and toward the interface. As that occurs, the transmitted beam grows weaker, and the reflected beam—the beam traveling back into the higher-index medium—becomes stronger. When a particular incident angle is reached, known as the **critical angle** (θ_c), all the light striking the interface will be reflected back; no light will be transmitted. This **total internal reflection** continues for all incident angles greater than the critical angle. If, for example, $n_i = 2$ while $n_t = 1$, Snell's Law maintains that $2 \sin \theta_i = \sin \theta_t$ and, pro-

The rays of light from the submerged portion of the pencil bend on leaving the water as they rise toward the viewer.

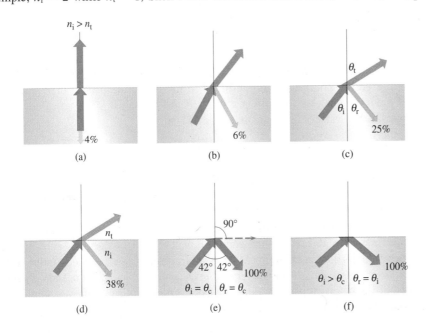

Figure 25.24 Internal reflection ($n_i > n_t$). As the incident angle increases, more and more light is reflected. (a) When $\theta_i = 0$, only about 4% of the light is reflected, while $\approx$96% is transmitted. (b) (c) (d) As θ_i increases, more light is reflected and less transmitted. (e) At the critical angle ($\theta_i = \theta_c$), all the light is internally reflected. (f) At values of θ_i greater than θ_c, the incident beam continues to be totally internally reflected.

The prism behaves like a mirror and reflects a portion of the pencil (reversing the lettering on it). The operating process is total internal reflection.

vided $\sin \theta_i \leq \frac{1}{2}$, the equation can be satisfied. But for incident angles greater than this critical value of $\theta_i = 30°$, the $\sin \theta_t$ would have to be greater than 1, and that cannot be—there is no transmitted beam.

Snell's Law yields

$$\sin \theta_i = \frac{n_t}{n_i} \sin \theta_t$$

but when $\theta_i = \theta_c$, we know that $\theta_t = 90°$, then $\sin 90° = 1$, and we get the defining equation

$$\sin \theta_c = \frac{n_t}{n_i} \qquad (25.5)$$

For incident angles equal to or greater than the critical angle, 100% of the energy in the incoming beam is reflected back into the incident medium, and no energy is transmitted. Recall that at the critical angle the transmission angle becomes 90°; the "transmitted" light propagates along the interface. Because that wave is limited to the boundary, its energy flows back and forth across the interface with no average transmission into the second medium.

Example 25.7 Imagine a beam of light traveling in a block of glass for which $n_g = 1.56$. Now suppose the light comes to the end of the block and impinges on an air-glass interface. What is the minimum incident angle that will result in all the light being reflected back into the glass?

Solution: [Given: $n_i = n_g = 1.56$, $n_t = n_a = 1.00$. Find: θ_c.] From Eq. (25.5), we have

$$\sin \theta_c = \frac{1.00}{1.56} = 0.641$$

and

$$\boxed{\theta_c = 39.9°}$$

▶ **Quick Check:** For $n_t = 1.5 = 3/2$, $\sin \theta_c = 2/3$ and $\theta_c = 42°$, so the above result is certainly reasonable.

The fact that the critical angle for air-glass typically ranges from 36° to 43° (depending on the kind of glass) is utilized in the reflecting prisms of Fig. 25.25. This is a convenient way to redirect a beam, and it's used in cameras, binoculars, telescopes, and a variety of other optical devices.

Fiberoptics

Easily the most important application of total internal reflection is **fiberoptics**. Techniques have evolved in recent times for efficiently conducting light and near-IR (1300 nm) energy along thin, transparent, dielectric fibers. Figure 25.26 depicts a glass (or plastic) fiber having a diameter of, say, 50 μm, just about the thickness of a human head-hair. Because of the small diameter, most of the light entering one face goes on to strike the cylindrical wall at a large angle and is subsequently totally internally reflected. This occurs over and over again, typically thousands of times per foot, as the beam propagates along the fiber. The smooth surface of a single filament must be kept clean of contamination, or the boundary conditions change and

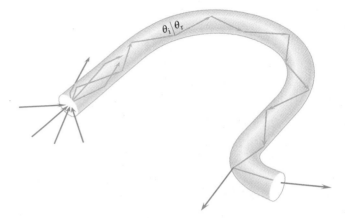

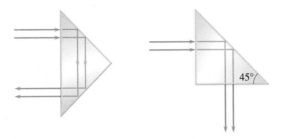

Figure 25.25 The rays strike the walls of these prisms at 45°, which is in excess of the critical angle ($36° \lesssim \theta_c \lesssim 43°$). The rays are then totally internally reflected.

Figure 25.26 Light captured within a thin transparent fiber. Note that if a bend is too tight $\theta_i < \theta_c$, and some light leaks out. Similarly, if a ray enters at too large an angle, it will leak out.

light leaks out at those spots. Accordingly, each fiber core is usually enshrouded in a transparent sheath of a lower-index material called a *cladding*.

All such thin filaments are quite flexible and, if not bent too tightly, will transmit light through twists and turns with relatively little loss. Bundles of free fibers whose ends are bound together and then polished form flexible light guides. If no attempt is made to keep the fibers in an ordered array so that the two end-faces are each a different jumble, the thing is called an *incoherent bundle*. These are light carriers. Easy to produce and inexpensive, they are used for remote light sensing and illumination. They conduct light, for example, from a conveniently positioned bulb to some otherwise inaccessible place such as an instrument panel in an airplane or deep into a human body.

Conversely, when the fibers are arranged so that their terminations on both end-faces are exactly the same, the bundle is said to be *coherent,* and we have a flexible image carrier. Frequently, these bundles are tipped off with a small lens so that they need not be in contact with the surface being viewed. Today, it is commonplace to use fiberoptic apparatus to poke into all sorts of unlikely places from nuclear reactor cores and jet engines to stomachs and reproductive organs. When a device is used to examine internal body cavities, it's called an *endoscope,* and there are bronchoscopes, colonoscopes, gastroscopes, and so on. An additional incoherent bundle incorporated into the device usually supplies the illumination.

The world is now in the first stages of a new era of optical communications, of light flashing along fibers, replacing electricity moving in metal wires—not for transmitting power, but information. The much higher frequencies of light allow for an incredible increase in data-handling capacity. For example, using some sophisticated transmitting techniques, a pair of copper telephone wires can be made to carry up to about two dozen simultaneous conversations. To get a feel for how much information that is, consider the fact that a single ordinary TV transmission is equivalent to about 1300 simultaneous phone conversations, which, in turn, is roughly the equal of sending some 2500 typewritten pages each and every second! So, at present, it's quite impractical to attempt to send television over copper phone lines. By comparison, it's already possible to transmit in excess of 12 000 simulta-

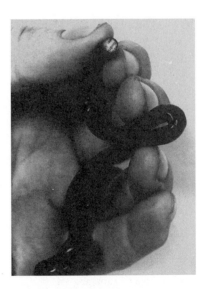

A bundle of thousands of carefully arranged thin glass fibers transmitting an image. One end of the bundle rests on a page of print, and the other end reveals the words below, despite the knots and bends.

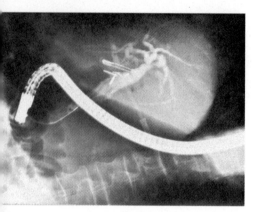

A colonoscope being used to examine a colon for cancer.

neous conversations over a single pair of fibers, which is more than nine TV channels—and this is only the beginning; the technology is in its infancy. Achieved capacities to date don't even begin to approach the theoretical limit. A pair of fibers will someday connect your home to a vast network of communications and computer facilities that will make the era of the copper wire seem charmingly primitive.

THE WORLD OF COLOR

We are now in a position to explore why the objects in our environment look the way they do. What makes paper white and coal black? Why is silver gray and grass green?

25.10 White, Black, and Gray

A reflecting surface is white when it diffusely scatters a broad range of frequencies under white-light illumination. This page is white. So are salt and snow. Clouds are white, as are powders and pills and cloth. Even the foam on a glass of beer is white, steam and soap bubbles and chalk and sugar and gray hair are white—and all this diversity shares a common structural similarity. All of these objects are composed of transparent grains or particles or fibers or bubbles that are large in size compared with the wavelength of light, but otherwise quite small. Each is much too large to produce Rayleigh scattering, too large to act as a single oscillator preferentially scattering at the higher frequencies. Despite the common opinion that things white are opaque, each grain or fiber is actually transparent. There is no such thing as a single particle of white pigment that by itself is opaque white. Materials that have no atomic resonances in the visible are transparent to light of all frequencies—they do not appear colored because they do not absorb any range of colors. Sprinkle a single layer of salt or sugar on this page, and the print will still be legible through the transparent crystals. Crush a piece of clear glass and the grains, like grains of sand, will appear white. Water is clear, but snow and steam and clouds are white.

Whiteness arises when the incident white light is reflected back out of the medium in all directions as if from countless point sources, from scatterers that show no preferential absorption. All colors of light come in, and all colors scatter out. When the medium is composed of many small transparent bodies, like the matted fibers of this page or the random jumble of sugar crystals on a spoon, each reflects some light, transmits some light, and reflects again (Fig. 25.27). And that process happens over and over, layer upon layer, with much of the light eventually reflected back into the general direction from which it came. Slightly lift a few pages of this book and peep under them—lots of light will be transmitted through the pages but, like a cloud, if the mass is thick enough, very little light will emerge and it will appear gray or black.

The fraction of light energy reflected at each surface depends on the difference between the indices of the two media—if there is no difference, the boundary essentially vanishes. *The closer the indices are to each other, the less light will be reflected,* and the less luminous will be the surface, Accordingly, to make white paint, we need only mix a powdered transparent material, the pigment, with a clear vehicle (such as oil or acrylic). If the refractive indices are equal, the pigment simply vanishes, and the paint is transparently useless. If the indices differ appreciably, there will be strong reflections at the countless surfaces and the paint will be a brilliant white. A really splendid white surface can reflect up to 98% of the incident light.

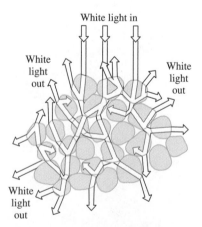

White light in

White light out

White light out

White light out

Figure 25.27 A number of transparent particles, which are large compared to the wavelengths of light. These might be cloth fibers, sugar crystals, talcum powder, or snow-flakes. White light enters, is reflected and refracted out.

Because of the low index of air, small particles such as salt, talc, and sugar, as well as all kinds of undyed fibers, surrounded by air appear bright white. That brightness will change tremendously if these fibers are immersed in a substance having an index between that of fiber and air, a substance like water. What happens to the bright whiteness of a tissue or a piece of cloth when it becomes wet?

A diffusely reflecting surface that absorbs somewhat uniformly right across the spectrum will reflect a bit less than a white surface and so appear matt gray. The less it reflects, the darker the gray, until it absorbs almost all the light and appears black. A surface that reflects perhaps 70% or 80%, but reflects specularly, will appear the familiar shiny gray of a typical metal. Metals possess tremendous numbers of free electrons that scatter light very effectively independent of frequency—the free electrons are not bound to the atoms and have no associated resonances. The amplitudes of the vibrations are an order-of-magnitude larger than they are for the bound electrons. The incident light cannot penetrate into the metal any more than a fraction of a wavelength or so before it's canceled completely. There is little or no refracted light—most of the energy is reflected out; the small remainder is absorbed.

If the metal is thin enough (only a few atom layers thick), it will transmit light and one can see through it. Look at a bright lamp through a CD. Nowadays, "two-way" mirrors that can be seen through are a common security device. A number of snack food products are packaged in shiny metal-coated plastic films. Hold one up to your eyes and look through it at a bright light. Note that the primary difference between a gray surface and a mirrored surface is one of diffuse versus specular reflection.

25.11 Colors

When the distribution of energy in a beam of light is fairly uniform across the spectrum, the light appears white; when it is not, the light usually appears colored. Figure 25.28 depicts typical frequency distributions for what would be perceived as red, green, and blue light. These curves show the dominant frequency regions, but beyond that there can be a great deal of variation in the distributions, and they will still provoke the eye-brain responses of red, green, and blue. Thomas Young, in the early 1800s, showed that a broad range of colors can be generated by mixing three beams of light, provided their frequencies were widely separated. When three such beams combine to produce white light, they are called **primary colors**. There is no single unique set of these primaries, nor do they have to be monochromatic. Since the widest range of colors can be created by mixing *light beams* of red (R), green (G), and blue (B), these tend to be the most commonly used. They are the three components (emitted by three phosphors) that generate the whole gamut of hues seen on a color TV set (p. 956). Keep in mind that at the moment we are talking about mixing light beams and not paint pigments.

Figure 25.29 summarizes the result of overlapping these three primaries in a number of different combinations: red plus blue is *magenta* (M), a reddish blue; blue plus green is *cyan* (C), a bluish green; and, most surprising, red plus green is *yellow* (Y). And the sum of all the three

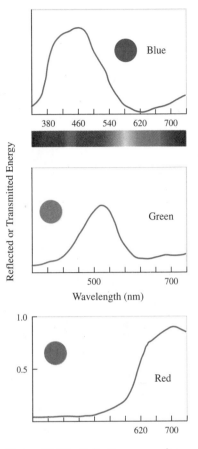

Figure 25.28 Reflection curves for blue, green, and red pigments are typical, but there is a great deal of variation within each color.

A white screen illuminated with red, blue, and green light appears white. Anyone standing between a lamp and the screen casts a shadow that has the complementary color.

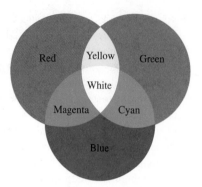

Figure 25.29 Three overlapping beams of colored light. A color TV set uses the same three primary light sources—red, green, and blue.

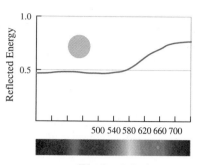

Figure 25.30 Spectral reflection of a pink pigment. The broad range of wavelengths produces a white background; adding the red peak yields pink. A shirt that reflects about half the light incident on it over most of the spectrum, but reflects even more in the red, looks pink.

primaries is white:

$$R + B + G = W$$
$$M + G = W, \quad \text{since} \quad R + B = M$$
$$C + R = W, \quad \text{since} \quad B + G = C$$
$$Y + B = W, \quad \text{since} \quad R + G = Y$$

Any two colored light beams that together produce white are said to be **complementary**, and the last three symbolic statements exemplify that situation. Now suppose we overlap a beam of magenta and a beam of yellow, represented symbolically by

$$M + Y = (R + B) + (R + G) = W + R$$

The result is pink, a combination of white and red. And that raises another point: we say that a color is **saturated**, that it is deep and intense, when it does not contain any white light. As can be seen in Fig. 25.30, pink is unsaturated red; red superimposed on a background of white.

Imagine a piece of yellow stained glass; that is, glass having a resonance in the blue, which it strongly absorbs. Looking through it at a white-light source composed of red, green, and blue, the glass would absorb blue, passing red and green, which is yellow (Fig. 25.31). Yellow cloth, paper, dye, paint, and ink all selectively absorb blue and reflect what remains—yellow—and that's why they appear yellow. And if you peer at something that is a pure blue through a yellow filter, the object will appear black. Here, the filter or the paint colors the light yellow by removing blue, and we speak of the process as *subtractive* coloration. On the other hand, *additive* coloration can only result from the overlapping of light beams. Red light and green light makes yellow; red paint and green paint makes brown.

A wide range of colors (including red, green, and blue) can be produced by passing light through various combinations of magenta, cyan, and yellow filters (Fig. 25.32). Magenta, cyan, and yellow are the primary colors of subtractive mix-

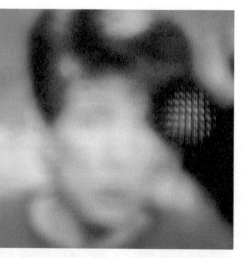

A color TV screen is made up of red, blue, and green regions that can be made visible with a magnifying glass.

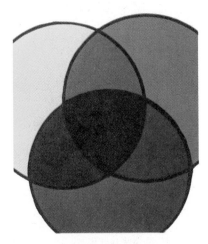

Filters of magenta, yellow, and cyan. These absorb certain frequencies and the process is called subtractive coloration. Where all three overlap, no light is passed.

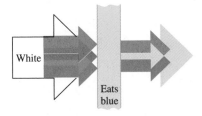

Figure 25.31 Yellow stained glass absorbs blue light, passing yellow. The incoming white beam corresponds to R + B + G. Subtracting blue yields R + G, which is yellow.

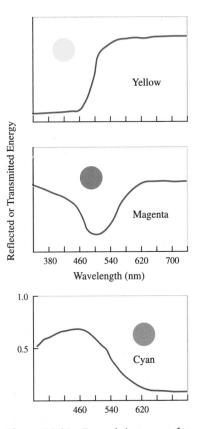

Figure 25.32 Transmission curves for colored filters. Or reflection curves for colored paints.

ing. (They are the primaries of the paint box, although they are very often mistakenly spoken of as red, blue, and yellow.) They are the basic colors of the dyes used to make photographs and the inks that printers use to print them. To color paint or cloth or anything else, we need only surround the transparent grains or fibers with a colored filter, a dye layer. White light passes through the film of dye, loses a select portion of the spectrum due to dissipative absorption, and emerges with the complementary color. Ideally, if you mix all the subtractive primaries together (either by combining paints or by stacking filters), you get no color, no light, just blackness. Each removes a region of the spectrum, and together they absorb it all. Selective absorption is the principal mechanism that colors the world from green grass to red roses to pink lips.

Core Material

The fundamental mechanism underlying the propagation of light through matter is **scattering**—the absorption and re-emission of radiant energy. *A ray is a line drawn in space corresponding to the direction of flow of radiant energy* (p. 942).

When a beam of light reflects from a smooth surface

$$\theta_i = \theta_r \qquad [25.1]$$

The angle of incidence equals the angle of reflection. This equation is the first part of the **Law of Reflection**. The second part maintains that *the incident ray, the perpendicular to the surface, and the reflected ray all lie in a plane* called the **plane-of-incidence**.

The ratio of the speed of an electromagnetic wave in vacuum to that in matter is the *index of refraction n*:

$$n = \frac{c}{v} \qquad [25.2]$$

The **Law of Refraction**, or **Snell's Law**, is

$$n_i \sin \theta_i = n_t \sin \theta_t \qquad [25.3]$$

where *the incident, reflected, and refracted rays all lie in the plane-of-incidence*. When light enters a refracting medium, its wavelength changes, and

$$\lambda = \frac{\lambda_0}{n} \qquad [25.4]$$

provides the wavelength in any medium of index *n*.

When the light impinges on an interface where $n_i > n_t$, there will be a special incident angle, the **critical angle** (θ_c), at which all the incoming light will be reflected back and none will be transmitted. This **total internal reflection** continues for all incident angles greater than the critical angle defined by

$$\sin \theta_c = \frac{n_t}{n_i} \qquad [25.5]$$

The widest range of colors can be created using a mix of light beams of red (R), green (G), and blue (B). Red plus blue is *magenta* (M); blue plus green is *cyan* (C); and red plus green is *yellow* (Y). And the sum of all three (p. 956) is white.

Suggestions on Problem Solving

1. The equation for the apparent depth of an object in a liquid (Fig. 25.23) can be remembered by keeping in mind that the ratio of indices equals the ratio of depths: smaller-to-larger, (or possibly, larger-to-smaller). Thus, for example, when air is above water, it's the index of air to the index of water as the apparent depth is to the real depth.

2. An error commonly made when dealing with total internal reflection is to substitute the wrong values in for the two indices—most often they are simply interchanged. Remember here

$\sin \theta_c = n_t/n_i$ and $n_i > n_t$. The index ratio must be less than one.

3. When dealing with problems involving plane mirrors, always draw a diagram. The central notion is that the angle of incidence equals the angle of reflection—all the rest is geometry. Often it is useful to construct rays that touch the very edges of the mirror and thereby define the limits of the system.

4. When treating a problem that pertains to wavelengths, remember to distinguish between the vacuum wavelength and the wavelength in the medium ($\lambda_0 > \lambda$). Keep in mind that $n = \lambda_0/\lambda$.

Discussion Questions

1. (a) Describe the kind of reflection that is taking place off this page. (b) Why is it that glossy paints can be deeper and richer in color than flat paints? (c) Why do the colors on a wet watercolor painting look so much more vivid than when they dry?

2. In what regard can we say that both reflection and refraction are manifestations of scattering?

3. Shaving lather is often substituted for meringue in pie-throwing fests. Why do they look so similar?

4. Suppose you have the bad luck to own a car whose exhaust has a bluish tint to it. What can you say about the effluvium?

5. (a) What is the angle of reflection for a ray incident normally on a smooth surface? (b) What is the angle of refraction for a beam striking an air-glass interface perpendicularly?

6. Modern picture frames sometimes come with "nonreflecting" glass (that is, glass whose surface has been deliberately made rough). What does it actually accomplish?

7. Is Venus in Velasquez's painting *The Toilet of Venus* (Fig. Q7) looking at her own face? Draw a ray diagram and in it drop a perpendicular from her nose to the plane of the mirror.

8. How is it that you can see your image in the surface of a polished *black* automobile? Explain what's happening. What allows you to see the gin in a glass, or the glass itself, for that matter?

9. The girl in Manet's painting, *The Bar at the Folies-Bergères* (Fig. Q9), is standing in front of a large plane mirror. We see reflected in it her back and the face of a man she seems to be talking to. From the Law of Reflection what, if anything, is amiss?

10. A properly cut diamond has a brilliant, lustrous appearance arising from the fact that most of the light entering the stone re-emerges. Explain how this happens and why diamond is especially well suited for the business.

11. Glass is often colored by adding to the silica, ions of the transition elements (Cr, Mn, Fe, Co, Ni, and Cu), usually in the

Figure Q7 *The Toilet of Venus* by Diego Rodriguez de Silva y Velasquez—National Gallery, London.

form of oxides—all have a partially filled electron shell. Ordinary window glass contains, as contaminants, traces of iron oxides (in both ferrous and ferric forms) that have weak absorptions in the red and blue and a strong resonance in the near-UV. What radiation can you expect will traverse a sheet of glass? What color do you expect the glass to have if seen edge-on? Explain your answers.

12. Can a magenta light beam always be separated by a prism into red and blue? Similarly, can a cyan light beam always be separated into green and blue?

13. (a) What "color" will a red apple appear under red-light illumination? (b) What "color" will a cyan ink appear under the

Figure Q9 *The Bar at the Folies-Bergères* by Edouard Manet—Courtauld Institute Galleries. London.

Figure Q16

Figure Q18

same conditions? (c) Why is a yellow filter often used when taking black-and-white photos of the clouds in the sky?

14. Which colored light appears the most saturated—the blue of the sky, the red of a traffic signal, or the pink of a rose?

15. (a) What "color" will a sheet of white paper appear under 650-nm illumination? (b) Given that you have a beam of cyan light shining on a white wall, what color beam should be added so that the reflected light is white?

16. Figure Q16 is a plot of the amount of light transmitted across the spectrum by a piece of stained glass. (a) What color does it appear? (b) Two beams of light are projected onto a white sheet of paper. If one passes through a cyan filter and the other through a yellow filter, what color results?

17. (a) A beam of cyan light passes into a yellow filter. What color emerges? (b) A beam of yellow light passes into a magenta filter. What color emerges? (c) What color results when two beams of light, one cyan and one magenta, are made to overlap on a white screen? (c) White light passes through a cyan filter followed by a magenta filter. What color emerges?

18. Figure Q18 shows the transmission spectrum of a sheet of colored plastic. What color is it? Redraw the curve so that it corresponds to the frequency distribution reflected from an orange.

Multiple Choice Questions

1. Standing on its surface looking upward, the Moon's sky is black because, unlike the Earth, the Moon (a) has no street lights (b) is very cold (c) has no atmosphere (d) has no oceans to reflect sunlight (e) none of these.

2. Cigarette smoke rising from the burning end has a bluish cast, while smoke exhaled is whitish. This difference happens because (a) the smoke contains a blue material that is left behind in the lungs (b) the body changes the chemistry so that the exhaled smoke is white (c) the smoke cools in the process and as a result turns white (d) water droplets from the body surround the inhaled smoke particles and, because they are large, they subsequently scatter white light (e) none of these.

3. Take a piece of ordinary window glass and crush it into a fine powder; holding the mound in hand, it will (a) appear black (b) become green (c) be as transparent as ever (d) appear white (e) none of these.

4. If you have ever scratched a piece of clear plastic or the varnished finish on a table top, you are familiar with the white marks that result. They have that characteristic appearance due to (a) changes in the chemistry of the surface (b) melting of the surface so it emits white light (c) electron bombardment of the surface via static charge that causes fluorescence (d) leaving behind a streak of powdered material that scatters a broad frequency band in the visible (e) none of these.

5. A material is transparent to a certain band of frequencies of electromagnetic radiation. Accordingly, we can say that the atoms in the material (a) are very small (b) have resonances in that band (c) are not interacting with each other (d) are not close together (e) none of these.

6. If the atmosphere were very much deeper than it is now, the Sun might appear (a) blue at sunset (b) red at noon (c) green at sunrise (d) violet at noon (e) none of these.

7. From our knowledge of the atom, it seems reasonable that all types of matter must have resonances somewhere in the electromagnetic spectrum and therefore must, if the matter is dense enough, (a) dissipatively absorb some radiant

energy (b) reflect across the spectrum (c) absorb broadly across the spectrum (d) be transparent at those resonances (e) none of these.

8. When light is incident on a smooth surface, the angle of incidence ordinarily (a) equals the angle of scattering (b) doesn't equal the angle of reflection (c) equals the angle of refraction (d) equals the polarization angle (e) none of these.

9. When light is incident on an interface, the angle(s) of (a) the reflected and refracted beams depend on the wavelength (b) the reflected and refracted beams are independent of frequency (c) the refracted beam is independent of wavelength (d) the reflected beam is independent of frequency (e) none of these.

10. Compared to an object in front of it, the image in a plane mirror is always (a) smaller (b) virtual (c) three times as far away (d) distorted (e) none of these.

11. A handful of white sandlike material is poured into a beaker of clear oil and vanishes from sight as it passes into the liquid. We can conclude that (a) the material dissolved (b) this phenomenon is quite impossible and could not have happened (c) the oil and material have mutually decomposed (d) the oil has the same index of refraction as the material (e) none of these.

12. A chicken is standing 1.0 m in front of a vertical plane mirror. A woman is standing 5.0 m from the mirror, behind and in line with the bird. How far from her will she see the image of the chicken? (a) 5.0 m (b) 1.0 m (c) 6.0 m (d) 4.0 m (e) none of these.

13. When a beam of light traveling in air enters a glass block, it ordinarily undergoes a change in (a) speed only (b) frequency only (c) wavelength only (d) speed and wavelength (e) none of these.

14. The bending or refraction of a beam of light as it enters a medium that has a greater index of refraction than the incident medium is due to a change in its (a) amplitude (b) effective speed (c) frequency (d) period (e) none of these.

15. What "color" will a yellow shirt appear under blue-light

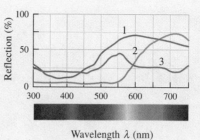

Figure MC20

illumination? (a) blue (b) yellow (c) white (d) black (e) none of these.

16. What combination of colored beams of light when overlapped on a white screen will produce black? (a) R + B + G (b) M + C + Y (c) M + R + C + Y + G (d) M + C + Y + R + B + G (e) none of these.

17. A beam of yellow light passes through a cyan filter. The color that emerges is (a) yellow (b) blue (c) green (d) red (e) none of these.

18. The color that results from the mixing of magenta and yellow paint is (a) red (b) blue (c) green (d) black (e) brown.

19. What color emerges when white light passes through a magenta filter followed by a green filter? (a) white (b) yellow (c) green (d) blue (e) none of these.

20. Figure MC20 shows three plots of the amount of light reflected over the visible region of the spectrum from a lemon, a ripe tomato, and a lettuce. These correspond, in turn, to curves (a) 1, 3, and 2 (b) 3, 2, and 1 (c) 1, 2, and 3 (d) 2, 3, and 1 (e) 3, 1, and 2.

21. A yellow surface is illuminated with magenta light. What color will it appear in reflected light? (a) blue (b) green (c) yellow (d) red (e) none of these.

Problems

SCATTERING
REFLECTION

1. [I] Two beams of light, one red ($\lambda_r = 780$ nm) and one violet ($\lambda_v = 390$ nm), pass through several hundred meters of air. What is the ratio of the amount of scattering of red to violet?

2. [I] A beam of white light crosses a large volume occupied by a tenuous molecular gas mixture of mostly oxygen and nitrogen. Compare the amount of scattering occurring for the yellow (580 nm) component with that of the violet (400 nm) component.

3. [I] A laser beam strikes a front-silvered mirror at an angle of 30° to the perpendicular. At what angle will it be reflected?

4. [I] A beam of parallel light strikes a flat polished metal surface perpendicularly. At what angle will light be reflected?

5. [I] Rays of light impinge on a smooth flat mirror at glancing incidence, making a tiny angle with respect to the surface. Approximately, what are the angles of incidence and reflection?

6. [I] A narrow beam of light impinges on a smooth glass plate at 25° measured from the perpendicular to the surface. What is the angle between the incident and reflected beams?

7. [I] Two rays leave a point source and strike a front-silvered mirror, making angles with the normal of 30° and 40°, respectively. What is the angle between the two reflected rays?

8. [I] A very narrow beam of light from a laser is incident on a horizontal mirror at an angle of 58°. The reflected beam strikes a wall at a spot 5.0 m away from the point of incidence where the beam hit the mirror. How far horizontally is the wall from that point of incidence?

9. [I] The tomb of FRED the Hero of Nod is a dark closed chamber with a small hole in a wall 3.0 m up from the floor. Once a year, on FRED's birthday, a beam of sunlight enters via the hole, strikes a small polished gold disk on the floor 4.0 m from the wall and reflects off it, lighting up a great diamond imbedded in the forehead of a glorious statue of FRED, 20 m from the wall. Roughly how tall is the statue?

10. [I] There are a number of practical devices that use rotating plane mirrors. Show that if a mirror rotates through an angle α, the reflected beam will move through an angle of 2α.

11. [I] A woman can see an object clearly when it's held at a distance of 25 cm from her eyes. How far should she hold a flat mirror to see her own image clearly?

12. [I] A man is walking at 1.0 m/s directly toward a flat mirror. At what speed is his image approaching him?

13. [I] Two upright 1.00-m-tall plane mirrors are placed parallel to each other 2.8 m apart. The top of the mirror on the right is then moved back a little so that its surface tilts away from the other mirror at an angle of 10.0° off the vertical. A narrow laser beam passes perpendicularly through a small hole in the very bottom of the mirror on the left. It subsequently strikes the tilted mirror from which it reflects. How many times will it reflect off the upright mirror on the left?

14. [I] Suppose both mirrors in Problem 13 were, say, 10 m high. What then would be the angle of reflection off the tilted mirror when the beam encounters it for the second time?

15. [I] A woman stands between a vertical mirror $\frac{1}{2}$ m tall and a distant tree whose height is H. She is 1.0 m from the mirror, and the tree is 11.0 m from the mirror. If she sees the tree just fill the mirror, how tall is the tree?

16. [III] 4.00% of the light energy incident normally on a microscope slide is reflected from the top surface, how much will be transmitted through two such slides?

17. [II] Suppose you are standing in front of a 4.0-ft-tall flat vertical mirror in which you can see some fraction of your body. What will happen to that fraction when you step farther from the mirror? Draw a ray diagram that proves your contention.

18. [II] A man 6 ft tall standing 10 ft from a 3-ft-high vertical mirror can see his entire image in it. If his eyes are 4 in. below the very top of his head, how high above the floor is the bottom edge of the mirror?

19. [II] We wish to set up an eye test in a rather modest-sized office and so must use a plane mirror to get the full required distance to the chart ($C_1 C_2$) as shown in Fig. P19. If the observer is a distance d from the front-silvered mirror (which is H meters tall) and the chart is h meters in height, write an expression for the minimum value of H that will suffice, in terms of h, d, and the object distance s_0.

20. [II] Figure P20 shows a ray striking one of two perpendicular mirrors at angle θ_{i1} and then striking the other at θ_{i2}. Find a relationship between these two angles and describe the paths taken by the incoming and outgoing rays.

21. [II] Return to Fig. P20, but this time imagine there is a frog sitting between the mirrors looking at himself. Draw a ray diagram showing the locations of all the possible images.

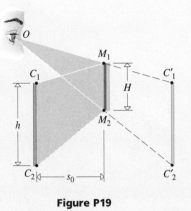

Figure P19

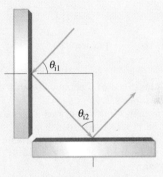

Figure P20

You should actually try this one with two ordinary mirrors held together. Look directly into the corner and hold up your right hand. Which hand will the image hold up? Notice where the seam is. Most people will see it run down the middle of their stronger eye. (A few with equal-strength eyes will see it down the middle of their faces.)

22. [II] Two vertical plane mirrors are brought together so that they make a wedge-angle of 35°, and a point source of light (S) is placed midway between them on a line bisecting the wedge-angle. Draw a simplified ray diagram locating all of the images of S by just using normal rays and making the object distance equal the image distance.

23. [III] Figure P23 shows an arrangement of mirrors forming a rangefinder for a camera. Mirror M_1 is partially silvered and fixed in position while M_2 is totally silvered and rotates about a vertical axis. Looking into M_1, you would see two images, and by rotating M_2, they are made to exactly overlap. Since d is known accurately and the angle through which the second mirror is turned can be easily measured, a scale can be computed that displays the distance L directly on the camera. Write an expression for L in terms of ϕ and d, taking the tilt of the fixed mirror to be 45°.

24. [III] Most mirrors in common use are back-silvered. Suppose you stand 1.000 0 m from such a mirror, which is made of 5.00-mm-thick plate glass ($n = 1.50 = 3/2$). How far behind the front surface will your image appear? Draw a ray diagram.

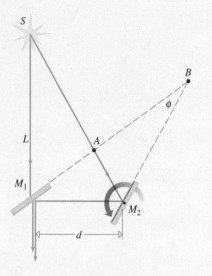

Figure P23

REFRACTION

25. [I] What is the speed of a beam of light in diamond if the index of refraction is 2.42?

26. [I] If the wavelength of a lightwave in vacuum is 540 nm, what will it be in water, where $n = 1.33$?

27. [I] What should be the index of refraction of a medium if it is to reduce the speed of light by 10% as compared to its speed in vacuum?

28. [I] If the speed of light (that is, the phase speed) in Fabulite (SrTiO$_3$) is 1.245×10^8 m/s, what is its index of refraction, to three significant figures?

29. [I] How far does yellow light travel in water in 1.00 s? What is the speed of yellow light in ethyl alcohol?

30. [I] If a narrow beam of light is incident on the surface of a tank of turpentine at 30°, at what angle will it be transmitted?

31. [I] A beam of light impinges on an air-liquid interface at an angle of 55°. The refracted ray is observed to be transmitted at 40°. What is the refractive index of the liquid?

32. [I] A swimmer shines a beam of light in water up toward the surface. It strikes the air-water interface at 35°. At what angle will it emerge into the air?

33. [I] Prove that to someone looking straight down into a swimming pool, the water will appear to be 3/4 of its true depth.

34. [I] A pool of water is 3.00 m deep and 4.00 m wide. Someone lying with his face close to the water is looking for a coin that is directly across the pool on the bottom far edge. Will he be able to see it?

35. [I] Radiant energy can be conveniently converted from IR (1.06 μm) to UV using nonlinear crystals. First, a portion of the IR is converted into its second harmonic on passing through an appropriate crystal. Determine the resulting frequency. Then, these two frequencies are mixed in a second crystal to generate the third harmonic. Find the wavelength of the third harmonic.

36. [I] What is the critical angle for diamond? Give your answer to two significant figures. What, if anything, does

the critical angle have to do with the luster of a well-cut diamond?

37. [I] What is the critical angle for an air-ice interface?

38. [I] Using a block of a transparent, unknown material, it is found that a beam of light inside the material is totally internally reflected at the air-block interface at an angle of 48.0°. What is its index of refraction?

39. [I] What is the minimum incident angle for internal reflection at the interface between two materials of indices 1.33 and 2.40, respectively?

40. [II] A 500-nm lightwave propagating in vacuum impinges normally on a glass sheet of index 1.60. How many waves span the glass if it's 1.00 cm thick?

41. [II] The photo in Fig. P41 shows a pulse of green light (530 nm, the second harmonic of 1.06 μm from a neodymium-doped glass laser) lasting about 10 picoseconds. The cell contains water ($n \approx 1.36$) and the scale is in millimeters. The exposure time was also about 10 ps. Explain what is shown in the picture. How far did the pulse travel during the exposure? How many wavelengths long (of green light) is the pulse?

Figure P41

42. [II] The American physicist R. W. Wood (1868–1955) pointed out the following: a pulse of red light entering a block of glass (with a measured value of $n = 1.52$) 12.00 miles thick will emerge 1.80 miles ahead of a blue pulse that entered at the same moment.   (a) How long will it take the red light to traverse the glass?   (b) How fast does the blue pulse travel?   (c) How much time will elapse between the emergence of the red and then the blue? Incidentally, Albert Michelson actually observed the effect in carbon disulfide.

43. [II] We wish to determine the index of refraction of a liquid filling a small tank as shown in Fig. P43. By moving up and down, it is found that the back edge of the container is just visible at an angle of 20.0° above the horizontal. Find the index.

44. [II] Yellow light from a sodium lamp ($\lambda_0 = 589$ nm) traverses a tank of glycerin (of index 1.47), which is 20.0 m long, in a time t_1. Now, if it takes a time t_2 for the light to pass through the same tank when filled

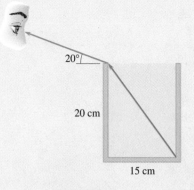

Figure P43

Figure P49

with carbon disulfide (of index 1.63), determine the value of $t_2 - t_1$.

45. [II] A beam of light impinges on the top surface of a 2.00-cm-thick parallel glass ($n = 1.50$) plate at an angle of 35°. How long is the actual path through the glass?

46. [II] Consider an air-glass interface with light incident in the air. If the index of refraction of the glass is 1.70, find the incident angle such that the transmission angle is to equal $\frac{1}{2}\theta_i$?

47. [II] Imagine that you focus a camera with a close-up bellows attachment on a letter printed on this page. Now, suppose the letter is covered with a 1.00-mm-thick microscope slide ($n = 1.55$). How high must the camera be raised in order to keep the letter in focus?

48. [II] A prism, ABC, is cut such that $\angle BCA = 90°$ and $\angle CBA = 45°$. What is the minimum value of its index of refraction if, while immersed in air, a beam traversing face AC is to be totally internally reflected from face BC?

49. [II] A fish looking straight upward toward the surface receives a cone of rays and sees a circle of light filled with the images of sky and ships and whatever else is up there (Fig. P49). This bright circular field is surrounded by darkness. Explain what is happening and compute the cone-angle.

50. [II] A block of glass with an index of 1.55 is covered with a layer of water of index 1.33. For light traveling in the glass, what is the critical angle at the interface?

51. [III] A lightwave propagates from point A to point B in vacuum. Suppose we introduce into its path a flat glass plate ($n_g = 1.50$) of thickness $L = 1.00$ mm. If the vacuum wavelength is 500 nm, how many waves span the space from A to B with and without the glass in place? What phase shift is introduced with the insertion of the plate?

52. [III] Refer to Fig. 25.22. If the plate has a thickness τ, show that the emerging beam is laterally displaced by a perpendicular distance d from the incident beam where

$$d = \frac{\tau \sin(\theta_{i1} - \theta_{t1})}{\cos \theta_{t1}}$$

53. [III] A coin rests on the bottom of a tank of water ($n_w = 1.33$) 1.00 m deep. On top of the water floats a layer of benzene ($n_b = 1.50$), which is 20.0 cm thick. Looking down nearly perpendicularly, how far beneath the topmost surface does the coin appear? Draw a ray diagram.

54. [III] With the results of Problem 52 in mind, by how much would a beam, incident at 30°, be displaced on traversing a plate of glass 5.00 cm thick having an index of 1.60?

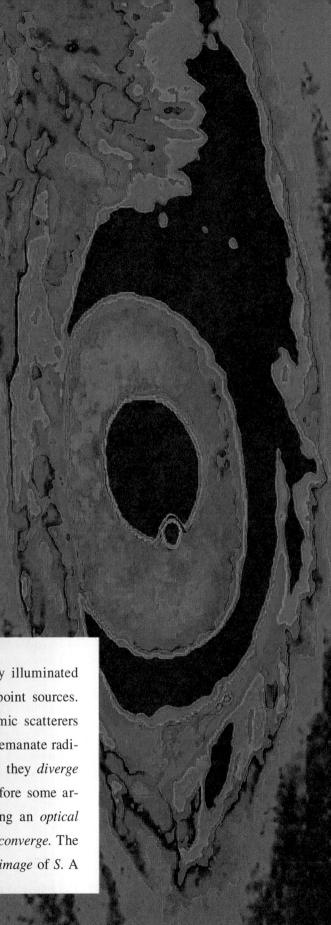

26

Geometrical Optics and Instruments

AN OBJECT THAT IS either self-luminous or externally illuminated can be imagined to have a surface covered with radiating point sources. Your face in the sunshine is a distribution of countless atomic scatterers each sending out little spherical wavelets. The associated rays emanate radially outward in the direction of the energy flow (Fig. 26.1); they *diverge* from each point source *S*. Now suppose the object stands before some arrangement of reflecting and/or refracting surfaces constituting an *optical system* whose function it is to collect the light and cause it to *converge*. The energy in the diverging cone arrives at *P*, which is called the *image* of *S*. A

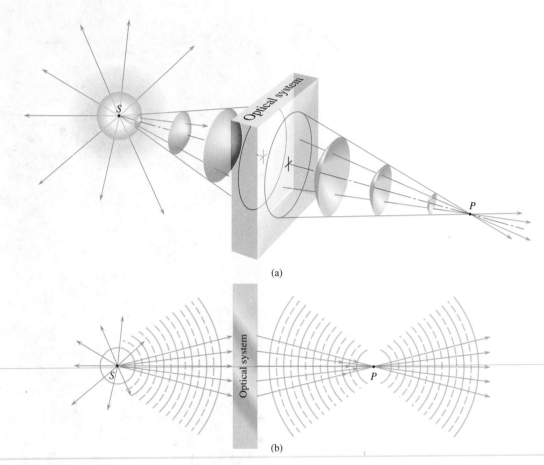

(a)

(b)

Figure 26.1 (a) A point source S sends out spherical waves. A cone of rays enters an optical system that inverts the wavefronts, causing them to converge on point P. (b) Rays diverge from S and a portion of them converge to P. If nothing stops the light at P, it continues on.

billion points on a face contribute light to a billion points making up the image of that face.

Real systems—cameras, telescopes, and eyeballs—cannot possibly collect *all* the light emitted from an object. When only a portion of any incident wavefront enters the system, there will inevitably be some diffraction. No matter how well made the optical system is, an object point will invariably be imaged as a somewhat enlarged *blur spot*. An otherwise perfect imaging system is thus said to be *diffraction limited*.

As the dimensions of the optical system become increasingly large in comparison to the wavelength, the effects of diffraction diminish. Accordingly, it is often reasonable to simply neglect diffraction and assume that all the rays travel in straight lines and go where they go via the Laws of Reflection and Refraction. As the wavelength becomes vanishingly small, diffraction disappears, rectilinear propagation obtains, and we have the idealized domain of **geometrical optics**. *That discipline deals with the manipulation of wavefronts (or rays) by means of interposing reflecting and/or refracting objects, neglecting any diffraction effects, and assuming rectilinear propagation.*

LENSES

The lens is no doubt the most widely used optical device, and that's not even considering the fact that we see the world through a pair of them. Human-made lenses date back at least to the *burning-glasses* of antiquity, which, as the name implies, were used to start fires long before the advent of matches—there's a reference to one of these in Aristophanes' play *The Clouds* (424 B.C.). From our perspective, *a lens is a refracting device (or discontinuity in the transmitting media) that reconfigures an incoming energy distribution.* And that much is true whether we are dealing with UV, lightwaves, IR, microwaves, radiowaves, or even sound waves (Fig. Q15, p. 523).

The configuration of a lens is determined by the reshaping of the wavefront it is to perform. Point sources are basic, and so it is often desirable to convert diverging spherical waves into a beam of plane waves; flashlights, projectors, and searchlights all do this in order to keep the beam from spreading out and weakening as it progresses. In just the reverse, it is frequently necessary to collect incoming parallel rays and bring them together at a point, thereby focusing the energy, as is done with a burning-glass or a telescope lens.

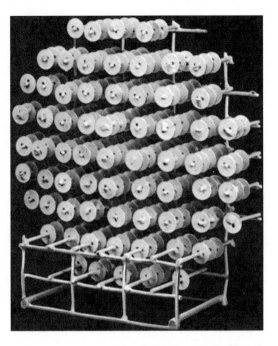

A lens for short-wavelength radio waves. The disks serve to refract these waves much as rows of atoms refract light.

26.1 Aspherical Surfaces

We can begin to understand how a lens works by interposing in the path of the wave a transparent substance in which the wave's speed is different than it was initially. Fig. 26.2a shows a diverging spherical wave traveling in an incident medium of index n_i, impinging on the curved interface of a transmitting medium of index n_t. When n_t is greater than n_i, the wave slows as it enters the new substance. The central area of the wavefront travels more slowly than its outer extremities, which are still moving through the incident medium. These extremities overtake the mid-region, continuously straightening out the wavefront. If the interface is properly configured, the spherical wavefront bends into a plane wave.

To find the required shape of the interface so we can make such a device, refer to Fig. 26.2c, wherein point A can lie anywhere on the boundary. *One wavefront is transformed into another provided the paths along which the energy propagates are all equal, thereby maintaining the phase of the wavefront.* A little spherical surface of constant phase emitted from S must evolve into a flat surface of constant phase at $\overline{DD'}$. That means that whatever path the light takes from S to $\overline{DD'}$, it must always be the same number of wavelengths long, so that the disturbance begins and ends in-phase. Radiant energy leaving S as a single wavefront must arrive at the plane $\overline{DD'}$, *having traveled for the same amount of time* to get there, no matter what the actual route taken by any particular ray. In other words, $\overline{F_1A}/\lambda_i$ (the number of wavelengths along the arbitrary ray from F_1 to A) plus $\overline{AD}/\lambda_t$ (the number of wavelengths along the ray from A to D) must be constant regardless of where on the interface A happens to be. Now, if we add these and then multiply by λ_0, we get

$$n_i(\overline{F_1A}) + n_t(\overline{AD}) = \text{constant} \tag{26.1}$$

Each term on the left is the length traveled in a medium multiplied by the index of that medium, and each represents what is called the **optical path length** (O.P.L.) traversed. The O.P.L. is the equivalent length in vacuum—if it is divided by c, we get the time it takes the light to travel the actual distance it did in the medium. If Eq. (26.1) is divided by c, the first term becomes the time it takes to travel from S to A and the second term, the time from A to D—the right side remains constant (not

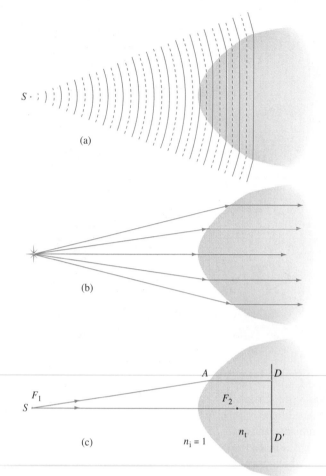

(a)

(b)

(c)

F_1
S

A

D

F_2

n_t

D'

$n_i = 1$

Figure 26.2 Light striking a hyperbolic interface between air and glass. (a) The wavefronts bend and straighten out. (b) The rays become parallel. (c) The hyperbola is such that the optical path from S to A to D is the same no matter where A is.

the same constant, but constant). Equation (26.1) is equivalent to saying that all paths from S to $\overline{DD'}$ must take the same amount of time to traverse.

Now, back to finding the shape of the interface. Dividing Eq. (26.1) by n_i, it becomes

$$\overline{F_1 A} + \left(\frac{n_t}{n_i}\right)\overline{AD} = \text{constant} \qquad (26.2)$$

This is the equation of a *hyperbola* where the eccentricity, which measures the bending of the curve, is given by $(n_t/n_i) > 1$; the greater the eccentricity, the flatter the hyperbola. When a point source is located at the focus F_1 and the interface between the two media is hyperbolic, plane waves will be transmitted into the higher-index material. Ellipsoidal and paraboloidal surfaces are also useful, but we'll limit our discussion here to hyperboloidal ones only (Fig. 26.3).

It's an easy matter now to construct lenses such that both the object and image points (or the incident and emerging light) will be outside of the medium of the lens. In Fig. 26.4a diverging incident spherical waves are made into plane waves at the first interface via the mechanism of Fig. 26.3b. These plane waves within the lens strike the back face perpendicularly and emerge unaltered; $\theta_i = 0$ and $\theta_t = 0$. And because the rays are reversible, *plane waves incoming from the right will converge to point F*, which is known as the **focal point** of the lens. Exposed to the parallel rays from the Sun, our rather sophisticated lens would serve nicely as a burning-glass.

In Fig. 26.4b, the plane waves within the lens are made to converge toward the axis by bending the second interface. Both of these lenses are thicker at their midpoints than at their edges and are therefore said to be **convex** (from the Latin *convexus*, meaning arched). Each lens causes the incoming beam to converge somewhat, to bend a bit more toward the central axis, and so they are referred to as **converging lenses**.

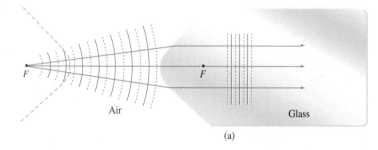

(a)

Air

F

F

Glass

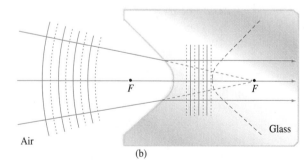

F

F

Air

Glass

(b)

Figure 26.3 Hyperbolic surfaces have two foci, *F*. In (a), light originating at *F* passes into the glass in the form of plane waves. In (b), light converging toward *F* passes into the glass as plane waves.

In both cases, the rays can be reversed so that they emerge into the air as either converging or diverging cones.

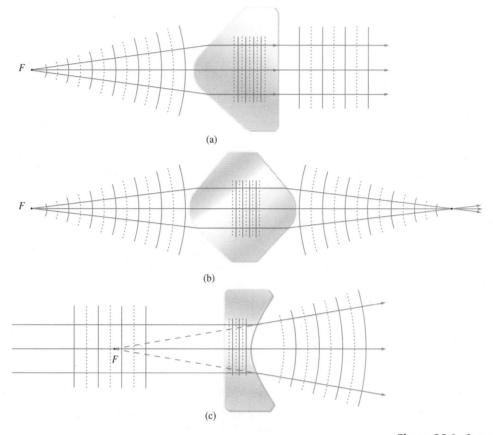

(a)

(b)

(c)

Figure 26.4 Several hyperbolic lenses.

In contrast, a **concave** lens (from the Latin *concavus,* meaning hollow—and most easily remembered because it contains the word *cave*) is thinner in the middle than at the edges, as is evident in Fig. 26.4c. It causes the rays that enter as a parallel bundle to diverge. All such devices that turn rays outward away from the central axis (that in so doing add divergence to the beam) are called **diverging lenses**. In Fig. 26.4c, parallel rays enter from the left and, on emerging, seem to diverge from *F*; still, that point is taken as a focal point. *When a parallel bundle of rays passes through a converging lens, the point to which it converges (or when passing through a diverging lens, the point from which it diverges) is a focal point of the lens.*

Optical elements (mirrors and lenses) having at least one curved surface that is not spherical are referred to as *aspherical.* Although they are easy to understand and they perform certain tasks exceedingly well, such elements are difficult to manufacture with great accuracy, Even so, many millions of aspherics have been made, and they can be found in instruments across the whole range of quality: in telescopes, projectors, cameras, and spy satellites.

26.2 Spherical Thin Lenses

In comparison to aspherics, spherical surfaces (actually, segments of spheres) are easy to fabricate. Unlike the hyperboloid, ellipsoid, or paraboloid, the sphere has no single central symmetry axis. All its diameters are alike, and the sphere can be generated by simply randomly rocking a rough-cut glass disk against an approximately spherical grinding tool. The two will wear each other away until both are perfectly spherical. The vast majority of lenses in use today have surfaces that are segments of spheres—despite the fact that those spherical surfaces are not ideal and will result in imaging errors known as *aberrations*. By using several components made of different materials to form compound lenses, these errors can be controlled so well that image quality need only be limited by diffraction.

As has already been seen, we cannot expect perfect imagery from spherical surfaces; hence, it will be necessary to place limitations on the way a spherical lens can be used so that it behaves appropriately. Thus, we only allow the lens to receive rays that strike it *not far from the central axis and enter only at shallow angles*—rays of this kind are said to be **paraxial**. No matter how the rays are drawn (and they will often be depicted making large angles for the sake of clarity), the rays are nonetheless paraxial. To simplify matters further, we will only deal with **thin lenses**; that is, *lenses for which the radii of curvature of the surfaces are large compared to the thickness*. Such lenses are quite common—most telescope and eyeglass lenses are thin.

The distances of objects and images are usually measured from the lens and can be on either side of it. It is important to associate a specific sign with each such distance so that it can be properly manipulated algebraically. *As a rule, light enters from the left*, and Fig. 26.5 shows how a typical ray is twice bent toward the central axis as it traverses a convex spherical lens. In Fig. 26.6a, we see two rays leaving the axial point source S and converging to the corresponding image point P. This is the basic geometry for which several possible sign conventions exist. We will take an **object distance** s_o to the *left* of the lens as positive and an **image distance** s_i to the *right* of the lens as positive. Furthermore, the radius of curvature of a lens surface is positive when its center point C is to the right of the surface. Here R_1, the radius of the first surface encountered, is positive while R_2, whose center C_2 is to the left, is negative. *All interfaces that bulge toward the left have positive radii, and all interfaces that bulge right have negative radii.*

To derive an equation for the operation of a spherical lens, note that the optical path lengths traversed by all rays from S to P are equal. All portions of the diverging wavefront take the same time to reach P, and all arrive in-phase. Let's examine two such paths, one along the central axis and one higher up, some arbitrary distance y. The O.P.L. from S to A to G to P must be the same as that from S to H to J to P. The wavefronts just in front of and behind the lens, Σ and Σ', have radii $\overline{SH} = \overline{SA}$ and $\overline{GP} = \overline{JP}$, respectively, and so if the O.P.L. from A to G equals that from H to J, the overall optical path lengths along the two routes will be equal. The paraxial limitation requires that y be small and since the lens is thin, we can approximate the path from A to G as a straight line (Fig. 26.6b). For simplicity, assume the lens has an index n_1 and it is immersed in air with an index of one. Thus, S will be imaged at P provided that

$$n_1\overline{HI} + n_1\overline{IJ} = \overline{AB} + \overline{BC} + n_1\overline{CD} + n_1\overline{DE} + \overline{EF} + \overline{FG} \qquad (26.3)$$

for all allowed values of y.

Since all the surfaces are spheres, the curves in the figure are circles, and there is a nice way to represent each of these little distances. Figure 26.6c shows a chord cutting a diameter ($2R$) of a circle perpendicularly. The piece σ is called the *saggita* (from the Latin for *arrow*) because it looks like an arrow resting on a bow. There is a theorem (proved in Problem 34) stating that for two such intersecting lines, the product of the two segments of one equals the

Figure 26.5 The radius drawn from C_1 is normal to the first surface, and as the ray enters the lens, it bends down *toward* that normal. The radius from C_2 is normal to the second surface, and as the ray emerges, since $n_1 > n_a$, the ray bends down *away* from that normal.

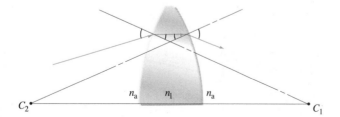

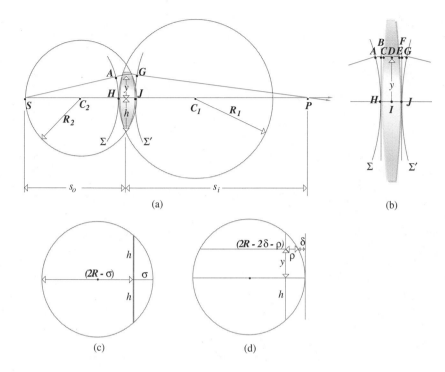

(a)

(b)

(c)

(d)

Figure 26.6 (a) The geometry of the thin lens. (b), (c), and (d) are associated with the approximations needed to determine the optical path lengths for rays from *S* to *P*.

product of the segments of the other:

$$\sigma(2R - \sigma) = (h)(h)$$

Hence

$$2R\sigma - \sigma^2 = h^2$$

But we are only interested in cases where $R \gg \sigma$ and there σ^2 is negligible compared to $2R\sigma$; hence, $2R\sigma \approx h^2$ and

$$\sigma \approx \frac{h^2}{2R} \qquad (26.4)$$

Figure 26.6d shows another chord, the segment ρ resembling a saggita, and the length δ extending beyond the circle. In the same way as above, it follows (Problem 35) that

$$\rho \approx \frac{(h^2 - y^2)}{2R} \qquad \text{and} \qquad \delta \approx \frac{y^2}{2R} \qquad (26.5)$$

These three approximations correspond to the segments in Fig. 26.6b where the surfaces of the lens form circles of radii R_1 and $-R_2$ and the radii of the wavefronts Σ and Σ' are represented by s_o and s_i, respectively. This last approximation is equivalent to saying that the lens is so thin we can measure the object and image distances from *either* its center or from its faces. Pressing on, we have

$$\overline{AB} = \delta_0 \approx y^2/2s_0 \qquad \overline{DE} = \rho_2 \approx (h^2 - y^2)/-2R_2$$
$$\overline{BC} = \delta_1 \approx y^2/2R_1 \qquad \overline{EF} = \delta_2 \approx y^2/-2R_2$$
$$\overline{CD} = \rho_1 \approx (h^2 - y^2)/2R_1 \qquad \overline{FG} = \delta_i \approx y^2/2s_i$$
$$\overline{HI} = \sigma_1 \approx h^2/2R_1 \qquad \overline{IJ} = \sigma_2 \approx h^2/-2R_2$$

Now, substituting all of these into Eq. (26.3) and shifting things around, we see that

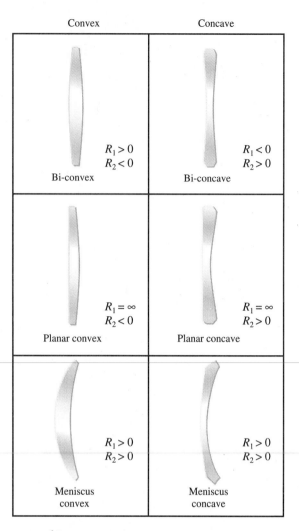

Figure 26.7 Cross sections of various centered spherical simple lenses. Take the surface on the left as number one, since it's encountered first by light coming from the left.

every term with h^2 vanishes; the $y^2/2$ factors out and cancels, leaving

$$\frac{1}{s_o} + \frac{1}{s_i} = (n_1 - 1)\left(\frac{1}{R_1} - \frac{1}{R_2}\right) \qquad (26.6)$$

This is the **Thin-Lens Equation**, often referred to as the **Lensmaker's Formula** because the left side (which treats what is going on external to the lens) is given in terms of the physical variables that would have to be selected to fabricate the lens (Fig. 26.7).

It can be shown (Problem 37) that, had the lens been immersed in a medium of index n_m rather than air, the Lensmaker's Formula would again result, but instead of the first term on the right being n_1, or equivalently $n_1/1$, it would now be n_1/n_m.

Example 26.1   A small point of light lies on the central axis 120 cm to the left of a thin bi-convex lens having radii of 60 cm and 30 cm. Given that the index of refraction of the lens is 1.50, find the location of the resulting image point. As far as the image distance is concerned, does it matter which way the lens is facing;

(continued)

(continued)

that is, which surface is toward the object? Unless otherwise informed, always assume the surrounding medium is air.

Solution: [Given: $s_o = 1.20$ m, $R_1 = 0.60$ m, $R_2 = -0.30$ m and $n_1 = 1.50$. Find: s_i.] Because the lens is bi-convex, the second radius encountered by incoming rays from the left is taken as a negative number. It follows from Eq. (26.6) that

$$\frac{1}{1.20 \text{ m}} + \frac{1}{s_i} = (1.50 - 1)\left(\frac{1}{0.60 \text{ m}} - \frac{1}{-0.30 \text{ m}}\right)$$

$$\frac{1}{s_i} = (0.50)\left(\frac{1}{0.20 \text{ m}}\right) - \frac{1}{1.20 \text{ m}} = \frac{2}{1.20 \text{ m}}$$

and $\boxed{s_i = 0.60 \text{ m}}$. The image distance is positive, and so the image lies to the right of the lens on the axis. Had we let $R_1 = 0.30$ m and $R_2 = -0.60$ m, nothing would have changed—*it doesn't matter which way a thin lens faces!*

▶ **Quick Check:** The right side of the Lensmaker's Formula equals $(0.50)/(0.20 \text{ m})$ and that quantity should equal the left side of the formula. Hence, $1/(1.20 \text{ m}) + 1/(0.60 \text{ m})$ should equal $1/(0.40 \text{ m})$, and it does.

26.3 Focal Points and Planes

Suppose that the object point S in Fig. 26.6a is moved far to the left. As $s_o \to \infty$, the rays enter the lens as a parallel bundle and are brought together at a specific image point known as the **image focal point**, F_i. The distance from the lens to this point is called the **image focal length**, f_i, where as $s_o \to \infty$, $s_i \to f_i$ as depicted in Fig. 26.8a. Similarly, the rays will emerge from the lens as a parallel bundle as $s_i \to \infty$ and the special object point for which this occurs is called the **object focal point**, F_o. The distance from the lens to this point is called the **object focal length**, f_o, where as $s_i \to \infty$, $s_o \to f_o$ as depicted in Fig. 26.8b. A thin lens surrounded by the same medium on both sides is a special case for which the image and object focal lengths are the same and the subscripts can be dropped altogether. Accordingly, go back to the Lensmaker's Formula and let $s_o \to \infty$, whereupon $(1/s_o) \to 0$ while $s_i \to f$, and we get

$$\frac{1}{f} = (n_1 - 1)\left(\frac{1}{R_1} - \frac{1}{R_2}\right) \tag{26.7}$$

Figure 26.8 Focal points for a converging lens. (a) A parallel bundle of rays passing through a thin lens is brought to convergence at the image focal point F_i. (b) A point of light at the object focal point F_o emits light that emerges from the lens as a parallel beam.

and the same thing results as $s_i \to \infty$. The focal length of a lens is determined by its physical makeup and can be positive or negative. In Figs. 26.8a and b, $R_1 > 0$ and $R_2 < 0$; hence, each focal length is positive. In Fig's. 26.9a and b, $R_1 < 0$ and $R_2 > 0$,

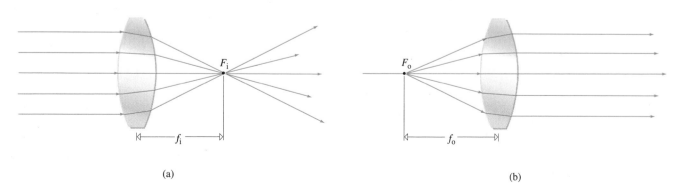

(a)   (b)

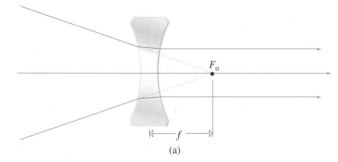

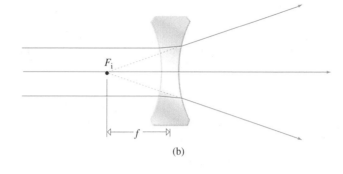

(a)

(b)

Figure 26.9 Focal points for a diverging lens. (a) Rays heading for F_o emerge parallel. (b) Rays entering parallel emerge as if from F_i.

and each focal length is negative. Going by the sign of its focal length, we commonly refer to a converging lens as a **positive lens** and a diverging one as a **negative lens**.

If for a bi-convex lens the radii are small (the lens has bulging faces), then the right side of the equation will be large, and the left side must also be large, meaning f must be small (the lens has a short focal length). The more "bent" the surfaces are, the more they bend the rays. On the contrary, when the surfaces are fairly flat (the radii are large), the focal length will be large. In fact, if both radii are equal and the lens is made of glass ($n_l \approx \frac{3}{2}$), it follows from Eq. (26.7) that $f \approx R$. For bi-convex and bi-concave lenses, f is ordinarily of the order-of-magnitude of the smaller radius, which is a good thing to keep in mind. This need not apply with meniscus lenses, where the magnitude of the focal length can be much greater than the radii.

Example 26.2 Determine the focal length in air of a thin spherical planar-convex lens having a radius of curvature of 50 mm and an index of 1.50. What, if anything, would happen to the focal length if the lens is placed in a tank of water?

Solution: [Given: the fact that the first surface is flat means that it has an infinite radius of curvature, $R_1 = \infty$; $R_2 = -0.050$ m and $n_l = 1.50$. Find: f when $n_m = 1.00$ and 1.33.] From Eq. (26.7),

$$\frac{1}{f} = (1.50 - 1)\left(\frac{1}{\infty} - \frac{1}{-0.050 \text{ m}}\right)$$

$1/\infty = 0$ and $\boxed{f = +0.10 \text{ m}}$. When the lens is surrounded by a medium of index n_m rather than air, n_l must be replaced by $n_l/n_m = 1.50/1.33$. The effect is to reduce the lens's ability to bring the rays into convergence and to increase the focal length to $+0.39$ m.

▶ **Quick Check:** f is positive as it should be for a convex lens. Moreover, $f = 2|R_2|$, which is also the right order-of-magnitude.

It is especially convenient to draw a ray along the central axis of a lens because it strikes each surface perpendicularly and passes straight through undeviated (Fig. 26.10). We now examine a tilted off-axis ray that enters the lens and emerges parallel to the incident direction. Such a ray must pass through a fixed point on the axis known as the **optical center** of the lens, O. It actually enters, bends a little, and emerges parallel to its incident direction. But because the lens is thin, the lateral displacement of the emerging ray is negligible. Thus, we can assume that *any paraxial ray heading toward the center of any thin lens will pass through O undeviated and may be drawn as a straight line.* It is customary when treating a thin lens to simply place the point O at the geometric center of the lens (Table 26.1).

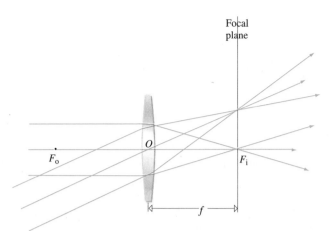

Focal plane

F_o

O

F_i

f

Figure 26.10 A ray headed for the center O of a thin lens passes straight through without bending. Ideally, all parallel bundles of rays focus on a plane, but actually only paraxial rays focus on that focal plane.

TABLE 26.1	Sign convention for spherical refracting surfaces and thin lenses*
s_o, f_o	+ left of O
s_i, f_i	+ right of O
R	+ if C is right of O
y_o, y_i	+ above optical axis

*Light enters from the left.

A bundle of parallel rays impinging on the lens at some angle with the central axis will be focused at a point on the central ray, the ray passing through O (Fig. 26.10). All such convergence points really lie on a curved surface, but if we limit the discussion to paraxial rays, that surface closely resembles a plane perpendicular to the central axis at the focal point. This is the **focal plane**, located a distance f from the lens, given by Eq. (26.7).

26.4 Finite Imagery: Lenses

Any point on an extended object sends out light in a great many directions, and if some of that light enters a positive lens, it can be made to converge to an image point. In that way, point-by-point, the image is formed somewhere in what is called the *image space*. Wherever two rays from an object point are made to converge, all rays from that point, passing through the lens, ideally converge. ***Find where any two rays from any object point cross, and you have found the corresponding image point***. And that is particularly easy to do using any two of the three special rays whose behavior we already know (Fig. 26.11). **Ray 1** heading for the center of any type of thin lens goes straight through. **Ray 2** entering a positive lens parallel to the central axis emerges passing through the image focus; a similar ray entering a negative lens emerges so that it can be extended back to pass through the image focus. **Ray 3** passing through the object focus of a positive lens emerges from the lens parallel to the central axis; a ray heading for the object focus of a negative lens emerges parallel to the central axis.

Since we know where these rays are going, we can simplify the drawings by constructing ray-paths with a single refraction taking place on a vertical line through the center of the lens. There really are two refractions, one at each face, but this way of proceeding will save us a lot of effort. Begin a ray diagram by setting down a horizontal axis and then sketching in a centered lens of arbitrary size and roughly the right shape. *The most important construction feature of the diagram is the location of the object and image focal points, which are at equal distances (f) on each side of the lens.* The vertical center line, the focal length, and the location of the object determine the entire geometry. As a rule, when not drawing everything to scale, make the lens about the same size as the focal length. Should you inadver-

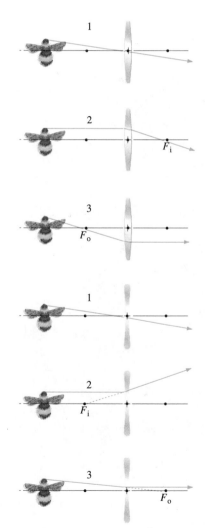

Figure 26.11 Tracing a few key rays through a positive and negative lens.

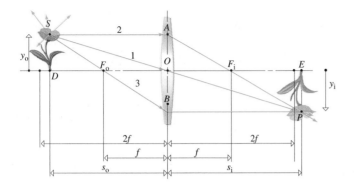

Figure 26.12 The geometry of image formation via a thin convex lens. For simplicity, refractions take place at the *A-O-B* plane.

A real image projected on the viewing screen of a 35-mm camera, much as the eye projects its image on the retina. Here a prism has been removed so you can see the image directly.

tantly draw a lens too small to accept a particular ray, it can still properly be drawn refracted from the vertical line as if the lens was pictured appropriately. At a distance of f on each side of the lens, put a mark on the axis to represent the focal points and then draw in another pair of points at distances of $2f$. These four points divide the object and image spaces into four distinctive regions.

Figure 26.12 depicts an object at some distance between f and $2f$ in front of a positive lens. To locate the resulting image, trace through the lens any two rays from each and every point on the flower—practically, a point on the very top and one on the bottom will do. From the topmost object point S, draw ray 1 straight through the center O of the lens. Now, we can either draw ray 2 or ray 3. If you like, draw all three; they, and indeed *all* rays entering the lens from that same object point, converge at the same conjugate image point P—that locates the top of the flower's image. The bottom of the image E can be found by doing the same thing over again for the bottom-most object point.

Here, ***the rays converge toward the image***, and we say that the image is **real**. The light actually arrives at each such point and if there is a viewing screen there (as in a movie theater), the image would appear on it, which is why it's called a *real* image. By comparison, when an image is formed of light diverging from it (as was the case with the plane mirror, p. 943), we say that the image is **virtual**—it cannot be projected onto a screen, but it certainly can be viewed directly. In much the same way, we refer to an object point as *real* when rays diverge from it and *virtual* when rays appear to converge toward it; the former situation is by far the most common, though the latter can occur (Fig. 26.9a).

The rays in Fig. 26.12 don't just end on the image, they go on past it to diverge again. Moving the viewing screen toward or away from the lens intercepts a more or less blurred version of the flower, whose sharpest image exists at the one location in space where the rays converge to points. Unless the rays are intercepted (for example, by a film-plate), nothing of the image can be seen from the side. There is no transverse propagation—the image is not visible laterally as the flower itself *is*. The lens forms an image of the flower, but it does not recreate the original lightwave coming from the flower; something is lost in the transformation (p. 1041).

The ray diagram can provide us with an analytic relationship between the object and image distances and the focal length. In Fig. 26.12, triangles $F_i EP$ and $F_i AO$ are similar, all their angles are equal, and so

$$\frac{\overline{PE}}{\overline{AO}} = \frac{s_i - f}{f}$$

wherein $\overline{AO} = \overline{SD}$. Triangles SOD and POE are also similar and therefore

$$\frac{\overline{PE}}{\overline{SD}} = \frac{s_i}{s_o} \tag{26.8}$$

Setting these two equations equal and rearranging terms, we get

$$\frac{1}{s_o} + \frac{1}{s_i} = \frac{1}{f} \tag{26.9}$$

which is the famous **Gaussian Lens Equation** first derived by C. F. Gauss in 1841.

Now that we see how Eq. (26.9) follows from the geometry of the lens and rays, it should be noted that the same expression results on comparing Eqs. (26.6) and (26.7).

Example 26.3 We wish to place an object 45 cm in front of a lens and have its image appear on a screen 90 cm behind the lens. What must be the focal length of the appropriate positive lens?

Solution: [Given: $s_o = 0.45$ m and $s_i = 0.90$ m. Find: f.] The problem does not mention the n or the radii of the lens—all of that is summarized in f, and so the analysis proceeds from Eq. (26.9), as

$$\frac{1}{0.45 \text{ m}} + \frac{1}{0.90 \text{ m}} = \frac{1}{f}$$

Using a calculator, this can easily be evaluated term by

term with the $(1/x)$ function

$$2.222 + 1.111 = \frac{1}{f}$$

and $\boxed{f = +0.30 \text{ m}}$

▶ **Quick Check:** Working without a calculator, recognize that Eq. (26.9) has the same form as that for two resistors adding in parallel; hence

$$f = \frac{s_i s_o}{s_i + s_o} = \frac{(0.45 \text{ m})(0.90 \text{ m})}{1.35 \text{ m}} \qquad (26.10)$$

and $f = +0.30$ m.

26.5 Magnification

The ratio of any transverse dimension of the image formed by an optical system to the corresponding dimension of the object is defined as the **transverse magnification** or, more often, simply the *magnification* M_T. In Fig. 26.12, the magnification is the height of the image divided by the height of the object:

$$M_T = \frac{y_i}{y_o} \qquad (26.11)$$

Here y_i is below the central axis and is traditionally taken to be a negative number. The image is upside down, meaning that *the magnification is negative whenever the image is inverted and positive when the image is right-side-up*. Bear in mind that the magnification refers to the ratio of image size to object size and need not only correspond to enlargement ($|M_T| > 1$). The image can certainly be minified ($|M_T| < 1$) or life-size ($|M_T| = 1$) as well (Table 26.2).

TABLE 26.2 Meanings associated with the signs of various thin lens parameters

Quantity	Sign +	Sign −
s_o	Real object	Virtual object
s_i	Real image	Virtual image
f	Converging lens	Diverging lens
y_o	Right-side-up object	Inverted object
y_i	Right-side-up image	Inverted image
M_T	Right-side-up image	Inverted image

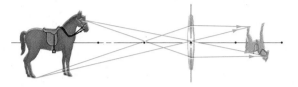

Figure 26.13 The image of a 3-dimensional object is itself 3-dimensional. The image exists in space and any portion of it can be viewed on a screen even though it cannot be seen from the side as drawn here.

From the similar triangles in Fig. 26.12, we got Eq. (26.8), which provides a convenient alternative statement for the magnification:

$$M_T = -\frac{s_i}{s_o} \qquad (26.12)$$

The minus sign here is necessary because both the object and image distances are positive and yet the image is inverted; that is, the magnification is negative.

It's easy to forget that the image exists in 3-dimensional space. As a reminder, Fig. 26.13 depicts a horse standing in front of a very big lens (a small one would do, but the ray diagram would be awkward to draw). The resulting minified image is real, and it exists in an extended region of space (although we generally view it in transverse slices on a flat screen).

Example 26.4 The horse in Fig. 26.13 is 2.25 m tall, and it stands with its face 15.0 m from the plane of the thin lens, whose focal length is 3.00 m. (a) Determine the location of the image of the equine nose. (b) What is the magnification? (c) How tall is the image? (d) If the horse's tail is 17.5 m from the lens, how long—nose-to-tail—is the image of the beast?

Solution: [Given: (a) $s_o = 15.0$ m, $f = +3.00$ m, $y_o = +2.25$ m; (d) $s_o = 17.5$ m. Find: s_i, front and rear; M_T; image height; and describe the image.] (a) From the Gaussian Lens Equation (26.9), we obtain

$$\frac{1}{15.0 \text{ m}} + \frac{1}{s_i} = \frac{1}{3.00 \text{ m}}$$

and $\boxed{s_i = +3.75 \text{ m}}$. (b) Computing the magnification from Eq. (26.12), we have

$$M_T = -\frac{s_i}{s_o} = -\frac{3.75 \text{ m}}{15.0 \text{ m}} = \boxed{-0.25}$$

(c) From the definition of magnification, Eq. (26.11), it follows that

$$y_i = M_T y_o = (-0.25)(2.25 \text{ m}) = \boxed{-0.563 \text{ m}}$$

where the minus sign tells us that the image is inverted. (d) Again from the Gaussian Equation, for the tail

$$\frac{1}{17.5 \text{ m}} + \frac{1}{s_i} = \frac{1}{3.00 \text{ m}}$$

and $s_i = +3.62$ m. The entire equine image is only $\boxed{0.13 \text{ m}}$ long.

▶ **Quick Check:** Because the image-distance is positive, the image is *real*. Because the magnification is negative, the image is *inverted*, and because the absolute value of the magnification is less than one the image is *minified*. Moreover, the image is between 1 and 2 focal lengths from the lens. All of which matches Fig. 26.13, where the object distance exceeds $2f$.

There are two important things to observe about the situation in Example 26.4 that apply to all real images formed by a positive lens. First, the image is evidently distorted, in that its length is reduced more than its height—the magnification transverse to the axis (M_T) is greater than the *longitudinal magnification* along the axial direction. This should not be surprising; the image of everything behind the horse, as far as the eye can see—out to infinity—must be compressed into the small space between the image-horse's rump and the lens (in fact, as we will see shortly, it occupies even less than that, ending at F_i).

The second important feature is that aside from being inverted, the image is oriented in an interesting way quite differently from that of the plane mirror. *The horse's nose, which is closer to the lens, is imaged farther away.* Were we to place a transverse observing screen far from the lens on the right and gradually move it to the left, closer in, we would first encounter the horse's face (with the saddle blurred). And then with the screen moved closer, the horse's head would be blurred, and the saddle would appear sharply "in focus." Nearest the lens, with horse and saddle fuzzy, only the tail might be clear.

26.6 A Single Lens

We are now in a position to understand the entire range of behavior of a single convex or concave lens. To that end, suppose that a distant point source sends out a cone of light that is intercepted by a positive lens (Fig. 26.14). If the source is at infinity (that is, so far away that it might just as well be infinity), rays coming from it entering the lens are essentially parallel and will be brought together at the focal point F_i. If the source point S_1 is closer, but still fairly far away, the cone of rays entering the lens is narrow, and the rays come in at shallow angles to the surface of the lens. Because the rays do not diverge greatly, the lens bends each one into convergence, and they arrive at point P_1. As the source moves closer, the entering rays diverge more and the resulting image point moves farther to the right. Finally, when the source point is at F_o, the rays are diverging so strongly that the lens can no longer bring them into convergence, and they emerge parallel to the central axis. Moving the source point closer results in rays that diverge so much on entering the lens that they still diverge on leaving. The image point is now virtual—***there are no real images of objects closer in than f.***

A positive lens operates with three distinct regions of object space. Suppose a man with an umbrella is standing near a tree somewhere in the most distant region, extending from ∞ to 2*f*, as indicated in Fig. 26.15a. His real, inverted image will be formed on the right of the lens between *f* and 2*f*; the farther away he is, the closer is his image to F_i (if he were at ∞, it would be at F_i). This situation corresponds to the way an eye or a camera works. The image on the retina is minified so that a panorama fits on the small screen; thus, we can see a whole oak tree at once rather than one acorn at a time. *As the man walks toward the lens, his image grows in size and slowly moves away from the lens.* This domain, from ∞ to 2*f*, is the first region of object space. It ends for the man at a distance of two focal lengths. When he stands at 2*f*, he is at the symmetry point of the system, where right and left sides (object and image) are identical. Figure 26.15b shows the situation: the triangles are now not just similar, they are congruent, and the real image is life-sized. This optical setup is that of a photocopy machine.

Figure 26.14 As the source moves closer, the rays diverge more and the image point moves out away from the lens. The emerging rays no longer converge once the object reaches the focal point; nearer in still, they diverge.

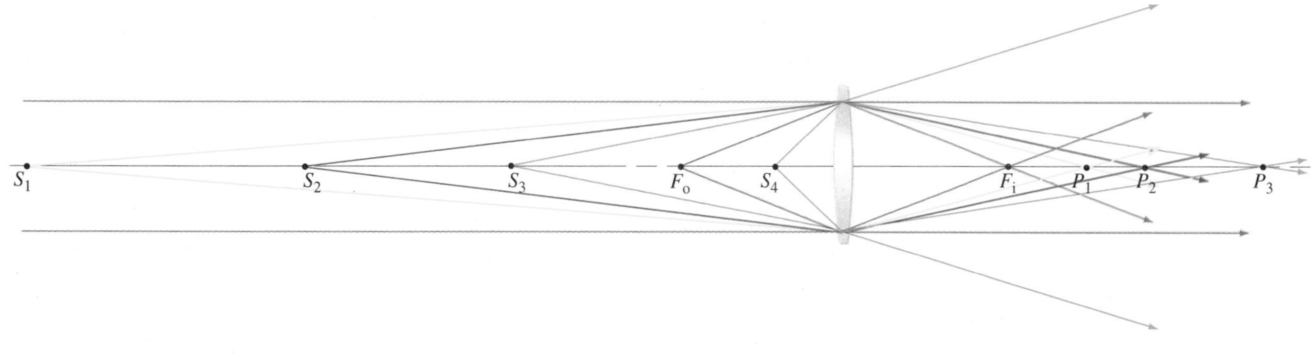

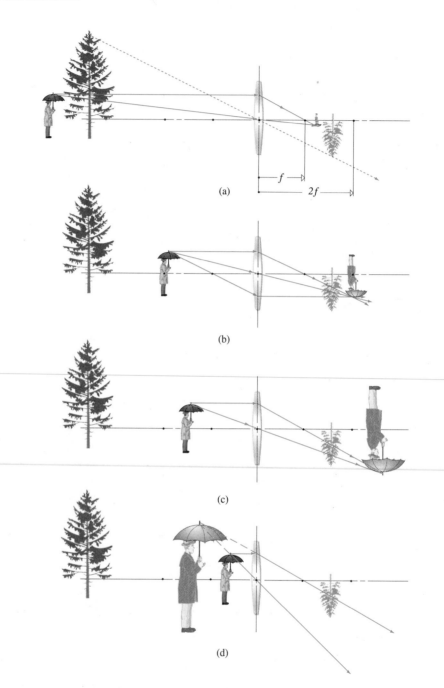

(a)

(b)

(c)

Figure 26.15 The operation of a thin positive lens.

(d)

As the man continues to walk toward the lens, moving from $2f$ toward f, his initially life-sized image gradually grows while it moves away from the lens, advancing to the right from $2f$ toward ∞. The second region of object space, from $2f$ to f, transforms into the image space from $2f$ to ∞. Notice that everything thus far has happened in a smooth, continuous way. As he walks in from very far away, his real image walks out; starting small, it reaches life-size at $2f$ and continues to grow, becoming immense as it moves away. This arrangement is that of a projector. The slide or film is the object located between $2f$ and f, each point of which will have a corresponding image point on the picture screen. Thereon will appear a real, enlarged, upside-down image of the illuminated portion of the slide. The projector is focused by just changing the slide-to-lens distance. And the inverted image is corrected by simply putting the slide or movie film in upside down.

It follows from Eq. (26.12) that the magnification approaches ∞ as the object nears f and s_i approaches ∞. When the object is exactly at the focal point, exactly at the boundary of the second region of object space, rays from each object point emerge from the lens parallel to one another. The image is no longer clearly observable on a screen, no matter how far away the screen is. Only a very large blur will be seen as if a real image were in focus somewhere far beyond the observing screen (namely, at ∞).

When the little man crosses beyond f into the third region of object space nearest the lens, he is so close and the rays so strongly diverging, that no real image is possible. An eye or a camera looking into the lens could bring the emerging diverging rays together. What one would see is a virtual, right-side-up image much like the image formed by a plane mirror (Fig. 26.15d), except enlarged. With the object at f, the image is an immense blur, but if the little man leans back a bit, his image forms—large, real and inverted; if he leans forward, his image blurs, vanishes, and then reappears—large, virtual, and right-side-up. It flips over, but that discontinuity happens when he is at f and, reasonably enough, cannot be observed because the image is gone. As he walks toward the lens, his image (as seen through the lens) will diminish ($M_T > 1$) until his face is flat up against the lens and he appears life-size. A positive lens operating on the light from an object located in this third region is known as a **magnifying glass**.

The concave lens (Fig. 26.16) operates in only one way and so is much easier to keep track of. It produces *only* virtual, right-side-up, minified images no matter where the object is located. Rays diverging from any object point are made to diverge even more. An object pressed up against the lens appears very nearly life-size, but one more distant is minified correspondingly. In all cases, the image distance is negative, and the image always appears on the left side of the lens—on the same side the light enters from. Table 26.3 summarizes these conclusions for both types of lens.

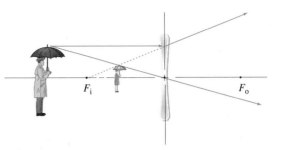

Figure 26.16 A concave lens forms a virtual, minified, right-side-up image.

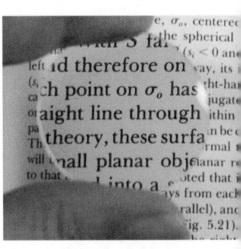

A positive lens serves as a magnifying glass when the object is closer to the lens than one focal length ($s_o < f$).

TABLE 26.3 Images of real objects formed by thin lenses

Convex				
Object		**Image**		
Location	**Type**	**Location**	**Orientation**	**Relative size**
$\infty > s_o > 2f$	Real	$f < s_i < 2f$	Inverted	Minified
$s_o = 2f$	Real	$s_i = 2f$	Inverted	Same size
$f < s_o < 2f$	Real	$\infty > s_i > 2f$	Inverted	Magnified
$s_o = f$		$\pm \infty$		
$s_o < f$	Virtual	$|s_i| > s_o$	Right-side-up	Magnified

Concave				
Object		**Image**		
Location	**Type**	**Location**	**Orientation**	**Relative size**
Anywhere	Virtual	$|s_i| < |f|$, $s_o > |s_i|$	Right-side-up	Minified

Example 26.5 A 5.00-cm-tall matchstick is standing 10 cm from a thin concave lens whose focal length is -30 cm. Determine the location and size of the image and describe it. Draw an appropriate ray diagram.

Solution: [Given: $y_o = 0.0500$ m, $s_o = 0.10$ m, and $f = -0.30$ m. Find: s_i and y_i.] From the Gaussian Lens Formula, Eq. (26.9),

$$\frac{1}{0.10 \text{ m}} + \frac{1}{s_i} = \frac{1}{-0.30 \text{ m}}$$

and $s_i = -1/13.3 = -0.075$ m $= \boxed{-7.5 \text{ cm}}$. The image distance is negative, meaning it is to the left of the lens, and therefore the image is virtual. Using Eq. (26.12), we can compute the magnification and from that the size of the image via Eq. (26.11):

$$M_T = -\frac{s_i}{s_o} = -\frac{-0.075 \text{ m}}{0.10 \text{ m}} = +0.75$$

and so

$$y_i = M_T y_o = 0.75(0.05 \text{ m}) = \boxed{0.038 \text{ m}}$$

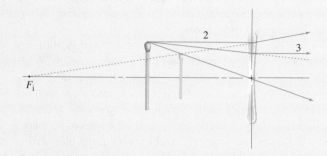

Figure 26.17 The image of a matchstick formed by a concave lens.

Figure 26.17 is the corresponding ray diagram. Notice how ray 3 heading for the object focus off to the right emerges parallel to the axis, as in Fig. 26.9a, while ray 2 entering parallel to the axis appears, on emerging, to be coming from the image focus.

▶ **Quick Check:** The image is right-side-up ($M_T > 0$), minified ($|M_T| < 1$), virtual ($s_i < 0$), and $s_o > |s_i|$, as Table 26.3 says it should be.

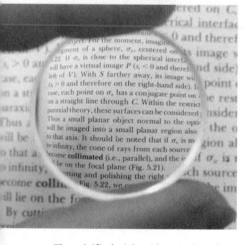

The minified, right-side-up, virtual image formed by a negative lens.

The Human Eye

The eye (Fig. 26.18) is an almost spherical (24 mm long by about 22 mm across) jellylike mass contained within a tough, flexible shell, the *sclera*. With the exception of the front portion or *cornea*, which is transparent, the sclera is white and opaque. The cornea is the first and strongest converging element of the eye system. Most of the bending impressed on a bundle of rays by the eye takes place at the air-cornea interface. The reason you can't see very well underwater ($n_w \approx 1.333$) is that the liquid's index is so close to that of the cornea ($n_C \approx 1.376$) that adequate refraction can no longer occur.

Having passed into the cornea, the light is only made slightly more convergent on emerging because it enters a chamber filled with a watery fluid known as the *aqueous humor* ($n_{ah} \approx 1.336$). Immersed in the aqueous is a variable diaphragm called the *iris*, which controls the amount of light entering the remaining portion of the eye by way of an aperature or *pupil*.

Just behind the iris is the *crystalline lens*. The lens (9 mm in diameter and 4 mm thick) is a complex, layered fibrous mass surrounded by an elastic membrane. In structure, it is somewhat like a small transparent onion, formed of roughly 22 000 very fine layers. As a whole, the lens is quite pliable, albeit less so with age. Its index of refraction varies from about 1.406 at the inner core to roughly 1.386 at the less dense cortex. The crystalline lens provides the needed fine-focusing mechanism via changes in its shape.

The cornea and crystalline lens can be treated as forming a double-element lens whose object focus is about 15.6 mm in front of the outer surface of the cornea and whose image focus is about 24.3 mm behind it on the retina. The combined lens has an optical center 17.1 mm in front of the retina, just at the rear edge of the crystalline lens. As a rule, $s_0 > 2f$.

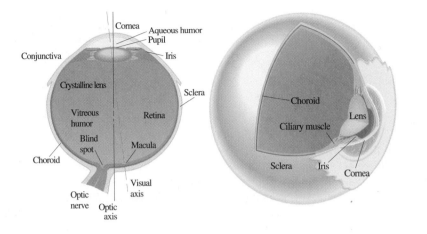

Figure 26.18 The human eye. The cornea is slightly flattened to control spherical aberration; the fact that rays from the edges of a lens usually do not focus at the same point as do rays near the center. Moreover, the spherical shape of the eyeball and retina eliminates problems associated with the fact that the image "plane" is not really flat. The retina contains 125 to 130 million photoreceptor cells.

Example 26.6   As we saw earlier (p. 271), the Moon subtends an angle of about 0.009 rad as seen from Earth. How big is its image formed by the human eye?

Solution: [Given: see Fig. 26.19. Find: y_i.] Since the Moon subtends an angle of 0.009 rad, it follows from Eq. (8.1) and the geometry that the image size is just $r\theta$ or, equivalently

$$|y_i| = s_i\theta = (17.1 \text{ mm})(0.009 \text{ rad}) = 0.15 \text{ mm}$$

and

$$\boxed{|y_i| = 0.2 \text{ mm}}$$

The absolute value is used because the image is inverted, and so y_i is negative by convention. The fact that the diameter is 0.15 mm means that the retinal image of the

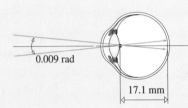
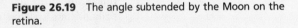

Figure 26.19 The angle subtended by the Moon on the retina.

face of the Moon is a dot smaller than the size of a period on this page!

▶ **Quick Check:** $|y_i|/s_i \approx (15 \times 10^{-5} \text{ m})/(17 \times 10^{-3} \text{ m}) \approx 1 \times 10^{-2} \text{ rad}$.

Behind the lens is another chamber filled with a transparent gelatinous substance, the *vitreous humor* ($n_{vh} \approx 1.337$). A thin, delicate, transparent multilayer of cells (from 0.5 mm to 0.1 mm thick) covers about 65% of the interior surface of that chamber. This structure is the light-sensitive *retina* (from the Latin *rete*, meaning net). The retina is the transducer that converts electromagnetic energy impinging on it into electrical nerve impulses that can be processed by the brain.

Accommodation.   The fine focusing or **accommodation** performed by the human eye is carried out by the crystalline lens. Since the image distance for the eye is fixed, the only way we can see things clearly at different object distances is if the focal length is changed. The lens is suspended by ligaments that are connected to a circular yoke of muscles. Ordinarily, these are relaxed and elongated, the aperture they encompass is large, and in that state they pull back on the network of fine fibers holding the rim of the lens. This draws the pliable lens into a fairly flat configuration, increasing its radii of curvature (especially of the anterior surface), which increases its focal length. With the muscles completely relaxed, the light from an object at infinity (which is practically speaking anywhere beyond about 5 m) is focused on the retina (Fig. 26.20). Not all eyes will do that well and so the **far-point**,

Figure 26.20 The human eye focuses on an object by contracting the ciliary muscles around the edge of the crystalline lens, which relaxes the tension on the lens so it contracts and bulges.

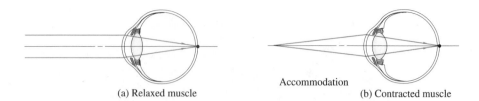

(a) Relaxed muscle

Accommodation (b) Contracted muscle

the point that is seen clearly by the unaccommodated eye, is frequently closer in than infinity (or even 5 m).

As the object moves closer, the muscles contract, the aperture encircling the lens gets smaller, and the lens bulges a little under its own elastic forces. In so doing, the focal length decreases, keeping the image on the retina. *The closest point that can be clearly seen with maximum accommodation is the* **near-point**. The lens doesn't change thickness by much more than about 0.5 mm over the whole range. A 10-year-old with a flexible lens may have a near-point as close as 7 cm, but that will typically move out to 12 cm by age 25, and 28 cm at around age 45—still a workable distance (11 inches) for reading without any inconvenience. But by 50 years, the near-point jumps to about 40 cm; by 60 years, it's out to 100 cm and by 70, it's at 400 cm.

The amount of accommodation required when an object is moved from infinity to 1 m is quite small; just enough to shift the image forward 0.06 mm. By contrast, moving the object from 1 m to 1/8 m requires enough accommodation to shift the image an additional 3.51 mm in order to keep it on the retina. Reading this page demands a disproportionate effort. Long study of things too close is one reason for eyestrain; we were apparently designed to look at, and out for, things afar.

The Camera

Figure 26.21 (a) Ideally, light entering the pinhole travels in a straight line, forming an image built up by countless little spots. (b) A pinhole camera photo (Science Building, Adelphi University). Hole diameter 0.5 mm, film plane distance 25 cm, A.S.A. 3000 shutter speed 0.25 s. Note depth of field.

The prototype of the modern photographic camera is the *camera obscura*, the earliest form of which was simply a dark room containing a small hole in one wall. Light entering the aperture formed an inverted image of the sunlit outside scene on an inside screen (Fig. 26.21). By replacing the viewing screen with a photosensitive sur-

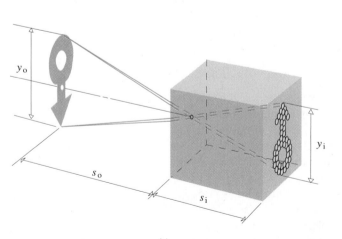

(a)

(b)

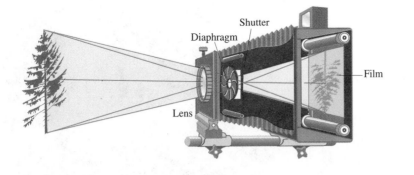

Figure 26.22 A large-format camera consists of a lens, followed by an adjustable diaphragm, a shutter that can open and close, and a sheet of film on which the image is formed.

face and filling the much-enlarged hole (which will let in a lot more light) with a lens, we have a photographic camera (Fig. 26.22), a TV camera, or a human eye, depending on the details.

Figure 26.23 depicts a typical camera lens with a focusing or distance scale that runs from ∞ to a fraction of a meter. When the camera is focused on infinity, the image distance takes on its smallest value; namely, f. At any lesser object distance, the image distance is greater than it was, and to keep the image on the film plane, the lens must be moved forward an amount δ such that $s_i - f = \delta$. From the lens equation, $s_i = s_o f/(s_o - f)$, and so $\delta = s_o f/(s_o - f) - f(s_o - f)/(s_o - f) = f^2/(s_o - f)$. For a camera lens where $s_o \gg f$, we have $\delta \approx f^2/s_o$.

The lens in the diagram has a focal length of 52 mm and to refocus it from infinity down to an object at, say, 10 m, the above equation tells us that the lens must be advanced a distance $\delta = 0.27$ mm. If it is moved another 0.27 mm from the film plane, it is now focused on an object at (10 m)/2 = 5 m; yet another movement of 0.27 mm sets it at (10 m)/3 = 3.3 m away, and so on. Rounded off, these are the numbers on the distance scale marked on the barrel of the lens.

Light enters the camera by way of a hole, and the size of that aperture and the duration over which it stays open together determine the amount of light allowed in. The time interval during which light is admitted is controlled with a shutter. The shutter is traditionally marked with a selection of settings designated 1000,

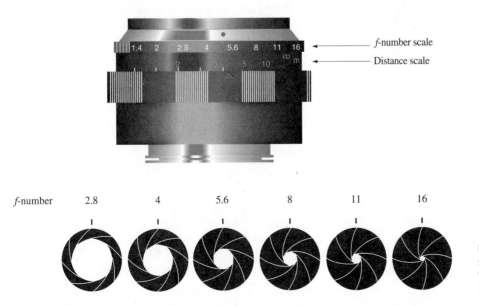

Figure 26.23 A camera lens showing possible settings of the variable diaphragm that is usually located within the lens.

A single-lens reflex camera. Light from the lens hits the mirror and goes up to the prism and out to the eye. When the shutter is released the mirror pops up, the light goes directly to the film, and then the mirror pops back down.

500, 250, 125, 60, 30, 15, 8, 4, 2, 1. These are fractional open-time intervals to be understood as 1/1000 s, 1/500 s, etc. up to 1 s, *each being twice as long as the one before it.*

The second light-control mechanism is a variable diaphragm, like the pupil of the eye (Fig. 26.23). The amount of light entering the camera from a broad source is proportional to the area of the open aperture, and that is proportional to its diameter squared (D^2). The concentration of light reaching the back of the camera, the *energy per unit area,* depends inversely on the area of the image (each side of which is dependent on the focal length) and, hence, is inversely proportional to f^2. The energy density on the image plane therefore goes as $(D/f)^2$. The ratio D/f is the *relative aperture,* whereas its inverse, f/D, is the **f-number**. *The smaller the f-number, the more light reaches the film.*

Since the energy density depends inversely on the square of the *f*-number, multiplying the *f*-number by $\sqrt{2} = 1.4$ *halves the amount of light reaching the film.* The seemingly strange values marked on the lens in Fig. 26.23 (1.4, 2, 2.8, 4, 5.6, 8, 11, 16, etc.) are consecutive *f*-numbers where each *stop,* as they are called, is a multiple of $\sqrt{2}$. As the lens is adjusted from $f/1.4$ to $f/2$ to $f/2.8$, etc., the diaphragm within it is set to close down by an amount that will halve the light passed by the previous opening (or double it if it's turned the other way).

Example 26.7   The lightmeter on a camera indicates that the proper amount of energy to expose a certain film will be delivered when the aperture is set at $f/2.8$ with a shutter speed of 1/120 s. The photographer wishing to increase the depth in space that will be in focus sets the aperture at $f/8$ instead. What should the shutter speed be?

(continued)

(continued)

Solution: [Given: *f*/2.8 and 1/120 s to pass same energy as *f*/8. Find: corresponding shutter speed.] *f*/8 is three stops down from *f*/2.8; that is, in going to *f*/8, the energy density will be halved three times. Accordingly, the energy must be doubled three times via the shutter if the total amount reaching the film is to be maintained. Rather than taking the photo at 1/120 s, the shutter time should be doubled three times—it should be left open eight times longer, that is, for 1/15 s .

The Magnifying Glass

To examine an object in detail, you simply bring it nearer the eye so that the retinal image increases in size. That procedure can continue until the object is at the near-point, beyond which the eye can no longer provide adequate accommodation because the rays diverge too much (Fig. 26.14). To enlarge the object further, a single positive lens can be used to add convergence to the visual system, allowing the object to be brought still closer. A lens so used is a *magnifying glass*. Its function is *to provide an image of a nearby object that is larger than that seen by the unaided eye*. It would be nice to have a right-side-up, magnified image where the rays entering the eye are not converging, and that's satisfied by placing the object within one focal length of the positive lens.

To deal with how large an object appears in some optical device, we must consider the size of its retinal image. Consequently, the *magnifying power*, or **angular magnification** M_A, of an instrument is defined as *the ratio of the size of the retinal image formed by the device to the size of the retinal image formed by the unaided eye at normal viewing distance*. The latter is taken as the distance to the near-point d_n. In Fig. 26.24, M_A is equivalent to the ratio of the angles α_a (aided) and α_u (unaided):

$$M_A = \frac{\alpha_a}{\alpha_u} \qquad (26.13)$$

Being restricted to the paraxial region, $\tan \alpha_a = y_i/L$ is very small and therefore approximately equal to α_a while $\tan \alpha_u = y_o/d_n \approx \alpha_u$. Since $-s_i/s_o = y_i/y_o$

$$M_A = \frac{y_i d_n}{y_o L} = -\frac{s_i d_n}{s_o L}$$

wherein both y_i and y_o are above the axis and positive. Taking all the previously unspecified distances in the diagram such as l, d_n, and L to be positive as indicated makes M_A positive as well. In the most commonly encountered application of the

Figure 26.24 A magnifying glass. (a) An object is examined directly by placing it at the near-point. The retinal image is then as large as possible. (b) With a magnifying glass, the same object results in a much larger retinal image.

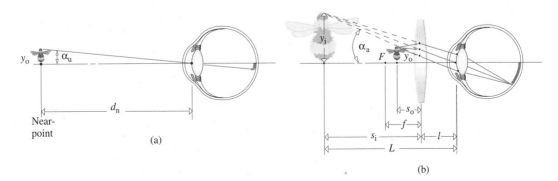

(a)

(b)

magnifying glass, the object is located at the focal point of the lens. In that case, the image is at infinity ($-s_i \approx L = \infty$; $s_o = f$) and

$$M_A = \frac{d_n}{f} \tag{26.14}$$

for all practical values of l. Rays emerging from the lens are parallel and can be viewed with a relaxed eye, which is a very important practical consideration.

Single lens magnifiers are usually limited by their aberrations to powers of about 2× or 3×. The famous Sherlock Holmes reading glass is an example of the type. More complicated multi-element magnifiers can be made in the range from 10× to 20×, and these are also used in microscopes and telescopes.

26.7 Thin-Lens Combinations

Optical systems are usually composed of several lenses and so we now examine a procedure for treating such arrangements. Figure 26.25 depicts two positive thin lenses L_1 and L_2 separated by a distance d, which here happens to be smaller than the focal length of either lens. The resulting image can be determined graphically by ray-tracing, using the following procedure. Imagine that L_2 is no longer there and

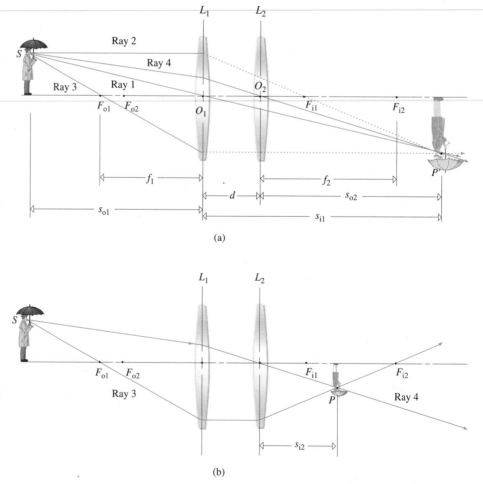

Figure 26.25 The image formed by two thin lenses. In (a) we imagine that lens L_2 is not there and we locate the image formed by L_1. Ray 4 is found passing through point O_2. It will be unchanged when lens L_2 is put back (b) and, along with ray 3, it locates the final image.

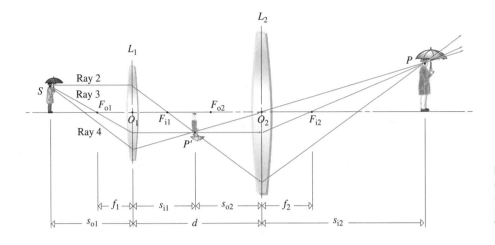

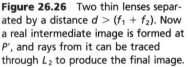

Figure 26.26 Two thin lenses separated by a distance $d > (f_1 + f_2)$. Now a real intermediate image is formed at P', and rays from it can be traced through L_2 to produce the final image.

find the image formed exclusively by L_1. This is easily done with ray 1 drawn through the center of L_1 and ray 2 entering parallel to the axis. These two rays fix point P' and the *intermediate image*. If L_2 was in place, a ray through its center would be undeviated; hence, construct ray 4 running backwards from P' through O_2 to L_1 and back to S. This ray is one of the two crucial rays that will locate the final image, since it passes properly through both lenses. The second necessary ray is just ray 3 through F_{o1}, which emerges from L_1 parallel to the central axis and therefore passes through F_{i2} on leaving the second lens. The point of intersection of these two rays determines the final image P of S. Notice that the intermediate image created by L_1 falls to the right of L_2 and can be thought of as a virtual object for L_2, which then goes on to create the final image.

Another example of the method is shown in Fig. 26.26 wherein the lenses are farther apart; in fact, $d > (f_1 + f_2)$. Again ray 1 and ray 2 fix the intermediate image at P'. And again ray 4 passing through P' and O_2 is drawn backwards through L_1 to S. This ray and ray 3 once more locate the final image P. Here, the intermediate image is real, and we could have proceeded by simply treating it as the real object for the second lens, forgetting about the first lens. Any two convenient rays from P' through the second lens will suffice to locate the final image.

The system can be treated analytically in much the same way—that is, by finding the intermediate image formed by the first lens and letting that function as the object for the second lens.

Example 26.8 Two positive lenses with focal lengths of 0.30 m and 0.50 m are separated by 0.20 m, as in Fig. 26.25. A red jellybean rests on the central axis 0.50 m in front of the first lens. Locate the resulting image with respect to the second lens.

Solution: [Given: $f_1 = +0.30$ m, $f_2 = +0.50$ m, $d = 0.20$ m, and $s_{o1} = 0.50$ m. Find: s_{i2}.] Applying the lens formula to L_1 only yields

$$\frac{1}{s_{i1}} = \frac{1}{f_1} - \frac{1}{s_{o1}} = \frac{1}{0.30 \text{ m}} - \frac{1}{0.50 \text{ m}} \quad (26.15)$$

and $s_{i1} = 0.75$ m. The intermediate image falls $(0.75 \text{ m} - 0.20 \text{ m}) = 0.55$ m to the right of the second lens. This value, then, is the virtual object distance for L_2, and because it is to the right, it is negative ($s_{o2} = -0.55$ m). Again, the lens formula for the second lens with this object yields

(continued)

(continued)

$$\frac{1}{s_{i2}} = \frac{1}{f_2} - \frac{1}{s_{o2}} = \frac{1}{0.50 \text{ m}} - \frac{1}{-0.55 \text{ m}} \quad (26.16)$$

and $\boxed{s_{i2} = +0.26 \text{ m}}$. The image is real and to the right of the last lens.

▶ **Quick Check:** The second lens adds more convergence, pulling the image in closer to the lens as if the object was farther away than it is. The image is real, inverted, and minified. See Example 26.9.

Example 26.9 Considering two thin lenses separated by a distance d, derive an expression for s_{i2} in terms of s_{o1}, d, and the focal lengths.

Solution: [Given: Fig. 26.25. Find: s_{i2}.] For L_1, using Eq. (26.15)

$$s_{i1} = \frac{s_{o1}f_1}{s_{o1} - f_1} \quad (26.17)$$

From Eq. (26.16), for L_2

$$s_{i2} = \frac{s_{o2}f_2}{s_{o2} - f_2}$$

Substituting in the fact that $s_{o2} = d - s_{i1}$, s_{i2} becomes

$$s_{i2} = \frac{(d - s_{i1})f_2}{(d - s_{i1} - f_2)}$$

Getting rid of s_{i1} using Eq. (26.17), we have

$$\boxed{s_{i2} = \frac{f_2 d - [f_2 s_{o1} f_1 / (s_{o1} - f_1)]}{d - f_2 - [s_{o1} f_1 / (s_{o1} - f_1)]}} \quad (26.18)$$

▶ **Quick Check:** Using the numbers from Example 26.8

$$s_{i2} =$$

$$\frac{(0.50 \text{ m})(0.20 \text{ m}) - (0.50 \text{ m})(0.50 \text{ m})(0.30 \text{ m})/(0.50 \text{ m} - 0.30 \text{ m})}{0.20 \text{ m} - 0.50 \text{ m} - (0.50 \text{ m})(0.30 \text{ m})/(0.50 \text{ m} - 0.30 \text{ m})}$$

and $s_{i2} = (-0.275 \text{ m}^2)/(-1.05 \text{ m}) = +0.26 \text{ m}$.

Suppose that the individual lenses are now brought close enough to touch one another, as is often done in compound systems. When $d = 0$, in Eq. (26.18) we can find the focal length of the combination by letting $s_{o1} \to \infty$, whereupon $s_{i2} = f$. The terms with d vanish, $(s_{o1} - f) \to s_{o1}$ and

$$f = \frac{f_1 f_2}{f_1 + f_2}$$

or

$$\frac{1}{f} = \frac{1}{f_1} + \frac{1}{f_2} \quad (26.19)$$

The focal lengths add like resistors in parallel.

Eyeglasses

Spectacles were probably invented some time in the late thirteenth century, possibly in Italy or China. A Florentine manuscript (1299), which no longer exists, spoke of "spectacles recently invented for the convenience of old men whose sight has begun to fail." In 1804, Wollaston, recognizing that traditional (fairly flat bi-convex and concave) eyeglasses provided good vision only while looking through their centers, patented a new deeply curved lens. These were the forerunners of modern meniscus lenses that allow the turning eyeball to see through them from center to margin without significant distortion.

It is customary in physiological optics to speak about the **dioptric power** $\mathscr{D}$ of a lens, which is simply the *reciprocal of the focal length*. A lens possesses great power when it strongly bends rays, which happens when it has a *short* focal length. Power has the units of inverse meters, or *diopters* (D): $1 \text{ m}^{-1} = 1 \text{ D}$. For instance, a converging lens with a focal length of $+10$ m has a power of 0.10 D, while a diverging lens with a focal length of -2 m has a power of $-\frac{1}{2}$ D. It follows from Eq. (26.7) that

$$\mathscr{D} = (n_1 - 1)\left(\frac{1}{R_1} - \frac{1}{R_2}\right) \qquad (26.20)$$

The combined focal length of two lenses in contact is given by Eq. (26.19), and so their total power is the sum of the individual powers

$$\mathscr{D} = \mathscr{D}_1 + \mathscr{D}_2 \qquad (26.21)$$

Example 26.10   Two lenses with focal lengths of $+0.100$ m and -0.333 m are held close together on a common centerline. Compute both the focal length and the power of the combination.

Solution: [Given: $f_1 = +0.100$ m and $f_2 = -0.333$ m. Find: f and $\mathscr{D}$.] The focal length is obtained from Eq. (26.19), as

$$\frac{1}{f} = \frac{1}{f_1} + \frac{1}{f_2} = \frac{1}{+0.100 \text{ m}} + \frac{1}{-0.333 \text{ m}}$$

and $\boxed{f = 0.143 \text{ m}}$. Equation (26.21) provides the power; thus

$$\mathscr{D} = \mathscr{D}_1 + \mathscr{D}_2 = (10.0 \text{ D}) + (-3.0 \text{ D}) = \boxed{7.0 \text{ D}}$$

▶ **Quick Check:** $1/f = 1/(0.143 \text{ m}) = 7.0$ D. A negative lens combined with a stronger positive lens yields a positive lens.

The human eye has a total power of roughly $+59$ D for the unaccommodated state (of which the cornea provides about $+43$ D). In the normal eye, that's just the refractive power needed to focus a parallel bundle of rays onto the retina. All too commonly, however, the image focus does not lie on the retina. This condition can arise either because of abnormal changes in the refracting mechanism (cornea, lens, and so on) or because of alterations in the length of the eyeball that upset the lens-retina distance. The latter is by far the more common cause. About 25% of the young adult population falls in the class of requiring as little as about ± 0.5 D or less of eyeglass correction, and perhaps as many as 65% need only ± 1.0 D or less.

Farsightedness, or *hyperopia,* is the defect that causes the image focus of the unaccommodated eye to fall behind the retina (Fig. 26.27). It is most often (perhaps 90% of the time) due to a shortening of the anteroposterior axis of the eye—the lens is too close to the retina. As a result, the image on the photoreceptors is formed of overlapping blotches of light and the picture is somewhat blurred. The *relaxed* hyperopic eye cannot bend the rays enough because it lacks the needed convergence and cannot see anything, near or far, clearly. But it can accommodate, thereby increasing its power and bringing into focus light from far away, which isn't very divergent to begin with (Fig. 26.27b). By accommodating, the farsighted eye can see clearly everything from infinity inward to some **near-point**. This near-point will be a lot farther away than it is in the normal eye. Any closer and the rays diverge too much; strain as the eye may, the image will be blurred.

To increase the power of the hyperopic visual system, a positive spectacle lens can be placed in front of the eye. This lens will allow the unaccommodated eye to see very distant objects clearly (Fig. 26.27d) while effectively pulling in the near-point so that it is at some close, convenient distance. Another way to appreciate this

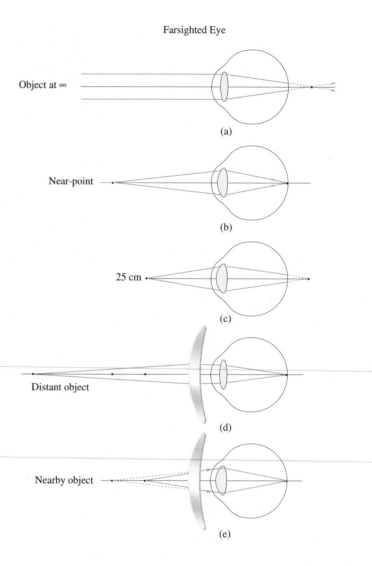

Farsighted Eye

Object at ∞

(a)

Near-point

(b)

25 cm

(c)

Distant object

(d)

Nearby object

(e)

Figure 26.27   Correction of the farsighted eye (a), which focuses parallel light beyond the retina. (b) The near-point is now farther away than 25 cm, which is normal (c). By adding more convergence to the eye, via a positive eyeglass lens, distant objects can be viewed in a relaxed state (d). Moreover, objects at 25 cm, which were blurred, are now clearly seen at the near-point (e).

is to realize that a nearby object (closer in than the focal length of the corrective lens) results in a distant right-side-up, virtual image. That image is located farther out than is the eye's unaided near-point and so can be seen clearly. These spectacles will cast real images—try it if you happen to be hyperopic.

Example 26.11   It is determined that a patient has a near-point at 50 cm. Approximate the eye to be 2.0 cm long. (a) Roughly how much power does the refracting system have when focused on an object at infinity? (b) When focused at 50 cm? (c) How much accommodation is required to see an object at 50-cm distance? (d) What power must the eye have to see clearly an object at the standard near-point distance of 25 cm?

(e) How much power should be added to the patient's vision system via reading glasses?

Solution: [Given: actual near-point is 50 cm, $s_i =$ 2.0 cm, standard near-point at 25 cm. Find: (a) power focused at ∞; (b) power focused at 50 cm; (c) accommodation for 50-cm vision; (d) power focused at 25 cm; (e) prescription for correction.] (a) When focused at in-

(continued)

(continued)

finity with the image falling on the retina 0.020 m beyond, the eye has a power of

$$\mathscr{D} = \frac{1}{f} = \frac{1}{\infty} + \frac{1}{0.020 \text{ m}} = \boxed{50 \text{ D}}$$

(b) Similarly, with $s_o = 0.50$ m and $s_i = 0.020$ m, the power is

$$\mathscr{D} = \frac{1}{f} = \frac{1}{0.50 \text{ m}} + \frac{1}{0.020 \text{ m}} = \boxed{52 \text{ D}}$$

(c) Thus, the eye adds 2 D of power via accommodation when changing focus from infinity to 0.50 m. This particular patient cannot provide any more than $\boxed{+2 \text{ D}}$

of accommodation. (d) When $s_o = 0.25$ m, the eye must bring to bear a power of

$$\mathscr{D} = \frac{1}{f} = \frac{1}{0.25 \text{ m}} + \frac{1}{0.020 \text{ m}} = \boxed{54 \text{ D}}$$

(e) Accordingly, this person is lacking $\boxed{+2 \text{ D}}$ of power that can be provided by correction lenses.

▶ **Quick Check:** (a) $f = \infty(0.02 \text{ m})/(\infty + 0.02 \text{ m}) = 0.02$ m. (b) $f = (0.50 \text{ m})(0.02 \text{ m})/0.52 \text{ m} = 1.9$ cm. (d) $f = (0.25 \text{ m})(0.02 \text{ m})/0.27 \text{ m} = 1.85 \text{ cm} = 1/(54 \text{ D})$.

Example 26.12 An optometrist finds that a farsighted person has a near-point at 125 cm. What power contact lenses will be required if they are to effectively move that point inward to a more workable distance of 25 cm? Use the fact that if the object is imaged at the near-point, it can be seen clearly.

Solution: [Given: $s_o = 0.25$ m and $d_n = 1.25$ m. Find: $\mathscr{D}_c$.] The eye can see the near-point clearly so we want 1.25 m to be the image distance of the correction lens. This s_i must be on the left of the lens and so it is negative, and the lens is being used as a magnifying glass. Accordingly, $s_i = -1.25$ m and $s_o = 0.25$ m. Both

the focal length and power of the contact lens can be found via Eq. (26.9); accordingly

$$\frac{1}{s_o} + \frac{1}{s_i} = \frac{1}{f_c} = \mathscr{D}_c = \frac{1}{0.25 \text{ m}} + \frac{1}{-1.25 \text{ m}} = \boxed{+3.2 \text{ D}}$$

The contact lens will form a virtual image of the book that appears at the near-point of the eye. By adding 3.2 D of power to the eye, the near-point of the corrected system becomes 0.25 m instead of 1.25 m.

▶ **Quick Check:** s_o is less than $f = 0.31$ m. $1/(1/4) - 1/(5/4) = 4.0 \text{ D} - 0.8 \text{ D} = 3.2 \text{ D}$.

Nearsightedness or *myopia* is the condition where parallel rays are brought to a focus in front of the retina; the power of the eye's refractive system is too large for the anterior-posterior axial length (Fig. 26.28). The problem occurs primarily because the eye elongates or the cornea changes shape. It is a situation that usually becomes noticeable in the teens and then levels off in severity at about age 25 or so. Interestingly, myopia hardly exists in "primitive" populations, whereas it is exceedingly common in so-called "advanced" civilizations (there are perhaps 40 million myopes in the United States).

With the myopic eye, images of faraway objects fall in front of the retina. And that's true for object distances from infinity inward to the so-called **far-point**, where the rays diverge enough so that the image is finally right on the retina and clearly visible. It is the farthest point that can be seen sharply by the unaided myopic eye. Depending on the degree of the problem, the far-point can be very much closer in than infinity, and all objects beyond it in space appear blurred. Moreover, the near-point is also closer than normal, which is a convenience for doing detailed work because it provides a bit more magnification.

In effect, the myopic eye has too much convergence; its positive power is too great. To correct the symptoms, we need only place a negative lens in front of the

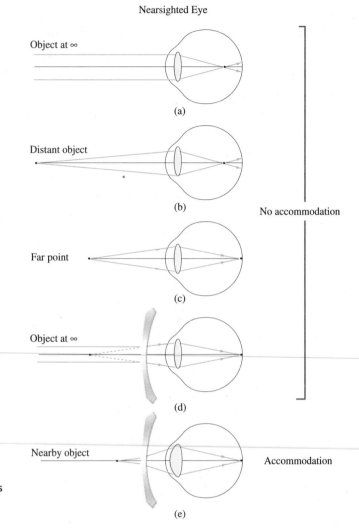

Figure 26.28 Correction of the nearsighted eye (a), which focuses parallel light in front of the retina. By adding some divergence to the eye, via a negative eyeglass lens, distant objects can be viewed (d) in a relaxed state. The far-point (c) is brought in closer (e).

eye. The image focus of the combined lens-eye system must fall on the retina. In other words, *parallel light is made to diverge just enough so that it appears to come from the far-point,* which can then be seen clearly by the unaccommodated eye. The intermediate image formed by the spectacle lens is at the far-point, which is the location of the object for the eye. *The far-point distance from the correction lens equals its focal length.* The eye views the right-side-up virtual images of all objects formed by the correction lens, and those images are located between its far- and near-points. The near-point also moves away a little, which is why myopes will often prefer to remove their spectacles when reading small print; they can then bring the material closer to the eye, increasing the magnification. If you are wearing glasses to correct myopia, try casting a real image with them—it can't be done.

Example 26.13 A person sees objects beyond a distance of 2.0 m to be blurred, but otherwise everything seems fine. What kind of contact lenses, with what power, should they wear?

(continued)

(continued)

Solution: [Given: a far-point of 2.0 m. Find: $\mathcal{D}$.] We want the image of an object at infinity to appear at the far-point of the eye ($s_o = \infty$, $s_i = -2.0$ m): Again, the far-point is on the left side of the lens and therefore the image distance is *negative*. Every distant object will then be imaged closer to the eye than the far-point and will be seen clearly. Hence, a lens must be added to the eye that has a power given by

$$\frac{1}{s_o} + \frac{1}{s_i} = \frac{1}{f} = \mathcal{D} = \frac{1}{\infty} + \frac{1}{-2.0 \text{ m}} = \boxed{-\tfrac{1}{2}\text{ D}}$$

▶ **Quick Check:** $f = \infty(-2.0 \text{ m})/(\infty - 2.0 \text{ m}) = -2.0$ m.

The Compound Microscope

The compound microscope goes the next step beyond the simple magnifier, providing still higher angular magnification (from 15× to around 1200×), achieved this time with a two-step arrangement. A simple version is illustrated in Fig. 26.29. The lens system closest to the object is the **objective**. It forms a real, inverted, magnified image of the object that is then viewed by the **eyepiece**. The latter is essentially a magnifying glass that looks at and enlarges the image created by the objective. Rays diverging from this intermediate image emerge from the eyepiece as a parallel bundle that can comfortably be viewed by a relaxed eye.

The problem is to examine an object that is close at hand, and since $M_T = -s_i/s_o$, the objective should have as small an s_o and as large an s_i as possible, which means, first, that the objective must be close to the object. Rearranging the lens equation yields $s_i = fs_o/(s_o - f)$, suggesting that the image distance will be appropriately large when $s_o \approx f$: *the objective must have a short focal length,* and the object must be positioned just beyond it so that the intermediate image is real. This image is then viewed by the eyepiece, which also must have a short focal length (f_E) since its magnification is $M_{AE} = d_n/f_E$. The intermediate image falls near the focal plane of the eyepiece so that the rays emerge parallel or almost so.

The eyepiece magnifies the intermediate image, which is a magnified version of the object; in other words, the total angular magnification of the system is the product of the magnifications of the objective (M_{TO}) and the eyepiece:

$$M_A = M_{TO} M_{AE}$$

Return to Fig. 26.12 and notice that since triangles AOF_i and PEF_i are similar, $y_i/y_o = -(s_i - f)/f = M_T$. Applied to the objective in Fig. 26.29, this expression becomes $M_{TO} = -L/f_O$, where the image distance minus the focal length of the objective is symbolized by L and is known as the **tube length**. Many manufacturers design their microscopes such that L is standardized at a length of about 160 mm. Using $M_{AE} = d_n/f_E$ and the fact that *it is customary to take the near-point d_n at 254 mm (10 in.)*, we have

$$M_A = \frac{-L}{f_O}\frac{d_n}{f_E} = \left(-\frac{160 \text{ mm}}{f_O}\right)\left(\frac{254 \text{ mm}}{f_E}\right) \tag{26.22}$$

where the focal lengths on the right are in *millimeters.*

The barrel of an objective with a focal length of, say, 32 mm, is engraved with the mark 5×, indicating a *magnification* of 5 = 160 mm/32 mm. Combined with a 10× eyepiece ($f_E = 25.4$ mm), the microscope then has a magnification of 50×—the apparent size is 50 times the actual size.

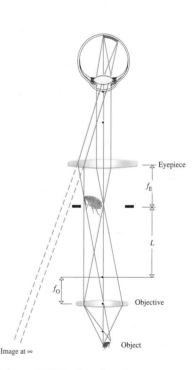

Figure 26.29 A rudimentary compound microscope. The objective forms a real magnified image of the object. That image is further magnified by the eyepiece. Because the intermediate image is at the focal point of the eyepiece, rays enter the eye in parallel bundles, and the eye is relaxed. Notice the large retinal image, which is, of course, what we are after.

Example 26.14 Suppose we wish to make a microscope (which can be used with a relaxed eye) out of two positive lenses both of focal length 25 mm. Assuming the object is positioned 27 mm from the objective, (a) how far apart should the lenses be, and (b) what magnification can we expect?

Solution: [Given: $f_O = f_E = 25$ mm and $s_o = 27$ mm. Find: (a) lens separation; (b) M_A.] (a) The intermediate image distance is obtained from the lens formula applied to the objective; thus

$$\frac{1}{27 \text{ mm}} + \frac{1}{s_i} = \frac{1}{25 \text{ mm}}$$

and $s_i = 3.38 \times 10^2$ mm. This value is the distance from the objective to the intermediate image, to which

must be added the focal length of the eyepiece to get the lens separation, and so

$$3.38 \times 10^2 \text{ mm} + 25 \text{ mm} = \boxed{3.6 \times 10^2 \text{ mm}}$$

(b) $M_{TO} = -s_i/s_o = -(3.38 \times 10^2 \text{ mm})/(27 \text{ mm}) = -12.5\times$, while the eyepiece has a magnification of $d_n \mathscr{D}_E = (254 \text{ mm})(1/25 \text{ mm}) = 10.2\times$. Since $M_A = M_{TO} M_{AE}$, the total magnification is $M_A = (-12.5) \times (10.2) = \boxed{-1.3 \times 10^2}$; the minus sign just means the image is inverted.

▶ **Quick Check:** $L = s_i - f_O = 338$ mm $- 25$ mm $= 313$ mm; $M_{TO} = -L/f_O = -(313 \text{ mm})/(25 \text{ mm}) = -12.5\times$. $M_{AE} = (254 \text{ mm})/f_E = 10.2\times$.

The Refracting Telescope

The primary function of the telescope is to enlarge the image of a *distant* object. The device shown in Fig. 26.30 has an objective and an eyepiece just like a microscope but because the job to be done is different, the structure is also different. The object is at a finite far distance from the device so that the intermediate image is located beyond the image focus of the objective. As with the microscope, this real, inverted image serves as the object for the eyepiece, which functions as a magnifier. Hence, the intermediate image is made to fall within one focal length (f_E) of the eyepiece so that the resulting final image it creates is virtual, enlarged, and remains inverted. In practice, *the position of the intermediate image is fixed and only the eyepiece is moved in order to focus the instrument.*

The central piece of design information is that the object is far away; that is, the object distance for the objective is very large in comparison to all the other distances in the system. The Gaussian Lens Equation then tells us that since $1/s_o \approx 0$, it follows that $s_i \approx f_O$. Unlike the microscope, where the intermediate image was magnified, here it must be *minified*. That might seem strange at first, but realize how

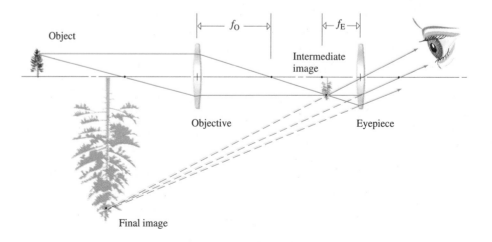

Figure 26.30 Kepler's astronomical telescope. It's used in astronomy where it doesn't matter if the image of the Moon or a star is upside down.

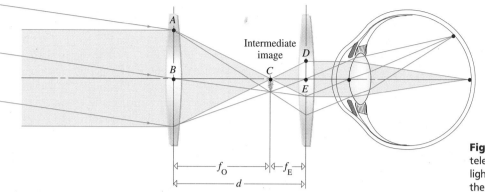

Figure 26.31 An astronomical telescope operating so that parallel light enters the objective and leaves the eyepiece.

awkward it would be to try to fit a larger-than-life image of, say, the Moon inside the tube of a telescope. The important feature of the real image is that it is now so close by that it can easily be examined with a magnifier.

Suppose the object focus of the eyepiece, which lies in front of it, overlaps the image focus of the objective, which lies behind it, as it does in Fig. 26.31. Then the separation between the two lenses equals the sum of their focal lengths. Parallel rays entering the scope from a very distant object exit in parallel bundles that can be viewed comfortably by the relaxed eye. That concern is an important one, and this arrangement is most often used.

In general, we want a minified intermediate image that is as large as is practical. Since $M_{TO} = -s_i/s_o \approx -f_O/s_o$, we need an objective lens with *as long a focal length as possible*. Again, the eyepiece views the intermediate image and magnifies it. Since the magnification varies inversely with f_E, the eyepiece should have a *short focal length*. In fact, the magnifying power of a telescope, adjusted so parallel rays emerge, is

$$M_A = -\frac{f_O}{f_E}$$

This equation is the reason why high-power refracting telescopes usually have long tubes into which one inserts a short focal-length eyepiece. When you look through the back end of a telescope, everything appears minified. The roles of eyepiece and objective are reversed, and because their focal lengths are quite different, the effect is striking. Reversing the telescope reduces the image size by the same factor it was previously increased by.

MIRRORS

Mirror systems are finding increasingly more extensive and important applications, particularly in the infrared, ultraviolet, and X-ray regions of the spectrum. It is relatively easy to construct a reflecting device that performs satisfactorily across a broad range of frequencies; the same cannot be said for refracting systems. Today, mirrors play a significant role in all sorts of devices, from spy satellites and copy machines to cameras, microscopes, and lasers.

26.8 Aspherical Mirrors

Curved mirrors that form images very much like those of lenses have been known since the ancient Greeks. Fortunately, we have already developed much of the con-

The 2.4-m-diameter primary mirror of the Hubble Space Telescope.

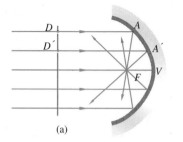
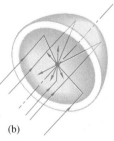

Figure 26.32   A paraboloidal mirror. (a) Parallel axial rays are brought to a focus at point *F*. (b) The configuration is 3-dimensional and symmetric about the central axis.

ceptual basis for analyzing curved mirrors and will be able to evolve the subject quickly without introducing many new ideas.

We again ask for the kind of surface that reshapes an incoming plane wave, this time via reflection, into an outgoing converging spherical wave. Figure 26.32 shows the geometry. We require that the optical path length from any point *D* on the plane wave to point *A* directly opposite it on the surface, and thence to the fixed point *F*, must be constant. Since there is now only one medium involved

$$\overline{DA} + \overline{AF} = \text{constant} = \overline{D'A'} + \overline{A'F'}$$

In two dimensions, this expression corresponds to the equation of a parabola (of eccentricity $n_t/n_i = 1$, p. 968) with its focus at *F*. A bundle of parallel rays reflecting off a paraboloidal mirror will be brought to a focus at *F*, a distance *f* from the **vertex** *V* of the mirror. Most of the world's older astronomical reflecting telescopes have energy-gathering mirrors that are paraboloidal.

Figure 26.33 depicts the behavior of several other aspherics. In recent years, these have been used to form images, via reflected X-rays, of a variety of phenomena from solar emission to laser-induced fusion. Today, the hyperboloid is the overwhelming choice for large telescopes, including the Hubble Space Telescope.

26.9   Spherical Mirrors

A spherical mirror has no one particular symmetry axis, and that can be a great advantage, especially when the device is not movable. In many applications restricted to paraxial optics, spherical mirrors perform quite well. Problem 76 deals with the proof that a sphere and a paraboloid coincide in the region close to the symmetry axis provided that $|f| = |R|/2$, as shown in Fig. 26.34. Thus, insofar as the rays are paraxial, they will encounter a region where sphere and paraboloid are nearly identical and the sphere will behave like the mirror in Fig. 26.32.

Absolute values, $|f|$ and $|R|$, are used because we have not yet agreed upon signs for mirror quantities. Using the previous convention, *R* in Fig. 26.35 is negative be-

Figure 26.33   Two aspherical mirrors.

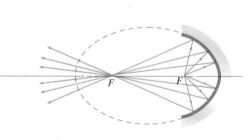

Concave elliptical

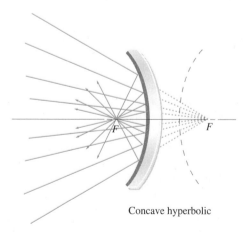

Concave hyperbolic

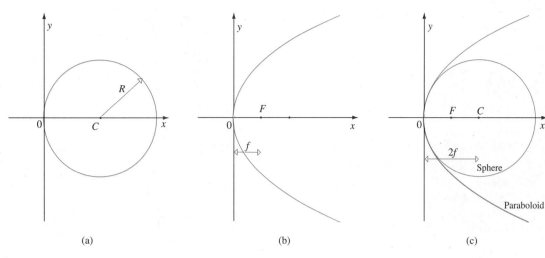

(a)   (b)   (c)

Figure 26.34 A sphere (a) and a paraboloid (b) coincide (c) in the region near the axis, such that $|R| = 2|f|$.

cause C is to the left of V. A parallel bundle of axial rays will converge to F, and so we take f to be positive, despite the fact that it is now measured to the left of the mirror (as compared with to the right of the lens). That being the case

$$f = -\frac{R}{2} \qquad (26.23)$$

The focal length of a spherical mirror equals one-half its radius.

The image P of S in Fig. 26.35 is real, but it's *on the left of the mirror,* unlike the situation with the positive lens. If we continue to require that real images have positive image distances, then we must henceforth take s_i to be *positive when left of the vertex,* exactly the same way that we took the focal length to be positive. These differences in the signs of f and s_i will be the only deviations from the convention for lenses, and with them all the appropriate equations will turn out the same as before. The image-formation geometry for the concave mirror is identical to that of the convex lens, and we need not rederive all the equations. Suffice it to say that

The 1000-ft radiotelescope at Arecibo, Puerto Rico, operates at 21 cm. The spherical bowl reflects radiant energy up to the focal-point detector suspended above it on cables attached to three towers.

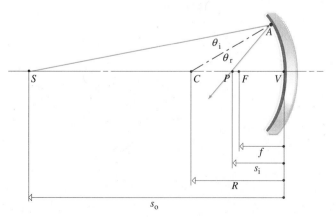

Figure 26.35 The geometry of a spherical mirror in the paraxial region.

Just like a convex lens, a concave mirror can form a real image. Here the mirror in the foreground casts a real, minified, inverted image of the candle flame.

paraxial rays in Fig. 26.35 obey the relationship

$$\frac{1}{s_o} + \frac{1}{s_i} = -\frac{2}{R} \qquad (26.24)$$

which is called the **mirror formula**. If $s_o \to \infty$, $s_i \to f_i = -R/2$ while if $s_i \to \infty$, $s_o \to f_o = -R/2$. In other words, both the image and object focal lengths equal f and

$$\frac{1}{s_o} + \frac{1}{s_i} = \frac{1}{f} \qquad (26.25)$$

Observe that f is positive for concave mirrors ($R < 0$) and negative for convex mirrors ($R > 0$). In the latter case, the image is formed behind the mirror in diverging light and cannot be projected upon a screen—it is virtual (Fig. 26.36a).

Example 26.15 A point source lies on the central axis 1.00 m in front of a concave spherical mirror having a radius of 20 cm. Locate and describe the resulting image.

Solution: [Given: $s_o = 1.00$ m and $R = -0.20$ m. Find: s_i.] From Eq. (26.24)

$$\frac{1}{s_i} = -\frac{2}{R} - \frac{1}{s_o} = -\frac{2}{-0.20 \text{ m}} - \frac{1}{1.00 \text{ m}}$$

and $\boxed{s_i = 0.11 \text{ m}}$. The image distance is positive and so the image itself must be real.

▶ **Quick Check:** $f = -R/2 = 0.10$ m = (1.00 m) × (0.11 m)/(1.00 m + 0.11 m) = 0.10 m.

Figure 26.36 The focal point, F, of a convex spherical mirror.

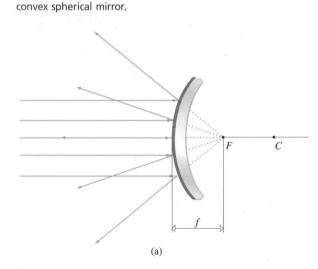

(a)

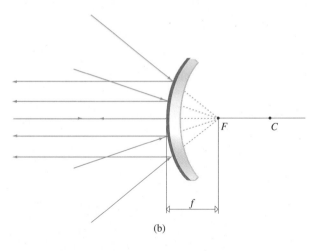

(b)

TABLE 26.4 Sign convention for spherical mirrors

Quantity	Sign	
	+	**−**
s_o	Left of V, real object	Right of V, virtual object
s_i	Left of V, real image	Right of V, virtual image
f	Concave mirror	Convex mirror
R	C right of V, convex	C left of V, concave
y_o	Above axis, right-side-up object	Below axis, inverted object
y_i	Above axis, right-side-up image	Below axis, inverted image

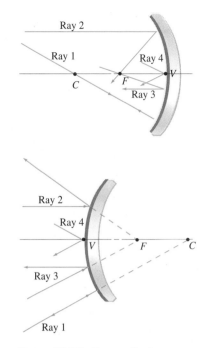

26.10 Finite Imagery: Mirrors

The remaining mirror properties are so similar to those of lenses that we need only mention them briefly without repeating the development. Accordingly, within the restrictions of paraxial theory, *any off-axis parallel bundle of rays will be focused to a point on the focal plane* a distance f from V. The image of any point on an object can again be located using any two easily drawn rays (Fig. 26.37). Unlike the lens, which is transparent and has separate object and image foci (one on each side), the mirror has only one focus. Table 26.4 summarizes the sign convention.

The ray diagram (Fig. 26.38) shows how the image of an extended object is created by a concave mirror. Because triangles SDV and PEV are similar (look at Fig. 26.12), their sides are proportional. Hence, taking distance measured down from the axis to be negative, $y_i/y_o = -s_i/s_o$, which, of course, is the transverse magnification M_T as defined earlier (p. 977). The striking similarity between the behavior of a concave mirror and a convex lens on one hand, and a convex mirror and a concave lens on the other, is evident on comparing Tables 26.3 and 26.5. Figure 26.39 graphically illustrates the entire range of responses of the concave mirror. Except for the fact that the mirror folds the rays over (so that object and image space overlap when the image is real and do not when it's virtual), the diagram is the same as Fig. 26.15.

Figure 26.37 Four easily drawn rays. Ray 1 heads toward C and reflects back along itself. Ray 2 comes in parallel to the central axis and reflects toward (or away from) F. Ray 3 passes through (or heads toward) F and reflects off parallel to the axis. Ray 4 strikes point V and reflects such that $\theta_i = \theta_r$.

A convex spherical mirror forming a virtual, right-side-up, minified image.

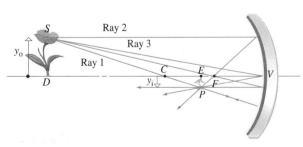

Figure 26.38 An extended image formed by a spherical concave mirror.

TABLE 26.5 Images of real objects formed by spherical mirrors

		Concave		
Object		Image		
Location	Type	Location	Orientation	Relative size
$\infty > s_o > 2f$	Real	$f < s_i < 2f$	Inverted	Minified
$s_o = 2f$	Real	$s_i = 2f$	Inverted	Same size
$f < s_o < 2f$	Real	$\infty > s_i > 2f$	Inverted	Magnified
$s_o = f$		$\pm\infty$		
$s_o < f$	Virtual	$|s_i| > s_o$	Right-side-up	Magnified

		Convex		
Object		Image		
Location	Type	Location	Orientation	Relative size
Anywhere	Virtual	$|s_i| < |f|, s_o > |s_i|$	Right-side-up	Minified

Example 26.16 A youngster looking into the shiny convex back of a spoon with a spherical bowl sees herself reflected therein. The bowl has a radius of 3.00 cm, and her nose is 25.0 cm from its surface. Where will the image of her nose appear? Describe the image completely.

Solution: [Given: $R = +0.030$ m and $s_o = 0.250$ m. Find: s_i and M_T.] To locate the image, use the mirror formula

$$\frac{1}{0.250 \text{ m}} + \frac{1}{s_i} = -\frac{2}{0.030 \text{ m}}$$

and $\boxed{s_i = -0.014 \text{ m}}$. The image distance is negative and so the virtual image is located to the right, behind the mirror. The magnification is

$$M_T = -\frac{s_i}{s_o} = -\frac{-0.014 \text{ m}}{0.250 \text{ m}} = \boxed{+0.056}$$

The image is minified and right-side up.

▶ **Quick Check:** $f = -R/2 = -0.015$ m $= (0.250 \text{ m})(-0.014 \text{ m})/(0.250 \text{ m} - 0.014 \text{ m}) = -0.015$ m.

The Reflecting Telescope

In addition to providing magnification, serious astronomical telescopes must meet another demand: they must gather in as much light as possible because the objects being viewed are generally extremely faint. As with the camera, the energy entering the system is proportional to the diameter of the objective—the bigger, the better. But there is a real difficulty in making big lenses. The largest such instrument in the world is the 40-inch diameter Yerkes refracting telescope in Wisconsin as compared to the 200-inch Palomar reflector in California. The problems are evident: a lens has to be perfectly transparent and free of internal flaws. A front-silvered mirror need not even be transparent. A lens can only be supported by its rim and may sag under its own weight; a mirror can be supported over its entire back. For these and other reasons (better frequency response, better aberration control, and so on), reflectors predominate in the domain of large telescopes, from observatories to spy satellites.

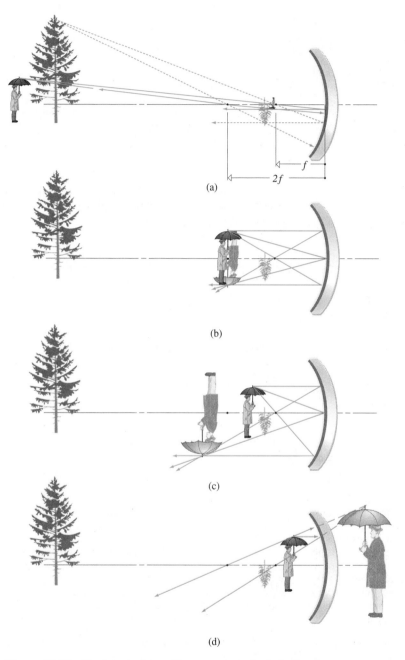

Figure 26.39 The images formed by a concave spherical mirror.

Invented by the Scotsman James Gregory in 1661, the reflecting telescope was first successfully constructed by Newton in 1668 (primarily because he had mistakenly concluded that the aberrations suffered by lenses were unavoidable). Two common reflectors are shown in Fig. 26.40. A plane mirror or prism brings the beam out to the side in the Newtonian version. The traditional Cassegrainian arrangement uses a convex hyperboloidal secondary mirror to increase the effective focal length of a paraboloidal primary. A modernized Cassegrainian with a hyperboloidal primary is now the leading telescope configuration.

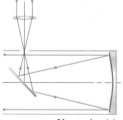

Newtonian (a)

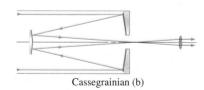

Cassegrainian (b)

Figure 26.40 Reflecting telescopes.

Core Material

As the wavelength of the radiant energy being processed becomes vanishingly small, diffraction disappears, rectilinear propagation obtains, and we have the idealized domain of **geometrical optics**. For thin spherical lenses, there is the **Lensmaker's Formula**

$$\frac{1}{s_o} + \frac{1}{s_i} = (n_l - 1)\left(\frac{1}{R_1} - \frac{1}{R_2}\right) \qquad [26.6]$$

where the **focal length** f is given as

$$\frac{1}{f} = (n_l - 1)\left(\frac{1}{R_1} - \frac{1}{R_2}\right) \qquad [26.7]$$

Comparing these two expressions yields the **Gaussian Lens Equation**

$$\frac{1}{s_o} + \frac{1}{s_i} = \frac{1}{f} \qquad [26.9]$$

The ratio of any transverse dimension of the image to the corresponding dimension of the object is the **transverse magnification** or, more often, simply the **magnification**, M_T:

$$M_T = \frac{y_i}{y_o} \qquad [26.11]$$

Alternatively $\qquad M_T = -\dfrac{s_i}{s_o} \qquad [26.12]$

If the object is at the focal point of a **magnifying glass**, its image is at infinity and

$$M_A = \frac{d_n}{f} \qquad [26.14]$$

When two lenses are being used at once, the image of the first is the object of the second, and the analysis can be carried out one lens at a time (p. 988). The focal length of two thin lenses in contact is given by

$$\frac{1}{f} = \frac{1}{f_1} + \frac{1}{f_2} \qquad [26.19]$$

The **dioptric power** $\mathscr{D}$ of a lens is the reciprocal of the focal length, and its units are inverse meters or *diopters* (D). The total power of two lenses in contact is

$$\mathscr{D} = \mathscr{D}_1 + \mathscr{D}_2 \qquad [26.21]$$

A spherical mirror behaves very much like a thin lens. It has a focal length given by

$$f = -\frac{R}{2} \qquad [26.23]$$

But we take f to be positive to the left of the mirror. The relationship between object distance, image distance, and focal length is represented by

$$\frac{1}{s_o} + \frac{1}{s_i} = -\frac{2}{R} \qquad [26.24]$$

and this is the **mirror formula**.

Suggestions on Problem Solving

1. Read the problem carefully for *given* information that is subtly stated. If the image created by a single thin lens appears on a wall, the lens is positive, its focal length is positive, the image is real, the image distance is positive, the image is inverted, and the magnification and image height are both negative. All of that detail may *not* be explicitly spelled out. Make a sketch of the lens or mirror, the object and image, putting everything in roughly where it belongs, with all the numbers included so the setup can be visualized as a whole. Worry about the ray diagram later.

2. When working problems, keep in mind the sign convention as you substitute in the numbers. Read your diagram to confirm the signs. If you know the image formed by the lens is inverted, y_i must go in as a negative number, just as M_T must be negative. If that image is 3 m tall, $y_i = -3$ m. Watch out for the signs of the radii of curvature of lenses and mirrors—these are often messed up. The sign convention requires that the *light enter from the left*. If this is not the case in a particular situation, redraw the figure.

3. Several of the equations in this chapter, such as the Lensmaker's Formula and the Gaussian Lens Equation, are written in terms of the reciprocals of quantities that are of interest. Among the most common errors made is to neglect to invert the final result. For example, computing the numerical value for $1/f$ and

then giving that number as your answer for f. Furthermore, remember that you *cannot* take an equation like $1/s_o + 1/s_i = 1/f$ and invert both sides!

4. In most cases, especially with only one lens or mirror involved, you should already have a good idea of what the image will be from Table 26.3 and/or Table 26.5. *Whenever possible, check the results of your calculations with reality; that is, with those tables.* If the object is just a little bit farther from the positive lens than one focal length and the image is computed to appear roughly one focal length beyond the lens, then you should know something is terribly wrong!

5. When drawing ray diagrams, remember that objects nearer to a positive lens produce real images that are farther away. Also keep in mind that all images formed by a convex mirror lie between the vertex and the focal point behind the mirror. If you have a rough idea of what to expect, your ray diagrams will not go wild.

6. Notice that the angular magnification of both the compound microscope and the astronomical telescope are negative because the images are inverted. That means that you will have to be a little careful about signs when doing problems dealing with these devices. A 10× telescope means $M_A = -f_O/f_E = -10$; both focal lengths are positive.

Discussion Questions

1. What happens to the focal length of a glass lens when it's taken from the air and placed in water? Explain your answer.

2. If a lens of glass surrounded by air is negative, what can be said about the identically shaped lens made of air surrounded by glass? A bubble in a glass of beer is a tiny lens. What kind?

3. Explain why it is that the focal length of a lens actually depends on the color of the light being transmitted.

4. How do goggles work to allow an underwater swimmer to see clearly?

5. What is a quick physical means of determining the approximate focal length of a converging lens? How might you use a known strong positive lens in order to find the focal length of a negative lens?

6. If a horse stands facing a positive lens, which part of the beast will be closest the lens in the real image? In the virtual image? Draw the appropriate ray diagrams.

7. Imagine a converging glass lens in a chamber filled with a gas under a few atmospheres of pressure. Suppose the lens is illuminated by parallel light. What, if anything, will happen to the point at which the beam converges if the gas is gradually pumped out?

8. If a real image is formed by a large plane mirror, what can be said about the incoming rays? Incidentally, a tiny flat mirror can form a real image just as a pinhole does.

9. What is the focal length of a plane mirror? What does Eq. (26.24), the mirror formula, say about the image distance? What then is the magnification of a flat mirror according to the equations?

10. Figure Q10 depicts a hyperboloidal mirror and its accompanying geometry. Explain what is happening in the diagram. Describe the wavefronts before and after reflection from the hyperboloidal mirror.

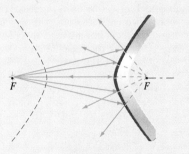

Figure Q10

11. Legend has it that Archimedes (ca. 287 B.C.–212 B.C.) burned the invading Roman fleet by focusing sunlight onto the sails. One version of the story has him on a mountainside lining up soldiers holding brightly polished shields. How could he have arranged the soldiers to accomplish this task?

12. Suppose we take two positive lenses and place them in contact with one another. In what sense is the combination more powerful than either lens separately?

13. Figure Q13 is a diagram of an X-ray camera used for diagnosing the implosion of tiny laser-fusion targets (Sect. 32.10) at the Lawrence Livermore Laboratory in California. Explain how it works to form an image of the target. (You should be able to figure out what's happening even though some aspects were not discussed in the text explicitly.)

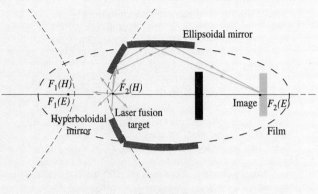

Figure Q13

14. A person with normal vision puts on a pair of eyeglasses made for a hyperope. What will things look like near and far with the eye relaxed? Not relaxed?

15. Microscopes operating in the visible are limited in magnification to about 1200×. How big an object would produce an image seen in the eyepiece to be $\frac{1}{2}$ mm across? What does that suggest about the limitations of the instrument? Will a 12 000× light microscope resolve any finer details of the object?

16. The spotlight in Fig. Q16 is a rather common, very simple one used in theaters in the United States. Explain what it does and how it works. Where is the filament located with respect to both the lens and the mirror? Why?

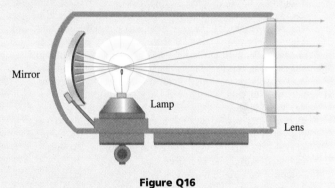

Figure Q16

17. Figure Q17 shows an ellipsoidal reflector spotlight, which is probably the most widely used form of the device. How does it work?

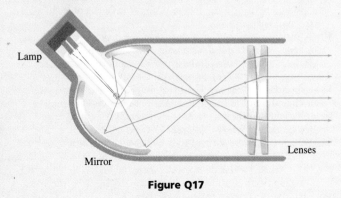

Figure Q17

Multiple Choice Questions

1. A converging lens is mounted on an optical bench, and a candle is placed in front of it so that an image of the flame appears on a screen 1.5f beyond the lens. The object is (a) between f and 2f (b) beyond 2f (c) closer than f (d) near infinity (e) none of these.

2. Which cannot be created by a negative lens? An image that is (a) virtual, right-side-up, and closer than f (b) right-side-up and smaller than life (c) minified, inverted, and virtual (d) minified, closer than f, and right-side-up (e) none of these.

3. Which cannot be formed by a positive lens? An image that is (a) virtual, right-side-up, and larger than life (b) virtual, inverted, and minified (c) real, inverted, and minified (d) real, inverted, and magnified (e) none of these.

4. A convex lens with a very small focal length is placed in contact with a convex lens with a very large focal length. The combined focal length will be (a) much larger than either (b) much smaller than either (c) approximately equal to the smaller (d) approximately equal to the larger (e) none of these.

5. A lens made of glass ($n = 1.50$) has a convex first surface with a radius of curvature of 2.0 m and a second concave surface with a radius of curvature of 1.0 m. Its focal length is (a) 3.0 m (b) $\frac{1}{2}$ m (c) $-\frac{1}{4}$ m (d) -4.0 m (e) none of these.

6. When an object is placed between the vertex and focal point of a negative lens, its image is (a) virtual, enlarged, and right-side-up (b) real, minified, and right-side-up (c) real, enlarged, and inverted (d) virtual, minified, and right-side-up (e) none of these.

7. A frog sitting 40 cm from a positive lens of focal length 20 cm hops out to 1.00 m. Its image on the right of the lens moves from (a) 40 cm to 1.00 m (b) 40 cm to 25 cm (c) 20 cm to 40 cm (d) 40 cm to 20 cm (e) none of these.

8. A bird is sitting on a perch peering into a small convex mirror in its cage. The image it sees is (a) always real (b) always virtual (c) always magnified (d) always inverted (e) none of these.

9. Suppose you make a calculation of the magnification of the image of a real object located somewhere in front of a convex spherical mirror and it turns out to be -1.5. You can then conclude that (a) the calculation was wrong (b) the image is smaller than life (c) the object was upside down (d) the image is larger than the object (e) none of these.

10. A little mirror used by a dentist to see inside the mouth is a concave spherical device with a focal length of 25 mm. Held 2.0 cm from a tooth, it provides a magnification of (a) 0.5 (b) 10 (c) 5 (d) 0.1 (e) none of these.

11. Suppose that you carry out the analysis of a mirror system and find that the image distance is negative. This finding tells you that the image is (a) magnified (b) inverted (c) black and white (d) real (e) none of these.

12. Which cannot be formed by a concave spherical mirror? An image that is (a) real, inverted, and minified (b) virtual, right-side-up, and larger than life (c) magnified, upside down, and beyond 2f (d) life-size, real, and inverted (e) none of these.

13. A concave makeup mirror magnifies, by a factor of 2, the face of anyone looking into it a distance of 25 cm. (The image is, of course, right-side-up.) The mirror's focal length must therefore be (a) 5 cm (b) 50 cm (c) 500 cm (d) 5000 cm (e) none of these.

14. The off-axis mirror forming a perfect point image depicted in Fig. MC14 is (a) a segment of a paraboloid (b) planar (c) a segment of an ellipsoid (d) a segment of a hyperboloid (e) none of these.

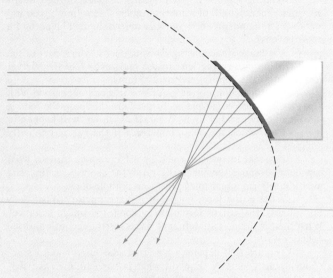

Figure MC14

15. Can a paraboloidal mirror be used to create a reflected diverging spherical wave? (a) no (b) yes, by sending a parallel beam into a concave mirror (c) yes, by sending a beam converging toward the focus of a convex mirror (d) yes, by sending a parallel beam toward a convex mirror (e) none of these.

16. George is nearsighted. George wears negative corrective lenses. The image focal point of either one of George's eyeglass lenses (a) lies at infinity (b) lies at his near-point (c) lies in the middle of his eye (d) lies at his far-point (e) none of these.

17. If the human eye, simply modeled, has a power of about 59 D, its focal length is (a) $+17$ mm (b) -5.9 mm (c) $+5.9$ mm (d) -17 mm (e) none of these.

18. A human eye is about 2 cm long and at night has a maximum pupil diameter of 8 mm. The f-number of the system is (a) 0.25 (b) 4 (c) 16 (d) 2.5 (e) none of these.

19. Reducing the radius of the diaphragm on a camera by a factor of 2 (a) decreases the f-number by a factor of 2 (b) increases the f-number by 2 (c) increases the f-number by a factor of 2 (d) decreases the f-number by 2 (e) none of these.

20. Suppose you slip out the eyepiece from a microscope and replace it with one having twice the focal length. The magnification of the instrument will (a) double (b) halve (c) remain unchanged (d) quadruple (e) none of these.

21. Suppose we come upon a bag of lenses that are marked +1.0 D, −4.0 D, +10 D, +0.10 D, +95 D, and −0.50 D. Which would you use to build a microscope with the largest possible magnification?   (a) +0.10 D as objective and +95 D as eyepiece   (b) −4.0 D as objective and −0.50 D as eyepiece   (c) +95 D as objective and +0.10 D as eyepiece   (d) +10 D as objective and +0.10 D as eyepiece   (e) none of these.

22. Suppose we come upon a bag of lenses that are marked +1.0 D, −4.0 D, +10 D, +0.10 D, +95 D, and −0.50 D. Which would you use to build an astronomical telescope with the largest possible magnification?   (a) +0.10 D as objective and +95 D as eyepiece   (b) −4.0 D as objective and −0.50 D as eyepiece   (c) +95 D as objective and +0.10 D as eyepiece   (d) +10 D as objective and +0.10 D as eyepiece   (e) none of these.

Problems

LENSES

1. [I] What is the radius of curvature of the spherical surface of a planar-convex lens that has a diameter of 20 cm, is $\frac{1}{2}$ cm thick at its center, and curves gradually to a sharp edge all around its periphery?

2. [I] A thin glass lens ($n = 1.50$), which is fatter in the middle than at its edges, has one flat face and one face with a radius of curvature of 1.00 m. How far from the lens will it focus sunlight?

3. [I] A bi-convex lens made of plastic ($n = 1.58$) has radii of curvature of +1.00 m and −1.00 m. What is its focal length?

4. [I] A glass ($n = 1.5$) bi-convex thin lens has radii of curvature having magnitudes of 2.0 m and 5.0 m. Determine its focal length in air.

5. [I] We measure the radius of curvature of the face of a planar-convex lens to be 2.00 m and find its focal length to be 1.60 m. What is its index of refraction?

6. [I] A cat is 30.0 cm in front of a positive lens that has a 10.0-cm focal length. Locate and describe the image.

7. [I] A convex lens having a 60.0-cm focal length is placed 100.0 cm from a frog. Where will the image of the frog be located?

8. [I] A window on a spaceship in the Mongoian Royal Fleet is flat on one side and slightly concave on the other. As a result, it has a focal length of −10 m. What does the Universe look like through that porthole? A beacon light is 50 m from the window. How far away will it appear to someone looking out?

9. [I] A light bulb is 0.75 m in front of a thin positive lens that has a focal length of 0.25 m.   (a) Completely describe the image.   (b) Draw a ray diagram.

10. [I] A sharp image of a bus appears on a piece of paper held 2.5 m behind a positive eyeglass lens that has a focal length of 2.0 m. How far away is the vehicle?

11. [I] A light bulb is 80.0 mm in front of a positive lens with a focal length of 120 mm. Determine the location and type of image formed.

12. [I] The lens in a camera has a focal length of 60.0 mm and is 100 mm from the film plane. How far away should the bug be that is having its picture taken?

13. [I] A camera lens has a focal length of 50 mm and an aperture with a radius of 12.5 mm when its diaphragm is wide open. What is the speed of the lens—that is, what is its minimum f-number?

14. [I] A 52.0-mm focal-length camera lens is focused on a very distant motorcycle heading toward the photographer. How far must the lens be advanced from the film plane if the driver is to be in focus when 5.00 m away?

15. [I] A 35-mm camera lens has a focal length of 52 mm and is marked $f/1.4$. What is the maximum diameter of the lens aperture?

16. [I] A photo of a racehorse is perfectly exposed but somewhat blurry when taken at 1/30 s and $f/16$. To "stop" the motion, the shutter speed is raised to 1/500 s. What must now be the new f-number?

17. [I] A lens used as a magnifying glass is found to provide an angular magnification of 2.0× when viewed by a normal eye in the relaxed state. What is its focal length?

18. [I] Holding the magnifying glass close to his eye, Dr. Watson brought the bloodstained scrap of cloth up toward it until his eye was relaxed and the image clear. If the glass has a focal length of 10 cm and the good doctor has a near-point of 25 cm, what was the magnification?

19. [I] A little magnifier marked 8× is for sale in a camera store. It is designed for examining photographs and allows light to enter from the sides through a clear cone-shaped stand that keeps the lens one focal length above whatever flat surface the thing sits on. How high above the surface should the lens be?

20. [I] The largest refracting telescope in the world at the Yerkes Observatory in Wisconsin has a 40-in.-diameter lens with a focal length of 63 ft. The reflector on Mount Palomar in California has a 200-in.-diameter parabolic mirror with a focal length of 55.5 ft. Determine the f-number for both. Which is the faster? (That is, all else being equal, which would require the shorter exposure time?) Roughly how many times faster is one than the other?

21. [I] The 20-cm-diameter image of the face of a clock appears on a screen 1.0 m from a positive lens. A negative lens is then placed in the path 90 cm behind the first lens, and the image moves out an additional 10 cm beyond where it was. Determine the focal length of the second lens. How big is the final image?

22. [I] A negative lens with a focal length of magnitude 15.0 cm is in contact with a positive lens having a focal length of 24.0 cm. What is the focal length of the combination?

23. [I] We wish to spread a narrow parallel laser beam out into a diverging cone of light so that it makes a large blotch on a nearby screen. At our disposal are two positive lenses, each with a focal length of 20 cm. If one lens is placed in the beam and the resulting blotch is still too

small, will it help any to put the other lens in right up against the first? What is the combined focal length?

24. [I] An optical system consists of three lenses with focal lengths of +10 cm, +20 cm, and +5.0 cm. The first and second of these are separated by 30 cm, the third and fourth by 5.0 cm. If parallel light is shone into the first lens, how far from the third will it be brought to a focus by the system?

25. [I] A rather expensive well-corrected (for aberrations) lens consists of three simple lenses of focal lengths +10 cm, −20 cm, and +5.0 cm all glued together. What is the combined focal length? Can it form real images?

26. [I] A lens positioned 1.00 m from a light bulb produces a sharply focused image of the filament on a screen 33 cm beyond the lens. What is the refractive power of the lens?

27. [I] The near-point of a person's eye is 100 cm away rather than a more desirable 25.4 cm. What contact lens should be prescribed?

28. [I] The far-point of a person's eye is 100 cm away rather than a more desirable ∞. What contact lens should be prescribed?

29. [I] Someone wearing corrective contact lenses having a focal length of −5.0 m can see quite normally. Determine this person's unaided far-point.

30. [I] A compound microscope is made with a 20× objective and a 5× eyepiece. What is the total magnification of the device?

31. [I] Two lenses used to form a homemade microscope are mounted in a tube with a separation of 10.0 cm. If the objective has a focal length of 10 mm and the eyepiece has a focal length of 30 mm, what is the so-called tube length?

32. [I] Typically, a good eyepiece will have a focal length of around an inch. Suppose we have one with a focal length of 2.5 cm and we wish to make a 10× astronomical telescope with it. What objective shall we use and how long will the scope end up?

33. [I] Suppose we wanted to make a telescope to look at the stars. At our disposal is an old eyeglass lens (+1.00 D) donated by hyperopic Aunt Jane and a little magnifier having a focal length of 3.0 cm that came with a stamp collection we got from someone. How long should the tube be for relaxed viewing? What magnification can be achieved?

34. [II] Prove that the products of the segments of two intersecting chords of a circle are equal. (See p. 971.)

35. [II] Referring to Fig. 26.6, prove that

$$\rho \approx \frac{(h^2 - y^2)}{2R} \quad \text{and} \quad \delta \approx \frac{y^2}{2R} \qquad [26.5]$$

36. [II] The image of a face is to be projected life-sized onto a screen via a bi-convex lens whose both radii equal 0.60 m. The lens is made of glass ($n = 1.5$), and the whole thing is taking place in air. (a) Compute the necessary location of the object. (b) How far must the lens be from the screen? (c) Draw a ray diagram.

37. [II] Starting with Eq. (26.3), show that, had the lens been immersed in a medium of index n_m rather than air, the Lensmaker's Formula would result, but instead of the first term on the right being n_1, there would now be an n_1/n_m term.

38. [II] Design the lens for a simple 35-mm slide projector that will cast an image on a screen 10 m away that is enlarged 100 times.

39. [II] A grasshopper sitting 10 cm to the left of a convex lens sees its image on a screen 30 cm to the right of the lens. It then jumps 7.5 cm toward the lens. (a) Where will its image be now? (b) Describe the image in both instances. (c) Draw a ray diagram of the situation after the jump.

40. [II] It's a clear cold winter's day in Nebraska, and a youngster playing outside decides to warm herself with an ice-burning glass. She takes a flat sheet of ice 5.0 cm thick, cuts it into a 50-cm-diameter disk, and then shapes one surface so that it's spherical, gradually thinning out to a sharp edge. If it's 5.0 cm thick at its center, determine the focal length of her lens.

41. [II] A photographer wishes to take a picture of his pet chicken Fred, who happens to have a fine face 5.0 cm high. While standing 2.0 m away, he selects a lens that will fill the film (24 mm top-to-bottom) with Fred's poultry physiognomy. What lens should be used?

42. [II] Suppose you wanted to take a picture of the Moon (which has a diameter of 0.273 times that of Earth and is 3.84×10^8 m away) using a 35-mm camera with a normal lens having a 50-mm focal length. How large will the image be on the film? The diameter of the Earth is about 1.27×10^7 m.

43. [II] A person who is nearsighted has a far-point at 1.00 m and a near-point at 25.0 cm. (a) What corrective lens should this individual wear if it's to be mounted 15.0 mm from the eye? (b) Where is the near-point when glasses are worn?

44. [II] George has been nearsighted since he was eighteen and now that he's fifty-five, his far-point has moved in to 3.00 m. But even more annoying, his near-point has recently migrated out to 45 cm. Assuming that both his eyes are the same, prescribe bifocals with the tops for distance and the bottoms for reading to be worn at 2.0 cm from the cornea. Incidentally, it was Ben Franklin who made the first pair of bifocals (by sawing two lenses in half).

45. [II] Using Fig. 26.21, show that the equation for the transverse magnification produced by a pinhole is the same as for a lens. What does this similarity tell us about the pinhole-film plane distance and the magnification?

46. [II] A combination lens consists of a positive lens with a focal length of +0.30 m located 10 cm in front of a negative lens with a focal length of −0.20 m. Determine the location and magnification of the image of an object 30 cm in front of the first lens by finding the effect of each lens in turn.

47. [II] A compound microscope formed of two lenses, a 20× objective and a 10× eyepiece, is adjusted for viewing by a relaxed eye. It has a standard tube length of 160 mm. (a) What is the total magnification of the device? (b) What is the focal length of each lens? (c) Compute the object distance.

48. [II] Suppose we have two lenses with focal lengths of 2.0 cm and 2.0 mm. How should they be arranged to make a microscope to view, with a relaxed eye, an object 2.5 mm from the front lens? What is the separation between lenses? What is the magnification of the device?

49. [III] The image projected by an equiconvex lens ($n = 1.50$) of a 5.0-cm-tall frog standing 0.60 m from a screen is to be 25 cm high. Compute the necessary radii of the lens.

50. [III] A thin double convex glass lens (with an index of 1.56) surrounded by air has a 10-cm focal length. If this lens is placed underwater (having an index of 1.33) 100 cm beyond a small fish, where will the guppy's image be formed?

51. [III] Write an expression for the focal length (f_w) of a thin lens that is immersed in water ($n_w = \frac{4}{3}$) in terms of the focal length it had when it was surrounded by air (f_a). Take the index of the lens to be 1.5.

52. [III] An equiconvex thin lens L_1 is placed in close contact with a thin negative lens L_2, whereupon the combination has a focal length of 0.50 m in air. If the lenses are made of glass with indices of 1.50 and 1.55, respectively, and if the focal length of L_2 is -0.50 m, compute the radius of each surface. Note that a convex surface of L_1 fits precisely against a concave surface of L_2.

53. [III] A homemade TV projection system uses a large positive lens to cast the image of the screen onto a wall. The final picture is enlarged 3 times and although rather dim, it's nice and clear. If the lens has a focal length of 60 cm, what should be the distance between the screen and the wall? Why use a large lens? How should we mount the set with respect to the lens?

54. [III] A nearsighted person has a far-point of 200 cm. What power eyeglass lens should be worn 2.0 cm from the cornea? What contact lens is equivalent to this? Show that both have the same far-point. Verify that Eq. (26.3) agrees with your results.

55. [III] Mary Lou got her first pair of reading glasses ($+2.0$ D) when she was forty-eight, in 1979. Now, in order to peruse her mail (still wearing those spectacles 2.0 cm down on her nose), she finds she must hold it 80 cm away from her eyes. Prescribe new glasses for her.

MIRRORS

56. [I] A convex mirror has a radius of curvature whose magnitude is 0.50 m. What is its focal length?

57. [I] A polished 50-cm-diameter steel ball in the hands of a statue reflects the scene around it. Determine the ball's focal length.

58. [I] We wish to convert light from a candle flame into a parallel beam using an inexpensive mirror 30 cm away from it. Design the appropriate arrangement and draw a ray diagram.

59. [I] If an object 200 cm from the vertex of a spherical concave mirror is imaged 400 cm in front of the mirror, what is the latter's focal length?

60. [I] A 15-cm-tall Teddy bear stands 60 cm from the vertex of a concave mirror having a radius of curvature of 60 cm. Determine the resulting image and describe it in detail. Draw a ray diagram illustrating the phenomenon.

61. [I] A display lamp having a bright 5.0-cm-long vertical filament is positioned 30 cm from a concave mirror that projects the bulb's image onto a wall 9.0 m from the vertex. (a) What is the radius of curvature of the mirror? (b) How big is the image?

62. [I] Suppose you had a spherical mirror with a 0.20-m

radius. If you wished to project the image of a candle flame onto a piece of paper 1.10 m away, where should the candle be located? Describe the image, giving the magnification as well. Draw a ray diagram.

63. [I] If your nose is 20 cm from a convex spherical mirror having a radius of 100 cm, what will the image look like? Draw a ray diagram.

64. [I] The cornea of the human eye behaves like a little convex spherical mirror when seen from nearby. Suppose you look at yourself reflected in someone's eye a distance of 20 cm away. Taking the radius of curvature of the cornea to be 8.0 mm, what will your image look like? Where will it be located?

65. [II] An Earth satellite in the shape of a 2.0-m ball is orbiting at an altitude of 500 km. It is being tracked by a telescope having a spherical concave mirror with a 1.0-m radius of curvature. How big is the image of the craft formed by the mirror?

66. [II] We wish to design an eye for a robot using a concave spherical mirror such that the image of a 1.0-m-tall object 10 m away fills its 1.0-cm-square photosensitive detector (which is movable for focusing purposes). Where should this detector be located with respect to the mirror? What should be the focal length of the mirror? Draw a ray diagram.

67. [II] You are herewith requested to design a little dentist's mirror to be fixed at the end of a shaft for use in someone's mouth. The requirements are (1) that the image be right-side-up as seen by the dentist, and (2) that when held 1.5 cm from a tooth, the mirror should produce an image twice life-size.

68. [II] Prove that with a spherical mirror of radius R, an object at a distance s_o will result in an image that is magnified by an amount

$$M_T = \frac{R}{2s_o + R}$$

69. [II] With the results of Problem 68 in mind, suppose that we have a concave mirror with a radius of curvature of 60 cm, which is forming an image of an object 2.4 m away. What will be the resulting magnification? Describe the image. Draw a ray diagram.

70. [II] A keratometer is a device used to measure the radius of curvature of the cornea of the eye, which is useful information when fitting contact lenses. In effect, an illuminated object is placed a known distance from the eye, and the reflected image off the cornea is observed. The instrument allows the operator to measure the size of that virtual image. Suppose, then, that the magnification is found to be $0.037\times$ when the object distance is set at 100 mm. What is the radius of curvature?

71. [II] Consider a spherical mirror. Show that the locations of the object and image are given by

$$s_o = \frac{f(M_T - 1)}{M_T} \quad \text{and} \quad s_i = -f(M_T - 1)$$

72. [II] Looking into the bowl of a spherical soup spoon, someone 25 cm away sees their image reflected with a

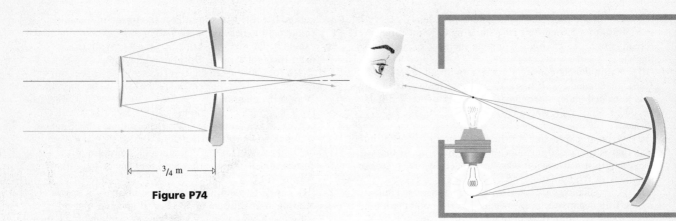

Figure P74

Figure P75

magnification of −0.064. Determine the radius of curvature of the spoon.

73. [III] A large, upright, convex spherical mirror in an amusement park is facing a plane mirror 10.0 m away. A youngster 1.0 m tall standing midway between the two sees herself twice as tall in the plane mirror as in the spherical one. In other words, the angle subtended at the observer by the image in the plane mirror is twice the angle subtended by the image in the spherical mirror. What is the focal length of the latter?

74. [III] The telescope depicted in Fig. P74 consists of two spherical mirrors. The larger (which has a hole through its center) has a radius of curvature of 2.0 m and the smaller of 60 cm. How far from the smaller mirror should the film plane be located if the object is a star? What is the effective focal length of the system?

75. [III] Figure P75 shows the arrangement of a classic illusion that works surprisingly well. A person looks through a window into the box and sees a glowing bulb in a socket. The bulb is turned off and vanishes, leaving only the socket seen in roomlight entering the window. Explain what is happening. If the bulb-mirror distance is 1.0 m, compute the mirror's radius of curvature.

76. [III] Starting with the equation $y^2 + (x - R)^2 = R^2$ for a circle whose center is displaced by a distance R from the coordinate origin, solve for x and expand the appropriate term in a binomial series. Compare the first term with the equation of a parabola having its vertex at the origin, $y^2 = 4fx$.

27

Physical Optics

LIGHT IS ELECTROMAGNETIC, AND it is wavelike, and (if we are careful to remember that what we usually observe are the macroscopic manifestations of torrents of photons) we can say that *light is an electromagnetic wave*. Because the wavelengths associated with the visible region of the spectrum are quite small, the wave nature of light can sometimes be neglected. Geometrical optics can be developed using only rays to indicate the direction light travels, without saying anything about the nature of what's traveling. Still, there are a variety of phenomena—*polarization, interfer-*

Crystals of potassium chloride, calcium carbonate (calcite), and sodium chloride (table salt). Only the calcite produces a double image.

ence, and *diffraction*—that reveal the underlying wave characteristics of radiant energy, and these form the study of **physical optics**.

POLARIZATION

A curious discovery made in 1669 led to one of the great insights into the nature of light. A newly found crystal, now called calcite, was observed to have the remarkable ability to produce double images of everything seen through it. That strange separation, or **polarization** of light, was first studied in depth by Huygens, who explained some aspects of it via his wave theory. In the early 1800s, Augustin Fresnel, joined by Arago, conducted a series of experiments to see if polarization had any effect on the process of interference. Their positive results were utterly bewildering because they, like Huygens, believed light was a longitudinal wave in the aether. For the next few years, Fresnel in France, allied with Thomas Young in England, wrestled with that stubborn problem until finally Young (in 1817) had the crucial revelation: **light is a transverse wave** (something Robert Hooke had suggested a century before).

We can wiggle a length of rope up and down in a vertical plane while the wave progresses horizontally (Fig. 27.1). The disturbance remains in that fixed plane, and we say that the wave is plane-polarized. When viewed end-on, the vibration appears to be along a line perpendicular to the propagation direction, and the situation is also referred to as *linear polarization.* The plane in which such a transverse wave oscillates is the *plane of polarization,* and it certainly need not be vertical—any orientation is equally possible. When the **E**-field oscillates in a plane, the TEM wave is **plane-polarized**.

It's even possible for the **E**-field to swing around in a circle, in which case the wave is said to be **circularly polarized**. If the advancing wave revolves clockwise (looking toward the source), then it's said to be *right-circularly polarized;* if counterclockwise, it's *left-circularly polarized.* The ordinary mode in which a jump rope is operated is a familiar example of the kind. With circular light, the electric-field vector remains constant in magnitude while it revolves once around with every advance of one wavelength (Fig. 27.2). Each sustained configuration is known as a *state of polarization,* and so there are $\mathscr{P}$-state (plane-polarized), $\mathscr{R}$-state (right-circular), and $\mathscr{L}$-state (left-circular) lightwaves.

We needn't concern ourselves with circular light other than to realize that photons can exist in two states, $\mathscr{R}$- and $\mathscr{L}$-, which carry oppositely directed angular momenta. All forms of light, all observed states, are a mix of these two photon states.

Figure 27.1 Plane-polarized waves. The rope oscillates in a plane. Notice how the end point vibrates along a line as the wave advances. As a result, the wave is also said to be linearly polarized.

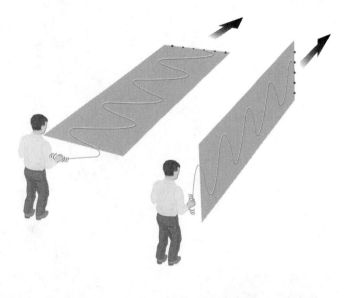

27.1 Natural Light

The light coming from an ordinary (nonlaser) source is a polychromatic jumble of overlapping emissions from a tremendous number of more or less independent atoms. Each atom radiates a tiny polarized photon-wavetrain lasting less

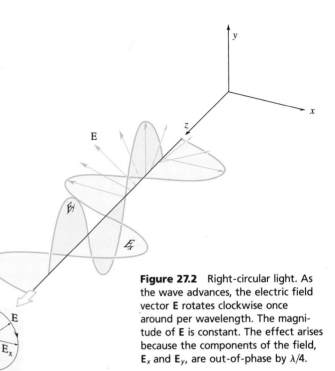

Figure 27.2 Right-circular light. As the wave advances, the electric field vector **E** rotates clockwise once around per wavelength. The magnitude of **E** is constant. The effect arises because the components of the field, **E**$_x$ and **E**$_y$, are out-of-phase by $\lambda/4$.

than $\approx 10^{-8}$ s, and these all arrive with no particular phase relationship or orientation, one to the other. The resultant field is the sum of all the separate wavetrain contributions at that location and is constantly changing as new photons arrive and others sweep by. Thus, the net optical field has some polarization at every point, at every instant, but that polarization varies exceedingly rapidly in a totally random fashion. Because the state of polarization of the light is sustained for less than $\approx 10^{-8}$ s, that state is essentially undetectable, and we refer to the light as **unpolarized**, or **natural**. This is the case even though, strictly speaking, there is no such thing as unpolarized light. Often, the light we experience in the environment is a blend of polarized and unpolarized known as *partially polarized*. Ordinary sky light, the light reflected from objects, and the light transmitted through windows are all partially polarized.

There are several ways to conceptually visualize unpolarized light. One of the easiest is to think of it as a superposition of many differently oriented plane-polarized waves of different amplitudes, all changing rapidly and randomly (Fig. 27.3a). In time, the resultant E-field points every which way, swiftly jumping from one orientation and amplitude to another.

Figure 27.3 (a) Natural light is a jumble of random, rapidly changing fields. (b) Looking toward the source, one sees a resultant field oscillating, first in one direction, then another, each lasting for a fraction of a period before **E** jumps abruptly to a new random orientation.

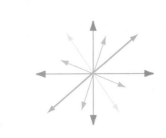

(a)   (b)

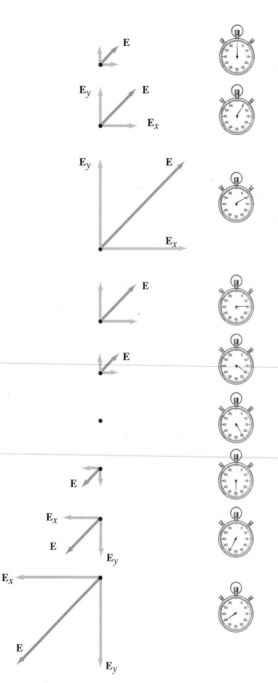

Figure 27.4 A sequence in time showing the oscillation of a linearly polarized light field. **E** is always along a 45° line in the first and third quadrants. The two component fields E_x and E_y are in-phase—when one is positive, the other is positive. As the wave sweeps past a fixed point in space, the field oscillates in time, as shown.

Any electric field vector can be resolved into two mutually perpendicular components (which we take for convenience as horizontal and vertical). For example, if at some instant the electric field vector points up and to the right, it can be thought of as the resultant of two smaller constituent fields, one to the right and one up (Fig. 27.4). As the tilted field oscillates, it can be resolved into—and replaced by—the two perpendicular fields that oscillate at the same rate in-step with each other (Fig. 27.5).

The natural-light field vector depicted in Fig. 27.3b can similarly be resolved into horizontal and vertical components that change from moment to moment as the vector changes. The result at any instant is a net horizontal field and a net vertical field. *We can think of the unpolarized jumble as equivalent to two independent, equal-amplitude $\mathcal{P}$-states varying rapidly and randomly along any two perpendicular axes.* Each of these carries $\frac{1}{2}$ the total irradiance (I_0) of the light.

27.2 Polarizers

A device that takes an input of natural light and transforms it into an output of polarized light is a **polarizer**. Natural light is a superposition of two independent, perpendicular $\mathcal{P}$-states. Anything that separates these two equal-irradiance components, discarding one and passing on the other, will produce a beam of light that has its **E**-field fixed in a plane, a beam that is plane-polarized. Such a device is a *linear polarizer*. **Ideally, if natural light of irradiance I_0 is incident on a linear polarizer, $\mathcal{P}$-state light of irradiance $\frac{1}{2}I_0$ will be transmitted.**

Before we go on, we must establish a means of experimentally verifying that a device is in fact a linear polarizer. When natural light is incident on the ideal linear polarizer of Fig. 27.6, only $\mathcal{P}$-state light emerges. It has an orientation parallel to a specific direction called the **transmission axis** of the polarizer. This is not literally a single axis but a direction—any line on the polarizer parallel to that axis is also a transmission axis. Only the component of the incident electric field parallel to the transmission axis will emerge as the transmitted light. An oscillating electric field **E** tilted over at an angle θ will come out of the polarizer.

If the polarizer is revolved in its own plane about the central axis, changing θ in the process, the transmission axis will rotate along with it, as will the emerging **E**-field. Nonetheless, the reading of the detector (for example, a photocell or a light meter on a camera) that corresponds to the transmitted irradiance will be constant because of the symmetry of natural light—each transmitted $\mathcal{P}$-state carries the same average amount of energy. Similarly, the human eye cannot distinguish the various polarization states and will see only a constant irradiance as the polarizer is revolved.

Now, suppose we introduce a second identical polarizer called an **analyzer**, whose transmission axis is vertical (Fig. 27.7). The first polarizer transmits a wave of amplitude E_{01} tilted over at an angle θ. Only the vertical component of that wave,

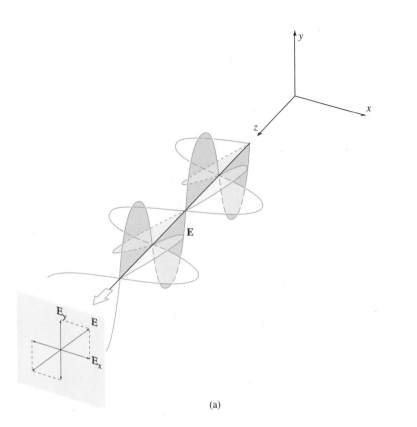

(a)

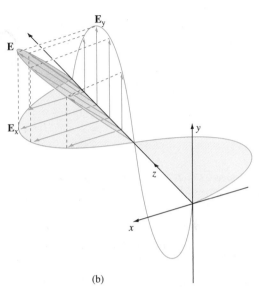

(b)

Figure 27.5 A linearly polarized wave. In Fig. 27.4, we examined the disturbance at a fixed point in space as time progressed. (a) Here, we examine the disturbance spread out in space at a given instant in time. (b) The resultant field at any instant is **E**, the sum of $\mathbf{E}_x$ and $\mathbf{E}_y$.

namely $E_{01} \cos \theta$, will be passed by the analyzer because its transmission axis is vertical. (Note that θ is the angle between the transmission axes of the polarizer and analyzer.) According to Eq. (24.8) the irradiance reaching the detector is proportional to the square of the amplitude ($E_{02} = E_{01} \cos \theta$) of the wave coming out of the analyzer; thus

$$I = \tfrac{1}{2} c\varepsilon_0 (E_{01} \cos \theta)^2 = (\tfrac{1}{2} c\varepsilon_0 E_{01}^2) \cos^2 \theta = I_1 \cos^2 \theta$$

where the irradiance leaving the first polarizer is I_1. Slowly revolving the analyzer about the central axis varies θ, and the transmitted irradiance changes as the cosine-

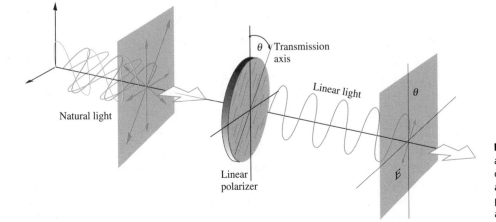

Figure 27.6 Natural light incident on a linear polarizer. The transmission axis of the polarizer is tilted at an angle θ and only the **E**-field parallel to θ is passed. In this case, the polarizer absorbs half the light that enters it.

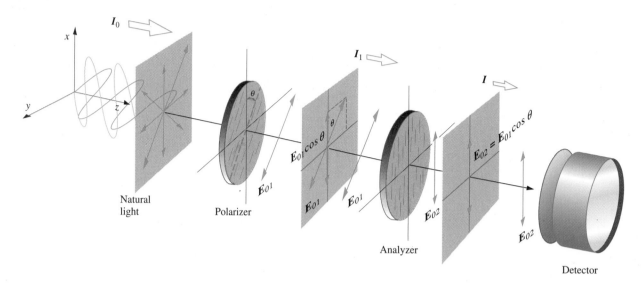

Figure 27.7 A linear polarizer and analyzer—Malus's Law. Natural light enters the polarizer and emerges in a $\mathcal{P}$-state at an angle θ. The analyzer passes a fraction of that light, such that $E_{02} = E_{01} \cos \theta$. The irradiance of the natural light incident on the polarizer is I_0, and it, in turn, transmits an irradiance $I_1 = \frac{1}{2} I_0$. This is incident on the analyzer, and it transmits an irradiance $I = I_1 \cos^2 \theta$.

squared. That is, it varies from its maximum value ($I = I_1$ when $\theta = 0$ and the two transmission axes are parallel) to its minimum value ($I = 0$ when $\theta = 90°$ and the axes are said to be **crossed**). The expression

$$I = I_1 \cos^2 \theta \qquad (27.1)$$

is known as **Malus's Law**, having first been published in 1809 by Étienne Malus, military engineer and captain in the army of Napoleon. Using the setup of Fig. 27.7 and Malus's Law, we can test polarizers to see if they are linear by simply revolving one in front of the other. In fact, if we analyze a beam of light through a revolving polarizer and find that it varies in accord with Eq. (27.1), going completely dark and then bright every 90°, the beam must be $\mathcal{P}$-state light.

Example 27.1 A beam of unpolarized light with an irradiance of 1000 W/m² impinges on an ideal linear polarizer whose transmission axis is vertical. The light is studied using a second ideal linear polarizer, and it is found that the final emerging beam has an irradiance of 250 W/m². (a) How much light leaves the first polarizer? (b) What is the orientation of the second polarizer?

Solution: [Given: $I_0 = 1000$ W/m² incident and $I = I_2 = 250$ W/m². Find: (a) I_1 and (b) θ.] (a) If natural light is incident with an irradiance I_0, half of that will be "eaten" by the first polarizer, which must discard half the light if it is to pass only a $\mathcal{P}$-state. Hence, $I_1 = \frac{1}{2} I_0 = \boxed{500 \text{ W/m}^2}$. (b) Since the second polarizer passes $I_2 = 250$ W/m² $= \frac{1}{2} I_1$, it follows from Malus's Law that $\cos^2 \theta = \frac{1}{2}$. Consequently, $\cos \theta = \sqrt{\frac{1}{2}}$ and $\boxed{\theta = 45°}$; the second polarizer is 45° from the vertical.

▶ **Quick Check:** Substituting back into Eq. (27.1), $I_2 = (500 \text{ W/m}^2) \cos^2 45° = 250 \text{ W/m}^2$.

There are many kinds of linear polarizers, among them several that operate by the dissipative absorption of one of the two perpendicular $\mathcal{P}$-state components of the incident light, a process broadly known as *dichroism*. Of these, we will examine two types—wire grids and Polaroids—the first only because it explains the second.

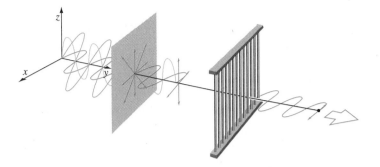

Figure 27.8   Natural light incident on a wire-grid polarizer. The unpolarized light can be represented as two uncorrelated $\mathscr{P}$-states, one horizontal and one vertical. The grid absorbs the vertical E-field and passes the horizontal $\mathscr{P}$-state.

The Wire-Grid Polarizer

The **wire-grid polarizer** consists of a grid of parallel conducting wires each spaced less than a wavelength apart. The electric field of an incident unpolarized electromagnetic wave can again be imagined resolved into two independent, perpendicular components—one parallel to the wires and the other transverse to them (Fig. 27.8). The vertical component of the field effectively drives the free electrons within the wires. Oscillating, up-and-down currents result along the length of each conductor. These currents dissipate some of the wave's energy via joule heating, thereby diminishing the vertical field. At the same time, the accelerating charges reradiate like tiny dipoles. But these emissions are out-of-phase with the incoming wave, thus canceling most of the remaining vertical field in the forward direction. The reradiation in the backward direction constitutes the reflected wave and carries off the remainder of the energy. By contrast, the currents oscillating across the wires are very restricted, and there is hardly any effect on the transverse electric field, which passes through the grid almost the same as it arrived.

The transmission axis of the grid is perpendicular to the wires. It is an extremely common error to assume that the field somehow slips between the wires and is vertically polarized—it doesn't and it isn't! Such grids have been made to operate in the visible (one had 2160 wires per mm) and infrared, though they are much easier to fabricate for microwaves.

Polaroids

In 1928, Edwin H. Land, then a nineteen-year-old undergraduate at Harvard College, invented the first dichroic *sheet polarizer.* That was the humble start of the now-famous Polaroid Corporation. In 1938, Land devised the much improved *H-sheet Polaroid,* which is now easily the most widely used linear polarizer. A sheet of clear polyvinyl alcohol is heated and stretched, thereby aligning—in nearly parallel rows—its long-chain hydrocarbon molecules. The sheet is then dipped into a dye solution rich in iodine. The iodine atoms impregnate the plastic and attach to each polymeric molecule, coating it and effectively forming a conducting chain along it. The conduction electrons associated with the iodine can now move down each chain, from one atom to the next, as if the coated molecule was a long, thin microscopic wire. The result is essentially a wire-grid polarizer with its transmission axis transverse to the aligned molecules. In natural light, each sheet looks gray because it absorbs roughly half the incident light.

A pair of crossed Polaroids. The electric field passed by the first polarizer is perpendicular to the transmission axis of the second polarizer.

Example 27.2 An unpolarized light beam of 800 W/m^2 is incident on an ideal pair of crossed linear polarizers. Now a third such polarizer is inserted between the other two with its transmission axis at $45°$ to that of each of the others. Determine the emerging irradiance before and after the insertion of the third polarizer and explain what's happening.

Solution: [Given: polarizers at $\theta = 0°$, $\theta = 90°$, and $\theta = 45°$, and $I_0 = 800$ W/m^2. Find: I_2 and I_3.] Before the third polarizer is inserted, no light passes through the crossed pair. It might seem that inserting yet another polarizer could have no effect, but that's wrong! If the first filter has a vertical transmission axis, then $I_1 = \frac{1}{2} I_0 = 400$ W/m^2 of vertical $\mathcal{P}$-state light will emerge from it. The second polarizer makes an angle of $45°$ with the first; hence, by Malus's Law

$$I_2 = I_1 \cos^2 45° = \boxed{200 \text{ W/m}^2}$$

The transmitted electric field is now tilted at $45°$—*the presence of the second polarizer has changed the field reaching the third polarizer.* The field is at $45°$ with respect to the last polarizer whose transmission axis is horizontal. Again from Malus's Law

$$I_3 = I_2 \cos^2 45° = \boxed{100 \text{ W/m}^2}$$

which is the amount of horizontal $\mathcal{P}$-state light finally emerging through all three filters.

▶ **Quick Check:** $E_{01} \propto \sqrt{I_1}$; $E_{02} = E_{01} \cos 45°$; $I_2 \propto E_{02}^2 = E_{01}^2 \cos^2 45°$; $I_2 = I_1 (1/\sqrt{2})^2 = \frac{1}{2} I_0 (1/\sqrt{2})^2 = \frac{1}{4} 800$ W/m^2.

27.3 Polarizing Processes

There are several important natural processes that produce polarized light. One of the most commonplace is the reflection from dielectric media. The glare spread across a window pane, a sheet of paper, or a classroom desk; the sheen on a balding head or a shiny nose are all partially polarized.

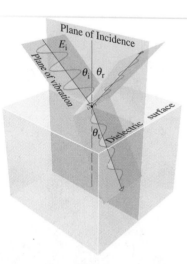

Figure 27.9 A wave with its *E*-field perpendicular to the plane-of-incidence reflecting and refracting at an interface. Electrons then oscillate perpendicular to the plane-of-incidence and reradiate light perpendicular to that plane.

Polarization by Reflection

We can understand what's happening in a simple way by again considering the electron-oscillator model of the atom, although the description is a little simplistic. Figure 27.9 shows an incident wave plane polarized *perpendicular to the plane-of-incidence*. The atoms in the surface are driven by the *E*-field parallel to the interface and they reradiate, producing the usual reflected and refracted waves, both of which are similarly polarized. On the other hand, if the incident *E*-field is linearly polarized *in the plane-of-incidence,* something rather different happens to the reflected wave. The atoms at and near the surface are again driven into oscillation by the *E*-field of the transmitted wave (Fig. 27.10), but this time that's not parallel to the interface. The oscillations are all perpendicular to the refracted ray, making a small angle θ with the reflected ray. Because these dipoles do not radiate along their oscillatory axes, the amount of light reflected is relatively small. In fact, if we could arrange things so that $\theta = 0$, whereupon the reflected and transmitted rays would be perpendicular ($\theta_r + \theta_t = 90°$), no light would be reflected at all.

The special angle of incidence for which this occurs is designated by θ_p and is called the **polarization angle**, or *Brewster's angle,* where $\theta_p + \theta_t = 90°$. It follows from Snell's Law and the fact that $\theta_t = 90° - \theta_p$ that

$$n_i \sin \theta_p = n_t \sin \theta_t = n_t \sin(90° - \theta_p) = n_t \cos \theta_p \qquad (27.2)$$

where shifting the sine by 90° makes it a cosine. Dividing both sides by cos θ_p and again by n_i yields

$$\tan \theta_p = \frac{n_t}{n_i} \qquad (27.3)$$

This equation is known as **Brewster's Law** after the man who discovered it empirically, Sir David Brewster, the inventor of the kaleidoscope.

When an incoming unpolarized beam is incident on a dielectric surface at an angle θ_p, *only the component polarized normal to the plane-of-incidence will be reflected*: ***the reflected light will be totally plane-polarized parallel to the interface*** (Fig. 27.11). At any other angle, the reflected light will be partially polarized. This provides a handy way to find the transmission axis of a linear polarizer. Locate the reflected glare of some light source on a flat, horizontal, nonconducting surface (at an angle around 50°, which is a typical value for θ_p). Now, slowly revolving the polarizer, view the glare through it and when the reflected light essentially vanishes, the transmission axis is vertical (see the photos on p. 1020).

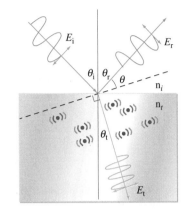

Figure 27.10 Electron-oscillators and Brewster's Law. Here, the incident E-field is in the plane-of-incidence. When $\theta = 0$, the electrons near the surface will oscillate parallel to what would ordinarily be the direction of the reflected beam. But they do not radiate along their axis of vibration, so there will be no reflected wave; $E_r = 0$.

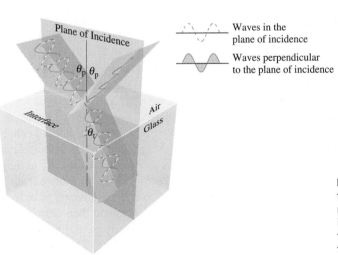

Figure 27.11 The polarization of light that occurs on reflection from a dielectric, such as glass, water, or plastic. At θ_p, the reflected beam is a $\mathcal{P}$-state perpendicular to the plane-of-incidence. The transmitted beam is strong in $\mathcal{P}$-state light parallel to the plane-of-incidence and weak in $\mathcal{P}$-state light perpendicular to the plane-of-incidence—it's partially polarized.

Example 27.3 The image of a child reflects off a wet city street. At what angle should the reflection be viewed if it is to be seen in totally polarized, linear light? Give your answer to three significant figures.

Solution: [Given: $n_i = 1.00$ and $n_t = 1.33$. Find: θ_p.] The relationship that governs polarized reflections is

Brewster's Law:

$$\tan \theta_p = \frac{n_t}{n_i} = \frac{1.33}{1.00} = 1.33$$

and so $\boxed{\theta_p = 53.1°}$.

▶ **Quick Check:** $\tan 53.1° = 1.33$.

(a)

(b)

Light reflecting off the puddle is partially polarized. When viewed through a Polaroid filter whose transmission axis is parallel to the ground, the glare is passed and visible (a). When the Polaroid's transmission axis is perpendicular to the water's surface, most of the glare vanishes (b).

Polarization by Scattering

When light is scattered by molecular-sized particles, as in Rayleigh scattering (p. 936), the particles behave like tiny dipole oscillators. Because the dipole radiation pattern is not the same in all directions, the scattered light will be polarized. Figure 27.12a depicts an incident vertically polarized plane wave driving a molecule, just as sunlight drives the molecules of the atmosphere. Light is reradiated in all directions, except along the axis of vibration—straight up and straight down. In Fig. 27.12b, the incident beam is horizontally polarized, and again it is scattered into all directions, except along the axis of vibration. Now, if the incident beam is unpolarized, it can be envisioned as equivalent to two perpendicular, uncorrelated $\mathcal{P}$-states. Figure 27.12c shows the scattering of such a disturbance, and it's nothing more than the superposition of the two previous situations (a) and (b). Notice that the light coming off perpendicular to the original beam is completely linearly polarized.

If you happen to have a piece of Polaroid, locate the Sun and then examine the sky roughly at right angles to the solar rays. It will appear only partially polarized, mainly because of the depolarizing effects of multiple scattering. The light is scattered and re-scattered several times before reaching you. That mechanism can be illustrated by putting a piece of waxed paper between crossed Polaroids. The multiple reflections within the paper jumble up the E-field orientations and depolarize the transmitted light. As a last example, return to the "put a few drops of milk in a tank of water" experiment of Chapter 25 (p. 939) where suspended, minute fat particles do the scattering. This time examine the scattered light transverse to the beam through a Polaroid—it will be polarized, just as is the sky light. Apparently, bees and pigeons can distinguish polarized sunlight and navigate using it even when the Sun cannot be seen during overcast days.

Birefringence

Many crystalline substances, such as ice, quartz, and calcite, have optical properties that are not the same in all directions. Because the atoms of these substances are not

A piece of waxed paper between crossed polarizers. Light, having passed through the first filter, is plane-polarized, but the hodgepodge of fibers depolarizes the beam. Try it with an ordinary piece of wet paper.

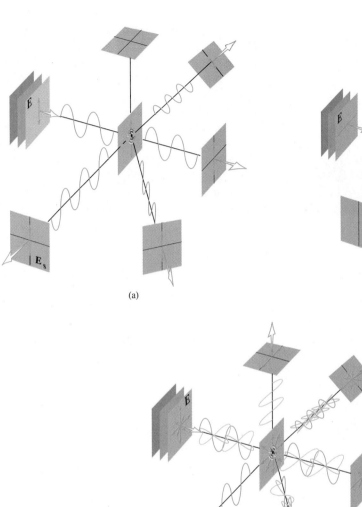

(a)

(b)

(c)

Figure 27.12 Scattering of polarized light by a molecule. Vertically polarized light scatters, as shown in (a); horizontally polarized light scatters as in (b). Since natural light (c) can be imagined as the sum of two perpendicular $\mathscr{P}$-states, it scatters as the superposition of (a) and (b). Notice how the light scattered perpendicular to the propagation direction in (c) is linearly polarized.

A pair of crossed polarizers. The lower one is noticeably darker than the upper one, indicating that scattered sky light is partially polarized.

uniformly arrayed, the forces on their electron clouds are different in different directions. Thus, the response of the atomic oscillators depends on the direction of the incoming E-field. As a result, such crystals display two different indices of refraction. The word *refringence* used to be used instead of our more modern term *refraction,* hence the name **birefringence**, meaning doubly refractive.

When unpolarized light enters a calcite crystal, each of the two constituent, independent, perpendicular $\mathscr{P}$-states encounters an essentially different medium, and each travels with a different speed. The two beams refract at different angles (ac-

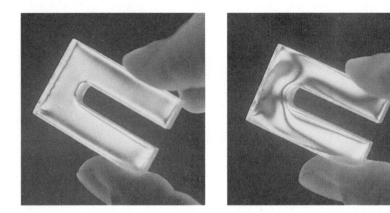

Stress within a material can cause shifts among its molecules, which then affect the transmission of polarized light. Here a piece of clear plastic between crossed Polaroids is being squeezed to increase that stress. Notice how dust, deliberately left between the crossed Polaroids scatters and depolarizes light, thereby becoming quite visible.

A piece of crumpled cellophane placed between crossed Polaroids. Depending on its thickness and the frequency of the light, the cellophane rotates the **E**-field by different amounts. Rotating one of the Polaroids will shift the colors to their complements.

cording to Snell's Law) and therefore separate, producing the double image discussed earlier (see the photo on p. 1012). By viewing both images through a linear polarizer, we can easily see that the two are plane-polarized in perpendicular directions. Many human-made substances, such as certain transparent food wraps, clear plastic tapes, and plastic microscope slides, are birefringent because of the alignment of their long-chain molecules—the best of all of these is cellophane. Crumple up a sheet of cellophane (still to be found wrapped around things like imported cigars) and insert it between crossed Polaroids. A profusion of multicolored regions will appear, resulting from the variations in thickness of the birefringent material—the indices of refraction of all substances are frequency dependent.

Some transparent isotropic materials that ordinarily show no polarization effects can have their molecules shifted by an applied force, producing an anisotropy and what is known as *stress birefringence*. Most clear plastic implements—rulers, spoons, cups, etc.—have this stress birefringence frozen in as they solidify. Between crossed polarizers, they reveal the internal stress pattern. To study the stresses within a machine part or an architectural element, we need only make a model of the piece in plastic and examine it in polarized light. Glass lenses when stressed in manufacture or by improper mounting will also show birefringence. Federal regulations require that eyeglasses be heat-treated to remove internal stresses that make them more likely to break. A simple test with Polaroids is used to evaluate the annealing. The back windows of automobiles are often deliberately stressed so that they will shatter into small harmless pieces on impact, and that induced birefringence is often visible, even without polarizers, at Brewster's angle.

INTERFERENCE

The basic idea of interference was introduced earlier (p. 514): in the region of overlap, two waves of the same frequency can combine constructively or destructively depending on their relative phase (Fig. 25.3, p. 937). Thus, if two identical point sources S_1 and S_2 emit spherical waves with the same wavelength, we can expect that the space surrounding them will contain some pattern of interference. Assuming the two sources continue to pump out light in-step with each other, the waves reaching an arbitrary point P in Fig. 27.13 will interfere in a stable observable fashion. Their relative phase (δ) at P will determine exactly how they interfere, and δ

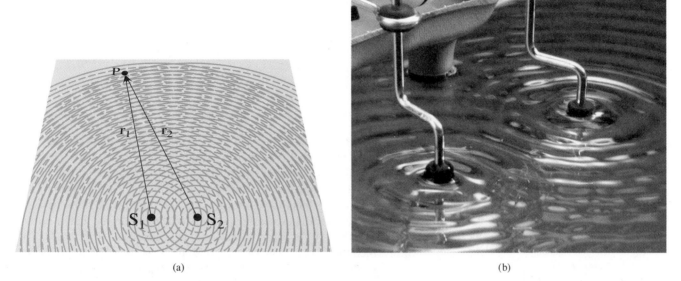

(a)   (b)

will depend on the difference between the two optical path lengths (that is, the index of the medium times the actual path length) traveled by the waves. Here, we will assume the index of refraction is one. Thus, if one route (r_1) is greater than the other (r_2) by half a wavelength ($r_1 - r_2 = \frac{1}{2}\lambda$), then $\delta = 180°$ and destructive interference occurs—the waves arrive out-of-step; trough overlays peak, there is cancellation, and P appears as a dark spot. The same thing would happen if the path difference was $1\frac{1}{2}\lambda$ or $2\frac{1}{2}\lambda$ or $3\frac{1}{2}\lambda$ and so, quite generally, a **minimum** in the irradiance occurs when

$$(r_1 - r_2) = m'\frac{1}{2}\lambda \qquad (27.4)$$

where $m' = \pm1, \pm3, \pm5, \ldots$. Similarly, if the path difference corresponds to λ or 2λ or 3λ, the overlapping waves will arrive in-phase, $\delta = 0$ or, equivalently, 2π or 4π or 6π, etc. In general then, a **maximum** occurs when

$$(r_1 - r_2) = m\lambda \qquad (27.5)$$

where $m = 0, \pm1, \pm2, \ldots$.

Suppose that the right side of either of these two expressions is held fixed (that is, $m' = $ constant or $m = $ constant). The left side then represents all the possible locations of P for which $(r_1 - r_2) = $ constant. But if the difference in the distances from any point P to two fixed points S_1 and S_2 is a constant, as it is here, then the locations of P, which satisfy Eqs. (27.4) and (27.5), form a family of hyperbolas; that's the definition of a hyperbola. In three dimensions, a vertical viewing screen placed at P, perpendicular to the plane of S_1 and S_2, shown in Fig. 27.13, will be covered with vertical bright and dark bands known as **interference fringes**. The energy, instead of being uniformly distributed, is redirected out of certain areas and into others—what is missing at minima appears at maxima: energy is conserved.

27.4 Coherence

It should be noted that the two sources producing interference need not really be in-phase with each other. A somewhat shifted but otherwise identical pattern will oc-

Figure 27.13 (a) Two point sources S_1 and S_2 sending out waves in-phase. Here we examine what's happening in a plane and draw the waves as circular. Where peak overlaps peak (two solid lines meet) and trough overlaps trough (two dashed lines meet), there is constructive interference and a maximum. Where peak overlaps trough (a solid line and a dashed line meet), there is destructive interference and a minimum. (b) Water waves in a ripple tank. (See the photo on p. 515.)

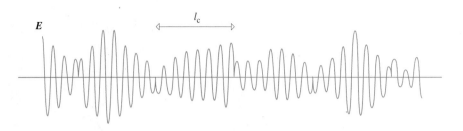

Figure 27.14 A light beam will be nicely sinusoidal for a time t_c, extending in space a length l_c, before it changes randomly. The more nearly monochromatic the light is, the longer is l_c. This is a schematic representation—the phase changes actually occur more smoothly.

cur even if there is some *initial phase difference* between the sources, so long as it remains constant. Such sources (which may or may not be in-step, but are always marching together) are said to be **coherent**. Remember that because of the granular nature of emission from atoms, a conventional quasimonochromatic source produces light that is a mix of photon wavetrains. At each illuminated point in space, there is a net field that oscillates nearly sinusoidally through fewer than a million cycles or so for less than 10^{-8} s before it randomly changes phase (Fig. 27.14). This interval (over which the wave resembles a sinusoid) is a measure of its **temporal coherence**. *The average time interval during which the lightwave oscillates in a predictable way* is known as the **coherence time** (t_c). The longer the coherence time, the greater the temporal coherence of the light.

The corresponding spatial extent over which the lightwave oscillates in a regular, predictable way is called the **coherence length**, l_c, where in vacuum $l_c = ct_c$. It is often convenient to picture the light beam as a progression of well-defined, more or less sinusoidal wavegroups of average length l_c whose phases are quite uncorrelated one to the other. Note that *temporal coherence is a manifestation of spectral purity*. If the light were ideally monochromatic, the wave would be a perfect sinusoid with an infinite coherence length. By comparison, a good laboratory discharge lamp has a coherence length of only several millimeters, whereas certain special lasers provide coherence lengths of tens of kilometers.

Two ordinary sources, two light bulbs or candle flames, can maintain a constant relative phase for a time no greater than t_c (which is $<10^{-8}$ s), and so the interference pattern they produce will randomly shift around in space at an exceedingly rapid rate, averaging out and making it quite impossible to observe. Until the advent of the laser, no two different sources could produce an observable interference pattern. Several decades ago interference was electronically detected using independent lasers, and recently (1993), fringe patterns produced by two separate lasers have even been photographed directly.

27.5 Young's Experiment

The main problem in producing interference is the sources: they must be *coherent*. And yet separate, independent, adequately coherent sources, other than the modern stabilized laser, don't exist! That dilemma was first solved two hundred years ago by Thomas Young in his classic double-slit experiment. He brilliantly took a single wavefront, split off from it two coherent portions, and had them interfere. In effect, he produced two coherent sources using a single wavefront. The technique is straightforward (Fig. 27.15a)—an opaque screen σ_a containing two identical small apertures is illuminated by a symmetrical wave that impinges on the two holes in the same way. That wave can be planar, spherical, or cylindrical provided that the phases of the two oscillating E-fields across the holes are identical. The light streaming from the apertures S_1 and S_2 pours out (diffracts) as if from two coherent identical sources.

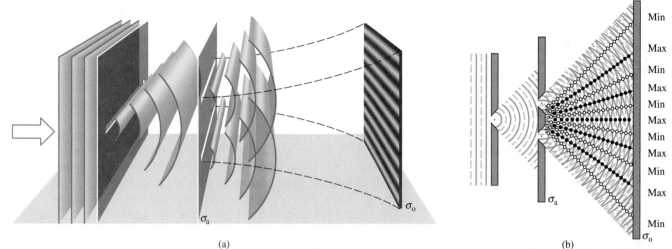

(a)

(b)

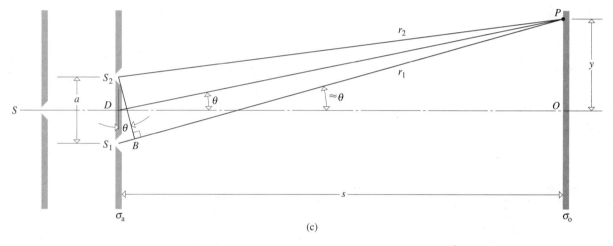

(c)

Figure 27.15 Young's Experiment. S_1 and S_2 are two coherent sources— they can be slits or holes (slits provide more light and a brighter pattern). The waves from them overlap, as in Fig. 27.13, and fill the space with fringes. The fringes that appear on a screen σ_o are horizontal in (a), (b), and (c). (b) is an edge-view of (a). With a laser beam, σ_a can be illuminated directly without the need for the single slit at far left. (c) Usually $s \gg a$ and $s \gg y$, in which case angle $S_1 S_2 B \approx \theta$.

The apertures could be pinholes, in which case the setup closely resembles Fig. 27.13. In Fig. 27.15a, at every point where trough lies on trough (dashed line crosses dashed line) or peak overlaps peak (solid line crosses solid line), the waves are in-phase and a maximum or bright spot appears. The lines connecting all of these maxima are rather straight hyperbolas. Similarly, where trough overlaps peak (dashed line crosses solid line), the waves are out-of-phase and a minimum or dark spot appears. These, too, lie on fairly straight hyperbolas. As the wavefronts sweep out and away from the apertures, the points of overlap—the maxima and minima— travel out along the respective hyperbolas.

Put two pinholes (about the size of a period) one diameter apart in an opaque piece of paper. Hold it very close to your eye and look through them at a distant streetlamp at night. The characteristic fringe pattern will appear on your retina.

A double-slit fringe pattern. When white light is used in Young's Experiment, each wavelength produces its own fringe pattern slightly shifted from the others. The result is a central white band ($m = 0$ for all λ) surrounded by fringes that are increasingly colored.

A double-slit fringe pattern. When the separation between slits, a, is decreased, the distances the fringes are from the central axis, $y_m = sm\lambda/a$, increases. Similarly, the fringes broaden since $\Delta y = s\lambda/a$. All of this is easily seen with an ordinary long-filament display light bulb.

In order to pass a greater amount of light and get a brighter pattern, narrow slits are used rather than pinholes. An observation screen σ_o would then be covered with fairly straight dark and light bands running parallel to the slits, as in Fig. 27.15b. There we see plane waves impinging on a slit to produce a cylindrical wave, but a laser beam could just as well illuminate σ_a with plane waves directly. In fact, if you place σ_a (a 3×5 index card will do) close to your eye and look through the slits (cut with a razor) at an ordinary straight-filament display bulb, you'll see a wonderful fringe pattern cast on your retina.

Generally, the slit separation a is very much smaller than the distance to the observation screen s. The former is usually less than a millimeter, the latter is typically several thousand times that. Consequently, some simplifying assumptions can be made in Fig. 27.15c. First, approximate the path length difference $(r_1 - r_2)$ from the two apertures to any point P on the observation screen as the distance $\overline{S_1 B}$ where $\overline{S_2 B}$ is perpendicular to $\overline{S_1 P}$. Then $a(\sin \theta) = \overline{S_1 B}$ and

$$(r_1 - r_2) = a \sin \theta \tag{27.6}$$

But θ is very small and so $\sin \theta \approx \theta$, in which case

$$(r_1 - r_2) \approx a\theta$$

Inasmuch as $\angle PDO = \theta$, $\tan \angle PDO = y/s \approx \theta$ and

$$(r_1 - r_2) \approx \frac{ay}{s}$$

Yet, for maxima

$$(r_1 - r_2) = m\lambda \tag{27.5}$$

which means that the height of the mth bright band (counting the central one as the zeroth) above the axis is

$$y_m \approx \frac{s}{a} m\lambda \tag{27.7}$$

These maxima correspond to locations of P where, as shown in Fig. 27.16, the optical path difference is either 0 (where $m = 0$), $\pm\lambda$ (where $m = \pm 1$), $\pm 2\lambda$ (where $m = \pm 2$), and so forth. The height of the fringes is λ-dependent, which means that with incident white light we can expect to see a white central band ($m = 0$) where all wavelengths overlap but that all other fringes will show some coloration.

The angular height of the mth maximum is obtained by comparing Eqs. (27.5) and (27.6); thus

$$a \sin \theta_m = m\lambda \tag{27.8}$$

or, since the angles are small, $\theta_m \approx m\lambda/a$.

The spacing between fringes on the screen, Δy, is just the difference between the locations of consecutive maxima. From Eq. (27.7) it follows that $\Delta y = y_{m+1} - y_m$ is

$$\Delta y \approx \frac{s}{a} \lambda \tag{27.9}$$

For a particular wavelength of light, the spacing of the fringes is constant and inversely proportional to the slit

Figure 27.16 The plane waves from a laser illuminate the double-slit experiment. The $m = \pm 1$ maxima occur where the optical path length difference $(r_1 - r_2)$ equals $\pm\lambda$.

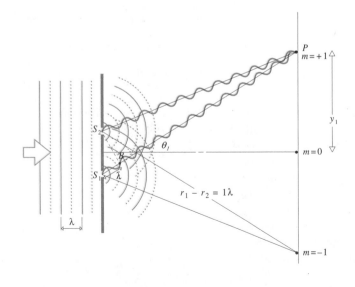

separation: the wider apart the slits, the finer the fringes. Ideally, the irradiance would vary as a constant-amplitude cosine-squared, but because of diffraction, that's only approximated in reality.

As P gets farther from the axis, $\overline{S_1 B}$ (which is less than or equal to $\overline{S_1 S_2}$) increases. If the primary source has a short coherence length, as the optical path difference increases, identically paired wavegroups will no longer be able to arrive at P exactly together—there will be an increasing amount of overlap of portions of uncorrelated wavegroups, and the contrast of the fringes will degrade (Fig. 27.17). It is possible for $l_c < \overline{S_1 B}$. Then, instead of two correlated portions of the same wave-

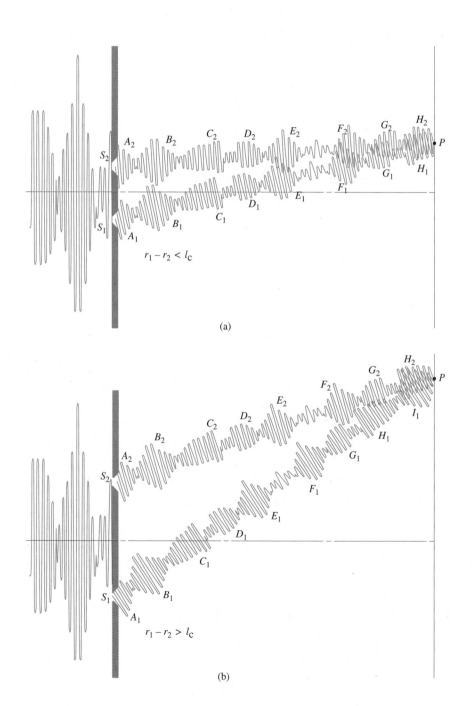

Figure 27.17 A schematic representation of interfering lightwaves. (a) Where $l_c > (r_1 - r_2)$, the interference will be sustained and observable. In (b), wavegroup E_1 from source S_1 arrives at P with wavegroup D_2 from S_2, and there is interference, but it lasts only for a tiny time before the pattern shifts when wavegroup D_1 begins to overlap wavegroup C_2, since the relative phases are different. Had either the l_c been larger or $(r_1 - r_2)$ smaller, wavegroup D_1 would more or less interact with its clone wavegroup D_2, and so on for each pair. The phases would then be correlated and the interference pattern stable, as in (a).

group arriving at *P*, only segments of different wavegroups will overlap, and the fringes will vanish.

Since a white-light source has a coherence length of only about 1 wavelength or so, it follows from Eq. (27.5) that very few fringes will be seen on either side of the central maximum; the more nearly monochromatic the light, the more fringes.

Example 27.4 Red light from a He-Ne laser ($\lambda = 632.8$ nm) is incident on a screen containing two very narrow horizontal slits separated by 0.200 mm. A fringe pattern appears on a white piece of paper held 1.00 m away. (a) Approximately how far (in radians and mm) above and below the central axis are the first zeros of irradiance? (b) How far (in mm) from the axis is the fifth bright band?

Solution: [Given: $\lambda = 632.8$ nm, $a = 0.200$ mm, and $s = 1.00$ m. Find: the angle for the first minimum, and y_5, the height of the fifth maximum.] (a) According to Eq. (27.4), the first minimum occurs when $m' = \pm 1$ and

$$(r_1 - r_2) = \pm\tfrac{1}{2}\lambda$$

Hence

$$a \sin \theta_1 = \pm\tfrac{1}{2}\lambda$$

and

$$\theta_1 \approx \frac{\pm\tfrac{1}{2}\lambda}{a} \approx \pm\tfrac{1}{2}\frac{(632.8 \times 10^{-9}\text{ m})}{(0.200 \times 10^{-3}\text{ m})}$$

$$\boxed{\theta_1 \approx \pm 1.58 \times 10^{-3}\text{ rad}}$$

or $y = s\theta_1 = (1.00\text{ m})(\pm 1.58 \times 10^{-3}\text{ rad}) = \boxed{\pm 1.58\text{ mm}}$. (b) From Eq. (27.7)

$$y_5 \approx \frac{s5\lambda}{a} \approx \frac{(1.00\text{ m})5(632.8 \times 10^{-9}\text{ m})}{(0.200 \times 10^{-3}\text{ m})}$$

$$\boxed{y_5 \approx 1.58 \times 10^{-2}\text{ m}}$$

▶ **Quick Check:** The fringe irradiance varies as cosine-squared and the answer to (a) is half a fringe width. The answer to (b) is a distance of 5 full fringes and therefore it is 10 times larger.

27.6 Thin Film Interference

In addition to the triumph of the double-slit experiment, Young was able to explain the colors arising from *thin films*: the colors of soap bubbles and oil slicks on a wet pavement; the rainbow of hues that appears on oxidized metal surfaces; the iridescence of peacock feathers and mother-of-pearl.

When light impinges on the first surface of a transparent film, a portion of the incident energy re-emerges as a reflected wave while the remainder is transmitted. The refracted energy subsequently encounters the second surface where, again, a portion is reflected and the remainder transmitted out of the film (Fig. 27.18). The film has the effect of dividing the wave into three segments: one reflected from the top surface, one reflected from the bottom, and one transmitted. Rather than utilizing two separate regions of the incident wavefront (each with the same amplitude as the incoming wave) as was done in Young's Experiment, the film shears the entire wavefront. These two reflected waves come off in the same direction and can be made to overlap at some point on the viewer's retina. The same thing would happen if this situation involved a thick block of glass, but there is a crucial difference. *If the film is thin in comparison to the coherence length of the light traversing it, the two waves will be correlated; that is, they will be fairly coherent and capable of interfering in a sustained fashion.* Because they travel different routes, depending on the film's thickness, they will ultimately interfere in some way that depends on that thickness.

Consider a nonuniform oil film floating on a puddle so that it thins out to a thickness of only a few wavelengths (and the dark surface beneath absorbs any transmitted light, keeping it from scattering back to the viewer). Wherever the film is exactly the right thickness for the two waves of emerging red light to undergo

The vivid iridescence of a peacock feather is due to interference of the light reflected from its complex layered surface.

constructive interference, the film will appear to reflect a spot of red light (and so on across the spectrum and across the film). These so-called *fringes of equal thickness* create a kind of colored topographical map of the film.

Let's analyze the situation for the simplified case of nearly perpendicularly incident light. Figure 27.19 shows a thin film of index n_f between media with indices n_1 and n_2. Ray 2 travels through the film, down and back up, crossing it twice. If the thickness at that point is d, then ray 2 traverses an additional optical path length of $2dn_f$ before rejoining ray 1, which means that ray 2 traveled an additional number of waves $2dn_f/\lambda_0$ farther. To see this, notice that the extra distance $\approx 2d$ in the film corresponds to an extra number of wavelengths equal to $2d/\lambda_f$, which, because $\lambda_f = \lambda_0/n_f$, equals $2dn_f/\lambda_0$. Since each wavelength is equivalent to a phase change of 2π rad, the two waves will have a relative phase difference δ of

$$\delta = \frac{4\pi dn_f}{\lambda_0} \qquad (27.10)$$

When this quantity equals a whole number (m) multiple of 2π, the two waves will be back in-phase, and that particular wavelength of light will undergo constructive interference:

$$\delta = 2\pi m = \frac{4\pi dn_f}{\lambda_0}$$

and therefore (CASE 1: *no phase shift on reflection*)

[*maxima*] $$d = \frac{m\lambda_0}{2n_f} = \frac{m\lambda_f}{2} \qquad m = 1, 2, 3, \ldots \qquad (27.11)$$

Maxima in normally reflected light occur when the film thickness is a whole number multiple of half the wavelength. And in the same way, *minima occur when the thickness is an odd multiple of one-quarter of the wavelength.*

As the film becomes thicker, exceeding several wavelengths, it becomes possible for two different colors to have maxima at the same spot. For example, a film 1000 nm thick will produce maxima for light with wavelengths within it of both 400 nm and 500 nm. At still greater thicknesses, where many wavelengths can interfere constructively at the same time, the reflected color becomes increasingly unsaturated, the fringe contrast decreases, and the pattern ultimately vanishes. Monochromatic illumination with its infinite coherence length will encounter no such limitations, and a sheet of window glass is still effectively a thin film for laser light.

Actually, Eq. (27.11) provides the thickness of the film for an interference maximum *only* when $n_1 > n_f > n_2$ or $n_1 < n_f < n_2$. In the first case, the reflections are both *internal* and, in the second, they're both *external*. As we saw earlier (p. 940), nearly normally incident light will undergo a relative phase shift of π rad between its internally and externally reflected components. That's not relevant in either of the above cases, which is why Eq. (27.11) applies. Notice that as the film gets vanishingly thin, all wavelengths will interfere more or less constructively everywhere, to gradually produce the uniform reflection from the interface (between the two surrounding media) that must exist when the film disappears altogether.

An even more commonly occurring situation has $n_1 < n_f > n_2$ or $n_1 > n_f < n_2$, as with a soap film in air in the first case, and an air film between two sheets of glass in the second. Now, one reflection is internal and the other is external, and there will be an additional $\pm\pi$ phase shift. Which sign we select is irrelevant, accordingly, using the minus

$$\delta = \frac{4\pi dn_f}{\lambda_0} - \pi$$

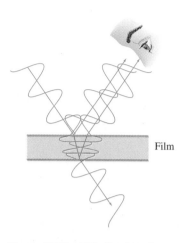

Figure 27.18 Thin film interference. Light reflected from the top and bottom of the film interferes to create a fringe pattern.

A wedge-shaped film made of liquid dishwashing soap, showing interference fringes. The top part is drained thin.

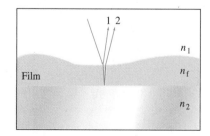

Figure 27.19 The reflection of light from the top and bottom of a thin film of index n_f.

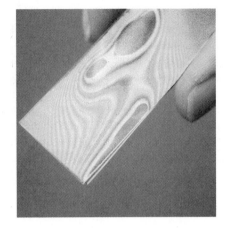

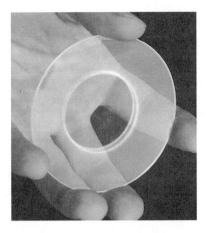

The interference pattern produced by a thin film of air between two microscope slides.

An antireflection coating in the shape of a circle, applied at the center of each side of a glass disk.

Interference from the thin air film between a convex lens and the flat sheet of glass it rests on. The illumination was quasimonocromatic, which is why the fringes aren't more colorful. These fringes were first studied in depth by Newton and are known as Newton's rings. See Problem 42, p. 1048.

Hence, constructive interference in reflected light now occurs when

$$\delta = 2\pi m = \frac{4\pi d n_f}{\lambda_0} - \pi$$

and therefore (CASE 2: 180° *phase shift on reflection*)

[*maxima*] $$d = \frac{(m + \frac{1}{2})\lambda_0}{2n_f} = \frac{(m + \frac{1}{2})\lambda_f}{2} \qquad m = 0, 1, 2, \ldots \qquad (27.12)$$

As the film becomes vanishingly thin ($d \to 0$, $d \ll \lambda_f$), the total phase difference, δ, approaches $-\pi$, there is total destructive interference, and the film appears uniformly black in reflected light and uniformly transparent in transmitted light. That's reasonable enough—since it must disappear altogether when $d = 0$, it should reflect less and less and transmit more and more as it gets there. Biological bimolecular lipid films (5 nm to 10 nm thick) appear black in reflected light, as does an ordinary liquid-soap film when it drains and becomes a small fraction of a wavelength thick (see the black band at the very top of the film in the photo on p. 1029).

Example 27.5 A soap film in air has an index of refraction of 1.34. If a region of the film appears bright red ($\lambda_0 = 633$ nm) in normally reflected light, what is its minimum thickness there?

Solution: [Given: $\lambda_0 = 633$ nm and $n_f = 1.34$. Find: d.] This situation corresponds to CASE 2, $1.00 < 1.34 > 1.00$; hence, using Eq. (27.12) with $m = 0$, which corre-

sponds to the minimum thickness, we have

$$d = \frac{(0 + \frac{1}{2})(633 \text{ nm})}{2(1.34)} = \boxed{118 \text{ nm}}$$

▶ **Quick Check:** O.P.L. $= 2(1.34)(118$ nm$) = 3.16 \times 10^{-7}$ nm; O.P.L.$/\lambda_0 = 0.5$, which, correctly, is an odd whole number multiple of $\frac{1}{2}$.

One of the most important practical applications of these ideas is the ***antireflection coating***. Whenever light reflects off an interface separating two dielectrics (such as glass and air), some fraction is reflected, and that can become very problematic in complex instruments where there might easily be a dozen or more such surfaces. As a solution, each optical element is coated with a thin, transparent, solid, dielectric film such that it does not reflect a specific range of wavelengths. Often that range is chosen to be in the yellow-green region of the spectrum, where the eye is most sensitive. Lenses coated for the yellow-green reflect in the blues and reds, giving the surface a familiar purple color.

Example 27.6 A glass microscope lens with an index of 1.55 is to be coated with a magnesium fluoride ($n = 1.38$) film to increase the transmission of normally incident yellow light ($\lambda_0 = 550$ nm). What minimum thickness should be deposited on the lens?

Solution: [Given: $n_f = 1.38$, $n_g = 1.55$, and $\lambda_0 = 550$ nm. Find: d for a minimum.] The refracted wave will traverse the film twice, and there will be no relative phase shift on reflection. Hence, we want a film one-quarter wavelength thick if the two waves are to be out-of-phase and interfere destructively; consequently

$$d = \frac{\lambda_0}{4n_f} = \frac{(550 \text{ nm})}{4(1.38)} = \boxed{99.6 \text{ nm}}$$

▶ **Quick Check:** O.P.L. $= n_f d = (1.38)(99.6 \text{ nm})$; O.P.L./$\lambda_0 = 0.25$, as it should; this result equals the number of waves, or $\frac{1}{4}$.

27.7 The Michelson Interferometer

There are a number of practical devices known as **interferometers** that produce fringe patterns very much in the nature of the thin film effects just considered. The most important of these, both historically and practically, is the *Michelson Interferometer*, shown in Fig. 27.20. Here, an extended source emits a wave that enters from the left. This could be an expanded laser beam, the light from a discharge lamp, or an ordinary tungsten bulb. The wave is then sheared into two equal-amplitude parts by a beamsplitter (M_S), which is usually a half-silvered mirror. The two waves are subsequently reflected by mirrors M_1 and M_2 and return to the beamsplitter. The

Figure 27.20 The Michelson Interferometer. Light enters at the left and is split into two equal beams by the beamsplitter M_S. Part goes on to mirror M_1 and part to mirror M_2. These beams are reflected back to M_S, and both pass on to the detector. When M_1 and M_2 are perpendicular, the fringes are circular and centered on the axis of the observer's eye. (b) is a simplified schematic representation of the wavefronts ignoring refraction.

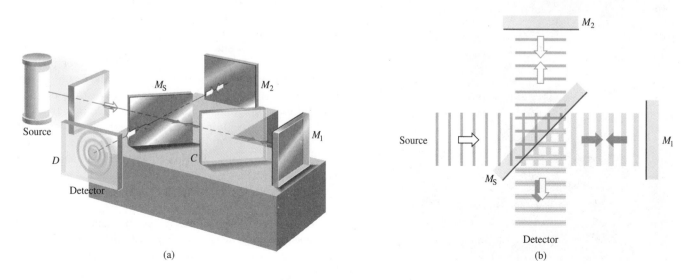

(a)

(b)

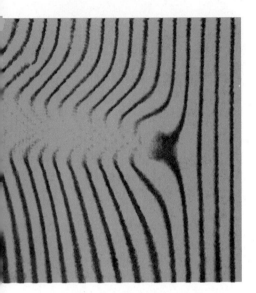

Wedge fringes in a Michelson Interferometer. The distortions are caused by a hot soldering-iron placed in one arm.

Diffraction pattern of a hand-held paper clip. This is simply the shadow cast on a wall using a He-Ne laser beam (a distant point source would do as well).

wave from M_1 reflects toward the detector, and the wave from M_2 traverses M_S, passing on to the detector. The waves are reunited, and we can expect to see an interference pattern that depends on the path length differences traversed as well as whatever phase shifts are introduced via reflections from the beamsplitter.

Since one beam passes through M_S three times and the other only once, there will be a good deal of difference in the optical path lengths, even when the two arms are of equal actual length. Moreover, the optical path length will depend on λ because of dispersion (p. 946) in the glass. The inclusion of a compensator plate, C, which is a duplicate of the beamsplitter (without the silvering), equals out the number of thicknesses of glass traversed and negates the effects of dispersion. With the compensator in place even white light can be used; without it, the illumination must be quasimonochromatic.

What we see when we look into the device at the position of the detector is the image of both mirrors superimposed. We see the beamsplitter and, at the same time, mirrored in it M_1 and, through it, M_2. In effect, we are looking at M_1 and M_2, one behind the other, with a layer of air between them whose thickness corresponds to the difference in the two path lengths.

If M_2 and the image of M_1 are not parallel, fringes of equal thickness appear. Thus, when the mirrors are set to form a triangular-shaped air film, a system of parallel, equal-spaced straight fringes aligned with the edge of the wedge (corresponding to lines of equal film thickness) will be seen, as in the photo. And here we see another fundamental advantage of this kind of interferometer: there is space within the arms that we have access to. A hot soldering iron inserted into one of the arms will change the index of refraction of the air, shift the fringes, and reveal an otherwise invisible phenomenon. There are several variations on Michelson's instrument that exploit this marvelous ability to study the structure of transparent systems (such as gases, plasmas, lenses, and so forth).

DIFFRACTION

An object placed between a point source and a screen casts an intricate shadow made up of bright and dark regions quite unlike anything one might expect from the tenets of geometrical optics. This deviation from rectilinear propagation, known as **diffraction**, *is a general characteristic of all wave phenomena occurring whenever a portion of a wavefront is obstructed in any way.* When in the course of encountering an obstacle, either transparent or opaque, a region of the wavefront is altered in amplitude or phase, diffraction occurs. Anyone who has ever walked at night in the rain wearing eyeglasses has probably seen the effect. Just put a drop of water on a glass plate, hold it near your eye, and look through it at a distant streetlight, and you'll see a complex system of bright and dark diffraction fringes. Similarly, the amoeba-like floaters that can be seen within your own eye when you squint at a bright broad source are diffraction patterns of drifting cellular debris cast on your retina.

Nowadays (because these fringes are characteristic of the objects that give rise to them, and because they can be analyzed by computer) all sorts of automatic processors

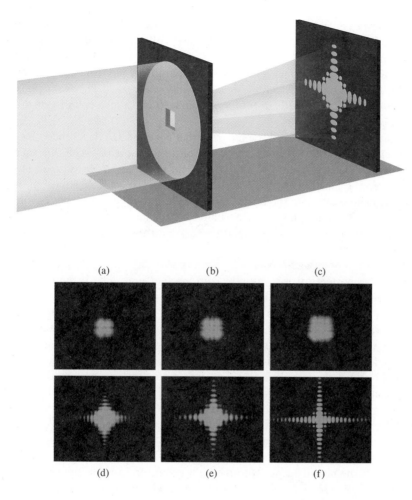

Figure 27.21 The shadow or diffraction patterns arising from a single square aperture. The hole was sequentially decreased, producing Fresnel diffraction (a), which gradually transformed (b), (c), (d), (e), into Fraunhofer diffraction (f). All apertures were illuminated by the plane waves from a He-Ne laser and the viewing screen was fixed in place.

utilize diffraction, for things as diverse as spotting tanks in aerial photos and checking fingerprints to measuring blood cell sizes.

The various unobstructed segments of a wavefront that propagate beyond an obstacle (be it sound or light) interfere to produce the particular energy-density distribution referred to as the diffraction pattern. There is no profound physical difference between interference and diffraction—in fact, the two are inseparable, though we tend to talk about them individually because that's the way the notions developed historically.

As far as the analysis is concerned, we recognize two distinct kinds of diffraction, though that, too, is more for practical reasons than fundamental ones. Figure 27.21 will help make the point. It shows a succession of diffraction patterns produced by a square hole illuminated by a laser. When the hole is large and the film very close to it by comparison, the pattern is complicated internally but the overall shape is recognizable—that's what is seen in (a), (b), and (c). But as the hole is made still smaller, *the pattern increasingly spreads out in directions perpendicular to the hole's edges* until it becomes quite unrecognizable. Finally, as in (f), a point will be reached where making the hole smaller simply makes the pattern grow larger without changing shape.

Since the effect depends on the relative size of the hole compared to its distance from the film, the same set of photos could just as well have been gotten by leaving the aperture fixed and moving the film plane farther and farther away. The first several pictures, from (a) to (e), represent what is called *near-field* or **Fresnel diffraction**, while (f) corresponds to *far-field* or **Fraunhofer diffraction**. The latter is a special case of the former, one that is comparatively easy to deal with mathematically. We will only consider *far-field diffraction where the incident light is planar and the film is essentially at infinity* (though 10 or 15 meters will generally do nicely in practice).

Consider a thin glass photographic plate that is completely blackened except for a small, clear aperture, perhaps in the shape of a little square. When this square is illuminated by a succession of normally incident plane waves, layer upon layer of atoms within the glass will scatter the light until the last sheet of atoms on the far side of the plate emits into the air in all directions. In effect, the light beyond the screen—the diffracted light—is the result of a tremendous number of point sources distributed uniformly across the aperture, all emitting secondary, more or less spherical, wavelets.

To make the point again from a different perspective, imagine the obstructing screen to be a perfect mirror, initially with no apertures. The incoming light reflects backward and none is transmitted. Whatever radiation comes from the atoms of the mirror cancels the forward-moving primary wave and simply produces the reflected wave. Now, cut out the same hole as before and remove the square plug. The back-scattered radiation from the screen no longer completely cancels the wave, and light emerges from the region of the hole. Still, if the plug were reinserted, the light would be exactly canceled. As a first approximation, let's assume that the mutual interaction of all the atomic oscillators is negligible; that is, the atoms in the screen are unaffected by the removal of the atoms in the plug. The field in the region beyond the screen will be that which existed there prior to the removal of the plug, namely zero, minus the contribution of the plug alone. All of which suggests that, except for being 180° out-of-phase, the light emitted by all the atoms spread across the surface of the plug is identical to the light emerging from the aperture.

The diffraction field, in this approximation, can be pictured as arising from a set of fictitious (noninteracting) oscillators distributed uniformly over the unobstructed area of the aperture. These imagined secondary point sources emit wavelets beyond the obstruction that mutually interfere to create the diffraction pattern.

Figure 27.22 illustrates how the notion is applied to the diffraction of a wave by an aperture. In (a), we envision the unobstructed wavefront replaced by a very large number of point sources emitting essentially spherical waves in the forward direction. This is equivalent (b) to rays traveling out from each secondary source-point in all directions. And this, in turn, is equivalent (c) to plane waves being diffracted in all directions. Since we are limiting the discussion to Fraunhofer diffraction, it is

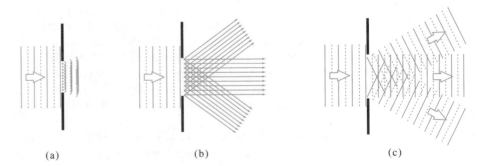

Figure 27.22 The diffraction of light through a narrow slit, represented by (a) spherical wavelets, (b) rays, and (c) plane waves.

(a)   (b)   (c)

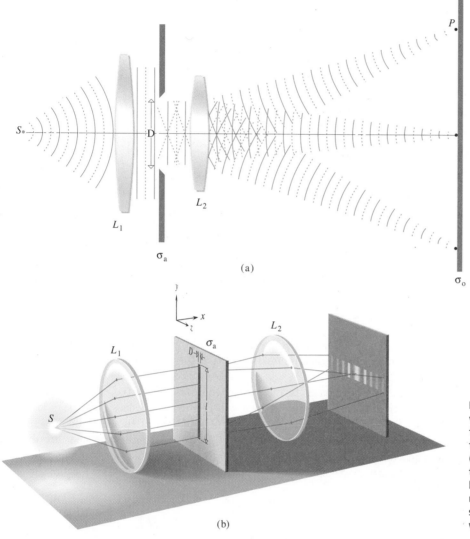

Figure 27.23 (a) The diffracted waves (seen in a horizontal plane) from the aperture are brought to a focus on a nearby screen by lens L_2. (b) When the aperture is a single slit, the diffraction pattern is a series of bands. With a point source S or a narrow laser beam illuminating the slit, the fringe pattern is narrow as well.

more convenient to use a few lenses to compress the setup rather than having to observe the pattern at infinity (Fig. 27.23).

27.8 Single-Slit Diffraction

Suppose that the aperture in Fig. 27.23a is a long narrow slit of width D, perpendicular to the plane of the diagram. Under monochromatic plane-wave illumination, we envision every point in the aperture emitting rays in all directions. The light that continues to propagate directly forward (Fig. 27.24a) is the undiffracted beam, all the waves arrive on the viewing screen in-phase, and a central bright region is formed. Figure 27.24b shows the specific bundle of rays coming off at an angle θ_1 where the path length difference between the rays from the very top and bottom, $D \sin \theta_1$, is made equal to one wavelength. A ray from the middle of the slit will then lag $\frac{1}{2}\lambda$ behind a ray from the top and exactly cancel it. Similarly, a ray from just below center will cancel a ray from just below the top and so on; all across the aper-

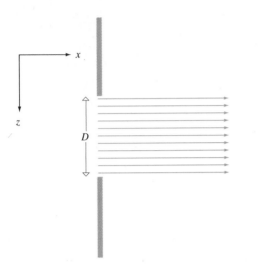

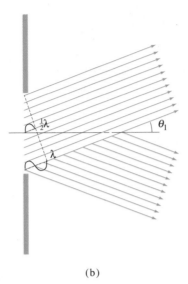

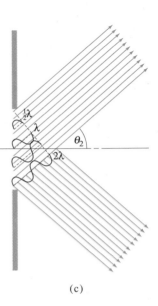

(a)   (b)   (c)

Figure 27.24 The zeros in the irradiance distribution produced by a single slit. (a) The central maximum is produced by undiffracted light. (b) The first pair of minima ($m' = \pm 1$) occur when the waves from the edge and center of the slit are out-of-phase by $\lambda/2$. (c) The second pair of minima ($m' = \pm 2$) occur when waves from each quarter of the slit cancel.

ture ray-pairs will cancel, yielding a black minimum. The irradiance has dropped from its high central maximum to the first zero on either side (Fig. 27.25) at $\sin \theta_1 = \pm \lambda / D$.

As the angle increases further, some small fraction of the rays will again interfere constructively, and the irradiance will rise to form a subsidiary peak less than 5% of the central maximum. Increasing the angle still further produces another minimum, Fig. 27.24c, when $D \sin \theta_2 = 2\lambda$. Now, imagine the aperture divided into quarters. Ray by ray, the top quarter will cancel the one beneath it and the next, the third quarter, will cancel the last quarter. Ray-pairs at the same locations in adja-

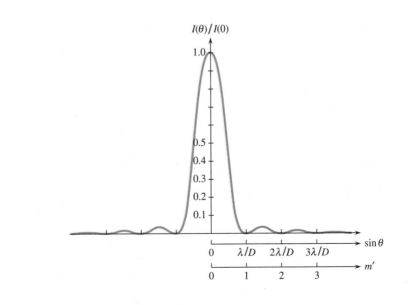

Figure 27.25 The Fraunhofer diffraction pattern of a single slit. This is a plot of the relative irradiance. Notice that the central maximum is twice the width of the little maxima.

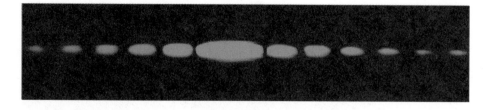

Diffraction pattern of a single vertical slit under narrow laser beam illumination.

cent segments are $\lambda/2$ out-of-phase and destructively interfere. In general, then, zeros of irradiance will occur when

$$D \sin \theta_{m'} = m'\lambda \qquad (27.13)$$

where $m' = \pm1, \pm2, \pm3, \ldots$.

Example 27.7 A single slit 0.10 mm wide is illuminated by plane waves from a helium-neon laser ($\lambda = 632.8$ nm). If the observing screen is 10 m away, determine the **width of the central maximum** (*defined as the distance between the centers of the adjoining minima*).

Solution: [Given: $D = 1.0 \times 10^{-4}$ m, $\lambda = 632.8 \times 10^{-9}$ m, and $s = 10$ m. Find: the width of the central maximum.] The two minima bounding the central maximum correspond to $m' = \pm1$ in Eq. (27.13); hence

$$\sin \theta_1 = \frac{\lambda}{D} = \frac{632.8 \times 10^{-9} \text{ m}}{1.0 \times 10^{-4} \text{ m}} = 632.8 \times 10^{-5}$$

and $\theta_1 = 0.36° = 6.3 \times 10^{-3}$ rad. The central band is $2(6.3 \times 10^{-3}$ rad) wide. Since the linear half-width is $s(\tan \theta_1)$, which in radians is approximately $s\theta_1$, the overall width of the central fringe is

$$s2\theta_1 = (10 \text{ m})(0.012\,6 \text{ rad}) = \boxed{0.13 \text{ m}}$$

▶ **Quick Check:** Using the above results, $2\theta_1 = (0.13 \text{ m})/s = 0.013 \approx 2 \sin \theta_1 = 2(632.8 \times 10^{-5}) = 0.013$. Alternatively, $\tan \theta_1 = y_1/s$; $2y_1 = 2s \tan \theta_1 = 2(10 \text{ m}) \tan 0.36° = 0.13$ m.

---

*As the slit narrows, and D gets smaller, the diffraction pattern spreads out and the central peak gets wider.** Again, the light fans out perpendicularly against the edges of the aperture just as it did in Young's Experiment. In fact, during that experiment, very narrow slits are used and each produces a single-slit diffraction pattern with a very wide central bright band. These two diffraction patterns overlap to create the familiar double-slit cosine-squared fringes (Fig. 27.26a and b).

27.9 The Diffraction Grating

Something rather interesting develops as the number of parallel slits (each spaced a distance a) is increased beyond two (Fig. 27.26). Identical single-slit diffraction patterns again overlap to produce maxima at the same locations as in Young's Experiment, but now these principal peaks are narrower, and faint subsidiary maxima appear between them. Three slits will produce one subsidiary maximum; four slits, two subsidiary maxima; and N slits, $(N - 2)$ subsidiary peaks between successive principal bright bands. As N increases, the secondary peaks become more numerous

---

*This interrelationship is very important for wave phenomena. It will appear over and over again in optics and later as well, when we consider the Heisenberg Uncertainty Principle (Sect. 31.9). The finer the details of the aperture, the more broadly is the light diffracted and vice versa.

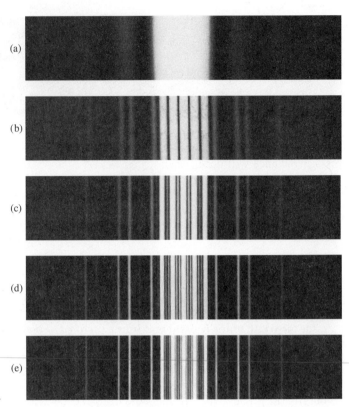

Figure 27.26 Diffraction patterns for slit systems composed of (a) 1, (b) 2, (c) 3, (d) 4, and (e) 5 identical, equally spaced, vertical slits. Notice that the maxima for several slits are where they were in the two-slit case. The more apertures, the finer the maxima become and the more faint the fringes appear between them. When there are thousands of identical slits, the various order maxima are quite fine and they appear separated by black regions.

and even fainter, with the effect that more of the diffracted light appears in the sharpened, widely spaced, principal bands. When there are thousands of slits, the bright bands are, for all practical purposes, separated by black regions where essentially no light arrives. It follows from Eq. (27.8) for Young's fringes that the principal maxima are to be found where

[*principal maxima*] $$a \sin \theta_m = m\lambda \qquad\qquad [27.8]$$

and $m = 0, \pm 1, \pm 2, \ldots$. A repetitive array of apertures or obstacles that alters the amplitude or phase of a wave is a **diffraction grating**, and Eq. (27.8) is known as

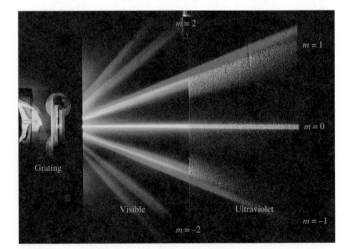

Light and UV passing through a transmission grating. The region on the left shows the visible spectrum, that on the right reveals the ultraviolet.

the *grating equation*. Master gratings with over 10 000 lines inscribed per centimeter are either ruled using a fine diamond point or generated holographically. Most gratings are plastic copies of master gratings, and they can be transparent *transmission gratings* or metal-coated *reflection gratings*.

Example 27.8 Light from a distant star enters a telescope and then passes through a diffraction grating. Each component of the emerging dispersed light is focused onto a curved strip of film. The grating is located on the central axis of the setup. It is found that a red beam (known as the hydrogen-α line) appears off axis, in the first order ($m = \pm 1$) spectrum, at an angle of 25.93°. If the lines of the grating are separated by 1.50×10^{-6} m, determine the wavelength of that light.

Solution: [Given: $\theta_1 = 25.93°$, $a = 1.50 \times 10^{-6}$ m, and $m = 1$. Find: λ.] From the grating equation, Eq. (27.8), we have

$$\lambda = \frac{a(\sin \theta_1)}{m} = \frac{(1.50 \times 10^{-6} \text{ m})(0.4373)}{1} = \boxed{656 \text{ nm}}$$

▶ **Quick Check:** $a \sin \theta_m = m\lambda$, $\theta_m = \sin^{-1} m\lambda/a = \sin^{-1} (0.437) = 25.9°$.

Gratings are used across much of the electromagnetic spectrum in research applications. Moreover, they are finding increasing commercial use as well. For example, one compact disk (CD) system uses a grating to split the laser beam that reads the disk into three ($m = 0$ and $m = \pm 1$) tracking beams.

Notice in Eq. (27.8) that it is only because a is greater than λ that more than the single $m = 0$ wave exists. The smaller a is, the fewer are the values of m that will satisfy the formula. Moreover, if $\lambda > a$, since the sine cannot exceed one, only $m = 0$ is possible. This is equally true for both transmission and reflection gratings, and it makes an important general point. When light scatters specularly from a smooth surface (p. 943), it does so such that one beam is reflected and the angle of incidence equals the angle of reflection. Because $\lambda \gg a$, where a is the spacing between atomic scatterers, only the $m = 0$ order can be sustained. By comparison, when the wavelength is reduced to the X-ray region where λ is comparable to a, the higher-order reflections emerge.

A CD has its information stored in the form of tiny raised areas that lie along a spiral path. The resulting $\frac{1}{4}\lambda$-high ridges cause the CD to behave like a reflection grating, as does an ordinary phonograph record held at glancing incidence. There are also a number of natural systems that possess the appropriate degree of order to display grating behavior. Certain beetles, wasps, and butterflies are decorated with diffraction-grating colors.

Look at a bright source through the regular array of ridges forming a bird's feather, and you'll see transmission-grating colors. Similarly, the fabric of an umbrella, a regular mesh nylon curtain, or an undergarment can serve as a 2-dimensional transmission grating. And the gemstone opal gets its shifting internal coloration from an ordered array of tiny silica (SiO_2) spheres that make it a 3-dimensional diffraction grating. As we'll see, for X-rays all crystals are 3-dimensional gratings (p. 1095).

A multi-aperture diffraction pattern. This is a picture of a white-light point source shot through a piece of tightly woven cloth.

Airy rings (1.0-mm-hole diameter). The center disk was allowed to get overexposed in order to make a few rings visible.

27.10 Circular Holes and Obstacles

When light from a distant point source (like a star or, more mundanely, an atom on a star's nose) is focused by a lens, the image formed is a small blotch rather than a perfect point. It must be so because the lens captures only a portion of the wavefront, and diffraction must occur—the blotch is the far-field diffraction pattern of the aperture of the lens. In the end, every image formed by a circular lens, whether in your

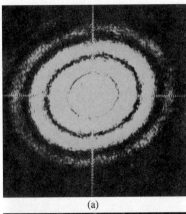

(a)

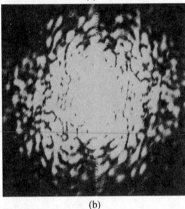

(b)

(a) The Fraunhofer diffraction pattern of a normal cervical cell. (b) The diffraction pattern of a malignant cervical cell is very different. Diffraction is being studied as a possible means of rapid automatic analysis of Pap tests for cancer.

Figure 27.27 Overlapping images of two point sources (separated by θ_a) that are just resolvable.

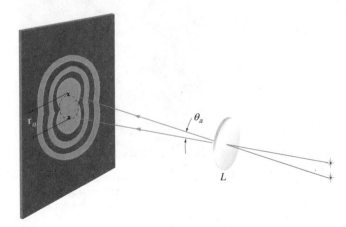

eye or a TV camera, is composed of countless overlapping circular-aperture Fraunhofer diffraction patterns. And that alone makes this particular configuration of the greatest importance. Unfortunately, even though a complete analysis was provided in the nineteenth century by Sir George B. Airy, Astronomer Royal of England, it's far too complicated for our purposes.

All Fraunhofer patterns, no matter how weirdly shaped the aperture, are symmetrical about the center point. Draw a line through that center, and the distribution of light outward along one-half of the line is the same as along the other. A circular hole must generate a circularly symmetric pattern. In fact, a circular aperture produces a circular disk of light, the so-called **Airy disk**, surrounded by a system of increasingly faint concentric rings. Actually, 84% of all the energy lies in the Airy disk. By the way, because there is so little energy in the higher-order fringes, circular lenses are better image-formers than rectangular ones.

A hole with a diameter D followed by a lens of focal length f will generate a diffraction pattern on a screen (in the lens's focal plane) having an Airy disk of radius r_a that can be shown to be

$$r_a \approx \frac{1.22 f \lambda}{D} \tag{27.14}$$

Alternatively, using the fact that $\theta_a \approx \sin \theta_a = r_a/f$, we can express the size of the central bright circle in terms of its angular half-width θ_a as

$$\theta_a \approx \frac{1.22 \lambda}{D} \tag{27.15}$$

in radians.

Suppose we wish to examine two equal-irradiance, incoherent, point sources—two close-together stars seen through a telescope, or two points on a virus viewed in a microscope. Because of diffraction, the images will spread out, and only if their angular separation exceeds this smearing will they appear distinct. As the two sources come closer together, their images come together, commingling into a single blend of fringes. Lord Rayleigh suggested that the criterion for *just being able to resolve the separate images* be that their center-to-center distance equal the radius of their Airy disks (Fig. 27.27). We can actually do a little better, but Rayleigh's criterion is both simple and standard. Equation (27.15) therefore specifies the **angular limit of resolution**, and the *resolving power* of an image-forming system is taken as the reciprocal of θ_a.

By decreasing the wavelength, we can increase the resolving power of an instrument. Ultraviolet microscopes can "see" finer details than light microscopes (which at best can distinguish two points as distinct that are no closer than about 0.12×10^{-6} m). Similarly, electron microscopes with wavelengths equivalent to roughly 10^{-4} to 10^{-5} times that of light have a limit of resolution of about 0.5 nm. By increasing the diameter of the objective lens or mirror of a telescope, it can collect more light, but the images it forms will be sharper as well. That's why spy satellites have large-diameter cameras.

Example 27.9 Compute the angular limit of resolution of the eye, assuming that it's determined only by diffraction. Take the pupil diameter to be 2.0 mm and $\lambda = 550$ nm. How far apart will two points be if they are just able to be distinguished at a distance of 25 cm from the eye?

Solution: [Given: $D = 2.0$ mm, $\lambda = 550$ nm, and $d_o = 25$ cm. Find: θ_a and y_o.] The limit of resolution follows from Eq. (27.15); thus

$$\theta_a \approx \frac{1.22\lambda}{D} = \frac{1.22(550 \times 10^{-9} \text{ m})}{2.0 \times 10^{-3}}$$

and

$$\boxed{\theta_a \approx 3.4 \times 10^{-4} \text{ rad}}$$

which is 1.9×10^{-2} degrees. At a distance of 25 cm, this angle corresponds to a linear distance of $y = d_o\theta_a$ or

$$y_o = (0.25 \text{ m})\theta_a = \boxed{8.4 \times 10^{-2} \text{ mm}}$$

which, at roughly 1/10 mm, is just about right for a normal eye.

▶ **Quick Check:** $\theta_a D/\lambda$ should equal 1.22, therefore $(0.34 \text{ mrad})D/\lambda = (0.34 \text{ mrad})(2.0 \text{ mm})/(550 \text{ nm}) \approx 680/550 \approx 1.2$.

In 1818, the young Fresnel entered his new wave theory in a competition sponsored by the French Academy. The judging committee consisted of such luminaries as Pierre Laplace, Jean Biot, Siméon Poisson, Dominique Arago, and Joseph Gay-Lussac. An ardent antagonist of the wave hypothesis, Poisson managed to deduce a remarkable and seemingly untenable conclusion from Fresnel's theory. He showed that the treatment predicted that a bright spot should appear at the very center of the shadow of a circular opaque obstacle; such an absurdity must surely disprove the entire theory! Arago, who was not one to accept anything on face value, went to the laboratory to learn the truth of Poisson's death blow. And there, wonder of wonders, at the center of the shadow he saw the "absurd"—a bright spot of light. Fresnel was right!

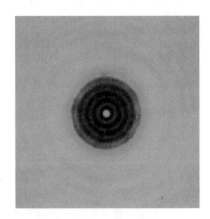

The shadow of an $\frac{1}{8}$-in.-diameter ball bearing illuminated with He-Ne laser light. To hold it in place, the ball bearing was glued to a microscope slide. The effect can easily be seen with a halogen lamp as the point source.

27.11 Holography

The technology of photography has been with us for a long time, and we're all accustomed to seeing the 3-dimensional world flattened onto a scrapbook page. A photograph is a record of the energy per unit area per unit time that once impinged on each of its surface points, and the light it scatters (the light we see when we look at it) is nothing more than a reflection of that frozen irradiance distribution. It is not an accurate reproduction of the original light field that came from the subject, but only a record of the square of the field's amplitude averaged over the exposure time. It reveals nothing about the phases of the waves that formed the image. On the other hand, if we could somehow reconstruct both the amplitude and phase of the wave coming from the object, the resulting light field would be indistinguishable from the original. One would then see the reformed image in perfect 3-dimensionality, exactly as if the object were there before us, generating the wave.

The technique of *image reconstruction* was invented by Dennis Gabor sometime around 1948. His research, which won him the 1971 Nobel Prize in physics, led to a practical means of recording and playing back the complete wave emanating from an object. The process is now called **holography** (from the Greek *holos* meaning whole). The crucial insight is that both the amplitude and phase information can be captured in a coded form via interference and preserved photographically. When two monochromatic waves generate an interference pattern, the shape,

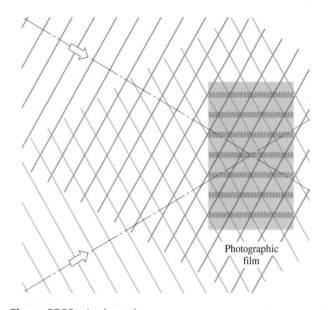

Photographic
film

Figure 27.28 A schematic representation of the interference of two plane waves traveling toward the same side of a photographic film, thereby creating a transmission hologram. (For simplicity, refraction has been ignored.)

contrast, and spacing of the fringes is a record of the physical characteristics of the overlapping waves themselves. If we wish to capture the vital features of an **object-wave**, we can cause it to interfere with a known, simply configured, **reference-wave**. The two should have roughly the same amplitude and, of course, be coherent. This last requirement kept Gabor's work in semi-unnoticed oblivion for about 15 years, until the laser was invented.

Consider the simple case of two plane waves traveling in the same general direction (Fig. 27.28) with a photographic plate inserted in the region of overlap. The wedge-shaped film viewed in a Michelson Interferometer produces the same configuration of two tilted plane waves, and so we can anticipate a fringe pattern of parallel bands across the emulsion. To appreciate this more clearly, just lay a pencil along each wavefront and move them both in their respective propagation directions at the same speed. The point of overlap, the maximum, will travel to the right along a straight line—along the fringe. As the shape of the object-wave deviates increasingly from a simple plane, the fringe pattern becomes more modulated and complex. Thus, if the object-wave is the light reflected from some one's face, the fringe system will be an unrecognizable tumult of fine bright and

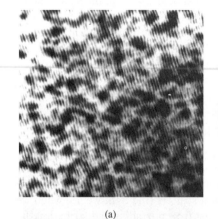

(a)

(b)

Figure 27.29 (a) A hologram. (b), (c), (d) Three different views photographed from the same holographic image generated by the hologram in (a). By moving your head (or in this case, a camera), you can see different regions of the scene. The lens in the hologram will magnify different objects depending on how you look through it.

(c)

(d)

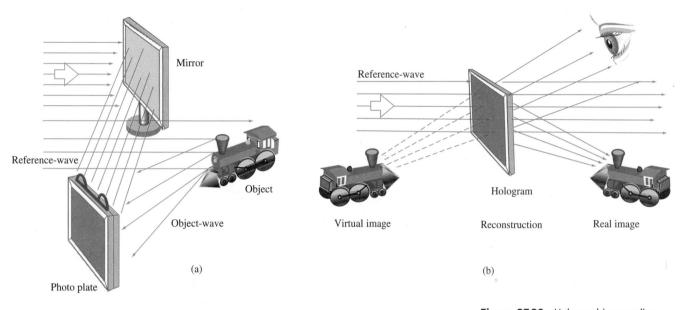

(a)

(b)

Figure 27.30 Holographic recording and reconstruction of an image. (a) The photo plate receives light from the object and a reference beam directly from the mirror. (b) To view the hologram, it is illuminated with a wave identical to the reference-wave.

dark regions. When developed, the silver atoms in the emulsion form a kind of 3-dimensional diffraction grating, which is the hologram (Fig. 27.29).

All we need do is illuminate a fine-grained film with both a scattered object-wave and a coherent reference-wave (Fig. 27.30). The developed film is a *transmission hologram,* and when it is illuminated from the back by a wave identical to the reference-wave (Fig. 27.30b), the original light field is reconstructed. The hologram during playback functions much like a diffraction grating that scatters the incoming plane waves into two off-axis beams. One of these gives rise to an extraneous real image, while the other corresponds to a duplicate of the original lightwave diverging from the object.

The light emerging from a hologram is identical to the light originally coming from the object (it's as if you were simply looking through a window). In the case of an extended scene, your eyes would have to refocus as you viewed different regions at various distances. And if you chose to photograph the vision, you would only have to point the camera through the "window," focus on the desired object, and shoot. Furthermore, since the light from any point on the subject spreads across much of the film plate, the information about that point is redundantly recorded all across the hologram. Consequently, just as you can peer out of a small opening in a window shade and see the entire scene beyond, you can peep through a small piece of the hologram and see everything as well. It's been suggested that the human brain functions in a similarly redundant way and that memory is, in that sense, holographic.

Core Material

When the electric field of a lightwave remains in a fixed plane the wave is **plane-polarized**, or *linearly polarized*. With *circular light,* the electric-field vector remains constant in magnitude while it revolves once around with every advance of one wavelength (p. 1012). Because the state of polarization of light from an ordinary source is sustained for less than 10^{-8} s, that state is essentially undetectable, and the light is said to be **unpolarized**, or **natural** (p. 1013).

Light that passes through two consecutive linear polarizers whose transmission axes make an angle θ, emerges with an irradiance given by

$$I = I_1 \cos^2 \theta \qquad [27.1]$$

where I_1 leaves the first filter. This equation is **Malus's Law** (p. 1016).

When an unpolarized beam is incident on a surface at an angle θ_p, known as **Brewster's angle**, the reflected light will be totally plane-polarized parallel to the interface. The **polarization angle** is given by

$$\tan \theta_p = \frac{n_t}{n_i} \qquad [27.3]$$

Light can also be linearly polarized via scattering (p. 1020).

In a region of overlap, two waves of the same frequency can combine constructively or destructively, depending on their relative phase, to produce a redistribution of energy in that area—this is the essence of **interference**. *The average time interval during which a lightwave oscillates in a predictable way is known as the* **coherence time** *(t_c) of the radiation (p. 1023). The corresponding spatial extent over which the lightwave oscillates in a regular, predictable way is called the* **coherence length** *(l_c), where in vacuum $l_c = ct_c$.*

The double-slit setup, known as Young's Experiment, is the premier demonstration of interference. Bright maxima appear (p. 1024) symmetrically about the centerline at distances of

$$y_m \approx \frac{s}{a} m\lambda \qquad [27.7]$$

The maxima are displaced from the central axis by the angles θ_m such that

$$a \sin \theta_m = m\lambda \qquad [27.8]$$

or since the angles are small, $\theta_m \approx m\lambda/a$. The spacing between fringes on the screen is

$$\Delta y \approx \frac{s}{a} \lambda \qquad [27.9]$$

Thin transparent films generate fringe patterns when the two waves reflected from the front and back surfaces interfere (p. 1028). Where the film is nonuniform, the pattern of so-called *fringes of equal thickness* creates a topographical map. When $n_1 > n_f > n_2$ or $n_1 < n_f < n_2$, reflected maxima occur where the thickness is

$$d = \frac{m\lambda_0}{2n_f} = \frac{m\lambda_f}{2} \qquad m = 1, 2, 3, \ldots \qquad [27.11]$$

An even more commonly occurring situation has $n_1 < n_f > n_2$ or $n_1 > n_f < n_2$. Now maxima occur where

$$d = \frac{(m + \frac{1}{2})\lambda_0}{2n_f} = \frac{(m + \frac{1}{2})\lambda_f}{2} \qquad m = 0, 1, 2, \ldots$$
$$[27.12]$$

The deviation of light from rectilinear propagation is **diffraction**. We concentrate on *Fraunhofer* diffraction, which effectively obtains *where the incident light is planar and the observation screen is at infinity*. In the case of a single slit of width D, zeros of irradiance will occur on both sides of the broad central maximum when

$$D \sin \theta_{m'} = m'\lambda \qquad [27.13]$$

where $m' = \pm 1, \pm 2, \pm 3, \ldots$. Similarly, when many parallel slits, each separated by a distance a, are present, narrow principal maxima will appear at the same locations as Young's fringes.

The angular half-width of the Airy disk is given, in radians, by

$$\theta_a \approx \frac{1.22\lambda}{D} \qquad [27.15]$$

Suggestions on Problem Solving

1. A common error associated with calculations of the polarization angle is to invert the n_t/n_i—watch out for that. Another frequent oversight is to forget that a linear polarizer reduces the irradiance of incident *unpolarized* light by 50%. Don't fail to square the cosine in Malus's Law.

2. When dealing with interference in thin films, be sure to determine whether or not there is a relative phase shift, due to internal and external reflection, before making any numerical calculations.

3. Many of the equations from the analysis of Young's Experiment carry over directly to the treatment of the diffraction grating. That means that Eqs. (27.7) through (27.9) apply to gratings, as does the expression $\theta_m = m\lambda/a$. Keep in mind that this last equation is always in radians!

4. The number density of lines on a grating is often given in lines per centimeter, *not* per meter, so on inverting this factor to compute a, the line spacing, you must first convert to lines/m if a is to be in meters.

Discussion Questions

1. Is it correct to say that ideally monochromatic light must be polarized? Explain.

2. Suppose that you had a beam of monochromatic light that you knew to be the in-phase superposition of two equal-amplitude, plane-polarized waves having their electric fields respectively horizontal and vertical. How might you verify this situation experimentally?

3. Can polychromatic light be polarized? In what sense is there no such thing as unpolarized light? Explain your answers.

4. Natural light is incident on a flat pane of glass at an angle of 45°. Describe the polarization of both the reflected and transmitted beams. How would that change if the light came in at Brewster's angle instead?

5. Suppose that you are looking through a linear polarizer, with its transmission axis vertical, at the surface of a pot of water, and at some angle of reflection find that the glare on the surface vanishes. What is happening? Would the vision change if you now replace the water with benzene, all else held constant? Explain.

6. What was the central problem that Young solved via his double-slit arrangement? How will the fringe pattern in his experiment change if the light source is gradually altered so that

its coherence time diminishes? Describe what will happen to the fringe system in some region where the optical path length difference from the two sources ultimately exceeds the coherence length.

7. An antireflection coating on glass has a thickness that corresponds to one-quarter of a wavelength of red light in the medium of the film. What color light will be reflected and what color transmitted when white light is incident normally? Assume $n_a > n_f > n_g$.

8. If you put a few drops of liquid soap in a cup with a little water and shake it around, you can make a great froth of bubbles. After several seconds, the bubbles will begin to show bright rainbow color patterns. Why must you wait before the colors appear?

9. Take a very close look at a worn surface under direct sunlight. A fingernail, an old coin, or the roof of an old car will all show a very fine granular pattern of tiny colored dots. This phenomenon is the so-called *speckle effect*. What might be its cause?

10. Fog up a piece of glass with your breath and look through the fogged plate in a darkened room at a white-light point source. You will see a system of concentric colored rings. What does this remind you of, and what do you think is causing the effect? Figure Q10 shows a point source seen through a glass plate covered with a layer of transparent, spherical lycopodium spores.

Figure Q10

11. Consider Young's Experiment where the apertures are two small circular holes, one next to the other, separated by a distance a, which is similarly small but still much larger than any of the incoming wavelengths. Assuming normally incident monochromatic plane-wave illumination, describe—qualitatively but in detail—the resulting fringe pattern on a very distant screen parallel to the plane of the apertures. Make a sketch of it.

12. Discuss the interrelationship between aperture size, illuminating wavelength, and *finite* distance to the observation screen (no lens after the hole) for near- and far-field diffraction. In other words, what will happen to the pattern on a screen as each of the above is varied?

13. How could you use a Michelson Interferometer to measure the coherence length of a discharge lamp?

14. Why do bright stars appear to the eye to be larger than faint ones?

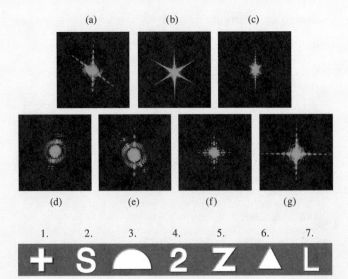

Figure Q17

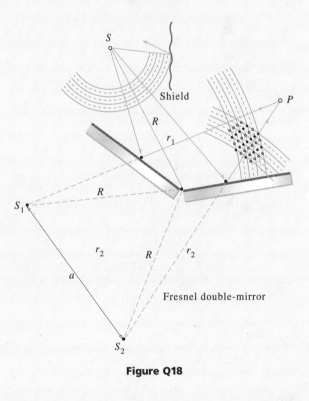

Fresnel double-mirror

Figure Q18

15. A nearsighted person looking through an adjustable circular aperture at a distant object can improve his vision by making the hole a bit smaller than his pupil—people squint to improve acuity. Still, if the aperture is made less than about 1.0 mm in diameter, the image will again blur. What is happening?

16. Using a TV or computer screen as a convenient broad source, stand a few feet from it and peer at it through a narrow slit (for example, the space between two straight fingers). Hold your hand about five or six inches from your eye and form a slit

between your fingers about a millimeter or less wide—you can rotate your hand slightly to effectively narrow the slit. Describe and explain the distribution of light within the slit.

17. Figure Q17 shows seven apertures and seven Fraunhofer diffraction patterns. Match up the patterns with the apertures that produced them.

18. Explain how the Fresnel double-mirror works to produce interference fringes (Fig. Q18). What will the pattern look like?

19. Imagine that you have a long-filament light bulb held vertically. If you look at it from several meters away through a vertical slit cut in a 3×5 card, you will see a lovely fringe system. Explain what's happening.

Multiple Choice Questions

1. The irradiance of a polarized lightwave is   (a) a function of the amplitude (to the first power) of the electric field   (b) independent of the electric field   (c) a function of the square of the amplitude of the electric field   (d) a function of one over the amplitude of the electric field   (e) none of these.

2. Natural light is light that is   (a) only found in nature   (b) linearly polarized   (c) completely unpolarized   (d) polarized but rapidly changing   (e) none of these.

3. The electric field of $\mathscr{R}$-state light   (a) has a constant magnitude and rotates   (b) has an oscillating magnitude and rotates clockwise   (c) has a sinusoidal amplitude and rotates counterclockwise   (d) has a constant amplitude and oscillates in a fixed plane   (e) none of these.

4. Unlike transverse waves, longitudinal waves cannot   (a) interfere   (b) diffract   (c) be reflected   (d) be polarized   (e) none of these.

5. If 100 W/m^2 of natural white light impinges on two ideal linear polarizers held one behind the other with their transmission axes parallel, the amount of light emerging will be   (a) 100 W/m^2 (b) 50 W/m^2 (c) 25 W/m^2 (d) 200 W/m^2 (e) none of these.

6. Light linearly polarized in the plane-of-incidence impinges on the surface of a glass plate in air at Brewster's angle. The transmitted beam is   (a) nonexistent   (b) linearly polarized in the plane-of-incidence   (c) linearly polarized perpendicular to the plane-of-incidence   (d) partially polarized   (e) none of these.

7. $\mathscr{P}$-state light parallel to the interface impinges on an air-water boundary at the polarization angle. The reflected beam will be   (a) linearly polarized perpendicular to the plane-of-incidence   (b) nonexistent   (c) partially polarized   (d) unpolarized   (e) none of these.

8. For two sources to be coherent, it is both necessary and sufficient that each   (a) be exactly in step, that is, in-phase   (b) have exactly the same amplitude   (c) be monochromatic   (d) be linearly polarized   (e) none of these.

9. Coherent lightwaves never arise from   (a) two lasers   (b) two pinholes   (c) two candles   (d) two slits   (e) any of these.

10. Increasing the coherence time of a source is equivalent to   (a) decreasing the coherence length   (b) increasing the speed   (c) making it more nearly monochromatic   (d) making it less monochromatic   (e) none of these.

11. Imagine that we cover each slit in Young's Experiment with one of two identical linear polarizers such that one transmission axis is parallel to the slits while the other is transverse to them. On the distant screen, we will observe   (a) no light at all   (b) a fairly uniform illumination   (c) cosine-squared fringes shifted $\frac{1}{2}\lambda$ from normal   (d) the usual fringe pattern   (e) none of these.

12. The middle of the first-order maximum, adjacent to the central bright fringe in the double-slit experiment, corresponds to a point where the optical path length difference from the two apertures is equal to   (a) λ (b) 0   (c) $\frac{1}{2}\lambda$ (d) $\frac{1}{4}\lambda$ (e) none of these.

13. Suppose you take a ring-shaped wire and dip it in liquid soap so that a circular film forms. Now, if you place it horizontally on a platform, which is then spun about a central vertical axis, and look at it under white light,   (a) a series of straight colored fringes will appear   (b) you will see a uniformly bright surface   (c) a series of concentric colored circular fringes surrounding a black central disk will appear   (d) a set of concentric circular colored bands surrounded by a black fringe will appear   (e) none of these.

14. A narrow slit of width D is illuminated by light coming from a monochromator so that the wavelength can be varied all across the visible spectrum. As λ decreases, the Fraunhofer diffraction pattern, viewed in the focal plane of a lens,   (a) shrinks with all the fringes getting narrower   (b) spreads out with all the fringes getting wider   (c) remains unchanged   (d) alters such that only the central maximum broadens   (e) none of these.

15. A small aperture is located some distance from an observation screen. As the wavelength of the illumination is gradually decreased, the far-field diffraction pattern seen on the screen   (a) remains unchanged   (b) changes to a near-field pattern   (c) changes to a Fraunhofer pattern   (d) stays as a far-field pattern but gradually expands   (e) none of these.

16. The effect of increasing the number of lines per centimeter of a grating is to   (a) increase the number of orders that can be seen   (b) allow for the use of longer wavelengths   (c) increase the spread of each spectral order   (d) produce no change in the diffracted light   (e) none of these.

17. A grating diffracts red light through an angle that is   (a) greater than the angle for blue light   (b) independent of frequency   (c) less than the angle for blue light   (d) the same as the angle for blue light   (e) none of these.

Problems

POLARIZATION

1. [I] What is the irradiance of a beam of linearly polarized electromagnetic radiation traveling through vacuum if the maximum value attained by its electric field is 1000 V/m?

2. [I] Plane-polarized light, oscillating in a vertical plane and having an energy flux density of 2000 W/m^2, impinges normally on a linear polarizer whose transmission axis is horizontal. What is the transmitted irradiance?

3. [I] A 100-W/m^2 beam of linearly polarized light with its electric field vertical impinges perpendicularly on an ideal linear polarizer with a vertical transmission axis. What is the irradiance of the transmitted beam?

4. [I] Light from an ordinary tungsten bulb arrives at an ideal linear polarizer with a radiant flux density of 100 W/m^2. What is its corresponding value on emerging?

5. [I] A beam of natural light with an irradiance of 500 W/m^2 impinges on the first of two consecutive ideal linear polarizers whose transmission axes are 30.0° apart. How much light emerges from the two?

6. [I] Many substances such as sugar and insulin are *optically active;* that is, they rotate the plane of polarization in proportion to both the path length and the concentration of the solution. A glass vessel is placed between crossed linear polarizers, and 50% of the natural light incident on the first polarizer is transmitted through the second polarizer. By how much did the sugar solution in the cell rotate the light passed by the first polarizer? This sort of arrangement can be used to determine such things as the amount of sugar in urine.

7. [I] A beam of white light linearly polarized with its electric field vertical and having an irradiance of 160 W/m^2 is incident normally on a linear polarizer whose transmission axis is 30° above the horizontal. How much light is transmitted?

8. [I] $\mathscr{P}$-state light aligned with its electric field vector at $+40°$ from the vertical impinges on an ideal sheet polarizer whose transmission axis is at $+10°$ from the vertical. What fraction of the incoming light emerges?

9. [I] The reflection of the sky coming off the surface of a pond ($n = 1.33$) is found to completely vanish when seen through a Polaroid filter. At what angle is the surface being examined? (Give the answer to two significant figures.)

10. [I] What is the polarization angle for reflection of light from the surface of a piece of glass ($n_g = 1.55$) immersed in water ($n_w = 1.33$)?

11. [II] Two ideal linear sheet polarizers are arranged with respect to the vertical with their transmission axes at 10° and 70°, respectively. If $\mathscr{P}$-state light at 40° enters the first polarizer, what fraction of its irradiance will emerge?

12. [II] Four ideal linear polarizers are stacked one behind the other with the transmission axes of the first vertical, the second at 30°, the third at 60°, and the fourth at 90°. What fraction of the incident unpolarized light emerges?

13. [II] A beam of light is reflected off the surface of a cup of benzene, and the light is examined with a linear sheet polarizer. It is found that when the central axis of the polarizer (that is, the perpendicular to the plane of the sheet) is tilted down from the vertical at an angle of 56.33°, the reflected light is completely passed, provided the transmission axis is parallel to the plane of the interface. From this information, compute the index of refraction of the liquid.

14. [II] Light reflected from a glass ($n_g = 1.65$) plate immersed in ethyl alcohol ($n_e = 1.36$) is found to be completely linearly polarized. At what angle will the partially polarized beam be transmitted into the plate?

15. [III] Two ideal linear polarizers are in place one behind the other. What angle should their transmission axes make if the incident unpolarized beam is to be reduced to 30% of its original irradiance?

INTERFERENCE

16. [I] A fine mercury arc lamp, emitting light at 546.1 nm, has a coherence length in vacuum of 0.60 m. Determine its coherence time.

17. [I] Red light from a He-Ne laser ($\lambda = 632.8$ nm) is incident in air on a screen containing two very narrow horizontal slits separated by 0.100 mm. A fringe pattern appears on a screen 2.00 m away. How far (in mm) above and below the central axis are the first zeros of irradiance?

18. [I] Red plane waves from a ruby laser ($\lambda = 694.3$ nm) in air impinge on two parallel slits in an opaque screen. A fringe pattern forms on a distant wall, and we see the fourth bright band 1.0° above the central axis. Calculate the separation between the slits.

19. [I] Two parallel slits 0.100 mm apart are illuminated by plane waves of quasimonochromatic light, and it is found that the fifth bright fringe is at an angle of 1.20°. Determine the wavelength of the light.

20. [I] A 3 × 5 card containing two pinholes, 0.08 mm in diameter separated center-to-center by 0.10 mm, is illuminated by red light from a He-Ne laser ($\lambda = 632.82$ nm). If the fringes on an observing screen are to be 10 mm apart, how far away should the screen be?

21. [I] Parallel rays of blue light from an argon ion laser ($\lambda = 487.99$ nm) are incident on a screen containing two very narrow slits separated by 0.200 mm. A fringe pattern appears on a sheet of film held 1.00 m away. How far (in mm) from the central axis is the fourth irradiance maximum?

22. [I] A collimated beam of light ($\lambda = 550$ nm) falls on a screen containing a pair of long, narrow slits separated by 0.10 mm. Determine the separation between the two third-order maxima on a screen 2.00 m from the apertures.

23. [I] A thin film of ethyl alcohol ($n_e = 1.36$) spread on a flat glass plate and illuminated with white light shows a lovely color pattern in reflection. If a region of the film reflects only green light (500 nm) strongly, how thick is it?

24. [I] A glass camera lens with an index of 1.55 is to be coated with a cryolite film ($n \approx 1.30$) to decrease the reflection of normally incident green light ($\lambda_0 = 500$ nm). What thickness should be deposited on the lens?

25. [I] A soap film of index 1.35 appears yellow (580 nm) when viewed from directly above. Compute several possible values of its thickness.

26. [II] White light falling on two long, narrow slits emerges and is observed on a distant screen. If red light ($\lambda_1 = 780$ nm) in the first-order fringe overlaps violet in the second-order fringe, what is the latter's wavelength?

27. [II] Considering the double-slit experiment, derive an equation for the distance $y_{m'}$ from the central axis to each irradiance *minimum*. (Hint: Consider Eq. (27.4). The first dark bands on either side of the central maximum correspond to $m' = \pm 1$.)

28. [II] A strip of photographic film is completely black except for two horizontal parallel slits, each 1.0 cm long by 0.10 mm wide, separated center-to-center by 0.50 mm. When illuminated by sunlight, all but the zeroth-order fringe show a spread of colors. If violet light appears on a screen 3.0 m away at a distance of 2.40 mm above and below the central axis in the first colored bands, what's its wavelength?

29. [II] Two 1.0-MHz radio antennas emitting in-phase are separated by 600 m along a north-south line. A radio placed 20 km east is equidistant from both transmitting antennas and picks up a fairly strong signal. How far north should that receiver be moved if it is again to detect a signal nearly as strong?

30. [II] Sunlight incident on a screen containing two long, narrow slits 0.20 mm apart casts a pattern on a white sheet of paper 2.0 m beyond. What is the distance separating the violet ($\lambda = 400$ nm) in the first-order band from the red ($\lambda = 600$ nm) in the second order?

31. [II] Two stereo speakers are 5.00 m apart, up against the east-west wall of someone's living room. They have been wired carelessly so that when one cone is displaced forward by the driving signal, the other is displaced backward. What's the main disadvantage of this arrangement to someone sitting on the midway line equidistant from the speakers? How far due east should the listener move if she is sitting up against the opposite wall 10.0 m away and she wishes to hear a peak in the power level of a 1000-Hz sound? Take the speed of sound to be 346 m/s.

32. [II] It's an easy matter to see fringes of equal thickness in a wedge-shaped film. Take two flat sheets of glass, one on top of the other, and separate them with a piece of paper (Fig. P32). If the wedge angle is α and the illumination is near normal at a wavelength in the film of λ_f, derive an expression for the distance (x_m) measured from the apex A to the successive maxima.

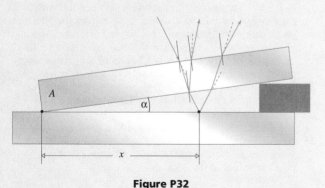

Figure P32

33. [II] Refer to the wedge-shaped film in Problem 32. What is the increase in film thickness as we move along from one maximum to the next? Derive an expression for the spacing Δx between consecutive bright fringes.

34. [II] With Fig. P32 in mind, suppose a wedge-shaped air film is made between two sheets of glass using a piece of paper 7.618×10^{-5} m thick as the spacer. If light of wavelength 550 nm comes down from directly above, determine the number of bright fringes that will be seen across the wedge.

35. [III] A Michelson Interferometer is illuminated with monochromatic light. One of its mirrors is then moved 2.53×10^{-5} m, and it is observed that 92 fringe-pairs, bright and dark, pass by in the process. Determine the wavelength of the incident beam.

36. [III] One of the mirrors of a Michelson Interferometer is moved, and 1000 fringe-pairs shift past the hairline in a viewing telescope during the process. If the device is illuminated with 500 nm light, how far was the mirror moved?

37. [III] Suppose we place a chamber 10.0 cm long (measured inside) with flat parallel windows in one arm of a Michelson Interferometer that is being illuminated by 600-nm light. If the refractive index of air is 1.000 29 and all the air is pumped out of the cell, how many fringe-pairs will shift by in the process?

38. [III] With regard to Young's Experiment, derive a general expression for the shift in the vertical position of the mth *maximum* as a result of placing a thin parallel sheet of glass of index n and thickness d directly over one of the slits. Identify your assumptions.

39. [III] Plane waves of monochromatic light impinge at an angle θ_i on a screen containing two narrow slits separated by a distance a. Derive an equation for the angle measured from the central axis that locates the mth maximum.

40. [III] Every source of light is composed of a range of frequencies distributed around some mean value. As a measure of the spectral purity of a source, we define the *ratio of the mean period to the coherence time* as the *frequency stability*. Accordingly, a typical He-Ne laser with a mean wavelength of 632.82 nm might have a frequency stability of 10^{-6}. Determine the coherence length of such a laser.

41. [III] A soap film of index 1.340 has a region where it is 550.0 nm thick. Determine the vacuum wavelengths of the radiation that is not reflected when the film is illuminated from above with sunlight.

42. [III] Figure P42 shows a setup for examining the shape of a lens. The lens is placed on an optical flat and illuminated at normal incidence by quasimonochromatic light. The gap between the lens and optical flat constitutes a circularly symmetric, wedge-shaped air film. The amount of uniformity in the resulting concentric system of circular fringes, known as Newton's rings, is a measure of the degree of perfection (see photo on p. 1030). Derive the expression

$$x_m = [(m + \tfrac{1}{2})\lambda_f R]^{\frac{1}{2}}$$

for the radius of the mth bright ring. Use the fact that $R \gg d$.

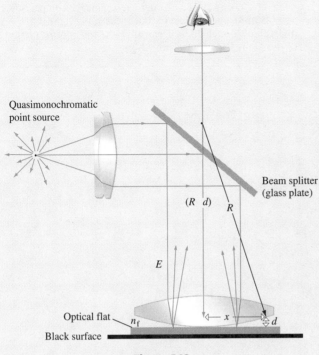

Quasimonochromatic
point source

Beam splitter
(glass plate)

$(R \; d)$ R

E

Optical flat n_f

$\leftarrow x \rightarrow$ d

Black surface

Figure P42

DIFFRACTION

43. [I] A single slit 0.10 mm wide is illuminated (in air) by plane waves from a krypton ion laser ($\lambda = 461.9$ nm). If the observing screen is 1.0 m away, determine the angular width of the central maximum as defined on p. 1037.

44. [I] A narrow single slit (in air) is illuminated by IR from a He–Ne laser at 1152.2 nm, and it is found that the center of the tenth dark band lies at an angle of 6.2° off the central axis. Please determine the width of the slit.

45. [I] A parallel beam of microwaves impinges on a metal screen that contains a 20-cm-wide, long horizontal slit. A detector moving parallel to the screen locates the first minimum at an angle of 36.87° above the central axis. Determine the wavelength of the radiation.

46. [I] A transmission grating whose lines are separated by 3.0×10^{-6} m is illuminated by a narrow beam of red light ($\lambda = 694.3$ nm) from a ruby laser. Spots of light, on both sides of the undeflected beam, appear on a screen 2.0 m away. How far from the central axis is either of the two nearest spots?

47. [I] We wish to study several different gratings by holding them each near one eye and looking through each at an ordinary light bulb, blocked off by two pieces of cardboard so that it forms a slit source. What will happen to the locations of the various-order spectra as we go from 200 lines/cm to 400 lines/cm to 800 lines/cm?

48. [I] A diffraction grating with slits 0.60×10^{-3} cm apart is illuminated by light with a wavelength of 500 nm. At what angle will the third-order maximum appear?

49. [I] A diffraction grating produces a second-order spectrum of yellow light ($\lambda = 550$ nm) at 25°. Determine the spacing between lines.

50. [I] If you peeped through a 0.75-mm-diameter hole at an eye chart, you might notice a decrease in visual acuity. Compute the angular limit of resolution, assuming that it's determined only by diffraction; take $\lambda = 550$ nm. Compare your results with the value of 1.7×10^{-4} rad that corresponds to a 4.0-mm pupil.

51. [I] The Mount Palomar telescope has a 508-cm-diameter objective mirror. Determine its angular limit of resolution at a wavelength of 550 nm, in radians, degrees, and seconds of arc.

52. [I] With Problem 51 in mind, how far apart must two objects be on the surface of the Moon if they are to be resolvable by the Palomar telescope? The Earth–Moon distance is 3.844×10^8 m; take $\lambda = 550$ nm.

53. [I] Compare the results of Problem 52 with the human eye; that is, how far apart must two objects be on the Moon if they are to be distinguished by eye? Assume a pupil diameter of 4.00 mm.

54. [II] Plane waves with a wavelength of 500 nm are incident on a long, narrow aperture 0.10 mm wide. How wide will the central maximum be on a screen 3.00 m from the slits? (Use the fringe-width defined in Example 27.7, p. 1037.)

55. [II] A beam of parallel rays of red light from a ruby laser ($\lambda = 694.3$ nm in air) incident perpendicularly on a single slit produces a central bright band that is 40 mm wide on an observing screen 3.50 m beyond. How wide is the slit? (See Problem 54.)

56. [II] Consider Problem 44. At what angle would the tenth minimum appear if the entire arrangement were immersed in water ($n_w = 1.33$) rather than air ($n_a = 1.000\,29$)?

57. [II] Bats use ultrahigh-frequency sound to reflect off targets. Estimate the typical frequency of this rodent sonar, presuming that bats eat moths. Take the speed of sound to be 330 m/s and a moth to be about 3.0 mm across.

58. [II] White light falls normally on a transmission grating that contains 1000 lines per centimeter. At what angle will we find red light ($\lambda = 650$ nm) in the first-order spectrum?

59. [II] Light from a sodium lamp has two strong yellow components at 589.592 3 nm and 588.995 3 nm. How far apart in the first-order spectrum will these two be on a screen 1.00 m from a grating having 10 000 lines/cm? (a) Assume Eq. (27.7) is accurate enough for our purposes. (b) Show that a more exact solution, arrived at using Eq. (27.8), yields a separation of 1.13 mm.

60. [II] Sunlight impinges on a transmission grating having 5000 lines per centimeter. Does the third-order spectrum overlap the second-order spectrum? Take red to be 780 nm and violet to be 390 nm.

61. [III] Two long slits 0.10 mm wide, separated (center-to-center) by 0.20 mm, in an opaque screen are illuminated by light of wavelength 500 nm. How many Young's fringes will be seen within the central bright band on a screen 2.0 m away? Determine a general rule for this situation in terms of the number N where $a = ND$. (Hint: Count fractions of fringes too.)

62. [III] Light with a frequency of 4.0×10^{14} Hz is incident on a grating having 10 000 lines per centimeter. What is the highest-order spectrum that can be seen with this device? Explain.

63. [III] A device used to measure the diffracted angle, and therefore the wavelength, of light passing through a grating is called a *spectrometer*. Suppose that such an instrument while in vacuum on Earth sends 500-nm light off at an angle of 20.0° in the first-order spectrum. By comparison, after landing on the planet Mongo, the same light is diffracted through 18.0°. Determine the index of refraction of the Mongoian atmosphere.

28

Special Relativity

THE SPECIAL THEORY OF Relativity represents the culmination of Classical Physics. It arises from a rethinking of the essence of space and time and the very nature of interaction. So profound are the ideas that they led to a flood of new conclusions, many of which were no less than shocking: time is relative and flows differently in systems that are in motion with respect to each other; the length of an object is not absolute, but depends on its relative motion with regard to each particular observer; the speed of light, c, is invariant and, moreover, it is the upper limit of all speeds in the Universe; the gravitational and electromagnetic interactions propagate at c rather than infinitely swiftly; except for Newton's Second Law (as he gave it, p. 129), most of the equations of classical dynamics are only approximations; mass and energy are equivalent in a way never conceived of before. These ideas and more follow from two simple postulates that Einstein introduced and then elegantly developed.

BEFORE THE SPECIAL THEORY

The research that laid the groundwork for the Special Theory was concerned with the behavior of light. Ironically, Relativity has nothing to do with light *per se,* although it has a lot to do with the constancy of the speed of light.

Prior to Einstein's work in 1905, it was widely believed that space was filled with aether and light was an electromagnetically induced vibration of that invisible elastic "goop." The common wisdom was that aether was a transparent solid that could support transverse lightwaves and at the same time a perfect fluid that could be parted effortlessly by the wandering planets.

28.1   The Michelson-Morley Experiment

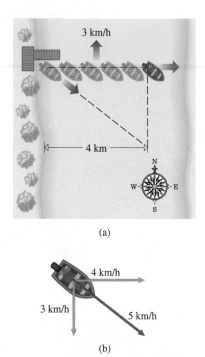

(a)

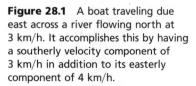

(b)

Figure 28.1 A boat traveling due east across a river flowing north at 3 km/h. It accomplishes this by having a southerly velocity component of 3 km/h in addition to its easterly component of 4 km/h.

The aether, even though it doesn't exist, played a crucial part in the development of the Special Theory. If we are to appreciate the context in which Relativity was conceived, it would be helpful to know a little about the aether as envisioned at the turn of the twentieth century. At that time, the most important single observation was the Michelson-Morley experiment, which had failed to detect any evidence of the aether and which seemed to suggest the invariance of the speed of light. It is arguably one of the most important experiments ever performed, and at the same time, it was a disappointing failure for the man who performed it.

James Clerk Maxwell, in 1879, the last year of his life (and the year Einstein was born), wrote a letter that was ultimately published. In it, he discussed a scheme for measuring the speed (v) at which the orbiting Earth plows through the aether. As the Earth moves, the ordinarily motionless aether presumably streams over the planet. The back-and-forth time of transit of a light beam in this aether wind should be different along or across the flow. To see this, imagine a wide river streaming north at 3 km/h and a motorboat capable of a single speed with respect to the water of 5 km/h (Fig. 28.1). Compare the transit times for two trips; 4 km upstream and back, versus 4 km across stream and back. Going upstream, bucking the current, the boat makes 5 km/h − 3 km/h = 2 km/h and reaches its destination in 2 h, whereupon it turns around and returns at 5 km/h + 3 km/h = 8 km/h, arriving back after an additional $\frac{1}{2}$ h; the total trip taking $2\frac{1}{2}$ h. By contrast, to cross the river directly, the boat must be headed south a bit so that it has a southerly velocity component equal to the river's northerly 3 km/h (p. 55). The two north-south velocities cancel and the boat travels due east. The 3-4-5 right-triangle geometry gives the boat a net easterly speed of 4 km/h. It travels out for 1 h and then returns (again heading somewhat south) after another 1 h—total transit time: 2 h. The time difference between the two journeys (here, $\frac{1}{2}$ h) reveals the presence of the moving medium. If v (the river's speed) was zero, the time difference would vanish.

Unfortunately, as Maxwell pointed out, in the case of an aether wind (instead of a river) flowing at v and a light beam (instead of a boat), the effect depends on $(v/c)^2$, and that "is quite too small to be observed." Even with the aether streaming past the planet at the Earth's orbital speed of $v = 3 \times 10^4$ m/s (that is, ≈66 000 mi/h), the quantity $(v/c)^2 \approx 10^{-8}$ and the transit-time difference for a 4 km-out-and-4 km-back journey is a mere tenth of a millionth of a millionth of a second. Not very promising.

Maxwell's letter came to the attention of a young instructor at the United States Naval Academy, Albert Abraham Michelson. Michelson took a leave of absence in 1880 to study in Europe, and while there, accepted Maxwell's challenge. In the past, several experimenters had used interferometers to examine the effects of motion on the transmission of light through the aether. But all of these efforts were limited in their precision, and though they failed to detect the aether, the results were unconvincing. Now Michelson designed an entirely new instrument that would allow him to make the decisive measurement precisely enough to reveal $(v/c)^2$ variations. His interferometer (funded by none other than Alexander Graham Bell) would surely detect an aether wind carrying along the lightwaves.

Figure 28.2 is a schematic drawing of the Michelson Interferometer; it's the same device we studied earlier (p. 1031). The difference between the along-the-aether and the across-the-aether transit times, Δt, is

$$\Delta t \approx \frac{L}{c}\beta^2 \tag{28.1}$$

where $\beta = v/c \approx 10^{-4}$. One wave travels a little longer than the other, and they arrive a bit out-of-phase. Hence, the fringes are shifted from where they would have been were the Earth at rest (Fig. 28.3). Since we can't stop the Earth, this shift is not immediately apparent, but that can be rectified by rotating the entire interferometer while observing the fringes. This will gradually interchange the two arms so that the one that was along the aether wind is now across it and vice versa. The cross-stream wave will go from leading by Δt to lagging by Δt as it becomes an upstream-downstream wave. That change of $2\Delta t$ in time is equivalent to an introduced path-length difference of $c2\Delta t$, which causes the fringe pattern to move. A shift of one wavelength results in a displacement of one bright-dark fringe-pair, and the observer (following around as the device is rotated through 90°) should see $N = c2\Delta t/\lambda \approx 2L\beta^2/\lambda$ fringe-pairs move past a cross hair in a viewing telescope.

When Michelson first performed the experiment in 1881, N should have been about four one-hundredths of a fringe, but he saw no shift at all. Confident of his measurements, he published the startling result that "there is no displacement of the interference bands." Still, his so-called *null result* was not very persuasive. With the urging of Lord Rayleigh and Lord Kelvin, Michelson decided to redo the experiment in an improved fashion and settle the matter. He was joined by E.W. Morley and they enlarged the apparatus to the point where the expected shift was now fourtenths of a fringe. Still no appreciable shift was observed—and this time the results could not be neglected—there was no detectable aether wind. *The speed of light is not influenced by the motion of the Earth.*

A modern version of the Michelson-Morley experiment performed in 1979 used stable lasers to improve the precision by a factor of 4000; it, too, found no sign of an aether wind.

The Lorentz-FitzGerald Contraction

In 1892, G. FitzGerald proposed a rather imaginative hypothesis to get around the Michelson-Morley result and still keep the aether wind blowing. The idea was elaborated by Lorentz and is today known as the **Lorentz-FitzGerald Contraction**. However *ad hoc* the notion was, it would re-emerge over a decade later as a natural consequence of Einstein's Special Theory of Relativity. What FitzGerald proposed was that the aether wind exerts a pressure on a body moving through it and the body

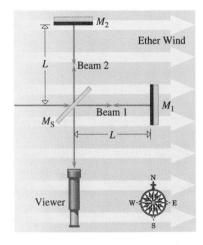

Figure 28.2 A simplified version of the Michelson Interferometer placed in the supposed aether wind.

Figure 28.3 With the interferometer set up to produce straight fringes, we would see a pattern like this one in the viewing telescope. It occurs when M_1 and M_2 are perpendicular to the plane of the instrument but not to each other.

The luminiferous aether, that is the only substance we are confident of in dynamics. . . . One thing we are sure of, and that is the reality and substantiality of the luminiferous aether.

LORD KELVIN
Popular Lectures and Addresses (1891)

compresses slightly: *every object moving at a speed v contracts along the direction of motion by a factor equal to* $\sqrt{1 - \beta^2}$. If the $M_S M_1$ arm of the interferometer shrinks so that L becomes $L\sqrt{1 - \beta^2}$, $\Delta t = 0$ (see Problem 1). There is no such aether pressure, but strange as it may seem, there is a Lorentz-FitzGerald Contraction.

Example 28.1 An object is moving at a speed of 0.200 0c. Determine the value of $1/\sqrt{1 - \beta^2}$ at that speed. Redo the calculation for a speed of 0.002 0c. Don't worry about significant figures.

Solution: [Given: speeds of 0.200 0c and 0.002 0c. Find: $1/\sqrt{1 - \beta^2}$.] $\beta = v/c = 0.200\,0$, $\sqrt{1 - \beta^2} = \sqrt{1 - 0.040\,0} = 0.979\,8$, and $1/\sqrt{1 - \beta^2} = \boxed{1.021}$. Now for $v = 0.002\,0c$, $\beta = 0.002\,0$, $\beta^2 = 4.0 \times 10^{-6}$, $(1 - \beta^2) = (1 - 0.000\,004\,0) = 0.999\,996\,0$ and $\sqrt{1 - \beta^2} = 0.999\,998\,0$. Thus $1/\sqrt{1 - \beta^2} = \boxed{1.000\,002}$.

▶ **Quick Check:** Knowing that we'll need it again (p. 1076), we now derive a useful approximation. From the binomial expansion with $x = -\beta^2$ and $n = -\frac{1}{2}$, we get

$$(1 + x)^n = 1 + nx + \frac{n(n - 1)x^2}{2} + \cdots$$

All the terms beyond the second one can be dropped since β^2 is tiny, and therefore $x^2 = (-\beta^2)^2$ is minuscule. Hence $1/\sqrt{1 - \beta^2} \approx (1 + nx) \approx 1 + \beta^2/2 = 1 + 0.000\,004\,0/2 = 1.000\,002$.

How Comfortable the Old Ideas.
The beginner will find it best to accept the aether theory, at least as a working hypothesis. . . . Even if future developments prove that the extreme relativists are right and that there is no aether, it is likely that the change will involve no serious readjustments so far as explanations of the ordinary phenomena are concerned.

A. A. KNOWLTON
Physics for College Students (1928)

There was a young fencer named Fisk,
Whose thrust was exceedingly brisk.
So fast was his action,
The Lorentz-FitzGerald contraction
Reduced his rapier to a disk.

ANONYMOUS

FitzGerald maintained that there was no fringe shift because the arm of the interferometer along the wind shrunk. And there's no easy way to measure that shrinkage because all rulers will likewise shrink. It was not until 1932, long after Einstein had put things right, that the notion of an aether-wind-induced contraction was laid to rest experimentally by R. Kennedy and E. Thorndike. In 1990, a modern laser version of their experiment accurate to within 70 parts per million confirmed their result.

Stimulated by Michelson's observation, physicists subsequently performed many elaborate optical, electrical, and magnetic experiments, all of which failed to detect the motion of the Earth with respect to the aether. In 1900, the brilliant French mathematician Jules Henri Poincaré wrote:

Our aether, does it really exist? I do not believe that more precise observations could ever reveal anything more than *relative* displacements.

THE SPECIAL THEORY OF RELATIVITY

Long after his billowing shock of hair had turned gray, Albert Einstein recalled how, when he was sixteen, he began to struggle with a troubling paradox. Light was understood to be electromagnetic, an intricate oscillatory web of linked time-varying electric and magnetic fields rippling through space. If we could travel out at speed c alongside a pulse of light, what would we see? Moving in imagination next to a wave peak, the disturbance would appear unchanging, motionless; yet electromagnetic theory does not allow such a situation. A stationary, nonvarying field-pulse cannot exist. In that regard, light is totally unlike a sound wave or even a stream of bullets, each of which can be followed and examined as if frozen in flight. Light is self-sustained by *change*. It is a thing of interwoven fields that, by alternation, generate each other—no variation in time, no existence. A stationary observer must see the pulse, even as the observer moving at c sees nothing—how strange. At sixteen,

Einstein had recognized a profound dilemma, a conflict between Newton's mechanics, which allowed travel at lightspeed and Maxwell's electrodynamics, which couldn't abide with it. One of these two formalisms was wrong.

28.2 The Two Postulates

In 1905, Einstein was an unknown clerk in the Bern Patent Office, Switzerland. Newly married, shy but friendly, the young man had plenty of time to think and create (reviewing patent applications wasn't very taxing). He often worked on his own ideas, hiding the calculations in a drawer whenever footsteps approached. That was the year he published five papers in the prestigious *Annalen der Physik,* representing three extraordinary new developments in physics. One of these was his first paper on Relativity, "*Zur Elektrodynamik bewegter Körper,*" ("On the Electrodynamics of Moving Bodies").

At twenty-six, Einstein had created the *special* or *restricted* theory—restricted in the sense of specialized, since it pertained only to *uniform motion.* The theory assumed the validity of two *postulates* that he believed were correct but could not otherwise prove. He then set out to derive the physical implications of those postulates, and in the process completely recast our understanding of space and time.

Albert Einstein (1879–1955).

The Principle of Relativity

The first of the two postulates is called the *Principle of Relativity.* It's a generalization of work done by Galileo and Newton. Both men recognized that uniform motion had no perceivable effect on mechanical systems. One can play pool or Ping Pong aboard a ship and never know the vessel is moving, regardless of the velocity, so long as it is constant. The woman juggling oranges in the lounge of a *747* can't tell from the behavior of the fruit in the air whether she is cruising at 600 mi/h or sitting at rest on the runway. This, then, is the *Classical Principle of Relativity: the laws of mechanics are the same for all observers in uniform motion.*

Newton had struggled with the distinction between absolute and relative motion. Was there something somewhere in the vast Universe that was totally stationary, something absolutely at rest from which all motion could be reckoned absolutely? "I hold space to be at rest" wrote Newton—space, unchanging and immovable, was his fixed reference frame. By the end of the 1880s, a motionless aether filled all space and provided a material backdrop for absolute rest throughout the Universe. A thing moves absolutely when it moves with respect to the aether. The aether had two theoretical functions that justified its existence in the face of untold contradictions: it provided the medium for lightwaves, and it was the signpost of absolute rest.

We speak of a uniformly moving observer as an **inertial observer**—someone standing still in a train, plane, or rocket ship that is itself moving at a constant velocity (p. 50). *A system that is moving at a constant velocity is an* **inertial system**. The name comes from the fact that the Law of Inertia holds in all inertial systems. A body at rest does not tend to stay at rest, nor does a body in motion tend to stay in uniform motion in a straight line if it's traveling in a system that is accelerating.

While in an inertial system, jet-plane commuters expect no novel experiences. Our battery-operated toothbrushes still hum along, computers compute, popcorn pops, cola tastes the same, and pizza smells the same. Experience suggests that not just the laws of mechanics are the same, but all the laws of physics are the same for inertial observers. All aspects of the physical environment are unaffected by uniform motion, and life goes on quite normally at 2 km/h or 2000 km/h. This conjecture Einstein raised to the status of his **First Postulate**, his **Principle of Relativity**:

All the laws of physics are the same for uniformly moving observers.

It is impossible, using experiments performed within inertial systems, to observe results that can distinguish between such systems, and no experiment whatsoever can establish whether a particular inertial system is uniformly moving or at "absolute rest." The concept of absolute rest thereby loses all significance; if it can never be determined, motion is relative, not absolute. The Principle of Relativity logically abolishes the concept of absolute rest and along with it the concept of absolute motion—**motion is relative**.

All the failed experiments that had vainly tried to measure the absolute motion of the Earth with respect to the aether had led Poincaré to the Principle of Relativity, and now Einstein embraced the same conclusion. He dismissed the aether wind, boldly dispensing with the concept of aether entirely:

> The introduction of a "luminiferous aether" will prove to be superfluous, inasmuch as the view here to be developed will not require an "absolutely stationary space."

The Constancy of the Speed of Light

Einstein's **Second Postulate** is the **Principle of the Constancy of the Speed of Light**:

> **Light propagates in free space with a speed c that is independent of the motion of the source.**

Now this much in itself is both orthodox and reasonable; after all, the speed of sound is independent of the motion of the source. A sound wave is launched into a medium, and the speed of the disturbance is only determined by the physical characteristics of the medium. The speed of the source is irrelevant. If light is a wave in the aether, this statement makes obvious sense. Still, there's more here than meets the eye; just three months before this paper, Einstein had submitted an article wherein he maintained that light was a stream of particles, in which case the Second Postulate is not so obvious.

Remember that the equations of Electromagnetic Theory—Maxwell's Equations—led to a wave equation that provided the speed of light in vacuum (p. 914). Assuming Maxwell's Equations are right and given the First Postulate, it must be that this same wave equation is applicable in all inertial systems—the vacuum speed of light measured on Earth or inside a rocket ship must be the same, independent of any uniform relative motion. In Maxwell's theory, the speed of light is a constant, not a motion-dependent variable. In other words, *the speed of light measured with respect to an inertial system must be the same for all such systems.*

The fact that the speed of light is independent of the motion of the source is not at all troublesome, but is it also independent of the motion of the detector (that is, the observer)? Certainly, the speed of sound is not; if the detector rushes toward the source, moving with respect to the air, the measured speed of sound increases. Just imagine two identical ships headed toward a motionless sound-emitting buoy, one steaming along at full speed and the other dead in the water. On both ships, the time it takes a blast of sound to sweep from bow to stern is measured, and the speed of the wave is computed in that inertial system. Clearly, for the ship moving toward the buoy, that time will be shorter (during the interval it takes the sound to traverse the ship, the stern will advance somewhat toward the pulse, shortening the effective length) and the wave speed will be determined to be faster.

Figure 28.4 depicts the same arrangement, this time with spaceships and light-waves. Both ships receive light from the outside source *S*, and both compare the

Figure 28.4 Two spaceships, one at rest, the other moving at speed *v* with respect to a source S. Both have on-board light beams, which they measure to travel at c. Both must measure the light from S to travel at c as well.

speed of that light to the speed of the light from their own sources. Ship 1 at rest with respect to the beacon must measure both its beam and the beacon's beam to have the same speed c. Ship 2 (moving at *v* with respect to *S*) must measure the beam from its own source to travel at c because of the Second Postulate. But it must also measure the light from the beacon to travel at c, as well. If it didn't, that would mean that the two beams could be used together as a motion detector to establish that ship 2 was actually moving, which is not allowed by the First Postulate. The conclusion is an astonishing one: *no matter how fast a light source moves toward or away from an inertial observer, and no matter how fast the observer moves toward or away from the source, the speed of the light passing from one to the other in vacuum will always be* c—**the speed of light is constant**; it *is* absolute. The Michelson-Morley experiment ends with a null result because the speed of light along each arm is identically the same—there is no fringe shift because there is no transit-time difference, $\Delta t = 0$.

The two postulates separately are innocent enough. It's when we mix them together, when we demand that they both apply at once, that things seem to become fantastic. The paradox of Einstein's youth is now no longer a paradox—one simply cannot travel next to a light beam and catch up to it. The spaceship in Fig. 28.5 could be rushing either toward or away from the source at any speed you like, say, 99% of c, and still the inertial observer aboard it will measure the speed of the beam to be c! This is one of the premier conclusions of the analysis; its disturbingly "illogical" nature suggests that our familiar, comfortable understanding of space and time requires revision.

To our best knowledge, neutrinos and gravitons both also travel at c, and there must equally well be a *Principle of the Constancy of the Speed of Neutrinos*. The important thing is the speed c, not the photons doing the traveling—they just happen to be the most convenient, easily detected, c-speed probe.

Experimental Confirmations of the Second Postulate

At the time Einstein proposed the Second Postulate, there had been no direct experimental evidence to confirm it—it was at birth a purely logical conjecture. That situation has changed considerably over the intervening decades, and the extensive confirmation that now exists has transformed the Second Postulate into what might better be called the *Law of the Constancy of c*.

In a most convincing contemporary experiment, subatomic particles known as neutral pions (p. 1214) were produced at a tremendous speed of 99.98% of c. Free neutral pions have a mean life of only 8.7×10^{-17} s before naturally decaying— vanishing into gamma rays. So here we have a pulse of pions traveling at 0.999 8c, and in a short while each of them transforms into two photons. Detectors in the forward direction record bursts of photons, and their speeds can be determined over the 31-m flight path. Instead of finding a speed of 0.999 8c + c ≈ 2c, which would have been expected from Newtonian kinematics, researchers found a speed of $2.997\,7 \pm 0.000\,4 \times 10^{8}$ m/s, in excellent agreement with the standard value of c measured with a stationary source.

28.3 Simultaneity and Time

Einstein was troubled by the fact that, on one hand, the speed of light must be invariant and, on the other, such invariance violates the customary addition rules for velocities. He spent most of a year struggling with the problem. Then it came to him: "Time cannot be absolutely defined, and there is an inseparable relation be-

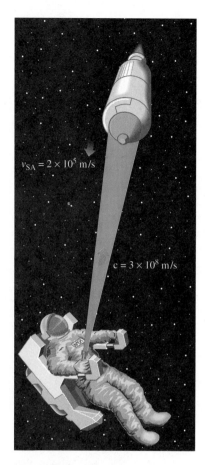

Figure 28.5 The speed of light in vacuum will be measured to be constant, regardless of the motion of either the source or the receiver. Here, a ship receives a light beam from an astronaut toward whom it is rushing at a speed v_{SA}. Regardless of the relative motion between the ship and the astronaut, a person on board the vessel will measure the light to arrive at c.

Aether Persisted. We have learned too that radiant heat energy is believed to be transmitted by a medium called the aether. At the present time, some scientists believe that other aether waves produce various other effects. . . . It is possible, then, that light waves are aether waves.

CHARLES E. DULL
Modern Physics (1939), a high
school text

tween time and signal velocity. With this new concept, I could resolve all the difficulties . . . . Within five weeks the Special Theory of Relativity was completed."

Light is the swiftest instrumentality at our disposal for communication and for
the perception of events (which are themselves the basis of time). Yet its speed is
finite, and that shapes our understanding. If c were infinite, the distinction between
Special Relativity and Newtonian theory would vanish. In fact, wherever c is so
large in comparison with the relevant motions that it can be considered effectively
infinite, Newtonian theory works very well. This is precisely why the discrepancies
went unnoticed so long—we live, for the most part, in a comparatively slow-moving
world. Running, driving, even spaceship-flying at thousands of miles per hour are
all mere crawling compared to 2.998×10^8 m/s (Table 28.1).

The link between space and time and c hardly intrudes in our everyday lives—
it doesn't matter that the people you are talking to 10 feet away are also ten-
thousandths of a millionth of a second away. Light takes about 10^{-9} s to travel 1 foot,
and the people are seen not as they are now, but as they were 10×10^{-9} s ago. When
you look out the window, the scene on your retina—the scene you see—is not all
happening at the same moment; the view of the Sun in the sky shows it as it was
8.3 minutes before Uncle George smiled in the foreground. The stars in a photo of
the night sky were not all there looking as they do at the same moment, even though
the light from them arrived on the film at the same instant. For that matter, the stars
you see might no longer exist even while you are "looking" at them: a star 100 light-
years away appears as it was 100 years ago when the light you see now first left it.
And if any of this seems strange, it is because you have always thought of the world
as if you saw things the instant they happened, as if $c = \infty$, but it doesn't.

TABLE 28.1 Various speeds and the corresponding β and γ values*

Object	Speed v	β v/c	γ $1/\sqrt{1 - (v/c)^2}$
Human walking	8 km/h	0.000 000 007	1.000 000 000
100-yard dash (max.)	10.0 m/s	0.000 000 033	1.000 000 000
Commercial automobile (max.)	62 m/s	0.000 000 21	1.000 000 000
Sound	333 m/s	0.000 001 11	1.000 000 000
SR-71 reconnaissance jet	980 m/s	0.000 003 27	1.000 000 000
Moon around Earth	1000 m/s	0.000 003 33	1.000 000 000
Apollo 10 (re-entry)	11.1 km/s	0.000 037	1.000 000 001
Escape speed (Earth)	11.2 km/s	0.000 037	1.000 000 001
Pioneer 10	14.4 km/s	0.000 048	1.000 000 001
Earth around Sun	29.6 km/s	0.000 099	1.000 000 005
Mercury orbital speed	47.9 km/s	0.000 16	1.000 000 013
Helios B solar probe	66.7 km/s	0.000 22	1.000 000 025
Earth-Sun around galaxy	2.1×10^5 m/s	0.000 70	1.000 000 245
Electrons in a TV tube	9×10^7 m/s	0.3	1.05
Muons at CERN	2.996×10^8 m/s	0.999 4	28.87
Electrons in Stanford Linear Accelerator (SLAC)	$2.997 9 \times 10^8$ m/s	0.999 999 999 7	4×10^4

*To better show the behavior of β and γ, little concern is given to significant figures.

The National Institute of Standards and Technology (NIST) sends out radio signals with which we can set our clocks—"When you hear the tone it will be . . . BEEP." Yet the farther you are away from the transmitter, the later the BEEP will arrive. An observer in the Andromeda galaxy will have to wait 2,200,000 years for the BEEP to arrive marking *now*. That's presumably no problem since we can make the necessary corrections for communication-time lags, knowing distance and c. Still, what does *now* mean for us in regard to our friend on Andromeda? That is, what does it mean to say that two events occur simultaneously, one here and one there, now? To be sure, if two events occur simultaneously at the same location, there's no problem—if two comets are seen to blow up immediately on crashing, we know the two events occurred absolutely simultaneously. Every other inertial observer in the Universe will see the same thing. Difficulties arise when the events occur at separate locations, and the greater their separation in space, the longer the period of inconclusiveness in time.

Figure 28.6 illustrates the problem quite simply. Part (a) is the view of a pulse of light as seen sequentially by an observer at rest inside the ship. She sees the expanding lightwave strike the front and back walls simultaneously. Note that *once the light is launched into space, it's on its own and it must travel at c in all directions as seen by any and all inertial observers*. Part (b) is the view as seen by an observer outside with respect to whom the ship is uniformly moving. He sees the lightwave, traveling at c in all directions, but now the rear wall advances on the source and is struck first while the front wall recedes and is struck second—the

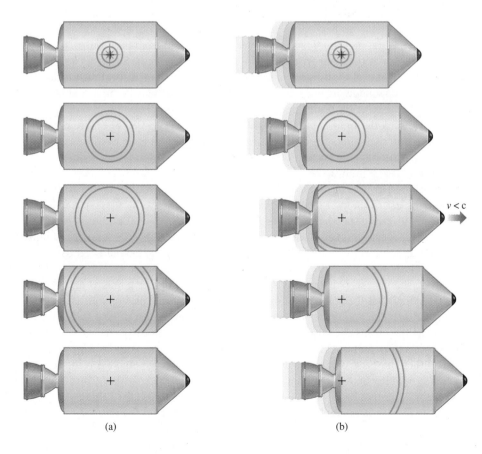

(a)   (b)

Figure 28.6 A flash of light in (a) a "stationary" and in (b) a "moving" ship as seen from outside. Or, equivalently, a flash in a ship seen (a) inside and (b) by an observer outside with respect to whom the ship is moving at a constant speed *v*.

events are *not simultaneous*. Notice that if the walls in the craft are mirrored, the waves will be reflected and meet in the center of the chamber in Fig. 28.6a. There's no ambiguity then; the same thing must happen in (a) and (b)—events that occur simultaneously at one place in space must be simultaneous for any inertial observer. What all of this suggests is that, like rest and motion, simultaneity is not absolute. You might say "who cares?" but if simultaneity, which is at the heart of the measurement of when events occur, is not absolute, time itself is not absolute.

Since that's really a big step to take, let's be a bit more rigorous about it. *Two spatially separated events are simultaneous if they are seen to occur at the same time by an observer located midway between the sites where the events happen.* This way, we will not have to introduce any corrections for different distances as was the case with the BEEPs from NIST. The question now is, do two events occurring simultaneously for one *midway inertial observer* occur simultaneously for all midway inertial observers regardless of their motion?

Imagine a rocket ship and a space station gliding past each other at any constant relative speed (Fig. 28.7). *We select the station commander,* a stubborn but honest fellow named Stan, *to be the one who will see the two events we are arranging as simultaneous.* Playing out his part, he switches on two explosive flares that just happen to be floating a few meters apart in space between him and the rocket. At the very moment they explode (as determined by him later), he finds himself lined up nose-to-nose with the rocket's pilot, a woman named Rosie no less willful than himself. Both are in the middle of their respective crafts. (It doesn't matter how these flares are themselves moving, or how they got there. We are only interested in them as light sources that will conveniently leave dent marks on each hull.) According to Stan, points A and A' are adjacent to each other at the same moment that B and B' are adjacent—the distances between dents on the station and on the ship are the same.

Stan in the station is a midway observer, and the light from both flares, fore and aft, reaches him at the same instant (we arranged it that way). So far as he is concerned, the two explosions are unquestionably simultaneous. Naturally enough, looking out the window he takes himself to be at rest and sees the rocket ship moving off to his rear. As a result, he sees Rosie advance on flare A and recede from flare B. He sees the light from flare A reach her first (Fig. 28.7b) before the light from flare B arrives (Fig. 28.7d). He expects Rosie to see the two events as not having occurred simultaneously, but he knows that he's right.

Rosie is also a midway observer, and the two dents on the rocket ship from the explosions are permanent proof of that. She looks out her window and sees the station moving off to the rear of her rocket ship (Fig. 28.8). We have arranged for the two beams to arrive together at Stan's location (Fig. 28.8c), and both observers see that happen because it takes place at a single point in space. Accordingly, Rosie must see the flash from flare A (Fig. 28.8b), which she records as having fired first, and then the flash from flare B, which she therefore knows must have fired second. Moreover, she observes that points A and A' coincided before points B and B' did. Understanding that the station is moving toward the later explosion and away from the earlier one, she anticipates Stan giving an erroneous report of simultaneity. She knows that the events she witnessed did not occur simultaneously, and all of the automatic detectors on board her ship will verify it.

If either the First or Second Postulates were wrong, we could determine who was really moving and would know which of the two observers was correct. But the discrepancies cannot be resolved—they are inherent in the nature of things; they are the reality. **There is no such thing as absolute simultaneity**—*events separated in*

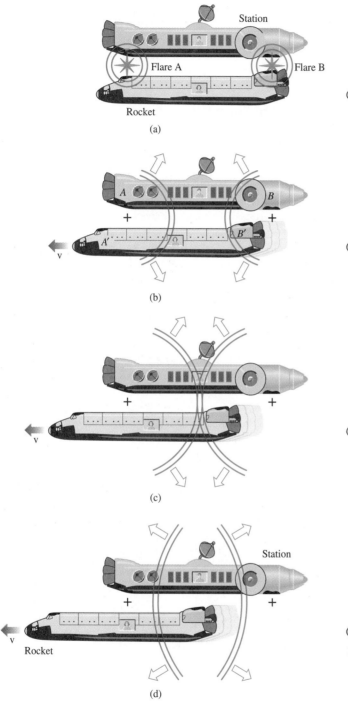

(a)

(b)

(c)

(d)

Figure 28.7 From the perspective of Stan, the Station Commander, the flares lit simultaneously and the rocket ship sailed off to the rear of his station. It follows from the Principle of Relativity that either inertial observer can assume that he or she is at rest and that it is the other person who is moving.

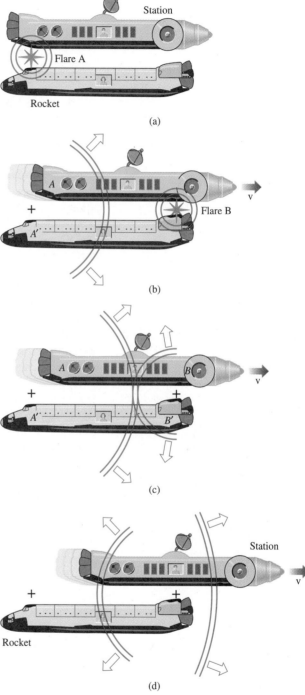

(a)

(b)

(c)

(d)

Figure 28.8 From the perspective of Rosie, the rocket's pilot, flare A fires first and the station sails off to the rear of her ship. The drawing neglects the fact that the station must appear to Rosie to be somewhat shrunken, a point we'll come back to later.

space that are simultaneous for midway observers in one inertial system are not necessarily simultaneous for midway observers in any other inertial system. Notice, too, that the lengths $\overline{AB}$ and $\overline{A'B'}$, which were seen by Stan in the station to be equal (Fig. 28.7), are not seen to be equal by Rosie in the rocket (Fig. 28.8); they will not agree on times or lengths, and both will be right.

28.4   The Hatter's Watch: Time Dilation

We reckon time by comparing simultaneous events—the moment the horse's nose touches the wire, and the placement of the hands on a stopwatch. Evidently, if the front flare in Fig. 28.7 was a flashing digital clock, each observer would see the other flare explode at a different time. Despite the centuries of thought to the contrary, **time is relative**, not absolute. There is no universal time that dances over us all equally, unalterably—the Universe doesn't grind away according to a single silent drumbeat.

Suppose we construct a light-clock, in which a pulse of light bounces back and forth between mirrors and some mechanism counts off the number of traversals, much like the familiar tick-tock. Figure 28.9a depicts such a device in a spaceship. With respect to the pilot on board, the clock is at rest, and she sees nothing extraordinary happening—everything aboard ship is normal. Quite the contrary is true according to an observer on the ground as the ship whisks by with a uniform relative velocity (Fig. 28.9b). He sees a light pulse in her clock behave as if it advanced diagonally as the vehicle carried the clock's mirrors past him. (Notice how the situation resembles the transverse arm of the Michelson Interferometer—indeed, the analysis will yield similar results.) While someone at rest with the clock sees the pulse travel to and fro between two stationary mirrors, the picture is very different when viewed from the ground. The pulse surely goes from one mirror to the other (both observers see the pulse strike either mirror—at such moments, the pulse and mirror are at the same point in space), but to the outside observer, the mirrors move along the flight path of the ship.

As seen by the pilot who is stationary with respect to the clock, the pulse travels a distance given by $c\,\Delta t_S$, where Δt_S is the elapsed interval on her clock (one tick's worth). To the observer on the ground with respect to whom the ship is moving at speed v, the pulse travels a longer diagonal path. Bound by the Second Postulate, the pulse (which must be seen by all inertial observers to progress at c) must take a longer time, Δt_M, to traverse the longer distance, $c\,\Delta t_M$. During that interval, as perceived from the ground, the ship advances a distance $v\Delta t_M$. It follows from the Pythagorean Theorem that

$$(c\,\Delta t_M)^2 = (v\Delta t_M)^2 + (c\,\Delta t_S)^2$$

Hence

$$(c\,\Delta t_M)^2 - (v\Delta t_M)^2 = (c\,\Delta t_S)^2$$

and

$$(\Delta t_M)^2 = \frac{c^2(\Delta t_S)^2}{c^2 - v^2} = \frac{(\Delta t_S)^2}{1 - \dfrac{v^2}{c^2}}$$

$$\Delta t_M = \frac{\Delta t_S}{\sqrt{1 - v^2/c^2}} \qquad (28.2)$$

Inasmuch as $\sqrt{1 - \beta^2}$ must be less than 1, the time interval seen by the outside observer with respect to whom the clock is moving (Δt_M) must be greater than the cor-

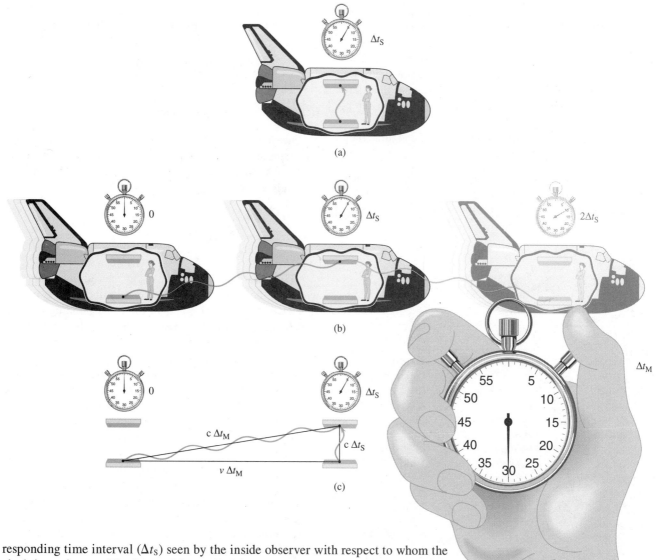

responding time interval (Δt_S) seen by the inside observer with respect to whom the clock is stationary. The interval between ticks and tocks is longer for the observer who sees the clock moving than for the one who doesn't. Let $\gamma = 1/\sqrt{1 - \beta^2}$, whereupon we can write

$$\Delta t_M = \gamma \Delta t_S \qquad (28.3)$$

with $\gamma > 1$. This slowing down of time is known as **time dilation**, and it's a very small effect. A clock aboard a commercial plane flying at top speed for $\approx$70,000 years would lose about 1 s compared to a clock on the ground.

    Time on a clock that is moving with respect to an observer is seen to run slower than time on a clock that is stationary with respect to that observer. And this is true for any clock (a wristwatch, a pendulum, a beating heart, a fertility cycle, or a dividing cell); all must slow down, all must match the light-clock. Otherwise we could easily learn from the difference who was actually moving, which is nonsense; no one is *actually* (or absolutely) moving—absolute motion violates the First Postulate.

Figure 28.9 (a) A person at rest relative to a light-clock sees what we might call an interval of stationary time Δt_S go by. (b) When that same clock is viewed by an observer who sees it moving, (c) that observer sees an interval Δt_M go by. He measures the moving clock to be running slow compared to his own stationary time. Only when an observer times two events occurring at the same place will he measure Δt_S. Here $\gamma = 6.0$.

Example 28.2 A college physics laboratory is under observation by aliens traveling on an asteroid. An undergrad seen measuring the period of a mass oscillating on a spring gets a value of 2.00 s. Given that the aliens are cruising by at a constant speed of 0.50c, and that they have nothing better to do, what period will they determine? Because they are moving rapidly, there can be substantial communication-time lags, so we assume they make any necessary corrections.

Solution: [Given: $v = 0.50c$ and $\Delta t_S = 2.00$ s. Find: the period determined by the aliens.] The period measured by an observer at rest with respect to the pendulum is 2.00 s. From Eq. (28.3), the aliens record an interval given by

$$\Delta t_M = \gamma \Delta t_S = \frac{1}{\sqrt{1 - \beta^2}} \Delta t_S$$

Here, $\sqrt{1 - \beta^2} = 0.866$ and $\boxed{\Delta t_M = 2.3 \text{ s}}$.

▶ **Quick Check:** The time interval is properly dilated; the aliens see the oscillations taking longer than the student does. Moreover, when $\beta = \frac{1}{2}$, Table 28.2 agrees with this value of γ.

TABLE 28.2 Values of β, $1/\gamma$, and γ

β	$1/\gamma$	γ
v/c	$\sqrt{1 - (v/c)^2}$	$1/\sqrt{1 - (v/c)^2}$
0.000 000	1.000 00	1.000 000
0.100 000	0.994 987	1.005 038
0.200 000	0.979 796	1.020 621
0.300 000	0.953 939	1.048 285
0.400 000	0.916 515	1.091 089
0.500 000	0.866 025	1.154 701
0.600 000	0.800 000	1.250 000
0.700 000	0.714 143	1.400 280
0.800 000	0.600 000	1.666 667
0.900 000	0.435 890	2.294 157
0.990 000	0.141 067	7.088 812
0.999 000	0.044 710	22.366 27
0.999 900	0.014 142	70.712 45
0.999 990	0.004 472	223.607
0.999 999	0.001 414	707.107

To keep things neat, the proper number of significant figures has not been kept consistently.

The duration of an event Δt_S, as measured by someone who sees the event to begin and end in one place is always shorter, by a factor of γ^{-1} than the corresponding interval Δt_M, as measured by an observer who sees the event to occur in a moving system.

Everyone at rest on the Earth experiences the same Earth-time, and *we call the time measured by an observer at rest with respect to the clock, the* **proper time**. Since none of us really rushes around very quickly, $c \gg v$, $\gamma \approx 1$, and quite generally $\Delta t_M \approx \Delta t_S$ for *everyone* on the planet, which is just what we expect. At its extremes, Relativity must yield the same results as our well-tested Classical Mechanics. As the Earth sails through space, it essentially has its own proper time. Anyone riding an asteroid past the planet will see our time running slower than their own time as seen on their "stationary" clocks. Someone else flashing by at a greater speed will see our "moving" clocks running still slower compared to their time. And the inverse is true: we will see their "moving" clocks run slow, as our own "stationary" clocks run at their normal rate. This reciprocity doesn't make the process less meaningful—time dilation has been measured. Certain nuclei vibrate and emit gamma-rays with very precise frequencies. When a sample of such a substance is heated, the gamma-ray frequency is reduced. The atoms move around more rapidly, and with respect to an observer at rest in the laboratory, their nuclear clocks run slower.

There have been many other experimental confirmations of time dilation over the years—the following is among the more interesting. Muons are subnuclear particles that are like heavy electrons. They are unstable, decaying into electrons and neutrinos. A muon at rest in the laboratory has a mean life of 2.2 μs, and this provides us with a convenient natural clock having a 2.2-μs tick-tock interval. In 1976, experimenters at the European Council for Nuclear Research (CERN) created a beam of muons traveling at 0.999 4c. These were injected into a large doughnut-shaped storage ring, where they were confined by powerful magnets and circulated until they decayed. Although a typical muon might, on the basis of Newtonian theory, be expected to survive 14 or 15 trips around the ring, most muons actually made in excess of 400 orbits. Electron detectors surrounding the ring established that the rapidly moving muons had a mean life about 30 times longer than when they were at rest ($\gamma = 28.87$). Equation (28.2) was confirmed to an accuracy of 0.2%.

More recently (1985), fast-moving neon atoms (excited by a laser so that they emitted light of a precise frequency) were used to confirm the time dilation to within an accuracy of 40 parts per million.

What then is time? If no one asks me, I know: if I wish to explain it to one that asketh, I know not.

ST. AUGUSTINE

28.5 Shrinking Alice

Neither time nor space is absolute in this Universe where the speed of light is constant. The fall of absolute simultaneity takes with it both absolute time and absolute length (or distance). *Only if both ends of a moving rod can be located at exactly the same instant can its length be measured accurately, and that cannot be done absolutely.* It's not much good in finding the length of a rod to say that the front end lined up with the 3-m mark of a ruler at 1:00 P.M., and 2 seconds later the rear end was next to the 2-m mark. We must watch the alignment with the ruler at both ends (separated in space) simultaneously, and different observers will not agree about that. If the flares in Fig. 28.7 are replaced by clocks, they could be synchronized to the same time by someone at rest with respect to them, but they will be seen to be unsynchronized by anyone moving relative to their inertial system. What is the distance between the exploding flares in Figs. 28.7 and 28.8? Remember there were two different distances seen by the two observers just because they couldn't agree on whether the flares fired off simultaneously or not. A rod has one **proper length** measured by any observer at rest with respect to it, but it can also have different shorter lengths measured by people who are in uniform motion with respect to it.

Imagine a meter stick in a rocket ship flying past you, and you also have a meter stick. *The sticks are aligned parallel to the relative velocity,* and along with appropriate clocks and sources, they are used to measure the speed of light. As seen by Rosie in the rocket, a pulse of light travels the length of her ruler (L_S) in a certain proper time at speed c. As seen by you, with respect to whom that experiment is moving, the pulse in the rocket traverses a length L_M (which may or may not equal L_S; that is, 1 m). You, looking into the rocket, see the pulse travel for fewer seconds on the rocket's clock, which you see running slow. You watch the pulse traverse the apparatus, in a room where time runs "slow," just as Rosie does, in a room where time runs "normally," and both of you must determine the speed to be c. For this outcome to be the case, the pulse in the rocket ship must travel along a path that to you is apparently shorter than it "ought to be" (shorter than L_S, shorter than 1 m) by a factor of γ; so that

$$L_M = \gamma^{-1} L_S \qquad (28.4)$$

or

$$L_M = L_S \sqrt{1 - v^2/c^2} \qquad (28.5)$$

Pilot Rosie moving along with her apparatus sees everything in the ship as properly normal and measures c. You, the outside viewer, see the rocket ship *shrunk along the line of motion*, the experiment on board it shrunk, and everything in the cabin happening slowly. And you understand why Rosie found the pulse to travel at c (her time was running slow by a factor of γ, but her experiment was shrunk by a factor of $1/\gamma$). With your proper length stick and proper time, your light pulses will also travel at c. And when Rosie views you on Earth, she sees everything of yours appropriately shrunk and your time running slow compared to her clock—the contractions and dilations are symmetrical via the First Postulate.

A moving observer measures an object to have a length (along the direction of motion) that is shorter than the length measured by an observer at rest with respect to the object (that is, shorter than the proper length). This is the **length contraction**, and it applies only to the direction of motion; *transverse distances are unaltered*. Equation (28.5) is mathematically identical to the Lorentz-FitzGerald Contraction (p. 1053), and it's often called by that name.

Example 28.3 A flying saucer descending straight down toward the Earth at 0.400 0c is first observed by an astronomer on the planet when it passes a satellite at an altitude of 3000 km. At that instant, what will be the ship's altitude as determined by its navigator?

Solution: [Given: $L_S = 3000$ km and $v = 0.400$ 0c. Find: the altitude.] The height measured by an observer with respect to whom the distance is stationary is $L_S = 3000$ km. Using Eq. (28.5), we have

$$L_M = L_S \sqrt{1 - v^2/c^2} = (3000 \text{ km})\sqrt{1 - 0.160\,0}$$

and

$$\boxed{L_M = 2750 \text{ km}}$$

▶ **Quick Check:** The speed is not very high so one expects a relatively small contraction. A glance at Table 28.2 confirms that $\sqrt{1 - v^2/c^2} = 0.92$.

There is no absolute distance between New York and Chicago; an atlas provides the proper distance as measured by people at rest with respect to the planet. But every traveler moving with regard to the surface sees his or her own version of that spatial interval, depending on his or her relative speed. Go fast enough, and London and San Francisco can be a meter or two apart. As for the reality of all of this (that is, is the spaceship actually squashed?), the business is similar to what happens with the Doppler Effect. Run toward a source and the sound pitch increases. The wave itself doesn't physically change, but the perception of it certainly does change. What you hear and measure—the reality of the experience—most assuredly depends on how you move with respect to the source.

You might find it satisfying to know that the electromagnetic field seen by differently moving inertial observers will be different. And that a detailed theory of the electromagnetic forces between the atoms within an object shows that the object must thereby contract by a factor of $1/\gamma$.

Example 28.4 A starship (some time in the very distant future) is headed for a galaxy that, according to human astronomy texts, is 200 light-years away from Earth. Flying a direct course, the ship reaches a cruising speed of 0.999c. What will be the Earth-galaxy distance as then determined by the navigator?

Solution: [Given: $L_S = 200$ ly and $v = 0.999c$. Find: the distance.] The first thing to settle is which distances are which in Eq. (28.5). L_M is the length as seen by someone moving with respect to the physical system in which the proper length is L_S. Thus

$$L_M = L_S \sqrt{1 - v^2/c^2} = (200 \text{ ly})\sqrt{1 - (0.999)^2}$$

and

$$\boxed{L_M = 8.94 \text{ ly}}$$

▶ **Quick Check:** The speed is very high so one expects a considerable contraction. A glance at Table 28.2 confirms that $\sqrt{1 - v^2/c^2} = 0.044\,7$. At these great speeds, the length is still a few percent of the proper length. (Get a feeling for the numbers you can expect by looking over the table.)

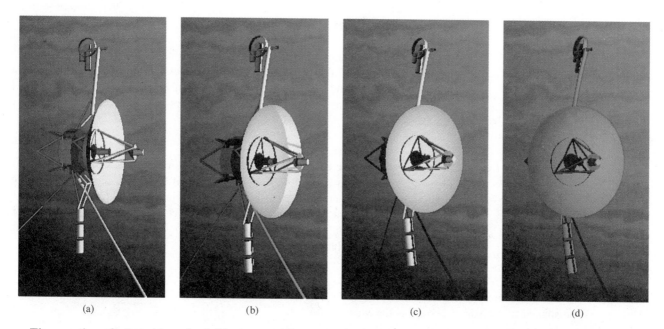

(a)   (b)   (c)   (d)

The question of what objects look like to a rapidly moving observer because of the length contraction is a complicated one. The semi-enlightened comic book renditions of things simply shrunk along the direction of motion—the long-skinny-people vision—is erroneous. Complications arise because light arriving at the retina (or on a piece of film) at some instant must have left different regions of the object at different times, depending how far away they are. Three-dimensional objects will therefore seem to be twisted and distorted, the more so the greater v (see photo above).

Computer images of a space probe as seen by an observer traveling with respect to it at four different speeds: (a) 0, (b) 0.25c, (c) 0.5c, and (d) 0.75c. Moving away at high speed, the light is red-shifted via the Doppler Effect (p. 469).

28.6 The Twin Effect

Someday, we may have rocket ships that can attain speeds high enough to experience significant time dilations and length contractions. This feat requires engines that can continue to exert thrust for very long periods so that the craft can accelerate at a humanly bearable rate (1 g would be nice) for years. Such an achievement is far beyond our present capabilities and may be for centuries. Practicalities aside, suppose we have such a starship. You and I meet at the launch pad, engage our identical stopwatches, shake hands, and you fly off to some star a distance $L_S = 50.00$ light-years away. The mission is to arrive at the star, plant the flag as it were, and promptly come home. To make the calculations simple, you quickly get the craft up to a wonderfully unrealistic cruising speed v of 0.999 8c and settle in for the journey. A glance out the window reveals to the passengers that the Earth-star distance is now $L_S\sqrt{1 - v^2/c^2}$. Since we both agree on *our* relative speed (and neglecting the comparatively short times it takes to negotiate the several accelerations), the trip out takes a proper on-board time of $(L_S\sqrt{1 - v^2/c^2})/v$. Thus, if T_S is the total round-trip flight time recorded by you, then

$$T_S = \frac{2L_S}{v}\sqrt{1 - v^2/c^2}$$

Similarly, if T_M is the total flight time I record while watching you through my telescope, then $vT_M = 2L_S$ and

$$T_M = \frac{2L_S}{v}$$

If we substitute this expression into the equation for T_S we're right back to the time dilation equation.

For this particular trip, $\gamma = 50.00$, $T_M = 2(50.00 \text{ ly})/(0.999\,8c) = 100.0$ y while $T_S = T_M/\gamma = 2.000$ y. (Note that in the expression for T_M, we could have entered L_S in meters and v in meters per second, but putting L_S in in light-years is equivalent to replacing it with the number-of-years-traveled-at-c multiplied by c; that is

$$1 \text{ light-year} = c \times (1 \text{ year})$$

in which case, the two c factors, regardless of their units, cancel yielding T_M in years directly). I watch you travel 100.00 ly total (out and back) at a speed nearly that of light, thus taking about 100 y. You see a contracted journey and travel only 1.000 ly out and 1.000 ly back taking about 2.000 y on your clock. I will greet you on the big day of your return in a wheelchair being around 100 y older than when you left, and you will saunter out having aged a mere 2 y.

Example 28.5 The nearest galaxy to ours in all the Universe is the shapeless star-island known as the Magellanic Cloud (located about 1.70×10^5 ly from midtown Manhattan). Assuming you could get up to a speed of $0.999\,99c$ in a negligible amount of time (which is sheer nonsense), how long would you say the trip to that galaxy will take? Incidentally, the fastest anyone has ever gone is only around $0.000\,037c$.

Solution: [Given: $v = 0.999\,99c$ and $L_S = 1.70 \times 10^5$ ly. Find: the flight time T_S on the traveler's clock.] You, the traveler, see a contracted distance $L_S\sqrt{1 - v^2/c^2}$, which is to be traversed at a speed $v = 0.999\,99c$. Hence, your proper time is

$$T_S = \frac{L_S\sqrt{1 - v^2/c^2}}{v}$$

and

$$T_S = \frac{(1.70 \times 10^5 \text{ ly})(4.472\,12 \times 10^{-3})}{0.999\,99c}$$

Thus

$$T_S = \frac{760 \text{ ly}}{c} = \boxed{760 \text{ y}}$$

It would seem we're not likely to travel to other galaxies using the technology we have at hand.

▶ **Quick Check:** Table 28.2 confirms the value of $\sqrt{1 - v^2/c^2}$. Multiplying 760 y by γ yields 1.7×10^5 y, which is the time in which we on Earth would see you make the 1.70×10^5-ly journey traveling at $0.999\,99c$.

This **twin effect** has often been called the *twin paradox* (usually enunciated with one twin staying and one traveling), but it's no paradox at all. It may be a startling result, but it's quite understandable within the context of the two postulates. Even so, Einstein pointed out that because of the accelerations on the part of the traveler, the analysis should more properly be done using General rather than Special Relativity. Until now, the situations we have treated have been symmetrical; inertial observers in relative motion see each other's clocks run slow. Here, however, there are accelerations, and we know from the inertial forces that the traveler is moving and not the stay-at-home. That determination can be made in an isolated laboratory aboard the vessel. It is those accelerations that change the path of the traveler through space and time, impressing on him the burden of being out of step with all the friends he left behind. In that sense, the time elapsed between two moments in a journey is route-dependent, just as is the distance traveled. As we will see (p. 1070), accelerations put bends in the space-time path and alter the time of the journey.

One of the most compelling confirmations of the reality of these conclusions was made in October of 1971. Then, four exceedingly accurate cesium-beam atomic clocks were flown around the world twice, on regularly scheduled commercial jet flights. The idea was to compare the clocks with those at the U. S. Naval Observatory before and after they had circumnavigated the Earth. Because of the planet's spin, there were two trips around, first eastward and then westward. Things were complicated by the presence of gravity, which affects what's happening via the General Theory of Relativity. (The problem is that time speeds up as the gravitational potential decreases with altitude. This was in addition to the speed-dependent slowdown of time expected from the Special Theory.) In any event, the eastward clocks should have lost 40 ± 23 ns in the journey, and the westward-flying clocks should have gained 275 ± 21 ns. After \$7600 was provided for airfare, it was found that, with respect to terrestrial reference standards, the eastward clocks lost 59 ± 10 ns, and the westward clocks had gained 273 ± 7 ns—in breathtaking agreement with theory!

28.7 Wonderland: Space-Time

> Gentlemen! The views of space and time which I wish to develop before you have sprung from the soil of experimental physics, and therein lies their strength.

Standing at the podium, Herman Minkowski delivered his lecture to the Eightieth Assembly of German Natural Scientists and Physicians, 1908. At that time, he was a renowned professor at Göttingen University. (Minkowski had once taught young Albert but was unimpressed—"in his student days Einstein had been a lazy dog.") Ironically, here he stood years later presenting a paper entitled "Space and Time" that represented an elegant reformulation of the Special Theory. Minkowski read on:

> Henceforth space by itself, and time by itself, are doomed to fade away into mere shadows, and only a kind of union of the two will preserve an independent reality.

He had recast Relativity in a 4-dimensional geometrical framework that would later prove invaluable to Einstein.

"Let's meet for lunch on the 30th floor of the hotel at the corner of 43rd and Park at 1:30 P.M." You know how to play the game—it takes three coordinates to locate anything in space. But in a real sense the event, our meeting, is not specified until we provide a fourth coordinate, 1:30 P.M. We can envision our own lives as a sequence of events in the 4-dimensional realm of **space-time**. We are swimming, as it were, through a space-time continuum.

The history of our lives, of any object in the Universe, can be imagined as a sequence of points sweeping out a smooth curve in 4-dimensional space-time known as a **world-line** (Fig. 28.10). Figure 28.11 is a simplified graphing of time versus a single spatial coordinate, where any motion is limited to running either way along the x-axis. The diagram is being drawn by you and me who are observers sitting together watching the passing scene. As with the more usual plots of x versus t, a straight line corresponds to a constant speed. Here, the slope $\Delta t/\Delta x$ equals $1/v$ so the faster an object moves, the less the tilt of its world-line. Indeed, an object at rest at $x = 0$ (namely, you or me) has a world-line corresponding to the t-axis. The coordinates are chosen so that the world-line of any pulse of light in the positive x-direction is a diagonal at $45°$. That is, if t is plotted in seconds, x is plotted in light-seconds (the distance light travels in one second). Since, as we'll see presently, c is the universal limiting speed, all world-lines must have greater slopes (lower

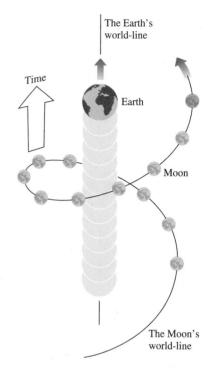

Figure 28.10 The Moon's unfolding world-line as it revolves about the Earth.

Dimension. I have already said that it is impossible to conceive more than three dimensions. A learned man of my acquaintance, however, believes that one might regard duration as a fourth dimension. . . . The idea may not be admitted, but it seems to be not without merit, if it be only the merit of originality.

DIDEROT
Encyclopédie (1777)

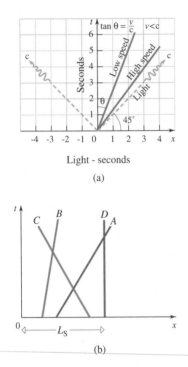

Light - seconds

(a)

(b)

Figure 28.11 World-lines for objects moving at different speeds. (a) A photon has the least-tilted world-line, traveling the greatest value of x in a given time interval. (b) With you and me at rest at $x = 0$, inertial observer A moves away from us, whereas inertial observer C moves toward us.

Billy Pilgrim says that the Universe does not look like a lot of bright little dots to the creatures from Tralfamadore. The creatures can see where each star has been and where it is going, so that the heavens are filled with rarefied, luminous spaghetti. And Tralfamadorians don't see human beings as two-legged creatures, either. They see them as great millipedes—"with babies' legs at one end and old people's legs at the other," says Billy Pilgrim.

KURT VONNEGUT, JR.
Slaughterhouse-Five (1968)

speeds) than this. In Fig. 28.11b, we see four inertial travelers: A and B moving away from you and me (who are still sitting at $x = 0$) at different speeds, traveler C coming toward us, and D just standing there a proper distance L_S away from us.

As something of a *tour de force*, Fig. 28.12 depicts a traveler who first accelerated away, then coasted out at a constant speed, slowed and stopped at a distance L_S, rested a while, and then came back to meet us after a time we measure to be T_M. This journey is essentially that of the traveler in the twin effect. The proper time measured by the traveler is marked out along his curved world-line. Note how the time intervals depend on the relative speed, matching those of the stay-at-home (played by you or me) only when $v = 0$. The larger is the relative speed between observer and observed, the more nearly the curve approaches the slope of the photon line, and therefore the larger is the difference between an interval of your proper time and an interval of the traveler's proper time. The total time elapsed from the start to the end of a journey in space-time is always less along a curved world-line than along a direct one. The slower the traveler moves out and back, the more nearly the route approaches the direct one, and the less difference there will be in the two proper times.

As the curve in Fig. 28.12 approaches the photon line, the interval of time stretches out. **For the photon, time does not pass.** In the frame traveling at c, time is stopped (p. 1075); photons don't age, on their clocks they cross the Universe in no time at all.

28.8 Addition of Velocities

Classical theory maintains that velocities add vectorially (p. 51), and at everyday speeds that certainly seems to be true. Accordingly, consider Fig. 28.13, which shows a coordinate system S' moving with respect to another such system S. The speed of O' relative to O is $v_{O'O}$. Now consider some object at point P moving in the space of both systems. For simplicity, we limit its motion to be in the $\pm x$-directions. Suppose that P travels at a speed relative to O of v_{PO}, and at a speed relative to O' of $v_{PO'}$. Classically, we learned that

$$v_{PO} = v_{PO'} + v_{O'O} \tag{28.6}$$

for all objects in uniform motion. But this certainly couldn't be true for a photon, for which it follows from the Second Postulate that $v_{PO} = v_{PO'} = c$ regardless of $v_{O'O}$. This dilemma can be sorted out by carefully applying the two postulates to the broader question of how coordinates in one system relate to those in another. We will skip the derivation—which isn't nearly as difficult as it is long—and simply state the *one-dimensional relativistic formula for the addition of velocities*:

$$v_{PO} = \frac{v_{PO'} + v_{O'O}}{1 + \dfrac{v_{PO'} v_{O'O}}{c^2}} \tag{28.7}$$

Note that the individual speeds can be either positive or negative. Equation (28.7) differs from Eq. (28.6) only because of the term $v_{PO'} v_{O'O}/c^2$. Thus, if either the object moves slowly with respect to S' (that is, $v_{PO'} \ll c$), or S' moves slowly with respect to S (that is, $v_{O'O} \ll c$), it follows that $v_{PO'} v_{O'O}/c^2 \ll 1$. In that case, the relativistic expression for v_{PO} becomes identical to the classical one. That's why Eq. (28.6) was used for about two centuries before there was even a hint of a problem.

Now for the true test: let's find the speed of light emitted from a moving source. Observer O', carrying a flashlight, is moving at a speed $v_{O'O}$ in the positive x-direction

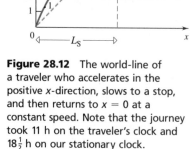

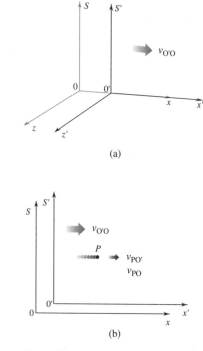

Figure 28.12 The world-line of a traveler who accelerates in the positive x-direction, slows to a stop, and then returns to $x = 0$ at a constant speed. Note that the journey took 11 h on the traveler's clock and $18\frac{1}{2}$ h on our stationary clock.

Figure 28.13 Two inertial systems S and S' moving with respect to each other at speed $v_{O'O}$. (a) Here, S is at rest and S' advances in the x-direction. (b) A particle P moves at a speed v_{PO} relative to S, and $v_{PO'}$ relative to S'.

relative to observer O. A stream of photons is sent out in the positive x-direction at a speed $v_{PO'} = c$, and we want to find its speed v_{PO} with respect to O, namely,

$$v_{PO} = \frac{v_{PO'} + v_{O'O}}{1 + \dfrac{v_{PO'} v_{O'O}}{c^2}} = \frac{c + v_{O'O}}{1 + \dfrac{c v_{O'O}}{c^2}} = \frac{c(c + v_{O'O})}{c + v_{O'O}} = c$$

Wonderful! No matter what value $v_{O'O}$ has, each observer sees the speed of light to be c.

As another example, consider the rocket ship (S') in Fig. 28.14. It is flying away from us (S) at speed $v_{O'O} = \frac{1}{2}c$ when it emits a pulse of light. A traveler aboard the ship sees the pulse moving at $v_{PO'} = -c$; the minus sign shows that it's advancing in the negative x-direction. The pulse's speed with respect to us is

$$v_{PO} = \frac{-c + \frac{1}{2}c}{1 + \dfrac{(-c)(\frac{1}{2}c)}{c^2}} = -c$$

The light comes toward us at c even though it was emitted from a platform receding at a speed of $\frac{1}{2}$c.

Figure 28.14 A rocket ship in system S' moving at speed $v_{O'O} = \frac{1}{2}$c emits a pulse of light toward the axis of system S.

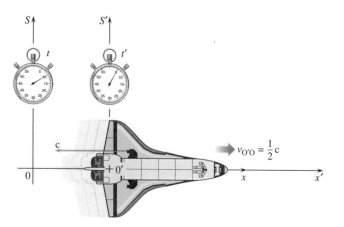

Example 28.6   Two galaxies are speeding away from the Earth along a line in opposite directions, each with a speed of 0.75c with respect to the planet. At what speed are they moving apart with respect to each other?

Solution: [Given: galactic speeds of 0.75c. Find: relative speed.] We have three moving bodies, and in Fig. 28.15, we let one galaxy be S, the Earth be S', and the other galaxy be P. Thus, S' moves forward with respect to S at $v_{O'O} = 3c/4$; P moves forward with respect to S' at $v_{PO'} = 3c/4$, and we want to find v_{PO}, the speed of P with respect to S. From Eq. (28.7)

$$v_{PO} = \frac{v_{PO'} + v_{O'O}}{1 + \frac{v_{PO'}v_{O'O}}{c^2}} = \frac{0.75c + 0.75c}{1 + \frac{(0.75c)(0.75c)}{c^2}} = \boxed{0.96c}$$

The galaxies separate at 96% of c.

▶ **Quick Check:** This result is at least reasonable

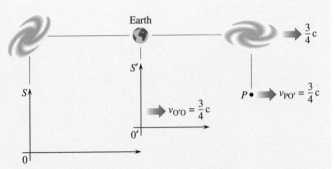

Figure 28.15   The galaxy on the right at point P is moving away from O' in S' at $v_{PO'} = \frac{3}{4}$c. With respect to the galaxy on the left (at rest in S), the Earth (at rest in S') is moving to the right at $v_{O'O} = \frac{3}{4}$c.

because the galaxies could not separate from each other at a speed equal to or greater than c since Eq. (28.7) says that even photons traveling in opposite directions ($v_{PO'} = c$ and $v_{O'O} = c$) only separate at c.

The velocity transformations are such that **the speed of one entity with respect to another is always less than or equal to** c. As far as the two galaxies in Example 28.6 are concerned, they separate from each other at 0.96c and can intercommunicate directly (via light beams traveling at c). An important point here is that *any two entities, regardless of their motions, can interact;* they can transport energy (at speed c) from one to the other. Thus, even though the gravitational and electromagnetic interactions propagate at the finite speed c, two objects cannot outrun their mutual interactions.

RELATIVISTIC DYNAMICS

Classical dynamics developed along two conceptual paths illuminated by the fundamental principles of Conservation of Momentum and Conservation of Energy. And so, faced with the necessity to reformulate dynamics, Einstein was guided by these same principles. Rather than attempt a complete derivation of relativistic dynamics, we limit our study to a few main results, their origins, and implications.

28.9   Relativistic Momentum

Both the energy and momentum of an object depend on the inertial reference system in which they are determined: the passenger holding a bowling ball on her lap flying to Paris "sees" it to have neither momentum nor kinetic energy. Since we now know, via Eq. (28.7), that the speed of an object with respect to different inertial frames has to be treated relativistically, it should be no surprise that both momentum (p) and energy (E) will have to be reformulated. Here, we are guided by our knowledge that the classical definition of momentum $\mathbf{p} = m\mathbf{v}$ works well at low speeds ($v \ll c$) and must be an approximation of a more precise *relativistic momentum*. We can get a sense of how the momentum might be modified by examining the details of a simple collision.

Let's first establish that if momentum is to be conserved relativistically, it cannot be the classical notion given by $\mathbf{p} = m\mathbf{v}$. To that end, Fig. 28.16 depicts an elastic collision between two identical balls. There is a "stationary" frame S and a person therein, observer 1, holding ball 1. Passing nearby is a uniformly "moving" frame S' (traveling at speed $v_{O'O}$), in which there is a second person, observer 2, holding ball 2. Just at the right moment, both observers throw their balls vertically toward one another at identical speeds equal to u. Classically, the balls have momenta of $p_y = mu$ that are equal in magnitude and opposite in direction, and the initial net vertical momentum equals zero. After the head-on collision, the balls are both reversed; the net vertical momentum is still zero, and classical momentum is conserved.

Now, let's run through this situation again, taking into account that the relative speed of S' with respect to S is appreciable. Observer 2 sees his ball travel a vertical round-trip distance of $2d$ at a speed u in a time $2d/u$. The same is true of observer 1, who also sees her ball travel a vertical distance of $2d$ at a speed u in a time $2d/u$. But observer 1 sees ball 2 to be moving and therefore to be experiencing a time dilation (by a factor of γ). Thus, observer 1 must see ball 2 take a longer time to make the round trip from, and back to, the hand of observer 2. Observer 1 sees ball 2 travel a vertical distance of $2d$ in a longer time, and therefore at a slower speed, u/γ, than his own ball moves. And yet the collision reverses the motion of his ball—classical momentum is not conserved. So we need a new statement of what is conserved and, although we might guess that it look like γmv, we certainly haven't proven it.

Nonetheless, the **relativistic linear momentum**, which is conserved, is indeed given by

$$\mathbf{p} = \gamma m\mathbf{v} = \frac{1}{\sqrt{1 - v^2/c^2}} m\mathbf{v} \tag{28.8}$$

As v/c becomes negligible, $\beta \to 0$, $\gamma \to 1$, and $\mathbf{p} \to m\mathbf{v}$, which is the classical value. Fig. 28.17 is a plot of p/mc, which classically equals β and relativistically equals $\gamma\beta$. As the speed of a body increases beyond $\beta \approx \frac{1}{2}$, the relativistic momentum climbs away from the classical value, approaching infinity as $v/c \to 1$.

Example 28.7 Electrons have a mass of 9.1094×10^{-31} kg. If one is traveling at 0.99c, what is its momentum?

Solution: [Given: $m = 9.1094 \times 10^{-31}$ kg and $v = 0.99c$. Find: p.] The momentum follows from Eq. (28.8), where we first compute γ to be 7.0888. Hence

$$p = \gamma mv$$

$$p = (7.0888)(9.1094 \times 10^{-31} \text{ kg})(2.96779 \times 10^8 \text{ m/s})$$

and $p = 1.91655 \times 10^{-21}$ kg·m/s $= \boxed{1.9 \times 10^{-21} \text{ kg·m/s}}$.

▶ **Quick Check:** From Table 28.2, γ looks okay at 7. The classical momentum is roughly $(9.1 \times 10^{-31})(3 \times 10^8) = 27 \times 10^{-23}$ kg·m/s, so our answer should be about $7(27 \times 10^{-23})$, which it is.

Though experiments have verified the validity of Eq. (28.8), the interpretation of $p = \gamma mv$ is unsettled at this moment. We can associate the γ with mass and suppose that there is a *relativistic mass* $m_R = \gamma m$, which is a function of speed. Then $p = m_R v$, which is nice. Einstein's early work seems ambivalent about the idea of a speed-dependent mass. Nonetheless, in 1948, he made it very clear that introducing the concept of relativistic mass was "not good." Experimentally, what we

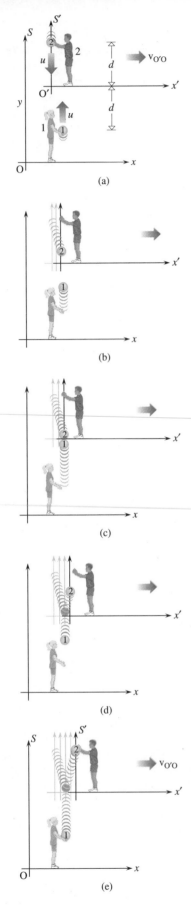

(a)

(b)

(c)

(d)

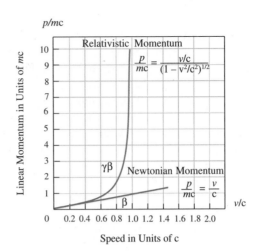

$$\frac{p}{mc} = \frac{v/c}{(1 - v^2/c^2)^{1/2}}$$

$$\frac{p}{mc} = \frac{v}{c}$$

Figure 28.17 Linear momentum in units of mc plotted against speed in units of c; that is, p/mc versus v/c. Here, we compare relativistic momentum against Newtonian momentum. Observe that although they both increase with v, the former approaches ∞ as v approaches c.

measure is a change in momentum with speed; there is no way to measure the mass of a moving body directly. Many physicists prefer to think of mass as speed dependent. A growing number of others find it more appealing to hold that mass is an invariant property of matter like charge. Both views lead to the same measured results; the difference is a matter of interpretation. *We shall take m to be the invariant mass*—the speed independent, reference-frame-independent, almost Newtonian mass (m)—though that idea will need some more discussion (p. 1078).

Relativistic Force

Newton's Second Law now provides a new perspective:

$$\mathbf{F} = \frac{d\mathbf{p}}{dt} = \frac{d}{dt}(\gamma m\mathbf{v}) = m\frac{d}{dt}(\gamma\mathbf{v})$$

wherein **p** is the relativistic momentum given by Eq. (28.8). Inasmuch as **p** is no longer directly proportional to **v**, **F** is no longer directly proportional to $d\mathbf{v}/dt$; that is, **F** does not equal $m\mathbf{a}$! Problem 64 examines the simple case of straight-line motion under the influence of a constant force and establishes that

[**F** *and* **v** *colinear*] $F = \gamma^3 ma$ (28.9)

In general, however, **F** and $d\mathbf{v}/dt$, which is **a**, are not even parallel to one another (as shown in Problem 67).

Now, suppose a constant force is applied to a body that is initially at rest. What happens to v? In that case, $F = p/t$, where p is the momentum after the elapse of a time t. Solving Eq. (28.8) for v^2, we obtain

$$v^2 = \frac{p^2/m^2}{1 + p^2/m^2c^2}$$

Figure 28.16 An elastic collision as viewed by observer 1 at rest in inertial system S. Observer 2 is at rest in system S', which moves relative to S at a speed $v_{O'O}$. Observer 1 sees time running slow for ball 2 by a factor of γ, which depends on its net relative speed.

(e)

Dividing by c^2 and taking the square root gives

$$\beta = \frac{v}{c} = \frac{Ft/mc}{\sqrt{1 + (Ft/mc)^2}}$$

Now consider the possible motion of an object. If the force and time are such that Ft is small and Ft/mc is therefore much less than 1, Ft/mc is negligible in the denominator and $\beta \approx Ft/mc$, whereupon $mv = Ft$, and we have the classical change-in-momentum-equals-impulse expression, Eq. (4.3). On the other hand, if Ft is large (as, for instance, if a large force is applied for a very long time), $(Ft/mc)^2 \gg 1$ and $\sqrt{1 + (Ft/mc)^2} \to Ft/mc$, and so as $Ft \to \infty$, $\beta \to 1$. What this means is that *regardless of how great the force or how long it is applied, the body will neither reach nor exceed lightspeed. As a body moves faster and faster under the influence of a force, it takes longer and longer (because of the time dilation) for the speed to further increase.* In effect, an inertial observer will see the acceleration decrease (much as if the body's mass had actually increased).

As v approaches c, time slows [Eq. (28.3)] until at $v = c$, $\beta = 1$, $\sqrt{1 - \beta^2} = 0$, $\gamma = \infty$, and a second stretches out into infinity. Nor is the Lorentz-FitzGerald Contraction any more comprehensible when $v = c$ and everything shrinks to nothingness. For speeds beyond c, all this becomes unreal, literally and mathematically. Only one conclusion is evident: **the speed of light is an upper limit on the rate of propagation of objects that have mass.** Nothing with mass can be accelerated to c (Question 11). To be precise, this conclusion does not preclude the existence of particles that have somehow been created with speeds in excess of c. These fantastic hypothetical entities, known as *tachyons,* have never been observed and may well never be. They are a product of the theoretical school that maintains that whatever is not explicitly forbidden must exist. But since we are not possessed of all the theories we are ever going to have, such a notion seems a trifle premature.

28.10 Relativistic Energy

We want now to arrive at an expression for the relativistic kinetic energy (KE) of a particle. The rigorous, though mathematically arduous, way to proceed is to go back to the idea of the work done by a force in the process of changing the body's speed (in a zero-PE situation). We'll sketch such a treatment here, leaving a more exhaustive analysis for Problem 68. Consider the case of one-dimensional motion for which Eq. (28.9) applies. Then

$$W = \int F\,dx = \int \gamma^3 ma\,dx$$

To rewrite $a\,dx$ in terms of v (which is what's inside γ) make use of the definitions $v = dx/dt$ and $a = dv/dt$:

$$a\,dx = \frac{dv}{dt}\,dx = \frac{dx}{dt}\,dv = v\,dv$$

hence

$$W = \int_0^v \gamma^3 mv\,dv$$

The work done goes into kinetic energy, and evaluating the integral leads to

$$KE = \frac{mc^2}{\sqrt{1 - v^2/c^2}} - mc^2 = \gamma mc^2 - mc^2 \qquad (28.10)$$

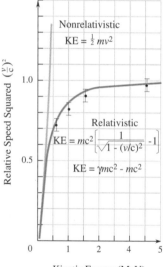

Figure 28.18 Here, electrons were given various values of kinetic energy by accelerating them across appropriate potential differences. Their speeds were then measured by finding the flight times over a fixed path. The dots represent the measured values, for which the vertical bars indicate the expected range of experimental error. Clearly, the results at high speeds confirm the validity of the relativistic expression for KE (the purple curve). The classical formula for KE corresponds to the straight dashed line, which is accurate at low speed. [Adapted from W. Bertozzi, *Am. J. Phys.* 32, 551 (1964).]

To see that this is consistent with our previous notions recall (p. 1054) that when $\beta \ll 1$, $\gamma = 1/\sqrt{1 - \beta^2} \approx 1 + \beta^2/2$, and so $\gamma m \approx m(1 + \beta^2/2)$. Multiplying this by c^2 gives

$$\gamma mc^2 \approx mc^2 + \tfrac{1}{2}mv^2$$

The term on the far right is the familiar low-speed $KE = \tfrac{1}{2}mv^2$: as expected

$$KE \approx \gamma mc^2 - mc^2$$

This is the **relativistic kinetic energy**, and it's no longer simply $\tfrac{1}{2}mv^2$. (See Fig. 28.18 for a lovely experimental confirmation.) Consequently

$$\gamma mc^2 = KE + mc^2 \qquad (28.11)$$

and each term has the units of energy. The second quantity on the right is independent of speed and is known as the **rest energy** (E_0). All of this suggests that we interpret the term on the left as the **total energy** of the body ($E = \gamma mc^2$):

Total energy = kinetic energy + rest energy

$$E = KE + E_0 \qquad (28.12)$$

Example 28.8 Every electron has a rest energy of 0.511 MeV. If one in particular is traveling at a speed of 0.900c, determine its total energy and its kinetic energy. Give your answers in MeV.

Solution: [Given: $E_0 = 0.511$ MeV and $v = 0.900c$. Find: E and KE.] Since $E = \gamma mc^2$, let's first determine γ by using

$$\gamma = \frac{1}{\sqrt{1 - \beta^2}} = \frac{1}{\sqrt{1 - (0.900)^2}} = 2.294\,2$$

Then realize that $mc^2 = 0.511$ MeV and so

$$E = 2.294\,2(0.511 \text{ MeV}) = \boxed{1.17 \text{ MeV}}$$

From Eq. (28.12)

$$KE = E - E_0 = 1.172 \text{ MeV} - 0.511 \text{ MeV}$$

and $\qquad \boxed{KE = 0.661 \text{ MeV}}$

▶ **Quick Check:** γ is greater than 1, which is a good sign, and it also checks with Table 28.2. As we will see in Problem 61, rest energy for a particle equals kinetic energy when $\beta = 0.866$, so this result where $\beta = 0.900$ is certainly the right magnitude.

When the body is at rest ($\gamma = 1$), $KE = 0$, and the total energy ($E = \gamma mc^2$) equals the rest energy, and so

$$E_0 = mc^2 \qquad (28.13)$$

This is perhaps the most famous outcome of the Special Theory. Confirmed experimentally, it stands beyond doubt. If there can be such a thing as the premier equation of the twentieth century, this is it. It has revolutionized Modern Physics and hurled us all into the age of nuclear weapons. Even so, the precise meaning of the relationship is still being argued in the scientific literature, primarily because we have not yet satisfactorily defined matter, mass, and energy. For example, is mass a congealed form of energy, or are mass and energy very different concepts that are simply proportional to each other and only interrelated via Eq. (28.13), just as F and a are interrelated by $F = ma$? Still, it has been established experimentally that matter possessing mass (for example, electrons and positrons) can be transformed into electromagnetic radiation and vice versa. At the level of our discussion, it's probably best to ignore some of these subtleties and follow the most common usage.

TABLE 28.3   Masses and rest energies for some particles and atoms

Object	Mass (kg)	Rest energy (MeV)
Photon	0	0
Neutrino	0	0
Electron (or positron)	$9.109\,389\,7 \times 10^{-31}$	0.510999
Proton	$1.672\,623\,1 \times 10^{-27}$	938.272
Neutron	$1.674\,929 \times 10^{-27}$	939.566
Muon	$1.883\,54 \times 10^{-28}$	105.659
Pion (+)	$2.416\,5 \times 10^{-28}$	135.56
Deuteron	$3.343\,584 \times 10^{-27}$	1875.612
Triton	$5.007\,357 \times 10^{-27}$	2808.920
Alpha	$6.644\,72 \times 10^{-27}$	3727.41
Hydrogen atom ($_1^1$H)	$1.673\,534 \times 10^{-27}$	938.783
Deuterium atom ($_1^2$H)	$3.344\,497 \times 10^{-27}$	1876.12
Tritium atom ($_1^3$H)	$5.008\,270 \times 10^{-27}$	2809.43
Helium-3 atom ($_2^3$He)	$5.008\,237 \times 10^{-27}$	2809.41
Helium atom ($_2^4$He)	$6.646\,482 \times 10^{-27}$	3728.40

Mass can be transformed into energy, and energy can be transformed into mass; hence

$$1\text{ kg} \leftrightarrow 8.987 \times 10^{16}\text{ J}$$
$$1\text{ kg} \leftrightarrow 5.609 \times 10^{29}\text{ MeV}$$

There is one unified concept: *mass-energy.* Had we known that a few centuries ago, we would not now have joules, Btus, calories, and kilowatt-hours to fuss with: the kilogram would do for all forms of mass-energy. The constant c^2 is a scale factor that numerically relates mass to energy. It's an immense number, $\approx 9 \times 10^{16}\text{ m}^2/\text{s}^2$, so even a tiny change in mass corresponds to an enormous change in energy. In Modern Physics, mass is often given in units of MeV/c^2; thus

$$1\text{ MeV/c} = 5.344\,29 \times 10^{-22}\text{ kg·m/s}$$
$$1\text{ MeV/c}^2 = 1.782\,663 \times 10^{-30}\text{ kg}$$

whereupon, for example, the mass of an electron is 0.510999 MeV/c^2 (Table 28.3).

Example 28.9 A 1.00-kg chicken is placed on the transporter of a fictitious starship whereupon it is converted directly into electromagnetic energy—a process that is theoretically possible, though technologically quite beyond our poor powers—so that it could be "beamed" down to the galley. What is the equivalent energy of the chicken? How much is that in kilowatt-hours?

Solution: [Given: $m = 1.00$ kg. Find: E_0.] From Eq. (28.13)

$$E_0 = mc^2 = (1.00\text{ kg})(2.998 \times 10^8\text{ m/s})^2$$

and

$$\boxed{E_0 = 8.99 \times 10^{16}\text{ J}}$$

1 kilowatt-hour = 1000 J/s × 60 s/min × 60 min = 3.60 MJ; hence, dividing this result into the energy yields $\boxed{E_0 = 2.50 \times 10^{10}\text{ kW·h}}$. That's the equivalent of running ten 100-W bulbs for 2.5×10^{10} hours, about 3 million years.

▶ **Quick Check:** Since 1 kg → 5.6×10^{29} MeV and 1 MeV = 1.6×10^{-13} J, 1 kg → 9×10^{16} J.

A single material particle at rest, by virtue of its very existence, has a rest energy. Similarly, a body composed of several particles possessing internal energy (thermal, potential, whatever), when taken as a whole, also has a net rest energy and a net mass. As Einstein put it, "the mass of a body is a measure of its energy-content." A hot apple pie has more rest energy and more mass than an otherwise identical cold apple pie. The greater mass is due purely to its greater energy content—**energy possesses inertia**. Thus, a flashlight emitting energy (ΔE) decreases in mass (by $\Delta E/c^2$), just as a plant absorbing that light gains in mass. A spring must weigh more after *elastic*-PE is stored in it, than before. No one has ever measured these minuscule variations in mass, but there is ample confirmation elsewhere. In the final analysis, all the familiar occurrences that liberate energy—from burning marshmallows to exploding dynamite—transform a minute amount of mass into energy. Ultimately, that is the source of the reaction energy, and this is as true for chemical energy as it is for energy liberated by a nuclear weapon.

In Chapter 9 we talked about Conservation of Mechanical Energy, and in Chapter 16 we generalized that concept to include thermal energy. The result was the First Law of Thermodynamics. Now we come to the final generalization: there is one all-encompassing law of **Conservation of Energy**:

The total energy of an isolated system always remains constant although any portion of it can be converted from one form to another, including rest energy.

Example 28.10   There are several fusion reactions that convert mass directly into energy and can power stars and drive hydrogen bombs. One such process fuses two nuclei of heavy hydrogen (deuterium) together, resulting in a still heavier hydrogen (tritium) nucleus, an ordinary hydrogen nucleus (proton), and the KE they fly off with. It's customary to write such a reaction in terms of the neutral atoms involved (neglecting the tiny amounts of energy holding the electrons to each atom, ≈ 10 eV). Thus

$$^2_1\text{H} + {}^2_1\text{H} \rightarrow {}^3_1\text{H} + {}^1_1\text{H} + \text{energy}$$

Determine the energy liberated per fusion.

Solution: [Given: the reaction. Find: energy liberated.] Energy is conserved; the total energy on the right in the reaction equals the total energy on the left. In other words, the difference in the rest energy before and after is the liberated KE. From Table 28.3, 1876.12 MeV + 1876.12 MeV = 2809.43 MeV + 938.783 MeV + energy

$$\text{energy} = 4.03 \text{ MeV} = \boxed{6.45 \times 10^{-13} \text{ J}}$$

where 1 MeV = 1.602×10^{-13} J.

▶ **Quick Check:** Nuclear reactions typically involve several MeV, so this result is the right order-of-magnitude.

It is not always possible practically to convert a quantity of mass completely into energy, or vice versa, although both effects are now commonplace. Still, one single gram of mass is equivalent to 9×10^{13} J, which is enough energy to raise 200 000 000 kg of water from 0°C to 100°C. That corresponds to the peak power output of Boulder Dam operating for 19 hours; namely, 25 million kilowatt-hours. An exploding kilogram of TNT liberates about 5 million joules, which is certainly formidable, though it represents a mass loss of only about 6×10^{-11} kg—far too little to measure. Chemical reactions, which are at heart relatively weak electrical interactions, release energies of the order of a few eV. By comparison, nuclear transmutations involving the far more powerful *strong force* (p. 1181) correspond to mass

TABLE 28.4 The fractional change of mass for various kinds of processes

Process releasing energy $\Delta E = \Delta(mc^2)$	$\Delta m/m$
Chemical	$\approx 1.5 \times 10^{-8}\%$ to $\approx 10^{-7}\%$
Nuclear fission	$\approx 0.1\%$
Nuclear fusion	$\approx 0.6\%$
Electron-positron annihilation	100%
Neutral pion decay into two photons	100%

changes of about 0.1%. This is roughly a million times more than the mass change in a chemical reaction (see Table 28.4) and, hence, directly observable.

The total energy can be written in terms of p without any explicit reference to v, and that provides some insights about photons. Starting with $E = \gamma mc^2$, square both sides and write it as

$$E^2 = \gamma^2 m^2 c^2(c^2 + v^2 - v^2)$$

Now
$$E^2 = \gamma^2 m^2 c^2(c^2 - v^2) + \gamma^2 m^2 c^2 v^2$$

Putting in the expression for γ, we have

$$E^2 = \frac{m^2 c^2(c^2 - v^2)}{(1 - v^2/c^2)} + \gamma^2 m^2 c^2 v^2$$

and using $p = \gamma mv$, substitute in $p^2 = \gamma^2 m^2 v^2$ to get

$$E^2 = m^2 c^4 + p^2 c^2 \qquad (28.14)$$

or
$$E^2 = E_0^2 + (pc)^2 \qquad (28.15)$$

Figure 28.19 shows the kind of striking confirmation of these conclusions that, today, is available experimentally.

Notice that inasmuch as $E_0^2 = (mc^2)^2$ is a constant independent of any reference frame, $E^2 - (pc)^2$ must also be invariant, even though E and p separately depend on v and are thus relative quantities. We shall not explore the implications of this, other than to point out that the invariance of $E^2 - (pc)^2$ implies that *in space-time there is a single notion that ties together what until now were the two separate ideas of energy and momentum.*

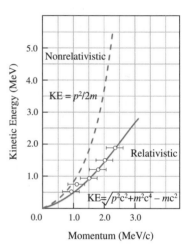

Figure 28.19 A comparison of the classical and relativistic relationships between KE and p. Momentum and energy are measured using electrons emitted via radioactive decay. Note how nicely the data fit the predicted relativistic curve. [Adapted from R. Kollarits, *Am. J. Phys* 40, 1125(1972).]

Example 28.11 A proton (of mass 938.3 MeV/c^2) is accelerated across a potential difference of 202.0 MV so that it has a kinetic energy of 202.0 MeV. Determine its total energy (in MeV) and momentum (in MeV/c). What is the speed of the particle?

Solution: [Given: $m = 938.3$ MeV/c^2, KE $= 202.0$ MeV. Find: E and p.] $E = KE + E_0 = KE + mc^2 = 202.0$ MeV $+ (938.3$ MeV/c$^2)c^2 = \boxed{1140 \text{ MeV}}$. From Eq. (28.15)

$$p = \frac{\sqrt{E^2 - E_0^2}}{c} = \boxed{647.5 \text{ MeV/c}}$$

To find the speed, we use the fact that $E = \gamma E_0$; $E = E_0/\sqrt{1 - v^2/c^2}$ and so

$$v = c\sqrt{1 - \frac{E_0^2}{E^2}} = c\sqrt{1 - \left(\frac{938.3}{1140}\right)^2} = \boxed{0.568\,3c}$$

▶ **Quick Check:** $p = \gamma mv$, $v = p/\gamma m = p/(E/c^2) = (647.5$ MeV/c$)/(1140$ MeV/c$^2) = 0.568\,0c$.

In addition to the photon, physicists have found it necessary to deal with other zero-mass particles (p. 1215) such as the several different neutrinos (whose complete lack of mass is being questioned) and the graviton. The existence of neutrinos is well confirmed experimentally, although the graviton is still a matter of speculation. When $m = 0$ in Eq. (28.14)

[*zero-mass*]
$$E = pc \qquad\qquad (28.16)$$

Moreover, because $E = \gamma mc^2$, $E/\gamma = mc^2$ and when $m = 0$, $E/\gamma = 0$. Since E is nonzero, $1/\gamma = 0 = \sqrt{1 - v^2/c^2}$, and it follows that for particles of zero mass $v = c$. **Particles of zero mass exist only at the speed** c.

Example 28.12 A neutral pion has a mass 264 times larger than the electron mass and a rest energy of 135 MeV. It is unstable and decays into two oppositely directed γ-ray photons. Assume the pion to be at rest and determine the energy and momentum of the photons.

Solution: [Given: $E_0 = 135$ MeV. Find: E and p for photons.] Since the pion is at rest when it decays, initially $p = 0$, and the momenta of the photons are equal and opposite. The total energy of the two photons (2E) must equal the energy of the pion: 135 MeV = 2E, and

for each photon $\boxed{E = 67.5 \text{ MeV}}$. From Eq. (28.16) for massless particles,

$$p = \frac{E}{c} = \boxed{67.5 \text{ MeV/c}}$$

and since 1 MeV/c = 5.3443×10^{-22} kg·m/s, $p = 3.61 \times 10^{-20}$ kg·m/s.

▶ **Quick Check:** $m = 264(9.11 \times 10^{-31}$ kg$) = 2.4 \times 10^{-28}$ kg; the photon energy is half the pion energy $E = \frac{1}{2}mc^2 = 1.08 \times 10^{-11}$ J = 67.5 MeV.

Though we shall not elaborate the argument, Relativity establishes that if an interaction is to be *causal* (in the sense that the act of initiation of an event must precede the event in time for all inertial observers), then the speed of propagation of the interaction (that is, the signal) cannot exceed c. The uncovering of a radioactive source sets a distant counter clattering; the emitted γ-rays travel from source to detector at c, and the sequence of events is causal. The transport of information (matter-energy in a discriminated form) in a cause-and-effect sequence cannot occur at a speed in excess of c. That does not mean that there cannot be apparent motions that are greater than c. Shine a laser beam at observer A leaning on a very distant screen. Now quickly jerk the laser so the light spot moves across the screen to observer B. Clearly, if you shift the laser fast enough and if the target is far enough away, the light spot travels from A to B in excess of c (oscilloscopes actually perform this trick all the time). Still, the photons that get to B were never at A—new photons a little behind in the stream arrive at each successive point on the screen. Light never literally travels from A to B.

Relativity does not forbid noncausal interactions. In fact, the concept is at the center of a great deal of modern experimentation and interpretation as it relates to Quantum Mechanics. There are some wonderfully mind-boggling contemporary experiments that are challenging our familiar notions of reality.

Core Material

The Michelson-Morley experiment established that the motion of the Earth through the supposed surrounding aether could not be detected (p. 1052). The **Principle of Relativity**, the *First Postulate* of the Special Theory, states that *all the laws of physics are the same for nonaccelerating observers*. Einstein's *Second Postulate*, the **Principle of the Constancy of the Speed of Light**, states that *light propagates in free space at the invariant speed* c (p. 1055).

There is no such thing as **absolute simultaneity**—events separated in space that are simultaneous for midway observers in one inertial system are not necessarily simultaneous for midway observers in any other inertial system (p. 1060). Time on a clock that is moving with respect to an observer runs slower than time on a clock that is stationary with respect to that observer; thus

$$\Delta t_{\text{M}} = \frac{\Delta t_{\text{S}}}{\sqrt{1 - v^2/c^2}} \qquad [28.2]$$

A moving observer sees an object to have a length (along the direction of motion) that is shorter than the length seen by an observer at rest with respect to the object; thus

$$L_{\text{M}} = L_{\text{S}}\sqrt{1 - v^2/c^2} \qquad [28.5]$$

The one-dimensional relativistic velocity transformation is

$$v_{\text{PO}} = \frac{v_{\text{PO}'} + v_{\text{O'O}}}{1 + \dfrac{v_{\text{PO}'} v_{\text{O'O}}}{c^2}} \qquad [28.7]$$

The relativistic momentum of a body of mass m is

$$\mathbf{p} = \gamma m\mathbf{v} = \frac{1}{\sqrt{1 - v^2/c^2}} m\mathbf{v} \qquad [28.8]$$

Here we take the approach that *m is not a function of v*. The relativistic energy of a body is given by

$$E = KE + E_0 \qquad [28.12]$$

where $$E = \gamma mc^2$$

and the rest energy is

$$E_0 = mc^2 \qquad [28.13]$$

The mass of a composite body changes if the rest energy of its component parts changes (p. 1079). In terms of momentum

$$E^2 = E_0^2 + (pc)^2 \qquad [28.15]$$

For zero-mass particles

$$E = pc \qquad [28.16]$$

Suggestions on Problem Solving

1. The factor $\gamma = 1/\sqrt{1 - \beta^2}$ is so ubiquitous in the work of this chapter that it merits a few words. It's common to forget either to square the β or to take the square root once having computed $1 - \beta^2$. Remember that $\gamma > 1$. A nice way to compute γ on an electronic calculator if v is given as a number (for example, $v = 2.43 \times 10^7$ m/s) with no reference to c is

$$v \; [\div] \; \text{c} \; [=] \; [x^2] \; [+/-] \; [+] \; 1 \; [=] \; [\sqrt{}] \; [1/x]$$

If you have v in terms of c (for example, $v = 0.9$c), then the calculation is more straightforward because you can enter v/c as a number (in this case, v/c = 0.9)

$$1 \; [-] \; v/\text{c} \; [x^2] \; [=] \; [\sqrt{}] \; [1/x]$$

Many calculators have constants stored in memory, and c is usually one of them.

2. When doing problems on time dilation and length contraction, the "hard part" is just figuring out which piece of information to call L_{M} and which L_{S}. For example: A kid on an asteroid holding a meter stick is going to measure the length of a space probe as it flies by at 0.9c parallel to the stick. As seen by the pilot, how long is the meter stick? "As seen by the pilot"—the stick moves past the probe at 0.9c, the stick is moving with respect to the frame we are interested in, the pilot's frame. Hence, the *proper length* of the stick measured by the kid is $L_{\text{S}} = 1$ m. We place ourselves in the frame of the pilot and see a moving stick of length L_{M}. The rest of the problem is just plug-in using Eq. (28.5).

Another useful approach is to *figure out who sees what shrunk or dilated before getting into the problem numerically*. In the above example, "As seen by the pilot" tells us whose perspective we are to determine. The pilot sees the ruler moving and *shrunk*. The answer must be less than 1 m. Keeping in mind that $\sqrt{1 - v^2/c^2} < 1$,

(Length seen by pilot) = (length seen by kid)$\sqrt{1 - v^2/c^2}$

The stick has only one proper length (measured at rest), and that length is the longest possible measure.

3. There are clever things that can be done regarding units in momentum and energy problems. For example, given rest energy in MeV, if you are asked to find momentum, determine it in terms of MeV/c. You needn't convert MeV to joules. Thus, if $v = \frac{1}{2}$c, $p = \gamma mv = \gamma m\frac{1}{2}$c; *now multiply top and bottom by* c; $p = \gamma\frac{1}{2}mc^2/c = \frac{1}{2}\gamma E_0/c$ and you enter E_0 in MeV and get p in MeV/c.

Discussion Questions

1. In terms of Classical Physics only, suppose you are sitting in a chair resting. (a) What will you see if you jump up, rush away at, say, a speed of 100c, spin around, and stop? (b) In a similar vein, how might you observe Lincoln's Gettysburg address? (c) Is all of this actually possible?

2. Prior to the Special Theory, Lorentz and others were faced with an interesting dilemma. Figure Q2 shows a beam of light and two observers moving along with it, one at a speed $v < c$ and the other at $v > c$. From a classical perspective, how would each see the light to move? What about the notion that the beam propagates in the direction of $\mathbf{E} \times \mathbf{B}$?

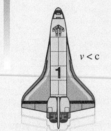

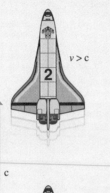

Light   c

Figure Q2

3. How might the history of physics been different if c were 3×10^2 m/s rather than 3×10^8 m/s?

4. Could we connect two distant spacecraft by a long rigid rod and communicate faster than we might with light by just wiggling the ends of the rod? Explain.

5. Imagine an ideally rigid long pair of scissors. As the two blades are closed, the cutting point moves out away from the pivot. The more nearly parallel the blades become, the more rapidly that point travels away. If the scissors are long enough and closed rapidly enough, the contact point will travel out faster than c. An ax slammed almost horizontally onto an equally large horizontal sheet of wood suggests the same effect. Think of the point of overlap of two crossed laser beams in the same plane that are moved toward being parallel. Does any of this violate Relativity? Explain. Can you think of another "thing" that moves faster than c?

6. Primary cosmic rays impacting on the upper atmosphere at altitudes of around 15 km create high-speed ($v = 0.999c$) muons. Despite their 1.5-μs half-life, about 200 muons are detected at ground level per square meter per second. That's surprising because half of them should disappear after 1.5 μs (or 450 m) and after 33.3 half-lives (15 km) a miniscule $10^{-8}\%$ of them should survive (much fewer than do). Explain what is happening from the perspective of the muon.

7. Suppose we are receiving a stream of photons of frequency f in vacuum and then decide to fly directly away from the source. Discuss what we will observe to happen to the photons' speed, momentum, total energy, and rest energy.

8. Astronaut Suzy inside her spaceship shines a flashlight beam on the front wall just as the craft passes the Earth. If the ship is traveling at or in excess of c, what would Suzy see? What would you on Earth see? Using the two postulates, what can you conclude?

9. As we have seen, when two parallel wires carry currents in the same direction, there is a "magnetic" attraction between them. Amazingly, it turns out that this effect is actually an electrostatic one and, more generally, that *magnetism is a relativistic manifestation of electrostatics.* There really is only one force—the electromagnetic force. Thus, envision two current-carrying wires and their moving electrons and stationary positive ions. What do things look like within wire 1 as seen by a moving electron in wire 2? Is there a Lorentz-FitzGerald Contraction and, if so, what is its apparent effect on the positive and negative charge densities?

10. For warm-ups, draw part of the world-line of a book at rest on your desk. Figure Q10 is a space-time diagram showing several world-lines. Describe the motion associated with each as completely as possible.

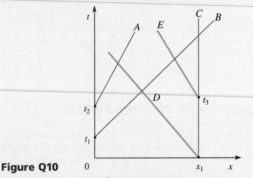

Figure Q10

11. Given that $E = \gamma mc^2$, what happens to a material object as it approaches the speed of light? What does this tell us about v compared to c?

12. Figure Q12 is a space-time diagram illustrating how light pulses can be used as time markers. Explain what's happening in detail. Why are the intervals seen by observer B longer than those seen by observer A? What does observer C see? Consider the flashes arriving at C to have emanated either from A or B. Each observer is in a different inertial reference frame.

13. Newton said "Absolute, true, and mathematical time, of itself, and from its own nature, flows equally without relation to anything external." Discuss the significance and validity of this statement.

14. The energy of a photon is entirely kinetic. Discuss the relationships between E, p, f, and λ for photons. How does the Doppler Effect influence these considerations? Can a light beam exert pressure on a target?

15. Starting with Eq. (28.7), show that either $v_{PO}/c = 1$ or $v_{OO}/c = 1$ implies $v_{PO}/c = 1$. When is v_{PO} less than the classical value? When is it greater?

16. You may have learned in chemistry class somewhere along the line that if you "weighed" all the reactants before and after a chemical reaction, the sum would be unchanged. That's

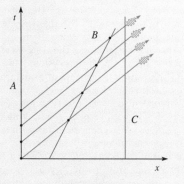

Figure Q12

the so-called Law of Conservation of Mass and it's confirmed to an accuracy of one part per ten million ($1/10^7$). Inasmuch as Relativity shows us that most chemical reactions involve mass changes ($\Delta m/m$) of a few parts per ten thousand million ($\approx 2/10^{10}$), what can be said about the Law of Conservation of Mass?

17. Poincaré challenged the idea of immediate action-at-a-distance, although Newton had accepted it. From Poincaré's perspective, for example, a change in the gravitational field of the Sun would take a finite time to reach the Earth. Generalizing this assumption, show that nothing can travel at an infinite speed. Assuming there is an upper-speed limit, discuss why it must be invariant with respect to all inertial reference frames. What does this suggest about the Michelson-Morley experiment and c?

18. You've no doubt heard in school the adage "Matter can neither be created nor destroyed" or its haunting corollary "Energy can neither be created nor destroyed." What can you say about these statements?

19. Here are two postulates: (1) All interactions propagate at a finite speed. (2) An object cannot interact with itself. Discuss how it follows that there must be an upper-speed limit.

Multiple Choice Questions

1. The Michelson-Morley experiment established   (a) that the Earth does not move with respect to the Sun   (b) that the aether moves at c as the Earth travels in its orbit   (c) that the aether is an elastic solid that streams over the Earth   (d) that there is no observable aether wind at the surface of the Earth   (e) none of these.

2. Simultaneity is   (a) dilated   (b) absolute   (c) invariant   (d) relative   (e) none of these.

3. A clock is moving at a uniform velocity with respect to an observer. The latter, comparing things to her clock, reports that the time on the moving clock is   (a) perfectly accurate   (b) fast   (c) slow   (d) running backwards   (e) none of these.

4. According to the Lorentz-FitzGerald Contraction   (a) a body contracts only when it accelerates   (b) a body contracts along the direction of motion   (c) the time it takes for a light clock to tick contracts   (d) a body contracts transverse to the direction of its motion   (e) none of these.

5. How long will a vertical meter stick appear to someone moving horizontally with respect to it at a speed of 0.99c?   (a) 99 cm   (b) 100 cm   (c) 0 cm   (d) 0.01 cm   (e) none of these.

6. Two inertial observers separating from each other at a speed of 99.99% of c each shine a light beam at one another.   (a) each will see the light arrive at a speed of c   (b) each will see the light arrive at a speed of 2c   (c) the light will not reach either of them   (d) the light will reach one at c and the other at 2c   (e) none of these.

7. A spaceship observes an asteroid coming directly toward it at a speed v. The ship fires its engines and travels away along the line of approach. It measures the asteroid's   (a) v and p to decrease, but E_0 is unaltered   (b) v, p, and E to be unchanged   (c) v only to decrease, p and E being unaltered   (d) p and E to increase, leaving v and E_0 unaltered   (e) none of these.

8. A long thin rocket takes off straight up from the ground. It is observed by three people: a man selling ice cream on the beach, a woman riding up an adjacent gantry elevator, and the pilot looking out a window. Referring to the rocket, ideally   (a) all three see the same length   (b) of the three, the pilot sees the shortest length   (c) of the three, the woman sees the shortest length   (d) of the three, the man with the ice cream sees the longest length   (e) none of these.

9. An astronaut heading out toward a star at a constant high speed can determine that he is in motion by   (a) the contraction of on-board meter sticks   (b) the slowing down of time on his clocks   (c) the increase of his mass   (d) the speeding up of his heart   (e) none of these.

10. An arrow 30 cm long is fired at very high speed directly toward a hollow cylinder 30 cm long. The arrow passes axially inside the cylinder. As seen by an observer at rest with respect to the cylinder,   (a) the arrow is shrunken and fits inside for an instant   (b) the arrow and cylinder are unchanged and stay exactly the same size   (c) the cylinder is shrunken and the arrow never fits completely inside   (d) the arrow and the cylinder are both shrunken and remain the same size   (e) none of these.

11. An arrow 30 cm long is fired at very high speed directly toward a hollow cylinder 30 cm long. The arrow passes axially inside the cylinder. As seen by an observer at rest with respect to the arrow,   (a) the arrow is shrunken and fits inside for an instant   (b) the arrow and cylinder are unchanged and stay exactly the same size   (c) the cylinder is shrunken and the arrow never fits completely inside   (d) the arrow and the cylinder are both shrunken and remain the same size   (e) none of these.

12. An observer in the laboratory section of a spacecraft performs a series of experiments to determine whether the ship is at rest or in uniform motion. He   (a) can succeed by making very precise time measurements   (b) can succeed by making very precise mass measurements   (c) can succeed by making very precise length and time measurements   (d) cannot succeed no matter what he does   (e) none of these.

13. When a battery is drained in ordinary use,   (a) its mass decreases because its internal rest energy decreases

(b) its mass decreases, but its rest energy remains unchanged   (c) its mass remains unchanged, but its internal rest energy decreases   (d) its mass decreases because its internal rest energy increases   (e) none of these.

14. When a photon is red-shifted via the Doppler Effect by having an observer moving away from the source,   (a) its energy is unchanged   (b) its momentum increases   (c) its energy increases   (d) both its momentum and energy decrease   (e) none of these.

15. An arrow 30 cm long is fired at very high speed v directly toward a hollow cylinder 30 cm long. The arrow passes axially inside the cylinder. An observer moving parallel to the cylinder in the direction of the arrow at a speed $\frac{1}{2}v$ sees the arrow moving forward at the same speed he sees the cylinder moving backward. He observes that   (a) the arrow is shrunken and fits inside for an instant   (b) the arrow and cylinder are unchanged and exactly the same size   (c) the cylinder is shrunken and the arrow never fits completely inside   (d) the arrow and the cylinder are both shrunken and are the same size   (e) none of these.

16. The proper time of a system is   (a) the time measured on clocks at rest in that system   (b) the time measured on clocks at rest in an inertial system moving properly with respect to the first system   (c) the time measured on clocks moving uniformly in that system   (d) the time that agrees with the National Institute of Standards and Technology   (e) none of these.

17. A photon's proper time   (a) is a function of its speed   (b) changes slowly with respect to a moving observer   (c) never changes   (d) changes faster with respect to systems at rest   (e) none of these.

18. Given that there are about 2.56×10^9 heartbeats in a statistically average lifetime of 70 years, people who are born and die on a spaceship moving at a constant speed of 0.600c can expect their hearts to beat a total of (a) $(0.600)(2.56 \times 10^9)$ times   (b) 2.56×10^9 times (c) $(0.800)(2.56 \times 10^9)$ times   (d) $(1.25)(2.56 \times 10^9)$ times   (e) none of these.

Problems

BEFORE THE SPECIAL THEORY
THE SPECIAL THEORY OF RELATIVITY

1. [I] In the Michelson-Morley experiment, the time it takes beam 1 to travel out and back, with and against the aether wind, is $t_{\parallel}$, and the time for beam 2 to go from M_S to M_2 to M_S is $t_{\perp}$, where

$$t_{\parallel} = \frac{2L}{c}\frac{1}{(1 - \beta^2)} \quad \text{and} \quad t_{\perp} = \frac{2L}{c}\frac{1}{\sqrt{1 - \beta^2}}$$

Suppose the $M_S M_1$ arm of the Michelson Interferometer shrinks as suggested by FitzGerald, so that L becomes $L\sqrt{1 - \beta^2}$. Show that $\Delta t = t_{\parallel} - t_{\perp}$ then equals zero.

2. [I] A light-minute is often symbolically written as 1 c·min. Define the concept and give its value in SI units.

3. [I] Show that $\beta^2 = (\gamma^2 - 1)/\gamma^2$.

4. [I] A computer in a spacecraft takes 2.00 μs to make a calculation, as measured by a co-moving observer on board. Someone traveling at a relative speed of 0.998c looks in the window. How long will that person determine the calculation to take?

5. [I] According to a biology text, it takes 45 s for blood to circulate around the human body. How long on his clocks will it take blood to circulate around the body of an astronaut traveling such that with respect to Earth $\gamma = 2$? How long will a ground-based observer see the astronaut's circulation take?

6. [I] A stopwatch in the hands of a terrestrial observer is seen by a passer-by in a flying saucer traveling relative to the planet with a γ of 2.00. If the earthling sees two events separated by a proper time interval of 60.0 s, how long will the interval be as seen by the traveler?

7. [I] The vaudeville team of Shultz and O'Hara, playing to a full house at Minsky's Burlesque, takes 60.0 s to tell their worst joke. How long will an observer in a rocket ship flying by at 0.995c have to wait to lip-read the punch line?

Will the observer start laughing after seeing the audience respond? Assume that the observer is equidistant from the stage when he perceives the start and finish of the joke so we needn't worry about communication-time lags.

8. [I] An astronaut on the way to Mars at 0.600c wishes to take a one-hour nap. She contacts Ground Control requesting a wake-up call. Neglecting any complications due to the time it takes the signal to reach the ship, how long should the flight controller let her sleep as measured on his clock?

9. [I] Suppose you look at the inhabitants of a passing asteroid and see all their clocks running such that 52.0 s goes by on them while 60.0 s elapse on your planetary clock. What is the relative speed of the asteroid and Earth?

10. [I] An astronaut is on the way directly to a star system along a straight line from Earth. Her instruments indicate that at the speed relative to the Earth at which she is traveling, her γ is 5/3 and the total distance she has to traverse is (6.0c)(1.0 y). What is the corresponding proper distance shown on Earth-based instruments?

11. [I] How large would a 1.00-m-long Italian bread look to a hungry observer (with very sharp eyes) if he sees the bread to have a speed of 0.900c in a direction parallel to the loaf?

12. [I] A long pepperoni is fired out of a new secret weapon developed by the Mediterranean arm of NATO. As it flashes past the reviewing stand, its length is measured to be 44.0% of what it was before launch. How fast is it moving?

13. [I] If an electron is hurtling down the two-mile long Stanford Linear Accelerator at 99.98% of c, how many meters long is the trip as seen by the electron?

14. [I] A 1.00-m-long rocket is to be fired past an observer who will locate both its ends simultaneously and determine its length. If that length is found to be 0.500 m, how fast was the rocket moving?

15. [I] A spacecraft passes a platform traveling parallel to it (at 0.600c) along its full length. The pilot of the spacecraft determines the platform to be 400.0 m long. You witness this event from the platform lounge where you happen to be having a drink. What is the proper length of the platform?

16. [I] A spaceship moving at a speed of 0.70c with respect to you fires a rocket in the forward direction at 0.40c relative to the ship. What is the speed of the rocket with respect to you? What would be expected classically?

17. [I] Two pulses of light are sent out in vacuum, one to the right and another colinearly to the left. What is the speed of either one with respect to the other? Show your work, please. What would be expected classically?

18. [I] A spaceship moving at a speed of 0.70c with respect to you fires a rocket in the backward direction at 0.40c relative to the ship. What is the speed of the rocket with respect to you? What would be expected classically?

19. [I] An electron traveling at 0.992c collides head-on with a positron traveling at 0.981c. Both speeds are measured with respect to the laboratory in which the experiment occurs. At what speed do they approach each other?

20. [II] Calculate the approximate time dilation that would be observed by someone on the ground watching a clock in a supersonic jet flying at 1800 mi/h. Use the binomial expansion (which is applicable at speeds less than around 0.3c). Try doing this calculation exactly—you'll see why the approximation is so useful.

21. [II] A bright yellow line painted along the length of a rocket is seen to be 1.000 yd long by an observer who is standing on the Earth as the vehicle sails by. Knowing that the proper length of the line is actually 1.000 m, how fast is the rocket traveling?

22. [II] Calculate the percentage length contraction for a high-speed jet plane traveling at 3600 mi/h. Try doing it exactly and then use the binomial expansion (which is okay for speeds less than around 0.3c).

23. [II] Two inertial observers pass each other at a relative speed v, and each notices a 10.0% length contraction of the other. Determine v.

24. [II] The navigator on a Federation starship traveling toward Earth along a path parallel to the Earth-Sun line sees the separation between these bodies to be 6.11 c·min. Consulting charts, the navigator finds the proper distance to be 8.33 c·min. How fast is the ship going?

25. [II] Two friends, Harvey (age 20) and Lisa (age 24), say their goodbyes. She boards a space shuttle for a round-trip flyby of the planet Mongo. Rapidly reaching cruising speed, she settles in for a long journey. Neglecting the short acceleration times (which is quite impractical), Lisa returns home 24 months later (as seen on her calendar). She is met by Harvey, who is 26 years old. At what speed was Lisa cruising?

26. [II] Mars is 80.00×10^6 km from Earth when a settler sends a laser beam message back home. As fate would have it, just as the flash of light goes out, a spacecraft flies by on the way toward Earth at 0.500 0c. How long will someone on Earth see the ship take to make the journey? How long will the message take to arrive on Earth? How long will the crew of the ship say the trip took? How long will they say the message took to get to Earth?

27. [II] A 1.000-m bar of steel travels at a constant speed of 0.600c with respect to an observer. The bar moves lengthwise along a line perpendicular to the observer's telescope. How long does it take the bar to fly past the cross hairs in the telescope?

28. [II] A starship passes a space station at a uniform 0.600c. The pilot looking into the gym area sees a runner traveling perpendicular to the flight path. The runner covers the straight 100-m gym track in 10.0 s by his watch. What was the runner's average speed as seen by the pilot? [Hint: How long is the track as seen by the pilot?]

29. [II] A laser beam fired from Earth is received by a spacecraft heading directly toward the source at a speed v. Show that a technician aboard the ship will measure the speed of the incoming beam to be c regardless of the value of v.

30. [II] Show that if the speed of an object in any one inertial frame is less than c, its speed in all other inertial frames (each moving with respect to the first at a speed less than c) will be less than c. [Hint: Let $v_{PO'} = ac$ and $v_{O'O} = bc$, where a and b are < 1.]

31. [III] Consider the passengers aboard a jet plane cruising at 1000 km/h with respect to an observer on the ground. Neglecting all other effects (for example, gravity), how long would the flight have to last before the clock aboard the plane was 1.00 s behind a clock on the ground with which it had been synchronized?

32. [III] With Problem 31 in mind, consider a satellite in orbit or a plane flying around the Earth at a uniform speed. We wish to determine how long the flight must last for there to be a discrepancy of 1.00 s between clocks on the ground and in the air. Show that this will indeed be the case when the proper time on the moving craft is

$$\Delta t_S = \frac{1.00 \text{ s}}{\gamma - 1}$$

33. [III] The TV mast on a spaceship passing Earth is fixed at 21.0° to the line of flight by the communications officer. If the ship zips by at 0.852c, at what angle will someone on Earth see the mast?

RELATIVISTIC DYNAMICS

34. [I] An electron ($m = 9.110 \times 10^{-31}$ kg) is traveling at 0.866c with respect to the face of a TV picture tube toward which it is heading. How much momentum does it have with respect to the tube?

35. [I] A neutron ($m = 1.675 \times 10^{-27}$ kg) is traveling at 0.50c with respect to a "stationary" target. How much momentum does it have with respect to the target?

36. [I] A proton has a mass of $1.672\,6 \times 10^{-27}$ kg. Determine its rest energy in joules and in MeV and compare your results with Table 28.3.

37. [I] A hypothetical particle is traveling with respect to an observer such that its total energy is seen to be twice its rest energy (which is 1000 MeV). How fast is it moving? What is its kinetic energy?

38. [I] An electron with a rest energy of 0.511 MeV has a kinetic energy of 0.089 MeV. What is its total energy?

39. [I] The total energy of a muon is 106.7 MeV. If its rest energy is 105.7 MeV, what is its kinetic energy?

40. [I] A proton ($m = 1.672\,6 \times 10^{-27}$ kg) moves with respect to the laboratory frame such that it has a $\gamma = 1.50$. What is its total energy in joules and electron-volts?

41. [I] A particle of mass 1.000×10^{-6} kg is fired from a device such that it has a muzzle "velocity" of $0.200\,0c$ with respect to an inertial observer. Determine the rest energy, total energy, and kinetic energy of the projectile with respect to the observer.

42. [I] An electron with a rest energy of 0.511 MeV travels with a γ of 5/3 with respect to a laboratory. What is its total energy?

43. [I] A deuteron is the heavy hydrogen nucleus $^2_1$H composed of a proton and a neutron. Its rest energy is 1875.6 MeV. How much energy must be supplied to rip it apart? That is, how much energy is liberated (as KE and as a γ-ray) when a deuteron is formed of a separate proton and neutron?

44. [I] A hypothetical particle has a rest energy of 1.500 MeV and a total energy of 3.000 MeV. What is its momentum?

45. [I] A clock, in the course of being wound, has 89.88 J of work done on its mainspring. By how much does its mass change?

46. [I] A proton and a neutron together form the nucleus of a deuterium atom that is completed with a single orbiting electron. The mass of such an atom is $3.344\,5 \times 10^{-27}$ kg. When the particles come together, mass is transformed, and the composite atom has less mass than the separate constituents; the equivalent energy difference is called the **binding energy**. Find the binding energy of deuterium in units of J and MeV to three figures.

47. [I] Compute the amount of energy liberated in the fusion reaction

$$^2\text{H} + {}^2\text{H} \rightarrow {}^3\text{He} + {}^1\text{n} + \text{energy}$$

Here, the neutron flies off with about 3 times the KE of the helium.

48. [I] Another important fusion reaction involving deuterium and tritium is

$$^2\text{H} + {}^3\text{H} \rightarrow {}^4\text{He} + {}^1\text{n} + \text{energy}$$

where the neutron flies off with about 4 times the KE of the helium. Determine the total amount of energy liberated.

49. [I] A photon is to have the same energy as a 1.000-MeV electron (that is, an electron with a KE of 1.000 MeV). What must be its frequency? [Hint: That's total energy.]

50. [II] Show that as the momentum of a body of mass m becomes very large, its speed approaches c, but does not exceed it.

51. [II] Suppose that an object traveling uniformly has a value of γ relative to an inertial observer. Derive an expression in terms of γ for the percent error introduced by using the classical momentum (p_c) instead of the relativistic momentum (p); that is, $(p - p_c)/p$. What is this error for a speed of 0.600c?

52. [II] An electron ($m = 9.109\,5 \times 10^{-31}$ kg) has a momentum of 8.603×10^{-23} N·s. What's its speed?

53. [II] If the ratio of the momentum to the mass of a proton is 0.101, determine its speed. Compare this ratio to its classically determined speed. When will the relativistic speed approximate the classical speed?

54. [II] An electron with rest energy 0.511 MeV travels at 0.80c with respect to the laboratory. What is its momentum in the lab frame in units of MeV/c?

55. [II] An object has a total energy equal to twice its rest energy. Show that $p = \sqrt{3}\ mc$.

56. [II] An electron has a kinetic energy of 0.200 0 MeV. Find its speed. Compare that with its classical speed.

57. [II] An electron at rest accelerates through a potential difference of 1.50 MV. Determine its speed on emerging and compare it with the erroneous classical value.

58. [II] We wish to accelerate an electron from rest up to 0.990c. Through what potential difference must it pass?

59. [II] A proton with a rest energy of 938.3 MeV has a momentum of 100.0 MeV/c. How fast is it moving?

60. [II] Show that for an object moving at speed v, the ratio of the relativistic KE to the classical KE, call it Γ, is

$$\Gamma = 2(\gamma - 1)/\beta^2$$

61. [II] Given a particle whose rest energy equals its kinetic energy, determine its β value.

62. [II] With Problem 60 in mind, show that

$$\gamma = \tfrac{1}{4}(\Gamma + \sqrt{\Gamma^2 + 8\Gamma})$$

63. [II] A photon is to have the same momentum as a 1.000-MeV electron (that is, an electron with a KE of 1.000 MeV). What must be its frequency?

64. [cc] A particle of mass m moves along the x-axis under the influence of a constant force F acting along that axis. Show that the resulting acceleration is given by

$$a = \frac{F}{m}\left[1 - \frac{v^2}{c^2}\right]^{3/2}$$

Assume $v = 0$ at $t = 0$.

65. [cc] A particle of mass m is acted on by a force F along the x-axis that is initially applied at $t = 0$. What is the speed of the particle as a function of time? [Hint: Make use of the results of the previous problem.]

66. [III] An object has an initial speed v and a corresponding momentum p. Show that if the momentum is to be doubled, the final speed must become

$$v_f = \frac{2vc}{\sqrt{c^2 + 3v^2}}$$

67. [ccc] Show that the force **F** and the acceleration $d\mathbf{v}/dt$ do not necessarily act in the same direction. [Hint: Begin with Newton's Second Law and show that the force depends on $d\mathbf{v}/dt$ and **v**, which need not always be parallel.]

68. [ccc] To derive the expression for the relativistic kinetic energy begin with

$$W = \int F\,dx = \int \frac{dp}{dt}\,dx$$

(a) Use the Chain Rule to show that

$$\frac{dp}{dt}\,dx = \frac{dp}{dv}\,v\,dv$$

(b) Then prove that

$$\frac{dp}{dv} = \gamma^3 m$$

Finally, establish that

$$W = \int_0^v \gamma^3 mv\,dv$$

which we know yields Eq. (28.10). Take the initial speed to be zero.

29

The Origins of Modern Physics

WE NOW BEGIN A sequence of five chapters dealing with Modern Physics. That discipline is a study of the submicroscopic world of atoms, and the particles that compose them, and the particles that compose those particles. It is essentially a creation of the twentieth century, but most of the work had its foundations in the late 1800s, a period when even the existence of the atom was not totally accepted. This chapter deals with the establishment of the atom as a real entity, and, beyond that, with the realization that it is a complex structure capable of coming apart into its constituent subatomic particles.

Here we study a series of important experiments that laid the groundwork for all that was to follow. We trace the discoveries of the electron, X-rays, radioactivity, the atomic nucleus, protons, and neutrons. The understanding of the atomic domain that evolved from these fundamental insights now guides much of the work done in contemporary physics, chemistry, and biology.

SUBATOMIC PARTICLES

The first subatomic particle to reveal itself was the electron, and that discovery was predicated on the study of electric currents in solids, liquids, and gases. We know now that charge is a property of substantial matter and that it is quantized; it comes in whole number multiples of a minimum amount corresponding to the magnitude of the charge on the electron. A little over a hundred years ago, physicists had no idea of what charge was; whether it was continuous or particulate, whether it was material or aethereal, like light.

29.1 The Quantum of Charge

Soon after the invention of the battery, Nicholson and Carlisle used one to decompose water via a process known as *electrolysis*. Their discovery was accidental; to improve the contact between a wire and their battery, they put some water on the connection and the liquid filled with gas bubbles. Thereafter, they inserted two wires into a beaker of water and passed a current through it. Bubbles of oxygen were released at the *anode,* and hydrogen was formed at the *cathode.* Humphry Davy and his assistant Michael Faraday used electrolysis to explore chemical interactions. When the conducting solution, the electrolyte, contains a dissolved salt of some metal like silver, that metal will be liberated at the cathode just as hydrogen is (Fig. 29.1). To account for the passage of current across the electrolyte, Faraday supposed that there was a flow of charged particles, which he called **ions**. Although he did not speculate as to their nature, we know that ions are atoms that have lost or gained one or more orbital electrons and therefore carry a net charge.

Faraday measured the mass of monovalent silver deposited on the cathode and the net charge provided (that is, current multiplied by the time it was applied) in the process. To liberate one *mole* of a single-valence element (singly charged $\pm$ ion), he found that a large quantity of charge F, now called a **faraday** and equal to 96 485 C, had to be provided. Remember that a mole of any element comprises $N_A = 6.022\,136\,7 \times 10^{23}$ atoms. Thus, for example, silver has an atomic mass of 107.87 units, and 1 faraday of charge liberates 6.02×10^{23} silver atoms with a net mass of 107.87 grams. Faraday teetered on the edge of electron theory, hinting that electricity was composed of particles of charge e. Given that there are N_A atoms per mole, and assuming each monovalent ion has a charge e, then a faraday of charge must correspond to an amount

$$F = N_A e \tag{29.1}$$

The striking thing is that the same quantity F is found for all monovalent elements regardless of their chemical properties; $e = F/N_A$ is a constant. For bivalent ions like copper, a charge of 2F must be supplied to liberate a mole of Cu, and each ion carries a charge of $2e$. Though F was directly measurable, Avogadro's number was not determined for several decades and that left e unknown. Still, 1 faraday of charge is transported by 1 mole of monovalent substance. It follows that $\frac{1}{2}$ faraday is

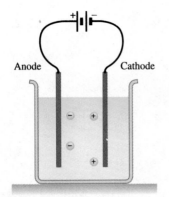

Figure 29.1 The passage of a current through an electrolyte. Negative ions move to the anode, positive ions move to the cathode.

Anode

Cathode

transported by $\frac{1}{2}$ mole, and so on down, *presumably* to the smallest unit of charge *e* associated with the smallest mass *m*; that is, the mass of a single atom of the material liberated. For example, we find from electrolysis that the **charge-to-mass ratio** for singly ionized hydrogen (H$^+$) is

$$\left(\frac{e}{m}\right)_{H^+} = 9.58 \times 10^7 \text{ C/kg}$$

Hydrogen is the lightest of the elements, and this value turns out to be the largest charge-to-mass ratio of any element. Although neither *e* nor *m* was known at the time, the corresponding ratios could be determined experimentally.

Example 29.1 During electrolysis, how much silver will be deposited by a current of 1.00 A applied for 1.00 s?

Solution: [Given: element silver, $I = 1.00$ A, and $t = 1.00$ s. Find: *m*.] We have the current and the time during which it flowed, so the net charge passed is

$$\Delta q = I\Delta t = (1.00 \text{ A})(1.00 \text{ s}) = 1.00 \text{ C}$$

Since 1 faraday of charge liberates 1 mole of silver (that is, 107.9 g), 1.00 C of charge liberates an amount *m*

where

$$\frac{96\ 485 \text{ C}}{107.9 \text{ g}} = \frac{1.00 \text{ C}}{m}$$

and

$$\boxed{m = 1.12 \text{ mg}}$$

▶ **Quick Check:** We are passing about 10^{-5} faraday, and that quantity should deposit about 10^{-5} of a mole of Ag, which equals 1.1 mg.

Remarkably, in 1874, the Irish physicist George Stoney used Eq. (29.1) along with the best available values of F and Avogadro's number to determine the basic unit of charge. His estimate of *e* was too small by a factor of about 20, but even that was still quite good, all things considered.

By 1881, Helmholtz asserted that, "If we accept the hypothesis that the elementary substances are composed of atoms, we cannot avoid concluding that electricity also, positive as well as negative, is divided into definite elementary portions, which behave like atoms of electricity." For a while, the German literature referred to *e* as "*das Helmholtzsche Elementarquantum*" (the Helmholtz elementary quantity, or **quantum**). In 1891, Stoney christened that fundamental unit of charge the *electron*.

29.2 Cathode Rays: Particles of Charge

A totally different route to the electron began one dark night in 1675 when the French astronomer J. Picard noticed in amazement that the barometer he was swinging as he walked began to give off an eerie flickering light. The effect was duplicated in the laboratory by Hauksbee using an electrostatic generator to provoke a rarefied gas in a bottle into the mysterious luminosity. Thus began the study of electrical discharges in low-pressure gases.

Outstanding among the many scientists in the field during the 1870s was Sir William Crookes, a rather unorthodox fellow who believed he could communicate with the dead. A "Crookes tube" with two sealed electrodes is illustrated in Fig. 29.2. (It's the forerunner of all the blazing bar and motel signs—the neon uglies, p. 1087.) When the tube was connected to the terminals of a high-voltage source, a glowing beam spread down its length. Different gases glowed with differ-

Figure 29.2 A cathode-ray or Crookes tube.

Figure 29.3 When the tube is bent, it becomes clear that the rays emanate from the negative terminal, or cathode.

A so-called neon sign is just a long tube filled with an appropriate gas, across which there is a sustained high-voltage discharge.

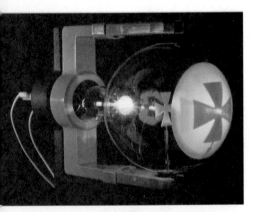

A cathode-ray tube. The metal cross intercepts the beam and casts a shadow on the face of the tube.

ent colors: mercury emitted greenish blue, while neon gave off a bright orange-red. By displacing the anode to the side (Fig. 29.3), Crookes verified that the emanations were actually streaming from the negative plate, not always making it to the anode. When these **cathode rays** struck the walls of the tube, the glass gave off a pale green fluorescent glow that betrayed the impact of the beam.

Objects that were inserted into the stream cast sharp shadows in the glow at the far end of the tube. This, too, suggested straight-line propagation. Crookes even put a little paddle wheel inside a tube and spun it around with the bombarding beam, which more and more appeared to be a stream of particles carrying energy and momentum. Later, he was able to deflect the beam with a magnet (p. 807) in the same way a current of negative charges would be deflected. What emerged from this experimentation was the "British" view that cathode rays were streams of negatively charged submicroscopic particles.

The opposing "German" view was most vigorously pressed by P. Lenard. Heinrich Hertz had discovered that cathode rays could pass through thin sheets of metal without leaving any holes, and that was enough to convince his assistant Lenard that the rays must be nonmaterial. Today, we know that bulk matter is mostly empty space and that some subatomic particles can sail through yards of concrete and steel as if they weren't there. To Lenard, cathode rays were without substance; they were waves like light—"phenomena in the aether."

There the matter stood, befogged by nationalism—corpuscle versus wave—a controversy that would last at least twenty years.

Discovering the Electron

When Joseph John Thomson began his research into the nature of cathode rays, he was the director of the famed Cavendish Laboratory at Cambridge University. In 1894, Thomson showed that cathode rays traveled much slower than light (another blow to the notion that they were "phenomena in the aether"). And then J. Perrin (1895) found that a metal obstacle located in the path of the beam acquired a negative charge. It remained for Thomson (his friends called him J. J.) to demonstrate that the cathode rays and the charge were one and the same.

To that end, Thomson constructed the device shown in Fig. 29.4, which is similar in many respects to a TV picture tube, a so-called CRT (cathode-ray tube). A high voltage between cathode (C) and anode (A) ionized the trace of residual gas, creating a cloud of positively charged atoms and negative electrons; the latter were the cathode rays. These rays were accelerated toward the positive anode reaching tremendous speeds (of as much as 60 000 mi/h). A small hole in the front of the anode allowed a narrow unobstructed stream to pass into the region beyond A. The speed v of the beam, which could be controlled via the cathode-anode voltage, was fairly uniform. Further along the tube, the beam traveled between two metal plates (P) that were attached to a variable dc voltage, creating a vertical electric field **E**. Coils straddling the tube produced a controllable horizontal magnetic field **B** crossing the same region.

The idea was to measure the charge-to-mass ratio in a two-step procedure. The beam was assumed to be composed of a stream of discrete particles of charge e and mass m_e that could be deflected in the vertical plane by either field. The first step was to determine v, which was done by adjusting E and B so that the forces produced were equal and opposite and the beam sailed through undeflected (the effect of gravity was negligible). This arrangement is the "velocity selector" we considered

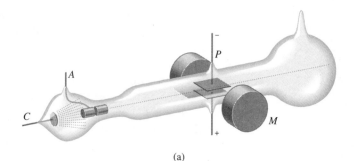

(a)

J. J. Thomson's original tube.

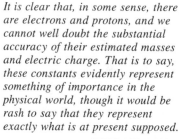

(b)

Figure 29.4 (a) J. J. Thomson's electron-beam device showing the horizontal electrical deflecting plates and the vertical magnetic coils. (b) Electrons experience a downward electric force and (with the B-field coming out of the page) an upward magnetic force.

earlier (Problem 57, p. 829), where it was shown that there's no deflection of the beam provided

$$v = \frac{E}{B}$$

Next, the B-field was shut off, and the beam was deflected by the remaining E-field. The cathode-ray particles experienced a downward Coulomb force (eE) and so accelerated down uniformly, via Newton's Second Law (eE/m_e). They each "fell," much as a stone would fall in a gravity field, for all the time (t) that they were between the plates. Since the horizontal speed v is constant, the traversal time is $L/v = L/(E/B)$. We want the distance each particle drops (y) in the course of passing through the plates. Thus, from Eq. (3.9), $y = \frac{1}{2}at^2$, or

$$y = \tfrac{1}{2}\left(\frac{eE}{m_e}\right)\left(\frac{LB}{E}\right)^2$$

Inasmuch as y can be determined directly from the point where the beam hits the screen (using the geometry of the tube, p. 1114), and since we know E from the plate voltage and B from the coil current, the only unknowns are e and m_e. Thus

$$\frac{e}{m_e} = \frac{2yE}{L^2B^2} \qquad (29.2)$$

The present-day accepted value of the ratio is

$$\frac{e}{m_e} = (1.758\,819\,62 \pm 0.000\,000\,53) \times 10^{11} \text{ C/kg}$$

It is clear that, in some sense, there are electrons and protons, and we cannot well doubt the substantial accuracy of their estimated masses and electric charge. That is to say, these constants evidently represent something of importance in the physical world, though it would be rash to say that they represent exactly what is at present supposed.

BERTRAND RUSSELL
The Analysis of Matter (1954)

The electrical matter consists of particles extremely subtle since it can permeate common matter, even the densest, with such freedom and ease as not to receive any appreciable resistance.

BENJAMIN FRANKLIN (ca. 1750)

J. J. Thomson at the Cavendish Laboratory.

Thomson suggested that the reason this ratio was so much larger (namely, 1836 times larger) than the corresponding one for hydrogen was that the mass of the electron was that much smaller than the mass of the H^+ ion. *The electron was a tiny fragment detached from a complex atom.* Not everyone in 1897 believed in atoms, and few were pleased with the notion of **subatomic particles**.

J. J. put the finishing touches on what today is generally called his "discovery of the electron" with two further experiments in 1899. First, he repeated the above measurement using a totally different source of electrons, namely the *photoelectric effect* (p. 1123), and got the same results for e/m_e. His second experiment was to measure e, again using a totally new procedure. His student C. T. R. Wilson (the inventor of the cloud chamber) had been working on the formation of clouds and found that droplets could form around charged particles. We shall not describe the methods used by Thomson and his colleagues other than to say that by 1901 he was able to arrive at a value for e that was only about 30% off. The work inspired a classic measurement by the American Robert Millikan, which we will examine.

The Oil-Drop Experiment

Millikan's famous oil-drop experiment is pictured in Fig. 29.5. A fine mist of oil is squirted from an atomizer above a small hole in the top plate of what can be thought of as a large parallel-plate capacitor. A few droplets descend through the hole into the region occupied by a variable E-field. Lit from the side, each minute droplet shines like a tiny star when seen through a viewing telescope. Most of the droplets

On Electrons. At first there were very few who believed in the existence of these bodies smaller than atoms. I was even told long afterwards by a distinguished physicist who had been present at my lecture at the Royal Institution that he thought I had been "pulling their legs."

J. J. THOMSON
English physicist

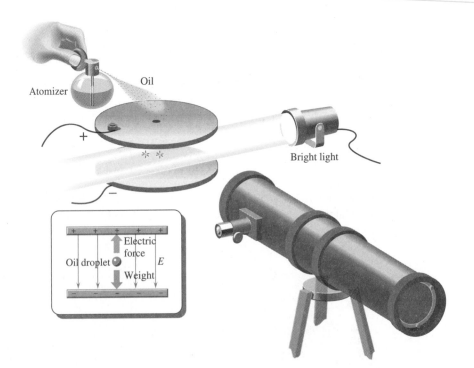

Figure 29.5 The Millikan oil-drop experiment. A mist of oil causes a few drops to enter between the plates of a capacitor. There, they experience a force due to the applied electric field.

get negatively charged on passing through the nozzle, picking up some small but unknown number of electrons. Once a drop is caught sight of, the voltage on the plates is carefully varied, thereby controlling a downward E and slowing the fall until that sphere of oil is suspended motionless midair in a balance between gravitational (mg) and electrical (qE) forces. At that moment $qE = mg$, and if the mass of the droplet (m) were known, the charge that it carried (q) would be known. The next procedure determines m.

The E-field is shut off and the same droplet watched as it again falls, soon reaching a constant terminal speed. By timing the motion of the droplet as it passes from one hairline down to another in the telescope, the terminal speed is measured. The theory of air resistance tells us that the terminal speed depends on the radius of the droplet, as well as on a number of other factors, all of which are known. Thus, from the terminal speed we have the radius, and from the radius we have the mass of the sphere. Millikan actually used a somewhat more complicated, though less arduous approach, but the idea is the same. His graduate student H. Fletcher methodically found the net charge on thousands of droplets, one by one. Millikan and Fletcher were then able to show that the droplets carried whole number multiples of a basic charge $q_e = e$, which was ascribed to the electron itself. They arrived at an average value of e of 1.592×10^{-19} C. Today, the best value of this fundamental quantum of charge (Table 29.1) is

$$e = (1.602\,177\,33 \pm 0.000\,000\,49) \times 10^{-19} \text{ C}$$

All charged subatomic matter observed to date, whether positive or negative, carries a net charge that is an integer multiple of e.

Combining e with the charge-to-mass ratio provides another remarkable number—the mass of the electron:

$$m_e = (9.109\,389\,7 \pm 0.000\,005\,4) \times 10^{-31} \text{ kg}$$

Using Eq. (29.1) and the measured value of the faraday, Millikan computed Avogadro's number to be 6.062×10^{23} molecules per mole (as compared to the present-day value of 6.022×10^{23}—not bad at all).

Given a charge-to-mass ratio for a hydrogen ion of 9.58×10^7 C/kg and the value of e, we can make a rough estimate of the size of an atom—silver, for instance. The relative atomic mass of hydrogen is 1.008, whereas that of silver is 107.9. The mass of a hydrogen ion is e divided by the charge-to-mass ratio: $(1.602 \times 10^{-19}$ C$)/(9.58 \times 10^7$ C/kg$) = 1.67 \times 10^{-27}$ kg. Thus, a silver atom has a mass of $(107.9/1.008)(1.67 \times 10^{-27}$ kg$) = 1.79 \times 10^{-25}$ kg. The density of silver is 10.4×10^3 kg/m^3, and so there must be $(10.4 \times 10^3$ kg/m$^3)/(1.79 \times 10^{-25}$ kg/atom$) = 5.8 \times 10^{28}$ atoms/m^3. The reciprocal of that is the number of cubic meters occupied per silver atom; namely, 1.72×10^{-29} m^3/atom. Each atom sits in a little imaginary cube of this volume. Assuming the silver atoms are tightly packed

TABLE 29.1 Some physical characteristics of the electron*

Mass m_e	$9.109\,389\,7(54) \times 10^{-31}$ kg
Charge e	$1.602\,177\,33(49) \times 10^{-19}$ C
Rest energy	$0.510\,999\,06(15)$ MeV
Charge-to-mass ratio e/m_e	$1.758\,819\,62(53) \times 10^{11}$ C/kg

*The parentheses show the uncertainty in the last two figures.

spheres, the cube root of 1.72×10^{-29} m^3 should be roughly the diameter of one such atom; that is, 0.26 nm.

29.3 X-Rays

Wilhelm Conrad Röntgen was fifty-five in 1895, a well-respected professor of physics at Würzburg. He was a private person who conducted his research quietly, almost secretively, so much so that his students nicknamed him *der Unzugängliche,* the unapproachable one. He had become interested in the work done on cathode rays by Hertz and Lenard, especially the latter's study of the passage of the rays though an aluminum-foil window and thence outside beyond the vacuum tube into the air. Lenard had surrounded his apparatus with lead and iron sheet to shield it from light and electric fields, and Röntgen also covered the tube, but luckily he used only thin black cardboard. In the darkened room, he noticed that one of his luminescent screens (a piece of paper coated with a barium salt) some distance away glowed brightly for no apparent reason. Astonished, he explored the effect until he was convinced that it was real. After a while, the cause seemed clear: some new invisible penetrating radiation was being emitted from the discharge tube. He called it **X-rays**, because *x* in mathematics represents an unknown.

Like light, X-rays expose photographic film, and because of their tremendous penetrating power, he was able to produce "photographs . . . of the shadows of the bones of the hand." Though the rays could ionize gases, they were similar to light in that they were not deflected by either electric or magnetic fields. To prove that X-rays were waves and not particles, Röntgen tried but failed to observe refraction, specular reflection, and polarization. Still, J. J. Thomson, among others, favored what we know to be the proper interpretation; namely, that X-rays are very short-wavelength electromagnetic radiation ($\lambda \approx 0.1$ nm). The correct wave picture was only slowly gaining acceptance when C. Barkla, in 1906, managed to observe the partial polarization of X-rays, thus establishing their transverse nature (see Discussion Question 7). Still, the wavelength of the radiation needed to be found.

X-Ray Diffraction

The question of the wavelength of X-rays was settled in 1912 by the German physicist Max von Laue (pronounced *fun* low*ùh*). He reasoned that X-rays could not be

Van Dyck's painting of *Saint Rosalie Interceding for the Plague-Stricken of Palermo,* as viewed in reflected light and in transmitted X-rays. Because the X-rays pass through the painting to reach the photographic film placed behind it, they reveal information about what is beneath the surface. (See photos on p. 1198.)

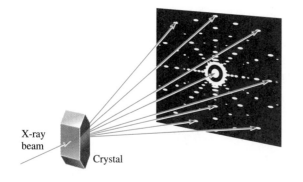

X-ray beam

Crystal

Figure 29.6   X-ray diffraction. A narrow X-ray beam enters a crystal, which diffracts it into many exiting beams. The process is akin to the operation of a 3-dimensional transmission grating.

diffracted by ordinary gratings because their wavelengths were much too small. As a rule, if an array of scatterers is to produce an observable diffraction pattern, their spacing must be of the order of the wavelength of the wave. Von Laue proposed that the ordered array of atoms in a crystal (which at the time were correctly believed to be separated by about 0.1 nm to 0.2 nm) might serve to produce diffraction. The atoms would scatter X-rays in all directions, but because of the regularity of the array, the waves traveling in certain directions would interfere constructively.

A crystal was placed in a narrow X-ray beam in front of a photographic plate (Fig. 29.6). If all went according to plan, the crystal would produce a number of off-axis diffracted beams, and an ordered pattern of dots would appear in the photo. The very first attempt was a resounding success, "proving" both the transverse nature of X-ray waves and the periodicity of the atoms in a crystal. The published accounts of this work caught the attention of W. H. Bragg, professor of physics at the University of Leeds, and his son, W. L. Bragg, a student at Cambridge. They devised a simpler diffraction process now known as Bragg scattering.

Figure 29.7 shows a regular array of atoms. We can imagine various sets of planes passing through the array in lots of different directions and containing different groups of atoms. Each such plane scatters rays in specific directions. For simplicity, let's deal with the set of horizontal planes. A parallel beam strikes the atoms and is scattered every which way. We have already seen that, for each plane, the reflected radiation will reinforce only in the direction such that it leaves the plane at the same angle at which it entered. That was what gave rise to the Law of Reflection (and it's true separately for each plane in the diagram). Thus, in Fig. 29.8 (p. 1096), ray 1 and ray 2 correspond to waves that will exactly reinforce each other. The question now is, What is the condition for rays from the second plane (and all others as well) to reinforce those from the first plane?

The waves associated with ray 1 and ray 3 will be in-phase, provided that their path-length difference is a whole number of wavelengths. Ray 3 travels farther than ray 1 by a distance equal to $2d \sin \theta$. For special values of θ; namely, θ_m, $2d \sin \theta_m$ equals $m\lambda$, where m is a whole number (1, 2, 3, . . .). It follows that constructive interference (Fig. 29.8b) occurs between waves from the various planes only when

$$2d \sin \theta_m = m\lambda \qquad (29.3)$$

This formula has come to be known as the **Bragg equation**. Consistent with the equation (for a fixed d and λ), there may be several angles at which diffraction can

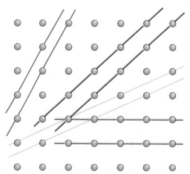

Figure 29.7   Several of the atomic planes off which X-rays may scatter. Notice that d is different in each case.

occur, each corresponding to an integer value of m known as the *order*. Although the angle of incidence equals the angle of reflection as with light, *only those incident angles satisfying the Bragg equation will result in reflection (or diffraction)*, and even then the exiting beam will be quite weak.

Since $\sin \theta$ must be equal to or less than 1

$$\frac{m\lambda}{2d} = \sin \theta_m \leq 1$$

Figure 29.8 (a) The scattering of X-rays from the planes of a crystal. (b) Waves interfering constructively as they scatter off an array of atoms. (c) The geometry of the scattering. Practically, one knows the direction of the incident beam and can measure the scattered beam. Thus, 2θ is determined experimentally.

hence with $m = 1$

$$\lambda \leq 2d$$

Generally d is at most 0.3 nm, which sets a practical limit of 0.6 nm on λ.

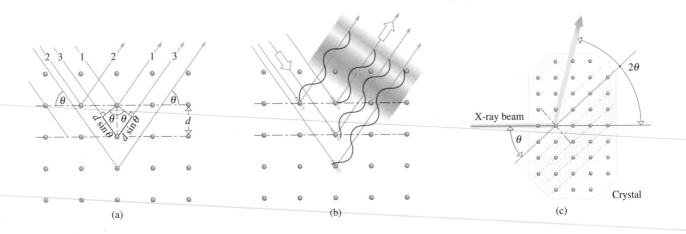

(a)   (b)   (c)

Example 29.2 A beam of quasimonochromatic X-rays with a wavelength of 0.15 nm is incident on a crystal. What is the space between atomic planes if the first-order diffraction maximum occurs at $\theta_1 = 16.0°$?

Solution: [Given: $\lambda = 0.15$ nm, $m = 1$, and $\theta_1 = 16.0°$. Find: d.] From the Bragg equation

$$d = \frac{m\lambda}{2 \sin \theta_m} = \frac{(1)(0.15 \text{ nm})}{2 \sin 16.0°} = \boxed{0.27 \text{ nm}}$$

▶ **Quick Check:** The answer is the right order-of-magnitude (that is, a few tenths of a nanometer). $2d \sin \theta_m = 1.49 \times 10^{-10}$ m $= m\lambda$.

X-rays are generated when high-speed electrons crash into a material target and rapidly decelerate, thereupon emitting radiant energy. A modern X-ray tube (Fig. 29.9) boils electrons out of a hot metal filament in a process called *thermionic emission* (discovered by Edison while he was working with light bulbs). These electrons are then accelerated across a potential difference of from 10^4 V to 10^6 V, in vacuum, whereupon they impact on a metal anode, decelerate, and radiate.

X-ray diffraction has become a routine technique for determining crystal structure. Equally as important, it has become a powerful tool for studying molecules, especially in the case of large repeating configurations such as proteins and DNA. Indeed, it was with the guidance of X-ray data that J. D. Watson and F. H. C. Crick were able (1953) to figure out the double-helix structure of DNA.

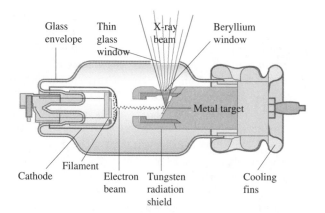

Figure 29.9 A modern X-ray tube. A beam of electrons crashes into a metal target, and X-rays are emitted.

29.4 The Discovery of Radioactivity

Röntgen's chance discovery of X-rays sent much of the scientific community into a flurry of activity. Henri Becquerel, in France, returned to his father's earlier work on fluorescent substances. These absorb and re-emit light, as does the pale green paint on the hands of a watch. The emission of X-rays had been related to the glowing region of the Crookes tube, and Becquerel wondered whether his glowing materials might also send out X-rays. As fate would have it, he began to work with one of his father's compounds, potassium uranylsulfate—a *uranium* salt. He exposed the salt to sunlight, and it actually did emit radiation that could penetrate the heavy paper wrapping on a photographic plate and fog the film.

On a sunless day in the winter of 1896, Becquerel placed the fluorescent compound atop a wrapped film plate as usual but, this time, he put it aside in the darkness of a drawer to wait for clearer skies. A few days later, almost as an afterthought, he developed the plate; there, amazingly, was the blackened outline of the piece of uranium salt! Prior exposure to light was apparently quite unnecessary— the radiation was emitted without it, without the fluorescence. By accident, he had discovered what Madame Curie would later name **radioactivity**, the property of certain substances to give out, all by themselves, penetrating radiation.

Becquerel studied the radiation and reported several of its properties (a few of them erroneously), including that it could discharge a charged electroscope, as could X-rays. By 1897, he turned to other things. Unlike X-rays, these new emanations couldn't produce those exciting pictures of bones, and they were generally greeted with indifference. This limbo of neglect lasted for about a year and a half until the problem was taken up by a promising young student at the Sorbonne in search of a Ph.D. problem. Marie (Manya) Sklodovska Curie accepted the challenge and began her life's work.

Mme. Curie determined that thorium, like uranium, was radioactive. She and her husband, Pierre, discovered an element about 400 times more active than uranium and called it *polonium* (the name deriving from Marie's homeland, Poland). No one knew at the time that the radiation was essentially debris flung out from the nucleus of an unstable atom as it spontaneously readjusted itself. Even so, Marie was the first (1898) to report "that radioactivity is an atomic property." Working for 45 months under the most dreadful conditions in an abandoned, leaking wooden shed, the Curies finally isolated yet another radioactive element: *radium*. After four years of incredible labor, they had distilled from over a ton of the uranium ore a mere 1/30 ounce of the pure radium salt. But this new element was two million

Entitled "Radium," this lithograph appeared in the December 22, 1904, issue of *Vanity Fair*. It depicts the Curies, Pierre holding a sample of the glowing element and Marie standing properly behind him.

Marie Sklodovska Curie (1867–1934). She is the only person to win Nobel prizes in physics and chemistry: in 1903 for her work on radioactivity, and in 1911 for her discovery of two new elements.

times more radioactive than uranium. Over the years, a single ounce of radium would spew out an amount of energy equal to that of burning ten tons of coal. That quietly blazing precious salt stays warm from its own internal heat and glows, self-luminous, like a firefly. The term *atomic energy* entered the vocabulary of science.

Marie received her Ph.D. at the Sorbonne on the basis of this epochal work. Within six months she, Pierre, and Becquerel shared the 1903 Nobel Prize in physics. Pierre was probably already suffering from radiation sickness (he used to keep the vials of radium in his pockets), and they were both too exhausted to even attend the ceremony in Stockholm. Three years later he was dead—run down in the street by a horsecart. In 1934, after a long illness, Marie Curie (the only person to ever win Nobel Prizes in both physics and chemistry) died of leukemia, a victim of many years of overexposure to radiation—too close to the firefly. Even the pages of her lab notebook were later found to be contaminated with radioactive fingerprints.

Alpha, Beta, and Gamma

Ernest Rutherford was twenty-four when he arrived at the Cavendish Laboratory in 1895. There, he began to study "Becquerel rays" and found that they were "complex, and that there are present at least two distinct types of radiation—one that is very readily absorbed, which will be termed for convenience α [**alpha**] radiation, and the other of a more penetrative character, which will be termed the β [**beta**] radiation." Only a short time before, Sir William Ramsay had discovered helium on Earth in a uranium-bearing mineral (and, along with F. Soddy, determined that it was liberated from radium). Thus, there was an early connection between α-rays and helium. (An alpha particle is a helium nucleus, two protons and two neutrons, but that would not be known for years.)

In 1903, Rutherford succeeded in deflecting α-rays using strong electric and magnetic fields, thus "proving" that they were positively charged particles (Fig. 29.10). Measurements of the charge-to-mass ratio of α-particles produced a value several thousand times smaller than that of the electron, again suggesting a large mass. Working with a young Ph.D., Hans Geiger (of Geiger counter fame, p. 1106), Rutherford concluded that the α-particle carried a charge of $+2e$, that is, twice the positive fundamental charge (twice the charge of the hydrogen ion).

In a beautiful experiment (1908), Rutherford and Royds captured alpha particles (Fig. 29.11). Alpha-emitting radon gas was collected above mercury in a thin-walled tube. After a week, the alpha particles that had passed into the surrounding vacuum region were pushed up into a capillary tube by raising the mercury level. On electrically exciting the accumulated gas, they found it to produce the characteristic emission spectrum of helium. It was not yet clear what atoms were, but "the α-particle, after it has lost its positive charge [by gaining two orbital electrons], is a helium atom." *An alpha particle is the nucleus of the helium atom.*

The work on β-rays progressed quickly in a number of European laboratories: they were deflected by magnetic fields (1899), found to have a negative charge (1900), revealed to possess a charge-to-mass ratio very near that of cathode rays (1900), and finally were determined to have a mass equal to that of the electron (1902). *Beta rays are electrons.*

P. Villard, in Paris, also studied the emissions from radium and, in 1900, reported the presence of highly penetrating rays. They were not even slightly deflected by strong magnetic fields, thus suggesting a kinship with light. Gradually, the evidence grew that this γ (**gamma**) radiation was electromagnetic; somewhat more energetic and shorter in wavelength, but otherwise identical to X-rays. The is-

Ernest Rutherford (1871–1937) came to the Cavendish in 1895. Fresh from New Zealand, this unpolished colonial was the first of the new research students and the first to work with J. J. Thomson himself. Rutherford received the Nobel Prize in chemistry, strangely enough, in 1908.

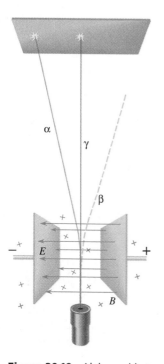

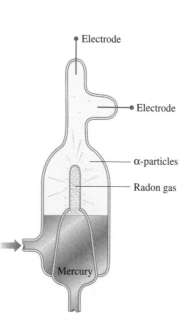

Figure 29.10 Alpha and beta rays, as charged particles, can be bent off course by both electric and magnetic means. Gamma rays, being electro-magnetic radiant energy, are unaffected by these fields.

Figure 29.11 Radioactive radon in the thin-walled chamber emits α-particles, which are captured within the thick outer vessel. When the mercury level is raised, the alphas are forced into the discharge tube at the top. There, a helium spectrum is produced.

sue was settled in 1914 when Rutherford and Andrade succeeded in reflecting γ-rays off the surface of a crystal.

Gamma photons (typically 0.01 MeV to 10 MeV) are highly penetrating. They can be completely absorbed only after passing through several feet of concrete or about one to five centimeters of lead. More potent (and more dangerous) than X-rays, gamma rays have no trouble whisking through a human body and destroying molecules along the way. Beta particles (0.025 MeV to 3.2 MeV), traveling at roughly 25%c to 99%c, can sail through upwards of 15 meters of air. A millimeter or so of aluminum will block them totally, and they will not burrow very deeply into people ($\approx$1 mm to 2 cm). They are about 100 times more penetrating than alphas, though far less so than gammas.

Alpha particles (4 MeV to 10 MeV) have a mass of 6.642×10^{-27} kg, roughly 7300 times that of an electron. They're ejected from atoms at formidable speeds (14×10^3 km/s to 22×10^3 km/s)—the alpha particles from radium are emitted at $\approx$15 190 km/s. Easily stopped, alphas will barely make it through a single sheet of paper, or 0.3 cm to 8.6 cm of air. Even so, breathing in radioactive dust can put alpha emitters inside the lungs, where they can be quite lethal. They are very highly ionizing (1000 times more so than β-rays). A 5-MeV alpha can create 40 000 ion pairs in the process of traversing 1 cm of dry air. As a result, alphas lose their energy rapidly and are especially hazardous to biological organisms (perhaps 20 times more damaging than beta or gamma radiation). Only very energetic alphas

A smoke alarm. A small amount of a radioactive α-emitter ionizes the air between two parallel metal plates. A voltage across the plates causes the ions to drift, creating a current. The presence of smoke cuts off the current and triggers the alarm.

TABLE 29.2   Ionizing radiation sources in the United States

Sources	Percent
Human activity	Total 18%
Medical and dental X-rays	11
Nuclear medicine	4
Consumer products	3
Occupational	0.3
Fallout	<0.3
Nuclear fuel cycle	0.1
Miscellaneous	0.1
Natural sources	Total 82%
Radon	55
Internal body emission	11
Terrestrial minerals	8
Cosmic rays	8

(>7.5 MeV) can penetrate human skin, and external radiation is usually not a problem (Table 29.2).

THE NUCLEAR ATOM

Today, we know beyond any doubt that an atom is composed of a central dense core of positive charge—the nucleus—surrounded by a fast-moving cloud of electrons. This nuclear model of the atom was advanced by Rutherford, but others before him paved the way. Earnshaw (1831) had argued that no system of charges interacting via an inverse-square-law force could be in a stable static equilibrium—if they're bound by a Coulomb force, *the particles that constitute the atom must be in motion.* It followed from the cathode-ray experiments that those particles "must" be electrons. J. Larmor (1900) proposed a vague scheme with *electrons attracted to a central positive charge* and, soon after that, J. Perrin (1901) put forth a Solar System model with *orbiting electrons.* But as we'll see, there were serious problems with the stability of all such systems—*orbiting electrons are accelerating and so should radiate,* lose energy, and collapse into the positive center.

Lord Kelvin in 1902 proposed that the atom might be imagined as a jellylike sphere of positive charge, embedded throughout with an equal negative charge in the form of electrons, like the raisins in a glob of pudding. The following year J. J. made a thorough investigation of the stability of such a scheme, and as a result, it came to be known as the "Thomson atom." This *raisin-pudding atom* was widely considered, for want of anything better.

29.5   Rutherford Scattering

One day in early 1909, Geiger went to Rutherford to ask if his student, an undergraduate named Marsden, might be allowed "to begin a small research." Being of a similar mind, Rutherford suggested a simple scattering experiment, whose results he was sure he knew beforehand. Rutherford had been the first to scatter α-particles from matter, and Geiger had done some work in the area as well. The idea was to study the way alpha particles were deflected as they traversed a thin foil, and from that, possibly learn something about the hidden structure of the atoms of the target.

A few milligrams of a radium compound in a hollow lead tube served as the gun, shooting α-particles in a well-defined beam. The target, an exceedingly delicate gold foil about 0.000 06 cm (about 20 millionths of an inch) thick, corresponded to only ≈1000 layers of atoms. Alphas that easily penetrated the foil would slam into a distant zinc sulphide screen (Fig. 29.12). With each impact, the screen would give off a minute flash of light that could be seen in a totally dark room and counted with great effort. (The experiment is nerve-wracking, to say the least.) Positioning the detector straight down in the forward direction, the experimenters had found that very few alphas were even slightly deflected away from the original beam, which for the most part went right through undeviated. That finding was reasonable enough; after all, the alpha particles were very massive and moving exceedingly fast. They would presumably sail right through the tenuous positive pudding and could hardly be pulled aside appreciably by the minute electron raisins

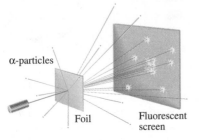

Figure 29.12 Rutherford's alpha particle scattering experiment.

α-particles

Foil

Fluorescent screen

(Fig. 29.13). So, when Rutherford suggested that Marsden look for alphas scattered through large angles (greater than 90°), he knew the eager undergraduate simply wouldn't find any.

Two or three days later, Geiger rather excitedly rushed back to "Papa," as they called the great man in those days, to tell him the unbelievable news: "We have been able to get some of the alpha particles coming backwards." Rutherford recalled, "It was almost as incredible as if you fired a 15-inch shell [that is, in diameter] at a piece of tissue paper and it came back and hit you." It took about two years for Rutherford to work out the details of a theory that would explain those extraordinary observations (Fig. 29.14).

It was clear from the start that each alpha particle could be blasted backward toward the source by a head-on collision with a concentrated, highly charged, positive, massive object—the **atomic nucleus**. We know from the study of elastic collisions (p. 344) that a projectile (the α-particle) will bounce back off a target (a nucleus) only when its mass is exceeded by that of the target (Fig. 9.22b). This kind of collision must be a very rare event (the nucleus must be exceedingly small, or else the undeviated penetration of the bulk of the beam would never occur). To be sure, the back-scattering happened only about once in every 10 000 or 20 000 impacts. Nonetheless, there were many millions of alphas being fired (a gram of radium undergoes about 4×10^{10} atomic disintegrations every second).

"Papa" saw the connection with the arching flight of a comet in the Sun's attractive gravitational field. In time, he was able to derive an equation for the scattering that would occur in a repulsive interaction between a target and projectile when both were positively charged. The α-particles sail off in hyperbolic orbits (Fig. 29.14). One morning in 1911, Rutherford happily sauntered into Geiger's lab to share one of the great secrets of the Universe: ***Each atom consists of a tiny massive concentration of positive charge, the nucleus*** (a term he introduced in 1912), *surrounded by a distribution of electrons*. Geiger immediately began to test the theoretical predictions—the specific dependence of the scattering on foil thickness, nuclear charge, alpha velocity, etc. Within a year, his measurements would convincingly bear out the power of the image, however fuzzy: the nuclear atom had arrived.

The Size of the Nucleus

Atoms were known to be about 10^{-10} m across (p. 362). Now Rutherford's scattering experiments provided a means of approximating the size of the nucleus. Consider a head-on collision between an α-particle and a nucleus, which we suppose to contain a number of positive charges Z, each of magnitude e. The α-particle initially sails in with a kinetic energy of

$$\mathrm{KE}_\alpha = \tfrac{1}{2} m_\alpha v_\alpha^2$$

and it gradually slows down as it approaches closer and closer to the repelling nucleus. At any distance r from the nucleus, the electric potential is given by Eq. (18.8); namely

$$V = \frac{kQ}{r}$$

where $Q = Ze$, the nuclear charge (assumed to be a point, which it isn't). Thus, at any distance r, the alpha has an electric potential energy (gravity is negligible) equal

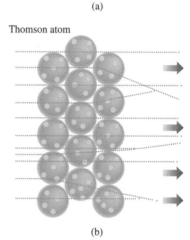

Figure 29.13 (a) The raisin-pudding atom—negative electrons embedded within and throughout a large positive fluff. In its simplest form, the electrons hover in circular patterns within fixed planes. (b) Bombarding α-rays should pass right through the raisin-pudding atom with very little deflection.

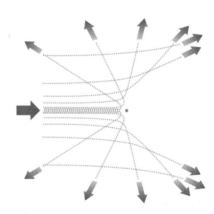

Figure 29.14 The scattering of a beam of α-particles by a positive massive nucleus.

to its charge $2e$ times V; that is

$$\text{PE}_\alpha = \frac{k(Ze)(2e)}{r}$$

This is only approximate since it neglects the cloud of orbital electrons (which will soon be penetrated anyway). The alpha will reach a point of closest approach to the nucleus at a distance $r = R$, where it momentarily stops (as if it had compressed a spring and is about to pop backward). At that point, its initial KE has been transformed into PE:

$$\text{KE}_\alpha = \text{PE}_\alpha$$

$$\tfrac{1}{2} m_\alpha v_\alpha^2 = \frac{k2Ze^2}{R}$$

and

$$R = \frac{4kZe^2}{m_\alpha v_\alpha^2} \tag{29.4}$$

This distance represents an upper limit on the nuclear radius. Thus, given $v_\alpha = 1.5 \times 10^7$ m/s and $m_\alpha = 6.6 \times 10^{-27}$ kg, we have to come up with a value of Z. At the time when Rutherford was doing these experiments, it was reasonable to guess that (as with the α-particles themselves) a nucleus would contain a number of charges equal to half the atomic weight; that is, $Z = \tfrac{1}{2}197 \approx 99$. Actually, for gold, $Z = 79$, but that knowledge would have to wait until the work of Moseley (p. 1139). We can use either number; the choice will not affect things much. Hence

$$R = \frac{4(9.0 \times 10^9 \text{ N·m}^2/\text{C}^2)(79)(1.60 \times 10^{-19} \text{ C})^2}{(6.62 \times 10^{-27} \text{ kg})(1.5 \times 10^7 \text{ m/s})^2} = 4.9 \times 10^{-14} \text{ m}$$

The nucleus is about ten thousand times smaller than the atom. Thus, if in imagination the nucleus of a typical atom is enlarged to the size of an apple (10 cm), the atom as a whole would be about 1 km across, most of it a vast emptiness.

In several regards Rutherford was quite lucky. He happened to be using both low-energy alphas (≈ 5 MeV) and targets with large nuclear charges. As a result, the alphas had enough kinetic energy to easily pierce the surrounding atomic electron clouds (which could then be neglected) rather than being appreciably scattered by them. Still, the alphas were not energetic enough to penetrate deeply into the so-called *nuclear Coulomb barrier*. In other words, working against the Coulomb repulsion, they never got much closer than about 5×10^{-14} m to the center of the nucleus. Had the α-particles been traveling with much more initial energy (say, around 25 MeV or so) Rutherford's results would have been very different; at shorter and shorter distances, the scattering deviates more and more from a purely Coulomb interaction (p. 1182). What he could not have known was that there is an additional nuclear force lurking at very short ranges and that its influence would have messed up his neat, simple understanding of what was happening; that challenge would come, but happily only later.

Another bit of good fortune for Rutherford was that the alphas were moving slowly enough to be in the classical rather than the relativistic regime. Given this case, the correct quantum-mechanical behavior for a Coulomb potential is identical to the predicted classical scattering. In short, Rutherford was lucky, and we, rather fortuitously, were given the nuclear atom.

Rutherford never satisfactorily addressed three outstanding issues concerning his model: the question of the stability of the atom, the necessity to explain atomic spectra, and the need to understand the Periodic Table within the context of atomic structure. That too would come later.

29.6 Atomic Spectra

In the late 1800s, there developed a body of research called *spectroscopy* that provided an accurate means of distinguishing between atomic species. The formulas describing the patterns of spectroscopic phenomena are of little interest to us in themselves. But they reflect, with extraordinary precision, the hidden structure of the atom. Ultimately, those formulas will serve as one of the most powerful tests of atomic theory (p. 1131).

Spectroscopy began with Newton and his experiments using white light and prisms, but it didn't become an analytic tool for exploring matter until the 1800s. A. J. Ångström, in 1853, used a discharge tube filled with various gases to study their spectra. Light from a specimen is made to pass through a slit in a screen and then through a prism or grating that separates the narrow beam into its constituent color bands, the **spectral lines** (Fig. 29.15). When a gas is excited, it emits specific wavelengths, and we see colored lines on a black background. This is known as the **emission spectrum** (Fig. 29.16). Inversely, when white light passes through the same gas, the atoms absorb those same specific wavelengths. This is the **absorption spectrum**, and we see black lines on a smoothly varying, bright-colored background. Ångström first measured the wavelengths of the four bright visible emission lines of hydrogen. Because hydrogen is the simplest of all the elements, these data would play a special role in the later development of theoretical atomic physics. J. Plücker gave the lines the names appearing in Table 29.3 and, by 1858, he had rightly suggested that the spectrum of any substance was a fingerprint that unambiguously specified its identity.

The very existence of spectral lines (and the knowledge that light was an oscillatory wave of some sort) suggested that atoms had a complex structure that could sustain many different internal vibrations. Maxwell pointed *that* fact out in 1875 but, even before then, George Stoney noticed that the hydrogen spectral wavelengths formed simple ratios:

$$H_\alpha : H_\beta : H_\delta = \frac{1}{20} : \frac{1}{27} : \frac{1}{32}$$

Stoney talked about, "the 32nd, 27th, and 20th harmonics of a fundamental vibration" of the atom. In 1885, Johann Balmer, having investigated the notion of atomic

Figure 29.15 (a) The formation of spectral lines by a prism. (b) The emission spectra of, from top to bottom, hydrogen, helium, and mercury.

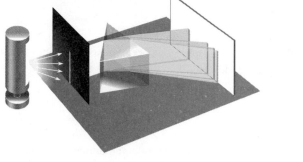

(a)

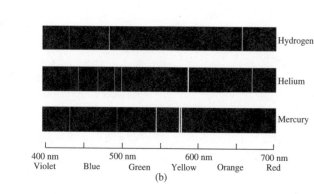

400 nm	500 nm	600 nm	700 nm
Violet	Blue Green	Yellow Orange	Red

(b)

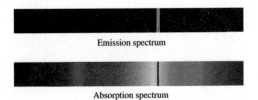

Emission spectrum

Absorption spectrum

Figure 29.16   Emission and absorption spectra of sodium.

TABLE 29.3   Hydrogen visible spectra

Line	Wavelength (nm)	Color
H_α	656.28	red
H_β	486.13	blue-green
H_γ	434.05	violet
H_δ	410.12	violet

harmonics, published a simple formula that rather remarkably yielded the observed hydrogen wavelengths. In its modern form, it is

$$\frac{1}{\lambda} = R\left[\frac{1}{2^2} - \frac{1}{n^2}\right], \qquad n = 3, 4, 5, \ldots \quad (29.5)$$

The equation generates the wavelengths of the various *visible* lines—of what has come to be known as the **Balmer series**—when we substitute in turn, $n = 3$ or 4 or 5, and so on. When λ is expressed in meters, R, which is called the **Rydberg constant**, is

$$R = 1.097\,373\,15 \times 10^7 \text{ m}^{-1}$$

which is quite close to the number Balmer came up with. Note that the series continues with the wavelengths getting shorter and shorter as $n \to \infty$ at the *series limit* (Problem 19); in time, this too, was confirmed experimentally.

Example 29.3   Use Balmer's formula to compute the wavelength of the red line in the hydrogen spectrum.

Solution: [Given: Balmer's formula. Find: λ for red line.] The red line has the longest wavelength and the lowest n; namely, $n = 3$. Thus

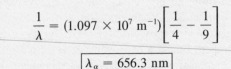

and

$$\boxed{\lambda_\alpha = 656.3 \text{ nm}}$$

▶ **Quick Check:** This λ is appropriate for red light and, of course, it agrees with Table 29.3.

The first term in the brackets of Eq. (29.5) contains a denominator of 2^2, and Balmer was so sure of himself that he proposed (off the top of his head, as it were) that there might well be other sets of lines for 1^2, 3^2, 4^2, and so on. As luck would have it, he was essentially right, and in time the **Lyman series**

$$\frac{1}{\lambda} = R\left[\frac{1}{1^2} - \frac{1}{n^2}\right], \qquad n = 2, 3, 4, \ldots \quad (29.6)$$

in the ultraviolet was observed. As were the **Paschen series**

$$\frac{1}{\lambda} = R\left[\frac{1}{3^2} - \frac{1}{n^2}\right], \qquad n = 4, 5, 6, \ldots \quad (29.7)$$

and two others (with $1/4^2$ and $1/5^2$) in the infrared (p. 1135).

It would take until 1913 before Niels Bohr could explain spectral lines in terms of atomic transitions between energy levels. We will deal with that discovery in the next chapter.

29.7   The Proton

Recall the cathode-ray experiments in which a cathode and anode were placed in a tube containing a low-pressure gas. A high potential across the electrodes caused the gas to become ionized, and a beam of electrons accelerated away from the cathode

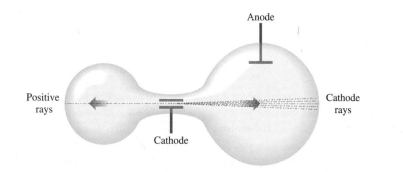

Figure 29.17 Cathode rays stream out toward the general location of the anode. Positive rays sail toward, through, and beyond the cathode.

out in the general direction of the anode (Fig. 29.17). It was noticed in 1886 that the glowing trail of a beam could also be projected in the opposite direction. The backward streaming of that emanation, along with the observation that it could be deflected by a magnetic field, was convincing evidence that these rays were positive particles; J. J. Thomson christened them *positive rays*. When their charge-to-mass ratio was measured, several rather low values were found (indicative of particles with large masses), on the order of those of the atoms themselves.

After the Rutherford model had been put forward, it was reasonable to suppose that positive rays were simply ionized atoms. Naturally enough, the e/m ratios depended on the particular trace gas in the tube. When Thomson performed the experiment with hydrogen (1907), he found two values of e/m, the larger for the light H^+ ion and the smaller for the ionized molecule H_2^+. The former compared nicely with the value arrived at via electrolysis. More and more it seemed that the positive nuclear particle (the thing many people called the H-particle) was the hydrogen atom stripped of its single orbital electron.

The clincher came when Rutherford, pounding the nuclei of various elements with alpha particles, discovered H-particles among the flying fragments. The apparatus (Fig. 29.18) consisted of an evacuated chamber containing an alpha source. The chamber was sealed with a piece of foil just thick enough to stop all the alphas. Nitrogen gas was then introduced into the chamber (later on, fluorine and other elements were used; in all cases, the results were the same). Scintillations were observed, indicating the presence of a more penetrating radiation that proved to have the same range and charge as the H-particle. The range in air was easily determined by moving the screen, and it was indicative of the kind of radiation being dealt with. Apparently, this positive particle must be a basic component of many if not all the elements. It wasn't until 1920 that Rutherford formally proposed the name **proton**

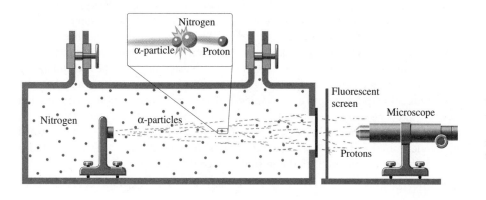

Figure 29.18 Alpha particles crashing into nitrogen atoms liberate protons, which strike the fluorescent screen, causing flashes that can be seen with a microscope. The chamber is sealed with a thin foil that will pass protons but not alpha particles.

(from the Greek *protos* for "first"), a word that had already quietly been in use for a dozen years.

With a mass of

$$m_\mathrm{p} = 1.672\,623\,1(10) \times 10^{-27} \text{ kg} = 938.272\,31(28) \text{ MeV/c}^2$$

the proton (p) is 1836 times the mass of the electron (e). The proton carries a quantum of charge as does the electron, and so these subatomic specks logically seem, if not twins, at least mates. Laboratory tests confirm that the proton and electron charges are indeed equal—in the most precise measurement ever carried out, the hydrogen atom has been found to be neutral to an accuracy of twenty-two decimal places. This equality of charge becomes even more remarkable when we learn that the proton is a complex structured thing; in fact, it remains one of the fundamental puzzles of Modern Physics (p. 1224).

29.8 The Neutron

We picture the hydrogen atom as a single nuclear proton coupled by an attractive Coulomb force to one orbital electron. The next element in the Periodic Table, helium, might then consist of two protons and two surrounding electrons; lithium might have three and three; and so on, all the way up to uranium with 92 protons and 92 electrons. Yet, the α-particle has a charge of $+2e$ and an e/m ratio of half that of the hydrogen nucleus; it has a mass of 4. And this is *not* quite right; something else must be happening in the nucleus. Such differences between **atomic number** (Z) and **atomic mass** (A) occur throughout the Table all the way up to uranium, which has 92 nuclear protons and a mass of 238. There is a good deal of mass unaccounted for when the uranium nucleus is considered to be a cluster of 92 protons. If everything is made up of the only two then-known particles— electrons and protons—we have a little problem here.

Rutherford, in 1920, proposed that all of these difficulties could be resolved by assuming the existence of tightly bound *proton-electron pairs,* neutral units that would add mass but not charge. The idea was wonderfully simple, but quite erroneous. Rutherford even suggested the name *neutron* for the pair-particle. And he promptly set a group of his "boys" to work to track down the neutral object; thus, James Chadwick began a 12-year search.

In 1930, it was discovered that when beryllium (Be) was bombarded with alpha particles, out streamed a flux of very energetic radiation that was assumed to be γ-emission, but which seemed a bit odd in some respects. It was soon determined that, as with γ-rays, the new radiation was both very penetrating and not deflected by a magnetic field. However, unlike γ-rays, it was not ionizing and could not discharge an electroscope. The Joliet-Curies, Irène (Mme Curie's daughter) and her husband, Frédéric Joliot-Curie, observed that the presence of the mystery emission (which they thought was electromagnetic) could be detected in a much enhanced way by making it impinge on a substance rich in hydrogen, that is, protons. When the beam struck a sheet of paraffin, protons were blasted out, and these could easily be picked up with a Geiger counter (Fig. 29.19). (The latter was simply a metal tube containing a low-pressure gas such as argon. A voltage of about 1000 V was applied between the central-wire anode and the tube. Ionizing radiation entered via a thin mica window and knocked electrons out of a few gas atoms. These rushed toward the anode, ionizing more atoms in the process. The resulting discharge created a current pulse that was amplified and sent to a counter.)

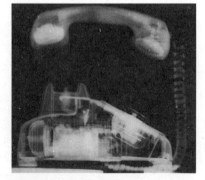

A radiograph of a telephone created with a beam of neutrons. These neutrophotos usually show considerably more of the subtle variations and details than do X-ray pictures, which they otherwise closely resemble.

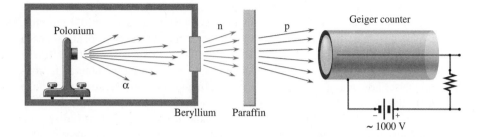

Figure 29.19 Polonium emits alphas, which knock neutrons out of the beryllium. These knock protons out of the paraffin.

Chadwick (1932) argued that a nonionizing particle capable of ejecting protons had to be the long-sought neutron (n). He could determine the speed of the protons from their range and was able to show from collision theory that all of his observations could be understood quantitatively by hypothesizing the presence of a neutral particle having about the mass of the proton. When the alpha bombardment of beryllium was finally understood, Chadwick wrote it out as

$$\,_2^4\alpha + \,_4^9\text{Be} \rightarrow \,_6^{12}\text{C} + \,_0^1\text{n}$$

(mass as the superscript, charge as the subscript). Beryllium absorbs an alpha and, with the emission of a neutron, is transformed into carbon.

Rutherford had been close (but no cigar): electrons don't live in the nucleus, and the neutron is as much a single particle as is the proton, though it is slightly more massive:

$$m_\text{n} = 1.674\,928\,6(10) \times 10^{-27}\ \text{kg} = 939.565\,63(28)\ \text{MeV/c}^2$$

Very soon after the announcement of the discovery of the neutron, Werner Heisenberg proposed the now accepted neutron-proton model of the atomic nucleus. *All nuclei are composed exclusively of neutrons and protons*, the total number of which (A) determines the atomic mass. Thus, uranium has 92 protons (an atomic number Z of 92) and 146 neutrons, yielding an atomic mass (A) of 238—everything works out neatly, at last.

Once removed from the nucleus, the neutron is unstable, decaying into an electron, a proton, and a neutrino (actually an antineutrino, p. 1190) at a rate such that half of the neutrons present at any moment will decay within roughly 10.4 minutes. This process is an example of β-decay, and it explains how electrons can be ejected from a nucleus wherein they do not exist—they are a byproduct of the transformation, created in the process. While nestled with the protons in the nucleus, the neutrons interact with (and seem to be more or less stabilized by) the others, so they don't decay in all nuclei. Moreover, neutrons and protons provide a glue (by way of the strong force) that helps keep the nucleus together. To underscore this intimate relationship (and because they are the exclusive components of all nuclei), the neutron and proton are called **nucleons**.

Neutrons are about 1.3 MeV/c^2 more massive than protons, and with this excess rest energy, it is the free neutron that decays "down" to the proton. Had the reverse been true, we could expect free protons to decay into neutrons. Happily for us, it's not the case, and hydrogen, water, and human beings are stable. (A proton can decay into a neutron but only in a fairly large nucleus that can provide the needed mass-energy.)

When nuclei decay, as when neutrons decay, the process is described statistically. We know that half the free neutrons in a box will decay after ≈ 10.4 minutes, but we have no idea which ones will be left. Each survivor seems identical, each seems unaffected by the passage of ≈ 10.4 min. The data suggests that the likelihood of a particular neutron decaying during any ≈ 10.4-minute interval is constant, no matter how many such intervals it has already survived. Neutrons don't age and, when they do decay, it is for no apparent reason. This strange result is the same one Rutherford observed for radioactive atoms, and it suggests that causality (and the accompanying predictability we are so fond of) has given way on some hidden level to the roulette wheel and the rule of probabilities. Today, it is widely believed that our inability to predict the individual behavior of a neutron or a nucleus is not the result of a correctable ignorance but is a fundamental manifestation of the quantum-mechanical nature of the Universe.

29.9 Radiation Damage: Dosimetry

When energetic radiation (generally in the form of positive ions, electrons and positrons, neutrons, and photons) passes through matter it can have a variety of transforming effects. With enough exposure (that is, with the deposition of large amounts of energy), interatomic bonds are broken and atoms are literally displaced; metals become brittle, living cells die, plastics decompose, and semiconductors (and their electronic progeny) deteriorate. Of course, this same ability to destroy can be harnessed to control cancer, sterilize food, introduce useful biological mutations, and produce sunglasses that darken automatically.

Among the most important effects of radiation on biological systems is the ionization of atoms and molecules. An atomic electron in the cloud surrounding the nucleus can be ripped away, leaving behind a positive ion. When that electron is subsequently captured by some other atom, a negative ion is created. The presence of these ion-pairs in living cells can alter the normal chemistry of the cell and have very deleterious effects, including the initiation of cancer.

The subject of *dosimetry* deals with the quantification and measurement of the "amount" of radiation, the so-called *dose* delivered to a system. There are four distinct radiation measurements that must be considered: the *source activity* (p. 1194), which deals with how much radiation is being generated at the place of origin; the *exposure,* which signifies the amount of ionizing radiation arriving at the system under study; the *absorbed dose,* which concerns the energy actually imparted to the system by the incident radiation; and the *biological equivalent dose,* which is associated with the effect the physical dose has on any particular biological system.

Exposure is defined specifically for photons with energies up to 3 MeV. It relates to the amount of ionization produced in a kilogram of dry air at STP. The earliest unit used was the *roentgen* (R), which corresponds to the generation of a specific number of ionization pairs or, equivalently, to the separation of a certain number of coulombs per kilogram of positive and negative charge: $1\ R = 2.58 \times 10^{-4}$ C/kg. Such a measure provides us with a sense of the ionizing intensity of the incident radiation. But it says nothing about the specific system being irradiated, which might even be totally transparent to the radiation.

The *absorbed dose* reflects the amount of the energy imparted to the system via the absorption of ionizing radiation incident upon it. If a one-roentgen photon beam impinges on soft biological tissue, an energy of approximately 0.01 J will be absorbed in each kilogram of material. On average a particle loses about 35 eV per ionization it produces. Accordingly, we define the *rad* (short for radiation absorbed dose) such that

$$1 \text{ rad} = 0.01 \text{ J/kg}$$

This is probably still the most widely used unit of absorbed dose, although the gray (Gy) is now the official SI unit: 1 Gy = 100 rad = 1.0 J/kg. The roentgen pertains only to photons, whereas the rad and gray apply to all particles.

Every different kind of radiation is likely to have a somewhat different effect on a particular biological system (cell, tissue, organ, or organism), and therefore it's not enough to simply say that the absorbed dose is so many rads. For instance, energetic neutrons are especially nasty at causing cataracts (the clouding of the lens in the eye that can lead to blindness). Thus, 1 rad of fast neutrons will inflict the same damage as 10 rads of X-rays. That's because the ionization caused by fast neutrons is concentrated in dense tracks along the path, which causes more serious alterations of the tissue.

To begin to take account of these differences in the *relative biological effectiveness* of the various forms of radiation, a multiplicative **quality factor** (QF) has been introduced. The damage from each form of radiation (which also depends on energy and the specifics of the target) is compared to that induced by 200-keV photons, taken as the standard; the resulting QF values are listed in Table 29.4. The **biological equivalent dose** is defined as the product of the absorbed dose (in rads) times the QF, and the corresponding unit is the rem (which stands for *rad equivalent man*). **One rem of any kind of radiation produces more or less the same amount of biological damage**, namely that of 1 rad of 200-keV photons. Thus 1 rad of 200-keV photons (QF = 1) corresponds to 1 rem and generates the same cataract damage as about 0.05 rad of fast neutrons (QF = 20), which also corresponds to 1 rem.

It is estimated that the average person receives an annual dose of radiation from all sources of about 0.40 rem (see Table 29.2). Of that perhaps 0.13 rem per year is the natural background due to cosmic rays and due to radioactivity in the immediate environment: in rocks, brick, soil, food, cigarette smoke, etc. Beyond that, another roughly 0.07 rem per year is picked up via medical X-rays.

Animals can be classified according to how radiosensitive they are. That's usually done by specifying the $LD_{50/30}$, that is, the single lethal whole-body dose that will result in death of 50% of the animals within a period of 30 days. (An animal surviving 30 days is not likely to expire soon thereafter.) Table 29.5 lists the $LD_{50/30}$

TABLE 29.4 Quality factors (QF) for several kinds of radiation

Radiation	Typical QF
Photons	1
Beta particles (> 30 keV)	1
Beta particles (< 30 keV)	2
Protons (1–10 MeV)	2
Neutrons (< 0.02 MeV)	3–5
Fast neutrons (1–10 MeV)	10 (body)–20 (eyes)
Protons (1–10 MeV)	10 (body)–20 (eyes)
Alpha particles	10–20
Heavy ions	up to 20

TABLE 29.5 Radiosensitivities of several organisms

Species	$LD_{50/30}$ (rem)
Dog	350
Mouse	400–600
Monkey	600
Man	500–700
Rat	600–1000
Frog	700
Newt	3000
Snail	8000–20 000
Virus	>100 000

TABLE 29.6	Immediate effects of whole-body single exposure to radiation—adults
Dose (rem)	**Effect**
0–10	No observable short-term effects.
10–100	Slight blood changes, decrease in white cell count. Temporary sterility at 35 rem for women and 50 rem for men.
100–200	Significant reduction in blood platelets and white cells (temporary). Some nausea, vomiting.
200–500	Severe blood damage, nausea, vomiting, hemorrhage, hair loss, death in many cases.
500–2000	Malfunction of small intestine and blood systems. Death in less than 2 months in most cases, but survival possible if treated.

for several species. The likelihood of death and the time it takes to occur both depend on the dose. The MST is the *mean survival time* for any whole-body dose. For humans, a whole-body dose of around 100 rem generally causes mild radiation sickness (Table 29.6). The few fatalities that occur have an MST of around 15 days. As the dose increases, the MST stays fairly constant, although the likelihood of demise gradually increases. Received in one dose, 250 rem will result in severe radiation sickness. A dose of 500 rem is likely to be lethal in 50% of the occurrences (with an MST that's still 15 days). All of these unfortunates succumb from damage to the bone marrow. As the dose increases further, the MST drops, reaching another plateau of three or four days at a dosage level of 1000 rem. Death now occurs in almost 100% of the untreated cases and is due to gastro-intestinal damage, which manifests itself more rapidly than bone marrow deterioration. With doses beyond 10 000 rem or so, the MST drops to a few hours and death results from the destruction of the central nervous system. At still higher doses, death occurs almost simultaneous to the irradiation from the direct destruction of molecules vital to the support of life.

Core Material

A faraday of charge (96 485 C) corresponds to

$$F = N_A e \qquad [29.1]$$

For single-valence elements, 9.65×10^4 C will liberate 1 mol, or equivalently, 9.65×10^7 C will liberate 1 kmol (p. 1088). The **charge-to-mass ratio** for singly ionized hydrogen (H^+) is

$$\left(\frac{e}{m}\right)_{H^+} = 9.58 \times 10^7 \text{ C/kg}$$

The charge-to-mass ratio of the electron is

$$\frac{e}{m_e} = \frac{2yE}{L^2 B^2} \qquad [29.2]$$

The accepted value of that ratio for electrons is

$$\frac{e}{m_e} = 1.7588 \times 10^{11} \text{ C/kg}$$

where $e = 1.602\,177\,33 \times 10^{-19}$ C and $m_e = 9.109\,389\,7 \times 10^{-31}$ kg. The **Bragg equation** (p. 1095) for X-ray diffraction is

$$2d \sin \theta_m = m\lambda \qquad [29.3]$$

Each atom consists of a tiny massive concentration of positive charge, the nucleus, surrounded by a distribution of electrons (p. 1100). *The nucleus is at least ten thousand times smaller than the atom.* The **Balmer series** for the spectrum of hydrogen is

$$\frac{1}{\lambda} = R\left[\frac{1}{2^2} - \frac{1}{n^2}\right] \qquad n = 3, 4, 5, \ldots \qquad [29.5]$$

wherein $R = 1.097\,373\,15 \times 10^7 \text{ m}^{-1}$

The masses of the proton and neutron are

$$m_p = 1.672\,623\,1(10) \times 10^{-27} \text{ kg} = 938.272\,31(28) \text{ MeV}/c^2$$

and

$$m_n = 1.674\,928\,6(10) \times 10^{-27} \text{ kg} = 939.565\,63(28) \text{ MeV}/c^2$$

All nuclei are composed exclusively of neutrons and protons.

Suggestions on Problem Solving

1. Return to Fig. 29.8c and notice that the angle between the transmitted and diffracted beams is 2θ. It is this angle that is usually measured experimentally, and it is called the *diffraction angle*. Among the first things to determine in a given X-ray diffraction problem is which angle is being provided or asked for.

2. When dealing with the spectral series for hydrogen, the order is Lyman (1), Balmer (2), Paschen (3), Brackett (4), Pfund (5); these are the numbers that are squared in the first term of the denominators for $1/\lambda$ [that is, Eq's. (29.5–29.7)]. The first spectral line in any particular series is the next number up;

for example, substituting $n = 3$ into Eq. (29.5) generates the first Balmer line. Remember to square those integers. Once again, we have one-over the quantity of interest (that is, $1/\lambda$), and it's common to forget to invert the result to get the final answer—be alert to this.

3. Remember that it's commonplace to specify the uniform electric field between parallel plates via $E = V/d$; that is, V and d are given, and it is assumed you therefore know E. Because d is small, it is often provided in centimeters; make sure to convert it into meters—failure to convert is a common error.

Discussion Questions

1. A molten bath of NaCl has a current passed through it. Chlorine gas bubbles off at the anode, and sodium is deposited at the cathode. Explain what's happening. Why did we start with molten NaCl rather than a water solution?

2. The chemical cell in Fig. Q2 depicts a silver nitrate solution, a silver anode, and a copper cathode. Explain what will happen when the switch is closed. How is it that the concentration of the solution remains unchanged? Describe the nature of the current both in the solution and in the external circuit. What, if anything, happens to the silver of the anode?

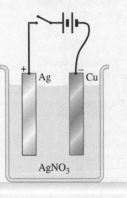

Figure Q2

3. Let's free our imaginations and speculate about the nature of the electron. Experiments to date have shown that if the electron has a "hard core," it's smaller than 10^{-16} cm. What are your feelings about what the electron "looks like"; that is, size, shape, charge distribution, etc? What questions immediately come to mind? Incidentally, no one knows what configuration it has or even if it has one. Nor do we know if it's finite or simply a point (whatever that means). We do believe that it is an elementary particle (that is, it has no smaller constituent parts). How does mass enter in your picture?

4. An equation for the mass M of material liberated at either electrode when a net charge of Q is passed across an electrolyte is

$$M = \frac{(Q)(\text{mass per mole})}{(\text{F})(\text{valence})}$$

Explain how this expression comes about.

5. Beyond their dynamical parameters (that is, velocity, momentum, etc.), is it reasonable to assume that all electrons are identical?

6. Compare X-ray diffraction from a crystal with the reflection of light from a smooth surface. How do the two processes differ? Although we speak of the "reflection" of X-rays, that description is somewhat misleading. Why?

7. Using detectors that were not themselves sensitive to polarization, Barkla (1906) performed a crucial experiment that showed that X-rays could be polarized. His arrangement consisted of two blocks of carbon as depicted in Fig. Q7 and an

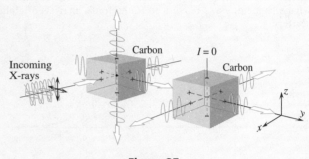

Figure Q7

incident unpolarized beam (entering at the left along the x-axis). From what you learned via Fig. 27.12, explain why the absence of scattered radiation from the second block outside the xy-plane proved the wave nature of the process.

8. Figure Q8 depicts a monochromatic beam of X-rays incident on a single crystal. Explain what is happening.

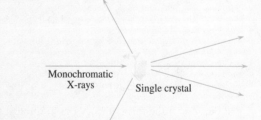

Figure Q8

9. Consider X-ray diffraction using both a single crystal and a chamber holding a monatomic gas. Figure Q9 shows the

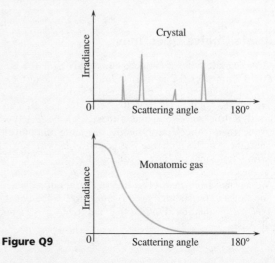

Figure Q9

irradiance reaching a detector as a function of the angle from the beam axis. Explain the features of the two curves and why they are so different. What would the curve look like for a liquid or amorphous solid?

10. Think of the nucleus as a uniformly charged positive sphere. Figure Q10a depicts the electric field for two different-sized spheres carrying the same total charge. Explain the diagram. Part (b) of the figure shows a head-on collision between each of these spheres and a small positive particle. Explain what's happening and relate it to Rutherford's analysis of alpha scattering.

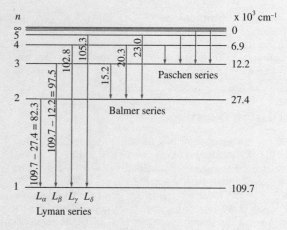

Figure Q11

97.5, 102.8, and 105.3. Thus, the difference between the first two wave numbers in the Lyman series is the H_α wave number in the Balmer series, which is

$$97.5 - 82.3 = 15.2$$

Figure Q11 displays several of these remarkable relationships and we see that the H_α line corresponds to what might be called a *transition* from the *3rd* to the *2nd* "level." **All the spectral lines can thus be seen as transitions from one level to another,** which is essentially the *Ritz Combination Principle,* first proposed in 1908. What is the significance of the lowest level (109 677), of the next lowest level (27 419), and so on? What does all of this suggest about what might be happening in the atoms giving off the light? (The Bohr atom, treated in Sect. 30.5, p. 1131, is predicated on similar reasoning.)

12. The electron, proton, and neutron all have magnetic properties; they each have a magnetic moment (p. 816) and each behaves like a tiny bar magnet. The electron's magnetic moment is roughly 666 times stronger than the proton's, which is about 1.5 times stronger than the neutron's. Experiments show that the magnetic moments of ordinary nuclei are comparable to those of the proton. Does this relationship suggest anything about electrons residing in the nucleus of an atom? Explain.

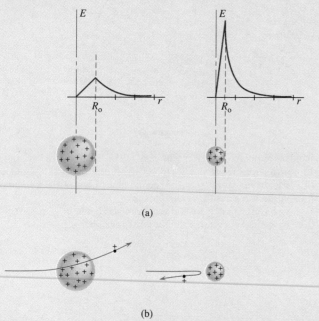

(a)

(b)

Figure Q10

11. Because of the form of Eq's. (29.5) through (29.7), it's useful to define $1/\lambda$ as the so-called *wave number.* It was noticed early on that the differences between certain wave numbers corresponded to other wave numbers. For example, the first four wave numbers in the Lyman series in units of $\times 10^3 \text{ cm}^{-1}$ are 82.3,

Multiple Choice Questions

1. A typical atom has a diameter of roughly   (a) 0.2 mm (b) 0.2 pm   (c) 0.2 m   (d) 0.2nm   (e) none of these.
2. A typical nucleus has a diameter of roughly (a) 1×10^{-14} mm   (b) 1×10^{-14} cm   (c) 1×10^{-14} m (d) 1×10^{-14} nm   (e) none of these.
3. We learned from electrolysis that one atom of a univalent substance carries a charge   (a) equal to F   (b) equal to F/N_A (c) equal to $2F/N_A$ (d) equal to N_A/F (e) none of these.
4. We learned from electrolysis that one atom of a bivalent substance carries a charge   (a) equal to F   (b) equal to F/N_A (c) equal to $2F/N_A$ (d) equal to N_A/F (e) none of these.

5. The magnitude of the charge of the electron is   (a) equal to F   (b) equal to F/N_A (c) equal to $2F/N_A$ (d) equal to N_A/F (e) none of these.
6. When a monochromatic X-ray beam impinges on a crystal (λ and d are fixed), there may be several values of the (a) angle θ at which diffraction will occur, and these correspond to different values of m (b) crystal spacing at which the reflected beam angle will exceed the incident angle   (c) energy at which the reflected beams emerges at an angle in excess of 90°   (d) reflection angle for each incident angle and these correspond to different values of f (e) none of these.
7. The largest Bragg angle through which a beam of X-rays

can be bent is   (a) $0°$ (b) $45°$ (c) $90°$ (d) $180°$
(e) none of these.

8. The light emitted via the excitation of low-pressure hydrogen gas is   (a) a continuous spectrum   (b) made up of a discrete spectrum of bright lines   (c) a mix of bright spectral lines superimposed on a continuous bright background   (d) composed exclusively of X-rays   (e) none of these.

9. Naturally occurring radioactive atoms can spontaneously emit   (a) δ-, γ-, and ξ-rays   (b) α-, β-, and γ-rays   (c) N-rays   (d) X-rays   (e) none of these.

10. Neutrons are highly penetrating because they   (a) are very thin and can slide between atoms   (b) never strike an atom because they twist as they advance   (c) are neutral and do not lose energy via Coulomb interactions with the atoms   (d) always travel at extremely high speeds in comparison to protons or electrons   (e) none of these.

11. Gamma-ray emissions can   (a) be distinguished from β-ray emissions using a magnetic field   (b) not be distinguished from β-ray emissions   (c) not be distinguished from α-ray emissions using a magnetic field   (d) be distinguished from a neutron beam by bending the latter's path via a B-field   (e) none of these.

12. An element emits spectral lines   (a) that are exactly the same as all other elements   (b) that are exactly the same as all other elements in its column of the Periodic Table   (c) that are characteristic of that element   (d) that are evenly spaced   (e) none of these.

13. In reference to the Balmer series, the longest wavelength line   (a) is associated with the smallest n, namely 1   (b) is associated with the smallest n, namely 3   (c) is associated with the largest n^2, namely 9   (d) is associated with the largest n namely ∞ (e) none of these.

14. In reference to the Balmer series, the shortest wavelength line   (a) is associated with the smallest n, namely 1   (b) is associated with the smallest n, namely 3   (c) is associated with the largest n^2, namely 9   (d) is associated with the largest n, namely ∞ (e) none of these.

15. The magnitude of the charge-to-mass ratio of the electron   (a) is zero   (b) is less than that for the proton   (c) is greater than that for the proton   (d) equals that of the neutron   (e) none of these.

16. The proton and the electron   (a) have the same size charge and mass   (b) have the same mass but differ in charge by a factor of about 2000   (c) have the same size charge and differ in mass by a factor of about 2000   (d) differ in both charge and mass by a factor of about 2000   (e) none of these.

17. The neutron   (a) has a mass slightly greater than that of the proton   (b) has a charge slightly greater than that of the proton   (c) has a mass slightly greater than that of the electron   (d) has a mass slightly less than that of the proton   (e) none of these.

Problems

SUBATOMIC PARTICLES
THE NUCLEAR ATOM

1. [I] How many electrons would you get if you bought a gram of them? How does that compare with the total number of stars in the entire Universe ($\approx 10^{22}$)?

2. [I] Determine the total mass of all the electrons in 2.00 g of hydrogen (H_2) gas at STP.

3. [I] A current of 965 A passes through a molten solution of NaCl for 100.0 s. How much chlorine gas will be liberated?

4. [I] For single-valence elements, verify that 1 faraday is equivalent to 9.65×10^7 C/kmole.

5. [I] What mass of metallic sodium will be deposited at the cathode if 50.0 A passes through a molten bath of NaCl for 5.00 minutes?

6. [I] How much charge must pass through a molten bath of NaCl in order to liberate 11.2 liters of chlorine gas (Cl_2) at STP?

7. [I] A current of 2.00 A is passed through molten $BaCl_2$ for 5.00 h. How much barium and chlorine will be liberated?

8. [I] In the Millikan oil-drop experiment, a droplet of mass 1.111×10^{-15} kg is held motionless by an electric field of 34.0 kV/m. How many extra electrons is it carrying?

9. [I] A tiny droplet of oil carries an extra charge of e and is held motionless between parallel plates 2.00 cm apart across which there is a potential difference of 60.0 kV. What is the mass of the droplet?

10. [I] A beam of electrons in a 45-kV X-ray tube bombards the target, producing 750 W of thermal energy per second. Nonetheless, only 1.00% of the beam's energy ends up as X-rays. Compute the average rate at which electrons strike the target. [Hint: Use the relationship between power, current, and voltage.]

11. [I] A beam of X-rays having a wavelength of 0.090 nm impinges on an unknown crystal. The strongest reflection maximum is observed to occur at an angle of $25°$. What is the spacing between the atomic planes doing the scattering?

12. [I] A crystal such as that in Fig. 29.8c can be used as a monochromator. It can select a single frequency (at a particular angle) out of a polychromatic incident beam. If the Bragg planes are known to be separated by 1.50 Å (that is, 0.150 nm), what wavelength will have a first-order peak at $30°$ from the beam axis?

13. [I] Determine the angle at which a narrow beam of monochromatic X-rays of wavelength 0.090 nm has a first-order reflection off a calcite crystal ($d = 0.303$ nm).

14. [I] A silver bromide crystal has atomic planes spaced by 0.288 nm. Given that an X-ray beam diffracts at an angle $2\theta = 35.0°$, producing a 1st-order maximum at a detector, what is the wavelength?

15. [I] X-rays reflect off a salt crystal whose atomic planes are 0.28 nm apart. If the first-order maximum is at $22°$ to the planes, what is the wavelength of the radiation?

16. [I] Confirm Stoney's notion that the hydrogen spectral wavelengths formed simple ratios, such as $H_\alpha : H_\beta := \frac{1}{20} : \frac{1}{27}$.

17. [I] Compute (to 4 significant figures) the wavelength of the second line of the hydrogen Balmer series.

18. [I] What is the frequency (to 4 significant figures) of the third line in the Balmer series?

19. [I] Compute (to 4 significant figures) the wavelength that corresponds to the series limit of the Balmer series.
20. [I] Compute (to 4 significant figures) the wavelength of the first line of the Paschen series.
21. [I] Determine (to 4 significant figures) the longest wavelength in the Balmer series for hydrogen.
22. [I] Compute (to 4 significant figures) the wavelength of the second line of the Paschen series.
23. [II] Imagine a block of pure metallic copper with a mass of 63.55 g and a density of 8.96 g/cm^3. Approximate the size of a copper atom.
24. [II] If 0.754 5 g of metallic silver are deposited onto an electrode using a current of 0.500 A for 22.5 minutes, what is the atomic mass of silver? The ions are singly charged. Refer to Discussion Question 2.
25. [II] Use Fig. P25 depicting Thomson's electron-beam apparatus to show that

$$\frac{e}{m_e} \approx \frac{E\theta}{B^2 L}$$

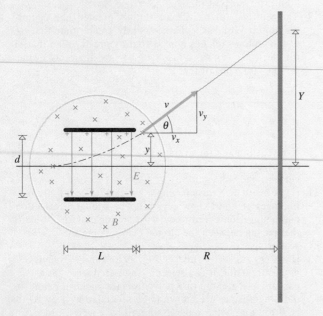

Figure P25

26. [II] If r is the radius of the path taken by an electron in the B-field region of Fig. P25, show that

$$\frac{e}{m_e} = \frac{E}{B^2 r}$$

and prove that this equation is equivalent to the result gotten in Problem 25.
27. [II] An electron in Thomson's apparatus moves under the influence of a B-field along a path with a radius of 15.00 cm. If an E-field of 20.0 kV/m makes the path straight and horizontal, find B.

28. [II] With the previous few problems in mind, knowing e/m for the electron, determine the strength of the B-field that will result in a deflection of 0.25 rad when $V = 180$ V, and $L = 4.5$ cm, and the plates are separated by 1.66 cm.
29. [II] An electron in a cathode-ray tube is emitted from a hot filament at a negligible speed, whereupon it accelerates across a potential difference of 1.00×10^6 V. Determine its final speed. [Hint: The correct answer is about half the classical value.]
30. [II] The continuous X-ray emission from a copper target impinges in a narrow beam on a calcite crystal with an atomic spacing of 0.303 nm. A detector picks up the first strong (first-order) maximum at 24.0° to the beam axis. What is the shortest wavelength present in the radiation?
31. [II] A narrow beam of X-rays of wavelength 0.200 nm impinges on a crystal as in Fig. 29.8c. The detector picks up the first-order maximum at an angle of 50.0° from the beam axis. What is the spacing and orientation of the atomic planes causing the reflection?
32. [II] X-rays of unknown wavelength are incident on a nickel single crystal having an interatomic separation of 0.215 nm, as shown in Fig. P32. The diffracted beam has a first-order maximum at an angle $\phi = 45°$. Determine the wavelength. [Hint: First compute d. Note that the planes *do not* run diagonally across each little square of atoms.]

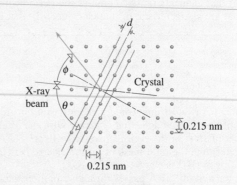

Figure P32

33. [II] A beam of 0.020 0-nm X-rays impinges on a crystal having an array of Bragg planes separated by 1.22×10^{-10} m. What is the highest-order reflected maximum?
34. [II] A polychromatic narrow beam of X-rays impinges on a crystal at 30° to a set of atomic planes. What wavelengths will be reflected at 30° if the atomic spacing is 0.050 0 nm?
35. [II] An alpha particle of mass 6.6×10^{-27} kg is fired at a speed of c/20 directly at an iron nucleus. About how close will it come to the center of the nucleus?
36. [II] What is the shortest wavelength of the Lyman series of hydrogen? Give your answer to 5 significant figures.
37. [II] If light is defined to correspond to the wavelength range from 390 nm to 780 nm, what is the shortest wavelength of the Balmer series that falls within that range?

38. [II] What is the longest wavelength of the Lyman series of hydrogen? Give your answer to 5 significant figures.

39. [III] Use Fig. P25 depicting Thomson's electron-beam apparatus to show that

$$Y = \left(\frac{e}{m_e}\right)\frac{B^2 L}{2E}(L + 2R)$$

40. [III] An oil droplet carrying a net charge of Q and having a mass m falls in air at a steady vertical terminal speed between two vertical parallel plates separated by a distance d. When a potential difference V is applied across the plates, the droplet moves uniformly at an angle θ with the vertical. Show that

$$\tan \theta = \frac{VQ}{mgd}$$

41. [III] An imperfect crystal has Bragg "planes" spaced on average by 0.100 nm, which, instead of being fixed everywhere, vary by ±0.001 nm from place to place. A polychromatic beam of X-rays diffracts from the crystal into a first-order peak at an angle of 30°. Determine the spread in the wavelengths of the emerging first-order beam at 30°.

42. [III] Rock salt (NaCl) is a cubic crystal with a molecular mass of 58.5 and a density of 2.16×10^3 kg/m³. Approximate the space between its atoms.

30

The Evolution of Quantum Theory

QUANTUM MECHANICS DEVELOPED IN two fairly distinct phases. First came the shocking basics: the appreciation of Planck's Constant, the quantum of action; the discovery of the quantization of energy and then, later, momentum; the conception of the photon; a crude but effective atomic model; and the wave-particle duality of matter. All these ideas form the body of knowledge known as the Old Quantum Theory. Building on that foundation, from 1925 to the present, physics moved into the contemporary, more theoretically abstract and mathematically driven stage of Quantum Mechanics, which will be discussed in Chapter 31.

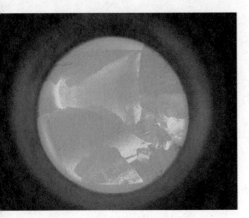

A red-hot furnace emits a spectrum of radiation that is independent of the furnace wall materials and is contingent only on the temperature.

THE OLD QUANTUM THEORY

It shouldn't be surprising that if physics was to be turned upside down, it would be done while trying to figure out what *light* (that is, radiant energy) was about. Quantum Mechanics had its earliest tentative beginnings in the theory of blackbody radiation, which itself began back in 1859. That year, Darwin published *The Origin of Species,* and it was the year that Gustav Robert Kirchhoff proffered an intellectual challenge that would lead to a revolution in physics.

30.1 Blackbody Radiation

Kirchhoff analyzed the way bodies in thermal equilibrium behave in the process of exchanging radiant energy. This *thermal radiation* (p. 586) is electromagnetic energy radiated by all objects, the source of which is their thermal energy. Suppose that the abilities of a body to emit and absorb electromagnetic energy are characterized by an **emission coefficient** ε_λ and an **absorption coefficient** α_λ. Epsilon is the energy per unit area per unit time emitted in a tiny wavelength range around λ (in units of $W/m^2/m$); any energy-measuring device admits a range of wavelengths. Alpha is the fraction of the incident energy absorbed per unit area per unit time in that wavelength range; it's unitless. These coefficients depend on both the nature of the surface of the body (color, texture, etc.) and the wavelength—a body that emits or absorbs well at one wavelength may emit or absorb poorly at another.

Now, imagine an isolated chamber of some sort in thermal equilibrium at a fixed temperature T. Clearly, it would be filled with radiant energy at lots of different wavelengths—just think of a glowing electric furnace. Assume that there is some formula, or **distribution function** I_λ, which depends on T and which tells us the intensity or amount of energy present at each wavelength. Put in numbers for T and λ, and the formula tells us the quantity of energy in the radiation within the cavity. Apparently, the *total* amount of energy at all wavelengths being absorbed by the walls versus the amount emitted by them must be the same, or else T will change. Kirchhoff argued that if the walls were made of different materials (which behave differently with T) that same balance would have to apply for *each* wavelength range individually. Thus, the energy absorbed at λ, namely, $\alpha_\lambda I_\lambda$, must equal the energy radiated, ε_λ, *and this is true for all materials no matter how different.* **Kirchhoff's Radiation Law** is thus

$$\frac{\varepsilon_\lambda}{\alpha_\lambda} = I_\lambda$$

wherein the distribution I_λ is a universal function the same for every type of cavity wall regardless of material, color, size, and shape and is only dependent on T and λ. That's extraordinary! Still, the famous British ceramist Thomas Wedgwood (1792) had long before noted that the objects in a fired kiln all turned glowing red together with the furnace walls regardless of their size, shape, or material constitution.

Although Kirchhoff did not provide the energy distribution function, he did point out that a perfectly absorbing body, one for which $\alpha_\lambda = 1$, will appear black and, in that special case, $I_\lambda = \varepsilon_\lambda$. The distribution function for a perfectly black object is the same as for an isolated chamber at the same temperature. This means that the radiant energy at equilibrium inside an isolated cavity is in every regard the same, "as if it came from a completely black body of the same temperature." The energy leaking from a small hole in the chamber should be identical to the radiation coming from a perfectly black object at the same temperature and that, as we will see, has important practical consequences.

Even though the scientific community accepted the challenge of determining I_λ, technical difficulties caused progress to be very slow. A simplified experimental setup is shown in Fig. 30.1. Radiant energy from a source passes through a slit and into a nonabsorbing prism. The wavelengths present are spread out into a continuous band that is sampled by a detector. Data must be extracted that is independent of the specifics of the detector. Thus, the best thing to plot is the radiant energy per unit time, which enters the detector per unit area (of the entrance window) per unit wavelength range (admitted by the detector). Figure 30.2 shows the kind of curves that were ultimately recorded (see also Fig. 15.9, p. 586), and each is a plot of I_λ at a specific temperature.

The Stefan-Boltzmann Law

In 1865, J. Tyndall published the result that the total energy emission of a heated platinum wire was 11.7 times as much operating at 1200°C (1473 K) as it was at 525°C (798 K). Amazingly, Josef Stefan (1879) noticed that the ratio of $(1473 \text{ K})^4$ to $(798 \text{ K})^4$ was 11.6, nearly 11.7, and he inferred that the rate at which energy is radiated is proportional to T^4. In this he was quite right (and quite lucky), because Tyndall's results were actually far from those of a blackbody. In any event, the conclusion was subsequently proven via a theoretical argument carried out by L. Boltzmann (1884). This was a traditional analysis of the radiation pressure exerted on a piston in a cylinder using the Laws of Thermodynamics and Kirchhoff's Law. The discussion progressed in much the same way one would treat a gas in a cylinder, but instead of atoms, the active agency was electromagnetic waves. The resulting **Stefan-Boltzmann Law** for blackbodies is

$$P = \sigma A T^4 \tag{30.1}$$

where P is the total power radiated at all wavelengths, A is the area of the radiating surface, T is the absolute temperature in kelvins, and σ is a universal constant now given as

$$\sigma = 5.670\,3 \times 10^{-8} \text{ W/m}^2 \cdot \text{K}^4$$

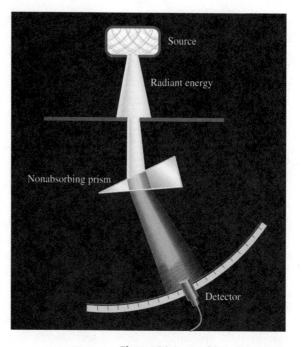

Figure 30.1 A schematic setup for measuring blackbody radiation.

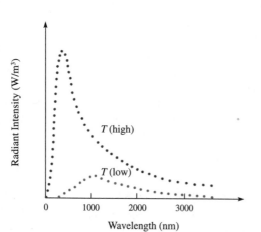

Figure 30.2 A plot of energy (J) per unit time (s), per unit area (m²), per unit wavelength interval (m), entering a detector successively centered at a number of different wavelengths. The source is a blackbody. Each curve represents I_λ at a given temperature.

TABLE 30.1 Representative values of total emissivity*

Material	ε
Aluminum foil	0.02
Copper, polished	0.03
Copper, oxidized	0.5
Carbon	0.8
White paint, flat	0.87
Red brick	0.9
Concrete	0.94
Black paint, flat	0.94
Soot	0.95

*T = 300 K, room temperature.

The total area under any one of the blackbody-radiation curves of Fig. 30.2 for a specific T is the power per unit area, and from Eq. (30.1) that's just $P/A = \sigma T^4$.

No real object is a perfect blackbody; carbon black has an absorptivity of nearly one, but only at certain frequencies (obviously including the visible). Its absorptivity is much lower than that in the far infrared. Still, most objects resemble a blackbody (at least at certain temperatures and wavelengths)—you, for instance, are nearly a blackbody for infrared. Thus, it's useful to write a similar expression for ordinary objects. To that end, we introduce a factor called the **total emissivity** (ε), which relates the radiated power to that of a blackbody for which $\varepsilon = 1$, at the same temperature, thus

$$P = \varepsilon \sigma A T^4 \qquad (30.2)$$

Table 30.1 provides a few values of ε (at room temperature), where $0 < \varepsilon < 1$. Note that emissivity is unitless.

Example 30.1 A cube of rough steel 10 cm on a side is heated in a furnace to a temperature of 400°C. Given that its total emissivity is 0.97, determine the rate at which it radiates energy from each face.

Solution: [Given: $L = 10$ cm, $T = 400°C$, and $\varepsilon = 0.97$. Find: P.] From the Stefan-Boltzmann Law, with $T = 673$ K and $A = 0.01$ m^2

$$P = \varepsilon \sigma A T^4$$
$$P = (0.97)(5.67 \times 10^{-8} \text{ W/m}^2\cdot\text{K}^4)(0.01 \text{ m}^2)(673 \text{ K})^4$$
and
$$\boxed{P = 1.1 \times 10^2 \text{ W}}$$

▶ **Quick Check:** $P \approx (1)(6 \times 10^{-8} \text{ W/m}^2\cdot\text{K}^4) \times (10^{-2} \text{ m}^2)(7 \times 10^2)^4 \approx 1.4 \times 10^2$ W.

Suppose a body with a **total absorptivity** of α is placed in an enclosure such as a cavity or a room having an emissivity ε_e and a temperature T_e. The body will radiate at a rate $\varepsilon \sigma A T^4$ and absorb energy inside the enclosure at a rate $\alpha(\varepsilon_e \sigma A T_e^4)$. But at any equilibrium temperature between body and enclosure (that is, $T = T_e$), these rates must be equal; hence, $\alpha \varepsilon_e = \varepsilon$ and that must be true for all temperatures. We can write an expression for the net power radiated (when $T > T_e$) or absorbed (when $T < T_e$) by the body:

$$P = \varepsilon \sigma A(T^4 - T_e^4) \qquad (30.3)$$

All bodies not at zero kelvin radiate, and the fact that T is raised to the fourth power makes the radiation sensitive to temperature increases. If a body at 0°C (273 K) is brought up to 100°C (373 K), it radiates about 3.5 times the previous power. Raising the temperature raises the net power radiated; that's why it gets harder and harder to increase the temperature of an object. (Just try heating a steel spoon to 1300°C.) Raising the temperature also shifts the distribution of energy among the various wavelengths present. When the filament of a light bulb "blows," the resistance, current, and temperature momentarily rise, and it goes from its normal operating reddish-white color to a bright flash of blue-white.

The Wien Displacement Law

In 1893, Wilhelm "Willy" Wien derived what has come to be known as the **Displacement Law**. Each blackbody curve reaches a maximum height at a value of

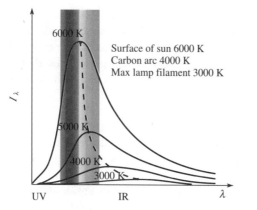

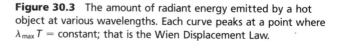

Figure 30.3 The amount of radiant energy emitted by a hot object at various wavelengths. Each curve peaks at a point where $\lambda_{max} T$ = constant; that is the Wien Displacement Law.

wavelength (λ_{max}), which is particular to it and therefore to the absolute temperature T. At that wavelength, the blackbody radiates the most energy. Wien correctly arrived at the fact that

$$\lambda_{max} T = \text{constant} \tag{30.4}$$

where the constant was found experimentally to be 0.002 898 m·K. The peak wavelength is inversely proportional to the temperature. *Raise the temperature, and the bulk of the radiation moves to shorter wavelengths and higher frequencies* (see the dashed curve in Fig. 30.3). As a glowing coal or a blazing star gets hotter, it goes from IR warm, to red-hot, to blue-white. A person or a piece of wood, both only approximating blackbodies, radiates mostly in the infrared and would only begin to glow faintly in the visible at around 600°C or 700°C, long after either had decomposed. The bright cherry red of a chunk of hot iron (p. 531) sets in at around 1300°C.

A hot filament and the spectra it emits. As the temperature rises from (a) to (b) to (c), the corresponding emission curves shift, as shown in Fig. 30.3. The peaks of the curves move toward the yellow, and the blue end of the spectrum increases in intensity as well. The result is that the filament shifts from cherry red to white-hot.

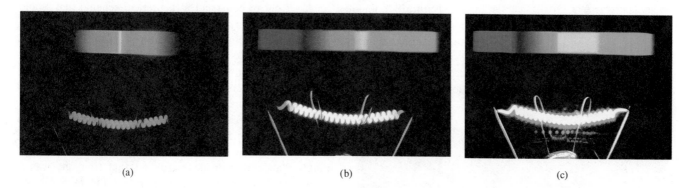

(a)   (b)   (c)

Example 30.2 On the average, your skin temperature is about 33°C. Assuming you radiate as does a blackbody at that temperature, at what wavelength do you emit the most energy?

Solution: [Given: $T = 33$°C. Find: λ_{max}.] Using Eq. (30.4), we obtain

$$\lambda_{max} = \frac{0.002\,898\ \text{m·K}}{306\ \text{K}} = \boxed{9.5\ \mu\text{m}}$$

or 9.5×10^3 nm.

▶ **Quick Check:** This result is well into the infrared and therefore reasonable. It's also independent of skin color.

Figure 30.4 Energy entering a small hole in a chamber rattles around until it's absorbed. In reverse, the aperture in a heated enclosure appears as a blackbody source.

In 1899, researchers tremendously advanced the state of experimentation by using, as a source of blackbody radiation, a small hole in a heated cavity (Fig. 30.4). Energy entering such an aperture reflects around inside until it's absorbed (the pupil of the eye appears black for precisely this reason). A near-perfect absorber is a near-perfect emitter, and the region of a small hole in an oven is a wonderful source of *blackbody radiation.*

Planck and His Energy Elements

Max Karl Ernst Ludwig Planck at forty-two was the reluctant father of Quantum Theory. Like so many other theoreticians at the turn of the century, he, too, was working on blackbody radiation. But Planck would succeed not only in producing Kirchhoff's distribution function, but he would turn physics upside-down in the process. We cannot follow the details of his derivation—they are far too complicated and, besides, the original derivation is wrong (Einstein corrected it years later). Still, it had such a powerful impact that it's worth looking at some of the features that are right.

Planck knew that if an arbitrary distribution of energetic molecules was injected into a constant-temperature chamber, it would ultimately rearrange itself into the Maxwell-Boltzmann distribution of speeds (p. 556) as it inevitably reached equilibrium. Presumably, if an arbitrary distribution of radiant energy is injected into a constant-temperature cavity, it, too, will ultimately rearrange itself into the Kirchhoff distribution of energies as it inevitably reaches equilibrium.

In October of 1900, Planck produced a distribution formula that was based on the latest experimental results. This mathematical contrivance, concocted "by happy guesswork," fit all the data available. It contained two fundamental constants, one of which (h) would come to be known as **Planck's Constant** (p. 918). That much by itself was quite a success, even if it didn't explain anything. Although Planck had no idea of it at the time, he was about to take a step that would inadvertently revolutionize our perception of the physical Universe.

Naturally enough, Planck set out to construct a theoretical scheme that would logically lead to the equation he had already devised. He assumed that the radiation in a chamber interacted with simple microscopic oscillators of some unspecified type. These vibrated on the surfaces of the cavity walls, absorbing and re-emitting radiant energy independent of the material. (In fact, the atoms of the walls do exactly that. Because of their tightly packed configuration in the solid walls, the atoms interact with a huge number of their neighbors. That completely blurs their usual characteristic sharp resonance vibrations, allowing them to oscillate over a broad range of frequencies and emit a continuous spectrum.) Try as he might, Planck was unsuccessful. At that time, he was a devotee of E. Mach, who had little regard for the reality of atoms, and yet the obstinate insolubility of the problem ultimately led Planck to "an act of desperation." He hesitantly turned to Boltzmann's "distasteful" statistical method, which had been designed to deal with the clouds of atoms that constitute a gas.

Boltzmann, the great proponent of the atom, and Planck were intellectual adversaries for a while. And now Planck was forced to use his rival's probabilistic formulation of entropy (p. 632), which—ironically—he would actually misapply. If Boltzmann's scheme for counting atoms

Most of the participants in the first Solvay Conference of 1911. Among the standees are Einstein (second from right), Planck (second from the left), and Rutherford (fourth from right). The young man next to him in the casual suit is Jeans. The lone woman is of course Madame Curie. J. J. Thomson missed the picture altogether.

was to be applied to something continuous, such as energy, some adjustments would have to be made in the procedure. Thus, according to Planck, the total energy of the oscillators had to be thought of, at least temporarily, as apportioned into "energy elements" so that they could be counted. These energy elements were given a value proportional to the frequency f of the resonators. Remember that he already had the formula he was after and in it there appeared the term hf. Planck's constant

$$6.626\,075\,5 \times 10^{-34} \text{ J·s} \qquad \text{or} \qquad 4.135\,669\,2 \times 10^{-15} \text{ eV·s}$$

is a very small number and so hf, which has the units of energy, is itself a very small quantity. Accordingly, he set the value of the energy elements equal to it:

$$\text{E} = hf \qquad (30.5)$$

This was a statistical analysis, and counting was central. Still, when the method was applied as Boltzmann intended, it naturally smoothed out energy, making it continuous as usual. Again, we needn't worry about the details; the amazing thing was that Planck had stumbled on a hidden mystery of nature: **energy is quantized**—it actually comes in tiny bursts as given by Eq. (30.5), but he certainly didn't realize it then.

The equation that finally resulted is presented here only so that you can see the answer to Kirchhoff's challenge, namely

$$I_\lambda = \frac{2\pi hc^2}{\lambda^5}\left[\frac{1}{e^{\frac{hc}{\lambda k_B T}} - 1}\right]$$

This is **Planck's Radiation Law**, and it fit blackbody data splendidly. Notice how the expression contains the speed of light, Boltzmann's Constant (p. 547), and Planck's Constant (h). It bridges Electromagnetic Theory to the domain of the atom. In 1901, Planck used blackbody data to arrive at a numerical value of k_B. With that and the universal gas constant R (p. 547), he determined N_A. And using Avogadro's number, the faraday, and Eq. (29.1), he calculated the charge on the electron! Remarkably, his answer was only 2% too small (which was much better than the result J. J. Thomson had gotten), and it was arrived at eight years before Millikan's measurements.

Although Eq. (30.5) represents a great departure from previous ideas, Planck did not mean to break with classical theory. It would have been unthinkable for him to even suggest that radiant energy was anything but continuous. "That energy is forced, at the outset, to remain together in certain quanta . . . ," Planck later remarked, "was purely a formal assumption and I really did not give it much thought." It was only around 1905, at the hands of a much bolder thinker, Albert Einstein, that we learned that the atomic oscillators were real, and that their energies were quantized. Each oscillator could only exist with an energy that was a whole number (n) multiple of hf (a little like the *gravitational*-PE of someone walking up a flight of stairs). Moreover, *radiant energy itself is quantized*, existing in localized blasts of an amount E = hf.

Max Karl Ernst Ludwig Planck (1858–1947). In 1889 he became Professor of Physics at the University of Berlin. There, he soon took up the problem of blackbody radiation first raised by his old teacher Kirchhoff, who had recently died (1887), and whose former position Planck now held.

30.2 Quantization of Energy: The Photoelectric Effect

The concept of the energy quantum went unnoticed for almost five years. It languished until Einstein completely recast blackbody theory. He had published three brilliant papers between 1902 and 1904 that laid out the principles of the discipline

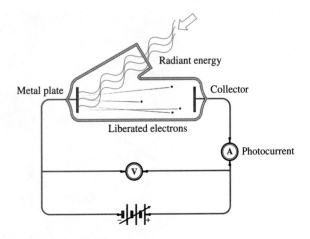

Figure 30.5 The Photoelectric Effect—radiant energy liberating electrons from a metal.

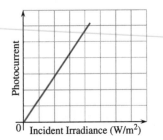

Figure 30.6 Photocurrent is directly proportional to the incident irradiance (that is, the energy per unit area per unit time).

known as *Statistical Mechanics*. Building on this work, Einstein in 1905 published a paper in which he showed that radiant energy behaved as if it was a collection of particles, light-quanta, each of energy *hf*. That same paper is most remembered for its analysis of the **Photoelectric Effect**.

The Photoelectric Effect was discovered by Heinrich Hertz, the man who had experimentally established the existence of electromagnetic waves. Ironically, the Photoelectric Effect, which he considered a *minor* observation, would ultimately lead to the overthrow of the classical understanding of electromagnetic waves. The Photoelectric Effect corresponds to the fact that *radiant energy (in the form of X-rays, ultraviolet, or light) impinging on various metals ejects electrons from their surfaces* (Fig. 30.5).

Classical theory suggests that the incident "light" arrives as an electromagnetic wave. If we use a uniform beam, its energy is presumably evenly spread across the entire wavefront (as it is with a water wave). The brighter the light, the greater its intensity, the larger the amplitudes of the *E*- and *B*-fields everywhere on a wavefront, and the more energy the wave delivers per second. These fields exert forces on the electrons in the metal and can even liberate some of them from the surface.

When the collector plate is made positive with respect to the emitter plate, *photoelectrons* easily traverse the tube, and that constitutes a *photocurrent* that is measurable with a microammeter. Classical theory predicts that increasing the brightness of the light beam provides more energy, thus liberating more electrons, and increasing the photocurrent (Fig. 30.6). This proportionality between photocurrent and irradiance (brightness) was shown to hold across a broad range from a low, where the eye cannot even see the light, to a high, where it cannot bear it.

If the collector's positive potential is gradually decreased, we can expect the photocurrent to slowly decrease as well. Figure 30.7 illustrates the effect for monochromatic light. Some electrons make it to the collector even at zero potential difference; and the brighter the light, the more current there will be. Evidently, if the collector is made negative, it will repel photoelectrons, decreasing the current even further. At a negative potential—specific to each metal—known as the **stopping potential** (V_S), all the electrons hurled out of the metal are turned back, and the photocurrent goes to zero. The existence of a stopping potential characteristic of the metal, but *independent of intensity,* was troubling. It would seem that V_S ought to

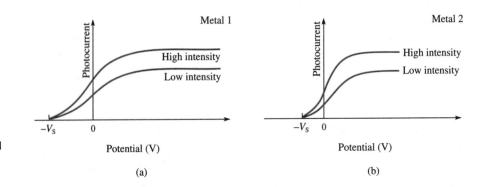

Figure 30.7 The variation of photocurrent with collecting potential and light intensity. Different metals have different stopping potentials.

depend on how bright, and therefore energetic, the light is. An electron has to traverse the region where the *E*-field is pushing against it and thus overcome a potential energy barrier in order to arrive at the collector. Reaching the maximum negative potential finally turns back even those electrons with a maximum KE, where

$$KE_{max} = eV_S \qquad (30.6)$$

The surprise is that KE_{max} is independent of the light intensity. With a low light level and very little energy coming in, the electron must somehow store up what it needs to pop out of the metal, and yet at high light levels, it must refrain from absorbing any more energy than KE_{max}—very strange.

However disconcerting these experimental results proved, when attention was turned to the frequency of the radiation, things became utterly bizzare. For any given metal, there is a specific threshold frequency below which photoemission does not occur, no matter how intense the incident radiant energy. That simply should not be; provided enough energy (via a bright beam), electrons should come pouring out. Moreover, the astounding facts were established (1903) that *the photoelectron's maximum kinetic energy was independent of beam intensity, and yet it was proportional to the frequency of the illumination.* Kinetic energy should only depend on the incident energy; it shouldn't have anything to do with frequency. How could frequency affect energy?

The time delay between the arrival of radiant energy and the emission of photoelectrons is now known to be less than 3×10^{-9} s; switching on even the dimmest source can "instantly" release electrons. That behavior is simply impossible within orthodox wave theory. Each incoming wave should spread uniformly over the surface of the illuminated metal, imparting only a minute amount of energy to each of the millions upon millions of atoms in its path. We can calculate, at least crudely, how long it would take an electron to absorb enough energy to attain the observed speeds (Problem 48). There is some ambiguity depending on how much absorbing "area" the electron presents to the beam, but all such calculations conclude that it will take from seconds to months of irradiation before an electron can be ejected—and yet there they were, flying off almost instantaneously. Something fundamental was obviously wrong with the accepted understanding of the way electromagnetic energy interacted with matter.

Remarkably, J. J. Thomson (1903) suggested that electromagnetic waves might well be radically different from other waves; perhaps a sort of concentration of radiant energy actually existed. After all, if one shone X-rays on a gas, only certain of the atoms, here and there, would be ionized, as if the beam had "hotspots" rather than being uniform. The notion was prophetic.

Einstein and the Photon

Planck had earlier introduced the concept of energy elements, and regardless of the inconsistencies, he had supposed that radiant energy was continuous and it behaved like a classical wave. Now Einstein, in 1905, stepped into the muddle. Audaciously breaking with the long-standing traditional view, he proposed that ***light itself is granular***, that it is actually composed of discrete bursts, particles of energy, or **photons**, as they came to be called.* The interaction between electromagnetic radiation

*The word *photon* was coined by G. N. Lewis in 1926.

and substantial matter occurs in jolts. Radiant energy is absorbed and emitted discontinuously because it itself is discontinuous!

Einstein's first paper on light-quanta, "On a Heuristic Point of View Concerning the Generation and Conversion of Light," was published in 1905 (before Relativity). And *heuristic* means something that serves as a guide in the solution of a problem but is otherwise itself unproved. In that spirit, he postulated that *every electromagnetic wave of frequency f is actually a stream of energy quanta, each with an energy*

$$E = hf = \frac{hc}{\lambda} \tag{30.7}$$

Example 30.3 In the atomic domain, energy is often measured in electron-volts. Accordingly, arrive at an expression for the energy of a light-quantum in eV when the wavelength is in nanometers. What is the energy of a quantum of 500-nm light?

Solution: [Given: λ in nm and particularly $\lambda = 500$ nm. Find: E in eV.] Using $E = hc/\lambda$ with $h = 4.135\,67 \times 10^{-15}$ eV·s (p. 1123), we get

$$E = \frac{1239.8 \text{ eV·nm}}{\lambda}$$

For $\lambda = 500$ nm, $\boxed{E = 2.48 \text{ eV}}$.

▶ **Quick Check:** The numerator here can be remembered as 1234.5, which is accurate to about 0.4%. The energy of a 500-nm quantum is in keeping with the energies associated with valence-electron processes, a few eV.

Nobody knows what a photon "looks like," but it is both localized and wavy (it has a frequency). The higher the frequency, the greater the energy of the individual photons. The irradiance of a monochromatic beam, the energy per unit area per unit time, is determined by the number of photons in the stream. The brighter the beam, the more photons.

A photon colliding with an electron in a metal can vanish, imparting essentially all of its energy to the electron. *The electron cannot be totally free*: because of the demands of momentum conservation, momentum must be transferred to the metal atoms, which nonetheless only pick up a negligible amount of energy. Even in the faintest beam, a single adequately energetic photon can kick loose an electron, and this can happen as soon as the illumination is turned on—there is no time delay. Raising the irradiance (intensity) of the beam increases the number of photons and therefore proportionately increases the current (Fig. 30.6). It takes energy to bring an electron up to the surface of the metal and subsequently liberate it. If electrons were not bound to the metal, they would be escaping all the time, leaving it positively charged, which doesn't happen. For an electron already up near the surface, this liberation energy is a minimum called the **work function** ϕ (Table 30.2). A photon's energy goes into freeing the electron, and whatever is left appears as KE. When the electron is at the surface, the liberating energy is a minimum, and the electron takes on a maximum KE given by

$$hf = KE_{max} + \phi \tag{30.8}$$

TABLE 30.2   Representative work function values

Metal	Work function (ϕ in eV)
Na	2.28
Co	3.90
Al	4.08
Pb	4.14
Zn	4.31
Fe	4.50
Cu	4.70
Ag	4.73
Pt	6.35

This wonderfully simple expression is known as **Einstein's Photoelectric Equation**, and it explains every aspect of the effect. Since $KE_{max} = hf - \phi$, increasing the intensity of the light leaves the maximum kinetic energy unchanged. Only by changing f is the KE_{max}, or equivalently, the stopping potential, changed for a given metal. The threshold frequency f_0 corresponds to the initiation of emission where $KE_{max} = 0$. Hence, $hf_0 = \phi$ and below a frequency of $f_0 = \phi/h$, the photocurrent will be zero.

It follows from Eq's. (30.8) and (30.6) that

$$eV_S = hf - \phi$$

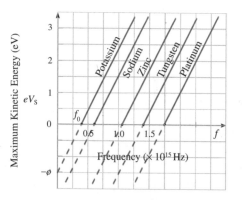

Figure 30.8 Each line follows the equation $eV_S = hf - \phi$. Thus, each intercepts the f-axis at f_0 and the energy axis at $-\phi$.

which has the familiar form of the equation of a straight line ($y = mx + b$). The theory predicted that for any and all metals, a plot of stopping-potential-multiplied-by-electron-charge (y) against frequency (x) will be a straight line of slope (m) equal to Planck's Constant, and y-intercept (b) equal to the negative of the work function. No such relationship was known prior to Einstein's paper of 1905 and, amazingly, exactly that relationship was subsequently observed in every respect! Not until 1914–1915 would Robert Millikan finally and conclusively determine that the Photoelectric Equation was in complete agreement with experiment. He had spent 10 years of meticulous labor intent on showing that Einstein was totally wrong, and in the end he had established "the exact validity" of the theory. And yet he was unwilling to accept the reality of the photon even as he accepted the Nobel Prize for his work. Figure 30.8 shows the kind of results Millikan got. Each metal produces a straight line that intersects the horizontal axis (where the voltage is zero) at its threshold frequency. And each intersects the vertical axis (where $f = 0$) at $-\phi$. Moreover, each line has a slope equal to Planck's Constant!

The above analysis is increasingly being called the *Single Photon Photoelectric Effect* because, nowadays, lasers can put out such a torrent of nearly monochromatic photons that it's possible to blast an electron with two or more light-quanta before it leaves the metal. The result is the *Multiple Photon Photoelectric Effect* (see Discussion Question 5).

Light propagates from one place to another as if it were a wave. Several centuries of work had established *that* with convincing credibility, and yet now it became equally as clear that *light interacts with matter in the processes of absorption and emission as if it were a stream of particles*. This, then, is the so-called **wave-particle duality,** the schizophrenia of light (and, as we will see, of matter in general). Radiant energy appears and disappears in minute localized blasts and is seemingly transported via spreading waves.

The idea of light-quanta was not received well at all, and it had very few early advocates. Paul Ehrenfest (who seems to have been independently thinking along similar lines in 1905), Max von Laue, and Johannes Stark (who in 1909 anticipated the Compton Effect, p. 1129) were the leading exceptions. Despite its clearly demonstrated theoretical power, it was so disconcerting to people educated in classical wave theory that it was especially slow to be accepted. When Planck recommended Einstein for membership in the Prussian Academy in 1913, he felt that, notwithstanding Einstein's demonstrated genius, he still had to apologize for him because "he may sometimes have missed the target in his speculations, as, for example, in his hypothesis of light-quanta."

A few frames of a motion picture showing the optical sound track at the right. Light passing through the track illuminates a photoelectric detector that converts the signal into a varying voltage.

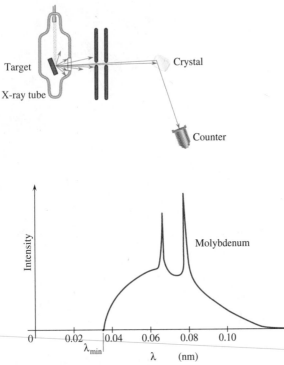

Figure 30.9 X-ray spectrum for a molybdenum target. The broad peak in intensity is due to Bremsstrahlung.

Despite the general indifference, Einstein continued to use his heuristic principle to make a number of splendid predictions, one of which (1911) concerned the generation of X-ray quanta. It would be another 10 years before the strange and wonderful notion that light was no longer a simple TEM wave became so potent that it had to be accepted, even if not quite "understood." Einstein received the 1921 Nobel Prize "for his services to Theoretical Physics, and especially for his discovery of the law of the photoelectric effect."

30.3 Bremsstrahlung

X-rays are generated (p. 1096) when high-speed electrons impacting on a dense target rapidly decelerate and thus radiate. The resulting broad continuous X-ray spectrum, illustrated in Fig. 30.9, is known as **Bremsstrahlung**. The spikes in the curve arise from the atomic structure of the specific target, and we'll deal with them later in this chapter. The process of X-ray production is something of an inverted Photoelectric Effect; in come electrons, out goes radiant energy. And both can only occur in the presence of heavy atoms that can take up some momentum. Using the notion of light-quanta, Einstein predicted there would be a high-frequency (low-wavelength) limit on the radiation.

Imagine an incident electron arriving with an energy eV after being accelerated across a potential of V. As it slows down within the target, it radiates one or more photons. The maximum photon frequency (f_{max}) occurs when all the electron's kinetic energy is radiated as a single photon, whereupon

$$E = hf_{max} = eV$$

Accordingly, there will be a cut-off frequency of

$$f_{max} = \frac{eV}{h}$$

or since $c = f\lambda$, a minimum wavelength of

$$\lambda_{min} = \frac{ch}{eV} \qquad (30.9)$$

In other words, the electron cannot emit more energy than it has, and that maximum amount corresponds to the maximum frequency (minimum wavelength) end of the Bremsstrahlung curve.

Example 30.4 Determine the smallest wavelength X-rays that can be emitted by an electron as it crashes into the metal mask on the front face of a color TV tube operating with an accelerating voltage of 20.0 kilovolts.

Solution: [Given: $V = 20.0$ kV. Find: λ_{min}.] From Eq. (30.9),

$$\lambda_{min} = \frac{ch}{eV}$$

(continued)

(continued)

$$\lambda_{\min} = \frac{(2.9979 \times 10^8 \text{ m/s})(4.13567 \times 10^{-15} \text{ eV·s})}{20.0 \times 10^3 \text{ eV}}$$

$$\boxed{\lambda_{\min} = 0.0620 \text{ nm}}$$

Notice how entering h in electron-volts allows us to enter the denominator in electron-volts. Finally

▶ **Quick Check:** An atom is about 0.1 nm across, and X-rays range in wavelength from there down to roughly 0.006 nm.

The existence of a maximum frequency was confirmed experimentally in 1915 by W. Duane and F. Hunt at Harvard. Using Eq. (30.8), they determined h to an accuracy of better than 4%.

30.4 The Compton Effect

Evidence of the reality of photons and the fact that they behave like particles with a well-defined energy and *momentum* was provided by the American Arthur H. Compton. He shone X-rays onto targets of low atomic number, like carbon. These have many loosely bound electrons that are essentially "free," and they scatter the radiant energy in a characteristic fashion. Classical theory is complicated by several factors, one of which is that the electrons recoil from the collision, and that introduces a Doppler Shift in the radiation they emit. Be that as it may, Compton reported in 1922 that the evidence at hand was in severe conflict with classical theory.

In 1923, Compton and Debye independently applied relativistic kinematics to the problem of a photon colliding with a "free" electron and thereby explained all the bewildering observations. Recall Eq. (28.16) for a massless particle, $E = pc$; hence, for photons for which $E = hf$

$$p = \frac{E}{c} = \frac{hf}{c}$$

and

$$p = \frac{h}{\lambda} \qquad (30.10)$$

Thus, both the momentum and the energy of a high-frequency (short-wavelength) photon, such as an X-ray quantum, exceed that of a low-frequency (long-wavelength) photon, such as a microwave quantum.

Imagine a beam of wavelength λ_i incident on a target. When photons collide with strongly bound electrons that stay put (and do not absorb energy), the photons are elastically scattered in all directions, and one sees this as radiation coming out with the incident wavelength unaltered. By contrast, picture an incident photon of momentum $\mathbf{p}_i$ (Fig. 30.10). If it strikes a free, essentially motionless, electron and imparts some of its momentum to that electron ($\mathbf{p}_e$), a new scattered photon with a longer wavelength and less momentum ($\mathbf{p}_s$) will come flying out at some angle θ.

This is an elastic-collision problem, and we analyze it just as we did earlier (p. 344). Applying Conservation of Momentum yields

$$\mathbf{p}_i = \mathbf{p}_s + \mathbf{p}_e \qquad (30.11)$$

Similarly, the initial energy of the system is associated with the incident photon

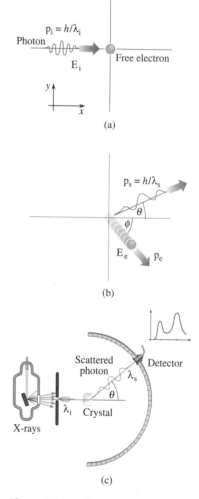

Figure 30.10 Compton scattering of X-rays. (a) A photon collides with a free electron and (b) both are scattered. (c) The scattered X-rays come off at an angle θ.

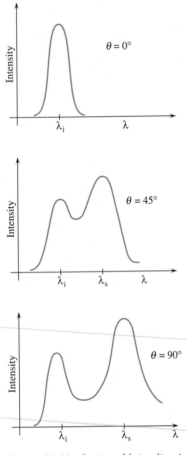

Figure 30.11 graphs, labeled $\theta = 0°$, $\theta = 45°$, and $\theta = 90°$, with axes Intensity versus λ, marked λ_i and λ_s.

Figure 30.11 Scattered intensity at several angles.

($E_i = hf_i$) and the rest energy of the electron ($m_e c^2$). After the collision, the energy is that of the scattered photon ($E_s = hf_s$) and the moving electron (E_e). Conservation of Energy states that

$$E_i + m_e c^2 = E_s + E_e \tag{30.12}$$

Equations (30.11) and (30.12) can be combined with the relativistic expression [Eq. (28.15), p. 1079] describing the energy of the electron

$$E_e^2 = c^2 p_e^2 + m_e^2 c^4$$

to yield (see Problem 47) a formula for the increase in wavelength

$$\Delta\lambda = \lambda_s - \lambda_i = \frac{h}{m_e c}(1 - \cos\theta) \tag{30.13}$$

Figure 30.11 shows the sort of results that are observed and that agree very nicely with Eq. (30.13). Note that the wavelength shift $\Delta\lambda$ only depends on the photon-scattering angle and is independent of the initial wavelength. It's also independent of the scattering material. The accuracy of the experiment was found to be within 1% by determining the multiplicative constant ($h/m_e c$), which is called the **Compton Wavelength** and which should theoretically equal 0.002 426 nm. In 1925, Bothe and Geiger, using counters, showed that the scattered electron and the scattered X-ray photon actually do fly off at the same time. More recent experiments (1950) verify that simultaneity to within 0.5 ns.

In the 1920s, the Compton Effect had the impact of eliminating the last vestiges of doubt: **whatever light is, it is a stream of particles capable of transferring energy and momentum that can also somehow behave as a wave**. Interestingly, though the contemporary consensus is that the Compton Effect is strong "proof" of the photon, people have nonetheless come up with alternative derivations in terms of the wave model—the business is far from settled.

Example 30.5 A 50.0-keV photon is Compton-scattered by a quasi-free electron. If the scattered photon comes off at 45.0°, what is its wavelength?

Solution: [Given: $E_i = 50.0$ keV and $\theta = 45.0°$. Find: λ_s.] To use Eq. (30.13), first compute λ_i using the results in Example 30.3, p. 1126:

$$\lambda_i = \frac{1240 \text{ eV·nm}}{50.0 \times 10^3 \text{ eV}} = 0.024\,8 \text{ nm}$$

Since

$$\lambda_s - \lambda_i = \frac{h}{m_e c}(1 - \cos\theta)$$

we have

$$\lambda_s = (0.024\,8 \text{ nm}) + (0.002\,426 \text{ nm})(1 - 0.707)$$

and

$$\boxed{\lambda_s = 0.025\,5 \text{ nm}}$$

▶ **Quick Check:** $\Delta\lambda = \lambda_s - \lambda_i = 0.025\,5$ nm $-$ 0.024 8 nm $= 7 \times 10^{-4}$ nm as compared to Eq. (30.13), where $(0.002\,426 \text{ nm})(1 - 0.707) = 7 \times 10^{-4}$ nm.

ATOMIC THEORY

Niels Henrik David Bohr was twenty-six when he arrived at Cambridge from Denmark in the autumn of 1911. He had come to work with J. J. Thomson, but they didn't hit it off well. As fate would have it, the guest speaker at the annual Cavendish dinner was Ernest Rutherford, who, only a few months before, had published his theory of the nuclear atom. Bohr met with Rutherford and, by the spring of 1912, he was happily immersed in the whirlwind of atomic research that churned at Rutherford's lab. "This young Dane," Rutherford once remarked, "is the most intelligent chap I've ever met." Bohr stayed only for four months before returning to Copenhagen, but he had already become convinced of the validity of the nuclear atom.

The atomic theory Bohr later created was a brilliant accomplishment at the time. It has, however, been surpassed by the far more powerful formulations of Quantum Mechanics. Still, we will study the Bohr Theory because it provides the basic conceptual vocabulary for modern atomic physics and because many of its conclusions are essentially valid.

30.5 The Bohr Atom

Bohr began his creation by guessing about the nature of the simplest atomic configuration, the single-electron atom; his first postulate assumed that *the electron sails around the nucleus in a circular orbit.* He then postulated that, unlike a planet, which can revolve permanently at any distance from the Sun, **atomic electrons exist in only certain stable, lasting orbits about the nucleus**. These are now called **stationary states**, and the one with the lowest energy is the **ground state** (p. 919). *While in such a stationary state, the atom does not radiate.* Everyone knew that something was wrong with the planetary model of the atom; electrons orbiting the nucleus, accelerating, should continuously radiate. Losing energy, they should wind inward, finally crashing into the nucleus. And it's easy to calculate that this death spiral will take about a hundred-millionth of a second. Our very existence is therefore an embarrassment to such a theory, which ridiculously insists that all the atoms in the Universe should have long ago collapsed. Why doesn't Bohr's orbiting electron radiate? Here, Bohr was refreshingly outrageous. He simply insisted it didn't, and that was that. Of course, his stance implied that Maxwell's Electromagnetic Theory somehow reaches profound limitations on the atomic level, but he didn't bother with that.

Bohr was not the only one groping toward an orbital model. J. W. Nicholson, an English astrophysicist, was perhaps his chief rival. Like Maxwell long before him, Nicholson raised the question of there being a relationship between the structure of an atom and the kind of spectrum it emits. Bohr believed that any such interdependence must be exceedingly complicated, and at first he avoided the notion altogether. But when a friend, H. M. Hansen, returned to Copenhagen after studying spectroscopy and the two began to discuss Bohr's work, the issue of spectral colors immediately came up. Hansen suggested that spectra (p. 1103) were not really so complex. "As soon as I saw Balmer's formula," Bohr recalled, "the whole thing was immediately clear to me."

Consider the simplest of all atoms, the hydrogen atom. It has a nuclear proton orbited by a single electron in its ground state. When the atom is appropriately stimulated (perhaps thermally via collisions, electrically, or even by absorbing light), the electron is excited into a higher energy orbit that is more distant from the nucleus. There it resides, ordinarily for about a nanosecond, before spontaneously descending to some inner orbit, ultimately dropping back to the ground state. During each drop (a moment when the theory goes impotent), the electron emits its excess energy as a burst of electromagnetic radiation—a photon. This is the now-famous **quantum jump**.

Niels Henrik David Bohr (1885–1962). Before Bohr left Denmark in 1943 with the Nazis on his trail, he dissolved in acid the gold Nobel Prize medals that Von Laue and Franck had given him for safekeeping. The bottle containing the gold solution was left on a shelf in Bohr's lab throughout the war. When he returned to Copenhagen, Bohr precipitated the gold and had the medals recast.

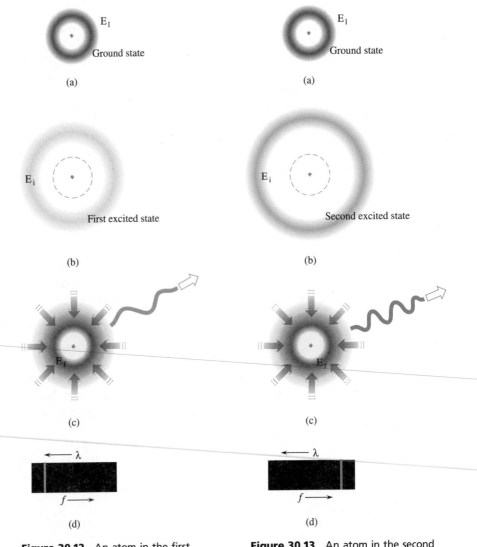

Figure 30.12 An atom in the first excited state drops back to the ground state with the emission of a long-wavelength, low-energy photon. The emission shows up as a long-wavelength, low-frequency spectral line.

Figure 30.13 An atom in the second excited state drops back to the ground state with the emission of a short-wavelength, high-energy photon.

The picture is a little like a Greek amphitheater, in which a ball revolving around the lowest tier is boosted up to some higher ring, where it orbits for a while, only to drop back to ground level in either one large or many small descents. The sequence of Balmer's frequencies became a ladder of orbits, of **energy levels**. When the electron in an excited atom drops from an initial (E_i) to a final (E_f) energy level, it emits the difference as a quantum hf; that is

$$E_i - E_f = hf \tag{30.14}$$

The larger the drop, the greater the photon frequency (Fig's. 30.12 and 30.13).

By the time Bohr began his work, the notion that an oscillator could have quantized energy levels that were whole number multiples (n) of hf was widely accepted;

E = *nhf* was an essential feature of the new blackbody theory. Planck recognized that *h* had the units of energy multiplied by time (J·s), or equivalently of momentum multiplied by distance. Since that product, known as *action*, was important in classical mechanics, he called *h* the **quantum of action**. The idea that action was quantized would be generalized into a guiding principle. For example, for an electron in the *n*th circular orbit of radius r_n, the distance traveled per revolution is $2\pi r_n$ and, since $p_n = m_e v_n$, we might guess that the action (momentum times distance) for each orbit would equal a whole number multiple of *h*: $(m_e v_n)(2\pi r_n) = nh$. This immediately leads to $m_e v_n r_n = nh/2\pi$, and that we recognize (p. 288) as the angular momentum (*L*) of the orbiting electron. In any event, Nicholson (1912) suggested that "the angular momentum of an atom can only rise and fall by discrete amounts." A year later, Bohr adopted the same idea as his second postulate; that is

$$L_n = m_e v_n r_n = n\frac{h}{2\pi}, \qquad n = 1, 2, 3, \ldots \qquad (30.15)$$

Here, each successive value of *n* corresponds to a higher orbit of larger radius and lower speed. The term $h/2\pi$ comes up so frequently that P. A. M. Dirac gave it its own symbol, $\hbar$, and we refer to it as "h-bar":

$$\hbar = 1.054 \times 10^{-34} \text{ J·s}$$

Angular momentum is quantized in whole number multiples of $\hbar$.

If we continue with this simple pictorial model (which we *will* have to abandon soon enough), we can derive an expression for the radii of the allowed orbits (Fig. 30.14) of a one-electron atom. Of course, hydrogen is the simplest such atom, but the theory works for such systems as singly ionized helium as well. The electron is attracted to a nucleus (which in general has *Z* protons and a net charge of $+Ze$) via a Coulomb interaction. Equating that force, $F_E = k_0(Ze)(e)/r^2$ with the centripetal force, $F_C = mv^2/r$, leads to

$$k_0\frac{Ze^2}{r_n^2} = \frac{m_e v_n^2}{r_n} \qquad (30.16)$$

which can be solved for

$$r_n = \frac{k_0 Ze^2}{m_e v_n^2}$$

The quantity v_n, which we do not know, can be eliminated from this expression by solving Eq. (30.15) for it. Whereupon $v_n = nh/2\pi m_e r_n$, and substituting this equation in the above and simplifying (canceling a factor of r_n) gives

$$r_n = n^2\frac{\hbar^2}{m_e k_0 Ze^2} = n^2 r_1 \qquad (30.17)$$

For hydrogen, $Z = 1$, and putting the appropriate numbers in Eq. (30.17) leads to $r_1 = 0.052\,917\,7$ nm. This is the radius of the ground state of the hydrogen atom,

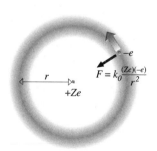

Figure 30.14 A single-electron atom showing the Coulomb interaction between electron and nucleus.

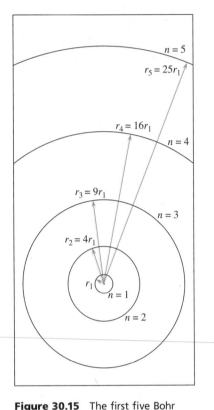

Figure 30.15 The first five Bohr orbits.

and it's called the **Bohr radius**. We know that an atom is about 0.1 nm across so this result is very encouraging (Fig. 30.15). The theory suggests (it only deals directly with single-electron systems) that for heavier atoms (larger Z), the increased nuclear charge draws more strongly on the surrounding electrons, shrinking their orbits. Thus, atomic diameters are all close to the same size. Uranium, which is 238 times more massive than hydrogen, has a diameter only about 3 times as great.

The model Bohr devised pictured the discrete spectral lines resulting from transitions between discrete energy levels. Clearly, energy is the next thing to consider. The total energy of an orbital electron (E_n) is its KE plus its *electrical*-PE = $qV = -eV$. Here, V is due to the nucleus and so $V = k_0 q/r = k_0(Ze)/r$. Thus

$$E_n = \tfrac{1}{2}m_e v_n^2 - \frac{k_0 Z e^2}{r_n}$$

which is the energy of the atom as a whole. It follows from Eq. (30.16) that $\tfrac{1}{2}m_e v_n^2 = kZe^2/2r_n$ and so

$$E_n = -\frac{k_0 Z e^2}{2r_n}$$

Notice that the total energy of the atom is negative, only becoming zero when the electron is removed to infinity, or the practical equivalent thereof (see Discussion Question 10). The dependence on r_n can be removed using Eq. (30.17), yielding

$$E_n = -\frac{2\pi^2 k_0^2 e^4 m_e}{h^2}\frac{Z^2}{n^2} \tag{30.18}$$

The integer n, which enumerates both the orbital radii and the energies, is known as the **principal quantum number**. Notice that as n gets larger, the radii increase as n^2, and the orbits get farther apart. On the other hand, the energy levels become less negative and approach zero as $n \to \infty$; they get closer together, ultimately becoming continuous when the energy is positive and the electron is unbound.

Equation (30.18) provides the energy of each orbit, and it can be simplified considerably by noting that the ground-state energy ($n = 1$) for hydrogen ($Z = 1$) turns out to equal -13.6 eV. This value is the hydrogen atom's largest negative energy and its most tightly bound configuration. In that case

[*for hydrogen*]
$$E_n = -\frac{13.6 \text{ eV}}{n^2} \tag{30.19}$$

The *first excited state* ($n = 2$) of the hydrogen atom has an energy of $-(13.6/2^2)$ eV, or -3.40 eV. The minus sign tells us that that much energy must be added to raise the electron to the zero level. We can think of it as a kind of energy well with the electron usually down -13.6 eV, at the bottom. Accordingly, the **ionization energy** needed to remove the electron from the atom setting it free (to raise it up and out of the potential well) is 13.6 eV, and this agrees precisely with the measured value! Being able to derive that from basic principles is an amazing accomplishment.

Example 30.6 How much energy must a hydrogen atom absorb if it is to be raised from the ground state to the first excited state? If that excitation energy is to come in the form of a photon, what must be its frequency?

(continued)

(continued)

Solution: [Given: H atom in ground state. Find: $E_2 - E_1$ and f.] We already know that $E_1 = -13.6$ eV and that $E_2 = -3.40$ eV; hence, the atom must receive an energy of

$$(-3.4 \text{ eV}) - (-13.6 \text{ eV}) = \boxed{10.2 \text{ eV}}$$

if it's to be raised into its first excited state. Hence, a photon for which

$$\Delta E = hf = 10.2 \text{ eV}$$

should have a frequency of

$$f = \frac{(10.2 \text{ eV})}{(4.136 \times 10^{-15} \text{ eV·s})} = \boxed{2.47 \times 10^{15} \text{ Hz}}$$

▶ **Quick Check:** $\lambda = 120$ nm, which is in the UV where it should be (see Fig. 30.16).

If the transition from an upper-excited state down to a lower state is accompanied by the emission of a photon as per Eq. (30.14), we should now be able to write an explicit expression for the resulting wavelength, or better yet, for $1/\lambda$. Since $hf = hc/\lambda$, rewrite Eq. (30.14) as

$$\frac{1}{\lambda} = \frac{1}{hc}(E_i - E_f)$$

or using Eq. (30.18)

$$\frac{1}{\lambda} = \frac{2\pi^2 k_0^2 e^4 m_e Z^2}{h^3 c}\left[\frac{1}{n_f^2} - \frac{1}{n_i^2}\right] \qquad (30.20)$$

where $n_f < n_i$, the atom is initially in a higher state than the one it drops down to. Now compare this equation to the Balmer series ($Z = 1$), for which $n_i = n$ and $n_f = 2$; that is

$$\frac{1}{\lambda} = R\left[\frac{1}{2^2} - \frac{1}{n^2}\right], \qquad n = 3, 4, 5, \ldots \qquad [29.5]$$

where all those constants in front of the bracketed term in Eq. (30.20) must equal RZ^2. When Bohr carried out the calculation in 1913 using the best values of the day, he got the measured R to within 1%—spectacular!

Each line in the Balmer series arises when the hydrogen atom in an excited state relaxes back in a single quantum jump to the first excited state ($n_f = 2$). Similarly, when the transition is down to the ground state ($n_f = 1$), the energies are greater and the resulting Lyman series, which wasn't discovered until 1916, is in the far-UV. The Paschen series (first observed in 1908) corresponds to $n_f = 3$. Brackett found a new series in 1922 in the IR for which $n_f = 4$, and Pfund, in 1924, located yet another in the IR for $n_f = 5$ (Fig. 30.16); all precisely as predicted by the Bohr Theory.

Energy Levels

An atom can be raised into any excited state by absorbing an amount of energy given by Eq. (30.14). This can happen, for instance, via collisions with other atoms or from bombardment by projectiles such as electrons (via an electric current) or photons. If an incoming photon does not have enough energy to raise the atom into its first excited state, the atom remains in its ground state, immediately

Figure 30.16 (a) Energy levels of the hydrogen atom and the transitions between them corresponding to the emission of radiation. Shown are the Lyman, Balmer, and Paschen series. (b) The spectral lines corresponding to the Balmer series. (Take a look at Fig. Q11, p. 1112.)

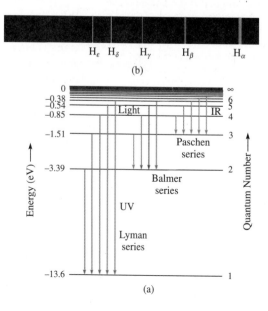

The first photograph of a solitary atom. The tiny blue-green dot at the center of this photo is a single barium ion. It was slowed or "cooled" by laser beams and is suspended in a radio frequency trap. (The large, white and red structure surrounding the ion is part of the trap.) The ion absorbs energy from the laser beam and re-emits at 493 nm.

elastically scattering away the energy (p. 920). Thus, a visible photon with an energy of around 3 eV cannot raise hydrogen into its first excited state and so hydrogen, normally in its ground state, is transparent to light; photons enter and leave the gas without any energy losses. Only if the photon (or any other impacting particle) has energy equal to, or in excess of, that needed for excitation will the atom "rise up" to a higher level. For example, consider a monatomic gas at room temperature. The atoms collide, but they don't give off light. Why not? According to Eq. (14.14), the average KE of such an atom is $\frac{3}{2}k_B T = \frac{3}{2}(1.38 \times 10^{-23} \text{ J/K})(300 \text{ K}) = 6.2 \times 10^{-21} \text{ J} = 0.04 \text{ eV}$. That's much too small to excite any of the atoms, and the collisions remain perfectly elastic (which is why the air doesn't glow on a warm day).

In recent times, it has become possible to examine the transitions made by single atoms from one specific energy level to another. For example, in 1986, an individual barium ion held, almost at rest, in a electromagnetic-field trap was stimulated using laser beams of precise frequency. The experiments (see Discussion Question 13) directly confirmed the existence of quantum jumps between energy levels.

30.6 Stimulated Emission: The Laser

In a conventional light source, such as a tungsten lamp or a neon sign, energy is usually *pumped* into the reacting atoms via collisions with an electrical current. An excited atom drops back to its ground state *spontaneously*, without any external inducement, emitting a randomly directed photon. The atoms are all essentially independent, and each photon in the emitted stream bears no particular phase relationship to any other.

Now, imagine a bunch of atoms somehow pumped up into an excited state. What happens next is not obvious. In 1917, Einstein pointed out that an excited atom can relax to a lower state via photon emission in two distinct ways. In one, the atom emits energy spontaneously, while in the other, it is triggered into emission by the presence of a photon of the proper frequency. The former process is known as **spontaneous emission**, the latter as **stimulated emission**. Suppose that the energy difference between an upper- and lower-energy state is $\Delta E = E_u - E_l$ such that if the transition down occurs, a photon hf_{ul} will be emitted. Let the atom be in the E_u-state with no concern as to how it got there. If the excited atom is exposed to a photon of frequency f_{ul}, it will immediately be stimulated into dropping down to the E_l state. A remarkable feature of the process is that *the emitted photon is in-phase with, has the polarization of, and propagates in the same direction as, the stimulating radiation*. This is a manifestation of the basic nature of photons to cluster in the same state (p. 1158). The incident lightwave is thereby increased in irradiance (Fig. 30.17).

Since most atoms are usually in their ground states, any incoming light is far more likely to be absorbed than to produce stimulated emission. This raises an interesting point: what happens if a substantial number of atoms could be excited into an upper state, leaving the lower state almost empty? Such a condition is called a

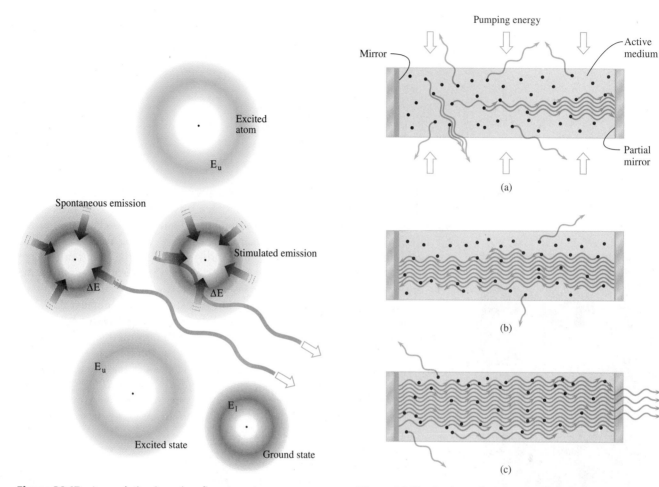

Figure 30.17 A population inversion: five atoms in a group where all but one are in an excited state. One atom (on the left) spontaneously returns to the ground state, emitting a photon, which causes a nearby atom to be stimulated into emitting an in-phase photon of its own.

Figure 30.18 A schematic representation of a laser. An active medium with most of its atoms in an excited state is located between two mirrors, one totally reflecting, the other 99% reflecting. Atoms spontaneously emit. Only photons along the axis stay within the cavity and stimulate additional emission. As the wave sweeps back and forth across the active medium, the wave builds until a portion leaks out as the laser beam. Energy can be pumped in continuously, in which case the beam can be continuous.

population inversion. An incident photon of the right frequency could then easily trigger an avalanche of stimulated photons—all in-phase, all perfectly in step. The initial triggering wave would continue to build as it swept across the active medium, as long as there were no dominant competitive processes (such as scattering) and provided the population inversion could be maintained. Moreover, the process could be enhanced by placing the active medium between two mirrors so that the light-wave passed back and forth through it many times. To allow some light to escape in the form of a beam, one of the mirrors could be made to be less than 100% reflecting; it would leak the beam (Fig. 30.18). In effect, energy (electrical, optical, chemical, whatever) would be pumped in to sustain the inversion, and a beam of light would be extracted. Such a device for **l**ight **a**mplification by **s**timulated **e**mission of **r**adiation is known as a **laser**.

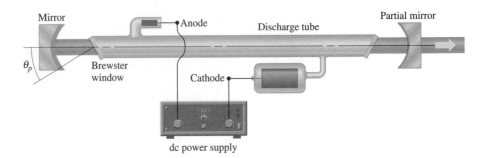

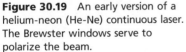

Figure 30.19 An early version of a helium-neon (He-Ne) continuous laser. The Brewster windows serve to polarize the beam.

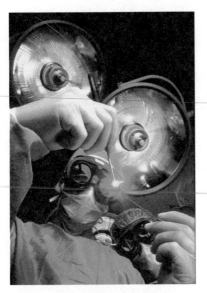

Surgery using a laser.

The two mirrors constitute a **resonant cavity**. Only certain wavelengths that "fit" (p. 510) can set up standing waves in the cavity, and only those wavelengths will be sustained. Consequently, the laser produces light of unmatched spectral purity; it is quasimonochromatic and therefore has a very long coherence length. Moreover, only waves sweeping back and forth parallel to the axis can undergo many reflections and thereby build in intensity. Thus, the rays emerging from the cavity are very nearly parallel, and the laser beam is highly directional.

The helium-neon (He-Ne) laser (Fig. 30.19) is still the most popular device of its kind. It puts out a continuous beam, usually of a few milliwatts of red light at 632.8 nm. The active medium, a mixture of about 10 Pa of neon (the active centers) to 100 Pa of helium, is placed in a gas-discharge tube. Pumping is accomplished through a high-voltage electrical discharge. What happens to the gases can be understood using the simplified energy-level diagram in Fig. 30.20. Many helium atoms are kicked up into a number of upper levels by the current. After dropping down, most accumulate in a long-lived state (E_{H2}), 20.61 eV above the ground level, from which there are no allowed radiative transitions.

The excited He atoms inelastically collide with and transfer energy to ground-state neon atoms, raising them into a long-lived energy state E_{N4}. This level is 20.66 eV above the ground level, the difference (0.05 eV) having been supplied from the KE of the colliding atoms. There then exists a population inversion, among the neon atoms, with respect to the lower E_{N3} level. Spontaneous photons initiate stimulated emission, and a chain reaction begins down from E_{N4} to E_{N3}. The result is the emission of bright red light at 632.8 nm. The E_{N3} level readily drains off to the E_{N2}

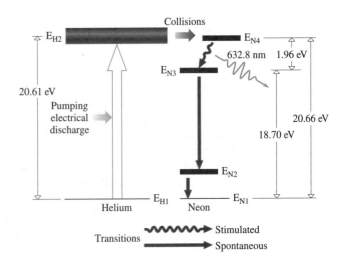

Figure 30.20 Simplified He-Ne laser energy levels. Here the lasing transition $E_{N4} \rightarrow E_{N3}$ is not down to the ground state. E_{N3} quickly dumps to E_{N2}, so the population inversion between E_{N3} and E_{N2} is constantly sustained.

level, sustaining the inversion. The latter is above the ground level, but that isn't significant—the important point is that lasing takes place between two upper levels. As a result, since the E_{N3} level is only sparsely occupied, the inversion is easy to maintain continuously, without having to half-empty the ground state.

Example 30.7 The helium-neon laser puts out a bright red beam at a wavelength of 632.8 nm. Please determine the difference in energy between the two states defining the transition.

Solution: [Given: $\lambda = 632.8$ nm. Find: ΔE.] There is a photon and an atomic energy transition, and that brings us back to Eq. (30.14): $E_i - E_f = hf$. Hence

$$\Delta E = \frac{hc}{\lambda} = \frac{(6.626\,2 \times 10^{-34}\ \text{J}\cdot\text{s})(2.997\,9 \times 10^8\ \text{m/s})}{632.8 \times 10^{-9}\ \text{m}}$$

and $\Delta E = \boxed{3.139 \times 10^{-19}\ \text{J}} = 1.959$ eV.

▶ **Quick Check:** The order-of-magnitude of the answer, a few eV, is appropriate for outer-electron transitions. The 1.96-eV transition corresponds to an energy in joules of $(1.96\ \text{eV})(1.601\,8 \times 10^{-19}\ \text{J}) = 3.14 \times 10^{-19}$ J, and since that value equals hf, $f = 4.738 \times 10^{14} = c/\lambda$ and $\lambda = 632.8$ nm.

30.7 Atomic Number

The **characteristic X-rays** that appear as sharp spikes superimposed upon the broad Bremsstrahlung spectrum (Fig. 30.9) come in groups much like the Balmer and Paschen series. Barkla (1911) named them the K, L, M, . . . and so forth series and, today, the spikes in each group are labeled (recall the hydrogen lines) K_α, K_β, K_γ, and so on. Just as the Balmer lines are an indication of the hidden outer electron structure of the atom, these series are equally as significant for the inner electrons. Bohr discussed this matter with his friend Henry Moseley, and Moseley, in 1913, brilliantly found the pattern. After a tremendous solitary effort, he was able to devise an empirical equation describing the frequencies of the K_α lines in terms of the particular atomic structure of the target; that is, in terms of something he called the *atomic number*.

Each element has an integer atomic number, and it increases, one unit at a time, from hydrogen (1), to helium (2), to lithium (3) and beryllium (4), up to uranium (92), just like a place number marking consecutive positions in the Periodic Table (inside back cover). Moseley astutely concluded: "This quantity [atomic number] can only be the charge on the central positive nucleus." *The **atomic number** is the number of units of nuclear charge, the number of protons in the nucleus (Z)*. The formula Moseley deduced was $f = C(Z - 1)^2$, where the constant C was subsequently shown to have the value $3cR/4$; hence

[*for K_α lines*] $$f = \left(\frac{3cR}{4}\right)(Z - 1)^2 \qquad (30.21)$$

Bohr conjectured that for all atoms, as the nuclear charge increased, the number of orbital electrons would also increase. One by one, they would build across the Periodic Table. Unhappily, the heavier atoms were beyond treating analytically with his simple theory. Despite that, W. Kossel (1914) suggested that if an inner tightly bound electron of one such atom was somehow removed from its orbit (for example, by bombarding it with cathode rays), all the other whirling electrons would cascade downward until that vacancy was filled. He called the innermost electron energy

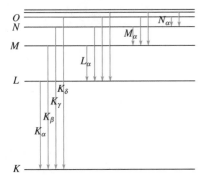

Figure 30.21   Inner-electron energy levels and the transitions that give rise to X-ray emission.

level K, the next L, and so on (Fig. 30.21), and that scheme led to a picture of electron transitions (much like the ones corresponding to the outer electrons) and the associated spectral lines. By comparison, these jumps were far greater in energy, and so emitted X-rays. For example, while 7.4 eV will remove the outermost electron from a lead atom, it takes 88 keV to remove either one of the K electrons.

Equation (30.21) can be rewritten for wavelengths, and it then takes on a familiar form:

$$\frac{1}{\lambda} = R(Z - 1)^2 \left[\frac{1}{1^2} - \frac{1}{2^2} \right] \tag{30.22}$$

This equation looks just like Eq. (30.20) except for the $(Z - 1)^2$ instead of Z^2. Ordinarily, the $n = 1$, K-level has two electrons in it, but now one has been removed, leaving one remaining. As an electron from an upper level plunges toward the hole, it sees an effective nuclear charge reduced (via Gauss's Law) by one negative unit because of the single nearby orbital electron; it sees $+e(Z - 1)$, not $+eZ$.

Example 30.8   A target emits K_α radiation of wavelength λ. Show that

$$Z = 1 + \sqrt{\frac{4}{3\lambda R}}$$

Then suppose that $\lambda = 0.2510$ nm. What metal is being used?

Solution: [Given: $\lambda = 0.2510$ nm. Prove above expression and find Z.] Starting with Eq. (30.22)

$$\frac{1}{\lambda} = R(Z - 1)^2 \left[\frac{1}{1^2} - \frac{1}{2^2} \right] = \frac{R(Z - 1)^2 3}{4}$$

and

$$(Z - 1)^2 = \frac{4}{3\lambda R}$$

Hence

$$Z = 1 + \sqrt{\frac{4}{3\lambda R}}$$

Thus

$$Z = 1 + \sqrt{\frac{4}{3(0.2510 \times 10^{-9} \text{ m})(1.097\,373 \times 10^7 \text{ m}^{-1})}}$$

and $Z = 23.0$, therefore the metal is $\boxed{\text{vanadium}}$.

▶ **Quick Check:** Using the expression on the left, $1/\lambda = R(22)^2 3/4$; $\lambda = 0.251$ nm.

The idea that the elements of the Periodic Table were to be ordered not by mass but by nuclear charge had been around for several years already. However, Moseley could directly observe the emitted X-rays and from them immediately tell whether or not the material under study was elemental and, if so, where in the Table it belonged! For instance, there was some controversy about argon (atomic weight 39.9) and potassium (atomic weight 39.1), since the former is an inert gas and should be located with the other inert gases (inside back cover), which would have it strangely preceding the lighter alkali metal. Moseley's X-rays revealed argon to have an Atomic Number of 18, whereas that of potassium was 19. Argon clearly belonged before potassium.

Unfortunately, Harry Moseley would solve no more of nature's mysteries; in 1915, at the age of twenty-eight, he was killed at Gallipoli in one of the most useless campaigns of World War I ("the war to end all wars").

Core Material

The total power radiated at all wavelengths by a blackbody is given by the **Stefan-Boltzmann Law**, namely

$$P = \sigma A T^4 \tag{30.1}$$

where A is the area, T is the absolute temperature in kelvins, and

$$\sigma = 5.6703 \times 10^{-8} \text{ W/m}^2 \cdot \text{K}^4$$

Given the **total emissivity** (ε), the power radiated by a surface is

$$P = \varepsilon\sigma AT^4 \qquad [30.2]$$

For a body in an enclosure, the net power radiated (when $T > T_e$) or absorbed (when $T < T_e$) is

$$P = \varepsilon\sigma A(T^4 - T_e^4) \qquad [30.3]$$

For blackbodies, **Wien's Displacement Law** is

$$\lambda_{max} T = \text{constant} \qquad [30.4]$$

where the constant is 0.002 898 m·K. **Planck's Radiation Law** is

$$I_\lambda = \frac{2\pi hc^2}{\lambda^5}\left[\frac{1}{e^{hc/\lambda k_B T} - 1}\right]$$

where $h = 6.626\,0755 \times 10^{-34}$ J·s (p. 1123).

In the Photoelectric Effect, the stopping potential is given by

$$KE_{max} = eV_s \qquad [30.6]$$

According to Einstein, **light is a stream of energy quanta**, each with an energy

$$E = hf = \frac{hc}{\lambda} \qquad [30.7]$$

The **Photoelectric Equation** is then

$$hf = KE_{max} + \phi \qquad [30.8]$$

The minimum wavelength of Bremsstrahlung (p. 1128) is given by

$$\lambda_{min} = \frac{ch}{eV} \qquad [30.9]$$

In the **Compton Effect**, a photon can scatter off a free electron with an accompanying wavelength shift given by

$$\Delta\lambda = \lambda_s - \lambda_i = \frac{h}{m_e c}(1 - \cos\theta) \qquad [30.13]$$

According to Bohr, for a hydrogen atom, a photon is emitted when the atom makes a transition from one energy level to another; that is

$$E_i - E_f = hf \qquad [30.14]$$

The resulting orbits have radii given by

$$r_n = n^2\frac{\hbar^2}{m_e k_0 Ze^2} = n^2 r_1 \qquad [30.17]$$

and energies given by

$$E_n = -\frac{2\pi^2 k_0^2 e^4 m_e}{h^2}\frac{Z^2}{n^2} \qquad [30.18]$$

Moseley showed (p. 1139) that

$$[\textit{for } K_\alpha \textit{ lines}] \qquad f = \left(\frac{3cR}{4}\right)(Z - 1)^2 \qquad [30.21]$$

Suggestions on Problem Solving

1. Some useful constants to keep in mind are

$$1 \text{ eV} = 1.602\,177 \times 10^{-19} \text{ J}$$
$$hc = 1.239\,842 \times 10^{-6} \text{ eV·m} \approx 1240 \text{ eV·nm}$$
$$h = 6.626\,08 \times 10^{-34} \text{ J·s} = 4.135\,67 \times 10^{-15} \text{ eV·s}$$
$$\hbar = 1.054\,573 \times 10^{-34} \text{ J·s} = 6.582\,12 \times 10^{-16} \text{ eV·s}$$

2. In problems involving electrons, we might be given the energy in eV or, equivalently, told the potential across which an electron moves. It's then convenient to use $m_e = 0.511$ MeV/c^2. The $1/c^2$ tells us we are dealing with mass, not energy (MeV). To get the electron's mass in kg, *multiply* 0.511 MeV/c^2 by 1.602×10^{-13} J/MeV and then *divide* by c^2 in (m/s)^2. The $1/c^2$ is not to be treated as if it were a unit like s in m/s, which we get

rid of by multiplying by s. The quantities $1/c^2$ or $1/c$ are literal divisions; they're not carried out numerically because we anticipate that they will soon cancel and it's convenient to leave them in this form, provided it doesn't confuse the issue.

3. When calculating things such as the minimum X-ray wavelength $\lambda_{min} = hc/eV$, the energy will often be encountered in electron-volts, whereupon it's helpful to use $hc = 1.239\,842 \times 10^{-6}$ eV·m.

4. Computations concerning the Compton Effect can be simplified a bit by utilizing the Compton wavelength: $(h/m_e c) = 0.002\,426$ nm.

5. Many modern calculators have built-in constants like c, h, e, k_B, m_e, etc. It will often be easier to use these built-in numbers and let them determine the form of the units (that is, eV versus J, etc.).

Discussion Questions

1. A candle or match flame usually has a distinct yellow color to it. One might guess that the color comes from sodium, but this is generally *not* the case. It's actually due to blackbody radiation from minute hot ($\approx$2000 K) particles of soot. How might you confirm this fact? (Incidentally, the bluish light at the base of the flame is from CH, C_2, and CO_2 molecular emission.)

2. Not very many solid materials can be heated to incandescence, and even fewer liquids can. Discuss and explain this observation. How do the following substances behave regarding incandescence: iron, water, glass, carbon, plastic, gasoline, banana, soap, wood, ceramics?

3. In order to test the photon hypothesis, Walther Bothe

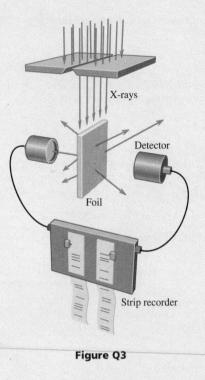

Figure Q3

conducted the following experiment (Fig. Q3). A low-intensity beam of X-rays was shone on a thin foil, which subsequently emitted X-rays (via X-ray fluorescence) toward two counters. The counts from the two detectors were recorded as they occurred and no correlation in time was observed. What does that "prove"? Explain.

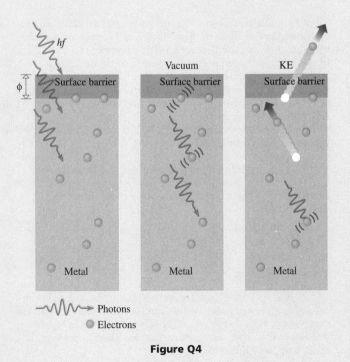

Figure Q4

4. Figure Q4 shows a sequence of drawings of three photons interacting with a metal. Explain each portion of the figure. Why does the liberated electron have a maximum kinetic energy?

5. With a bright source, such as a high-powered laser, it is possible to have a photoelectron absorb two or more photons before it leaves the metal (p. 1127). Given that an electron absorbs N identical photons of frequency f, how should Eq. (30.8) be modified, if at all? What will happen to the stopping potential as compared to the single photon effect? Does the work function change? Does the threshold frequency change? Explain your answers in detail.

6. Figure Q6 shows a current-versus-voltage curve for the Photoelectric Effect using polychromatic light. Explain every detail of the curve. Which frequency is higher?

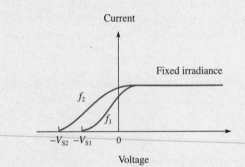

Figure Q6

7. Light absorbed by a substance may result in a chemical change of that substance via a process known as a *photochemical reaction* (which is why wine and aspirin are kept in dark-colored bottles). For each such reaction, there is a threshold frequency below which it will not occur. Assuming that the molecule must absorb an amount of energy that will trigger the process E_a, known as the *activation energy*, explain (as Einstein first did) the existence of a threshold frequency.

8. Figure Q8 shows some data for gamma-ray Compton scattering. Even though the gamma-ray wavelengths are much smaller than those of X-rays, the slope was found to be 2.4×10^{-12} m, in fine agreement with theory. Explain the curve in every regard.

9. A photon cannot be absorbed by a free electron—prove that statement by using Conservation of Energy and Momentum. Assume the electron is initially at rest. First, draw a diagram of the problem as seen from a coordinate system fixed with the center-of-mass—show it before and after the collision. Note that by definition the initial momentum in the center-of-mass system is zero. What is the final momentum? Write expressions for the initial and final energy. Now write the statement of Conservation of Energy (as seen from the center-of-mass). The fact that this implies that $m > \gamma m$ makes the initial assumption impossible.

10. Discuss the relationships between E, KE, and PE for any Bohr orbit of hydrogen. In the process explain Fig. Q10 and offer a definition of the binding energy of the electron. Which of these quantities is positive and which negative?

11. Figure Q11 shows a hypothetical 4-level laser. How does it work and what are the advantages of the arrangement?

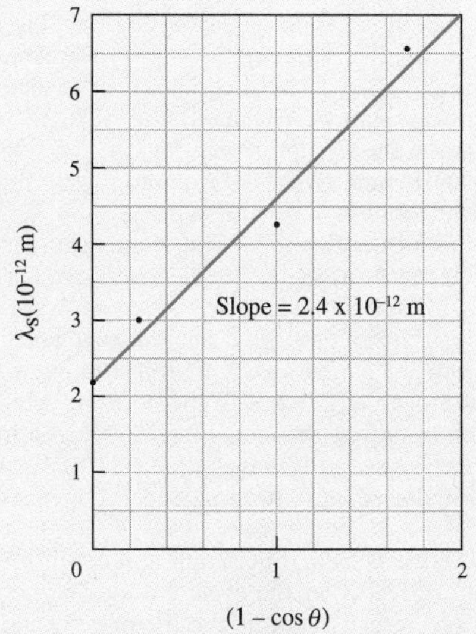

Figure Q8

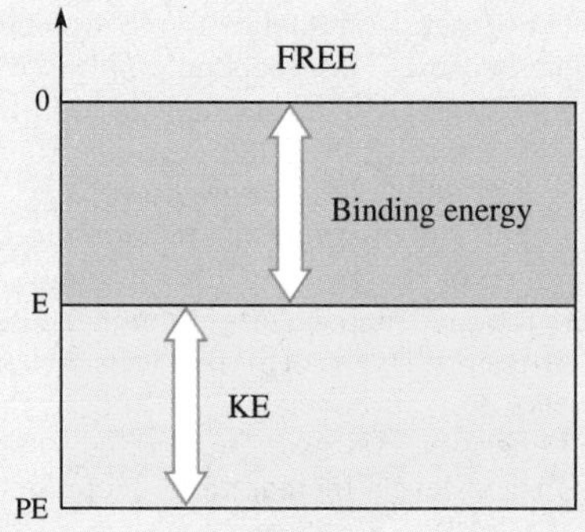

Figure Q10

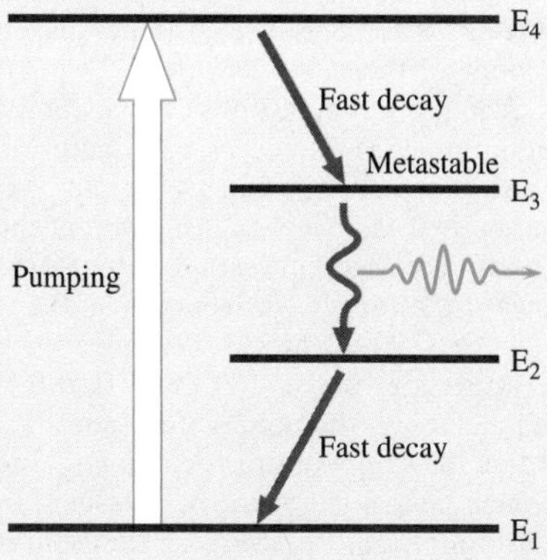

Figure Q11

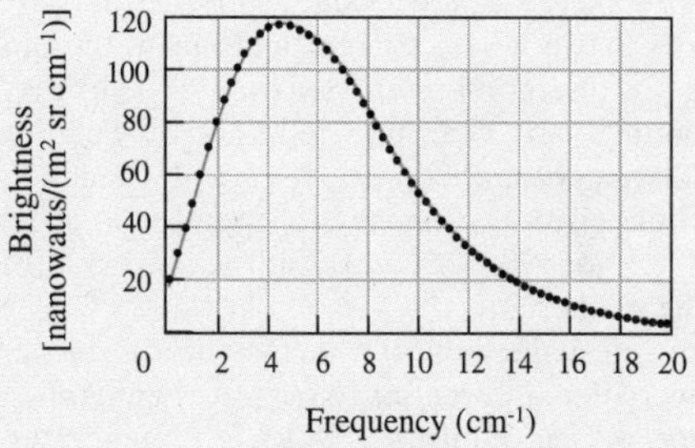

Figure Q12

12. It is theorized that the Universe was created roughly 17 thousand million years ago and is still expanding from that explosive moment. If that's the case, the original radiant fireball should have cooled considerably (via the Doppler Effect) as the Universe spread out to its present size. Less than a year after the so-called Big Bang, the radiation and matter should have reached equilibrium. By 300,000 years later, the radiation was red-shifted down to $\approx$4000 K and neutral atoms formed. Since then, the Universe has expanded about a thousandfold, and we can now expect isotropic cosmic background radiation with a blackbody temperature of about 3 K. Figure Q12 is a plot of brightness versus wavelength of the radiation detected by the *Cosmic Background Explorer* (*COBE*) satellite. The measured microwave spectrum corresponds to the black dots, the solid line is the Planck blackbody curve for a temperature of 2.735 $\pm$ 0.06 K. What can you conclude from all of this?

13. In 1986, a single barium ion (and in a different experiment, a mercury ion) was held in a "trap" and excited by laser light (p. 1136). The ion has two energy levels above the ground state that were utilized. The first was an ordinary short-lived state, and when bathed in light of the proper frequency, the atom absorbed and emitted 100 million times per second, putting out a seemingly continuous stream of light that was easily observed in a microscope. The other higher state was metastable and, once excited, the atom took several seconds to make the transition back down to the ground state via the emission of a lone photon (which went undetected). When stimulated into the first state, the atom shone with a "continuous" light, but whenever it was twinked into the metastable state, the shining light blinked off only to blink back on seconds later. Explain what happened.

Multiple Choice Questions

1. Pour hot water into a box whose outer walls are painted either black or white. Place thermal detectors at each face. Even though the entire surface of the box is at one temperature, (a) the black faces will be hotter (b) the white faces will be hotter (c) the black faces will radiate more energy (d) the white faces will radiate more energy (e) none of these.
2. It follows from the Stefan-Boltzmann Law that the energy per unit area per unit time radiated by a blackbody (a) depends on its total mass (b) depends on its color

(c) depends on its temperature   (d) depends on its total area   (e) none of these.

3. When the absolute temperature of a blackbody is doubled, its total radiation   (a) increases by a factor of 16   (b) stays constant   (c) increases by a factor of 4   (d) decreases by a factor of 2   (e) none of these.

4. All blackbody radiation curves have a peak that   (a) is shifted toward the longer wavelengths as the temperature goes up   (b) is shifted toward the longer wavelengths as the temperature goes down   (c) is shifted toward the higher frequencies as the temperature goes down   (d) is not shifted as the temperature goes up   (e) none of these.

5. Old-fashioned carbon filament incandescent light bulbs operated at around 2100 K. Modern tungsten lamps operate at around 2500 K. This more than triples the efficiency because   (a) it's always better to have the strength of metal   (b) at the higher temperature the blackbody curve puts more energy in the visible   (c) the Stefan-Boltzmann Law tells us we will get more light out of the tungsten   (d) we always get more energy from a metal than from a semiconductor like carbon   (e) none of these.

6. For us humans the theoretically ideal incandescent light bulb would have a filament that operated at around   (a) $20°$ C   (b) $98.6°$ F   (c) 2500 K   (d) 6000 K   (e) none of these.

7. In order to escape from the surface of a metal, an electron must overcome a potential barrier at the metal-vacuum interface, which requires   (a) a high temperature $T = hf/k_B$ (b) an energy equal to ϕ (c) a stopping potential   (d) a photon of energy $h\lambda$ (e) none of these.

8. The work function of a metal depends on   (a) the frequency of the incident radiation   (b) the voltage applied   (c) the maximum kinetic energy of the electrons   (d) the stopping potential   (e) none of these.

9. In the photoelectric effect, if $f > f_0$ and the irradiance of the incident beam is doubled, the photocurrent   (a) is unchanged   (b) decreases by a factor of 4   (c) doubles   (d) is halved   (e) none of these.

10. When dealing with photoelectrons, a plot of   (a) e times the stopping potential against photon frequency is a straight line for each metal   (b) photocurrent against potential is a straight line of slope h for each metal   (c) photocurrent against incident intensity is a straight line of slope h for each metal   (d) incident intensity against photon frequency is a straight line of slope ϕ for each metal   (e) none of these.

11. When a photon is Compton-scattered, the   (a) minimum shift $\Delta\lambda$ will equal twice the Compton wavelength   (b) minimum shift $\Delta\lambda$ will equal h times twice the Compton wavelength   (c) maximum shift $\Delta\lambda$ will equal h times twice the Compton wavelength   (d) maximum shift $\Delta\lambda$ will equal twice the Compton wavelength   (e) none of these.

12. Conservation of both energy and momentum lead to the conclusion that   (a) only free electrons can absorb a photon   (b) only bound electrons can absorb a photon   (c) no electrons can absorb a photon   (d) both free and bound electrons can absorb a photon   (e) none of these.

13. As the hydrogen atom goes from one Bohr orbit to another, increasing in radius, the   (a) electron's speed increases   (b) electron's speed decreases   (c) electron's speed remains unaltered   (d) proton's speed increases   (e) none of these.

14. The higher the principal quantum number of a Bohr orbit, the   (a) smaller the orbit and the less negative the energy   (b) larger the orbit and the more negative the energy   (c) larger the orbit and the less positive the energy   (d) larger the orbit and the more positive the energy   (e) none of these.

15. The classical planetary model of the hydrogen atom is unacceptable because   (a) the Coulomb force is too weak to hold the electron   (b) the nucleus is too powerful to allow the electron to orbit it   (c) the electron is accelerating and must radiate away its energy   (d) the electron's angular momentum is all wrong to be in such an atom   (e) none of these.

16. The greater the principal quantum number, the   (a) closer are adjacent energy levels   (b) farther apart are adjacent energy levels   (c) more nearly constant are the separations of adjacent energy levels   (d) more rapidly the separations between adjacent levels increase and decrease   (e) none of these.

17. A population inversion means that   (a) there are more atoms in one gas than in another   (b) there are more atoms in some excited state than in a lower state   (c) there are more states populated than unpopulated   (d) the lower states are filled rather than the higher ones   (e) none of these.

18. To pump a laser means to   (a) lower all its electron states   (b) remove some gas from it   (c) put pressure on the active medium   (d) excite its atoms   (e) none of these.

Problems

THE OLD QUANTUM THEORY
ATOMIC THEORY

1. [I] Suppose that a person has an average skin temperature of $33°$ and a total naked area of 1.4 m^2. If the person's total emissivity is 97%, find the net power radiated per unit area, the irradiance, when the environment is room temperature. How much energy does that person radiate per s?

2. [I] An object resembling a blackbody is raised from a temperature of 100 K to 1000 K. By how much does the

amount of energy it radiates increase?

3. [I] Assume you are a blackbody at 33°C (external temperature). What is the wavelength at which you radiate most energy? (See Problem 2).

4. [I] An object resembling a blackbody at room temperature (20°C) radiates energy into the environment. What is the wavelength that carries away the most energy?

5. [I] A class O blue-white star has a surface temperature of around 40×10^3 K. At what frequency will it radiate most of its energy?

6. [I] The Sun radiates the most energy at 470 nm. Taking it to be a blackbody, determine the temperature of its outer layer, the photosphere.

7. [I] Show that the units of action defined as either energy multiplied by time, or momentum multiplied by position, are equivalent.

8. [I] What is the energy in joules of a 0.100-nm photon?

9. [I] What is the wavelength of a 2.50-eV photon?

10. [I] What is the energy in keV of a 0.20-nm photon?

11. [I] A typical light level for good reading corresponds to about 2×10^{13} photons per second per square centimeter. If these have an average wavelength of 550 nm, what is the corresponding irradiance?

12. [I] What is the threshold frequency that will result in photoemission from zinc?

13. [I] Photons of wavelength 650 nm liberate photoelectrons with negligible kinetic energy from a metal target. What is the threshold frequency of the metal?

14. [I] Determine the cut-off wavelength for the emission of photoelectrons from tungsten, which has a work function of 4.52 eV.

15. [I] Ultraviolet radiation of wavelength 200 nm is shone on a zinc target. What's the maximum kinetic energy of the emitted electrons?

16. [I] In the photoelectric effect, what frequency of radiant energy should be shone on iron to produce electrons with a maximum kinetic energy of 1.50 eV?

17. [I] X-rays with a wavelength of 110 pm are scattered off free electrons at an angle of 20.0°. Find the change in λ.

18. [I] What is the shortest-wavelength X-ray emanation that can be expected from a tube operating at a voltage of 30.0 kV?

19. [I] At what scattering angle does the photon come off in the Compton Effect when the electron has a maximum kinetic energy?

20. [I] Compton shone 71.0-pm photons at a target containing loosely bound electrons and examined the scattered photon beam at 90.0°. What was the wavelength of the scattered photon?

21. [I] A 60-keV photon is Compton-scattered off an electron at rest. In the process, a 25-keV photon comes sailing off at 35°. What is the kinetic energy of the scattered electron?

22. [I] Determine (to 3 significant figures) the energy in electron-volts and in joules of the first excited state of the hydrogen atom.

23. [I] Compute the energy in eV of the second excited state of a hydrogen atom (to 3 significant figures).

24. [I] With Problems 22 and 23 in mind, what is the energy (in joules) of the quantum of light given out when a hydrogen atom drops from its second to its first excited state?

25. [I] With the last problem in mind, what is the frequency and wavelength of the light given out when a hydrogen atom drops from its second to its first excited state?

26. [I] The carbon dioxide laser usually emits at 10.6 μm in the IR. Typically such a device puts out a continuous emission of from a few watts to several kilowatts. Determine the energy difference in eV between the two laser levels for this wavelength.

27. [I] What is the radius of the second excited state of the hydrogen atom according to the Bohr Theory?

28. [I] A certain variation of He-Ne laser operates between

two levels that are 3.655×10^{-19} J apart. What color is the beam? Compute the wavelength.

29. [I] Compute the principal quantum number corresponding to a hydrogen atom with a 1-m orbital radius.

30. [I] The anode of an X-ray tube emits K_α radiation of wavelength 0.075 9 nm. What metal is the target made of?

31. [II] At what temperature will an object resembling a blackbody emit a maximum amount of energy per unit wavelength in the red end of the visible region of the spectrum ($\lambda = 650$ nm)?

32. [II] A blackbody is at a temperature of 6000 K. At what wavelength will it radiate the most energy per unit wavelength?

33. [II] The current in a Photoelectric Effect experiment decreases to zero when the retarding voltage is raised to 1.25 V. What is the maximum speed of the electrons?

34. [II] Suppose that 60-nm radiant energy is incident on a metal whose work function is negligibly small. Compute the maximum speed of the liberated photoelectrons.

35. [II] Photons of wavelength 220 nm impact on a metal target and liberate electrons with kinetic energies ranging from 0 to 61×10^{-20} J. Determine the threshold frequency and wavelength.

36. [II] In the Photoelectric Effect, the target has a stopping potential of 4.00 V when the incident energy has a momentum of 3.50×10^{-27} kg·m/s. What is the threshold frequency?

37. [II] An electron traveling at 1.00×10^8 m/s is fired at a metal target. On impact, it rapidly decelerates to half that speed emitting a photon in the process. What is the wavelength of the photon?

38. [II] In the Compton experiment, derive an expression for the kinetic energy of the scattered electron in terms of the incident and scattered photon wavelengths.

39. [II] On scattering via the Compton Effect, a photon undergoes a *fractional wavelength change* $(\lambda_s - \lambda_i)/\lambda_i$ equal to 6.00%. If the incident photon has a wavelength of 0.020 0 nm, at what angle is the detector to the incident beam?

40. [II] X-ray photons of wavelength 0.220 0 nm are Compton-scattered at 45°. What is the energy of the scattered photons (in eV)?

41. [II] What is the speed of the electron in the ground state of the hydrogen atom according to Bohr?

42. [II] Determine the force holding the electron in orbit in the ground state of the hydrogen atom.

43. [II] Compute the ground-state energy of singly ionized helium using the Bohr Theory.

44. [II] Use the Bohr Theory to prove that for hydrogen-like atoms

$$v_n = \frac{k_0 Z e^2}{n \hbar}$$

45. [II] Derive an expression for the frequency (f_n) of the electron revolving in the nth Bohr orbit of a hydrogen-like atom.

46. [II] With Problem 45 in mind, use the Bohr Theory to prove that

$$|E_n| = \tfrac{1}{2} n h f_n$$

for the various orbits (and not $E_n = n h f_n$).

47. [III] The Compton equation (30.13) is a formula in terms of photon parameters and the constant m_e. To derive it, start with Eq. (30.11) and write expressions for conservation of the x- and y-components of momentum. Show that

$$p_i^2 + p_s^2 - 2p_i p_s \cos \theta = p_e^2$$

Then, using Conservation of Energy, establish that

$$E_e = E_i + m_e c^2 - E_f$$

Substitute both of these formulas into the equation for the electron's energy

$$E_e^2 = c^2 p_e^2 + m_e^2 c^4$$

thereby getting an expression containing the electron's constant rest-energy and the energies and momenta of only the photons. Using $E = pc$ for photons, simplify, and Eq. (30.13) follows in short order.

48. [III] Approximate the time it takes for an electron in the Photoelectric Effect to absorb enough energy (≈ 1 eV or 2 eV) from a classical electromagnetic wave to be liberated. First assume a He-Ne laser source of 10 W/m². Compare this result with the time corresponding to an irradiance of about 1 μW/m², which is detectable by sodium. Assume the electron in a sodium atom absorbs over an area equal in radius (0.10 nm) to its orbit (which should be giving it the benefit of the doubt).

49. [III] When a photon is absorbed by an atom in the process of excitation, the atom must recoil if it is to conserve momentum. Consider a photon able to raise a 939-MeV/c² hydrogen atom into its first excited state. Determine the recoil energy of the atom and notice that it is negligible.

50. [III] Compute the energy of the third excited state of the helium atom in the Bohr Theory and compare it to the first excited state of hydrogen.

51. [ccc] According to Planck the energy per unit area per unit time per wavelength interval emitted by a blackbody at a temperature T is given by

$$I_\lambda = \frac{2\pi h c^2}{\lambda^5} \left[\frac{1}{e^{hc/\lambda k_B T} - 1} \right]$$

Accordingly, the total power radiated per unit area of the blackbody is equal to the area under the corresponding I_λ-versus-λ curve, at the specific temperature (see Fig. 30.2). Use all of this to derive the Stefan-Boltzmann Law. [Hint: To clean up the exponential, change variables in the integral so that $u = hc/\lambda k_B T$. Use the fact that $\int_0^\infty u^n \, du/(e^u - 1) = \Gamma(n + 1)\zeta(n + 1)$, where the gamma function is given by $\Gamma(n + 1) = n!$ and the Riemann zeta function for $n = 3$ is $\zeta(4) = \pi^4/90$.]

52. [ccc] We wish to establish that the stable orbits in the Bohr Theory correspond to those of minimum energy. Accordingly, show that the energy of an orbit as a function of r is given by

$$E(r) = \frac{n^2 h^2}{8\pi^2 m_e r^2} - \frac{k_0 Z e^2}{r}$$

Prove that for radii having values of

$$r = \frac{n^2 \hbar^2}{m_e k_0 Z e^2}$$

the energy is a minimum.

Quantum Mechanics

THE PACE OF THINGS intellectual picked up after the First World War (caution seemed less fitting in a battered world). The Jazz Era was largely a period of revolt and irreverence, and that selfsame turbulent mood gave form to the new physics that rose like a flash out of the Roaring Twenties. The hodgepodge of *ad hoc* ideas that constituted the Old Quantum Theory had gone about as far as it could. Quantum Mechanics emerged around 1925 as a complete theoretical structure that subsumed, and went far beyond, Classical Physics and the Old Quantum Theory.

Quantum Mechanics is far too mathematically complex for us to attempt to understand it here. Indeed, as Richard Feynman (1967) penetratingly remarked, "nobody understands quantum mechanics." But we can get a sense of it, see where it came from, learn a little about its strengths and weaknesses, and appreciate some of its accomplishments.

THE CONCEPTUAL BASIS OF QUANTUM MECHANICS

Amazingly, when physicists set about reformulating classical theory in the 1920s, they seemed at first to have come upon an overabundance of riches. Heisenberg created Matrix Mechanics, Dirac produced Transformation Theory, and Schrödinger devised Wave Mechanics, all almost simultaneously. Though their approaches were distinctive and their resulting formalisms seemed equally unique, it was soon shown that all three theories were essentially equivalent. These formulations are the basis of what is broadly called Quantum Mechanics. Our study will be cursory at best, and limited to only Wave Mechanics.

31.1 De Broglie Waves

In the summer of 1923, the French aristrocrat Prince Louis Victor Pierre Raymond de Broglie (pronounced to rhyme with *Troy*) proposed that the wave-particle duality (p. 1126) manifested by radiant energy might be a fundamental characteristic of *all* entities (for example, electrons, protons, and neutrons). Einstein had already shown that substantial matter (matter possessing mass) and radiant energy are interconvertible. Why should they not display similar properties? In particular, might not substantial matter have some sort of wave aspect associated with it? Suppose that Eq. (30.10), $p = h/\lambda$, which describes photons, applies to all matter. In that eventuality, with $p = mv$ for material entities, we have

$$\lambda = \frac{h}{mv} \qquad (31.1)$$

Prince Louis de Broglie (1892–1987) spent much of his life teaching physics at the Sorbonne and at the Institut Henri Poincaré, both in Paris.

Electron micrograph of a neuron (see Fig. Q1). With an accelerating voltage of around 10^5 V, the electrons have a de Broglie wavelength of about 0.004 nm. The photo of an insect shown on p. 1147 was also made using electrons.

Matter in motion has a wavelength. Not that de Broglie knew how to visualize such waves or even what was doing the waving. A particle moving with a constant momentum is associated with a monochromatic wave ($\lambda = h/p$) having a single frequency determined by its *total energy* ($f = E/h$) via Eq. (24.9). As with all waves, it moves with a phase speed given by $v_p = f\lambda$, which is usually different from, though related to, the speed of the particle (Problem 24).

However theoretical, the idea that electrons and protons were that much more like photons brought a pleasing symmetry to nature. This fascinating speculation was the basis of a Ph.D. thesis that de Broglie submitted to his physics professors in Paris. Still, the concept was considered outlandish, and a copy of the work was sent to Herr Professor Einstein for his opinion. The great man was enthusiastically supportive, and the degree was promptly granted. Six years later (1929), de Broglie received the Nobel Prize for the idea that particles are waves—but a remarkable accident happened first.

In April of 1925, the American C. J. Davisson was busy scattering electrons off a polycrystalline nickel target.

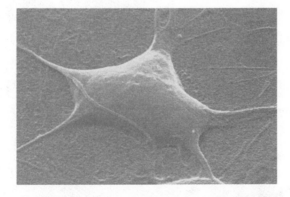

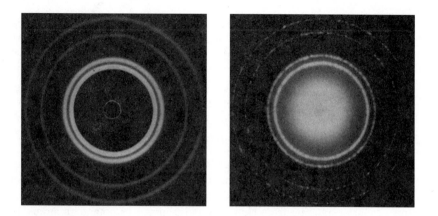

The concentric ring patterns produced when a beam of X-rays (left) and a beam of electrons (right) passes through the same thin polycrystalline aluminum foil. The two diffraction patterns are almost identical.

While collaborating with L. Germer, an explosion rocked the lab, but after putting the experiment back together, it was clear that something very strange had happened: the data "completely changed." Unbeknownst to them, while cleaning up the target through prolonged heating, the nickel sample had reformed into a few large crystals. When they subsequently directed an electron beam at the target, they saw a diffraction pattern identical to the one produced by X-rays (p. 1095), even though it's impossible for a stream of particles to do that. Another year passed before anyone realized that Davisson and Germer had verified de Broglie's hypothesis; Eq. (31.1) matched the data in every detail. *Electrons with comparable momenta to those of X-ray quanta are diffracted as if they were waves of the same wavelength.*

Example 31.1   Calculate the wavelength of an electron once it has been accelerated through a potential of 110 V and compare that to X-rays. Assume the speed is nonrelativistic.

Solution: [Given: $V = 110$ V for an electron. Find: λ.] To get λ, we will apply Eq. (31.1), and so we first need v. Since for the electron $KE_f = PE_i$, $\frac{1}{2}m_e v^2 = eV$

$$v = \sqrt{\frac{2eV}{m_e}} = 6.22 \times 10^6 \text{ m/s}$$

it follows that

$$\lambda = \frac{h}{m_e v} = 1.17 \times 10^{-10} \text{ m} = \boxed{0.117 \text{ nm}}$$

This is in the wavelength range of X-rays, just about the size of an atom.

▶ **Quick Check:**   We'll first derive a useful expression for λ: $KE = p^2/2m$, $\lambda = h/p = h/\sqrt{2m(KE)}$, multiplying top and bottom by c, and using the fact that $hc = 1239.8$ eV·nm (p. 1141), we have $\lambda = (1239.8 \text{ eV·nm})/\sqrt{2(mc^2)(KE)}$. For an electron, $mc^2 = 0.511$ MeV, and using the kinetic energy in eV, we have

[*for electrons*] $$\lambda = \frac{1.226}{\sqrt{KE}} \text{ nm} \qquad (31.2)$$

Here, where $KE = 110$ eV, $\lambda = 0.117$ nm.

Davisson shared the 1937 Nobel Prize with G. P. Thomson, J. J. Thomson's son, and therein lies another irony. The younger Thomson started out deliberately to test the idea of matter waves. By passing electrons through a thin metal foil, he confirmed that they behaved exactly as do X-rays of equal wavelength. And so J. J. Thomson "proved" that electrons were *particles,* and G. P. Thomson "proved" they were *waves.*

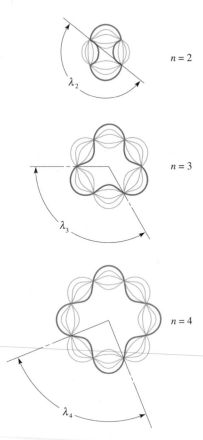

Figure 31.1 De Broglie standing waves for three excited states ($n = 2$, 3, 4) of hydrogen. Each Bohr orbit corresponds to a standing-wave configuration for the electron.

The rabbi spoke three times; the first talk was brilliant: clear and simple. I understood every word. The second was even better: deep and subtle. I didn't understand much, but the rabbi understood all of it. The third talk was a great and unforgettable experience. I understood nothing and the rabbi himself didn't understand much either.

A story told by Bohr about a young man reporting on his visit to a great rabbi as related by VICTOR WEISSKOPF

De Broglie's wave-particle theory offered another provocative interpretation of the Bohr atom. The quantization of angular momentum, $L = n\hbar$, means that $mvr = nh/2\pi$ and $mv = nh/2\pi r$. But if $mv = h/\lambda$, $n\lambda = 2\pi r$; in other words, the circumference of each of Bohr's allowed circular orbits ($2\pi r$) exactly equals an integer multiple of the electron wavelength ($n\lambda$). The orbits are electron standing-wave configurations (Fig. 31.1).

31.2 The Principle of Complementarity

When light propagates beyond an obstruction or through a hole, it casts a complex pattern in the process of diffracting, a pattern that can be analyzed in all its detail only if the light is taken as a wave. It is remarkable that researchers using all sorts of beams of matter, ranging from electrons and neutrons to potassium atoms, have now observed exactly the same complex interference and diffraction patterns. Classically, there is just no way to understand how beams of atoms can come together on a screen in clumps that exactly match optical interference patterns.

If the "logic" of Eq. (31.1) holds, and no one knows if it does, it applies equally well to oranges, baseballs, and fire engines as it does to electrons. The momentum of an ordinary macroscopic object in motion is so gigantic that λ is much too small to ever measure, and that even applies to something as small as a tiny dust mote gently falling through the air. Bullets may well have de Broglie wavelengths, though we'll never see them bending around obstacles. Unless you can pass a camel through the eye of a needle, observing diffraction (and thereby λ) is out of the question for structures made up of millions of atoms.

Microparticles (electrons, protons, photons, atoms, and so forth) propagate as if they were waves and exchange energy as if they were particles—that's the *wave-particle duality*. When we measure the arrival of one of them at a detector, we always measure the energy it delivers to a single point, even though that point may be part of a pattern that could only be created by a wave. In any experimental event, such entities will do one or the other—they cannot simultaneously manifest the properties of wave and particle. The imagery of particle and wave are two aspects of the complete picture of the photon, and they do not compete; rather, they complement each other. That notion (which does not immediately make things any less disconcerting) was first mentioned by Bohr in 1925 and came to be known as the **Principle of Complementarity**. Energy multiplied by temporal period ($T = 1/f$) equals momentum multiplied by spatial period (λ):

$$ET = p\lambda = h$$

The particle notions of E and p complement the wave notions of T and λ.

31.3 The Schrödinger Wave Equation

In 1657, Pierre Fermat proposed the intriguing idea that light, in going from one place to another, traveled along the route that took the least time. In a single simple statement, Fermat seemed to have gotten an overview of optics. His successes (for example, the derivation of Snell's Law) stimulated a great deal of effort to supersede Newton's Laws with a similar formulation. The resulting dynamical *Principle of Least Action* was introduced, albeit in a confused fashion, by Maupertuis (1747) and formalized by William Rowan Hamilton about a hundred years later. The principle asserts that a dynamical system moves in such a way as to keep its action ($\Sigma p \Delta q$) a minimum (here p is the momentum and q is the corresponding, or *conjugate*, space variable, for example, p_x and x or L and θ). The kinship between the propagation of light and the motion of a material particle is already there in this early work.

Hamilton found that there are mathematical surfaces over which the action of a system is constant, and that these *surfaces of constant action* propagate in a completely analogous fashion to the way surfaces of constant phase (that is, wavefronts) propagate in optics. The same equations apply to both! The angular temporal frequency ($\omega = 2\pi f$) of the light corresponds to the energy (E) of the particle; the angular spatial frequency ($k = 2\pi/\lambda$) of the light corresponds to the momentum (p) of the particle. Hamilton maintained that Newtonian Mechanics corresponded to the ray representation of Geometrical Optics. But the latter is only an approximation of the complete description of Wave Optics. What it lacks is interference, and what Hamilton's ray-dynamics for material entities lacked was also interference. Of course, this all happened almost a hundred years before de Broglie introduced the notion of the interference of electrons (the electron wasn't even discovered yet). For us, the question almost leaps from the page: *is Newtonian Mechanics an approximation of some as yet undiscovered Wave Mechanics?*

The Viennese physicist Erwin Schrödinger learned about de Broglie's wave hypothesis from a footnote in one of Einstein's papers—he was already quite familiar with Hamilton's work. Just before Christmas, 1925, he went off to the Swiss Alps for vacation, leaving his wife at home. He took along a copy of de Broglie's thesis and an old girlfriend (Erwin was a rather blatant philanderer). When he and the lady returned from their holiday two and a half weeks later, the great breakthrough had already been made. Starting essentially where Hamilton had left off, he produced an extraordinary formula now universally known as **Schrödinger's Wave Equation**:

Erwin Schrödinger (1887–1961). In 1925, he became interested in de Broglie waves, and by the end of that year he derived a correct relativistic version of Wave Mechanics. Thinking he was in error (because, without the concept of spin, the theory does not predict fine structure), he abandoned it and published the more limited nonrelativistic wave equation.

$$\frac{1}{2m}(i\hbar)^2\nabla^2\Psi + U\Psi = i\hbar\frac{\partial\Psi}{\partial t}$$

It was literally created by Schrödinger and presented as a postulate because it works. The equation cannot be derived directly from Classical Theory, although it has its roots there. Mathematically, it's well beyond what we can deal with here, but because it's one of the most important intellectual accomplishments of the twentieth century, it's worth examining. During the derivation, the expression had a constant K in it. Upon applying the analysis to the orbital electron of the hydrogen atom, Schrödinger immediately got the Bohr energy levels, provided K was set equal to $\hbar$. Interestingly, the theory naturally produced the quantum number n, but (like both Heisenberg's and Dirac's previous analyses), $\hbar$ had to be inserted into the formalism. The above equation is the full, time-dependent three-dimensional formulation. The symbol $\nabla^2\Psi$ stands for a set of partial second derivatives of Ψ in terms of the three space variables and the $\partial\Psi/\partial t$ is the partial derivative of Ψ with respect to time. Thus, the equation talks about the "motion" in space and time of a function Ψ (Greek letter capital *psi*) referred to as the **wavefunction**.

Schrödinger's Equation is the equation of motion of de Broglie waves, Ψ-waves, whatever they are. The symbol U represents the *potential energy* of the system under consideration. When U is constant in time, as it often is, the time-dependent part of the wave function can be canceled, leaving only the space-dependent part ψ. The one-dimensional, time-independent Schrödinger equation is then

$$\frac{1}{2m}(i\hbar)^2\frac{d^2\psi}{dx^2} + U\psi = E\psi$$

When we treat a particle, such as an electron in an atom, m is its mass, U is its potential energy, and E is its total energy. Recalling that $p^2/2m = $ KE, we see that the Schrödinger Equation is suggestive of the classical expression for the energy (KE + U = E) as Hamilton first wrote it, namely

$$\frac{1}{2m}p^2 + U = E$$

A particle constrained to a region by an interaction forms a *bound system*—for example, an electron in an atom. *The quantization of the energy of a bound system is a consequence of the wave equation and certain restrictions placed on the wavefunction.* Those "reasonable" restrictions simply require that ψ have only one value at each point, that it be continuous, and that it be finite everywhere. The quantum numbers that were introduced into the Bohr Theory in an *ad hoc* way now emerge, as Schrödinger put it, "in the same way as the integers specifying the number of modes in a vibrating string."

In general, Ψ is complex; it contains $i = \sqrt{-1}$ and thus is a so-called imaginary quantity. Accordingly, Ψ is not directly measurable; it has no physical existence. In other words, Ψ does not carry energy, as do all physical waves, and cannot be detected first-hand via that energy. Still, it has all the usual mathematical attributes of a wave: it has a frequency, amplitude, and phase; it obeys the superposition principle and so mathematically undergoes interference and diffraction. When two subatomic particles have identical physical characteristics (mass, charge, velocity, and so on), we say they are in the same *quantum state,* wherein they have identical wavefunctions. It is assumed that such particles are physically indistinguishable from one another.

New theories are rarely complete at the moment of presentation, and Schrödinger's was no exception. Although it proved to be remarkably effective, several features were quite troubling from the outset. In particular, what was the physical significance of Ψ? In the early days, Schrödinger assumed that the negative charge of the atom was actually spread out in space around the nucleus and that Ψ was related to the density of that charge-cloud. Curiously, although Schrödinger's theory became an immense success, the spread-out-charge interpretation of Ψ was a disappointing failure.

Probability Waves

Max Born, who was a professor of physics at Göttingen, was troubled by the idea that Ψ could undergo diffraction and be split up into separate portions—how could it characterize an electron that doesn't fragment? One can detect individual electrons with a Geiger counter and see their separate tracks in a cloud chamber. These are real manifestations of a highly localized entity.* In the spring of 1926, guided by a suggestion Einstein had made with regard to photons and the electric field, Born proposed what has now come to be the orthodox interpretation of Ψ. He suggested that the Ψ-wave associated with a particle was a probability function, an information wave that would tell us not where the particle is, but only where it is likely to be. Today, the first postulate of the orthodox interpretation of Quantum Mechanics is that **the state of a system is completely specified by its wavefunction**.

If we shine the image of some scene on a photon detector and then drastically lower the light level, we can watch the flash of each light-quantum arriving, one at a time. The recorded spots of light seem unpredictably random, but the image

---

*Electrons interact with each other over distances through their electric fields. Thus, two colliding beams scatter electrons via this field interaction. If the electrons have some size, some extent in space, as the beams come closer and closer, there should be a deviation from pure field scatter. As of this writing, distances down to 10^{-16} cm have been probed with no sign of a hard core—if the electron has a size, it's extremely small.

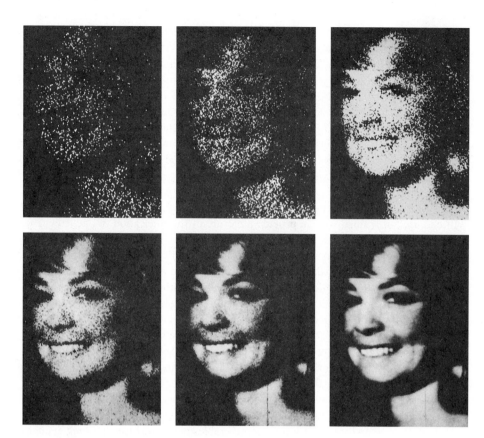

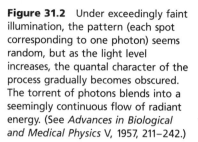

Figure 31.2 Under exceedingly faint illumination, the pattern (each spot corresponding to one photon) seems random, but as the light level increases, the quantal character of the process gradually becomes obscured. The torrent of photons blends into a seemingly continuous flow of radiant energy. (See *Advances in Biological and Medical Physics* V, 1957, 211–242.)

gradually builds until it is recognizable (Fig. 31.2). Under ordinary conditions, a great torrent of photons produces the picture almost all at once, obscuring the fundamental grainy nature of the process. Similarly, when light impinges on the double-slit arrangement of Young's Experiment (Fig. 31.3), it produces a flutter of flashes on the screen that blend into the familiar irradiance distribution of bright and dark bands. The irradiance (I) is proportional to the square of the amplitude of the electric-field wave (E_0^2) as per Eq. (24.8). The most likely place for a photon to

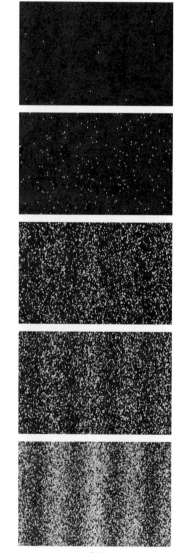

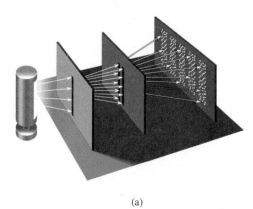

(a)

Figure 31.3 (a) Young's double-slit experiment. Flash by flash, photon by photon, the pattern gradually develops. (b) The double-slit interference pattern, but this time it's generated by a beam of electrons rather than photons. The number of electrons in each frame runs from 10 to 100 to 3000 to 20 000 and, finally, to 70 000.

(b)

arrive is where the pattern is brightest, where I is largest. Thus, the probability of a photon arriving at any particular location is given by the amplitude of the field squared. Naturally enough, Born suggested that if the experiment is carried out with electrons, or neutrons, or with any microparticle, the probability of finding such a particle at a specific location is given by the square of the amplitude of the wavefunction. Just as waves of electric fields interfere to produce complicated fringe patterns, Ψ-waves interfere to produce much the same patterns.

Schrödinger's Equation is quite as deterministic as is Newton's; that is, once the wavefunction is known, its future form can be computed precisely. Nonetheless, *the Ψ-wave can only tell us the probability of finding a particle at a given point in space.* Determinism as it applies to measurable quantities vanishes in the microworld; what is observable is probabilistic. We never know where a particular electron will end up, but instead we can provide a picture of the likelihood of occurrence of all its various possibilities. Large numbers of particles respond as a group whose overall behavior can be accurately predicted statistically. Even if the individual is lost in the crowd, the crowd *en masse* does its thing in a well-defined fashion that can be measured, understood, and anticipated. If we shine a beam of electrons at a pair of tiny slits, we can calculate the resulting fringe pattern precisely even though we know nothing of the specific motion of any one particle. Similarly, an actuarial table can tell you the probability of dying for any person in your age group. But it obviously cannot reveal how long you, a particular person, will live.

Still, Quantum Mechanics as we currently understand it challenges us at almost every turn. Suppose we again shine a beam of photons, electrons, or neutrons on a double slit. Now lower the beam intensity to the point where only one microparticle enters the apparatus at a time. It will pass through one of the slits (or not) and cause a flash on the screen (or not) at some unpredictable spot. But if we wait long enough, perhaps a few months, flash by flash, the traditional fringe system will appear. It seems that *each photon, electron, or neutron interferes with itself* (whether it interferes *only* with itself is another question). And it follows that the wavefunction, which undergoes the interference, must extend over both apertures and pass through both apertures. In that sense, the individual photon passes through both slits.

Envision a laser beam comprising a stream of photons. For what it's worth, one might picture the wavefunction of one such light-quantum as a thin, flat disk of progressing undulations—the photon has a probability of being anywhere within the confines of the beam and so the wavefunction is defined by the beam diameter. Some would say that the photon was a flying disk, while others keep their wavefunctions purely mathematical and think of the photon as a localized concentration of energy. Still others don't worry about what a photon looks like since it can't be seen in transit anyway. It's even possible in the case of light with a broad classical wavefront (starlight, for example) to separate the slits in Young's Experiment by 10 or 20 meters and still observe interference. If we buy the Ψ-wave picture, we must buy the idea that the photon's wavefunction then also extends over tens of meters.

The photon itself must somehow "see" both slits, however far apart they are, and yet arrive at the observation screen as a tiny localized blast of energy. If we cover either aperture, the interference vanishes. Indeed, if we place a particle detector behind either aperture so as to determine if the photon (or electron) has passed through it alone (as a particle might), the interference pattern also vanishes (see Discussion Question 5). And that is presumably true even if the beam is not obstructed as it is detected. Once the particle aspect is observed—once we know it has passed through one slit or the other—the wave aspect is obliterated. It's safe to say that many physicists are still not totally happy with the orthodox (or any other) explanation of these strange phenomena.

On the concept of quantum jumps. The whole idea of quantum jumps necessarily leads to nonsense. . . . If we are still going to have to put up with these damn quantum jumps, I am sorry that I ever had anything to do with quantum theory.

ERWIN SCHRÖDINGER

Just as surprising, if we set up several million identical double-slit experiments, shine a single identical photon (or electron or neutron) at each one at the same moment, and then superimpose the results from each, the total pattern should be the same fringe system as always. Each photon addresses the diffracting apertures individually and it progresses, guided by its wavefunction pattern, in a totally random way. Where on the theoretically computable, observably verifiable distribution it will land is indeterminant, but that it will land on the distribution is certain. This state of affairs is very peculiar! *It appears that identical particles in identical situations need not behave identically*—that's the essence of *quantum randomness,* which must be postulated in order to account for the observations. And again, not every physicist is happy wth these weird goings on, though most are pragmatic enough to keep on cranking out amazing solutions to real physical problems.

If we determine Ψ from Schrödinger's Equation for an electron moving in the Coulomb field of a proton, we get a whole range of potential electron locations, a cloud of probabilities. The most likely place to find the electron in the ground state is out at a distance identically equal to the Bohr radius, 0.052 9 nm (Fig. 31.4). There is no longer any reason to think of atomic electrons flying around in little orbits, though what they are doing from one moment to the next is a mystery. Figure 31.5 is a representation of the first few excited states of hydrogen. Keeping in mind that we are dealing with one rapidly moving electron, imagine a cloudlike afterimage around the nucleus, revealing where the electron has been and where it is likely to be.

The indeterministic substructure of the Universe was hinted at decades before Quantum Mechanics, when it was discovered that the decay of radioactive atoms was only predictable statistically (p. 1108). Generally, that realization was rationalized as simple classical ignorance, which we could hope to overcome, rather than profound quantum-mechanical indeterminism, which could never be eliminated. Even though it is now the consensus, the idea of a probabilistic world (wherein things like particle decay happen spontaneously, without apparent cause) does not sit well with all physicists. It certainly didn't sit well with Einstein, who complained that God does not play dice with the Universe.

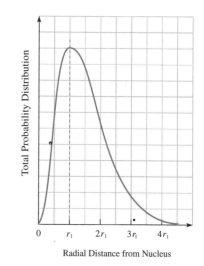

Figure 31.4 A plot of the probability of finding the electron at a distance from the nucleus of the hydrogen atom in its ground state ($n = 1$, $l = 0$). The most likely location is at the Bohr radius.

Figure 31.5 These are plots of the probability density, $|\Psi|^2$, for the electron in the ground state and in several excited states of the hydrogen atom. Each picture is a slice of the 3-dimensional electron cloud with the nucleus at the center. The brighter the region, the higher the probability of finding the electron. Each pattern corresponds to a specific set (p. 1159) of quantum numbers (n, l, m_l). For example, the ground state is specified by (1, 0, 0).

QUANTUM PHYSICS

Guided by challenging experimental observations and powered by a potent theoretical machinery, Quantum Mechanics developed rapidly in the 1920s and 30s. Attempts to understand the Zeeman Effect stimulated the introduction of the fundamental concept of *spin*. Another revelation, the *Pauli Exclusion Principle,* helped to uncover the detailed structure of the atomic electron cloud. The *Uncertainty Principle,* a profound overview of the microworld, was formulated by Heisenberg. A relativistic quantum theory, *Quantum Electrodynamics,* was introduced by Dirac, and it predicted the existence of *antimatter.* All of these milestones are explored in the remainder of this chapter.

31.4 Quantum Numbers

Modern Quantum Mechanics generates a set of fundamental quantum numbers in a mathematical way that flows naturally from the theory of differential equations. These numbers describe the state of a quantum-mechanical system and therefore specify the wavefunction. The

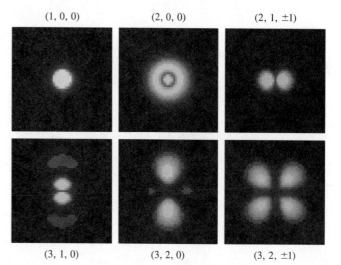

Schrödinger Equation has solutions that are energy values and, in one dimension, these are each associated with a value of the **principal quantum number** n. For a three-dimensional atomic system, there must be at least three quantum numbers (n, l, and m_l), and the rules for specifying these numbers come directly from the mathematics. These same rules for quantizing dynamical systems were more or less anticipated by the Old Quantum Theory, though in an *ad hoc* fashion.

Ever since Michelson's work (1891), it was known that each of the hydrogen spectral lines actually comprised several lines, very close together. With that in mind, Arnold Sommerfeld (1915) generalized the Bohr Theory to encompass this **fine structure**. What he needed was a very small additional energy term that would produce the observed multiplet structure. Accordingly, he assumed the orbits were elliptical and would precess about the nucleus. Surprisingly enough, when this variation on Bohr's theme was framed relativistically, it produced the desired effect in full agreement with observation. Thus, was the fine structure of hydrogen "explained." Sommerfeld's theory, which is now totally obsolete, was so accurate and visually appealing that the picture of interweaving ellipses persists even today. (Just ask any youngster to draw an atom).

Each possible motion of a system is known as a *degree of freedom* and, in effect, Sommerfeld had added another degree of freedom to the one circular orbital motion of the Bohr electron; elliptical orbits could swing around the nucleus. To the principal quantum number n, which essentially determined the orbital energy, he added a new quantum number—an important insight. To put things in a modern context, we call it the **orbital angular momentum quantum number**, symbolized by l and having integer values from 0 to $n - 1$ (that is, $n \neq l$). This quantum number determines the magnitude of the total orbital angular momentum (**L**), fixing it in steps of multiples of $\hbar$. Here the modern theory departs dramatically from Bohr's treatment in that it allows for an $l = 0$, zero angular momentum state. That possibility does great mischief to any thoughts of *orbiting* electrons!

31.5 The Zeeman Effect

The next problem to demand another degree of freedom (and a new quantum number) was remotely necessitated by Faraday, who was convinced that magnetism should affect light. Following his lead, Pieter Zeeman (1896), who had the advantages of a powerful electromagnet and a fine diffraction grating, placed a sodium flame in a B-field. Whenever "the current was put on, the two D lines were distinctly widened"—the magnetic field influences the way atoms emit light. The **Zeeman Effect** was quickly explained classically: the orbiting electron is a tiny current loop that has a magnetic moment and experiences a torque. As a result, the angular momentum vector revolves around **B**, and that alters the energy slightly, producing a splitting of the spectral lines (Fig. 31.6).

Debye and Sommerfeld were able to interpret the Zeeman Effect within the Old Quantum Theory by associating a new quantum number with the magnetic moment of the orbital electron. Today, this number is written as m_l and is known as the **orbital magnetic quantum number**. The energy of a bar magnet, a dipole equivalent to a current loop, in a B-field depends on its orientation with respect to the field. Accordingly, consider the magnetic moment of an orbital electron pointing in the opposite direction of its

Figure 31.6 An orbiting electron creates a magnetic moment that revolves around **B**.

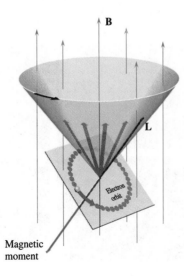

Magnetic
moment

angular momentum. When the electron is placed in a magnetic field, **L** can only assume certain orientations with respect to **B**. Sommerfeld called this *space quantization* ("spatial quantization" might be better). In other words, if **B** is in, say, the z-direction, L_z is quantized such that

$$L_z = m_l \hbar$$

wherein m_l goes from $-l$ to 0 to $+l$ in integer steps. For example, when $l = 2$, m_l can be -2, -1, 0, $+1$, or $+2$. And if $m_l = 0$, $L_z = 0$ and **L** will be perpendicular to **B**. For states other than the $n = 1$ ground state (for which $l = 0$ and $m_l = 0$), an applied magnetic field will split the energy levels into 3, 5, etc., sublevels whose separation depends on the strength of the field (Fig. 31.7).

31.6 Spin

Almost as soon as the problem of the Zeeman Effect was solved, it was complicated again. Although there were lines that behaved according to theory (the so-called Normal Zeeman Effect), it was discovered in 1897 that others, like those of sodium, actually split into a bewildering array of lines in a magnetic field (the Anomolous Zeeman Effect).

R. Kronig, from Columbia University, conceived the idea (1925) that the electron might be spinning about its own axis and therefore have a magnetic moment. Although Compton (1921) had already suggested that "the electron itself, spinning like a tiny gyroscope, is probably the ultimate magnetic particle," he never applied the idea. Kronig did, but there were relativistic problems with the notion of a finite spinning charged particle, and when several colleagues (among them, Pauli and Heisenberg) rejected the hypothesis, he decided to let it go unpublished. Ironically, S. A. Goudsmit and G. E. Uhlenbeck together independently introduced the idea of electron spin after reading a paper by Pauli in which he showed the necessity for associating four quantum numbers with the electron. They, too, hesitated, but fortunately they were working under Ehrenfest who told them "that it was either highly important or nonsense" and, in any event, he had already sent their paper off to be published.

Classically, an object such as the Earth can have any value of spin angular momentum—the quantity has always been assumed to be continuous. But that's not possible with particles in the atomic domain. To match experimental results, the **spin angular momentum quantum number** (s) for electrons must have only one value, $s = \frac{1}{2}$; because of this, electrons are referred to as *spin-$\frac{1}{2}$ particles*. Spatial quantization again applies because the spin magnetic moment orients itself with regard to an applied B-field. Thus, the component of the spin angular momentum vector in the direction of **B** is quantized and equals $m_s \hbar$ where m_s is the **spin magnetic quantum number**. Moreover, m_s can only change by steps of 1 going from $+s$ to $-s$; in other words, $m_s = \pm\frac{1}{2}$. The component of spin angular momentum of an electron in the direction of B can only equal $\pm\frac{1}{2}\hbar$. We speak of **spin-up** ($+\frac{1}{2}$) and **spin-down** ($-\frac{1}{2}$) electrons. The electron's energy in the magnetic field depends on its spin orientation, and so all of the levels are split by the applied B-field, thereby producing the Anomalous Zeeman Effect.

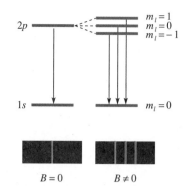

Figure 31.7 The splitting of energy levels in a magnetic field; the Zeeman Effect. The application of a magnetic field causes the atom to have three close-together excited states. The result is three emission lines where there ordinarily is only one.

A sunspot. These dark blotches on the face of the Sun are associated with powerful localized magnetic fields. The thin black vertical line in (a) corresponds to the location of the slit opening in a spectrograph that analyzes the light within that narrow region. In (b) we see three spectral lines running vertically. Where the slit opening passes into the strong magnetic field of the sunspot, the middle line clearly splits (b) into three lines as a result of the Zeeman Effect. By measuring the separations of these lines, the magnetic field ($\approx$4 T) can be determined.

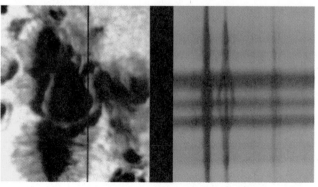

(a)   (b)

TABLE 31.1   Atomic quantum numbers

Name	Symbol	Values
Principal quantum number	n	$1, 2, 3, \ldots$
Orbital angular momentum quantum number	l	$0, 1, 2, \ldots, (n-1)$
Orbital magnetic quantum number	m_l	$0, \pm1, \pm2, \ldots, \pm l$
Spin angular momentum quantum number	m_s	$\pm\frac{1}{2}$

The notion of spin is not supposed to be taken literally. Particles do have angular momentum—a beam of light can twist a torsion pendulum—yet there are serious relativistic problems associated with picturing particles as tiny revolving entities. The idea of the electron being a minute spinning charged sphere is appealing but very problematic. What is said instead is that it has an inherent spin angular momentum (without the need for revolving), but that concept doesn't completely satisfy those who like their angular momentum to involve a clear angular motion. Still, in 1928, Dirac demonstrated that a relativistic quantum theory of the electron naturally provides intrinsic spin via an additional quantum number (Table 31.1).

31.7 The Pauli Exclusion Principle

The quantum numbers specify the state of a system and are therefore extraordinarily important. As Pauli showed, the state of any atomic electron can be determined via four quantum numbers: n, l, m_l, and m_s. Furthermore, **no two atomic electrons can occupy the same state**; that is, no two electrons in the same atom can have the same four quantum numbers. In a generalized form, this idea, known as the **Pauli Exclusion Principle**, is a basic precept of nature applicable to a wide range of situations.

We learn from Quantum Mechanics that the Exclusion Principle applies to particles that are known as **fermions**. These are particles (whether elementary or compound) that have spins that are odd integer multiples of $\frac{1}{2}$; particles such as the proton, neutron, and neutrino. It does not apply to the group of particles called **bosons**, which have zero or integer spins. Alpha particles, which are composed of even numbers of fermions, are bosons. Photons are spin-1 particles and therefore bosons.

The state of a photon is specified by its momentum vector and polarization. Thus, a planar monochromatic lightwave can be thought of as a stream of photons all in the same state; a circumstance that is possible because they are bosons. Indeed, bosons tend to accumulate in the lowest possible energy state. If that were the case for fermions, all atomic electrons would collect in the same lowest-energy level and the atom would compress. Moreover, when atoms are brought very close to one another (so close that their wavefunctions overlap), the Exclusion Principle obtains, which is why two hydrogen atoms can form a molecule only when the atoms have spin-up and spin-down electrons, respectively.

To date, tests of the Exclusion Principle have never found it wanting; researchers have seen no violations of it down to an accuracy of less than 2 parts in 10^{26}.

31.8 Electron Shells

During the 1920s, Bohr, Stoner, and others constructed a model of the electron structure of all the elements in the Periodic Table (inside back cover). Moseley's

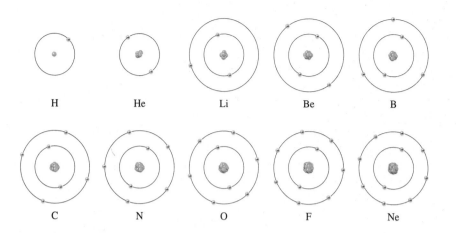

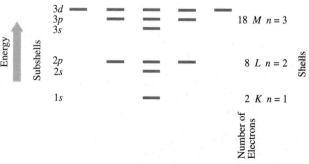

Figure 31.8 A schematic representation of the electronic structure of the first 10 atoms in the Periodic Table. (The sizes of the nuclei are tremendously exaggerated, as are the electrons.)

work provided a knowledge of the number of nuclear protons and, since the atom is neutral, that's the number of orbital electrons. It's no simple matter to sort out the electron structure (Fig. 31.8); for that analysis, chemical behavior and atomic spectra were the guides.

Electrons are ordered in shells and subshells about the various nuclei according to rules associated with their quantum numbers. Each shell corresponds to a specific value of n, and they are traditionally given letter names, where

$$\text{Shell} = K\ L\ M\ N\ O \ldots$$
$$n = 1\ 2\ 3\ 4\ 5 \ldots$$

Remember that l ranges from $(n - 1)$ to 0; that m_l goes from $-l$ to $+l$; and that m_s is either $+\frac{1}{2}$ or $-\frac{1}{2}$. Again, by tradition, the states of an electron are given letter names, where

$$\text{Subshell} = s\ p\ d\ f\ g\ h \ldots$$
$$l = 0\ 1\ 2\ 3\ 4\ 5 \ldots$$

A **shell** is a group of states that have the same principal quantum number. A **subshell** is a smaller group of states that has both the same value of n and l (Fig. 31.9). An **orbital** is specified by the three quantum numbers n, l, and m_l, and it can contain two electrons; one spin-up, one spin-down. (Each short horizontal line in Fig's. 31.9 and 31.10 represents an orbital.) And a **state** is specified by all four quantum numbers and contains one electron, as per the Exclusion Principle.

Hydrogen ($Z = 1$) has one electron and is chemically active. Its one unpaired electron gives it a valence of one and requires that it enter into covalent bonds in which the electron is shared with another atom. In the ground state, $n = 1$, $l = 0$, $m_l = 0$, and $m_s = \pm\frac{1}{2}$ (Fig. 31.11). The electron configuration is said to be a $1s$ orbital ($n = 1$, $l = 0$) of the K-shell (Fig. 31.9). Once two hydrogen atoms have combined (H_2) by sharing their spin-up, spin-down electrons, there are no longer any unpaired charges, and a third hydrogen cannot join the group (Table 31.2, p. 1161).

Helium ($Z = 2$) has two electrons, and because it's a stable *Noble Gas* (in the right-most column of the Periodic Table), we can conclude that two electrons must correspond to a completed system. Thus (even before the Exclusion Principle), it was presumed that the first (innermost) two

Figure 31.9 Ordering of atomic energy levels in terms of quantum numbers n, l, m_l. For example, the $3d$ subshell corresponds to $n = 3$ and $l = 2$. It comprises five orbitals corresponding to $m_l = -2$, -1, 0, $+1$, and $+2$.

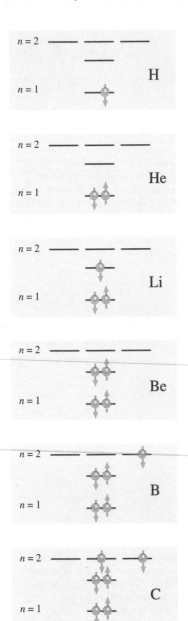

Figure 31.10 Distributions of electrons in the energy levels for the first six elements: hydrogen (H), helium (He), lithium (L), beryllium (Be), boron (B), and carbon (C).

electrons of any atom form a closed K-shell. In the ground state, the quantum numbers $(n, l, \mathrm{m}_l, \mathrm{m}_s)$ for one electron are $(1, 0, 0, -\frac{1}{2})$ and for the other are $(1, 0, 0, +\frac{1}{2})$. The K-shell has only two states and is filled. The 2-electron configuration is $1s^2$, where the superscript is the number of electrons (Table 31.3, p. 1162). Both electrons are close to the nucleus, and the atom has a high ionization energy of 24.6 eV (Fig. 31.12); it holds tightly to its electrons and does not interact effectively with other atoms.

Lithium ($Z = 3$) is next. It has three electrons, two in the first closed K-shell and the third rather far off in the second, or L-shell. With no room in the $n = 1$ shell, the third electron goes into the $l = 0$, s-subshell of the $n = 2$ shell; (n, l, m_l, m_s) for the third electron corresponds to $(2, 0, 0, -\frac{1}{2})$. The overall configuration is $1s^2 \cdot 2s^1$.

Beryllium ($Z = 4$) has four electrons, the last of which fills the $l = 0$, s-subshell and has quantum numbers $(2, 0, 0, +\frac{1}{2})$. The 4-electron ground-state configuration is $1s^2 \cdot 2s^2$; there are no unpaired electrons and the valence is zero. As it happens, the energy difference between the $(2, 0, 0, \frac{1}{2})$ state and the $(2, 1, 0, \frac{1}{2})$ state is very small and the beryllium atom can easily go into the $1s^2 \cdot 2s^1 2p^1$ configuration; that is, one electron can be raised up into the p-subshell. Accordingly, it can have a valence of either 0 or 2, but nothing else (Fig. 31.10).

Boron ($Z = 5$) has five electrons, the last of which goes into the next open level in the p-subshell; $(2, 1, +1, -\frac{1}{2})$. The 5-electron ground-state configuration is $1s^2 \cdot 2s^2 2p^1$, and a valence of 1 is to be expected. Here again, the configuration can be altered with the input of a small amount of energy, which is available to the atom via collisions even at room temperature. It then takes the form $1s^2 \cdot 2s^1 2p^2$.

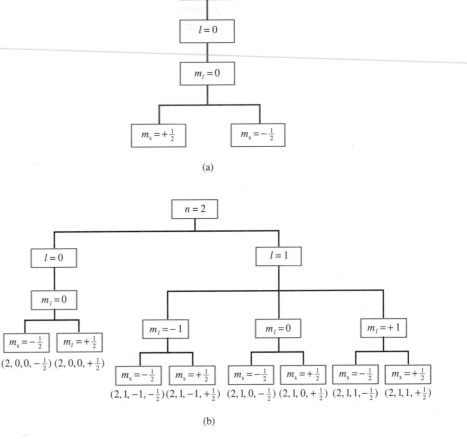

Figure 31.11 The quantum numbers and corresponding states (n, l, m_l, m_s) for the principal quantum numbers (a) $n = 1$ and (b) $n = 2$ for hydrogen.

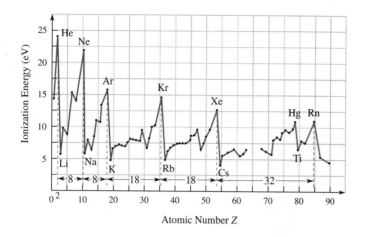

Figure 31.12 Ionization energy versus Atomic Number. Closed shells occur at $Z = 2, 10, 18, 36,$ and 54. An atom such as sodium has one electron beyond a closed shell ($Z = 11$), and it's shielded from the nucleus, so its binding energy is small.

TABLE 31.2   Ground-state configurations of several elements

Element	Symbol	Atomic number, Z	Electronic configuration
Hydrogen	H	1	$1s$
Helium	He	2	$1s^2.$
Lithium	Li	3	$1s^2.2s$
Beryllium	Be	4	$1s^2.2s^2$
Boron	B	5	$1s^2.2s^22p$
Carbon	C	6	$1s^2.2s^22p^2$
Nitrogen	N	7	$1s^2.2s^22p^3$
Oxygen	O	8	$1s^2.2s^22p^4$
Fluorine	F	9	$1s^2.2s^22p^5$
Neon	Ne	10	$1s^2.2s^22p^6.$
Sodium	Na	11	$1s^2.2s^22p^6.3s$
Magnesium	Mg	12	$1s^2.2s^22p^6.3s^2$
Aluminum	Al	13	$1s^2.2s^22p^6.3s^23p$
Silicon	Si	14	$1s^2.2s^22p^6.3s^23p^2$
Phosphorus	P	15	$1s^2.2s^22p^6.3s^23p^3$
Sulfur	S	16	$1s^2.2s^22p^6.3s^23p^4$
Chlorine	Cl	17	$1s^2.2s^22p^6.3s^23p^5$
Argon	Ar	18	$1s^2.2s^22p^6.3s^23p^6.$

Carbon ($Z = 6$) has six electrons, two in the $n = 1$ shell and four in the $n = 2$ shell. The last electron goes into the p-subshell; $(2, 1, 0, -\frac{1}{2})$. As a rule, *electrons filling a subshell do not double up in an orbital until each orbital has one.* Because of their mutual repulsion, the electrons get as far apart as possible by going into different orbitals. The 6-electron ground-state configuration is $1s^2.2s^22p^2$. Thus, we would expect carbon (Fig. 31.10) to participate in covalent bonds and have a valence of 2. And yet it almost always displays a valence of 4. This occurs because it takes a mere ≈ 2 eV to break the $2s^2$ pair, raising one of the electrons into the empty orbital of the p-subshell. There are then four unpaired electrons, yielding a valence of 4. Figure 31.13 shows how carbon can have valences of 0, 2, or 4, but nothing else. It's this range of valences that allows carbon to share electrons, forming single, double, and triple bonds (for example, C—O, C=C, C=O, C=N, C≡C, C≡N), and

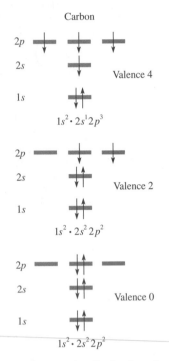

Figure 31.13 The distribution of electrons in carbon corresponding to valences of 0, 2, and 4. Each orbital fills up only when it contains two electrons. Accordingly, the valence corresponds to the number of single electrons that can be shared.

						K	L		M			N			
					n:	**1**	**2**		**3**			**4**			
Z	**Element**				*l:*	*s*	*s*	*p*	*s*	*p*	*d*	*s*	*p*	*d*	*f*
1	Hydrogen	H				1									
2	Helium	He				2									
3	Lithium	Li				2	1								
4	Beryllium	Be				2	2								
5	Boron	B				2	2	1							
6	Carbon	C				2	2	2							
7	Nitrogen	N				2	2	3							
8	Oxygen	O				2	2	4							
9	Fluorine	F				2	2	5							
10	Neon	Ne				2	2	6							
11	Sodium	Na				2	2	6	1						
12	Magnesium	Mg				2	2	6	2						
13	Aluminum	Al				2	2	6	2	1					
14	Silicon	Si				2	2	6	2	2					
15	Phosphorus	P				2	2	6	2	3					
16	Sulfur	S				2	2	6	2	4					
17	Chlorine	Cl				2	2	6	2	5					
18	Argon	Ar				2	2	6	2	6					
19	Potassium	K				2	2	6	2	6	·	1			
20	Calcium	Ca				2	2	6	2	6	·	2			
21	Scandium	Sc				2	2	6	2	6	1	2			
22	Titanium	Ti				2	2	6	2	6	2	2			
23	Vanadium	V				2	2	6	2	6	3	2			
24	Chromium	Cr				2	2	6	2	6	5	1			
25	Manganese	Mn				2	2	6	2	6	5	2			
26	Iron	Fe				2	2	6	2	6	6	2			
27	Cobalt	Co				2	2	6	2	6	7	2			
28	Nickel	Ni				2	2	6	2	6	8	2			
29	Copper	Cu				2	2	6	2	6	10	1			
30	Zinc	Zn				2	2	6	2	6	10	2			
31	Gallium	Ga				2	2	6	2	6	10	2	1		
32	Germanium	Ge				2	2	6	2	6	10	2	2		
33	Arsenic	As				2	2	6	2	6	10	2	3		

TABLE 31.3   Electron configurations of some atoms in their ground states

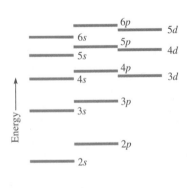

Figure 31.14 Relative positions of atomic orbitals.

thereby producing a wide range of complex molecules, including the ones that are the basis of all known life.

At the point where stable neon ($Z = 10$) is reached (Table 31.4), its ten electrons complete the first two shells, with two and eight electrons, respectively, ($1s^2 \cdot 2s^2 2p^6$). Argon ($Z = 18$) is the next Noble Gas, and it fills the $3p$-subshell in the *M*-shell with a configuration of $1s^2 \cdot 2s^2 2p^6 \cdot 3s^2 3p^6$. After argon, the order of several subshells changes because, for example, the $4s$ level is a little less energetic than the $3d$, so it fills first (Fig. 31.14). Krypton ($Z = 36$) fills the *M*-shell and has a configuration of $1s^2 \cdot 2s^2 2p^6 \cdot 3s^2 3p^6 \cdot 3d^{10} 4s^2 4p^6$. And so on up the Periodic Table.

With few exceptions, ***all the atoms in any one column of the Periodic Table have the same number of unpaired electrons***. The very active Alkali Metals all have only one such electron, which is easily given up or shared; the less active Alkaline

Earths (such as magnesium and calcium) have two; the boron, carbon, nitrogen, and oxygen families have three, four, five, and six, respectively; and the highly reactive Halogens each contain an almost complete shell of seven electrons.

Example 31.2 Determine the ground-state electron configuration for $_{12}$Mg. Start with neon, for which $(1s^2 \cdot 2s^2 2p^6)$.

Solution: [Given: $_{12}$Mg. Find: its electron configuration.] Neon has 10 electrons $(1s^2 \cdot 2s^2 2p^6)$ and a closed p-subshell. The $n = 2$ shell allows $l = 1$ or 0, so there is an s- and a p-subshell. The eleventh electron ($n = 3$) corresponds to $(1s^2 \cdot 2s^2 2p^6 \cdot 3s^1)$ and the twelfth to $\boxed{(1s^2 \cdot 2s^2 2p^6 \cdot 3s^2)}$.

▶ **Quick Check:** Look at Table 31.2, p. 1161.

31.9 The Uncertainty Principle

Born's statistical interpretation suggests that we can compute only the likelihood of individual events. This "blurriness" of vision, which is inherent in the formulation of Wave Mechanics, might seem peculiar. After all, if we know the precise location of a proton to begin with, as well as how fast and in what direction it's moving, why can't we follow it *exactly*? Perhaps that's not the right question, however; perhaps we should be bolder and ask whether or not we really can simultaneously measure both the position and velocity precisely in the first place. The product of position and momentum, or time and energy, brings us back to *action* and the quantum-of-action (h), which is at the heart of things quantum-mechanical. Questions about the measurability and definability of quantum concepts had occupied Werner Heisenberg in 1927 before he came up with the crucial insight known as the **Uncertainty Principle**.

Ordinarily, we observe the position of something by scattering something else off it: radar waves off a car; sound waves off a whale; lightwaves off your face; or the end of a cane off the curb. A stream of sunshine photons scattering from a flying baseball hardly affects its path, but one photon slamming into an electron drastically and unpredictably alters the electron's motion. Evidently, to measure anything in Atomland we will need a probe as well, but even the most delicate one possible will obtrude on the measurement. In order to observe the position of a microparticle precisely, we must—to cut down on diffraction—use a short-wavelength (and, therefore, a high-energy) probe. And that, in turn, will blast the particle away, making knowledge of its velocity and momentum even less precise. To approximate the effect, we use a photon of wavelength λ to locate a tiny object along the x-axis. Reasonably enough, assume the position can be measured to an accuracy of $\Delta x \approx \lambda$. At most, the photon can transfer all its momentum (h/λ) to the object, whose own momentum will then be uncertain by an amount $\Delta p_x \approx h/\lambda$. Hence

$$\Delta p_x \Delta x \approx h \qquad (31.3)$$

and this is a crude form of the Heisenberg Uncertainty Principle. It quantifies the "giveth and taketh away" interrelationship between so-called conjugate variables, such as position and momentum, and energy and time (p. 1150). Even though it's clear that the very act of measuring unavoidably obtrudes on the atomic level, most physicists now believe that the Uncertainty Principle has a yet more fundamental basis.

Consider a particle moving along the x-axis with a speed v and a momentum p_x. Momentum, according to de Broglie ($p = h/\lambda$), depends on wavelength, which, in

TABLE 31.4	Groups of elements	

Halogens	Number of electrons
Fluorine	9
Chlorine	17
Bromine	35
Iodine	53
Astatine	85

Noble Gases	Number of electrons
Helium	2
Neon	10
Argon	18
Krypton	36
Xenon	54
Radon	86

Alkali Metals	Number of electrons
Lithium	3
Sodium	11
Potassium	19
Rubidium	37
Cesium	55
Francium	87

Werner Karl Heisenberg (1901–1976) was awarded the Nobel Prize for physics in 1932 for his discovery of the Uncertainty Principle.

turn, corresponds to an extension in space. To specify λ, we must think of observing a cycle of the wave in space. The wave-mechanical concept of a precisely determined momentum is thus in direct conceptual conflict with the notion of a precisely defined conjugate position (x). Similarly, consider a particle of energy E moving at a time t. Energy (E = hf) depends on frequency, which, in turn, corresponds to an extension in time. To specify f, we must think of observing a cycle of the wave in time. The wave-mechanical concept of a precisely determined energy is thus incompatible with the notion of a precisely defined conjugate time (t).

Suppose we set out to simultaneously determine a pair of conjugate variables, say, p_x and x. Having made a number of measurements of any physical quantity, we would find a spread in the data about an average value. Accordingly, we record a momentum spread of Δp_x and a position spread of Δx. Classically, we would expect to be able to simultaneously sharpen up these values, making them both more and more precise, with no inherent conceptual limitations keeping us from improving things. We need only measure v over a vanishingly small track (Δx) centered at x. But the wave nature of the particle makes that process much less straightforward. To fix the momentum, we must fix the wavelength with precision, and to do so will necessitate allowing it some extension in space, thereby blurring x.

This linkage between conjugate variables is at the center of Heisenberg's discovery of the Uncertainty Principle, which is

$$\Delta p_x \Delta x \geq \tfrac{1}{2}\hbar \tag{31.4}$$

The product of the simultaneous uncertainties in position and momentum is at best equal to $\tfrac{1}{2}\hbar$. And the conceptually related expression

$$\Delta E \, \Delta t \geq \tfrac{1}{2}\hbar \tag{31.5}$$

holds as well. We cannot simultaneously know both the momentum and the position to any accuracy we wish. Homing in on one conjugate variable decreases its uncertainty and increases the uncertainty in the other variable.

Consider the implications of the above conclusions as they relate to complementarity; that is, to the wave-particle duality. The Uncertainty Principle sees to it that we will never be able to experimentally resolve the either/or issue of whether a photon is a particle or a wave. Similarly, precisely measuring the position of an electron fixes it with a sharp spatial localization, whereupon it may correctly be considered a particle. Yet when $\Delta x = 0$, it follows that $\Delta p = \infty$, and the electron has zero wavelength ($p = h/\lambda$); it has no wavelike attributes. The basis of the wave-particle duality thus seems to be the conceptual conflict between position and momentum as reflected in the Uncertainty Principle, and that sets the agenda for much of Quantum Mechanics. If we cannot know precisely an electron's present, we will have to rely on statistical means to tell its future. The very existence of the quantum of action smears certainty into probability and blurs cause and effect so that reality is not quite as sharp as we once thought it was. *Just as the unique character of Relativity arises from the fact that c is finite and not infinite, the unique character of Quantum Mechanics arises from the fact that h is finite and not zero.* The classical world behaves as if c = ∞ and h = 0. The real world behaves as if c = 3×10^8 m/s and $h = 7 \times 10^{-34}$ J·s. It's only on the macroscopic scale, where h is relatively negligible, that energy and momentum seem continuous and the world appears to be classical.

The more important fundamental laws and facts of physical science have all been discovered, and these are now so firmly established that the possibility of their ever being supplanted in consequence of new discoveries is exceedingly remote. Our future discoveries must be looked for in the sixth place of decimals.

ALBERT MICHELSON (1899)
American physicist

Δ*p* and Diffraction

Not surprisingly, a similar interrelationship to that which exists between momentum and position appears with light-waves: closing down a single slit (p. 1035) increases the spread in the diffraction pattern. Recall the equation for single-slit diffraction of light of wavelength λ; namely,

$$D \sin \theta_{m'} = m'\lambda \qquad [27.13]$$

Making *D*, the slit width, smaller increases θ. To see how the Uncertainty Principle can be used to provide more than just a numerical limit, consider the beam of electrons or photons impinging on a slit of width *D* in Fig. 31.15. The hole confines the beam, restricting it in the *y*-direction to a width *D* and making the uncertainty in the sidewise position of any electron Δ*y* = *D*. The electron must acquire a sidewise *y*-component of momentum Δ*p_y* on passing through the slit such that at best

$$\Delta p_y \, \Delta y \approx h$$

That's what is traditionally called diffraction; by passing through the slit, the electron was effectively localized in *y*, but its *y*-momentum component simultaneously spread out. From the diagram $\Delta p_y = p \tan \theta \approx p \sin \theta = (h/\lambda) \sin \theta$, and since $\Delta p_y \, \Delta y \approx h$

$$\left(\frac{h}{\lambda} \sin \theta \right) D \approx h$$

At best, the electron flies off at an angle somewhere between 0 and θ, the latter given by

$$\sin \theta \approx \frac{\lambda}{D}$$

From Eq. (27.13), that's the location of the first minimum (*m'* = 1). It effectively defines the central diffraction maximum, where most of the electrons will end up. If we knew nothing about diffraction, the Uncertainty Principle would tell us to expect the electrons to be spread out over the central region and beyond, rather than to cast a sharp image of the slit.

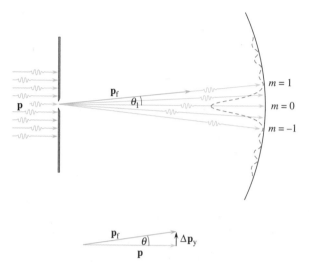

Figure 31.15 The Uncertainty Principle as applied to a stream of photons passing through a narrow slit of width *D*.

Uncertainty and Nuclear Electrons

It was pointed out earlier (p. 1106) that electrons do not reside in the nuclei of atoms, and a persuasive argument to that effect can be made using the Uncertainty Principle. An electron imprisoned in a nucleus must have a position uncertainty not greater than the size of the nucleus. The Uncertainty Principle then demands a corresponding indeterminancy in momentum, allowing values of *p* and therefore of energy quite inconsistent with observation.

Example 31.3 Assume that an atomic nucleus is 1.0 × 10⁻¹⁴ m in diameter and calculate the kinetic energy of an electron confined to such a space. Compare that to the energy range for beta rays (0.025 MeV to 3.2 MeV). Recall that an electron has a rest energy of 0.511 MeV.

(continued)

(continued)

Solution: [Given: nuclear diameter of 1.0×10^{-14} m. Find: KE.] The electron can be anywhere in the nucleus, and so $\Delta x \approx 10^{-14}$ m. Hence, the smallest uncertainty in momentum is

$$\Delta p_x \approx \frac{\frac{1}{2}\hbar}{\Delta x} = \frac{5.27 \times 10^{-35} \text{ J·s}}{1.0 \times 10^{-14} \text{ m}} = 5.27 \times 10^{-21} \text{ J·s/m}$$

The electron's momentum must be at least as large as this result, and so its total energy is

$$E^2 = E_0^2 + (pc)^2 \qquad [28.15]$$

For an electron, $E_0 = 8.19 \times 10^{-14}$ J and

$$E^2 = 6.70 \times 10^{-27} \text{ J}^2 + (5.27 \times 10^{-21} \text{ J·s/m})^2$$
$$\times (2.99 \times 10^8 \text{ m/s})^2$$

Thus $\qquad E = 1.58 \times 10^{-12}$ J $= 9.85$ MeV

and with $E_0 = 0.511$ MeV we have

$$\boxed{KE = E - E_0 \approx \boxed{9.3 \text{ MeV}}}$$

This is the smallest energy, and it's appreciably larger than the observed β-ray energies. That difference strongly suggests that there are ordinarily no electrons confined to the nucleus.

▶ **Quick Check:** To redo the calculation in MeV, use $h = 4.14 \times 10^{-15}$ eV·s in the expression $\Delta p \approx \frac{1}{2}\hbar/\Delta x$. Hence, $p = 0.0329$ eV·s/m and $p^2 c^2 = 9.73 \times 10^{13}$ eV^2. Since $E_0 = 0.511$ MeV, $E^2 = E_0^2 + p^2 c^2$ and $E \approx 9.9$ MeV.

Electrons can only exist inside the nucleus if there is a tremendous force restraining them there. A large collapsing star can have such immense pressures at its core that orbital electrons are crushed into the nuclei to create a giant ball of neutrons—a neutron star.

Determinism

Ultimately, we do not even know if a particle simultaneously has a perfectly defined position and momentum. There are many physicists who believe that it does not. In any case, we could not measure it if it did; not because we're momentarily without the proper method, but because no such method is theoretically possible. Heisenberg's original German word for his concept was *Unbestimmtheitsprinzip*, which better translates as "Indeterminacy Principle." "Uncertainty" suggests that a particle has definite simultaneous values of conjugate variables whose determination is uncertain, which was not Heisenberg's belief, though not everyone agrees.

There was a time when scientists like famed Laplace idly fancied that if they knew the initial location and momentum of all the particles in the Universe, they could (theoretically at least) foretell the future and retell the past. (This, despite free will and passion, as if your next warm kiss had been set in motion some 17 thousand million years ago when the Universe began.) That's the ultimate in *determinism,* and however beyond our actual computational abilities, the logic still disturbs. Though Newtonian theory weaves a deterministic plot where every sigh results from the inexorable confluence of physical law, Quantum Mechanics happily puts a touch of chance in the Hatter's clockwork. The future of even one lone electron is quite beyond prediction; the Uncertainty Principle makes it impossible to precisely know its initial conditions. So kiss whomever you like, change your mind however you will, and rest assured that physics, at least, knows that *it* cannot hope to calculate the future.

An intelligence which at a given moment knew all the forces that animate nature, and the respective positions of the beings that compose it, could condense into a single formula the movement of the greatest bodies of the Universe and that of the least atom: for such an intelligence nothing could be uncertain, the past and future would be before its eyes.

PIERRE SIMON, MARQUIS DE LAPLACE
(1749–1827)

I think I can safely say that nobody understands quantum mechanics.

RICHARD P. FEYNMAN
The Character of Physical Law (1967)

31.10 QED and Antimatter

In 1928, things were still simple, ominously simple: atoms were composed only of electrons and protons; the neutron had not yet been discovered; and there were no other microparticles in the zoo except for the photon. That was the year P. A. M. Dirac published "The Relativistic Theory of the Electron," for which he would share the 1933 Nobel Prize with Schrödinger. It was Dirac's intention to reformulate Quantum Mechanics so that it was consistent with Special Relativity. A relativistic quantum theory would naturally mate with Maxwell's Equations and, as such, is now known as **Quantum Electrodynamics** and affectionately called QED. The resulting *Dirac equation* was a triumph: among other things, electron spin came out naturally, Sommerfeld's fine structure formula was derived, and the magnetic moment of the electron was calculated. However splendid, the theory seemed to be flawed by one small aspect, one enigmatic mathematical quirk: it had twice as many energy states as were presumably called for.

Dirac traced the problem back to the relationship $E^2 = c^2p^2 + m^2c^4$; as with all square roots, it has two solutions; that is

$$E = \pm c\sqrt{p^2 + m^2c^2}$$

Paul Adrien Maurice Dirac (1902–1984). He shared the 1933 Nobel Prize in physics with Schrödinger.

Yet the energy of a free electron cannot be negative. It made no sense to keep the negative solution—its implications were too bizarre to be real. As is customary in such cases, Dirac simply put aside the ugly twin as a mathematical aberration. We do that all the time in Classical Physics. It soon became clear, however, that the negative portion could not be overlooked without seriously weakening the whole theory. Now there was a dilemma—a richly potent theory embarrassed by what seemed a bit of technical minutia.

It took Dirac until the end of 1929 to come up with a bold solution. The negative-energy states are real: many electrons have presumably radiated their energy and descended down into these lowest-possible states. Since the $\approx 10^{80}$ ordinary electrons in this Universe of ours are not observed to tumble into the oblivion of the negative Wonderland, it must be that all the negative states are occupied, in accordance with the Exclusion Principle. Wonderland has no vacancies. The situation was like a large hotel with half its floors below ground and half above. With most of the upper rooms empty, there could be lots of shifting around by the guests, but below ground level all the rooms were occupied. No one new could move down and in until someone moved up and out.

Instead of being empty, vacuum was now to be imagined filled with an invisible multitude of unseen electrons sailing around with negative energies. If one of them was somehow removed, pulled up into the ordinary positive world, it would leave behind an empty hole, a bubble in the negative sea of electrons. Dirac recognized that this hole would appear to us as a positive charge and proposed that it was the only known positive particle, the proton. J. Robert Oppenheimer quickly showed that this proposal was untenable. The hole had to behave as if it were the mirror image of the electron. "A hole, if there is one would be a new kind of particle, unknown to experimental physics, having the same mass and opposite charge to the electron." So wrote Dirac in 1931. "We may call such a particle an anti-electron."

This flight of creative imagination might have remained little more than a mathematical fantasy if not for a cosmic-ray study taking place at the California Institute of Technology. Carl Anderson, working with Millikan, was examining these particles (predominantly protons, possibly emitted from supernovas), which stream in on Earth from space. They were being observed using a cloud chamber placed in a uniform magnetic field so that positive and negative particles would follow oppositely curved paths. Among the thousands of photographs taken, Anderson noticed one that seemed curious. It clearly showed two oppositely curved tracks, one of which was certainly made by an electron; the other suggested an antielectron, but

All Nature is but Art, unknown to thee;
All Chance, Direction, which thou canst not see;
All Discord, Harmony not understood.

ALEXANDER POPE (1688–1744)

that conclusion was still "very radical at the time." Others had seen these positive tracks and either ignored them or dismissed them as "dirt." But Dirac's paper changed the world view, and the facts in the cloud chamber changed accordingly. By the summer of 1932, Anderson had clear evidence of the existence of the antielectron! It was Anderson who christened it the **positron**.

Today, it is commonplace to create electron-positron pairs—to have substantial matter materialize out of radiant energy (see the photo on p. 701). A blast of electromagnetic energy, a gamma ray with zero mass, can disappear and in its place an electron and a positron pop into existence. This ultimate alchemy occurs provided that the gamma-ray photon carries an energy equal to at least the total rest energy of the two particles (and provided there is a heavy object nearby to conserve momentum). Just as pair creation is possible, so is pair annihilation. A positron and *any* electron it approaches can come together and totally obliterate each other, vanishing in a puff of gamma rays (as if the electron had dropped back into the hole and disappeared).

Example 31.4 Determine the minimum energy (in joules and MeV) that a gamma photon must have to create an electron-positron pair.

Solution: [Given: an electron-positron pair. Find: minimum energy to create.] The minimum energy, E, will equal the rest energy of the pair; namely

$$E_0 = 2m_e c^2$$

$$E_0 = 2(9.11 \times 10^{-31} \text{ kg})(3.00 \times 10^8 \text{ m/s})^2$$

and so

$$\boxed{E = 1.64 \times 10^{-13} \text{ J}}$$

or

$$\boxed{E = 1.02 \text{ MeV}}$$

▶ **Quick Check:** Since $E = hf$

$$f = \frac{E}{h} = \frac{1.64 \times 10^{-13} \text{ J}}{6.63 \times 10^{-34} \text{ J·s}} = 2.48 \times 10^{20} \text{ Hz}$$

This result corresponds to 0.001 2 nm, which is a gamma ray.

Richard Phillips Feynman (1918–1988).

The proton is 1836 times more massive than the electron. Thus the creation of the **antiproton**, which would require that much more energy, could not be attempted until a new generation of accelerators was available. In 1955, the bevatron at the University of California, Berkeley, produced the first human-made antiproton—a negative speck with the same mass as the proton. Highly energetic beams of protons (p) and antiprotons (p̄) can slam together, creating all sorts of new subnuclear particles (p. 1218). On rare occasions, there is a cancellation of the equal and opposite charges and the creation of a neutron and an **antineutron** (1956). The similarity of neutrons (n) and protons is underscored by the fact that protons can similarly interact with antineutrons (n̄), as can antiprotons (often called p-bars) and neutrons.

Nowadays, antimatter is familiar in the laboratory; there are even artificial radioactive isotopes, such as ^{22}Na, that emit positrons on decaying and so serve as convenient sources. For example, by injecting someone with glucose laced with such a radioactive tracer, we can get a picture of the metabolism of various regions of the brain (Fig. 31.16). Antinucleons have been joined together to make compound structures such as antideuterons and antialpha particles. Researchers have made *positronium,* a hydrogen-like atom composed of a positron and an electron bound together in a somewhat stable form. And an exotic atom called *antiprotonic hydrogen,* a proton and antiproton orbiting each other, was created in 1978. Protons, electrons, neu-

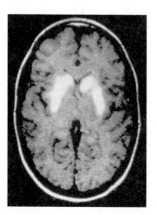

Photon
detector
array

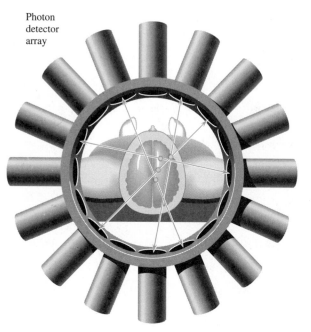

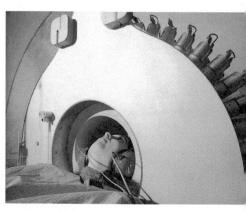

Figure 31.16 Positron Emission Tomography (PET). A patient receives an injection of some substance such as glucose that has been tagged with a radioactive element. The glucose soon goes to the brain, where it emits positrons. The positrons annihilate with electrons, producing pairs of gamma photons that are readily measured by a surrounding array of detectors. The images that result can reveal Alzheimer's disease, schizophrenia, and a wide range of other disorders of the brain.

trons, neutrinos (indeed, all the material subnuclear particles) have their antimatter twins, their mirror-image annihilators.

Dirac's theory worked well; in its first approximation, it agreed nicely with the known experiments of the day. His prediction of the electron's magnetic moment was accurate to two decimal places. But there gradually developed some very troublesome features: an atomic electron should interact both with the field of the nucleus and, to some extent, with its own field. When attempts were made to add in appropriate contributions, the theory went wild, producing totally weird results— things took on infinite values. In 1947, Willis Lamb at Columbia University used microwave techniques to discover that two quantum states of hydrogen actually had slightly different energies even though the Dirac theory predicted that they had to be *exactly* identical. This *Lamb shift* stimulated a modern-day reformulation of Quantum Electrodynamics, principally by Julian Schwinger and Richard Feynman of the United States and Sinitiro Tomonaga of Japan. They devised a mathematical scheme for making the theory workable known as *renormalization*. It got around the infinites, producing "normal" results.

Despite the fact that QED, the theory of the electron and photon, is certainly limited, it nonetheless is among the premier theoretical formalisms of physics. In the several instances where QED can be carried through, it provides an amazing degree of accuracy. For example, the magnetic moment of the electron as determined theoretically via QED has the numerical value 1.001 159 652 46, with an error of about 20 in the last two digits. The accepted measured value is 1.001 159 652 193, with a possible error of about 10 in the last two digits. A theory that can do that will engender a great deal of faith in its practitioners.

As we'll see (p. 1218), the structure of QED is very special in a way that few people were concerned about before 1954; it's a *gauge theory,* derivable from a symmetry principle. In time, QED would become a model for all the theoretical advances in High Energy Physics, the great work of the second half of the twentieth century.

Despite its enormous practical success, quantum theory is so contrary to intuition that, even after 45 years, the experts themselves still do not agree what to make of it.

BRYCE DEWITT
Physics Today (September 1970)

Core Material

According to de Broglie, the wavelength of a particle is

$$\lambda = \frac{h}{mv} \qquad [31.1]$$

and using the kinetic energy in eV, for nonrelativistic situations

[for electrons] $$\qquad \lambda = \frac{1.226\,4}{\sqrt{KE}}\text{ nm} \qquad [31.2]$$

Material particles that are moving have momentum and therefore a wavelength.

Heisenberg, Schrödinger, and Dirac each formulated a quantum-mechanical theory, all of which are closely related. **Schrödinger's Equation** in simple form is

$$-\frac{\hbar^2}{2m}\frac{d^2\psi}{dx^2} + U\psi = E\psi$$

and it is the basis of Wave Mechanics (p. 1150). The wavefunction, ψ, determines the probability of occurrences.

No two fermions in any system can occupy the same state; that is, have the same quantum numbers. This concept is the **Pauli Exclusion Principle** (p. 1155). The state of an atomic electron is fixed by four *quantum numbers: n, l, m_l, and m_s.*

One form of the **Heisenberg Uncertainty Principle** (p. 1163) is

$$\Delta p_x \Delta x \approx h \qquad [31.3]$$

or, more precisely

$$\Delta p_x \Delta x \geq \tfrac{1}{2}\hbar \qquad [31.4]$$

The position and linear momentum of an object cannot both be measured with unlimited precision at the same time. Similarly

$$\Delta E \Delta t \geq \tfrac{1}{2}\hbar \qquad [31.5]$$

Suggestions on Problem Solving

1. Some useful constants to keep in mind are

$$1\text{ eV} = 1.602\,177 \times 10^{-19}\text{ J}$$
$$hc = 1.239\,842 \times 10^{-6}\text{ eV·m} \approx 1240\text{ eV·nm}$$
$$h = 6.626\,08 \times 10^{-34}\text{ J·s} = 4.135\,67 \times 10^{-15}\text{ eV·s}$$
$$\hbar = 1.054\,573 \times 10^{-34}\text{ J·s} = 6.582\,12 \times 10^{-16}\text{ eV·s}$$
$$\tfrac{1}{2}\hbar = 3.29 \times 10^{-16}\text{ eV·s}$$

2. We often have to find such things as the wavelength of, say, an electron, and so need its momentum ($p = h/\lambda$). And here a decision has to be made as to whether to compute the classical value $p = \sqrt{(KE)2m}$ or the relativistic value $p = \sqrt{(E^2 - E_0^2)}/c$. When $E_0 \gg KE$, we can use the classical formulation. For ex-

ample, for an electron with a KE of 0.1 MeV (as compared to its rest energy of 0.5 MeV), using the classical momentum in calculating wavelength introduces a 5% error.

3. When calculating the de Broglie wavelength ($\lambda = h/p$) and frequency ($f = E/h$) for anything other than a photon, remember that λf is the phase speed of the wave, *not* the speed of the particle and *not* c.

4. The following summary of levels and quantum numbers is handy to have:

Principal	$n = 1, 2, 3, 4, \ldots$
Orbital	$l = 0, 1, 2, 3, \ldots, n-1$
Magnetic	$m_l = l, l-1, \ldots, 0, \ldots, -l+1, -l$
Spin	$m_s = -\tfrac{1}{2}, +\tfrac{1}{2}$

Discussion Questions

1. Given that the resolving power of a microscope is proportional to the wavelength of the illumination, why might an electron microscope be appealing? Figure Q1 shows a transmission electron microscope invented by Ruska in 1935. How does it work? Incidentally, electrons traveling within the axially symmetric magnetic field inside the gap of a current-carrying coil tend to be brought inward to a focus. Compare this tendency with the behavior of X-rays. While an optical instrument might have a resolution of 200 nm or so, a modern electron microscope can resolve objects as small as about 0.1 nm. (The limit on resolution is actually set by the spherical aberration of the lenses and not λ.)

2. A narrow, very sparse beam of monoenergetic protons is shone on an even narrower slit in an opaque screen. Make a set of drawings showing the pattern you would expect to observe

after several increasing intervals of time. Explain your conclusions. What if the beam were not monoenergetic, what would happen?

3. Draw a crude pictorial representation of the electron shell structure of sodium fluoride (NaF) and explain how the molecule is held together. Do the same for hydrogen chloride (HCl).

4. Considering the shell model of the atom, derive an expression for the number of electrons in the nth shell and also the number in the lth subshell. Explain your reasoning.

5. Imagine a screen with two narrow slits in it and suppose there is an incident beam of monochromatic electrons that encompasses both apertures (Fig. Q5). An interference pattern typical of Young's Experiment will be observed on a distant surface. Now suppose that each slit is surrounded by a coil of wire such that, as an electron passes through the opening and then the

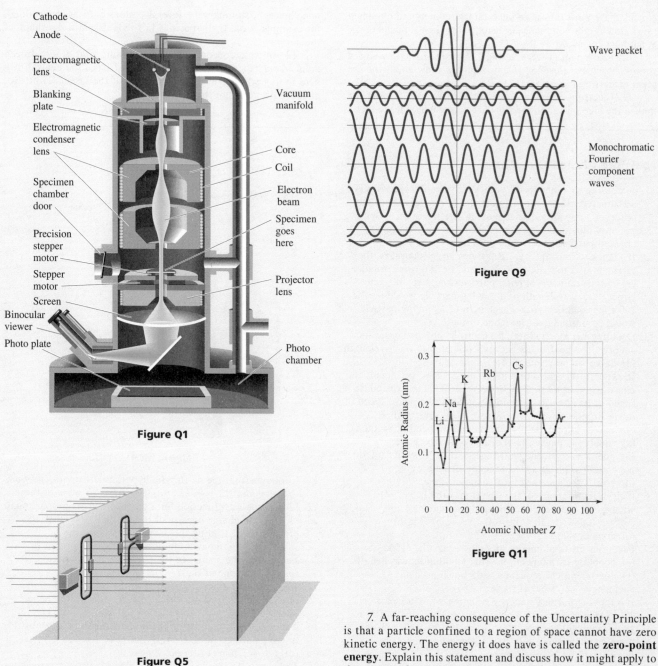

Cathode

Anode

Electromagnetic lens

Blanking plate

Vacuum manifold

Electromagnetic condenser lens

Core

Coil

Specimen chamber door

Electron beam

Specimen goes here

Precision stepper motor

Stepper motor

Projector lens

Screen

Binocular viewer

Photo plate

Photo chamber

Figure Q1

Wave packet

Monochromatic Fourier component waves

Figure Q9

Atomic Radius (nm)

Li

Na

K

Rb

Cs

Atomic Number Z

Figure Q11

Figure Q5

coil, a current will be induced and the passage of the particle appropriately recorded. Use the arguments of Quantum Mechanics to describe what, if anything, will happen to the interference pattern once the switches on the coils are closed. If only one coil was used, would that change things?

6. Like photons, electrons have a wavelength, frequency, spin, and phase. Might it be possible to build a laserlike device that will produce tremendously intense beams of coherent electrons? Explain.

7. A far-reaching consequence of the Uncertainty Principle is that a particle confined to a region of space cannot have zero kinetic energy. The energy it does have is called the **zero-point energy**. Explain this statement and discuss how it might apply to the energy of a material at a temperature of absolute zero.

8. In 1929, Otto Stern shone a beam of neutral helium atoms (with a de Broglie wavelength of about 0.13 nm) at an angle to the surface of a crystal. What do you think he saw when he examined the pattern of the reflected helium atoms? Explain.

9. A reasonable way to represent a particle mathematically is to use a localized wavefunction like that of Fig. Q9. Such a pulse is known as a *wave packet,* and it's equivalent to the super-position of a number of monochromatic waves, as indicated in the figure. The tighter (shorter in space) the packet, the more numerous the needed monochromatic contributions (remember Fourier's analysis). In what sense does this description of the

particle have built into it an uncertainty in energy and momentum? How might you make Δp zero? What then happens to Δx? What happens as $\Delta x \to 0$?

10. Can a photon create a single electron? Explain your reasoning. What conclusions, if any, can you draw about the creation of particles in general?

11. Electron clouds are statistical distributions, and that makes the size of an atom somewhat ambiguous. Figure Q11 is nonetheless a summary of several atomic radii as determined, for example, from their spacings in solids. Explain the shape of the curve.

12. In what conceptual sense is a Bohr orbit of a hydrogen atom like a vibrating circular steel hoop? Does the central idea here in any way relate to the structure of the laser? [Hint: Remember Kundt's tube?]

Multiple Choice Questions

1. According to contemporary physics, wavelike behavior is a characteristic of   (a) all particles at rest relative to the observer   (b) all particles moving relative to the observer   (c) only moving charged particles   (d) only stationary charged particles   (e) none of these.

2. If Planck's Constant were 100 times larger than it is, the   (a) mass of a moving particle would be 100 times smaller   (b) momentum of a moving particle would be 100 times larger   (c) wavelength of a moving particle would be 100 times smaller   (d) wavelength of a moving particle would be unchanged   (e) none of these.

3. Doubling the momentum of a neutron   (a) decreases its energy   (b) doubles its energy   (c) doubles its wavelength   (d) halves its wavelength   (e) none of these.

4. In general, the speed of a material particle   (a) equals the phase speed of its de Broglie wave   (b) equals the phase speed of its Ψ-wave   (c) does not equal the phase speed of its de Broglie wave   (d) equals c   (e) none of these.

5. For a monochromatic photon, the phase speed in vacuum   (a) equals the phase speed of its de Broglie wave   (b) does not equal the phase speed of its Ψ-wave   (c) does not equal the phase speed of its de Broglie wave   (d) equals the speed γc (e) none of these.

6. If the circumference of the first Bohr orbit is 3.3×10^{-10} m, what is the wavelength of the ground-state electron?   (a) 0   (b) ∞ (c) 3.3×10^{-10} m   (d) $(3.3 \times 10^{-10}$ m$)/h$ (e) none of these.

7. Doubling the total energy of a meson has the effect of   (a) doubling its momentum   (b) doubling its wavelength   (c) quartering its frequency   (d) doubling its frequency   (e) none of these.

8. If a moving particle's energy is increased by a factor of 10,   (a) its frequency increases by a factor of 10   (b) its frequency decreases by a factor of 10   (c) its frequency remains unchanged   (d) its wavelength increases by a factor of 10   (e) none of these.

9. Neutrons from a reactor are slowed down by passing them through graphite (Fig. MC9) so that they have a de Broglie wavelength of about 0.1 nm. The beam is directed at a crystal, and the neutrons   (a) because they are uncharged pass right through the crystal, totally unaffected by it   (b) reflect off at an angle equal to the incident angle, just like a stream of baseballs   (c) are totally absorbed   (d) reflect off in several beams, as determined by the Bragg equation   (e) none of these.

10. It follows from the de Broglie hypothesis that   (a) $E = \omega \hbar$ (b) $E = \omega h$ (c) $E = f \hbar$ (d) $E = \lambda \hbar$ (e) none of these.

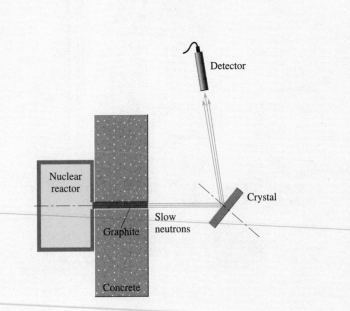

Figure MC9

11. It follows from the de Broglie hypothesis that   (a) $p = \omega \hbar$ (b) $p = \lambda h$ (c) $p = k \hbar$ (d) $p = \lambda \hbar$ (e) none of these.

12. If Ψ is the wavefunction for a particle, $|\Psi|^2$ is proportional to   (a) the charge density of the particle   (b) the probability of finding the particle at a point in space   (c) the momentum of the particle at a point in space   (d) the energy of the particle at the point in space   (e) none of these.

13. The principal quantum number of the 6th excited state of hydrogen is   (a) 6   (b) 5   (c) 4   (d) 7   (e) none of these.

14. The maximum number of electrons that can be contained in the f-subshell of an atom is   (a) 14   (b) 10   (c) 6   (d) 12   (e) none of these.

15. The size of the space in which an electron is confined determines the uncertainty in its   (a) linear momentum   (b) maximum angular momentum   (c) spin angular momentum   (d) its mean lifetime   (e) none of these.

16. We might say that an electron in the ground state cannot radiate because   (a) if it did, its wavelength would increase and it could not fit in any smaller lower-energy orbit   (b) it has no energy to radiate in the ground state   (c) it would decrease in wavelength and lose momentum   (d) it would not conserve both energy and momentum   (e) none of these.

Problems

THE CONCEPTUAL BASIS OF QUANTUM MECHANICS
QUANTUM PHYSICS

1. [I] Calculate the wavelength of a 60-kg person jogging along at 2.0 m/s.
2. [I] What is the de Broglie wavelength of a 10.00-g bullet traveling at 331 m/s?
3. [I] Determine the wavelength of an electron traveling at a speed of c/10.
4. [I] Determine the frequency of an electron traveling at a speed of c/10. Use relativistic considerations.
5. [I] A *thermal neutron* is one that is traveling at a speed comparable to that of a gas molecule at room temperature. In other words, one for which $(3/2)k_B T = KE$, and that (at 293 K) turns out to be 6.068×10^{-21} J. What is the wavelength of a thermal neutron?
6. [I] Use the results of Problems 3 and 4 to determine the phase speed of the de Broglie wave of an electron traveling at c/10. Note that v_p is greater than c.
7. [I] Suppose we are to build an electron microscope and we want it to operate at a wavelength of 0.10 nm. What accelerating voltage should be used?
8. [I] What are the values of the quantum numbers n and l for a 3d electron state?
9. [I] Consider the second excited state of the hydrogen atom. What are the values of the appropriate quantum numbers n, l, and m_l?
10. [I] Can two electrons in an atom have quantum number sets of $(2, 0, 0, +\frac{1}{2})$ and $(2, 0, 0, -\frac{1}{2})$, respectively? Explain.
11. [I] How many electrons can reside in the K, L, M, and N shells of an atom?
12. [I] How many electrons can reside in each subshell of the M-shell of an atom?
13. [I] Imagine a particle flying along the x-axis in a box of length L. What is the uncertainty in its momentum along the x-axis?
14. [I] A 100-g ball is confined to move in a 1.00-m-long frictionless tube lying along the x-axis. What is the minimum uncertainty in its speed? And how far will it move in a year at that speed?
15. [I] The position of a 0.001 00-kg particle along the x-axis is measured to be $1.243\,7 \pm 0.000\,5$ cm from the tip of a probe. What is the minimum uncertainty in its speed? (Hint: That's $\pm 0.000\,5 \times 10^{-2}$ m.)
16. [I] Since a charged pi meson at rest exists on average for only 26 ns, its energy will not be able to be measured with unlimited precision. Determine the minimum uncertainty in the meson's rest energy.
17. [I] A typical excited atomic state has a lifetime of 10^{-8} s. Determine the uncertainty in the energy of such a state in joules and electron-volts. (This is the unavoidable "blurriness" of the energy level and, rather than a sharp line, it produces an emission band known as the *natural linewidth*.)
18. [I] With Problem 17 in mind, determine the minimum uncertainty in the frequency of the emitted photon when there is a transition down to the ground state.
19. [I] A rho meson at rest has a mean lifetime of 4.4×10^{-24} s and an energy of 765 MeV. Determine

the minimum uncertainty in the energy and write it as a fraction of the rest energy.

20. [II] What is the wavelength of an electron whose KE is 4.00 MeV? Since this energy is comparable with the rest energy, use relativistic arguments.
21. [II] What is the wavelength of an electron with a KE of 20 eV?
22. [II] Determine the kinetic energy of an electron that has a wavelength of 1.00 m.
23. [II] Consider a particle that has a large KE compared to its rest energy. Show that its wavelength is approximately equal to the wavelength of an equal-energy photon.
24. [II] If v_p is the phase speed of the de Broglie wave of a particle traveling at a speed v, prove that $v_p v = c^2$. Use Special Relativity.
25. [II] For a lightwave, the phase speed is given by $c = E/p$. Assume the same form for a de Broglie wave and find a relationship between its phase speed (v_p) and the particle's speed v. Assume classical conditions; that is, let $E = KE$. Your result neglects the rest energy and will not compare very well with the relativistic expression $v v_p = c^2$.
26. [II] With Problem 25 in mind and assuming an electron in a hydrogen atom has an energy given by $E_n = -hf_n$, where f_n is the frequency of the electron in the nth "orbit," show that de Broglie's hypothesis leads to the same orbital energies as the Bohr Theory; namely,

$$ E_n = -\frac{n\hbar v_n}{2r_n} $$

provided $E = KE$.

27. [II] Using Bohr Theory, determine the energy of a photon that would excite the electron in a hydrogen atom from the K-shell to the M-shell.
28. [II] Make a table of all of the allowed four quantum numbers for the first three shells of the hydrogen atom. How many electrons can each shell accommodate?
29. [II] An excited hydrogen atom with its electron in the O-shell drops down to the L-shell, emitting a photon in the process. Determine the energy of that photon using Bohr Theory.
30. [II] What are the possible values of n, l, and m_l for a 5f atomic state? (First, check to see if it's allowed.)
31. [II] What are the possible values of n, l, and m_l for a 3f atomic state, and what can be said about that state?
32. [II] A 10.0-μg particle is traveling at 2.00 cm/s. Given that there is a 1.00% uncertainty in its speed, what is the least uncertainty in its position?
33. [II] Determine the ground-state electron configuration for $_{17}$Cl.
34. [II] Prove that a 10-g beetle whose position is known to within 1.00 nm can move with a speed uncertainty of 1.00 nm per year and still not violate the Uncertainty Principle.
35. [II] A 10.0-g particle is traveling at 20.0 cm/s. If the uncertainty in its momentum is 1 part in 1000, determine the uncertainty in its position.
36. [II] Approximate the minimum kinetic energy of an

electron confined to a region the size of an atom (0.10 nm).

37. [II] The position of an electron is measured within an uncertainty of 0.100 nm. What will be its minimum position uncertainty 2.00 s later?

38. [II] Angular momentum (L) and angle (θ), in radians, are conjugate variables. Accordingly, beginning with $\Delta p_x \Delta x \geq \frac{1}{2}\hbar$, apply it to circular motion and derive the uncertainty relationship for angular momentum.

39. [III] Gold is placed in an oven, melted, vaporized, and kept at a temperature of 1600 K. A parallel beam of atoms (each of mass 3.271×10^{-25} kg) emerges directly toward a circular hole of radius r in a screen (Fig. P39). Taking all the atoms to have the same speed, use the Uncertainty Principle to approximate the spot size formed by the gold on a screen that is 1.000 m away.

40. [III] In the Davisson-Germer experiment, 54-V electrons were scattered from a nickel crystal (in the first order) at 65°. The spacing of the crystal's atomic planes was found, using X-rays, to be 0.091 nm. Show that these results conform to de Broglie's hypothesis.

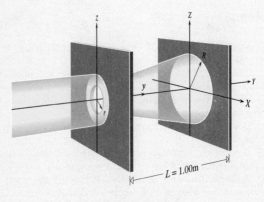

Figure P39

32

Nuclear Physics

AN ATOM, WITH A radius of about 10^{-10} m, consists of a cloud of electrons moving at great speeds around a positively charged nucleus comprised of protons and neutrons. That miniscule core, containing over 99.9% of the total mass, is roughly 10^4 times smaller than the atom as a whole. Like the atom, the nucleus is a bound system and can exist in a number of quantum states beyond its lowest-energy ground state.

We now turn our attention to the atomic nucleus. Though crucial to the structure of the atom, it plays an unobtrusive role hidden deep below the swirling cloud of electrons. On Earth, the atomic nucleus reveals itself in

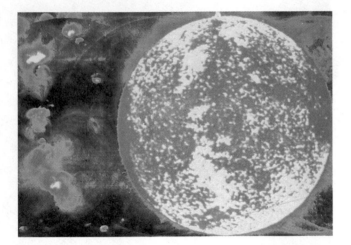

The Sun is a swirling sphere of plasma. At its center is a thermonuclear furnace driven by a gigantic fusion reaction.

the phenomenon of radioactivity and in nuclear weapons that unleash vast amounts of energy. On a much grander scale, nuclear fusion powers the stars (p. 1204).

NUCLEAR STRUCTURE

During the past several decades, experiments have provided reliable information about the size, shape, the distribution of charge within, and the magnetism of, the nucleus. Three principal techniques are used: (1) the nucleus can be probed with high-energy ($\approx$10 GeV), short-wavelength beams, usually of electrons; (2) orbital electrons interact with the nucleus, and the electromagnetic energy they emit provides information about nuclear structure; (3) a beam of positive particles can be scattered inelastically, transferring some of its energy to the nucleus, which becomes excited. The gamma radiation emitted as the nucleus returns to the ground state provides data on the nuclear configuration. It has been found that the nucleus is a complex shifting structure made of rapidly moving parts. The first clues to its composition came from studying isotopes.

32.1   Isotopes: Birds of a Feather

All atoms of a particular element were supposed to be identical. That assertion had been central to atomic theory since before Dalton's time, but it was wrong. There *are* different kinds of every one of the elements, often six or seven distinct variations. Rutherford and F. Soddy traced the decay of several heavy, naturally occurring radioactive elements, such as uranium and thorium. These decayed, spewing out particles and continuously transforming into different elements until they ended up as lead. Along the way, elements appeared that were chemically identical to others but that had different radioactive characteristics. Although ordinary lead has an atomic mass of 207.20 (we'll straighten out the units in a moment), lead present in uranium ore had a mass of only 206.05. Soddy named these variations of a given element **isotopes** (from the Greek *isos* meaning "same" and *topos* for "place"— having the same place in the Periodic Table).

In 1913, J. J. Thomson and F. W. Aston were the first to separate the isotopes of an element. They had put neon gas (atomic mass 20.2) into a positive ion tube and deflected the beam with charged plates and magnets to measure the atomic mass. At the face of the tube *two* distinct spots appeared. The conclusion was unmistakable: "Neon is not a single gas, but a mixture of two gases, one of which has an atomic weight [sic] of about 20, and the other of about 22."

A particular species of nucleus specified by a characteristic value of both **atomic number** (Z) and mass number, or **nucleon number** (A), is called a **nuclide**, and there are about 1500 nuclides. Every distinctly different nucleus is a specific nuclide. **Isotopes** of a given element are nuclides having the same Z (that is, the same nuclear charge), but having a different A; the total number of nucleons is different. Since $A - Z$ is the number of neutrons present (N), it follows that *isotopes of an element differ from one another only in their value of N.* Letting X be the chemical symbol for any element, we can specify any nuclide by writing it as

$$^A_Z X_N$$

The name of an element is associated with a specific value of Z, a specific place in the Periodic Table. Any atom containing 10 protons is neon no matter what its nucleon number. Moreover, 10 protons and 10 neutrons make neon ($^{20}_{10}\text{Ne}_{10}$), just as 10 protons and 12 neutrons make a different form of neon ($^{22}_{10}\text{Ne}_{12}$). Apparently, this notation is redundant: $Z = 10$ means neon, and vice versa. The distinction is also often made by referring to neon-20, or Ne-20.

When found in our atmosphere, 90% of natural neon is of the first and lighter variety while the other 10% is the heavier kind. The *chemical atomic mass* of an element referenced in the Periodic Table reflects the natural abundances of the isotopes in a typical sample. These values were arrived at before the knowledge of the existence of isotopes and correspond to a weighted average of all the isotopes naturally present in the environment. For neon that average turns out to be 20.179 7.

In the early part of the nineteenth century, it made sense to determine relative atomic masses. Thus, hydrogen was set at 1, and all the other elements were measured with respect to it. Today, masses in the atomic domain are often specified in **unified atomic mass units** (u) where a neutral carbon atom ($^{12}_{6}\text{C}$) is defined to have a mass of precisely 12.000 000 u (see Table 32.1). Because of the practical link to the laboratory, with its accelerating electric fields, the unit MeV/c^2 is also widely used:

$$1 \text{ u} = 1.660\,540 \times 10^{-27} \text{ kg} = 931.494 \text{ MeV/c}^2$$

As a rule, A for each isotope differs slightly, but significantly, from its mass in units of u. (The exception is ^{12}C.)

TABLE 32.1   Atomic mass of some representative nuclides

Element	Symbol	Mass (u)
Hydrogen	$^{1}_{1}\text{H}$	1.007 825
Deuterium	$^{2}_{1}\text{H}$ (D)	2.014 102
Tritium*	$^{3}_{1}\text{H}$ (T)	3.016 049
Helium	$^{3}_{2}\text{He}$	3.016 029
Helium	$^{4}_{2}\text{He}$	4.002 603
Lithium*	$^{5}_{3}\text{Li}$	5.012 54
Lithium	$^{6}_{3}\text{Li}$	6.015 121
Beryllium	$^{9}_{4}\text{Be}$	9.012 182
Nitrogen	$^{14}_{7}\text{N}$	14.003 074
Nitrogen	$^{15}_{7}\text{N}$	15.000 109
Nitrogen*	$^{16}_{7}\text{N}$	16.006 100
Oxygen	$^{16}_{8}\text{O}$	15.994 915
Oxygen	$^{17}_{8}\text{O}$	16.999 131
Oxygen	$^{18}_{8}\text{O}$	17.999 160
Lead	$^{204}_{82}\text{Pb}$	203.973 020
Lead*	$^{205}_{82}\text{Pb}$	204.974 458
Lead	$^{207}_{82}\text{Pb}$	206.975 872
Uranium*	$^{233}_{92}\text{U}$	233.039 628
Uranium*	$^{235}_{92}\text{U}$	235.043 924
Uranium*	$^{238}_{92}\text{U}$	238.050 785

*Radioactive. (Remember Table 29.3.)

Example 32.1   Show that the chemical atomic mass of neon should be about 20.18 u, given that ^{20}Ne and ^{22}Ne have natural abundances of about 90.51% and 9.22%, and masses of 19.99 u and 21.99 u, respectively. (Your error will be due to the fact that we have overlooked trace amounts of ^{21}Ne.)

Solution: [Given: isotopes with masses of 19.99 u and 21.99 u, and abundances of 90.51% and 9.22%. Find: the chemical atomic mass.] The lighter neon is almost 10 times more abundant and therefore 10 times more influential in determining the chemical atomic mass of any natural sample than is the heavier isotope. The weighted average mass is thus

$$90.51\% \,(19.99 \text{ u}) + 9.22\% \,(21.99 \text{ u})$$

or 18.09 u + 2.03 u = $\boxed{20.12 \text{ u}}$.

▶ **Quick Check:** As already given, the actual value is 20.179 7 u.

Some 280 isotopes of the naturally occurring elements are stable and presumably will last for all time (Fig. 32.1). Around 1200 others (human-made and natural) are radioactive and transient. All the elements beyond uranium, from $Z = 93$ to ≈ 110, are produced in the laboratory and are radioactive; if they ever existed in abundance in nature, most were short-lived enough to have decayed to unobservably low levels. Some elements, such as xenon and iodine, have more than a dozen known

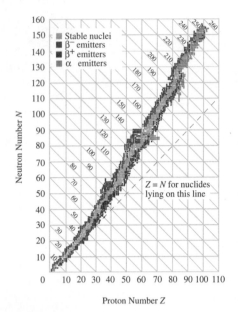

Figure 32.1 A plot of *N* versus *Z* for stable and unstable nuclides. The so-called *line of stability* runs up through the gold-colored band, which represents the stable nuclei. Each integer value of *N* and *Z* lying on a colored region corresponds to a nuclide.

The 60-inch cyclotron at the Argonne National Laboratory. A beam of deuterons (the nuclei of deuterium) is streaming out at a speed of ≈28 000 miles per second.

isotopes each. Nine of the xenon forms are stable, but iodine has only one stable isotope, as do aluminum and gold.

The three isotopes of hydrogen ($^1_1$H, $^2_1$H, $^3_1$H) are so different from each other and so important that, unlike all others, they have their own names (Fig. 32.2). Ordinary *hydrogen* with a single proton ($Z = 1$, $A = 1$) is the lightest and most common (99.985%). **Deuterium** (also written $^2_1$D) with an added neutron ($Z = 1$, $A = 2$) is quite rare (0.015%). For every 6500 ordinary hydrogen atoms in a natural sample, there is only 1 deuterium atom. Radioactive **tritium** ($^3_1$T) with still another neutron ($Z = 1$, $A = 3$) is by far the least abundant (for every 10^{18} atoms of ^{1}H there is 1 of ^{3}T). Since these isotopes differ enormously in mass, and since the nuclei must interact somewhat differently with the single orbital electron, it's not surprising that they are distinctive chemically. For example, living organisms respond differently to water formed of oxygen and deuterium, known as **heavy water**, than they do to the ordinary brew, which differs as well in its freezing and boiling points. Heavy-water ice cubes, though they look and taste ordinary enough, sink to the bottom of a glass of tap water.

Insofar as the nucleus has little influence on the outer electrons, isotopes behave identically chemically except for some minor effects. Different isotopes of the same element can be separated, but as a rule only by mechanical means. The fact that a uranium-based atomic bomb requires large amounts of ^{235}U, which must be separated from ^{238}U, makes their production quite difficult for all but the most technologically advanced nations (p. 1198).

32.2 Nuclear Size, Shape, and Spin

Experiments pioneered by R. Hofstadter in the 1950s have revealed that nuclei are fairly spherical, with most being slightly ellipsoidal, though they come elongated, flattened a little, and pear-shaped as well. Figure 32.3 is a plot of nuclear charge density measured out along a radial distance in femtometers. The same results hold for the mass-density, with neutrons and protons being distributed in much the same way. Recall that 1 fm = 10^{-15} m; it's a distance that's convenient in the nuclear domain just as the nanometer was convenient in the atomic domain. (Nowadays, 1 fm is often called a *fermi* in honor of the Italian-American physicist Enrico Fermi.) The density drops off gradually over an outer thickness of roughly 2.5 fm—the nucleus thins out across this one-nucleon-thick surface region.

The nuclear radius *R* is often taken as the distance from the center to the half-density point (Fig. 32.3b). Independent of *A*, the density of a nucleus is constant for much of its radius. That means that the number of nucleons contained in a nucleus (assumed to be spherical) simply depends on its volume $\frac{4}{3}\pi R^3$. Hence, $A \propto R^3$ and $R \propto A^{1/3}$. Using a proportionality constant R_0 to make this relationship into

an equality, we have

$$R = R_0 A^{1/3} \qquad (32.1)$$

R_0 turns out to be ≈ 1.2 fm (Fig. 32.4). This situation should remind us of a drop of water, which also has a volume proportionate to the number of molecules it contains.

Example 32.2 Determine the radius of the carbon nucleus ($A = 12.0$).

Solution: [Given: $A = 12.0$ u. Find: R.] Using Eq. (32.1), we have

$$R = (1.2 \text{ fm})A^{1/3} = (1.2 \text{ fm})(2.29) = \boxed{2.7 \text{ fm}}$$

▶ **Quick Check:** Compare this result with Fig. 32.3.

The density of *nuclear matter,* as it's called, is the mass over the volume, and since A is equivalent to the mass in units of u, where 1 u $= 1.66 \times 10^{-27}$ kg

$$\rho = \frac{m}{\frac{4}{3}\pi R^3} = \frac{A(1.66 \times 10^{-27} \text{ kg})}{\frac{4}{3}\pi R_0^3 A} = \frac{1.66 \times 10^{-27} \text{ kg}}{7.24 \times 10^{-45} \text{ m}^3} = 2.3 \times 10^{17} \text{ kg/m}^3$$

This value is tremendously large; by comparison, the density of water is a mere 10^3 kg/m³. Nuclear matter in bulk does not exist on Earth. You would know if it did because a cubic inch of it would weigh about 4×10^9 tons. Still, it is likely that certain celestial objects, such as neutron stars, consist of nuclear matter. Each such object is essentially a gigantic nucleus with gravity forcing it together.

Neutrons and protons are fermions with spins of $\frac{1}{2}$ and angular momenta of $\frac{1}{2}\hbar$. The *Shell Model* (p. 1184) assumes that the nucleus has energy levels very like the atom. The Exclusion Principle requires that only two protons (one spin-up, one spin-down) and two neutrons (one spin-up, one spin-down) can occupy any level. The total spin of a nucleus is the sum of the spins of its parts—*odd-A nuclides are fermions, even-A nuclides are bosons.* Thus, for an even-even nucleus (even Z, even N), such as ${}_2^4$He, ${}_6^{12}$C, or ${}_8^{16}$O, the total spin is zero. The nucleus of deuterium, the **deuteron** (${}_1^2$H), is typical of odd-odd nuclides; it has unfilled sublevels (one unpaired neutron and one unpaired proton) and a spin of 1. Even-odd and odd-even nuclides have one unpaired nucleon and therefore total spins that are odd-integer multiples of

Hydrogen

Deuterium

Tritium

Figure 32.2 The three hydrogen isotopes.

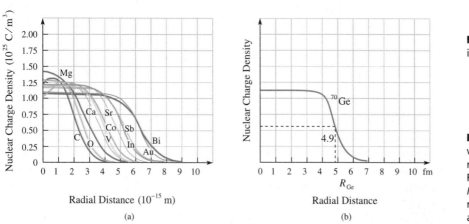

(a)

(b)

Figure 32.3 (a) The charge density versus distance from the center of the nucleus for several nuclei. [Data from R. Hofstadter, *Annual Reviews of Nuclear Science* **7**, 231 (1957).] (b) The radius of the germanium nucleus is about 4.9 fm.

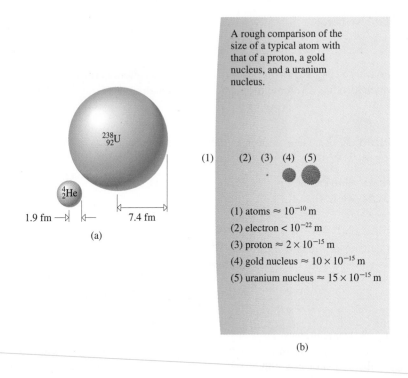

A rough comparison of the size of a typical atom with that of a proton, a gold nucleus, and a uranium nucleus.

$^{238}_{92}$U

$^4_2$He

1.9 fm → ⊢← ⊢←——— 7.4 fm ———⊣

(a)

(1)   (2) (3) (4) (5)

(1) atoms ≈ 10^{-10} m
(2) electron < 10^{-22} m
(3) proton ≈ 2×10^{-15} m
(4) gold nucleus ≈ 10×10^{-15} m
(5) uranium nucleus ≈ 15×10^{-15} m

(b)

Figure 32.4 (a) shows the relative sizes of helium and uranium nuclei. (b) gives a sense of the size difference between a nucleus and an atom.

$\frac{1}{2}$. Thus, the spin of $^7_3$Li (3 protons, 4 neutrons) arises from a group of two protons and two neutrons with spin zero plus the remaining 3 spin-$\frac{1}{2}$ nucleons, yielding a total of $\frac{3}{2}$.

The presence of spin suggests the possibility of a magnetic moment, which is certainly the case for a charged particle or an entity composed of charged particles. The magnetic moment of the proton is about 0.15% that of the electron and is in the same direction as its spin. This much is determined experimentally; as yet there is no complete theory of nuclear magnetism that accounts for the details. The magnetic moment of the proton ($+1.41 \times 10^{-26}$ J/T) is about 3 times larger than expected from its mass and the way the electron behaves. We can already anticipate that the proton will be a far more complicated entity than was first thought. The neutron, too, has a magnetic moment; it's smaller than the proton's and in the opposite direction to its own spin. The nonzero magnetic moment implies that the neutron, though it is neutral, has some internal nonuniform charge distribution. We now know that the proton and neutron are each composed of three still smaller charged quarks. A typical nucleus will have a net magnetic moment that depends on the number and arrangement of its nucleons.

By applying magnetic fields to an atom, it's possible to alter the spin state of the nucleus in a process discovered in 1946 called **nuclear magnetic resonance** (NMR). In the case of hydrogen, a proton can align its magnetic moment either parallel or antiparallel to an applied static B-field. The spin-up state has a slightly lower energy (ΔE) than the spin-down state. If a photon with an energy (hf) equal to the difference in energy between these two spin states (namely, ΔE) impinges on the proton, it can be absorbed. Thus, a pulse of radiofrequency electromagnetic energy of exactly the correct frequency can resonate the protons, exciting them into the higher-energy state. The same is true of more complex nuclei that have their characteristic resonant frequencies. When the nuclei relax back to their ground states, they each emit a photon of the same resonant frequency, which is easily detected. Either the absorption or re-emission of photons at different points in the specimen can be used to produce images of its internal structure. The human body is about 75% water and contains an abundance of hydrogen and therefore of protons. **Magnetic resonance imaging** (MRI) usually uses the NMR of protons to noninvasively form images of body tissue of living patients (Fig. 32.5).

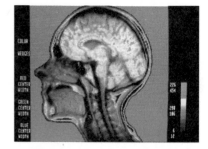

32.3 The Nuclear Force

No sooner had the nucleus been revealed, than there arose the obvious question: what holds it together? By 1925, there was a recognition of the need for a new type of force. A cluster of positively charged particles must repel one another with an electrostatic force (like charges repel). Moreover, we know from scattering experiments that the $1/r^2$ Coulomb force works right down to nuclear distances. A simple calculation of the repulsion between two protons separated by a distance that puts them just about in contact within a nucleus yields a value of around 50 N (about 11 lb). That interaction is enormous when considered in light of the tiny masses of the protons; no nuclei other than hydrogen should ever have formed, and those that exist should explode immediately. Evidently, there is another kind of force, the **nuclear force**, operating within the nucleus. *The nuclear force binds neutrons and protons together to form nuclei.* It had its conceptual origins in a proposed short-range, neutron-proton force first suggested (1932) by Heisenberg. Now known to have an effective range of only about 1 fm, the nuclear force is powerfully attractive, imparting potential energies to nucleons of as much as 100 MeV. It is repulsive at distances less than about 0.5 fm (two nucleons cannot occupy the same space), and it depends on the spins of the interacting particles.

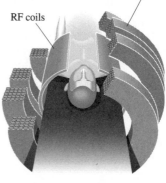

Magnetic field coils

RF coils

The nuclear force is a manifestation of the more fundamental and less restricted *strong force,* which affects a whole class of particles (hadrons), not just nucleons. The most potent of all known interactions, the strong force (at a proton-proton separation of 2 fm) is about 100 times stronger than the electromagnetic force and roughly 10^{34} times stronger than the gravitational force (Table 32.2). While it might take as much as 8 MeV to remove a nucleon from a nucleus, the electron in a hydrogen atom can be ionized with a mere 13.6 eV. For that reason, pound for pound, a nuclear reaction can liberate millions of times more energy than a chemical reaction. We will discuss the strong force at greater length in Chapter 33.

From experiments starting in 1936, we have found that the nuclear force exists between any two nucleons. The evidence for the n↔n attraction is inferential, but the p↔p and p↔n interactions can be measured, although indirectly. This is done using beams of neutrons or protons scattered from a target consisting mostly of hydrogen (that is, protons). Electrons are completely immune to the nuclear force, which is why they were so effective at probing the nuclear charge distribution.

Rutherford's scattering experiments of 1913 established, that down to distances of roughly 10^{-14} m, α-particles interacted with nuclei via the Coulomb force. In 1919, he fired 5-MeV α-particles at low-Z nuclei, thereby minimizing the Coulomb repulsion. What he found (using hydrogen nuclei as targets) was that at distances of ≈ 3.5 fm the resulting scattering markedly deviated from that predicted by electrodynamics (Fig. 32.6).

At relatively large distances, protons repel one another via the Coulomb force, but there is a change at roughly 3 fm, where the interaction becomes increasingly attractive. A monoenergetic neutron beam colliding with protons shows little or no interaction to about 2 fm, whereupon there is again an increasingly strong attractive

Figure 32.5 Magnetic resonance imaging (MRI). A human body is subjected to a powerful constant *nonuniform* magnetic field. The energy levels available to the hydrogen nuclei split in two (much like the Zeeman Effect), separating by $\Delta E \propto B$. A broad band radiofrequency (RF) pulse of electromagnetic energy then excites some of the nuclei ($hf = \Delta E \propto B$). When the nuclei subsequently relax, they emit RF photons ($f \propto B$), which are used to construct an image of the distribution of nuclei based on a knowledge of the spatial configuration of *B*.

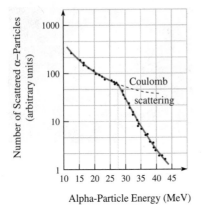

Figure 32.6 Alpha particles elastically scattered from a lead target. [Data from G.W. Farwell and H.E. Wegner, *Phys. Rev.* Vol. 95: 1212 (1954).] At around 27.5 MeV, the curve abruptly deviates from purely Coulomb scattering. The nuclear force now comes into play as the α-particles get close enough to the nucleus for that short-range force to be influential.

TABLE 32.2 The Four Forces of nature*

Force	Interacts between	Strength[1]	Effective range
Gravitational	All mass-energy[2]	10^{-34}	Unlimited
Weak	All material particles (quarks and leptons)	10^{-2}	$\approx 10^{-17}$ m
Electromagnetic	Electromagnetic charges	10^{2}	Unlimited
Strong	Many subnuclear particles (quarks and gluons)	10^{4}	$\approx 10^{-15}$ m

*At the temperatures that exist today, we see four distinct interactions. At much higher energies, these blend together (p. 1234).
[1]The strengths (in newtons) are for two protons separated, center-to-center, by 2 fm.
[2]Gravity and the strong force both act on their own field quanta.

force. At very small separations, the interaction quickly becomes repulsive. The nuclear force is so powerfully repulsive at very small distances that nucleons rarely get closer to one another than about 0.4 fm. Estimates of the nucleon radius range from about 0.3 fm to 1 fm.

Figure 32.7a depicts a crude potential-energy curve for a proton interacting with a nucleus of radius R. At large distances, the curve is positive and Coulombic—an approaching proton experiences a repulsion. The positive *electrical*-PE increases as $1/r$. By contrast, a neutron does not "feel" any electrical force—it approaches the nucleus along the zero-PE axis in Fig. 32.7b. At the surface of the nucleus, a nucleon is tremendously influenced by the attractive nuclear force. The resulting negative potential energy confines the nucleon to the tiny region of the nucleus. Bound neutrons and protons rattle around inside the nucleus like submicroscopic bees flying in a well as much as 50 MeV deep.

The nuclear force is *saturable:* each bound nucleon interacts with only a few of its nearest neighbors. If a nucleon is added to a nucleus, it will not interact with all the other particles via the nuclear force. That means that the more massive nuclei should have nearly the same density as small nuclei, which is borne out by experiment (Fig. 32.3). Remember that atoms, held together by the $1/r^2$ Coulomb force, are nearly all the same size regardless of A. Adding more charge to both the nucleus and the electron cloud increases the long-range interaction, and that essentially

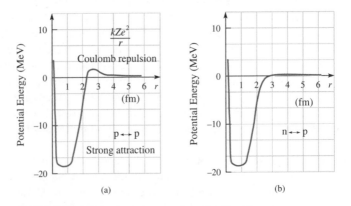

Figure 32.7 Approximate potential wells for (a) proton-proton and (b) neutron-proton interactions.

compresses the atom. By contrast, nuclei get larger with A via Eq. (32.1). The nucleus will prove to be more similar in behavior to a droplet of water than it is to an atom.

The nuclear force performs as do the forces between molecules—both fall off rapidly with separation. Moreover, two water molecules also repel one another when they come very close together and their electron clouds start to overlap. Nucleons and water molecules, while shifting around, will tend to stay separated by a distance such that their mutual attraction is maximized. The separation between nucleons is comparable to the range of the nuclear force, and the surface of the nucleus is about that thick as well. As with a liquid, the nucleus also experiences surface tension; nucleons at the surface will be pulled inward, which is why nuclei are essentially spherical. In a sense, the nucleus is a droplet of nuclear matter.

When a neutron and a proton come together to form a deuteron, they do it only if the two spins are parallel, suggesting yet another complication in the way the nuclear force works: it has a substantial spin dependence. When nucleon spins are antiparallel, the nuclear force between the two particles is about a factor of 2 weaker.

32.4 Nuclear Stability

Figure 32.1 reveals that stable light nuclei up to about $A = 20$ are almost all composed of equal numbers of neutrons and protons ($Z = N$). The hydrogen nucleus ($A = 1$) is a single proton and naturally stable (though, according to the Grand Unified Theories, protons should decay, no evidence has yet been found to support that conclusion—see p. 1234). Protons repel one another with the Coulomb force, and it takes some doing to get two of them close enough together to stick via the nuclear force. There is evidence that the di-proton has been created, but it lasts for less than 10^{-18} s. On the other hand, neutrons experience the nuclear force while being immune to electrostatic repulsion; they therefore serve as a source of nuclear glue, but that's a bit simplistic, since the di-neutron is unstable. The next heavier nucleus ($A = 2$) results when a neutron clings to a proton with parallel spin, forming a stable deuteron ($Z = 1$, $N = 1$). Two protons can be held together with the inclusion of a neutron, thereby making stable ^3_2He. Two deuterons can combine to create helium ($Z = 2$, $N = 2$).

Both the spin and the magnetic moment of the α-particle are zero. Since the magnetic moments of the neutron and proton are different, that tells us that the two neutrons (spin-up and spin-down) pair together, as do the two protons (spin-up and spin-down). The corresponding closed stable system (the α-particle) plays an important role in the scheme of the nuclides (p. 1186). That's borne out by the observation that, while there are only four stable odd-odd nuclei (^2_1H, ^6_3Li, $^{10}_5\text{B}$, and $^{14}_7\text{N}$, for which $Z = N$), there are 160 stable even-even nuclei. Furthermore, when a nuclide is blasted with a nucleon, it's much more likely that an α-particle will be emitted than a deuteron.

Most of the radioactive nuclides in Fig. 32.1 are of the *artificially induced* variety, made in the laboratory by bombarding other nuclides. There are only about 20 naturally occurring radioactive isotopes in the range up to lead ($Z = 82$). Beyond that, all nuclides are radioactive, though bismuth decays so slowly it might as well be considered stable.

The Shell Model

Nuclides with certain numbers of neutrons or protons are especially stable and abundant. These so-called **magic numbers** with N or Z equaling

2, 8, 20, 28, 50, 82, 126

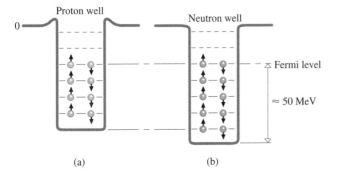

Figure 32.8 Potential-energy wells for (a) protons and (b) neutrons in the same nucleus. To bring the Fermi levels in line, protons can convert to neutrons by emitting positron-neutrino pairs. Note the unfilled higher-energy levels available when the nucleus is excited.

Maria Goeppert Mayer (1906–1972). German-born American physicist shared the 1963 Nobel Prize in physics with J. H. D. Jensen.

suggested a closed shell structure reminiscent of the configurations of the Noble Gases in atomic theory (p. 1159). A nucleus with either N or Z magic is tightly bound. One with both N and Z magic is very tightly bound, highly stable, and abundant. Such is the case with $^{4}_{2}\text{He}_2$, $^{16}_{8}\text{O}_8$, $^{40}_{20}\text{Ca}_{20}$, $^{48}_{20}\text{Ca}_{28}$, and $^{208}_{82}\text{Pb}_{126}$, the five known nuclei with double magic (see Discussion Question 1).

A nucleon within the nucleus is surrounded by other nucleons, and the nuclear forces they exert on it in opposing directions should tend to cancel; an interior nucleon will experience little local influence via the nuclear force. Don't picture the nucleus as a static cluster of tightly packed cannon balls! Instead, imagine it to be more like a cloud of flying bees, each of which sees and attracts only its nearest neighbors. Although there is a complex tumult of motion, the swarm hangs together as a single entity. In fact, if the bees fly around in pairs, the analogy would be still better. In any event, we can assume that a nucleon "sees" an average force that retains it within the nucleus. The Uncertainty Principle requires that particles confined to a small region of space have a correspondingly large uncertainty in their motion and so at least a minimum kinetic energy.

Suppose that each nucleon moves around in a complex independent way within the nucleus. Each is restrained by the surface of the nucleus, as if in a potential-energy well. Figure 32.8 depicts a crude representation of the average potential-energy wells for a proton and neutron in a medium- to large-sized nucleus. The difference in shape and depth is due to the Coulomb force. The proton's well is not as deep because of the repulsion.

Each nucleon is accompanied by a de Broglie wave, a standing wave, fitting into the nucleus, just as the electron's wave fit into the atom. Thus, each nucleon should correspond to a particular standing wave pattern or, equivalently, should occupy a particular quantum orbital. A group of energy levels that are close to one another is called a shell, and this scheme is the **shell model** of the nucleus. In the same way that an atomic level can contain at most two different electrons, each nuclear energy level can harbor two protons and two neutrons. No exclusion acts between neutrons and protons, and they can enter the same states. The levels fill up from the bottom. The highest occupied level corresponding to the greatest kinetic energy is called the **Fermi level**. Although we might expect a moving nucleon to undergo frequent collisions and be scattered every which way, the Exclusion Principle and the demand for filled levels will keep that from happening. There are very few available energy levels for a scattered nucleon to enter, and so they cannot easily be knocked off course.

Once in identical states, a proton and neutron experience a maximum attractive interaction, and so the simplest closed subshell system, that of the helium nucleus, is energetically favorable. When the energy levels were computed quantum-mechanically, it was found that the nucleus had a system of closely spaced subshells. These come in groups constituting major shells that are themselves much farther apart energetically. M. Goeppert Mayer and J. H. D. Jensen further showed (1947) that the magic numbers corresponded to the number of states in the major shells. When the major shells are filled, the corresponding nuclei were especially stable. For example, a nucleus with 50 neutrons or 50 protons has a filled shell. Thus, tin ($Z = 50$), which is relatively abundant in nature, has 10 stable isotopes.

Nuclides off the **line of stability** that runs through the stable nuclei spontaneously decay, often changing neutrons to protons or vice versa, and becoming more tightly bound and energetically stable. Observe how the line of stability in Fig. 32.1 increasingly bends toward the N-axis, indicating a larger percentage of neutrons. Because of the long range of the Coulomb force, the greater the number of protons, the more the nucleus tends to decay, and the more neutrons are needed to stabilize it. The progression of stable nuclei ends at lead; further increasing the percentage of neutrons will not keep things together. There are no pea-sized nuclei, and not until the mass becomes immense can gravity help to hold nuclear matter together on a large scale as in a neutron star.

Binding Energy

If we fire a neutron directly at a proton so that they come very close, the two can grab hold of each other via the strong force. Rammed together, the system will emit a burst of electromagnetic energy, a 2.224-MeV gamma-ray photon. Thus formed, the deuteron (2.013 553 u) has shed mass; its constituents have drawn tightly together and lost some of the plumpness they had apart. This transformed mass is known as the **mass defect** (Δm), and it indicates how tightly a nuclide is bound. In this case

$$m_p + m_n = (1.007\,276\,u) + (1.008\,665\,u) = 2.015\,941\,u$$

whereas the deuteron mass (m_d) is only 2.013 553 u. The difference between the mass of the separate components and the combined nucleus is the mass defect; namely

$$\Delta m = 0.002\,388\,u$$

Since 1 u is equal to 931.494 MeV/c^2, the mass defect corresponds to a **binding energy** ($E_B = \Delta m c^2$) of 2.224 MeV—exactly the energy ejected as a photon. In reverse, to split a deuteron into a neutron and a proton, 2.224 MeV (or 3.56×10^{-13} J) has to be supplied, for example, via a photon or a collision. Of course, 2.224 MeV is an immense amount of energy on an atomic scale. It takes only 2×10^{-18} J, one hundred thousand times less energy, to pull the electron out of a deuterium atom, and even that is 10 times the energy released when the "burning" of a carbon atom forms CO_2.

Example 32.3 When a neutron is removed from a $^{43}_{20}$Ca atom (of mass 42.958 766 u), it will be transformed into a $^{42}_{20}$Ca atom (of mass 41.958 618 u). What minimum energy must be provided to accomplish the removal?

Solution: [Given: atomic masses of 42.958 766 u and 41.958 618 u. Find: the energy to remove a neutron.] Let's first find the difference between the masses of the two nuclei. If that difference is less than the mass of a neutron, the deficiency (Δm) has to be made up by pumping energy in. We were given atomic masses because they are what's generally tabulated rather than nuclear masses. That fact doesn't matter here because taking the difference cancels the mass of the atomic electrons; thus

(mass of ^{43}Ca) − (mass of ^{42}Ca) = 1.000 148 u

Not surprisingly, this is less than the neutron mass (1.008 665 u); hence, the difference

$$\Delta m = (1.008\,665\,u) - (1.000\,148\,u) = 0.008\,517\,u$$

(at the least) will have to be supplied as energy. Consequently

$$E = (0.008\,517\,u)(931.494\,\text{MeV/u}) = \boxed{7.934\,\text{MeV}}$$

▶ **Quick Check:** As we'll see presently, 8 MeV is typical for the binding-energy-per-nucleon (which is an average value); for heavy nuclei, our answer is reasonable. (See Problem 29.)

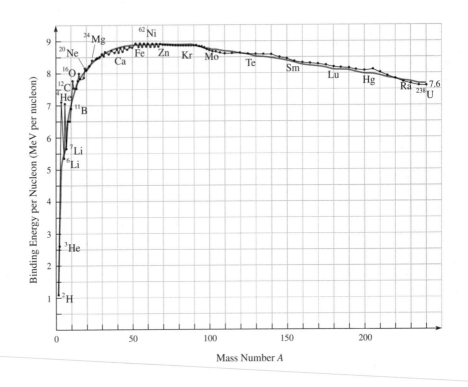

Figure 32.9 Average binding-energy-per-nucleon for the nuclides that occur naturally (including short-lived ^{8}Be, the second spike near He) versus mass number.

Planck first realized that *any bound system should have less mass than the sum of its constituents* and, in 1913, Langevin applied that insight to the nucleus. *The coming together of nucleons to form a nuclide is accompanied by a conversion of mass to energy.* The mass defect is equivalent to the total binding energy of the nuclide. Dividing that energy by the total number of nucleons yields an average measure of how strongly each nucleon is bound to the composite system; namely (E_B/A), the **binding-energy-per-nucleon**. (In a way, it's like our swarm of bees flying around in a well where we ask how much energy it will take on average to lift each of them up and out.)

A plot of binding-energy-per-nucleon for all the elements from hydrogen to uranium is given in Fig. 32.9. Hydrogen, with one proton, has no binding energy at all; from there, the curve rises to 1.1 MeV for the deuteron, 2.8 MeV for ^{3}H, 2.6 MeV for ^{3}He, then on up to a 7.074-MeV spike for tightly bound helium, falling back to 5.3 MeV for ^{6}Li, then rising to 5.6 MeV for ^{7}Li, and gradually continuing upward. The significance of the tightly bound α-particle structure is manifest by nuclides that can be thought of as comprising whole number multiples of ^{4}He. Thus, ^{8}Be (2 alphas), which decays almost immediately, nonetheless has a substantial binding energy of 7.06 MeV. Stable ^{12}C (3 alphas) and ^{16}O (4 alphas) have relatively high binding energies compared to their neighbors. Moreover, those nuclides, whose Z and A values correspond to whole number groupings of α-particles, also have zero spin and zero magnetic moment.

The binding-energy-per-nucleon curve reaches a maximum of 8.795 MeV at nickel-62, which is the most stable and most tightly bound of all nuclides. That fairly flat peak in the curve (around $Z = 60$) contributes to the abundance of iron in the Universe. From there on, the graph gradually drops, because of the Coulomb repulsion of the protons, all the way to uranium. Except for the lightest nuclei, E_B/A is fairly constant at about 8 MeV per nucleon. That this value is independent of A

again suggests that the nucleus is held together by a short-range force. With a long-ranged force such as gravity, the size of the composite body is crucial; it's much more difficult to remove a 10-kg stone from the surface of the Earth than from the surface of the Moon. To the contrary, where the force is short-ranged, as with the intermolecular force holding water together, there is the same independence of quantity; it takes the same amount of energy to evaporate 10 kg of water from a kiddy pool as from an ocean.

The binding-energy-per-nucleon is the energy above and beyond its kinetic energy (the Fermi level) that must be added to a nucleon to remove it from the nucleus (Fig. 32.8). For small nuclei, the well depth is small because there are fewer nucleons acting, and therefore the binding energy is also small. For large nuclei, the electrostatic potential energy raises the bottom of the well and decreases the binding energy at the top.

Notice that if we select a nuclide way off on either side of the peak in the binding-energy-per-nucleon curve and alter its structure so as to move up the curve toward Ni, a very large amount of energy could be liberated. Thus, if two light nuclei (say, of hydrogen) could be joined together (that is, fused), the resulting nuclide would reside further up the curve and would have a greater binding-energy-per-nucleon. Each of its nucleons would be more tightly squeezed and individually less massive than it was prior to the union. The resulting mass defect would appear as liberated energy in the well-known process of *nuclear fusion*. On the other hand, splitting a large nucleus (from the right side of the curve) into small fragments also transforms mass. The binding-energy-per-nucleon of the fragments is higher than it was for the original nucleus that broke up. This process of *nuclear fission* liberates copious amounts of energy. It's no accident that the key elements of the nuclear age, hydrogen and uranium, are the end points of the binding energy curve.

NUCLEAR TRANSFORMATION

The transformation of one nuclide into another can take place either spontaneously or as a result of an external stimulus. The remainder of this chapter deals with a variety of nuclear transformations. We examine the spontaneous processes of alpha, beta, and gamma decay; introduce the weak force; study the mathematical description of the rate of radioactive decay; and explore the phenomenon of induced radioactivity. The discussion ends with an introduction to fission and fusion as applied to bombs and stars.

32.5 Radioactive Decay

As we saw earlier, nuclei spontaneously transform themselves into more energetically favorable configurations via alpha, beta, and gamma emission in a process called **radioactive decay**. Usually, when a nuclide resides above the line of stability in Fig. 32.1, it decays by emitting an electron, thereby transforming a neutron into a proton and moving downward and to the right, coming closer to the line. Similarly, when a nuclide resides beneath the line of stability, it emits a positron or, occasionally, an alpha, transforming to the left toward the line.

A given radioactive decay can be a single step in a long sequence of transformations from one unstable nuclide to another, ultimately ending in a stable form. There are three *naturally occurring* radioactive series beginning with ^{238}U, ^{235}U, and ^{232}Th (Table 32.3), all ending in different isotopes of lead. The series shown in Fig. 32.10

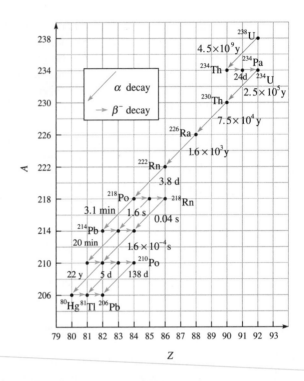

Figure 32.10 The naturally occurring decay series beginning with uranium-238.

TABLE 32.3   The three natural radioactive series

Series	Starting nuclide	Half-life (years)	Stable end nuclide
^{235}U-Actinium	$^{235}_{92}$U	7.04×10^8	$^{207}_{82}$Pb
^{232}Thorium	$^{232}_{90}$Th	1.41×10^{10}	$^{208}_{82}$Pb
^{238}U-Radium	$^{238}_{92}$U	4.47×10^9	$^{206}_{82}$Pb

begins with the slowly decaying nuclide uranium-238. A little less than half of all the U-238 present at the formation of the Solar System some 5×10^9 years ago still remains. By contrast, radium-226 decays away fairly rapidly (half of it is transformed every 1600 years), and all of Earth's original radium atoms have long ago disappeared. Yet, radium is still plentiful because, like so many other radioactive substances, it is continuously replenished (Fig. 32.10).

Alpha Decay

Alpha emission is rare in the light nuclides, although at around $Z = 60$, there is a small cluster of α-emitters. It is beyond $Z = 82$, where there are no stable nuclides, that alpha emission predominates. There, it drives the nuclides down in both Z and N, toward the line of stability. The emission of an α-particle ($^4_2$He, or $^4_2\alpha$) decreases N by 2, decreases Z by 2, and decreases A by 4. In general, the decay can be written as

$$^A_Z X \rightarrow {}^{A-4}_{Z-2}Y + {}^4_2He + Q \tag{32.2}$$

where X is called the **parent nucleus**, Y is the **daughter nucleus**, and Q is the **disintegration energy**. The rules of the game require that the process *conserve* *charge, nucleon number (A), angular and linear momentum,* and *mass-energy.* Alpha decay is a rearranging: it doesn't change the total number of nucleons present in the formula, nor does it alter the net charge, but it does alter the net mass. Whenever such a rearrangement happens by itself, energy is liberated in the amount Q such that

$$Q = (m_X - m_Y - m_\alpha)c^2 \qquad (32.3)$$

where m_X, m_Y, and m_α are the masses of the parent, daughter, and alpha, respectively. The disintegration energy (here given in joules) appears as the total kinetic energy of the daughter nuclide and the α-particle. A positive Q means that the process takes place spontaneously; a negative Q means it cannot. Using the binding energy data, it can be shown that for heavy nuclei, the average value of Q is around 5 MeV. Typical of the process is the *spontaneous decay* (Fig. 32.11) of uranium-238 into thorium-234:

$$^{238}_{92}U \rightarrow \ ^{234}_{90}Th + \ ^{4}_{2}\alpha + 4.3 \text{ MeV}$$

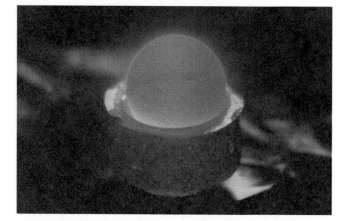

A self-luminous 1000°C radioactive sphere. At its core are six ounces of plutonium (^{238}Pu) dioxide, clad in an iridium shell, surrounded by a graphite casing. The α-particles emitted in the decay are absorbed in the surrounding shell, where they impart thermal energy at a rate of 100 W and will continue to do so for decades.

Example 32.4 A too-common hazard in many homes is the radioactive gas radon-222. It's produced in the ground by the alpha decay of radium-226 and often leaks into basements. Write out the transformation equation and determine (to 3 figures) the total kinetic energy of the decay products in MeV. The masses of the radium, radon, and helium *atoms* are 226.025 406 u, 222.017 574 u, and 4.002 603 u, respectively. As a rule, most of the KE is carried off by the light particle; here, the alpha. Very little ($\approx$0.1 MeV) goes to the massive recoiling daughter.

Solution: [Given: $m_X = 226.025\,406$ u, $m_Y = 222.017\,574$ u, and $m_\alpha = 4.002\,603$ u. Find: KE.] Using

$$^{A}_{Z}X \rightarrow \ ^{A-4}_{Z-2}Y + \ ^{4}_{2}He + Q \qquad [32.2]$$

we obtain

$$^{226}_{88}Ra \rightarrow \ ^{222}_{86}Rn + \ ^{4}_{2}He + Q$$

The KE equals Q, where

$$Q = (m_X - m_Y - m_\alpha)c^2 \qquad [32.3]$$

The atomic masses can be used because the electrons cancel. In MeV

$$Q = (226.025\,406 \text{ u} - 222.017\,574 \text{ u} \\ - 4.002\,603 \text{ u})(931.494 \text{ MeV/u})$$

and

$$Q = (0.005\,229 \text{ u})(931.494 \text{ MeV/u}) = 4.870\,8 \text{ MeV}$$

The net KE of the decay products is $\boxed{4.87 \text{ MeV}}$.

▶ **Quick Check:** Our answer is roughly 5 MeV, as we know it should be.

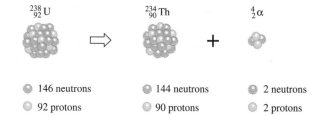

$^{238}_{92}U$ $^{234}_{90}Th$ $^{4}_{2}\alpha$

146 neutrons   144 neutrons   2 neutrons
92 protons   90 protons   2 protons

Figure 32.11 An unstable parent nuclide (U-238) decaying, via α emission, into a daughter nuclide (Th-234).

Beta Decay and the Neutrino

Three distinct processes are each a form of *beta decay*. In $\boldsymbol{\beta^-}$ **decay**, an electron ($_{-1}^{0}$e) is emitted from a nucleus as a neutron transforms into a proton. In $\boldsymbol{\beta^+}$ **decay**, a positron ($_{+1}^{0}$e) is emitted from a nucleus as a proton transforms into a neutron. In **electron capture**, one of the orbital electrons in an inner shell of the cloud is drawn into the nucleus, transforming a proton into a neutron.

By the end of the 1920s, experiments on electron beta decay had revealed some extremely puzzling aspects. A parent nucleus was assumed to decay into a daughter after the creation and immediate emission of an electron. One problem was an apparent violation of Conservation of Linear Momentum. When a parent nucleus (the best results are obtained with a monatomic gas) more or less at rest decays, the electron and recoiling daughter nucleus should move in opposite directions. That's the only way the net linear momentum before and after can be zero, and surprisingly, that didn't always happen. The basic process occurring was assumed to be the spontaneous transformation of a neutron into a proton and an electron, but each of these particles is a spin-$\frac{1}{2}$ fermion. The original neutron spin is $\frac{1}{2}$, whereas the resulting total proton-electron spin could be either 0 or 1. Clearly, this situation violated Conservation of Angular Momentum.

An even more confounding observation was made by Chadwick in 1914. Figure 32.12 shows the broad energy spectrum measured for emitted electrons from a typical beta emitter. Each parent atom is identical, as is each daughter, and so every escaping electron should be identical and have the same energy, but they do not. If the basic process corresponds to n $\rightarrow$ p + e$^-$, we can expect a disintegration energy of

$$Q = (m_n - m_p - m_e)c^2$$

which equals 0.783 MeV (see Problem 38). Except for a very small amount of energy appearing as the recoil KE of the proton, Q ought to be exactly the KE of every emitted electron. In the case of a heavy nucleus, even less energy is given to the recoil. In other words, all electrons should appear with the maximum kinetic energy (KE_{max}) terminating the curve in Fig. 32.12 for each emitter. That energy (Q) is equivalent to the parent-daughter-electron mass difference. Experimentally, we find that all electrons are ejected with less than this energy. To explain that evident violation of Conservation of Energy, it was suggested that the β-particles lost energy via collisions while emerging from the atom. That suggestion was rejected by C. D. Ellis (who, incidentally, became a physicist only after being interned in a German prison camp with Chadwick during World War I) and W. A. Wooster. Using a thick lead-walled calorimeter, they proved that the total amount of energy deposited corresponded to the average energy of the spectrum, which is about 40% of what it ought to be (either energy was not conserved, or 60% of it mysteriously escaped from the calorimeter).

Some physicists felt compelled to reject Conservation of Energy—Madame Curie and Niels Bohr were among those skeptics. Rather than accept the obvious facts that were there for all to see, Wolfgang Pauli remained steadfast. In 1930, he postulated that an invisible particle emitted along with each electron carried off precisely the amount of energy needed to sustain conservation. Sometime later, Enrico Fermi, while answering a question, almost jokingly referred to this phantom speck as a **neutrino** (which in Italian means "little neutral one"), and the name stuck. Today (for reasons that will become clear) we call it an **electron-antineutrino** ($\overline{\nu}_e$). Not surprisingly, Pauli proposed that it was uncharged and so immune from the electromagnetic force. Because this neutral particle could pass through the thick lead walls of the Ellis-Wooster calorimeter, it had to be unaffected by the strong

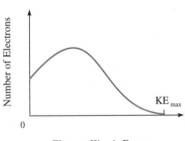

Figure 32.12 Energy spectrum of the electrons in a typical beta decay.

force. To conserve angular momentum, Pauli argued that his phantom particle had a spin of $\frac{1}{2}$. Today, it's known that neutrinos and antineutrinos do have spin-$\frac{1}{2}$. When an electron and an antineutrino are created as a pair, their spins are oppositely directed.

The kinetic energies of some emitted electrons approach KE_{max}, meaning that, in the extreme, the accompanying antineutrino can have a total energy that is vanishingly small. This total energy equals its rest energy (mc^2) plus its kinetic energy. The only way that sum can be vanishingly small is if m is zero or nearly zero, while $KE \approx 0$ as well. By 1933, Perrin had argued from the β-spectrum "that the neutrino has *zero intrinsic mass*." Like a photon, it would then travel at c, and like a photon with zero rest energy, it could carry ample kinetic energy to balance out the demands of conservation. Intrigued by these ideas, de Broglie (1934) suggested that, as with the electron, the neutrino should have a mirror image, the antineutrino.

Electron-beta decay arises when a neutron decays into a proton, an electron, and an electron-antineutrino,

$$\text{n} \rightarrow \text{p} + {}_{-1}^{0}\text{e} + \bar{\nu}_e \qquad (32.4)$$

In the case of a radioactive nuclide, the decay has the form

$$_Z^A\text{X} \rightarrow {}_{Z+1}^{A}\text{Y} + {}_{-1}^{0}\text{e} + \bar{\nu}_e + Q \qquad (32.5)$$

where Q is the kinetic energy of all three particles emerging from the transformation. Similarly, positron-beta decay arises when a proton decays into a neutron, a positron, and an electron-neutrino:

$$\text{p} \rightarrow \text{n} + {}_{+1}^{0}\text{e} + \nu_e \qquad (32.6)$$

In the case of a radioactive nuclide, the decay has the form

$$_Z^A\text{X} \rightarrow {}_{Z-1}^{A}\text{Y} + {}_{+1}^{0}\text{e} + \nu_e + Q \qquad (32.7)$$

Electron-capture, the process wherein a parent nucleus absorbs one of its own orbital electrons, competes with β^+ decay; that is

$$_Z^A\text{X} + {}_{-1}^{0}\text{e} \rightarrow {}_{Z-1}^{A}\text{Y} + \nu_e + Q \qquad (32.8)$$

For example, beryllium-7 captures an electron and transforms into lithium-7:

$$_4^7\text{Be} + {}_{-1}^{0}\text{e} \rightarrow {}_3^7\text{Li} + \nu_e$$

The distinction between the electron-neutrino (ν_e) and the electron-antineutrino ($\bar{\nu}_e$) is that the former is left-handed (its spin angular momentum vector is antiparallel to its linear momentum vector) and the latter is right-handed (its angular and linear momentum vectors are parallel). *The emission of a neutrino is equivalent in many regards to the absorption of an antineutrino, and vice versa.*

Because it doesn't "feel" either the strong or the electromagnetic force, a neutrino can sail through a thickness of over a light-year of solid lead before undergoing a single collision. In the 1930s, there was little hope of ever confirming their existence. Twenty-five years later, the feat was accomplished. Using a nuclear reactor (which is a copious source of antineutrinos, most arising from the decay of free neutrons), C. L. Cowan and F. Reines bombarded the protons in a huge quantity of water. If antineutrinos existed, the inverse process to that of Eq. (32.6)—add an antineutrino to both sides—should produce observable effects. Roughly 1 in 10^{12} antineutrinos struck a proton head-on, converting it into a neutron and a positron, both of which were then detected. The electron-antineutrino exists. More recently (1986), it has been determined that its mass (if it has mass) is less than a mere 27 eV/c^2 (which is about 19 000 times less than the electron). The issue of a nonzero neutrino mass is crucial in astrophysics. Even if the neutrino corresponded to just a few eV, they are so numerous that much of the mass of the Universe would be invisible neutrinos. That hidden mass might well determine the future fate of the entire Universe.

At 7:35 A.M. on February 23, 1987, counters on two huge particle detectors deep below the surface of the Earth, one near Cleveland and the other near Tokyo, began flashing for several seconds (p. 1234). The cause of that unprecedented event was revealed several hours later when the light from an exploding star in a nearby galaxy (170 000 ly away) arrived at Earth. The cataclysmic eruption of Supernova 1987A emitted a blast of neutrinos that immediately flew out into space at or very near the speed of light. Several hours later, the shock wave reached the surface of the star, and it began emitting a blaze of light a million times brighter than the Sun. The pulse of neutrinos and flood of photons had flown toward Earth for 170,000 years, arriving about two hours apart.

32.6 The Weak Force

In 1934, Fermi introduced a highly successful theory of beta decay predicated on the existence of a new interaction in nature, the **weak force**. The very fact that neutrinos were so penetrating suggested that they were uninfluenced by both the strong and the electromagnetic forces, and yet something (other than gravity) both blasted them into existence and promoted their occasional interaction with matter.

In its broadest meaning, a force is an agent of change; it may manifest itself as a push or a pull (a changer of motion), or it may be a changer of some aspect of the state of a system, a transformer. The electromagnetic force can transform mass-energy in an atomic system, thereby creating a photon (or it can transform a stick of dynamite into a puff of smoke). In the case of beta decay, it was necessary to conceive of a new force that could transmute a neutron into a proton, or vice versa.

A free neutron has a half-life of 10.4 minutes (a very long time in Atomland), and that implies that the operative force is very weak—it's argued that the stronger the force, the faster the processes it drives. As we'll see in Chapter 33, there are hundreds of unstable exotic subnuclear particles. Some of these decay via the strong force (producing strongly interacting particles) and have lifetimes of only $\approx 10^{-23}$ s or so. By contrast, a decay resulting in a photon and so driven by the electromagnetic force might take $\approx 10^{-20}$ s. Weak decays involving neutrinos happen much more slowly, with lifetimes of perhaps $\approx 10^{-8}$ s.

The weak force is a million times more feeble than the strong force (Table 32.2), though it is immensely ($\approx 10^{32}$ times) more powerful than gravity. It reveals its presence in the macroscopic world primarily through radioactive transformation (there being hundreds of β-decaying nuclides) and also via the results of fission and fusion occasionally rather spectacularly in a supernova. Most subnuclear particles (the vast majority of them are unstable) interact via the weak force. The spontaneous transmutation of many subnuclear particles occurs through the weak force. It is extremely short-ranged, so much so that it was

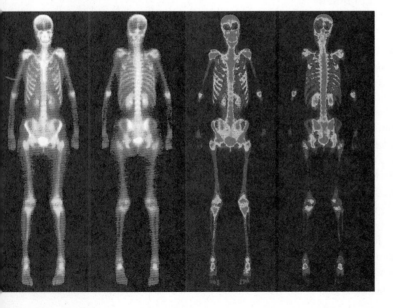

Full-body bone scans. Gamma-ray images of a patient three hours after being injected with 25 millicuries (see Problem 40) of technetium-99, a γ-ray emitter with a half-life of 6 h.

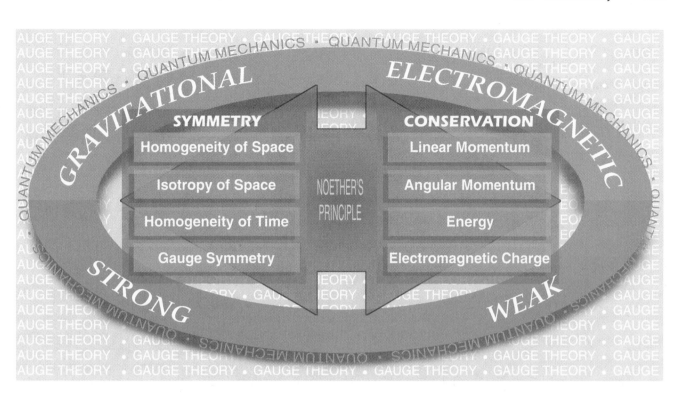

Figure 32.13 The gravitational, electromagnetic, strong, and weak forces are the four fundamental interactions that effectuate all change.

only in the 1980s that accurate determinations of its extent could be attempted. Beyond about 10^{-17} m (10^{-2} fm), the weak force becomes exceedingly small. It is for the most part an inside-the-particles force rather than a between-the-particles force. Still, as we become more sophisticated experimentally, we are able to detect the minute influence of the weak force reaching from the nucleus to the electron cloud of an atom (Fig. 32.13).

32.7 Gamma Decay

After an alpha or beta decay, a daughter nucleus may momentarily end up in an excited state; that is, with a nucleon in a higher energy level than the ground state (Fig. 32.8). The nucleus then quickly relaxes, inevitably reaching the lowest available energy configuration. The energy difference (≈ 1 keV to ≈ 1 MeV) is emitted as one or more gamma photons (hf). As with an atom, it is possible to excite a nucleus by pumping in energy. This might be accomplished via the absorption of a gamma photon of the proper frequency, or it might happen as the result of a collision with a massive particle. For example, a slow-moving neutron can be absorbed by a ^{238}U nucleus, whereupon it becomes an excited ^{239}U* nucleus (the asterisk indicates an excited state). Returning to its ground state, it emits a gamma photon; that is

$$_{0}^{1}\text{n} + {}^{238}\text{U} \rightarrow {}^{239}\text{U*} \rightarrow {}^{239}\text{U} + \gamma$$

When observed independent of its nuclear source, a gamma photon is identical in all respects to a photon of the same energy arising from any other means. It's only because most nuclear energy levels are widely separated that gamma rays are usually more energetic than photons emitted from atomic electron transitions.

Example 32.5 Nitrogen-12 beta decays into carbon-12. The latter, $^{12}C^*$, then emits a 4.43-MeV gamma photon during de-excitation. Compute the mass of the $^{12}C^*$ atom.

Solution: [Given: gamma energy of 4.43 MeV and ^{12}C system. Find: $^{12}C^*$ atomic mass.] The mass of the ^{12}C atom is 12.000 000 u. The equivalent mass of the photon is gotten from $E = mc^2$, or the fact that 1.000 00 u = 931.494 MeV/c^2:

$$m_\gamma = \frac{4.43 \text{ MeV}}{931.494 \text{ MeV/c}^2} = 0.004\,76 \text{ u}$$

Hence, the total mass of the excited carbon-12 atom is

$$12.000\,00 \text{ u} + 0.004\,76 \text{ u} = \boxed{12.004\,76 \text{ u}}$$

▶ **Quick Check:** Let's redo the calculation using kilograms. $(4.43 \text{ MeV})(1.782\,663 \times 10^{-30} \text{ kg/MeV}) = 7.897\,197 \times 10^{-30} \text{ kg}$; $(12.000\,00 \text{ u})(1.660\,540 \times 10^{-27} \text{ kg/u}) = 1.992\,648 \times 10^{-26} \text{ kg}$; hence, the total mass is $1.993\,438 \times 10^{-26} \text{ kg}$; or 12.004 76 u.

32.8 Half-Life

While working with highly radioactive ^{220}Rn, Rutherford observed that the intensity of the emissions decreased with time in a precise and predictable way. The amount of radiation emerging from a sample of a radioactive element is virtually independent of the surrounding environment (that is, the chemical compound it's in, the temperature, pressure, etc.). The quantitative measure of radioactive intensity is the *number of disintegrations per second,* known also as the **decay rate,** or the **activity.** The SI unit of decay rate is the **becquerel** (Bq), which is *one disintegration per second.* Activity seems to be dependent only on the type and amount of radioactive material present at that moment, which is both a common and a very special behavior. Many processes unfold at a rate that depends on the amount of participating ingredient present at that instant (for example, the decay of the charge on a capacitor, Fig. 20.19).

Letting N stand for the number of radioactive atoms present at a given instant, $|\Delta N/\Delta t|$ represents the decay rate (R). This rate can be made independent of the size of the particular sample by dividing it by N. Thus, $|\Delta N/\Delta t|/N$ is the fraction of the atoms of a given species disintegrating per unit time, regardless of the sample size. This quantity has been found experimentally to be constant over a wide range of sources and for long periods of time. It's called the **decay constant** and is usually symbolized, rather unfortunately, by λ, which is *not* to be confused with wavelength (sorry about that):

$$\lambda = \frac{|\Delta N/\Delta t|}{N} = \text{constant} \tag{32.9}$$

The unit of λ is s^{-1}, and it should be thought of as the fractional disintegration rate.

Suppose a detector is set up next to the sample, and the counting rate per convenient interval of time is recorded. Start at $t = 0$ with a rate R_0. From Eq. (32.9), $R_0 = \lambda N_0$ where N_0 is the number of atoms present initially. Figure 32.14 is a typical plot of R against time. The measured decay rate drops to $\frac{1}{2}R_0$ after a time interval Rutherford called the **half-life** ($t_{1/2}$). The known radioactive nuclides have half-lives ranging from roughly 10^{-22} s to about 10^{21} y. Upon waiting two half-lives, the decay rate will be one-quarter its original value; after three half-lives, it will be half of that or one-eighth its original value, and so on, with R approaching, but never reaching, zero. Thus, after n half-lives, the decay rate is

Figure 32.14 The exponential decay of the activity R of a radioactive nuclide.

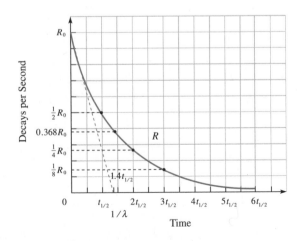

$$R = (\tfrac{1}{2})^n R_0 \qquad (32.10)$$

Each radioactive element has its own decay curve, some declining more steeply than others (just like the two curves in Fig. 22.21), but all will have the same basic shape. A straight line having the initial slope of each curve will intersect the time axis at the *time constant,* or **mean lifetime** (τ), at which moment $R = 0.368\,R_0$. The greater the activity ($R = |\Delta N/\Delta t|$), the greater λ is, the steeper the slope, and the smaller is τ. In fact, $\tau = 1/\lambda$. The red curve in Fig. 32.15 is a replot of Fig. 32.14, with R_0 set equal to 1 and τ set equal to 1; the fundamental decay curve. Let's try to fit this nice smooth graph with a simple mathematical function: 2^{-t} or $1/2^t$ is too slow descending, and 3^{-t} is a little too fast. After some routine calculation, it can be determined that $(2.718\,2818\ldots)^{-t}$ fits the decay exceedingly well. It's traditional to call $2.718\,2818\ldots$ simply e (after the mathematician Euler), whereupon the basic decay curve is e^{-t} (Appendix A-3).

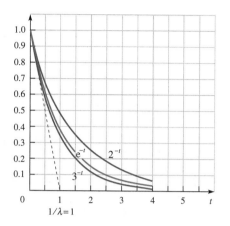

Figure 32.15 A comparison of 2^{-t}, e^{-t}, and 3^{-t}.

All decay curves drop to $1/e = 0.368$ of their initial values at $t = \tau = 1/\lambda$, and so the basic curve can be particularized to a specific decay rate by writing it as $1e^{-\lambda t}$. Moreover, letting the initial activity be some value R_0, the decay rate for any radioactive element becomes

$$R = R_0 e^{-\lambda t} \qquad (32.11)$$

At $t = 0$, $e^0 = 1$, and $R = R_0$, as it should be. Now, let's consider the half-life. When $t = t_{1/2}$, $R = R_0 e^{-\lambda t_{1/2}} = \tfrac{1}{2}R_0$, thus $e^{-\lambda t_{1/2}} = \tfrac{1}{2}$; yet as an electronic calculator can show $e^{-0.693} = \tfrac{1}{2}$, hence

$$\lambda t_{1/2} = 0.693 \qquad (32.12)$$

or $t_{1/2} = 0.693\tau$. The average value of the exponential is reached at $t = \tau$, which is why it's called the *mean lifetime.* If we know the half-life or the lifetime, we can calculate the decay constant or vice versa. A large decay constant means a short half-life.

It follows from Eq's. (32.9) and (32.11) that

$$R = \lambda N_0 e^{-\lambda t} = \lambda N \qquad (32.13)$$

hence the number of atoms remaining after a time t is

$$N = N_0 e^{-\lambda t} \qquad (32.14)$$

Notice that both sides of this equation could be multiplied by the mass of the nuclide, thereby yielding an expression for the total mass of the remaining undecayed element.

Equation (32.14) was deduced theoretically in 1905 by E. von Schweidler, who accomplished the task without any concern for nuclear structure. He simply assumed that the decay process was subject to the laws of chance: The probability of an atom disintegrating in any time interval is independent of its past history and is the same for all atoms of the same kind. The decay law, Eq. (32.14), is actually a statistical relationship that is the result of a very large number of events occurring according to the laws of probability. On average the number of atoms decaying per second is indeed λN, but there is always some fluctuation around that number. (The usual derivation of the decay law, which involves the calculus and assumes N is a continuous function of t, is therefore a bit simplistic; it's left for Problem 67.)

Example 32.6 Radium-226 is found to have a decay constant of 1.36×10^{-11} Bq. Determine its half-life in years. Given that the Curies had roughly 200 g of radium in 1898, how much of it will remain 100 years later?

Solution: [Given: $\lambda = 1.36 \times 10^{-11}$ decays/s, $m_0 = 200$ g, and $t = 100$ y. Find: $t_{1/2}$ and m.] Using Eq. (32.12), we have

$$t_{1/2} = \frac{0.693}{1.36 \times 10^{-11} \text{ s}^{-1}} = 5.096 \times 10^{10} \text{ s}$$

$$\boxed{t_{1/2} = 1.62 \times 10^3 \text{ y}}$$

Since $t = 100$ y $= 3.15 \times 10^9$ s, and $\lambda = 0.693/t_{1/2} = 1.36 \times 10^{-11}$ s^{-1}

$$m = m_0 e^{-(1.36 \times 10^{-11} \text{ s}^{-1})(3.15 \times 10^9 \text{ s})}$$

and

$$m = (200 \text{ g})(0.958) = \boxed{192 \text{ g}}$$

▶ **Quick Check:** The factor $(-\lambda t)$ to which the exponential is raised must be a pure number; its units must cancel as they do here. With a half-life of 1620 y, $\lambda = 0.693/(1620 \text{ y}) = 4.28 \times 10^{-4}$ y^{-1} and $e^{-\lambda t} = e^{-0.0428} = 0.958$.

The half-life of radioactive carbon-14 is 5730 years (see Table 32.4), and it serves as an archeological clock. The ratio of the amount of ^{14}C to stable ^{12}C in the atmosphere is fairly constant at about 1.3×10^{-12}. Thus, any living organism takes in both isotopes in that fixed proportion as it assimilates carbon. Once dead, the organism takes in no new carbon, the ^{14}C decays, and the ratio changes in a way that marks the passage of time. For instance, if a piece of charcoal from an ancient fireplace contains one-quarter of the original ratio of ^{14}C to ^{12}C, then a time of $2t_{1/2} = 11\,460$ years has passed since the wood died.

Radioactivity is a quantum-mechanical phenomenon and Einstein (1916) first realized that Eq. (32.13) was a manifestation of the inherent statistical nature of the process. It is now the common wisdom that an unstable nucleus decays spontaneously (just as an atom in an excited state emits a photon spontaneously); that

TABLE 32.4   Half-lives of some radioisotopes

Isotope	Decay mode	Half-life
Rubidium-87	e^-	4.7×10^{10} y
Uranium-238	α	4.5×10^9 y
Plutonium-239	α	2.4×10^4 y
Carbon-14	e^-	5730 y
Radium-226	α	1600 y
Strontium-90	e^-	28 y
Tritium-3	e^-	12.26 y
Cobalt-60	e^-	5.24 y
Iodine-131	e^-	8 d
Radon-222	α	3.82 d
Technetium-104	e^-	18 min
Fluorine-17	e^+	66 s
Polonium-213	α	4×10^{-6} s
Beryllium-8	α	1×10^{-16} s

there is no knowable external trigger that fires them off in a way that can be anticipated; that identical atoms behave, all by themselves, in nonidentical ways that conform *en masse* to the rules of probability. It would seem that radioactivity provides a glimpse of the quantum-mechanical microworld playing havoc with classical cause and effect. An atom that is ten thousand years old is supposedly identical to an atom of the same species that is ten seconds old; from this moment on, one of these may live for ten thousand years and the other for ten seconds, and yet we don't know which will do what. Whether that ignorance is a profound one, forced on us by the probabilistic nature of the Universe, or just a practical one inflicted by our own limitations remains to be seen.

32.9 Induced Radioactivity

Irène and Jean Frédéric Joliot-Curie were not having much luck in the early 1930s. Chadwick's announcement of his discovery of the neutron in 1932 was a shock. They had often produced these new particles in their own laboratory but had thought they were gamma rays. Hot on the trail of the positron, they lost that glory to Anderson in 1933. Still, their triumph (and the Nobel Prize in chemistry) would soon come (1934) out of a series of experiments bombarding light elements with α-particles. In particular, they transmuted aluminum into phosphorus:

$$^{4}_{2}\text{He} + ^{27}_{13}\text{Al} \rightarrow ^{31}_{15}\text{P*} \rightarrow ^{30}_{15}\text{P} + ^{1}_{0}\text{n}$$

The α-particle was captured into the Al-27 nucleus, forming a highly unstable *compound nucleus*, P-31, that immediately decayed into P-30. The surprising thing was that the phosphorus went on emitting positrons even after the alpha bombardment ceased, as

$$^{30}_{15}\text{P} \rightarrow ^{30}_{14}\text{Si} + ^{0}_{+1}\text{e}$$

They had **artificially induced radioactivity** and, within a year, created a whole group of **radioisotopes**. The stage was set for our present-day medical, biological, and industrial use of these materials—even then the implications were enormous.

Enrico Fermi in Rome was particularly fascinated by the work of the Joliot-Curies, and he set out to test the novel idea he had of using neutrons to create radioisotopes. Since these neutral bullets experienced no electrical repulsion, he rightly reasoned they would easily approach and enter a target nucleus. Within weeks, Fermi published his first positive results. Systematically, he bombarded every element he could get his hands on, all the way up to uranium. Along the way, he realized that neutrons that had been slowed down in their passage through certain materials were even more potent at instigating transmutations than were ordinary fast neutrons. These **thermal neutrons** (moving at the speeds of room temperature air molecules) spent more time in the vicinity of a nucleus as they traveled by and could be captured more effectively. "The Pope" (that was Fermi's nickname because of his strong advocacy of the "new faith" of Quantum Mechanics) concluded that the light hydrogen atoms (protons) in a sheet of paraffin would be very efficient at slowing the neutrons. Colliding objects that are about the same mass transfer the most momentum (p. 345). To everyone's amazement, inserting a sheet of paraffin into the neutron beam increased its ability to produce transmutations a hundredfold.

Uranium bombarded by thermal neutrons exhibited some surprising behavior in that it subsequently emitted beta rays. The natural conclusion was that a nucleus had swallowed up a neutron, emitted an electron, and ended up with an additional proton. Fermi believed that uranium had thus been transformed into a new element one

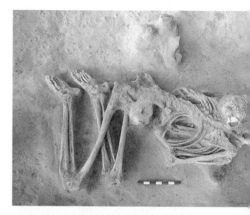

Archeological remains, like this skeleton, can be dated using carbon-14.

Enrico Fermi (1901–1954), Italian-American physicist, was one of the leading figures in the development of the atomic bomb. After years of working with radiation, he died from cancer at the age of 53.

box higher in the Periodic Table, the first *transuranic* element. Although that transformation does occasionally take place, the predominant effect was something far more spectacular: nuclear fission.

32.10   Fission and Fusion

Shortly after Fermi's announcement, Ida Noddack, a German chemist, published an alternative scenario that at the time, though it was correct, seemed absolutely crazy. She proposed that the incoming neutron had ruptured the uranium nucleus "into several big fragments which are really isotopes of already known elements." Everyone, including the world's foremost radiochemist, Professor Otto Hahn at the Kaiser Wilhelm Institute in Berlin, completely dismissed the idea as "impossible." After all, how could a small, low-energy neutron burst the largest known nucleus?

Hahn and his colleague Lise Meitner, joined by F. Strassmann, repeated the Italian experiments with uranium. The chemical analysis was extremely difficult, and the results remained inconclusive, but that wasn't their only worry. Dr. Meitner was an Austrian Jew, and despite Planck's personal appeal to Hitler, she was dismissed from the Institute where she had worked for 30 years. That summer (1938), she fled to Stockholm. Just before Christmas, Hahn and Strassmann finally realized that they were dealing with isotopes of the light element barium. It was inescapable: the uranium nucleus had indeed split in two (Fig. 32.16). They dashed off a letter to the editor of the journal *Naturwissenschaften* and sent a copy to their exiled friend Meitner.

Just then Meitner was being visited by her nephew, a young physicist named Otto Frisch who had fled the Nazis and was working at Bohr's Institute in Copen-

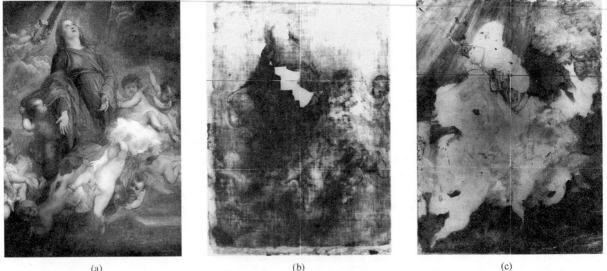

(a)   (b)   (c)

By bombarding a painting (a), in this case one by Van Dyck, with neutrons, different elements in the paint and varnish can be made radioactive. For example, manganese, once common in the dark brown pigment umber, absorbs neutrons with the subsequent emission of electrons. (b) A special film sensitive to electrons reveals the original layers of umber. The white region free of manganese is a modern repair. (c) Four days later, after the emission from the umber faded away, radiation from phosphorus in charcoal revealed a whole new aspect to the painting. Notice the inverted charcoal sketch of the head of a person, now clearly visible. (See p. 1094.)

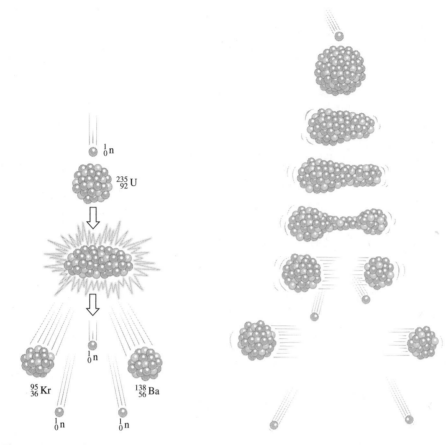

Figure 32.16   The fission of U-235.   **Figure 32.17**   Nuclear fission.

hagen. With Hahn's letter in hand, the two went for a walk in the woods. Frisch was well versed in Bohr's new model, which maintained that the nucleus resembled a droplet of liquid bound by surface tension. They sat down on a tree stump and began making calculations on scraps of paper. A very large nucleus such as uranium with its many protons would experience an internal electrostatic repulsion that would all but cancel the cohesive surface tension. Even the slight nudge of an incoming neutron might cause it to oscillate and elongate (Fig. 32.17), whereupon the Coulomb repulsion would tend to increase the distortion until, swamped by the short-range strong force, the two pieces would snap into a separate existence and fly apart (all in $\approx 10^{-22}$ s).

Remembering the binding energy curve (p. 1186), Meitner could assume that the fragment nuclei might have a binding-energy-per-nucleon of about 8.5 MeV compared to about 7.6 MeV for a heavy nucleus. The energy released would be the difference $(8.5 - 7.6)$ MeV for each of the roughly 240 nucleons, or about 216 MeV. That's a lot of energy for a single atomic event (roughly a million times the amount liberated by the combustion of one molecule of gasoline). About 1% of the original nuclear mass would be converted into energy, most of it (168 MeV) going into the KE of the two fragments.

A few days later, Frisch caught up with Bohr as he was about to sail off to America to visit Einstein. "Oh, what fools we have been!" Bohr blurted out, "We ought to have seen that before." Hurriedly, Frisch and Meitner wrote a paper an-

Lise Meitner (1878–1968).

nouncing their results and, borrowing a term from the biology of cell division, Frisch named the process **nuclear fission**.

Naturally occurring uranium is composed of three isotopes, U-234, U-235, and U-238. By far the most abundant at 99.27% is U-238, followed by U-235 at only 0.72%, and a mere trace of U-234. On the basis of theoretical considerations, Bohr and his former student J. A. Wheeler concluded that it was the rare U-235 that had undergone slow-neutron fission. In fact, it's the only naturally occurring element that can do that little trick. (The common, and far more stable, isotope U-238 can be split, but only by fast-neutron bombardment.) A slow neutron captured by U-235 results in the highly unstable compound nucleus U-236, which immediately ruptures into two large pieces. One of the common fragment pairs is barium-138 and krypton-95:

$$\ _{0}^{1}\text{n} + \ _{92}^{235}\text{U} \rightarrow \ _{92}^{236}\text{U}^{*} \rightarrow \ _{56}^{141}\text{Ba} + \ _{36}^{92}\text{Kr} + 3\ _{0}^{1}\text{n}$$

The Chain Reaction: On the Way to the Bomb

In the 1930s, most people (including Einstein) shared Rutherford's view: "Anyone who expects a source of power from the transformation of these atoms is talking moonshine." Leo Szilard, a brash young Hungarian physicist, was incensed by Rutherford's remark, and in short order he conceived the concept of the **chain reaction**. Szilard speculated that if a nuclide could be found that, when struck by a neutron, would emit *two* or more neutrons, the resulting fission process could continue on its own, building up into a horrendous cascade (Fig. 32.18). Sensing that Europe would soon be engulfed in war, Szilard accepted a post at Columbia University and arrived in America in January, 1939. Fermi came to Columbia that same month. His mother's antecedents were Jews, as was his wife, and so after stopping in Stockholm to pick up the 1938 Nobel Prize, the family continued on to New York.

Only weeks after Fermi's arrival, Bohr reached New York with the news of the discovery of fission. "The Pope" immediately appreciated the possibilities of mak-

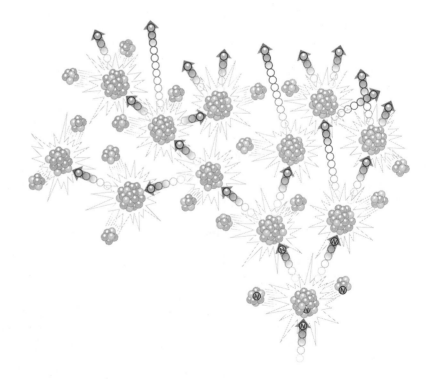

Figure 32.18 A branching chain reaction arising from uranium fission.

The article in the right column reports Einstein's opinion (1934) that attempts at loosing the "energy of the atom" would be fruitless—an opinion he soon changed.

ing a nuclear bomb, but there were troubling questions that had to be resolved. Only if each fission event emitted an average of at least two neutrons could an explosive device be developed. In March, Szilard and Zinn determined experimentally that, on average, between two and three fast neutrons were indeed emitted per fission. A single such event liberates about 3.2×10^{-11} J, which is a great deal of energy in Atomland, but not very much on the human scale of things. That was why the chain reaction—and, in particular, a chain reaction that fanned out and grew—was so crucial if a vast amount of energy was to be liberated. If the first nucleus to split fired out two neutrons, those two could split two other nuclei in the second generation, and so on. Ideally, after, say, 80 generations, an incredible 1.2×10^{24} atoms would have shattered in a fraction of a millisecond. About 0.5 kg of uranium would vanish, releasing 3.8×10^{13} J, or the equivalent of roughly 10 kilotons of TNT.

When Hitler embargoed the export of uranium from Czechoslovakian mines, Szilard was convinced the Germans were developing nuclear weapons.* In August of 1939, he went to Einstein with a letter to President Roosevelt, which the highly respected scientist signed. It advised "that extremely powerful bombs of a new type may thus be constructed." The all-out push to develop the atomic bomb began on December 6, 1941, just one day before Pearl Harbor was attacked by the Japanese.

Nuclear Reactors

A chunk of natural uranium is mostly U-238, and these nuclei don't easily split. Moreover, fast neutrons from a fissioning nucleus can be captured equally well by U-238 and U-235. Thus, pure natural uranium will not support a chain reaction. But U-238 has much less appetite for slow neutrons than does U-235, and that difference was exploited in the design of the first **fission reactor**, the controlled chain-reaction machine. If uranium was interspersed within a *moderator* (some light substance that slowed neutrons down), it would have almost the same effect as removing the U-238 altogether, leaving only the U-235 to interact. The most promising candidates for moderators were beryllium, heavy water (regular hydrogen absorbs too many neu-

*Szilard was right. Germany had an active nuclear weapons program led by Heisenberg. Because he seems never to have actually believed in the possibility of making a bomb, Heisenberg's efforts were rather half-hearted. In 1943, the Japanese joined the race to build an atomic bomb. A lack of uranium made things difficult for them, and when a submarine from Hitler carrying two tons of it was sunk by the Allies, their project faltered.

The time will come when atomic energy will take the place of coal as a source of power. . . . I hope that the human race will not discover how to use this energy until it has brains enough to use it properly.

SIR OLIVER LODGE (1920)

We are justified in reflecting that scientists who can construct and demolish elements at will may also be capable of causing nuclear transformations of an explosive character.

FRITZ HOUTERMANS
Berlin (1932)

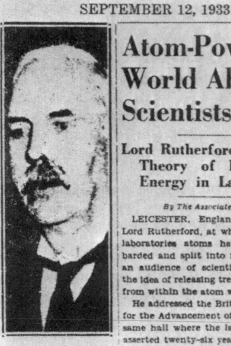

SEPTEMBER 12, 1933

Atom-Powered World Absurd, Scientists Told

Lord Rutherford Scoffs at Theory of Harnessing Energy in Laboratories

By The Associated Press

LEICESTER, England, Sept. 11.— Lord Rutherford, at whose Cambridge laboratories atoms have been bombarded and split into fragments, told an audience of scientists today that the idea of releasing tremendous power from within the atom was absurd.

He addressed the British Association for the Advancement of Science in the same hall where the late Lord Kelvin asserted twenty-six years ago that the atom was indestructible.

Describing the shattering of atoms by use of 5,000,000 volts of electricity, Lord Rutherford discounted hopes advanced by some scientists that profitable power could be thus extracted.

"The energy produced by the breaking down of the atom is a very poor kind of thing," he said. "Any one who expects a source of power from the transformation of these atoms is talking moonshine. . . . We hope in the next few years to get some idea of what these atoms are, how they are made and the way they are worked."

©1933, N.Y. HERALD TRIBUNE CO.

Lord Rutherford

trons), and carbon. Beryllium was crossed off the list because it is rare and toxic. Heavy water would have been perfect, but there were only a few quarts of it in the United States, and there wasn't time to prepare the several tons needed. The choice fell to carbon in the form of pure graphite.

Fermi and his group moved the top-secret operation to the University of Chicago. They set up shop in the squash courts (where the ceilings were more than 25 feet high) under the abandoned football stands of Stagg Field. There, they stacked 40 000 graphite bricks, into which were inserted chunks of uranium oxide. Several *control rods* of cadmium, which is a voracious absorber of neutrons, were built into the reactor to keep a leash on the chain reaction. On December 2, 1942, the cadmium rods were carefully withdrawn, and the world's first self-sustained controlled nuclear chain reaction occurred.

The absorption by U-238 of a neutron (Fig. 32.19) within the core of the reactor produced a new radioactive transuranic element called neptunium. With a half-life of $2\frac{1}{2}$ days, neptunium decays into yet another new element called *plutonium*. It was predicted that Pu-239 would undergo slow-neutron fission even easier than U-235, and that made it a prime candidate for use in a bomb. With that as the driving force, the multimillion-dollar Plutonium Project got underway almost immediately. Three giant water-cooled production reactors were constructed in 1943 on the Columbia River near Hanford, Washington. After cooking for a few months, intensely radioactive slugs of U-238 were taken out of the reactor and the $^{239}_{94}$Pu extracted. Each reactor could produce about half a pound of plutonium a day from otherwise plentiful and comparatively worthless U-238.

The modern power reactor is very similar in principle to these early machines. It usually uses natural uranium *enriched* to contain a few percent U-235, with ordinary water as the moderator (Fig. 32.20). The reactor is used essentially as a furnace to supply "heat" (via steam) that drives a traditional turbine and ultimately generates electricity. The

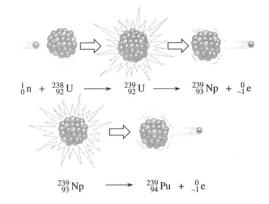

$$^{1}_{0}n + {}^{238}_{92}U \longrightarrow {}^{239}_{92}U \longrightarrow {}^{239}_{93}Np + {}^{0}_{-1}e$$

$$^{239}_{93}Np \longrightarrow {}^{239}_{94}Pu + {}^{0}_{-1}e$$

Figure 32.19 Plutonium-239 produced from uranium-238 by neutron bombardment.

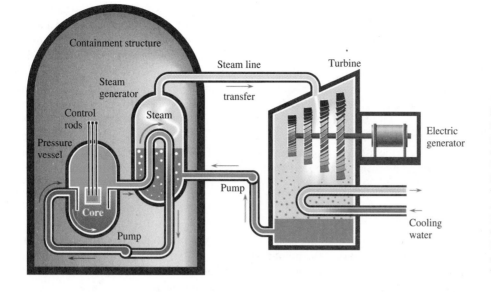

Figure 32.20 A pressurized water reactor system. A modern power reactor usually uses natural uranium enriched to contain a few percent U-235, with ordinary water as the moderator. The kinetic energy of the fission fragments and neutrons absorbed in the reactor transforms into thermal energy. This energy is carried out of the core by a circulating liquid coolant. The reactor is essentially a furnace to supply heat that drives a traditional turbine.

kinetic energy of fission fragments and neutrons absorbed in the reactor transforms into copious amounts of thermal energy. A liquid coolant circulating around the fuel rods carries this energy out of the core. A modern 1000-million-watt reactor is charged with about 90 000 kilograms of fuel, a third of which is replenished each year. It's because a rather innocent-looking power reactor can generate plutonium that it becomes very difficult to control the spread of nuclear weapons (which is why the United States, in 1991, bombed Iraq's purported power reactors in the first few days of the Gulf War).

The Atomic Bomb

The actual design of the bomb was carried out at Los Alamos, New Mexico, under the direction of J. Robert Oppenheimer ("Oppie" to almost everyone). The design team included, among many others, Bohr, Chadwick, Frisch, Fermi, and Feynman. General Groves, who headed the entire bomb program (code-named the Manhattan Project), once remarked "At great expense, we have gathered here the largest collection of crackpots ever seen."

The principle underlying an A-bomb is simple. A small spherical chunk of fissionable material under bombardment by neutrons can support only a limited chain reaction; too many neutrons released from the fissioning nuclei escape from the sample through its surface. If a sphere of U-235 or Pu-239 is made larger, its volume increases faster than its surface area, more neutrons are produced within it, and proportionately fewer escape from its surface. There is therefore a *critical size* at which point the average number of neutrons from each fission that will cause another fission (rather than being lost somehow) equals one. Under such circumstances, the chain reaction is just initiated. Any increase in the amount of fissionable material beyond that produces a branching, rapidly growing chain reaction—such a mass is *supercritical*. A solid sphere of U-235 about the size of a softball, struck by a stray neutron, will roar into a furious, multimillion-degree plasma in a mil-

A depiction of the historic start-up of the world's first nuclear reactor under the direction of Fermi.

Technicians loading the fuel into a modern reactor at a nuclear power plant.

lionth of a second. That's all it takes: one chunk in excess of the *critical mass* (a few kilograms) and off it goes. Of course, the U-235 or Pu-239 has to be kept in a sub-critical configuration so it can be safely transported. The hard part is to assemble a supercritical mass fast enough so that it doesn't just melt and fizzle. That was especially difficult to do with plutonium, which is very unstable.

The first bomb to be designed was a brute force device so simple it was used in war without having been previously tested. Two pieces of U-235, each too small to ignite on its own, were rapidly brought together into a *supercritical* mass. The bomb was little more than a short cannon that fired a subcritical projectile of U-235 into another subcritical target screwed into the end of the muzzle. On August 6, 1945, at $8:15\frac{1}{2}$ in the morning, it dropped from the belly of a B-29 over Hiroshima, Japan. Seconds later, 1500 feet above the ground, it burst into a 15-kiloton fireball.

Starfire: Thermonuclear Fusion

Stars and hydrogen bombs are both powered by a process called **thermonuclear fusion**. Figure 32.9 makes the point that fusing two light nuclei together produces a heavier nucleus higher up the binding-energy-per-nucleon curve. On bonding more tightly, a fraction of each nucleon's mass is converted to energy in accord with Einstein's $E_0 = mc^2$.

Stars are born, live, and die—some explode and disperse; others, latecomers, rise out of their ashes. First-generation stars are composed of primordial hydrogen laced with helium. They are formed when a great interstellar cloud of gas slowly collapses under its own gravitational force, heating up and growing more dense as the material plunges inward. At densities 100 times that of water and temperatures of 15 to 20 million K, an almost-star is a seething plasma ball where bare nuclei rush about at tremendous speeds. Two protons, each with a charge $+e$, can overcome their Coulomb repulsion and come close enough ($r < 10^{-14}$ m) for the strong force to lock them together. That happens provided they approach one another with a kinetic energy that can override the repulsion; a net KE equal to their electrical potential energy at a center-to-center separation r:

$$\mathrm{PE_E} = k_0 \frac{e^2}{r} = (9 \times 10^9 \text{ N·m}^2/\text{C}^2)\frac{(1.6 \times 10^{-19} \text{ C})^2}{10^{-14} \text{ m}}$$

or
$$\mathrm{PE_E} = 2.3 \times 10^{-14} \text{ J} = 0.14 \text{ MeV}$$

which follows from Eq. (18.8). A gas of protons with an average KE of 0.14 MeV has a temperature given by Eq. (14.14); accordingly

$$(\mathrm{KE})_{\mathrm{av}} = \tfrac{3}{2} k_{\mathrm{B}} T$$

and so

$$T = \frac{(0.66)(2.3 \times 10^{-14} \text{ J})}{1.38 \times 10^{-23} \text{ J/K}} = 1 \times 10^9 \text{ K}$$

Fusion can take place at temperatures roughly 50 times lower than this for two reasons. First, this value is an average KE, and many particles have much higher energies. Second, there are quantum-mechanical effects that allow the particles to penetrate the Coulomb barrier. In any event, high speeds and high temperatures are called for, which is why the process is known as *thermo*nuclear fusion. Once fusion begins, the furnace at the center of a star puffs it up with an outrushing of released

On the Atomic Bomb. That is the biggest fool thing we have ever done. The bomb will never go off, and I speak as an expert in explosives.

ADM. WILLIAM LEAHY (1945)

A nuclear explosion.

There is another form of temptation even more fraught with danger. This is the disease of curiosity. . . . It is this which drives us on to try to discover the secrets of nature, those secrets which are beyond our understanding, which can avail us nothing, and which men should not wish to learn. . . . In this immense forest, full of pitfalls and perils, I have drawn myself back, and pulled myself away from these thorns. In the midst of all these things which float unceasingly around me in my everyday life, I am never surprised at any of them, and never captivated by my genuine desire to study them. . . . I no longer dream of the stars.

ST. AUGUSTINE
Confessions

energy, and the gravitational collapse stops; the plasma ball stabilizes and a star is born.

Many thermonuclear fusion processes can drive a star, depending on its physical circumstances (temperature, mass, available fuel, and so on). Each process generates energy out of matter, cooking up heavier elements along the way. Except for the hydrogen in your body (which goes back roughly 17 thousand million years to the first moments of creation), you are star dust, the ash created in the belly of some supernova long before our Sun was born. We believe the Sun (a second-generation star) is predominantly powered by a process known as the *proton-proton chain*. It begins with two protons coming together to form a deuteron ($^2_1$H), a positron, and a neutrino:

$$^1_1\text{H} + {^1_1}\text{H} \rightarrow {^2_1}\text{H} + {^{\ 0}_{+1}}\text{e} + \nu_\text{e}$$

The deuteron then captures another proton, producing helium:

$$^2_1\text{H} + {^1_1}\text{H} \rightarrow {^3_2}\text{He} + \gamma$$

Finally, two helium-3 nuclei fuse in the transformation

$$^3_2\text{He} + {^3_2}\text{He} \rightarrow {^4_2}\text{He} + {^1_1}\text{H} + {^1_1}\text{H}$$

forming He-4 ash, two protons, and a great deal of energy. In effect, the total process is

$$4\text{p} \rightarrow \alpha + 2\,{^{\ 0}_{+1}}\text{e} + 2\nu_\text{e} + \gamma$$

The liberated energy (6.2 MeV per proton) is carried off mostly by gamma rays, positrons, and neutrinos. The gamma photons are absorbed by the plasma. The

J. R. Oppenheimer and General Groves in front of the remains of the vaporized 100-foot steel tower that had held the first atomic bomb (a plutonium device) ever exploded. The desert sand was seared into a glassy jade green crust. One wonders at the wisdom of visiting such a place so soon after an explosion.

positrons are annihilated when they come in contact with electrons generating more photons (which are also absorbed and add more energy). About 2% of the energy liberated comes off with the escaping neutrinos. The Sun unleashes 3.8×10^{26} J each second; each second it pours into space the energy equivalent of burning three million million million gallons of gasoline. Every second roughly 660 000 000 tons of hydrogen are transmuted into helium. In the process, about 4 600 000 tons (4.2×10^{9} kg) of matter transform into radiant energy. And so it has raged for at least five thousand million years.

Core Material

Masses in the atomic domain are often specified in *unified atomic mass units* (u) where a neutral $^{12}_{6}$C atom is defined to be precisely 12.000 000 u:

$$1 \text{ u} = 1.660\,540 \times 10^{-27} \text{ kg} = 931.494 \text{ MeV/c}^2$$

The nuclear radius R is

$$R = R_0 A^{1/3} \qquad [32.1]$$

R_0 turns out to be ≈ 1.2 fm (p. 1179).

When nucleons combine to form a nucleus, an amount of mass Δm known as the **mass defect** is converted into **binding energy** $E_B = \Delta mc^2$. Alpha decay has the form

$$^{A}_{Z}X \rightarrow {}^{A-4}_{Z-2}Y + {}^{4}_{2}He + Q \qquad [32.2]$$

where X is called the **parent nucleus**, Y is the **daughter nucleus**, and Q is the **disintegration energy** (p. 1189),

$$Q = (m_X - m_Y - m_\alpha)c^2 \qquad [32.3]$$

Here m_X, m_Y, and m_α are the masses of the parent, daughter, and alpha, respectively. Electron-beta decay arises when (p. 1191)

$$n \rightarrow p + e^- + \bar{\nu}_e \qquad [32.4]$$

In the case of a radioactive nuclide, the decay has the form

$$^{A}_{Z}X \rightarrow {}^{A}_{Z+1}Y + {}^{0}_{-1}e + \bar{\nu}_e + Q \qquad [32.5]$$

Positron-beta decay arises when

$$p \rightarrow n + e^+ + \nu_e \qquad [32.6]$$

In the case of a radioactive nuclide, the decay has the form

$$^{A}_{Z}X \rightarrow {}^{A}_{Z-1}Y + {}^{0}_{+1}e + \nu_e + Q \qquad [32.7]$$

Electron-capture, the process wherein a parent nucleus absorbs one of its own orbital electrons, competes with β^+ decay (p. 1191):

$$^{A}_{Z}X + {}^{0}_{-1}e \rightarrow {}^{A}_{Z-1}Y + \nu_e + Q \qquad [32.8]$$

The SI unit of decay rate is the *becquerel* (Bq), which is equivalent to one disintegration per second. After n half-lives, the decay rate R will be

$$R = \left(\tfrac{1}{2}\right)^n R_0 \qquad [32.10]$$

and (p. 1195)

$$R = R_0 e^{-\lambda t} \qquad [32.11]$$

where λ is the decay constant. The half-life is related to λ by

$$\lambda t_{1/2} = 0.693 \qquad [32.12]$$

Since

$$R = \lambda N \qquad [32.13]$$

where N is the number of atoms remaining, it follows that

$$N = N_0 e^{-\lambda t} \qquad [32.14]$$

Suggestions on Problem Solving

1. Some useful numerical quantities for this chapter are

$1 \text{ u} = 1.660\,540 \times 10^{-27} \text{ kg} = 931.494 \text{ MeV/c}^2$
$1 \text{ eV/c}^2 = 1.782\,663 \times 10^{-36} \text{ kg}$
$1 \text{ eV} = 1.602\,177 \times 10^{-19} \text{ J}$
$m_e = 9.109\,389\,7(54) \times 10^{-31} \text{ kg}$
$\quad = 0.510\,999\,06(15) \text{ MeV/c}^2 = 5.485\,80 \times 10^{-4} \text{ u}$

$m_p = 1.672\,623\,1(10) \times 10^{-27} \text{ kg}$
$\quad = 938.272\,31(28) \text{ MeV/c}^2 = 1.007\,276 \text{ u}$
$m_n = 1.674\,928\,6(10) \times 10^{-27} \text{ kg}$
$\quad = 939.565\,63(28) \text{ MeV/c}^2 = 1.008\,665 \text{ u}$
$m_\alpha = 4.001\,506 \text{ u}$

2. Most tabulations of masses for the various isotopes are *atomic* mass listings, not nuclear mass. When you are given that the atomic mass for carbon-10 is 10.016 856 u, it is for the neutral atom, and it includes the six electrons. One way to handle this situation is to use atomic masses throughout any calculations (for example, of mass defect). Just *keep track of the electrons.* There will be a difference in the electron binding energies, but that's generally totally negligible. Accordingly, it may often be more convenient to use the mass of the hydrogen atom (1.007 825 u) in place of the proton mass—the difference is the electron.

3. A typical problem concerning radioactive decay might involve one or two of Eq's. (32.10) through (32.13). If you need the activity (or decay rate) R, that's Eq. (32.11). Any question about the mass, or fractional amount, number of atoms, or percentage of a substance remaining is addressed to Eq. (32.13). You may be given τ, λ, or $t_{1/2}$ and be required to find one of the others. Similarly, the equations for R and N are written in terms of λ, though you may be provided τ or $t_{1/2}$. *Don't forget the minus signs* in those two exponential expressions. Review Appendix A-6 on exponentials and logarithms. You can be asked to solve for λ in either Eq. (32.11) or (32.13), and that will require undoing the exponential by taking the natural log. Watch the time units in the exponential equations: you can use λ in decays/year provided you enter t in years—just don't mix years with days or seconds.

4. The advent of the electronic calculator has made it just as easy to use Eq. (32.10) $R = \left(\frac{1}{2}\right)^n R_0$ directly as to use Eq. (32.11), $R = R_0 e^{-\lambda t}$. Accordingly, you'll be able to check your work and are advised to always do so—everyone makes numerical errors, we just have to catch them.

5. To calculate $R = R_0 A^{1/3}$ use the $x^{1/y}$ key on your electronic calculator. Enter the number for A, which is x, hit the $x^{1/y}$ key, enter 3, and hit the $=$ key.

Discussion Questions

1. Figure Q1 is a plot of the excitation energy necessary to raise a nuclide with a given number of neutrons up to the next highest energy level. What's the significance of the data in light of the magic numbers?

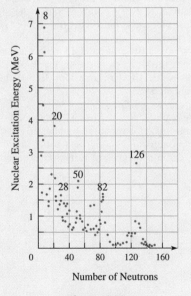

Figure Q1

2. The two nuclides ^{235}U and ^{239}Pu undergo fission via bombardment with neutrons of very little KE (even less than 0.01 eV). By contrast, ^{238}U and ^{232}Th will fission, provided the neutrons have kinetic energies in excess of 1 MeV. How is this fact used in a fission reactor? What does it suggest about fuels for bombs? Incidentally, it's been proposed that small atomic explosions could be used to propel a spaceship up to tremendous speeds. (Perhaps some good will come of this yet.)

3. Your standard H-bomb (what a thought) is essentially an ellipsoidal chunk of lithium-6 deuteride, an opalescent white solid, surrounded by an atom-bomb trigger. The whole thing is often encased in a U-238 shell, which initially reflects energy back onto the ^{6}LiD. The cheap U-238 also undergoes fission, allowing one to increase the output even more. (The Russians, with their big missiles and poor guidance systems, prefer large bombs.) The United States tested the first lithium-deuteride fission-fusion-fission device in 1953, ushering in a new era of human folly with a 15-megaton roar. Given the reactions

$$^2_1D + {}^3_1T \rightarrow {}^4_2He + n + 17.6 \text{ MeV}$$

and

$$^6_3Li + n \rightarrow {}^4_2He + {}^3_1T + energy$$

describe as much of what happens in an H-bomb as you can. Comment on the limitations, if any, on the mass of the U-238 shell. How might tritium be produced in a factory situation?

4. During World War II, the only large producer of heavy water in the world was the Norsk Hydro plant at Vemork, Norway. Toward the end of 1942, a group of Anglo-Norwegian commandos carried out two fairly successful raids on the facility. The plant was in Nazi hands, and the Nazis were shipping all of its output to Germany. In the fall of 1943, after the plant had been repaired, American planes attacked it again. At that point, it should have been clear that we knew and that they knew that we knew. Why did we attack the plant? In the United States, we often use enriched U-238 and ordinary water in reactors. By contrast, the Canadians generally use D_2O and natural uranium. Explain.

5. Figure Q5 depicts a vacuum chamber into which are pointed a number of powerful laser beams. A tiny pellet of deuterium and tritium is dropped inside and blasted from all sides by the beams. This is a so-called *inertial confinement* reactor. What does it do and why? Knowing that neutrons are a likely by-product, see if you can figure out the formula for a possible nuclear reaction that might take place.

6. Figure Q6 shows a Tokamak Fusion Reactor. A high-temperature deuterium-tritium plasma is confined by powerful magnets so that it stays suspended within the vacuum chamber.

Figure Q5

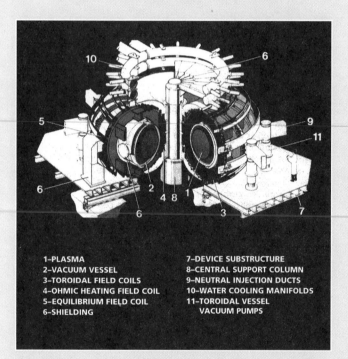

1–PLASMA
2–VACUUM VESSEL
3–TOROIDAL FIELD COILS
4–OHMIC HEATING FIELD COIL
5–EQUILIBRIUM FIELD COIL
6–SHIELDING

7–DEVICE SUBSTRUCTURE
8–CENTRAL SUPPORT COLUMN
9–NEUTRAL INJECTION DUCTS
10–WATER COOLING MANIFOLDS
11–TOROIDAL VESSEL
      VACUUM PUMPS

Figure Q6

By pumping exceedingly large currents through the plasma, temperatures of tens of millions of degrees can be attained. What do you think this thing is supposed to do? What are the high temperatures for? Why is fusion power appealing?

7. Discuss the difference between the half-life and the mean life of a radioactive substance. Relate these ideas to the life of a human being. Given a typical sample of Americans, 90% will make it to age 30, 80% to age 50, 50% to age 68, 25% to age 77, 10% to age 84, 1% to age 92, 0.1% to age 97, 0.01% to age 100, and 0% to age 200.

8. Hans Bethe at Cornell proposed a high-temperature fusion process known as the *carbon-nitrogen cycle* that may well power some stars:

$$
\begin{array}{lr}
 & \text{MeV} \\
{}^{12}\text{C} + p \rightarrow {}^{13}\text{N} + \gamma & 2.0 \\
{}^{13}\text{N} \rightarrow {}^{13}\text{C} + e^+ + \nu_e & 2.2 \\
{}^{13}\text{C} + p \rightarrow {}^{14}\text{N} + \gamma & 7.5 \\
{}^{14}\text{N} + p \rightarrow {}^{15}\text{O} + \gamma & 7.3 \\
{}^{15}\text{O} \rightarrow {}^{15}\text{N} + e^+ + \nu_e & 2.7 \\
{}^{15}\text{N} + p \rightarrow {}^{12}\text{C} + {}^{4}\text{He} & \underline{5.0} \\
 & 26.7
\end{array}
$$

Explain what's happening. What form does the energy produced take? What is the net result of running through the cycle?

9. Return to Fig. 32.14, the exponential curve, which is a rather special function. For example, consider the slope of the curve. To do that qualitatively, imagine short line segments tangent to the curve at every point. What is the significance of the slope? Now plot the slope versus time. What is the shape of the curve?

10. Let's have a little fun and do a bit of speculating. What would happen to the Universe if the nuclear force vanished, *now*?

11. What would the Universe look like if the electromagnetic force vanished, *now*?

12. Suppose we produce the molecules of ${}^{12}_{6}\text{C}\,{}^{16}_{8}\text{O}$ and $({}^{14}_{7}\text{N})_2$. Then we singly ionize them (that is, remove one electron) and accelerate them inside a vacuum chamber so we can determine their masses. Each molecule has a total of 14 protons and 14 neutrons. Explain in general terms what we would observe.

Multiple Choice Questions

1. A nuclide is specified completely by   (a) its atomic number   (b) its proton number   (c) the number of nucleons it has   (d) giving both Z and A (e) none of these.

2. The quantity A is the   (a) proton number   (b) nucleon number   (c) neutron number   (d) atomic number   (e) none of these.

3. Isotopes are nuclides having   (a) the same A and Z y values   (b) the same Z, but different values of A (c) the same A, but different values of Z (d) different values of both A and Z (e) none of these.

4. When a nucleus undergoes alpha decay, the daughter as compared to the parent has   (a) the same Z and an N reduced by 4   (b) Z reduced by 4 and A reduced by 2

(c) A reduced by 4 and Z reduced by 2   (d) Z increased by 2 and N reduced by 2   (e) none of these.

5. Beta particles are   (a) protons emitted from a nucleus   (b) electrons emitted from a nucleus   (c) neutrons emitted from the nucleus   (d) electrons emitted from the atomic cloud   (e) none of these.

6. The mass of a nucleus composed of several nucleons is   (a) always less than the sum of masses of its constituents   (b) sometimes less than the sum of masses of its constituents   (c) always more than the sum of masses of its constituents   (d) always equal to the sum of masses of its constituents   (e) none of these.

7. When a nucleus undergoes electron-beta decay, the daughter as compared to the parent has   (a) the same Z and an N reduced by 1   (b) Z reduced by 1 and A reduced by 1   (c) A reduced by 1 and Z increased by 1   (d) Z increased by 1 and N reduced by 1   (e) none of these.

8. The nucleus $^{234}_{90}$Th β decays into   (a) $^{234}_{89}$Ac   (b) $^{235}_{91}$Pa   (c) $^{234}_{91}$Th   (d) $^{234}_{91}$Pa   (e) none of these.

9. The missing particle in the reaction

$$^{2}_{1}\text{D} + ^{199}_{80}\text{Hg} \rightarrow ^{197}_{79}\text{Au} + (?)$$

is a(n)   (a) gamma   (b) alpha   (c) neutron   (d) proton   (e) electron.

10. The missing particle in the reaction

$$^{2}_{1}\text{D} + ^{196}_{78}\text{Pt} \rightarrow ^{197}_{79}\text{Au} + (?)$$

is a(n)   (a) deuteron   (b) alpha   (c) neutron   (d) proton   (e) gamma.

11. The missing particle in the reaction

$$^{1}_{0}\text{n} + ^{198}_{80}\text{Hg} \rightarrow ^{197}_{79}\text{Au} + (?)$$

is a(n)   (a) deuteron   (b) alpha   (c) neutron   (d) proton   (e) electron.

12. The missing particle in the reaction

$$^{1}_{0}\text{n} + ^{196}_{78}\text{Pt} \rightarrow ^{197}_{78}\text{Pt} + (?)$$

is a(n)   (a) deuteron   (b) alpha   (c) neutron   (d) proton   (e) gamma.

13. The missing particle in the reaction

$$\gamma + ^{198}_{80}\text{Pt} \rightarrow ^{197}_{79}\text{Au} + (?)$$

is a(n)   (a) electron   (b) alpha   (c) neutron   (d) proton   (e) gamma.

14. An antineutrino always accompanies   (a) α decay   (b) β^- decay   (c) gamma emission   (d) neutron emission   (e) none of these.

15. In a nuclear reaction, the value of Q (a) is always negative   (b) is positive in a spontaneous decay   (c) is positive only for gamma emission   (d) is always negative for proton emission   (e) none of these.

16. The process whereby a nucleus splits into two very roughly equal parts is known as   (a) fusion   (b) beta decay   (c) fission   (d) nucleation   (e) none of these.

17. A nuclide with a half-life of 10 days has a decay constant of   (a) 0.069 decays/day   (b) 6.9 decays/day   (c) 1/10 decays/day   (d) 10 decays/day   (e) none of these.

18. Given that a freshly prepared nuclide has a half-life of 10 days, the percentage of it remaining after 30 days is   (a) 30%   (b) 10%   (c) 12.5%   (d) 72.5%   (e) none of these.

19. As a naturally occurring radionuclide decays,   (a) its λ increases   (b) its τ decreases   (c) its $t_{1/2}$ decreases   (d) its R remains constant   (e) none of these.

20. The Sun is powered by   (a) nuclear beta decay   (b) nuclear fission   (c) thermonuclear fusion   (d) nuclear gamma decay   (e) none of these.

Problems

NUCLEAR STRUCTURE

1. [I] How many nucleons does the nuclide $^{111}_{50}$Sn possess?

2. [I] What is the proper symbol for a nucleus having 14 protons and 15 neutrons?

3. [I] How many neutrons and how many protons does the nuclide $^{15}_{7}$N possess?

4. [I] How many neutrons are in a ^{15}O nucleus?

5. [I] Identify each of the following nuclides: $^{211}_{87}$X, $^{202}_{82}$X, $^{105}_{47}$X, and $^{142}_{59}$X.

6. [I] What is the difference structurally between $^{183}_{76}$Os and $^{193}_{76}$Os?

7. [I] The stable isotope of nitrogen, $^{14}_{7}$N, has an atomic mass of 14.003 074 u. How much is that in MeV/c^2 and GeV/c^2? Give the answers to six significant figures.

8. [I] Inasmuch as nuclear matter has a density of 2.3×10^{17} kg/m^3, how many cubic meters of water would have to be compressed into a cubic centimeter to match that density?

9. [I] The mass of the lithium-6 isotope is 5.603 051 GeV/c^2. How much is that in kg? Give your answer to seven significant figures.

10. [I] There are two stable isotopes of chlorine, Cl-35 (with an atomic mass of 34.968 853 u) and Cl-37 (with an atomic mass of 36.965 903 u). They are found with relative abundances of 75.77% and 24.23%, respectively. Determine the atomic mass of chlorine as it would be listed in the Periodic Table.

11. [I] Niobium has several isotopes, but only Nb-93 is stable.

Its atomic mass is $1.542\,748 \times 10^{-25}$ kg. Determine its mass in unified atomic mass units and compare it with the atomic mass listed for niobium in the Periodic Table. Explain your results.

12. [I] What is the radius of a gold nucleus? Gold has a nucleon number of 197.

13. [I] Determine the diameter of the uranium isotope ^{235}U.

14. [I] The radius of a nucleus doubles whenever the number of nucleons increases by a multiplicative factor of how much?

15. [I] If a nucleus is determined to be 7.2 fm in diameter, what mass number does it correspond to?

16. [I] A boron atom ($^{10}_{5}$B) has a mass of 10.012 937 u. What is the mass defect of its nucleus?

17. [I] Using the results of Problem 16, determine the binding energy of the boron nucleus.

18. [I] Use Fig. 32.9 to approximate the binding-energy-per-nucleon for boron-10. What is the total binding energy of this nuclide? How does that compare with the computed value?

19. [I] Using the results of Problem 17, determine the average binding-energy-per-nucleon for boron-10. Compare your answer with that of Problem 18.

20. [I] Use Fig. 32.9 to determine an approximate value for the total binding energy of a U-238 nucleus.

21. [I] Find the binding energy of the last neutron in the nucleus of oxygen-16. The masses of the atoms ^{16}O and ^{15}O are 15.994 915 u and 15.003 065 u, respectively.

22. [I] Calculate the binding energy of the last neutron in $^{13}_{6}$C. The atomic mass of carbon-13 is 13.003 355 u.

23. [II] Bromine has at least two stable isotopes: Br-79 (with an atomic mass of 78.918 336 u) and Br-81 (with an atomic mass of 80.916 289 u). Given that the relative abundance of Br-79 is 50.7% and the Periodic Table lists its atomic mass as 79.909 u, is it likely to have any other long-lived isotopes? Explain via numerical analysis.

24. [II] Approximate the ratio of the density of an atom to the density of its nucleus. Take the radius of the atom to be 0.05 nm and that of the nucleus to be 1.2 fm.

25. [II] How much bigger is the radius of the nucleus of U-238 than that of He-4?

26. [II] What is the atomic number of the nuclide that has a diameter one-quarter that of tellurium-128?

27. [II] Use Avogadro's Number and the definition of the atomic mass unit to show that 1 u $= 1.66 \times 10^{-27}$ kg.

28. [II] Determine the binding-energy-per-nucleon of helium to four figures. [Hint: Watch out for the electrons.]

29. [II] Iron-54 has an atomic mass of 53.939 613 u. Determine its nuclear mass defect in atomic mass units. Find the binding-energy-per-nucleon in MeV (four figures will do).

30. [II] With Problem 29 in mind, find the binding-energy-per-nucleon for iron-55 (atomic mass 54.938 296 u) and compare results.

31. [II] The neutrons in an isotope tend to pair and thereby bind strongly. For example, compute the minimum energy needed to remove a neutron from calcium-41 (atomic mass 40.962 278 u) as compared to calcium-42 (atomic mass 41.958 618 u). The former has 21 neutrons, the latter 22. Calcium-40 has an atomic mass of 39.962 591 u.

32. [II] What is the minimum amount of energy necessary to remove a proton from the nucleus of a $^{42}_{20}$Ca atom, thereby converting it into a $^{41}_{19}$K atom? The former has a mass of 41.958 618 u, the latter 40.961 825 u, and a hydrogen atom has a mass of 1.007 825 u.

33. [II] Uranium-232 (with an atomic mass of 232.037 13 u) is radioactive. It emits an α-particle and transforms into thorium-228 (with an atomic mass of 228.028 73 u). Determine the KE available to the decay products. Assuming the U-232 was at rest, this KE must be shared by the Th-228 and the alpha so as to conserve momentum. [Hint: The above are the masses of the atoms.]

NUCLEAR TRANSFORMATION

34. [I] When boron is struck by an alpha, it is transmuted into nitrogen-13, the reaction being

$$^{10}_{5}\text{B} + ^{4}_{2}\text{He} \rightarrow ^{13}_{7}\text{N} + ^{1}_{0}\text{n}$$

which then decays (with a half-life of $\approx$10 minutes) via positron emission. Write out that decay reaction.

35. [I] Irène and Pierre Joliot-Curie bombarded a foil of aluminum with α-particles, transmuting some of the nuclei and producing neutrons in the process. Write out the transformation formula.

36. [I] Rubidium-87 undergoes electron-β decay. What is the daughter nuclide? Write out that decay reaction.

37. [I] Samarium-147 decays via alpha emission. What is the resulting daughter nuclide? Write out that decay reaction.

38. [I] Calculate the maximum kinetic energy available (in MeV) to the electron and electron-antineutrino created by the decay of a neutron.

39. [I] Consider the reaction in which lithium-7 is struck by a proton; that is

$$\text{p} + ^{7}_{3}\text{Li} \rightarrow \alpha + \alpha$$

Compute the difference in mass before and after the collision. How much KE will the alphas have (in excess of the KE the proton delivered)? The mass of the ^{1}H, ^{7}Li, and ^{4}He *atoms* are 1.007 825 u, 7.016 003 u, 4.002 603 u, respectively.

40. [I] An old unit of activity that approximated the decay rate of radium is the *curie* (Ci) where 1 Ci = 3.7 × 10^{10} decays/s. What is the equivalent activity of one microcurie (μCi) in becquerels?

41. [I] A cobalt atom is at the core of the vitamin B_{12} molecule. Accordingly, radioactive Co-60 has been used as a tracer to study B_{12} absorption defect in pernicious anemia. Cobalt-60 has a half-life of 5.3 years. What is its decay constant in disintegrations per second?

42. [I] With Problem 40 in mind, given that the level of radioactive activity in the human body is naturally 10 nCi, how many nuclear disintegrations occur per second inside you?

43. [I] A radioactive sample is studied for a period of 36 hours, during which time its decay rate decreases to one-eighth its original value. What is its half-life?

44. [I] In humans, iodine is readily taken up by the thyroid,

which requires it in the making of the hormone thyroxine. To study metabolism and treat thyroid disease, the isotope $^{131}_{53}$I is often introduced into the body. Given that it has a half-life of 8 days, what fraction of the original activity remains after 8 weeks? Assume none is excreted. Take a look at Problem 46.

45. [I] Radium-226 emits an α-particle and decays into radon-222 with a half-life of 1620 years. What is its decay constant in decays/year?

46. [I] Radionuclides that have entered the human body are sometimes biologically excreted in an approximately exponential way. The effective half-life is then

$$1/t_{1/2}(\text{eff}) = 1/t_{1/2}(\text{bio}) + 1/t_{1/2}$$

Given that iodine-131 has a half-life of 8.0 d and a biological half-life of 138 d, what is its effective half-life? How much of it will remain after 8.0 weeks? (See Problem 44.)

47. [I] Given that radon has a half-life of 3.8 days, what is its mean life?

48. [I] Radon undergoes α decay with a half-life of 3.8 days. Given some original amount, how much of it will remain after 11.4 days?

49. [I] A radioactive sample has a decay constant of 7.69×10^{-3} decays/s. What is its average lifetime and its half-life?

50. [I] A sample of radioactive oxygen-15 has a half-life of 2.1 minutes and a decay rate of 5.5×10^{-3} decays/s. By how much will the amount of this isotope be diminished after 4.0 s?

51. [I] Protactinium-234 has a half-life of 1.18 minutes. If 1.00 mg of it is freshly created, what fraction will be left in 1.00 h?

52. [II] Polonium-210 is radioactive. Confirm that it is energetically possible for it to decay via alpha emission in the following way

$$^{210}_{84}\text{Po} \rightarrow {}^{206}_{82}\text{Pb} + \alpha$$

What is the net kinetic energy of the decay particles, assuming the polonium nucleus to be at rest? The masses of the polonium and lead atoms are 209.982 848 u and 205.974 440 u, respectively. [Hint: Watch out for the electrons.]

53. [II] By bombarding beryllium with alphas, we get the following reaction

$$^{9}_{4}\text{Be} + {}^{4}_{2}\alpha \rightarrow {}^{12}_{6}\text{C} + {}^{1}_{0}\text{n} + Q$$

The mass of the Be atom is 9.012 182 u, the mass of the He atom is 4.002 603 u. Please compute the value of Q in excess of the alpha's incoming KE.

54. [II] Given the reaction

$$^{6}_{3}\text{Li} + \text{p} \rightarrow \alpha + {}^{3}_{2}\text{He} + Q$$

If $Q = 4.0185$ MeV in excess of the proton's KE and the lithium was at rest, find the mass of the helium-3 atom.

55. [II] Determine the value of Q, which is required in order to initiate the following reaction

$$\alpha + {}^{14}\text{N} \rightarrow {}^{17}\text{O} + \text{p}$$

This energy must come by way of the KE of the alpha, which actually has to be a bit higher than Q because some KE must be given to the other particles if momentum is to be conserved.

56. [II] Can lead-206 (of atomic mass 205.974 440 u) spontaneously decay via alpha emission? Explain your answer completely. The atomic mass of mercury-202 is 201.970 617 u.

57. [II] Calculate the activity of one milligram of radon-222, which has a decay constant of 2.1×10^{-6} decays/s.

58. [II] Uranium-238 has a half-life of 4.5×10^{9} years. The Earth was formed about 5×10^{9} years ago. What fraction of the ^{238}U present then is still around today?

59. [II] Radioactive phosphorus-32 has been used to study bone metabolism and for the treatment of blood diseases. What is the activity of a sample of 5×10^{16} atoms if it has a half-life of 14.3 days?

60. [II] Carbon-11 is radioactive with a half-life of 20.4 minutes. If a sample initially has 1.0×10^{17} carbon-11 atoms in it, what is its activity after 10 minutes?

61. [II] Oxygen-15 has a half-life of 2.1 min. How many ^{15}O atoms are present in a source with an activity of 4.1 mCi? (See Problem 40.)

62. [II] Consider a sample of freshly cut wood that has been reduced to pure carbon. What is the activity per gram resulting from the decay of ^{14}C (half-life of 5730 y) in this material?

63. [II] Determine the maximum permissible concentration in the air of tritium in the workplace, given that the maximum permissible annual intake of tritium is 444 MBq. A typical person inhales ≈ 10 m^3 of air per working day. Use a 50 work-week year.

64. [II] A 50-g chunk of charcoal is found in the buried remains of an ancient city destroyed by invaders. The carbon-14 activity of the sample is 200 decays/min. Roughly when was the city destroyed (more accurately, when was the tree felled from which the charcoal came)? Refer back to Problem 62.

65. [II] Determine the amount of energy liberated in the fusion reaction

$$4\text{p} \rightarrow \alpha + 2{}^{0}_{+1}\text{e} + 2\nu_{e} + \gamma$$

The masses of the proton, electron, and alpha particle are 1.007 276 u, 0.000 548 580 u, and 4.001 506 u, respectively.

66. [II] Radioisotopes are sometimes used to produce electricity to power such things as interplanetary probes and pacemakers. Consider the radioactive nuclide Po-210, which has a half-life of ≈ 140 d, emitting 5.30-MeV alphas. If all the KE of the alphas is absorbed in a radioisotope thermal generator (known as an RTG, in the trade), what's the average thermal power (in watts) developed per gram during the first 140 days of operation?

67. [cc] Assume that N, the number of radioactive atoms present in a sample, is a continuous function of t; but, of course, it is not. The quantity N is some whole number and not a continuous thing like displacement, speed, or time.

It's very different from these ideas that can be divided into infinitesimal pieces. Still when N is very large and the decay is rapid the process can be approximated as if it were continuous, whereupon $dN/dt = -\lambda N$. Using that and the calculus derive Eq. (32.14).

68. [III] A radioactive sample shows a decrease by a factor of 10 in activity over a period of 5.0 minutes. What is its decay constant?

69. [III] Show that $t_{1/2} = (\ln 2)/\lambda$.

33

High-Energy Physics

AS THE TWENTIETH CENTURY unfolded, the number of observed subnuclear particles gradually increased. The electron, proton, and neutron were joined by the muon, pion, positron, and so forth, until by the 1970s, several hundred distinct particles were identified. The obvious problem was to determine which of these tiny specks of matter was *elementary* in the sense of being a single homogeneous entity having no internal structure. It now seems certain that most of these subnuclear particles are clusters of two or more fundamental entities called *quarks* (p. 1224). These are the primary building blocks that fuse together via the strong force to create neutrons,

TABLE 33.1 Some fairly long-lived elementary particles

Category	Name	Particle	Antiparticle	Mass (MeV/c²)	B	L_e	L_μ	L_τ	S	Lifetime (s)
Leptons	Electron	e^-	e^+	0.51100	0	±1	0	0	0	Stable
	Neutrino (e)	ν_e	$\bar{\nu}_e$	0 ($<14 \times 10^{-6}$)	0	±1	0	0	0	Stable
	Muon	μ^-	μ^+	105.659	0	0	±1	0	0	2.197×10^{-6}
	Neutrino (μ)	ν_μ	$\bar{\nu}_\mu$	0 (<0.25)	0	0	±1	0	0	Stable
	Tau	τ^-	τ^+	1784.2 ± 3.2	0	0	0	±1	0	$(4.6 \pm 1.9) \times 10^{-13}$
	Neutrino (τ)	ν_τ	$\bar{\nu}_\tau$	0 (<35)	0	0	0	±1	0	Stable
Hadrons										
Mesons	Pion	π^+	π^-	139.567	0	0	0	0	0	2.60×10^{-8}
		π^0	π^0	134.96	0	0	0	0	0	0.87×10^{-16}
	Kaon	K^+	K^-	493.7	0	0	0	0	±1	1.24×10^{-8}
		K^0	$\bar{K}^0$	497.7	0	0	0	0	±1	0.9×10^{-10}
Baryons	Proton	p	$\bar{p}$	938.3	±1	0	0	0	0	Stable
	Neutron	n	$\bar{n}$	939.6	±1	0	0	0	0	898
	Lambda	Λ^0	$\bar{\Lambda}^0$	1115.6	±1	0	0	0	∓1	2.6×10^{-10}
	Sigma	Σ^+	$\bar{\Sigma}^+$	1189.4	±1	0	0	0	∓1	0.80×10^{-10}
		Σ^0	$\bar{\Sigma}^0$	1192.5	±1	0	0	0	∓1	5.8×10^{-20}
		Σ^-	$\bar{\Sigma}^-$	1197.3	±1	0	0	0	∓1	1.5×10^{-10}
	Xi	Ξ^0	$\bar{\Xi}^0$	1315	±1	0	0	0	∓2	2.9×10^{-10}
		Ξ^-	$\bar{\Xi}^-$	1321	±1	0	0	0	∓2	1.64×10^{-10}
	Omega	Ω^-	$\bar{\Omega}^-$	1672	±1	0	0	0	∓3	0.82×10^{-10}

protons, pions, and so forth. The complementary group of elementary entities, the *leptons,* do not experience the strong force and do not combine to form composite subnuclear particles. They exist only individually and, to within the limits of our most powerful techniques, each (the electron, neutrino, and so on) appears to be a distinct structureless object.

Table 33.1 provides a representative sample of the more long-lived subnuclear particles. They are listed primarily according to their modes of interaction. All of them experience or *couple to* the gravitational interaction, which nonetheless plays a negligible role in particle physics because it's so feeble. The more dominant influences arise from the electromagnetic, strong, and weak interactions, and these are transmitted by a set of elementary virtual particles known as the *gauge bosons* (p. 1228).

Our low-energy world of trees and rocks and politicians is made up almost entirely of three fundamental material particle types: the electron (a lepton), the u-quark, and the d-quark. Along with the neutrino, these constitute the *first generation* of matter, the ordinary stuff still remaining in the cool Universe as it presently exists. At higher temperatures and greater energies, we can produce more exotic forms of matter—the *second* and *third generations* (p. 1227). These highly unstable particles once existed in abundance in the blazing early moments of the primordial Universe (p. 1232). Today, we can just begin to recreate some of those incredibly violent conditions. In studying the subnuclear domain, we gain a view, however darkly, of the first moments of Creation.

ELEMENTARY PARTICLES

The hundreds of subnuclear particles discovered in the second half of the twentieth century offered physicists an opportunity to begin to sort out the fundamental nature of matter and its interactions. The process was similar to finding the electron-cloud structure of the atom beginning with the elements of the Periodic Table. First, they (elements or particles) were grouped by way of shared behavior. Then they were arranged according to measurable characteristics (such as mass and charge) underlying that behavior. And, finally, the hidden structure that gave rise to all the patterns of experimental data was hypothesized and essentially confirmed.

33.1 Leptons

The first group of twelve particles in Table 33.1 constitutes the **leptons** and **antileptons** (from the Greek *leptos,* meaning "slight"). These are the particles that couple to the weak force, and if they are electrically charged, they react to the electromagnetic force as well. *All are immune to the strong force.* The most familiar and the lightest of the charged leptons is the electron. The muon (μ^-), found among the products of cosmic-ray bombardment (1936), was the first unstable subnuclear particle to be discovered. Muons are commonplace, rushing toward the surface of the planet in a flow of roughly 1 per cm^2 per minute. Even as you read these lines, muons are more or less harmlessly streaming downward through your body like a gentle penetrating cosmic rain. The muon decays (in about 2.2 μs via the weak interaction) into an electron and two neutrinos:

$$\mu^- \rightarrow e^- + \nu_\mu + \bar{\nu}_e$$

Subnuclear particle tracks left in a bubble chamber. (See Discussion Question 7.)

Otherwise, it behaves in every way like an overweight electron. The same can be said about the tau (τ), although, since it was only discovered in the 1970s, we have less experience with it. The tau is almost twice as massive as the proton; nonetheless, it is believed to be truly elementary. Both muons and taus play no known significant role in the scheme of things, and their very existence is puzzling in a Universe that seems to have a preference for economy.

At our current limits of observation, leptons appear to be structureless, point-like entities. All are spin-$\frac{1}{2}$ fermions. Each lepton having substantial mass is created in conjunction with its own variety of essentially massless neutrino. Thus, there are electron-neutrinos, muon-neutrinos, and tau-neutrinos. In total, there are probably more neutrinos than anything else—they outnumber electrons and protons by a factor of perhaps a thousand million. The Universe is awash with neutrinos, some of which are harmlessly passing through your body even now. Each lepton comes in both particle and antiparticle varieties, making a total of six electrically charged and six neutral leptons. These form three—electron, muon, and tau—particle-neutrino couplets: e and ν_e; μ and ν_μ; and τ and ν_τ.

$$\begin{bmatrix} e \\ \nu_e \end{bmatrix} \quad \begin{bmatrix} \mu \\ \nu_\mu \end{bmatrix} \quad \begin{bmatrix} \tau \\ \nu_\tau \end{bmatrix}$$

The systematic patterns of production and decay of leptons suggested that an underlying conservation law might be at work. Remember the special feature that leptons are created in couplets. It is believed that **lepton number** (L) is conserved in all presently attainable processes involving any members of this family. Thus, each of the three subsets of leptons (electron-type, muon-type, and tau-type) conserves its own lepton number (L_e, L_μ, L_τ). In each subset, the particle (for example, e$^-$) has a lepton number (L_e) of +1, whereas the antiparticle (e$^+$) has a lepton number of -1, and the same is true of the associated neutrino and antineutrino. Thus the decay reaction

$$\mu^- \rightarrow e^- + \nu_\mu + \bar{\nu}_e$$

for which $\quad\quad L_e: \quad 0 \rightarrow 1 + 0 - 1$

and $\quad\quad\quad L_\mu: \quad 1 \rightarrow 0 + 1 + 0$

separately conserves both muon- and electron-lepton numbers.

33.2 Hadrons

Hadrons are the strongly interacting composite particles. They are all fairly massive; ergo, the name, which derives from the Greek *hadros,* meaning "bulky." A succession of hadrons was discovered between 1947 and 1954 in cosmic-ray studies and at the Brookhaven Cosmotron in New York. That machine, a 3-GeV accelerator, made it possible for the first time to bring to bear the needed amounts of energy to create heavy particles via $E_0 = mc^2$. These conjured hadrons were christened the kaon (K: $\approx$500 MeV/c^2), the lambda (Λ: $\approx$1100 MeV/c^2), the sigma (Σ: $\approx$1200 MeV/c^2), and the xi (Ξ: $\approx$1300 MeV/c^2). All were observed via photographs of the several-centimeter-long tracks they left in visual detectors. These trails, marking their passage between creation and decay, corresponded to lifetimes of around 0.1 ns to 10 ns. In contrast to the twelve leptons, there are hundreds of hadrons, which increasingly suggested that hadrons might not be elementary. Every hadron couples to the strong, weak, and gravitational interactions, and some also couple to the electromagnetic interaction. All of them, with the possible exception of the proton, decay—*some rapidly, via the strong interaction; some less rapidly, via the electromagnetic interaction; and others still more slowly, via the weak interaction.* Hadrons form two distinct subgroups (defined by their spins): *mesons* and *baryons.*

Mesons

The Greek root *meso* means "middle," and it was applied in 1939 to particles whose mass was between that of the electron and proton. Today, the term **meson** refers to hadrons that are bosons (they have spins of 0, 1, 2, . . .). Table 33.1 lists only the several pions (or π-mesons) and kaons (or K-mesons). Although there are dozens of other known mesons, these serve our purposes for the time being.

The three members of the *pion* family (π^+, π^-, π^0) are the lightest of the mesons. Proton-proton collisions carrying enough energy create pions:

$$p + p \rightarrow p + p + \pi^+ + \pi^-$$
$$p + p \rightarrow p + n + \pi^+$$

and pion-proton collisions produce more of them, for example

$$p + \pi^- \rightarrow p + \pi^- + \pi^0$$
$$p + \pi^- \rightarrow n + \pi^0$$

Each process conserves angular momentum (that is, spin; **the pion has zero spin**), electromagnetic charge, and nucleon number (that is, the number of nucleons). Whereas π^+ and π^- can be thought of as particle and antiparticle, neutral π^0 is its own antiparticle. The π^0 has a fleeting lifetime (0.87×10^{-16} s) usually decaying (via the electromagnetic interaction) into two photons

$$\pi^0 \rightarrow \gamma + \gamma$$

Notice that there is *no* conservation of meson number. The most common decay route for the charged pions is fixed by the fact that there are no hadrons lighter than themselves; they must decay into leptons (muons) via the weak interaction; thus

$$\pi^+ \rightarrow \mu^+ + \nu_\mu \qquad \pi^- \rightarrow \mu^- + \bar{\nu}_\mu$$

Pions play a very special role in nature: they make a major contribution to mediating the nuclear force between nucleons (p. 1181).

The kaon (K-meson) is produced in collisions via the strong interaction. There are four zero-spin kaons: K^+, K^-, K^0, and its antiparticle $\overline{K}^0$. They are unstable and undergo a number of different decay reactions. For example, each can decay into a pair of pions:

$$K^0 \rightarrow \pi^+ + \pi^-$$
$$\overline{K}^0 \rightarrow \pi^0 + \pi^0$$
$$K^+ \rightarrow \pi^+ + \pi^0$$

Two surprising things were observed here and with a group of other newly discovered particles as well. First, the reaction products were always created in twos, even though there was no known reason for it. For example, every time a K^0 was produced, a Λ^0 was also produced. Second, instead of a decay time typical of the strong interaction ($\approx 10^{-23}$ s), these transformations sauntered along with lifetimes of from 0.1 ns to 10 ns. It was as if the particles, born out of the strong force, died by the weak force.

That mysterious behavior was denoted by calling these objects **strange particles**. To explain what was happening, Murray Gell-Mann, in the 1950s, introduced a new quantum number that reflects a new conserved quantity. Like electromagnetic charge or spin, he proposed that these particles uniquely possess a quality of *strangeness* (as if they carried a strangeness charge), and he assigned a numerical value of it (S) to each hadron, according to its observed behavior. *The strong and electromagnetic interactions both conserve strangeness. The weak interaction does not, and the strangeness may change, but by no more than one unit.* Pions and nucleons, which are not at all strange, have a strangeness of 0. The kaon (K^+ and K^0) has a strangeness of $+1$, whereas the strange baryon known as lambda has a strangeness of -1.

The strong interaction could create a pair of strange particles out of more ordinary matter (with zero strangeness), provided the resulting pair had canceling values of strangeness. Thus, a pion and a proton (net strangeness, 0) can interact strongly to produce a K-zero and a lambda-zero of strangeness $+1$ and -1, respectively: $\pi^- + p \rightarrow K^0 + \Lambda^0$. The logic forbids strange particles (like K^0), once created, from decaying to lighter nonstrange particles ($K^0 \rightarrow \pi^+ + \pi^-$) by the rapid route of the strong interaction. Strange particles can accomplish such decay only via the weak interaction and only slowly. This kind of *ad hoc* conceptual fine-tuning reveals the way physics develops; in the absence of a proven theoretical formalism, phenomenological relationships are derived, or guessed at, from observed patterns. The idea, however well it worked then, really makes sense only when viewed from the perspective of the quark theory (p. 1224). Strangeness simply depends on the number of strange and antistrange quarks composing the particles.

Baryons

Baryons derive their name from the Greek *barys,* meaning "heavy." They are the heavy hadrons but, more importantly, they are all fermions. Neutrons and protons, the nucleons, are the most well-known baryons. Table 33.1 displays a representative selection of baryons. Generalizing from the concept of nucleon number, it was proposed that the number of baryons and antibaryons in a reaction is conserved, and so they are assigned a ***baryon number*** (B). For baryons $B = +1$, whereas antibaryons have $B = -1$, and all nonbaryons have $B = 0$ (see Discussion Question 11). This system has the effect of "explaining" why the proton does not decay into still lighter particles, even though such a decay is not otherwise in violation of any known principle. In all particle reactions (at the energies we can attain), the total baryon number must, it is assumed, be conserved. Thus, the strong reaction

$$\pi^- + p \rightarrow K^0 + \Lambda^0$$

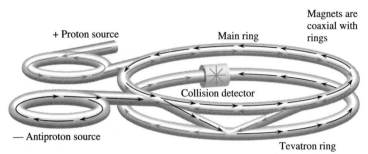

The world's most powerful particle accelerator at the Fermi National Accelerator Laboratory. When it actually has the funds to operate, it accelerates protons in one ring and antiprotons in the other. These beams are then made to collide with a total energy of ≈2000 GeV. Superconducting magnets force the particles to stay within the rings.

conserves electromagnetic charge (zero on both sides), strangeness (zero on both sides), angular momentum ($\frac{1}{2}$ on both sides), and baryon number ($+1$ on both sides), and therefore can and does take place. The heavier strange baryons (for example, Λ, Σ, Ξ, Ω) decay to lighter baryons and ultimately to the proton.

Whereas there are conservation laws for leptons and baryons, there are no restrictions on the number of photons or mesons entering and leaving an encounter. These particles can appear and vanish at will. In fact, the photon and the pion have $L = S = B = 0$. In the case of the photon and the neutral pion, each is its own antiparticle. With all of these quantum numbers equaling zero, including the electromagnetic charge, it is impossible to distinguish particle from antiparticle. As we'll see presently, photons and pions play a very special role in Quantum Field Theory.

QUANTUM FIELD THEORY

Prior to the introduction of Quantum Mechanics, particles and fields were considered interrelated though distinct entities; particles possessed intrinsic features (for example, mass and/or electromagnetic charge) that gave rise to external fields (for example, gravitational and/or electromagnetic). Force fields emanated from particles and filled the surrounding space. They carried energy and were, in a sense, real continuous media that interconnected all interacting particles and mediated their interactions. Particles were composed of matter, fields were composed of energy. The force field was the nineteenth century's answer to the age-old mystery of action-at-a-distance. Still, particles that do not react to any force fields are unobservable and physically meaningless. Force fields that do not act upon any particles are equally without significance. The ideas of particle and field take meaning from their interrelationship.

The concept of field began to change drastically with the introduction of Einstein's photon. The electromagnetic field does not, after all, have its energy continuously spread out in space. **The photon is the quantum of the electromagnetic field, and it carries the energy and momentum of the field**. The interaction of two charged particles corresponding to the electromagnetic force, transmitting energy and momentum from one to the other, must take place through the exchange of electromagnetic energy quanta—photons. Quantum Electrodynamics (QED), the theory of such interactions, was the first successful application of these ideas (p. 1166).

Figure 33.1 is a representation of two electrons as they undergo an elastic collision. It is called a **Feynman diagram**. Suppose each electron is initially traveling at the same speed. The electrons first approach and then recede from one another along a line in space that is projected upward in the increasing time direction in the diagram. The electron on the left emits a photon (the wiggly line), and for a moment (Δt) there are two electrons and one photon. The electron on the right absorbs the photon, and the interaction is momentarily over; other photons will subsequently go back and forth between the electrons. The average force is proportional to the rate of transfer of momentum mediated by the exchange of the photons. The measure of the probability of both emission and absorption of photons is the charge. Hence, the force must be proportional to both interacting charges (recall Coulomb's Law). Think of the repulsive interaction between two astronauts floating in space, throwing a ball back and forth (p. 141).

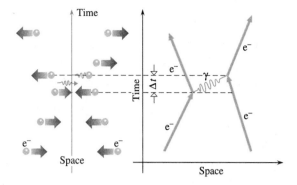

Figure 33.1 A Feynman diagram showing the scattering of two electrons via the exchange of a virtual photon.

The electrons' exchange interaction is a quantum effect and cannot be visualized in classical terms. Still, repulsion by way of an exchange force can be thought of via the astronaut-ball analogy. However, attraction between an electron and a proton through exchange is unvisualizable, unless you resort to nonsense like the astronauts facing away from each other catching boomerangs backwards so that they're pushed together. Figure 33.2 depicts the attraction, but in a way that does not attempt to be faithful to the kinematics. Feynman diagrams are symbolic—they're computational devices in QED and are not concerned with accurately picturing the particle trajectories. Thus, the horizontal distances are of little significance, and the arrangement of Fig. 33.1 is often used for both attraction and repulsion. The important part is the interaction.

The collision in Fig. 33.1 is elastic; the energy of either electron is unchanged throughout, and yet during the time Δt, the system contains an additional amount of energy hf corresponding to the photon. For a time Δt, Conservation of Energy is seemingly violated! Can this situation be tolerated? One answer offered by modern physics is *yes,* provided it can never be observed. In other words, there is always some uncertainty (ΔE) in the measured value of the energy of a system. The Heisenberg Uncertainty Principle (p. 1163) tells us that

$$\Delta E \, \Delta t \geq \tfrac{1}{2} \hbar$$

Nonconservation of energy up to an amount ΔE will be hidden by the ever-present energy uncertainty, provided the time available to make the observation (Δt) is restrictively small; namely

$$\Delta t \leq \tfrac{1}{2} \frac{\hbar}{\Delta E}$$

(If a moment of nonconservation is totally unobservable, is Conservation of Energy actually violated?) The energy uncertainty will exceed the photon energy hf if the photon exists for a time less than

$$\Delta t = \tfrac{1}{2} \frac{\hbar}{hf} = \frac{1}{4\pi f}$$

This unobservable photon can travel a maximum distance of

$$R = c \, \Delta t = \frac{c}{4\pi f} \tag{33.1}$$

and since its frequency can be arbitrarily small, the range of the force transmitted

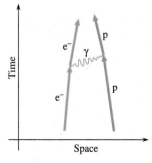

Figure 33.2 A Feynman space-time diagram of the electron-proton interaction.

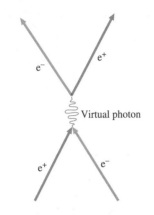

Figure 33.3 The annihilation and creation of electron-positron pairs.

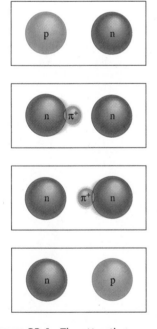

Figure 33.4 The attractive interaction between a neutron and proton through the exchange of a positive pion.

by the massless photon is unlimited. Such unobservable exchange quanta are called **virtual photons**.

In contrast to *real* quanta, these *virtual* quanta are the messengers of the interaction. In the Feynman diagrams, they are the internal segments that begin and end within the figure. They effectively "tell" the material particles what's happening. A photon that is observable in the sense that it is detected by an eyeball or a Geiger counter is real enough. A photon that never leaves the region of interaction between charges (Fig. 33.3) and vanishes in the process of communicating the electromagnetic force—a photon that cannot therefore be seen by a detector (so that we never observe any violation of the basic conservation laws)—is a virtual photon. It need not obey the equation for the total energy of a real particle: $E^2 = m^2c^4 + p^2c^2$.

The distinction between real and virtual photons is not always obvious. At the extreme of short range and short existence, messenger photons can have tremendous energies (modern accelerators can generate virtual quanta of upwards of 100 GeV). At the other extreme, because the photon is massless and moves at the ultimate speed, there is no limitation on the time it can travel with its message (its clocks don't appreciate time anyway, p. 1070). The range of virtual photons is limitless, and the range of the electromagnetic interaction is limitless. If Δt is tiny, as it usually is in particle physics, there's no issue; if Δt is large, as it might be in astronomy, the imagery can get a little murky. The question of whether a particular photon that has traveled through space for a million years is real or virtual loses significance: ΔE is vanishingly small.

By the late 1920s, it was recognized that the known material particles (protons and electrons) could each be considered as the quantum of a specific particle field. From this perspective, there are electron fields, proton fields, and so on; the Universe is a set of quantized fields. Reality (a cup of coffee and a hamburger) is the totality of observable manifestations of field quanta. And from that perspective, the problem of the wave-particle duality no longer exists: matter is field.

When Fermi formulated his theory of the weak interaction in 1932, he founded it on the principles of QED. Not long after that, the Japanese physicist Hideki Yukawa proposed (1934) that the strong interaction was mediated by the exchange of a massive virtual boson. At the time, the only known particles were the electron, proton, and neutron. As we'll see, he was able to predict that the mass of this new messenger would be between m_e and m_n and thus it came to be called the meson. It has since been identified specifically as the π-meson, or pion. Strong (hadronic) interactions occur with equal strength between electrically positive, negative, and neutral particles, and so the idea was extended to include three exchange particles: one positive (π^+), one negative (π^-), and one neutral (π^0). It was proposed that virtual particles, emitted and absorbed, constantly fly back and forth between nucleons which, in turn, are transformed—the proton and neutron are two alternative states of the nucleon (Fig. 33.4). The *virtual meson* of mass m_π and rest energy $m_\pi c^2$ traveling at nearly the speed of light has a maximum range provided by Eq. (33.1), of

$$R \approx c\,\Delta t \approx c\left(\frac{\frac{1}{2}\hbar}{\Delta E}\right) \approx c\left(\frac{\frac{1}{2}\hbar}{m_\pi c^2}\right) \approx \frac{h}{4\pi m_\pi c}$$

Given the known range of the nuclear force (≈ 1 fm), the predicted particle mass comes out about $200m_e$ or ≈ 100 MeV/c². As a rule, ***when enough energy is present such that*** $E_0 = mc^2$, ***a real particle corresponding to the virtual one can be created***. In 1934, the only source of sufficient energy was cosmic radiation. But not until 1947 was it possible to study high-energy cosmic-ray collisions using photographic emulsions. And only then was the Yukawa particle finally found: the **pion** was discovered (Fig. 33.5).

Contemporary Quantum Field Theory operates under several assumptions: (1) the essential reality is the set of quantum fields—nothing else exists; (2) these fields obey the rules of Special Relativity and Quantum Mechanics; (3) the intensity of a field at some location is a measure of the likelihood of finding an associated particle at that location; (4) the field quanta interact as the fields interact. **All particle interactions (all forces) are mediated by field quanta**. We now believe that pion exchange between hadrons is actually a low-strength residual manifestation of a still more basic and more powerful interaction involving the exchange of gluons among quarks (p. 1224).

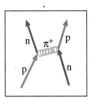

Figure 33.5 Nucleon-nucleon interactions through pion exchange.

33.3 Gauge Theory

It is now widely accepted that all field theories that accurately portray nature must possess a particular type of mathematical structure known as **gauge symmetry**. The meaning of the term is subtle and depends on several subsidiary ideas, and so we will develop it slowly. The discovery of the significance of gauge symmetry, one of the greatest of the century, was prompted by the realization that the General Theory of Relativity was gauge symmetric. Einstein's analysis of mass (or gravitational charge) in terms of the curvature of space and time held the secret to treating all forms of charge, all theories of the very essence of the cosmos. That insight was underscored by the awareness that renormalized QED (p. 1169) happened also to be gauge symmetric. Today, it is believed that we have a mathematical test of the legitimacy of any new field theory; even better, we have a potent guide to formulating such a theory—that's an amazing thing to be able to say! Physicists maintain that the hidden structure of the Universe, whatever detailed form it takes, is gauge invariant.

The philosophical insight that sprang from Relativity Theory was that all the laws of physics are the same for all observers—anyone anywhere must experience nature to behave in the same way. And yet a thousand scientists on a thousand planets across the Universe will surely have their own definitions and formulations, their own arbitrary sets of units, arbitrary base levels for things such as zero speed, zero voltage, and zero potential energy. If one body of physical law rules the cosmos, different local constructs must be equivalent—the mindset of the scientist cannot alter the underlying physical reality. A correct theory must be expressible in different local languages in terms of local conceptions and still carry the same truth. All such theories should be translatable from one foreign mathematical language to any other.

The question of units isn't a problem: we can *transform* units from one system to another—an inch is as good as 2.54 centimeters. There are no completely natural units; all are dreamed up in the minds of scientists and cannot possibly affect physical law. Nor can the arbitrarily assigned base levels of relative concepts such as potential energy have any effect on law. There is no natural zero of voltage, and however it may be assigned, we can again transform from one level to another with no problem (that's why there are zero reset knobs on meters). This freedom to assign base levels of certain important physical quantities is a manifestation of an underlying symmetry.

In the broadest terms, a physical system possesses symmetry if something happens (whereupon there is a change of some kind) and no observable change in the

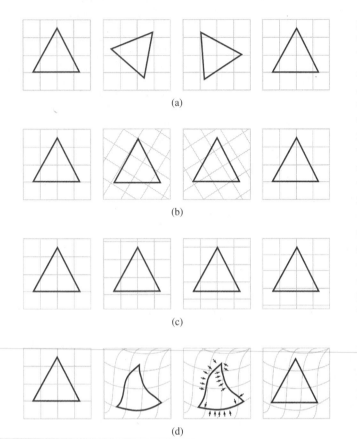

(a)

(b)

(c)

(d)

Figure 33.6 (a) The rotational symmetry of an equilateral triangle. (b) Rotating the entire coordinate system has no effect on the shape of the triangle. (c) Translating the entire coordinate system downward has no effect on the shape of the triangle. (d) Altering the coordinate system in a point-to-point, local way distorts the triangle. Even so, overlaying a force field, a gauge field, will reestablish the original configuration of the triangle.

final state of the system results. We've already talked about the geometrical symmetry of a snowflake (p. 7), which can be rotated through 60° and come up unaltered. Here, however, we are concerned with *abstract,* or *internal, symmetry* such that the alteration of some aspect of a system described by a set of equations does not change the solution of those equations. An equilateral triangle looks the same if we rotate it through 120° (Fig. 33.6a). The angle is continuously variable, and the symmetry operation is said to be *continuous.* By comparison, the triangle is also symmetric via a reflection through any one of its altitudes, but this is a jump, a discontinuous symmetry operation. Recall (p. 6) the discussion of **Noether's Principle**: *for every continuous symmetry, there is a corresponding conservation law and vice versa* (Fig. 33.7). Accordingly, we will only be concerned with continuous symmetries and indeed only with ones of a very restricted kind; namely, *local* rather than *global* symmetries. It is this last stricture that is the defining feature of gauge invariance.

Think of an object, a wire triangle, and imagine a mathematical coordinate grid extending in all directions against which the shape of the figure is to be described mathematically. Instead of an object, we might transform a physical quantity represented by a mathematical function (for example, the potential), but the triangle is easier to visualize. Every point on the wire is transformed to a point on the coordinate grid. Rotating the grid (Fig. 33.6b) will not alter the lengths of the sides or the angles of the projected triangle, nor will translating the grid (Fig. 33.6c). These are global transformations; they happen everywhere identically and do not affect our system (the shape of the triangle). When a feature of the system is invariant under such transformation, we say the system has a *global symmetry.*

To give the idea a more subtle application, consider yourself coasting around on a bicycle on top of a hill. It's a nice smooth mound, so the equipotential lines (the lines of equal height) are smooth concentric closed curves. If the entire hill is now transformed upward 100 m, you'll not notice a thing as you roll up and down its trails; assuming gravity is constant, the potential energy is independent of absolute height, and the forces on the bike only depend on the slopes of the surfaces. Similarly, the gravitational potential contour map, the potential field, is unchanged— this is a global symmetry.

By contrast, if the grid of Fig. 33.6d is altered independently from point to point so that it's twisted every which way, the shape of the triangle as formulated in this distorted space changes, and the symmetry is lost. But suppose we bend the wire here and there applying a force (call it a gauge force) at every point to counter the distortion and re-symmetrize this local transformation. We then have imparted **local symmetry** in the presence of an added compensating **gauge force field**. Go back up the hill on that bike. With a local height transformation that changes the potential from point to point, the hillside will be distorted, covered with humps and potholes; the contour lines will be full of wiggles. But suppose we somehow introduce a re-symmetrizing force field at every point that miraculously readjusts the net force field acting on you to its original configuration; we make a transformation of

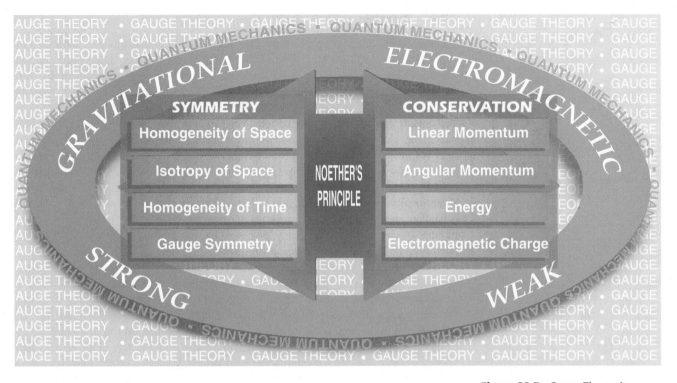

Figure 33.7 Gauge Theory is predicated on the notion of local symmetry and the concept of field.

the potential field. You would then happily bike around as if nothing had changed. The compensating gauge field establishes a local symmetry, a gauge symmetry.

Alternatively, if we place a mapmaker at "every" point on the hill and ask that they all independently establish a potential for just the one location at which each resides, we'll again get contours that are a wiggly mess—the cartographers choose the zero altitudes wherever they like, making the height of each point quite unrelated to the next. Now, let each person make up a complete map (leaving out the numerical values of the contours): they'll all be identical—the hill is the hill. If we add the proper field, the gauge field, to the wiggly-mess field, the mess will be transformed into a correct contour map of the hill. *A correct theory must be independent of all the arbitrary definitions that went into it; it must be gauge invariant.* It must be transformable from one measuring framework, or gauge, to another without affecting the conclusions of the formalism.

Classical electromagnetism is gauge invariant; Conservation of Charge is associated with symmetry in the electromagnetic potentials (p. 700), and for that symmetry to be local, one must introduce a compensating gauge field that turns out to be Maxwell's familiar electromagnetic field. The conservation of electromagnetic charge inexorably leads to classical Electromagnetic Theory. This gauge invariance carries over into the quantized domain of QED. Here, the phase of the de Broglie wave of a charged particle is made locally symmetric. To accomplish that, a new compensating field interacting with the charged particles is required, and it turns out to be the photon field. ***The photon is the quantum of the gauge field associated with electromagnetic charge***.

Both the gravitational and Coulomb force fields are long-range gauge fields. These are mediated by massless gauge quanta traveling at the speed of light (gravitons and photons). Moreover, the interaction is proportional to the source's quantum number; that is, its "charge" (mass, electromagnetic charge, etc.).

33.4 Quarks

The theoretical quandary produced by the proliferation of hadrons was addressed independently in 1963 by Murray Gell-Mann and George Zweig. They proposed that all known hadrons were composite particles, each made up of a cluster of a two or more truly elementary particles. These, Gell-Mann lightheartedly called **quarks**, from the phrase "Three quarks for Muster Mark" in James Joyce's novel *Finnegans Wake*. How many quarks make up a baryon? Baryons are fermions, and if they have parts, the parts must be fermions—any cluster of bosons is a boson. Thus, all baryons must be made up of an odd number of quarks. The simplest thing is to assume quarks are spin-$\frac{1}{2}$ fermions. Accordingly, Gell-Mann and Zweig proposed that *all baryons are composed of three quarks and all antibaryons are composed of three antiquarks*. Each quark has a baryon number of $+1/3$ and each antiquark a baryon number of $-1/3$. Similarly, mesons are bosons and must have an even number of quarks. Moreover, they have zero baryon number, and so *all mesons are composed of one quark and one antiquark*.

To account for every then-known hadron, Gell-Mann and Zweig required three varieties, or **flavors**, of quark. With a whimsical flare, these quark types came to be called **up** (u), **down** (d), and **strange** (s). Their most off-putting characteristic was the requirement that quarks have fractional electromagnetic charge, either $\pm 1/3$ or $\pm 2/3$ of the fundamental electron charge (Table 33.2). The proton and neutron, the ordinary matter of our Universe, are quark clusters uud and udd (Fig. 33.8), which suggests that the u- and d-quarks are also ordinary matter rather than the exotic matter created in high-energy collisions. Table 33.3 shows the quark combinations forming a number of hadrons. Note that if a meson is composed of a quark and antiquark of the same type or flavor (as is π^0), the constituents can rapidly annihilate one another and the lifetime is very brief. *The strong interaction cannot*

TABLE 33.2   Characteristics of quarks and antiquarks

				Quarks				
Flavor	Symbol	Charge	Spin	Baryon number	Strangeness	Charm	Bottomness	Topness
Up	u	$+\frac{2}{3}e$	$\frac{1}{2}\hbar$	$\frac{1}{3}$	0	0	0	0
Down	d	$-\frac{1}{3}e$	$\frac{1}{2}\hbar$	$\frac{1}{3}$	0	0	0	0
Strange	s	$-\frac{1}{3}e$	$\frac{1}{2}\hbar$	$\frac{1}{3}$	-1	0	0	0
Charmed	c	$+\frac{2}{3}e$	$\frac{1}{2}\hbar$	$\frac{1}{3}$	0	$+1$	0	0
Bottom	b	$-\frac{1}{3}e$	$\frac{1}{2}\hbar$	$\frac{1}{3}$	0	0	$+1$	0
Top	t	$+\frac{2}{3}e$	$\frac{1}{2}\hbar$	$\frac{1}{3}$	0	0	0	$+1$

				Antiquarks				
Flavor	Symbol	Charge	Spin	Baryon number	Strangeness	Charm	Bottomness	Topness
Up	$\bar{u}$	$-\frac{2}{3}e$	$\frac{1}{2}\hbar$	$-\frac{1}{3}$	0	0	0	0
Down	$\bar{d}$	$+\frac{1}{3}e$	$\frac{1}{2}\hbar$	$-\frac{1}{3}$	0	0	0	0
Strange	$\bar{s}$	$+\frac{1}{3}e$	$\frac{1}{2}\hbar$	$-\frac{1}{3}$	$+1$	0	0	0
Charmed	$\bar{c}$	$-\frac{2}{3}e$	$\frac{1}{2}\hbar$	$-\frac{1}{3}$	0	-1	0	0
Bottom	$\bar{b}$	$+\frac{1}{3}e$	$\frac{1}{2}\hbar$	$-\frac{1}{3}$	0	0	-1	0
Top	$\bar{t}$	$-\frac{2}{3}e$	$\frac{1}{2}\hbar$	$-\frac{1}{3}$	0	0	0	-1

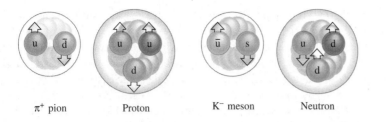

Figure 33.8 The quark composition of the pion, proton, meson, and neutron. As we'll see, each hadron also contains a swarm of gluons (p. 1150) that bind the quarks together. These gluons contribute significantly to the hadron's physical characteristics. For example, much of a neutron's or proton's spin is carried by its gluons.

transform flavor. If the flavors are different, the quark and antiquark will eventually annihilate and the meson decay, but only via the weak interaction and only relatively slowly. *The weak interaction can transform flavor*.

One might expect that blasting two protons together would produce a shower of quarks—they ought to be easy to generate and easy to identify, and yet their shyness proved a continuing embarrassment and impediment to the acceptance of the theory. No free quark has ever been observed. Accordingly, theoreticians have made a case for **quark confinement**, the notion that free quarks cannot exist, but that contention is not entirely convincing (it certainly wasn't in the early 1970s). Still, a series of experiments begun in 1969 at the Stanford Linear Accelerator Center (SLAC) in California and repeated using neutrinos at CERN gave the theory a needed boost. Probing with high-energy electrons (20 GeV), the SLAC group established that the proton and neutron were made up of three small, hard lumps of charge. The nucleon is composed of three pointlike charges that move around quite freely, like three bees inside a small balloon. But even that picture is now known to be an oversimplification—nucleons seem to be much more complicated.

A theoretical objection to the quark model was raised in the early 1960s: there are several hadrons composed of the same flavor quarks, and their existence violates the Pauli Exclusion Principle. For example, the Δ^{++} baryon corresponds to uuu, three u-quarks (three fermions in the same state is a no no no). A way out of that quark quandary developed that has since proven to be remarkably fruitful for a number of other reasons. It was proposed that each flavor of quark (at the time u, d, and s) actually came in three varieties, or **colors**: red, green, and blue. Each quark carries one of three types of *color charge*. The u-quarks in Δ^{++}—namely, ($u_R u_G u_B$)—were not really identical, and therefore there is no problem with the Exclusion Principle (Fig. 33.9).

Of course, nothing is colored in the ordinary sense of the word, but the analogy with light is convenient and makes the details easy to remember. Quarks carry color charges—they possess redness, blueness, or greenness. Antiquarks have anticolor charges of antired (think of it as cyan), antiblue (think of it as yellow), and antigreen (think of it as magenta). Red, blue, and green color are like positive electromagnetic charge; antired, antiblue, and antigreen anticolor are like negative electromagnetic charge—*it is these qualities that give rise to the strong force fields*.

Since no hadron displays an observable property that can be associated with color, it follows that color is an internal characteristic—*all hadrons are color neutral* (just as the neutron is electromagnetic-charge neutral). This means that either the total amount of each color is zero (as with mesons that are color-anticolor pairs, for example, $q_R \bar{q}_R$), or that all the colors are present in equal amounts (as with baryons, $q_R q_B q_G$). The latter is analogous to the fact that red, blue, and green light beams add to make white. This scheme demands that no observable particle be composed of either two (qq) or four (qqqq) quarks—nor can free quarks exist. We might say that an atom is electromagnetically white and that an ion is colored. And because it attracts opposite charges, an ion has a higher energy and tends to be-

TABLE 33.3	The quark composition of several hadrons*
Particle	**Quarks**
Mesons	
π^0	$u\bar{u}$, $d\bar{d}$ mix
π^+	$u\bar{d}$
π^-	$\bar{u}d$
η	$d\bar{d}$, $u\bar{u}$ mix
η'	$s\bar{s}$
K^0	$d\bar{s}$
$\bar{K}^0$	$\bar{d}s$
K^+	$u\bar{s}$
K^-	$\bar{u}s$
J/ψ	$c\bar{c}$
Υ	$b\bar{b}$
Baryons	
p	uud
n	udd
Δ^0	udd
Δ^{++}	uuu
Δ^+	uud
Δ^-	ddd
Σ^+	uus
Σ^-	dds
Σ^0	uds
Ξ^0	uss
Ξ^-	dss
Λ^0	uds
Ω^-	sss

*Where the quark compositions are the same, their spin alignments are different.

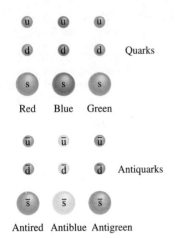

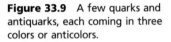

Figure 33.9 A few quarks and antiquarks, each coming in three colors or anticolors.

come white by picking up an electron. Similarly, quarks tend to form white neutral composites.

As fate would have it, quark theory, which was none too popular, was revivified by a chance discovery. In 1974, two teams of scientists—one at Brookhaven National Laboratory under C. C. Ting, and the other at SLAC under B. Richter—almost simultaneously discovered a new hadron, called J by one and ψ (psi) by the other. The J/ψ meson was three times more massive than the proton and, remarkably, lived for 10^{-20} s before decaying. That's 10^3 times longer than is normal for a hadron of that mass—there was some fundamentally new physics at work here. Moreover, the quark theory as it stood was no help; it was all filled up, there were no more particles that could be accounted for with three quarks.

The difficulty was soon settled by appealing to an idea suggested by Sheldon L. Glashow that had been around for some time but had found little support. Because there were then four known leptons, it had been proposed (on the grounds of natural symmetry among elementary particles) that there ought to be four quarks. The new addition had already been named **charm** (c), though up until the discovery of J-psi, there was no reason to take it seriously. It is now accepted that J/ψ corresponds to the two-quark bound state ($c\bar{c}$). A whole clutch of very massive charmed mesons and baryons, particles containing the heavy charm quark, have since been discovered.

This story played itself out again in 1975 when the tau (τ) lepton was discovered. Assuming there was also a τ-neutrino to be found, there were then six leptons and four quarks. By 1977, a new meson, the upsilon, was observed. It is 75 times more massive than the pion. With little hesitation, a fifth heavy quark flavor—**bottom**, or *beauty* (b)—was conjured up, and the upsilon was recognized to be ($b\bar{b}$). The mesons ($b\bar{d}$) and ($b\bar{u}$) were found in 1983. Evidence of discovery of the sixth heavy quark flavor, named **top**, or *truth* (t), was announced at the Fermi National Accelerator Laboratory in 1994. Experiments (1990) strongly suggest that nature probably cannot accommodate more than three kinds of neutrinos, and therefore there will be a total of six leptons and presumably no more than six quark flavors forming three generations of elementary particles (Fig. 33.10).

One might wonder what kind of interquark force is at work. Could some theoretical Coulomb-like interaction be devised that is proportional to color charge, one that would match all the diversity of strong reactions? In 1954, the first powerful step in that direction was taken by C. N. ("Frank") Yang and R. L. Mills. They produced the mathematical framework on which would ultimately rest modern Quantum Field Theory. Their work would make possible a satisfactory description in terms of the quantum of the gauge field of color: the gluon.

33.5 Quantum Chromodynamics

The interaction between two point-like quarks is mediated by the exchange of a boson messenger amusingly called the **gluon**. This is the basis of the fundamental *strong interaction,* or **color force** (Fig. 33.11). The quantum of the color field, the gluon, is a spin-1, electromagnetically neutral, massless particle referred to as a *vector boson*. Spin-1 particles are bosons that have wavefunctions in the form of a 4-dimensional vector; hence, the name. Because the quarks come in three colors and the absorption or emission of a gluon can change the quark's color (though not its flavor), it turns out that there are eight possible different couplings and Color Gauge Theory postulates eight massless gluons (Fig. 33.12). These differ significantly from the photon, which is chargeless, in that six of them carry color charge.

Family	Particle				Charge	Mass (GeV/c^2)
First Generation	Quarks	Up	• • •		$\frac{2}{3}$	0.330
		Down	• • •		$-\frac{1}{3}$	0.333
	Leptons	Electron	•		-1	5.11×10^{-4}
		e-neutrino	○		0	$< 1.4 \times 10^{-8}$
Second Generation	Quarks	Charm	● ● ●		$\frac{2}{3}$	1.65
		Strange	● ● ●		$-\frac{1}{3}$	0.486
	Leptons	Muon	●		-1	0.106
		μ-neutrino	○		0	$< 2.5 \times 10^{-4}$
Third Generation	Quarks	Top	● ● ●		$\frac{2}{3}$	≈ 188
		Bottom	● ● ●		$-\frac{1}{3}$	4.5
	Leptons	Tau	●		-1	1.78
		τ-neutrino	○		0	< 0.035

Figure 33.10 The three generations of elementary fermion (the associated antiparticles are not shown). Both the top quark and the τ-neutrino are presumed to exist, although they remain experimentally unconfirmed.

Each of these transports color and anticolor. Figure 33.13 shows how a red quark radiating a red-antigreen gluon loses red and has left behind green—it becomes charged green.

The strong interaction acts via color (for example, as with the meson depicted in Fig. 33.14). Gluon exchange also holds the hadrons together (Fig. 33.15) as composite entities (in twos and threes), but the transfer of individual gluons between separate hadrons is essentially precluded. Hadrons interacting with other hadrons experience the effects of the strong force mostly via the exchange of quark-antiquark composite particles (Yukawa's mesons). Figure 33.16 depicts the quark picture of the strong force proton-proton interaction. The creation and annihilation of quark pairs allows the transfer of a mediating pion, but the basic interaction is between quarks

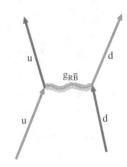

Figure 33.11 The interaction of an up-quark and a down-quark mediated by the exchange of a red-antiblue gluon.

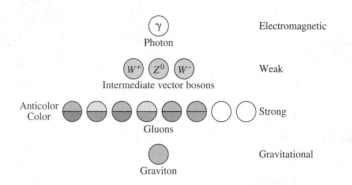

Figure 33.12 Boson force mediators.

and gluons. The strong force holding the nucleus of an atom together is the rather feeble remnant of the interquark color force.

In summary then, quarks are bound to one another (via the strong or color force) to form hadrons through the exchange of gluons. Unlike the electromagnetic force, which is communicated by chargeless photons, most gluons carry color charge, and their absorption or emission changes the color of the participating quark. The strong interaction leaves the flavor (u, d, s, c, t, b) of a quark unaffected, although its color is generally altered. Only quark-antiquark pairs of the same flavor can annihilate each other in a strong interaction (that's not the case for a weak interaction). Moreover, quarks cannot decay via the strong force, although they can via the weak force.

The gauge theory of the color field, the strong interaction, was rather colorfully christened **Quantum Chromodynamics** (QCD)—*chroma* in Greek means color—by Gell-Mann. This is a mathematically sophisticated, renormalizable gauge theory modeled after QED. To date, QCD has been successful in dealing with experimental findings, and although it's certainly incomplete, it may even be the right and true theory (or more likely, part thereof), but that remains to be seen.

33.6 The Electroweak Force

Fermi's nonrenormalizable theory of the weak interaction (1932) pictured the process of neutron decay at a single point in space-time (Fig. 33.17a, p. 1230). He had oversimplified things, avoiding the question of the carrier of his new force. Two years later, Yukawa suggested that the weak force (like the electromagnetic interaction of QED) was mediated by a massive messenger particle. After decades of neglect, the idea was revived by Julian Schwinger (1956), who attempted to describe the weak force in terms of Gauge Theory. He called the intermediary the W-particle (for *weak*). Like the photon, it had to have a spin of 1 if angular momentum was to be conserved; as a result, it came to be known as the *intermediate vector boson*. The extremely short range (≈ 0.01 fm) of the weak force required that its mediator be quite massive and quite small.

Schwinger postulated the existence of two charged vector bosons. The neutron decays into a proton and a virtual W-particle, which must therefore carry a negative electromagnetic charge (Fig. 33.17b). The W^- then decays into an electron and an electron-antineutrino. Similarly, he argued that the creation of a positron and a neutrino must be the result of the decay of a positively electromagnetically charged W-particle, the W^+ (Fig. 33.18a). The emission or absorption of a charged intermediate vector boson by a fundamental particle results in the prompt transformation of

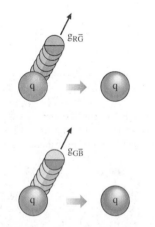

Figure 33.13 The transformation of a quark from one color to another. A $g_{R\overline{G}}$ gluon carries away red and antigreen (or magenta), leaving behind only green, and the quark becomes charged green. The emission of a $g_{G\overline{B}}$ gluon carries away green and antiblue (or yellow), leaving behind a blue quark.

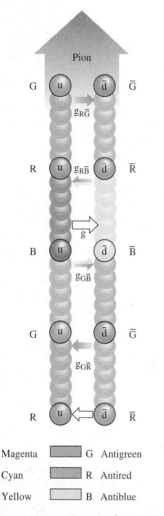

Magenta ▭ G Antigreen
Cyan ▭ R Antired
Yellow ▭ B Antiblue

Figure 33.14 The color force. A schematic representation of the interaction between the two quarks (u, d̄) comprising a pion. The attraction is mediated by the exchange of color-charged and uncharged gluons.

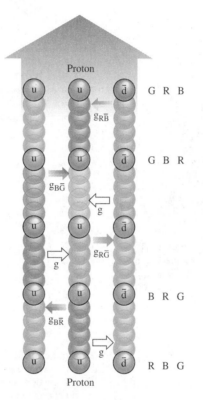

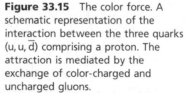

Figure 33.15 The color force. A schematic representation of the interaction between the three quarks (u, u, d̄) comprising a proton. The attraction is mediated by the exchange of color-charged and uncharged gluons.

that particle. That's something new among the forces we have studied. The emission or absorption of a photon alters the phase of the de Broglie wave, but it doesn't transform the emitting particle. Similarly, gluon exchange alters color, not flavor, so there's no change of particle type there either.

The weak force acts differently, depending on the handedness of the participants. Thus, the effect of the weak force on an electron changes depending on the particle's motion; that is, on the alignment or antialignment of its linear and spin angular momenta (a property we shall not discuss further). The weak force operates between fermions, and so leptons and quarks (that is, left-handed particles and right-handed antiparticles) couple to it; they possess weak charge. The W's carry both electromagnetic charge and weak charge.

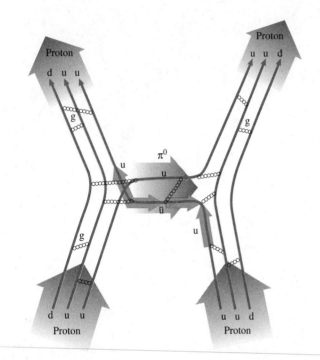

Figure 33.16 The strong interaction between protons. Two protons exchanging a neutral pion as understood via the quark model. The creation of one u-ū pair and the annihilation of another allows the transfer of a π^0. Gluons exchanged between quarks hold the hadrons together.

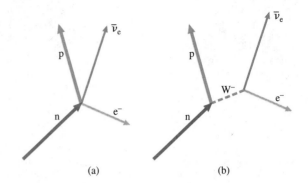

Figure 33.17 The decay of a neutron into a proton, electron, and electron-antineutrino. (a) The decay occurring at a single point. (b) The decay as mediated by a negative vector boson, which itself transforms into an electron and neutrino.

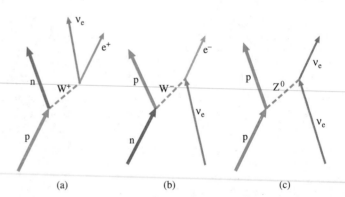

Figure 33.18 (a) Weak interactions as mediated by the three intermediate vector bosons. (b) A change in the charge of any of the particles creates a *charged current*. When, as in (c), there is no change in charge, we have a so-called *neutral current*. Note that if we slide the incoming neutrinos up so that they are created by the W^- or Z^0, they become outgoing antineutrinos.

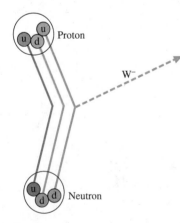

Figure 33.19 The transformation of a down-quark into an up-quark via the emission of a W^- intermediate vector boson. The result is the transformation of a neutron into a proton.

If, as a result of emitting a W^-, a neutron (udd) is transformed into a proton (udu), it must be that a d-quark is transformed into a u-quark (Fig. 33.19). The weak interaction can alter the flavor of a quark without altering its color. In fact, the emission or absorption of a $W^{\pm}$ results in the prompt transformation of one type of lepton (for example, e^-) into another (for example, $\overline{\nu}_e$), as shown in Fig. 33.20. *All the various nuclear decays that take place via the weak force involve quark flavor transformations.*

The fact that the W-particles are electromagnetically charged suggested to Schwinger that there is an intimate relationship between the weak and electromagnetic interactions; the W-particles carry both weak charge and electromagnetic charge. In the late 1950s, Schwinger turned to other concerns, but before he did, he asked one of his Ph.D. students, "Shelly" Glashow, to think about a possible connection between the weak and electromagnetic interactions. Years later (1961), Glashow published a seminal paper introducing a third messenger, the neutral Z boson, but few people paid it much attention. Like the $W^{\pm}$, these messengers were quite massive and yet gauge bosons presumably had to be massless—that, along with the fact that Glashow's model was nonrenormalizable, left much to be desired. Amazingly, Steven Weinberg, an old high school rival of Glashow, was independently working on the same problem at the Massachusetts Institute of Technology. He devised a scheme (1967) by which the vector bosons could take on mass and still fit into a gauge theory but again, no one, not even Glashow, seemed to notice it. A year later, Abdus Salam published a very similar unifying treatment that met with the same neglect. In 1969, a young graduate student in Holland, Gerhard 't Hooft, demonstrated how Yang-Mills gauge fields (p. 1226) could be made renormalizable, and the whole area of study came alive. The Glashow-Weinberg-Salam synthesis, the **Electroweak Theory**, won these three physicists the Nobel Prize in 1979. They had devised a quantum gauge theory that naturally resulted in four gauge quanta: three massive vector bosons, W^-, W^+, and a new neutral one, Z^0 (zee-zero), along with a fourth massless force-carrier, the familiar, though no less mysterious, photon (Table 33.4).

Gauge quanta are supposed to be massless. How did the theory produce massive gauge bosons? How did the symmetry demanded by Gauge Theory get suspended, or "broken"? The idea of a symmetry spontaneously being disrupted or broken is not new. Think of a sample of ferromagnetic material (p. 794) above its Curie temperature; its atoms have spins pointing every which way. A microscopic observer would encounter magnetic uniformity in all directions. Indeed, Maxwell's Equations are symmetric in space; there are no preferred directions for the theory. Yet, if the sample is cooled, that symmetry is spontaneously broken as the atoms align themselves arbitrarily into tiny domains—the energy of an aligned set of interacting magnets is less than the energy of a random grouping. The system (without any external asymmetrical influences) spontaneously breaks the symmetry of the laws of electromagnetism without violating those laws. In a quantum-mechanical Universe, where subatomic particles shift around constantly, this sort of energy-driven descent to the ground state is quite reasonable.

The W- and Z-particles "should" be massless like the photon. According to Electroweak Theory, they acquire their mass as a result of the *ad hoc* presence of an additional field known (after the person who first studied it) as a *Higgs field*. The latter is a spinless, directionless scalar field (as yet undetected experimentally). Space, at the low temperature at which it exists today, is permeated by the Higgs

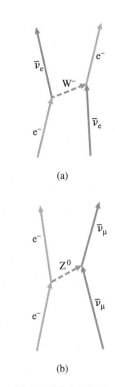

Figure 33.20 (a) An electron scattering with an electron-antineutrino via the weak interaction mediated by a W^- boson. The W carries charge, and we see an electron transformed into a neutrino. (b) An e^- and $\bar{\nu}_\mu$ come in and go out, but now the scattering is mediated by a Z^0 boson. The Z^0 is neutral and there is no change in electromagnetic charge. Note how the interaction resembles that mediated by a photon.

TABLE 33.4 The Four Forces of nature*

Force	Applicable to	Strength[1]	Effective range	Mediator
Gravity	All leptons and hadrons[2]	6×10^{-39}	Unlimited	Graviton
Weak	All leptons and hadrons	10^{-5}	$\approx 10^{-17}$ m	$W^\pm$ and Z^0
Electromagnetic	Charged leptons and hadrons	1/137	Unlimited	Photon γ
Strong	All hadrons	1	$\approx 10^{-15}$ m	Gluons g

*At the temperatures that exist today, we see four distinct interactions. At much higher energies, these blend together, presumably becoming one.
[1]Here is a slightly different scheme from the one we saw before for comparing the strengths relative to the strong force as 1.
[2]Gravity and the strong force both act on their own field quanta.

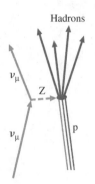

Figure 33.21 Inelastic scattering corresponding to a neutral current.

The Higgs boson is like a toilet. Everyone has one, but it's not necessarily an object of pride and beauty.

SHELDON LEE GLASHOW (1984)

field, which drags on the W- and Z-particles like cold, thick honey. In other words, vast numbers of hypothetical spin-0 Higgs bosons condense in the ground state, producing a classical field that acts on the Ws and Zs. No longer able to travel through space at lightspeed, the W and Z bosons behave like massive particles compared to the photon, which doesn't interact with the Higgs "honey" field. At high temperatures (that is to say, at high energies), the Higgs "honey" thins out and does not affect the W- and Z-particles. They supposedly become massless (at about 1000 GeV), thereby unifying in a blazing bliss with the ever-constant photon. The weak and the electromagnetic interactions are one and the same force, mediated by the exchange of these four field quanta.

All of this cannot help but seem bizarre, and yet the Electroweak Theory made several striking predictions that were subsequently confirmed in every regard. First was the existence of the *neutral-current process,* which is a weak interaction without any resulting change in electromagnetic charge. For example, the theory proposed that a very energetic neutrino might collide with a proton. Much of the neutrino's energy ($\approx$100 GeV) could be given up to the creation of a neutral Z that could pass over to a quark in the proton (Fig. 33.21). The neutrino could scatter off physically unaltered while the energy transformed into a burst of quark-antiquark pairs that would form a spray of hadrons with a net electromagnetic charge of plus one. That bit of high-energy alchemy was confirmed at CERN and later at Fermilab in 1973.

Second, the theory predicted that the masses of the intermediate vector bosons would be 80 GeV/c^2 for W$^\pm$ and 90 GeV/c^2 for Z. These values were way beyond the energy of any existing accelerators, and the direct observation of the real incarnations of these bosons had to wait for the construction of the proton-antiproton collider at CERN. Finally, in 1983, a team of 130 physicists led by Carlo Rubbia and Simon Van der Meer (both of whom shared the 1984 Nobel Prize for their work) succeeded in producing and detecting the intermediate vector bosons. One out of roughly five million p-$\bar{p}$ collisions fused together a quark from a proton and an antiquark from an antiproton, creating a vector boson that then disintegrated in less than 10^{-24} s. From the tracks made by the debris, the researchers could unambiguously identify the bosons. The masses determined for W$^\pm$ (81 GeV/c^2) and Z^0 (91 GeV/c^2) were in splendid accord with the predictions.

A recent test of the Electroweak Theory used laser beams to detect tiny distortions of heavy atoms such as cesium. The weak force between the nucleus and the orbital electrons of an atom has a minute but measurable effect. To date, all the results are in excellent agreement with theory. Still, the whole idea of the Higgs mechanism (which seems inelegant, but necessary) remains to be resolved, and until it is, few physicists will be completely content with the existing formalism.

33.7 GUTs and Beyond: The Creation of the Universe

Just as the electric and magnetic forces were unified by Maxwell's Theory into a single electromagnetic field, contemporary theory has produced a two-fold unification, a merger of the weak and the electromagnetic fields into a single **electroweak field**. What we see as the separate weak and electromagnetic interactions is a result of the cool environment of today's Universe—the unity is hidden. Only in the largest machines ever made can we even begin to simulate the inferno of primordial creation. Our Universe, space and time, began some seventeen thousand million

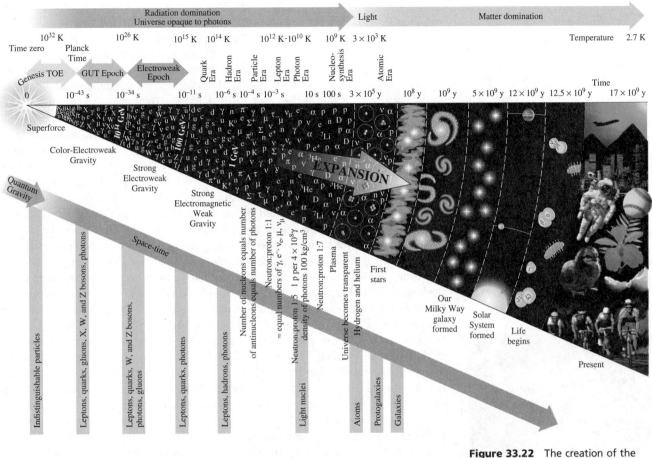

Figure 33.22 The creation of the Universe.

years ago. The interval from $\approx 10^{-34}$ s to $\approx 10^{-11}$ s after the initiation of Genesis (Time Zero) can be called the *electroweak epoch* (Fig. 33.22). Ending at an incredible temperature of around 3×10^{15} K, all there was was a dense seething soup of leptons, quarks, and massless gluons, photons, Ws and Zs. During the epoch (at temperatures above 3×10^{15} K), the weak force and the electromagnetic force shared the same inverse-square behavior and were of equal strength (Fig. 33.23). As the Universe cooled further, the electroweak symmetry shattered, Ws and Zs took on mass and vanished, and the cosmos became ruled by the Four Forces we know.

As the Universe continued to expand, it cooled to around 10^{14} K in less than its first microsecond. Free quarks and gluons vanished, coming together to form all the hadrons physicists have worked so hard to recreate. Then the heavy hadrons also annihilated, leaving (at $\approx 10^{-4}$ s) a dense quantum brew of leptons and light hadrons (protons, neutrons, pions, and so on). Neutrons and protons in almost equal number were transmuted into one another via weak interactions: $p + e^- \leftrightarrow n + \nu$ and $n + e^+ \leftrightarrow p + \bar{\nu}$. Because neutrons are heavier and require more energy to create, more protons resulted as the Universe cooled. After 1 s, the temperature was $\approx 10^{10}$ K. The unstable heavy leptons, the muons and taus, disappeared leaving neutrons, protons, electrons, positrons, and neutrinos—by now, there were five protons to every neutron.

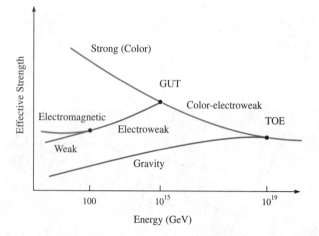

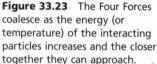

Figure 33.23 The Four Forces coalesce as the energy (or temperature) of the interacting particles increases and the closer together they can approach.

A 60 × 70 × 80-ft tank containing 8000 tons of ultra-pure water (a rich supply of protons) forms the body of a giant particle detector. Here a diver inspects the photomultiplier tubes. Although designed to detect proton-decay debris predicted by GUTs, in February of 1987 it recorded a flux of neutrinos from supernova 1987A.

The world was pervaded by a fireball of radiant energy; photons ruled the cosmos for the next ≈100 s. The Universe continued to expand, the particles separated; collisions and nucleon transformations ceased. After several minutes, appreciable numbers of free neutrons decayed into protons. But, at around 100 s, the temperature had dropped to the point ($\approx 10^9$ K) where light nuclei (deuterons and α-particles) could form via fusion and remain bound—all was plasma. Alpha production stopped when the available neutrons were locked away. After a few hundred thousand years, the Universe cooled enough for atoms of hydrogen (75%) and helium (25%) to form without being blasted apart by electromagnetic radiation. The cosmos was then dominated by matter, and in a billion years or so, proto-galaxies came into being, first-generation stars flashed on, lived, and died. Five thousand million years ago, the Sun was born, and the Earth came into being. A moment later life began, and you turned the page.

Today, we understand all the marvels of nature in terms of three primary gauge fields: the electroweak, the strong, and the gravitational fields. The first two, treated quantum mechanically via the Electroweak Theory and QCD, form the basis of what is generally referred to as the **Standard Model**. Driven by the aesthetic of *unification*, gauge theories have been proposed that attempt to unite the strong and electroweak interactions. These so-called **GUT**, or Grand Unification Theories, seek to find a relationship that will allow a threefold unification bringing together the electroweak gauge quanta and the gluons. The six quark flavors and the six lepton types then dissolve into a single lepton-quark fermion field.

One such renormalizable GUT supposes that quarks and leptons brought together at a distance of around 10^{-31} m will transform one into the other. The gauge bosons mediating such transformations, as well as transformations among quarks, are known as X particles. They would have to be incredibly massive, perhaps 10^{15} times the mass of the proton ($\approx 10^{15}$ GeV/c^2)—well beyond our laboratory capabilities for the foreseeable future.

A prominent prediction of the simplest form of GUT is the ultimate decay of the proton. In the unlikely event that two quarks inside a proton happen to come closer than 10^{-31} m, they might exchange an X gauge boson, converting the proton into a positron and a pion. Rather than being timeless, the theoretical proton lifespan is $\approx 10^{31}$ years (which, albeit finite, exceeds the age of the Universe by a factor of 10^{21}). To date, despite considerable experimental effort, no sign of proton decay has been observed. The proton's minimum life expectancy is now set at 10^{32} years, and climbing.

The GUT epoch of the Universe lasted over the miniscule interval from $\approx 10^{-43}$ s to $\approx 10^{-34}$ s after Creation. Within the continuum of space-time, a dense mass of leptons, quarks, and gauge bosons interacted, paying no attention to color, flavor, or electromagnetic charge. Only two forces operated: the color-electroweak and gravity. Unlike the Standard Model, GUTs predict that neutrinos have mass, and here experiments are as yet inconclusive. Studies of supernova 1987A (p. 117) suggest that if the electron-neutrino has any mass at all, it's less than a mere 16 eV. Though certainly unconfirmed, GUTs nonetheless remain compelling.

The Universe came into existence as simple as it could be, becoming increasingly diverse only as it cooled enough for complexity to crystallize—for symmetries to spontaneously break. Already by $\approx 10^{-43}$ s beyond Time Zero, gravity had settled out from the once single *superforce* that had ruled from the beginning. Even more ambitious than GUTs are the attempts to bring a quantum theory of gravity (with its mediator, the graviton) into a fourfold unification. That ultimate goal is amusingly called the *Theory of Everything* (TOE). It is estimated that the effects of quantum gravity will become significant at a particle separation of 10^{-35} m, which corresponds to a phenomenal energy of 10^{19} GeV. This value is totally beyond direct experimental observation—our most powerful machines will not likely exceed 10^5 GeV for decades. Since gravity acts on both fermions and bosons, it would seem that the TOE should account for the transformation of these two particle types into one another.

During the TOE epoch prior to $\approx 10^{-43}$ s, the entire Universe was subnuclear in size and surely quantum-mechanical in disposition. The ambient temperature ($\approx 10^{32}$ K) and density ($\approx 10^{92}$ times greater than water) were unimaginable. Every point in the inferno could spontaneously become a Black Hole dropping through the fabric of the cosmos and then evaporating back. That disconnected seething foam of space and time was quantized. It is here in the primal whirlwind, even before space-time congealed into a gravitational continuum, that our physics is most challenged. It is here, in the distant reaches of imagination, that we are most bold. Yet who could doubt that the whirlwind was a gauge whirlwind?

All there is today, stars, wind, baseballs, lovers, all there is, was there in the foam of beginning. Shattered symmetries, epoch upon epoch, fanned out diversity. Quarkstuff, you and I, for a moment contemplating Creation—how naive, how wonderful.

Core Material

Leptons are the particles that couple to the weak force and, if they are electrically charged, to the electromagnetic force, but leptons *are immune to the strong force*. **Hadrons** are the strongly interacting composite particles that form two distinct subgroups: *mesons* and *baryons*. Mesons are bosons; baryons are fermions (p. 1216). On a nonfundamental level, the nuclear force arises from the exchange of mesons (p. 1220). With a mass m, these have a maximum range of

$$R \approx c\,\Delta t \approx c\left(\frac{\frac{1}{2}\hbar}{m_\pi c^2}\right) \approx \frac{h}{4\pi m_\pi c}$$

All field theories must possess a particular type of mathematical structure known as gauge symmetry (p. 1221).

Baryons are composed of three quarks; antibaryons are composed of three antiquarks. Mesons are composed of one quark and one antiquark. There are six flavors of quark: *up* (u), *down* (d), *strange* (s), *charm* (c), *bottom* (b), and *top* (t) (p. 1224). Each flavor comes in three varieties or *colors*—red, blue, or green. Antiquarks have anticolor charges of antired, antiblue, and antigreen. Hadrons are color neutral (p. 1225). The interaction between quarks is mediated by the exchange of eight massless gluons (p. 1226). The gauge theory of the color field is called Quantum Chromodynamics (QCD).

The weak interaction is mediated by three *intermediate vector bosons*: W^+, W^-, and Z^0 (p. 1228). The weak interaction can alter the flavor of a quark without altering its color. *Electroweak Theory* combines the electromagnetic and weak forces (p. 1232). GUTs, or Grand Unification Theories, seek a threefold unification, bringing together the electroweak gauge quanta and the gluons.

Suggestions on Problem Solving

1. Some useful numerical quantities for this chapter:

$$1\ \text{eV} = 1.602\,177 \times 10^{-19}\ \text{J}$$

$$hc = 1.239\,842 \times 10^{-6}\ \text{eV·m}$$

$$k_B = 1.380\,66 \times 10^{-23}\ \text{J/K} = 8.617\,4 \times 10^{-5}\ \text{eV/K}$$

2. How would you go about determining if the reaction

$$\mu^+ + \nu_\mu \rightarrow \pi^+$$

is possible? First, you might check the electromagnetic charge; $+1 + 0 \rightarrow +1$ (it's okay). Then check the spin; $\pm\frac{1}{2} + \pm\frac{1}{2} \rightarrow 0$

(it's okay). Then check the baryon number; $0 \rightarrow 0$ (it's okay). Then check the lepton number. There are only μ-type leptons; hence, on the left, we have -1 for the μ^+ antimuon and $+1$ for the μ-neutrino, yielding a total of zero on both sides. These are not strange particles, so we needn't worry about S. Finally, checking the energy, the muon has less mass than the pion, and the neutrino can balance the formula by supplying the energy difference. Conclusion: the reaction is allowed. Moreover, the inverse reaction

$$\pi^+ \rightarrow \mu^+ + \nu_\mu$$

must also be allowed.

3. Notice that adding the same particle or antiparticle to both sides doesn't change any of the requirements and must result in allowed processes as well. Thus, adding an antiparticle to both sides, we see that the reaction

$$\pi^+ + \bar{\nu}_\mu \rightarrow \mu^+ + \nu_\mu + \bar{\nu}_\mu$$

results in

$$\pi^+ + \bar{\nu}_\mu \rightarrow \mu^+$$

which is fine, provided the μ has the right amount of KE. This is equivalent to the notion that *carrying a particle from one side of the formula to the other changes it to an antiparticle,* and vice versa. And *that's equivalent in a Feynman diagram to changing an incoming particle to an outgoing antiparticle,* or vice versa.

4. Remember that *the production of strange particles via the strong force conserves strangeness;* only weak-force decays do not conserve strangeness.

Discussion Questions

1. Figure Q1a is a drawing of the Stanford Linear Collider (SLC) showing its three-kilometer straightaway. The cathode fires two successive bunches of electrons. The damping rings condense the beams, which are focused down to a diameter of a few millionths of a meter later on. Discuss, in general terms, what this 100-GeV machine does and how it does it. What might this process have to do with the Z^0 intermediate vector boson? These are the heaviest known real particles and are of great interest because, among other reasons, they have many possible decay modes. Figure Q1b depicts the simplest linac, the drift-tube accelerator. How does it work? [Hint: The particles spend the same amount of time drifting along inside each cylinder.]

2. In many respects, the Z^0 behaves like a heavy photon. Discuss this notion, using the diagrams of Fig. Q2. Here, the generic label *quark* or *lepton* means that any such particle coming in also sails out. Explain what's happening in each illustration. An interaction between quarks can take place via Zs or Ws. How do these differ?

3. The μ^- and μ^+ are antiparticles of each other, and because the μ^- decays to an e^-, we take it as the matter and the other as the antimatter. By contrast, π mesons are not so easy to categorize; π^0 seems to be its own antiparticle, but the π^+ and π^- always decay into a lepton and an antilepton. Discuss this situation as it relates to the quark picture. Is the classification of matter and antimatter unambiguous? Does it matter which pion we call the antiparticle?

4. It is known that a Feynman diagram can be turned on its side to yield an entirely new physical insight, provided that particles moving backward in time are interpreted as antiparticles. Is this attribute true of Fig. Q2? Explain.

5. Every hadron interaction must be able to be analyzed in terms of the quark model. Accordingly, explain in detail how the reaction

$$\pi^- + p \rightarrow \Lambda^0 + K^0$$

takes place.

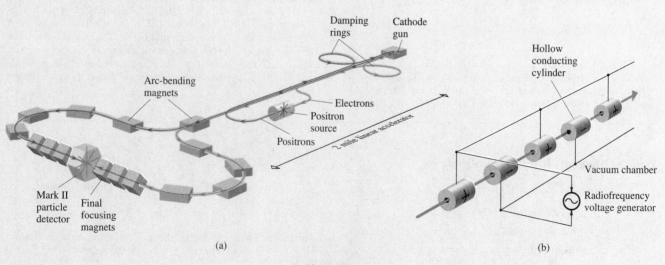

(a)

(b)

Figure Q1

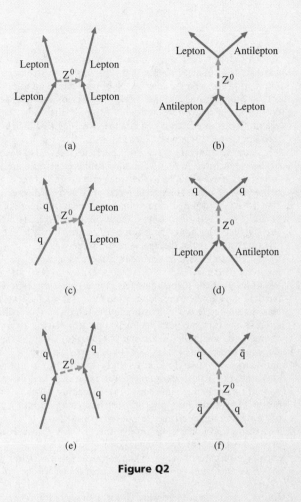

Figure Q2

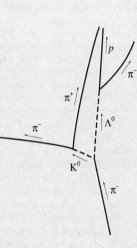

Figure Q7

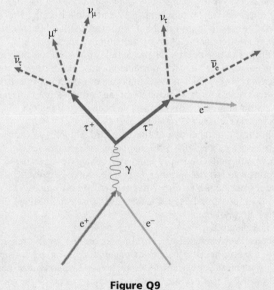

Figure Q9

6. With Question 5 in mind, realize that the Λ^0 may rapidly decay into a negative pion and a proton while the K^0 may decay into a positive and a negative pion. Write out the complete reaction starting with the proton. In what sense do we end up with more matter than we start with? How is that manifest in the quark model? Where does the matter come from? Draw a Feynman diagram of the whole process.

7. Figure Q7 shows a bubble chamber photo of some particle tracks. The chamber is filled with supercooled liquid hydrogen. When a particle ionizes the hydrogen atoms, tiny bubbles form, revealing the path. Compare the photo and the sketch, and explain what happened. Why did a few of the participants remain invisible? The tracks are bent by a known magnetic field—what can we tell from the radius of curvature of each track? Write out a formula for what took place.

8. Is it correct to say that *every observable subnuclear event occurs via creation and annihilation?* Explain.

9. In 1975, Martin Perl reported the existence of some unexpected muon-electron events associated with high-energy electron-positron collisions: $e^+ + e^- \rightarrow e^- + \mu^+ + ?$. He suggested that the observations were the result of the decay of a new particle-antiparticle pair of very massive leptons he called *tau*. Discuss the reactions depicted in Fig. Q9. The effect first appeared when the two colliding particles had a total combined energy of about 3.6 GeV. Why is that significant?

10. Neutrons decay when they are free—why might they be stable when bound to protons in a nucleus? [Hint: Consider the situation in terms of energy.] In light of GUTs, might they be less than totally stable, even in the nucleus?

11. The idea of baryon number as a conserved quantity was introduced by Weyl and Stückelberg (1930) to supposedly explain the proton's stability. Let's reconsider the notion. First, does it *explain* anything? Discuss the nature of the "law"—is it arbitrary? Is it applicable beyond particle physics, as is, for example, Conservation of Angular Momentum? What does the fact that the Universe contains far more protons than antiprotons suggest? How does the probable electrical neutrality of the Universe fit into this discussion? What about lepton number? What do GUTs say about all of this?

Multiple Choice Questions

1. A proton and an electron that are close together   (a) will sometimes interact through the strong force   (b) will never interact through the gravity force   (c) will never interact through the weak force   (d) will never interact through the strong force   (e) none of these.

2. Mesons are   (a) always composed of either two u-quarks or two d-quarks   (b) always composed of two quarks of different flavor   (c) always composed of a quark and an antiquark   (d) never composed of a quark and an antiquark of the same flavor   (e) none of these.

3. A meson and a baryon that come very close together   (a) will only sometimes interact through the gravity force   (b) will never interact through the gravity force   (c) will never interact through the weak force   (d) will never interact through the strong force   (e) none of these.

4. Decays that last in the range from 10^{-20} s to 10^{-23} s are   (a) driven by the weak interaction   (b) driven by gravity   (c) driven by the strong force   (d) driven by the electromagnetic force   (e) none of these.

5. The reaction $e^- + p \rightarrow n + \bar{\nu}_e$ is   (a) possible because it conserves everything   (b) impossible because it does not conserve baryon number   (c) impossible because it does not conserve lepton number   (d) impossible because it does not conserve charge   (e) none of these.

6. From the perspective of the Standard Model, the reaction $p \rightarrow \pi^+ + \pi^0$ is   (a) possible because it conserves everything   (b) impossible because it does not conserve strangeness   (c) impossible because it does not conserve lepton number   (d) impossible because it does not conserve charge   (e) none of these.

7. From the perspective of the Standard Model, the reaction $p \rightarrow e^+ + \gamma$ is   (a) possible because it conserves everything   (b) impossible only because it does not conserve baryon number   (c) impossible because it does not conserve either baryon or lepton number   (d) impossible because it does not conserve charge   (e) none of these.

8. The decay $K^+ \rightarrow \gamma + \gamma$ is   (a) possible because it conserves everything   (b) impossible for several reasons, one being because it is electromagnetic and must conserve strangeness   (c) impossible for several reasons, one being because it is electromagnetic and does not conserve meson number   (d) impossible for several reasons, one being because it does not conserve baryon number   (e) none of these.

9. The reaction $\nu_e + n \rightarrow e^- + p$ is   (a) possible because it conserves everything   (b) impossible because it does not conserve baryon number   (c) impossible because it does not conserve lepton number   (d) impossible because it does not conserve charge   (e) none of these.

10. Which is true?   (a) only the weak interaction can change one flavor of quark into another   (b) only the strong interaction can change one flavor of quark into another   (c) both the strong and the weak interactions can change one flavor of quark into another   (d) only the gravitational interaction can change one flavor of quark into another   (e) none of these.

11. All strongly interacting material particles   (a) also couple to the weak force   (b) also couple to the electromagnetic force   (c) never couple to the weak force   (d) never couple to the electromagnetic force   (e) none of these.

12. The presence of a neutrino among the decay products of a particle   (a) tells us nothing about the force responsible   (b) suggests that the strong force was at work   (c) suggests that the electromagnetic force was at work   (d) tells us that the weak force was at work   (e) none of these.

13. We know that the electromagnetic force has controlled a particle decay when the decay products   (a) include at least one particle with electromagnetic charge   (b) include at least one intermediate vector boson   (c) have a zero net charge   (d) include at least one photon   (e) none of these.

14. Baryons are composed of   (a) two quarks and an antiquark   (b) two antiquarks and a quark   (c) a lepton and a quark   (d) three quarks   (e) none of these.

15. A quark can change flavor by   (a) absorbing a W boson   (b) emitting a photon   (c) absorbing a gluon   (d) emitting an electron   (e) none of these.

16. The thing that holds a meson together is   (a) the exchange of W^- bosons   (b) the exchange of gluons   (c) the exchange of gauge bosons of the electromagnetic field   (d) the absorption and emission of quarks   (e) none of these.

17. Gluons can attract each other to form so-called *glueballs* because   (a) gluons are sticky   (b) gluons exchange photons   (c) gluons carry color charge   (d) gluons emit and absorb $W^\pm$ bosons   (e) none of these.

18. An antineutron can be distinguished from a neutron by   (a) its opposite electromagnetic charge   (b) its opposite magnetic moment   (c) its smaller mass   (d) its greater tendency to self-annihilate   (e) none of these.

19. Prior to 10^{-43} s after Time Zero, we believe that the Universe   (a) was the size of the Moon   (b) was ruled by a single superforce   (c) was in the so-called GUT era   (d) did not exist   (e) none of these.

20. We believe that a correct fundamental field theory must   (a) be simple   (b) predict that the proton is unstable   (c) have global symmetry   (d) be gauge invariant   (e) none of these.

Problems

ELEMENTARY PARTICLES
QUANTUM FIELD THEORY

1. [I] Knowing that Δ^* is a baryon with spin $\frac{3}{2}$ and 0 strangeness and that η and η' are mesons with 0 spin, 0 electromagnetic charge, and 0 strangeness, indicate which forces determine the following reactions: (a) $\eta \rightarrow \gamma + \gamma$ (b) $\Delta^* \rightarrow p + \pi$ (c) $\Lambda^0 \rightarrow \pi^- + p$ (d) $\eta' \rightarrow \eta + \pi + \pi$. Explain your answers fully.

2. [I] Determine the forces that are responsible for each of the following reactions: (a) $\pi^+ \rightarrow \mu^+ + \nu_\mu$

(b) $\pi^0 \to \gamma + \gamma$ (c) $\pi^- + p \to K^0 + \Lambda^0$. Explain your answers fully.

3. [I] State some of the conservation laws that are violated in each of the following reactions:
   (a) $\Lambda^0 \to \pi^+ + \pi^- + e^- + e^+$
   (b) $\pi^+ \to \mu^+ + \gamma$ (c) $\Lambda^0 \to \pi^- + \pi^+ + p$
   (d) $p + p \to \pi^+ + \Sigma^+$

4. [I] Give at least one conservation law violated in each of the following reactions?   (a) $n + \nu_e \to e^+ + \pi^-$
   (b) $p + \bar{\nu}_e + e^- \to \pi^0 + \gamma$
   (c) $p + p \to \pi^+ + K^+ + n + n$
   (d) $p + n \to K^- + K^+ + \pi^0$

5. [I] Does the reaction
$$p + n \to p + \bar{p} + p$$
occur and if not, why not?

6. [I] Is the reaction
$$\mu^- \to e^- + \bar{\nu}_e$$
possible and if not, why not?

7. [I] Is the reaction
$$\mu^- \to e^- + \nu_e + \nu_\mu$$
possible and if not, why not?

8. [I] Is the reaction
$$p + \pi^- \to p + \pi^- + \pi^0$$
possible and if not, why not? By what interaction could it proceed?

9. [I] Is the reaction
$$K^0 \to \pi^+ + \pi^- + \pi^0$$
possible and if not, why not? By which interaction will it proceed?

10. [I] Is the reaction
$$\pi^- + p \to K^0 + n$$
likely to be observed?

11. [I] Is the reaction
$$K^- \to \mu^- + \bar{\nu}_\mu$$
forbidden by any conservation law? How does it decay?

12. [I] Is the reaction
$$\pi^- + p \to K^+ + K^-$$
forbidden by any conservation law?

13. [I] Is the reaction
$$\pi^+ \to \mu^+ + \nu_\mu$$
forbidden by any conservation law? If not, how much energy will be released as KE as a result of the decay? (Assume the neutrino to have negligible mass and the pion to be at rest initially.)

14. [I] Determine the particle that is missing in each of the following reactions: (a) $p + (\ ) \to n + \mu^+$
   (b) $p + p \to p + K^+ + (\ )$
   (c) $p + p \to p + n + \pi^+ + (\ )$.

15. [I] Consider the decay
$$\Lambda^0 \to \pi^- + p$$
which takes a rather long 10^{-10} s. Assuming the lambda to be at rest, what is the net KE of the resulting particles? What conservation law does the reaction not have to obey?

16. [I] If the reaction
$$\mu^- \to e^- + \bar{\nu}_e + \nu_\mu$$
can occur, what is the maximum energy carried by the two essentially massless neutrinos assuming the muon to be at rest?

17. [I] If a proton and an antiproton, both essentially at rest, were to completely annihilate each other, how much energy could be liberated?

18. [I] What is the maximum total kinetic energy shared by the pion and electron in the reaction
$$K^+ \to \pi^0 + e^+ + \nu_e$$
assuming the kaon to be at rest?

19. [I] Approximately how much energy must be provided in order to create a separated u-quark and $\bar{u}$-antiquark pair?

20. [I] What is the quark composition of an antiproton?

21. [I] If the mean lifetime of a proton were a mere 10^{30} y, how many protons would we have to put in a tank in order to observe an average of one decay per year?

22. [II] Is the decay $\Xi^0 \to n + \pi^0$ forbidden and if so, why?

23. [II] Consider a mediating particle of mass m; derive an expression for its rest energy in terms of the range of the associated force.

24. [II] Two protons moving head-on toward each other at the same speed collide, causing the reaction
$$p + p \to p + p + \pi^0$$
Determine the minimum kinetic energy of each proton.

25. [II] Two protons moving toward one another at the same speed collide to produce the reaction
$$p + p \to n + p + \pi^+$$
What is the minimum amount of KE each must have in order for the reaction to take place?

26. [II] Consider the reaction
$$p + p \to p + \Lambda^0 + K^0 + \pi^+$$
Determine the minimum KE the protons must have if this event is to occur. For simplicity, we take the protons to be moving toward one another at the same speed.

27. [II] Determine the Q value of the reaction
$$\pi^- + p \to K^0 + \Lambda^0$$
Could this reaction take place if the pion and proton were moving slowly toward each other just before impact?

28. [II] How much energy would appear in the form of gamma rays if a positron with a KE of 10.0 MeV was to annihilate an electron with a KE of 20.0 MeV?

29. [II] A neutrino has an energy of 100 MeV. Assuming it has zero mass, what is its frequency, wavelength, and momentum? Use SI units.

30. [II] A D^+ meson has quantum numbers $B = 0$, $S = 0$, $C = +1$, and $Q = +1$ (and no topness or bottomness). What is its quark configuration?

31. [II] A D^0 meson has quantum numbers $B = 0$, $S = 0$, $C = +1$, and $Q = 0$ (and no topness or bottomness). What is its quark configuration?

32. [II] A K^- meson has quantum numbers $B = 0$, $S = -1$, $C = 0$, and $Q = -1$ (and no topness or bottomness). What is its quark configuration? What is the configuration of K^+? Explain your answer in detail.

33. [II] A Λ^0 baryon has quantum numbers $B = +1$, $S = -1$, $C = 0$, $Q = 0$, spin $= \frac{1}{2}$ (and no topness or bottomness). What is its quark configuration? Explain your answer in detail.

34. [II] Explain how, according to the quark model, this reaction takes place

$$\pi^0 \rightarrow \gamma + \gamma$$

35. [II] Explain how, according to the quark model, this reaction takes place

$$K^0 \rightarrow \pi^+ + \pi^-$$

(First write it out in terms of quarks and then discuss what happens.)

36. [II] Explain how, according to the quark model, this reaction takes place

$$\Omega^- \rightarrow \Lambda^0 + K^-$$

(First write it out in terms of quarks and then discuss what happens.)

37. [II] Suppose an electron and a positron, each very nearly at rest, annihilate forming a single *virtual* photon. What would be its frequency?

38. [II] The energy at which the weak and electromagnetic forces unify into the electroweak force is around 100 GeV (Fig. 33.23). What temperature does that correspond to? (Check your answer against Fig. 33.22.)

39. [II] One GUT formulation supposes that quarks and leptons brought together at a distance of around 10^{-31} m will transform one into the other. Determine the mass of the gauge boson that would have such a range. Give your answer in GeV/c^2.

40. [II] Assume the body of a typical human being contains roughly 10^{28} protons. If the proton had a half-life of only 10^{10} years, what would be its disintegration rate in decays per second?

41. [II] With Problem 39 in mind, at what temperature will this unification occur? (Check your answer against Fig. 33.22.)

42. [II] GUTs predict that both neutrons and protons can decay in the nucleus via routes that do not conserve baryon number. Roughly how many nucleons are in a liter of water? If the mean lifetime of a nucleon was 10^{20} y, how many decays would occur in the liter of water per year? What does this result suggest about the lifetime of a proton?

Appendixes: A Mathematical Review

APPENDIX A   ALGEBRA

Algebra is essentially a body of rules and procedures for logically exploring the relationships between concepts using symbols to provide a concise and easily read format. Our concern is usually with equations describing interdependencies among physical quantities. Thus, the length traveled (l) in a time (t) by a bird moving at a speed (v) is given by $l = vt$. **When we have one equation we can solve for only one unknown quantity**. Thus, suppose the speed is known to be 2 m/s and the length traveled is 500 m (ignoring units for the moment), the equation becomes $500 = 2t$; one equation, one unknown (namely, t). How long did it take the bird to make the trip? Solve for t; *get the unknown all by itself on one side of the equation*. To do that, remember the following three logical rules:

1. The same quantity (constant or variable) can be added to or subtracted from both sides of an equation without changing the equality: $100 = 100$; $100 - 2 = 100 - 2$.
2. Both sides of an equation can be multiplied or divided by the same quantity (constant or variable) without changing the equality: $100 = 100$; $100/2 = 100/2$.
3. Both sides of an equation can be raised to the same power (squared or square rooted, cubed or cube rooted) without changing the equality: $100 = 100$; $\sqrt{100} = \sqrt{100}$.

Wherever possible, apply rule 1 *first followed by rule* 2. *Isolate the variable to whatever power it is raised, and if that power is other than one, use rule* 3 *to make it one.*

Back to the bird: $2t = 500$. To get the t alone, remove the 2 (using rule 2) by dividing both sides by 2; hence, $t = 250$ seconds.

Let's solve $8x + 2 = 42$ for x. To get the x alone, first remove the 2 via rule 1, $8x = 42 - 2 = 40$. Next, use rule 2 to remove the 8; $x = 40/8 = 5$.

Now solve $5x^2 + 12 = 57$ for x. Apply rule 1: $5x^2 = 57 - 12 = 45$. Next isolate x^2 using rule 2 to divide both sides by 5: $x^2 = 9$. Now use rule 3, taking the square root of both sides: $x = \pm 3$. Note that there are two solutions: $+3$ and -3.

Solve $\frac{3}{4}t^2 - 6 = 0$ for t. Use rule 1 to move the 6, yielding $\frac{3}{4}t^2 = 6$. Now, using rule 2, multiple both sides by 4 to get $3t^2 = 24$. Use it again, dividing both sides by 3 to get $t^2 = 8$. Now, using rule 3, take the square root of both sides: $t = \sqrt{8}$.

Some Trouble Spots

1. Dividing by fractions can be troublesome. $8/\frac{1}{4}$ is *not* 2. **A fraction is not changed when the top and bottom are multiplied by the same quantity**. Here, multiply top and bottom by 4 to get 32/1. Similarly, given $\frac{1}{4}/\frac{1}{2}$ multiply top and bottom by 2 to get $\frac{1}{2}/1 = \frac{1}{2}$.
2. $(a + b)^2$ is *not* equal to $a^2 + b^2$! $(a + b)^2 = (a + b)(a + b) = aa + ab + ba + bb = a^2 + 2ab + b^2$.

3. $\frac{1}{a} + \frac{1}{b}$ is *not* equal to $\frac{1}{a+b}$. We can only add terms with the same denominators. To get a common denominator (ab), multiply top and bottom by whatever is needed: $\frac{b}{ba} + \frac{a}{ab}$ and add. Thus

$$\frac{1}{a} + \frac{1}{b} = \frac{a+b}{ab}$$

Similarly

$$\frac{1}{2} + \frac{2}{5} = \frac{5}{10} + \frac{4}{10} = \frac{9}{10}$$

4. Remember that $\sqrt{a}\sqrt{b} = \sqrt{ab}$; thus, $\sqrt{4}\sqrt{9} = \sqrt{36}$, or $2 \times 3 = 6$. Similarly, $\sqrt{8} = \sqrt{4}\sqrt{2} = 2\sqrt{2}$; hence, $\sqrt{25gt^2} = 5t\sqrt{g}$.

5. Given the relationship $F = GmM/R^2 = 2GmM \neq R^2$, how does F change if m is doubled? The new F is $G(2m)M/R^2$, which is twice the original F. How does F change when R is doubled? The new F is $GmM/(2R)^2 = GmM/4R^2$, which is one-quarter the original F.

A-1 Exponents

A square with sides of length a has an area of $a \times a$, which is more concisely written as a^2 (a raised to the second power). Because of that, a^2 is read "a squared." In the same way, a cube with sides of a has a volume of $a \times a \times a = a^3$ or "a cubed." In general, a raised to the nth **power** is a^n, where the **exponent** n can be any number, fractional or whole, positive or negative. Quantities raised to various powers are often multiplied and divided. Thus, $a^2 \times a^3 = (a \times a) \times (a \times a \times a) = a^2 a^3 = a^5$ and, in general

$$(a^n)(a^m) = a^{n+m} \qquad \text{(A.1)}$$

The base numbers (a) being raised to the powers n and m are the same. Alternatively, if the base numbers (a and b) are different and the powers are the same, then

$$(a^n)(b^n) = (ab)^n \qquad \text{(A.2)}$$

Now, for division: $2^3/2^2 = 8/4 = 2 = 2^1$; the two 2's on the bottom cancel two 2's on the top or, equivalently, the exponent on the bottom subtracts from that on the top. More generally

$$\frac{a^3}{a^2} = \frac{a \times a \times a}{a \times a} = a^{3-2} = a$$

As a rule

$$\frac{a^n}{a^m} = a^{n-m} \qquad \text{(A.3)}$$

which suggests that we define

$$\frac{1}{a^m} = a^{-m} \qquad \text{(A.4)}$$

Hence, $1/5 = 5^{-1}$, $1/3^2 = 3^{-2}$, and $1/4^{-2} = 4^2$. Inasmuch as $a/a = 1$, it follows from Eq. (A.3) that $a^{1-1} = a^0 = 1$. **Any quantity raised to the zero power is 1.**

Fractional exponents correspond to roots, thus

$$a^{1/n} = \sqrt[n]{a} \qquad \text{(A.5)}$$

Thus, using Eq. (A.1), we have

$$\sqrt{a}\sqrt{a} = a^{\frac{1}{2}}a^{\frac{1}{2}} = a^1 = a$$

Since powers undo roots and vice versa

$$(a^{1/n})^n = 1$$

For example, $(25^{\frac{1}{2}})^2 = 5^2 = 25$. In general

$$(a^n)^m = a^{nm} \qquad (A.6)$$

where n and m can be either whole or fractional numbers.

A-2   Powers of Ten: Scientific Notation

We deal with numbers that are extremely small (like the mass of an electron, 0.000 000 000 000 000 000 000 000 000 000 911 kg) and extremely large (like the number of stars in the Universe, $\approx 10\,000\,000\,000\,000\,000\,000\,000$). The *scientific notation* is a shorthand way of writing numbers in terms of powers of ten:

$10^0 = 1$ $10^0 = 1$
$10^1 = 10$ $10^{-1} = 1/10 = 0.1$
$10^2 = (10 \times 10) = 100$ $10^{-2} = 1/(10 \times 10) = 0.01$
$10^3 = (10 \times 10 \times 10) = 1000$ $10^{-3} = 1/(10 \times 10 \times 10) = 0.001$
$10^4 = (10 \times 10 \times 10 \times 10) = 10\,000$ $10^{-4} = 1/(10 \times 10 \times 10 \times 10) = 0.0001$

and so forth. For positive exponents, *the number of zeros corresponds to the power of ten*. Thus, the number of stars in the Universe is a 1 with 22 zeros, or 1×10^{22}, or just 10^{22}.

Suppose we want to express a number greater than 1.0 given in ordinary form (such as the number of seconds in a year—31 560 000) in scientific notation. Start by indicating the decimal at its reference position (31 560 000.0). Then decide to where you would like to move it (for example, 31*560 000.0). Here the decimal is to be relocated six places to the left yielding 31.56; multiplying by 10^6 will move it back six places to the right, where it started. Thus, $31\,560\,000 = 31.560 \times 1\,000\,000 = 31.56 \times 10^6$, or 3.156×10^7, or 0.3156×10^8. To write a number less than 1.0 in scientific notation (for example, 0.000 000 000 000 000 000 000 000 000 000 911 kg), again locate where you would like the decimal to appear (0.000 000 000 000 000 000 000 000 000 9*11 kg). That's 31 places to the right, so multiplying 9.11 by 10^{-31} will shift the decimal back 31 places to the left where it started. The mass of an electron is 9.11×10^{-31} kg.

When multiplying or dividing numbers in scientific notation, process the numerical terms separately from the exponents; as

$$(1.1 \times 10^{12})(5.0 \times 10^{17}) = (1.1 \times 5.0)(10^{12} \times 10^{17}) = 5.5 \times 10^{29}$$

$$\frac{(1.1 \times 10^{12})}{(5.0 \times 10^{17})} = \frac{(1.1)}{(5.0)} \times \frac{(10^{12})}{(10^{17})} = 0.22 \times 10^{-5}$$

Scientific calculators have an "exp" key that allows you to enter exponents, and they will keep track of the decimal automatically.

A-3   Logarithms

Suppose we have a positive number y expressed as a power of b where $b > 0$ and $b \neq 1$; accordingly

$$y = b^x$$

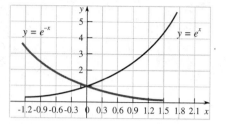

Figure A1

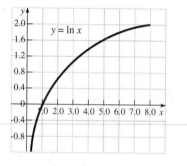

Figure A2

Let's define x to be the logarithm of y, to the base b; thus

$$x = \log_b y$$

The logarithm x is the exponent, and the two equations are just different ways of saying the same thing. For example, if $2^3 = 8$, then $3 = \log_2 8$; if $8^{-4/3} = 1/16$, then $-4/3 = \log_8 1/16$. Because logarithms are exponents, the laws governing the two are very similar. Thus, independent of the base

$$\log(ab) = \log a + \log b$$

$$\log \frac{a}{b} = \log a - \log b$$

$$\log a^n = n \log a$$

where

$$\log_b b = 1$$

$$\log 1 = 0$$

and

$$b^{\log_b a} = a$$

There are two widely used bases, 10 and $e = 2.718\,281\,828 \cdots$. The **common logarithms** arise when $y = 10^x$, whereupon $x = \log_{10} y$. Before the advent of the electronic calculator, common logs were used to carry out complicated calculations. Our interest in them arises out of the idea of intensity-level (p. 455) and the dB. The **natural logarithms** arise when $y = e^x$, whereupon $x = \ln y$ (it's customary to write $\ln y$ rather than $\log_e y$). The exponential function e^x (Fig. A1) and the natural log (Fig. A2) occur in a great many physical situations in which the rate-of-change of some quantity depends on that quantity. We can see this dependence in Fig. A1: the rate-of-change of the curve $y = e^x$ equals e^x. When the curve has small values, the slope is small; when the curve has large values, the slope is large. This property is very special and it means that if some quantity N changes in time exponentially, then $\Delta N/\Delta t \propto N$.

It follows from the above that

$$\ln e^x = x \qquad \text{and} \qquad e^{\ln x} = x$$

Notice that with $y = e^x$ when $y = 1$, $x = 0$; hence, $y = 2$ when $2 = e^x$, whereupon $x = \ln 2 = 0.693$. Doubling y yields $y = 4$ when $x = \ln 4 = 1.386 = (0.693 + 0.693)$, and again $y = 8$ when $x = \ln 8 = 2.079 = (0.693 + 1.386)$. Every time x increases by 0.693, y doubles. Indeed, any time x increases by a fixed amount, y will increase by a fixed multiplicative factor—y increases by the same factor in equal intervals of x. When x increases by $\ln 10 = 2.302\,6$, y increases by a factor of 10.

A-4 Proportionalities and Equations

We frequently find that one physical quantity depends on another: the force exerted by a spring depends on the amount it's stretched; the current through a resistor depends on the voltage across it. Suppose the quantities A and B depend on one another such that doubling either doubles the other. They are said to be **directly proportional** to each other or, symbolically, $A \propto B$. On the other hand, it's possible that doubling one could halve the other, in which case $A \propto 1/B$, and they are **inversely proportional**. Table A1 gives a list of values of A and B. At first, there may seem no relationship between them, but form the ratio B/A and we see that there is a

TABLE A1

A	B
3.1	9.3
18.0	54
5.2	15.6
71.0	213

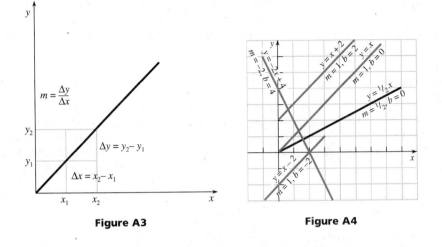

Figure A3

Figure A4

consistent pattern: $B/A = 3$. Thus, $B = 3A$ and 3 is known as a **constant-of-proportionality**. When two quantities are proportional, we can always turn the proportionality into an equality using an appropriate constant-of-proportionality. The circumference of a circle (C) is directly proportional to the diameter (D), where the constant of proportionality is π: $C = \pi D$. The area of a sphere (A) is directly proportional to the square of the radius (r), where the constant-of-proportionality is 4π: $A = 4\pi r^2$.

When the variables in an equation occur only to the first power, it's called a **linear equation**; for example, $y = mx$, where m is a constant. Here, when $x = 0$, $y = 0$, and the curve (Fig. A3) is a straight line passing through the origin. For any two points on the line, $\Delta y/\Delta x$ is its **slope**, or tilt, and it equals m. If a constant (b) is added to y such that $y = mx + b$, the line shifts parallel to itself; b is the place where the line crosses the y-axis, the y-*intercept* (Fig. A4).

The Quadratic Equation

When an object is uniformly accelerating, it travels a distance s given by Eq. (3.9); namely, $s = v_0 t + \frac{1}{2}at^2$. This expression is typical of a class of equations that contains a variable (t) raised to the second power, and it's called a **quadratic equation**. The standard form has everything on the left set equal to zero on the right:

$$ax^2 + bx + c = 0$$

where a, b, and c are constants. This equation can be solved directly, but it's common practice to memorize the solution in the form

$$x_\pm = \frac{-b \pm \sqrt{b^2 - 4ac}}{2a}$$

where, provided $b^2 > 4ac$, there are two real solutions. For example, $2x^2 + 6x - 20 = 0$ corresponds to $a = 2$, $b = 6$, and $c = -20$; therefore

$$x_+ = \frac{-6 + \sqrt{36 - 4(2)(-20)}}{2(2)} = \frac{-6 + \sqrt{196}}{4} = +2$$

$$x_- = \frac{-6 + \sqrt{36 - 4(2)(-20)}}{2(2)} = \frac{-6 - \sqrt{196}}{4} = -5$$

It may happen in a physical analysis that only one of the two solutions corresponds to a possible situation; we then simply ignore the other. For example, we might solve Eq. (3.9) for the time and find one of the solutions to be negative—it works mathematically but not physically.

Simultaneous Equations

For every unknown, we must have an independent equation (one that isn't equivalent to any other equation in the set). Situations with one and two unknowns are common throughout the discipline. Indeed, we will encounter problems in circuit analysis where there are as many unknowns as there are loops in the circuit, but three will be our limit. Suppose there are two unknowns, x and y, and two equations

$$4x - 2y = 16$$
$$3x + 4y = 23$$

Lots of numbers satisfy each of these equations (for example, in the first one, $x = 4$, $y = 0$ or $x = 0$, $y = -8$). We want to solve these *simultaneously* so that the solutions satisfy both equations at the same time. There are two ways to accomplish this goal. (1) *Solve either equation for either unknown in terms of the other unknown, and substitute that into the remaining equation.* From the first equation, $4x = 16 + 2y$, $x = 4 + \frac{1}{2}y$. Now put this result into the second equation so it has only one unknown $3(4 + \frac{1}{2}y) + 4y = 23$ and so $12 + 3y/2 + 4y = 23$, $3y/2 + 8y/2 = 23 - 12$, $11y/2 = 11$, $y/2 = 1$, $y = 2$. Put that back into the first equation: $4x - 2(2) = 16$, $4x = 20$, and $x = 5$. (2) Alternatively, *multiply either or both of the equations by whatever numbers it takes to make the same coefficient appear for either one of the unknowns. Then add or subtract the two, thereby removing one unknown.* Here, to get rid of y, multiply the first equation $4x - 2y = 16$ by 2

getting $8x - 4y = 32$

and add $\underline{3x + 4y = 23}$

$11x \qquad = 55$

or $x = 5$. Substitute this result into either equation and solve for y.

A-5 Approximations

It's often desirable to approximate the solution to a problem, either as a quick check or because the exact solution is too elaborate to deal with. Newton's **Binomial Theorem**, expressed as

$$(a + b)^n = a^n + na^{n-1}b + \frac{n(n - 1)}{2 \times 1}a^{n-3}b^2 + \frac{n(n - 1)(n - 2)}{3 \times 2 \times 1}a^{n-3}b^3 + \cdots$$

$$(A.7)$$

produces some very helpful approximations. In general, if n is a positive integer, the series ends in a finite number of terms, otherwise it has an infinite number. Expressions of the form $(1 + x)^n$ arise frequently and can be approximated using the Binomial Theorem ($a = 1$, $b = x$); thus

$$(1 + x)^n = 1 + nx + \frac{n(n - 1)}{2}x^2 + \cdots$$

$$(A.8)$$

When x is very small ($x \ll 1$), the x^2 and higher terms will be negligibly small and

$[x \ll 1]$	$(1 + x)^n \approx 1 + nx$
with $n = 2$	$(1 + x)^2 \approx 1 + 2x$
with $n = 3$	$(1 + x)^3 \approx 1 + 3x$
with $n = \frac{1}{2}$	$(1 + x)^{\frac{1}{2}} = \sqrt{1 + x} \approx 1 + \frac{1}{2}x$
with $n = -\frac{1}{2}$	$(1 + x)^{-\frac{1}{2}} = \dfrac{1}{\sqrt{1 + x}} \approx 1 - \frac{1}{2}x$
with $n = -1$	$(1 + x)^{-1} = \dfrac{1}{(1 + n)} \approx 1 - x$
with $n = \frac{1}{2}$	$(1 - x)^{\frac{1}{2}} = \sqrt{(1 - x)} \approx 1 - \frac{1}{2}x$

Try taking the square root of $1.000\,000\,0010$ on your calculator. Now, let $x = 0.000\,000\,0010$ and the square root of $(1 + x)$ is $\approx 1.000\,000\,0005$. For $a \gg b$

$$(a + b)^n = a^n \left[1 + \frac{b}{a} \right] \approx a^n \left[1 + n\frac{b}{a} \right] \qquad \text{(A.9)}$$

What is the value of $a/(a + b)^2$ when $a \gg b$? Since b is negligible compared to a, $(a + b)^2 \approx (a)^2$ and $a/(a + b)^2 \approx 1/a$. Check this expression with Eq. (A.9), $n = 2$ and $(a + b)^2 \approx a^2(1 + 2b/a) \approx a^2$. Suppose $b \gg a$, $(a + b)^2 \approx (b)^2$, and $a/(a + b)^2 \approx a/b^2$. We frequently know how a system behaves at its extremes and can confirm the analysis at those extremes when a key quantity is very large or very small. For example, any equation for the force between two magnets must go to zero as the separation between them gets very large.

A useful means of improving the likelihood that an analysis is correct is to simplify the numbers by rounding them off and then to run quickly through the calculation to get a crude answer. When a number is rounded off to the nearest power of 10, we say it's an **order-of-magnitude** figure. Thus, the order-of-magnitude of the acceleration due to gravity on Earth (9.81 m/s^2) is 10 m/s^2; the great mountains have heights of the order-of-magnitude of 10 km; the electron's mass ($9.109\,3897 \times 10^{-31}$ kg) is of the order-of-magnitude of 10^{-30} kg; the number of atoms per cubic centimeter of a solid is of the order-of-magnitude of 10^{23}.

APPENDIX B   GEOMETRY

Figure B1 provides a number of useful relationships for the angles formed by intersecting lines and triangles. Of special interest is the **Pythagorean Theorem**, which relates the sides of any right triangle; consequently

$$C = \sqrt{A^2 + B^2}$$

as in Fig. B2. There are a number of right triangles with whole number sides—the 3–4–5 and 5–12–13 are the most commonly encountered (Fig. B3).

Figure B4 displays the areas and volumes of shapes that are frequently used in this text. Notice that the area of the curved surface of a circular cylinder ($2\pi Rh$) is that swept out by a line of height h moved around the circumference of the base ($2\pi R$). In a similar way, the volume of a cylinder is the area of the base times the height (that is, the volume swept out by a vertical line of height h moved to every point on the base area).

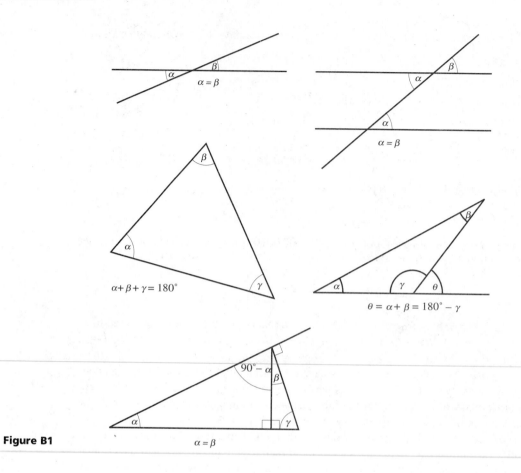

$\alpha = \beta$

$\alpha = \beta$

$\alpha + \beta + \gamma = 180°$

$\theta = \alpha + \beta = 180° - \gamma$

$90° - \alpha$

$\alpha = \beta$

Figure B1

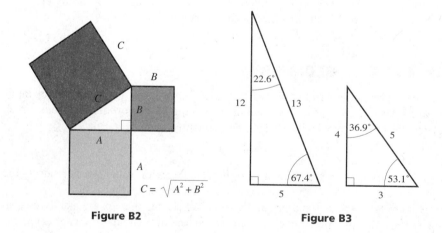

$C = \sqrt{A^2 + B^2}$

Figure B2

Figure B3

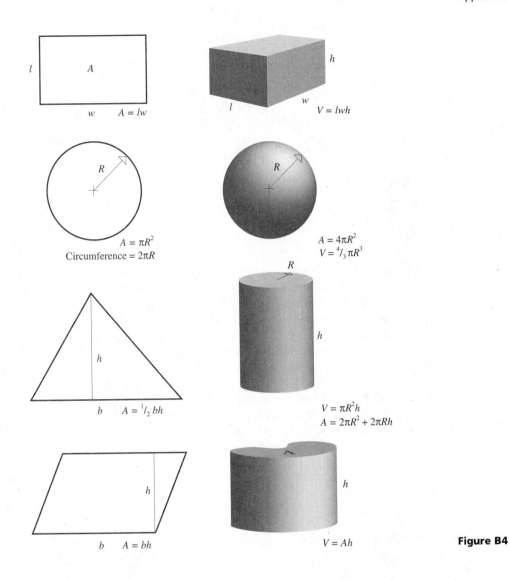

$A = lw$

$V = lwh$

$A = \pi R^2$
Circumference $= 2\pi R$

$A = 4\pi R^2$
$V = {}^4/_3 \pi R^3$

$A = {}^1/_2 bh$

$V = \pi R^2 h$
$A = 2\pi R^2 + 2\pi Rh$

$A = bh$

$V = Ah$

Figure B4

APPENDIX C   TRIGONOMETRY

Figure C1a shows a right triangle with an acute angle θ and the sides (A) adjacent and (B) opposite to it. A particular value of θ fixes a set of triangles, a few of which are shown in Fig. C1b. Although the triangles are different, the ratios of the sides in any one of them is the same for all of them. In other words, the ratio of the opposite over the adjacent is the same for each triangle and is indicative of θ. The various possible ratios are given names, and electronic calculators are programmed to compute them. Knowing θ we can determine any ratio and vice versa. Thus

$$\sin \theta = \frac{\text{opposite}}{\text{hypotenuse}} \qquad \cos \theta = \frac{\text{adjacent}}{\text{hypotenuse}} \qquad \tan \theta = \frac{\text{opposite}}{\text{adjacent}}$$

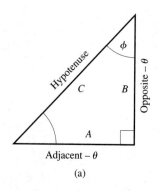

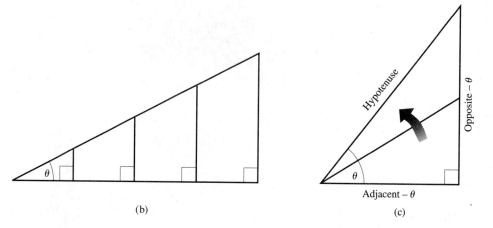

(a)   (b)   (c)

Figure C1

For example, for an angle of 30°, the ratio of the opposite to the adjacent side will always be 0.5774. To determine as much on a calculator, first make sure it's in the *degree* mode, then punch in [3][0][tan]. Alternatively, knowing the ratio, you can find the angle by entering [0][·][5][7][7][4][tan⁻¹].

Referring to Fig. C1a, observe that $\sin \theta = B/C = \cos \phi$ and that $\cos \theta = A/C = \sin \phi$; don't confuse these relationships. In Fig. C1c, increasing θ, with the adjacent side constant, increases both the opposite side and the hypotenuse. When $\theta \approx 0$, $B \approx 0$, $A \approx C$ and so $\sin \theta = 0$, $\cos \theta = 1$, and $\tan \theta = 0$. Similarly, when $\theta \approx 90°$, $C \approx B$ and both are tremendously large. Thus, $\sin 90° = 1$, $\cos 90° = 0$, and $\tan 90° = 1$. These results are summarized in Fig. C2, which provides $\sin \theta$, $\cos \theta$, and $\tan \theta$ for all values of θ between 0 and 360°.

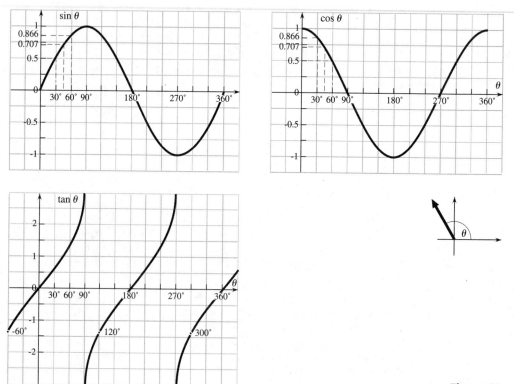

Figure C2

Some useful relationships are

$$\tan \theta = \frac{\sin \theta}{\cos \theta}$$

$$\sin^2 \theta + \cos^2 \theta = 1$$

$$\sin 2\theta = 2 \sin \theta \cos \theta$$

For all triangles (Fig. C3)

Law of Sines $$\frac{\sin \alpha}{A} = \frac{\sin \beta}{B} = \frac{\sin \gamma}{C}$$

Law of Cosines $$C^2 = A^2 + B^2 - 2AB \cos \gamma.$$

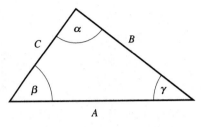

Figure C3

APPENDIX D   VECTORS

Vectors are discussed at length in Chapter 2. Here, we'll go over some of the trouble spots often encountered when dealing with them. Figure D1 shows a vector **C** in each of the four quadrants and the corresponding x and y components. In each case

$$C_x = C \cos \theta \quad \text{and} \quad C_y = C \sin \theta$$

where θ is measured either up or down from the x-axis and is always acute. Now, suppose we add several vectors along the x-axis and get $\mathbf{C}_x$ and add several along the y-axis and get $\mathbf{C}_y$. The next step in determining **C** is to use $\tan \theta = C_y/C_x$ to find θ. It's best to write the tangent as $\tan \theta = |C_y|/|C_x|$, using the absolute values of the components so that $\tan \theta$ is always positive and θ is always between 0° and 90°. Figure C2c is a plot of $\tan \theta$ against θ from 0° to 360°, where θ is always measured from the positive x-axis. Thus, the inverse tangent of -1.732 is both 120° and 300°,

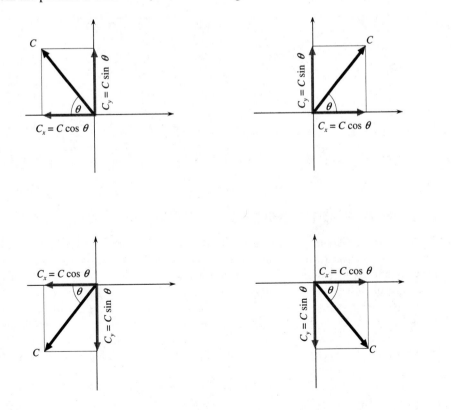

Figure D1

lying in either the second or fourth quadrant, as expected from the figure. A calculator provides another form of the answer—namely, $-60°$ (equivalent to $300°$). It neglects $120°$ altogether, which means that if we take the ratio of the components when finding the tangent and one or both of them are negative, there will be some ambiguity in the resulting angle. Use the ratio of absolute values; determine which quadrant the resultant is in by drawing C_x and C_y. Then measure $\theta < 90°$ up or down from the horizontal axis, as in Fig. D1.

We now wish to analytically add the two vectors **A** and **B** in the four situations depicted in Fig. D2. In each case, find $C_x = A_x + B_x$ and $C_y = A_y + B_y$. Since the angles are between 0 and 90°, the sines and cosines will be positive. Thus, oppositely directed components must be assigned plus and minus signs. Enter any component as plus $(+)$ if it is in the positive x- or y-direction and as minus $(-)$ if it is in the negative x- or y-direction. The appropriate expressions are shown in each case in Fig. D2.

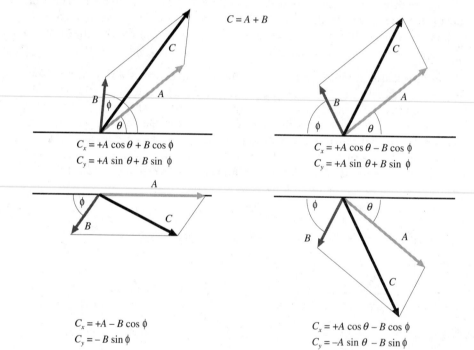

$$C = A + B$$

$$C_x = +A\cos\theta + B\cos\phi$$
$$C_y = +A\sin\theta + B\sin\phi$$

$$C_x = +A\cos\theta - B\cos\phi$$
$$C_y = +A\sin\theta + B\sin\phi$$

Figure D2

$$C_x = +A - B\cos\phi$$
$$C_y = -B\sin\phi$$

$$C_x = +A\cos\theta - B\cos\phi$$
$$C_y = -A\sin\theta - B\sin\phi$$

APPENDIX E   DIMENSIONS

To determine a distance, we must specify the units in which the measurement will be made:—feet, yards, light-years, meters, and so forth. But all such measurements have something in common—they are all lengths, regardless of the specific unit. The type of unit is the *dimension* of the quantity; in this case, the dimension of length indicated as $[L]$. Area is a distance times another distance and so has the dimensions of $[L][L] = [L^2]$. We can assign dimensions to all the quantities in mechanics using the basic dimensions of *length* $[L]$, *time* $[T]$, and *mass* $[M]$. Thus, speed, which is measured in units of m/s, ft/s, mi/h, km/h, and so on, has the dimensions of length/time or $[L]/[T]$.

The volume of a sphere of radius R is $\frac{4}{3}\pi R^3$, whereas that of a cube of side a is

a^3. Each has units of m^3 or ft^3 and each has the dimensions of volume [L^3], suggesting that a technique of dimensional analysis, with appropriate logical rules, can be used to check our equations. To that end, consider the equation representing three physical quantities A, B, and C, as

$$C = A + B$$

There are three rules. The first is: *dimensions can be treated as algebraic quantities*—added, subtracted, multiplied, and divided accordingly. Furthermore, A and B must have the same units (we can't add seconds to meters). Our second rule is: **only quantities having the same dimensions can be added or subtracted**. Whatever it might be, it's clear that C must have the same units as $A + B$. More generally, our third rule is: *the dimensions on either side of an equal sign must be the same.* All physical equations must obey these rules.

If you come up with an expression that obeys the rules, it *may* be correct; if it doesn't obey them, it's *definitely* incorrect. For example, in Chapter 3 we find that the distance traveled s by an object that was initially moving at speed v_i after it accelerates at a constant rate a for a time t is given by $s = v_i t + \frac{1}{2}at^2$. Knowing that the dimensions of speed and acceleration are [L]/[T] and [L]/[T^2] respectively, apply dimensional analysis. The expression in dimensional form is

$$[L] = \frac{[L]}{[T]}[T] + \frac{[L]}{[T^2]}[T^2]$$

Cancelling time on the right, the equation obeys the second rule since each term being added has the dimension [L]. And it obeys the first rule as well. Conclusion: the equation may be correct.

APPENDIX F   CALCULUS

F-1   The Derivative

Physics is the study of change in the physical universe, and *calculus is the mathematics of change.* Any occurrence, any physical process, that is described in terms of one or more parameters will change in a specific way when any of those parameters change. Such a mathematical description is called a function, and calculus is all about how functions change.

A **function**, call it f, is a logical procedure—embodied in a mathematical expression—that provides a specific output value, associated with each input value of some quantity on which f depends. Let f depend on a variable quantity x, which is the input; for each value of x there is one and only one value of f, the output. The function*, $f(x)$, might be an expression that yields the weight of a chicken in terms of the size of its feet (x) or the distance your car can travel in terms of the volume of its gas tank.

Figure F1 depicts a portion of a plot of a representative function, $f(x)$, against the variable x. The function has the value $f(x)$ at x, and when the variable changes to ($x + h$) the function becomes $f(x + h)$. Between the values of x and ($x + h$) the curve rises by an amount $f(x + h) - f(x)$. This occurs over the so-called *run* of the

*To help distinguish a function from a product, the parentheses associated with a symbol for a function, and everything therein, will be set in italic.

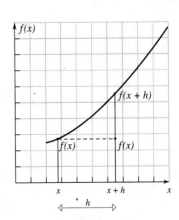

Figure F1

variable: $(x + h) - x = h$. The rise over the run,

$$\frac{f(x + h) - f(x)}{h}$$

is the *Difference Quotient* and it corresponds to the slope of the straight line between the two points $[x, f(x)]$ and $[(x + h), f(x + h)]$. As h becomes smaller, the interval diminishes and the straight line more and more approaches the tangent to the curve at $f(x)$. That special value of the difference quotient that is approached in the limit as $h \to 0$ is the slope of the tangent to the curve at the point $[x, f(x)]$. It is called the **derivative** of $f(x)$ and is symbolically represented as df/dx where

$$\frac{df}{dx} = \lim_{h \to 0} \frac{f(x + h) - f(x)}{h} \tag{F.1}$$

The derivative is the instantaneous rate-of-change of $f(x)$ with respect to x, and it is read "dee eff, dee ex." The derivative tells us how a function changes as the variable on which it depends changes. In physics the function usually describes some physical quantity, the speed of a rocket or the weight of water in a tank, and the variable might be the thrust of the engine or the size of a hole in the side of the tank. Changing the thrust changes the speed in a well-defined fashion that can be determined using calculus. The variable—which we simply called x—might be *any* physical parameter on which the process depends: length, time, mass, speed, density, voltage, whatever.

Derivatives are computed directly via Eq. (F.1), but as a rule, one simply memorizes and applies one or more of the following procedures.

The Derivative of a Constant

If f is a constant function, $f(x) = C$, then $df/dx = 0$: *the derivative of a constant is zero.* There is no rate-of-change of a constant.

The Power Rule

If $f(x) = x^n$, where n is *any* real number,

$$\frac{d}{dx}(x^n) = nx^{n-1} \tag{F.2}$$

For example, the derivative of $f(x) = x^3$ with respect to x is $df/dx = 3x^2$.

The Product Rule

The derivative of the product of two functions $f(x)$ and $g(x)$ is given by

$$\frac{d}{dx}[f(x)g(x)] = g\frac{df}{dx} + f\frac{dg}{dx} \tag{F.3}$$

When one of the functions in Eq. (F.3) is a constant C, because $dC/dx = 0$,

$$\frac{d}{dx}[Cf(x)] = C\frac{df}{dx} \tag{F.4}$$

For example, given $h(x) = 3x^3$, find dh/dx. To apply the Product Rule, take $f = 3$ and $g = x^3$ and hence $dh/dx = (x^3)(0) + 3(3x^2) = 9x^2$. Now for a somewhat more complicated example, determine dh/dx when $h(x) = (x^2 - 1)(3x^3 + 2x)$. Let $f(x) = (x^2 - 1)$ and $g(x) = (3x^3 + 2x)$, whereupon it follows that $dh/dx = (3x^3 + 2x)(2x) + (x^2 - 1)(9x^2 + 2)$. Notice that *the derivative of the product is not usually equal to the product of the derivatives.*

The Derivative of a Sum of Functions

Given two differentiable functions $f(x)$ and $g(x)$, the derivative of their sum is given by the sum of their derivatives, that is,

$$\frac{d}{dx}[f(x) + g(x)] = \frac{df}{dx} + \frac{dg}{dx} \tag{F.5}$$

The Chain Rule

If f is a differentiable function of u and if u is a differentiable function of x, then

$$\frac{df}{dx} = \frac{df}{du}\frac{du}{dx} \tag{F.6}$$

This relationship is especially important when dealing with interdependent quantities such as speed ($v = dx/dt$) and acceleration ($a = dv/dt$), in which case

$$a = \frac{dv}{dt} = \frac{dv}{dx}\frac{dx}{dt} = \frac{dv}{dx}v \tag{F.7}$$

The Chain and Power Rules Combined

If n is any real number and $u = f(x)$ is differentiable, then

$$\frac{d}{dx}(u^n) = nu^{n-1}\frac{du}{dx} \tag{F.8}$$

For example, if we want the derivative of $y = (x^3 - 1)^2$, take $u = (x^3 - 1)$, in which case $n = 2$ and $\frac{dy}{dx} = \frac{d}{dx}(x^3 - 1)^2 = 2(x^3 - 1)3x^2 = 6x^2(x^3 - 1)$.

Some Common Derivatives

$$\frac{d}{dx}\sin x = \cos x \tag{F.9}$$

More generally, when the argument of the sine function is itself some differentiable expression $f(x)$,

$$\frac{d}{dx}[\sin f(x)] = [\cos f(x)]\frac{df}{dx} \tag{F.10}$$

or equivalently using $u = f(x)$

$$\frac{d}{dx}[\sin u] = [\cos u]\frac{du}{dx}$$

For example, with $\omega = $ constant, we find the derivative of $\sin \omega t$ with respect to t, that is, $\frac{d}{dt}\sin \omega t = (\cos \omega t)\frac{d}{dt}\omega t = \omega \cos \omega t$. Similarly,

$$\frac{d}{dx}\cos x = -\sin x \tag{F.11}$$

More generally, when the argument of the cosine function is itself some differentiable expression $f(x)$,

$$\frac{d}{dx}[\cos f(x)] = [-\sin f(x)]\frac{df}{dx} \tag{F.12}$$

or equivalently using $u = f(x)$

$$\frac{d}{dx}[\cos u] = [-\sin u]\frac{du}{dx}$$

Inasmuch as $\tan x = (\sin x)/(\cos x)$,

$$\frac{d}{dx}\tan x = \frac{1}{\cos^2 x} \tag{F.13}$$

The distinguishing characteristic of the function e^x is that the slope of the curve corresponding to e^x equals e^x, that is,

$$\frac{d}{dx}e^x = e^x \tag{F.14}$$

More generally, when $u = g(x)$ is some differentiable function,

$$\frac{d}{dx}e^{[g(x)]} = e^{[g(x)]}\frac{d}{dx}g(x) \tag{F.15}$$

or

$$\frac{d}{dx}e^u = e^u\frac{d}{dx}u$$

For example, $\dfrac{d}{dx}e^{x^2} = e^{x^2}\dfrac{d}{dx}x^2 = 2xe^{x^2}$. The derivative of a natural logarithm is

$$\frac{d}{dx}\ln x = \frac{1}{x} \tag{F.16}$$

More generally, when $u = g(x)$ is some differentiable function,

$$\frac{d}{dx}\ln[g(x)] = \frac{1}{g(x)}\frac{d}{dx}g(x) \tag{F.17}$$

or

$$\frac{d}{dx}\ln u = \frac{1}{u}\frac{d}{dx}u \tag{F.18}$$

For example, $\dfrac{d}{dx}\ln x^2 = \dfrac{1}{x^2}2x = \dfrac{2}{x}$.

Differentials

Given $y = f(x)$ we can take dy/dx to be a single entity, the derivative of y with respect to x, which is itself a function of x. Thus let $dy/dx = F(x)$. In the spirit of the difference quotient considered above, the derivative can also be interpreted as a ratio of two tiny quantities called **differentials** where

$$dy = F(x)\,dx$$

and dx is an independent variable whereas dy is a dependent variable. Accordingly, dy is the change in y resulting when there is an associated change dx in x. For example, if $y = f(x) = x^3$, then $dy/dx = F(x) = 3x^2$ and $dy = 3x^2\,dx$.

The Second Derivative

The process of differentiation can be carried on beyond the first derivative to the second, third, fourth, and so on, simply by taking successive derivatives. Each differentiation produces a new function whose derivative can be taken in turn. Thus the second derivative of $f(x)$ is

$$\frac{d^2f}{dx^2} = \frac{d}{dx}\left(\frac{df}{dx}\right) \qquad \text{(F.19)}$$

For example, given $f(x) = x^3 - 7x^2 + 1$, the first derivative is $df/dx = 3x^2 - 14x$ and the second derivative is $d^2f/dx^2 = 6x - 14$. We will have little reason to go beyond the second derivative, although the third derivative of distance with respect to time does have physical significance.

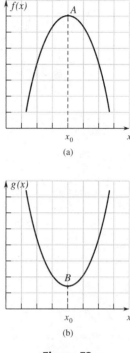

(a)

(b)

Figure F2

Maxima and Minima

At point A in Fig. F2a the curve of $f(x)$ is higher than at any neighboring points, and that function is said to have a *maximum value* at $x = x_0$. Similarly, at point B in Fig. F2b the curve of $g(x)$ has a *minimum value* at $x = x_0$. Both points, which are called **extrema**, share the feature that the tangent to the curve there at x_0 is horizontal. It follows that if $f(x)$ has an extremum at x_0 and if the derivative of $f(x)$ evaluated at $x = x_0$, that is $[df/dx]_{x=x_0}$, exists, then $[df/dx]_{x=x_0} = 0$. The slope of the curve is zero at the extrema—which is Fermat's Theorem. The proviso that the derivative exists arises because it is possible to have a function that has an extremum at some point but no derivative there—as is the case for $f(x) = |x|$, which has a minimum at $x = 0$, but no derivative there.

It is also possible for a function to have a horizontal tangent, $[df/dx]_{x=x_0} = 0$, and yet not have a maximum or minimum at that location (Fig. F3). The function $f(x) = x^3$ has zero slope at $x = 0$, but that's a **point of inflection**, also called a *stationary point*, where the slope changes from decreasing to increasing (or vice versa) as x increases.

The function

$$f(x) = x^3 - 3x^2 + 1 \qquad -\frac{1}{2} \le x \le 4$$

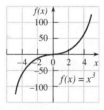

Figure F3

is plotted in Fig. F4. To find its extrema (a) take the derivative, (b) set it equal to zero, (c) solve for the values of x that cause the derivative to be zero, and (d) check the values of $f(x)$ at the end points of the interval to see if these happen to be either a maximum or minimum. Thus, (a) $df/dx = 3x^2 - 6x = 3x(x - 2)$, (b) $3x(x - 2) = 0$, and so (c) extrema are at $x = 0$ and $x = 2$. The point $x = 0$, where $f(x) = 1$, corresponds to a **local maximum**. The **absolute maximum**, which is the largest value of $f(x)$ in the interval, occurs at the end point $x = 4$ where substitution into $f(x)$ yields $f(4) = 17$. The absolute minimum occurs at $x = 2$.

It is obvious from a plot of $f(x)$ where the maxima and minima are, but a graph may not always be handy. Accomplishing the same thing analytically requires that we examine how the slope changes. If the slope of the curve (df/dx) decreases, changing from positive to negative as x increases across x_0, then $f(x_0)$ is a maximum. Similarly, if the slope increases, changing from negative to positive as x increases across x_0, then $f(x_0)$ is a minimum. That's equivalent to saying that if $d^2f/dx^2 < 0$ at $x = x_0$, $f(x_0)$ is a maximum and if $d^2f/dx^2 > 0$ at $x = x_0$, $f(x_0)$ is a minimum. (That procedure fails when the second derivative equals zero.) In the

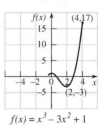

Figure F4

above example, $df/dx = 3x^2 - 6x$ and $d^2f/dx^2 = 6x - 6$. At the extremum $x = 0$, $d^2f/dx^2 = -6$, and the point corresponds to a maximum. At the other extremum $x = 2$, $d^2f/dx^2 = +6$ and the point marks a minimum.

Partial Derivatives

If f is a function of two variables x and y, that is $f(x, y)$, the first-order **partial derivative** of $f(x, y)$ with respect to x is written as $\partial f/\partial x$. It represents the rate-of-change of f as x changes and is found by treating y as a constant and taking the derivative of $f(x, y)$ with respect to x. Similarly, $\partial f/\partial y$ is found by treating x as a constant and taking the derivative of $f(x, y)$ with respect to y. For example, given $f(x, y) = 3x^3 - xy$, $\partial f/\partial x = 9x^2 - y$, whereas $\partial f/\partial y = -x$.

F-2   The Integral

The concept of integration has two different but related meanings in calculus. One has to do with determining the sum or total of something, such as the area under a curve or the volume of a three-dimensional figure or the length of a path. The other refers to the process for finding a function from its derivative. The two conceptions are usually distinguished, respectively, as *definite* and *indefinite* integration. We will deal with the latter first and, after both have been considered, show how the two are intimately related.

Start with the notion that integration is the inverse process to differentiation, just as multiplication is the inverse of division; if you multiply any number by 6 and then divide by 6 you get back to the original number. If you differentiate a function and then integrate the result you get back the original function—or at least something very close to it.

Antiderivatives

Suppose we have a function $f(x)$ and its derivative $F(x) = df/dx$. The function $f(x)$ is said to be the **antiderivative** of $F(x)$ and it can be found by running the procedures that led to the derivative, backward. Each of the above rules for forming the derivative can be used, where appropriate, to compute an antiderivative; obviously, if $F(x)$ is a sine or cosine function you use Eq. (F.12) or (F.10), and so forth. For example, when $f(x) = x^n$, we know that

$$\frac{d}{dx}(x^n) = nx^{n-1} = F(x) \qquad \text{[F.2]}$$

and so if we have $F(x)$ in the form of x^N, it follows that $N = n - 1$. Dividing $F(x)$ by $n = (N + 1)$ yields x^{n-1} and increasing the power by 1 brings us back to $f(x) = x^n$, the antiderivative. For example, given $F(x) = 3x^2 = df/dx$, we have $N = 2 = n - 1$, therefore $n = 3$ and the antiderivative is $f(x) = x^3$.

Notice that because the derivative of a constant is zero, $f(x) = x^2 + 5$ and $f(x) = x^2 - 12$ both have the same derivative, namely $F(x) = 2x$. That means that *the antiderivative is not unique* and we can, at best, construct it to within some additive constant C. In this particular case, the antiderivative is then $f(x) = 2x + C$. The constant C, which must be included, can be determined when we know the appropriate initial conditions of the problem.

The Indefinite Integral

The antiderivative $f(x)$ has come to be known as the **indefinite integral** of $F(x)$.

The indefinite integral is generally a function of the variable and it is written as

$$f(x) = \int F(x)\, dx \qquad (F.20)$$

Equation (F.20) simply means that $df/dx = F(x)$ and so

$$\frac{d}{dx}\left[\int F(x)\, dx\right] = F(x) \qquad (F.21)$$

differentiation undoes integration, and vice versa.

It's helpful to consider $df/dx = F(x)$ in the form of differentials, namely,

$$df(x) = F(x)\, dx$$

and integrating both sides yields

$$\int df(x) = \int F(x)\, dx$$

We have to be careful here because *the antiderivative is not unique* and

$$\int df(x) = f(x) + C$$

In general,

$$\int du = u + C$$

Integrating the differential of a function yields the function plus an arbitrary constant. For example, given $dy/dx = 3x^2$, $dy = 3x^2\, dx$, but $dy = dx^3 = 3x^2\, dx$. Hence

$$y = \int dy = \int d(x^3) = \int 3x^2\, dx = x^3 + C$$

It follows from Eq. (F.4) that if K is any constant

$$\int Kf(x)\, dx = K\int f(x)\, dx \qquad (F.22)$$

Only constants can be moved out in front of an integral.

Similarly, it follows from Eq. (F.5) that

$$\int [f(x) \pm g(x)]\, dx = \int f(x)\, dx \pm \int g(x)\, dx \qquad (F.23)$$

Some Indefinite Integrals

$$\int K\, dx = Kx + C \qquad\qquad \int x^n\, dx = \frac{x^{n+1}}{n+1} + C \quad (n \neq -1)$$

$$\int x^{-1}\, dx = \ln|x| + C \qquad\qquad \int \sin Kx\, dx = -\frac{\cos Kx}{K} + C$$

$$\int \cos Kx\, dx = \frac{\sin Kx}{K} + C \qquad \int e^{Kx}\, dx = \frac{1}{K}e^{Kx} + C$$

$$\int e^{-Kx}\, dx = -\frac{1}{K}e^{-Kx} + C \qquad \int xe^{-Kx}\, dx = -\frac{1}{K^2}(Kx + 1)e^{-Kx} + C$$

$$\int \frac{1}{x^2 + K^2}\,dx = \frac{1}{K}\tan^{-1}\left(\frac{x}{K}\right) + C \qquad \int \frac{1}{\sqrt{K^2 - x^2}}\,dx = \sin^{-1}\left(\frac{x}{K}\right) + C$$

$$\int \frac{1}{\sqrt{K^2 + x^2}}\,dx = \ln(x + \sqrt{K^2 + x^2}) + C \qquad \int \frac{x}{\sqrt{K^2 + x^2}}\,dx = \sqrt{K^2 + x^2} + C$$

$$\int \frac{x}{(K^2 + x^2)^{3/2}}\,dx = -\frac{1}{\sqrt{K^2 + x^2}} + C \qquad \int \frac{K^2}{(K^2 + x^2)^{3/2}}\,dx = \frac{x}{\sqrt{K^2 + x^2}} + C$$

Initial Conditions

The arbitrary constant, C, present after indefinite integration, can be determined uniquely by using whatever constraints there are on the function, provided such information is available. These constraints are called **initial conditions**, and such terminology is especially appropriate for physical situations where we might know, for example, a body's initial position or speed or the initial charge on a capacitor.

As an example let's find the function whose slope at point (x, y) is $3x^2$, provided the curve goes through point $(2, 0)$. There is a whole family of curves with that slope, but only one goes through point $(2, 0)$ and that is the initial condition that distinguishes the specific function and sets the value of C. Given $dy/dx = 3x^2$, $dy = 3x^2\,dx$; integrating both sides leads to

$$\int dy = \int 3x^2\,dx$$

and $y = x^3 + C$ where we combined the two arbitrary constants, one from each integration, into C. But we are given that $y = 0$ when $x = 2$, which means that $C = -8$ and so the desired function is $y(x) = 3x^2 - 8$.

The Definite Integral

Historically, differential calculus began with the problem of finding the tangent to a curve and, using the concept of the limit, arrived at the derivative. By contrast, integral calculus began with the problem of finding the area under a curve and, using the concept of the limit, arrived at the definite integral (see p. 101).

Consider the area under the curve corresponding to $F(x)$, bounded by the vertical lines $x = a$ and $x = b$ as shown in Fig. F.5. Divide that region into n narrow rectangular segments each of width Δx and height determined by $F(x)$ at that location. The area of the jth rectangle is then $A_j = F(x_j)\Delta x$, and the total area is approximated by the sum of all such contributions, namely, $\sum_{j=1}^{n} f(x_j)\Delta x$. The more rectangular segments, the better the sum approximates the area. If now $n \to \infty$ as Δx approaches 0, the limiting value of the summation precisely equals the area and that limiting value is called the **definite integral**

$$\lim_{n \to \infty} \sum_{j=1}^{n} F(x_j)\,\Delta x = \int_a^b F(x)\,dx$$

This definition along with the definition of the antiderivative provides an easy way to evaluate definite integrals rather than using limits of sums.

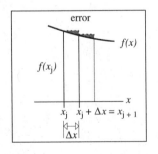

(a)

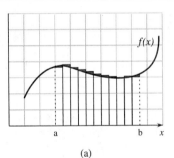

(b)

Figure F5

Given that $F(x)$ is continuous on the interval from $x = a$ to $x = b$ and that $f(x)$ is its antiderivative, it follows that

$$\int_a^b F(x)\, dx = f(x)\,|_a^b = f(b) - f(a) \tag{F.24}$$

We substitute the extreme values of $x = b$ and $x = a$ into $f(x)$ to get $f(b)$ and $f(a)$. This is the **Fundamental Theorem of Calculus**, so called because it unites differential and integral calculus. *To compute the definite integral of $F(x)$ one need only take the difference between the antiderivative of $F(x)$ evaluated at the upper and lower limits*. For example,

$$\int_1^4 2x\, dx = \frac{2x^2}{2}\,\Big|_1^4 = 4^2 - 1^2 = 15$$

where the C's in the antiderivative always cancel so we need not bother writing them down. Generally, ***the definite integral results in a numerical value and not a function***.

Limits and the Indefinite Integral

In physics, where we often know the initial conditions of a problem, it is very common to specify, at least partially, the limits of the indefinite integration in a way that automatically deals with C. For example, once again find the function whose slope at point (x, y) is $3x^2$, where the curve goes through point $(2, 0)$. This time we set the lower limits at the known initial conditions. When x has its initial value of $x_0 = 2$, the corresponding initial value of y is $y_0 = 0$. Similarly, with no particular upper limit on x, y is correspondingly unspecified and we can write the indefinite integrals as

$$\int_{y_0}^y dy = \int_{x_0}^x 3x^2\, dx$$

or more particularly

$$\int_0^y dy = \int_2^x 3x^2\, dx$$

This leads to

$$y - y_0 = x^3\,|_{x_0}^x \tag{F.25}$$

and

$$y - 0 = x^3\,|_2^x = x^3 - 8 \tag{F.26}$$

which is identical to the function determined above.

Compare this procedure using limits to the one discussed earlier that resulted in a constant of integration C. Each indefinite integration produces just such a constant. The integral on the left is $\int dy = y + C_1$. Comparing this with the left side of Eqs. (F.25) and (F.26) tells us that here $C_1 = -y_0 = 0$. The integral on the right is $\int 3x^2\, dx = x^3 + C_2$ and $C_2 = -(x_0)^3 = -(2)^3 = -8$. Combining the two integration constants into one, $y = x^3 + C$ where $C = y_0 - (x_0)^3 = -8$ and $y = x^3 - 8$. The two methods are equivalent.

The Average Value of a Function

The average value of any continuous function $f(x)$ on the interval from $x = a$ to $x = b$ is given by

$$f_{av} = \frac{1}{b - a} \int_a^b f(x)\, dx \tag{F.27}$$

Often we average over time and the variable (x) becomes t, but there are also situations in which a function must be averaged over a distance, area, or volume of space.

Answer Section

CHAPTER 1

Answers to Odd-Numbered Multiple Choice Questions
1. a 3. a 5. c 7. d 9. d 11. d 13. d
15. b 17. a 19. c

Answers to Odd-Numbered Problems
1. $10\,000\,000\,000 = 10^{10}$ 3. (a) 0.010 s
(b) 0.001 s (c) 10 000 s (d) 100 000 000 s
(e) 0.000 001 s 5. 1.00×10^7 mm
7. 76.2 mm 9. 5×10^2 nm 11. 2×10^2
13. 86 400 s 15. 9.46×10^{15} m 17. 8.4×10^{10} bacteria 19. 13 times greater
21. (a) 0.000 001 00 g (b) 0.000 000 000 001 g
(c) 0.100 0 g (d) 0.010 000 g (e) 10 000 g
23. 3×10^8 m/s, 3.00×10^8 m/s, 2.998×10^8 m/s, $2.997\,924\,6 \times 10^8$ m/s
25. $(1.602\,177\,33 \pm 0.000\,000\,48) \times 10^{-19}$ C
27. 0.092 90 m^2 29. $V = 2.2 \times 10^{19}$ m^3
35. 0.393 7 in./cm 37. 1.6×10^3 m
39. 0.092 90 m^2 41. 237×10^{-6} m^3
43. 1×10^{-7} g; 0.3×10^2 kg 45. 1806.8 m
47. 1.00 kg 49. 6.1×10^{-3} g^3

Solutions to Selected Problems
3. (a) 0.010 s; (b) 0.001 s; (c) 10 000 s;
(d) 100 000 000 s; (e) 0.000 001 s.
7. 1.00 in. = 2.54 cm = 25.4 mm;
3.00 in. = 76.2 mm. 11. 1 kg = 1000 g;
(1000 g)/(5 g) = 200 = 2×10^2. 15. (5.88 $\times$ 10^{12} mi)(1.609 km/mi) = 9.46×10^{15} m.
21. (a) 0.000 001 00 g; (b) 0.000 000 000 001 g;
(c) 0.100 0 g; (d) 0.010 000 g; (e) 10 000 g.
25. $(1.602\,177\,33 \pm 0.000\,000\,48) \times$
10^{-19} C. 29. $V = (4/3)\pi R_{\mathbb{C}}^3 = 2.199\,1 \times$
10^{19} m$^3 = 2.2 \times 10^{19}$ m^3.
33.

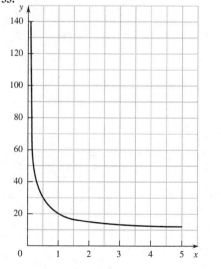

37. (45 hairs/d) (365 1/4 d/y) =
16 436 hairs/y; (16 436) (0.10 m) =
1.6×10^3 m. 41. 1 liter = 1000 cm^3 =
10^{-3} m^3; hence, 1.00 cup = 237×10^{-6} m^3.
45. These add up to 1806.818 m, and the least number of decimal places is one, so the answer is 1806.8 m. 49. (0.0021 g) $\times$
$(655.1 \times 10^3$ g)$(4.41 \times 10^{-6}$ g) =
6.1×10^{-3} g^3.
53.

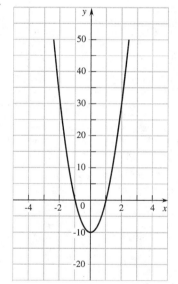

$y_{min} = -10$ $y = 0$ at $x = \pm 1$

$y = 10x^2 - 10$

59.

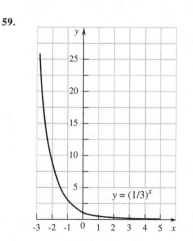

$y = (1/3)^x$

CHAPTER 2

Answers to Odd-Numbered Multiple Choice Questions
1. d 3. b 5. c 7. c 9. c 11. d 13. b
15. a 17. c

Answers to Odd-Numbered Problems
1. 29.5 km/s 3. (a) 1.0×10^{-9} s
(b) 3.336×10^{-6} s 5. 1.000 m/s =
3.600 km/h; 1.000 m/s = 2.237 mi/h;
1.000 m/s = 3.281 ft/s 7. 2.0 m; $v_{max} =$
4.0 m/s, $v_{min} = 0$; from 1.5 s to 3.3 s and from
4.0 s to 5.0 s 9. (a) 60 km/h, 20 km/h
(b) 30 km/h 11. 15 m 13. 0.33 m/s
15. (a) $16t^3$ (b) $-t^{14}$ (c) $-t^{-2}$ (d) t^{-6}
17. $v = 4.8$ m/s, $y = 9.6$ m 19. (a) m/s^2
(b) $y = 10.0$ m (c) $t = 1.43$ s (d) curve is an inverted parabola (e) at $t = 0$ the slope is zero, at $t = 1.00$ s the slope is -9.81 m/s
(f) object is moving increasingly fast in the negative direction. 21. $-2\pi Rv$ 23. 1.7 km
25. 120 km/h 27. 5.65 m; 4.20 m/s
29. $2CBt = v(t)$; at $t = 4.0$ s, $v = 8.0CB$,
$v = 0$ at $t = 0$. 31. $v = 2.0$ s, at 70 m
33. $dV/dt = 4\pi v^3 t^2 = 0.085$ m^3/s.
35. 6.7 km/h 37. $v(t) = 2Ct$ 39. (a) $dy/dt =$
$-(x/y)\,dx/dt$ (b) -0.15 m/s (c) It speeds up as it descends 41. 1.4 m 43. 15 km
45. $D = 10$ m 47. zero, 36 km 49. $s =$
24 m–straight up 51. $\mathbf{C} = 5$ units–due N
53. $v_{av} = 4.3$ m/s, $\mathbf{v}_{av} = 0.30$ m/s–south
55. $\mathbf{s} = 63$ m–72° up from the horizontal
57. (a) zero (b) $\mathbf{v} = 0.4$ m/s–south
59. $v_{av} = 150$ m/s; $\mathbf{v}_{av} = 150$ m/s–upward: round trip $v_{av} = 150$ m/s; $\mathbf{v}_{av} = 0$
61. $\mathbf{v}_{av} = 2.1$ m/s–southeast 63. $t = 3.75$ h; separation is 70.7 km 65. (a) $2 \times$
10^2 m/s–north (b) 2×10^2 m/s–north
67. $\dfrac{ds}{dt} = \dfrac{s_y}{\sqrt{s_x^2 + s_y^2}}\dfrac{ds_y}{dt}$; (b) 4.47 m/s
69. $\dfrac{ds}{dt} = \dfrac{(500.0\text{ m}^2/\text{s}^2)t}{\sqrt{(500.0\text{ m}^2/\text{s}^2)t^2 + 100\text{ m}^2}}$;
22.3 m/s 71. 100 s 73. 10×10^2 m
75. $\theta_1 = 67.38°$; $\theta_2 = 22.62°$ 77. 10 m
79. 12 m 81. 74 km/h 83. (a) $v_H =$
254 m/s (b) $v_V = 159$ m/s (c) $v = 300$ m/s
85. $v_{av} = 1.3 \times 10^3$ km/h 87. $\theta = 11°$;
$v = 10$ m/s 89. (a) no (b) $\theta = 42°$
(c) 35 km/h 91. 22 km/h–3.0° W of N
93. (a) $v_{RS} = 1470$ km/h; $v_{RC} = 0$ (b) $v_{RC} =$
2940 km/h (c) eastward (d) Florida is closer to the equator

Solutions to Selected Problems
1. $v_{av} = l/t = (885$ km)/(30.0 s) = 29.5 km/s.

5. 1.000 m/s $=$

$$\frac{(1.000 \text{ m})/(1000 \text{ m/km})}{(1.000 \text{ s})/(60.00 \text{ s/min})(60.00 \text{ min/h})}$$

$= 3.600$ km/h: 1.000 m/s $=$

$$\frac{(1.000 \text{ m})(3.281 \text{ ft/m})/(5280 \text{ ft/mi})}{(1.000 \text{ s})/(60.00 \text{ s/min})(60.00 \text{ min/h})}$$

$= 2.237$ mi/h: 1.000 m/s $=$

$$\frac{(1.000 \text{ m})(3.281 \text{ ft/m})}{(1.000 \text{ s})} = 3.281$$ ft/s.

9. (a) $v_{av} = l/t = (15 \text{ km})/(0.25 \text{ h}) = 60$ km/h; $v_{av} = l/t = (15 \text{ km})/(0.75 \text{ h}) = 20$ km/h; (b) $v_{av} = l/t = (30 \text{ km})/(1.0 \text{ h}) = 30$ km/h. **15.** (a) $dy/dt = 16t^3$; (b) $dx/dt = -(15t^{14})/15 = -t^{14}$; (c) $dx/dt = -t^{-2}$ (d) $dy/dt = -(-5t^{-6})/5 = t^{-6}$. **23.** Assuming the time for the light to reach you is negligible, $l = (3.3 \times 10^2 \text{ m/s}) \times (5.0 \text{ s}) = 1.7$ km. The error incurred by not including the time for light to travel is a few millimeters. **29.** $dz/dt = 2CBt = v(t)$; at $t = 4.0$ s, $v = 8.0CB$; $v = 0$ at $t = 0$. **35.** L is the one-way distance; $v_{av} = l/t = 2L/[L/(10 \text{ km/h}) + L/(5.0 \text{ km/h})] = 2/[1/(10 \text{ km/h}) + 1/(5.0 \text{ km/h})] = 20/(3.0 \text{ km/h}) = 6.7$ km/h. **37.** $l = Ct^2$; $(l + \Delta l) = C(t + \Delta t)^2 = C[t^2 + 2t\Delta t + (\Delta t)^2]$; $\Delta l = Ct^2 + 2Ct\Delta t + C(\Delta t)^2 - 1$; $v_{av} = \Delta l/\Delta t = [2Ct\Delta t + C(\Delta t)^2]/\Delta t = 2Ct + C\Delta t$. Letting $\Delta t \rightarrow 0$, in $v_{av} = 2Ct + C\Delta t$, the second term vanishes and $v = 2Ct$. **43.** This is a 3-4-5 right triangle, so the displacement is 15 km. **47.** Zero, 36 km. **53.** $v_{av} = l/t = (43 \text{ m})/(10 \text{ s}) = 4.3$ m/s; $\mathbf{v}_{av} = s/t = (3.0 \text{ m–south})/(10 \text{ s}) = 0.30$ m/s–south. **57.** (a) The slopes are zero, as are both velocities. (b) The slope is -2.0 m in 5.0 s or $v = (-2.0 \text{ m})/(5.0 \text{ s}) = -0.4$ m/s and so $\mathbf{v} = 0.4$ m/s–south. **61.** From Eq. (2.6), $\mathbf{v}_{av} = (\mathbf{s}_f - \mathbf{s}_i)/\Delta t$; $\Delta s = \sqrt{(3.0 \text{ m})^2 + (3.0 \text{ m})^2}$–southeast $= 4.24$ m–southeast; $\mathbf{v}_{av} = 2.1$ m/s–southeast. **65.** (a) $\mathbf{v}_{av} = (600 \text{ m})/(3 \text{ s})$–north $= 2 \times 10^2$ m/s–north; (b) also 2×10^2 m/s–north. **71.** Each runs 500 m in time t at speed 5.00 m/s; $(500 \text{ m})/(5.00 \text{ m/s}) = 100$ s. Alternatively, together they cover the track at a speed of 5.00 m/s $+ 5.00$ m/s $= 10.0$ m/s; hence, $(1000 \text{ m})/(10.0 \text{ m/s}) = 100$ s. **75.** $\tan \theta_1 = 12/5 = 2.400$, $\theta_1 = 67.38°$; $\tan \theta_2 = 5/12 = 0.4167$, $\theta_2 = 22.62°$. **79.** $\tan 25° = h/(25 \text{ m}) = 0.466$; $h = 11.66$ m, or 12 m. **85.** $\cos \theta = R_\theta/R$; $R_\theta = R \cos \theta = (6400 \text{ km}) \cos 41° = 4830$ km; the circumference is $2\pi R_\theta = 3.0 \times 10^4$ km; $v_{av} = (3.0 \times 10^4 \text{ km})/(24 \text{ h}) = 1.3 \times 10^3$ km/h, or 7.9×10^2 mi/h. **89.** (a) No. Those velocity vectors cannot form a right triangle because the hypotenuse ($\mathbf{v}_{BW}$) must be larger than either of the other sides, and here $v_{BW} = v_{WE}$. (b) $\mathbf{v}_{BE}$ cuts directly across the river. Head at $\mathbf{v}_{BW}$ somewhat upstream at an angle θ such that $\mathbf{v}_{BE} = \mathbf{v}_{BW} + \mathbf{v}_{WE}$, $\sin \theta = (20 \text{ mi/h})/(30 \text{ mi/h}) = 0.667$, $\theta = 42°$. (c) $v_{BE} = [(30 \text{ mi/h})^2 - (20 \text{ mi/h})^2]^{1/2} = 22$ mi/h $= 35$ km/h. **93.** (a) $v_{RS} = 1470$ km/h; $\mathbf{v}_{RC} = \mathbf{v}_{RS} + \mathbf{v}_{SC} = 1470$ km/h–west $+ 1470$ km/h–

east $= 0$. (b) $\mathbf{v}_{RC} = 1470$ km/h–east $+ 1470$ km/h–east $= 2940$ km/h–east. (c) Eastward. (d) Florida is closer to the equator and is moving faster than Maine.

CHAPTER 3
Answers to Odd-Numbered Multiple Choice Questions
1. a **3.** c **5.** b **7.** b **9.** e **11.** a **13.** d **15.** c **17.** d

Answers to Odd-Numbered Problems
1. 10 m/s² **3.** 0.17 m/s²–north **5.** 0.23 m/s² **7.** 18 m/s² **9.** 5.0 m/s² **11.** 20 m/s² **13.** At $t = 0$, $v = 25.0$ m/s; at $t = 1.0$ s, $v = 20.0$ m/s, $a = -5.00$ m/s², the acceleration is constant **15.** $a_T = 10.0$ m/s² **17.** $a_T(t) = 4.0$ m/s² $- 2t(1.0 \text{ m/s}^3)$ **19.** 3.5 m/s² **21.** (a) $a_{av} = 0$ (b) $a = -1.1$ m/s² **23.** -17 km/s² **25.** $a = A + 2Bt$ **27.** (a) $v(t) = 0.30t^2 - 1.20t + 0.90$ (b) 0, -0.3 m/s, and 0.90 m/s at $t = 1.0$ s and 3.0 s **29.** $a = g$ **31.** $a_T = (12 \text{ m/s}^3)t - 10$ m/s²; 26 m/s² **33.** zero at $t = (1/6)$s **35.** $a = 10$ cm/s², $\theta = 37°$ **37.** $a = (0.040 \text{ s}^{-2})x + 8.0$ m/s² **39.** 1×10^2 m/s² **41.** $v = 4.7$ m/s **43.** $(a_T)_{av} = -1.09$ m/s² **45.** $v_0 = 48$ m/s **47.** $a = -0.9 \times 10^6$ m/s² **49.** 0.2 s **51.** $s_3 - s_2 = 1.6$ m **53.** 244.5 m and they don't collide **55.** (a) 70 m (b) 90 m **57.** 0.1×10^2 m **59.** 10 m/s² **61.** 74 m **63.** $s = 125$ m **65.** 0.12 km **67.** 3.1 m/s; 10 ft/s; 7.0 mi/h **69.** -9.8 m/s, -20 m/s; -49 m/s, -98 m/s; -4.9 m, -20 m, -0.12 km, -0.49 km **71.** $v = 423$ m/s $= 947$ mi/h **73.** s versus t graph is a parabola. s versus t^2 graph is a straight line **77.** bullet falls 4.90 cm below target center **79.** 20 m **81.** the equation must be wrong **83.** 4.7 m **85.** -4.5 m **87.** $v(t) = \frac{1}{2}Ct^2 + 12$ m/s **89.** $v_0 = 4.9$ m/s **91.** 127.5 m **93.** $a = -gs_{max}/s_a$ **97.** 11 m/s **99.** 24.6 m, $v = 50.8$ m/s **101.** proof **103.** 125 m/s: 208 m **107.** 10.2 m/s **109.** $s_B = 6.1$ m **111.** $24t^4 - 255t^3 + 901t^2 - 400 = 0$

Solutions to Selected Problems
3. Taking north as positive, $a_{av} = \Delta v/\Delta t = [(+10 \text{ m/s}) - (-10 \text{ m/s})]/(120 \text{ s}) = 0.17$ m/s²–north. **7.** $a_{av} = \Delta v/\Delta t = (244 \times 0.4470 \text{ m/s})/(6.2 \text{ s}) = 18$ m/s². **11.** Draw a line on the curve from $t = 0$ to $t = 3$ s: the rise is 6(10 m/s); the run is 3.0 s; the slope is (60 m/s)/(3.0 s) $= 20$ m/s². **15.** $l = 10t + 5t^2$; $v = 10 + 10t$; $a_T = 10.0$ m/s². **23.** $a_{av} = \Delta v/\Delta t = (-60 \text{ km/h} \times 1000 \text{ m/km} \div 3600 \text{ s/h})(1 \text{ s}/1000) = -17$ km/s². **25.** $v = At + Bt^2$; $(v + \Delta v) = A(t + \Delta t) + B(t + \Delta t)^2 = At + A\Delta t + Bt^2 + B(\Delta t)^2 + B2t\Delta t$; $\Delta v = A\Delta t + B(\Delta t)^2 + 2Bt\Delta t$; $a_{av} = \Delta v/\Delta t = A + 2Bt + B\Delta t$; as $\Delta t \rightarrow 0$, $a = A + 2Bt$. **29.** $dv/dt = (\frac{1}{2}2g \, ds/dt) \times (k + 2gs)^{-1/2} = g$. **37.** $a = (dx/dt)(dv/dx) = v \, dv/dx = (0.20 \text{ s}^{-1})v = (0.040 \text{ s}^{-2})x + 8.0$ m/s². **41.** $v^2 = v_0^2 + 2as = (1.5 \text{ m/s})^2 + 2(1.0 \text{ m/s}^2)(10 \text{ m}) = 22$ m²/s²; $v = 4.7$ m/s. **49.** From Eq. (3.12), $s =$

$v_0 t_R - v_0^2/2a$; $t_R = (23.3 \text{ m})/(16.7 \text{ m/s}) + (16.7 \text{ m/s})/2(-7 \text{ m/s}^2) = 0.2$ s. **53.** To stop, the Express needs $s_E = -v_0^2/2a = -(26.67 \text{ m/s})^2/2(-4 \text{ m/s}^2) = 88.9$ m; the Flyer needs $s_F = -v_0^2/2a = -(30.56 \text{ m/s})^2/2(-3 \text{ m/s}^2) = 155.6$ m; for a total of 244.5 m and they don't collide. **59.** It covered 20 m in 2.0 s; $s = v_i t + \frac{1}{2}at^2$; the initial relative speed was zero; $a = 2s/t^2 = 10$ m/s². **65.** His initial speed with respect to the train is 50 km/h, and he passes it as if it were standing still. Hence $s = v_0 t + \frac{1}{2}at^2$, 500 m $= (13.89 \text{ m/s})t + \frac{1}{2}(10 \text{ m/s}^2)t^2$, and using the quadratic formula, $t = [-(13.89 \text{ m/s}) \pm \sqrt{(13.89 \text{ m/s})^2 - 4(5.0 \text{ m/s}^2)(-500 \text{ m})}]/2(5.0 \text{ m/s}^2) = (-13.89 \text{ m/s} \pm 100.96 \text{ m/s})/(10 \text{ m/s}^2) = 8.7$ s; the train travels $s = (13.89 \text{ m/s})(8.7 \text{ s}) = 0.12$ km. **71.** Taking down as positive, $v^2 = v_0^2 + 2as = 0 + 2(9.81 \text{ m/s}^2)(9144 \text{ m})$; $v = 423$ m/s $= 947$ mi/h. **75.** $v_{0H} = v_0 \cos \theta$, $v_{0V} = v_0 \sin \theta$, hence Eq. (3.16) yields the desired expression. **79.** (Taking down as positive) $s_V = \frac{1}{2}gt^2 = \frac{1}{2}(9.81 \text{ m/s}^2)(2.0 \text{ s})^2 = 19.6$ m, or to two figures, 20 m. **89.** $s = v_0 t + \frac{1}{2}at^2$; $0 = v_0(1.0) + \frac{1}{2}(-9.81 \text{ m/s}^2)(1.0 \text{ s})^2$; $v_0 = 4.9$ m/s. **93.** The lift-off speed is v_l, where $v_l^2 = 2as_a$. During the deceleration, $0 = v_l^2 + 2gs_{max}$ or $v_l^2 = -2gs_{max} = 2as_a$; $a = -gs_{max}/s_a$. **99.** To avoid the quadratic, use Eq's. (3.10) and (3.8) applied to the vertical motion, where $v_{0V} = (40.0 \text{ m/s}) \times \sin 60.0° = 34.64$ m/s. Taking down as positive, $v_V^2 = (34.64 \text{ m/s})^2 + 2(9.81 \text{ m/s}^2)(50.0 \text{ m})$; and $v_V = 46.70$ m/s. $s_V = 50.0$ m $= \frac{1}{2}(v_{0V} + v_V)t = (40.67 \text{ m/s})t$ and $t = 1.23$ s. It strikes the water a distance downrange of $v_H t = (40.0 \text{ m/s})(\cos 60.0°) \times (1.23 \text{ s}) = 24.6$ m. It hits at a speed of $v_f = (v_H^2 + v_V^2)^{1/2} = 50.8$ m/s. **105.** $ds/dt = v$ and so $a = v \, dv/ds$; $\int_0^v v \, dv = \int_0^s a \, ds$; $\frac{1}{2}(v^2 - v_0^2) = a(s - 0)$. **109.** Down is positive; over his height $s = v_0 t + \frac{1}{2}at^2$; 2 m $= v_0(0.20 \text{ s}) + \frac{1}{2}(9.81 \text{ m/s}^2)(0.20 \text{ s})^2$; $v_0 = 9.02$ m/s at top of his head; at ground $v = v_0 + gt = 9.02$ m/s $+ (9.81 \text{ m/s}^2) \times (0.20 \text{ s}) = 10.98$ m/s; for total fall, $v^2 = v_0^2 + 2as_B$; $(10.98 \text{ m/s})^2 = 0 + 2(9.81 \text{ m/s}^2)s_B$; $s_B = 6.1$ m. **111.** Using the Pythagorean Theorem, we obtain $(20 \text{ m})^2 = (v_H t)^2 + (v_{0V}t + \frac{1}{2}gt^2)^2$, $v_{0H} = v_0 \cos \theta = (30 \text{ m/s}) \times \cos 60°$ and $v_{0V} = v_0 \sin \theta = (30 \text{ m/s}) \sin 60°$. $24t^4 - 255t^3 + 901t^2 - 400 = 0$.

CHAPTER 4
Answers to Odd-Numbered Multiple Choice Questions
1. a **3.** a **5.** c **7.** c **9.** e **11.** c **13.** b **15.** c **17.** a **19.** c

Answers to Odd-Numbered Problems
1. $\mathbf{v}_M = 10$ m/s–horizontal **3.** 1.0000 m **5.** 8.36 km compared to 110 m **7.** 16.7 km **9.** 100 N **11.** 100 N on each person **13.** $F = 298$ N at $\theta = +35°$ **15.** 0.12 kN at $\theta = 54°$ **17.** (a) 24.0 m (b) 96.0 m

21. 500 N at 76.9° up from x-axis
23. one due east; the other two, ±60°
25. 400 N–straight up   27. 37 m
29. -10 N·s   31. unchanged, 10 kg·m/s
33. -2 MN   35. 0.012 s   37. 3.3 m/s–south
39. -0.12 m/s   41. $\mathbf{F}(t) = (16t + 4.0)$–north
43. 52.5 m/s in the positive x-direction
45. 270 m/s   47. 1.53 kN   49. 0.3 N
51. 203 N   55. $s_h = 1.7$ m   57. 42 MN
59. 1.0 kN   61. -1.00 kN   63. 18 m/s
65. 0.020 2 m/s opposite to snowball
67. 83.9 kN   69. 16.25 kN·s   71. $\Delta p = 10 \times 10^4$ kg·m/s   73. proof   75. -0.12 MN
77. $v_{Nf} = -0.100$ m/s, $v'_{Sf} = 0.198$ m/s, $v''_{Sf} = 0.400$ m/s

Solutions to Selected Problems

1. At maximum altitude, $v_V = 0$, hence $\mathbf{v}_M = 10$ m/s–horizontal.   5.   Using Eq. (3.15), $s_p = -v^2/2g = 8.36$ km, which is tremendously different from 110 m! The difference is due to drag, which is large because of the high speed and effective because of the small mass.   11. The force is 100 N on each person.   13. $F_{1x} = (100$ N$) \times \cos 45° = 70.7$ N; $F_{2x} = (200$ N$) \cos 30° = 173.2$ N; $F_{1y} = (100$ N$) \sin 45° = 70.7$ N; $F_{2y} = (200$ N$) \sin 30° = 100$ N; $F^2 = (243.9$ N$)^2 + (170.7$ N$)^2$; $F = 298$ N; at $\theta = \tan^{-1} 170.7/243.9 = +35°$.   17. Just to use something other than Eq. (3.15), $v_{iV} = (30.7$ m/s$) \sin 45° = 21.71$ m/s, take up as positive; $v_{fV} = v_{iV} + gt$, $0 = 21.71$ m/s $+ (-9.81$ m/s$^2)t_p$; $t_p = 2.21$ s. $s_{Hp} = v_H t_p = (21.7$ m/s$)(2.21$ s$) = 48.0$ m. (a) $s_p = \frac{1}{2}(v_{iV} + v_{fV})t_p = \frac{1}{2}(21.7$ m/s$)(2.21$ s$) = 24.0$ m above the launch height. (b) $s_R = 2s_{Hp} = 96.0$ m.   21. There is 90° between the two vectors, which form a 3-4-5 right triangle, so the resultant is 500 N at an angle of 36.9° + 40° up from the x-axis.   25. By symmetry, the horizontal components of the forces cancel, leaving only 4(200 N) $\times \cos 60° = 400$ N acting straight up.
29. Taking the direction of motion as positive, $\Delta p = m \Delta v = (1.0$ kg$) \times (-10$ m/s$) = -10$ N·s.   33. To find F_{av}, the force exerted on the gas, take the direction of the exhaust as positive and use $F_{av} \Delta t = m \Delta v$; $F_{av} = (1000$ kg$)(2$ km/s$)/(1.000$ s$) = 2$ MN. Thrust is -2 MN.   37. The initial momentum of the boat-person system is zero, $\mathbf{p}_i = \mathbf{p}_f = 0$, and $m_p v_{pf} + m_b v_{bf} = 0$; taking motion north as positive, $(50$ kg$)(+10$ m/s$) = -(150$ kg$)v_{bf}$; $v_{bf} = -3.3$ m/s; or 3.3 m/s–south.
43. $\Delta p_x = \int F_x dt = (2.00$ N/s$^2)t^3$; over the interval from 0 to $t = 5.00$ s, $\Delta p_x = 250$ N·s, $\Delta p_x/m = \Delta v_x = 12.5$ m/s; and the final velocity is 52.5 m/s in the positive x-direction.   49. 2 oz = 2/16 lb; $m = (2/16$ lb$) \times (0.453\,6$ kg/lb$) = 0.056\,7$ kg; $F_{av} = m \Delta v/\Delta t = (0.056\,7$ kg$)(0.50$ m/s$)/(0.1$ s$) = 0.284$ N, or 0.3 N.   53. $F_{av} = m \Delta v/\Delta t$; impact speed from Eq. (3.10), $v_I = \sqrt{2gs_h}$. The duration of the collision follows from Eq. (3.8), $s_c = \frac{1}{2}v_I \Delta t$, and so $\Delta t = 2s_c/v_I$; hence, $F_{av} = m \Delta v/\Delta t = mv_I/(2s_c/v_I) = mv_I^2/2s_c = mgs_h/s_c$.

59. Taking the initial direction as positive, $F_{av} = m \Delta v/\Delta t$; from Eq. (3.8), $s = \frac{1}{2}v_i \Delta t$, $\Delta t = 2(0.20$ m$)/(200$ m/s$) = 2.0$ ms; $F_{av} = (0.010$ kg$)(200$ m/s$)/(0.002\,0$ s$) = 1.0$ kN.   63. $p_i = (10\,000$ kg$)(25$ m/s$) = 2.0 \times 10^5$ kg·m/s $= p_f = (10\,000$ kg $+ 900$ kg$)v_f$; $v_f = 18.3$ m/s, or 18 m/s.   67. Let's see what a is first, then t, and then F. $v^2 = v_0^2 + 2as$; $a = (69.5$ m/s$)^2/(914.4$ m$) = 2.64$ m/s^2; $v = v_0 + at$; $t = (69.5$ m/s$)/(2.64$ m/s$^2) = 26.3$ s; m $= 31\,751$ kg; $F = m \Delta v/\Delta t = (31\,751$ kg$)(69.5$ m/s$)/(26.3$ s$) = 83.9$ kN. We could have started with Eq. (3.8).
71. $\Delta p = \int F(t) dt = (10$ kN$)t \vert_{10.0\,s}^{20.0\,s} + 2\pi(4.0$ kN$) \cos 2\pi t \vert_{10.0\,s}^{20.0\,s} = 10 \times 10^4$ kg·m/s.
75. 120 mi/h $= 53.64$ m/s; if v_T is his terminal speed and s_D the depth of the crater, Eq. (3.8) yields $\Delta t = 2s_D/v_T = 0.039\,8$ s. $F_{av} = m \Delta v/\Delta t = (90$ kg$)(-53.64$ m/s$)/(0.039\,8$ s$) = -0.12$ MN.   77. $p_{Ni} + p_{ai} = 0 = p_{Nf} + p_{af} = (100$ kg$)v_{Nf} + (0.500$ kg$) \times (20.0$ m/s$)$; $v_{Nf} = -0.100$ m/s—he moves in the opposite direction to the asteroid, whose direction is taken as positive; she catches it and so $p'_{ai} = p'_{af} + p'_{Sf}$; $(0.500$ kg$) \times (20.0$ m/s$) = v'_{Sf}(m_a + m_S) = v'_{Sf}(50.5$ kg$)$; $v'_{Sf} = 0.198$ m/s. Now she throws it back, $p''_{af} + p''_{Sf} = p''_{Si} + p''_{ai}$; $(50.5$ kg$) \times (0.198$ m/s$) = (0.500$ kg$)(-20.0$ m/s$) + (50.0$ kg$)v''_{Sf}$; $v''_{Sf} = 0.400$ m/s.

CHAPTER 5

Answers to Odd-Numbered Multiple Choice Questions

1. c   3. d   5. e   7. c   9. b   11. c   13. d
15. d   17. a   19. c

Answers to Odd-Numbered Problems

1. -71 N   3. >4.9 N   5. 1.5 F$_w$ 7. 592 N, 392 N   9. 0.586 g 11. F_i; $F_i/2$; $5F_i$; $a(t) = F(t)/m$ 13. $F(t) = m(6t - 12)$ 15. $F = 9.8$ N
17. 508 N, upward   19. 4.81 m/s^2 21. $\theta = 22°$ 23. $F_{T1} = 1.00$ kN, $F_{T2} = 475$ N, $F_f = 125$ N   25. $v = 12$ m/s   27. $a = 1.7$ km/s^2, 7.8×10^{-4} N   29. $v(\tau) = -\tau(F_i/m)$
31. $F = (1.20$ kN/s$)t + (1.00$ kN$)$, $a(10.0$ s$) = 130$ m/s^2 33. $v(t) = (60.0$ m/s$^3)t^2 + (40.0$ m/s$^2)t + 2.00$ m/s   35. $F_{T1} = 0.03$ kN, $F_{T3} = 0.02$ kN, $a = \frac{1}{4}g$, the tensions are internal to the 3-body system, the weights are external   37. $g_M = 3.4$ m/s^2 39. proof
41. $v = [(mg/K) + (v_0^2 - mg/K)e^{-2(K/m)y}]^{1/2}$
43. 17 m/s   45. $F_C = 6.6 \times 10^2$ N, friction
47. $F_C = 2.1$ kN   49. $R = 9.5 \times 10^2$ m
51. $a_C = 1.8 \times 10^6$ m/s^2 53. 0.12 kN
55. $F_N = m(v^2/r + g)$ 57. 19.2 m/s
59. (a) time to go once around, $T = 55$ s, or 0.018 rotations per second   (b) $a_C = 4\pi^2 r/T^2$ 61. 99.7 N   63. $\mu_s = 0.53$
65. $a = \mu_s g$ 67. 0.31   69. crate slides toward back of truck   71. 0.10   73. 20 m
75. 3 s   77. $\mu_k = 0.070$ 79. $\mu_k = 0.30$
81. (a) $a = 0$ (b) $a = 1$ m/s^2 83. $\theta = 0.2 \times 10^2$ degrees   85. $\mu_k = 0.11$

Solutions to Selected Problems

1. $F = ma = 0$; $F_f = -F \cos 45° = -(100$ N$)(0.707) = -71$ N.   7. $+\uparrow \Sigma F_V$

$= ma = F_N - F_W$; $F_N = ma + F_W = m(a + g) = (40.0$ kg$)(14.81$ m/s$^2) = 592$ N. $F_W = mg = 392$ N.   11. At $t = 0$, $F = F_i$; at $t = (\tau/4)$, $F = F_i/2$; at $t = \tau$, $F = 5F_i$; $a(t) = F(t)/m$, $v(\tau)$ can be gotten by integrating $a(t)$ from 0 to τ.   17. $+\downarrow \Sigma F_V = ma = F_W + F_f$; $m(2.00$ m/s$^2) = mg + F_f$; $F_f = -(7.81$ m/s$^2)(65.0$ kg$) = -508$ N, or 508 N, upward.   23. The tension in the front rope is 1.00 kN. For the first barge (taking the pull of the tug to be positive and to the right), $\overset{+}{\rightarrow}\Sigma F_{H1} = m_1 a = 1.00$ kN $- F_{T2} - F_f = (4.00 \times 10^3$ kg$)(0.100$ m/s$^2)$; for the second barge, $\overset{+}{\rightarrow}\Sigma F_{H2} = m_2 a = F_{T2} - F_f = (3.50 \times 10^3$ kg$)(0.100$ m/s$^2)$. Adding the two equations, $-2F_f = (7.50 \times 10^3$ kg$) \times (0.100$ m/s$^2) - 1.00$ kN; $F_f = +125$ N, opposing the motion; $F_{T2} = 475$ N.
27. $s_R = 12$ in. $= 0.304\,8$ m $= [2v_i^2/(9.8$ m/s$^2)] \cos \theta \sin \theta$; $v_i = 1.729$ m/s; (a) $a = \Delta v/\Delta t = (1.729$ m/s$)/(1.0 \times 10^{-3}$ s$) = 1.7$ km/s^2; (b) $F = ma = (4.5 \times 10^{-7}$ kg$) \times (1.73$ km/s$^2) = 7.8 \times 10^{-4}$ N.   29. $v(\tau) = (1/m)\int_0^\tau F(t) dt = [F_i t/m - (2.0t^2/\tau)(F_i/m)]_0^\tau = (F_i \tau - 2.0\tau F_i)(1/m) = -\tau F_i/m$.   35. For mass-1, $+\downarrow\Sigma F_{V1} = m_1 a = F_{W1} - F_{T1}$; for mass-2, $\overset{+}{\rightarrow}\Sigma F_{H2} = m_2 a = F_{T1} - F_{T2}$; for mass-3, $+\uparrow\Sigma F_{V3} = m_3 a = F_{T2} - F_{W3}$; adding all three of these, $a(m_1 + m_2 + m_3) = F_{W1} - F_{W3} = g(m_1 - m_3)$; (b) $a = g(2$ kg$)/(8$ kg$) = \frac{1}{4}g$; (c) the tensions are internal to the 3-body system, the weights are external; (a) $m_1 a = m_1 g - F_{T1}$; $F_{T1} = m_1(g - a) = 29.4$ N $= 0.03$ kN. $m_3 a = F_{T3} - F_{W3}$, $F_{T3} = m_3(a + g) = 24.5$ N $= 0.02$ kN.
39. $mg - Kv = m \, dv/dt$; but $mg = Kv_T$ and so $v_T = gm/K$; hence $g - gv/v_T = dv/dt$ and $\int_0^t g \, dt = \int_0^v \dfrac{dv}{(1 - v/v_T)}$; using the fact that $\int \dfrac{dx}{ax + b} = \dfrac{1}{a}\ln(ax + b)$; $gt = -v_T \ln(1 - v/v_T)$; $mg = Kv_T$; $-Kt/m = \ln(1 - v/v_T)$; $e^{-Kt/m} = 1 - v/v_T$; and the required answer follows immediately.
43. $F_C = mv^2/r$, $v = \sqrt{(20$ N$)(0.355\,6$ m$)/(0.025$ kg$)} = 17$ m/s.
47. The distance traveled once around is $2\pi r$, hence $v = 2(2\pi r)/(1.0$ s$)$; $r = 1.828$ m; $v = 22.98$ m/s; $F_C = mv^2/r = (7.257$ kg$)v^2/r = 2096.8$ N, or 2.1 kN.
51. $a_C = v^2/r$; $r = (0.100$ m$) \sin 30° + 0.050$ m $= 0.10$ m; $v = 2\pi r(40\,000)/60 = 418.88$ m/s; $a_C = 1.8 \times 10^6$ m/s^2.   55. Using Fig. 5.24 with F_T replaced by F_N, we get $F_N = m(v^2/r + g \cos \theta)$; at the bottom, $\theta = 0$, $\cos \theta = 1$; $F_N = m(v^2/r + g)$.   59. (a) $a_C = v^2/r = 1.0g$; $v^2 = \frac{1}{2}(1500$ m$)g$; $v = 85.76$ m/s; the cylinder must go once around a distance $(2\pi r)$ in a time $(2\pi r)/v = T = 54.9$ s, or 55 s; (b) $a_C = v^2/r = (2\pi r/T)^2/r = 4\pi^2 r/T^2$, where the time to go once around is the same throughout the station. Hence, the simulated gravity varies linearly with r, decreasing with distance "up" from the floor.   61. $+\downarrow\Sigma F_V = ma = F_G - F_N$; F_N is equal and opposite to the effective weight; F_G is the actual gravitation force downward, namely, 100 N.

$F_N = F_G - ma_C = F_G - (F_G/g)v^2/R_\oplus = 100 \text{ N} - 0.345 \text{ N} = 99.7 \text{ N}$.
65. $F = ma$; $a = F_f/m = mg\mu_s/m = \mu_s g$. **69.** For no slipping, $+\swarrow\Sigma F_\parallel = 0 = F_W \sin\theta - F_f(\text{max})$; and $\mu_s = \tan\theta$, 0.3 = $\tan\theta$; $\theta = 16.7°$; hence, the crate slides toward the back of the truck. **73.** $v = 50.0 \text{ km/h} = 13.89 \text{ m/s}$; to find the stopping distance, $v^2 = v_0^2 + 2as$; $s = -v_0^2/2a$; where $a < 0$ is determined from $F_f(\text{max}) = \mu_s mg = ma$; hence, $a = \mu_s g$ and so $s = -v_0^2/2(-\mu_s g) = 20 \text{ m}$. **77.** $+\swarrow\Sigma F_\parallel = 0 = F_W \sin\theta - F_f$; $\tan\theta = \mu_k = 0.070$. At $-10°\text{C}$, the water film that usually forms via melting under the skis would not. **81.** (a) $F_H = F\cos\theta = (300 \text{ N})(0.866) = 259.8 \text{ N}$, which must exceed $F_f(\text{max}) = 0.5[(50 \text{ kg})g + (300 \text{ N}) \times \sin 30°] = 320 \text{ N}$, but it does not, so there is no acceleration; (b) $F_f(\text{max}) = 256 \text{ N}$—it moves! $\overset{+}{\rightarrow}\Sigma F_H = ma = 259.8 \text{ N} - 0.3(640.3 \text{ N}) = 67.7 \text{ N}$ and $a = 1.4 \text{ m/s}^2$, or because the friction coefficients are known to only one figure, $a = 1 \text{ m/s}^2$. **85.** $v^2 = 2as$, $a = (8.33 \text{ m/s})^2/2(30.0 \text{ m}) = 1.158 \text{ m/s}^2$; $v = at$, $t = (8.33 \text{ m/s})/(1.158 \text{ m/s}^2) = 7.196 \text{ s}$; the box accelerates with respect to the truck at a_b, where $s_b = \frac{1}{2}a_b t^2 = 1.00 \text{ m}$, and $a_b = 0.0386 \text{ m/s}^2$. If there were no friction, it would have accelerated back at 1.158 m/s^2; hence, $(1.158 \text{ m/s}^2 - 0.0386 \text{ m/s}^2)m = F_f = \mu_k mg$, $\mu_k = 0.11$.

CHAPTER 6

Answers to Odd-Numbered Multiple Choice Questions

1. d **3.** d **5.** c **7.** e **9.** b **11.** a **13.** e **15.** d **17.** d **19.** c

Answers to Odd-Numbered Problems

1. (a) 250 N (b) 200 N **3.** (a) 30 N (b) 38 N, 2.0 N **5.** they slide equally **7.** $F_{Tb} = 0$, $F_{Tl} = 98$ N **9.** $\theta = 38°$, $F_T = 16$ N **11.** 39.7 N **13.** $F_T = 50.0$ N, 100 N **15.** $F_{BC} = 0$, $F_{DE} = 0$, $F_{FG} = 0$ **17.** 6.7 N **19.** $F_T = 3.1$ kN **21.** 75°, 193 N **23.** $\mu_s = 0.93$ **25.** $F_{RB} = 2.00$ kN, $F_{RA} = 1.73$ kN $= F_{RC}$, $F_{RD} = 3.00$ kN **27.** 100 N·m **29.** 140 N **31.** $F_R = 355.9$ N, $x = 0.7317$ m **33.** $F_{RB} = 6000$ N, $F_{RA} = 2000$ N **35.** $F_{N1} = F_{N2} = 69.3$ N **37.** 1.0 kg **39.** $F_{RD} = 360$ N, $F_{RC} = 80.0$ N, $F_{RA} = 40$ N **41.** $F_n = 0.83$ kN **43.** $F_f = 11$ N **45.** $F_{AH} = -1.5$ kN (to the left); $F_{AV} = -0.50$ kN (down) **47.** midface in the glue 8 cm from the end **49.** $x_{cg} = 0.063$ m, $y_{cg} = 0.063$ m **51.** $F_{T2} = 100$ N, $F_W = 173$ N **53.** yes **55.** yes **59.** $F_{RG} = 0.31$ kN at 74° **61.** $F_{Nh} = 0.21$ kN, $F_{Nt} = 0.11$ kN **63.** $x_{cg} = F_{N2}h/(F_{N1} + F_{N2})$ **65.** $F_T = 90.4$ N

Solutions to Selected Problems

3. (a) $(2.0 \text{ N/m})(20 \text{ m})(3/4) = 30$ N. (b) 1.0 m from the top, the tension is 38 N; 1.0 m from the bottom, it's 2.0 N. **5.** For each one,

$+\swarrow\Sigma F_\parallel = F_W \sin\theta - \mu_s F_W \cos\theta = 0$; and $\mu_s = \tan\theta$ independent of weight, so they tend to slide equally. **9.** $+\uparrow\Sigma F_V = (20 \text{ N})\cos 30° + F_T \cos 30° - 30 \text{ N} = 0$; $\overset{+}{\rightarrow}\Sigma F_H = F_T \sin\theta - (20 \text{ N})\sin 30° = 0$; hence, $F_T \cos\theta = 12.679$ N and $F_T \sin\theta = 10$ N. We could divide these and get $\sin\theta/\cos\theta = 0.7887 = \tan\theta$ and $\theta = 38°$, or square and add them, since $\sin^2\theta + \cos^2\theta = 1$, and get $F_T^2 = (12.697 \text{ N})^2 + (10 \text{ N})^2$; $F_T = 16$ N. **15.** Since BC is perpendicular to AC and CE at joint C, there can be no vertical force in BC; hence, force in BC is zero. The same is true at joint D (so $F_{DE} = 0$) and at joint G (so $F_{FG} = 0$). **19.** $+\uparrow\Sigma F_V = 2F_T \sin 5.0° - 533.8 \text{ N} = 0$; $F_T = 3.1$ kN. **25.** On the small ball, $\overset{+}{\rightarrow}\Sigma F_H = F_{RA} - F_{RB}\cos 30° = 0$ and $+\uparrow\Sigma F_V = F_{RB}\sin 30.0° - 1.00 \text{ kN} = 0$; $F_{RB} = 2.00$ kN. $F_{RA} = 1.73$ kN $= F_{RC}$; $F_{RD} =$ the net weight $= 3.00$ kN. **31.** $+\uparrow\Sigma F_V = 533.8 \text{ N} - 889.66 \text{ N} + F_R = 0$; $F_R = 355.9$ N; $\overset{\curvearrowleft+}{}\Sigma\tau_S = (355.9 \text{ N}) \times (1.829 \text{ m}) - (889.66 \text{ N})x = 0$; $x = 0.7317$ m from Selma. **35.** The reaction forces are each normal to the surfaces because there is no friction. $F_{N1} = F_W \sin 45.0° = F_W \cos 45.0° = F_{N2} = 69.3$ N. **41.** $\overset{\curvearrowleft+}{}\Sigma\tau = (300 \text{ N})(0.25 \text{ m}) - F_n(0.09 \text{ m}) = 0$; $F_n = 0.83$ kN. **45.** Take the horizontal bar out and draw a free-body diagram; $\overset{\curvearrowleft+}{}\Sigma\tau_A = (1000 \text{ N})(3.0 \text{ m}) - (2.0 \text{ m})F_C \sin 45° = 0$; $F_C = 2121.6$ N. At A, the horizontal force is $F_{AH} = -F_{CH} = -1.5$ kN (force is to the left). $+\uparrow\Sigma F_V = F_C \sin 45° - 1000 \text{ N} - F_{AV} = 0$; $F_{AV} = -0.50$ kN (force is down). **49.** Taking the upright side as $y = 0$, $\Sigma mx = (0.50 \text{ kg/m})(0.25 \text{ m})(0.125 \text{ m}) = 0.0156 \text{ kg·m}$ and $x_{cg} = (0.0156 \text{ kg·m})/(0.25 \text{ kg}) = 0.063$ m, and, by symmetry, $y_{cg} = 0.063$ m. **55.** If each is of length L, the c.g. of the topmost block is at the edge of the one beneath it. Measured from the left end, for the top two blocks, $x_{cg} = (mL + mL/2)/2m = 3L/4$, and that is located just above the edge of the third block. For the top three blocks, $x_{cg} = (2mL + mL/2)/3m = 5L/6$. For the top four blocks, $x_{cg} = (3mL + mL/2)/4m = 7L/8$. The overhangs are $L/8$, $L/6$, and $L/4$, which is more than $L/2$. **59.** $+\uparrow\Sigma F_V = F_N - g(30 \text{ kg}) = 0$; $F_N = 0.29$ kN. $\overset{\curvearrowleft+}{}\Sigma\tau = 0 = F_N(8.0 \text{ m})\cos 60° - (F_f)(8.0 \text{ m})\sin 60° - g(30 \text{ kg})(4.0 \text{ m})\cos 60°$; $F_f = 85$ N, $F_{RG} = 0.31$ kN at 74°. **63.** $\overset{\curvearrowleft+}{}\Sigma\tau_1 = F_W x_{cg} - F_{N2}h = 0$; $\overset{\curvearrowleft+}{}\Sigma\tau_2 = F_{N1}h - F_W(h - x_{cg}) = 0$; $F_W = F_{N2}h/x_{cg}$; $F_{N1}h - F_{N2}h(h - x_{cg})/x_{cg} = 0$; $x_{cg} = F_{N2}h/(F_{N1} + F_{N2})$. **65.** Taking O at the base of the wall, $\overset{\curvearrowleft+}{}\Sigma\tau_o = (4.00 \text{ m})F_{N2} - (3.00 \text{ m})F_W(2.00 \text{ m}) = 0$; and $\overset{+}{\rightarrow}\Sigma F_H = F_T \cos\theta - F_{N1} = 0$, where θ is the angle the rope makes with the ground. From the geometry (we are dealing here with a 3-4-5 right triangle), $\theta = 10.62°$. $+\uparrow\Sigma F_V = F_{N2} - F_W - F_T \sin\theta = 0$. Eliminating F_{N1} from the first and second equations, $-(3.00 \text{ m}) \times F_T \cos\theta - F_W(2.00 \text{ m}) + (4.00 \text{ m})F_{N2} = 0$; and substituting F_{N2} into the third equation, $(0.75 F_T \cos\theta + 0.50 F_W) - F_W - F_T \sin\theta =$

0. Hence, $F_T(0.75 \cos\theta - \sin\theta) = 0.50 F_W$; $F_T = 90.4$ N.

CHAPTER 7

Answers to Odd-Numbered Multiple Choice Questions

1. c **3.** d **5.** c **7.** c **9.** b **11.** d **13.** c **15.** b

Answers to Odd-Numbered Problems

1. would be halved **3.** m = 1.22×10^5 kg **5.** 4 times larger **7.** 2.5×10^{16} N **9.** $s_Q = 1.1$ m **11.** $F_E = 2 \times 10^{39} F_G$ **13.** $0.11 M_\oplus$ **15.** $F_z \approx 2\pi Gm_\bullet\rho\tau$ **17.** 4.4 m **21.** 0.379 g_0 **23.** $g_n = 1 \times 10^{12}$ m/s^2 **27.** $v = [2GM(1/r - 1/r_0)]^{1/2}$ **29.** $dg_\oplus/dr = -2GM_\oplus/IR_\oplus$; -3.083×10^{-6} (m/s^2)/m **31.** $F_G = GmMr/R^3$ **33.** $T = 1 \times 10^{-3}$ s **37.** 1.62 m/s^2 **39.** $F_x = GmM/[x_0\sqrt{(L/2)^2 + x_0^2}]$; $F_x = 2Gm\lambda_m/x_0$ **41.** 1.65×10^3 m/s **45.** $T_2 = 2.8 T_1$ **47.** 3.32×10^{-5} m/s^2 **49.** 5.9×10^{-3} m/s^2 **51.** 8.000 Earth-years **55.** 1.078 AU, 1.12 Earth-years **51.** $GM/2R^2$ **59.** 1.05×10^3 kg/m^3 **61.** 9.7 d, 1.440 km/s

Solutions to Selected Problems

3. $F_G = Gm^2/r^2$; $1.00 \text{ N} = (6.67 \times 10^{-11} \text{ N·m}^2/\text{kg}^2)m^2/(1.00 \text{ m})^2$; $m = 1.22 \times 10^5$ kg. **9.** $v^2 = 0 = v_0^2 - 2gs$; $s = v_0^2/2g$, $s_\oplus/s_Q = g_Q/g_\oplus$, $(1.00 \text{ m})/s_Q = 0.88$, $s_Q = 1.1$ m. **13.** $g_\delta = GM_\delta/R_\delta^2$, $M_\delta = (3.7 \text{ m/s}^2)(\frac{1}{2}6.8 \times 10^6 \text{ m})^2/(6.67 \times 10^{-11} \text{ N·m}^2/\text{kg}^2) = 6.4 \times 10^{23}$ kg $= 0.11 M_\oplus$. **17.** To find g_Q we need the mass; $M_Q = \frac{4}{3}\pi r^3 \rho = 4.82 \times 10^{24}$ kg; $g_Q = GM_Q/R_Q^2 = 8.79$ m/s^2; $s = \frac{1}{2}g_Q t^2 = 4.4$ m. **21.** $g_\delta = GM_\delta R_\delta^2 = G(0.108 M_\oplus)/R_\oplus^2(0.534)^2 = 0.379 g_0$. **27.** Taking upward as positive, $m \, dv/dt = -GmM/r^2$. Multiply both sides by $v \, dt = dr$; $v \, dv = -GM dr/r^2$ and integrate between r and r_0 to get $v^2 = 2GM(1/r - 1/r_0)$ and $v = [2GM(1/r - 1/r_0)]^{1/2}$. When $r_0 \gg R$, $v = [2GM(1/r)]^{1/2} = (2R_\oplus g_0)^{1/2}$. **31.** $F_G = GmM_r/r^2$, where $M_r = \frac{4}{3}\pi r^3 \rho$, and $\rho = M/(4\pi R^3/3)$; hence, $M_r = Mr^3/R^3$ and $F_G = GmMr/R^3$. **35.** $g_p = GM/(R + h)^2 = GM/R^2(1 + h/R)^2 = (GM/R^2)(1 + h/R)^{-2}$. From the binomial expansion, the bracketed term becomes $1 + (-2)(1)h/R + \frac{1}{2}(-2)(-3) \times (1)(h/R)^2 + \ldots$, which is $\approx(1 - \frac{1}{2}h/R)$ since the $(h/R)^2$ term is very small. **37.** $g_\mathbb{C} \approx GM_\mathbb{C}(1 - 2h/R)/R^2 \approx GM_\mathbb{C}[1 - (200 \text{ m})/(1.74 \times 10^6 \text{ m})]/R_\mathbb{C}^2 \approx 0.999885 \, GM_\mathbb{C}/R_\mathbb{C}^2 \approx 1.62$ m/s^2. The difference is negligible. **43.** From Eq. (7.7), $T^2 = 4\pi^2 r_\oplus^3/M_\oplus G = 9.90 \times 10^{-14} r_\oplus^3$; $T = 3.15 \times 10^{-7} r_\oplus^{3/2}$. **47.** $g_\mathbb{C} = GM_\mathbb{C}/r_{\mathbb{C}\oplus}^2 = 3.32 \times 10^{-5}$ m/s^2. **51.** From Eq. (7.6), $T^2 = r_\odot^3/C_\oplus = (4.000 \text{ AU})^3/(1 \text{ AU}^3/\text{Earth-year}^2) = 64.00$ Earth-years^2; $T = 8.000$ Earth-years. **55.** $r_{\odot I} = \frac{1}{2}(0.186 \text{ AU} + 1.97 \text{ AU}) = 1.078$ AU. $r_{\odot I}^3/T_I^2 = 1 = (1.078 \text{ AU})^3/T_I^2$; $T_I = 1.12$ Earth-years. **61.** From Eqs. (7.6) and (7.7), you find that $C_\oplus = GM_\oplus/4\pi^2 =$

$1.010 \times 10^{13} \text{ m}^3/\text{s}^2 = r_{\oplus s}^3/T_s^2 = (\frac{1}{3} 3.844 \times 10^8 \text{ m})^3/T_s^2$ and $T_s = 0.8385 \times 10^6 \text{ s} = 9.7$ days; $v = 2\pi r_{\oplus s}/T_s = 1.440$ km/s. Check this answer using Eq. (7.8).

CHAPTER 8

Answers to Odd-Numbered Multiple Choice Questions

1. b **3.** e **5.** b **7.** c **9.** b **11.** b **13.** d
15. a **17.** d **19.** b

Answers to Odd-Numbered Problems

1. 0.017453 rad **3.** 1.6 rad, 0.02 rad, 0.02 rad, 2.0 rad **5.** $D = 3 \times 10^5$ m
7. 2.0 m **9.** 69 rev **11.** 12.2 rad/s
13. 2.3 m **15.** 105 rad/s² **17.** 4.2 s
19. (a) $50.0A$ (b) $\omega = 2At$ (c) $\omega = 4.00A$
21. $\alpha(5) = 1.4$ rad/s². **23.** 0.70 m
25. 2.97 × 10⁴ rad/s, 1.99 × 10⁻¹ rad/s
27. 0.2 m **29.** 0.029 rad/s, 0.26 m/s
31. 2.7 m/s **33.** $v_E = 0.1$ m/s, upward
35. $\alpha = 9.82$ rad/s², 16.0 s
37. 4.84×10^3 km, $\theta = 40.6°$
39. $\omega_4 = 1200$ rpm, $\omega_5 = 400$ rpm, $\omega_n = \omega_1 N_1/N_n$; if n is odd, the direction is unchanged **43.** 1.4 m **45.** 0.75 m/s–upward
47. 25 rev **49.** -2.4×10^{-22} rad/s², 6.51×10^{-5} rad/s, 27 h **51.** MR^2
53. 450% decrease **55.** $\frac{1}{3}ml^2$
57. 3×10^{-3} kg·m²/s **59.** 0.20 rad/s²
61. $I = R^2M$ **63.** $I = \frac{1}{2}MR^2$ **65.** $x_{cm} = \frac{2}{3}L$
67. 426 rad/s² **69.** $a = mg/(m + I/R^2)$
71. $a = 1.5$ m/s² **73.** (a) 1.6×10^{-4} kg·m²
(b) 8.0×10^{-2} kg·m² (c) ≈0.080 kg·m²
(d) 0.010 kg·m/s (e) 4.0×10^{-3} kg·m²/s
(f) 5.0×10^{-2} rad/s **75.** $a = \frac{1}{2}g \sin \theta = g/4$
79. $I = \frac{1}{2}M(R_i^2 + R_0^2)$ **81.** 1.2 m/s, 1.8 × 10⁴ kg·m²/s, 24 m/s, 17 × 10³ N
85. $I \approx \frac{1}{3}Ml^2$

Solutions to Selected Problems

3. (ii) 90° = 1.6 rad; tan $\alpha = 1/100$, $\alpha = 5.7 \times 10^{-1}$, $2\alpha = 1.15° = 0.0201$ rad; $\theta = \ell/r = (2.00$ m$)/(100$ m$) = 0.0200$ rad.
(i) $\theta = \ell/r = 2.00$ rad. **5.** $\ell = r\theta = (3.8 \times 10^8$ m$)(8 \times 10^{-4}) = 3 \times 10^5$ m, which is the diameter. **9.** $2\pi r = 2\pi(0.15$ m$) = 0.942$ m, $(65$ m$)/2\pi r = 69$ rev.
13. $\omega = 2\pi(1800$ rpm$)/(60$ s/m$) = 188.5$ rad/s, $v = R\omega = (0.12$ m$) \times (188.5$ rad/s$) = 22.62$ m/s; $\ell = vt = (22.62$ m/s$)(0.10$ s$) = 2.3$ m. **17.** $\omega_0 = 200(2\pi)/(60$ rad/s$) = 20.94$ rad/s; $0 = \omega_0 + (-5$ rad/s²$)t$; $t = 4.2$ s. **21.** $d\omega/dt = \alpha(t) = 0.40$ rad/s² + $(0.20$ rad/s³$)t$; at $t = 5.00$ s, $\alpha(5) = 1.4$ rad/s².
27. $\omega^2 = 0 + 2\alpha\theta$, $(20 \times 2\pi/60)^2 = 2(5$ rad/s²$)\theta$, $\theta = 0.4$ rad, or 0.07 turns. $(0.4$ rad$)(0.5$ m$) = 0.2$ m.
31. $v = 2\pi R(100$ rpm$)/(60$ s/min$) = 8.7$ ft/s = 2.7 m/s. **35.** $\omega^2 = \omega_0^2 + 2\alpha\theta$, $(1500 \times 2\pi/60)^2 = 2\alpha(200 \times 2\pi)$, $\alpha = 9.82$ rad/s². $\omega = \omega_0 + \alpha t$, $t = (1500 \times 2\pi/60)/(9.82$ rad/s²$) = 16.0$ s.
39. 1500 rpm × 20 = $\omega_4 25$; $\omega_4 = 1200$ rpm; $\omega_5 = 400$ rpm. $\omega_n = \omega_1 N_1/N_n$, where N is the number of teeth and n is the number of gears. If n is odd,

the direction is unchanged. **43.** $a_C = R\omega^2$, $a_T = R(4.0$ rad/s²$)$, $a = \sqrt{a_C^2 + a_T^2}$, $(8.0$ m/s²$)^2 = R^2\omega^4 + R^2(4.0$ rad/s²$)^2$; $R^2 = 2.0$ m², $R = 1.4$ m. **45.** $\ell_3 = (0.24$ m$)\theta_1$, $\ell_5 = (0.12$ m$)$ $\theta_1 = (0.20$ m$)\theta_2$, $\ell_4 = (0.15$ m$)\theta_2$, so $\theta_2 = 0.60\theta_1$ and $\ell_4 = (0.15$ m$)(0.60 \theta_1) = (0.90 \times 10^{-1}) \times \theta_1 = 0.90 \times 10^{-1}$ $(\ell_3/0.24) = 0.375\ell_3$, $v_4 = 0.375v_3 = 0.375(2.0$ m/s$) = 0.75$ m/s, upward. **53.** 450% decrease; $L_i = L_f$.
57. $L = I\omega$; $I \approx mr^2 = (2 \times 10^{-3}$ kg$) \times (0.5$ m$)^2 = 5 \times 10^{-4}$ kg·m²; $L \approx (5 \times 10^{-4}$ kg·m²$)(2\pi$ rad/s$) = 3 \times 10^{-3}$ kg·m²/s.
61. Envision a tiny mass element dm a perpendicular distance R from the axis; $I = \int R^2 dm = R^2M$, since R is constant. **67.** $\Sigma \tau = 50.0$ ft·lb $= I\alpha$, $I = mR^2 = (F_W/g)R^2$, 50.0 ft·lb $= [(3.22$ lb$)/(32.2$ ft/s²$)] \times [(13$ in.$)/(12$ in./ft$)]^2\alpha$; $\alpha = 426$ rad/s².
71. $+\downarrow\Sigma F_v = ma = F_W - F_T$; $\overset{\curvearrowright}{+}\Sigma\tau = I\alpha = F_T R - 0.060$ N·m; $a = R\alpha$; $F_T = (I\alpha + 0.060$ N·m$)/R = Ia/R^2 + (0.060$ N·m$)/R$; $F_W - Ia/R^2 - (0.060$ N·m$)/R = ma$, $a(m + I/R^2) = Mg - (0.060$ N·m$)/R$, $a = 9.2/6.0 = 1.5$ m/s², as compared to 1.6 m/s² previously. **75.** $+\searrow\Sigma F_{\parallel} = F_W \sin \theta - F_f = ma$; $\Sigma \tau_{cm} = F_f R = I_{cm}\alpha = mR^2\alpha$; $a = R\alpha$; $F_f R = mRa$; $F_W = mg$; $a = \frac{1}{2}g \sin \theta = g/4$. **81.** $v = a_T t = 1.0 \times 10^{-3}g(120$ s$) = 1.2$ m/s; $L = rmv = 1.8 \times 10^4$ kg·m²/s. $L = (5.0$ m$)(150$ kg$)v = 1.8 \times 10^4$ kg·m²/s, $v = 24$ m/s, or 54 mi/h. $F_C = mv^2/r = 17 \times 10^3$ N, or 4×10^3 lb! **83.** $+\searrow\Sigma F_{\parallel} = mg \sin \theta - F_f = ma$; $\overset{\curvearrowright}{+}\Sigma\tau_{cm} = F_f R = I\alpha$, $\alpha = a/R$, $I_{cm} = \frac{1}{2}mR^2$, $F_f = \frac{1}{2}ma$; $mg \sin \theta = (3/2)ma$, $a = (2/3)g \sin \theta$; $s = h/\sin \theta$, $v^2 = 2as = 2(2/3)(g \sin \theta)h/\sin \theta = (4/3)gh$. $v = 2\sqrt{gh/3}$.

CHAPTER 9

Answers to Odd-Numbered Multiple Choice Questions

1. a **3.** a **5.** a **7.** c **9.** c **11.** a **13.** a
15. c **17.** e **19.** c **21.** d

Answers to Odd-Numbered Problems

1. 23 J **3.** 2.0 J **5.** 20 MJ **7.** 4.00 N
9. 5.0 W **11.** 10 m/s **13.** 5.0 m
15. $W = 1.88$ kJ **17.** $P = \tau\omega$
19. 0.50 kN·m **21.** 1.2×10^2 J
23. (a) 0.20 kN·m (b) 2.5 N·m/cm²
(c) 0.20 kN·m **25.** 5.3 s **27.** 60 W/m²
29. 74 W/m² **31.** 90 J **33.** $W = 4.20$ kJ
35. 0.232 kJ **37.** $P_{av} = 20$ W **39.** $W = mgh$
41. 10^{15} W **43.** 0.98 m/s, $KE_G = 0.95$ J, $KE_B = 0.29$ kJ **45.** 6×10^2 W **47.** 1.2×10^6 J **53.** 3.62 J **55.** (a) 2.9×10^5 J
(b) 7.2×10^4 J/s **57.** 8 min **59.** (a) 100 J
(b) no rope (c) +100 J **61.** (a) 0.16 $g_\oplus$
(b) 1.6×10^5 J **63.** (a) $v = 0$ when $t = 4.00$ s; (b) KE = 481 J; (c) $0 < t < 4.00$ s. **65.** KE = 1.13 kJ **67.** 1.7×10^3 m **69.** 5.2 m **71.** both **73.** 4.85×10^3 J, $h = 9.00$ m or 10.0 m above ground
75. 0.46 km/s **77.** -6.2×10^7 J, -3.1×10^7 J, -2.1×10^7 J, -1.6×10^7 J,

-1.2×10^7 J, -0.62×10^7 J, E = -1.0×10^7 J **79.** 2.4 km/s
81. $v_{esc}(\text{Moon})/v_{esc}(\text{Earth}) = 0.046$
83. 2121 m/s **85.** 0.070 m/s
87. $v_{2f} = 15$ m/s, $v_{1f} = -10$ m/s
89. $v = \sqrt{2(m_2 - m_1)gy/(m_1 + m_2)}$
91. $v_B = [(m_B + m_C)/m_B]\sqrt{2gh}$

Solutions to Selected Problems

3. $W = Fl = (20$ N$)(0.10$ m$) = 2.0$ J.
5. The thrust is constant at 10×10^3 N; hence, $W = (2.0 \times 10^3$ m$)(10^4$ N$) = 20$ MJ.
11. $P = Fv = 1.0 \times 10^4$ J/s = $(1.0 \times 10^3$ N$)v$; $v = 10$ m/s, or 22 mi/h. **19.** $\sin \theta = 5/13$, $W = (F_W \sin \theta)l = (100$ N$)(5/13)(13$ m$) = 500$ N·m. Straight up, $W = (100$ N$)(5.0$ m$) = 0.50$ kN·m. **23.** (a) $W = (100$ N$)(2.0$ m$) = 0.20$ kN·m; (b) (1.0 cm)(1.0 cm) = $(10$ N$)(0.25$ m$) = 2.5$ N·m/cm²; (c) 10 cm × 8.0 cm = 80 cm²; $(80$ cm²$)(2.5$ N·m/cm²$) = 0.20$ kN·m. **27.** $\Delta W/\Delta t = (10^7$ J$)/(24$ h × 3600 s/h$) = 1.2 \times 10^2$ W. BMR = $(90$ W$)/(1.5$ m²$) = 60$ W/m². **33.** $W = \int_0^{10.0} (24.0x + 9.00x^2)\, dx = (12.0x^2 + 3.00x^3)|_0^{10.0} = 4.20$ kJ. **39.** $dW = FL\, d\theta$, solve the two equilibrium equations to get $F = mg \sin \theta$; the integral yields $W = mgL(-\cos \theta)|_0^{\theta(\max)}$ but $\cos \theta(\max) = (L - h)/L$ and so $W = mgh$.
41. KE = $\frac{1}{2}mv^2 = \frac{1}{2}(0.5 \times 10^{-3}$ kg$) \times (200 \times 10^3$ m$)^2 = 1 \times 10^7$ J. $P = \Delta KE/\Delta t = (1 \times 10^7$ J$)/(10^{-8}$ s$) = 10^{15}$ W. **45.** $PE_G = mgh$; $h = (6 \times 10^9$ J$)/(10^6$ kg$) \times (9.8$ m/s²$) = 6 \times 10^2$ m. **53.** $v(t) = dh/dt = 9.810t + 6.02$ at $t = 0$, $v = 6.02$ m/s and KE = $\frac{1}{2}mv^2 = 3.62$ J. **57.** $Pt = W = \Delta KE = \frac{1}{2}(2 \times 10^3$ kg$)(268$ m/s$)^2 = 7.18 \times 10^7$ J; $t = 8$ min. **61.** (a) $g_{\mathbb{C}} = GM_{\mathbb{C}}/R_{\mathbb{C}}^2 = (1/100)/(1/4)^2 g_\oplus = 0.16\ g_\oplus$.
(b) $\Delta PE = mg\Delta h = (1000$ kg$)(0.16) \times (9.8$ m/s²$)(100$ m$) = 1.6 \times 10^5$ J. **67.** Let θ be the incline angle; $\tan \theta \approx \sin \theta = \frac{1}{2}\%$; the net work done by all forces acting on the train equals its ΔKE, $W = [22.7 \times 10^4$ N $- (35$ N$)(1.8 \times 10^3) - (1.8 \times 10^7$ N$) \sin \theta]l = (7.4 \times 10^4$ N$)l$; $\Delta KE = \frac{1}{2}(45)[(4.0 \times 10^5$ N$)/(9.8$ m/s²$)][(13.4$ m/s$)^2 - (6.7$ m/s$)^2] = 1.237 \times 10^8$ J; hence, $W = \Delta KE$ and $l = (1.237 \times 10^8$ J$)/(7.4 \times 10^4$ N$) = 1.7 \times 10^3$ m. **71.** For one car, KE = $\frac{1}{2}mv^2 = \frac{1}{2}[(7.12$ kN$)/(9.81$ m/s²$)] \times (26.67$ m/s$)^2 = 258$ kN·m; and for the other, KE = twice that; viz., 516 kN·m. For one car on top of the mountain, PE = $mgh = (7.12$ kN$)(33.5$ m$) = 238.5$ kN·m; and for the other, PE = twice that; viz., 477 kN·m. Thus, both cars have enough energy to coast over the top. **75.** $p_i = (1.00$ kg$)v_i + 0 = p_f = (91$ kg$)(5.0$ m/s$)$; $v_i = 0.46$ km/s.
79. $v_{esc} = (2GM/R)^{1/2} = [2(6.67 \times 10^{-11}$ N·m²/kg²$)(7.4 \times 10^{22}$ kg$)/(1.74 \times 10^6$ m$)]^{1/2} = 2.4$ km/s. It can be launched in any direction that does not intersect the Moon.
85. Taking east as positive, $p_i = (+8.0$ kg$) \times (0.15$ m/s$) - (2.0$ kg$)(0.25$ m/s$) = 0.70$ kg·m/s = $p_f = (10.0$ kg$)v_f$; $v_f = 0.07$ m/s. **87.** $p_i = m(15.0$ m/s$) +$

$m(-10 \text{ m/s}) = mv_{1f} + mv_{2f}, +5.0 = v_{1f} + v_{2f}.$
But $v_{2f} - v_{1f} = v_{1i} - v_{2i} = 25 \text{ m/s}$; hence, adding the last two equations, $2v_{2f} = 30 \text{ m/s}$, $v_{2f} = 15 \text{ m/s}$, and $v_{1f} = -10 \text{ m/s}$. **89.** $PE_i = 0$; $E_i = 0$; hence, $\frac{1}{2}m_1 v^2 + m_1 gy + \frac{1}{2}m_2 v^2 - m_2 gy = 0$ a moment after release. $v = \sqrt{2(m_2 - m_1)gy/(m_1 + m_2)}.$
91. $m_B v_B = (m_B + m_C)v_C.$ After impact, $\frac{1}{2}(m_B + m_C)v_C^2 = (m_B + m_C)gh, v_C^2 = 2gh,$ $v_B = [(m_B + m_C)/m_B]\sqrt{2gh}.$

CHAPTER 10

Answers to Odd-Numbered Multiple Choice Questions

1. d **3.** d **5.** b **7.** d **9.** c **11.** b **13.** b
15. a **17.** c **19.** b

Answers to Odd-Numbered Problems

1. 6.02×10^{23} **3.** $6 \times 10^7 \text{ kg/m}^3$ **5.** 49 kg
7. 19 GJ **9.** 12.0×10^{23} **11.** 303 u **13.** 6×10^{77} **15.** $m = 510 \text{ kg}$; 51.0 kg/m **17.** 3.3×10^{28} molecules/m^3 **19.** 3.14 kg **21.** 0.46 nm
23. $W = 1.10 \text{ kJ}$ **25.** 3.3 nm **27.** 1.0 kN/m
29. 2 MPa **31.** 0.15% **33.** 0.399 8 m
35. same **37.** 1×10^{-2} m **39.** $1.7 \times 10^2 \text{ N/m}^2$ **41.** $\dfrac{dV}{d\rho} = -\dfrac{m}{\rho^2}; B = \dfrac{F/A}{\Delta\rho/\rho}$
43. 5.0×10^2 N/m **45.** 16 J/m^2
47. 0.77 mm **49.** 29.5 kN **51.** 0.22%
53. $R = 1.01$ cm, 0.173% **55.** 0.911 MN
59. $g\rho L_0^2/2Y$ **61.** $KE_{max} = \frac{1}{2}(mg)^2/k + mgh$

Solutions to Selected Problems

1. 6.02×10^{23} **5.** (13 gallons)(3.785 liters/gallon) = 49.2 liters; 1 liter = 1000 cm^3; hence, the human body contains $\approx 49 \times 10^3$ cm^3, or $\approx 49 \times 10^3$ g, or 49 kg. **9.** $C_{12}H_{22}O_{11}$: 12(12) + 22(1) + 11(16) = 342; hence, 684 g is 2 gram-moles and, therefore, 12.0×10^{23} molecules. **13.** $(10^{33}$ g per star) $\times$ $(10^{11}$ stars per galaxy)$(10^{11}$ galaxies)/ $(10$ g per mole) = number of moles = 1×10^{54} moles; $(10^{54}$ moles)$(6.02 \times 10^{23}$ atoms/mole) = 6×10^{77} atoms. **19.** The mass of a molecule is 18 u, of which 2 u is hydrogen; that is, 11.1%; hence, 11.1% of 62.4 lb is 6.93 lb $\rightarrow$ 3.14 kg. **25.** 22.4 liters contain 6.0×10^{23} molecules; hence, the volume "occupied" by each one is $(22.4 \times 10^3$ cm$^3)$ $\times$ $(10^{-6}$ m^3 cm$^3)/6.0 \times 10^{23}$ = 3.73×10^{-26} m^3; and the cube root of this is 3.3×10^{-9} m, or 3.3 nm, which is roughly 10 times the nucleus-to-nucleus separation in a solid. **27.** $F = ks$; $k = F/s = (50 \text{ N})/$ $(0.05 \text{ m}) = 1.0 \text{ kN/m}$. **29.** $\sigma = F/A =$ $(200 \text{ N})/(10^{-4} \text{ m}^2) = 2 \times 10^6 \text{ N/m}^2 = 2\text{MPa}$. **33.** $\Delta L/L_0 = \varepsilon$; $\Delta L = (0.400 \text{ 0 m})(5.000 \times 10^{-4}) = 2.000 \times 10^{-4}$ m and $L = L_0 - \Delta L = 0.399$ 8 m. **37.** $Y = (F/A)/(\Delta L/L_0)$; $\Delta L = (F/A)(Y/L_0) = FL_0/YA$; the mike is already on the rod and so only the weight of the *sumatori* enters; $\Delta L = (9.8 \text{ m/s}^2) \times$ $(181.4 \text{ kg})(4 \text{ m})/(200 \times 10^9 \text{ Pa})\pi(\frac{1}{2}2 \times 10^{-3})^2 = 1 \times 10^{-4}$ m. **43.** $k = F/s =$ $(10 \text{ N})/(2.0 \times 10^{-2} \text{ m}) = 5.0 \times 10^2 \text{ N/m}$.
47. $F/A < 435$ MPa; $(200 \text{ N})/(435 \text{ MPa}) < A$; $A > 4.598 \times 10^{-7}$ m^2; $\pi R^2 > 4.598 \times 10^{-7}$

m^2; $R > 3.8 \times 10^{-4}$ m; minimum diameter is 0.77 mm. **51.** $Y = 50$ GPa = (110 MPa)/ε_R; $\varepsilon_R = (110 \text{ MPa})/(50 \text{ GPa}) = 2.2 \times 10^{-3} = 0.22\%$. **57.** $PE_e = F^2\frac{1}{2}L_0/2AY +$ $F^2\frac{1}{2}L_0/2(2A)Y = 3F^2L_0/8AY$. **63.** $k = YA/L_0$, $Y = (F/A)/(\Delta L/L_0) = \sigma_R/(\Delta L/L_0)$; $PE_e =$ $\frac{1}{2}k(\Delta L)^2 = \frac{1}{2}(YA/L_0)(\sigma_R L_0/Y)^2 = \frac{1}{2}AL_0\sigma_R^2/Y$.

CHAPTER 11

Answers to Odd-Numbered Multiple Choice Questions

1. b **3.** d **5.** a **7.** a **9.** a **11.** d **13.** b
15. c **17.** c **19.** d

Answers to Odd-Numbered Problems

1. 3×10^4 Pa **3.** 50.7 N/cm^2 **7.** 1.0×10^4 Pa **9.** 10.3 m **11.** 1.1×10^8 Pa **13.** 4.1×10^5 Pa **15.** 10.5 **17.** 13 mN **19.** $dP/dz = \rho g$
25. 3.9×10^3 Pa **27.** -1.1×10^4 Pa
29. 1.1×10^5 Pa **31.** 21.5, platinum, $V = 3.00 \times 10^{-8}$ m^3 **33.** 10% is visible, 8% is above **35.** 300 cm^3 **37.** 1.1×10^5 N
39. 72.7 kg **41.** 4.6 kN, 0.15 m
43. 2.7×10^{-2} N **51.** 2×10^5 N
53. $\rho_1 = (F_{Wa} - F_{Wl})\rho_w/(F_{Wa} - F_{Ww})$
55. 16×10^{-3} m^3/s **57.** -27 kPa
59. 8.8 kPa **61.** $v = \sqrt{2P_G/\rho}$ **63.** $x = 2\sqrt{yh}$ **65.** 12 m/s **67.** $v = \sqrt{2\Delta P/\rho}$
69. $v_p = \sqrt{2g\Delta y/[(A_p^2/A_t^2) - 1]}$ **71.** 82 m/s
73. $v_2 = \sqrt{2gh}, v_3 = \sqrt{2g(h + Y)}$
75. $v_2 = \sqrt{2gH}$ **79.** $\dfrac{dv_x}{dr} = \dfrac{\Delta P}{2\eta l}r$
81. $y = h \sin^2\theta$

Solutions to Selected Problems

3. $P_i = 5(1.013 \times 10^5 \text{ N/m}^2) = 5.07 \times 10^5$ N/m^2 = $(5.07 \times 10^5 \text{ N})/(10^4 \text{ cm}^2) = 50.7$ N/cm^2. **7.** (3.0 in.)(2.54 cm/in.) = 7.62 cm; $\Delta P = \rho g \Delta h = (13.6 \times 10^3 \text{ kg/m}^3) \times$ $(9.8 \text{ m/s}^2)(7.62 \times 10^{-2} \text{ m}) = 1.0 \times 10^4$ Pa. **11.** $P = \rho gh =$ $(1.025 \times 10^3 \text{ kg/m}^3)(9.8 \text{ m/s}^2) \times$ $(11 \times 10^3 \text{ m}) = 1.1 \times 10^8$ Pa, or 1.6×10^4 psi. **15.** From Eq. (11.9), (10.0 g)/(0.952 g) = 10.5, and it's likely silver. **21.** $\displaystyle\int_{P_0}^{P} dP =$ $-B\displaystyle\int_{V_0}^{V} \frac{1}{V}dV, P - P_0 = -B \ln\frac{V}{V_0}$, rearranging terms and raising each side to the eth power yields the desired equation. **25.** $P = \rho gh =$ $(1.0 \times 10^3 \text{ kg/m}^3)(9.8 \text{ m/s}^2)(0.40 \text{ m}) =$ 3.9×10^3 Pa. **29.** At the level of the lower surface, $P = P_A + \rho g\Delta h = (1.013 \times 10^5 \text{ Pa}) +$ $(1.00 \times 10^3 \text{ kg/m}^3)(9.8 \text{ m/s}^2)(0.61 \text{ m}) =$ 1.1×10^5 Pa, or 1.1 atm. **31.** From Eq. (11.9), $gm/(gm - gm_r) = (6.327 \times 10^{-3} \text{ N})/(6.327 \times 10^{-3} \text{ N} - 6.033 \times 10^{-3} \text{ N}) = 21.5$; platinum; weight of water displaced is $(6.327 \times 10^{-3} \text{ N} - 6.033 \times 10^{-3} \text{ N}) = 2.94 \times 10^{-4}$ N, equivalent to 3.00×10^{-5} kg, or 3.00×10^{-8} m^3. **35.** $P_i V_i = P_f V_f$; (228 cm Hg)(100 cm^3) = (76 cm Hg)V_f; $V_f = 300$ cm^3. **39.** mass of air displaced = mass of helium + mass of balloon; $V(1.29 \text{ kg/m}^3) = V(0.178 \text{ kg/m}^3) +$ 454 kg; $V = 408.3$ m^3, for a mass of $\rho V =$ $(0.178 \text{ kg/m}^3)(408.3 \text{ m}^3) = 72.7$ kg.

43. There are two surfaces here, so $F = 2\gamma L = 2\gamma\pi D = 2\pi(72.8 \times 10^{-3} \text{ N/m})(6.0 \times 10^{-2} \text{ m}) = 2.7 \times 10^{-2}$ N, which is the weight of a 2.8-g mass. **45.** $l = (b/a)y$; $dA = l\,dy = (b/a)y\,dy; F = \displaystyle\int_0^a \rho g(y + h_0) \times$ $\left(\dfrac{b}{a}y\right)dy = \dfrac{\rho gb}{a}\left[\dfrac{y^3}{3} + \dfrac{h_0 y^2}{2}\right]_0^a$ and the required equation follows immediately. **49.** The area of the left face is $A_a = L(L \sin\phi)$, the area of the top face is $A_c = L^2$, and that of the bottom face is $A_b = L^2 \cos\phi$. So the corresponding forces are $F_a = P_a L^2 \sin\phi$, $F_c = P_c L^2$, and $F_b = P_b L^2 \cos\phi$. Taking the sum of the horizontal forces equal to zero, we have $F_a = F_c \sin\phi$; $P_a L^2 \sin\phi = P_c L^2 \sin\phi$. Thus, $P_a = P_c$. The weight of the fluid wedge is the volume, $\frac{1}{2}L(L \sin\phi)(L \cos\phi)$, times ρg; hence, the sum of the vertical forces yields $F_b = F_c \cos\phi + F_W$ or $P_b L^2 \cos\phi = P_c L^2 \times$ $\cos\phi + \frac{1}{2}\rho g\,L^3 \sin\phi\cos\phi$ and $P_b = P_c + \frac{1}{2}\rho gL \sin\phi.$ As $L\rightarrow 0$, we get $P_b = P_c$, which equals P_a. **51.** Because the pressure varies uniformly with depth, the average pressure on the wall is $\frac{1}{2}$ the pressure at the bottom. From Problem 11.1, $P_1 = 3 \times 10^4$ Pa; hence, $F = (1.5 \times 10^4 \text{ Pa}) \times$ $(5 \text{ m} \times 3 \text{ m}) = 2 \times 10^5$ N. **53.** $F_{Wa} - F_{Ww} = \rho_w gV; V = (F_{Wa} - F_{Ww})/\rho_w g;$ $V = (F_{Wa} - F_{Wl})/\rho_1 g;$ setting these equal, $(F_{Wa} - F_{Ww})/\rho_w g = (F_{Wa} - F_{Wl})/\rho_1 g;$ and $\rho_1 = \dfrac{(F_{Wa} - F_{Wl})}{(F_{Wa} - F_{Ww})}\rho_w.$ **57.** From Eq. (11.18), $P_1 - P_2 = \rho g(y_2 - y_1) = (0.68 \times 10^3 \text{ kg/m}^3)(9.8 \text{ m/s}^2)(-4.0 \text{ m}) = -27$ kPa. $P_2 > P_1$. **61.** $v = \sqrt{2gh}, P_G = \rho gh$; hence, $v = \sqrt{2P_G/\rho}$. **65.** Consider a slug of gas of volume $V = AL$, where the length $L = v\Delta t$. The mass per unit time (Δt) flowing is 1.0 kg/s $= V\rho/\Delta t = Av\Delta t\rho/\Delta t = \frac{1}{4}\pi D^2\rho v; v =$ $4(1.0 \text{ kg/s})/\pi D^2\rho = 12$ m/s. **73.** At the surface and the spigot using gauge pressure, $0 + \rho g(h + Y) + 0 = 0 + \rho gY +$ $\frac{1}{2}\rho v_2^2$, hence, $v_2 = \sqrt{2gh}$. Then, between points 2 and 3, $0 + \rho gY + \frac{1}{2}\rho(2gh) = 0 +$ $0 + \frac{1}{2}\rho v_3^2$; $v_3 = \sqrt{2g(h + Y)}$. These are just what we expect if the water free-falls. **77.** From Eq. (11.20), $v_2^2 = v_1^2 + 2gh$ but $A_1 v_1 = A_2 v_2$ and so $v_2^2(1 - A_2^2/A_1^2) = 2gh,$ the desired equation follows. (b) $\dfrac{dV}{dt} =$ $\dfrac{dV}{dh}\cdot\dfrac{dh}{dt} = A_1\dfrac{dh}{dt} = A_2 v_2; \dfrac{dh}{dt} = \dfrac{A_2}{A_1}v_2$; and the desired result follows on substituting in v_2. **81.** $P_2 + \frac{1}{2}\rho v_2^2 + \rho gy_2 = P_3 + \frac{1}{2}\rho v_3^2 + \rho gy_3$; $0 + \frac{1}{2}\rho v_2^2 + 0 = 0 + \frac{1}{2}\rho v_3^2 + \rho gy$; since there is no horizontal acceleration, $v_3 = v_2 \cos\theta$, $y = \frac{1}{2}(v_2^2 - v_2^2 \cos^2\theta)/g$, but $\cos^2\theta +$ $\sin^2\theta = 1$; hence, $y = (\frac{1}{2}v_2^2 \sin^2\theta)/g$ and, since $v_2^2 = 2gh$, $y = h \sin^2\theta$ and it cannot exceed h. **83.** Imagine a flow tube from some distant point 1 in front of the plane at the level of the top of the wing; then $P_1 + \frac{1}{2}\rho v_1^2 + \rho gy_1 = P_\alpha + \frac{1}{2}\rho v_\alpha^2 + \rho gy_\alpha.$

Now do the same thing for a tube starting at point 2 just below 1 and running to the bottom of the wing as $P_2 + \frac{1}{2}\rho v_2^2 + \rho g y_2 = P_\beta + \frac{1}{2}\rho v_\beta^2 + \rho g y_\beta$. Since $P_1 \approx P_2$, $v_1 \approx v_2$, $y_1 \approx y_2$, $P_\alpha + \frac{1}{2}\rho v_\alpha^2 + \rho g y_\alpha = P_\beta + \frac{1}{2}\rho v_\beta^2 + \rho g y_\beta$, and with $y_\alpha \approx y_\beta$, $F_L = A(P_\beta - P_\alpha) = \frac{1}{2}\rho(v_\alpha^2 - v_\beta^2)A$.

CHAPTER 12

Answers to Odd-Numbered Multiple Choice Questions
1. b **3.** a **5.** c **7.** c **9.** d **11.** c **13.** a **15.** d **17.** d **19.** b **21.** c

Answers to Odd-Numbered Problems
1. 1.8 s **3.** 1.3 Hz, 2.6π rad/s **5.** (a) 5.0 m (b) 0.064 Hz (c) 0.10 rad (d) $x = 3.1$ m **9.** 4.9×10^2 m/s² **11.** T = 4.4 s **13.** $\varepsilon = \pi$ **15.** $z = +0.50$ m, $z = -0.50$ m, $z = 0$ **17.** $k = 2.6$ N/m **19.** $k = 784$ N/m, $f = 3.15$ Hz **21.** $f_0 = (1/2\pi)\sqrt{(k_1 + k_2)/m}$ **23.** 0.16 Hz **25.** L = 24.8 m **31.** $m\,d^2x/dt^2 + b\,dx/dt + kx = 0$ **33.** (a) $\omega t = 0, \pi, 2\pi, 3\pi, \ldots$ (b) touches top when $\omega t = \pi/2, 5\pi/2, 9\pi/2, \ldots$ and bottom when $\omega t = 3\pi/2, 7\pi/2, 11\pi/2, \ldots$. (c) $dy/dt = A_0[\exp(-\frac{1}{2}Ct)]\,[\omega \cos \omega t - \frac{1}{2}C \sin \omega t]$ **37.** $g = 4\pi^2\Delta L/T^2$ **39.** 3.5 Hz **41.** $T = 2\pi\sqrt{b\rho/\rho_w g}$ **43.** $v_b = 0.86$ km/s **45.** 1.2 T **47.** $T = 2\pi\sqrt{m/k}$ **49.** $dg = -\dfrac{g\sqrt{g}\,dT}{\pi\sqrt{L}}$ **59.** 500 nm **61.** $\lambda = 3.40$ m **65.** 12 m **67.** 1.3×10^2 m/s **69.** 1.08×10^3 N **75.** $v_y = -\omega A \cos(kx - \omega t)$ **83.** 6.3 m/s² **85.** 100 m/s, 314 rad/s, 0.020 0 s, 2.00 m **87.** $v = 1.0 \times 10^2$ m/s **89.** $F_T = 1.9 \times 10^2$ N **91.** $y = -0.7$ A, $y = -0.7$ A **93.** $v_{max} = \sqrt{gL}$, $v = \sqrt{g\Delta y + MgL/m}$

Solutions to Selected Problems
1. $T = 1/f = 1/[(33\frac{1}{3}\text{ rpm})/(60 \text{ s/min})] = 1.8$ s. **5.** By comparison with Eq. (13.3), (a) A = 5.0 m (b) $\omega = 2\pi f = 0.40$ rad/s; $f = 0.064$ Hz (c) $\varepsilon = 0.10$ rad (d) $x = 3.1$ m. **9.** $a_0 = -A(2\pi f)^2$; dropping the sign, $a_0 = (0.005 \text{ m})4\pi^2(50 \text{ Hz})^2 = 4.9 \times 10^2$ m/s². **13.** $x = A \cos \theta = A \cos(\omega t + \varepsilon) = A \cos(2\pi ft + \varepsilon)$; hence, at $t = 0$, $x = -3 = A \cos \varepsilon$, and since $A = 3$, $\cos \varepsilon = -1$ and $\varepsilon = \pi$. **17.** $f = (1/2\pi)\sqrt{k/m}$; $k = (2\pi f)^2 m = 2.6$ N/m. **21.** Since $x = x_1 = x_2$ and $F = F_1 + F_2$; hence, $kx = k_1x_1 + k_2x_2$ and $k = k_1 + k_2$. Now, simply substitute k into Eq. (12.12). **25.** From Eq. (12.17), $T = 2\pi\sqrt{L/g}$; hence, $L = gT^2/4\pi^2 = 24.8$ m. **29.** $F = ma = -kx$; therefore $a = d^2x/dt^2 = -kx/m$. **33.** (a) $y = 0$ when $\sin \omega t = 0$, i.e., when $\omega t = 0, \pi, 2\pi, 3\pi, \ldots$ (b) It touches the top curve when $\sin \omega t = 1$ at $\omega t = \pi/2, 5\pi/2, 9\pi/2, \ldots$ and it touches the bottom curve when $\sin \omega t = -1$ at $\omega t = 3\pi/2, 7\pi/2, 11\pi/2, \ldots$. (c) taking the derivative of y yields the speed, $dy/dt = A_0[\exp(-\frac{1}{2}Ct)]\,[\omega \cos \omega t - \frac{1}{2}C \sin \omega t]$. **37.** $k\Delta L = mg$; hence, $m/k = \Delta L/g$ and, from Eq. (12.13), $T = 2\pi\sqrt{\Delta L/g}$; hence, $g = 4\pi^2\Delta L/T^2$.

41. The additional buoyant force if the block is pushed down a distance y is $-acy\rho_w g$, which is the restoring force $-ky$, so the motion is SHM, since $F \propto -y$ and $k = ac\rho_w g$. Here, $m = abc\rho$, so $T = 2\pi\sqrt{(abc\rho)/(ac\rho_w g)} = 2\pi\sqrt{(b\rho)/(\rho_w g)}$. **45.** From Eq. (12.17), $T = 2\pi\sqrt{L/g}$, and if $L' = L + 50\% L = 1.50 L$, $T' = 2\pi\sqrt{L'/g} = \sqrt{1.50}\,T = 1.2\,T$. **49.** $T = 2\pi\sqrt{L/g} = 2\pi\sqrt{L}\,g^{-1/2}$; $dT = 2\pi\sqrt{L}\,(-\frac{1}{2})g^{-3/2}dg$; $dg = -\dfrac{g\sqrt{g}\,dT}{\pi\sqrt{L}}$. **53.** $PE = mgL(1 - \cos \theta)$; $KE = \frac{1}{2}m[L\,d\theta/dt]^2$; $E = PE + KE$, which must be constant, and so $dE/dt = 0 = mgL \sin \theta(d\theta/dt) + mL^2(d\theta/dt)(d^2\theta/dt^2)$; hence $(d^2\theta/dt^2) = -(g \sin \theta)/L = -(g/L)\theta$ for small angles. **57.** $x = L \cos \phi + R \cos \theta$; $\cos \phi = \sqrt{1 - \sin^2 \phi}$, but $L \sin \phi = R \sin \theta$; $x = R \cos \theta + L\sqrt{1 - (R/L)^2 \sin^2 \theta}$. Although the first term corresponds to SHM, the presence of the second term negates that. **61.** $v = f\lambda$; $\lambda = (1498 \text{ m/s})/(440 \text{ Hz}) = 3.40$ m. **65.** The round-trip distance is $2x = vt = (330 \text{ m/s})(70 \text{ ms}) = 23.1$ m, $x = 12$ m. **67.** $v = \sqrt{F_T/\mu} = \sqrt{(10 \text{ N})/(0.59 \times 10^{-3} \text{ kg/m})} = 1.3 \times 10^2$ m/s. **69.** $F_T = v^2\mu = 1.08 \times 10^3$ N. **73.**

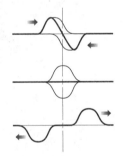

75. $v_y = [dy/dt]_{x\,constant} = -\omega A \cos(kx - \omega t)$ **81.** Using the trigonometric identity $\sin(\alpha \mp \beta) = \sin \alpha \cos \beta \mp \cos \alpha \sin \beta$, where $\alpha = k(x - vt)$ and $\beta = \mp\frac{1}{2}\pi$, we get $y = \mp A \cos \frac{2\pi}{\lambda}(x - vt)$. **85.** $v = \sqrt{F_T/\mu} = 100$ m/s; $\omega = 2\pi f = 314$ rad/s; $v = f\lambda$; $\lambda = 2.00$ m; $T = 1/f = 0.020\,0$ s. **89.** $m = F_W/g = (0.20 \text{ N})/(9.81 \text{ m/s}^2) = 0.020\,4$ kg; $\mu = 1.7 \times 10^{-3}$ kg/m; $v = \sqrt{F_T/\mu}$; $F_T = v^2\mu = 1.9 \times 10^2$ N, or 42 lb. **91.** The phase is $2\pi[(t/0.01) - (5.0/40)] = 2\pi[(t/0.01) - (1/8)] = (200\pi t - \frac{1}{4}\pi)$; hence, $y = A \sin(200\pi t - \frac{1}{4}\pi)$ is the displacement of the bead. At $t = 0$, $y = A \sin(-\frac{1}{4}\pi) = -0.7$ A, and at $t = 0.01$ s, which is one period later, $y = A \sin(2\pi - \frac{1}{4}\pi) = -0.7$ A.

CHAPTER 13

Answers to Odd-Numbered Multiple Choice Questions
1. a **3.** d **5.** e **7.** d **9.** d **11.** b **13.** c **15.** a **17.** b **19.** a

Answers to Odd-Numbered Problems
1. $T = 2.27$ ms, $\lambda = 0.780$ m **3.** 3.3×10^{-4} m **5.** 2.91 **7.** 4.0×10^{-2} W/m², E = 4.0 MJ **9.** 0.79 m **11.** 1.9 ms **13.** 40 dB **15.** 10 dB **17.** 0 dB **19.** 4.8 dB **21.** 5.0×10^{-5} W/m² **23.** 51 dB **25.** $4 \times 10^{-3}v$ **27.** 89.8 dB **29.** 87 dB, 75 dB **31.** $R = 2.2$ km **33.** 200 m **35.** $\beta = 20 \log_{10}(P/P_0)$ **37.** $v = \sqrt{\gamma P/\rho}$ **41.** (b) displacement and pressure are $\frac{1}{4}\lambda$ out-of-phase **43.** 25.8 dB **45.** 996 Hz or 1004 Hz **47.** 1.0 Hz **49.** $\lambda_6 = \frac{1}{6}$ m **51.** $f_1 = 100$ Hz **55.** zero **57.** intensity increases by a factor of 4 times **59.** 466 Hz **61.** 0.782 m/s **63.** For $N = 1$, nodes when $x = 0$ and $x = L$. Antinodes are at $x = L/4$ and $3L/4$. $v_y = -2\pi f_N A_N \sin kx \sin(2\pi f_N t)$ **65.** air, $\lambda = 1.4$ m; string, 66 cm **67.** 2 Hz **69.** 1.76 kHz **71.** $\Delta f = 14$ Hz **73.** 210 N **75.** 1.8 kHz **77.** 344 m/s **79.** $v = 340$ m/s **83.** $f_0 - f_s = 2v_t f_s/(v - v_t)$ **85.** $2A \cos[\frac{1}{2}(\omega_1 - \omega_2)t]$ is a cosinusoidally modulated amplitude

Solutions to Selected Problems
1. $T = 1/f = 1/440$ Hz = 2.27 ms. $\lambda = v/f = (343 \text{ m/s})/(440 \text{ Hz}) = 0.780$ m. **5.** $v = f\lambda = (1000 \text{ Hz})\lambda = 344$ m/s; $\lambda = 0.344$ m; 2.91 waves per meter. **11.** $\tau = \frac{1}{2}T = 1/2f = 1/2(261.6 \text{ Hz}) = 1.9$ ms. **15.** Ten times the power means 10 times the intensity; therefore $\Delta\beta = 10 \log_{10} 10 = 10$ dB. **19.** $\Delta\beta = 10 \log_{10} 3 = 4.8$ dB. **23.** The increase over one voice is $\Delta\beta = 10 \log_{10} 25 = 14$ dB; hence, each singer puts out 65 dB − 14 dB = 51 dB. **27.** $I = (1.2 \text{ W})/4\pi R^2 = 0.955$ mW/m²; $\beta = 10 \log_{10} 0.955 \times 10^9 = 89.8$ dB. **29.** $\beta = 10 \log_{10}(56 \times 10^{-5} \text{ W/m}^2)/(10^{-12} \text{ W/m}^2) = 87$ dB. The intensity drops with the inverse square of the distance; $(5.0 \text{ m})^2/(20 \text{ m})^2 = I/(56 \times 10^{-5} \text{ W/m}^2)$; at 20.0 m, $I = 3.5 \times 10^{-5}$ W/m²; and there $\beta = 75$ dB. **33.** −40 dB means a change of −4 bel or 10^{-4}; hence, from the Inverse Square Law, $R^2/(2.0 \text{ m})^2 = 1/10^{-4}$ of whatever the original intensity was. $R = 200$ m. **39.** $ds/dt = \omega s_0 \sin(kx - \omega t)$, so $v(max) = \omega s_0$. (b) If a thin layer of air of cross-sectional area A and mass Δm is pumped forward by a speaker its average energy $\Delta E_{av} = \Delta E = \Delta KE(max) = \frac{1}{2}\Delta m(\omega s_0)^2$ and if the volume of air displaced is $\Delta V = A \Delta x$, $\Delta m = \rho A \Delta x$: hence $\Delta E = \frac{1}{2}\rho A \Delta x(\omega s_0)^2$; $dE/dx = \frac{1}{2}\rho A(\omega s_0)^2$. **41.** (a) N/m² = (N/m²)(1/m)(m); (b) the displacement and pressure are $\frac{1}{4}\lambda$, or 90°, out-of-phase; (c) $P_0 = \frac{2\pi}{\lambda}BA$; $v = \sqrt{B/\rho}$; $P_0 = \frac{2\pi}{\lambda}v^2\rho A$. The Bulk Modulus relates pressure change to volume change in the gas, and that volume change, in turn, is related to the displacement of the molecules. P is a pressure change from ambient, and so we can expect it to be associated with B and displacement. **45.** The beat frequency is 4 Hz; hence, the wire is vibrating at either 996 Hz or 1004 Hz. **49.** $\frac{1}{2}\lambda = L = \frac{1}{2}$ m,

$\lambda = 1.0$ m; $\lambda_6 = 1/6$ m. **53.** $f_N = (N/2L) \times \sqrt{F_T/\mu}$; but $\mu = m/L$; hence, $f_N = (\frac{1}{2}N) \times \sqrt{F_T/L^2(m/L)}$, which equals the desired expression. **57.** The addition of a 180°, or $\frac{1}{2}\lambda$, phase shift will bring the waves into phase and they will now reinforce one another, creating a peak pressure 2 times as great and an intensity 4 times that of either speaker alone. **61.** $f_0 = f_s(v + v_0)/(v + v_s)$; here $v_s = 0$ and $v_0 = v(f_0 - f_s)/f_s = (344 \text{ m/s})(1/440) = 0.782$ m/s. **67.** $f \propto \sqrt{F_T}$; hence, the increased frequency is given by $f/(440 \text{ Hz}) = \sqrt{1.010 \, F_T/F_T}$; $f = 442$ Hz and the beat frequency is 2 Hz. **71.** $f_1/f_1' = 200 \text{ Hz}/f_1' = (331.5 \text{ m/s})/(331.5 + 0.60 \times 40) = 0.932$; $f_1' = 214$ Hz; $\Delta f = 14$ Hz. **75.** $v = \sqrt{Y/\rho} = 3516$ m/s; $v = f\lambda$, $\lambda = 2L = 2.00$ m; $f = v/2L = (3516 \text{ m/s})/(2.00 \text{ m}) = 1.8$ kHz. **79.** There is a tiny residual signal because the paths are different; hence, the amplitudes are slightly different and the direct and reflected waves cancel, but not completely. The angle between these two beams is $\tan \theta = 1.146/2.5$, $\theta = 24.62°$; the speaker-table distance is $1.146/\sin 24.62° = 2.75$ m. The direct beam travels 5.00 m, the reflected beam travels 5.50 m, and there is cancellation; the path difference (0.50 m) must be an odd multiple of $\frac{1}{2}\lambda$. Speed is ≈ 344 m/s, and since $f = 340$ Hz, $\lambda \approx 1$ m; hence, 0.50 m cannot be $3\lambda/2$ or $5\lambda/2$ and must be $\frac{1}{2}\lambda$, ergo $\lambda = 1.00$ m and $v = 340$ m/s. **85.** $y = y_1 + y_2 = A \sin(\omega_1 t) + A \sin(\omega_2 t)$. Using the identity $\sin \alpha + \sin \beta = 2 \sin \frac{1}{2}(\alpha + \beta) \cos \frac{1}{2}(\alpha - \beta)$, we get the desired result. Remember $\omega_1 \approx \omega_2$; there is a relatively high-frequency sinusoidal carrier oscillating at the average frequency $\frac{1}{2}(\omega_1 + \omega_2)$, and this is multiplied by an amplitude term $2A \cos[\frac{1}{2}(\omega_1 - \omega_2)t]$. This amplitude varies slowly in time, and so the modulated carrier has maxima whenever $\cos[\frac{1}{2}(\omega_1 - \omega_2)t] = \pm 1$; hence, beats occur at twice that frequency, namely, at $(\omega_1 - \omega_2)$ or $(f_1 - f_2)$.

CHAPTER 14

Answers to Odd-Numbered Multiple Choice Questions

1. b **3.** a **5.** c **7.** d **9.** a **11.** b **13.** a **15.** a **17.** b **19.** e **21.** a

Answers to Odd-Numbered Problems

1. 37°C **3.** 55.6 K **5.** 10×10^{6}°C $\approx 18 \times 10^{6}$°F **7.** 927°F **9.** $\Delta L = 25 \times 10^{-6}$ m **11.** 5.0 mm **13.** 0.54 m **15.** 1.008 cm² **19.** 10.002 m **21.** 0.51 cm³ **23.** $\Delta H = -1.5$ mm **25.** 0.79×10^{-6} m³ **27.** 7.9×10^3 kg/m³ **29.** $T = 4$°C **31.** 2.006 s, ran fast **35.** 9.3 MPa **37.** 1.0 MPa **39.** 1.1×10^3 cm³ **41.** $T_f = 199$ K **43.** (a) vapor (b) liquid (c) vapor (d) vapor (e) vapor **45.** 2.992×10^{-26} kg **47.** 3.01×10^{26} molecules **49.** 174 K **51.** $N = 27 \times 10^{18}$ **53.** 3.0 m/s **55.** 478.03 m/s

57. $dV/dP = -C/P^2$ **59.** 3.93×10^{-6} m³/s **61.** (a) $P = mg/A$ (b) $\Delta F = -mgA \Delta y/V$ **63.** 1.19×10^{22} **65.** 1.1 m³ **67.** 43 kg **71.** 76 kJ **73.** $v_{rms} = (3P/\rho)^{1/2}$ **77.** 189 cm³ **79.** 4.3 atm **81.** 0.427% **83.** $\ln(P/P_i) = -(Mg/RC) \ln[T_0/(T_0 - Cy)]$

Solutions to Selected Problems

1. 37°C **5.** 10 000 000 K $\to$ 10 000 273°C $\approx 10 \times 10^{6}$°C $\approx 18 \times 10^{6}$°F. **9.** $\alpha = 25 \times 10^{-6}$ K^{-1}; hence, $\Delta L = 25 \times 10^{-6}$ m. **13.** $\Delta L = \alpha L_0 \Delta T = (12 \times 10^{-6} \text{ K}^{-1}) \times (1280 \text{ m})(35 \text{ K}) = 0.54$ m. **19.** $\Delta L = \alpha L_0 \Delta T = (12.15 \times 10^{-6} \text{ K}^{-1})(10.000 \text{ m}) \times (16.0 \text{ K}) = 1.9 \times 10^{-3}$ m. $L = 10.002$ m. 2 mm is an appreciable error, but it's not likely to trouble a carpenter. **23.** $\Delta V = \beta V_0 \Delta T = (182 \times 10^{-6} \text{ K}^{-1})(800.0 \times 10^{-6} \text{ m}^3) \times (-95 \text{ K}) = -1.383 \times 10^{-5}$ m³ for the mercury. $\Delta V = \beta V_0 \Delta T = (26 \times 10^{-6} \text{ K}^{-1}) \times (800.0 \times 10^{-6} \text{ m}^3)(-95 \text{ K}) = -1.98 \times 10^{-6}$ m³ for the glass. Hence, the total $\Delta V = -1.185 \times 10^{-5}$ m³. $\Delta V = A \, \Delta H = \pi r^2 \Delta H$; $\Delta H = -1.5$ mm. **27.** $\rho_i = m/V_i$; $\rho_f = m/(V_i + \Delta V)$; $\rho_i V_i = \rho_f(V_i + \Delta V) = \rho_f V_i(1 + \beta \Delta T)$; $\rho_i/(1 + \beta \Delta T) = \rho_f = (7.85 \times 10^3 \text{ kg/m}^3)/[(1 + (36 \times 10^{-6} \text{ K}^{-1}) \times (-20 \text{ K})] = 7.9 \times 10^3$ kg/m³. **33.** $\rho_0 = m/V_0$; $\rho = m/(V_0 + \Delta V)$; $\rho_0 V_0 = \rho(V_0 + \Delta V) = \rho V_0(1 + \beta \Delta T)$; $\rho_0/(1 + \beta \Delta T) = \rho$: from the Binomial Theorem, $(1 + x)^n \approx (1 + nx)$ for $x \ll 1$; here, $n = -1$, and so $\rho \approx \rho_0(1 - \beta \Delta T)$. **35.** Stress $= Y\alpha \Delta T = (25 \text{ GPa})(10 \times 10^{-6} \text{ K}^{-1})(37 \text{ K}) = 9.3$ MPa. Concrete is strong in compression (Table 10.3), so it's more likely to buckle than break. **37.** $P_i V_i = P_f V_f$; $P_f = 10 P_i = 1.0$ MPa. **41.** No change in P; $V_i/T_i = V_f/T_f$; (300 cm³)/(298 K) = (200 cm³)/T_f; $T_f = 199$ K. **45.** $2H \to 2(1.008 \text{ u})$; 0 $\to$ 15.999 u; (18.015 u)(1.660 6 $\times 10^{-27}$ kg/u) $= 2.992 \times 10^{-26}$ kg. **49.** $PV = nRT$; $(202.6 \times 10^3 \text{ Pa})(10.0 \times 10^{-3} \text{ m}^3) = (1.40 \text{ mol}) \times (8.314 \text{ J/mol·K})T$; $T = 174$ K. **53.** $v_{av} = (1.0 \text{ m/s} + 2.0 \text{ m/s} + 3.0 \text{ m/s} + 4.0 \text{ m/s} + 5.0 \text{ m/s})/5 = 15/5$ m/s $= 3.0$ m/s. **59.** $PV = C$; $P(dV/dt) + V(dP/dt) = 0$; $dV/dt = -(V/P)dP/dt = -[(550 \times 10^{-6} \text{ m}^3)/(140 \text{ kPa})](1.00 \text{ kPa/s}) = 3.93 \times 10^{-6}$ m³/s. **63.** $PV = Nk_B T$; $T = 98.6$°F $= 37$°C $\to 310$ K; $PV/k_B T = (101.46 \text{ kPa})(500 \times 10^{-6} \text{ m}^3)/(1.380 \, 7 \times 10^{-23} \text{ J/K})(310 \text{ K}) = N = 1.19 \times 10^{22}$ molecules, which is the same as the number of stars there are in the entire Universe. **67.** $n = PV/RT = (19.4 \times 10^3 \text{ Pa})(2000 \text{ m}^3)/(8.314 \text{ J/mol·K})(218 \text{ K}) = 21.4 \times 10^3$ moles. Each mole has a mass of 0.002 kg; hence, the total mass is $(0.002 \text{ kg/mol})(21.4 \times 10^3 \text{ mol}) = 43$ kg. **71.** $(KE)_{av} = (3/2)k_B T = (3/2)(1.380 \, 662 \times 10^{-23} \text{ J/K})(273.15 \text{ K}) = 5.66 \times 10^{-21}$ J/molecule. 0.50 m³ = 500 liter; (500 liter)/(22.4 liter/mol) = 22.3 mol; (22.3 mol)(6.022 $\times 10^{23}$ molecules/mol) $\times (5.66 \times 10^{-21}$ J/molecule) $= 76$ kJ. **77.** To find the pressure on the gas, 3.90 cm Hg =

5.199 6 kPa, $P_i = (101.05 - 5.199 \, 6)$ kPa $= 95.85$ kPa; $P_i V_i/T_i = P_f V_f/T_f$; (95.85 kPa) $\times (214 \text{ cm}^3)/(293.15 \text{ K}) = (101.32 \text{ kPa})V_f/(273 \text{ K})$; $V_f = 189$ cm³. **81.** $v_{rms} = \sqrt{3k_B T/m}$; hence, the ratio of the rms-speeds varies as the inverse of the square roots of the masses. The masses are $238 \text{ u} + 6(18.998 \text{ u}) = 351.99$ u and $235 \text{ u} + 6(18.998 \text{ u}) = 348.99$ u. $(v_{rms}^{235} - v_{rms}^{238})/v_{rms}^{235} = [(348.99 \text{ u})^{-1/2} - (351.99 \text{ u})^{-1/2}]/(348.99 \text{ u})^{-1/2} = 0.427\%$.

CHAPTER 15

Answers to Odd-Numbered Multiple Choice Questions

1. e **3.** c **5.** c **7.** b **9.** a **11.** c **13.** a **15.** b **17.** b **19.** c

Answers to Odd-Numbered Problems

1. 63 kJ **3.** 116 W **5.** 9×10^5 J **7.** 17 K **9.** 0.68 kJ **11.** 0.68 kg **13.** 1.2×10^8 J **15.** 113 g **17.** 0.921 kg **19.** $W = 4.148$ J ≈ 1 cal **21.** 58°C **23.** +0.43 K **25.** 1.9 h **27.** 27 kcal **29.** 0.74 MJ **31.** $c = 0.65$ kJ/kg·K **33.** 1.8 h **35.** 12.7 g **37.** $Q = m[K_1(T_f - T_i) - \frac{1}{2}K_2(T_f^2 - T_i^2) + \frac{1}{3}K_3(T_f^3 - T_i^3)]$ **39.** 43 kJ **41.** $c = 443$ J/kg·K **43.** 41.0 kJ **45.** 3.5 MJ **47.** 0.230 kg vaporizes **49.** 0.17 MJ **51.** 0.029 kg **53.** 0.125 kg **55.** $T_f = 24$°C **57.** 2.1×10^5 J **59.** 2.65 h **61.** $T = T_s + (T_0 - T_s)e^{-Kt}$ **63.** $H = -(A/L)[a(T_0 - T_1) + (b/2)(T_0^2 - T_1^2)]$ **65.** 7.4×10^{-20} J **67.** 0.068 kg **69.** 354 m/s **71.** 1×10^2 g **73.** $T_f = 0$°C **75.** 0°C **77.** 2.3 m **81.** 0.36 MJ, $T_f = 1234$ K

Solutions to Selected Problems

1. 15 kcal $= 63$ kJ. **5.** (7 kcal/g)(30 g) $= 210$ kcal $= 879$ kJ, or to 1 significant figure, 9×10^5 J. **9.** $Q = cm\Delta T = (0.060) \times (1.0 \text{ kcal/kg})(0.030 \text{ kg})(-90 \text{ K}) = 0.162$ kcal $= 162$ cal $= 0.68$ kJ. **13.** $m = (3785 \times 10^{-6} \text{ m}^3)(0.68 \times 10^3 \text{ kg/m}^3) = 2.574$ kg; (2.574 kg)(48 MJ/kg) $= 124$ MJ $= 1.2 \times 10^8$ J. **17.** $-(910 \text{ J/kg·K})m(-173 \text{ K}) = (4186 \text{ J/kg·K})(4.95 \text{ kg})(7.00 \text{ K})$; $m(157 \, 430 \text{ J/kg}) = 145 \, 044.9$ J; $m = 0.921$ kg. **21.** $Q/t = cm\Delta T/t$; 1.5 liters/min $= 1.5$ kg/min $= 0.025$ kg/s; 4000 W $= (4186 \text{ J/kg·K})(0.025 \text{ kg/s})\Delta T$; $\Delta T = 38$ K; $T_f = 58$°C. **25.** $Q = cm\Delta T = (4186 \text{ J/kg·K})(1.00 \text{ kg})(80 \text{ K}) = 334 \, 880$ J; $(334 \, 880 \text{ J})/(50 \text{ J/s}) = 6697.6$ s $= 1.9$ h. **29.** 150 lb $\to 68.04$ kg; $Q = cm\Delta T$; $\Delta T = 33.89$°C $- 37$°C $= -3.11$ K; $Q = (3474.4 \text{ J/kg·K})(68.04 \text{ kg})(3.11 \text{ K}) = 0.74$ MJ. **33.** The average specific heat capacity of the body is 3.5 kJ/kg·K (0.83 kcal/kg·C°), and normal temperature is 37°C. From Eq. (19.1), $Q = cm\Delta T = (3.5 \text{ kJ/kg·K})(70 \text{ kg})(6 \text{ K}) = 1470$ kJ will raise the temperature 6 K. Since 200 kcal/h $= 837.2$ kJ/h, it will take $(1470 \text{ kJ})/(837.2 \text{ kJ/h}) = 1.8$ h.

37. $Q = m \int_{T_i}^{T_f} (K_1 - K_2 T + K_3 T^2)\, dT =$
$m[K_1 T - \frac{1}{2} K_2 T^2 + \frac{1}{3} K_3 T^3]_{T_i}^{T_f} =$
$m[K_1(T_f - T_i) - K_2 T_f^2/2 + K_2 T_i^2/2 +$
$K_3 T_f^3/3 - K_3 T_i^3/3]$ and $Q = m[K_1(T_f - T_i) - \frac{1}{2} K_2(T_f^2 - T_i^2) + \frac{1}{3} K_3(T_f^3 - T_i^3)]$

41. The tension in the rope is $g(6 \text{ kg}) = 58.8$ N, and this is the force exerted (via friction) on the rope by the rod, and it does it over a distance of $2\pi R(240) = 22.62$ m. The work thus done is $(58.8 \text{ N})(22.62 \text{ m}) = 1330$ J, and this result equals the heat-in: $Q = cm\Delta T$; $1330 \text{ J} = c(0.250 \text{ kg})(12.0 \text{ K})$; $c = 443$ J/kg·K, which is what we might expect for some sort of steel. **45.** $L_v = 2336 \times 10^3$ J/kg; $Q = mL_v = (1.5 \text{ kg}) \times (2336 \times 10^3 \text{ J/kg}) = 3.5$ MJ. **49.** $Q = cm\Delta T + mL_v = (138 \text{ J/kg·K}) \times (0.50 \text{ kg})(630 \text{ K} - 240 \text{ K}) + (0.50 \text{ kg}) \times (296 \times 10^3 \text{ J/kg}) = 26\,910 \text{ J} + 148\,000 = 0.17$ MJ. **53.** $-m(334\,000 \text{ J/kg}) = (0.50 \text{ kg})(4186 \text{ J/kg·K})(0° - 20°\text{C})$; $m = 0.125$ kg. **57.** $Q/t = \lambda_T A \Delta T/d = (18 \text{ Cal·cm/}$ m²·h·C°$)(1.4 \text{ m}^2)(4°\text{C})/(2.0 \text{ cm}) = 50.4$ Cal/h $= 0.014$ kcal/s $= 0.059$ kJ/s $= 0.059$ kW. Therefore, 2.1×10^5 J per hour.

61. $\dfrac{dT}{(T - T_s)} = -K \, dt$; $\displaystyle\int_{T_0}^{T} \dfrac{dT}{(T - T_s)} =$
$\displaystyle\int_0^t (-K)\, dt$; $\ln(T - T_s) - \ln(T_0 - T_s) = -Kt$
and $T = T_s + (T_0 - T_s)e^{-Kt}$.
65. 1 mol = 18 g; at STP there are 6.02×10^{23} molecules/mol; to vaporize 1 mol requires $Q = mL_v = (0.018 \text{ kg})(2492 \text{ kJ/kg}) = 44.86$ kJ. Hence, per molecule the energy is $(44.86 \times 10^3 \text{ J})/(6.02 \times 10^{23}$ molecules/mol$) = 7.4 \times 10^{-20}$ J. **69.** $Q = mL_f$; $mL_f = (23 \times 10^3 \text{ J/kg})m$; to which must be added $Q = cm\Delta T = (130 \text{ J/kg·K})m \times (327 \text{ K} - 23 \text{ K}) = (39\,520 \text{ J/kg})m$; the total is then $(62\,520 \text{ J/kg})m = \frac{1}{2}mv^2$; $v = 354$ m/s. **73.** $-Q_{out} = Q_{in}$; if it dropped to zero, the calorimeter would provide an amount of heat equal to $(1.00 \text{ kcal/kg·K})(0.398 \text{ kg})(5.1 \text{ K}) + (0.093 \text{ kcal/kg·K})(0.102 \text{ kg})(5.1 \text{ K}) = 2.078$ kcal; the amount of heat needed to melt the ice is $(0.040\,5 \text{ kg})(80 \text{ kcal/kg}) = 3.24$ kcal; hence, not all the ice melts (!), and $T_f = 0°\text{C}$. **77.** $Q/t = k_{TB} A \Delta T/d_B = k_{TA} A \Delta T/d_A$; $k_{TB}/d_B = k_{TA}/d_A$; $k_{TB}/k_{TA} = 0.60/0.026 = 23 = d_B/d_A$; $d_B = 23$ m. **83.** In the first set-up, $\mathrm{P}t = c_w m\Delta T + \mathrm{P}_l t$ and, in the second, $\mathrm{P}'t = c_w m'\Delta T + \mathrm{P}_l t$; note that the mean temperature is the same in both cases, and so P_l is the same. Thus, solve for P_l in both equations and set them equal. $\mathrm{P} - c_w m\Delta T/t = \mathrm{P}' - c_w m'\Delta T/t$; and $(\mathrm{P} - \mathrm{P}')t = c_w \Delta T(m - m')$.

CHAPTER 16
Answers to Odd-Numbered Multiple Choice Questions
1. b **3.** d **5.** d **7.** b **9.** e **11.** b **13.** c
15. a **17.** d **19.** b

Answers to Odd-Numbered Problems
1. +250 J **3.** −450 J **5.** +500 J **7.** 519 J
9. $W = \frac{1}{4}P_i V_i$ **11.** $W = \frac{1}{2}P_i V_i$ **13.** +12 J
15. 5.5% **17.** decrease T_L **19.** 3.5 kW
21. 2.68 J/K **25.** +75.6 J/K **27.** $\Delta S_f =$
1.22 kJ/K, $\Delta S = 12$ J/K **31.** $W = K(V_f^3 - V_i^3)/3$ **33.** +9.8 kJ **35.** 0.22 g
37. $T_f = -78°\text{C}$ **39.** (a) 0.038 4 MPa
(b) 207 K **41.** $W = nRT \ln(P_i/P_f)$
45. impossible **47.** 399 W **49.** $Q_H/W_i =$
6.5, pay for 1 unit of work and get 6.5 units of heat **51.** 0.69 W **53.** $T_H = 547$ K **55.** $e =$
62.5%, irreversible **57.** +0.014 kJ/K
59. $W = 6.0$ MJ, $e = 67\%$
61. $W = nRT \ln \dfrac{P_i}{P_f}$ **65.** $W = \dfrac{nRP_0}{a} \times$
$\ln \dfrac{(P_0 + aT_f)}{(P_0 + aT_i)}$ **67.** $\Delta S = nR \ln \dfrac{V_f}{V_i}$; change in
entropy is positive. **71.** $T_i/T_f = 0.49$
73. 5.8 J/K

Solutions to Selected Problems
1. From Eq. (16.2), $\Delta U = Q - W = (+500 \text{ J}) - (+250 \text{ J}) = +250$ J.
5. Since $\Delta V = 0$, $W = 0$, and from Eq. (16.1), $Q = \Delta U = +500$ J. **9.** $W = \frac{1}{2}(\frac{1}{2}P_i) \times (2V_i - V_i) = \frac{1}{4}P_i V_i$. **13.** The work done is the area under the curve, which is $\frac{1}{2}(3.0 \text{ MPa}) \times (2.0 \times 10^{-6} \text{ m}^3) + (4.0 \times 10^{-6} \text{ m}^3) \times (2.0 \text{ MPa}) + (1.0 \text{ MPa})(1.0 \times 10^{-6} \text{ m}^3) = 3.0 \text{ J} + 8.0 \text{ J} + 1.0 \text{ J} = 12$ J, and this is positive, since the gas expanded and it did work on the surroundings. **17.** $e_c = 1 - (T_L/T_H) = (T_H - T_L)/T_H$; increasing T_H increases both the numerator and the denominator; decreasing T_L increases the numerator only and therefore has the more profound effect. **21.** $\Delta S = (1.00 \text{ kJ})/(373 \text{ K}) = 2.68$ J/K. **25.** $\Delta S = Q/T = mL_f/T = (0.500 \text{ kg})(205 \text{ kJ/kg})/(1356 \text{ K}) = +75.6$ J/K. **29.** $B = -V dP/dV$; for an adiabatic process $PV^\gamma = $ constant; $d(PV^\gamma)/dV = V^\gamma dP/dV + P\gamma V^{\gamma-1} = 0$ and so $B = P\gamma$ and $K = 1/P\gamma$. **33.** $W = P\Delta V = (0.30 \text{ MPa})(0.50 \text{ m}^2)(+0.065 \text{ m}) = +9.8$ kJ.
37. $P_i V_i^\gamma = P_f V_f^\gamma$, where we assume the rapid expansion means an adiabatic process; final pressure is 1 atm; we don't have initial volume, so let's work per-unit-volume. $(4.5 \text{ atm})(1 \text{ m}^3) = (1 \text{ atm}) V_f^\gamma$; $V_f = 4.5^{1/\gamma} = 2.93 \text{ m}^3$ per m³, where $\gamma = 1.4$. To find T_f, use Ideal Gas Law; $P_i V_i/T_i = P_f V_f/T_f$; (4.5 atm) $\times (1 \text{ m}^3)/(300 \text{ K}) = (1.0 \text{ atm})(2.93 \text{ m}^3)/T_f$; $T_f = 195$ K, or $-78°\text{C}$. Notice that we didn't need to know the actual initial volume.
41. From Eq. (16.5), $W = nRT \ln(V_f/V_i)$, but $P_i V_i = P_f V_f$; hence, $W = nRT \ln(P_i/P_f)$.
45. The maximum efficiency from Eq. (16.9) requires the temperatures in kelvins; $T_H \to 144°\text{F} = 62.2°\text{C}$ or 355.2 K; $T_L \to 5°\text{F} = -15°\text{C}$ or 258 K; $e_c = 1 - (258/335.2) = 23\%$. As compared to the general expression for efficiency, Eq. (16.10), $e = (26 \text{ kJ})/(102 \text{ kJ}) = 25.4\%$. The engine exceeds the efficiency of a Carnot engine and is therefore impossible. **49.** $\eta = $ heat-in/

work-in $= Q_L/W_i = 5.5$. Moreover, $Q_H = Q_L + W_i = 5.5 W_i + W_i = 6.5 W_i$; hence, the ratio we want $Q_H/W_i = 6.5$. You pay for 1 unit of work and get 6.5 units of heat into the house, which compares very well with electrical heating (6.5 times better), where 100% of the electrical energy is converted 'into heat and you have to pay for each unit of energy that comes into the house.
53. $e_c = 1 - T_L/T_H = 42.2\%$ and $T_L = 273.4$ K. $T_H = 547$ K. **57.** Since the masses are equal, $\Delta T = \pm 4.0$ K; thus, $Q = mc\Delta T = (20 \text{ kg})(4.186 \text{ kJ/kg·K})(\pm 4.0 \text{ K}) = \pm 334.88$ kJ. Average temperatures are 38°C and 34°C; $\Delta S = (-334.88 \text{ kJ})/(311 \text{ K}) + (+334.88 \text{ kJ})/(307 \text{ K}) = (-1.077 \text{ kJ/K}) + (+1.091 \text{ kJ/K}) = +0.014$ kJ/K. **63.** (a) The internal energy of the gas of N molecules is all KE; $U = 3Nk_B T/2$, and so a change in temperature produces a change in internal energy $dU = 3Nk_B dT/2$; (b) $PV = Nk_B T$ and $dW = P dV = [(Nk_B T)/V]dV$; $dU + dW = 0$, and so $(dT/T) + (2/3)(dV/V) = 0$; $\ln T + (2/3) \ln V = $ constant; $TV^{2/3} = $ constant $\times T$; $PV^{5/3} = $ constant.
69. Using Eq. (16.7), $P_i V_i^\gamma = P_f V_f^\gamma$, we have $P_i/P_f = (V_f/V_i)^\gamma$; $P_i V_i = nRT_i$ and $P_f V_f = nRT_f$ and, dividing these, $P_i/P_f = T_i V_i/T_f V_i$; getting rid of the pressures, $(V_f/V_i)^\gamma = T_i V_f/T_f V_i$, from which Eq. (16.19) follows immediately.
73. From Problem 60, $\Delta S = nR \ln(V_f/V_i) = (1.0 \text{ mol})(8.314\,4 \text{ J/mol·K}) \ln 2 = 5.8$ J/K.

CHAPTER 17
Answers to Odd-Numbered Multiple Choice Questions
1. c **3.** b **5.** c **7.** a **9.** a **11.** d **13.** b
15. c **17.** d **19.** e

Answers to Odd-Numbered Problems
1. 5.7×10^{-18} kg **3.** $\pm 1.1 \times 10^{-5}$ C
5. 2.3 N **7.** 5.3 μC **9.** zero **11.** 8.0 μN
13. zero **15.** $q = \pi C R^4$ **17.** 0.58 N
19. 8.3 N, $\theta = 49°$ **21.** 6.6 N, $\theta = 74°$
23. 0.11 μC **27.** $F_x = kqQx/r^3$
29. $q = -0.34Q$, $a = 0.41$, where Q and q are separated by ad **31.** 8.0×10^2 N/C
33. −1/10 N/C due east **35.** 5.6×10^{-11} N/C straight down **37.** 5.1×10^{11} N/C
39. 7.8×10^4 N/C **41.** 6.4×10^2 N/C straight upward **43.** $E = \sigma/\varepsilon_0$ **45.** 4.4 nC
47. $E_x = -\lambda k/x_0$ **49.** 9.6×10^{10} m/s²
51. $E = R\sigma/r\varepsilon$ **53.** $E = \sigma/\varepsilon_0$
55. 2.6×10^{15} m/s² **57.** $E = Q/4\pi R^2 \varepsilon_0$ within the space, zero beyond the shells
59. $E = 2k\lambda/x$ **63.** $E_y = -2k\lambda/R$
65. $E(r) = (kQ/r^2)(1 - 5r^3/2R^3 + 3r^5/2R^5)$. When $r = R$, $E = 0$. **67.** $E = rQ/4\pi\varepsilon_0 R^3$
69. 1.4×10^{21} N/C

Solutions to Selected Problems
1. $N = (10^{-6} \text{ C})/(-1.6 \times 10^{-19} \text{ C}) = 6.2 \times 10^{12}$ electrons. Each has a mass of 9.109×10^{-31} kg; hence, the mass increase is 5.7×10^{-18} kg. **5.** $F = kqq/r^2 = (9.0 \times 10^9 \text{ N·m}^2/\text{C}^2)(1.6 \times 10^{-19} \text{ C})^2/(1.0 \times$

10^{-14} m$)^2 = 2.3$ N.   **9.** By symmetry, the net force is zero.   **17.** $F_3 = F_{31} + F_{32} =$ $(9 \times 10^9$ N·m^2/C$^2)(-12.5 \times 10^{-6}$ C) $\times$ $(-10.0 \times 10^{-6}$ C)/(3.0 m)$^2 + (9 \times 10^9$ N· m^2/C$^2)(-5.0 \times 10^{-6}$ C)$(-10.0 \times 10^{-6}$ C)/ $(1.0$ m)$^2 = 0.125$ N + 0.450 N = 0.58 N, in the increasing x-direction.   **19.** See Fig. AN 19. $F_{12} = kq_1q_2/r_{12}^2 = 3.6$ N; $F_{13} =$ $kq_1q_3/r_{13}^2 = 4.5$ N; $F_{14} = kq_1q_4/r_{14}^2 = 1.8$ N; $F_{x1} = (1.8$ N) + (4.5 N)4/5 = 5.4 N; $F_{y1} = -(3.6$ N) $-$ (4.5 N)3/5 = -6.3 N; $F_1 = \sqrt{5.4^2 + 6.3^2} = 8.3$ N; $\theta =$ $\tan^{-1}(6.3/5.4) = 49°$.

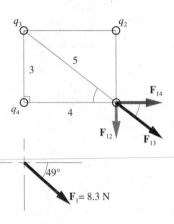

Fig. AN 19.

25. $dF = kq(Q/L)(dx/r^2)$; only the y-component doesn't cancel and so $F_y = kq\lambda \int (\cos \theta/r^2)\, dx$; $r^2 = h^2 + x^2$; $x = h \tan \theta$; $dx = h\sec^2\theta\, d\theta$; and $\sec^2\theta = 1 + \tan^2 \theta$;   **29.** The repulsive force between the two known charges is $F =$ $kQ2Q/d^2$; put a negative charge $-q$ on the line connecting the two positive charges and between them at a distance ad from $+Q$ and $(1 - a)d$ from $+2Q$. The attraction between $+Q$ and $-q$ will cancel the repulsive force on $+Q$ when $k2Q^2/d^2 = kqQ/(ad)^2$ or when $a^2 = q/2Q$; similarly $+2Q$ will be in equilibrium when $k2Q^2/d^2 = kq2Q/$ $[(1 - a)d]^2$, that is, $(1 - a)^2 = q/Q$; hence, $(1 - \sqrt{q/2Q}) = \sqrt{q/Q}$; $1 - \sqrt{q}/$ $\sqrt{2Q} = \sqrt{q}/\sqrt{Q}$; $1 = \sqrt{q}(1 + \sqrt{2})/$ $(\sqrt{2}\sqrt{Q})$; $q = 2Q/(1 + \sqrt{2})^2 = -0.34\ Q$ and $a = 0.41$.   **33.** $E = F/q = (2.0$ nN)/ $(-20$ nC) $= -1/10$ N/C; due east.   **37.** $E =$ $kq_e/r^2 = (9.0 \times 10^9$ N·m^2/C$^2)(1.6 \times 10^{-19}$ C)/ $(5.3 \times 10^{-11}$ m)$^2 = 5.1 \times 10^{11}$ N/C. The horizontal field cancels, leaving only twice the vertical contribution from each charge: $E = 2(kq/r^2) \cos \theta = 2[(9.0 \times 10^9$ N·m^2/ C$^2)(+50 \times 10^{-9}$ C)/(1.0 m)$^2](\cos 45°) =$ 6.4×10^2 N/C, straight upward.   **43.** See Fig. AN 43. The field is parallel to the curved side and perpendicular to the endface; since $E = 0$ inside the metal, $EA = \sigma A/\varepsilon_0$ and $E = \sigma/\varepsilon_0$.

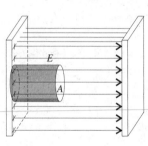

Fig. AN 41.

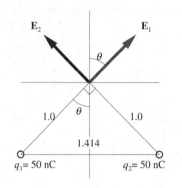

Fig. AN 43.

49. $F = qE = ma$; $a = qE/m = (1.6 \times 10^{-19}$ C)$(20 \times 10^3$ N/C)/$(3.35 \times 10^{-26}$ kg) $=$ 9.6×10^{10} m/s^2.   **53.** See Fig. AN 53. Use a very short perpendicular Gaussian cylindrical surface. Assume that E and σ are constant over the tiny area of the cylinder (both may vary from region to region, depending on the shape of the conductor). Embed one endface within the conductor, where $E = 0$. Hence, $EA = \sigma A/\varepsilon_0$ and $E = \sigma/\varepsilon_0$.

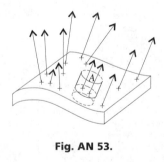

Fig. AN 53.

57. See Fig. AN 57. Surround the inner sphere with a spherical Gaussian surface of radius R ranging in the gap between the shells. The field is radial and $EA = E4\pi R^2 = Q/\varepsilon_0$; hence, $E = Q/4\pi R^2\varepsilon_0$, within the space. Now draw a Gaussian surface encompassing the

large sphere anywhere beyond it. Again, if there is a field, by symmetry it would have to be radial, but now the net charge enclosed is zero and the field is therefore zero.

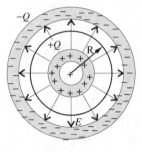

Fig. AN 57.

63. $dE = (k\lambda/R^2)\, dl$; for every element on the right of the y-axis there is an element on the left of it and so all horizontal field contributions cancel, leaving only ones in the direction of $-y$.

$$E_y = -\int_{-\pi/2}^{\pi/2} (k\lambda/R^2) \cos \theta\, dl; \quad dl = R\, d\theta;$$

and so

$$E_y = -\int_{-\pi/2}^{\pi/2} (k\lambda/R) \cos \theta\, d\theta = -2k\lambda/R.$$

69. $E = kq_e/r^2 = (9 \times 10^9$ N·m^2/C$^2) \times$ $(1.6 \times 10^{-19}$ C)/$(1.0 \times 10^{-15}$ m)$^2 = 1.4 \times$ 10^{21} N/C.

CHAPTER 18

Answers to Odd-Numbered Multiple Choice Questions

1. b   **3.** c   **5.** b   **7.** d   **9.** d   **11.** d   **13.** c **15.** d   **17.** b   **19.** b

Answers to Odd-Numbered Problems

1. -25.0×10^{-7} J   **3.** ±0.10 V   **5.** V = 0 **7.** -3.6×10^5 V   **9.** ΔKE = +500 eV, ΔPE$_E = -500$ eV   **11.** $V_R - V_r =$ $-(kQ/2R)(1 - r^2/R^2)$ **13.** $V_O = kQ/r$; $V_R = kQ/R$ **17.** $E_x = kQx(x^2 + R^2)^{-3/2}$ **19.** ΔKE $= 4.8 \times 10^{-15}$ J, 1.0×10^8 m/s **21.** 1.4×10^{-17} J   **23.** 1.3×10^2 V **25.** 5.58×10^{-12} V   **27.** $+0.23$ J **29.** 11 MV/m   **33.** $V = k\frac{Q}{L} \ln \frac{\sqrt{(L/2)^2 + h^2} + L/2}{\sqrt{(L/2)^2 + h^2} - L/2}$ **35.** $E_x = 2kQ/R^2$; $E_x = \sigma/2\varepsilon_0$ **37.** $E_x = -kp(y^2 - 2x^2)/(x^2 + y^2)^{5/2}$; $E_y = kp3xy/(x^2 + y^2)^{5/2}$. **39.** $V = -8.3$ kV, $E = 0$ **41.** $Q = 1.5 \times 10^{-10}$ C   **43.** $Q = 1 \times 10^{-5}$ C **45.** $R = 90$ mm   **47.** 5.0 μF **49.** 8.9×10^{-10} F   **51.** 2.5 pF   **53.** 110 pF **55.** 0.99 μF   **57.** 10 pF   **59.** 1.4 nJ

61. 0.01 F/m^2 **63.** 7×10^{-4} C/m^2
65. 1.5 nJ/m   **67.** 9 pF   **69.** 13 μF
71. 3.6×10^{-10} C   **73.** 1.0 V   **75.** 2.2 nJ
77. $C = L/[2k \ln(R_b/R_a)]$ **79.** 24 nF
81. 1.4 nJ

Solutions to Selected Problems

1. $\Delta W = \Delta PE_E = q\Delta V = (-25.0 \times 10^{-9}$ C$)(100$ V$) = -25.0 \times 10^{-7}$ J. This is the work done against the field. The negative sign means the charge has work done on it by the field. It moves from 0 to 100 V and so in the opposite direction to the field. That's the direction a negative charge moves in spontaneously, propelled by the field.   **5.** A charge of $-Q$ is induced on the inner surface of the outer sphere. At a point outside, the potential is equivalent to that due to a point charge of $+Q$ and a point charge of $-Q$, both at the center. Hence, at P, $V = 0$.
13. $E_r = kQ/r^2$; $V_O = -\int_\infty^r E_r \, dr = -\int_\infty^r (kQ/r^2) \, dr = kQ/r$. Because the potential is continuous $V_R = kQ/R$.   **21.** At the surface, $V = k_0Q/R = (8.99 \times 10^9$ N·m^2/C$) \times (1.00 \times 10^{-9}$ C$)/(0.10$ m$) = 90$ V. $W = q_e\Delta V = (1.6 \times 10^{-19}$ C$) \times (90$ V$) = 1.4 \times 10^{-17}$ J.   **27.** The potential at the point is $V = k_0Q_1/r_1 + k_0Q_2/r_2 = (8.99 \times 10^9$ N·m^2/C$^2)(+10 \mu$C$)/(4.0$ m$) + (8.99 \times 10^9$ N·m^2/C$^2)(-25 \mu$C$)/(5.0$ m$) = -22.5$ kV. $W = q\Delta V = (-10 \mu$C$) \times (-22.5$ kV $- 0) = +0.23$ J.
33. $V = k\frac{Q}{L}\int_{-L/2}^{L/2} \frac{dx}{\sqrt{x^2 + h^2}} = k\frac{Q}{L}\ln\frac{\sqrt{(L/2)^2 + h^2} + L/2}{\sqrt{(L/2)^2 + h^2} - L/2}$ **39.** Each face has a diagonal of $\sqrt{2}$ m, and so there is a tilted rectangle $\sqrt{2}$ by 1.0 m passing through the center and having four spheres at its corners. The diagonal of that square is $\sqrt{3}$ m long and half of that, $\frac{1}{2}\sqrt{3}$, is the corner-to-center distance. Hence, $V = 8kQ/r = -8.3$ kV. And $E = 0$.   **41.** $Q = CV = (100 \times 10^{-12}$ F$)(1.5$ V$) = 1.5 \times 10^{-10}$ C.   **47.** $C = Q/V = (20 \mu$C$)/(4.0$ V$) = 5.0 \mu$F.   **51.** $C = \varepsilon A/d$; $A = \pi R^2 = \pi(0.25 \times 10^{-2}$ m$)^2 = 1.96 \times 10^{-5}$ m^2; $C = (4.8)(8.85 \times 10^{-12}$ F/m$) \times (1.96 \times 10^{-5}$ m$^2)/(0.33 \times 10^{-3}$ m$) = 2.5$ pF.
57. The 12-pF and 4.0-pF capacitors are in series and so equivalent to 3.0 pF, which, in turn, is in parallel with 7.0 pF. Hence, $C = 10$ pF.   **61.** The membrane is like a rolled-up parallel-plate capacitor, so $C = \varepsilon A/d$; $C/A = \varepsilon/d = 7\varepsilon_0/d = 7(8.85 \times 10^{-12}$ C^2/N·m$^2)/(6$ nm$) = 0.01$ F/m^2.
65. $\Delta PE_E/L = \frac{1}{2}CV^2/L = \frac{1}{2}(3 \times 10^{-7}$ F/m$) \times (0.1$ V$)^2 = 1.5$ nJ/m.   **69.** The two 6.0-μF capacitors are in series (yielding 3.0 μF), as are the two 5.0-μF capacitors (yielding 2.5 μF). Now everything is in parallel, so 3.0μF $+ 7.5 \mu$F $+ 2.5 \mu$F $= 13 \mu$F.
71. The 9.0-pF capacitor is in parallel with the battery, and so the voltage is 12 V. All the capacitors are in parallel; hence, $C = 30$ pF. $Q = CV = (30$ pF$)(12$ V$) = 3.6 \times 10^{-10}$ C.

81. The diagram shows successive simplifications of the circuit. The equivalent capacitance is 20 pF; hence, $PE_E = \frac{1}{2}CV^2 = \frac{1}{2}(20$ pF$)(12$ V$)^2 = 1.4$ nJ.

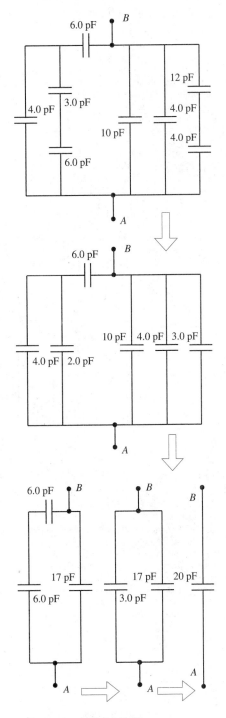

Fig. AN 81.

CHAPTER 19

Answers to Odd-Numbered Multiple Choice Questions
1. c   **3.** d   **5.** c   **7.** e   **9.** b   **11.** b   **13.** a
15. c   **17.** d   **19.** c   **21.** a

Answers to Odd-Numbered Problems
1. 10 μA   **3.** 96.5 A   **5.** 3.6×10^2 C
7. 6.2×10^{14} **9.** 120 C   **11.** 1.5×10^3 cells
13. 36 kC   **15.** $I = (6.00$ A/s$^2)t^2 + (8.00$ A/s$)t$ **17.** $q = 7.00$ C   **19.** $J = (6.00 \times 10^6$ C/s^3·m$^2)t^2 - (2.00 \times 10^6$ C/s·m$^2)$; 4.00×10^6 A/m^2.   **21.** 27 h   **23.** (a) A and B (b) A and E (c) A and D and C   (d) A and D   (e) A and F when the headlights are on
25. 20 cells in series form a row, and there are 20 rows in parallel   **27.** 27 h
31. 0.11 MC   **33.** 10 Ω **35.** 50×10^9 Ω
37. if it's 1 m on a side
41. 1.6×10^{-8} Ω·m   **43.** 1.9 mm
45. (a) 0.12 Ω (b) 0.10 Ω **47.** 0.1 A
49. 0.1 W   **51.** 0.16 A   **53.** 2.2 kW
55. 0.017 V/m   **57.** 1×10^3 °C
59. 26 Ω **61.** 220° C
63. 87 kW   **65.** (a) 2.0×10^{-4} Ω
(b) 2.4×10^{-5} V (c) 2.9×10^{-7} W
67. 5.2 kW   **69.** 30 W   **71.** 1.3 W, 12 Ω
73. 12 W   **75.** $R = (\rho/4\pi)[(1/r_i) - (1/r_o)]$
77. $\Delta R/R_0 = 2.7$, poor: 12
81. $E = V/[r \ln(r_o/r_i)]$; $I = 2\pi LV/[\rho \ln(r_o/r_i)]$; $R = [\rho \ln(r_o/r_i)]/(2\pi L)$.

Solutions to Selected Problems
1. $I = \Delta Q/\Delta t = (10 \mu$C$)/(1.0$ s$) = 10 \mu$A.
7. $I = \Delta Q/\Delta t$; $\Delta Q = (1.0$ mA$)(0.10$ s$) = 0.10$ mC. The number of protons is $(0.10 \times 10^{-3}$ C$)/(1.602 \times 10^{-19}$ C$) = 6.2 \times 10^{14}$.
11. $(220$ V$)/(0.15$ V per cell$) = 1466.7$ cells $= 1.5 \times 10^3$ cells. Salt water, because of the presence of ions, is a much better conductor.   **15.** $I = dq/dt = 3(2.00$ C/s$^3)t^2 + 2(4.00$ C/s$^2)t = (6.00$ A/s$^2)t^2 + (8.00$ A/s$)t$.
21. Number of coulombs passing per second is $(I/A)A = (1.0$ MA/m$^2)(1.00 \times 10^{-6}$ m$^2) = 1.0$ C/s; or 6.24×10^{18} electrons/s; hence, $(6.02 \times 10^{23}$ electrons$)/(6.24 \times 10^{18}$ electrons/s$) = 96\,507$ s $= 27$ h.   **25.** First, do the voltage, $(9.0$ V$)/(0.45$ V$) = 20$; we need 20 cells in series forming one row. To get the current, we must have $(440$ mA$)/(22$ mA$) = 20$ such rows in parallel.
29. $I = JA = \sigma E$; therefore $dq/dt = \sigma AE = -\sigma A \, dV/dx$.   **33.** $R = V/I = (100$ V$)/(10$ A$) = 10$ Ω.   **37.** If the cube is 1 m on a side, $R = \rho$.   **41.** $R = \rho L/A$; $\rho = AR/L = 1.6 \times 10^{-8}$ Ω·m.
45. (a) From Eq. (19.8), $R = 0.10$ Ω $+ (0.10$ $\Omega)(0.005\,0$ K$^{-1}) \times (30$ K$) = 0.12$ Ω (b) since the coefficient is essentially zero, $R = 0.10$ Ω.   **49.** $P = IV = (16 \times 10^{-3}$ A$)(9$ V$) = 0.144$ W $= 0.1$ W.   **55.** From Eq. (19.6) for a uniform E-field, $V = Ed$, so we must find V; since $V = IR$, we need R; $R = \rho L/A = (1.7 \times 10^{-8}$ Ω·m$)(1.0$ m$)/(1.0 \times 10^{-6}$ m$^2) = 0.017$ Ω; $V = (1.0$ A$)(0.017$ $\Omega) = 0.017$ V;

$E = V/d = (0.017 \text{ V})/(1.0 \text{ m}) = 0.017 \text{ V/m}$.
59. $R_f = R_0(1 + \alpha_0\Delta T) = (10 \ \Omega) \times (1 + 0.003 \ 9 \text{ K}^{-1} \times 400 \text{ K}) = 25.6 \ \Omega = 26 \ \Omega$. **63.** The current entering the factory must be $I = P/V = (45 \times 10^3 \text{ W})/(110 \text{ V}) = 409.09$ A; the total power loss in the cables is $I^2R = (409.09 \ \Omega)^2(2 \times 0.25 \ \Omega/\text{mile} \times 0.50 \text{ mile}) = 41.84$ kW; hence, the net power supplied must be 45 kW + 41.8 kW = 87 kW. **67.** P = (12 A)(500 V) = 6.0 kW. 86%(6.0 kW) = 5.2 kW = 6.9 hp. **71.** P = $IV = (1/3 \text{ A})(3.9 \text{ V}) = 1.3$ W. P = $I^2R = 1.3$ W = $(1/3 \text{ A})^2R$; R = 12 Ω. **79.** See Fig. AN 79. From the diagram $R/(T' + T) = R_0/T'$; and so $R = R_0(1 + T/T')$; comparing this expression with Eq. (19.8), we find that T is equivalent to ΔT and $\alpha_0 = 1/T'$.

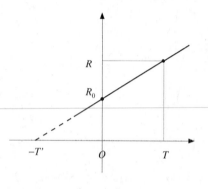

Fig. AN 79.

CHAPTER 20

Answers to Odd-Numbered Multiple Choice Questions

1. d **3.** d **5.** c **7.** c **9.** a **11.** b **13.** d
15. c **17.** a **19.** b **21.** b

Answers to Odd-Numbered Problems

1. 1.5 A, goes out **3.** 0.15 A, P doubles 1.8 W to 3.6 W **5.** 1 Ω **7.** 5.0 A **11.** 2 Ω **13.** 5.5 Ω **15.** 5.0 Ω **17.** 5 Ω **19.** 9 A through 17 Ω, 6 A through 3 Ω, 3 A through 4 Ω and 2 Ω **21.** 1 A through 4 Ω, 6/9 A through 3 Ω, 3/9 A through 6 Ω **23.** 10 Ω **25.** R_s =0.05 Ω **27.** 4.0 s **29.** 6.0 μA, 2.0 s **31.** R_e = 1.8 Ω **33.** 2 A through 3 Ω and 1 Ω, 5 A through 2 Ω, 2 A through 4 Ω and 1 Ω **35.** 4 Ω, 95 W **37.** 24 W, C is 1.5 V above A **39.** 0.89 A, 8.6 V **41.** 27 W **43.** R_s = 0.100 Ω **45.** 0.12 A, 0.60 ms **47.** 1.6 μA **49.** $I = I_ie^{-t/RC}$; $I_i = \mathcal{E}/R$ **51.** E = $\frac{1}{2}C\mathcal{E}^2$ **53.** 3 A, 3 A, 6 A **55.** $I_1 =$ 2 A, $I_2 = 2/3$ A, $I_3 = 4/3$ A **57.** 4 V **59.** 32 V **61.** $I_1 = 2.0$ A, $I_2 = 1.0$ A, $I_3 = 1.0$ A **63.** 8 μC **65.** $R_1 = 46 \ \Omega$, $\mathcal{E}_1 = 14$ V **67.** 28 Ω **69.** (a) zero (b) 36 μC **71.** R = 4.0 Ω, V = 30 V, $I_1 = 8.0$ A, $I_2 = I_6 = 2.0$ A, $I_3 = 1.0$ A, $I_4 = I_5 = 5.0$ A **73.** A at +14 V, E at +29 V, F at +12 V, D, C, B, and G at zero. **75.** 42/5 Ω.

Solutions to Selected Problems

3. $R_e = 80 \ \Omega$; $I = V/R_e = 0.15$ A. After S is closed $R_e = 40 \ \Omega$, and $I = 0.30$ A. P = IV, so the power doubles from 1.8 W to 3.6 W. **7.** Each lamp draws $I = P/V = (100 \text{ W})/(100 \text{ V}) = 1.0$ A; hence, the source must provide 5.0 A. **11.** 6 Ω in parallel with 3 Ω is 2 Ω. **17.** The only resistance between A and B is 5 Ω. **21.** The 3.0-Ω and 6.0-Ω resistors in parallel equal 2.0 Ω, and that's in series with the 4.0-Ω resistor; hence, $R_e = 6.0 \ \Omega$. $I = V/R_e = (6.0 \text{ V})/(6.0 \ \Omega) = 1.0$ A. This current passes through the 4.0-Ω resistor and splits 6 parts out of 9, that is, 6/9 A, through the 3.0-Ω resistor and 3/9 A through the 6.0-Ω resistor. **25.** $(50 \ \Omega)(0.001 \ I) = (0.999 \ I)R_s$; $R_s = 0.05 \ \Omega$. **29.** $I_i = V/R = (12.0 \text{ V})/(2.0 \text{ M}\Omega) = 6.0 \ \mu$A; $RC = (2.0 \text{ M}\Omega)(1.0 \ \mu\text{F}) = 2.0$ s. **31.** 3 Ω and 9 Ω are in parallel, and that's in parallel with the 18-Ω resistor yielding 2 Ω, which is in parallel with 3 Ω and 6 Ω for a total of 1 Ω; hence, $R_e = 1.8 \ \Omega$. **37.** See Fig. AN 37. The equivalent resistance is 6.0 Ω; the battery provides 2.0 A and 24 W. Working backwards, the currents split as shown. Going from A to B, there is a rise of $\frac{1}{2}$ V, and from B to C, another rise of 1 V; hence, C is $1\frac{1}{2}$ V above A.

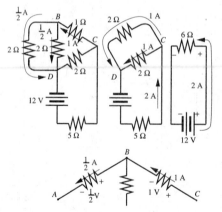

Fig. AN 37.

39. See Fig. AN 39. The equivalent resistance is 10.14 Ω. The current from the battery is 0.89 A, and terminal voltage is 9.0 V − (0.89 A)(0.50 Ω) = 8.6 V. **43.** The coil carries 1.00 mA when the meter carries 1.00 A, and so the shunt carries a maximum of 0.999 A; $(1.00 \times 10^{-3} \text{ A})r = (0.999 \text{ A})R_s$, $R_s = 0.100 \ \Omega$. **45.** See Fig. AN 45. $I_i = V/R = (12 \text{ V})/(100 \ \Omega) = 0.12$ A; $RC = (100 \ \Omega)(6.0 \ \mu\text{F}) = 0.60$ ms.

49. $\dfrac{d}{dt}\left(\mathcal{E} - \dfrac{Q}{C} - IR\right) = 0 - \dfrac{1}{C}\dfrac{dQ}{dt} - R\dfrac{dI}{dt} = 0$; $R\dfrac{dI}{dt} = -\dfrac{1}{C}I$; $\dfrac{dI}{I} = -\dfrac{1}{RC}dt$;
$\displaystyle\int_{I_i}^{I}\dfrac{dI}{I} = -\int_0^t \dfrac{1}{RC}dt$ and $\ln(I/I_i) = -t/RC$ and

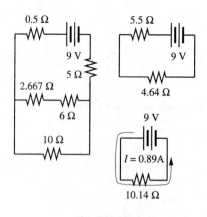

Fig. AN 39.

raising both sides to the eth power, $I = I_ie^{-t/RC}$. At $t = 0$, $I_i = \mathcal{E}/R$ because the charge on the capacitor is zero all the voltage is across the resistor.

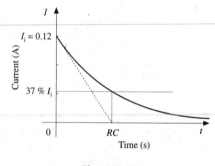

Fig. AN 45.

55. See Fig. AN 55. At A, $I_1 = I_2 + I_3$ around loop A-B-C-D-A going clockwise: $-3I_2 - 3I_2 + 20 \text{ V} - 8I_1 = 0$. Around loop A-B-C-A, clockwise: $-3I_2 - 3I_2 + 3I_3 = 0$. Around loop A-C-D-A, clockwise: $-3I_3 + 20 \text{ V} - 8I_1 = 0$. Hence, $I_1 = 2$ A, $I_2 = 2/3$ A, $I_3 = 4/3$ A.

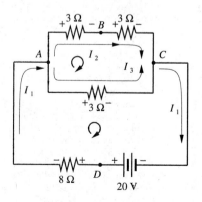

Fig. AN 55.

59. The drop across the top branch is 16 V; with that across the middle branch the current through it must be 1.0 A; hence, the current in the bottom branch, from the Node Rule, is 1.0 A. The drop across the two bottom resistors is 16 V; hence, the voltmeter must read 32 V.   **63.** The equivalent resistance in the steady state is 12 Ω; hence, the current is 1 A. The voltage across C is (1 A)(4 Ω) = 4 V, and since $CV = Q$, $Q = 8\ \mu C$.   **67.** See Fig. AN 67. $I_1 = 200$ mA, $V_{AB} = +2.4$ V; hence, around the first loop clockwise: $-0.200R + 12 - 2.4 - 4 = 0$, $R = 28\ \Omega$.

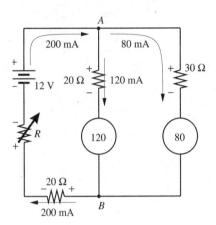

Fig. AN 67.

75. Place a source of, say, 18 V, across the group. Then solve for the six branch currents: here, 6/7 A goes through both 12-Ω resistors, 3/7 A goes through the central 6-Ω resistor, and 9/7 A passes through each outer 6-Ω resistor. Thus, the source provides 15/7 A, and so the equivalent resistance is 42/5 Ω, independent of the source.

CHAPTER 21

Answers to Odd-Numbered Multiple Choice Questions
1. d   **3.** c   **5.** c   **7.** b   **9.** b   **11.** a   **13.** a
15. b   **17.** c   **19.** c   **21.** a

Answers to Odd-Numbered Problems
3. 4.0×10^{-6} T   **5.** 4.0 mT   **7.** 0.10 mT
9. 63 μT   **11.** 0.63 nT   **13.** 0.20 A
15. 2.5 mT   **17.** $B = \mu_0 I/2R$; $B = N\mu_0 I/2R$
19. 1.3×10^{-11} T clockwise looking toward the source   **21.** 500 A   **23.** 24×10^{-7} T
25. 31° W of N   **27.** $B_z = \mu_0 I R^2/2z^3$
29. $\approx 6 \times 10^9$ A   **31.** $B = \mu_0 NI/2\pi r$, same for single wire   **37.** $B = \mu_0 I/2\pi R$
39. $B = \mu_0 I R^2/2(x^2 + R^2)^{3/2}$
41. $B_z = (\mu_0 I/4)[(1/R_b) - (1/R_a)]$
43. $B = \dfrac{\mu_0}{4\pi} \dfrac{IL}{h\sqrt{L^2 + h^2}} - \dfrac{\mu_0}{24} \dfrac{I}{R}$

47. $B = \mu_0 i$ inside, $B = 0$ outside
49. +x-direction   **51.** 1.3 T   **55.** 1.8×10^7 m/s   **59.** $-y$-direction   **61.** 1.00 A
63. 0.31 mN   **65.** 0.019 N·m   **67.** 3.8×10^{18} m/s^2 **69.** circle, $R = 14$ mm
71. $p = qBR$ **75.** 0.5 m/s   **77.** $\tau_f = 0.75\ \tau_i$
79. 2.0×10^{-5} N, repulsive
81. $F = (IC/h)\,[L/\sqrt{(L/2)^2 + h^2}]$
83. $f = qB/2\pi m$ **85.** $\mu_M = \omega Q R^2/3$

Solutions to Selected Problems
3. From Eq. (21.2), $B = (1.26 \times 10^{-6}$ T·m/A$)(10$ A$)/2\pi(0.50$ m$) = 4.0 \times 10^{-6}$ T.
7. $V = IR$; $I = (12$ V$)/(1.2\ \Omega) = 10$ A; $B = \mu_0 I/2\pi r = 0.10$ mT.   **11.** $B_z = \mu_0 I/2R = 0.63$ nT.   **15.** $V = IR$, $I = 0.20$ A; $B_z \approx \mu_0 nI = 2.5$ mT.   **21.** $B = \mu_0 I/2\pi r$; $I = 500$ A.   **25.** $B = \mu_0 I/2\pi r = 0.30 \times 10^{-4}$ T, west—due to the current; the net field is at an angle of $\tan^{-1} 0.30/0.50$ or 31° west of north.   **29.** We take a single circular equatorial current loop circulating with a radius of 2.3×10^3 km—it would be more realistic to assume several such loops distributed over the spherical core. We know that the field drops off as in Fig. 21.19b and the formula given. At a distance of $2.78R$, it's down by a factor of about 26. $B = \mu_0 I/2R$; 0.6×10^{-4} T = $(1.26 \times 10^{-6}$ T·m/A$)I/2(2.3 \times 10^6$ m$)$; $I = 0.22 \times 10^9$ A, but this distance is to be 6.3×10^6 m away along the z-axis, so the current should be roughly 26 times larger, or $I \approx 6 \times 10^9$ A. Since there is a south magnetic pole in the north, the current circulates from east to west.   **31.** Using Ampère's Law around a circular path of radius R within the torus, $\Sigma B_\parallel \Delta l = \mu_0 \Sigma I$. If we take the field to be axial (a solenoid bent around on itself), $B2\pi r = \mu_0 NI$; $B = \mu_0 NI/2\pi r$. The field varies inversely with r—the density of the windings is not the same on the inner and outer surfaces. For a single wire, $B = \mu_0 (NI)/2\pi r$, which is identical to the coil.   **35.** See Fig. AN 35. For a finite sheet, the field lines would loop around the sheet counterclockwise looking into the current (along the minus z-axis). The sheet is infinite so there cannot be a preferred direction other than parallel to it—no $\pm y$-component to the field. Moreover, there's no reason for the field to be any different at different distances from the sheet because, being infinite, the sheet "looks" the same as seen from any point above it. In effect, the notion of "distance from" loses its traditional meaning in this idealized situation. Going around the Ampèrean loop in the figure, $\Sigma B_\parallel \Delta l = \mu_0 \Sigma I$; $Bl + Bl = \mu_0 il$; and $B = \frac{1}{2}\mu_0 i$.   **39.** $dB = \dfrac{\mu_0}{4\pi} \dfrac{I\,dl \sin \theta}{r^2}$ where $\theta = 90°$ and so $dB = \dfrac{\mu_0}{4\pi} \dfrac{I\,dl}{(R^2 + x^2)}$; only the field along the x-axis contributes. $B = \int dB \cos \phi = \int \dfrac{R}{r} dB = \int \dfrac{R}{\sqrt{x^2 + R^2}} dB$;

$B = \int \dfrac{R}{\sqrt{x^2 + R^2}} \dfrac{\mu_0}{4\pi} \dfrac{I\,dl}{(R^2 + x^2)} = \dfrac{\mu_0}{4\pi} \dfrac{IR}{(x^2 + R^2)^{3/2}} \int dl$; the integral around the loop equals $2\pi R$ and so $B = \dfrac{\mu_0}{2} \dfrac{IR^2}{(x^2 + R^2)^{3/2}}$.

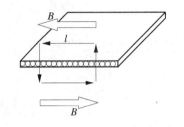

Fig. AN 35.

49. In the +x-direction.   **53.** Using $F = qvB$; $B = F/qv$; 1 T = 1 N/C·m/s = 1 (N·m)s/C·m^2 = 1 (J/C)s/m^2 = 1 V·s/m^2.   **57.** $qE = qvB$; $v = E/B$.   **61.** $F = IlB$; $I = F/lB = 1.00$ A.   **65.** $\tau = NIAB \sin \phi = (20)(1.5$ A$)(1.3 \times 10^{-3}$ m$^2) \times (0.90$ T$)(0.529\ 9) = 0.019$ N·m   **67.** $F = qvB \sin \theta = ma$; $a = (1.60 \times 10^{-19}$ C$)(5.0 \times 10^6$ m/s$)(5.0$ T$)0.866/(9.11 \times 10^{-31}$ kg$) = 3.8 \times 10^{18}$ m/s^2.   **71.** $qvB = mv^2/R$; $qB = mv/R$; $p = qBR$.   **75.** $V = vBl$; $v = V/Bl = (1.0\ \mu$V$)/(0.50$ mT$)(4$ mm$) = 0.5$ m/s.   **77.** $\tau = NIAB$; hence, $\tau_i/\tau_f = I_i/I_f = I_i/0.75I_i$; $\tau_f = 0.75\tau_i$.   **83.** $qvB = mv^2/R$, $qB = mv/R$, $v = qRB/m$; if f is the frequency, $1/f$ is the period, which equals $2\pi R/v$; hence, $f = v/2\pi R$ and $f = qB/2\pi m$.

CHAPTER 22

Answers to Odd-Numbered Multiple Choice Questions
1. e   **3.** c   **5.** a   **7.** a   **9.** c   **11.** c   **13.** b
15. d   **17.** d   **19.** b   **21.** b

Answers to Odd-Numbered Problems
1. 3.0 μWb   **3.** 1.2 Wb/m^2 **5.** 0.1 T, south   **7.** 0.50 V   **9.** 75 ms   **11.** 8.0 V
13. 0.60 V   **15.** bulb would not light
17. 90 μV   **19.** $\mathcal{E} = \pi R^2 \omega B \sin \omega t$
21. $\mathcal{E} = NAB_0 Ce^{-Ct}$ **23.** $\mathcal{E} = 2\pi f NBA \sin 2\pi ft$ **25.** $I = Blv/R$, clockwise, constant for a time l/v
27. 4.1 μC   **29.** 3.5 mV, counterclockwise   **31.** P = $(Blv)^2/R$
33. $B = RK\theta/NA$ **35.** 0.83 mT
37. $I = (NAB_0/R)\omega \cos \phi \sin \omega t$
39. $\Delta V = (\mu_0 Iv/2\pi) \ln[(y_0 + L)/y_0]$
43. $E = -CR^2/2r$ **45.** 1.6 V   **47.** 39 V
49. 100 V, $f = 60.000$ Hz   **51.** $L = 0.26$ H
53. 0.80 mWb   **55.** $N = 2.5 \times 10^3$
57. 8.4×10^2 **59.** 5.0 H   **61.** 2.5 H

63. 0.632 **65.** 1.2 H **67.** 21 mT
69. 12 mV **71.** $\mathscr{E} = (1.3\text{ kV})\sin 200\pi t$
73. 0.15 s **75.** 79 mJ/m **77.** (a) 6.0 A
(b) 12 ms (c) 3.8 A **79.** (a) 2.0 A
(b) 4.0 s (c) $\approx$20 s **83.** 7×10^{16} J,
7×10^8 gallons of gasoline
85. $I = -(V_0/2\pi fL)\sin 2\pi ft$; $V_0/2\pi fL$
89. $PE_M = (\mu_0 I^2/4\pi)l \ln(r_b/r_a)$ **91.** $\Delta I/\Delta t =$
$(V - IR)/L$, 0.89 kA/s never reaches 1.0 A
93. 21 V

Solutions to Selected Problems
1. $\Phi_M = B_\perp A = (1.2\text{ mT})(0.002\,5\text{ m}^2) =$
3.0 μWb. **5.** The final field minus the
initial field is the change in the field and
that's 0.1 T, south. **9.** From Eq.(22.3),
$\Delta t = 15\theta \times (0.030\text{ Wb})/(60\text{ V}) = 75$ ms.
13. $\mathscr{E} = vBl = (200\text{ m/s})(0.050\text{ mT})(60\text{ m}) =$
0.60 V. **21.** $\Phi_M = BA = A(B_0 e^{-Ct})$;
$\mathscr{E} = -Nd\Phi_M/dt = -NAB_0(-Ce^{-Ct}) =$
$NAB_0 Ce^{-Ct}$ **25.** As soon as the leading
edge of the loop starts cutting field lines,
there is an emf $= Blv$, which produces a
counterclockwise current $I = Blv/R$. The emf
is constant (for a time l/v) until the trailing
edge of the loop enters the field. It produces
an oppositely directed emf (the flux no longer
changes), and the net emf goes to zero. It
remains zero until the leading edge emerges
from the field at a time $3l/v$ after it entered,
whereupon the emf jumps to Blv in the
opposite direction. A current $I = Blv/R$
appears clockwise and remains constant for a
time l/v. **29.** From Eq. (22.1), $N\Delta\Phi_M =$
$1(0.500\text{ T})\cos 30° \times (1.6 \times 10^{-3}\text{ m}^2) =$
6.928×10^{-4} T·m^2; from Eq.(22.3), emf $=$
3.5 mV. Counterclockwise. **33.** $I = V/R$;
$\Delta Q = N\Delta\Phi_M/R = NBA/R = K\theta$, $B =$
$RK\theta/NA$. **39.** From Eq. (22.4) $\Delta V =$
$\int_{y_0}^{y_0+L} Bv\,dy = \frac{\mu_0 I}{2\pi}v \int_{y_0}^{y_0+L}\frac{dy}{y} =$
$(\mu_0 Iv/2\pi) \ln[(y_0 + L)/y_0]$. **47.** For
each length $\mathscr{E} = vBl =$
$(22\text{ m/s})(0.35\text{ T})(0.10\text{ m}) = 0.77$ V. Since
there are 2 lengths per turn and 25 turns,
$50(0.77\text{ V}) = 39$ V. **51.** $N\Phi_M = LI$;
$L = (500)(0.002\,0\text{ Wb})/(3.8\text{ A}) = 0.26$ H.
55. $N\Phi_M = LI$; $N = (2.5\text{ H})(1.80\text{ A})/$
$(1.80\text{ mWb}) = 2.5 \times 10^3$. **59.** $\mathscr{E} = -L\dfrac{\Delta I}{\Delta t}$,
$L = -(10\text{ V})/(2.0\text{ A/s}) = 5.0$ H (forget the
sign, since we don't know if the current is
increasing or decreasing). **63.** $1 - 0.368 =$
0.632. **69.** $\mathscr{E} = \frac{1}{2}Br^2 2\pi f = \frac{1}{2}(0.050\text{ mT} \times$
$\cos 40°)(5.0\text{ m})^2 2\pi(240/60) = 12$ mV.
73. $\mathscr{E} = -L\dfrac{\Delta I}{\Delta t}$; $\Delta t = (1.5\text{ H})(10\text{ A})/$
$(100\text{ V}) = 0.15$ s. **77.** (a) $I = V/R =$
$(150\text{ V})/(25\ \Omega) = 6.0$ A; (b) $L/R = (300\text{ mH})/$
$(25\ \Omega) = 12$ ms; (c) $0.632(6.0\text{ A}) = 3.8$ A.
81. From Eq. (22.9), $L \approx \mu N^2 A/l = An^2 l\mu$
where An^2 is a constant, so ΔL depends on the
change in μl. Once the shaft is inserted,
part of the region of the coil (d long) changes

from μ_0 to μ; hence, $\mu d + \mu_0(l - d)$ is the
new value, $\mu_0 l$ the original value. The
difference $d(\mu - \mu_0)$ is what we want; hence

$$\Delta L \approx d(\mu - \mu_0)\frac{N^2 A}{l^2}$$

85. $\mathscr{E} = -L\,dI/dt$; $V = V_0 \cos 2\pi ft =$
$-L\,dI/dt$; $dI = -(V_0/L) \cos 2\pi ft\,dt$;

$$I = \int -(V_0/L) \cos 2\pi ft\,dt =$$

$$-(V_0/2\pi fL)\int \cos 2\pi ft\,dt =$$

$-(V_0/2\pi fL) \sin 2\pi ft$ and the maximum
current is $V_0/2\pi fL$. **91.** From Eq. (22.11),
$\Delta I/\Delta t = (V - IR)/L$, which (in the limit as
$\Delta t \to 0$) is also the instantaneous rate of
change of current, or the slope of the curve in
Fig. 22.21 at any instant. At $t = 0$, $I = 0$ and
$\Delta I/\Delta t = V/L = 2.4$ kA/s. At $t = L/R$, $I =$
$0.63V/R$ and $\Delta I/\Delta t = (0.37\,V)/L = 0.89$ kA/s.
The maximum current is $V/R = 0.60$ A, so it
never reaches 1.0 A.

CHAPTER 23
Answers to Odd-Numbered Multiple Choice Questions
1. b **3.** d **5.** e **7.** c **9.** b **11.** e **13.** a
15. b

Answers to Odd-Numbered Problems
1. (a) zero (b) 84.9 V **3.** $i = (14\text{ A}) \times$
$\sin 2\pi(50\text{ Hz})t$ **5.** 141 V **7.** 0.833 A
9. 113 W **11.** 0.13 W **13.** 2.00 Ω
15. 22 μF **17.** $f = 11$ Hz **19.** 94 Ω
21. $q = CV_m \cos(\omega t)$; $i(t) = -\omega CV_m \sin(\omega t)$
23. $v = (I_m/\omega C)\sin(\omega t - \pi/2)$ **25.** $v =$
$\omega LI_m \sin(\omega t + \pi/2)$ **27.** $i(t) = I_0 +$
$(V_m/\omega L) \sin \omega t$ **29.** $f = 100$ Hz **31.** one
resistor is probably shorted internally
33. 20 A, 120 V **35.** 15 W, 240 W
39. 2.12 A **41.** (a) something is wrong with
it (b) new capacitor is shorted internally
43. $V_{rms} = V_m/\sqrt{3}$ **45.** $V_{rms} = (0.730)V_0$
47. $P_m = 2P_{av}$ **49.** 10.1 Ω **51.** 219 Ω
53. 1.30 kΩ **55.** $\theta = 26.6°$, voltage leads
current **57.** 0.25 A **59.** $-63°$
61. $R = 1.7$ kΩ **63.** 500 Ω
65. 5.1×10^{-9} F **67.** 52.8 mH
69. 100:1 **71.** $V_s = 30$ V, $I_s = 4.0$ A
73. 5.0 A, $\theta = 51°$ **75.** $V_L = 302$ V,
$V = 477$ V **77.** 27 μF **81.** (a) 20:1
(b) $I_p = 0.11$ A **83.** 98.3% **85.** $L\dfrac{d^2 i}{dt^2} +$
$R\dfrac{di}{dt} + \dfrac{i}{C} = -V_m\omega \sin \omega t$ **87.** $I_{rms} = \dfrac{I_m}{2}$
89. 20:1

Solutions to Selected Problems
3. $I_m = 1.414(10\text{ A}) = 14$ A; $i =$
$(14\text{ A}) \sin 2\pi(50\text{ Hz})t$. **7.** $P_{av} = 100$ W $= IV$;
$I = (100\text{ W})/(120\text{ V}) = 0.833$ A.
9. $P_{av} = IV = (1.50\text{ A})(75.0\text{ V}) = 113$ W.

13. $X_C = 1/2\pi fC$; $X_{Ci}/X_{Cf} = f_f/f_i = 100$;
hence, the final reactance is 2.00 Ω.
17. $X_L = 2\pi fL$; $f = (10\ \Omega)/2\pi(0.15\text{ H}) =$
10.6 Hz $= 11$ Hz. **23.** $v = q/C =$
$-(I_m/\omega C) \cos \omega t = (I_m/\omega C) \sin(\omega t - \pi/2)$
29. $v = V_m 0.951$; $0.951 = \sin 2\pi f(0.002\,00\text{ s})$;
$f = 100$ Hz. **33.** $1/R_e = 1/(10\ \Omega) +$
$1/(15\ \Omega)$, $R_e = 6.0\ \Omega$; $I = V/R_e = 20$ A. The
voltmeter reads 120 V. **37.** $X_L = 2\pi fL$;
$V = -L\Delta i/\Delta t$, so $L = -V\Delta t/\Delta i$ and 1 H $=$
1 V·s/A; hence, reactance has units of
$(1/s)(H) = V/A = \Omega$. **41.** (a) $X_C =$
$1/2\pi fC = 132.6\ \Omega$; hence, $I = V/X_C =$
$(120\text{ V})/(132.6\ \Omega) = 0.90$ A. The current
should be about 0.9 A and, instead, the
ammeter reads 0.25 A. Something is wrong
with that capacitor. (b) The fuse would blow
if the new capacitor was shorted internally—
it would short out the first capacitor.

45. $[v^2]_{av} = \dfrac{1}{T}\int v^2\,dt =$

$\dfrac{1}{T}\int_{-T/2}^{T/2} \{V_0[1 - 4(t/T)^2]\}^2\,dt =$

$\dfrac{V_0^2}{T}\int_{-T/2}^{T/2} [1 - 8(t/T)^2 + 16(t/T)^4]\,dt =$

$\dfrac{V_0^2}{T}\left[t - \dfrac{8}{T^2}\dfrac{t^3}{3} + \dfrac{16}{T^4}\dfrac{t^5}{5}\right]_{-T/2}^{T/2} =$

$\dfrac{V_0^2}{T}\left[T - \dfrac{16}{3T^2}\left(\dfrac{T}{2}\right)^3 + \dfrac{32}{5T^4}\left(\dfrac{T}{2}\right)^5\right] =$

$V_0^2\left[1 - \dfrac{2}{3} + \dfrac{1}{5}\right] = (0.533)V_0^2$; $V_{rms} =$

$\sqrt{[v^2]_{av}} = \sqrt{(0.533)V_0^2} = (0.731)V_0$
49. $Z = \sqrt{R^2 + X^2} = 10.1\ \Omega$. **51.** $Z =$
$\sqrt{R^2 + X^2} = 219\ \Omega$. **55.** $\tan \theta = X/R$;
$\theta = 26.6°$, and voltage leads current.
59. $\tan \theta = (200\ \Omega)/(100\ \Omega) = 2.0$;
$\theta = -63°$, this is a capacitive reactance.
63. $Z = 2500\ \Omega - 2000\ \Omega = 500\ \Omega$.
65. $C = 1/4\pi^2 f_0^2 L = 5.1 \times 10^{-9}$ F—such
values are not unusual. **69.** 100:1.
73. $2\pi f = 1257$ s^{-1}; $X_L = 2\pi fL = 37.7\ \Omega$;
$Z = \sqrt{R^2 + X^2} = 48.18\ \Omega$; $I = V/Z = 5.0$ A;
$\tan \theta = X_L/R = 1.257$, $\theta = 51°$, voltage leads
current. **75.** $2\pi f = 377$ s^{-1}; $X_L = 2\pi fL =$
150.8 Ω; $Z = \sqrt{R^2 + X^2} = 238.7\ \Omega$; $I =$
$V_R/R = (370\text{ V})/(185\ \Omega) = 2.00$ A. $V_L =$
$IX_L = 302$ V. $\tan \theta = X_L/R$; $\theta = 39.18°$;
$V = V_R/\cos \theta = 477$ V. **79.** $Z = \sqrt{R^2 + X_C^2}$;
$I = V_i/Z$; $V_0 = IX_C$; $V_0 = V_i X_C/Z$;
$V_0/V_i = X_C/Z = (1/2\pi fC)/\sqrt{R^2 + X_C^2} =$
$1/\sqrt{1 + (2\pi fRC)^2}$. **81.** (a) $V_p N_s = V_s N_p$;
$N_p/N_s = V_p/V_s = 20:1$; (b) $2\pi f = 377$ s^{-1};
$X_L = 2\pi fL = 1.131$ kΩ; $I_p = V/X_L = 0.11$ A.
85. $L\dfrac{di}{dt} + Ri + q/C = V_m \cos(\omega t)$ and taking
the derivative $L\dfrac{d^2 i}{dt^2} + R\dfrac{di}{dt} + \dfrac{i}{C} =$
$-V_m\omega \sin \omega t$. **89.** $Z_p = V_p/I_p$; $Z_s = V_s/I_s$;
$Z_p/Z_s = (V_p/V_s) \times (I_s/I_p)$ but, as we saw
earlier, $V_p N_s = V_s N_p$ and $N_p I_p = N_s I_s$; hence,
$Z_p = Z_s(N_p/N_s)^2$. $\sqrt{Z_p/Z_s} = N_p/N_s = 20:1$.

CHAPTER 24

Answers to Odd-Numbered Multiple Choice Questions
1. c **3.** b **5.** a **7.** b **9.** a **11.** d **13.** c
15. d **17.** b **19.** e **21.** c

Answers to Odd-Numbered Problems
1. 12.6×10^6 m^{-1} **3.** 10.00×10^3 m
7. 20 V/m **11.** radiowaves, 100 s
13. 1.50×10^7 m **15.** 5 μH
17. 1.50×10^6 Hz **19.** 1.2×10^{-9} J/m^2·s
21. 6.30×10^4 J **23.** 620 nm, 484 THz,
3.20×10^{-19} J, orange-red light **27.** -20 V/m
29. $B_0 = 6.7 \times 10^{-7}$ T **31.** 14.5
33. 53 W/m^2 **35.** $E_{max} = 1.00 \times 10^4$ eV,
1.24×10^{-10} m **37.** (a) 628 nm (b) 4.78 $\times$
10^{14} Hz (c) positive x-direction
39. (a) 0.600 m (b) 1.0×10^6 J/m^3

Solutions to Selected Problems
1. $k = 2\pi/\lambda = 2\pi/(500 \text{ nm}) = 12.6 \times 10^6$ m^{-1}.
5. See Fig. AN 5. At $x = \lambda/4$, $kx = (2\pi/\lambda) \times (\lambda/4) = \pi/2$.

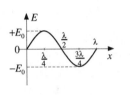

Fig. AN 5.

9. See Fig. AN 9. $E = 10 \sin (kx + \varepsilon)$ mark off the axis in intervals of $\lambda/4$. Notice that shifting the phase by 90° results in a cosine function.

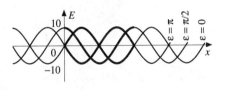

Fig. AN 9.

13. $c = f\lambda_0$, $\lambda_0 = c/f = (3.00 \times 10^8$ m/s$)/(20.0$ Hz$) = 1.50 \times 10^7$ m. **17.** $f = c/\lambda = (3.00 \times 10^8$ m/s$)/(200$ m$) = 1.50 \times 10^6$ Hz. **21.** Irradiance is energy divided by area divided by time: $E/At = I$, $E = (1.05 \times 10^3$ W/m$^2)(1.00$ m$^2)(60.0$ s$) = 6.30 \times 10^4$ J. **23.** 1 eV $= 1.602 \times 10^{-19}$ J; $E = hf = 3.20 \times 10^{-19}$ J; $f = 484$ THz; $\lambda_0 = c/(484$ THz$) = 620$ nm; orange-red light. **27.** $E(0,0) = 20 \cos (0 - 0 + \pi) = -20$; $E(0, T/4) = 20 \cos (0 - 2\pi fT/4 + \pi) = 20 \cos (+\pi/2) = 0$; $E(0, T) = 20 \cos (0 - 2\pi fT + \pi) = 20 \cos (-\pi) = -20$ V/m.

31. $\Delta l = c \, \Delta t = (3.00 \times 10^8$ m/s$) \times (30.0 \times 10^{-15}$ s$) = 9.00 \times 10^{-6}$ m. Now dividing this result by the wavelength yields the number of waves: $(9.00 \times 10^{-6}$ m$)/(620 \times 10^{-9}$ m$) = 14.5$. **37.** (a) $k = 10^7 = 2\pi/\lambda$, $\lambda = 628$ nm. (b) $f = c/\lambda = (3.00 \times 10^8$ m/s$)/(628 \times 10^{-9}$ m$) = 4.78 \times 10^{14}$ Hz. (c) It moves in the positive x-direction.
39. (a) $l = c\Delta t = (3.00 \times 10^8$ m/s$)(2.00 \times 10^{-9}$ s$) = 0.600$ m. (b) The volume of one pulse is $(0.600$ m$)(\pi R^2) = 2.945 \times 10^{-6}$ m^3; hence, $(3.0$ J$)/(2.945 \times 10^{-6}$ m$^3) = 1.0 \times 10^6$ J/m^3.

CHAPTER 25

Answers to Odd-Numbered Multiple Choice Questions
1. c **3.** d **5.** e **7.** a **9.** d. **11.** d **13.** d
15. d **17.** c **19.** e **21.** d

Answers to Odd-Numbered Problems
1. 6.25% **3.** 30° **5.** $\approx$90° **7.** 10°
9. 12 m **11.** 12.5 cm **13.** it misses the mirror **15.** 6.0 m **17.** unchanged **19.** $H = hd/(d + s_0)$ **23.** $L = d/\tan 2\phi$ **25.** 1.24×10^8 m/s **27.** 1.1 **29.** 2.25×10^8 m, 2.21×10^8 m/s **31.** 1.3 **35.** 566 THz, 353 nm **37.** $\theta_c = 49.8°$ **39.** $\theta_c = 33.7°$ **41.** 2.2 mm, 4.2×10^3 waves **43.** $n_i = 1.57$ **45.** 2.2 cm **47.** 0.355 mm **49.** 98° **51.** $(\overline{AB} - L)/\lambda_0$, $(\overline{AB}/\lambda_0) + L(1/\lambda - 1/\lambda_0)$, 2000π **53.** 0.885 m

Solutions to Selected Problems
1. $(1/\lambda_r)^4/(1/\lambda_v)^4 = (\lambda_v/\lambda_r)^4 = 6.25\%$.
9. The statue is 16 m from the point of incidence, and since the ray-triangles are similar, 4 m:16 m as 3 m:Y and $Y = 12$ m.
15. See Fig. AN 15. $h = \frac{1}{2}$m, $d = 1.0$ m, $D = 11.0$ m, find H. $h/d = H/(d + D)$, $h = H/(1.0 + D)$, $\frac{1}{2}(1.0 + D) = H = \frac{1}{2}(1.0 + 11.0) = 6.0$ m.

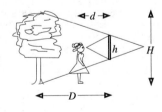

Fig. AN 15.

17. You still see the same fraction of your total body. **21.** See Fig. AN 21. There will be two images arising from single reflections from each of the single mirrors, and there will be a third image due to the double reflections, one from each mirror. The central image is *not* reversed in its handedness—you hold up a right hand, and diagonally across from it, the image will hold up its right hand.

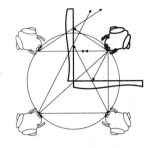

Fig. AN 21.

23. See Fig. AN 23. Construct a normal at M_2; the angle it makes with $\overline{SM_2}$ call α. $\angle M_1 A M_2 = 45° + 2\phi$, hence $(45° + 2\alpha) + (45° + 2\phi) = 180°$, and so $\phi = 45° - \alpha$ and $\angle M_1 S M_2 = 90° - 2\alpha = 2\phi$, thus $\tan 2\phi = d/L$; hence, $L = d/\tan 2\phi$.

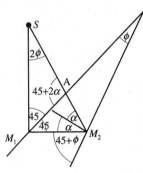

Fig. AN 23.

29. $l = vt = (c/n)t = (3.00 \times 10^8$ m/s$) \times (1.00$ s$)/1.333 = 2.25 \times 10^8$ m. $n = 1.36 = c/v$; $v = c/n = 2.21 \times 10^8$ m/s.
31. $1.00 \sin 55° = n \sin 40°$; $n = 1.27$ or 1.3.
35. $\lambda = (1.06 \ \mu\text{m})/2 = 0.530 \ \mu\text{m} = 530$ nm, hence $f = 566$ THz. $(1.06 \ \mu\text{m})/3 = 0.353 \ \mu\text{m} = 353$ nm in the near-UV.
41. The photo shows the trail of the pulse as it moves through the water during the 10-ps exposure time. $v = c/n = (3.0 \times 10^8$ m/s$)/1.36 = 2.2 \times 10^8$ m/s; the pulse moves $v \Delta t = (2.2 \times 10^8$ m/s$)(10 \times 10^{-12}$ s$) = 2.2$ mm during the exposure and is about 2.2 mm long to begin with. $(2.2 \times 10^{-3}$ m$)/(530 \times 10^{-9}$ m$) = 4.2 \times 10^3$ waves.
43. $\tan \theta_i = 15/20$, $\theta_i = 36.87°$; from Snell's Law $n_i \sin 36.87° = 1.00 \sin 70°$, $n_i = 1.57$.
47. The glass will change the depth of the object from d_R to d_A, where $d_A/d_R = 1.00/1.55$; but $d_R = 1.00$ mm; hence, $d_A = 0.645$ mm, and the camera must be raised 1.00 mm $- 0.645$ mm $= 0.355$ mm.
51. The number of waves in vacuum is $\overline{AB}/\lambda_0$. With the glass in place, there are $(\overline{AB} - L)/\lambda_0$ waves in vacuum and an additional L/λ waves in glass for a total of $(\overline{AB}/\lambda_0) + L(1/\lambda - 1/\lambda_0)$. The difference in number is $L(1/\lambda - 1/\lambda_0)$, giving a phase shift $\Delta\phi$ of 2π for each wave;

hence, $2\pi L(1/\lambda - 1/\lambda_0) = 2\pi L(n/\lambda_0 - 1/\lambda_0) = 2\pi L/2\lambda_0 = 2000\pi$.   **53.** See Fig. AN 53. $d_{A1}/d_{R1} = 1.50/1.33$; $d_{R1} = 1.00$ m; $d_{A1} = 1.1278$ m; $d_{R2} = d_{A1} + 0.20$ m; $d_{A2}/d_{R2} = 1.00/1.50$; $d_{A2} = (1.327\ 8)(1.00/1.50) = 0.885$ m.

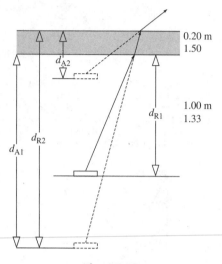

Fig. AN 53.

CHAPTER 26

Answers to Odd-Numbered Multiple Choice Questions
1. b   **3.** b   **5.** d   **7.** b   **9.** a   **11.** e   **13.** b   **15.** d   **17.** a   **19.** c   **21.** e

Answers to Odd-Numbered Problems
1. 100 cm   **3.** 0.86 m   **5.** 2.25   **7.** $s_i = 150$ cm   **9.** $s_i = 0.38$ m; real, inverted, and minified   **11.** $s_i = -240$ mm, virtual   **13.** $f/2$ **15.** 37 mm   **17.** 0.13 m   **19.** 3 cm   **21.** $f = -20$ cm, 40 cm in dia.   **23.** $f = 10$ cm, will help   **25.** $f = +4.0$ cm, yes   **27.** $\mathscr{D} = +2.94$ D   **29.** $s_i = -5.0$ m   **31.** $L = 6.0$ cm   **33.** 1.03 m, 33×   **39.** (a) -3.8 cm (b) 1st—real, inverted, magnified; after jump—virtual, right-side-up, magnified   **41.** $f = 6.5 \times 10^2$ mm   **43.** (a) $\mathscr{D} = -1.02$ D (b) $s_0 = 30.9$ cm   **45.** The greater is s_i, the greater is M_T **47.** (a) 200× (b) $f_E = 25.4$ mm, $f_0 = 8.0$ mm (c) $s_0 = 8.4$ mm   **49.** R = 0.083 m   **51.** $f_w = 4f_a$ **53.** 3.2 m, set should be upside down   **55.** $\mathscr{D} = 5.0$ D   **57.** -13 cm   **59.** 1.33 m   **61.** (a) -0.58 m (b) 1.5 m   **63.** virtual, right-side-up, minified (0.7 times), 0.14 m behind mirror   **65.** 2.0×10^{-6} m   **67.** R = -0.060 m   **69.** $M_T = -0.14$, real, inverted, and minified   **73.** $f = -2.5$ m   **75.** -1.0 m

Solutions to Selected Problems
1. From Eq. (26.4), $\sigma \approx h^2/2R$, $\frac{1}{2}$ cm $\approx$ $(10 \text{ cm})^2/2R$; R = 100 cm.   **3.** $1/f = 0.58 \times (2 \text{ m}^{-1})$; $f = 0.86$ m.   **5.** $1/f = (n-1) \times [1/(2.00 \text{ m}) - 1/\infty]$; $1/(1.60 \text{ m}) =$

$(0.500 \text{ m}^{-1})(n-1)$; $n = 2.25$.   **9.** Given the focal length and the object-distance, we want the image-distance that follows from the Gaussian lens formula: $1/s_0 + 1/s_i = 1/f$; $1/s_i = 1/f - 1/s_0 = 2.66$ m^{-1}; $s_i = 0.38$ m. The object is three focal lengths away from the lens and the image is real, inverted, and minified.   **13.** The diameter is 25 mm; hence, f-number $= (50 \text{ mm})/(25 \text{ mm}) = f/2$.   **17.** $M_A = d_n \mathscr{D}$, $2.0 = (0.254 \text{ m}) \mathscr{D}$, $\mathscr{D} = 7.87$ D and $f = 0.13$ m.   **21.** For the second lens $s_0 = -10$ cm and $s_i = +20$ cm; hence, $1/f = 1/(-10 \text{ cm}) + 1/(20 \text{ cm})$; $f = -20$ cm. $M_T = -(20 \text{ cm})/(-10 \text{ cm}) = +2$, and the image is 40 cm in diameter.   **25.** Applying Eq. (26.19) to two lenses at a time, $1/f = 1/10 + 1/(-20) + 1/5.0$; $f = +4.0$ cm. Since it's positive, it can form real images.   **29.** The lens makes an object at a distance s_0, corresponding to infinity, appear to be at the far-point (s_i), where it can be seen clearly. $1/(-5.0 \text{ m}) = 1/\infty + 1/s_i$; $s_i = -5.0$ m.   **33.** Putting the long focal length ($+1.00$ m) first, we'll need a tube of length $f_0 + f_E = 1.00$ m $+ 0.030$ m $= 1.03$ m. $M_A = -(1.00 \text{ m})/(0.030 \text{ m}) = 33\times$.   **37.** $n_1\overline{HI} + n_1\overline{IJ} = n_m\overline{AB} + n_m\overline{BC} + n_1\overline{CD} + n_1\overline{DE} + n_m\overline{EF} + n_m\overline{FG}$, now divide both sides by n_m and you are back to the form of the first derivation provided $n_1/n_m \to n_1$.   **41.** $M_T = y_i/y_0 = -s_i/s_0$; $s_i = (-2.0 \text{ m}) \times (-0.024 \text{ m})/(0.05 \text{ m}) = 0.96$ m. $1/(2.0 \text{ m}) + 1/(0.96 \text{ m}) = 1/f$; $f = 6.5 \times 10^2$ mm.   **47.** (a) $20 \times 10 = 200\times$. (b) $f_E = (254 \text{ mm})/M_{AE} = 25.4$ mm, $f_0 = (160 \text{ mm})/(-20) = 8.0$ mm. (c) $1/s_0 = 1/f_0 - 1/s_i = 1/(8.0 \text{ mm}) - 1/(L + f_0) = 1/(8.0 \text{ mm}) - 1/(168 \text{ mm})$; $s_0 = 8.4$ mm.   **51.** From Eq. (26.7) $1/f_w = (n_l/n_w - 1)/(n_l - 1)f_a = (1.5/1.33 - 1)/(1.5 - 1)f_a = 0.125/0.5f_a$; $f_w = 4f_a$.   **55.** With her glasses on, 2.0 D $= 1/(0.80 \text{ m} - 0.020 \text{ m}) + 1/s_i$, which gives the distance from the eyeglass plane at which she can see clearly; that is, $(s_i + 2.0 \text{ cm})$ is her new near-point. The prescribed focal length of her new spectacles must satisfy the expression $1/f = 1/(0.254 \text{ m} - 0.020 \text{ m}) + 1/s_i$. Combining these two equations; $1/f - 4.274$ D $= 2.0$ D $- 1.282$ D; and $\mathscr{D} = 5.0$ D.   **59.** $1/(2.00 \text{ m}) + 1/(4.00 \text{ m}) = 1/f$; $f = 1.33$ m.   **63.** $1/(0.20 \text{ m}) + 1/s_i = -2/(1.00 \text{ m})$; $s_i = -0.14$ m. $M_T = -(-0.14 \text{ m})/(0.20 \text{ m}) = +0.7$. The image is virtual, minified (0.7 times life-size), and 0.14 m behind the mirror.   **67.** For the image to be magnified and erect, the mirror must be concave and the image virtual; $s_0 = 0.015$ m; $M_T = 2 = -s_i/(+0.015 \text{ m})$; $s_i = -0.030$ m and, hence, $1/f = 1/(0.015 \text{ m}) - 1/(0.030 \text{ m})$; $f = 0.030$ m and $f = -R/2$; R = -0.060 m.   **71.** Take another look at your solution of Problem 68; $M_T(s_0 - f) = -f$; hence, $s_0 = (-f + fM_T)/M_T$, which equals what we want. Furthermore, $s_i = fs_0/(s_0 - f)$, and on substituting in for s_0, we do indeed get the desired expression.   **75.** The mirror projects a real, inverted, life-size image onto the empty

socket. $s_0 = 2f = s_i = 1.0$ m $= 2(-R/2)$; R = -1.0 m.

CHAPTER 27

Answers to Odd-Numbered Multiple Choice Questions
1. c   **3.** a   **5.** b   **7.** a   **9.** c   **11.** b   **13.** c   **15.** b   **17.** a

Answers to Odd-Numbered Problems
1. 1327 W/m^2 **3.** 100 W/m^2 **5.** 188 W/m^2 **7.** 40 W/m^2 **9.** 53°   **11.** 0.19   **13.** 1.501   **15.** 39°   **17.** ± 6.33 mm   **19.** 419 nm   **21.** 9.76 mm   **23.** 1.84×10^{-7} m   **25.** 107 nm, 322 nm, 537 nm, etc.   **27.** $y_{m'} = sm\frac{1}{2}\lambda/a$ **29.** 10 km   **31.** 0.346 m   **33.** $\Delta x = \lambda_f/2\alpha$ **35.** 550 nm   **37.** 96.6   **39.** $\theta_m = \sin^{-1}(\sin \theta_i \pm m\lambda/a)$ **41.** 1474 nm, 737.0 nm, 491.3 nm, 368.5 nm   **43.** 0.53°   **45.** 12 cm   **47.** Δy doubles   **49.** 2.6 μm   **51.** 1.32×10^{-7} rad, 7.57×10^{-6} degrees, 2.72×10^{-2} s   **53.** 64.5 km   **55.** 0.12 mm   **57.** 11×10^4 Hz   **59.** 0.597 mm   **61.** 4 fringes, 2N   **63.** 1.11

Solutions to Selected Problems
1. $I = \frac{1}{2}c\varepsilon_0 E_0^2 = \frac{1}{2}(2.998 \times 10^8 \text{ m/s}) \times (8.854 \times 10^{-12} \text{ C}^2/\text{N·m}^2)(1000 \text{ V/m})^2$; hence, $I = 1327$ W/m^2.   **5.** $I = I_1 \cos^2\theta = (250 \text{ W/m}^2)\cos^2 30 = 188$ W/m^2.   **9.** $\tan \theta_p = n_t/n_i = 1.3/1.00$, $\theta_p = 53°$.   **13.** $\tan \theta_p = n_t/n_i$; $n_t = \tan 56.33° = 1.501$.   **17.** According to Eq. (27.4), the first minimum occurs when $m' = \pm 1$. Hence, $a \sin \theta_1 = \pm\frac{1}{2}\lambda$; $\theta_1 \approx \pm\frac{1}{2}\lambda/a = \pm\frac{1}{2} \times (632.8 \times 10^{-9} \text{ m})/(0.100 \times 10^{-3} \text{ m}) = \pm 3.16 \times 10^{-3}$ rad, since $y = s\theta_1 = (2.00 \text{ m})(\pm 3.16 \times 10^{-3} \text{ rad}) = \pm 6.33$ mm.   **21.** $y_4 = s4\lambda/a = (1.00 \text{ m})4(487.99 \times 10^{-9} \text{ m})/(0.200 \times 10^{-3} \text{ m}) = 9.76 \times 10^{-3}$ m.   **25.** There will be a 180° phase shift due to the reflections, and so the film must be a quarter-wavelength thick or odd multiples thereof. $d = (m + \frac{1}{2})\lambda_0/2n_f$; for $m = 0$, $d = \lambda_0/4n_f = (580 \text{ nm})/4(1.35) = 107$ nm; for $m = 1$, $d = 322$ nm; for $m = 2$, $d = 537$ nm; etc.   **29.** From Eq. (27.7), $y_1 = s\lambda/a = sc/fa = (20 \times 10^3 \text{ m})(3.0 \times 10^8 \text{ m/s})/(1.0 \times 10^6 \text{ Hz})(600 \text{ m}) = 1.0 \times 10^4$ m.   **33.** The film increases half a wavelength, $\lambda_f/2$. Using the results of the previous problem, we obtain $x_{m+1} - x_m = \Delta x = \lambda_f/2\alpha$.   **35.** Each arm is traversed twice, so a motion of $\lambda/2$ causes a single fringe-pair to shift past; hence, $92\lambda/2 = 2.53 \times 10^{-5}$ m and $\lambda = 550$ nm.   **37.** If the cell has a length L and is traversed twice, then the change in optical path length with and without air is $2Ln_a - 2L$, and the number of waves changed is $2L(n_a - 1)/\lambda_0$, each corresponding to a fringe-pair, and so there are 96.6 of them.   **41.** There will be a relative phase shift on reflection of $\frac{1}{2}\lambda$; hence, the longest wave not reflected is 550.0 nm $= \frac{1}{2}\lambda_f$, or $\lambda_f = 1100$ nm, and so $\lambda_0 = (1100 \text{ nm})1.340 = 1474$ nm in the IR. When 550.0 nm $= \lambda_f$, $\lambda_0 = (550.0 \text{ nm}) \times 1.340 = 737.0$ nm. Similarly, when

550.0 nm = $3\lambda_f/2$ = 366.6 nm or λ_0 = 491.3 nm, there is no reflection. And, finally, when $2\lambda_f$ = 550.0 nm, λ_f = 275.0 nm; and λ_0 = 368.5 nm in the UV, which is not reflected. **45.** From Eq. (27.13), λ = (20 cm) sin 36.87° = 12 cm. **49.** From Eq. (27.8), a = $2(550 \times 10^{-9}$ m)/0.422 6 = 2.6×10^{-6} m. **53.** θ_a = $1.22\lambda/D$ = $1.22(550 \times 10^{-9}$ m)/(4.00 × 10^{-3} m) = 1.68×10^{-4} rad; hence, $L/(3.844 \times 10^8$ m) = 1.68×10^{-4} m; L = 64.5 km. **57.** The wavelength should be less than the size of the target or else diffraction will be troublesome. v = $f\lambda$, f > (330 m/s)/(0.003 0 m) = 11 × 10^4 Hz. **61.** The angular width of the diffraction maximum is obtained from Eq. (27.13), sin θ_1 = $1\lambda/D$ = (500 × 10^{-9} m)/ (0.10 × 10^{-3} m) = 0.005; θ_1 = 0.286 5°, and twice this value is 0.573 0°. The angular width of one of Young's fringes is θ = λ/a = (500 × 10^{-9} m)/(0.20 × 10^{-3} m) = 2.5 × 10^{-3} rad = 0.143 2°; hence, 4.0 fringes will exist. And that might have been obvious from the fact that a = $2D$ and the number of fringes is $2N$.

CHAPTER 28

Answers to Odd-Numbered Multiple Choice Questions
1. d **3.** c **5.** b **7.** a **9.** e **11.** e **13.** a
15. d **17.** c

Answers to Odd-Numbered Problems
5. 90 s, Earth-based observer; 45 s, astronaut **7.** 601 s **9.** 0.499c **11.** 0.436 m **13.** 64.34 m **15.** 500 m **17.** c, classically 2c **19.** 0.999 9c ≈ c **21.** v = 0.404 8c **23.** 0.436c **25.** v = 0.9c **27.** 4.45 ns **31.** 73.8 × 10^3 y **33.** 36.3° **35.** 2.9 × 10^{-19} kg·m/s **37.** KE = 1000 MeV, v = 0.866 0c **39.** 1.0 MeV **41.** E_0 = 8.988 × 10^{10} J, E = 9.173 × 10^{10} J, KE = 1.853 × 10^9 J **43.** 2.2 MeV **45.** 1.000 × 10^{-15} kg **47.** 3.26 MeV **49.** f = 3.653 × 10^{20} Hz **51.** $(\gamma - 1)/\gamma$, 20% **53.** v = 0.101 m/s for both, when $\gamma \approx 1$ **57.** v = 0.967c, v = 2.4c (classically) **59.** 0.106 0c **61.** 0.866 **63.** 3.438 × 10^{20} Hz **65.** v = $Ftc/\sqrt{m^2c^2 + F^2t^2}$

Solutions to Selected Problems
1. $t_\parallel$ = $\dfrac{2L}{c}\dfrac{1}{(1 - \beta^2)} \to \dfrac{2L\sqrt{1 - \beta^2}}{c(1 - \beta^2)}$ = $\dfrac{2L}{c}\dfrac{1}{\sqrt{1 - \beta^2}}$. **5.** Δt_M = $\Delta t_S/\sqrt{1 - v^2/c^2}$ = 2(45 s) = 90 s for the Earth-based observer; 45 s for the astronaut. **9.** Δt_M = $\Delta t_S/\sqrt{1 - v^2/c^2}$; 60.0 s = (52.0 s)/$\sqrt{1 - v^2/c^2}$; $\sqrt{1 - v^2/c^2}$ = 0.866 7, v = 0.499c. **13.** L_M = $L_S\sqrt{1 - v^2/c^2}$ = (2 mi) × (5280 ft/mi)(0.020 00) = 211.1 ft = 64.34 m. **17.** Let S' be the frame of our laboratory, such that one photon moves right at $v_{PO'}$ = c. The other photon in S moves to the left; that is, S' moves right (+x-direction) with respect to S at $v_{O'O}$ = c. The speed of one photon with

respect to the other is v_{PO} = (c + c)/(1 + cc/c^2) = c. Classically: 2c. **19.** Put the electron at rest in S, and have the proton move in S', which is approaching S at −0.992c (that is, moving left). The proton travels at −0.981c (also moving left). v_{PO} = [(−0.981c) + (−0.992c)]/[1 + (0.981c)(0.992c)/c^2] = −1.973c/1.973 2 ≈ −c. **23.** L_M = $0.900L_S$ = L_S/γ, γ = 1/0.900 = 1.111, 1.2345 = 1/(1 − β^2), β^2 = $(\gamma^2 - 1)/\gamma^2$ = 0.190, v = 0.436c. **31.** $\gamma \approx 1 + \frac{1}{2}\beta^2$ = 1 + $\frac{1}{2}$8.585 × 10^{-13}; Δt_M = $\gamma\Delta t_S$ = (1 + $\frac{1}{2}$8.585 × 10^{-13})Δt_S = Δt_S + ($\frac{1}{2}$8.585 × 10^{-13})Δt_S, and we want ($\frac{1}{2}$8.585 × 10^{-13}) Δt_S = 1.00 s; hence, Δt_S = 2.330 × 10^{12} s ≈ 73.8 × 10^3 y. **35.** p = γmv = 1.155(1.675 × 10^{-27} kg) × (0.50c) = 2.9 × 10^{-19} kg·m/s. **37.** E = KE + E_0 = $2E_0$, KE = E_0 = 1000 MeV; E = 2000 MeV = γmc^2 = γ(1000 MeV), γ = 2 = 1/$\sqrt{1 - v^2/c^2}$, v = 0.866 0c. **41.** E_0 = mc^2 = 8.988 × 10^{10} J; E = γmc^2 = 1.021(8.988 × 10^{10} J) = 9.173 × 10^{10} J; E = KE + E_0, KE = E − E_0 = 1.853 × 10^9 J. **43.** 938.272 MeV + 939.566 MeV = 1877.83 MeV; this result minus 1875.6 MeV is 2.2 MeV. **47.** 2(1876.12 MeV) − 2809.41 MeV − 939.566 MeV = 3.26 MeV. **51.** $(\gamma mv − mv)/\gamma mv$ = $(\gamma − 1)/\gamma$, γ = 1.25, $(\gamma − 1)/\gamma$ = 20%. **55.** E^2 = $E_0^2 + (pc)^2$, $4E_0^2$ = $E_0^2 + (pc)^2$, $3E_0^2$ = p^2c^2, p = $\sqrt{3}$ mc. **57.** KE = 1.50 Mev, E = (1.50 MeV + 0.511 MeV) = γmc^2 = γE_0 = γ(0.511 MeV), γ = 3.935, β^2 = $(\gamma^2 − 1)/\gamma^2$ = 0.935, v = 0.967c. KE_c = $\frac{1}{2}mv^2$ = 2.40 × 10^{-13} J, v = 2.4c. **61.** KE = E_0, E = KE + E_0 = $2E_0$, γmc^2 = $2mc^2$, γ = 2 = 1/$\sqrt{1 − v^2/c^2}$, 4 − 1 = $4\beta^2$, β = $\sqrt{3}/2$ = 0.866. **63.** E^2 = $E_0^2 + (pc)^2$, $(pc)^2$ = (1.000 MeV + 0.511 MeV)^2 − (0.511 MeV)^2, p = 1.422 MeV/c, which is the momentum of the electron. For the photon E = pc = 1.422 MeV = hf, f = 3.438 × 10^{20} Hz.

CHAPTER 29

Answers to Odd-Numbered Multiple Choice Questions
1. d **3.** b **5.** b **7.** c **9.** b **11.** a **13.** b
15. c **17.** a

Answers to Odd-Numbered Problems
1. 1.0978 × 10^{27} **3.** 35.5 g of Cl **5.** 3.55 g **7.** Ba, 25.6 g; Cl, 13.2 g **9.** 4.90 × 10^{-14} kg **11.** 0.11 nm **13.** θ = 8.54° **15.** 0.21 nm **17.** λ = 486.2 nm **19.** λ_∞ = 364.6 nm **21.** λ_α = 656.3 nm **23.** 2.28 × 10^{-10} m **27.** B = 8.71 × 10^{-4} T **29.** 2.82 × 10^8 m/s **31.** 0.237 nm at 25.0° **33.** 12 **35.** 1.6 × 10^{-14} m **37.** 397 nm **41.** $\Delta\lambda$ = ±0.001 nm

Solutions to Selected Problems
3. (965 A)(100.0 s) = 96.5 kC = 1 faraday; hence, 1 mole of each is liberated; 35.5 g of Cl. **7.** The valence of barium is 2 and that of chlorine is 1. Hence, for Ba, m = (2.00 A) ×

(5.00 h)(3600 s/h)(137 g)/(96.5 kC)2 = 25.6 g; for Cl, m = (2.00 A)(5.00 h) × (3600 s/h)(35.5 g)/(96.5 kC)1 = 13.2 g. **11.** d = 1(0.090 nm)/(2 sin 25°) = 0.11 nm. **15.** λ = 2(0.28 nm) sin 22° = 0.21 nm. **19.** $1/\lambda$ = (1.097 × 10^7 m^{-1})(1/4 − 1/∞); and λ_∞ = 364.6 nm. **21.** $1/\lambda$ = (1.097 × 10^7 m^{-1})(1/4 − 1/9); and λ_α = 656.3 nm. **25.** tan $\theta \approx \theta$ = v_y/v_x = a_yt/v_x = $(eE/m_e)t/$ (E/B) where t = L/v_x; hence, $\theta \approx eLB^2/Eme$. **31.** d = $1\lambda/(2$ sin 25.0°) = 0.237 nm at 25.0°. **33.** The largest angle is 90°; 2(1.22 × 10^{-10} m) sin 90° = m × (0.020 0 × 10^{-9} m); m = 12.2, so to nearest integer m = 12. **37.** 1/(390 nm) = (1.097 × 10^7 m^{-1})(1/2^2 − 1/n^2); and n = 7.8, so we must use n = 7; $1/\lambda$ = (1.097 × 10^7 m^{-1})(1/2^2 − 1/7^2), and λ = 397 nm. **39.** Y = $y + R$ tan θ; tan θ = v_y/v_x = eLB^2/Em_e. Using Eq. (29.2), we have y = $eL^2B^2/2m_e E$ and Y = $(e/m_e)\dfrac{B^2L}{2E}(L + 2R)$. **41.** $m\lambda$ = $2d$ sin θ; $m\Delta\lambda$ = $2 \Delta d$ sin θ; $\Delta\lambda$ = ±0.001 nm.

CHAPTER 30

Answers to Odd-Numbered Multiple Choice Questions
1. c **3.** a **5.** b **7.** b **9.** c **11.** d **13.** b
15. c **17.** b

Answers to Odd-Numbered Problems
1. I = 76.9 W/m^2, P = 108 W **3.** 9.4 μm, IR **5.** 4.1 × 10^{15} Hz **9.** 496 nm **11.** 0.07 W/m^2 **13.** 4.61 × 10^{14} Hz **15.** 1.89 eV **17.** 1.46 × 10^{-4} nm **19.** 180° **21.** 35 keV **23.** −1.51 eV **25.** f = 4.57 × 10^{14} Hz, λ = 656 nm **27.** 476 pm **29.** n = 137 467 **31.** T = 4.46 × 10^3 K **33.** 0.663 × 10^6 m/s **35.** 4.4 × 10^{14} Hz, 6.8 × 10^{-7} m **37.** 5.82 × 10^{-11} m **39.** 59.6° **41.** 2.19 × 10^6 m/s **43.** −54.4 eV **45.** f_n = $m_ek^2Z^2e^4/$ $2\pi n^3\hbar^3$ **49.** 5.54 × 10^{-8} eV

Solutions to Selected Problems
1. P/A = $\varepsilon\sigma(T^4 − T_e^4)$ = (0.97)(5.670 3 × 10^{-8} W/m^2·K^4)(306^4 − 293^4) = I = 76.9 W/m^2. P = 108 W. **5.** From Eq. (30.4), $\lambda_{max}T$ = 0.002 898 m·K; λ_{max} = 72 nm; f_{max} = 4.1 × 10^{15} Hz. **9.** E = hf = hc/λ, λ = hc/E = 496 nm. **13.** KE ≈ 0 is the threshold condition; f_0 = c/λ = 4.61 × 10^{14} Hz.

17. $\Delta\lambda$ = $\lambda_s − \lambda_i$ = $\dfrac{h}{m_ec}$(1 − cos θ) = (0.002 426 nm)(1 − cos 20°) = 1.46 × 10^{-4} nm. **23.** n = 1 is the ground state, n = 2 the first excited state, n = 3 the second excited state; E_3 = −(13.6 eV)/3^2 = −1.51 eV. **27.** n = 3; from Eq. (30.17), r = 3^2(0.052 9 nm) = 476 pm. **31.** Taking $\lambda_{max} \approx 650$ nm, $\lambda_{max}T$ = 0.002 898 m·K; T = 4.46 × 10^3 K. **35.** In order to use hf_0 = ϕ, we first need to find ϕ; hf = $KE_{max} + \phi$; ϕ = 2.929 × 10^{-19} J; f_0 = ϕ/h = 4.4 × 10^{14} Hz; λ_0 = 6.8 × 10^{-7} m. **39.** 0.060 0(20.0 pm) = (2.426 pm)(1 − cos θ); cos θ = 0.505;

$\theta = 59.6°$.   **43.** From Eq's. (30.18) and (30.19), $E_1 = (-13.6 \text{ eV})Z^2 = -54.4$ eV.   **47.** For x-component, $p_i - p_s \cos\theta = p_e \times \cos\phi$; for y-component, $p_s \sin\theta = p_e \sin\phi$; square and add these two, noting that $\cos^2\phi + \sin^2\phi = 1$; (A) $p_i^2 + p_s^2 - 2p_ip_s\cos\theta = p_e^2$. For the electron (B), $E_e^2 = c^2p_e^2 + m_e^2c^4$. Conservation of Energy is (C) $E_i + m_ec^2 = E_s + E_e$. Now rewrite this equation as $E_e = E_i + m_ec^2 - E_s$ and square it and plug it into (B); also multiply (A) by c^2 and plug it into (B) to get $(E_i + m_ec^2 - E_s)^2 = c^2(p_i^2 + p_s^2 - 2p_ip_s\cos\theta) + m_e^2c^4$. Using $E = pc$ for photons and simplifying, this equation becomes $1/E_s - 1/E_i = (1/m_ec^2)(1 - \cos\theta)$, and since $E = hf = hc/\lambda$ for photons

$$\lambda_s - \lambda_i = \frac{h}{m_ec}(1 - \cos\theta) \qquad [30.13]$$

49. $\Delta E = (-3.40 \text{ eV}) - (-13.6 \text{ eV}) = 10.2$ eV; the photon of momentum $p = E/c$, Eq. (28.16), imparts that momentum to the atom; the atom's recoil energy is then $p^2/2M = E^2/2Mc^2 = (10.2 \text{ eV})^2/2(939 \times 10^6 \text{ eV/c}^2)c^2 = 5.54 \times 10^{-8}$ eV.

CHAPTER 31

Answers to Odd-Numbered Multiple Choice Questions

1. b   **3.** d   **5.** a   **7.** d   **9.** d   **11.** c   **13.** d   **15.** a

Answers to Odd-Numbered Problems

1. 5.5×10^{-36} m   **3.** 2.4×10^{-11} m   **5.** 0.147 nm   **7.** 150 V   **9.** $l = 0, m_l = 0$; $l = 1, m_l = -1, 0, +1$; $l = 2, m_l = -2, -1, 0, +1, +2$ **11.** K has 2, L has 8, M has 18, N has 32   **13.** $\hbar/2L$ **15.** 5×10^{-27} m/s   **17.** 5×10^{-27} J, 3×10^{-8} eV   **19.** 75 MeV, 9.8%   **21.** 0.27 nm   **23.** $p \approx E/c$ **25.** $v_p = \frac{1}{2}v$ **27.** 12.1 eV   **29.** 2.86 eV   **31.** cannot exist   **33.** $1s^2 \cdot 2s^2 2p^6 \cdot 3s^2 3p^5$ **35.** 2.64×10^{-29} m   **37.** 1.16×10^6 m   **39.** $R = r + (1.79 \times 10^{-13} \text{ m})/r$

Solutions to Selected Problems

1. $p = h/\lambda$, $\lambda = h/p = h/mv = 5.5 \times 10^{-36}$ m.   **5.** $\lambda = h/p = h/\sqrt{2m\,\text{KE}} = 0.147$ nm.   **9.** $n = 3$; $n - 1 = 2$; hence, $l = 2, 1,$ or 0. m_l goes from -1 to $+1$; that is, when

$l = 0, m_l = 0$
$l = 1, m_l = -1, 0, +1$
$l = 2, m_l = -2, -1, 0, +1, +2$

13. $\Delta x = L$, $\Delta p_x \approx \frac{1}{2}\hbar/\Delta x = \frac{1}{2}\hbar/L$.   **17.** $\Delta t = 10$ ns; $\Delta E = \frac{1}{2}\hbar/\Delta t = (5.272\,9 \times 10^{-35} \text{ J·s})/(10 \text{ ns}) = 5 \times 10^{-27}$ J $= 3 \times 10^{-8}$ eV.   **21.** $E_0 \gg$ KE; hence, classical form is okay; $p = \sqrt{2m_e\,\text{KE}}$, KE $= (20 \text{ eV})(1.60 \times 10^{-19} \text{ J/eV})$; $p = 2.414\,6 \times 10^{-24}$ kg·m/s; $\lambda = h/p = 2.7 \times 10^{-10}$ m $= 0.27$ nm.   **25.** $v_p = E/p = (p^2/2m)/p = p/2m$; $v_p = \frac{1}{2}v$.   **29.** From Eq. (30.19), $\Delta E = (-13.6 \text{ eV})/5^2 - (-13.6 \text{ eV})/2^2 = 2.86$ eV.   **33.** $1s^2 \cdot 2s^2 2p^6 \cdot 3s^2 3p^5$ **37.** $\Delta z \Delta p_z \geq \frac{1}{2}\hbar$;

$m\Delta v_z \approx \frac{1}{2}\hbar/\Delta z$; $\Delta v_z \approx 0.578\,8 \times 10^6$ m/s. After 2.00 s, $\Delta z = \Delta v_z t = 1.16 \times 10^6$ m.   **39.** $\Delta z = 2r$; $\Delta p_z \approx \frac{1}{2}\hbar/\Delta z \approx \frac{1}{2}\hbar/2r$; $m\Delta v_z \approx \frac{1}{2}\hbar/2r$; $v_z \approx \frac{1}{2}\hbar/2mr$; in the time τ it takes to reach the observation screen, the beam spreads a distance $Z' = v_z\tau$. To find τ, we find v_y; KE $= (3/2)k_BT$; $\frac{1}{2}mv_y^2 = (3/2)k_B(1600 \text{ K})$; $v_y^2 = 3k_B(1600 \text{ K})/m$; $v_y = 450.1$ m/s. $\tau = (1.000 \text{ m})/(450.1 \text{ m/s}) = 2.221$ ms. $Z' = (\frac{1}{2}\hbar/2mr) \times (2.221 \text{ ms}) = (1.79 \times 10^{-13} \text{ m})/r$; and $R = r + Z' = r + (1.79 \times 10^{-13} \text{ m})/r$.

CHAPTER 32

Answers to Odd-Numbered Multiple Choice Questions

1. d   **3.** b   **5.** b   **7.** d   **9.** b   **11.** a   **13.** d   **15.** b   **17.** a   **19.** e

Answers to Odd-Numbered Problems

1. 111   **3.** 7 protons, 8 neutrons   **5.** Fr, Pb, Ag, Pr   **7.** 13 043.8 MeV/c², 13.043 8 GeV/c²   **9.** $9.988\,352 \times 10^{-27}$ kg   **11.** 92.906 38 u   **13.** 15 fm   **15.** 27   **17.** 64.749 MeV   **19.** 6.474 9 MeV/nucleon   **21.** 15.663 MeV   **23.** no   **25.** 3.9 times   **29.** 8.736 MeV   **31.** Ca-41, 8.36 MeV; Ca-42, 11.48 MeV   **33.** 5.40 MeV   **35.** $^{27}_{13}\text{Al} + {}^4_2\alpha \rightarrow {}^{30}_{15}\text{P} + {}^1_0\text{n}$ **37.** $^{147}_{62}\text{Sm} \rightarrow {}^{143}_{60}\text{Nd} + {}^4_2\alpha$ **39.** 17.346 MeV   **41.** 4.1×10^{-9} s^{-1} **43.** $t_{1/2} = 12$ h   **45.** 4.28×10^{-4} decays/y   **47.** 5.5 d   **49.** 90.1 s   **51.** $N/N_0 = 4.97 \times 10^{-16}$ **53.** 5.700 74 MeV   **55.** 1.191 MeV   **57.** 5.7×10^{12} Bq   **59.** 3×10^{10} Bq   **61.** 2.8×10^{10} atoms   **63.** 0.18 MBq/m³   **65.** 24.685 4 MeV

Solutions to Selected Problems

5. Fr, Pb, Ag, and Pr.   **9.** $(5.603\,051 \text{ GeV/c}^2) \times (1.782\,663 \times 10^{-36} \text{ kg/eV})(1.000\,000 \times 10^9 \text{ eV/GeV}) = 9.988\,352 \times 10^{-27}$ kg.   **13.** $R = R_0A^{1/3} = (1.2 \text{ fm})(6.17) = 7.4$ fm. $2R = 15$ fm.   **19.** If $E_B = 64.749$ MeV, $E_B/A = 6.474\,9$ MeV/nucleon.   **21.** $\Delta m = 1.008\,665$ u $+ 15.003\,065$ u $- 15.994\,915$ u $= 0.016\,815$ u $\rightarrow 15.663$ MeV.   **27.** Mass of C-12 atom is (12 kg)/$(6.022 \times 10^{26} \text{ atoms/kmole}) = 1.99 \times 10^{-26}$ kg; dividing by 12 yields 1 u $= 1.66 \times 10^{-27}$ kg.   **31.** For Ca-41: $(40.962\,278 \text{ u}) - (1.008\,665 \text{ u}) - (39.962\,591 \text{ u}) = -0.008\,978$ u. For Ca-42: $(41.958\,618 \text{ u}) - (1.008\,665 \text{ u}) - (40.962\,278 \text{ u}) = -0.012\,325$ u. One must add 8.36 MeV to remove a neutron from Ca-41 and 11.48 MeV to remove one from Ca-42.   **37.** $^{147}_{62}\text{Sm} \rightarrow {}^4_2\alpha + X$; $X = {}^{143}_{60}X$, which is neodymium-143.   **41.** $\lambda = 0.693/t_{1/2} = 0.693/(167 \times 10^6 \text{ s}) = 4.1 \times 10^{-9}$ s^{-1}.   **45.** $\lambda = 0.693/t_{1/2} = 4.28 \times 10^{-4}$ y^{-1}.   **49.** $\tau = 1/\lambda = 130$ s; $t_{1/2} = 0.693/\lambda = 90.1$ s.   **51.** $\lambda = 0.693/(1.18 \text{ min}) = 0.587 \text{ min}^{-1}$. $N/N_0 = \exp[(-0.587 \text{ min}^{-1})(60.0 \text{ min})] = 4.97 \times 10^{-16}$.   **55.** Using atomic masses; initial $-$ final mass; $(4.002\,603 \text{ u}) + (14.003\,074 \text{ u}) - (16.999\,131 \text{ u}) - (1.007\,825 \text{ u}) = -0.001\,279$ u; α must supply

1.191 MeV.   **59.** $t_{1/2} = 1.236 \times 10^6$ s; from Eq. (32.12), $\lambda = 5.609 \times 10^{-7}$ s^{-1}; $R_0 = \lambda N_0 = 3 \times 10^{10}$ Bq.   **63.** (444 MBq)/(50 weeks)(5 d/week)(10 m³/d) = 0.18 MBq/m³.   **65.** 4(1.007 276 u) $-$ (4.001 506 u) $-$ 2(0.000 548 580 u) $= 0.026\,501$ u; or 24.685 4 MeV.   **69.** $N = N_0 \exp(-\lambda t)$; $N_0/2 = N_0 \exp(-\lambda t_{1/2})$; $\exp(\lambda t_{1/2}) = 2$; $\ln 2 = \lambda t_{1/2}$.

CHAPTER 33

Answers to Odd-Numbered Multiple Choice Questions

1. d   **3.** e   **5.** c   **7.** c   **9.** a   **11.** a   **13.** d   **15.** a   **17.** c   **19.** b

Answers to Odd-Numbered Problems

1. (a) electromagnetic (b) strong (c) weak (d) strong   **3.** (a) baryon number and strangeness (b) lepton number (c) charge and strangeness (d) strangeness and baryon number   **5.** no   **7.** no, electron-lepton number   **9.** yes   **11.** no, weakly   **13.** not forbidden, 33.9 MeV   **15.** 37.7 MeV, strangeness   **17.** 1877 MeV   **19.** 0.660 GeV   **21.** 10^{30} **23.** $E_0 = hc/4\pi R$ **25.** 70.45 MeV   **27.** -535.4 MeV, no—must have appreciable KE.   **29.** 2.42×10^{22} Hz, 1.24×10^{-14} m, 5.34×10^{-20} kg·m/s   **31.** c$\bar{\text{u}}$ **33.** $\Lambda° = $ uds   **35.** d$\bar{\text{s}} \rightarrow$ u$\bar{\text{d}}$ + d$\bar{\text{u}}$ **37.** 2.47×10^{20} Hz   **39.** $\approx 10^{15}$ GeV/c²   **41.** 8×10^{27} K

Solutions to Selected Problems

1. (a) Only the electromagnetic force affects photons. (b) Since there's no strangeness, this must be a strong-force decay. (c) $\Lambda°$ has strangeness, so this must be a weak-force decay. (d) With no strangeness, this must be a strong-force decay.   **5.** No. It does not conserve baryon number—the p$\bar{\text{p}}$ pair can cancel, leaving only one nucleon on the right. Also, it doesn't conserve angular momentum.   **9.** Yes. It conserves electromagnetic charge, spin, and B. There is no meson number to worry about. Because of the strangeness of K°, its decay will be controlled by the weak force.   **13.** Not forbidden. (139.6 MeV) $-$ (105.7 MeV) = 33.9 MeV.   **17.** $2(m_pc^2) = 2(938.3 \text{ MeV}) = 1877$ MeV.   **21.** If 1 proton takes an average of 10^{30} y to decay, in a group of $N = 10^{30}$ protons one should decay within 1 y. $\tau = 10^{30}$ y $= 1/\lambda$; $\lambda = 10^{-30}$ fractional decays/y; $\lambda N = 1$.   **25.** 2(938.3 MeV) $-$ (939.6 MeV) $-$ (938.3 MeV) $-$ (139.6 MeV) $= -140.9$ MeV; each proton must bring in 70.45 MeV.   **29.** E $= hf$; $f = E/h = 2.417\,99 \times 10^{22}$ Hz $= 2.42 \times 10^{22}$ Hz; $\lambda = c/f = 1.239\,8 \times 10^{-14}$ m $= 1.24 \times 10^{-14}$ m; $p = h/\lambda = 5.34 \times 10^{-20}$ kg·m/s.   **33.** It's made up of three quarks (no antiquarks). To get the strangeness, we need $S = -1$ or one s-quark that has a $Q = -1/3$, so we must add a net charge of $+1/3$. One u and one d will do it. $\Lambda° = $ uds.   **37.** E $= hf = 2m_ec^2 = 1.637 \times 10^{-13}$ J; $f = 2.47 \times 10^{20}$ Hz.   **41.** E $= (3/2)k_BT$; $T = 2E/3k_B = 8 \times 10^{27}$ K.

Credits

This page constitutes an extension of the copyright page. We have made every effort to trace the ownership of all copyrighted material and to secure permission from copyright holders. In the event of any question arising as to the use of any material, we will be pleased to make the necessary corrections in future printings. Thanks are due to the following authors, publishers, and agents for permission to use the material indicated.

Chapter 2: 32: Figure 2.2 adapted from Eugene Hecht, *Physics in Perspective.* © 1980 by Addison-Wesley Publishing Company, Inc. **Chapter 3: 73**: Figure 3.3 from *Road & Track,* December 1977, p. 49. Reprinted by permission. **85**: Excerpt: From Associated Press News Release, July 1971. Reprinted by permission. **94**: Figure 3.12a adapted from Eugene Hecht, *Physics in Perspective,* © 1980 by Addison-Wesley Publishing Company, Inc. **Chapter 4: 121**: Figure 4.4 adapted from Eugene Hecht, *Physics in Perspective.* © 1980 by Addison-Wesley Publishing Company, Inc. **146**: Figure Q19 Courtesy of PASCO Scientific. **Chapter 5: 155**: Figure 5.2 adapted from Eugene Hecht, *Physics in Perspective.* © 1980 by Addison-Wesley Publishing Company, Inc. **165**: Figure 5.15 from Eugene Hecht, *Physics in Perspective.* © 1980 by Addison-Wesley Publishing Company, Inc. **Chapter 7: 244**: Table 7.1: Adapted from *Scientific American,* August 1992. **Chapter 9: 342**: Figure 9.21: Adapted from Eugene Hecht, *Physics in Perspective.* (Adapted from pg. 100). © 1980 by Addison-Wesley Publishing Company, Inc. **Chapter 10: 365**: Figure 10.3: From Eugene Hecht, *Physics in Perspective.* © 1980 by Addison-Wesley Publishing Company, Inc. **Chapter 11: 410**: Figure 11.22: From Eugene Hecht, *Physics in Perspective.* © 1980 by Addison-Wesley Publishing Company, Inc. **411**: Figure 11.23: From Eugene Hecht, *Physics in Perspective,* © 1980 by Addison-Wesley Publishing Company, Inc. **435**: Figure Q10: Adapted from Eugene Hecht, *Physics in Perspective.* © 1980 by Addison-Wesley Publishing Company, Inc. **Chapter 16: 618**: Figure 16.19 from Eugene Hecht, *Physics in Perspective.* © 1980 by Addison-Wesley Publishing Company, Inc. **Chapter 18: 687**: Figure 18.2 from Eugene Hecht, *Physics in Perspective.* © 1980 by Addison-Wesley Publishing Company, Inc. **Chapter 19: 747**: Figure Q15 from *Newsday,* Wednesday, July 29, 1987. Reprinted by permission of Los Angeles Times Syndicate International. **Chapter 21: 793**: Figure 21.11 adapted from Eugene Hecht, *Physics in Perspective.* © 1980 by Addison-Wesley Publishing Company, Inc. **793**: Figure 21.13 adapted from Eugene Hecht, *Physics in Perspective.* © 1980 by Addison-Wesley Publishing Company, Inc. **817**: Figure 21.39 adapted from Eugene Hecht, *Physics in Perspective.* © 1980 by Addison-Wesley Publishing Company, Inc. **Chapter 23: 898**: Figure Q15 based on Figure 4.1, from *49 Easy-to-Build Electronic Projects,* by Robert M. Brown and Tom Kneitel. Copyright 1981 by TAB Books, a Division of McGraw-Hill Inc., Blue Ridge Summit, PA 17294-0850, (1-800-233-1128). **Chapter 28: 1075**: Figure 28.18 adapted from W. Bertozzi, 1964, *American Journal of Physics, 32,* p. 551. Adapted by permission. **1079**: Figure 28.19 adapted from R. Kollarits, 1972, *American Journal of Physics, 40,* p. 1125. Adapted by permission. **Chapter 32: 1178**: Figure 32.1 from A. P. Arya, *Elementary Modern Physics* (Figure 11.5), © 1974 by Addison-Wesley Publishing Company, Inc. Reprinted by permission. **1179**: Figure 32.3 data from R. Hofstadter, *Annual Reviews of Nuclear Science, 7,* 231, 1957. **1182**: Figure 32.6 data from G. W. Farwell and H. E. Wegner, 1954, *Physical Review, 95,* 1212.

Photo Credits

Chapter 1: 1, J. Wray; **5**, National Optical Astronomy Observatories; **7**, Brooks/Cole; **8**, The Bryn Mawr College Archives; **9**, (top) Fotomas Index, London; (bottom) E. H.; **10**, (two at top) Stock, Boston, (two at bottom) Photo Researchers, Inc.; **12**, (top) Focus on Sports, (bottom) National Bureau of Standards; **13**, (two at top), Stock, Boston, (two at bottom) Photo Researchers, Inc.; **14**, (top) Stock, Boston, (bottom) Photo Researchers, Inc.; **16**, E. H.; **18**, Stephen Frisch/Stock, Boston; **21**, J. Anthony Tyson/AT&T Bell Labs

Chapter 2: 27, Brooks/Cole; **28**, Gift of Collection Société Anonyme/Yale University Art Gallery; **32**, Ana S. Arias; **36**, © The Harold E. Edgerton 1992 Trust; **50**, Canadian Forces

Chapter 3: 69, Brooks/Cole; **71**, Duomo Photography; **86**, Vandystadt/Photo Researchers, Inc.; **88**, Biblioteca Ambosiana, Milan, Italy; **90**, Peticolas/Megna/Fundamental Photographs; **96**, Dawson Jones/Stock, Boston

Chapter 4: 115, Brooks/Cole; **116**, Burndy Library; **117**, NASA; **118**, Roger-Viollet; **119**, Biblioteca Ambrosiana, Milan, Italy; **122**, E. H.; **129**, © The Harold E. Edgerton 1992 Trust; **130**, Rick Stewart/Allsport USA; **135**, E. H.; **136**, Richards/PhotoEdit; **137**, Hewlett-Packard; **138**, Mickey Gibson/Animals, Animals; **139**, Yoram Lehmann/Peter Arnold, Inc.; **145**, U.S. Air Force; **152**, Tracy Lee Didas/U.S. Navy

Chapter 5: 153, Brooks/Cole; **158**, Gerard Vandystadt/Photo Researchers, Inc.; **160**, McDonough/Focus on Sports; **161**, Joseph Sohm/Stock, Boston; **165**, Bob Martin/Allsport; **170**, (top) James Sugar/Black Star; (bottom) Gerard Vandystadt/Photo Researchers, Inc.; **172**, David Yarrow/Allsport; **174**, Six Flags Great Adventure; **175**, Leonard Harris/Stock, Boston; **176**, Brooks/Cole; **179**, (top) David Madison/Duomo Photography, (bottom) Gerard Vandystadt/Allsport USA; **180**, Gerard Vandystadt/Allsport USA; **183**, (left) E. H., (right) E. H.; **184**, E. H.

Chapter 6: 193, Brooks/Cole; **195**, Bill Beaver; **199**, Larry Molmud/Mucking Otis Press; **200**, Armed Forces Institute of Pathology; **201**, SKA; **203**, E. H.; **209**, Vanguard Racing Sailboats; **214**, Bob Daemmrich/Stock, Boston; **217**, (left) Bob Daemmrich/Stock, Boston, (right) E. H.; **220**, Hsinhai News Agency; **222**, E. H.; **223**, (top) E. H., (bottom) Central Scientific Company

Chapter 7: 237, Brooks/Cole; **238**, (top) SKA, (bottom) Jet Propulsion Laboratory; **239**, Burndy Library; **243**, Central Scientific Company; **244**, James Sugar/Black Star; **247**, James Sugar; **251**, **252**, Uwe Fink/Dept. of Planetary Sciences, University of Arizona; **256**, NASA; **260**, Raymond E. Arvidson/Washington University

Chapter 8: 269, Brooks/Cole; **286**, Central Scientific Company; **287**, Richard Megna/Fundamental Photographs; **293**, (left) NASA, (right) National Optical Astronomy Observatories, (middle) National Optical Astronomy Observatories, **294**, (top) Shigeru Hashimoto, (bottom) Thomas Eakins; **297**, (top) Jet Propulsion Laboratory, (bottom) Gordon Garradd/Photo Researchers, Inc.; **298**, Sargent Welch

Chapter 9: 311, Brooks/Cole; **322**, James Sugar/Black Star; **325**, (top) Smithsonian Institution, (bottom) Rick Rockman/Duomo Photography; **327**, Marty Stouffer/Animals, Animals; **330**, Dorothy Littell/Stock, Boston; **332**, NASA; **335**, Robert Bloomberg; **338**, The Harold E. Edgerton 1992 Trust; **344**, (left) Erik Anderson/Stock, Boston, (right) The Harold E. Edgerton 1992 Trust, (bottom) The Harold E. Edgerton 1992 Trust; **346**, Central Scientific Company

bridge; **1092**, Cavendish Laboratory, University of Cambridge; **1094**, (top, bottom) The Metropolitan Museum of Art, Purchase, 1871; **1097**, E. H.; **1098**, (top, bottom) Smithsonian Institution; **1099**, E. H.; **1103**, Stuart Kenter Assoc.; **1104**, Stuart Kenter Assoc.; **1106**, Courtesy General Electric Co.

Chapter 30:   **1117**, Brooks/Cole; **1118**, PhotoDisc; **1121**, (a, b, c) Kodansha, Ltd.; **1122**, Agency International de Physique Solvag/AIP Neils Bohr Library; **1123**, Courtesy, German Information Center; **1127**, Linda J. LaRosa; **1131**, AIP Emilio Segre Visual Archives; **1136**, Warren Nagourney; **1138**, Laserscope

Chapter 31:   **1147**, Brooks/Cole; **1148**, (top) French Embassy, Press & Information Division, (bottom) Photo Researchers, Inc.; **1149**, (left, right) Educational Development Center, Inc.; **1151**, AIP Neils Bohr Library; **1153**, (left) Courtesy, David Sarnoff Research Center, (right) A. Tonomura/Hitachi Advanced Research Center; **1155**, A. J. G. Hey/University of Southampton; **1157**, National Optical Astronomy Observatories; **1164**, AIP Neils Bohr Library, Archives for History of Quantum Physics; **1167**, AIP Neils Bohr Library, Fankuchen Collection; **1168**, AIP Emilio Segre Visual Archives; **1169**, (left) McConnell Brain Imaging Centre, Montreal Neurological Institute, (right) Dan McCoy/Rainbow

Chapter 32:   **1175**, Brooks/Cole; **1176**, Photo Researchers, Inc.; **1178**, Argonne National Laboratory; **1181**, (top, bottom) Dan McCoy/Rainbow; **1184**, AIP Meggers Gallery of Nobel Laureates; **1189**, EG&G Mound Applied Technology; **1192**, S. M. Larson and S. J. Goldsmith/Nuclear Medicine Service, Memorial Sloan-Kettering Cancer Center; **1197**, A. Agelarakis/Adelphi University; **1198**, (top) Argonne National Laboratory; **1199**, from Armin Hermann, *The New Physics,* Munich: Heinz Moos Verlag, 1979; **1201**, *Pittsburgh Post-Gazette,* AIP Emilio Segre Visual Archives; **1202**, *N. Y. Herald Tribune;* **1203**, painting by Gary Sheahan, photo courtesy Argonne National Laboratory; **1204**, U. S. Council for Energy Awareness; **1205**, (top) U. S. Navy, (bottom) UPI/Bettmann Archive; **1208**, Lawrence Livermore National Laboratory

Chapter 33:   **1213**, Brooks/Cole; **1215**, Fermilab Visual Media Services; **1218**, Fermilab Visual Media Services; **1233**, Brooks/Cole; **1234**, University of Michigan, Physics; **1237**, E. H.

Index

For entries on pp. 643–1240, please see Volume 2.

Need help solving the text's problems?

Introducing *Brooks/Cole Exerciser (BCX) v 2.0 for DOS, Windows, and Macintosh*

BCX is an easy-to-use software program that helps you solve the types of problems that appear in Hecht's *Physics: Calculus*

Features

- Each section in *BCX* begins with a problem from the text. *BCX* walks you through solving this problem.

- For each *BCX* problem, you are asked to indicate the correct solution. Choosing a correct answer prompts another problem, but choosing an incorrect answer repeats the question and offers a hint.

- A second wrong answer prompts *BCX* to provide the complete solution and an illustration of the concepts behind the problem.

- *BCX*'s record-keeping capabilities allow you to track how you are doing and to rework problems to improve your score.

- You can print your score or view it on the screen.

- The "Bookmark" feature allows you to quit the program halfway through a section and to return easily to that exact point.

Price: $20.25
DOS: ISBN: 0-534-33987-5. **Mac**: ISBN: 0-534-34171-3. **Windows**: ISBN: 0-534-34168-3.

System Requirements
<u>DOS</u>: 80x86 or later. DOS 3.2 or later. EGA, VGA, or SVGA monitor. 400 KB free RAM. 2.5 MB available on hard drive.

<u>Macintosh</u>: Mac SE or later, but not native PowerMac. System 6.0.5 or later. Monochrome or color monitor. 1 MB available on hard drive.

<u>Windows</u>: 80x86 or later. Windows 3.1 in Enhanced Mode or Windows 95. VGA or better graphics card. 1 MB free memory. 2.5 MB available on hard disk.

To order, please use the attached coupon or call 1-800-354-9706.

ORDER FORM

To order, simply fill out this coupon and return it to Brooks/Cole along with your check, money order, or credit card information.

Yes! I would like to order *Brooks/Cole Exerciser (BCX) v 2.0* for Hecht's *Physics: Calculus:*

_____ **DOS**: $20.25. (ISBN: 0-534-33987-5).
_____ **Mac**: $20.25. (ISBN: 0-534-34171-3).
_____ **Windows**: $20.25. (ISBN: 0-534-34168-3).

Residents of: AL, AZ, CA, CT, CO, FL, GA, IL, IN, KS, KY, LA, MA, MD, MI, MN, MO, NC, NJ, NY, OH, PA, RI, SC, TN, TX, UT, VA, WA, WI must add appropriate state sales tax.

Subtotal _____
Tax _____
Handling $2.00 _____
Total _____

Payment Options

_____ Check or money order enclosed

or bill my _____VISA _____MasterCard _____American Express

Card Number: _____

Expiration Date: _____

Signature: _____

Note: Credit card billing and shipping addresses must be the same.

Please ship my order to: (please print)
Name _____
Street Address_____
City_____ State_____ Zip+4_____
Telephone (   ) _____

Mail to:
Brooks/Cole Publishing Company
Dept. BCX-Studentinbk
511 Forest Lodge Road
Pacific Grove, California 93950-5098
Phone: (408) 373-0728; Fax: (408) 375-6414

Prices subject to change without notice.

12/95

Now your students can get tutorial help with a program that contains physics problems and solutions similar to those in Hecht's *Physics: Calculus*

Introducing *Brooks/Cole Exerciser (BCX) v 2.0 for DOS, Windows, and Macintosh*

BCX is an easy-to-use software program that helps your students solve the types of problems that appear in Hecht's *Physics: Calculus*

Features

- Each section in *BCX* begins with a problem from the text. *BCX* walks your students through solving this problem.

- For each *BCX* problem, your students are asked to indicate the correct solution. Choosing a correct answer prompts another problem, but choosing an incorrect answer repeats the question and offers a hint.

- A second wrong answer prompts *BCX* to provide the complete solution and an illustration of the concepts behind the problem.

- *BCX*'s record-keeping capabilities allow your students to track how they are doing and to rework problems to improve their scores.

- Students can print their scores or view them on the screen.

- The "Bookmark" feature allows students to quit the program halfway through a section and to return easily to that exact point.

Price: $20.25
DOS: ISBN: 0-534-33987-5. **Mac**: ISBN: 0-534-34171-3. **Windows**: ISBN: 0-534-34168-3.

Demos: Free
DOS Demo: ISBN: 0-534-33988-3. **Mac Demo**: ISBN: 0-534-341721-1.
Windows Demo: ISBN: 0-534-34169-1.

Site License Available: $37.50 + $5.00/CPU
DOS Site License: ISBN: 0-534-34167-5. **Macintosh Site License**: ISBN: 0-534-34182-9.
Windows Site License: ISBN: 0-534-34170-5.

Bundles Available
For more information on bundles, please call: 800-245-6724.

System Requirements
DOS: 80x86 or later. DOS 3.2 or later. EGA, VGA, or SVGA monitor. 400 KB free RAM. 2.5 MB available on hard drive.
Macintosh: Mac SE or later, but not native PowerMac. System 6.0.5 or later. Monochrome or color monitor. 1 MB available on hard drive.
Windows: 80x86 or later. Windows 3.1 in Enhanced Mode or Windows 95. VGA or better graphics card. 1 MB free memory. 2.5 MB available on hard disk.

For a complimentary demonstration copy, please use the attached coupon or call 1-800-354-9706.

REVIEW COPY REQUEST

I wish to review a complimentary demonstration copy of *Brooks/Cole Exerciser (BCX) v 2.0* **for Hecht's** *Physics: Calculus* :

_____ **DOS** (ISBN: 0-534-33988-3).
_____ **Mac** (ISBN: 0-534-341721-1).
_____ **Windows** (ISBN: 0-534-34169-1).

for adoption consideration as _____ required _____recommended software

for my course:_____
                        (course number)   *(course title)*

Review copy needed: _____ Text decision _____Course begins_____
                        mo/day/year   mo/day/year   mo/day/year

Number of students you teach_____per yr/qtr/sem *(circle one)*

Is this the first time you have taught this course? _____yes _____no

If not, how often do you teach this course?_____

The adoption decision is made by ___ me alone ___ me as part of a committee___ other

Others involved in the decision who should receive copies:

_____

_____

Book(s) in use *(author/title)*:   *Please Check One*

_____ ___required ___recommended

_____ ___required ___recommended

I am ___ very likely ___somewhat likely ____ not all likely to change my text at this time.

(To expedite the processing of your review copy request, please provide all requested information. Your answers to the questions will not affect the approval of your request.)

Name_____

Department_____ School_____

Street Address_____

City_____ State_____ Zip+4 _____

Office phone number(_____)_____

Office hours *(please circle)* M T W Th F / Time_____

Detach and mail to:

Brooks/Cole Publishing Company
Dept. BCX-FCinbk
511 Forest Lodge Road
Pacific Grove, California 93950-5098
Phone: (408) 373-0728; Fax: (408) 375-6414

In Canada, contact Nelson Canada, College Division, 1-800-268-2222   12/95

USEFUL PHYSICAL DATA

Standard Temperature and Pressure (STP)		$0°C = 273.15$ K 1 atm $\equiv 101.325$ kPa
Water		
Density, relative (4°C)	ρ_w	1.000×10^3 kg/m^3
Heat of fusion	L_f	333.7 kJ/kg
Heat of vaporization	L_v	2259 kJ/kg
Specific heat capacity	c	4.186 kJ/kg·K
Index of refraction	n_w	1.33
Standard acceleration due to Earth's gravity		$9.806\,65$ m/s^2
Speed of sound in air (20°C)		343 m/s
Speed of sound in air (STP)		331 m/s
Density of dry air (STP)		1.29 kg/m^3
Molecular mass of air		28.98 g/mol

ASTROPHYSICAL DATA

	Earth	Moon	Sun
Mass	5.975×10^{24} kg	7.35×10^{22} kg	1.987×10^{30} kg
Mean radius	6.371×10^6 m	1.74×10^6 m	6.96×10^8 m
Mean density	5.52×10^3 kg/m^3	3.33×10^3 kg/m^3	1.41×10^3 kg/m^3
Orbital period about galactic center			200×10^6 y
Orbital period	365 d 5 h 48 min	27.3 d	
Mean distance from Sun	1.50×10^{11} m		
Mean distance from Earth		3.85×10^8 m	
Surface gravitational acceleration	9.81 m/s^2	1.62 m/s^2	274 m/s^2
Surface pressure	1.013×10^5 Pa		
Magnetic moment	8.0×10^{22} A·m^2		
Surface temperature	≈ 287 K	125 K–375 K	5.8×10^3 K
Power output			3.85×10^{26} W
Period of rotation	23 h 56 min 4.1 s	27.3 d	~26 d

THE GREEK ALPHABET

Alpha	A	α	Nu	N	ν
Beta	B	β	Xi	Ξ	ξ
Gamma	Γ	γ	Omicron	O	o
Delta	Δ	δ	Pi	Π	π
Epsilon	E	ϵ	Rho	P	ρ
Zeta	Z	ζ	Sigma	Σ	σ
Eta	H	η	Tau	T	τ
Theta	Θ	θ	Upsilon	Υ	υ
Iota	I	ι	Phi	Φ	ϕ
Kappa	K	κ	Chi	X	χ
Lambda	Λ	λ	Psi	Ψ	ψ
Mu	M	μ	Omega	Ω	ω